Springer Monographs in Mathematics

This series publishes advanced monographs giving well-written presentations of the "state-of-the-art" in fields of mathematical research that have acquired the maturity needed for such a treatment. They are sufficiently self-contained to be accessible to more than just the intimate specialists of the subject, and sufficiently comprehensive to remain valuable references for many years. Besides the current state of knowledge in its field, an SMM volume should ideally describe its relevance to and interaction with neighbouring fields of mathematics, and give pointers to future directions of research.

Michael Stiebitz • Thomas Schweser
Bjarne Toft

Brooks' Theorem

Graph Coloring and Critical Graphs

Springer

Michael Stiebitz
Fakultät Mathematik & Naturwissenschaften
Technische Universität Ilmenau
Ilmenau, Germany

Thomas Schweser
Actian Germany GmbH
Ilmenau, Germany

Bjarne Toft
Department of Mathematics & Computer Science
The University of Southern Denmark
Odense M, Denmark

ISSN 1439-7382 ISSN 2196-9922 (electronic)
Springer Monographs in Mathematics
ISBN 978-3-031-50064-0 ISBN 978-3-031-50065-7 (eBook)
https://doi.org/10.1007/978-3-031-50065-7

This Springer imprint is published by the registered company Springer Nature Switzerland AG
The registered company address is: Gewerbestrasse 11, 6330 Cham, Switzerland

Paper in this product is recyclable.

In memory of Gabriel Andrew Dirac, Tibor Gallai and Horst Sachs

Preface

Brooks' Theorem from 1941 is a cornerstone in graph theory. Until then graph coloring theory was centered around planar graphs and the four color problem. Brooks' Theorem was the first nontrivial theorem to relate the chromatic number of general graphs to other graph parameters, in this case to the maximum degree.

Rowland Leonard Brooks (1916–1993) studied mathematics at Cambridge University, where he made friends with Cedric Austin Bardell Smith (1912–2002), Arthur Harold Stone (1916–2000) and William Thomas Tutte (1917–2002). The four together solved the problem of dividing a square into a finite number of smaller squares, all of different sizes, relating it to flow theory for graphs. But it was Brooks alone who took up abstract graph coloring and found the theorem that bears his name. Brooks' paper was communicated to the Cambridge Philosophical Society by Tutte and published in their proceedings in 1941.

The first monograph on abstract graph theory, written in German by the Hungarian Dénes König (1884–1944), had appeared already in 1936. In the academic year 1904/05 König – together with several other brilliant students – attended Minkowski's lectures on Analysis Situs in Göttingen, where Minkowski attempted to solve the four color problem. The book by König did not contain coloring theory for abstract graphs – it did not yet exist. But the lectures in Budapest by König made graph theory – together with number theory – a favorite topic of his students, among them György Hajós (1912–1972), Paul Erdős (1913–1996), Paul Turán (1910–1976) and Tibor Gallai (1912–1992). After the war a gifted young Hungarian student Peter Ungar came to London, where he teamed up with another Hungarian student Gabriel Andrew Dirac (1925–1984), who had moved to England from Budapest in 1937 as a 12 year old boy with his sister, when their mother Margit Wigner (1904–2002) and the world famous physicist Paul Adrian Maurice Dirac (1902–1984) married. Ungar told the young Dirac about graph theory, and Dirac took it as a topic of his ph.d. studies, supervised by Richard Rado (1906–1989), resulting in his thesis *On the colouring of graphs* in 1951. In his studies Dirac rediscovered Brooks' Theorem, without knowing the 1941 paper. He was informed about it by the external examiner Cedric A. B. Smith, resulting in some last minute changes in the thesis. The notion of critical graphs was introduced and studied by Dirac as a key concept in the thesis. During the 1950s Dirac continued to develop the theory, and the study was also taken up by others, in particular by Paul Erdős and Tibor Gallai in Hungary and by Horst Sachs (1927–2016) in East-Germany. In 1963 Dirac spent a year with Sachs as professor at the Technical University of Ilmenau.

In the period 1966–1970 Bjarne Toft had the good fortune to study graph theory with Dirac as supervisor, resulting in his thesis *Some contributions to the theory of colour-critical graphs*. In 1969 he spent a semester in Budapest, where he met Gallai, who in weekly meetings told about his results and problems, in particular relating to Dirac's critical graphs. One such problem asked if the number of connected components in the low-vertex subgraph of a k-critical graph is at least that number for the high-vertex subgraph (where the high vertices are those of degree at least k, and the low vertices those of degree $k-1$). This problem was transferred by Toft to Sachs and his students at the Technical University in Ilmenau, East Germany, where it was solved by Michael Stiebitz and first published in 1985 in his habilitation thesis *Beiträge zur Theorie der Färbungskritischen Graphen*. In 1985 Stiebitz spent a semester in Budapest and also had conversations with Gallai about critical graphs.

We are deeply grateful for the including and open atmosphere we have always met in the graph theory community, to a large extend inspired and fostered through the openness of Erdős and the Hungarian mathematicians in general. Toft is grateful to his fellow student Ivan Tafteberg Jakobsen, who for many years was an important partner, in Denmark, England and Hungary. Also in England, Germany and Denmark we met openness in our relations with Dirac and Sachs. In England Toft got in close contact with Cedric A. B. Smith and via him met Brooks, who worked as a tax-inspector in London.

Our project to write about Brooks' Theorem started in 2015, where Stiebitz and Toft published a book-chapter about Brooks' Theorem. This was a starting point of the project - to dig deeper and create a whole monograph, and we were joined by Stiebitz's ph.d.-student Thomas Schweser. But the literature related to graph coloring has really exploded, and it is impossible, even in a lengthy monograph, to cover everything related to Brooks' Theorem. We had to be selective and have concentrated in particular on three central aspects: the various proofs of Brooks' Theorem, the various extensions of it, and similar theorems for other graph parameters. Our book presents a comprehensive overview of the development and see it in context. It describes results, both early and recent, and explains relations. We hope that it will serve as a reference to a wealth of information, now scattered in journals, proceedings and dissertations. The reader gets easy access to this information, including best known proofs of the results described.

We are grateful to Cambridge University Press for the permission to reproduce the fundamental paper of R. L. Brooks [On colouring the Nodes of a Network, *Proceedings of the Cambridge Philosophical Society* 37 (1941), pp. 194–197]. It is the content of Appendix A1.

Our thanks also go to Tommy R. Jensen, who made us aware of W. T. Tutte's lecture *Fifty Years of Graph Colouring* at the University of Waterloo, June 26, 1992, and who gave us access to Tutte's original manuscript. In this context our gratitude also goes to Daniel Younger, who acted as executor of Tutte's Will, for the permission to reproduce Tutte's manuscript as Appendix B1. In an e-mail to us in March 2023 Younger wrote: *If I have any authority in this matter, it is my judgement that Tutte would have been pleased to have this lecture-document made public, in a way such as your Monograph.*

We have benefitted from the continued support from our work places, the Mathematical Institute of the Technical University of Ilmenau and the Department of Mathematics and Computer Science and its excellent library at the University of Southern Denmark. Several other individuals provided much help and support, among them Alexander Bock, Thomas Böhme, Thomas Fischer, Florian Hörsch, Christian Klaue, and Till Preuster. And Elena Griniari, Francesca Bonadei, and Francesca Ferrari at Springer have been indispensable in their continued efficiency and support in bringing this monograph to life.

Ilmenau and Odense,
October 2023

Michael Stiebitz
Thomas Schweser
Bjarne Toft

Contents

1 Degree Bounds for the Chromatic Number 1
1.1 Coloring Preliminaries 1
1.2 Brooks' Theorem 3
1.3 List Coloring of Graphs 10
1.4 DP-Coloring of Multigraphs 15
1.5 Equitable Coloring of Graphs 27
1.6 Critical Graphs and Gallai's Theorem 32
1.7 Complete Graphs, Cycles and Trees 36
1.8 Line Graphs and Total Graphs 37
1.9 Coloring of Weighted Graphs 40
1.10 Perfect Graphs 44
1.11 Exercises 48
1.12 Notes 53

2 Degeneracy and Colorings 61
2.1 Coloring Tree-like Graphs 61
2.2 Coloring and Subtrees of Graphs 68
2.3 Coloring and Girth 71
2.4 Coloring, Cycle Lengths and DFS Trees 76
2.5 Partition into Degenerate Subgraphs 81
2.6 Generalized Graph Coloring 90
2.7 Exercises 95
2.8 Notes 100

3 Colorings and Orientations of Graphs 115
3.1 The Gallai–Roy Theorem 115
3.2 Kernel Perfect Digraphs 118
3.3 Maximum Out-Degree 121
3.4 Normal Digraphs 124
3.5 The Graph Polynomial 128
3.6 Eulerian Digraphs 135

3.7 Number of Colorings 140
3.8 Orientation vs. Brooks 145
3.9 Exercises 149
3.10 Notes 154

4 Properties of Critical Graphs 167
4.1 Construction of Critical Graphs 167
4.2 Critical Graphs of Low Connectivity 175
4.3 Decomposable Critical Graphs 181
4.4 Subgraphs of Critical Graphs 186
4.5 Independent Sets and Degrees in Critical Graphs 197
4.6 Circumference in Critical Graphs 206
4.7 Other Types of Color Critical Graphs 209
4.8 Exercises 216
4.9 Notes 221

5 Critical Graphs with few Edges 231
5.1 Preliminaries and Ore's Conjecture 231
5.2 List-Critical Graphs with Few Edges 234
5.3 Extremal Graphs whose Order is Close to χ 239
5.4 A Bound for 4-Critical Graphs 242
5.5 A Bound for k-Critical Graphs 247
5.6 Triangle-Free Graphs 261
5.7 Exercises 272
5.8 Notes 275

6 Bounding χ by Δ and ω 293
6.1 Conjectures by Borodin, Kostochka and Reed 293
6.2 Hitting Sets and Independent Transversals 295
6.3 A Weakening of Reed's Conjecture 303
6.4 Independent Sets 310
6.5 Mozhan Partitions 314
6.6 Graphs with χ Close to Δ 327
6.7 Graphs with Bounded Clique Number 334
6.8 Exercises 341
6.9 Notes 343

7 Coloring of Hypergraphs 355
7.1 Coloring Preliminaries 355
7.2 Brooks' Theorem for Hypergraphs 357
7.3 Properties of Critical Hypergraphs 360
7.4 Constructions of Critical Hypergraphs 363
7.5 Critical Hypergraphs with Low Connectivity 370
7.6 Decomposable Critical Hypergraphs 375
7.7 Critical Hypergraphs with Few Edges 380
7.8 Exercises 384

7.9 Notes ... 388

8 Homomorphisms and Colorings ... 401
8.1 Homomorphism Universal Graph Properties ... 401
8.2 Fractional Colorings and Kneser Graphs ... 404
8.3 Circular Colorings and Circular Graphs ... 413
8.4 Odd Girth and Mycielski Graphs ... 418
8.5 Odd Colorings and Gyárfás Graphs ... 426
8.6 The Rise and Fall of the Product Conjecture ... 439
8.7 Topological Lower Bounds for χ ... 452
8.8 Exercises ... 462
8.9 Notes ... 465

9 Coloring Graphs on Surfaces ... 483
9.1 Map Color Theorems for Arbitrary Surfaces ... 483
9.2 Map Color Theorems for the Plane ... 488
9.3 Map Color Theorems for the Projective Plane ... 502
9.4 Exercises ... 513
9.5 Notes ... 515

A Brooks' Fundamental Paper ... 537
A.1 On Colouring the Nodes of a Network ... 537
A.2 Rowland Leonard Brooks 1916–1993 ... 540
A.3 Brooks' Theorem and Beyond ... 542

B Tutte's Lecture from 1992 ... 545
B.1 Fifty Years of Graph Colouring ... 545
B.2 William Thomas Tutte 1917–2002 ... 550
B.3 Tutte's Flow Conjectures ... 551

C Basic Graph Theory Concepts ... 553
C.1 Sets, Mappings, and Permutations ... 553
C.2 Graphs and Subgraphs ... 554
C.3 Path, Cycles and Complete Graphs ... 555
C.4 Connectivity and Blocks ... 556
C.5 Distance, Girth and Odd Girth ... 558
C.6 Trees and Bipartite Graphs ... 560
C.7 Graph Operations and Graph Modifications ... 562
C.8 Minors and Subdivisions ... 563
C.9 Multipartite Graphs and Turán Graphs ... 564
C.10 Automorphism Group of Graphs ... 566
C.11 Graph Properties and Graph Invariants ... 567
C.12 Multigraphs and Directed Multigraphs ... 570
C.13 Graphs on Surfaces ... 573
C.14 Hypergraphs ... 582
C.15 Lovász Local Lemma ... 586

C.16 Fekete's Lemma 590

References 593

Graph and Hypergraph Parameters 633

Graph and Hypergraph Families 635

Operations and Relations 637

Other Notation 639

Index 643

Chapter 1
Degree Bounds for the Chromatic Number

Until the 1940s, coloring theory focused almost exclusively on coloring of maps. However, already in 1879 A. B. Kempe suggested coloring of abstract graphs as a possible topic. Around 1930 H. A. Whitney considered chromatic polynomials of graphs, rather than maps, in his Harvard thesis and in the resulting paper of 1932 about coloring of graphs. But it was only with the seminal paper by R. L. Brooks in 1941 that coloring of abstract graphs emerged as a topic of study in its own right. Brooks' result, which has become known as Brooks' theorem, relates the chromatic number to the maximum degree of a given graph. Over the years, graph coloring theory has developed into a rich theory and, as emphasized by B. Reed in his extensive paper in 1998 about ω, Δ, and χ, Brooks' theorem is just the tip of the iceberg.

1.1 Coloring Preliminaries

Let G be an arbitrary graph, and let C be a set. A **coloring** of G with **color set** C is a mapping $\varphi : V(G) \to C$ that assigns to each vertex v of G a **color** $\varphi(v) \in C$ such that $\varphi(u) \neq \varphi(v)$ whenever $uv \in E(G)$. If $|C| = k$ with $k \in \mathbb{N}_0$, then we also say that φ is a k**-coloring** of G. We say that G is k**-colorable** if G has a k-coloring. Obviously, G is k-colorable if and only if G has a coloring with color set $C = [1, k]$. The least integer $k \geq 0$ for which G has a k-coloring is called the **chromatic number** of G, written $\chi(G)$. A graph with chromatic number k is also said to be k**-chromatic**. So $\chi(G) \leq k$ if and only if G is k-colorable, and $\chi(G) = k$ if and only if G is k-colorable, but not $(k-1)$-colorable. Note that any coloring of G induces a coloring with the same color set of each of its subgraphs. Consequently, the chromatic number is a monotone graph parameter, that is,

$$H \subseteq G \Rightarrow \chi(H) \leq \chi(G). \tag{1.1}$$

M. Stiebitz et al., *Brooks' Theorem*, Springer Monographs in Mathematics,
https://doi.org/10.1007/978-3-031-50065-7_1

Obviously, a mapping $\varphi : V(G) \to C$ is a coloring of G with color set C if and only if, for every color $c \in C$, the preimage

$$\varphi^{-1}(c) = \{v \in V(G) \mid \varphi(v) = c\}$$

is an **independent set** of G (possibly empty), that is, no two vertices of $\varphi^{-1}(c)$ are joined by an edge of G. These preimages of a coloring are also referred to as **color classes**. So there is a one-to-one correspondence between colorings of G with a set of k colors and sequences $(I_1, I_2, \dots, I_k)$ of disjoint independent sets of G whose union is $V(G)$. The maximum cardinality of an independent set of G is the **independence number** of G, written $\alpha(G)$. Thus any color class has at most $\alpha(G)$ vertices, which implies that any coloring of G with a set of k colors satisfies $|G| \leq k\alpha(G)$ and so

$$|G| \leq \chi(G)\alpha(G). \tag{1.2}$$

Since each vertex of G forms an independent set, trivially $\chi(G) \leq |G|$. By (1.2), we then conclude that

$$\chi(G) = |G| \Leftrightarrow \alpha(G) \leq 1 \Leftrightarrow G \text{ is complete.} \tag{1.3}$$

The largest order of a complete graph contained in G as a subgraph is the **clique number** of G, written $\omega(G)$. By (1.1) and (1.3), it then follows that

$$\omega(G) \leq \chi(G). \tag{1.4}$$

The chromatic number $\chi(G)$ is the least integer k for which $V(G)$ can be partitioned into k independent sets. Consequently, $\chi(G) = 0$ if and only if $G = \varnothing$. Furthermore, $\chi(G) \leq 1$ if and only if G has no edges; and $\chi(G) \leq 2$ if and only if G is bipartite. To decide whether $\chi(G) \leq 2$, we can use König's theorem (Theorem C.7). The answer is positive if $|G| \leq 2$. Furthermore, unless $G = \varnothing$,

$$\chi(G) = \max\{\chi(H) \mid H \text{ is a component of } G\}, \tag{1.5}$$

since the components of a graph can be colored independently of each other. So we may assume that G is connected. Now we may construct a spanning tree T of G and fix a vertex r of T. Then the sets $X = \{v \in V(T) \mid \text{dist}_T(r,v) \text{ is even}\}$ and $Y = \{v \in V(T) \mid \text{dist}_T(r,v) \text{ is odd}\}$ form a bipartition (X,Y) of T. If this is also a bipartition of G, we obtain $\chi(G) \leq 2$. Otherwise, two vertices v and w of the same part are joined by an edge of G, and this edge together with the path vTw forms an odd cycle C contained in G, implying that $\chi(G) \geq \chi(C) = 3$. This shows that there is a polynomial time algorithm to decide whether a given graph is 2-colorable, that is, this decision problem belongs to the complexity class P. However, as proved by Stockmeyer [981], for every fixed integer $k \geq 3$, the decision problem whether a given graph is k-colorable is NP-complete. So the determination of the chromatic number is an NP-hard optimization problem. Note that NP-complete problems form a class of decision problems, no member of which is known to have a polynomial time algorithm, but such that if any one of the problems does have such an algorithm, then they all do.

König's theorem implies, in particular, that a graph is 2-colorable or it contains a cycle C whose length $\ell(C)$ is an odd number. In 1982 Neumann-Lara [780, Theorem 7] proved the following generalization of this result.

Theorem 1.1 *Let G be a graph, let $k \geq 2$ be an integer, and let $p \in [2,k]$. Then $\chi(G) \leq k$ or G contains at least as many cycles of length $1 \pmod p$ as an edge of K_{k+1} is contained in a cycle of length $1+p$, i.e., the number of cycles of length $1 \pmod p$ contained in G is at least $(k-1)(k-2)\cdots(k-p+1)$.*

Proof Assume $\chi(G) \geq k+1$. Since G is finite, G contains a minimal subgraph G' having chromatic number at least $k+1$. Then $\chi(G') \geq k+1$ and $\chi(G'-e) \leq k$ for every edge e of G'. Obviously, G' contains an edge $e=vv'$, since $k \geq 2$ and so $\chi(G') \geq 3$. Then there exists a coloring φ of $G'-e$ with color set $C=[1,k]$. Since $\chi(G') \geq k+1$, we obtain that $\varphi(v)=\varphi(v')$. Consider a cyclic permutation π of p colors of C, where $\pi = (\varphi(v), \pi(\varphi(v)), \pi^2(\varphi(v)), \ldots, \pi^{p-1}(\varphi(v)))$ and $\pi^p(\varphi(v)) = \varphi(v)$. Let $N_0 = \{v\}$ and, for $i \geq 1$, let N_i be the set of vertices of $V(G') \setminus (N_0 \cup N_1 \cup \cdots \cup N_{i-1})$ having color $\pi^i(\varphi(v))$ and being neighbors in $G'-e$ of vertices belonging to N_{i-1}. Changing the color of all vertices u in $N = N_0 \cup N_1 \cup N_2 \cup \ldots$ from $\varphi(u)$ to $\pi(\varphi(u))$ results in a new coloring φ' of $G'-e$ with color set $C=[1,k]$. Since $\chi(G) \geq k+1$, it follows that $\varphi'(v) = \varphi'(v')$. Since $p \geq 2$, $\varphi'(v') \neq \varphi(v')$. Thus $v' \in N$, implying that $G'-e$ contains a v-v' path of length $0 \pmod p$. Adding vv' gives a cycle of length $1 \pmod p$. Since π may be chosen in as many different ways as an edge of K_{k+1} is contained in a cycle of length $1+p$, the theorem follows. □

1.2 Brooks' Theorem

In modern terminology Brooks' theorem, published in his influential paper [173] in 1941 and reprinted here in Appendix A.1, is as follows:

Let G be a graph with maximum degree Δ, where $\Delta > 2$. Suppose that no component of G is a complete graph $K_{\Delta+1}$. Then it is possible to color the vertices of G with Δ colors so that no two vertices of the same color are joined by an edge in G, and hence G has chromatic number at most Δ.

Brooks observed that it suffices to prove the result for connected graphs (by (1.5)). He also noticed that a coloring of a graph G can be obtained by giving to each vertex in turn a color different from all those already assigned to vertices to which it is adjacent. This sequential coloring algorithm obviously leads to a $(\Delta+1)$-coloring of G where Δ is the maximum degree of G. Consequently, any graph G satisfies

$$\chi(G) \leq \Delta(G) + 1. \tag{1.6}$$

The three missing cases, where $\Delta = 0, 1, 2$, can easily be included in Brooks' theorem. A connected graph with maximum degree $\Delta = 0$ or 1 is a $K_{\Delta+1}$ and has therefore chromatic number $\Delta+1$. A connected graph G with maximum degree $\Delta = 2$ is a path or cycle with at least one edge, which yields $\chi(G) = 2$, unless G is an odd cycle and so $\chi(G) = 3$. So another possible way of formulating Brooks' fundamental result is as follows.

Theorem 1.2 (BROOKS' THEOREM) *Let G be a connected graph of maximum degree Δ. Then $\chi(G) \leq \Delta+1$, where equality holds if and only if either $\Delta = 2$ and G is an odd cycle, or $\Delta \neq 2$ and $G = K_{\Delta+1}$.*

To prove Brooks' result it suffices to show that every connected graph G with maximum degree $\Delta \geq 3$ admits a Δ-coloring, provided that G is different from $K_{\Delta+1}$. Apart from Brooks' original proof, which is reprinted in Appendix A.1, many different proofs of this fundamental result have been presented over the years in the graph theory literature. In principle there are four basic approaches for proving Brooks' theorem.

A sequential coloring algorithm

The most popular proof of Brooks' result is due to Lovász [694]. Lovász's proof from 1975 provides an efficient algorithm that produces a Δ-coloring of a connected graph G with maximum degree $\Delta \geq 3$, provided that G is different from $K_{\Delta+1}$. To find such a coloring of G we use the following **sequential coloring algorithm**:

Starting from a fixed vertex order $\ell = (v_1, v_2, \dots, v_n)$ of G, we consider the vertices in turn and color each vertex v_i with the smallest positive integer not already used to color any neighbor of v_i among $v_1, v_2, \dots, v_{i-1}$.

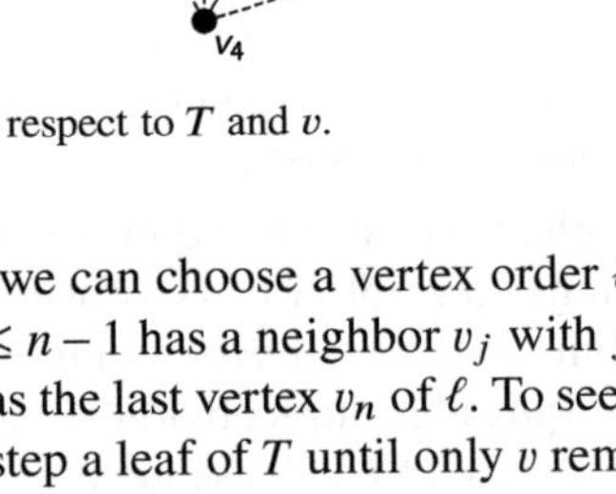

Fig. 1.1 A good labeling with respect to T and v.

Since G is connected, we can choose a vertex order $\ell = (v_1, v_2, \dots, v_n)$ such that each vertex v_i with $1 \leq i \leq n-1$ has a neighbor v_j with $j > i$. Furthermore, we may choose any vertex v of G as the last vertex v_n of ℓ. To see this, take a spanning tree T of G and remove step by step a leaf of T until only v remains (see Figure 1.1). If we apply the sequential coloring algorithm to this vertex order ℓ, then for each vertex v_i with $1 \leq i \leq n-1$ we use a color from the color set $C = [1, \Delta]$, since v_i has at most $\Delta - 1$ neighbors among $v_1, v_2, \dots, v_{i-1}$. This shows, in particular, that $G - v$ admits a coloring with color set C. This coloring can be extended to a coloring of G without using a new color if v has at most $\Delta - 1$ neighbors, or if v has two neighbors colored with the same color. So we are done if $\delta(G) \leq \Delta - 1$. We are also done if G has a separating vertex v. Then G is the union of two connected graphs having only vertex v in common and v has degree at most $\Delta - 1$ in these two graphs. Hence both graphs can be colored with colors from the set C in such a way that v receives the same color (permute the colors in one graph if necessary), thus yielding a coloring of G with color set C. It remains to consider the case when G is Δ-regular and 2-connected. Since G is not a complete graph and $\Delta \geq 3$, there are three vertices u, v and w for which $uv, vw \in E(G)$, $uw \notin E(G)$, and $G - u - w$ is connected. If G is 3-connected, this is obvious. If G is not 3-connected, then let v belong to a separating set of two vertices. Then $G - v$ is connected, but not 2-connected, so there are two end-blocks B_1 and B_2 of $G - v$. For $i = 1, 2$, let u_i be the only separating vertex of $G - v$ contained in B_i. Since G is 2-connected, v has a neighbor u in $B_1 - u_1$ as well as a neighbor w in $B_2 - u_2$. Clearly, $uw \notin E(G)$ and, since $d_G(v) \geq 3$, $G - u - w$ is connected. We can thus choose the vertex order ℓ in such a way that $v_1 = u$, $v_2 = w$ and $v_n = v$. The sequential coloring algorithm applied to this ordering ℓ then assigns color 1 to both u and w, and hence it terminates with a coloring of G with color set C. This completes the proof of Brooks' theorem.

A Kempe-change argument

The Kempe-change argument was introduced by Kempe [568] in his attempt to prove the four color theorem. Let G be a graph and let φ be a coloring of G with color set C. For two distinct colors $c, c' \in C$, a component of the graph $G[\varphi^{-1}(c, c')]$ is called a **(c, c')-chain** of G with respect to φ. Now choose a (c, c')-chain H of G with respect to φ. If we interchange the colors c and c' on the vertices of H, then we obtain a new coloring φ' of G with color set C. We say that φ' is obtained from φ by **recoloring** H, and we write $\varphi' = \varphi/H$. This operation is often called a **Kempe-change**.

The first proof of Brooks' theorem by means of Kempe-changes was presented in 1969 by Melnikov and Vizing [731]. Their proof uses induction on the number of vertices. However, one may also use induction on the number of edges. So let G be a connected graph with maximum degree at most Δ ($\Delta \geq 3$), and suppose that G is not a complete graph $K_{\Delta+1}$. We prove by induction on the number m of edges of G that G has a coloring with color set $C = [1, \Delta]$. This is evident if $m = 0$, so assume that $m \geq 1$. Let $e = vw$ be an arbitrary edge of G. Then $G' = G - e$ has maximum degree at most Δ and no component of G' is a $K_{\Delta+1}$. The induction hypothesis then implies that we can color each component of G' with color set C, thus yielding a coloring of G' with color set C. Let $\mathcal{C}$ be the set of all colorings of G' with color set C. If there is a coloring $\varphi \in \mathcal{C}$ with $\varphi(v) \neq \varphi(w)$, then φ is also a coloring of G and we are done. So assume that $\varphi(v) = \varphi(w)$ for all $\varphi \in \mathcal{C}$. For a coloring $\varphi \in \mathcal{C}$, let $c_\varphi = \varphi(v) (= \varphi(w))$ and $C_\varphi = C \setminus \{c_\varphi\}$. Furthermore, for a color $c \in C_\varphi$, let $H_{\varphi,c}$ be the (c_φ, c)-chain of G' with respect to φ containing the vertex v. The graph $H_{\varphi,c}$ contains the vertex w for all colors $c \in C_\varphi$; for otherwise, $\varphi' = \varphi/H_{\varphi,c}$ belongs to $\mathcal{C}$ and $\varphi'(v) \neq \varphi'(w)$, contradicting the assumption. Since $|C_\varphi| = \Delta - 1$ and $d_{G'}(u) \leq \Delta - 1$ for $u \in \{v, w\}$, this implies that both vertices v and w have degree 1 in $H_{\varphi,c}$ for all $c \in C_\varphi$. Next we claim that $H_{\varphi,c}$ is a v-w path of G' for all $c \in C_\varphi$. Indeed, there is a path P from v to w in $H_{\varphi,c}$. If $H_{\varphi,c} \neq P$, then P has an inner vertex with three identically colored neighbors in $H_{\varphi,c}$; let u be the first such vertex on P (see Figure 1.2). Then at most $\Delta - 2$ colors are used on the neighbors of u in G' and we may recolor u. This results in a coloring $\varphi' \in C$ with $c_{\varphi'} = c_\varphi$ such that w does not belong to $H_{\varphi',c}$, giving a contradiction. Next, we claim that, for any two distinct colors c, c' in C_φ, the v-w paths $H_{\varphi,c}$ and $H_{\varphi,c'}$ are internally disjoint. Otherwise there is a vertex u in $H_{\varphi,c} \cap H_{\varphi,c'}$ with $u \notin \{v, w\}$. Then u has two neighbors colored c and two neighbors colored c'. Hence $\Delta \geq 4$ and we may recolor u with a color from $C \setminus \{c_\varphi, c, c'\}$. The new coloring φ' belongs to C and $c_{\varphi'} = c_\varphi$, but w does not belong to $H_{\varphi',c}$, giving a contradiction.

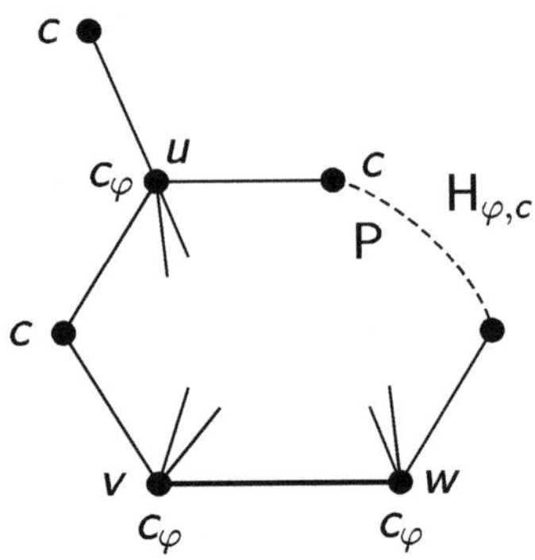

Fig. 1.2 The Kempe chain $H_{\varphi,c}$ is a v-w path.

Finally, we claim that, for all colors $c \in C_\varphi$, the v-w path $H_{\varphi,c}$ has length two. For otherwise, there is a color $c \in C_\varphi$, such that $H_{\varphi,c} = (v, v', \ldots, u, w)$ with $v' \neq u$. Clearly, $\varphi(v') = \varphi(u) = c$. Since $\Delta \geq 3$, there is a color $c' \in C_\varphi \setminus \{c\}$. Now, recolor $H_{\varphi,c'}$ by swapping the colors c_φ and c' on $H_{\varphi,c'} - v$ and recoloring v by c. This results in a coloring φ' of $G - vv'$ with $\varphi'(v) = \varphi'(v') = c$. Moreover, u belongs to the (c, c_φ)-chain from v', and u also belongs to the (c, c')-chain from v. This contradicts the two chains being internally disjoint, which follows from above, replacing vw by vv'. This proves the claim that the v-w path $H_{\varphi,c}$ has length two for all colors $c \in C_\varphi$. As a consequence we obtain that v and w have $\Delta - 1$ common neighbors in G. Since vw is an arbitrary edge of G, we then conclude that $G = K_{\Delta+1}$, contradicting the hypothesis. This completes the proof.

Gallai trees and independent sets

Brooks' theorem (Theorem 1.2) has the following corollary.

(∗) *If G is a connected graph of maximum degree Δ and G contains no $K_{\Delta+1}$ (and no odd cycle when $\Delta = 2$), then G contains an independent set I such that $G - I$ contains no K_Δ (and no odd cycle when $\Delta = 3$).*

On the other hand, it is easy to see that (∗) implies Theorem 1.2. Let I^* be a maximal independent set containing I. Then each component of $G - I^*$ satisfies the condition of Brooks' theorem with Δ replaced by $\Delta - 1$, and hence Theorem 1.2 follows by induction on the maximum degree. The statement (∗) may thus be seen as another formulation of Brooks' theorem.

In fact, the independent set I in (∗) may be assumed not only to be maximal, but maximum. Let namely, $\tilde{I}$ be a maximum independent set of G with largest possible intersection with the independent set I given in (∗). Suppose that $G - \tilde{I}$ contains a K_Δ (or an odd cycle when $\Delta = 3$). Then there is a vertex v of the K_Δ (or the odd cycle when $\Delta = 3$) contained in I, and v is joined to $\Delta - 1$ vertices out of I and $\tilde{I}$. Since $\tilde{I}$ is maximum, v is joined also to a vertex w of $\tilde{I}$, and therefore v has degree Δ in G and the vertex w is unique. Then $(\tilde{I} \setminus \{w\}) \cup \{v\}$ is a maximum independent set of G with a larger intersection with I than $\tilde{I}$. This contradiction shows that the independent set I in (∗) may be chosen to be maximum. Catlin [195] and independently Gerencsér [413] first proved that if G is a connected graph with maximum degree Δ with $\Delta \geq 1$, which is neither a complete graph nor an odd cycle, then G has a Δ-coloring in which one color class is a maximum independent set. This result motivated Tverberg [1043] to give a proof of Brooks' theorem by describing directly how to obtain the independent set I in (∗). First we introduce some notation.

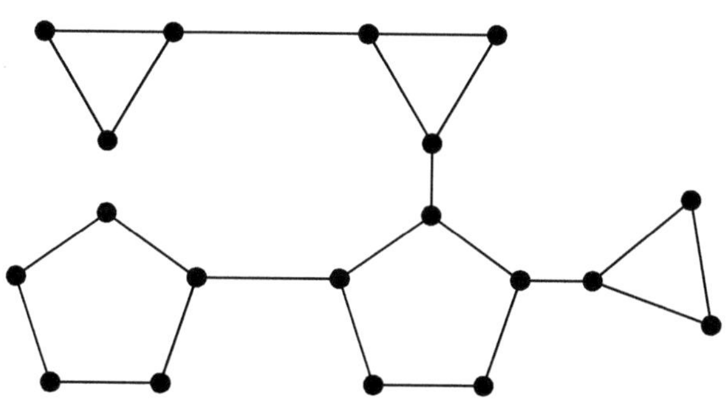

Fig. 1.3 An ε_3-graph.

A nonempty graph is called a **brick** if it is a complete graph or an odd cycle. If B is a brick, then B is regular and we say that B is a k-**brick** if B is $(k-1)$-regular. So if B is a k-brick, then $k \geq 1$ and $B = K_k$ or $k = 3$ and $B = C_{2q+1}$ for some $q \geq 1$. A graph all of whose blocks are bricks is called a **Gallai tree** when it is connected and a **Gallai forest** in general. So any component of a Gallai forest is a Gallai tree. Gallai trees occur first in a paper by Gallai [397] about the structure of critical graphs. Following Gallai, an ε_k-**graph** with $k \geq 2$ is a Gallai tree defined recursively as follows: Every k-brick is an ε_k-graph; and every graph obtained from an ε_k-graph G and a vertex disjoint k-brick B by adding an edge between a vertex v of G with $d_G(v) = k - 1$ and a vertex of B is an ε_k-graph (see Figure 1.3). Tverberg's proof of Brooks' result is based on the following lemma:

Let G be a connected graph with maximum degree Δ, where $\Delta \geq 2$. If G is neither a $(\Delta+1)$-brick nor an ε_Δ-graph, then there is a vertex v of degree Δ in G such that no component of $G-v$ is an ε_Δ-graph.

Before we prove this lemma, let us see how it implies Brooks' theorem. So let G be a connected graph with maximum degree Δ, and suppose that G is not a (Δ+1)-brick. We prove by induction on Δ that G has a Δ-coloring. This is obvious if $\Delta \leq 2$, so assume that $\Delta \geq 3$. If G is an ε_Δ-graph, then we easily find a Δ-coloring of G. If G is not an ε_Δ-graph, we can apply the lemma. So we can delete step by step vertices $v_1, v_2, \ldots, v_p$ such that v_i has degree Δ in $G_i = G - \{v_1, v_2, \ldots, v_{i-1}\}$ ($i = 1 \ldots, p$), no component of G_i is an ε_Δ-graph, and G_{p+1} has maximum degree at most $\Delta - 1$. Consequently, no component of G_{p+1} is a Δ-brick. The induction hypothesis then implies that each component of G_{p+1} has a $(\Delta-1)$-coloring. Thus G_{p+1} has a $(\Delta-1)$-coloring and this coloring can be extended to a Δ-coloring of G, since $X = \{v_1, v_2, \ldots, v_p\}$ is an independent set of G.

So it remains to prove Tverberg's lemma. The lemma is evident if G contains no Δ-brick. So assume that G contains a Δ-brick. Then it also contains an induced Δ-brick. For an induced Δ-brick B, let D_B be the set of vertices of B having degree Δ in G. Assume that there is an induced Δ-brick B for which D_B consists of exactly one vertex v. Then no component of $G-v$ is an ε_Δ-graph, since otherwise G would be an ε_Δ-graph, contradicting the hypothesis. So it remains to consider the case when $|D_B| \geq 2$ for each induced Δ-brick B of G. If there is an induced Δ-brick B with $V(B) \setminus D_B \neq \varnothing$, then we may choose a vertex $u \in V(B) \setminus D_B$ and a neighbor v of u in B belonging to D_B. If one component of $G-v$ would be an ε_Δ-graph, then u would not belong to that component and we would find an induced Δ-brick B' of G such that $|D_{B'}| = 1$, giving a contradiction. If each induced Δ-brick B of G satisfies $V(B) = D_B$, then let v be any vertex of degree Δ in G. If a component H of $G-v$ is a Δ-brick, then each vertex of H must be adjacent to v, which yields a (Δ+1)-brick, contradicting the hypothesis. If a component H of $G-v$ is a larger ε_Δ-graph, then H has at least two Δ-bricks containing at most one vertex of degree Δ in H, so at least $2(\Delta-1)$ of its vertices are adjacent to v in G. But $2(\Delta-1) > \Delta$ for $\Delta \geq 3$. For $\Delta = 2$ the larger ε_2-graph is an odd path and hence G is an odd cycle, that is, a 3-brick. This contradiction completes the proof of the lemma.

Precoloring extension

The following proof by Rabern [839] from 2013 is similar to Tverberg's proof from 1981. Rabern's proof is also based on a reduction, but only to cubic graphs and not to 2-regular graphs. This makes the reduction much simpler. However, the basic case becomes more complicated; cubic graphs are treated by applying a precoloring extension argument.

For $\Delta \geq 3$, we prove that if G is a graph with maximum degree at most Δ and with $\omega(G) \leq \Delta$, then $\chi(G) \leq \Delta$. Suppose this is false and let G be a counterexample whose order is minimum. Then $\chi(G) \geq \Delta + 1$, and the minimality of G implies that $\chi(G-v) \leq \Delta$ for all $v \in V(G)$. From this we deduce that $\delta(G) \geq \Delta$ (for otherwise, if $d_G(v) \le \Delta - 1$, then a Δ-coloring of $G - v$ can be extended to a Δ-coloring of G, giving a contradiction), and so G is Δ-regular.

Let v be an arbitrary vertex of G. Then $G - v$ has a Δ-coloring φ. Since $d_G(v) = \Delta \ge 3$ and $\omega(G) \le \Delta < \chi(G)$, the Δ neighbours of v in G belong to different color classes in φ and two of these neighbours are not adjacent in G. Furthermore, each of the Δ colors must be used on each K_Δ contained in $G - v$. As a consequence we obtain that there is a color class M in φ which shares a vertex with each K_Δ contained in G. Let I be a maximal independent set of G containing M, and let $H = G - I$. Then $\chi(H) \ge \chi(G) - 1 \ge \Delta$, $\omega(H) \le \Delta - 1$, and $\Delta(H) \le \Delta - 1$, reducing the proof to the case where Δ is one less. By repetition, the proof is reduced to the case $\Delta = 3$.

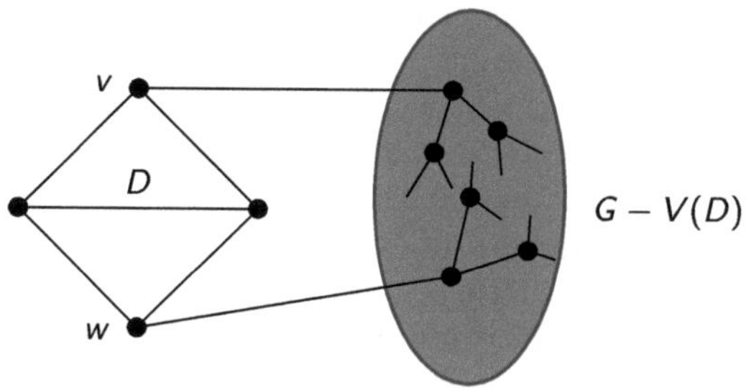

Fig. 1.4 Extending a 3-coloring of $G - V(D)$.

Suppose therefore that $\Delta = 3$. If G contains a **diamond** D, that is, a K_4 less an edge vw, then $G' = G - D$ has a 3-coloring. Since both v and w have only one neighbor in G', we can extend this 3-coloring by first coloring v and w the same and then coloring the two adjacent vertices of D (see Figure 1.4). This results in a 3-coloring of G, giving a contradiction. Now assume that G contains no diamond. Since G is 3-regular, there is an induced cycle $C \subseteq G$ (take a shortest cycle of G). Since G contains no diamond and no K_4, we deduce that the set $N = N_G(C) \setminus V(C)$ has at least two vertices, say x and y. Then $H = G - C + xy$ has maximum degree at most 3 and $K_4 \nsubseteq H$. Thus, because of the minimality of G, there is a coloring φ of H with color set $\{1,2,3\}$. For a vertex v of C, let $L(v) = \{1,2,3\} \setminus \{\varphi(w)\}$, where w is the only neighbor of v in H. Clearly, $|L(v)| = 2$ for all $v \in V(C)$ and, since $\varphi(x) \ne \varphi(y)$, we have $L(v) \ne L(v')$ for two distinct vertices $v, v' \in V(C)$, where we may assume that vv' is an edge of C and that there is a color $c \in L(v) \setminus L(v')$. Let $C = (v_1, v_2, \ldots, v_p, v_1)$ with $v_1 = v$ and $v_p = v'$. First we color v_1 with c, and then, for $i \in [2, p]$, we color v_i with a color $c_i \in L(v_i)$ different from the color of v_{i-1}. This results in a 3-coloring φ' of C such that $\varphi'(v) \in L(v)$ for all $v \in V(C)$, and so $\varphi \cup \varphi'$ is a 3-coloring of G. This contradiction completes the proof.

1.3 List Coloring of Graphs

Brooks' theorem tells us that Δ colors suffices to color all vertices of a connected graph with maximum degree Δ, unless the graph is a brick, that is, a complete graph or an odd cycle. However, if we assign an individual list $L(v)$ of colors to each vertex v of G and require to choose the color of v from its list $L(v)$, then $d(v)$ colors in each list $L(v)$ suffices to color all vertices of G, unless G is a Gallai tree. This list version of Brooks' result was proved by Vizing [1051] and, independently, by Erdős, Rubin, and Taylor [355].

Let G be a graph, let $f : V(G) \to \mathbb{N}_0$ be a function, and let $k \geq 0$ be an integer. A **list-assignment** L of G is a map that assigns to every vertex v of G a set (list) $L(v)$ of colors. We say that L is an f**-assignment** if $|L(v)| = f(v)$ for all $v \in V$, and a k**-assignment** if $|L(v)| = k$ for all $v \in V$, respectively. An L**-coloring** of G is a coloring φ of G with color set $C = \bigcup_{v \in V(G)} L(v)$ such that $\varphi(v) \in L(v)$ for all $v \in V$. If G admits an L-coloring, then G is said to be L**-colorable** or **list-colorable** if it is clear to which list-assignment we refer. The graph G is defined to be f**-list-colorable** or f**-choosable** if G is L-colorable for every f-assignment L of G. When $f(v) = k$ for all $v \in V$, the corresponding term becomes k**-list-colorable** or k**-choosable**. The **list chromatic number** or **choice number** of G, denoted by $\chi_\ell(G)$, is the least number $k \geq 0$ for which G is k-list-colorable.

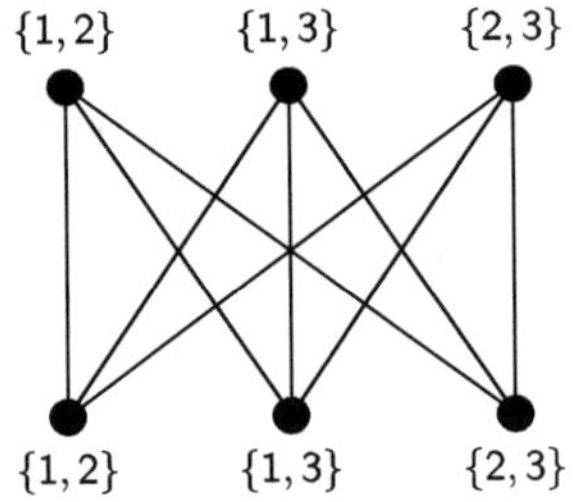

Fig. 1.5 A 2-assignment L for $K_{3,3}$ with no L-coloring.

The list coloring concept was introduced by Vizing [1051] and, independently, by Erdős, Rubin, and Taylor [355]. Note that an ordinary coloring of G with color set C may be regarded as a list coloring for the constant list-assignment L, where $L(v) = C$ for all $v \in V(G)$. Also, every k-list-colorable graph is k-colorable and, therefore, every graph G satisfies $\chi(G) \leq \chi_\ell(G)$. At the first glance it may seem surprising that the converse statement is not true. The more different the color lists are the easier it should be to find a list coloring. However, this intuition is wrong. Even for bipartite graphs, i.e., graphs with $\chi \leq 2$, the list chromatic number can be arbitrarily large. This was observed by Vizing [1051] as well as by Erdős, Rubin, and Taylor [355]. To construct a bipartite graph with large list chromatic number take the complete bipartite graph $G = K_{m,m}$ with $m = \binom{2r-1}{r}$ for some integer $r \geq 2$ and take the color set $C = [1, 2r-1]$. Then there is a list-assignment L of G such that each r-subset of C is assigned to exactly one vertex in each part of the bipartition (A, B)

of G, see Figure 1.5. If G admits an L-coloring φ, then $X = \{\varphi(a) \mid a \in A\}$ and $Y = \{\varphi(b) \mid b \in B\}$ are disjoint subsets of C, implying that one of the two sets, say X, has at most $r-1$ elements. But then there is a vertex $a \in A$ such that $L(a) \cap X = \varnothing$, contradicting $\varphi(a) \in L(a)$. So G admits no L-coloring and hence $\chi_\ell(G) \geq r+1$.

For a graph G of order $n \geq 1$, we can apply the sequential coloring algorithm to some vertex order $\ell = (v_1, v_2, \dots, v_n)$ of G. So we assign to each vertex v_i a color, if any, from a list $L(v_i)$, where the color is different from all those colors already assigned to the neighbors of v_i among $v_1, v_2, \dots, v_{i-1}$. This procedure obviously terminates with an L-coloring of G, provided that

$$|L(v_i)| \geq d_{G[v_1, v_2, \dots, v_i]}(v_i) + 1 \text{ for all } i \in [1, n].$$

Consequently, G is f-list-colorable, when $f(v) = d_G(v) + 1$ for all $v \in V(G)$. The least integer $p \geq 1$ for which there exists a vertex order $\ell = (v_1, v_2, \dots, v_n)$ of G satisfying $d_{G[v_1, v_2, \dots, v_i]}(v_i) + 1 \leq p$ for all $i \in [1, n]$ is called the **coloring number** of G, written $\operatorname{col}(G)$. The coloring number is an upper bound for the list chromatic number. Observe that $\operatorname{col}(G) \leq \Delta(G) + 1$, where equality holds if and only if G has a Δ-regular component (see Exercise 1.16). Summarizing, we obtain the following (using also (1.4)).

Proposition 1.3 *Every graph G is f-list-colorable, where $f(v) = d_G(v) + 1$ for all $v \in V(G)$. Furthermore, every graph G satisfies*

$$\omega(G) \leq \chi(G) \leq \chi_\ell(G) \leq \operatorname{col}(G) \leq \Delta(G) + 1. \tag{1.7}$$

Remark 1.4 For convenience, we allow a graph to be empty, that is, $V(G) = E(G) = \varnothing$, briefly $G = \varnothing$. By definition, we put $\alpha(\varnothing) = \omega(\varnothing) = \chi(\varnothing) = \chi_\ell(\varnothing) = \Delta(\varnothing) = \delta(\varnothing) = 0$ and $\operatorname{col}(\varnothing) = 1$.

The coloring number was first defined and investigated by Erdős and Hajnal [347] in 1966. Two years later Szekeres and Wilf [985] established another upper bound for the chromatic number that turned out to be equal to the coloring number, namely 1 plus the maximum minimum degree of the subgraphs of G. That the coloring number can be computed by a polynomial time algorithm was observed independently by various researchers including Finck and Sachs [372], Matula [725], and possibly others.

Let G be a nonempty graph of order n. Then there is a vertex order $\ell = (v_1, v_2, \dots, v_n)$ of G such that $d_{G[v_1, v_2, \dots, v_i]}(v_i) + 1 \leq \operatorname{col}(G)$ for all $i \in [1, n]$. Let H' be a nonempty subgraph of G with largest minimum degree, and let v_i be the vertex of H' with largest index. Then $H' \subseteq G[v_1, v_2, \dots, v_i]$ which yields

$$\max_{\varnothing \neq H \subseteq G} \delta(H) + 1 = \delta(H') + 1 \leq d_{G[v_1, v_2, \dots, v_i]}(v_i) + 1 \leq \operatorname{col}(G).$$

A **smallest last order** $v_n, v_{n-1}, \dots, v_1$ of the vertices of G is obtained by letting v_i be a vertex of minimum degree in the subgraph $G_i = G - \{v_{i+1}, v_{i+2}, \dots, v_n\}$ for $i = n, n-1, \dots, 1$, where $G_n = G$. It follows that

$$\mathrm{col}(G) \leq \max_{1\leq i\leq n} d_{G[v_1,v_2,\ldots,v_i]}(v_i)+1 = \max_{1\leq i\leq n} \delta(G_i)+1 \leq \max_{\varnothing\neq H\subseteq G} \delta(H)+1.$$

Hence to obtain an order $(v_1, v_2, \ldots, v_n)$ of G that achieves the coloring number of G, we first choose v_n with $d_G(v_n) = \delta(G)$, then choose as v_{n-1} a vertex of minimum degree in $G - v_n$, and so on. Summarizing we have obtained the following.

Proposition 1.5 *Every nonempty graph G of order n satisfies*

$$\mathrm{col}(G) = \max_{\varnothing\neq H\subseteq G} \delta(H)+1 = \max_{1\leq i\leq n} d_{G[v_1,v_2,\ldots,v_i]}(v_i)+1,$$

where $v_n, v_{n-1}, \ldots, v_1$ is a smallest last order of G. Furthermore,

$$\mathrm{col}(G) = \min_{\pi\in S_n} d_{G[v_{\pi(1)},v_{\pi(2)},\ldots,v_{\pi(i)}]}(v_{\pi(i)})+1.$$

A natural upper bound for the coloring number of a graph G is its **maximum average degree** $\mathrm{mad}(G)$. If G is nonempty, then

$$\mathrm{mad}(G) = \max_{\varnothing\neq H\subseteq G} \frac{2|E(H)|}{|V(H)|}$$

else $\mathrm{mad}(G) = 0$. By Proposition 1.5, $\mathrm{col}(G) \leq \lfloor \mathrm{mad}(G) \rfloor + 1 \leq \Delta(G)+1$ for each graph G. It is also easy to see that every graph G satisfies $\mathrm{mad}(G) \leq 2\mathrm{col}(G) - 2$ (see Exercise 1.17). In Section 3.3 we explain how $\mathrm{mad}(G)$ can be computed by a polynomial time algorithm. Several results about colorings were originally proved using $\mathrm{mad}(G)$. The most famous example is Heawood's bound for the chromatic number of graphs embedded in a surface (see Theorem C.21)

In 1956 Nordhaus and Gaddum [790] proved that a graph G and its complement $\overline{G}$ satisfy $\chi(G)+\chi(\overline{G}) \leq n+1$, where $n = |G|$. As proved by Finck and Sachs [372], this inequality can be strengthened considerably.

Proposition 1.6 *Every graph G of order $n \geq 1$ satisfies* $\mathrm{col}(G)+\mathrm{col}(\overline{G}) \leq n+1$.

Proof Let H be a nonempty subgraph of G with largest minimum degree, and let H' be a nonempty subgraph of $\overline{G}$ with largest minimum degree. If H and H' are vertex disjoint, then Proposition 1.5 implies

$$\mathrm{col}(G)+\mathrm{col}(\overline{G}) = \delta(H)+\delta(H')+2 \leq (|H|-1)+(|H'|-1)+2 \leq n.$$

If H and H' share a vertex v, then Proposition 1.5 implies

$$\mathrm{col}(G)+\mathrm{col}(\overline{G}) = \delta(H)+\delta(H')+2 \leq d_H(v)+d_{H'}(v)+2 \leq n+1.$$

Thus the proof is complete. □

As noticed above, the sequential coloring argument shows that every nonempty graph G is f-list-colorable if $f(v) = d_G(v)+1$ for all $v \in V(G)$. In general, though, this is rather generous. We call (G, L) an **uncolorable pair** if G is a connected graph, L is a list-assignment of G such that $|L(v)| \geq d_G(v)$ for all $v \in V(G)$, and G is not L-colorable.

To characterize all uncolorable pairs, we shall use the following reduction. Let (G, L) be an uncolorable pair with $|G| \geq 2$, let v be a nonseparating vertex of G, and let $c \in L(v)$ be a color. Furthermore, let $G' = G - v$ and, for $u \in V(G')$, let $L'(u) = L(u) \setminus \{c\}$ if $uv \in E(G)$ and $L'(u) = L(u)$ otherwise. Then (G', L') is an uncolorable pair and we write $(G', L') = (G, L)/(v, c)$. The following two theorems were obtained, independently, by Borodin [148, 149] and by Erdős, Rubin, and Taylor [355].

Theorem 1.7 *If (G, L) is an uncolorable pair, then the following statements hold:*

(a) $|L(v)| = d_G(v)$ *for all* $v \in V(G)$.
(b) *If G contains no separating vertex, then the list $L(v)$ is the same for each vertex v of G and G is regular.*
(c) *G is a Gallai tree.*

Proof The proof is by induction on the order n of G. The statements are evident if $n = 1$, so assume that $n \geq 2$. In order to prove (a), let v be an arbitrary vertex of G. Since G is connected and $n \geq 2$, there is a nonseparating vertex $u \neq v$ in G and $L(u) \neq \varnothing$. If $c \in L(u)$ is an arbitrary color, then $(G', L') = (G, L)/(u, c)$ is an uncolorable pair and the induction hypothesis implies that $d_{G'}(v) = |L'(v)|$ and hence $d_G(v) = |L(v)|$.

In order to prove (b), suppose this is false, that is, G contains no separating vertex, but there are two vertices v and w with $L(v) \neq L(w)$. Since G is connected, we may assume that $vw \in E(G)$. Furthermore, we may assume that there is a color $c \in L(v) \setminus L(w)$. However, then the uncolorable pair $(G', L') = (G, L)/(v, c)$ satisfies $d_{G'}(w) < |L'(w)|$, contradicting (a). Hence $L(v)$ is the same set for all $v \in V(G)$, and so G is regular (by (a)).

In order to prove (c), we consider two cases. First, assume that G contains a separating vertex v. Then G contains at least two end-blocks B_1 and B_2. The end-block B_i contains a vertex v_i, which is a nonseparating vertex of G. The induction hypothesis applied to the uncolorable pair $(G_i, L_i) = (G, L)/(v_i, c_i)$ then implies that $G_i = G - v_i$ is a Gallai tree (for $i = 1, 2$). Since each block $B \neq B_i$ of G is a block of $G - v_i$, this shows that G is a Gallai tree. It remains to consider the case that G contains no separating vertex, that is, G itself is a block. Then (b) implies that G is r-regular with $r \geq 1$, and there is a set C of r colors such that $L(v) = C$ for all $v \in V(G)$. Let v be an arbitrary vertex of G. The induction hypothesis applied to $(G', L') = (G, L)/(v, c)$ implies that $G - v$ is a Gallai tree, that is, each block of $G - v$ is a complete graph or an odd cycle. If $G' = G - v$ is a block, then it follows from (b) that G' is $(r-1)$-regular, which yields that both G' and G are complete graphs. If $G' = G - v$ is not a block, then G' has at least two end-blocks and every end-block must be $(r-1)$-regular, which yields $2(r-1) \leq d_G(v) = r$ and hence $r = 2$. Consequently, G is a cycle and the same two colors are available at each vertex. Since G is not L-colorable, this implies that G is an odd cycle. □

Borodin as well as Erdős, Rubin, and Taylor also provided a characterization of the lists that occur in uncolorable pairs. Let G be a graph. Let $\mathfrak{B}(G)$ be the set of blocks of G, and, for $v \in V(G)$, let $\mathfrak{B}_v(G) = \{B \in \mathfrak{B}(G) \mid v \in V(B)\}$. Define $\Lambda(G)$ to be the set of all mappings λ that assign to each block $B \in \mathfrak{B}(G)$ a set $\lambda(B)$ of $\Delta(B)$ colors such that $\lambda(B) \cap \lambda(B') = \varnothing$ whenever $B, B' \in \mathfrak{B}(G)$ are distinct blocks which have a common vertex. Note that two blocks of G can have only one vertex in common (see also Appendix C.4). For a mapping $\lambda \in \Lambda(G)$, let L_λ denote the list assignment of G with

$$L_\lambda(v) = \bigcup_{B \in \mathfrak{B}_v(G)} \lambda(B)$$

for every $v \in V(G)$. If G is a Gallai tree, then each block of G is a brick and hence regular. This implies that $|L_\lambda(v)| = d_G(v)$ for all $v \in V(G)$.

Theorem 1.8 *Let G be a Gallai tree and let L be a list-assignment of G. Then (G, L) is an uncolorable pair if and only if $L = L_\lambda$ for some $\lambda \in \Lambda(G)$.*

Proof The proof is by induction on the number of blocks of G. If G has only one block, the statement follows from Theorem 1.7(b) and from the fact that any brick B is regular and satisfies $\chi(B) = \Delta(B) + 1$. So assume that G has at least two blocks. Then G has an end-block G_1, and there exists exactly one separating vertex v of G contained in G_1. Let $G_2 = G - (G_1 - v)$. Then $G = G_1 \cup G_2$, $G_1 \cap G_2 = (\{v\}, \varnothing)$ and $\mathfrak{B}(G_2) = \mathfrak{B}(G) \setminus \{G_1\}$. Clearly, both G_1 and G_2 are Gallai trees.

First assume that $L = L_\lambda$ for some $\lambda \in \Lambda(G)$. For $i \in \{1, 2\}$, let λ_i be the restriction of λ to $\mathfrak{B}(G_i)$ and let $L_i = L_{\lambda_i}$. Then $\lambda_i \in \Lambda(G_i)$ and the induction hypothesis implies that (G_i, L_i) is an uncolorable pair. Clearly, $L = L_1 \cup L_2$ and $L_1(v) \cap L_2(v) = \varnothing$. Since G_i has no L_i-coloring for $i \in \{1, 2\}$, we then conclude that G has no L-coloring. Furthermore, we have $d_G(v) = d_{G_1}(v) + d_{G_2}(v) = |L_1(v)| + |L_2(v)| = |L(v)|$. This implies that (G, L) is an uncolorable pair.

Now assume that (G, L) is an uncolorable pair. For $i \in \{1, 2\}$, let C_i be the set of colors $c \in L(v)$ such that no L-coloring φ_i of G_i satisfies $\varphi_i(v) = c$. If $L(v) \setminus (C_1 \cup C_2)$ contains a color c, then there is an L-coloring φ_i of G_i with $\varphi_i(v) = c$ $(i \in \{1, 2\})$, and so $\varphi = \varphi_1 \cup \varphi_2$ is an L-coloring of G. This contradiction shows that $L(v) = C_1 \cup C_2$. For the Gallai tree G_i (with $i \in \{1, 2\}$) define the list-assignment L_i by

$$L_i(w) = \begin{cases} L(w) & \text{if } w \in V(G_i) \setminus \{v\}, \\ C_i & \text{if } w = v. \end{cases}$$

By definition of C_i, G_i is not L_i-colorable. Since (G, L) is an uncolorable pair, $|L_i(w)| = |L(w)| \geq d_G(w) = d_{G_i}(w)$ for all $w \in V(G_i) \setminus \{v\}$. Since G_i is not L_i-colorable, it then follows from Theorem 1.7(a) that $|C_i| = |L_i(v)| \leq d_{G_i}(v)$. Since $L(v) = C_1 \cup C_2$,

$$|C_1| + |C_2| \geq |C_1 \cup C_2| = |L(v)| \geq d_G(v) = d_{G_1}(v) + d_{G_2}(v),$$

which yields $|L_i(v)| = |C_i| = d_{G_i}(v)$ and $C_1 \cap C_2 = \varnothing$. Consequently, (G_i, L_i) is an uncolorable pair and the induction hypothesis implies that $L_i = L_{\lambda_i}$ for some map $\lambda_i \in \Lambda(G_i)$. Then the mapping λ with $\lambda(B) = \lambda_2(B)$ for $B \in \mathfrak{B}(G) \setminus \{G_1\}$ and $\lambda(G_1) = \lambda_1(G_1)$ belongs to $\Lambda(G)$ and $L = L_\lambda$. This completes the proof. □

Corollary 1.9 (Borodin / Erdős, Rubin, Taylor) *Let G be a connected graph and let L be a list-assignment of G. Then (G, L) is an uncolorable pair if and only if G is a Gallai tree and $L = L_\lambda$ for some $\lambda \in \Lambda(G)$.*

A graph G is said to be **degree-choosable** if G is d_G-list-colorable, that is, G is L-colorable for every list-assignment L of G satisfying $|L(v)| \geq d_G(v)$ for all $v \in V(G)$. Another immediate consequence of Theorem 1.8 is the following result.

Corollary 1.10 (Borodin / Erdős, Rubin, Taylor) *A connected graph G is not degree choosable if and only if G is a Gallai tree.*

That Brooks' theorem remains true when replacing the chromatic number by the list chromatic number follows immediately from Theorem 1.7 and was proved by Vizing [1051] and, independently, by Erdős, Rubin, and Taylor [355].

Theorem 1.11 (Vizing / Erdős, Rubin, Taylor) *Let G be a connected graph of maximum degree $\Delta \geq 0$. Then $\chi_\ell(G) \leq \Delta + 1$, where equality holds if and only if either $\Delta = 2$ and G is an odd cycle, or $\Delta \neq 2$ and $G = K_{\Delta+1}$.*

Proof For the proof it suffices to show that if $\chi_\ell(G) \geq \Delta + 1$, then G is a brick. So there is a Δ-assignment L of G for which G has no L-coloring, and so (G, L) is an uncolorable pair. It follows from Theorem 1.7 that G is a Δ-regular Gallai tree, which implies that G has only one block and therefore is a brick. □

1.4 DP-Coloring of Multigraphs

In this section we consider **multigraphs**, that is, graphs that may have multiple edges, but no loops. Terminology and notation for graphs are used similarly for multigraphs, see also Appendix C.12. Let G be an arbitrary multigraph. For distinct vertices $v, w \in V(G)$, let $\mu_G(v, w) = |E_G(v, w)|$ be the **multiplicity** of the vertex pair $\{v, w\}$. A **cover** of G is a pair (X, H) consisting of a map X and a graph H satisfying the following conditions:

(C1) $X : V(G) \to 2^{V(H)}$ is a mapping that assigns to each vertex $v \in V(G)$ a vertex set $X_v = X(v) \subseteq V(H)$, such that the sets X_v with $v \in V(G)$ are pairwise disjoint.

(C2) H is a graph with vertex set $V(H) = \bigcup_{v \in V(G)} X_v$ such that X_v is an independent set of H for each $v \in V(G)$, and for any two distinct vertices $v, w \in V(G)$ the set $E_H(X_v, X_w)$ is the union of $\mu_G(v, w)$ (possibly empty) matchings.

We now introduce the **DP-coloring problem** for the multigraph G. Let (X, H) be a cover of G. A vertex set $T \subseteq V(H)$ is called a **transversal** of (X, H) if $|T \cap X_v| = 1$ for each vertex $v \in V(G)$. An **independent transversal** of (X, H) is a transversal of

(X,H) which is an independent set of H. An independent transversal of (X,H) is also called an **(X,H)-coloring** of G, and the vertices of H are referred to as **colors**. We say that G is **(X,H)-colorable** if G admits an (X,H)-coloring. Let $f: V(G) \to \mathbb{N}_0$ be a function. The multigraph G is called **DP-f-colorable**, if G is (X,H)-colorable whenever (X,H) is a cover of G and $|X_v| \geq f(v)$ for all $v \in V(G)$. When $f(v) = k$ for all $v \in V(H)$, the term becomes **DP-k-colorable**. The **DP-chromatic number** of G, written $\chi_{\mathrm{DP}}(G)$, is the least integer $k \geq 0$ for which G is DP-k-colorable.

DP-colorings are closely related to list colorings. The set X_v is the list of colors available for v, and a matching edge between X_v and X_w indicates that the ends of that edge are considered the same color, so cannot both be chosen in a coloring of G. Let us show more formally how the list coloring problem can be reduced to the DP-coloring problem. Given a graph G and a list-assignment L of G, let (X,H) be the cover of G such that $X_v = \{v\} \times L(v)$ for all $v \in V(G)$, and two distinct vertices (v,c) and (v',c') are adjacent in H if and only if $c = c'$ and $vv' \in E(G)$. It is easy to check that (X,H) is indeed a cover of G. We claim that G is L-colorable if and only if G is (X,H)-colorable. If φ is an L-coloring of G, then $T = \{(v,\varphi(v)) \mid v \in V(G)\}$ is an independent transversal of H, and so G is (X,H)-colorable. If conversely T is an independent transversal of H, then there is a mapping φ from the vertex set of G into a color set such that $\varphi(v) \in L(v)$ for all $v \in V(G)$ and $T = \{(v,\varphi(v)) \mid v \in V(G)\}$ is an independent set of H, that is, φ is an L-coloring of G. This proves the claim that G is L-colorable if and only if G is (X,H)-colorable. Clearly, $|X_v| = |L(v)|$ for all $v \in V(G)$. Consequently, if G is DP-f-colorable for a function f, then G is f-list-colorable. In particular, $\chi(G) \leq \chi_\ell(G) \leq \chi_{\mathrm{DP}}(G)$. As we shall see below, the second inequality can be strict, even for simple graphs.

To obtain an upper bound for the DP-chromatic number of a multigraph G, we use a similar procedure as for the list coloring problem. So let (X,H) be a cover of G. To construct an independent transversal T of (X,H) we use a sequential coloring algorithm with respect to a given vertex order $\ell = (v_1, v_2, \ldots, v_n)$ of G. So we assign to each vertex v_i a color (vertex), if any, from the set X_{v_i}, where the color is in H not adjacent to all those colors already assigned to the neighbors of v_i among $v_1, v_2, \ldots, v_{i-1}$. This procedure obviously terminates with an (X,H)-coloring of G, provided that

$$|X_{v_i}| \geq d_{G[v_1,v_2,\ldots,v_i]}(v_i) + 1 \text{ for all } i \in [1,n].$$

This follows from the fact that if v, w are distinct vertices of G, then for $x \in X_w$ we have $|N_H(x) \cap X_v| \leq \mu_G(v,w)$. So at most $d_{G[v_1,v_2,\ldots,v_i]}(v_i)$ colors are forbidden in the set X_{v_i} by the already chosen colors. Consequently, G is DP-f-colorable, when $f(v) = d_G(v) + 1$ for all $v \in V(G)$ and, moreover, if G is a graph, then $\chi_{\mathrm{DP}}(G) \leq \mathrm{col}(G)$. The definition of the coloring number can be easily extended to multigraphs; in particular, the **coloring number** of a multigraph G is the least integer $k \geq 0$ for which every nonempty submultigraph of G has a vertex of degree at most k. Summarizing, we obtain the following sequence of inequalities for every multigraph G:

$$\omega(G) \leq \chi(G) \leq \chi_\ell(G) \leq \chi_{\mathrm{DP}}(G) \leq \mathrm{col}(G) \leq \Delta(G) + 1. \tag{1.8}$$

If G is a cycle, then $\chi_{\rm DP}(G) \leq 3$ and we have equality if G is an odd cycle. To see that we also have equality if G is an even cycle, we have to construct an appropriate cover of G. Assume that $V(G) = [1,n]$ and $E(G) = \{vw \mid v,w \in V(G) \text{ and } v-w \equiv 1 \pmod{n}\}$ with $n \geq 2$ even. Let (X,H) be the cover of G with $X_v = \{v\} \times \{1,2\}$ for all $v \in V(G)$ and $E(H) = \{(v,i)(w,j) \mid |v-w| = 1 \text{ and } i = j; \text{or } \{v,w\} = \{1,n\} \text{ and } i-j \equiv 1 \pmod{1}\}$. Note that (X,H) is a cover of G with $|X_v| = 2$ for all $v \in V(G)$. Clearly, $H = C_{2n}$ and (X,H) has no independent transversal. This shows that $\chi_{\rm DP}(G) \geq 3$ and hence $\chi_{\rm DP}(G) = 3$. The fact that $\chi_{\rm DP}(C_n) = 3$ for all $n \geq 3$ and not only for odd n, as for list-coloring, marks an important difference between the DP-chromatic number and the list-chromatic number.

A multigraph G is called **DP-degree-colorable** if G is (X,H)-colorable whenever (X,H) is a cover of G such that $|X_v| \geq d_G(v)$ for all $v \in V(H)$. Our aim is not only to characterize DP-degree-colorable multigraphs, but also the corresponding bad covers for the non DP-degree-colorable cases. Clearly, it suffices to do this for connected multigraphs.

For a multigraph H and an integer $t \geq 1$, let $G = tH$ denote the multigraph obtained from H by replacing each edge in H by a set of t parallel edges so that distinct vertices v,w of G satisfy $\mu_G(v,w) = t\mu_H(v,w)$, In particular, $H = 1H$. A multigraph G is called a **DP-brick** if either G is a tK_n for some integers $t,n \geq 1$ or G is a tC_n for some integers $t \geq 1$ and $n \geq 4$. For integers $t,n \geq 1$, let $K_{(n,t)} = K_{t,t,\ldots,t}$ denote the complete n-partite graph all of whose partite sets have t vertices. If G is a multigraph, then we denote by $A(G)$ the set of all 2-subsets $\{v,w\}$ of $V(G)$ such that v,w are adjacent in G; the graph $H = (V(G),A(G))$ is said to be the **underlying graph** of G.

A **feasible configuration** is a triple (G,X,H) consisting of a connected multigraph G and a cover (X,H) of G. A feasible configuration (G,X,H) is said to be **degree-feasible** if $|X_v| \geq d_G(v)$ for all $v \in V(H)$. Furthermore (G,X,H) is called **colorable** if G is (X,H)-colorable, otherwise it is called **uncolorable**. The next proposition lists some basic properties of feasible configurations; the proofs are straightforward and left to the reader.

Proposition 1.12 *Let (G,X,H) be a feasible configuration. Then the following statements hold:*

(a) *For distinct vertices v,w of G, $H[X_v \cup X_w]$ is a bipartite graph with parts X_v and X_w whose maximum degree is at most $\mu_G(v,w)$. Furthermore, $d_H(x) \leq d_G(v)$ for every vertex $v \in V(G)$ and every color $x \in X_v$.*

(b) *Let H' be a spanning subgraph of H. Then (G,X,H') is a feasible configuration. If (G,X,H) is colorable, then (G,X,H') is colorable, too. Furthermore, (G,X,H) is degree-feasible if and only if (G,X,H') is degree feasible.*

Proposition 1.12(b) leads to the following concept. A feasible configuration (G,X,H) is **minimal uncolorable** if (G,X,H) is uncolorable, but $(G,X,H-e)$ is colorable for every edge $e \in E(H)$. Clearly, if (G,X,H) is an uncolorable feasible configuration, then there is a spanning subgraph H' of H such that (G,X,H') is a minimal uncolorable feasible configuration. This follows from Proposition 1.12(b) and the fact that if $|G| \geq 2$ and $\tilde{H}$ is the edgeless spanning subgraph of H, then $(G, X, \tilde{H})$ is a colorable feasible configuration.

In order to characterize the class of minimal uncolorable degree-feasible configurations, we introduce two basic types of degree-feasible configurations and a construction.

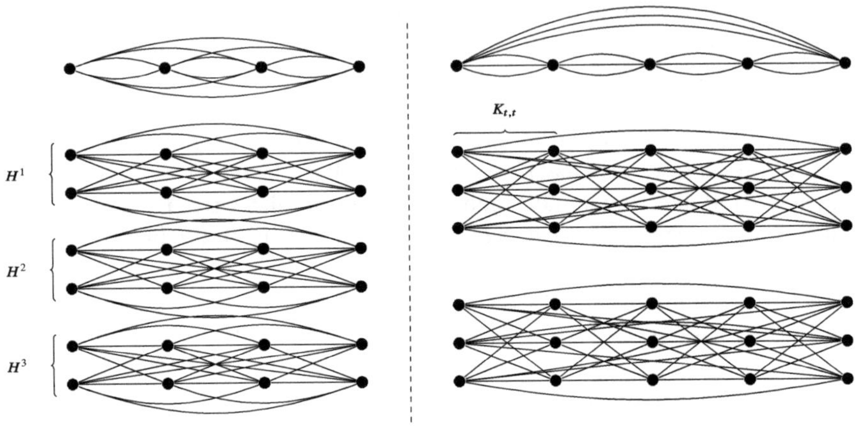

Fig. 1.6 A K-configuration with $G = 2K_4$ and an odd C-configuration with $G = 3C_5$.

We call (G, X, H) a **K-configuration** if $G = tK_n$ for some integers $t, n \geq 1$, and (X, H) is a cover of G such that, for every vertex $v \in V(G)$, there is a partition $(X_v^1, X_v^2, \ldots, X_v^{n-1})$ of X_v satisfying the following conditions:

- For every $i \in [1, n-1]$, the graph $H^i = H[\bigcup_{v \in V(G)} X_v^i]$ is a $K_{(n,t)}$ whose partite sets are the sets X_v^i with $v \in V(G)$, and
- $H = H^1 \cup H^2 \cup \cdots \cup H^{n-1}$.

It is easy to check that any K-configuration is a minimal uncolorable degree-feasible configuration. Note that in case of $n = 1$, we get $G = K_1$, $X = \varnothing$ and $H = \varnothing$.

We call (G, X, H) an **odd C-configuration** if $G = tC_n$ for some integers $t \geq 1$, $n \geq 5$ odd, and (X, H) is a cover of G such that for every vertex $v \in V(G)$, there is a partition (X_v^1, X_v^2) of X_v satisfying the following conditions:

- For every $i \in \{1, 2\}$ and every set $\{v, w\} \in A(G)$, the graph $H^i_{\{v,w\}} = H[X_v^i \cup X_w^i]$ is a $K_{t,t}$ whose partite sets are X_v^i and X_w^i, and
- H is the union of all graphs $H^i_{\{v,w\}}$ with $i \in \{1, 2\}$ and $\{v, w\} \in A(G)$

It is easy to check that any odd C-configuration is a minimal uncolorable degree-feasible configuration.

We call (G, X, H) an **even C-configuration** if $G = tC_n$ for some integers $t \geq 1$, $n \geq 4$ even, and (X, H) is a cover of G such that for every vertex $v \in V(G)$, there is a partition (X_v^1, X_v^2) of X_v, and there is a set $\{u, u'\} \in A(G)$ satisfying the following conditions:

- For every $i \in \{1,2\}$ and for every set $\{v,w\} \in A(G)$ different from $\{u,u'\}$, the graph $H^i_{\{v,w\}} = H[X^i_v \cup X^i_w]$ is a $K_{t,t}$ whose partite sets are X^i_v and X^i_w,
- $H^1_{\{u,u'\}} = H[X^1_u \cup X^2_{u'}]$ is a $K_{t,t}$ whose partite sets are X^1_u and $X^2_{u'}$,
- $H^2_{\{u,u'\}} = H[X^2_u \cup X^1_{u'}]$ is a $K_{t,t}$ whose partite sets are X^2_u and $X^1_{u'}$, and
- H is the union of all graphs $H^i_{\{v,w\}}$ with $i \in \{1,2\}$ and $\{v,w\} \in A(G)$

It is easy to check that any even C-configuration is a minimal uncolorable degree-feasible configuration. We call (G,X,H) a **C-configuration** if (G,X,H) is either an even or an odd C-configuration. Figure 1.6 shows a K-configuration with $G = 2K_4$ and an odd C-configuration with $G = 3C_5$.

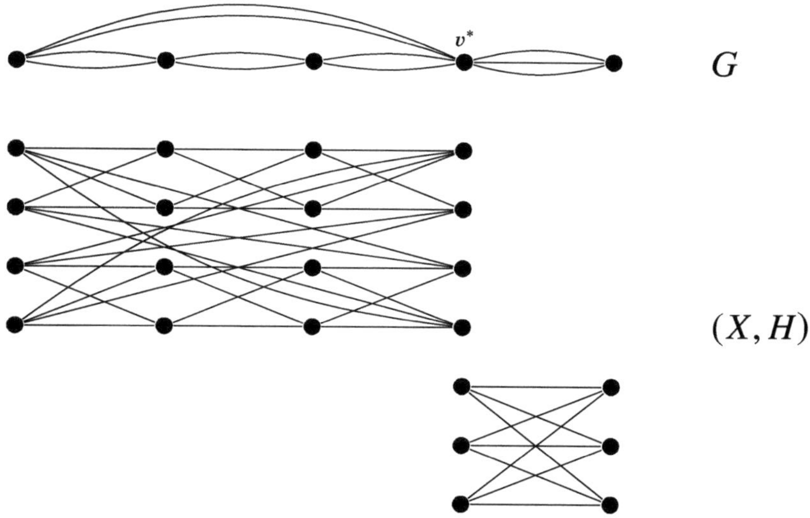

Fig. 1.7 A constructible configuration obtained by the merging operation.

Let (G^1,X^1,H^1) and (G^2,X^2,H^2) be two feasible configurations, which are **disjoint**, that is, $V(G^1) \cap V(G^2) = \varnothing$ and $V(H^1) \cap V(H^2) = \varnothing$. Furthermore, let $v^1 \in V(G^1)$ and $v^2 \in V(G^2)$, and let G be the graph obtained from G^1 and G^2 by identifying v^1 and v^2 to a new vertex $v^* \notin V(G^1) \cup V(G^2)$. Let $H = H^1 \cup H^2$, and let $X : V(G) \to 2^{V(H)}$ be the mapping such that

$$X_v = \begin{cases} X^1_{v^1} \cup X^2_{v^2} & \text{if } v = v^*, \\ X^i_v & \text{if } v \in V(G^i) \setminus \{v^i\} \text{ and } i \in \{1,2\}. \end{cases}$$

for $v \in V(H)$. Clearly, (G,X,H) is a feasible configuration and we say that (G,X,H) is obtained from (G^1,X^1,H^1) and (G^2,X^2,H^2) by **merging** v^1 and v^2 to v^*. Since $d_G(v^*) = d_{G^1}(v^1) + d_{G^2}(v^2)$ and $|X_{v^*}| = |X_{v^1}| + |X_{v^2}|$, it follows that (G,X,H) is degree-feasible if both (G^1,X^1,H^1) and (G^2,X^2,H^2) are degree-feasible.

Now we define the class of **constructible configurations** as the smallest class of feasible configurations that contains each K-configuration as well as each C-

configuration and that is closed under the above described merging operation. The proof of the next proposition can be done by induction on the number of blocks and is straightforward. Figure 1.7 shows a constructible configuration obtained by the merging operation applied to a C-configuration with $G^1 = 2C_4$ and a K-configuration with $G^2 = 3K_2$.

Proposition 1.13 *Let (G,X,H) be a constructible configuration. Then for each block $B \in \mathfrak{B}(G)$, there is a uniquely determined cover (X^B,H^B) of B such that the following statements hold:*

(a) *For every block $B \in \mathfrak{B}(G)$, (B,X^B,H^B) is either a K-configuration or a C-configuration.*
(b) *The graphs H^B with $B \in \mathfrak{B}(G)$ are pairwise disjoint and $H = \bigcup_{B \in \mathfrak{B}(G)} H^B$.*
(c) *Every vertex v of G satisfies $X_v = \bigcup_{B \in \mathfrak{B}_v(G)} X_v^B$.*

Our aim ist to show that the class of minimal uncolorable degree-feasible configurations coincides with the class of constructible configurations. To this end, we shall apply the following reduction.

Proposition 1.14 *Let (G,X,H) be a feasible configuration such that $|G| \geq 2$, let v be a nonseparating vertex of G, and let $x \in X_v$ be a color. For the multigraph $G' = G - v$ define a cover (X',H') as follows. For a vertex $w \in V(G')$,*

$$\text{let } X'_w = X_w \setminus N_H(x), \text{ and let } H' = H - (X_v \cup N_H(x)).$$

Then (G',X',H') is a feasible configuration, and in what follows we briefly write $(G',X',H') = (G,X,H)/(v,x)$. Furthermore, the following statements hold:

(a) *If (G,X,H) is degree-feasible, then (G',X',H') is degree-feasible, too.*
(b) *If (G,X,H) is uncolorable, then (G',X',H') is uncolorable, too.*

Proof That (X',H') is indeed a cover of G' is obvious, and so (G',X',H') is a feasible configuration. For $w \in V(G')$ we have $d_G(w) = d_{G'}(w) + \mu_G(v,w)$ and $|N_H(x) \cap X_w| \leq \mu_G(v,w)$ (see Proposition 1.12(a)), which leads to

$$|X'_w| = |X_w| - |N_H(x) \cap X_w| \geq |X_w| - \mu_G(v,w).$$

So $|X_w| \geq d_G(w)$ implies $|X'_w| \geq d_{G'}(w)$. This proves (a). Clearly, if T' is an independent transversal of (X',H'), then $T = T' \cup \{x\}$ is an independent transversal of (X,H), which proves (b). □

Proposition 1.15 *Let (G,X,H) be an uncolorable degree-feasible configuration. Then the following statements hold:*

(a) *$|X_v| = d_G(v)$ for all $v \in V(G)$.*
(b) *If u is a nonseparating vertex of G and v is a vertex of $G-u$, then $|N_H(x) \cap X_v| = \mu_G(u,v)$ for all colors $x \in X_u$.*

(c) *If G is a block, then G is regular, and for distinct vertices v, w of G, the graph $H[X_v \cup X_w]$ is a $\mu_G(v,w)$-regular bipartite graph whose partite sets are X_v and X_w.*

(d) *For every vertex $v \in V(G)$ there exists an independent set T of $H - v$ such that $|T \cap X_w| = 1$ for all $w \in V(G) \setminus \{v\}$.*

Proof The proof of (a) is by induction on the order of G. If G consists of one vertex v, then $X_v = \varnothing$ and $H = \varnothing$, so (a) holds. So assume that $|G| \geq 2$, and let v be an arbitrary vertex of G. Since G is connected, there is a nonseparating vertex $u \neq v$ in G and $X_u \neq \varnothing$. Hence there is a color $x \in X_u$ and $(G', X', H') = (G, X, H)/(u, x)$ is an uncolorable degree-feasible configuration. The induction hypothesis leads to $|X'_v| = d_{G'}(v)$. By Proposition 1.12(a), we then obtain that

$$\begin{aligned} d_{G'}(v) = |X'_v| &= |X_v| - |N_H(x) \cap X_v| \\ &\geq |X_v| - \mu_G(u,v) \geq d_G(v) - \mu_G(u,v) = d_{G'}(v) \end{aligned}$$

which implies $|X_v| = d_G(v)$ and $|N_H(x) \cap X_v| = \mu_G(u,v)$. This proves (a) and the same argument can be used to prove (b).

To prove (c), assume that G is a block. Let v, w be distinct vertices of G. Then $H[X_v \cup X_w]$ is a $\mu_G(v,w)$-regular bipartite graph with parts X_v, X_w (by (b)). Hence, if $vw \in E(G)$, then $|X_v| = |X_w|$ and so $d_G(v) = d_G(w)$ (by (a)). This proves (c).

For the proof of (d), let v be an arbitrary vertex of G. Let U be the vertex set of a component of $G - v$, let X' be the restriction of X to U, and let $H' = H[\bigcup_{u \in U} X_u]$. Then $(G[U], X', H')$ is a degree feasible configuration, and $|X_u| = d_G(u) > d_{G[U]}(u)$ for at least one vertex $u \in U$. It then follows from (a) that there is an independent transversal T_U of (X', H'). Clearly, if T is the union of the sets T_U over all components $G[U]$ of $G - v$, we obtain an independent set of $H - v$ such that $|T \cap X_w| = 1$ for all $w \in V(G) \setminus \{v\}$. This proves (d). □

Proposition 1.16 *Let (G, X, H) be obtained from two disjoint degree-feasible configurations (G^1, X^1, H^1) and (G^2, X^2, H^2) by merging $v^1 \in V(G^1)$ and $v^2 \in V(G^2)$ to a new vertex v^*. Then (G, X, H) is a degree-feasible configuration, and the following conditions are equivalent:*

(a) *Both (G^1, X^1, H^1) and (G^2, X^2, H^2) are minimal uncolorable.*

(b) *(G, X, H) is minimal uncolorable.*

Proof First, assume that (a) holds. If (G, X, H) is colorable, then there exists an independent transversal T of (X, H). Then T is an independent set of H containing exactly one vertex from each set X_v with $v \in V(H)$. Since $X_{v^*} = X^1_{v^1} \cup X^2_{v^2}$, this implies (by symmetry) that $|T \cap X^1_{v^1}| = 1$. Hence, $T^1 = T \cap V(H^1)$ is an independent transversal of (X^1, H^1), and so (G^1, X^1, H^1) is colorable. This contradiction to (a) shows that (G, X, H) is uncolorable. Now let $e \in E(H)$ be an arbitrary edge. Since $H = H^1 \cup H^2$, this implies (by symmetry) that $e \in E(H^1)$. Since (G^1, X^1, H^1) is minimal uncolorable, the cover $(X^1, H^1 - e)$ has an independent transversal T^1. Since

(G^2, X^2, H^2) is minimal uncolorable and G^2 is connected, there is an independent set T^2 in H^2 such that $|T^2 \cap X_v^2| = 1$ for all $v \in V(H^2) \setminus \{v^2\}$ (by Proposition 1.15(d)). Since $H = H^1 \cup H^2$ and $H_1 \cap H_2 = \varnothing$, the set $T = T^1 \cup T^2$ is an independent transversal of $(X, H - e)$, and so $(G, X, H - e)$ is colorable. This shows that (b) holds, that is, (G, X, H) is minimal uncolorable.

Now, assume that (b) holds. In order to prove that (a) holds, it suffices to show that (G^1, X^1, H^1) is minimal uncolorable (by symmetry). Suppose that (G^1, X^1, H^1) is colorable, that is, (X^1, H^1) has an independent transversal T^1. Since (G, X, H) is minimal uncolorable and G is connected, there is an independent set T^2 in H such that $|T^2 \cap X_v^2| = 1$ for all $v \in V(G^2) \setminus \{v^*\}$ (by Proposition 1.15(d)). But then, $T = T^1 \cup T^2$ is an independent transversal of (X, H) and so (G, X, H) is colorable, a contradiction to (b). This shows that (G^1, X^1, H^1) is uncolorable. To complete the proof, let $e \in E(H^1)$ be an arbitrary edge. Since (G, X, H) is minimal uncolorable and $e \in E(H)$, there is an independent transversal T of $(X, H - e)$. Then $T^1 = T \cap V(H^1)$ is an independent transversal of $(X^1, H^1 - e)$, and so $(G^1, X^1, H^1 - e)$ is colorable. Thus (G^1, X^1, H^1) is minimal uncolorable. □

Theorem 1.17 *Let (G, X, H) be a degree-feasible configuration. Then (G, X, H) is minimal uncolorable if and only if (G, X, H) is constructible.*

Proof If (G, X, H) is constructible, then (G, X, H) is minimal uncolorable; this follows from Proposition 1.16 and the fact that each K-configuration as well as each C-configuration is a minimal uncolorable degree-feasible configuration. Hence we only need to show the reverse statement. So assume that (G, X, H) is a degree-feasible configuration that is minimal uncolorable. We show by induction on the order of G that (G, X, H) is constructible. If $|G| = 1$, then $X = \varnothing$, $H = \varnothing$ and so (G, X, H) is a K-configuration, and we are done. Assume that $|G| \geq 2$. By Proposition 1.15(a), every vertex v of G satisfies

$$|X_v| = d_G(v) \tag{1.9}$$

To complete the proof, we distinguish two cases.

Case 1: *G contains a separating vertex v^*.* Then G is the union of two connected induced submultigraphs G^1 and G^2 with $V(G^1) \cap V(G^2) = \{v^*\}$ and $|G^j| < |G|$ for $j \in \{1, 2\}$. For $j \in \{1, 2\}$, let $\mathcal{T}^j$ denote the set of all independent sets T of H such that $|T \cap X_v| = 1$ for all $v \in V(G^j)$. It then follows from Proposition 1.15(d) that both sets $\mathcal{T}^1$ and $\mathcal{T}^2$ are nonempty. For $j \in \{1, 2\}$, let X_j be the set of all vertices of X_{v^*} that do not occur in any independent set belonging to $\mathcal{T}^j$. We claim that the set $M = X_{v^*} \setminus (X_1 \cup X_2)$ is empty. Otherwise, there would be two sets, say $T^1 \in \mathcal{T}^1$ and $T^2 \in \mathcal{T}^2$, which both contain a vertex $x \in X_{v^*}$. But then $T = T^1 \cup T^2$ would be an independent transversal of (X, H). This is due to the fact that if $v \in V(G^1) \setminus \{v^*\}$ and $w \in V(G^2) \setminus \{v^*\}$, then $\mu_G(v, w) = 0$ and so $E_H(X_v, X_w) = \varnothing$ (by Proposition 1.12(a)). Consequently, (G, X, H) would be colorable. This contradiction proves the claim that $M = \varnothing$, and so $X_{v^*} = X_1 \cup X_2$. For $j \in \{1, 2\}$, we now define a cover (X^j, H^j) of G^j as follows. For $v \in V(G^j)$, let

$$X_v^j = \begin{cases} X_v & \text{if } v \neq v^*, \\ X_j & \text{if } v = v^*, \end{cases}$$

and let $H^j = H[\bigcup_{v\in V(G^j)} X^j_v]$. Clearly, (G^j, X^j, H^j) is a feasible configuration, and from the definition of the set X_j it follows that (X^j, H^j) has no independent transversal. Consequently, (G^j, X^j, H^j) is uncolorable. Furthermore, every vertex $v \in V(G^j) \setminus \{v^*\}$ satisfies $|X_v| = d_G(v) = d_{G^j}(v)$ (by (1.9)). Since (G^j, X^j, H^j) is uncolorable, Proposition 1.15(a) implies that $|X^j_{v^*}| \le d_{G^j}(v^*)$. Since we have $X_{v^*} = X_1 \cup X_2 = X^1_{v^*} \cup X^2_{v^*}$, it follows from (1.9) that

$$|X^1_{v^*}| + |X^2_{v^*}| \ge |X^1_{v^*} \cup X^2_{v^*}| = |X_{v^*}| = d_G(v^*) = d_{G^1}(v^*) + d_{G^2}(v^*),$$

which yields $|X^j_{v^*}| = d_{G^j}(v^*)$ and $X^1_{v^*} \cap X^2_{v^*} = \varnothing$. Consequently, (G^j, X^j, H^j) is a degree-feasible configuration. Furthermore, $H' = H^1 \cup H^2$ is a spanning subgraph of H and $V(H^1) \cap V(H^2) = \varnothing$. Hence, (G, X, H') is a degree-feasible configuration that is obtained from two isomorphic copies of (G^1, X^1, H^1) and (G^2, X^2, H^2) by the merging operation. Clearly, (G, X, H') is uncolorable. Otherwise, there would exist an independent transversal T of (X, H') and, by symmetry, T would contain a vertex of $X^1_{v^*}$, but then $T^1 = T \cap V(H^1)$ would be an independent transversal of (X^1, H^1), which is impossible. Since (G, X, H) is minimal uncolorable and H' is a spanning subgraph of H, we then conclude that $H' = H$. Consequently, (G, X, H) is obtained from two isomorphic copies of (G^1, X^1, H^1) and (G^2, X^2, H^2) by the merging operation, and Proposition 1.16 then implies that both (G^1, X^1, H^1) and (G^2, X^2, H^2) are minimal uncolorable (and also degree-feasible). By the induction hypothesis it then follows that both (G^1, X^1, H^1) and (G^2, X^2, H^2) are constructible, and so (G, X, H) is constructible. This settles the first case.

Case 2: *G is a block.* Our aim is to show that G is a DP-brick and (G, X, H) is either a K-configuration or a C-configuration. To this end, we shall prove a sequence of claims.

Claim 1.17.1 *Let v be an arbitrary vertex of G, let $x \in X_v$ be an arbitrary color, and let $(G', X', H') = (G, X, H)/(v,x)$. Then there is a spanning subgraph $\tilde{H}$ of H' such that $(G', X', \tilde{H})$ is minimal uncolorable. Furthermore, $(G', X', \tilde{H})$ is constructible, and so each block of $G' = G - v$ is a DP-brick.*

Proof: Since $|G| \ge 2$ and G is connected, $X_v \ne \varnothing$ (by (1.9)). Hence, by Proposition 1.14, $(G', X', H') = (G, X, H)/(v,x)$ is an uncolorable degree-feasible configuration, and, therefore, there is a spanning subgraph $\tilde{H}$ of H' such that $(G', X', \tilde{H})$ is minimal uncolorable. The induction hypothesis then implies that $(G', X', \tilde{H})$ is constructible. In particular, $G' = G - v$ and each block of G' is a DP-brick. △

By a **multicycle** or a **multipath** we mean a multigraph that can be obtained from a cycle, respectively a path, by replacing each edge e of the cycle or path by a set of t_e parallel edges, where $t_e \ge 1$. For integers $s, t \ge 1$, we say that a multigraph H is an (s,t)**-multicycle**, if H can be obtained from an even cycle C by replacing each edge of a perfect matching of C by a set of s parallel edges and any other edge of C by a set of t parallel edges. Note that an (s,t)-multicycle is r-regular with $r = s + t$. Furthermore if H is a regular multicycle, then either $H = tC_n$ for some integers $t \ge 1$ and $n \ge 3$ or H is an (s,t)-multicycle for some integers $s, t \ge 1$.

Claim 1.17.2 *The multigraph G is a DP-brick.*

Proof: Since G is a block, it follows from Proposition 1.15(c) that G is r-regular for some integer $r \geq 1$. If v is a vertex of G, then each block of $G-v$ is a DP-brick (by Claim 1.17.1). Let S be the set of all vertices v of G such that $G-v$ is a block. If $S \neq \varnothing$, then for every vertex $v \in S$, $G-v$ is a DB-brick and hence regular. Since G is regular, too, for $v \in S$ there is an integer $t_v \geq 1$ such that $\mu_G(v,w) = t_v$ for all $w \in V(G) \setminus \{v\}$. But then $S = V(G)$ and we easily conclude that $t_v = t$ for all $v \in V(G)$, and so $G = tK_n$ where n is the order of G.

It remains to consider the case when $S = \varnothing$. Let v be an arbitrary vertex of G. Then each block of $G-v$ is a DP-brick and hence regular. Furthermore, $G-v$ has at least two end-blocks. Now let B be an arbitrary end-block of $G-v$. Then B is t_B-regular for an integer $t_B \geq 1$, and B contains exactly one separating vertex v_B of $G-v$. Since G is r-regular, there is an integer s_B such that $\mu_G(v,w) = s_B$ for all vertices $w \in V(B) \setminus \{v_B\}$. This implies that $|B| = 2$, since otherwise every vertex of $B - v_B$ would belong to S, and so S would be nonempty, which is impossible. Consequently, $B = t_B K_2$ and $r = t_B + s_B$. So $V(B) = \{v', v_B\}$ and $N_G(v') = \{v, v_B\}$. If we repeat the argument with v' instead of v, we conclude that G is a multicycle. Since G is regular, this implies that either $G = tC_n$ with $t \geq 1$ and $n \geq 3$, or G is an (s,t)-multicycle with $s \neq t$, say $1 \leq s < t$. In the former case we are done. To exclude the latter case, assume that G is an (s,t)-multicycle with $1 \leq s < t$. Then each vertex v of G satisfies $|X_v| = s+t$ (by (1.9)). Let v be any vertex of G. Then $G-v$ is a multipath and one end-block of $G-v$, say B, is a tK_2. Hence B consists of two vertices, say u and w, such that $d_{G-v}(u) = t$ and $d_{G-v}(w) = s+t$. Now let $x \in X_v$ be a color, and let $(G', X', H') = (G, X, H)/(v,x)$. Then there is a spanning subgraph $\tilde{H}$ of H' such that $(G', X', \tilde{H})$ is constructible (by Claim 1.17.1). It then follows from (1.9) and Proposition 1.13 that $|X'_u| = t$, $|X'_w| = s+t$ and there is a subset X^1_w of X'_w such that $|X^1_w| = t$ and $H^1 = \tilde{H}[X'_u \cup X^1_w]$ is a $K_{t,t}$ with parts X'_u, X^1_w. The graph H^1 is a subgraph of $H^2 = H[X_u \cup X_w]$, and H^2 is a t-regular bipartite graph with parts X_u, X_w (by Proposition 1.15(c)). Since $|X_u| = |X_w| = s+t$ and $1 \leq s < t$, this is impossible. Thus the claim is proved. △

By Claim 1.17.2, there are only two possibilities for the multigraph G. Either $G = tK_n$ with $t \geq 1$ and $n \geq 2$, or $G = tC_n$ with $t \geq 1$ and $n \geq 4$.

Claim 1.17.3 *Suppose that $G = tK_n$ for integers $t \geq 1$ and $n \geq 2$. Then (G, X, H) is a K-configuration.*

Proof: Since (G, X, H) is minimal uncolorable, for each vertex v of G and each pair u, w of distinct vertices of G, we have

(a) $|X_v| = t(n-1)$ and $H[X_u \cup X_w]$ is a t-regular bipartite graph with parts X_u, X_w

(by (1.9) and Proposition 1.15(c)). If $n = 2$, then G has exactly two vertices, say u and w, and $H[X_u \cup X_w]$ is a $K_{t,t}$ (by (a)), and so (G, X, H) is a K-configuration as claimed.

Now assume that $n \geq 3$. Let v be an arbitrary vertex of G, and let $x \in X_v$ be an arbitrary color. Furthermore, let $(G', X', H') = (G, X, H)/(v,x)$. Then there is a spanning subgraph $\tilde{H}$ of $H' = H - (X_v \cup N_H(x))$ such that $(G', X', \tilde{H})$ is a constructible configuration (by Claim 1.17.1). Since $G' = G - v = tK_{n-1}$, $(G', X', \tilde{H})$ is a K-configuration. Consequently, for every vertex $w \in V(G)$, there is a partition $(X_w^1, X_w^2, \dots, X_w^{n-2})$ of $X'_w = X_w \setminus N_H(x)$ such that, for $i \in [1, n-2]$,

(b) the graph $H^i = \tilde{H}[\bigcup_{w \in V(G')} X_w^i]$ is a $K_{(n-1,t)}$ whose partite sets are the sets X_w^i with $w \in V(G')$, and $\tilde{H} = H^1 \cup H^2 \cup \dots \cup H^{n-2}$.

For $w \in V(G')$ let $X_w^{n-1} = X_w \setminus X'_w$. Then, for every vertex $w \in V(G')$, $|X_w^{n-1}| = t$ and $(X_w^1, X_w^2, \dots, X_w^{n-1})$ is a partition of X_w. Since $\tilde{H}$ is a spanning subgraph of H', it follows from (a) and (b) that H^i is an induced subgraph of H (for $i \in [1, n-2]$) and the graph

$$H^{n-1} = H[\bigcup_{w \in V(G')} X_w^{n-1}]$$

is a $K_{(n-1,t)}$ whose partite sets are the sets X_w^{n-1} with $w \in V(G')$. Furthermore, $H - X_v = H^1 \cup H^2 \cup \dots \cup H^{n-1}$, and $N_H(x) = V(H^{n-1})$. Since the color $x \in X_v$ was chosen arbitrarily, it then follows from (a) and (b) that there is a partition $(X_v^1, X_v^2, \dots, X_v^{n-1})$ of X_v such that, for $i \in [1, n-1]$, we have $|X_v^i| = t$ and $N_H(x) = V(H^i)$ for all $x \in X_v^i$. Consequently, for $i \in [1, n-1]$, the graph

$$H_i = H[\bigcup_{w \in V(G)} X_w^i]$$

is a $K_{(n,t)}$ whose partite sets are the sets X_w^i with $w \in V(G)$ and, moreover, $H = H_1 \cup H_2 \cup \dots \cup H_{n-1}$. Hence (G, X, H) is a K-configuration. △

Claim 1.17.4 *Suppose that $G = tC_n$ for integers $t \ge 1$ and $n \ge 4$. Then (G, X, H) is a C-configuration.*

Proof: Since (G, X, H) is minimal uncolorable, for each vertex $v \in V(G)$ and each 2-set $\{u, w\} \in A(G)$, we have

(a) $|X_v| = 2t$ and $H[X_u \cup X_w]$ is a t-regular bipartite graph with parts X_u, X_w

(by (1.9) and Proposition 1.15(c)). Let v be an arbitrary vertex of G, and let $x \in X_v$ be an arbitrary color. Furthermore, let $(G', X', H') = (G, X, H)/(v, x)$. Then there is a spanning subgraph $\tilde{H}$ of $H' = H - (X_v \cup N_H(x))$ such that $(G', X', \tilde{H})$ is a constructible configuration (by Claim 1.17.1). Since $G' = G - v = tP_{n-1}$, the vertices of G' can be arranged in a sequence, say $v_1, v_2, \dots, v_{n-1}$, such that two vertices are adjacent in G' if and only if they are consecutive in the sequence. Note that $N_G(v) = \{v_1, v_{n-1}\}$ and each block of G' is a tK_2. We claim that for every vertex w of G' there is a partition (X_w^1, X_w^2) of X_w such that the following conditions hold:

(b) For every $i \in \{1, 2\}$ and every $k \in [1, n-2]$ the graph $H_k^i = H[X_{v_k}^i \cup X_{v_{k+1}}^i]$ is a $K_{t,t}$ whose partite sets are $X_{v_k}^i$ and $X_{v_{k+1}}^i$.
(c) The graph $H - X_v$ is the union of all graphs H_k^i with $i \in \{1, 2\}$ and $k \in [1, n-2]$.
(d) If n is even, then $N_H(x) = X_{v_1}^1 \cup X_{v_{n-1}}^2$ or $N_H(x) = X_{v_1}^2 \cup X_{v_{n-1}}^1$.
(e) If n is odd, then $N_H(x) = X_{v_1}^1 \cup X_{v_{n-1}}^1$ or $N_H(x) = X_{v_1}^2 \cup X_{v_{n-1}}^2$.

For $k \in [1, n-2]$, the multigraph $B^k = G[\{v_k, v_{k+1}\}]$ is a block of G'. Clearly, $\mathfrak{B}(G') = \{B^1, B^2, \ldots, B^{n-2}\}$ and the only end-blocks of G' are B^1 and B^{n-2}. Since $(G', X', \tilde{H})$ is a constructible configuration and each block of G' is a tK_2, we obtain from Proposition 1.13 that for every $k \in [1, n-2]$ there is a uniquely determined cover $(\tilde{X}^k, \tilde{H}^k)$ of B^k such that

- $\tilde{H}^k$ is a $K_{t,t}$ with parts $\tilde{X}^k_{v_k}, \tilde{X}^k_{v_{k+1}}$,
- $\tilde{H}$ is the disjoint union of the graphs $\tilde{H}^1, \tilde{H}^2, \ldots, \tilde{H}^{n-2}$,
- $X'_{v_1} = \tilde{X}^1_{v_1}$, $X'_{v_k} = \tilde{X}^{k-1}_{v_k} \cup \tilde{X}^k_{v_k}$ $(k \in [2, n-2])$, and $X'_{v_{n-1}} = \tilde{X}^{n-2}_{v_{n-1}}$.

Since $\{v_k, v_{k+1}\} \in A(G)$ for $k \in [1, n-2]$, it follows from (a) that $\tilde{H}^k$ is an induced subgraph of H. Let $\tilde{X}^0_{v_1} = X_{v_1} \setminus X'_{v_1}$ and $\tilde{X}^{n-1}_{v_{n-1}} = X_{v_{n-1}} \setminus X'_{v_{n-1}}$. Then both sets $\tilde{X}^0_{v_1}$ and $\tilde{X}^{n-1}_{v_{n-1}}$ have exactly t elements, and $N_H(x) = \tilde{X}^0_{v_1} \cup \tilde{X}^{n-1}_{v_{n-1}}$. Furthermore we conclude from (a) that, for $k \in [1, n-2]$,

- the graph $H[\tilde{X}^{k-1}_{v_k} \cup \tilde{X}^{k+1}_{v_{k+1}}]$ is a $K_{t,t}$ with parts $\tilde{X}^{k-1}_{v_k}$ and $\tilde{X}^{k+1}_{v_{k+1}}$.

If n is even, then let

$$(X^1_{v_1}, X^1_{v_2}, \ldots, X^1_{v_{n-1}}) = (\tilde{X}^1_{v_1}, \tilde{X}^1_{v_2}, \tilde{X}^3_{v_3}, \tilde{X}^3_{v_4}, \ldots, \tilde{X}^{n-3}_{v_{n-3}}, \tilde{X}^{n-3}_{v_{n-2}}, \tilde{X}^{n-1}_{v_{n-1}})$$

and

$$(X^2_{v_1}, X^2_{v_2}, \ldots, X^2_{v_{n-1}}) = (\tilde{X}^0_{v_1}, \tilde{X}^2_{v_2}, \tilde{X}^2_{v_3}, \tilde{X}^4_{v_4}, \tilde{X}^4_{v_5} \ldots, \tilde{X}^{n-2}_{v_{n-2}}, \tilde{X}^{n-2}_{v_{n-1}}).$$

If n is odd, then let

$$(X^1_{v_1}, X^1_{v_2}, \ldots, X^1_{v_{n-1}}) = (\tilde{X}^1_{v_1}, \tilde{X}^1_{v_2}, \tilde{X}^3_{v_3}, \tilde{X}^3_{v_4}, \ldots, \tilde{X}^{n-2}_{v_{n-2}}, \tilde{X}^{n-2}_{v_{n-1}})$$

and

$$(X^2_{v_1}, X^2_{v_2}, \ldots, X^2_{v_{n-1}}) = (\tilde{X}^0_{v_1}, \tilde{X}^2_{v_2}, \tilde{X}^2_{v_3}, \ldots, \tilde{X}^{n-3}_{v_{n-3}}, \tilde{X}^{n-3}_{v_{n-2}}, \tilde{X}^{n-1}_{v_{n-1}}).$$

Taking (a) and Proposition 1.15(b) into account, it is easy to check that, for every vertex w of G', (X^1_w, X^2_w) is a partition of X_w such that the conditions (b), (c), (d), and (e) are satisfied. Since the color $x \in X_v$ was chosen arbitrarily, it then follows from (a) and Proposition 1.15(b) that there is a partition (X^1_v, X^2_v) of X_v such that $|X^1_v| = |X^2_v| = t$ and the following conditions hold:

- If n is even, then $N_H(x) = X^1_{v_1} \cup X^2_{v_{n-1}}$ for all $x \in X^1_v$ and $N_H(x) = X^2_{v_1} \cup X^1_{v_{n-1}}$ for all $x \in X^2_v$.
- If n is odd, then $N_H(x) = X^1_{v_1} \cup X^1_{v_{n-1}}$ for all $x \in X^1_v$ and $N_H(x) = X^2_{v_1} \cup X^2_{v_{n-1}}$ for all $x \in X^2_v$.

Clearly, this implies that (G, X, H) is a C-configuration, and the claim is proved. △

This settles Case 2. Hence in both cases we proved that (G, X, H) is a constructible configuration. This completes the proof of Theorem 1.17. □

Corollary 1.18 *Let (G, X, H) be a degree-feasible configuration. If (G, X, H) is minimal uncolorable, then for each block $B \in \mathfrak{B}(G)$, there is a uniquely determined cover (X^B, H^B) of B such that the following statements hold:*

(a) *For every block* $B \in \mathfrak{B}(G)$, (B, X^B, H^B) *is either a* K*-configuration or a* C*-configuration.*
(b) *The graphs* H^B *with* $B \in \mathfrak{B}(G)$ *are pairwise disjoint and* $H = \bigcup_{B\in\mathfrak{B}(G)} H^B$.
(c) *For every vertex* v *of* G, $X_v = \bigcup_{B\in\mathfrak{B}_v(G)} X_v^B$.

Corollary 1.19 (Bernshteyn, Kostochka, and Pron) *A connected multigraph* G *is not DP-degree-colorable if and only if each block of* G *is a DP-brick.*

The following result is a Brooks-type theorem for the DP-chromatic number, it clearly implies Brooks' theorem for the chromatic number.

Theorem 1.20 (Bernshteyn, Kostochka, and Pron) *Let* G *be a connected multigraph. Then* $\chi_{\rm DP}(G) \le \Delta(G)+1$, *where equality holds if and only if* G *is a DP-brick.*

Proof That $\chi_{\rm DP}(G) \le \Delta(G)+1$ follows from (1.8). That every DP-brick G satisfies $\chi_{\rm DP}(G) = \Delta(G)+1$ is obvious: consider a K-configuration respectively a C-configuration. Now assume that $\chi_{\rm DP}(G) = \Delta(G)+1$. Then there is a cover (X,H) of G such that $|X_v| \ge \Delta(G)$ for all $v \in V(G)$ and G is not (X,H)-colorable, and so (G,X,H) is an uncolorable degree-feasible configuration. Hence there is a spanning subgraph H' of H such that (G,X,H') is minimal uncolorable. Then G is regular (by Proposition 1.15(a)) and each block of G is a DP-brick (by Theorem1.17). Since any DP-brick is also regular, this implies that G has only one block and is therefore a DP-brick. This completes the proof. □

1.5 Equitable Coloring of Graphs

An **equitable** k**-coloring** of a graph G is a k-coloring of G, in which the sizes of any two color classes differ by at most 1. Let G be a graph and $k \ge 0$ be an integer. If $|G| \le k$, then G has an equitable k-coloring. If $|G| \ge k \ge 1$, then G has an equitable k-coloring if and only if $V(G)$ has a partition $\mathcal{X}$ into k independent sets of G such that $\lfloor |G|/k \rfloor \le |I| \le \lceil |G|/k \rceil$ for $I \in \mathcal{X}$.

In 1964 Erdős [333] proposed the conjecture that any graph G with $\Delta(G) \le k$ admits an equitable $(k+1)$-coloring. That this is indeed the case was proved in 1970 by Hajnal and Szemerédi [458] with a long and complicated argument. A shorter proof for the Hajnal–Szemerédi theorem was given in 2008 by Kierstead and Kostochka [574]. A refinement of this proof was obtained two years later by Kierstead, Kostochka, Mydlarz, and Szemerédi [577].

Theorem 1.21 (Hajnal and Szemerédi) *Let* G *be a graph with* $\Delta(G) \le k$ *for an integer* $k \ge 0$. *Then* G *has an equitable* $(k+1)$*-coloring.*

Before we discuss the proof of the above result, we need some preparation. In the following let G be an arbitrary graph, let $p \in \mathbb{N}$, and let φ be a coloring of G with (a finite) color set C. For a subset $A \subseteq C$ and a vertex $v \in V(G)$, let $G_\varphi(A) = G[\varphi^{-1}(A)]$,

$n_\varphi(A) = |G_\varphi(A)|$, $N_{G,\varphi}(v:A) = N_G(v) \cap \varphi^{-1}(A)$, and $d_{G,\varphi}(v:A) = |N_{G,\varphi}(v:A)|$. We omit the index φ if it is clear to which coloring we refer, and we identify the set $A = \{c\}$ with the element c. We call φ a **p-uniform C-coloring** if $n(c) = p$ for all colors $c \in C$. We call φ a **nearly p-uniform C-coloring** if there are two colors, say $c^+, c^- \in C$ such that $n(c^+) = p+1$, $n(c^-) = p-1$ and $n(c) = p$ for all colors $c \in C \setminus \{c^+, c^-\}$, and we call c^+ the **positive color** and c^- the **negative color** under φ. Note that in both cases we have $|G| = p|C|$. Our goal is to show that a nearly p-uniform C-coloring of G can be transformed into a p-uniform C-coloring of G, using a sequence of elementary Kempe changes.

To control the recoloring process, we use a simple multidigraph associated with (G,φ), denoted by $D = D(G,\varphi)$ (see Appendix C.12 for the notation). The vertex set of D is the color set C, and two distinct colors $c, c' \in C$ are joined in D by an edge cc' (directed from c to c') if there is a vertex $v \in V(G)$ with $\varphi(v) = c$ and $N_{G,\varphi}(v:c') = \varnothing$; in this case we say that v is **(c,c')-movable** under φ. Let $M_\varphi(c,c')$ denote the set of vertices $v \in V(G)$ that are (c,c')-movable under φ. Hence, $M_\varphi(c,c') \neq \varnothing$ if and only if $cc' \in E(D)$. The following is worth noting. If $P = (c_1, c_2, \ldots, c_h)$ is a directed path of D and $v_i \in M_\varphi(c_i, c_{i+1})$ for $i \in [1, h-1]$, then we obtain a new coloring φ' with color set C from φ by recoloring v_i with c_{i+1} for $i \in [1, h-1]$; in this case we write $\varphi' = \varphi/P$ or $\varphi' = \varphi/(P, v_1, v_2, \ldots, v_{h-1})$ and say that φ' is obtained by **recoloring** φ along P. Then for a vertex $v \in V(G)$ we have

$$\varphi'(v) = \varphi(v) \text{ when } v \neq v_i, \text{ and } \varphi'(v_i) = \varphi(v_{i+1}) = c_{i+1} \text{ for } i \in [1, h-1], \tag{1.10}$$

and, moreover, for a color $c \in C$ we have

$$n_{\varphi'}(c) = n_\varphi(c) \text{ if } c \notin \{c_1, c_h\}, n_{\varphi'}(c_1) = n_\varphi(c_1) - 1, \text{ and } n_{\varphi'}(c_h) = n_\varphi(c_h) + 1. \tag{1.11}$$

For a vertex $c \in V(D)$ and a set $X \subseteq V(D)$, we denote by $N_D^+(c)$ the set of out-neighbors of c in D and set $d_D^+(c:X) = |N_D^+(c) \cap X|$.

Remark 1.22 Let G be a graph, let φ be a nearly p-uniform C-coloring of G with $p \geq 1$, and let $D = D(G,\varphi)$. Then we use the following notation. We denote by c^+ and c^- the positive and the negative color under φ, respectively. Furthermore, let A denote the set of colors $c \in C$ such that D has a directed path from c to c^-, and put $B = C \setminus A$. We say that a color $c' \in A$ is **avoidable** under φ if for every color $c \in A \setminus \{c'\}$ there exists in D a directed path P from c to c^- such that $c' \notin V(P)$. Let A' be the set of colors $c' \in A$ that are avoidable under φ. Let B' denote the set of colors $c \in B$ such that D contains a directed path from c^+ to c. Note that $c^- \in A$ and that $c^- \in A'$ only if $A = A' = \{c^-\}$. Also note that $c^+ \in A \cup B'$. The cardinalities of the sets A, A', B and B' are denoted by a, a', b and b', respectively. If we want to make clear that we refer to the coloring φ, we also write $A(\varphi)$, $a(\varphi)$ and so on.

Lemma 1.23 *Let G be a graph, let φ be a nearly p-uniform C-coloring of G with $|C| = k+1 \geq 2$ and $p \geq 1$, and let $D = D(G,\varphi)$. If $d_G(v) \leq k$ for every vertex v with $\varphi(v) \in A' \cup B$, then the following statements hold:*

(a) *If $v \in V(G)$ and $\varphi(v) \in B$, then $d_G(v:c) \geq 1$ for every color $c \in A$, $d_G(v:A) \geq a$, and $d_G(v:B) \leq b-1$.*
(b) *If $v \in V(G)$ and $\varphi(v) \in B'$, then $d_G(v:c) \geq 1$ for every color $c \in B \setminus B'$.*
(c) *If $c^+ \in A$, then G has a p-uniform C-coloring.*
(d) *If $c^+ \in B$, then $n_\varphi(A) = ap-1$, $n_\varphi(B) = bp+1$, $c^- \in A \setminus A'$, and $1 \leq a' < a$.*
(e) *If $c^+ \in B$, then $b \leq a'$ or $d^+_D(c:A) < b$ for some color $c \in A'$.*

Proof Let $c' = \varphi(v)$. If $c' \in B$, then for all colors $c \in A$, we have $M_\varphi(c',c) = \varnothing$ and hence $d_G(v:c) \geq 1$ (for otherwise there would be a directed path in D from c' to c^-, which is impossible). Therefore, $d_G(v:A) \geq |A| = a$, which implies that $d_G(v:B) \leq |B|-1 = b-1$, since $d_G(v:A)+d_G(v:B) = d_G(v) \leq k = a+b-1$. If $c' \in B'$, then for all colors $c \in B \setminus B'$, we have $M_\varphi(c',c) = \varnothing$ and hence $d_G(v:c) \geq 1$ (for otherwise there would be a directed path in D from c^+ to c, which is impossible). If $c^+ \in A$, then D has a directed path P from c^+ to c^-, and it follows from (1.11) that $\varphi' = \varphi/P$ is a p-uniform C-coloring of G. This proves (a), (b), and (c).

For the proof of (d) and (e), suppose that $c^+ \in B$. Note that $c^- \in A$ and $|C| = a+b = k+1$. Since φ is a nearly p-uniform C-coloring of G, we obtain that $n_\varphi(A) = ap-1$ and $n_\varphi(B) = bp+1$. If $c^- \in A'$, then $A = A' = \{c^-\}$, which implies that $n_\varphi(A') = n_\varphi(A) = p-1$ and $n_\varphi(B) = kp+1$. Since $d_G(v) \leq k$ for all $v \in \varphi^{-1}(A')$, this gives $|E_G(\varphi^{-1}(A'), \varphi^{-1}(B))| \leq kn_\varphi(A') = k(p-1) < n_\varphi(B)$, a contradiction to (a). So $c^- \in A \setminus A'$ and hence $a' < a$. The first vertex of a longest directed path in D ending in c^- is in A', hence $a' \geq 1$. This proves (d). For each color $c \in A$ there is a directed path in D from c to c^-. Hence there is an ordering of the a colors in A, say $c_1 = c^-, c_2, \ldots, c_a$ such that each color c_i with $i \in [2,a]$ has an out-neighbor c_j in D with $j < i$. Let ℓ be the largest index such that c_ℓ is nonavoidable under φ, i.e. $c_\ell \in A \setminus A'$. Then $a' \geq a-\ell$. Since $c_\ell \in A \setminus A'$, there is a color $c \in A$ such that $D - c_\ell$ has no directed path from c to $c^- = c_1$. This implies that $c = c_j$ with $j > \ell$ and $N^+_D(c) \cap \{c_1, c_2, \ldots, c_{\ell-1}\} = \varnothing$, which gives $d^+_D(c:A) \leq a-\ell$. This implies (e), since $a-\ell < b$ or $b \leq a-\ell \leq a'$. □

Lemma 1.24 *Let G be a graph, let φ be a nearly p-uniform C-coloring of G with $|C| = k+1 \geq 2$ and $p \geq 1$, and let $D = D(G,\varphi)$. If $d_G(v) \leq k$ for every vertex v with $\varphi(v) \in A' \cup B$, then G has a p-uniform C-coloring.*

Proof The proof is by contradiction. So let (G,C,φ) be a counterexample for which $b = b(\varphi)$ is minimum. We consider the set $L \subseteq V(G) \times V(G)$ defined by

$$L = \{(v,w) \mid \varphi(v) \in B, \varphi(w) \in A' \text{ and } N_G(v:\varphi(w)) = \{w\}\}.$$

We need the following statement.

(a) If $(v,w) \in L$, then $d_G(w:c) \geq 1$ for all colors $c \in A \setminus \{\varphi(w)\}$, and as a consequence $d_G(w:A) \geq a-1$ and $d_G(w:B) \leq b$.

Proof of (a): Suppose this is false. Let $c_0 = \varphi(v)$ and $c_1 = \varphi(w)$. Then there is a color, say $c_2 \in A \setminus \{c_1\}$ such that $d_G(w:c_2) = 0$, which implies that $w \in M_\varphi(c_1,c_2)$. Since $c_1 \in A'$, there is a directed path in D from c_2 to c^- avoiding c_1. Hence, there is a directed path $P = (c_1, c_2, \ldots, c_h = c^-)$ and a vertex $v_i \in M_\varphi(c_i, c_{i+1})$ for $i \in [1, h-1]$ with $v_1 = w$. Note that $V(P) \subseteq A$ and $N_{G,\varphi}(v : c_1) = \{w\}$. Then $\varphi_1 = \varphi/(P, v_1, v_2, \ldots, v_{h-1})$ is a nearly p-uniform C-coloring of G with $N_{G,\varphi_1}(v : c_1) = \varnothing$ and $c^-(\varphi_1) = c_1$. So if we recolor v with c_1, we get a coloring φ' of G, such that $n_{f'}(c) = p$ for all colors $c \in A$. Hence φ' restricted to $G_1 = G_{\varphi'}(A)$ is a p-uniform A-coloring φ_1 of G_1. Let φ_2 be the restriction of φ' to $G_2 = G_{\varphi'}(B)(= G_\varphi(B) - v)$. If $c_0 = c^+(\varphi)$, then φ_2 is a p-uniform B-coloring of G_2. Then $\varphi_1 \cup \varphi_2$ is a p-uniform C-coloring of G, a contradiction. If $c_0 \neq c^+(\varphi)$, then φ_2 is a nearly p-uniform B-coloring of G_2 with $b(\varphi_2) < b = b(\varphi)$. From Lemma 1.23(a) it follows that $|G_2| = pb$ and and $\Delta(G_2) \leq b-1$ (note that $b = |B|$ and $G_2 = G_\varphi(B) - v$). By the choice of (G, C, φ), this implies that there is a p-uniform B-coloring φ_2' of G_2. Then $\varphi_1 \cup \varphi_2'$ is a p-uniform C-coloring of G, a contradiction. This proves (a). ◇

If $(v, w) \in L$, then $e = vw$ is an edge of G and we call e an L-edge; note that $\varphi(v) \in B$ and $\varphi(w) \in A'$. The ends of an L-edge are called L-vertices and L-neighbors of each other. Note that we have $a + b = k + 1$, $d_G(v) = d_G(v : A) + d_G(v : B)$ for all $v \in V(G)$, and $|G| = p(k+1)$. Moreover, by assumption and Lemma 1.23(c), we get $c^+ \in B$ and $d_G(u) \leq k$ whenever $\varphi(u) \in A \cup B'$. To complete the proof of Lemma 1.24 we consider two cases according to Lemma 1.23(e).

Case 1: $d_D^+(c : A) < b$ *for some color* $c \in A'$. Since $|A| = a$, this implies that we have $|A \setminus N_D^+(c : A)| \geq a - b$. Hence, every vertex $w \in \varphi^{-1}(c)$ satisfies $d_G(w : A) \geq a - b$ and $d_G(w : B) \leq k - d_G(w : A) \leq 2b - 1$. Let S be the set of L-vertices belonging to $\varphi^{-1}(c)$, let $T = \varphi^{-1}(c) \setminus S$, and let $N = N_G(S) \cap \varphi^{-1}(B)$. If $v \in \varphi^{-1}(B) \setminus N$, then v has no L-neighbor in $\varphi^{-1}(c)$, which gives $|N_G(v) \cap T| \geq 2$ (by Lemma 1.23(a)). Hence we obtain

$$2|\varphi^{-1}(B) \setminus N| \leq |E_G(T, \varphi^{-1}(B) \setminus N)| \leq (2b-1)|T| \leq 2b|T|.$$

Since $d_G(w) \leq k$ for all $w \in S$, we obtain that $|E_G(S, \varphi^{-1}(A))| + |N| \leq k|S|$. We have $c \notin \{c^-, c^+\}$ and $|\varphi^{-1}(B)| = n_\varphi(B) = pb + 1$ (by Lemma 1.23(d)), which gives, in particular, that $|S| + |T| = |\varphi^{-1}(c)| = p$. Summarizing, it follows that

$$\begin{aligned} bp + |E_G(S, \varphi^{-1}(A))| &< |\varphi^{-1}(B) \setminus N| + |N| + |E_G(S, \varphi^{-1}(A))| \\ &\leq b|T| + k|S| = bp + (a-1)|S|, \end{aligned}$$

which gives $|E_G(S, \varphi^{-1}(A))| < (a-1)|S|$. This implies that $d_G(w : A) \leq a - 2$ for some L-vertex $w \in S$, contradicting statement (a).

Case 2: $b \leq a'$. Let $v \in \varphi^{-1}(B')$ be an arbitrary vertex. Denote by $\ell(v)$ the number of L-neighbors of v (which all belong to $\varphi^{-1}(A')$ and have different colors). From Lemma 1.23(a)(b) it follows that we have

$$d_G(v : A) \geq a + a' - \ell(v) \text{ and } d_G(v : B) - d_G(v : B') \geq b - b',$$

which gives $\ell(v) \geq a' - b' + d_G(v : B') + 1$ (note that $d_G(v) \leq k = a + b - 1$). The graph $G_\varphi(B')$ has a maximal independent set I containing $\varphi^{-1}(c^+)$. Then we get $\sum_{v \in I}(d_G(v : B') + 1) \geq n_\varphi(B') = b'p + 1$. Since $a' \geq b \geq b'$ (by the assumption of the case), $|I| \geq n_\varphi(c^+) = p+1$ and $c^- \in A \setminus A'$ (by Lemma 1.23(d)), we obtain that

$$\sum_{v \in I} \ell(v) \geq \sum_{v \in I} (a' - b' + d_G(v : B') + 1)$$
$$\geq (a' - b')(p+1) + b'p + 1 > a'p = n_\varphi(A').$$

Consequently, there are two distinct vertices in $I \subseteq \varphi^{-1}(B')$, say v_1 and v_2, and a vertex $w \in \varphi^{-1}(A')$ such that $(v_i, w) \in L$ for $i \in \{1,2\}$. Let $c_1 = \varphi(w)$, $A_1 = A \setminus \{c_1\}$ and $B_1 = B \cup \{c_1\}$. The multidigraph $D[A]$ has a directed path $P = (c_1, c_2, \ldots, c_h = c^-)$. Hence there is a vertex $w_1 \in M_\varphi(c_1, c_2)$ such that recoloring φ along P results in p-uniform A_1-coloring φ_1 of $G_1 = G_\varphi(A) - W_1$ with $W_1 = (\varphi^{-1}(c_1) \setminus \{w_1\})$. From (a) it follows that $w_1 \neq w$. Since $\varphi(v_1) \in B'$, $D[B]$ contains a directed P from $c^+(\varphi)$ to $\varphi(v_1)$. Recoloring φ along P leads to a p-uniform B-coloring φ' of $G_\varphi(B) - v_1$. Note that $N_G(v_i) \cap W_1 = \{w\}$ for $i \in \{1,2\}$. From (a) it follows that there is a color $c' \in B$ such that every vertex $v \in N_G(w) \setminus \{v_1\}$ satisfies $\varphi'(v) \neq c'$. If we color w with c' and v_1 with c_1, we get a nearly p-uniform B_1-coloring ϕ of $G_2 = G - V(G_1)(= G[\varphi^{-1}(B_1)] - w_1)$, where c_1 is the negative color of ϕ and $\phi^{-1}(c_1) = W'$ with $W' = (W_1 \setminus \{w\}) \cup \{v_1\}$. Since v_1 and v_2 belong to the independent set I of G, we have $N_G(v_2) \cap W' = \emptyset$. Hence $\phi(v_2)c_1$ is a directed edge of $D(G, \phi)$, which gives $c_1 = c^-(\phi) \in A(\phi) \setminus A'(\phi)$, $\phi(v_2) \in A(\phi)$, $b(\phi) < b = |B_1| - 1$, $U = A'(\phi) \cup B(\phi) \subseteq B(\varphi)$, and $d_{G_2}(v) \leq b$ for $v \in U$ (using Lemma 1.23(a) and $d_G(v) \leq k$). By the choice of (G, C, φ), there exists a p-uniform B_1-coloring φ_2 of G_2. Then $\varphi_1 \cup \varphi_2$ is a p-uniform C-coloring of G, which is impossible. This completes the proof of Lemma 1.24. □

Proof of Theorem 1.21: Let G be a graph with $\Delta(G) \leq k$ and $k \geq 0$. To show that G has an equitable $(k+1)$-coloring, we may assume that $|G|$ is divisible by $k+1$. For otherwise, we have $|G| = p(k+1) - r$, where $1 \leq r \leq k$. If G' is the disjoint union of G and K_r, then $|G'|$ is divisible by $k+1$, $\Delta(G') \leq k$, and the restriction of any equitable $(k+1)$-coloring of G' to G is an equitable $(k+1)$-coloring of G.

To prove that every graph with maximum degree at most k and order divisible by $k+1$ has an equitable $(k+1)$-coloring, we apply induction on the number m of edges of the graph. The base step $m = 0$ is trivial. So consider the induction step $m \geq 1$. Let G be a graph with m edges such that $\Delta(G) \leq k$ and $|G| = p(k+1)$. Then $k, p \geq 1$. Let $e = uv$ be an edge of G. Since $\Delta(G - e) \leq k$, the induction hypothesis implies that there exists an equitable $(k+1)$-coloring $\tilde{\varphi}$ of $G - e$, say with color set $C = [1, k+1]$. Then, for each color $c \in C$, the color class $\tilde{\varphi}^{-1}(c)$ contains p vertices. If $\tilde{\varphi}(u) \neq \tilde{\varphi}(v)$, then $\tilde{\varphi}$ is also an equitable $(k+1)$-coloring of G and we are done. Otherwise, one color class $\tilde{\varphi}^{-1}(c)$ contains both u and v. Since u has at most $k-1$ neighbors in $G - e$ and $|C| = k+1$, there is another color class $\tilde{\varphi}^{-1}(c')$ with $c' \in C \setminus \{c\}$ such that no neighbor of u in $G - e$ belongs to $\tilde{\varphi}^{-1}(c')$. Then we can recolor u with c'. This results in a coloring φ of G with color set C such that $|\varphi^{-1}(c)| = p - 1$, $|\varphi^{-1}(c')| = p + 1$ and $|\varphi^{-1}(c^*)| = p$ for all colors $c^* \in C \setminus \{c, c'\}$, that is, φ is a nearly p-unifom C-coloring of G. Then Lemma 1.24 implies that G has an equitable $(k+1)$-coloring. □

The **equitable chromatic number** $\chi^=(G)$ of a graph G is the least integer k for which G admits an equitable k-coloring. The equitable chromatic number is a graph parameter and any graph G satisfies

$$\chi(G) \leq \chi^{=}(G) \leq \Delta(G) + 1, \tag{1.12}$$

where the latter inequality is a consequence of Theorem 1.21 and the former inequality follows from the fact that every equitable k-coloring is a particular k-coloring. Unlike the chromatic number, the equitable chromatic number is not monotone. If G is the disjoint union of two disjoint copies of $K_{1,n}$, then $\chi(G) = \chi^{=}(G) = 2$, but $\chi^{=}(K_{1,n}) = 1 + \lceil \frac{n}{2} \rceil$. The complete bipartite graph $K_{2p+1,2p+1}$ has an equitable 2-coloring, but no equitable $(2p+1)$-coloring.

In view of Brooks' theorem and (1.12), it would be of interest to know whether the complete graphs and the odd cycles are the only connected graphs G with $\chi^{=}(G) = \Delta(G) + 1$. The question whether such a Brooks-type result holds for the equitable chromatic number was first raised by Meyer [734]. Lih and Wu [677] have shown that the answer is affirmative for all connected bipartite graphs; they actually proved that if G is a connected bipartite graph with maximum degree Δ different from any complete bipartite graph $K_{2n+1,2n+1}$ with $n \geq 1$, then G has an equitable Δ-coloring. This result led Chen, Lih, and Wu [207] to propose the following conjecture.

Conjecture 1.25 (The Equitable Δ-Coloring Conjecture) *Let G be a connected graph with maximum degree Δ, where $\Delta \geq 1$. Then G has no equitable Δ-coloring if and only if $G = K_{\Delta+1}$, or $\Delta = 2$ and G is an odd cycle, or Δ is odd and $G = K_{\Delta,\Delta}$.*

A graph has a k-coloring if and only if each of its component has one. However, for equitable k-coloring this need not be the case as we saw above for $k = 2$. Kierstead and Kostochka [576] extended the equitable Δ-coloring conjecture to arbitrary graphs.

Conjecture 1.26 (Kierstead and Kostochka) *Suppose that $\Delta \geq 6$ and G is a Δ-colorable graph with maximum degree at most Δ. Then G has no equitable Δ-coloring if and only if Δ is odd and the vertex set of G can be partitioned into sets $V_0, V_1, \ldots, V_t$ such that $G[V_0] = K_{\Delta,\Delta}$ and $G[V_i] = K_\Delta$ for $i \in [1,t]$.*

1.6 Critical Graphs and Gallai's Theorem

As previously mentioned, the chromatic number is a monotone graph parameter. The chromatic number decreases by at most one if an edge or a vertex is deleted, that is, if G is a graph and $t \in V(G) \cup E(G)$, then

$$\chi(G) - 1 \leq \chi(G - \{t\}) \leq \chi(G). \tag{1.13}$$

An element $t \in V(G) \cup E(G)$ is said to be **critical** if $\chi(G - \{t\}) \leq \chi(G) - 1$. A graph G is defined to be **critical**, respectively k**-critical**, if $\chi(H) < \chi(G) = k$ for every proper subgraph H of G. By (1.1) it follows that a graph G is critical if and only if all edges and vertices of G are critical.

Let us first discuss why critical graphs form a useful concept. Suppose we are interested in a graph property $\mathcal{P}$, which is **monotone**, that is, $H \subseteq G \in \mathcal{P}$ implies $H \in \mathcal{P}$. There are many examples for such graph properties, including the class of all graphs, the class of planar graphs, and the class of triangle-free graphs. Suppose also that ρ is a graph parameter for $\mathcal{P}$, that is, a mapping that assigns to each graph of $\mathcal{P}$ a real number such that $\rho(G) = \rho(H)$ whenever G and H are isomorphic graphs of $\mathcal{P}$. If the aim is to bound the chromatic number for the graphs of $\mathcal{P}$ from above by the parameter ρ, then we can apply the critical graph method, provided that ρ is **monotone**, that is, $H \subseteq G \in \mathcal{P}$ implies $\rho(H) \leq \rho(G)$.

Proposition 1.27 (Critical Graph Method) *Let $\mathcal{P}$ be a monotone graph property and let ρ be a monotone graph parameter defined for $\mathcal{P}$. Then the following statements hold:*

(a) *For every graph $G \in \mathcal{P}$ there exists a critical graph $H \in \mathcal{P}$ such that $H \subseteq G$ and $\chi(H) = \chi(G)$.*
(b) *If $\chi(H) \leq \rho(H)$ for every critical graph $H \in \mathcal{P}$, then $\chi(G) \leq \rho(G)$ for every graph $G \in \mathcal{P}$.*

Proof Every graph G contains a minimal subgraph H with $\chi(H) = \chi(G)$. Since χ is a monotone graph parameter and $\mathcal{P}$ is a monotone graph property, H is critical (with respect to χ) and $H \in \mathcal{P}$. This proves (a). For the proof of (b), let $G \in \mathcal{P}$ be an arbitrary graph. By (a), G contains a critical subgraph $H \in \mathcal{P}$ with $\chi(H) = \chi(G)$. Then $\chi(H) \leq \rho(H)$; and since ρ is monotone, we obtain $\chi(G) = \chi(H) \leq \rho(H) \leq \rho(G)$, as required. □

Proposition 1.28 *The following statements hold:*

(a) *If G is k-critical, H is k-chromatic and $H \subseteq G$, then $H = G$.*
(b) *$\chi(G) \geq k$ if and only if there is a k-critical graph H with $H \subseteq G$.*

Proof By definition, a k-critical graph cannot contain a k-chromatic graph as a proper subgraph. This proves (a). If a graph G contains a k-critical graph H as a subgraph, then $\chi(G) \geq \chi(H) = k$. Conversely, if $\chi(G) \geq k$, then it follows from (1.13) that there is a subgraph G' of G with $\chi(G') = k$. By Lemma 1.27(a), G' and hence G contains a k-critical subgraph H. □

For an integer $k \geq 0$, let $\text{Crit}(k)$ denote the class of all k-critical graphs. Proposition 1.28(b) says that the class $\mathcal{G}(\chi \leq k-1)$ of graphs with chromatic number at most $k-1$ can be characterized by a family of forbidden subgraphs, where the family consists of all k-critical graphs, that is,

$$\mathcal{G}(\chi \leq k-1) = \text{Forb}_{\subseteq}(\text{Crit}(k))$$

(see Appendix C.11). Clearly, $\text{Crit}(0) = \{\varnothing\}$, $\text{Crit}(1) = \{K_1\}$, and $\text{Crit}(2) = \{K_2\}$. From König's characterization of bipartite graphs (Theorem C.7) we deduce that

$$\text{Crit}(3) = \{C_n \mid n \equiv 1 \pmod 2\}.$$

However, no reasonable characterization of 4-critical graphs, or equivalently of 3-colorability, seems possible. If we want to check whether a connected graph is k-critical it suffices to investigate all edge deleted subgraphs.

Proposition 1.29 *For an integer $k \geq 2$, a graph G is k-critical if and only if G is connected, $\chi(G) \geq k$ and $\chi(G-e) \leq k-1$ for all $e \in E(G)$.*

Proof If G is k-critical, then $\chi(G) = k$ and $\chi(G-e) \leq k-1$ for all $e \in E(G)$, since $G - e \subset G$. Furthermore, it follows from (1.5) that G is connected. This proves the "only if" implication. To see the "if" implication, we first note that $\chi(G) \geq k \geq 2$ implies that G has at least one edge e. Since $\chi(G-e) \leq k-1 < \chi(G)$, it follows from (1.13) that $\chi(G) = k$. Let $H \subset G$. Since G is connected, we deduce that there is an edge $e \in E(G)$ such that $H \subseteq G - e$, which gives $\chi(H) \leq \chi(G-e) \leq k-1$. This shows that G is k-critical. □

Proposition 1.30 *Let G be a graph. Assume that $G = G_1 \cup G_2$ is the union of two induced subgraphs G_1 and G_2 of G. If $G_1 \cap G_2$ is a complete graph, then $\chi(G) = \max\{\chi(G_1), \chi(G_2)\}$.*

Proof By symmetry, we may assume that $\chi(G_1) \leq \chi(G_2)$. Then, for $i = 1, 2$, there is a coloring φ_i of G_i with a set C_i of $\chi(G_i)$ colors, where the color sets are chosen so that $C_1 \subseteq C_2$. Since $X = V(G_1) \cap V(G_2)$ is a clique of G (and hence of G_i), we have $|\varphi_1(X)| = |\varphi_2(X)| = |X|$. Thus we can permute the colors of G_1 so that $\varphi_1(v) = \varphi_2(v)$ for all $v \in X$. Since $E_G(V(G_1) \setminus X, V(G_2) \setminus X) = \varnothing$, it then follows that $\varphi = \varphi_1 \cup \varphi_2$ is a coloring of G with color set C_2 and so $\chi(G) \leq \chi(G_2)$. Since $G_2 \subseteq G$, we have $\chi(G_2) \leq \chi(G)$. Thus $\chi(G) = \chi(G_2) = \max\{\chi(G_1), \chi(G_2)\}$. □

Corollary 1.31 *If G is a k-critical graph with $k \geq 1$, then no separating set of G is a clique, and as a consequence G is 2-connected, unless $k \leq 2$.*

Theorem 1.32 (Dirac) *If G is a k-critical graph with $k \geq 1$, then G is $(k-1)$-edge-connected, in particular $\delta(G) \geq k-1$.*

Proof Since G is k-critical, $|G| \geq k$. If $k = 1, 2$, then G is a K_k and the statement is evident. So assume $k \geq 3$. Then there is a minimum cut $F = \partial_G(X)$, i.e., $\varnothing \neq X \subset V(G)$ and $|F| = \kappa'(G)$. Since G is critical, G is connected and so $F \neq \varnothing$. Now let $uv \in F$ be an arbitrary edge. Then $\chi(G-uv) \leq k-1$ and in any $(k-1)$-coloring of $G - uv$ the vertices u and v receive the same color c. For any of the remaining $k-2$ colors c', the vertices u and v are joined by a path whose vertices are colored alternately with c and c'. Otherwise, by a Kempe-change, we would get a $(k-1)$-coloring of $G - uv$, where u and v receive different colors, which would be a $(k-1)$-coloring of G, giving a contradiction. Hence, together with the edge uv, there are $k-1$ edge disjoint paths between u and v. This implies that $\kappa'(G) = |F| \geq k-1$, as required. □

Theorem 1.32 leads to a natural way of classifying the vertices of a critical graph into two classes. The vertices of a k-critical G having degree $k-1$ are called **low vertices** of G, and the remaining vertices are called **high vertices** of G. So any high vertex of G has degree at least k in G. Furthermore, let G_L be the subgraph of G induced by the low vertices, and let G_H be the subgraph of G induced by the high vertices. We call G_L the **low vertex subgraph** of G and G_H the **high vertex subgraph** of G. This classification is due to Gallai [397] who proved the following extension of Brooks' theorem.

Theorem 1.33 (Gallai) *The low vertex subgraph of any critical graph is a Gallai forest.*

Proof Let G be a k-critical graph. If $k \leq 2$, then $G_L = G = K_k$ and so the statement trivially holds. So assume that $k \geq 3$. If $G_L = \varnothing$, then there is nothing to prove. Otherwise, let H be a component of G_L. Since G is critical, there is a coloring φ of $G - V(H)$ with a set C of $k-1$ colors. For $v \in V(H)$, let $L(v) = C \setminus \varphi(N_G(v) \setminus V(H))$. Then, for all $v \in V(H)$, $|L(v)| \geq d_H(v)$, since $d_G(v) = |C|$. But $\chi(G) = k$, so H is not L-colorable. Consequently, (H, L) is an uncolorable pair, which implies that H is a Gallai tree (by Theorem 1.7). □

As an immediate consequence of Gallai's characterization of the low vertex subgraph of a critical graph, we deduce that any nonempty critical graph G satisfies the following version of Brooks' theorem

$$G_H = \varnothing \Leftrightarrow \chi(G) = \Delta(G) + 1 \Leftrightarrow G \text{ is a brick.}$$

Proposition 1.34 *Let G be a k-chromatic graph with $k \geq 2$, and let φ be a coloring of G with a set C of k colors. Then the following statements hold:*

(a) *For every color $c \in C$, the color class $\varphi^{-1}(c)$ contains a vertex v which has a neighbor in each color class $\varphi^{-1}(c')$ whenever $c' \in C \setminus \{c\}$.*
(b) *G has at least $\binom{k}{2}$ edges.*
(c) *If $C' \subseteq C$ and $X = \bigcup_{c \in C'} \varphi^{-1}(c)$, then $\chi(G[X]) = |C'|$.*

Proof In order to prove (a), assume that it is false. So there is a color class $\varphi^{-1}(c)$ such that for each vertex $v \in \varphi^{-1}(c)$ there is a color $c_v \neq c$ with $N_G(v) \cap \varphi^{-1}(c_v) = \varnothing$. Then we can recolor each vertex $v \in \varphi^{-1}(c)$ with color c_v. This results in a coloring of G with color set $C' = C \setminus \{c\}$ and so $\chi(G) \leq |C'| = k - 1$, which is a contradiction.

Statement (a) implies, in particular, that if φ is a k-coloring of G, then there is at least one edge between any two distinct color classes, which gives $|E(G)| \geq \binom{k}{2}$. Thus (b) is proved.

If (c) is false, then $X = \bigcup_{c \in C'} \varphi^{-1}(c)$ satisfies $\chi(G[X]) \leq |C'| - 1$ and, therefore, there exists a coloring φ_1 of $G[X]$ with a set C_1 of at most $|C'| - 1$ colors, where $C_1 \subseteq C'$. Clearly, $\varphi_2 = \varphi|_{V(G) \setminus X}$ is a coloring of $G - X$ with color set $C_2 = C \setminus C'$. Then $\varphi^+ = \varphi_1 \cup \varphi_2$ is a coloring of G with color set $C^+ = C_1 \cup C_2$ and so $\chi(G) \leq |C^+| = |C_1| + |C_2| = |C| - 1 = k - 1$, which is a contradiction. □

1.7 Complete Graphs, Cycles and Trees

Let G be a graph, and let H be a subgraph of G. A **chord** of H in G is an edge $uv \in E(G) \setminus E(H)$ such that $u, v \in V(H)$. If H has no chord in G, then H is said to be **chordless**. Obviously, H is a chordless subgraph of G if and only if H is an induced subgraph of G.

Clearly, any even cycle of a Gallai forest has at least two chords. That this property characterizes Gallai forests was proved by Gallai [397, Satz (1.9)] who used this result for proving Theorem 1.33; this argument was also used by Carsten Thomassen (see Theorem 5.39).

Theorem 1.35 (GALLAI) *Let G be a graph in which every even cycle has at least two chords. Then any block of G is a brick, and so G is a Gallai forest.*

Proof For the proof it suffices to show that if G is a block, but not a brick, then G contains an even cycle with at most one chord. To this end, choose a cycle C of G of minimum length. Then C is without chords, so if C is even we are done.

If C is a K_3, then let K be a maximal complete subgraph in G. Since G is a block, but not a brick, there are vertices outside K, hence there is a shortest path P of length at least 2 joining two vertices u and v of K with all interior vertices outside K. If P has length 2, then let w be a vertex of K not joined to the interior vertex x of P. Such a vertex exists by the maximality of K. Then $\{u, v, w, x\}$ induces a 4-cycle of G with exactly one chord. If P has length at least 3, then P together with a path of length 1 or 2 in K joining u and v is an even cycle with at most one chord, since otherwise there would be a shorter path P.

Finally, if C is odd and of length at least 5, then again let P be a shortest path of length at least 2 joining two vertices u and v of C and with all interior vertices outside C. The vertices u and v divide C into two paths P_1 and P_2, one of which, say P_1, has the same parity as P. Then $P \cup P_1$ is an even cycle without at a chord, since C and P are shortest. This completes the proof. □

A basic fact in graph theory says that any connected graph G contains a **spanning tree**, that is, a tree $T \subseteq G$ with $V(T) = V(G)$. A spanning tree of a connected graph G is called an **independence tree** of G if the leaves of T form an independent set in G. It is easy to check that the following types of graphs have no independence tree: cycles, complete graphs, and complete bipartite graphs $K_{p,p}$ with $p \geq 1$. That this list is complete was proved by Böhme et al. [124]. As we shall see in Section 2.1, this result can be used to give an alternative proof of Brooks' theorem.

Theorem 1.36 *Let G be a connected graph of order $n \geq 3$. Then the following statements are equivalent:*

(a) *G is C_n, K_n, or $K_{n/2,n/2}$ for n even.*
(b) *G has no independence tree.*
(c) *G has no depth-first-search tree in which all leaves are pairwise nonadjacent in G.*

(d) *G has a Hamilton path, and every Hamilton path of G is contained in a Hamilton cycle of G.*

Proof First we show that (a) implies (b). Clearly, neither C_n nor K_n has an independence tree. So suppose, to the contrary, that $K_{p,p}$ with $p \geq 2$ has an independence tree T. Then all leaves of T would lie in the same part of the complete bipartite graph $K_{p,p}$. Hence all vertices of the other part would have degree at least two in T, which give $|E(T)| \geq 2p = |V(T)|$, a contradiction. This shows that (a) implies (b). Obviously, (b) implies (c).

To prove that (c) implies (d), suppose that G does not satisfies (d). Then G contains a path P whose ends are nonadjacent and have no neighbors in $V(G) \setminus V(P)$, for example a longest path has this property. If u and v are the ends of P, then there exists a depth-first-search tree T with root u for which $P = uTv$, which implies that $V^{(1)}(T)$ is an independent set (see Proposition C.6). So (c) implies (d).

To prove that (d) implies (a), suppose that G satisfies (d). Clearly, G contains a Hamilton cycle C. Choose a cyclic orientation of C so that each vertex v of C has a successor v^+ and a predecessor v^-. For a vertex v, let $v^{+1} = v^+$ and $v^{+(1+i)} = v^{(+i)+}$. For two distinct vertices v and w of C, let vCw (respectively, $vC^{-1}w$) denote the v-w path of C that contains v^+ (respectively, v^-), so that $C = vCw \cup vC^{-1}w$.

If C has no chord, then $G = C$ and we are done. So let vw be an arbitrary chord of C. Then $v^+CwvC^{-1}w^+$ is a Hamilton path of G with ends v^+ and w^+, which implies that that v^+w^+ is a chord of C (by (d)). Likewise, v^-w^- is a chord of C. If $\ell(vCw) \geq 4$, then $v^{++}w$ and v^+w^- are also chords of C. The first follows from (d) and the fact that $v^{++}Cw^-v^-C^{-1}w^+v^+vw$ is a Hamilton path, and the second we get in the same way starting from the chord v^-w^-.

Now assume that C has a chord vw such that $\ell(vCw) = 2$. Let $x = v^+(= w^-)$. Then xw^+ is an edge of G and $xw^{+i} \in E(G)$ implies $xw^{+(i+1)} \in E(G)$, since then $w^{+(i+1)}CvwCw^{+i}x$ is a Hamilton path of G. It then follows that v^+ is adjacent to every vertex of G. Since $v^{+i}w^{+i}$ is a chord of C with $\ell(v^{+i}Cw^{+i}) = 2$, we deduce that $G = K_n$. If no chord vw of C satisfies $\ell(vCw) = 2$, then $\ell(vCw)$ is odd for all chords vw of C; moreover, all chords vw of C with $\ell(vCw) \equiv 1 \pmod 2$ are present in C. Thus $G = K_{p,p}$ with $2p = n$. This completes the proof. □

1.8 Line Graphs and Total Graphs

For a graph or multigraph G, the **line graph** of G, denoted by $\mathrm{L}(G)$, is the graph whose vertex set corresponds to the edge set of G and in which two vertices are adjacent if the corresponding edges of G have a common end. Two distinct edges of G having a common end are called **adjacent edges** of G. A coloring of $\mathrm{L}(G)$ is therefore a map $\varphi : E(G) \to C$ such that $\varphi(e) \neq \varphi(e')$ whenever e and e' are adjacent edges of G. A coloring of $\mathrm{L}(G)$ is also called an **edge coloring** of G. The chromatic number of $\mathrm{L}(G)$ is called the **chromatic index** of G, written $\chi'(G)$; and the list-chromatic number of $\mathrm{L}(G)$ is called the **list-chromatic index** of G, written $\chi'_\ell(G)$. So every graph G satisfies $\chi'(G) = \chi(\mathrm{L}(G))$ and $\chi'_\ell(G) = \chi'_\ell(\mathrm{L}(G))$.

Let G be a multigraph with at least one edge. If v is a vertex of G, then $E_G(v)$ is a clique in $\mathrm{L}(G)$ and so $\omega(\mathrm{L}(G)) \geq \Delta(G)$. Furthermore, $\Delta(\mathrm{L}(G)) \leq 2\Delta(G) - 2$. From (1.7) it then follows that

$$\Delta(G) \leq \chi'(G) \leq \chi'_\ell(G) \leq 2\Delta(G) - 1. \tag{1.14}$$

That the lower bound for the chromatic index is sharp for the class of bipartite multigraphs was proved in 1916 by König [609].

Theorem 1.37 (KÖNIG) *For any bipartite multigraph G, $\chi'(G) = \Delta(G)$.*

In contrast to the chromatic number, multiple edges greatly affect the chromatic index. Given a multigraph G, the maximum number of edges of G with the same two ends is called the **multiplicity** of G, denoted by $\mu(G)$. If G is a multigraph consisting of three vertices that are pairwise joined by m parallel edges, then $\mathrm{L}(G) = K_{3m}$ and so $\chi'(G) = 3m$; moreover, we have $\Delta(G) = 2m$ and $\mu(G) = m$. This graph family shows that the following two upper bounds for the chromatic index of a multigraph are best possible. The first upper bound for χ' in terms of Δ alone was established in 1949 by Shannon [934]; the second upper bound of χ' in terms of Δ and μ was established in 1964 by Vizing [1049] and, independently, by Gupta [438].

Theorem 1.38 (SHANNON) *For any multigraph G, $\chi'(G) \leq \lfloor \frac{3}{2}\Delta(G) \rfloor$.*

Theorem 1.39 (VIZING / GUPTA) *For any multigraph G, $\chi'(G) \leq \Delta(G) + \mu(G)$.*

Another lower bound for the chromatic number is in terms of the independence number. Given a multigraph G, an independent set of $\mathrm{L}(G)$ is a set of pairwise non-adjacent edges of G; such an edge set is called a **matching** of G. The independence number of $L(G)$ is called the **matching number** of G, denoted by $\alpha'(G)$. It follows from (1.2) that $|E(G)| \leq \alpha'(G)\chi'(G)$ for any multigraph G. Since $\alpha'(G) \leq \lfloor \frac{1}{2}|G| \rfloor$, it follows that $|E(G)| \leq \lfloor \frac{1}{2}|G| \rfloor \chi'(G)$. Clearly, this inequality holds for every subgraph of G, too. For a multigraph G, define

$$\mathcal{W}(G) = \max_{H \subseteq G, |H| \geq 2} \left\lceil \frac{|E(H)|}{\lfloor \frac{1}{2}|H| \rfloor} \right\rceil$$

when $|G| \geq 2$, and $\mathcal{W}(G) = 0$ otherwise; the parameter $\mathcal{W}(G)$ is called the **edge density** of G. As Scheinerman and Ullman [898] proved, the maximum can always be achieved for an induced submultigraph H of G having odd order, provided that $|G| \geq 3$. Hence, every multigraph G with $|G| \geq 3$ satisfies

$$\mathcal{W}(G) = \max_{X \subseteq V(G), |X| \geq 3\ odd} \left\lceil \frac{2|E(G[X])|}{|X| - 1} \right\rceil$$

Every multigraph G with $|G| \leq 2$ satisfies $\mathcal{W}(G) = \Delta(G) = \chi'(G)$. Clearly, $\mathcal{W}(G) \leq \chi'(G)$ for any multigraph G. There is no upper bound of the chromatic index in terms of the edge density. For the star $K_{1,n}$ ($n \geq 3$), we have $\mathcal{W}(K_{1,n}) = 2$ and $\chi'(K_{1,n}) = n$. On the other hand, the chromatic index can be bounded from above in terms of the maximum degree and the edge density. In 2018, Chen, Jing, and Zang [209] proposed a solution to a longstanding open problem, known as the Goldberg-Seymour Conjecture; this conjecture was proposed, independently, by Goldberg [420] and [421], Seymour [927] and [928], Andersen [61], and Gupta [439].

Theorem 1.40 (Chen, Jing, and Zang) *Every multigraph G satisfies $\chi'(G) \leq \max\{\Delta(G)+1, \mathcal{W}(G)\}$.*

The inequality in Theorem 1.40 is equivalent to the statement that $\chi'(G) \in \{\Delta(G), \Delta(G)+1, \mathcal{W}(G)\}$. This is a far reaching extension of the Vizing–Gupta bound, which implies that the chromatic index of any (simple) graph G can have only two values: either $\chi'(G) = \Delta(G)$ or $\chi'(G) = \Delta(G)+1$.

We have observed that the gap between the chromatic number and the list-chromatic number can be arbitrarily large, even within the class of bipartite graphs. In view of this, the following conjecture, suggested independently by various researchers, including Vizing, Albertson, Collins, Tucker, Gupta, Bollobás and Harris, is surprising. This conjecture is used to be known as the **list-colouring conjecture**, but a more specific name was suggested by Ellingham and Goddyn [324], Woodall [1074], and possibly others.

Conjecture 1.41 (The List-Edge-Coloring Conjecture) $\chi'_\ell(G) = \chi'(G)$ *for any multigraph G.*

This conjecture, abbreviated LECC, asserts that for line graphs of multigraphs there is no gap between the list-chromatic number and the chromatic number. The conjecture seems difficult to attack and might even be false for the class of all multigraphs. Therefore, it is reasonable to search for interesting subclasses for which the list coloring conjecture can be proved. An appropriate candidate for such a subclass was already proposed by Dinitz, namely the class of complete bipartite graphs $K_{n,n}$. The most spectacular result about edge coloring is Galvin's proof [402] that the list chromatic index of any bipartite multigraph is equal to its chromatic index (see Theorem 3.6 and Corollary 3.7).

Given a multigraph G, the **total graph** of G, denoted by $\mathrm{T}(G)$, is the graph whose vertex set is $V(G) \cup E(G)$ and in which two vertices are adjacent if the corresponding elements of G are adjacent or incident. So if $H = \mathrm{T}(G)$, then $V(H) = V(G) \cup E(G)$, $H[V(G)] = G$, $H[E(G)] = \mathrm{L}(G)$, and $N_H(v) = E_G(v)$ for all $v \in V(G)$. A coloring of $\mathrm{T}(G)$ is called a **total coloring** of G; a total coloring of G assigns a color to each vertex and to each edge of G so that colored objects have different colors when they are adjacent or incident. Furthermore, $\chi''(G) = \chi(\mathrm{T}(G))$ is called the **total chromatic number** of G, and $\chi''_\ell(G) = \chi_\ell(\mathrm{T}(G))$ is called the **total list-chromatic number** of G.

Let G be a multigraph. If v is a vertex of G, then $E_G(v) \cup \{v\}$ is a clique in $\mathrm{T}(G)$ and so $\omega(\mathrm{T}(G)) \geq \Delta(G)+1$. Furthermore, $\Delta(\mathrm{L}(G)) \leq 2\Delta(G)$. From (1.7) it then follows that

$$\Delta(G)+1 \le \chi''(G) \le \chi''_\ell(G) \le 2\Delta(G)+1.$$

The following result is folklore; it relates the total list chromatic number to the list chromatic index.

Proposition 1.42 $\chi''(G) \le \chi''_\ell(G) \le \chi'_\ell(G)+2$ *for any multigraph* G.

Proof Let $H = \mathrm{T}(G)$ and let L be a k-assignment for H with $k = \chi'_\ell(G)+2$. Since $\chi_\ell(G) \le \Delta(G)+1 \le k$, there is an L-coloring φ of $G = H[V(G)]$. Define a list-assignment L' of $\mathrm{L}(G) = H[E(G)]$ by $L'(e) = L(e) \setminus \{\varphi(v), \varphi(w)\}$ for every edge $e \in E_G(v,w)$ $(v,w \in V(G))$. Then $|L'(e)| = |L(e)|-2 = \chi'_\ell(G)$ for all $e \in E(G)$ and, therefore, there is an L'-coloring φ' of $\mathrm{L}(G) = H[E(G)]$. But then $\varphi \cup \varphi'$ is an L-coloring of H, which proves the proposition. □

In 1965 Behzad [89] proposed the conjecture that $\chi''(G) \le \Delta(G)+2$ for every graph G. In 1968 Vizing [1050] published a paper containing a wealth of graph theory problems including the conjecture that $\chi''(G) \le \Delta(G)+\mu(G)+1$ for any multigraph G. As a further extension of this conjecture, we offer the following conjecture.

Conjecture 1.43 (TOTAL GRAPH CONJECTURE) $\chi''_\ell(G) \le \chi'_\ell(G)+1$ *for any multigraph* G.

That the class of total graphs of multigraphs forms another class of graphs for which the chromatic number and list chromatic number always coincide was conjectured about the same time and independently of each other by by Borodin, Kostochka, and Woodall [160], by Juvan, Mohar, and Škrekovski [542], and by Hilton and Johnson [498].

Conjecture 1.44 (THE LIST-TOTAL-COLORING CONJECTURE) $\chi''_\ell(G) = \chi''(G)$ *for any multigraph* G.

1.9 Coloring of Weighted Graphs

A **weighted graph** is a pair (G,g) consisting of a graph G and a vertex function $g: V(G) \to \mathbb{N}_0$. When $g \equiv p$ is a constant function, that is, $g(v) = p$ for all $v \in V(G)$, where $p \ge 0$ is an integer, we let the value p denote the function g and we write (G,p) instead of (G,g).

Given a weighted graph (G,g), a **coloring** of (G,g) with **color set** C is a map φ which assigns to each vertex v of G a set $\varphi(v) \subseteq C$ of $g(v)$ colors so that $\varphi(v) \cap \varphi(w) = \varnothing$ whenever $vw \in E(G)$. When $|C| = k$, we also say that φ is a k**-coloring** of (G,g). A weighted graph is k**-colorable** if it admits a k-coloring. The **chromatic number** of a weighted graph (G,g), denoted by $\chi(G,g)$, is the least integer k for which (G,g) has an k-coloring. Observe that $\chi(G,1) = \chi(G)$. If L is a list-assignment for G, then an L**-coloring** of (G,g) is a coloring φ of (G,g) with color set $C = \bigcup_{v \in V(G)} L(v)$ such that $\varphi(v) \subseteq L(v)$ for all $v \in V(G)$. If (G,g) admits an L-coloring, then (G,g)

is L-**colorable**. The weighted graph (G,g) is said to be f-**list-colorable**, where $f : V(G) \to \mathbb{N}_0$ is a vertex function, if (G,g) is L-colorable for every f-assignment L of G. When $f(v) = k$ for all $v \in V(G)$, the corresponding term becomes k-**list-colorable**. The **list chromatic number** of (G,g), denoted by $\chi_\ell(G,g)$, is the least integer k for which (G,g) is k-list-colorable. Clearly, $\chi_\ell(G,1) = \chi_\ell(G)$ and every weighted graph (G,g) satisfies $\chi(G,g) \leq \chi_\ell(G,g)$.

The coloring problem for weighted graphs can be reduced to the conventional coloring problem. With a weighted graph (G,g) we can associate a graph $G[g]$ with vertex set

$$V(G[g]) = \{(v,i) \mid v \in V(G), i \in [1,g(v)], g(v) \geq 1\}$$

and edge set

$$E(G[g]) = \{(v,i)(w,j) \mid vw \in E(G) \vee (v = w \wedge i \neq j)\}.$$

The graph $G[g]$ is called an **inflation** of G. The set $X_v = \{v\} \times [1,g(v)]$ is a clique of $G[g]$ satisfying $|X_v| = g(v)$, we call this set the v-**clique** of $G[g]$. For two vertices $v \neq w$ of G, the set $E_{G[g]}(X_v, X_w)$ consists of all edges between X_v and X_w if $vw \in E(G)$, otherwise it is the empty set.

Now consider a coloring φ of the weighted graph (G,g) with color set C. Since $\varphi(v)$ is a set of $g(v)$ colors for each vertex v of G, there exists a mapping $\phi : V(G[g]) \to C$ such that $\phi|_{X_v}$ is a bijection between X_v and $\varphi(v)$ for all $v \in V(G)$. Since $\varphi(v) \cap \varphi(w) = \varnothing$ whenever $vw \in E(G)$, ϕ is a coloring of $G[g]$ with color set C. Conversely, if ϕ is a coloring of $G[g]$ with color set C, then the mapping $\varphi : V(G) \to C$ with $\varphi(v) = \phi(X_v)$ for all $v \in V(G)$ is a coloring of the weighted graph (G,g) with color set C. This shows that every weighted graph (G,g) satisfies

$$\chi(G,g) = \chi(G[g]). \tag{1.15}$$

The chromatic number of weighted graphs is subadditive in the following sense. If (G,g_1) and (G,g_2) are two weighted graphs, then

$$\chi(G,g_1+g_2) \leq \chi(G,g_1) + \chi(G,g_2). \tag{1.16}$$

This follows from the fact that if φ_i is a coloring of (G,g_i) with color set C_i, where $C_1 \cap C_2 = \varnothing$, then the mapping φ with $\varphi(v) = \varphi_1(v) \cup \varphi_2(v)$ for all $v \in V(G)$ is a coloring of (G,g_1+g_2).

Weighted graphs with a constant weight function are particularly interesting. Given a graph G and a constant weight function $g \equiv p$ with $p \in \mathbb{N}$, the graph $G[g]$ is the lexicographic product $G[K_p]$ of G and K_p (see Appendix C.7). Clearly, $G[K_0]$ is the empty graph and $G[K_1]$ is isomorphic to G.

Proposition 1.45 *Let G be a nonempty graph. Then the following statements hold:*

(a) $\chi(G,p) = \chi(G[K_p])$ *for all* $p \in \mathbb{N}$.
(b) $\chi(G[K_{p+q}]) \leq \chi(G[K_p]) + \chi(G[K_q])$ *for all* $p,q \in \mathbb{N}$.
(c) $\chi(G[K_p]) \leq p\chi(G)$ *for all* $p \in \mathbb{N}$.

(d) *The limit* $\lim_{p\to\infty}\frac{\chi(G[K_p])}{p}$ *exists and satisfies*

$$\lim_{p\to\infty}\frac{\chi(G[K_p])}{p}=\inf_{p\in\mathbb{N}}\frac{\chi(G[K_p])}{p}\le\chi(G).$$

Proof Statements (a) and (b) are immediate consequences of (1.15) and (1.16). Statement (c) follows from (b) and the fact that $G[K_1]\cong G$. Statement (d) is a consequence of (b), (c) and Fekete's lemma (Lemma C.43). □

Proposition 1.46 *Let* H *be a graph with* $\chi(H)=n$. *Then* $\chi(G[H])=\chi(G[K_n])$ *for any graph* G.

Proof Since $\chi(H)=n$, there is a coloring φ of H with color set $V(K_n)$. Then it is easy to check that if ϕ is a coloring of $G[K_n]$ with color set C, then the mapping ϕ' with $\phi'(v,w)=\phi(v,\varphi(w))$ for all $(v,w)\in V(G[H])$ is a coloring of $G[H]$ with color set C. It follows that $\chi(G[H])\le\chi(G[K_n])$.

Conversely, let φ be a coloring of $G[H]$ with a set C of $\chi(G[H])$ colors. Then the subgraph of $G[H]$ induced by $X_v=\{v\}\times V(H)$ is isomorphic to H for all $v\in V(H)$. Since $\chi(H)=n$, the set $\varphi(X_v)$ contains at least n colors and $\varphi(X_v)\cap\varphi(X_w)=\varnothing$ whenever $vw\in E(H)$. Thus there is a set $C_v\subset\varphi(X_v)$ of n colors for all $v\in V(G)$ and the mapping φ' with $\varphi'(v)=C_v$ for $v\in V(G)$ is a coloring of (G,n) with color set $C'=\bigcup_v C_v$. Since $|C'|\le|C|$, it follows that $\chi(G[K_n])=\chi(G,n)\le\chi(G[H])$.□

Let (G,g) be a weighted graph. For $X\subseteq V(G)$, let $g(X)=\sum_{v\in X}g(v)$, where $g(\varnothing)=0$. For the clique number and the maximum degree of the inflation graph $G[g]$ we have

$$\omega(G[g])=\max\{g(X)\mid X\subseteq V(G)\text{ is a clique of }G\},$$

and

$$\Delta(G[g])=\max\{g(N_G(v)\cup\{v\})\mid v\in V(G)\}-1$$

Proposition 1.47 *If* G *is the inflation of a bipartite graph, then* $\chi(G)=\omega(G)$.

Proof Suppose that $G=B[g]$ is the inflation of a bipartite graph B. Since $\omega(G)\le\chi(G)$, it suffices to show that $\chi(G)\le\omega(G)$. The proof is by double induction, first on the order of B and then on $g(V(B))$. For $|B|\le 1$, this is obviously true, since then G is a complete graph. So assume that $|B|\ge 2$.

First consider the case that there is a vertex $v\in V(B)$ such that $d_B(v)=0$ or $g(v)=0$. Let $B'=B-v$ and $g'=g|_{V(B')}$. Then the induction hypothesis implies that $\chi(B'[g'])=\omega(B'[g'])$. In booth cases, we obtain that $\chi(G)=\max\{\chi(B'[g']),g(v)\}=\max\{\omega(B'[g']),g(v)\}=\omega(B[g])=\omega(G)$. So we are done. It remains to consider the case when $g(v)\ge 1$ for all $v\in V(B)$ and $\delta(B)\ge 1$. Define the function $g':V(B)\to\mathbb{N}_0$ by $g'(v)=g(v)-1$ for all $v\in V(G)$. Since $\delta(B)\ge 1$, we have $\omega(B[g])=\omega(B[g'])+2$ and $\chi(B)=2$. By the induction hypothesis applied to $B[g']$, and by (1.15) and (1.16), it then follows that

$$\begin{aligned}\chi(G) = \chi(B[g]) = \chi(B,g) &\le \chi(B,g') + \chi(B,1) = \chi(B[g']) + \chi(B)\\ &\le \omega(B[g']) + 2 = \omega(B[g]) = \omega(G).\end{aligned}$$

Thus the proof is complete. □

Theorem 1.48 *If G is the inflation graph of a cycle C, then*

$$\chi(G) = \max\left\{\omega(G), \left\lceil \frac{|G|}{\lfloor \frac{1}{2}|C| \rfloor} \right\rceil\right\}.$$

Proof Let us denote the right hand side of the above equation by $R(G)$. Since G is the inflation of a cycle C, it follows that $\alpha(G) \le \lfloor \frac{1}{2}|C| \rfloor$. Thus it follows from (1.2) and (1.7) that $\chi(G) \ge R(G)$. So it remains to prove that $\chi(G) \le R(G)$. We prove this inequality by induction on $|G|$. If $\chi(G) \le \omega(G)$, then there is nothing to prove. So assume that $\chi(G) \ge \omega(G) + 1$. By Proposition 1.47, it then follows that C is an odd cycle with $|C| \ge 5$ and $G = C[g]$, where $g(v) \ge 1$ for all $v \in V(C)$. For a vertex v of C, let $X_v = \{v\} \times [1, g(v)]$. Then X_v is a clique in G with $|X_v| = g(v)$ and $\omega(G) = \max\{g(v) + g(w) \mid vw \in E(C)\}$. In what follows, we use the fact that each independent set X of G satisfies $|X \cap X_v| \le 1$ for all $v \in V(C)$.

First consider the case that there is an edge vw in C such that $g(v) + g(w) < \omega(G)$. There is an independent set X in G such that $|X| = \lfloor \frac{1}{2}|C| \rfloor$ and $X \cap X_z = \varnothing$ for $z \in \{v, w\}$. Then $G' = G - X$ is a inflation of the cycle C, where $|G'| = |G| - \lfloor \frac{1}{2}|C| \rfloor$ and $\omega(G') \le \omega(G) - 1$. It follows from the induction hypothesis that $\chi(G') \le R(G') \le R(G) - 1$. Since X is an independent set of G, this implies that $\chi(G) \le \chi(G') + 1 \le R(G)$, as required.

Now consider the case that $g(v) + g(w) = \omega(G)$ for all edges vw of C. Since $|C|$ is odd and ≥ 5, it follows that there is an integer $p \ge 1$ and $g(v) = p$ for all $v \in V(G)$. This implies that $\omega(G) = 2p$. As in the first case, there is an independent set X in G with $|X| = \lfloor \frac{1}{2}|C| \rfloor$. Then $G' = G - X$ is an inflation of C, where $|G'| = |G| - \lfloor \frac{1}{2}|C| \rfloor$ and $\omega(G') = \omega(G)$. Then the induction hypothesis implies that

$$\chi(G') \le R(G') = \max\left\{\omega(G), \left\lceil \frac{|G|}{\lfloor \frac{1}{2}|C| \rfloor} \right\rceil - 1\right\}.$$

Since $|C|$ is odd and $|G| = p|C|$, we obtain that

$$\omega(G) + 1 = 2p + 1 \le \left\lceil \frac{2p|C|}{|C| - 1} \right\rceil = \left\lceil \frac{|G|}{\lfloor \frac{1}{2}|C| \rfloor} \right\rceil,$$

which implies that

$$\chi(G) \le R(G') + 1 = \max\left\{\omega(G) + 1, \left\lceil \frac{|G|}{\lfloor \frac{1}{2}|C| \rfloor} \right\rceil\right\} = R(G),$$

as required. □

Corollary 1.49 $\chi(C_{2n+1}[K_p]) = 2p + \lceil \frac{p}{n} \rceil$ *for any* $n, p \in \mathbb{N}$.

While a weighted graph (G,g) and its associated inflation graph $G[g]$ have the same chromatic number, for the list chromatic number we only have $\chi_\ell(G,g) \leq \chi_\ell(G[g])$. To see this, let L be an arbitrary k-assignment for G, where $k = \chi_\ell(G[g])$. Define a k-assignment L' for $G[g]$ by $L'(v,i) = L(v)$ for all vertices (v,i) of $G[g]$. Then there is an L'-coloring φ' of $G[g]$, and so the mapping φ that assigns to each vertex v of G the color set $\varphi(v) = \varphi'(X_v)$, where X_v is the v-clique of $G[g]$, is an L-coloring of (G,g). This shows that $\chi_\ell(G,g) \leq k$, as claimed. As a consequence of (1.7) we obtain the following sequence of inequalities for any weighted graph (G,g):

$$\omega(G[g]) \leq \chi(G,g) \leq \chi_\ell(G,g) \leq \chi_\ell(G[g]) \leq \operatorname{col}(G[g]) \leq \Delta(G[g]) + 1. \quad (1.17)$$

A weighted graph (G,g) is called **connected** if G is a connected graph. If $g(v) \geq 1$ for all $v \in V(G)$, then we also write $g \geq 1$. Clearly, if (G,g) is a connected weighted graph with $g \geq 1$, then $G[g]$ is a complete graph if and only if G itself is a complete graph. As a consequence of Theorem 1.11 we obtain that any connected weighted graph (G,g) with $g \geq 1$ satisfies $\chi_\ell(G,g) \leq \Delta(G[g]) + 1$, where equality holds if and only if G is a complete graph or $g \equiv 1$ and G is an odd cycle. In Section 3.8, we shall use Galvin's kernel method to prove the following extension of this result obtained in 2013 by Stiebitz, Tuza and Voigt [979].

Theorem 1.50 (Stiebitz, Tuza, and Voigt) *Every connected weighted graph* (G,g) *with* $g \geq 1$ *satisfies*

$$\chi_\ell(G,g) \leq \Delta(G[g]) + 1 - \min_{v \in V(G)} g(v),$$

unless G *is a complete graph or an odd cycle.*

Erdős, Rubin, and Taylor [355] conjectured that the list chromatic number of a weighted graph is subadditive like the chromatic number.

Conjecture 1.51 (The $(a{:}b)$-choosability conjecture) $\chi_\ell(G, a \cdot b) \leq a \cdot \chi_\ell(G,b)$ *for any graph* G *and any integers* $a, b \geq 0$.

In 2019, Dvořák, Hu, and Sereni [311] have disproved this conjecture, by constructing a graph G with $\chi_\ell(G) = \chi_\ell(G,1) = 4$, but with $\chi_\ell(G,2) > 8$.

1.10 Perfect Graphs

The simple observation that $\chi \geq \omega$ led Claude Berge in the 1960s to introduce the class of perfect graphs. A graph G is called **perfect** if any induced subgraph H of G satisfies $\chi(H) = \omega(H)$. A graph that is not perfect is called **imperfect**. Clearly, any induced subgraph of a perfect graph is perfect, too. A graph G is

minimally imperfect if it is imperfect but each proper induced subgraph of G is perfect. Examples of perfect graphs are bipartite graphs and their complements, while examples of imperfect graphs are odd cycles of length at least 5 and their complements. The odd cycles, with the exception of C_3, are minimally imperfect. A graph is perfect if and only if it contains no minimally imperfect graph as an induced subgraph.

Proposition 1.52 *Let G be a minimally imperfect graph, and let $I \subseteq V(G)$ be an independent set of G. Then $\omega(G - I) = \omega(G)$.*

Proof The statement is evident if $I = \varnothing$. So assume that $I \neq \varnothing$. Since G is minimal imperfect, we have $\omega(G-I) = \chi(G-I) \geq \chi(G) - 1 \geq \omega(G) \geq \omega(G-I)$, implying that $\omega(G-I) = \omega(G)$. □

Lemma 1.53 *Let G be a minimally imperfect graph with independence number α and clique number ω. Then there exists a set $\mathcal{I}$ of $\alpha\omega + 1$ independent sets of G such that each vertex of G belongs to precisely α sets of $\mathcal{I}$. Furthermore, for each independent set $I \in \mathcal{I}$, G has a clique $C = C(I)$ of ω vertices such that $C(I) \cap I = \varnothing$ and $|C(I) \cap I'| = 1$ for all $I' \in \mathcal{I} \setminus \{I\}$. Moreover, $|G| = \alpha\omega + 1$ and the independents sets in $\mathcal{I}$ have all size equal to α.*

Proof Since G has independence number α, there is an independent set I_0 in G with α vertices. Note that $\alpha \geq 1$ and so $I_0 \neq \varnothing$. Let $v \in I_0$ be an arbitrary vertex. Since G is minimally imperfect, $G - v$ is perfect and hence $\chi(G - v) = \omega(G - v) = \omega$ (by Proposition 1.52). Thus $V(G) \setminus \{v\}$ has a partition $\mathcal{I}^v$ into ω independent sets, and thus $|G| \leq \alpha\omega + 1$. Then $\mathcal{I} = \{I_0\} \cup \bigcup_{v \in I_0} \mathcal{I}^v$ is a set of $\alpha\omega + 1$ independent sets. They are all different: if a set I is in both $\mathcal{I}^v$ and $\mathcal{I}^w$, then $G - I$ contains a K_ω (by Proposition 1.52) and it would contain both v and w, but they are both in I_0 and not joined by an edge. By the construction of $\mathcal{I}$, it follows that each vertex of G belongs to precisely α independent sets in $\mathcal{I}$. Let $I \in \mathcal{I}$ be an arbitrary independent set. Since $\omega(G - I) = \omega$ (by Proposition 1.52), there is a clique $C = C(I)$ of $G - I$ with ω vertices. Then $C = \{C(I) \mid I \in \mathcal{I}\}$ is a set of $\alpha\omega + 1$ different cliques each with ω vertices. Clearly, $C(I) \cap I = \varnothing$. Because each of the ω vertices in $C(I)$ lies in α of the independent sets in $\mathcal{I}$, and because each independent set contains at most one vertex of the clique $C(I)$, we deduce that $|C(I) \cap I'| = 1$ for all $I' \in \mathcal{I} \setminus \{I\}$.

Summarizing G contains $\alpha\omega + 1$ independent sets $I^0, I^1, \ldots, I^{\alpha\omega}$ and $\alpha\omega + 1$ cliques $C^0, C^1, \ldots, C^{\alpha\omega}$ such that (a) each vertex of G belongs to precisely α sets I^i, (b) each clique C^j has ω elements, and (c) if $i, j \in [0, \alpha\omega]$, then

$$|I^i \cap C^j| = \begin{cases} 0 \text{ if } i = j, \\ 1 \text{ if } i \neq j. \end{cases}$$

Let $v_1, v_2, \ldots, v_n$ be an enumeration of the n vertices of G. Now let A, B be the corresponding $n \times (\alpha\omega + 1)$ incidence matrices associated with the independent sets respectively with the cliques. So if $v_i \in I^j$, then $A(i, j) = 1$ else $A(i, j) = 0$; and if $v_i \in C^j$, then $B(i, j) = 1$ else $B(i, j) = 0$. It then follows from (a), (b) and (c) that the $(\alpha\omega + 1) \times (\alpha\omega + 1)$ matrix $M = A^T B$ satisfies the equations $M(i, j) = 1$ if $i \neq j$ and

$M(i,i) = 0$. Then $\det(M) \neq 0$ and, therefore, M has full rang $\alpha\omega + 1$. This implies that both A and B are also of rang $\alpha\omega + 1$. Since these two matrices have n rows, we have $|G| = n \geq \alpha\omega + 1$. Hence $|G| = \alpha\omega + 1$. This proves the lemma. □

The following characterization of perfect graphs was proposed by A. Hajnal; it was confirmed in 1972 by Lovász [690].

Theorem 1.54 (LOVÁSZ) *A graph G is perfect if and only if $\alpha(H)\omega(H) \geq |H|$ for every induced subgraph H of G.*

Proof First assume that G is perfect. If H is an induced subgraph of G, then H is perfect, too. Using (1.2), we deduce that $|H| \leq \alpha(H)\chi(H) \leq \alpha(H)\omega(H)$. This proves the "only if" implication.

For the proof of the "if" implication, suppose that it is false and let G be a counterexample with $|G|$ is minimum. Then G is imperfect and $|H| \leq \alpha(H)\omega(H)$ for every induced subgraph H of G. By the choice of G, we deduce that any proper induced subgraph of G is perfect, so G is minimally imperfect. Then Lemma 1.53 implies that $|G| \geq \alpha(G)\omega(G) + 1$, a contradiction. □

The inequality in the above theorem is invariant under complementation. Given a graph G, H is an induced subgraph of G if and only if $\overline{H}$ is an induced subgraph of $\overline{G}$. Furthermore, for any induced subgraph H of G, we have $|\overline{H}| = |H|$, $\omega(\overline{H}) = \alpha(H)$ and $\alpha(\overline{H}) = \omega(H)$. Theorem 1.54 thus implies the following result obtained in 1972 by Lovász [691]. This result was conjectured by Claude Berge.

Theorem 1.55 (THE WEAK PERFECT GRAPH THEOREM) *A graph is perfect if and only if its complement is perfect.*

The original proof of the weak perfect graph theorem given by Lovász [691] was based on the following result.

Theorem 1.56 *Any inflation of a perfect graph is again a perfect graph.*

Proof Let G be a perfect graph. Expanding a vertex v of G to an edge vv' produces a new graph G' by adding a vertex v' and joining v' to v and all neighbors of v. Clearly, any inflation of G can be obtained by successive expanding a vertex to an edge. Hence, in order to prove the theorem it suffices to show that if G' is obtained from G by expanding a vertex v of G to an edge vv' then G' is perfect. So let H' be an induced subgraph of G'. Either H' is isomorphic to an induced subgraph of G or $v, v' \in V(H')$ and H' is obtained from $H = H' - v'$ by expanding v to the edge vv'. In the former case, H' is a perfect graph and so $\chi(H') = \omega(H')$. In the latter case we distinguish two subcases. Let $\omega(H) = \omega$. Then $\omega \leq \omega(H') \leq \omega + 1$ and $\chi(H') \leq \chi(H) + 1$. Clearly, H is a perfect graph and so $\chi(H) = \omega$. So if $\omega(H') = \omega + 1$, then $\omega(H') \leq \chi(H') \leq \chi(H) + 1 = \omega + 1 = \omega(H')$ and we are done. It remains to consider the case when $\omega(H') = \omega$. Then no $K_\omega \subseteq H$ contains v, since together with v' this would yield a $K_{\omega+1}$ in H'. Consider a coloring of H with ω colors and let I denote the color class containing v. Then each $K_\omega \subseteq H$ mets I, but not v, and so $\tilde{H} = H - (I \setminus \{v\})$ has clique number $\omega(\tilde{H}) \leq \omega - 1$. Clearly, $\tilde{H}$ is a perfect graph and we can color $\tilde{H}$ with $\omega - 1$ colors. Since I is an independent set of H and $v \in I$, the set $I' = (I \setminus \{v\}) \cup \{v'\}$ is an independent set of H' and $I' = V(H') \setminus V(\tilde{H})$. Consequently, we can extend our coloring of $\tilde{H}$ with $\omega - 1$ colors to a coloring of H' with ω colors, which gives $\chi(H') \leq \omega = \omega(H')$ as required. □

The class of perfect graphs contains several interesting subclasses. Here we shall study the class of chordal graphs. A graph is **chordal** if each of its cycles of length greater than three has a chord, or equivalently, if the graph contains no induced cycle of length at least four. Thus any induced subgraph of a chordal graph is again a chordal graph. Clearly, complete graphs and trees are chordal. As we shall see, chordal graphs have a treelike structure and can be decomposed into complete graphs. This follows from the following result of Dirac [300].

Theorem 1.57 *Let G be a connected chordal graph which is not complete. Then any minimal separator of G is a clique.*

Proof Let $S \subseteq V(G)$ be a minimal separator of G. Suppose, on the contrary, that there are two nonadjacent vertices v and w in S. Since S is a separator, $G - S$ has two components G_1 and G_2. Since neither $S \setminus \{v\}$ nor $S \setminus \{w\}$ is a separator of G, we deduce that v as well as w have neighbors in both G_1 and G_2. Thus there exists a shortest v-w path P_i all of whose internal vertices lie in G_i (for $i = 1, 2$). Then $P_1 \cup P_2$ is an induced cycle of length at least four, giving a contradiction. □

A graph G is the **clique sum** of two graphs G_1 and G_2 if both graphs are proper induced subgraphs of G such that $G = G_1 \cup G_2$ and $G_1 \cap G_2$ is a complete graph. A simple consequence of the above theorem is that every chordal graph is either a complete graph, or it is the clique sum of two smaller chordal graphs. Conversely, the clique sum of two chordal graphs is again a chordal graph. We shall establish a refinement of this decomposition result. A vertex of a graph G is **simplicial** if its neighborhood is a clique in G. Another consequence of Theorem 1.57 is the following result of Dirac [300].

Theorem 1.58 *Any connected chordal graph G contains a simplicial vertex.*

Proof The statement is evident if G is complete. If G is a chordal graph which is not complete, we prove by induction on the order of G that G contains two nonadjacent simplicial vertices. Since G is not complete, it contains a minimal separator S. By Theorem 1.57, S is a clique in G. Hence G contains two proper induced subgraphs G_1 and G_2 such that $G = G_1 \cup G_2$ and $G_1 \cap G_2 = G[S]$. To complete the proof, we claim that G_i contains a simplicial vertex v_i $(i = 1, 2)$ which is not contained in S. Then v_1 and v_2 are simplicial vertices of G and $v_1 v_2 \notin E(G)$, so we are done. The claim is evident if G_i is a complete graph, since then any vertex of G_i is simplicial. If G_i is not complete, then the induction hypothesis implies that G_i contains two nonadjacent simplicial vertices and, since S is a clique of G_i, one of these two vertices is not contained in S. □

That chordal graphs are perfect is an immediate consequence of Theorem 1.57 and Proposition 1.30. However, Theorem 1.58 implies the following slightly stronger result.

Theorem 1.59 *For any chordal graph G, $\chi(G) = \chi_\ell(G) = \omega(G) = \mathrm{col}(G)$.*

Proof By Proposition 1.3, it suffices to show that $\mathrm{col}(G) \le \omega(G)$. To show this, let H be a nonempty subgraph of G and let G' be the subgraph of G induced by $V(H)$. Then G' is chordal and Theorem 1.58 implies that G' contains a simplicial vertex v. Then $\delta(H) \le d_{G'}(v) \le \omega(G') - 1 \le \omega(G) - 1$. By Proposition 1.5, this implies that $\mathrm{col}(G) \le \omega(G)$. □

As mentioned above, a graph is perfect if and only if it contains no minimally imperfect graph as an induced subgraph. Two types of minimally imperfect graphs are odd cycles of length at least 5 and their complements (by the weak perfect graph theorem). That there are no other minimally imperfect graphs was conjectured in the 1960s by Berge; this conjecture became known as the strong perfect graph conjecture. In 2002 Chudnovsky, Robertson, Seymour, and Thomas [230] succeeded to verify this conjecture; in this way, the strong perfect graph conjecture has become the strong perfect graph theorem. The proof is long and technical; even to present an outline of the proof is beyond the scope of this book.

Theorem 1.60 (The Strong Perfect Graph Theorem) *A graph G is perfect if and only if neither G nor $\overline{G}$ contains an odd cycle of length at least 5 as an induced subgraph.*

1.11 Exercises

1.1 Let $G = \overline{C_n}$ be the complement of a cycle of order n. Determine $\alpha(G)$, $\omega(G)$, $\Delta(G)$, and $\chi(G)$.

1.2 Prove that if G is a graph with n vertices and m edges, then

$$\frac{n^2}{n^2 - 2m} \le \chi(G) \le \tfrac{1}{2} + \sqrt{2m + \tfrac{1}{4}}.$$

1.3 Prove that every graph G satisfies $\chi(G) \le |G| - \alpha(G) + 1$.

1.4 Let G be a connected graph, let $k \ge 2$ be an integer, and let $p \in [2, k]$. If $\chi(G) \ge k + 1$, then Theorem 1.1 implies that G contains many cycles of length $1 \pmod{p}$. Construct such a cycle by means of a DFS tree T and a labeling of the vertex set of T. (Tuza [1039])

1.5 Show that if G is a graph and $vw \in E(\overline{G})$, then

$$\chi(G) = \min\{\chi(G + vw), \chi(G/\{v, w\})\},$$

where $G/\{v, w\}$ is obtained from G by identifying v and w.

1.6 Given a graph G and a vertex order ℓ of G, let $A(G, \ell)$ be the number of colors used by the sequential coloring algorithm.

1. Prove that every graph G admits a vertex order ℓ for which $A(G,\ell) = \chi(G)$.
2. Construct a bipartite graph G of order $2n$ and a vertex order ℓ of G for which $A(G,\ell) = n$.

1.7 Prove that if a connected graph G with maximum degree Δ is neither a complete graph nor an odd cycle, then G has a Δ-coloring with color classes $I_1, I_2, \ldots, I_\Delta$ such that I_1 is a maximum independent set of G and for each $i \in [2,\Delta]$, the set I_i is a maximum independent set of $G - (I_1 \cup I_2 \cup \cdots \cup I_i)$.

1.8 Prove that for every 2-connected graph G the following conditions are equivalent:

1. G is a complete graph or an odd cycle.
2. Any two nonadjacent vertices of G form a separating set of G.
3. Any two vertices of distance two in G form a separating set of G.

(Bryant [183])

1.9 Prove that if every depth-first-search tree of a connected graph G is a Hamilton path, then G is a cycle, a complete graph, or a complete bipartite graph $K_{n,n}$. Use this to prove Brooks' theorem. (Bondy [141])

1.10 Deduce from Brooks' theorem that any graph G with maximum degree $\Delta \geq 3$ and $\omega(G) \leq \Delta$ satisfies $\alpha(G) \geq \frac{|G|}{\Delta}$. Albertson, Bollobás, and Tucker [32] proved that any graph G with maximum degree $\Delta \geq 3$ and $\omega(G) \leq \Delta - 1$ satisfies $\alpha(G) > \frac{|G|}{\Delta}$, except when $G = C_8^2$ or $G = C_5[K_2]$. Here H^2 denotes the **square** of the graph H, that is, the graph with the same vertex set as H in which two vertices are joined by an edge if and only if their distance in H is one or two.

1.11 Determine $\chi_\ell(K_{3,3})$.

1.12 Let $p \geq 2$ and $q \geq p^p$. Show that $\chi_\ell(K_{p,q}) = p+1$. (Hint: Figure 1.9 shows a bad list-assignment when $p = 2$.)

1.13 Let G be a complete graph of order n, where $n \geq 1$, and let L be a list-assignment of G. Deduce from Hall's theorem (Theorem C.11) that G is L-colorable if and only if the lists $L(v)$ satisfy Hall's condition, that is, $|\bigcup_{v \in S} L(v)| \geq |S|$ for all $S \subseteq V(G)$. (Vizing [1051])

1.14 List coloring vs. independence number: Let G be a nonempty graph and let L be a list-assignment. Denote by $G \square L$ the graph with vertex set $V(G \square L) = \{(v,c) \mid v \in V(G), c \in L(v)\}$, where two vertices (v,c) and (w,c') are joined by an edge if and only if $v = w$ and $c \neq c'$, or $vw \in E(G)$ and $c = c'$. Prove that G is L-colorable if and only if $\alpha(G \square L) = |G|$. (Vizing [1051])

1.15 List coloring vs. ordinary coloring: Let G be a nonempty graph, let L be a list-assignment, and let C be the set of colors used in the union of all the lists. For each color c, take a new vertex v_c and join a vertex v of G with a vertex v_c if and only if $c \notin L(v)$. Let $G(L)$ denote the resulting graph and let $|C| = k$. Show that G is L-colorable if and only if $G(L)$ is k-colorable. (Biró, Hujter, and Tuza [119])

1.16 Show that, for every graph G, $\mathrm{col}(G) = \Delta(G)+1$ if and only if G has a Δ-regular component.

1.17 Prove that every graph G satisfies $\mathrm{mad}(G) \le 2\mathrm{col}(G) - 2$.

1.18 A **signed multigraph** is a pair (G,σ) consisting of a multigraph G and a mapping $\sigma : E(G) \to \{+1,-1\}$; the value $\sigma(e)$ is called the **sign** of the edge e. If $\sigma(e) = +1$ for all edges $e \in E(G)$, we call (G,σ) an **all positive signed multigraph**, and write $(G,\sigma) = (G,+)$. If $X \subseteq V(G)$, then a new signed multigraph (G,σ') can be obtained by reversing the sign of all edges belonging to $\partial_G(X)$, that is, $\sigma'(e) = -\sigma(e)$ if $e \in \partial_G(X)$ else $\sigma'(e) = \sigma(e)$. We then say that (G,σ') is obtained from (G,σ) by **switching** at X, and write $(G,\sigma') = (G,\sigma)^X$. Two signed multigraphs (G,σ) and (G,σ') are called **switching equivalent**, if there is a vertex set $X \subseteq V(G)$ such that $(G,\sigma') = (G,\sigma)^X$. A signed multigraph (G,σ) is called **balanced** if for every cycle C of G the **sign product** $\prod_{e\in E(C)} \sigma(e)$ is positive, otherwise it is called **unbalanced**. For a simple graph H, let (G,σ) be the signed graph obtained from H by replacing each edge e of H by two parallel edges e_1, e_2 with $\sigma(e_1) = +1$ and $\sigma(e_2) = -1$; we then write $(G,\sigma) = (2H,\pm)$.

A **coloring** of a signed multigraph (G,σ) is a mapping $\varphi : V(G) \to \mathbb{Z}$ such that every edge $e \in E_G(v,w)$ satisfies $\varphi(v) \ne \sigma(e)\varphi(w)$, that is, if $\sigma(e) = +1$, then $\varphi(v) \ne \varphi(w)$, else $\varphi(v) \ne -\varphi(w)$. For an integer $h \ge 0$, let $Z_{2h} = \{\pm 1, \pm 2, \ldots, \pm h\}$ and $Z_{2h+1} = Z_{2h} \cup \{0\}$. The **chromatic number** $\chi(G,\sigma)$ is the least integer k such that (G,σ) admits a coloring φ with $\mathrm{im}(\varphi) \subseteq Z_k$. If L is a list-assignment of G with color set $\mathbb{Z}$, then (G,σ) is said to be L**-colorable** if (G,σ) admits a coloring φ such that $\varphi(v) \in L(v)$ for all $v \in V(G)$. The **list chromatic number** $\chi_\ell(G,\sigma)$ is the least integer $k \ge 0$ such that (G,L) is L-colorable for all list-assignments L of G with color set $\mathbb{Z}$ such that $|L(v)| \ge k$ for all $v \in V(G)$. A signed graph (G,σ) is called **degree-choosable** if (G,σ) is L-colorable for all list-assignments of G with color set $\mathbb{Z}$ such that $|L(v)| \ge d_G(v)$ for all $v \in V(G)$.

A signed multigraph (G,σ) is called a **signed brick** if (1) G is a complete graph or an odd cycle and (G,σ) is balanced, or (2) G is an even cycle and (G,σ) is unbalanced, or (3) $(G,\sigma) = (2K_n,\pm)$ with $n \ge 2$, or (4) $(G,\sigma) = (2C_n,\pm)$ with $n \ge 5$ odd.

1. Show that switching equivalent defines an equivalence relation for the class of signed multigraphs.
2. Show that a signed multigraph (G,σ) is balanced if there is a vertex set $X \subseteq V(G)$ such that $\partial_G(X) = \{e \in E(G) \mid \sigma(e) = -1\}$. (Harary [468])
3. Show that a signed multigraph (G,σ) is balanced if and only if (G,σ) is switching equivalent to $(G,+)$. (Zaslavsky [1090])
4. Show that if (G,σ) and (G,σ') are switching equivalent, then $\chi(G,\sigma) = \chi(G,\sigma')$ and $\chi_\ell(G,\sigma) = \chi_\ell(G,\sigma')$.
5. Show that $\chi(G,+) = \chi(G)$ and $\chi_\ell(G,+) = \chi_\ell(G)$.
6. Show that if H is a simple graph, then $\chi(2H,\pm) = 2\chi(H) - 1$.
7. Show that if (G,σ) is a signed multigraph, then $\chi_\ell(G,\sigma) \le \chi_{\mathrm{DP}}(G)$.

8. Let (G,σ) be a signed connected multigraph. Deduce from Theorem 1.17 that (G,σ) is not degree-choosable if and only if each block of (G,σ) is a signed brick. (Schweser and Stiebitz [908])
9. Let (G,σ) be a signed connected multigraph. Show that $\chi_\ell(G,\sigma) \leq \Delta(G)+1$, where equality holds if and only if (G,σ) is a signed brick. (Schweser and Stiebitz [908])

1.19 Let (G,σ) be a signed simple graph, and $(\mathbb{Z}_n,+)$ be the cyclic group of integers modulo n. A $\mathbb{Z}_n$-coloring of (G,σ) is a mapping $\varphi : V(G) \to \mathbb{Z}_n$ such that every edge $e = vw$ satisfies $\varphi(v) \neq \varphi(w)$ if $\sigma(e) = +1$ and $\varphi(v) \neq -\varphi(w)$ if $\sigma(e) = -1$. The **signed chromatic number** $\chi_\pm(G,\sigma)$ is the least integer n such that (G,σ) has a $\mathbb{Z}_n$-coloring. Use Theorem 1.17 to show that if G is not an odd cycle and (G,σ) is not a balanced complete graph, then $\chi_\pm(G,\sigma) \leq \Delta(G)$. (Kang [553])

1.20 Prove that a complete bipartite graph $K_{n,n}$ has an equitable k-coloring if and only if $\lceil n/\lfloor k/2\rfloor\rceil - \lceil n/\lceil k/2\rceil\rceil \leq 1$. (Lih and Wu [677])

1.21 For $k \geq 3$, let $\mathcal{DG}(k)$ denote the family of all graphs G whose vertex set consists of three nonempty pairwise disjoint sets X, Y_1 and Y_2 with

$$|Y_1| + |Y_2| = |X| + 1 = k-1$$

and two additional vertices v_1 and v_2 such that X and $Y_1 \cup Y_2$ are cliques in G not joined by any edge, and $N_G(v_i) = X \cup Y_i$ for $i = 1,2$. Prove that every graph $G \in \mathcal{DG}(k)$ is k-critical. (Dirac [303], Gallai [397])

1.22 For $k \geq 4$, let $\mathcal{EG}(k)$ denote the family of all graphs G whose vertex set consists of four nonempty pairwise disjoint sets X, Y_1, Y_2 and Y_3 with

$$|Y_1| + |Y_2| + |Y_3| = |X| + 1 = k-1$$

and three additional vertices v_1, v_2 and v_3 such that X and $Y_1 \cup Y_2 \cup Y_3$ are cliques in G not joined by any edge, and $N_G(v_i) = X \cup Y_i$ for $i = 1,2,3$. Prove that every graph $G \in \mathcal{EG}(k)$ is k-critical. (Kostochka and Stiebitz [630])

1.23 Prove that if G is the line graph of a graph, then $\chi(G) \in \{\omega(G), \omega(G)+1\}$.

1.24 Prove that if G is a multigraph, then

$$\omega(\mathrm{L}(G)) = \max\{\Delta(G), \max\{|E(G[X])| \mid X \in [V(G)]^3\}\}.$$

1.25 Let G be a bipartite multigraph with $\Delta(G) \leq k$

1. Prove König's theorem that G has an edge coloring with a set of k colors and so $\chi'(G) \leq k$. (Hint: Use induction on $|E(G)|$ and a Kempe-change argument.)
2. Show that the line graph of G is a perfect graph.
3. Prove that G has an equitable edge coloring with a set of k colors and so $\chi^=(\mathrm{L}(G)) \leq k$. (Hint: Choose an edge coloring φ with a set C of k colors such that $\sum_{c\in C} |\varphi^{-1}(c)|$ is minimum.)

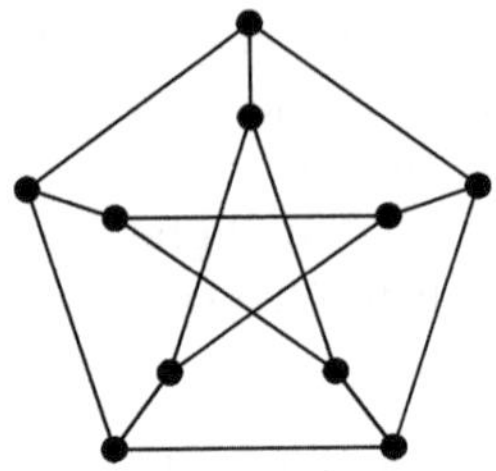

Fig. 1.8 The Petersen graph.

1.26 Let G be the **Petersen graph** (see Figure 1.8). Prove that $\chi'(G) = 4$.

1.27 CATLIN GRAPH

1. Prove that $\chi(G[H]) \leq \chi(G)\chi(H)$
2. The graph $G = C_5[K_n]$ is known as the **Catlin graph**. Using Corollary 1.49, deduce that $\chi(G) = \lceil \frac{5n}{2} \rceil = \lceil \frac{1}{2}(\omega(G) + \Delta(G) + 1) \rceil$. (A.V. Kostochka)
3. Show that if $G = C_5[K_3]$, then $\chi(G) = 8$ and K_9 is a minor of G, but K_8 is no topological minor of G. (Catlin [194])

1.28 Show that if $k \geq 3$ and $G = C[g]$ is an inflation of an odd cycle $C = C_{2p+1}$ with $p \geq 2$ and $g(v) \geq 1$ for all $v \in V(C)$, then G is k-critical if and only if $\omega(G) \leq k-1$ and $|G| = (k-1)p+1$. (Hint: Use Theorem 1.48 and consider first $G - v$.) (T. Gallai, see Krusenstjerna-Hafstrøm and Toft [659])

1.29 Let $\mathcal{T}$ be a set of subtrees of a tree T.

1. Prove that if any two members of $\mathcal{T}$ have a common vertex, then all members of $\mathcal{T}$ have a common vertex. (Hint: Apply induction on $|T|$ respectively on $|\mathcal{T}|$, or choose a root in T.)
2. Prove that if G is the intersection graph of $\mathcal{T}$, that is, $V(G) = \mathcal{T}$ and two distinct members of $\mathcal{T}$ are adjacent in G if and only if they have a common vertex, then G is a chordal graph.

1.30 Let G be an **interval graph**, that is, there is a closed real interval I_v for each vertex v of G such that $vw \in E(G)$ if and only if $I_v \cap I_w \neq \emptyset$.

1. Show that G is a chordal graph.
2. For $t \in \mathbb{R}$, let $\mathrm{int}(t)$ denote the maximum number of vertices v of G such that $t \in I_v$. Show that $\omega(G) = \max\{\mathrm{int}(t) \mid t \in \mathbb{R}\}$.
3. Prove (without using Theorem 1.59) that $\chi(G) = \omega(G)$.

1.31 Duplicating a vertex v of a graph produces a new graph G' by adding a new vertex v' and joining v' to all neighbors of v (but not to v). Show that if G' is obtained from a perfect graph G by duplicating a vertex, then G' is perfect, too.

1.32 Prove that a graph G is perfect if and only if, for every induced subgraph H of G, the independence number of H is equal to the minimum number of cliques into which the vertex set of H can be partitioned.

1.33 Let G be a graph whose vertex set is the union of two disjoint p-cliques X and Y with $p \geq 1$ such that $|E_G(X,Y)| \leq p$. Show that G is a perfect graph with $\chi(G) = \omega(G) \leq p+1$. Furthermore, show that $\chi(G) = p+1$ if and only if $E_G(X,Y) \subseteq E_G(v)$ for some vertex v of G and $|E_G(X,Y)| = p$. (Hint: Apply Theorem 1.55 and use the fact that the function $f(s) = s(p-s)$ is concave.)

1.12 Notes

The classical **graph coloring problem** is to find an **optimal coloring** of a graph G, that is, a coloring of G with $\chi(G)$ colors. By a result of Karp [556], the determination of the chromatic number is an NP-hard optimization problem. The NP-hardness of the graph coloring problem gives rise to the necessity of heuristic algorithms. A **graph coloring algorithm** A is an algorithm that applied to any graph G (respectively, to any graph belonging to a given graph property) produces a coloring of G. The number of colors A uses on G is denoted by $A(G)$. Every graph coloring algorithm A satisfies $\chi(G) \leq A(G)$ for every graph G, so A defines a graph parameter which is an upper bound for the chromatic number. A graph coloring algorithm A is said to be **optimal** if $A(G) = \chi(G)$ for every graph G and **suboptimal** otherwise. Furthermore, A is said to be **polynomial** if its time complexity or running time $t = t(G)$ is bounded from above by a polynomial in $n = |G|$ and $m = |E(G)|$, that is, $t(G) \leq p(n,m)$ for some polynomial $p = p(x,y)$ over the real numbers in two variables. A graph parameter ρ is said to be an **efficiently realizable upper bound** for χ if there exists a polynomial graph coloring algorithm A such that every graph G satisfies $A(G) \leq \rho(G)$.

Brooks' theorem from 1941 provided the first efficiently realizable upper bound for the chromatic number. For a graph G, let $br(G) = \Delta(G)+1$ if some component of G is a complete graph of order $\Delta(G)+1$ or if $\Delta(G) = 2$ and some component of G is an odd cycle, otherwise let $br(G) = \Delta(G)$. It is straightforward to show that the parameter br can be computed by a polynomial time algorithm. On the one hand, Brooks' theorem tells us that br is an upper bound for the chromatic number. On the other hand, Brooks' proof, which is based on sequential coloring and color interchange, leads to a polynomial coloring algorithm A such that $A(G) \leq br(G)$ for every graph G.

There are two basic types of graph coloring algorithms: **sequential coloring** and **maximal independent set coloring**. Any sequential coloring algorithm operates in two stages: in the first stage the algorithm fixes a vertex order $\ell = (v_1, v_2, \ldots, v_n)$ of the input graph G, either an arbitrary one or one that satisfies a certain property; in the second stage the vertices are colored in this order by colors from the set $\mathbb{N}$ assigning to each vertex v_i the least color $c(v_i)$ possible, so

$$c(v_i) = \min(\mathbb{N} \setminus \{c(v_j) \mid j < i, v_iv_j \in E(G)\}).$$

This procedure of color assignment is also known as a **greedy coloring**.

In a maximal independent set coloring the vertices of G are also given in some order. For a given color $c \in \mathbb{N}$ an uncolored vertex v_i is assigned color c if v_i is not adjacent to any vertex already colored with c. Having assigned color c to all possible vertices, the algorithm deletes them and repeats the same procedure for the remaining subgraph of G and color $c+1$. Hence the vertices of color k form a maximal independent set in $G - \bigcup_{j=1}^{k-1} V_j$, where V_j are the set of vertices having color j. The number of colors used by these two heuristics depends very much on the particular vertex order; in fact for both methods there are vertex orders yielding an optimal coloring, but no polynomial time algorithms are known to find such a vertex order.

Brooks noticed that a sequential coloring algorithm leads to a coloring of a given graph using at most $\Delta+1$ colors, regardless of the chosen vertex order. In 1975 Lovász [694] gave a new proof of Brooks' theorem which yields a polynomial sequential coloring algorithm A satisfying $A(G) \leq br(G)$ for every graph G; the idea of this algorithm is to choose the initial vertex order more carefully. A related idea was used in 1969 by Ponstein [817]. In 2003 Bondy [141] showed that it is possible to implement Lovász's sequential coloring algorithm to run in $O(n+m)$ time using a depth-first-search tree.

Color interchange is another simple method of improving the effectiveness of any sequential coloring algorithm. If the next greedy coloring step requires the assignment of a new color to vertex v_i, then it may be possible to perform a sequence of Kempe-changes, and thus enable the assignment of a not new color to vertex v_i. The first proof of Brooks' theorem using the color interchange method was given in 1969 by Melnikov and Vizing [731]. Kostochka, and Nakprasit [625] used the color interchange method to prove the following strengthening of Brooks' theorem.

Theorem 1.61 (Kostochka and Nakprasit) *Let $\Delta \geq 3$ and let G be a graph with $\omega(G) \leq \Delta$ and $\Delta(G) \leq \Delta$. Then for every vertex v of G and every coloring φ of $G - v$ with color set $C = [1, \Delta]$, there is a coloring ϕ of G with color set C such that $|\phi^{-1}(c)| \geq |\varphi^{-1}(c)|$ for all colors $c \in C$.*

Maximal independent set graph coloring algorithms were first investigated by Johnson [537]. Tverberg's proof of Brooks' theorem discussed in Section 1.2 can be transformed into a polynomial maximal independent set coloring algorithm A for which $A(G) \leq br(G)$ for all G; the algorithm A can be implemented to run in $O(mn)$ time.

There are two important features to take into account when selecting a (sub-optimal) graph coloring algorithm A, the **worst-case running time** of A and the **performance guarantee** of A defined by

$$A(n) = \max\left\{\frac{A(G)}{\chi(G)} \mid G \text{ is a graph of order } n\right\}.$$

A graph coloring algorithm A is called a **t-relative approximation algorithm** if $A(G) \leq t\chi(G)$ for all G, i.e. $A(n) \leq t$. The algorithm A is called a **t-absolute approximation algorithm** if $|A(G) - \chi(G)| \leq t$ for all G. The performance guarantee of most polynomial graph coloring algorithms A is only linear, i.e., $A(n) = O(n)$. Johnson [537] proved that there is a polynomial maximal independent set coloring algorithm A whose performance guarantee is sublinear, i.e., $A(n) = O(n/\log n)$. Lund and Yanakakis [706] proved the following remarkable result:

Theorem 1.62 (LUND AND YANAKAKIS) *There is a constant $\varepsilon > 0$ such that: if there is a polynomial graph coloring algorithm A with performance guarantee $A(n) \leq n^{\varepsilon}$, then there is an optimal polynomial graph coloring algorithm. In particular, this is the case if there is a polynomial graph coloring algorithm A with $A(G) \leq c \cdot \chi(G) + d$ for all G and for some constants c and d.*

As a consequence of the above result, we conclude that even the task of estimating the value of the chromatic number by means of any t-relative or t-absolute approximation algorithm is NP-hard. Using new techniques, inspired by zero-knowledge proof systems, Feige and Kilian [368] have shown that the chromatic number is hard to approximate by a polynomial algorithm within a factor of $n^{1-\varepsilon}$ for any constant $\varepsilon > 0$, assuming $\mathsf{NP} \not\subseteq \mathsf{ZPP}$.

List coloring is a very natural extension of the classical coloring model. We still pick a single color for each vertex v, but with the restriction that the color must come from a particular list $L(v)$ of available colors and not from a common color set C. This coloring model was introduced in the second half of the 1970s, in two papers, by Vizing [1051] and, independently, by Erdős, Rubin, and Taylor [355], under the name **choosability**. In the 1980s only a few papers appeared on this subject (see e.g. [132], [133], [217], [449]). However, the situation changed in the early 1990s with the publication of the pioneering works of Alon and Tarsi [58], Alon [44], [45], Voigt [1052], Thomassen [1008] and Galvin [402]. Since then the list coloring concept has been actively investigated and the number of publications in this field has increased immensely. Several survey papers have helped readers to keep track of the subject. The surveys of Tuza [1040], Kratochvíl, Tuza and Voigt [648], Woodall [1074], and Stiebitz and Voigt [980] all provides interesting information on the subject.

A result about graph coloring whose proof mainly relies on the sequential coloring method can quite often be transformed into a result about list coloring of graphs. Conversely, list coloring arguments provide useful tools for solving coloring problems. So it is often an advantage to study list coloring of graphs even if one is primarily interested in ordinary coloring.

The most special case of the list coloring problem is obviously the coloring problem, with all lists are identical. As pointed out by Biro, Hujter, and Tuza [119] the three problems: list coloring, coloring, and precoloring extension, are easily reducible to each other (see Exercise 1.15). Here a **precoloring** of a graph G means a coloring of a subgraph of G. The problem whether a coloring φ of an induced subgraph H of G with color set C can be extended to a coloring of G with color set C leads to a list coloring problem for the graph $G' = G - V(H)$. So if L is the list-assignment for G' with $L(v) = C \setminus \varphi(N_G(v) \cap V(H))$, then φ can be extended to a coloring of G with color set C if and only if G' is L-colorable. In this case we only need to investigate a specific list-assignment. However, if we want to show that a given graph is k-list-colorable, we have to argue why there exist list colorings for all possible k-assignments. This is one of the reason why the list chromatic number is more difficult to compute than the chromatic number. This is the case even for small graphs with a simple structure. Obviously, a graph is 1-list-colorable if and only if it is edgeless. A characterization of 2-list-colorable graphs was given by Erdős, Rubin, and Taylor [355]. If G is a connected graph, the **core** of G is the graph obtained from G by repeatedly deleting vertices of degree at most 1 until there is no such vertex. Then we have the following result from [355].

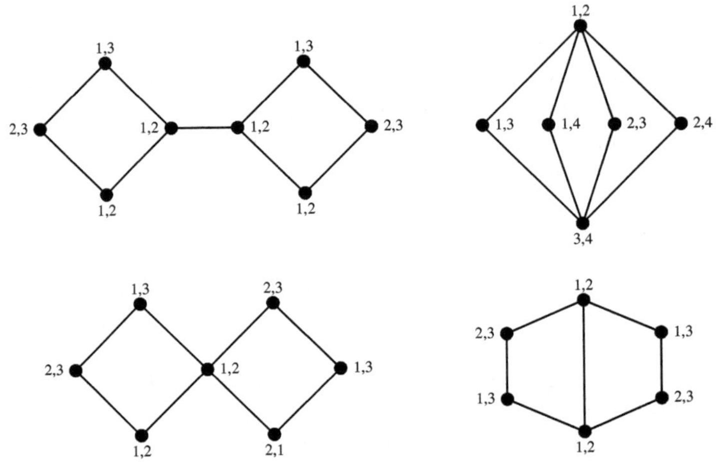

Fig. 1.9 Bipartite graphs with an uncolorable 2-assignment.

Theorem 1.63 (Erdős, Rubin, and Taylor) *A graph is 2-list-colorable if and only if the core of each of its components is either the empty graph, or an even cycle, or a graph consisting of two vertices with three internally disjoint even paths between them, where the length of at least two of the paths is exactly 2.*

How can we prove that a certain graph G is not k-list-colorable, that is, that $\chi_\ell(G) > k$? One obvious way would be to find an appropriate k-assignment L for G such that there is no L-coloring of G. A list-assignment L for G such that G has no L-coloring is also referred to as an **uncolorable list-assignment** or **bad list-assignment** of G. In many proofs we implicity use the fact that a graph G is f-list colorable if and only if G is L-colorable for every list-assignment L for G satisfying $|L(v)| \geq f(v)$ for all $v \in V(G)$, i.e., that = can be replaced by $\geq$. Figure 1.9 shows four bipartite graphs each with a bad 2-assignment, proving that all these four graphs are not 2-list-colorable. Consider the pair (G, L), where G is the complete bipartite graph $K_{2,4}$. To show that there is no L-coloring of G, consider the two vertices v, w of the first color class with $L(v) = \{1, 2\}$, and let a, b, c, d be the four vertices in the second color class with $L(a) = \{1, 3\}$ and so on. If we color v with 1, then there is only one possibility for coloring a and b and hence no choice for w. If we color v

with 2, then there is only one possibility for coloring c and d and hence no choice for w. This type of arguments works in all five cases.

The proof of Theorem 1.7 which combines the sequential coloring method with a reduction of the lists is due to Kostochka, Stiebitz, and Wirth [636]; see also [976]. The original proof of Erdős, Rubin, and Taylor [355] combines the sequential coloring method with a structural result (Theorem 1.35) characterizing Gallai forests.

While solving a problem on list coloring of planar graphs, Dvořák and Postle [318] introduced the notion of DP-coloring under the name **correspondence coloring**; the name was later changed by Bernshteyn, Kostochka, and Pron [110]. The DP-coloring of a graph G reduces the problem of finding a list coloring of G to the problem of finding an independent transversal in an appropriate cover (X,H) assigned to G. This is similar to a former reduction by Plesnevič and Vizing [810] of the k-coloring problem of G to the problem of finding an independent set of size $|G|$ in the cartesian product $G\square K_k$, that is, $\chi(G) \leq k$ if and only if $\alpha(G\square K_k) = |G|$ (see Exercise 1.14).

The advantage of DP-coloring is that it not only generalizes list coloring, but allows for reductions previously only possible for ordinary colorings. For ordinary coloring, the reductions commonly involve identifications of vertices. For list coloring it is in general not possible to identify vertices, since they might have different lists. However, for DP-coloring it is possible to identify vertices as long as we do not create multiple edges. Dvořák and Postle [318] used this fact to prove the following result, answering a question raised by Borodin [152].

Theorem 1.64 (Dvořák and Postle) *Every planar graph G without cycles of length 4 to 8 satisfies $\chi_\ell(G) \leq 3$, and so G is 3-list-colorable.*

The characterization of DP-degree-colorable multigraphs (Corollary 1.19) and the Brooks-type result for the DP-chromatic number (Theorem 1.20) are due to Bernshteyn, Kostochka, and Pron [110]. Proposition 1.14 and Proposition 1.15 are taken from [905]; that the merging operation preserves uncolorability was already observed in [110]. When dealing with DP-coloring, it is interesting to characterize not only the non-DP-colorable multigraphs, but also the corresponding uncolorable covers (see Theorem 1.17); this was first done by Kim and Ozeki [591]. However, the proof of Theorem 1.17 given by Kim and Ozeki is in two steps. First they use the characterization of DP-degree-colorable multigraphs due to Bernshteyn et al., and based on this they characterize the uncolorable covers. Our proof of Theorem 1.17 is similar to the proof of a result by Schweser [905] dealing with DP-degree-colorable hypergraphs with multiple edges. As noticed by Kim and Ozeki Theorem 1.17 can be used to characterize degree-choosable signed multigraphs (see Exercise 1.18.9); this shows that it is reasonable to consider DP-coloring for multigraphs and not only for graphs.

Equitable colorings form a subclass of all colorings. Meyer [734] introduced the equitable chromatic number $\chi^=$ as a new graph parameter. Detailed information and further references concerning equitable colorings of graphs can be found in [574], [576], [577], and [676]. As pointed out by Kierstead and Kostochka [574], [576], equitable colorings arise naturally in several applications of graph coloring and Theorem 1.21 has interesting applications in extremal and probabilistic combinatorics. Our proof of Theorem 1.21 is based on ideas of Kierstead and Kostochka [574] as well as Kierstead, Kostochka, Mydlarz, and Szemerédi [577]. Without doubt, the equitable Δ-coloring conjecture (Conjecture 1.25) remains one of the most important unsolved problems in this area. As pointed out by Kostochka (private communication), Theorem 1.21 combined with Theorem 1.61 yields the following result, supporting the equitable Δ-coloring conjecture.

Theorem 1.65 (Kostochka) *Let $\Delta \geq 3$ and let G be a graph with $\omega(G) \leq \Delta$ and $\Delta(G) \leq \Delta$. Then G has a Δ-coloring φ such that $|\varphi^{-1}(c)| \geq \lfloor |G|/(\Delta+1) \rfloor$ for all colors $c \in [1, \Delta]$.*

In 1951 Gabriel A. Dirac submitted a Ph.D thesis, supervised by Richard Rado and entitled *On the Colouring of Graphs*, to the University of London. In this thesis Dirac continued the study of coloring abstract graphs initiated by Brooks, and he introduced the notion of critical graphs as an important method. The initial inspiration for Dirac to take up graph coloring probably came from Peter Ungar, who came to London from Budapest in the late 1940's. Dirac himself came to England from Hungary in the 1930's, when his mother married the physicists P. A. M. Dirac. That critical graphs form a useful concept relies on the fact that many problems concerning the chromatic number of graphs can be reduced to critical graphs. As emphasized by Dirac in his thesis, *every general feature of k-chromatic graphs is possessed also by critical k-chromatic graphs, on the other hand a critical graph is more sharply defined and less arbitrary than a noncritical graph.* Dirac was aware of the critical graph method (Proposition 1.27) and he established the basic properties of critical graphs (Propositions 1.28, 1.29, 1.30, Corollary 1.31, and Theorem 1.32). Dirac published his results about the structure of critical graphs in the early 1950s in a sequence of papers [289], [291], [292], and [293].

The two structural results in Section 1.7 have proven useful; applications of both results are discussed in the following chapters. Theorem 1.35 first appeared in the pioneering work of Gallai [397]. The result was rediscovered by Erdős, Rubin, and Taylor [355] and is sometimes called Rubin's Block Lemma. A proof going back to lecture notes from 2007 by Robin Thomas at Georgia Institute of Technology was published by Hladký, Král' and Schauz [500] and by Cranston and Rabern [260]. Our proof is a simplification of Thomas' proof. The proof of Theorem 1.36 is from [124]. Chartrand and Kronk [205] showed that the graphs in Theorem 1.36(a) are those graphs in which every depth-first-search tree is a Hamilton path. Similar questions have been discussed by Thomassen [1000]. As discussed in [124], from the proof of Theorem 1.36 one easily extract a polynomial algorithm that either finds an independence tree or shows that no such tree exists (and recognizes that the graph is a cycle, a complete graph, or a balanced complete bipartite graph).

The **edge coloring problem** is to find an optimal edge coloring of a given multigraph G, that is, an edge coloring of G with a set of $\chi'(G)$ colors, where

$\chi'(G) = \chi(\mathrm{L}(G))$. On the one hand, edge coloring may be considered as (vertex) coloring restricted to line graphs. On the other hand, edge coloring has grown large and distinct enough, with its own problems, to be regarded as a subject in its own right. We refer the reader to the monograph on graph edge coloring by Stiebitz, Scheide, Toft, and Favrholdt [973]. That the coloring problem restricted to line graphs remains NP-complete was proved by Hoyler [504]. So the determination of the chromatic index χ' is an NP-hard optimization problem. However, the proof of Shannon's bound (Theorem 1.38) yields a polynomial time $\frac{3}{2}$-relative approximation algorithm for the chromatic index.

The announced proof of the Goldberg–Seymour conjecture by Chen, Jing, and Zang [209] (see Theorem 1.40) is a major breakthrough in the study of edge coloring. To the history of this conjecture and to the numerous partial results supporting the conjecture, the reader is referred to the book [973, Chapter 6]. Based on Tashkinovs [994] fundamental recoloring method Scheide [897] proved in 2010 that

$$\chi'(G) \leq \max\left\{ \mathcal{w}(G), \Delta(G) + \sqrt{\frac{1}{2}(\Delta(G)-1)} \right\}$$

for every multigraph G. A similar bound was established by Haxell and McDonald [484], namely

$$\chi'(G) \leq \max\left\{ \mathcal{w}(G), \Delta(G) + 2\sqrt{\mu(G)\ln \Delta(G)} \right\}.$$

The **total coloring problem** is to find an optimal total coloring of a given multigraph G, that is, a total coloring of G with a set of $\chi''(G)$ colors, where $\chi''(G) = \chi(\mathrm{T}(G))$. Compared with edge coloring the theory of total coloring has received less attention, and there is still no monograph on this subject. However, some results on total colorings are covered in the book on graph coloring and the probabilistic method by Molloy and Reed [756]. For further information on total coloring the reader is referred to the 2023 survey paper by Geetha, Narayanan, and Somasundaram [411]. The delightful 2009 book [950] by Soifer contains historical facts on this and many other coloring problems.

Results about colorings of weighted graphs can be found in the literature under a variety of names such as **multi-colorings** (Hilton, Rado, and Scott [499]), **tuple-colorings** (Geller [412] and Stahl [956]) and **set-colorings** (Bollobás and Thomason [135]). The first result dealing with set colorings of planar maps was obtained in 1972 by Meyer [733]. Hilton, Rado, and Scott [499] introduced the graph parameter

$$\chi^*(G) = \inf_{p \in \mathbb{N}} \frac{\chi(G,p)}{p};$$

they showed, using Meyer's result, that every planar graph G satisfies $\chi^*(G) < 5$. The parameter χ^* is known under the name **set chromatic number** and **fractional chromatic number** (see Section 8.2). Stahl [956] proved several basic results about the chromatic number of weighted graphs with constant weight functions (Propositions 1.45, 1.46, 1.47, and Corollary 1.49). The formula for the chromatic number of weighted cycles in Theorem 1.48 was obtained in the mid 1960's by T. Gallai (see Krusenstjerna-Hafstrøm and Toft [659] for an account of Gallai's result). Any inflation of a cycle is the line graph of a multigraph whose underlying graph is a cycle. Such multigraphs are also referred to as ring graphs; the edge chromatic number of ring graphs was established by B. Rothschild and J. Stemple (see Ore's book [794] and the book [973, Theorem 6.3]).

List colorings of weighted graphs were considered already by Erdős, Rubin, and Taylor [355] under the notion $(a:b)$-**choosability**. A graph G is said to be $(n:p)$-**choosable** if $\chi_\ell(G,p) \leq n$. For an overview of results about list coloring of weighted graphs we refer to the surveys [648], [980], [1040] and [1074]. The concept of DP-colorings of weighted graphs has attracted less interest; until present we only know two papers: one by Bernshteyn, Kostochka, and Zhu [111] and another one by Kaul and Mudrock [558]. For later use (see Section 8.9) we shall briefly introduce the concept here. Let (G,g) be a weighted graph, and let (X,H) be a cover of G. An (X,H)-**coloring** of (G,g) is an independent set I of H such that $|I \cap X_v| \geq g(v)$ for all $v \in V(G)$. We say that (G,g) is (X,H)-**colorable** if (G,g) admits an (X,H) coloring. The **DP-chromatic number** of (G,g), written $\chi_{\mathrm{DP}}(G,p)$, is the least integer $k \geq 0$ such that (G,g) is (X,H)-colorable whenever (X,H) is a cover of G and $|X_v| \geq k$ for all $v \in V(G)$.

Perfect graphs play an important role in combinatorial optimization (see the book [903] by Schrijver), in algorithmic graph theory (see Golumbic's book [424]), and in game theory (see the survey by Boros and Gurvich [161]). The significance of perfect graphs relies on the fact that that each of their induced subgraphs H fulfils a min–max equality, $\chi(H) = \omega(H)$. Many fundamental classes of graphs are perfect (see the book of Ramirez-Alfonsin and Reed [846]). Our proof of the weak perfect graph theorem is due to Gasparian [409]; his proof illustrates that simple linear algebra methods are powerful tools for obtaining combinatorial results. Lovász's inflation Theorem 1.56 shows that the so-called pluperfect graphs of Fulkerson [395] are in fact perfect. Using LP duality Fulkerson [394] had earlier proved that complements of pluperfect graphs are pluperfect. That chordal graphs are perfect was proved around the same time and independently by Hajnal and Surányi [457], by Berge [97], and by Dirac [300]. We also recommend Berge's personal recollections of the history of perfect graphs in [102], and the paper by Roussel, Rusu, and Thuillier [878].

In 2009 Alexander Soifer published a very personal and interesting monograph [950] on graph coloring problems from the boundary of geometry, combinatorics and number theory, thus from a different angle than our present book.

Chapter 2
Degeneracy and Colorings

As discussed in the previous chapter, every graph G satisfies $\chi(G) \le \text{col}(G) \le \Delta(G) + 1$. However, for many graph classes, the difference between the coloring number and the maximum degree can be arbitrarily large. For example, planar graphs have unbounded maximum degree, but their coloring number is at most 6. While Brooks' theorem provides a characterization of graphs satisfying $\chi = \Delta + 1$, a characterization of graphs satisfying $\chi = \text{col}$ seems to be unattainable. A graph satisfies $\text{col} = \Delta + 1$ if and only if it has a Δ-regular component.

2.1 Coloring Tree-like Graphs

A graph G satisfying $\text{col}(G) \le k + 1$ for an integer $k \ge 0$ is also referred to as a **k-degenerate graph**. By Proposition 1.5, a graph G is k-degenerate if and only if every nonempty subgraph of G has minimum degree at most k. Thus a graph is 0-degenerate if and only if it is edgeless, and 1-degenerate if and only if it is a forest. By (1.7), a k-degenerate graph has chromatic number at most $k + 1$. In the sequel we shall discuss an extension of this simple result.

Given a graph G and a color set C, we denote by $\mathcal{CO}(G, C)$ the set of all colorings of G with color set C, and we write $\mathcal{CO}(G, k)$ when $C = [1, k]$. If $\varphi \in \mathcal{CO}(G, C)$, then $\pi \circ \varphi \in \mathcal{CO}(G, C)$ for all permutations $\pi \in \text{Sym}(C)$. Two colorings $\varphi, \varphi' \in \mathcal{CO}(G, C)$ are **equivalent**, written $\varphi \sim \varphi'$, if there exists a permutation $\pi \in \text{Sym}(C)$ such that $\varphi' = \pi \circ \varphi$; this obviously defines an equivalence relation. A coloring $\varphi \in \mathcal{CO}(G, C)$ induces a partition $\mathcal{X} = \{\varphi^{-1}(c) \mid c \in C \wedge \varphi^{-1}(c) \neq \varnothing\}$ of $V(G)$ into independent sets, where $|\mathcal{X}| = |\text{im}(\varphi)|$. Then two colorings in $\mathcal{CO}(G, C)$ are equivalent if and only if they induce the same partition.

A graph G is called **uniquely C-colorable** if any two colorings of $\mathcal{CO}(G, C)$ are equivalent. If $|C| = k$, the corresponding term is **uniquely k-colorable**. A nonempty graph G is uniquely 1-colorable if and only if G is edgeless; and uniquely 2-colorable if and only if G is a connected bipartite graph. If $k \ge |G|$, then G is uniquely k-

M. Stiebitz et al., *Brooks' Theorem*, Springer Monographs in Mathematics,
https://doi.org/10.1007/978-3-031-50065-7_2

colorable if and only if G is a complete graph. On the other hand, every uniquely k-colorable graph G with $|G| \geq k$ has chromatic number k.

Let G be a graph, let H be a subgraph of G, let $\phi \in CO(H,C)$, and let $\varphi \in CO(G,C)$. We call ϕ a **restriction** of φ to H, or a φ an **extension** of ϕ onto G, if $\varphi(v) = \phi(v)$ for all $v \in V(H)$. In this case, we write $\phi = \varphi|_H$.

The following result, due to Stiebitz [970], gives a sufficient condition for extending a given k-coloring of an induced subgraph to the entire graph by means of an appropriately chosen spanning forest. As we shall discuss below this result implies Brooks' theorem. Given a graph G and a subgraph H of G, let $G - H = G - V(H)$.

Theorem 2.1 *Let G be a graph, and let H be an induced subgraph of G satisfying $\chi(H) \leq k$ for some integer $k \geq 3$. Suppose that G has a spanning forest F such that (1) $F[V(K)]$ is a tree for every component K of H, and (2) $d_G(v) \leq d_F(v) + k - 2$ for every vertex v of $G - H$. Then $\chi(G) \leq k$.*

Proof The proof is by reductio ad absurdum. So suppose that the theorem is false and let G, H and F be a counterexample minimizing the order $|G - H|$. Then H is k-colorable, but not G, and F is a spanning forest of G satisfying (1) and (2).

Clearly, $|G - H| > 0$, for otherwise $G = H$ and so $\chi(G) = \chi(H) \leq k$, a contradiction. Furthermore, $H \neq \varnothing$, for otherwise (2) implies that $G - E(F)$ has maximum degree at most $k-2$ and so G is $(k-1)$-degenerate, since the spanning forest F is 1-degenerate. But then $\chi(G) \leq k$, a contradiction. Consequently, $G - H$ has vertices and for the number of components of H we have $\mathrm{com}(H) \geq 1$.

For a vertex v of $G - H$, let $\mathcal{K}_v$ denote the set of components of H containing a vertex of $N_G(v)$. First, we claim that $\mathcal{K}_v \neq \varnothing$ for all vertices v of $G - H$. For otherwise we could add such a vertex v to H and would obtain a counterexample G, $H' = G[V(H) \cup \{v\}]$ and F, with $|G - H'| < |G - H|$, a contradiction.

Claim 2.1.1 *Let v be a vertex of $G - H$, and let $\mathcal{K} \subseteq \mathcal{K}_v$ be a set of p components, where $p \geq 1$. If $d_G(v) \leq p + k - 2$, then there is a path in $F - v$ whose ends belong to different components of $\mathcal{K}$.*

Proof: Suppose to the contrary that this is false. For $K \in \mathcal{K}$, choose a vertex $v_K \in N_G(v) \cap V(K)$. Let $I = \{v_K \mid K \in \mathcal{K}\}$ be the set of all these neighbors of v, and let $\hat{H} = \bigcup \mathcal{K}$ be the union of the components in $\mathcal{K}$. Observe that I is an independent set of H and hence also of G, F and $\hat{H}$. Form G', H', F' and $\hat{H}'$ from G, H, F and $\hat{H}$, respectively, by first deleting v and then by identifying I to a new vertex u (see Appendix C.7). Then $|G' - H'| < |G - H|$ and, moreover, K is a component of H' if and only if $K = \hat{H}'$ or K is a component of H different from the components in $\mathcal{K}$. Since $\chi(H) \leq k$, there is a k-coloring of H. By permuting colors in the components of H, we obtain a k-coloring of H in which all vertices of the independent set I have the same color. This coloring then induces a k-coloring of H', that is, $\chi(H') \leq k$. Since there is no path in $F - v$ between two different components of $\mathcal{K}$, it is easy to check that F' is a spanning forest of G' satisfying condition (1) and (2) of the theorem with respect to G' and H'. The minimality of $|G - H|$ then implies that G' has a k-coloring. This leads to a k-coloring of $G - v$ in which all vertices of

I receive the same color. Then the number h of colors used on $N_G(v)$ satisfies $h \leq |N_G(v)| - |I| + 1 = d_G(v) - p + 1 \leq k - 1$, from which we conclude that one of the k colors remains free for v and so $\chi(G) \leq k$, a contradiction. △

For a vertex v of $G - H$, let $\mathcal{F}_v$ denote the set of components of H containing a vertex of $N_F(v)$. Clearly, $\mathcal{F}_v \subseteq \mathcal{K}_v$ and, by condition (1), no component of $\mathcal{F}_v$ contains two vertices of $N_F(v)$ implying that $|\mathcal{F}_v| = |N_F(v) \cap V(H)|$.

Claim 2.1.2 *If v is a vertex of $G - H$, then $d_F(v) \geq 2$ and $d_{F-H}(v) \geq 1$.*

Proof: Since $\mathcal{K}_v \neq \emptyset$, it follows from Claim 2.1.1 with $p = 1$ that $d_G(v) \geq k$ and so $d_F(v) \geq d_G(v) - k + 2 \geq 2$ (by condition (2)). It remains to show that $d_{F-H}(v) \geq 1$. Suppose this is false and so $d_{F-H}(v) = 0$. This implies that $|\mathcal{F}_v| = |N_F(v) \cap V(H)| = d_F(v) \geq 2$. Clearly, there is no path in $F - v$ between two components of $\mathcal{F}_v$. Since $\mathcal{F}_v \subseteq \mathcal{K}_v$ and $d_G(v) \leq d_F(v) + k - 2$ (by condition (2)), we then obtain a contradiction to Claim 2.1.1 with $p = d_F(v)$. △

To finish the proof, first, we show that there is a vertex v in $G - H$ such that $d_{F-H}(v) = 1$ and $\mathcal{F}_v \cap \mathcal{F}_u = \emptyset$ for all vertices u of $G - H$ with $u \neq v$. To this end, let F' and H' be the graphs obtained from F and H, respectively, by identifying the vertices of each component of H to a single vertex. By (1), F' remains a forest such that $F' - H' = F - H$ and $F'' = F' - (F' - H')$ is edgeless. By Claim 2.1.2, each vertex of $F - H$ satisfies $d_{F'}(v) = d_F(v) \geq 2$ and $d_{F-H}(v) \geq 1$. Consequently $X = \{v \in V(F') \mid d_{F'}(v) \leq 1\}$ is a subset of $V(F'')$. Since F'' is edgeless, the forest $F' - X$ has a vertex v such that $d_{F'-X}(v) = 1$ and v belongs to $F - H$. Then v is a vertex of $G - H$ such that $d_{F-H}(v) = 1$ and each vertex u of $G - H$ with $u \neq v$ satisfies $\mathcal{F}_v \cap \mathcal{F}_u = \emptyset$.

Let $H_1, H_2, \ldots, H_s$ be the components of $\mathcal{F}_v$. Then $s \geq 1$ (by Claim 2.1.2) and, for every $i \in [1, s]$, there is a unique vertex v_i of H_i belonging to $N_F(v)$, and $d_F(v) = s + d_{F-H}(v) = s + 1$. Next, we claim that $\mathcal{K}_v = \mathcal{F}_v$. For otherwise, there would be a component $H_{s+1} \in \mathcal{K}_v \setminus \mathcal{F}_v$. Since v is the only vertex of $G - H$ that is adjacent in F to any of $H_1, H_2, \ldots, H_s$, there can be no path in $F - v$ between any pair among $H_1, H_2, \ldots, H_s, H_{s+1}$. Since $d_G(v) \leq d_F(v) + k - 2 = s + 1 + k - 2$ (by condition (2)), this leads to a contradiction to Claim 2.1.1 with $p = s + 1$. Hence $\mathcal{K}_v = \mathcal{F}_v$ as claimed. This implies that $F[V(K)]$ is a tree for every component K of $H' = G[V(H) \cup \{v\}]$. So the spanning forest F of G satisfies condition (1) and (2) with respect to the spanning subgraph H' of G. Since $\chi(H) \leq k$, there is a k-coloring of H. By permuting colors in the components of H, this leads to a k-coloring of H such that the vertices $v_1, v_2, \ldots, v_s$ get the same color. Then for the number h of colors used on $N_G(v) \cap V(H)$ we obtain $h \leq |N_G(v) \cap V(H)| - s + 1$. Since $d_{F-H}(v) = 1$ and $d_F(v) = s + 1$, it then follows from (2) that $h \leq d_G(v) - 1 - s + 1 \leq d_F(v) - s + k - 2 = k - 1$. Hence the k-coloring of $H = H' - v$ can be extended to H' and so $\chi(H') \leq k$. Clearly, $|G - H'| < |G - H|$. But then the minimality of $|G - H|$ implies that $\chi(G) \leq k$. This contradiction completes the proof of Theorem 2.1. □

Let us note that the proof of Theorem 2.1 yields a polynomial time algorithm to construct a k-coloring ϕ of G from any given k-coloring φ of H by permuting colors in the components of H so that $\phi|_K \sim \varphi|_K$ for every component K of H (see Exercise 2.6).

Let us discuss two applications of the above theorem. First we show that it implies Brooks' theorem. So let G be a connected graph with maximum degree $\Delta \geq 3$, different from $K_{\Delta+1}$. If G is bipartite, then $\chi(G) \leq 2$ and we are done. If G is not bipartite, then Theorem 1.36 implies that G contains a spanning tree T such that the leaves of T form an independent set I of G. Now we apply Theorem 2.1 with $H = G[I]$, $F = T$ and $k = \Delta$. Clearly, $\chi(H) = 1 \leq k$ and each component of H is an isolated vertex, which is a tree. Finally, each vertex v of $G - H$ satisfies $d_F(v) \geq 2$ and so $d_G(v) \leq \Delta = k \leq d_F(v) + k - 2$. Consequently, Theorem 2.1 implies that $\chi(G) \leq k = \Delta$.

The second application deals with the union of graphs. The following result about the chromatic number of the union of two graphs is folklore. If G_1 and G_2 are two graphs, then

$$\chi(G_1 \cup G_2) \leq \chi(G_1) \cdot \chi(G_2). \tag{2.1}$$

To prove this inequality, we may suppose that $V(G_1) = V(G_2)$, since otherwise we may add isolated vertices. The result then follows from the fact that if φ_i is a coloring of G_i with color set C_i for $i \in \{1,2\}$, then the map φ with $\varphi(v) = (\varphi_1(v), \varphi_2(v))$ for all $v \in V(G)$ is a coloring of $G_1 \cup G_2$ with color set $C = C_1 \times C_2$.

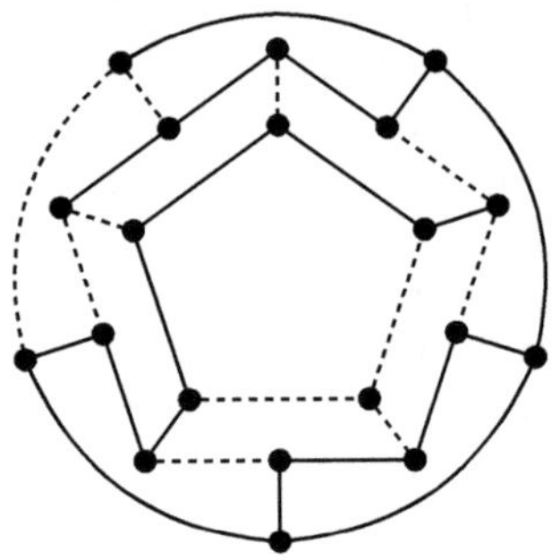

Fig. 2.1 The dodecahedron is the union of a forest and a star forest (dashed lines).

So if G is the union of two forests, then (2.1) implies that $\chi(G) \leq 4$, and the complete graph K_4, which is the union of two paths of length 3, shows that equality can hold. A **star forest** is a forest not containing a path of length 3, that is, each component of a star forest is a star $K_{1,n}$ for some integer $n \geq 0$. Suppose that a graph $G = F \cup F'$ is the union of a forest F and a star forest F'. We may assume that $V(F) = V(F') = V(G)$, otherwise we can add isolated vertices to F or F'. For each component T of F', choose the unique vertex of T having maximum degree if $|T| \geq 3$, else chose an arbitrary vertex of T. Let X denote the set of all these vertices and let $H = G[X]$. Then $H = F[X]$ and hence $\chi(H) \leq 2$. If v is a vertex of $G - H$, then $d_G(v) \leq d_F(v) + 1$. So the spanning forest F of G satisfies the conditions (1) and (2) of Theorem 2.1 with respect to G, H and $k = 3$. Consequently, we obtain from Theorem 2.1 that $\chi(G) \leq 3$. This result provides an affirmative answer to a coloring problem proposed by Sauer [890] at the graph theory conference in Keszthely (Hungary) in 1993, see also [970].

Whether inequality (2.1) remains true if we replace the chromatic number by the list chromatic number is unknown. That a graph which is the union of a forest and a star forest is not only 3-colorable, but also 3-list-colorable follows from the following result, see also Conjecture 2.3 and Figure 2.1.

Theorem 2.2 *Let $G = F \cup B$ be the union of a forest F and a bipartite graph B with parts X and Y such that $d_B(y) \le t$ for every vertex $y \in Y$, where $t \ge 0$ is an integer. Then $\chi_\ell(G) \le t+2$.*

Proof We may assume that $V(G) = V(F) = V(B) = X \cup Y$, otherwise we add some isolated vertices to F or B. A list-assignment L of G is called *correct* if

$$|L(v)| \ge \begin{cases} d_B(v)+2 & \text{if } v \in Y \\ t+2 & \text{if } v \in X. \end{cases}$$

Clearly, for the proof of the theorem it suffices to show that G is L-colorable for every correct list-assignment L of G. Hence, in what follows let L be an arbitrary correct list-assignment for G.

A subset M of X (possibly empty) is called an X-*transversal* of F if $|M \cap V(T)| = 1$ for every component T of F satisfying $X \cap V(T) \ne \varnothing$.

Consider an arbitrary X-transversal M of F and choose for each vertex $x \in M$ a color $c_x \in L(x)$. We shall prove by induction on $|E(G)| + |V(G)$ that there is an L-coloring φ of G satisfying $\varphi(x) = c_x$ for all $x \in M$. Such a coloring φ is said to be *compatible* with the precoloring $(c_x)_{x \in M}$ of M. Note that M is an independent set in G.

For $X = \varnothing$, we have $G = F$, $M = \varnothing$ and $|L(v)| \ge 2$ for all $v \in V(G)$. Since $G = F$ is 1-degenerate, there is an L-coloring φ of G. Whence, we may assume that $X \ne \varnothing$. Then $M \ne \varnothing$. Choose an arbitrary vertex $x' \in M$ and let $c' = c_{x'}$.

First, assume that $d_B(x') \ge 1$. Then there is an edge $x'y_0 \in E(B)$ for some vertex $y_0 \in Y$. Let $B' = B - x'y_0$ and $G' = F \cup B' = G - x'y_0$. Define L' to be the list-assignment with

$$L'(v) = \begin{cases} L(y_0) \setminus \{c'\} & \text{if } v = y_0 \\ L(v) & \text{otherwise.} \end{cases}$$

Clearly, L' is a correct list-assignment for G'. From the induction hypothesis it then follows that there is an L'-coloring φ of G' that is compatible with the precoloring of M. Since $\varphi(y_0) \in L'(y_0) = L(y_0) \setminus \{c'\}$, we conclude that φ is an L-coloring of G that is compatible with the precoloring of M.

Now, assume that $d_B(x') = 0$. If $d_F(x') = 0$, then by the induction hypothesis there is an L-coloring φ' of $G' = G - x'$ that is compatible with the precoloring of $M' = M \setminus \{x'\}$. Clearly, the map φ defined by $\varphi(x') = c'$ and $\varphi(v) = \varphi'(v)$ for all $v \in V(G') = V(G) \setminus \{x'\}$ is an L-coloring of G that is compatible with the precoloring of M. Whence, we may assume that $d_F(x') \ge 1$. Let z denote an arbitrary neighbor of x' in F. We consider two cases.

Case 1: $z \in X$. Then we argue as follows. First, we delete the edge $x'z$ from F. Then the forest $F' = F - x'z$ has one more component than F and $M' = M \cup \{z\}$ is an X-transversal of F'. Since $|L(z)| \geq d_B(z)+2 \geq 2$, there is a color $c_z \in L(z) \setminus \{c'\}$. By the induction hypothesis, there is an L-coloring φ of $G' = G - x'z$ that is compatible with the precoloring of M'. In particular, we have $\varphi(z) = c_z \neq c' = \varphi(x')$. This implies that φ is an L-coloring of G that is compatible with the precoloring of M.

Case 2: $z \in Y$. If $d_B(z) = 0$, then B is a bipartite graph with parts $X \cup \{z\}$ and $Y \setminus \{z\}$ and we can argue as in Case 1, since $|L(z)| \geq t+2 \geq 2$. Whence, we may assume that $d_B(z) \geq 1$. Let x_0 be an arbitrary neighbor of z in B. Clearly, we have $x_0 \in X$.

Let $F' = F - zx'$ and $G' = G - zx'$. Moreover, let T denote the component of F containing x' and hence also z. Then $T' = T - zx'$ is the union of two components, say T_1 and T_2, where $x' \in V(T_1)$ and $z \in V(T_2)$. Obviously, $\tilde{T}$ is a component of F' if and only if $\tilde{T}$ is a component of F different from T or $\tilde{T} \in \{T_1, T_2\}$. If T_2 contains a vertex of X, then we can choose some vertex $\tilde{x} \in X \cap V(T_2)$ and a color $\tilde{c} \in L(\tilde{x})$ to obtain a X-transversal M' of F' and a precoloring of M'. Otherwise, $M' = M$ is a X-transversal of F'.

If $c' = c_{x'}$ is not contained in the list $L(z)$, then we argue as follows. By the induction hypothesis, there is an L-coloring φ of $G' = G - zx'$ that is compatible with the precoloring of M'. Then $\varphi(x') = c' \neq \varphi(z)$ and, therefore, φ is an L-coloring of G that is compatible with the precoloring of M. Therefore, in what follows we may assume that $c' \in L(z)$. To complete the proof, we consider two cases.

Case A: $x_0 \notin V(T_2)$. Let T_0 denote the component of $F' = F - zx'$ containing x_0. Then $T_0 \neq T_2$ and, therefore, $F^0 = F' + zy_0$ is a forest and M is a X-transversal of F^0. Then $B^0 = B - zx_0$ is a bipartite graph with parts X and Y and $d_{B^0}(z) = d_B(z) - 1$. Let L^0 be the list-assignment of $G' = G - zx'$ defined by

$$L^0(v) = \begin{cases} L(z) \setminus \{c'\} & \text{if } v = z \\ L(v) & \text{otherwise.} \end{cases}$$

Clearly, $G' = G - zx' = F^0 \cup B^0$ and L^0 is a correct list-assignment of G'. By the induction hypothesis, there is an L^0-coloring φ of G' that is compatible with the precoloring of M. Then $\varphi(z) \neq c' = \varphi(x')$ and, therefore, φ is an L-coloring of G that is compatible with the precoloring of M.

Case B: $x_0 \in V(T_2)$. Hence $x_0 \in V(T)$ and, since $x' \in V(T) \cap M$, we have $x_0 \notin M$. Since $x_0 \in X$, we have $|L(x_0)| \geq t+2$. Since $z \in Y$, we have $|L(z)| \geq d_B(z)+2$. We may assume that equality holds, otherwise we remove some colors from the list $L(z)$. Since $d_B(z) \leq t$, this implies that $|L(x_0)| \geq |L(z)|$. Since $c' \in L(z)$, this implies that there is a color $\gamma \in L(x_0) \setminus (L(z)\ \{c'\})$. Now, let $G^0 = G' - zx_0 = G - zx' - zx_0$, $M^0 = M \cup \{x_0\}$, $c_{x_0} = \gamma$, and

$$L^0(v) = \begin{cases} L(z) \setminus \{c'\} & \text{if } v = z \\ L(v) & \text{otherwise.} \end{cases}$$

Clearly, $G^0 = F' \cup (B - zx_0)$ and L^0 is a correct list-assignment of G^0. Moreover, M^0 is a X-transversal of F' with a well defined precoloring. By the induction hypothesis, there is an L^0-coloring φ of G^0 that is compatible with the precoloring of M^0. In particular, we then have $\varphi(x') = c' \neq \varphi(z)$ and $\varphi(x_0) = \gamma \neq \varphi(z)$. Therefore, φ is an L-coloring of G that is compatible with the precoloring of M. □

Conjecture 2.3 (THE FOREST-PLUS-STARFOREST CONJECTURE) *Every planar graph with girth at least 5 is the union of a forest and a star forest.*

If $G = G_1 \cup G_2 \cup \cdots \cup G_p$ is the union of p graphs, then it follows from (2.1) that

$$\chi(G) \leq \chi(G_1) \cdot \chi(G_2) \cdot \ldots \cdot \chi(G_p).$$

However, if $\chi(G_i) = \mathrm{col}(G_i)$ for $i \in [1, p]$, then the bound can be improved as proved by Klein and Schönheim [601].

Theorem 2.4 *If $G = G_1 \cup G_2 \cup \cdots \cup G_p$ is the union of p graphs such that the ith graph satisfies $\mathrm{col}(G_i) \leq d_i + 1$ for a positive integer d_i, then*

$$\chi(G) \leq \mathrm{col}(G) \leq 2(d_1 + d_2 + \cdots + d_p)$$

and

$$\omega(G) \leq \sum_{i=1}^{p} d_i + \left\lfloor \frac{1}{2}\left(1 + \sqrt{1 + 8 \sum_{1 \leq i < j \leq p} d_i d_j}\right) \right\rfloor.$$

Proof The class $\mathcal{G}(\mathrm{col} \leq d)$ of graphs with coloring number at most d is a monotone and additive graph property. Hence we may assume that $V(G_i) = V(G)$ for $i \in [1, p]$, otherwise we can add some isolated vertices to G_i. Since $\chi(G) \leq \mathrm{col}(G)$ (by Proposition 1.7), it suffices to show that $\mathrm{col}(G) \leq d$ for $d = 2(d_1 + d_2 + \ldots + d_p)$. To this end, let H be a nonempty subgraph of G. Since $\delta(H) \leq |H| - 1$, there is nothing to prove if $|H| \leq d$, So assume that $|H| \geq d + 1$. Then, for $i \in [1, p]$, $G_i[V(H)]$ is a d_i-degenerate graph of order at least d_i and, therefore,

$$|E(G_i[V(H)]| \leq d_i|H| - d_i(d_i + 1)/2$$

(see Exercise 2.35). Consequently,

$$2|E(H)| \leq d|H| - \{d_1(d_1 + 1) + d_2(d_2 + 1) + \ldots + d_p(d_p + 1)\},$$

which implies that $\delta(H) \leq d - 1$. It remains to establish the upper bound for the clique number of G. So assume that $\omega(G) = n$. If $n \leq d_1 + d_2 + \ldots + d_n$, we are done. Otherwise, let X be an n-clique of G. Then $G[X]$ is the union of the graphs $H_i = G_i[X]$ $(i = 1, 2, \ldots, p)$. Since H_i is a d_i-degenerate graph of order at least d_i, we have $|E(H_i)| \leq d_i n - d_i(d_i + 1)/2$. Consequently, we obtain

$$\frac{n(n-1)}{2} = |E(G[X])| \leq \sum_{i=1}^{p} \left(d_i n - \frac{d_i(d_i + 1)}{2}\right),$$

which is equivalent to

$$n^2 - n(1+2(d_1+\cdots d_n)) + (d_1(d_1+1)+\cdots+d_p(d_p+1)) \leq 0.$$

This quadratic inequality in n yields the required bound for the clique number n. □

That the bound for the clique number in the above theorem is sharp was proved by Klein [600] (see also [601]). So if n is the right hand side of this bound, then K_n is the union of p graphs, where the ith graph is d_i-degenerate. Let us consider the special case where G is the union of p forests. Then the above theorem implies that $\chi(G) \leq 2p$ and $\omega(G) \leq 2p$. That the complete graph K_{2p} is the union of p forests follows from a result of Nash-Williams [769] (see also Edmonds [321], Nash-Williams [768] and Tutte [1034]); this result says that a multigraph G is the union of p forests if and only if $|E_G(U,U)| \leq p(|U|-1)$ for every nonempty set $U \subseteq V(G)$. Consequently, if G is the union of p forests, then the bound for the chromatic number in the above theorem is sharp; and this bound is the same as the bound for the clique number. In all other cases, the bound for the clique number is smaller than the bound for the chromatic number and Klein and Schönheim [601] proposed the conjecture that the bound for the chromatic number can be replaced by the bound for the clique number.

Conjecture 2.5 (THE KLEIN–SCHÖNHEIM CONJECTURE) *Assume that G is the union of p graphs such that the ith graph is d_i-degenerated for a positive integer d_i. Then*

$$\chi(G) \leq \sum_{i=1}^{p} d_i + \left\lfloor \frac{1}{2}\left(1+\sqrt{1+8\sum_{1\leq i<j\leq p} d_i d_j}\right)\right\rfloor.$$

2.2 Coloring and Subtrees of Graphs

Here is a formulation of Brooks' theorem in terms of forbidden subgraphs:

Every connected graph $G \neq K_{k+1}$ $(k \geq 3)$ not containing $K_{1,k+1}$ as a subgraph satisfies $\chi(G) \leq k$.

Taking this formulation into account, it becomes natural to exclude arbitrary trees of order $k+2$, and in fact Mihók [736] proved the following result.

Theorem 2.6 (MIHÓK) *Let T be a tree of order $k+2$, where $k \geq 3$. Then every connected graph $G \neq K_{k+1}$ not containing T as a subgraph satisfies $\chi(G) \leq k$.*

Let $\text{Forb}_{\subseteq}(T)$ denote the class of graphs not containing any copy of T as a subgraph. So Theorem 2.6 says that if T is a tree of order $k+2$ and $k \geq 3$, then every connected graph of $\text{Forb}_{\subseteq}(T)$, except K_{k+1}, has chromatic number at most k. However, unless $T = K_{1,k+1}$, every such graph has coloring number at most k, as proved by Mihók. So Theorem 2.6 is a consequence of Brooks' theorem and the following result.

Lemma 2.7 *Let G be a connected graph, let k be a positive integer, and let T be a tree of order $k+2$ different from $K_{1,k+1}$. If $G \neq K_{k+1}$ and $\delta(G) \geq k$, then G contains a copy of T as a subgraph.*

Proof Note that if X is a vertex set of G with $|X| \leq k$, then each vertex in X has a neighbor outside X in G. Let x be a leaf of T, let y be the only neighbor of x in T, and let $T' = T - x$. Since $T \neq K_{1,k+1}$, we can choose x such that T' contains a path with three vertices starting at y. Since G is not complete, there is a path $P = (u, v, w)$ in G such that $uw \notin E(G)$. Starting with P we can construct, step by step, a copy $\tilde{T}'$ of T' such that P is a path in $\tilde{T}'$ and u is a copy of y. Since $d_G(u) \geq k$ and $uw \notin E(G)$, the vertex u has a neighbor u' in G not belonging to $\tilde{T}'$. Hence $\tilde{T} = \tilde{T}' + uu'$ is a copy of T in G as claimed. □

Mihók's result about the chromatic number of graphs not containing a specific tree as a subgraph shows that the most difficult case is when T is a star. In all other cases, the bound for the chromatic number is in fact a bound for the coloring number. However, if we forbid a specific tree T as an induced subgraph, then the corresponding problem becomes much more difficult. Recall that a graph is **T-free** if no induced subgraph of G is isomorphic to T, and that Forb(T) denotes the class of T-free graphs. In 1987 Gyárfás [443] published an interesting (and entertaining) paper about *Problems from the world surrounding perfect graphs*. One of the many interesting problems and conjectures presented in this paper is the following conjecture, first proposed by Gyárfas [442], and which was independently proposed by Sumner [983].

Conjecture 2.8 (The Gyárfás–Sumner Tree Conjecture) *For every tree T there is a function f_T such that every T-free graph G satisfies $\chi(G) \leq f_T(\omega(G))$.*

Let us rephrase this conjecture, using the notation of Gyárfás [443]. A class of graphs $\mathcal{P}$ (usually a hereditary graph property) is called **χ-bounded** if there is a function $f : \mathbb{N} \to \mathbb{N}$ such that $\chi(G) \leq f(\omega(G))$ for every nonempty graph $G \in \mathcal{P}$; the function f is then called a **χ-bounding function** of $\mathcal{P}$. For instance, the class of perfect graphs is χ-bounded with $f(x) = x$ as a χ-bounding function, and the class of line graphs of graphs is χ-bounded with $f(x) = x + 1$ as a χ-bounding function (by Theorem 1.39, see also Exercise 1.23).

Let $\mathcal{X}$ be a class of graphs. A graph G is called **$\mathcal{X}$-free** if no induced subgraph of G is isomorphic to a member in $\mathcal{X}$. The class of all $\mathcal{X}$-free graphs is denoted by Forb($\mathcal{X}$). If $\mathcal{X} = \{H\}$ is a singleton, then we write Forb(H) rather than Forb($\mathcal{X}$).

In 1959 Erdős [328] used the probabilistic method to show the existence of graphs with arbitrarily large girth and chromatic number. Recall that the **girth** $g(G)$ of a graph G is the length of the shortest cycle contained in G; if G does not contain any cycle (i.e., G is a forest), we set $g(G) = \infty$ (see Appendix C.5).

Theorem 2.9 (Erdős) *For all positive integers k and ℓ there exists a graph G with chromatic number $\chi(G) \geq k$ and girth $g(G) \geq \ell$.*

We will present a constructive proof of Erdős' theorem in next section of this chapter. An immediate consequence of Theorem 2.9 is that if a graph H contains a cycle, then the class Forb(H) is not χ-bounded. Hence the conjecture of Gyárfás and Sumner is equivalent to the assertion that Forb(H) is χ-bounded if and only if H is a forest (see Exercise 2.18). Not much progress has been made since the conjecture was posed by Gyárfás and Sumner. That the class Forb$(K_{1,n})$ is χ-bounded follows from Ramsey's theorem (see Exercise 2.12). That the conjecture is true for paths was proved by Gyárfás [443].

Theorem 2.10 (GYÁRFÁS) *Let $n \geq 1$ be an integer. Then then graph class* Forb(P_n) *is χ-bounded and $f_n(x) = (n-1)^{x-1}$ is a χ-bounding function, that is, every graph $G \in$* Forb(P_n) *satisfies $\chi(G) \leq (n-1)^{\omega(G)-1}$.*

Proof It suffices to show that every P_n-free graph G with $\omega(G) \leq p+1$ satisfies $\chi(G) \leq (n-1)^p$. The proof is by induction on $p \geq 0$. If $p = 0$, then the statement is evident as G is edgeless and $\chi(G) \leq 1$. Assume that $p \geq 1$ and let G be a P_n-free graph with $\omega(G) \leq p+1$. Suppose to the contrary that $\chi(G) > (n-1)^p$. We shall reach a contradiction by construction an induced path P of G having order n. To this end, we shall construct a sequence $(G_0, G_1, \ldots, G_n)$ of graphs with $G_0 = G$ and a sequence $(v_1, v_2, \ldots, v_n)$ of vertices such that for every $i \in [1,n]$ the following statements hold:

(a) G_i is a connected induced subgraph of G_{i-1} with $v_i \in V(G_i)$,
(b) if $1 \leq j < i$, then $v_j \notin V(G_i)$ and, for $v \in V(G_i)$, we have $vv_j \in E(G)$ if and only if $j = i-1$ and $v = v_i$, and
(c) $\chi(G_i) > (n-i)(n-1)^{p-1}$.

Let G_1 be a component of $G_0 = G$ with $\chi(G_1) = \chi(G) > (n-1)^p$ (see (1.5)), and let v_1 be any vertex of G_1. Then (G_0, G_1) and v_1 satisfy (a), (b), and (c) for $i = 1$.

Assume that $(G_0, G_1, \ldots, G_k)$ and $(v_1, v_2, \ldots, v_k)$ are already defined for some $k \in [1, n-1]$ such that (a), (b), (c) are satisfied for $i \in [1,k]$.

Construct G_{k+1} and v_{k+1} as follows. Let $A = N_G(v_k) \cap V(G_k)$ and $B = V(G_k) \setminus (A \cup \{v_k\})$. Then $A \cap \{v_j \mid j \in [1, k-1]\} = \varnothing$ (by (b)). Furthermore,

$$\omega(G[A]) \leq \omega(G_k) - 1 \leq \omega(G) - 1 \leq p,$$

and so the induction hypothesis implies that we have

(+) $\chi(G[A]) \leq (n-1)^{p-1}$.

First, assume that $B \neq \varnothing$. Then $\chi(G_k) \leq \chi(G[A]) + \chi(G[B])$ since a coloring of $G[A]$ with $\chi(G[A])$ colors, a coloring of $G[B]$ with $\chi(G[B]$ new colors and an assignment of any color used on B to v_k yields a coloring of G_k. By (c) and (+), this implies that

$$\begin{aligned}\chi(G[B]) &\geq \chi(G_k) - \chi(G[A]) > (n-k)(n-1)^{p-1} - (n-1)^{p-1} \\ &= (n-(k+1))(n-1)^{p-1}.\end{aligned}$$

Consequently, by (1.5), there is a component H of $G[B]$ with $\chi(H) = \chi(G[B]) > (n-(k+1))(n-1)^{p-1}$. Since G_k is connected (by (a)), there is a vertex $v \in A$ such that $H' = G_k[V(H) \cup \{v\}]$ is connected. Now, let $G_{k+1} = H'$ and $v_{k+1} = v$. Then it is easy to check that $(G_0, G_1, \ldots, G_{k+1})$ and $(v_1, v_2, \ldots, v_{k+1})$ satisfy the requirements (a), (b), and (c) for all $i \in [1, k+1]$.

Next, assume that $B = \varnothing$. Then $V(G_k) = A \cup \{v_k\}$ and so $\chi(G_k) \leq \chi(G[A]) + 1$. By (c) and (+) this leads to $(n-k)(n-1)^{p-1} < (n-1)^{p-1} + 1$, which implies $k = n-1$. Since $A \neq \varnothing$ (by properties (a) and (c) for G_k), we can chose v_n as any vertex of A and $G_n = (\{v_n\}, \varnothing)$.

This shows the existence of the sequence $(G_0, G_1, \ldots, G_n)$ and the corresponding sequence $(v_1, v_2, \ldots, v_n)$ satisfying the conditions (a), (b), and (c) for all $i \in [1, n]$. Consequently, $P = (v_1, v_2, \ldots, v_n)$ is a path of order n contained in G as an induced subgraph. This contradiction completes the induction step and hence the proof. □

Corollary 2.11 *Every graph in* $\mathrm{Forb}(P_n, K_3)$ *with* $n \geq 2$ *satisfies* $\chi(G) \leq n-1$.

2.3 Coloring and Girth

Following an idea of Alon, Kostochka, Reiniger, West, and Zhu [52], we give an explicit construction of sparse graphs having large girth and large chromatic number. We use the notation introduced in [52]. Readers who are not familiar with the basic terminology of rooted trees may consult Appendix C.6. A graph G is called an **r-augmented tree**, if there is a rooted tree (T, v) and G is obtained from T by adding r edges from each leaf to its ancestors. We then call (T, v) the **underlying rooted tree** of G; and the edges of the set $E(G) \setminus E(T)$ are called **augmenting edges** of G. A **complete d-ary tree of heigth m** is a rooted tree whose internal vertices have d children and whose leaves have distance m from the root. For $d, r, \ell \in \mathbb{N}$, we call G a **(d, r, ℓ)-graph** if G is a bipartite r-augmented tree whose underlying tree is a complete d-ary tree and the girth of G satisfies $g(G) \geq \ell$. First, we need the following result.

Theorem 2.12 *For* $d, r, \ell \in \mathbb{N}$ *with* ℓ *even and* $\ell \geq 4$, *there is a* (d, r, ℓ)*-graph.*

Proof If there exists a (d, r, ℓ)-graph, then let $m(d, r, \ell)$ denote the least height of the underlying rooted tree of such a graph; otherwise, let $m(d, r, \ell) = \infty$. To prove the theorem it suffices to show that $m(d, r, \ell)$ is finite whenever $d, r, \ell \in \mathbb{N}$ and $\ell \geq 4$ is even. We prove this by double induction, using the following three claims, assuming that $d, r, \ell \in \mathbb{N}$ and $\ell \geq 4$ is even.

Claim 2.12.1 $m(d, r, 4) \leq 2r + 1$.

Proof: Chose a complete d-ary tree of height $2r+1$, and join each leaf to its r nonparent ancestors having odd distance from it in the tree. Then it is easy to check that the resulting graph is a $(d, r, 4)$-graph, which proves the claim. △

Claim 2.12.2 $m(d,1,\ell+2) \le 2+m(d,d^2,\ell)$.

Proof: We may assume that $m(d,d^2,\ell)$ is finite, since otherwise there is nothing to prove. Then let G' be a (d,d^2,ℓ)-graph whose underlying rooted tree (T',v') has height $m = m(d,d^2,\ell)$. For each leaf v of (T',v'), let (T_v,v) be a complete d-ary tree of height 2 such that T_v and T have only vertex v in common, and let f_v be a bijection between the d^2 augmenting edges of G' incident with v and the d^2 leaves of T_v. Assume that the trees T_v are disjoint. Now let T be the union of T' and the trees T_v taken over all leaves v of (T',v'). Then (T,v') is a complete d-ary tree of height $m+2$. For each leaf v of (T',v') replace in each augmenting edge $e \in E_{G'}(v)$ the vertex v by $f_v(e)$, and let F denote the set of all these new edges. Then the graph G with vertex set $V(T)$ and edge set $E(T) \cup F$ is an 1-augmented tree whose underlying rooted tree is (T,v'). Since for each leaf v of (T',v') and each augmenting edge $e \in E_{G'}(v)$ we have $d_T(v',f_v(e)) = d_{T'}(v',v)+2$, the graph G is bipartite as G' is bipartite. We claim that $g(G) \ge \ell+2$. Suppose this is false. Then G contains a cycle C with length $\ell(C) < \ell+2$. Since G is bipartite and ℓ is even, this implies $\ell(C) \le \ell$. Clearly, C must contain an augmenting edge of G, say xy with y being a leaf of (T,v'). Then there is a leaf v of (T',v') such that y belongs to T_v. Since $d_G(y) = 2$, the cycle C contains the edge yz of T_v incident with y. Contracting in G the trees T_w of height 2 into leaves of (T',v') contracts C to a closed walk C' of G' with $\ell(C') < \ell$. Since C' traverses edge vx only once, the remaining walk from x to v along C' contains a path that with vx completes a cycle of G' having length less than ℓ, contradicting $g(G') \ge \ell$. This proves the claim that $g(G) \ge \ell+2$. Consequently G' is a $(d,1,\ell+2)$ graph and $m(d,1,\ell+2) \le 2+m(d,d^2,\ell)$. △

Claim 2.12.3 $m(d,r+1,\ell) \le m_1+m_2-1$, *where* $m_1 = 2\left\lfloor \frac{m(d,1,\ell)}{2} \right\rfloor + 1$ *and* $m_2 = m(d^{m_1},r,\ell)$.

Proof: We may assume that m_1 and m_2 are finite, since otherwise there is nothing to prove. Then there is a $(d,1,\ell)$-graph H_1 whose underlying rooted tree has hight $m(d,1,\ell)$. If $m(d,1,\ell)$ is odd, then $m_1 = m(d,1,\ell)$ and we take as G_1 the graph H_1. If $m(d,1,\ell)$ is even, then $m_1 = m(d,1,\ell)+1$ and we take as G_1 the graph obtained from d disjoint copies of H_1 by adding a new vertex and joining it to the roots of those graphs. Hence, in both cases G_1 is a $(d,1,\ell)$-graph whose underlying rooted tree (T_1,v_1) has hight m_1.

Let $d' = d^{m_1}$. Then the rooted tree (T_1,v_1) has d' leaves. Furthermore, there is a (d',r,ℓ)-graph H_2 whose underlying rooted tree (T_2,v_2) has hight $m_2 = m(d',r,\ell)$. Let (T',v_2) be the rooted tree formed from (T_2,v_2) by starting from the root v_2 and keeping only d children of each included vertex, except that all all d' children are kept at the last level. Hence, (T',v_2) has $d^{m_2-1}d'$ leaves and deleting all these leaves from T' results in a complete d-ary tree of hight m_2-1. Now, let $G_2 = H_2[V(T_2)]$. Since all ancestors in T_2 of a leaf belonging to T' appear in T', each leaf of T' has r ancestors as neighbors in G_2. Consequently, G_2 is a r-augmented tree and his underlying rooted tree is (T',v_2).

Eventually, we construct a graph G from G_1 and G_2 as follows. Let U be the set of vertices having level m_2-1 in T'. For each vertex $u \in U$, let X_u be the set of children of u in T'. Note that $|X_u| = d'$ and each vertex of X_u is a leaf of T'. The sets of leaves of T' is the disjoint union of the sets X_u with $u \in U$. For each vertex $u \in U$, let G_1^u be a copy of G_1 and let (T_1^u, u) be the corresponding rooted tree of G_1^u. Furthermore, let f_u be a bijection between the set X_u and the set of d' leaves of T_1^u. Assume that all the the copies G_1^u and the graph G_2 are disjoint. Now let (T, v_2) be the rooted tree obtained from (T', v_2) by deleting the leaves of T' and by adding the trees T_1^u with $u \in U$, that is,

$$T = (T' - \bigcup_{u \in U} X_u) \cup \bigcup_{u \in U} T_1^u.$$

Clearly, (T, v_2) is a complete d-ary rooted tree of hight $m_1 + m_2 - 1$. Let G' be the graph obtained from G_2 by deleting all leaves of T' and adding the graphs G_1^u with $u \in L$. We call an augmenting edge of G_1^u a *short edge*. For each augmenting edge e of G_2 we add to G' a new augmenting edge e' formed as follows. The augmenting edge e of G_2 joins a leaf v of T' with another vertex w of T' (which is not a leaf). Then there is a unique vertex $u \in U$ such that $v \in X_u$. Now let e' be the edge joining w with the leaf vertex $f_u(v)$ of T_1^u (respectively of T); we say that e' is a *long edge* belonging to e. Let G be the resulting graph obtained from G' by adding all long edges. Then G is a $(r+1)$-augmented tree and (T, v_2) is the underlying rooted tree.

Since G_1^u is bipartite for all $u \in U$, it follows that G' is bipartite. If $e' = f_u(v)w$ is a long edge belonging to the augmenting edge $e = vw$, where $v \in X_u$ and $u \in U$, then $d_T(f_u(v), w) = d_{T'}(v, w) + m_1 - 1$ and $m_1 - 1$ is even. Consequently, G is bipartite as G_2 is bipartite.

It remains to show that $g(G) \geq \ell$. To this end, let C be an arbitrary cycle of G. If C contains no long edge, then C is a cycle in a copy of G_1 and so $\ell(C) \geq g(G_1) \geq \ell$. If C contains a long edge, then we contract in G the rooted tree (T_1^u, u) into the rooted tree (S^u, u) with $S^u = T'[X_u \cup \{u\}]$ (for $u \in U$). This contracts G into G_2 and C into an closed walk C' of G_2 containing an augmented edge $e = xy$ of G_2. Since f_u is a bijection between the leaves of (T_1^u, u) and the leaves of (S^u, u), the edge e is not repeated in C'. Hence the remaining walk from x to y along C' contains a path that with e completes a cycle $\tilde{C}$ of G_2. Then $\ell(C) \geq \ell(\tilde{C}) \geq g(G_2) \geq \ell$. This proves that $g(G) \geq \ell$, and hence G is a $(d, r+1, \ell)$-graph and $m(d, r+1, \ell) \leq m_1 + m_2 - 1$. Hence the claim is proved. △

To complete the proof of the theorem, we claim that $m(d, r, \ell)$ is finite for all $\ell, r, d \in \mathbb{N}$ with ℓ even and $\ell \geq 4$. We use induction on ℓ. As the base step, $m(d, r, 4)$ is finite for all $r, d \in \mathbb{N}$ by Claim 2.12.1. For the induction step assume that for even $\ell \geq 4$, $m(d, r, \ell)$ is finite for all $r, d \in \mathbb{N}$. We shall show that $m(d, r, \ell+2)$ is finite for all $r, d \in \mathbb{N}$. To this end, we use induction on r. That $m(d, 1, \ell+2)$ is finite for all $d \in \mathbb{N}$ follows from Claim 2.12.2 as $m(d, r, \ell)$ is finite for all $r, d \in \mathbb{N}$. Now assume that $r \geq 1$ and $m(d, r, \ell+2)$ is finite for all $d \in \mathbb{N}$. Then Claim 2.12.3 implies that $m(d, r+1, \ell+2)$ is finite for all $d \in \mathbb{N}$ as $m(d, 1, \ell+2)$ is finite for all $d \in \mathbb{N}$. This completes the proof of the claim that $m(d, r, \ell)$ is finite for all $\ell, r, d \in \mathbb{N}$ with ℓ even and $\ell \geq 4$. Thus the theorem is proved. □

Before we explain how (d,r,ℓ)-graphs can be used to construct graphs with arbitrarily large girth and chromatic number, we need some further notation. Let (T,v_T) be an arbitrary complete k-ary tree rooted at v_T. Let L denote the set of leaves of T. For a vertex v of T, let $E_T^+(v)$ denote the set of **out-going edges** of v in T, that is, the set of edges of the form vw such that w is a child of v. Note that for $v \in V(T)$ we have $|E_T^+(v)| = k$ if $v \notin L$ and $|E_T^+(v)| = 0$ otherwise. A path P of T between the root and a leaf of T is called a **full path** of T. If P is a full path, then for each nonleaf vertex v of P exactly one out-going edge of v in T belongs to P; we call this edge the **out-going edge** of v **along** P. An **ordering** of (T,v_T) is a mapping $f : E(T) \to [1,k]$ which induces for every nonleaf vertex v of T a bijection between $E_T^+(v)$ and $[1,k]$. So an ordering of (T,v_T) induces, indeed, a linear order on the out-going edges of each nonleaf vertex. Now fix an ordering f of (T,v_T) and let $\varphi : V(T) \to [1,k]$ be a mapping. A full path P of T is called a φ**-path** with respect to f, if every nonleaf vertex v belonging to P satisfies $\varphi(v) = f(e)$ where e is the out-going edge of v along P. Whenever $\varphi : V(T) \to [1,k]$ is a mapping, there is a unique φ path with respect to f; just start from the root and repeatedly follow the out-going edge whose f-value is equal to the φ-value of the current vertex.

Proof of Theorem 2.9: Let k and ℓ be integers with $k \geq 2$ and $\ell \geq 3$. For the proof of the theorem it suffices to show that there is a graph H with $\chi(H) \geq k+1$ and $g(H) \geq \ell$. By Theorem 2.12 there is a $(k,k+1,2\ell)$-graph G. Let (T,v_T) be the underlying rooted tree of G, and let L be the set of leaves of T. Note that (T,v_T) is a complete k-ary tree rooted at v_T. We fix an ordering f of (T,v_T). For $v \in L$, let $P = v_T T v$ be the full path of T between v_T and v. Then there is a set $X_v \subseteq V(P) \setminus \{v\}$ of $k+1$ vertices each of which is joined with v by an augmenting edge of G. The pigeonhole principle then yields a 2-set $e_v = \{x,y\} \subseteq X_v$ such that that the outgoing edge of x along P and the outgoing edge of y along P have the same f-value. Let H be the graph with vertex set $V(H) = V(T) \setminus L$ and edge set $E(H) = \{e_v \mid v \in L\}$.

First, we claim that $\chi(H) \geq k+1$. To this end, let $\varphi : V(T) \to [1,k]$ be an arbitrary mapping. Then, in T, there is a unique φ-path with respect to f, joining the root v_T with some leaf $v \in L$. Then the corresponding edge $e_v = \{x,y\}$ satisfies $\varphi(x) = \varphi(y)$. So φ is not a k-coloring of H, which implies that $\chi(H) \geq k+1$.

Next, we claim that $g(H) \geq \ell$. To see this, let C be a cycle of H, say with edge set $e_1, e_2, \ldots, e_p$. For each edge e_i there is a leaf $v_i \in L$ such that v_i is in G adjacent to both ends of e_i, say $e_i = \{x_i, y_i\}$. Now form graph C' in G by replacing each edge e_i of C by the path $P_i = (x_i, v_i, y_i)$ contained in G. Since for each leaf v of T we formed exactly one edge e_v in H, the leaves $v_1, v_2, \ldots, v_p$ are distinct. Hence C' is a cycle of G and we have $\ell(C') = 2\ell(C) \geq 2\ell$ as G is a $(k,k+1,2\ell)$-graph. Consequently, $\ell(C) \geq \ell$, which implies that $g(H) \geq \ell$. □

If we choose the $(k,k+1,2\ell)$-graph in the above construction of the graph H so that the underlying rooted tree has height h, say $h = m(k,k+1,2\ell)$, then H satisfies

$$|E(H)| = |L| = k^h \text{ and } |H| = |V(T)| - |L| = \frac{k^h - 1}{k-1},$$

which leads to $|E(H)| = (k-1)|H| + 1$. Therefore, the average degree of H satisfies

$$ad(H) = \frac{2|E(H)|}{|H|} = 2(k-1) + \frac{2}{|H|} \leq 2k - 1.$$

However, if we also want to bound the maximum average degree, we need to modify the construction.

Theorem 2.13 *For $k, \ell \in \mathbb{N}$ there is a graph G satisfying $\chi(G) \geq k+1$, $g(G) \geq \ell$ and* $\text{mad}(G) \leq 2(k-1)$.

Proof Fix an even integer $\ell \in \mathbb{N}$. For the proof of the theorem it suffices to construct a sequence of graphs, say $(G_k)_{k=1}^{\infty}$, such that $\chi(G_k) \geq k+1$, $g(G_k) \geq \ell$ and $\text{mad}(G_k) \leq 2(k-1)$. We construct this sequence recursively. For the basis step, let $G_1 = K_2$ and let G_2 be an odd cycle of length at least ℓ. Given G_{k-1} with $k \geq 3$, let $r = |G_{k-1}|$.

To construct G_k, we first choose a $(k, (r-1)k+1, \ell)$-graph G and specify an ordering f of the underlying rooted tree (T, v_T) of G; note that such a graph exists by Theorem 2.12.

First, we construct a subtree T' of T as follows. Let L be the set of leaves of T, and let $U = V(T) \setminus (L \cup \{v_T\})$. Then, for every vertex $u \in U$, there is exactly one outgoing edge of u in T, say e_u, such that the f-value of e_u is equal to the f-value of the incoming edge of u in T; let T_u denote the component of $T - e_u$ not containing u. Now T' is obtained from T by deleting all subtrees T_u with $u \in U$. Then (T', v_T) is a rooted tree such that $L' = L \cap V(T')$ is the set of leaves of T', and each nonleaf of T' has degree k in T'. Furthermore, the ordering f (more precisely, the restriction of f to $E(T')$) is an edge coloring of T' with color set $C = [1, k]$. For $v \in L'$, let $P = v_T T v$ be the full path of T between v_T and v. Note that $P = v_T T' v$ and there is a set $X_v \subseteq V(P) \setminus \{v\}$ of $(r-1)k+1$ vertices each of which is joined to v by an augmenting edge of G. The pigeonhole principle then yields a set $Y_v \subseteq X_v$ with $|Y_v| = r$ such that the outgoing edges of the vertices in Y_v along P have the same f-value.

Now, we are ready to construct G_k. For $v \in L'$, let G_v be a copy of G_{k-1}, and let $\tilde{T} = T' - L'$. We may assume that the graphs $\tilde{T}$ and G_u with $u \in L'$ are disjoint. Then G_k is obtained from the disjoint union of $\tilde{T}$ and G_v with $v \in L$ by adding, for each vertex $v \in L$, a perfect matching between G_v and Y_v. This can be done as $|G_v| = |Y_v| = r$.

First, we claim that $\chi(G_k) \geq k+1$. Suppose this is false. Then there is a coloring φ of G_k with color set $C = [1, k]$. Clearly, φ is a coloring of the tree $\tilde{T}$ (=$T' - L'$). Now, start from the root v_T and repeatedly follow the outgoing edge in T' whose f-value is equal to the φ-value of the current vertex. Since f is an edge coloring of T' with color set C, this leads to an φ-path P with respect to f, say $P = v_T T' v$ with $v \in L'$. Then Y_v is a monochromatic set with respect to φ, i.e., the color set $\varphi(Y_v)$ consists of a single color $c \in C$. Since Y_v and G_v are joined by a perfect matching, this implies that φ induces a coloring of G_v with color set $C \setminus \{c\}$. Hence the copy G_v of G_{k-1} satisfies $\chi(G_v) \leq k-1$, contradicting $\chi(G_{k-1}) \geq k$. This proves the claim that $\chi(G_k) \geq k+1$.

Next, we prove that $g(G_k) \geq \ell$. To this end, let C be an arbitrary cycle of G_k. Our aim ist to show that $\ell(C) \geq \ell$. If C lies in some copy G_v of G_{k-1}, then $\ell(C) \geq g(G_{k-1}) \geq \ell$ and we are done. Otherwise, C contains an edge joining a copy of G_{k-1}

and the tree $\tilde{T}$. Then, contracting each copy G_v with $v \in L'$ to a single vertex, say v, yields a closed walk C' of $G' = G[V(T')]$ using some augmenting edge of G' (respectively G). Since each vertex in a copy G_v inherits only one augmenting edge, each augmenting edge occurs only once in C'. Hence as in the proof of Claim 2.12.3, C' contains a cycle $\tilde{C}$ of G', and, therefore, $\ell(C) \geq \ell(\tilde{C}) \geq g(G) \geq \ell$ as G is a $(k,(r-1)k+1,\ell)$-graph. So we are done, too.

In order to bound the maximum average degree of G_k, it suffices to consider induced subgraphs of G_k. For a vertex set $U \subseteq V(G_k)$, let

$$d(U) = \sum_{u \in U} d_{G_k[U]}(u).$$

Let X be an arbitrary nonempty subset of $V(G_k)$. We claim that

$$d(X) \leq 2(k-1)|X|.$$

To this end, we split X into two disjoint subsets, namely $Y = X \cap V(\tilde{T})$ and $Z = X \setminus Y$. Then $G_k[Z]$ is a subgraph of the disjoint union of G_v with $v \in L'$. Since each G_v is a copy of G_{k-1} and $\text{mad}(G_{k-1}) \leq 2(k-2)$, this leads to $d(Z) \leq 2(k-2)|Z|$. Each vertex of Z has in G_k at most one neighbor in Y, which leads to $d(X) \leq d(Y) + d(Z) + 2|Z|$. Since $G_k[Y]$ is a subgraph of the tree $\tilde{T}$, we obtain $d(Y) \leq 2|Y|$. Summarizing, we get

$$d(X) \leq d(Y) + d(Z) + 2|Z| \leq 2|Y| + 2(k-2)|Z| + 2|Z| \leq 2(k-1)|X|,$$

as claimed. Clearly, this shows that $\text{mad}(G_k) \leq 2(k-1)$. Hence, the proof is complete. □

2.4 Coloring, Cycle Lengths and DFS Trees

In this section we shall establish some relation between the chromatic number and the cycle spectrum of a graph. The set of cycle lengths of a graph G is denoted by $\text{CL}(G)$, that is,

$$CL(G) = \{\ell(C) \mid C \text{ is a cycle of } G\}.$$

Then $|CL(G)|$ is the number of different cycle length of G. Furthermore, let $\text{CL}_e(G)$ be the even numbers of $CL(G)$, and let $\text{CL}_o(G)$ be the odd numbers of $CL(G)$. With this notation, G is a bipartite graph if and only if $|CL_o(G)| = 0$. The next result is therefore an extension of Königs' theorem characterizing bipartite graphs. The result was conjectured by Bollobás and Erdős, see [340], and proved by Gyárfás [444].

Theorem 2.14 (GYÁRFÁS) *Every graph G satisfies $\chi(G) \leq 2|CL_o(G)| + 2$.*

Proof Let G be a graph with $|CL_o(G)| = r$. By (1.5), we may assume that G is connected graph. To show that $\chi(G) \leq 2r+2$, fix a DFS tree (T, v_0) of G, say of hight h (see Appendix C.6). For $i \in [0,h]$, let $L_i = \{v \in V(G) \mid d_T(v_0, v) = i\}$ be the set of vertices having level i in T. Then each set L_i is an independent set of G (see Proposition C.6). If $uv \in E(G) \setminus E(T)$ such that $u \in L_i$, $v \in L_j$, and $i < j$, then $C = uTv + uv$ is a cycle of length $\ell = j - i + 1$, where ℓ is odd if and only if $i \equiv j \pmod 2$. Since G has only r different odd cycle lengths, this implies that

$$d_j = |\{i \mid i < j, i \equiv j \pmod 2, \text{ and } E_G(L_i, L_j) \neq \emptyset\}|$$

satisfies $d_j \leq r$ for all $j \in [1, h]$. For $p \in \{0, 1\}$, let L^p be the union of the sets L_j with $j \equiv p \pmod 2$. Then $V(G)$ is the disjoint union of L^0 and L^1, and $\chi(G) \leq \chi(G[L^0]) + \chi(G[L^1])$.

To find a coloring of $G[L^p]$ we use a sequential coloring algorithm where we color each vertex of the set L_j with the same color. Since $d_j \leq r$, this can be done with a set of $r+1$ colors, and so $\chi(G[L^p]) \leq r+1$ for $p \in \{0, 1\}$. Consequently, $\chi(G) \leq 2r+2$, which proves the theorem. □

Gyárfás proved a stronger result, relating the coloring number to the number of different odd cycle length. He also proved that $\chi(G) = 2|CL_o(G)| + 2$ implies that G contains a complete graph of order $2|CL_o(G)| + 2$, as conjectured by Bollobás and Erdős. Mihók and Schiermeyer [737] proved a counterpart to Gyárfás' result for the number of even cycle lengths.

Theorem 2.15 *Let G be a graph with $|CL_e(G)| = s$. Then $\mathrm{col}(G) \leq 2s+3$ and equality holds if and only if $s = 0$ and G contains an odd cycle, or $s \geq 1$ and G contains a K_{2s+3}.*

Proof Let G be a graph with $|CL_e(G)| = s$. For a path P of G and an end v of P, let $D_v(P) = \{\ell(vPu) \mid u \in V(P) \text{ and } uv \in E(G)\}$. Furthermore, for $r \in \{0, 1\}$, let $D_v^r(P) = \{\ell \in D_v(P) \mid \ell \equiv r \pmod 2\}$. If $u \in V(P)$ and $\ell(vPu) = p$, we say that u is the p-vertex of (P, v). Assume that $p, q \in D_v(P)$ with $2 \leq p < q$, u is the p-vertex of (P, v), and w is the q-vertex of (P, v). Then $C_p = vPu + uv$ is a cycle of G with $\ell(C_p) = p+1$, and $C_{p,q} = vu + uPw + wv$ is a cycle of G with $\ell(C_{p,q}) = q - p + 2$.

Clearly, $1 \in D_v^1(P)$ unless $|P| = 1$. If $p \in D_v^1(P)$ and $p \geq 3$, then $p+1 \in CL_e(G)$. Consequently, $|D_v^1(P)| \leq s+1$. If $p, q \in D_v^0(P)$ and $p < q$, then $q - p + 2 \in CL_e(G)$. Therefore, $|D_v^0(P)| \leq s+1$. Summarizing, we obtain that $|D_v(P)| \leq 2s+2$.

First, we claim that $\mathrm{col}(G) \leq 2s+3$. To this end, let H be an nonempty subgraph of G. Then there is a longest path P in H. If v is an end of P, then all neighbors of v in H belong to P, and, therefore, $d_H(v) = |D_v(P)| \leq 2s+2$. This implies that $\delta(H) \leq 2s+2$. Hence, $\mathrm{col}(G) \leq 2s+3$ as claimed. If G contains a subgraph H_s such that H_0 is an odd cycle and $H_s = K_{2s+3}$ if $s \geq 1$, then $2s+3 = \mathrm{col}(H_s) \leq \mathrm{col}(G) \leq 2s+3$, which yields $\mathrm{col}(G) = 2s+3$.

Now assume that $\mathrm{col}(G) \geq 2s+3$. Our aim is to show that $s = 0$ and G contains an odd cycle, or $s \geq 1$ and G contains a K_{2s+3}. Since $\mathrm{col}(G) \geq 2s+3$, there is an induced subgraph H of G such that $\delta(H) \geq 2s+2 \geq 2$. Let P be an arbitrary longest path of H, and let v be an end of P. Since $|D_v^r(P)| \leq s+1$ for $r \in \{0, 1\}$, we obtain

$$2s+2 \leq d_H(v) = |D_v(P)| = |D_v^0(P)| + |D_v^1(P)| \leq 2s+2,$$

which leads to $|D_v^0(P)| = |D_v^1(P)| = s+1$. If $s = 0$, then G contains an odd cycle of length $p+1$ with $p \in D_v^0(P)$ and we are done. So we may assume that $s \geq 1$. Then

$$D_v^0(G) = \{p_1, p_2, \ldots, p_{s+1}\} \text{ with } 2 \leq p_1 < p_2 < \cdots < p_{s+1}.$$

If $CL_e = CL_e(G)$, then $|CL_e| = s$ and $CL_e = \{p_{i+1} - p_1 + 2 \mid 1 \leq i \leq s\}$. For $i \in [1,s]$, let $\ell_i = p_{i+1} - p_i$. Then the set CL_e consists of the s values

$$\ell_1 + 2, \ell_1 + \ell_2 + 2, \ldots, \ell_1 + \ell_2 + \cdots + \ell_s + 2$$

listed in increasing order. For $i \in [1,s]$ the values

$$\ell_i + 2, \ell_{i-1} + \ell_i + 2, \ldots, \ell_1 + \ell_2 + \cdots + \ell_i + 2$$

are also even cycle lengths in increasing order belonging to CL_e. Consequently, $\ell_i = \ell_1 = \ell$ for $i \in [1,s]$, and so $CL_e = \{\ell+2, 2\ell+2, \ldots, s\ell+2\}$, which leads to

$$D_v^0(P) = \{p_1, p_1+\ell, p_1+2\ell, \ldots, p_1+s\ell\}.$$

On the other hand, $CL_e(G) = \{p+1 \mid p \in D_v^1(P), p \geq 3\}$, which implies that

$$D_v^1(P) = \{1, \ell+1, 2\ell+1, \ldots, s\ell+1\}$$

We now apply the exchange methods for longest paths. Let w_i be the $(i\ell+1)$-vertex of (P,v), let v' be the ℓ-vertex of (P,v), and let w be the end of P different from v. Then $P' = v'Pv + vw_1 + w_1Pw$ is also a longest path of H, the vertex v' is an end of P', $v'w_1 \in E(H)$, and $d_{P'}(v', w_1) = \ell + 1$. Hence $D_{v'}^1(P') = D_v^1(P)$, which implies, that $v'w_i \in E(H)$ for $i \in [1,s]$, and so (v', w_1, v, w_2, v') is a cycle of length 4 in G. Consequently, $\ell = 2$, and $D_v^1(P) = D_{v'}^1(P') = \{1,3,\ldots,2s+1\}$. Furthermore, $D_v^0(P) = \{p_1, p_1+2, p_1+4, \ldots, p_1+2s\}$ and, hence, $D_v^0(P') = \{p_1', p_1'+2, p_1+4, \ldots, p_1'+2s\}$. If $vv' \in E(H)$, then $p_1 = p_1' = 2$. Otherwise, both p_1 and p_1' are even and at least 4. By symmetry, we may assume that $p_1 < p_1'$. Then let u_1 be the p_1-vertex of (P,v) and let u_1' be the $(p_1'+2s)$-vertex of (P',v'). Note that u_1' is also the $(p_1'+2s)$-vertex of (P,v). Then $C = u_1Pu_1' + u_1v' + v'Pv + vu_1$ is a cycle with $\ell(C) = (p_1' - p_1) + 2s + 4$, and so $\ell(C)$ is even and $\ell(C) \geq 2s+4$, which is impossible as $CL_e = \{4,6,\ldots,2s+2\}$. This shows that $p_1 = 2$, and so $D_v(P) = [1, 2s+2]$. If v_i is the i-vertex of (P,v), then v_i for $1 \leq i \leq$ is an end of the longest path $P_i = v_iPv + v_{i+1}Pw$, and so $D_{v_i}(P_i) = [1, 2s+2]$. Consequently $G[v, v_1, v_2, \ldots, v_{2s+2}]$ is a K_{2s+3}. This completes the proof of the theorem. □

Corollary 2.16 (Mihók and Schiermeyer) *Let G be a graph with $|CL_e(G)| = s$. Then $\chi(G) \leq 2s+3$ and equality holds if and only if $s = 0$ and G contains an odd cycle, or $s \geq 1$ and G contains a K_{2s+3}.*

Given a graph G, define

$$\mathrm{c}(G) = \max CL(G), \mathrm{c_o}(G)(G) = \max CL_o(G) \text{ and } \mathrm{c_e}(G)(G) = \max CL_e(G),$$

where we set $\max\varnothing = -\infty$; these parameters are called the **circumference** of G, the **odd circumference** of G, and the **even circumference** of G, respectively. If G is a nonbipartite graph, then $2|CL_o(G)|+1 \le \mathrm{c_o}(G)$, and from Theorem 2.14 it then follows that $\chi(G) \le \mathrm{c_o}(G)+1$. The complete graph of order $\mathrm{c_o}(G)+1$ shows that this inequality is tight. On the other hand, at the 2nd Cantania Combinatorial Conference (September 1992) Erdős [341, Problem 2.7] proposed the conjecture, that for every $\varepsilon > 0$ there exists $k_0(\varepsilon)$ such that for $k \ge k_0(\varepsilon)$, every triangle-free k-chromatic graph G satisfies $|CL_o(G)| \ge k^{2-\varepsilon}$. That this is true was proved in 2017 by Kostochka, Sudakov and Verstraëte [637]. We will not present a proof of this result. However, we shall show that graphs with girth at least 5 satisfies $\mathrm{c_o} \ge O(\chi^2)$. The proof of this result, which is due to Diwan, Kenkre, and Vishwanathan [304], demonstrates the usefulness of online coloring algorithms combined with DFS trees.

An **online coloring algorithm** $\mathcal{A}$ takes as input a graph G and a linear order $\preceq$ on its vertex set, that is, $\preceq$ is a binary relation on V(G), which is reflexive, transitive, antisymmetric and total. The pair $(G,\preceq)$ is called an **online graph**. For $v \in V(G)$, let $G[\preceq v]$ denote the subgraph of G induced by the vertex set $\{w \in V(G) \mid w \preceq v\}$. The output of the algorithm $\mathcal{A}$ is a coloring of G, where the color of a vertex v is determined solely by the isomorphism types of $G[\preceq v]$. Intuitively, based on the vertex order $(v_1,v_2,\dots,v_n)$ of G with $v_1 \preceq v_2 \preceq \cdots \preceq v_n$ the algorithm $\mathcal{A}$ colors the vertices in turn and at the time a color is irrevocably assigned to v_i, the algorithm can only see $G[v_1,v_2,\dots,v_i]$.

Let $\mathcal{A}$ be an online coloring algorithm. For an input $(G,\preceq)$ of $\mathcal{A}$, we denote by $\chi_{\mathcal{A}}(G,\preceq)$ the number of colors that $\mathcal{A}$ uses to color $(G,\preceq)$. For a graph G, the maximum of $\chi_{\mathcal{A}}(G,\preceq)$ over all linear orders $\preceq$ on V(G) is denoted by $\chi_{\mathcal{A}}(G)$. If $\mathcal{P}$ is a **hereditary graph property**, that is, a class of graphs that is closed under isomorphism and under taking induced subgraphs (see Appendix C.11), then we denote by $\chi_{\mathcal{A},\mathcal{P}}(n)$ the maximum of $\chi_{\mathcal{A}}(G)$ taken over all graphs $G \in \mathcal{P}$ with $|G| \le n$. Hence, the function $\chi_{\mathcal{A},\mathcal{P}}(n)$ is monotone increasing in n.

The most popular online coloring algorithm is the algorithm **First-Fit** (FF), that is, the sequential coloring algorithm combined with the greedy strategy. The algorithm FF uses the color set $\mathbb{N}$. If $(G,\preceq)$ is an input of FF and $(v_1,v_2,\dots,v_n)$ is the vertex order of G with $v_1 \preceq v_2 \preceq \cdots \preceq v_n$, then FF colors the vertices in turn and color each vertex v_i with the smallest positive integer not already used to color any neighbor of v_i in $G[v_1,v_2,\dots,v_i]$.

That the arithmetic mean of the order and the clique number of a graph is an upper bound of its chromatic number is folklore.

Proposition 2.17 *Let ω be a positive integer, and let $\mathcal{P}$ denote the class of graphs with clique number at most ω. Then $\chi_{FF,\mathcal{P}}(n) \le \frac{n+\omega}{2}$.*

Proof Let $(G,\preceq)$ be an input of FF with $G \in \mathcal{P}$ and $|G| \le n$. Let $k = \chi_{FF}(G)$ be the number of colors used by FF to color $(G,\preceq)$, and let p be the number of color classes in this coloring consisting of just one vertex, and let $u_1,u_2,\dots,u_p$ be the corresponding vertices with $u_1 \preceq u_2 \preceq \cdots \preceq u_p$. Since we use FF, it then follows that u_i is adjacent in $G[\preceq u_i]$ to the vertices $u_1,u_2,\dots u_{i-1}$ for $i \in [2,p]$.

Consequently, $G[u_1,u_2,\dots,u_p]$ is a complete graph and so $p \le \omega$. Since exactly k colors are used for the vertices of G, we obtain $p+2(k-p) \le |G| \le n$, which yields $k \le (n+p)/2 \le (n+\omega)/2$. □

Proposition 2.18 *Let $\mathcal{P}$ denote the class of graphs with girth at least 5. Then $\chi_{FF,\mathcal{P}}(n) \le 2\sqrt{n}$.*

Proof Let $(G,\preceq)$ be an input of FF with $G \in \mathcal{P}$ and $|G| \le n$. Let $k = \chi_{FF}(G)$ be the number of colors used by FF to color $(G,\preceq)$ and let φ be the corresponding coloring of G. Then there is a vertex v with $\varphi(v) = k$. Consequently, in $G[\preceq v]$ the vertex v has a set X of $k-1$ neighbours such that $\varphi(X) = [1,k-1]$. If $u \in X$ and $\varphi(u) = i$, then in $G[\preceq u]$ there is a set X_i of $i-1$ neighbors of u such that $\varphi(X_i) = [1,i-1]$. Since $g(G) \ge 5$, the sets $\{v\}, X, X_1, \dots, X_{k-1}$ are pairwise disjoint subsets of $V(G)$. Consequently, $1+1+2+3+\dots+(k-1) \le |G| \le n$, which yields $k \le 2\sqrt{n}$. □

If P is a path and v_0 is an end of P, then there is a natural linear order $\preceq$ of $V(P)$ satisfying $u \preceq w$ if $d_P(v_0,u) \le d_P(v_0,w)$. We say that $\preceq$ is the **distance order** of (P,v_0). If $\mathcal{P}$ is a hereditary graph property and $\mathcal{A}$ is an online coloring algorithm that can be applied to any online graph $(G,\preceq)$ with $G \in \mathcal{P}$, then we say that $\mathcal{A}$ is an online coloring algorithm for $\mathcal{P}$.

Theorem 2.19 *Let $\mathcal{P}$ be a hereditary graph property, and let $\mathcal{A}$ be an online coloring algorithm for $\mathcal{P}$. If $G \in \mathcal{P}$ has a DFS tree of hight h, then $\chi(G) \le \chi_{\mathcal{A},\mathcal{P}}(h+1)$.*

Proof Let $G \in \mathcal{P}$ be a graph and let (T,v_0) be a DFS tree of G with hight h. For each leaf u of (T,v_0), let $P_u = v_0Tu$, let $G_u = G[V(P_u)]$, and let $\preceq_u$ be the distance order of (P_u,v_0). Then $|P_u| \le h+1$ for every leaf u. Hence, the algorithm $\mathcal{A}$ applied to the online graph $(G_u,\preceq_u)$ yields a coloring φ_u of G_u using at most $\chi_{\mathcal{A},\mathcal{P}}(h+1)$ colors. Next we show that the union φ of the colorings φ_u defines a mapping that assigns to each vertex of G exactly one color. To this end, consider a vertex w that belongs to G_u as well as to $G_{u'}$ where u and u' are distinct leaves of (T,v_0). Then w belongs to P_u as well as to $P_{u'}$, and hence $G_u[\preceq_u w] = G_{u'}[\preceq_{u'} w](= G')$. Consequently, the two colorings φ_u and $\varphi_{u'}$ restricted to the graph G' are the same, and so $\varphi_u(w) = \varphi_{u'}(w)$. So φ is a mapping from $V(G)$ to a set of at most $\chi_{\mathcal{A},\mathcal{P}}(h+1)$ colors. Since G is the union of the graphs G_u (see Proposition C.6), φ is a coloring of G, and so $\chi(G) \le \chi_{\mathcal{A},\mathcal{P}}(h+1)$. □

Corollary 2.20 *Let G be a graph with clique number at most ω, and let p be the length of a longest length of a path in G. Then $\chi(G) \le \frac{\omega+p+1}{2}$.*

Proof By (1.5), we may assume that G is a connected graph. Then G has a DFS tree and the height h of this DFS tree is at most p. Then the statement follows from Proposition 2.17 and Theorem 2.19. □

Theorem 2.21 *Let $\mathcal{P}$ be a hereditary graph property, and let $\mathcal{A}$ be an online coloring algorithm for $\mathcal{P}$. If $G \in \mathcal{P}$ and $\mathrm{c}(G) \le 2\ell$ with $\ell \ge 2$, then $\chi(G) \le 3\chi_{\mathcal{A},\mathcal{P}}(\ell)$.*

Proof By (1.5), we may assume that G is a connected graph. Then G has a DFS tree, say (T, v_0). For each leaf u of (T, v_0), let $P_u = v_0 T u$, let $G_u = G[V(P_u)]$, and let $\preceq_u$ be the distance order of (P_u, v_0). For $i \in \mathbb{N}$, let

$$L_i = \{v \in V(T) \mid (i-1)\ell \le d_T(v, v_0) \le i\ell - 1\}.$$

For each leaf u, let $P_u^i = P \cap T[L_i]$ and $G_u^i = G[V(P_u^i)]$. If $n_u = |P_u|$, then $n_u = p_u \ell + r_u$ with $0 \le r_u \le \ell - 1$, and the path P_u is partitioned into p_u pathes $P_u^1, P_u^2, \dots, P_u^{p_u}$, where each of the first $p_u - 1$ paths has length $\ell - 1$ and the last path has length r_u. Hence, the algorithm $\mathcal{A}$ applied to the online graph $(G_u^i, \preceq_u)$ yields a coloring φ_u^i of G_u^i using at most $k = \chi_{\mathcal{A},\mathcal{P}}(\ell)$ colors. We may assume that we use the color set $C = [1, k]$. As in the proof of Theorem 2.19 it follows that the union φ of the colorings φ_u^i is a mapping from the vertex set $V(G)$ into the set C. We now construct a new mapping φ' from $V(G)$ to the color set $C' = [1, 3k]$ as follows. Note that $V(G)$ is the disjoint union of the sets L_i. If a vertex v of G belongs to L_i, then $i \equiv r \pmod 3$ with $r \in \{0, 1, 2\}$ and we set $\varphi'(v) = \varphi(v) + rk$. We claim that φ' is a coloring of G. To this end, let vw be an arbitrary edge of G. Then there is a leaf u of (T, v_0) and vw is an edge of G_u (by Proposition C.6). Since $\mathrm{c}(G) \le 2\ell$, we obtain that $d_{P_u}(v, w) = d_T(v, w) \le 2\ell$. Hence if $v \preceq_u w$ and v belongs to L_i, then w belongs to $L_i \cup L_{i+1} \cup L_{i+2}$ and so $\varphi'(v) \ne \varphi'(w)$. This proves the claim that φ' is a coloring of G with color set C', and thus $\chi(G) \le 3k$. □

Corollary 2.22 *Every graph G with $g(G) \ge 5$ and $\mathrm{c}(G) \le 2\ell$ ($\ell \ge 2$) satisfies $\chi(G) \le 6\sqrt{\ell}$.*

Proof Let $\mathcal{P}$ be the class of graphs with girth at least 5. Clearly, $\mathcal{P}$ is a hereditary graph property and the result follows from Proposition 2.18 and Theorem 2.21 □

As proved by Voss [1055, Theorem 2] in 1977 and rediscovered by Kenkre and Vishwanathan [569, Lemma 3] every 2-connected nonbipartite G satisfies $\mathrm{c}(G) \le 2\mathrm{c_o}(G) - 2$. Therefore, Corollary 2.22 implies the following result.

Corollary 2.23 *Every graph G with $g(G) \ge 5$ and $\mathrm{c_o}(G) \le \ell + 1$ ($\ell \ge 2$) satisfies $\chi(G) \le 6\sqrt{\ell}$.*

Proof Supose $\chi(G) > 6\sqrt{\ell} \ge 6$. Then G contains a critical graph H with $\chi(H) = \chi(G)$ (by Proposition 1.27(a)). Then H is 2-connected (by Corollary 1.31), and $g(H) \ge g(G) \ge 5$. As H is nonbipartite, $\mathrm{c}(H) \le 2\mathrm{c_o}(H) - 2 \le 2\mathrm{c_o}(G) - 2 \le 2\ell$. But then Corollary 2.22 implies $\chi(G) = \chi(H) \le 6\sqrt{\ell}$, which is impossible. □

2.5 Partition into Degenerate Subgraphs

A **partition** and p-**partition** of a multigraph G is a sequence $(G_1, G_2, \dots, G_p)$ of p induced submultigraphs of G (possibly empty) such that each vertex of G belongs to exactly one multigraph in the sequence. A coloring of G with a set of p colors may be viewed as a partition of G into p edgeless multigraphs or, equivalently, into p multigraphs each of which has maximum degree zero. In this section we are interested in partitions of multigraphs into a fixed number of induced submultigraphs satisfying certain degree constraints. One of the earliest results in this direction is due to Lovász [686].

Theorem 2.24 (LOVÁSZ) *Let $p \in \mathbb{N}$, and let $d_1, d_2, \ldots, d_p \in \mathbb{N}_0$. If a multigraph G satisfies*

$$\Delta(G) \leq d_1 + d_2 + \cdots d_p + p - 1,$$

then G has a partition $(G_1, G_2, \ldots, G_p)$ such that $\Delta(G_i) \leq d_i$ for each $i \in [1, p]$.

Proof The result is evident if $p = 1$. For $p \geq 2$ the result easily follows from the basic case $p = 2$ by induction. For $p = 2$, a partition (G_1, G_2) for which $|E(G_1)| + |E(G_2)| + d_2|G_1| + d_1|G_2|$ is minimum has the desired property, since otherwise a vertex of too large a degree could be moved to the other side to obtain a better partition. □

On the one hand, Lovász's partition result only gives the trivial inequality $\chi \leq \Delta + 1$. On the other hand, combining this partition result with Brooks' theorem leads to an improvement of that theorem. Brooks's theorem implies that graphs with maximum degree $\Delta \geq 3$ and clique number ω satisfy $\chi \leq \max\{\Delta, \omega\}$. So Brooks' theorem provides the first result bounding the chromatic number of a graph from above in terms of the maximum degree and the clique number. Improvements of Brooks's result in this direction were first obtained independently around the same time by Borodin and Kostochka [157], Catlin [193] and Lawrence [668]

Theorem 2.25 *Let ω and Δ be integers with $3 \leq \omega \leq \Delta$. Then every graph G with maximum degree Δ and clique number at most ω satisfies*

$$\chi(G) \leq \Delta + 1 - \left\lfloor \frac{\Delta + 1}{\omega + 1} \right\rfloor. \tag{2.2}$$

Proof We apply Lovász's partition theorem with $p = \lfloor (\Delta + 1)/(\omega + 1) \rfloor$. Now we can choose $d_i = \omega$ for $1 \leq i \leq p - 1$ and $d_p \geq \omega$, so that $d_1 + d_2 + \ldots + d_p = \Delta - p + 1$. By Theorem 2.24, there is a partition $(G_1, G_2, \ldots, G_p)$ of G such that $\Delta(G_i) \leq d_i$, and hence $\chi(G_i) \leq d_i$ by Brooks's theorem when $\omega \geq 3$. Consequently, $\chi(G) \leq \chi(G_1) + \chi(G_2) + \ldots + \chi(G_p) \leq d_1 + d_2 + \ldots + d_p = \Delta - p + 1$. □

Next we are interested in partitions of multigraphs into induced submultigraphs with bounded degeneracy. So let G be a multigraph, let $d \in \mathbb{N}$ be an integer, and let $h : V(G) \to \mathbb{N}_0$ be a function. Then G is said to be **strictly h-degenerate** if every nonempty submultigraph H of G contains a vertex v such that $d_H(v) < h(v)$. So if G is a graph and $h \equiv d$ is the constant function, that is, $h(v) = d$ for all $v \in V(G)$, then G is strictly h-degenerate if and only if G is $(d-1)$-degenerate, which is equivalent to $\text{col}(G) \leq d$. A function $f : V(G) \to \mathbb{N}_0^p$ with $p \in \mathbb{N}$ is called a **vector function** of G. For such a vector function f of G, we denote by f_i the ith coordinate of f, so that

$f = (f_1, f_2, \ldots, f_p)$. Let $\mathcal{V}_p(G)$ denote the set of all vector functions of G with p coordinates. If $f \in \mathcal{V}_p(G)$, then an f**-partition** of G is a partition $(G_1, G_2, \ldots, G_p)$ of G such that G_i is strictly f_i-degenerate for every $i \in [1, p]$. If G has an f-partition, then G is said to be f**-partitionable**.

A vector function $f \in \mathcal{V}_p(G)$ may also be considered as vector function for every submultigraph of G. So we denote the restriction of f to a submultigraph H of G also by f and say that f is a vector function of H. If G is f-partitionable, then each of its submultigraphs is f-partitionable, too. In this sense, f-partitionability is a monotone multigraph property. The property is also additive, that is, G is f-partitionable if and only if each component of G is f-partitionable.

Note that the list coloring problem for a given graph G can be encoded as a partition problem for G. To see this, let L be a list-assignment of G, and let C be the set of colors used in the union of all the lists. Assume that $|C| = p$. Then we can rename the colors so that $C = [1, p]$. Now let $f \in \mathcal{V}_p(G)$ be the vector function such that $f_i(v) = 1$ if $i \in L(v)$ and $f_i(v) = 0$ if $i \notin L(v)$. If φ is an L-coloring of G, then $(G_1, G_2, \ldots, G_p)$ with $G_i = G[\varphi^{-1}(i)]$ for $i \in [1, p]$ is an f-partition of G. Conversely, if $(G_1, G_2, \ldots, G_p)$ is an f-partition of G, then G_i is edgeless for all $i \in [1, p]$ and, moreover, $v \in V(G_i)$ implies that the subgraph $H = (\{v\}, \varnothing)$ satisfies $0 = d_H(v) < f_i(v) \leq 1$, and so $f_i(v) = 1$, i.e., $i \in L(v)$. So φ with $\varphi(v) = i$ if $v \in V(G_i)$ is an L-coloring of G. Therefore, G is L-colorable if and only if G is f-partitionable. Furthermore, every vertex v of G satisfies $|L(v)| = f_1(v) + f_2(v) + \cdots + f_p(v)$. So if we consider only binary vector functions $f \in \mathcal{V}_p(G)$ satisfying $f_1(v) + f_2(v) + \cdots + f_p(v) \geq d_G(v)$ for all $v \in V(G)$, then based on the proof of Corollary 1.9 we obtain a good characterization of all such vector functions f for which G is f-partitionable. We shall extend this characterization to nonbinary vector functions.

We now define a class of pairs (G, f) called **hard pairs** (alternatively, we say that G is f**-hard**), see Figure 2.2. Recall that for a multigraph H and an integer $t \geq 1$, we denote by $G = tH$ the multigraph obtained from H by replacing each edge in H by a set of t parallel edges. Let G be a connected multigraph and let $f \in \mathcal{V}_p(G)$. Then G is f-hard if one of the following four conditions hold:

(H1) G is a block and there is an $j \in [1, p]$ such that

$$f_i(v) = \begin{cases} d_G(v) & \text{if } i = j, \\ 0 & \text{if } i \in [1, p] \setminus \{j\} \end{cases}$$

for all $v \in V(G)$. In this case, we also say that (G, f) is a **mono-block**.

(H2) $G = tK_n$ for some integers $t, n \geq 1$ and there are integers $n_1, n_2, \ldots, n_p$ such that $n_1 + n_2 + \cdots + n_p = n - 1$ and $f(v) = (tn_1, tn_2, \ldots, tn_p)$ for all $v \in V(G)$.

(H3) $G = tC_n$ for some integers t, n, where $t \geq 1$ and $n \geq 3$ is odd, and there are two integers $k, \ell \in [1, p]$ such that

$$f_i(v) = \begin{cases} t & \text{if } i \in \{k, \ell\}, \\ 0 & \text{if } i \in [1, p] \setminus \{k, \ell\} \end{cases}$$

for all $v \in V(G)$.

(H4) There are two hard pairs, say (G^1, f^1) and (G^2, f^2) with $f^1 \in \mathcal{V}_p(G^1)$, $f^2 \in \mathcal{V}_p(G^2)$, and with $V(G^1) \cap V(G^2) = \varnothing$ and $E(G^1) \cap E(G^2) = \varnothing$ such that G is obtained from $G^1 \cup G^1$ by identifying a vertex $v^1 \in V(G^1)$ with a vertex $v^2 \in V(G^2)$ to a new vertex v^* and f satisfies

$$f(v) = \begin{cases} f^1(v) & \text{if } v \in V(G^1 - v^1), \\ f^2(v) & \text{if } v \in V(G^2 - v^2), \\ f^1(v^1) + f^2(v^2) & \text{if } v = v^* \end{cases}$$

for all $v \in V(G)$. In this case we say that (G, f) is obtained from (G^1, f^1) and (G^2, f^2) by **merging** v^1 and v^2 to v^*.

The following result was obtained in 2000 by Borodin, Kostochka, and Toft [159]; this result, originally only proved for graphs, provides a far reaching extension of Brooks' theorem and several other results (see also Exercises 2.30–2.33).

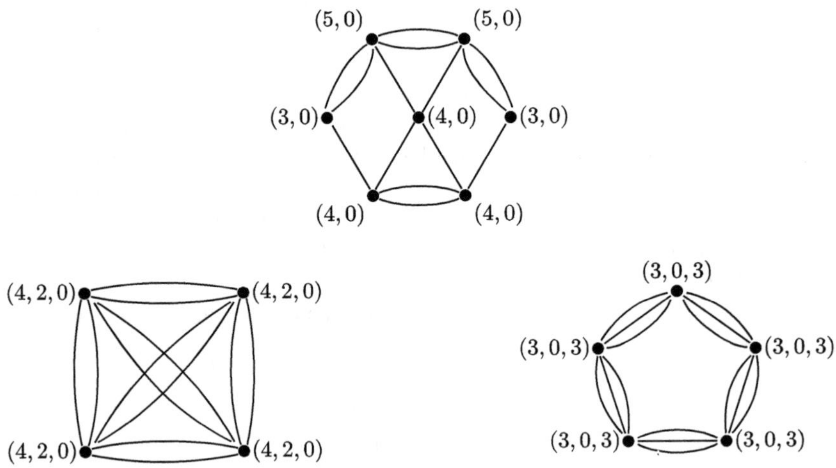

Fig. 2.2 Some examples of hard pairs.

Theorem 2.26 (Borodin, Kostochka, and Toft) *Let G be a multigraph, and let $f \in \mathcal{V}_p(G)$ be a vector function with $p \geq 1$ such that $f_1(v) + f_2(v) + \cdots + f_p(v) \geq d_G(v)$ for all $v \in V(G)$. Then G is not f-partitionable if and only if G has a component which is f-hard.*

Since a multigraph has an f-partition if and only if each of its components has one, it suffices to prove Theorem 2.26 for connected multigraphs. First we prove the "if" implication of that theorem.

Proposition 2.27 *Let G be a multigraph, and let $f \in \mathcal{V}_p(G)$ be a vector function with $p \geq 1$. If G is f-hard, then the following statements hold:*

(a) $f_1(v)+f_2(v)+\cdots+f_p(v)=d_G(v)$ *for all* $v\in V(G)$.
(b) *If* u *and* u' *are two nonseparating vertices of* G *contained in the same block of* G*, then* $f(u)=f(u')$ *or* $f_i(u)=f_i(u')=0$ *for all but one index* $i\in[1,p]$.
(c) G *is not* f*-partitionable.*

Proof Statement (a) is evident, and statement (b) follows simply by induction on the number of blocks of G. The proof of statement (c) is by reductio ad absurdum. So we suppose that G is f-hard, G has an f-partition $(G_1,G_2,\dots,G_p)$, and $|G|$ is minimum with respect to these conditions. If f_i is the zero-function, then $G_i=\varnothing$. So if (G,f) satisfies (H1), then $G_i=\varnothing$ for every $i\in[1,p]\setminus\{j\}$, implying that $G_j=G$. However $G_j=G$ is not strictly f_j-degenerate, since each vertex v of G satisfies $d_G(v)=f_j(v)$, a contradiction. If (G,f) satisfies (H2), then $G=tK_n$ for some integer $t\ge 1$ and $f(v)=(tn_1,tn_2,\dots,tn_p)$ for all $v\in V(G)$, where $n_1,n_2,\dots,n_p\in\mathbb{N}_0$ are integers such that $n_1+n_2+\cdots+n_p=n-1$. Then $G_i=\varnothing$, or there is a vertex v in G_i such that $d_{G_i}(v)=t(|G_i|-1)<f_i(v)=tn_i$. This implies that $|G_i|\le n_i$ for all $i\in[1,p]$, and so $|G|=|G_1|+|G_2|+\cdots+|G_p|\le n_1+n_2+\cdots+n_p=n-1$, a contradiction. If (G,f) satisfies (H3), then $G=tC_n$ for $n\ge 3$ odd and $t\ge 1$, and (G_k,G_ℓ) is a partition of G. As n is odd, one of the parts, say G_k, contains a tK_2. As $f_k\equiv t$, the multigraph G_k is not strictly f_k-degenerate, a contradiction. It remains to consider the case that (G,f) is obtained from two hard pairs (G^1,f^1) and (G^2,f^2) by merging v^1 and v^2 to v^* (so we may assume that $v^*=v_1=v_2$). By the minimality of G, it follows that for $j\in\{1,2\}$ the multigraph G^j is not f^j-partitionable. For $i\in[1,p]$ and $j\in\{1,2\}$ let $G_i^j=G_i\cap G^j$. By symmetry we may assume that v^* belongs to G_1. Hence, if $i\in[2,p]$, then G_i^1 is strictly f_i^1-degenerate and G_i^2 is strictly f_i^2-degenerate. Consequently, for $j\in\{1,2\}$ the multigraph G_1^j is not strictly f_1^j-degenerate and, therefore, there is a nonempty submultigraph H^j of G_1^j such that $d_{H^j}(v)\ge f_1^j(v)$ for all $v\in V(H^j)$. It then follows from (H4) that $H=H^1\cup H^2$ is a nonempty submultigraph of G_1 such that $d_H(v)\ge f_1(v)$ for all $v\in V(H)$, contradicting the assumption that G_1 is strictly f_1-degenerate. □

To prove the "only if" implication of Theorem 2.26 (see Proposition 2.30), we need some preparation. We call (G,f) an **nonpartitionable pair of dimension** p if G is a connected multigraph, $f\in\mathcal{V}_p(G)$ is a vector function such that

$$f_1(v)+f_2(v)+\cdots+f_p(v)\ge d_G(v)$$

for all $v\in V(G)$, and G is not f-partitionable. If H is an induced submultigraph of a multigraph G and $v\in V(G)\setminus V(H)$, then let $H+v=G[V(H)\cup\{v\}]$. To show that any nonpartitionable pair is hard, we shall use the following reduction method.

Proposition 2.28 *Let* (G,f) *be a nonpartitionable pair of dimension* $p\ge 1$*, let* z *be a nonseparating vertex of* G*, and let* $j\in[1,p]$ *be an index such that* $f_j(z)\ne 0$*. For the graph* $G'=G-z$*, define* $f'\in\mathcal{V}_p(G')$ *to be the vector function satisfying*

$$f_i'(v)=\begin{cases}\max\{0,f_j(v)-\mu_G(z,v)\} & \text{if } v\in V(G-z) \text{ and } i=j,\\ f_i(v) & \text{otherwise}\end{cases}$$

for all $v \in V(G')$ *and all* $i \in [1, p]$. *Then* (G', f') *is a nonpartitionable pair of dimension* p, *and in what follows, we write* $(G', f') = (G, f)/(z, j)$.

Proof By symmetry, we may suppose that $j = 1$. By assumption, $f_1(z) \geq 1$ and G has no f-partition. This implies that $|G| \geq 2$, and so G' is a connected multigraph. Suppose, to the contrary, that G' has an f'-partition, say $(G_1, G_2, \ldots, G_p)$. Clearly, for every $i \in [2, p]$, the multigraph G_i is strictly f_i-degenerate. We claim that $G_1' = G_1 + z$ is strictly f_1-degenerate. To this end, let H be a nonempty submultigraph of G_1'. If $H' = H \cap G_1$ is a nonempty submultigraph of G_1, then there is a vertex $v \in V(H')$ such that $d_{H'}(v) < f_1'(v)$. Since $f_1'(v) > 0$, this leads to

$$d_H(v) = d_{H'}(v) + \mu_H(v, z) < f_1'(v) + \mu_G(v, z) = f_1(v)$$

and we are done. Otherwise H only consists of the vertex z and $d_H(z) = 0 < f_1(z)$. This shows that G_1' is strictly f_1-degenerate. Then $(G_1', G_2, \ldots, G_p)$ is an f-partition of G, which is impossible. □

Proposition 2.29 *Let* (G, f) *be a nonpartitionable pair of dimension* $p \geq 1$. *Then the following statements hold:*

(a) $f_1(v) + f_2(v) + \cdots f_p(v) = d_G(v)$ *for all* $v \in V(G)$.
(b) *If* z *is a nonseparating vertex of* G *such that* $f_j(z) \neq 0$ *for some index* $j \in [1, p]$, *then* $f_j(v) \geq \mu_G(z, v)$ *for all* $v \in V(G - z)$.
(c) *If* $|G| \geq 2$ *and* u *is an arbitrary vertex of* G, *then* $G - u$ *has an* f*-partition and any* f*-partition* $(G_1, G_2, \ldots, G_p)$ *of* $G - u$ *satisfies* $f_i(u) = d_{G_i + u}(u)$ *for every* $i \in [1, p]$.

Proof The proof of (a) is by induction on the order n of G. The statement is evident if $n = 1$, so assume that $n \geq 2$. Let v be an arbitrary vertex of G. Then there is a nonseparating vertex $z \neq v$ in G. Since $f_1(z) + f_2(z) + \cdots + f_p(z) \geq d_G(z) \geq 1$, there is an index $j \in [1, p]$ such that $f_j(z) \geq 1$. Then $(G', f') = (G, f)/(z, j)$ is a nonpartitionable pair (by Proposition 2.28). From the induction hypothesis we then obtain that

$$\begin{aligned} d_{G'}(v) &= f_1'(v) + f_2'(v) + \cdots + f_p'(v) \\ &\geq f_1(v) + f_2(v) + \cdots + f_p(v) - \mu_G(z, v) \\ &\geq d_G(v) - \mu_G(z, v) = d_{G'}(v), \end{aligned}$$

which implies that $d_G(v) = f_1(v) + f_2(v) + \cdots + f_p(v)$. This proves (a).

In order to prove (b), suppose this is false. Then G has a nonseparating vertex z and there is a vertex v and an index $j \in [1, p]$ such that $f_j(z) \neq 0$, and $f_j(v) < \mu_G(z, v)$. By symmetry, we may assume that $j = 1$. Then $(G', f') = (G, f)/(z, 1)$ is a nonpartitionable pair (by Proposition 2.28) such that $0 = f_1'(v) > f_j(v) - \mu_G(z, v)$ and $f_i'(v) = f_i(v)$ for all $i \in [2, p]$. By (a), this yields

$$\begin{aligned} d_G(v) - \mu_G(z,v) = d_{G'}(v) &= f_1'(v) + f_2'(v) + \cdots + f_p'(v) \\ &> f_1(v) + f_2(v) + \cdots + f_p(v) - \mu_G(z,v) \\ &= d_G(v) - \mu_G(z,v), \end{aligned}$$

which is impossible. This proves (b).

In order to prove (c), let u be an arbitrary vertex of G and put $G' = G - u$. Then each component H of G' contains a vertex $v \in N_G(u)$, and this vertex satisfies $f_1(v) + f_2(v) + \cdots + f_p(v) > d_H(v)$. Then (a) implies that each component of G' is f-partitionable, and hence G' is f-partitionable. Now let $(G_1, G_2, \ldots, G_p)$ be an arbitrary f-partition of G'. Let $i \in [1, p]$ be an arbitrary index. Since G has no f-partition, $G_i' = G_i + u$ is not strictly f_i-degenerate, and hence there is a nonempty submultigraph H of G_i' such that $d_H(v) \geq f_i(v)$ for all $v \in V(H)$. Since G_i is strictly f_i-degenerate, we obtain that u belongs to H, which leads to $f_i(u) \leq d_H(u) \leq d_{G_i'}(u)$. Since $f_1(u) + f_2(u) + \cdots + f_p(u) = d_G(u)$ (by (a)) and $d_G(u) = d_{G_1'}(u) + d_{G_2'}(u) + \cdots + d_{G_p'}(u)$, we obtain that $f_i(u) = d_{G_i'}(u)$ as required. □

Proposition 2.30 *If (G, f) is a nonpartitionable pair, then G is f-hard.*

Proof The proof is by reductio ad absurdum. So let (G, f) be a smallest counterexample, that is,

(1) (G, f) is a nonpartitionable pair, say of dimension $p \geq 1$,
(2) (G, f) is no hard pair, and
(3) $|G|$ is minimum subject to (1) and (2).

By Proposition 2.29(a), each vertex v of G satisfies

$$f_1(v) + f_2(v) + \cdots + f_p(v) = d_G(v). \tag{2.3}$$

Furthermore, $|G| \geq 2$, for otherwise, (G, f) would be a hard pair of type (H1), contradicting (2).

Claim 2.30.1 *G is a block, that is, G has no separating vertex.*

Proof: Suppose, to the contrary, that G contains a separating vertex, say v^*. Then G is the union of two connected induced submultigraphs G^1 and G^2 having only vertex v^* in common such that $|G^j| < |G|$ for $j \in \{1, 2\}$. By Proposition 2.29(c), $G - v^*$ has an f-partition $(G_1, G_2, \ldots, G_p)$ and $f_i(v^*) = d_{G_i + v^*}(v^*)$ for every $i \in [1, p]$. For $i \in [1, p]$, let $G_i^1 = G_i \cap G^1$ and $G_i^2 = G_i \cap G^2$. Note that G_i is the disjoint union of G_i^1 and G_i^2, and

$$f_i(v^*) = d_{G_i + v^*}(v^*) = d_{G_i^1 + v^*}(v^*) + d_{G_i^2 + v^*}(v^*). \tag{2.4}$$

For $j \in \{1, 2\}$, let $f^j \in \mathcal{V}_p(G^j)$ be the vector function satisfying

$$f_i^j(v) = \begin{cases} f_i(v) & \text{if } v \in V(G^j - v^*), \\ d_{G_i^j + v^*}(v^*) & \text{if } v = v^*. \end{cases}$$

for all $v \in V(G^j)$ and all $i \in [1,p]$. Taking (2.3) and (2.4) into account, we conclude that

$$f_1^j(v) + f_2^j(v) + \cdots + f_p^j(v) = d_{G^j}(v)$$

for $j \in \{1,2\}$ and for every vertex v of G^j. First assume that G^j is not f^j-partitionable for $j \in \{1,2\}$. Then, as (G^j, f^j) satisfies (1) and $|G^j| < |G|$, it follows from (3) that G^j is f^j-hard. So booth (G^1, f^1) and (G^2, f^2) are hard pairs, and so G is f-hard (by (H4)), a contradiction to (2). Now assume, by symmetry, that G^1 has an f^1-partition $(G'_1, G'_2, \ldots, G'_p)$. We shall combine this partition with the f-partition $(G_1^2, G_2^2, \ldots, G_p^2)$ of $G^2 - v^*$. By symmetry, we may assume that v^* belongs to G'_1. Now let $(H_1, H_2, \ldots, H_p)$ be a partition of G with $H_1 = G'_1 \cup G_1^2 + v^*$ and $H_i = G'_i \cup G_i^2$ for $i \in [2,p]$. Obviously, for every $i \in [2,p]$ the multigraph H_i is strictly f_i-degenerate. We claim, that H_1 is strictly f_1-degenerate. To show this, let H be a nonempty submultigraph of H_1. If $H \subseteq G_1^2$, then there is a vertex v in H such that $d_H(v) < f(v)$, since G_1^2 is strictly f_1-degenerate. Otherwise, $H' = H \cap G'_1$ is a nonempty submultigraph of G'_1 and, since G'_1 is strictly f_1^1-degenerate, there is a vertex v in H' such that $d_{H'}(v) < f_1^1(v)$. If $v \neq v^*$, then $d_H(v) = d_{H'}(v) < f_1^1(v) = f_1(v)$ and we are done. If $v = v^*$, then (2.4) and the definition of f_1^j implies that

$$d_H(v^*) \le d_{H'}(v^*) + d_{G_1^2 + v^*}(v^*) < f_1^1(v^*) + f_1^2(v^*) = f_1(v^*)$$

and we are done, too. This proves the claim that H_1 is strictly f_1-degenerate. Consequently, G has an f-partition, contradicting (1). This proves the claim. △

Claim 2.30.2 *Suppose that there is a vertex $z \in V(G)$ and an index $j \in [1,p]$ such that $f_j(z) \neq 0$. Then $(G', f') = (G, f)/(z, j)$ is a nonpartitionable pair and the following statements hold:*

(a) *(G', f') is a hard pair.*
(b) *$f_j(v) \ge \mu_G(z,v)$ and $f'_j(v) = f_j(v) - \mu_G(z,v)$ for all $v \in V(G)$.*

Proof: Since G is a block (by Claim 2.30.1) and $|G| \ge 2$, z is a nonseparating vertex of G and $G' = G - z \neq \varnothing$. Since (G, f) is a nonpartitonable pair (by (1)), (G', f') is a nonpartitionable pair (by Proposition 2.28). From (3) it then follows that G' is f'-hard. This proves (a). Statement (b) follows from Proposition 2.29(b). This proves the claim. △

As a consequence of Claim 2.30.2(b), we obtain that each coordinate f_i of the vector function f is either constant zero or nowhere zero. Let I denote the set of all indices $i \in [1,p]$ such that f_i is nowhere zero. Clearly, $|I| \ge 1$ and , by symmetry, we may assume that $1 \in I$. If $|I| = 1$, then (2.3) implies that

$$f(v) = (d_G(v), 0, \ldots, 0)$$

for all $v \in V(G)$ and hence G is f-hard (by (H1)), contradicting (2). Hence $|I| \ge 2$ and, again by symmetry, we may assume that $2 \in I$, and so both f_1 and f_2 are nowhere zero. If $|G| = 2$, say $V(G) = \{u, v\}$, then $(G - u, G - v, \varnothing, \ldots, \varnothing)$ is an f-partition of

G, contradiction (1). Hence, $|G| \geq 3$. As G is a block and $|G| \geq 3$, $|N_G(v)| \geq 2$ for every vertex v of G.

Claim 2.30.3 *The underlying graph of G is not a cycle.*

Proof: Suppose this is false. Let v be an arbitrary vertex of G. Then $|N_G(v)| = 2$, say $N_G(v) = \{z_1, z_2\}$, and so $d_G(v) = \mu_G(z_1, v) + \mu_G(z_2, v)$. By Claim 2.30.2(b), $f_j(v) \geq \mu_G(z_i, v)$ for all $j \in I$ and for $i \in \{1,2\}$. It then follows from (2.3) that $\mu_G(v, z_1) = \mu_G(v, z_2) = m_v$ and $f(v) = (m_v, m_v, 0, \ldots 0)$. This obviously implies that there is an integer $t \geq 1$ such that $G = tC_n$ and $f(v) = (t, t, 0, \ldots, 0)$ for all $v \in V(G)$. As G has no f-partition, n is odd and so (G, f) is a hard pair (by (H3)), contradicting (2). This proves the claim. △

Claim 2.30.4 *The underlying graph of G is not complete.*

Proof: Suppose this is false. Let n be the order of G. As $n \geq 3$, it then follows from Claim 2.30.3 that $n \geq 4$. First, assume that $G = tK_n$. Let u and v be two arbitrary vertices of G. By Proposition 2.29(c), $G - u$ has an f-partition $(G_1, G_2, \ldots, G_p)$ and $f_i(u) = d_{G_i+u}(u) = t|G_i|$ for $i \in [1, p]$. Then v belongs to exactly one multigraph G_j for some $j \in [1, p]$, say $j = 1$. Then $f_1(u) = d_{G_1+u}(u) > d_{(G_1+u)-v}(u)$ and so $G_1' = (G_1 + u) - v$ is strictly f_1-degenerate. Consequently, $(G_1', G_2, \ldots, G_p)$ is an f-partition of $G - v$ with $|G_1'| = |G_1|$. By Proposition 2.29(c), $f_i(v) = t|G_i|$ for $i \in [1, p]$. This shows that every vertex w of G satisfies $f_i(w) = t|G_i|$ for all $i \in [1, p]$, and hence (G, f) is a hard pair (by (H2)), a contradiction to (2).

Now assume that G is different from tK_n for every $t \geq 1$. Then, as $n \geq 4$, there is a vertex z such that $G - z$ is different from sK_{n-1} for every s. Now let $(G', f') = (G, f)/(z, 1)$. As $1 \in I$, it follows from Claim 2.30.2 that (G', f') is a hard pair and

$$f'(v) = (f_1(v) - \mu_G(z, v), f_2(v), f_3(v), \ldots, f_p(v))$$

for all $v \in V(G')$. As the underlying graph of $G' = G - z$ is complete, but not a sK_{n-1}, we conclude that (G', f') is a mono-block. As $2 \in I$, this implies that every vertex v of G' satisfies $f_1(v) = \mu_G(z, v)$, $f_2(v) = d_{G'}(v)$, and $f_j(v) = 0$ for $j \in [3, p]$. Consequently, $I = \{1,2\}$. Let v be an arbitrary vertex of G'. As G' is connected and $f_2 = d_{G'}$, $G' - v$ is strictly f_2-degenerate and, therefore, $(G[v], G' - v, \varnothing, \ldots, \varnothing)$ is an f-partition of $G - z$. Then Proposition 2.29(c) implies that $f_1(z) = \mu_G(z, v)$ and $f_2(z) = d_{G'-v}(z)$. As v was chosen arbitrarily, this implies that there is an integer t_z such that $f_1(u) = t_z = \mu_G(z, u)$ for all $u \in V(G - z)$, and $f_1(z) = t_z$. As $I = \{1,2\}$, we can apply the same argument to the second coordinate f_2. Then we obtain that $f_2(u) = t_z$ for all $u \in V(G - z)$, and $f_2(z) = t_z$. Consequently, $f(v) = (t_z, t_z, 0, \ldots, 0)$ for every vertex v of G. This implies, in particular, that $d_G(z) = 2t_z$ and hence $|G| = 3$, a contradiction to $|G| \geq 4$. This proves the claim. △

Claim 2.30.5 *Let z be an arbitrary vertex of G, let $G' = G - z$, and let u and u' be two nonseparating vertices of G' contained in the same block of G'. Then either $\{u, u'\} \subseteq N_G(z)$ or $\{u, u'\} \cap N_G(z) = \varnothing$.*

Proof: Suppose this is false. By symmetry, this implies that $u \in N_G(z)$ and $u' \notin N_G(z)$. Now put $(G', f') = (G, f)/(z, 1)$ and $(G', f'') = (G, f)/(z, 2)$. By Claim 2.30.2 it follows that G' is both f'-hard and f''-hard. Since $u' \notin N_G(z)$, we obtain that $f(u') = f'(u') = f''(u')$. Then Proposition 2.27(b) implies that $f'(u) = f'(u')$ and $f''(u) = f''(u')$. Using Claim 2.30.2(b), the first equation leads to $f_1(u) - \mu_G(z, u) = f_1(u')$ and $f_2(u) = f_2(u')$, and the second equation leads to $f_1(u) = f_1(u')$ and $f_2(u) - \mu_G(z, u) = f_2(u')$. As $\mu_G(z, u) \geq 1$, this is impossible. Thus the claim is proved. △

To complete the proof of the proposition, let z be an arbitrary vertex of G. First we claim that $G - z$ is not a block. Suppose this is false. Then z is adjacent to all vertices of $G - z$ (by Claim 2.30.5). If z' is an arbitrary vertex of $G - z$, then $G - z'$ is a block, too, and so $N_G(z') = V(G - z')$. This shows that the underlying graph of G is complete, a contradiction to Claim 2.30.4. This proves the claim that $G - z$ is not a block. Hence $G - z$ has at least two end-blocks. Now let B be an arbitrary end-block of $G' = G - z$. Then B contains exactly one separating vertex of G', say v_B. It follows from Claim 2.30.5 that $V(B) \setminus \{v_B\} \subseteq N_G(z)$. Now we claim that $|B| = 2$. For otherwise, $|B| \geq 3$ and if $z' \in V(B - v_B)$, then it is easy to see that $G - z'$ is a block, a contradiction. This proves the claim that $|B| = 2$, say $V(B) = \{v_B, z'\}$. Then $N_G(z') = \{v_B, z\}$ and so $G - z'$ has exactly two end-blocks each of order 2, one containing z and the other one containing v_B. Hence $|N_G(z)| = 2$, too. Since z was chosen arbitrarily, we conclude that the underlying graph of G is a cycle, a contradiction to Claim 2.30.3. This completes the proof of the proposition. □

2.6 Generalized Graph Coloring

If we drop the condition $\varphi(v) \neq \varphi(w)$ for all edges $vw \in E(G)$ from the definition of coloring, then φ is called an **improper coloring** of G; accordingly, the term **proper coloring** is used in the graph theory literature in order to emphasize that the condition holds. So any mapping from the vertex set of a graph G to a set C is an improper coloring of G with color set C. However, improper colorings become a subject of interest only when some restrictions are imposed. One possibility would be to replace the usual coloring condition that each color class forms an independent set by some other condition. So let $\mathcal{P}$ be a **graph property**, that is, a class of graphs that is closed under isomorphism (see Appendix C.11). An improper coloring $\varphi : V(G) \to C$ is called a **$\mathcal{P}$-coloring** of G with color set C if $G[\varphi^{-1}(c)] \in \mathcal{P}$ for all colors $c \in C$. The **$\mathcal{P}$-chromatic number** of G, denoted by $\chi(G : \mathcal{P})$, is the least integer k for which there exists a $\mathcal{P}$-coloring of G with a set of k colors. Clearly, if $\mathcal{P}$ is the class of all edgeless graphs, then $\chi(G : \mathcal{P}) = \chi(G)$. There is also a list version for this generalized coloring concept. If L is a list-assignment for G, then a **$(\mathcal{P}, L)$-coloring** of G is a $\mathcal{P}$-coloring φ of G with color set $C = \bigcup_{v \in V(G)} L(v)$ such that $\varphi(v) \in L(v)$ for all $v \in V(G)$. We say that G is **$(\mathcal{P}, L)$-colorable** if G admits a $(\mathcal{P}, L)$-coloring. The **$\mathcal{P}$-list-chromatic number** of G, written $\chi_\ell(G : \mathcal{P})$, is the smallest integer k such that G is $(\mathcal{P}, L)$-colorable for every k-assignment L of G.

In order to establish a Brooks-type bound for the $\mathcal{P}$-list chromatic number, we shall restrict our attention to special graph properties. A graph property $\mathcal{P}$ is called **smooth** if the following two conditions hold:

(P1) $\mathcal{P}$ is **hereditary**, that is, $\mathcal{P}$ is closed under taking induced subgraphs.
(P2) $\mathcal{P}$ is **nontrivial**, that is, $\mathcal{P}$ contains a nonempty graph, but not all graphs.

For a smooth graph property $\mathcal{P}$, define

$$\mathcal{F}(\mathcal{P}) = \{H \mid H \notin \mathcal{P}, \forall v \in V(H) : H - v \in \mathcal{P}\}.$$

and

$$d(\mathcal{P}) = \min\{\delta(H) \mid H \in \mathcal{F}(\mathcal{P})\}.$$

Note that every graph in $\mathcal{F}(\mathcal{P})$ is nonempty and $\mathcal{F}(\mathcal{P}) = \mathrm{Crit}_{\leqslant}(\mathcal{P})$ (see Appendix C.11). If $\mathcal{P}$ is the class of edgeless graphs, then $\mathcal{P}$ is a smooth graph property for which $\mathcal{F}(\mathcal{P}) = \{K_2\}$ and $d(\mathcal{P}) = 1$.

Proposition 2.31 *If $\mathcal{P}$ is a smooth graph property, then the following statements hold:*

(a) $K_0, K_1 \in \mathcal{P}$.
(b) *A graph H belongs to $\mathcal{F}(\mathcal{P})$ if and only if any proper induced subgraph of H belongs to $\mathcal{P}$, but H itself does not belong to $\mathcal{P}$.*
(c) *$G \notin \mathcal{P}$ if and only if G contains an induced subgraph H with $H \in \mathcal{F}(\mathcal{P})$.*
(d) *$\mathcal{F}(\mathcal{P}) \neq \varnothing$ and $d(\mathcal{P}) \in \mathbb{N}_0$.*
(e) *If $G \notin \mathcal{P}$, but $G - v \in \mathcal{P}$ for some vertex v of G, then $d_G(v) \geq d(\mathcal{P})$.*

Proof By (P2), $\mathcal{P}$ contains a nonempty graph G. Since any induced subgraph of G belongs to $\mathcal{P}$ (by (P1)), we have $K_0, K_1 \in \mathcal{P}$, which proves (a). Since $H - v$ is a proper induced subgraph of H for every vertex v of H, statement (b) is an immediate consequence of (P1) and the definition of $\mathcal{F}(\mathcal{P})$.

To proof (c), let G be a graph. If G contains an induced subgraph H with $H \in \mathcal{F}(\mathcal{P})$, then $H \notin \mathcal{P}$ and so $G \notin \mathcal{P}$ (by (P1)). Now suppose that $G \notin \mathcal{P}$. Among all induced subgraphs of G not contained in $\mathcal{P}$, let H be one whose order is minimum. Then $|H| \geq 2$ (by (a)) and any proper induced subgraph of H belongs to $\mathcal{P}$. So it follows from (b) that $H \in \mathcal{F}(\mathcal{P})$. This complete the proof of (c). Since $\mathcal{P}$ does not contain all graphs (by (P2)), (c) implies (d).

For the proof of (e), suppose that $G \notin \mathcal{P}$, but $G - v \in \mathcal{P}$ for some vertex v of G. By (c), G contains an induced subgraph H with $H \in \mathcal{F}(\mathcal{P})$. Then $v \in V(H)$, since otherwise H would be an induced subgraph of $G - v$ and so H would belong to $\mathcal{P}$ as $G - v$ belongs to $\mathcal{P}$, giving a contradiction. Then we deduce that $d_G(v) \geq d_H(v) \geq \delta(H) \geq d(\mathcal{P})$. This proves (e). □

Let $\mathcal{P}$ be a smooth graph property, and let G be a graph with $n \geq 1$ vertices. If C is a set of at least n colors and $\varphi : V(G) \to C$ is a injective mapping, then φ is a $\mathcal{P}$-coloring of G, because $G[\varphi^{-1}(c)] \in \{K_0, K_1\}$ for all $c \in C$ and $K_0, K_1 \in \mathcal{P}$ (by Proposition 2.31(a)). Consequently, we have

$$\chi(G:\mathcal{P}) \le \chi_\ell(G:\mathcal{P}) \le |G|.$$

If φ is a $\mathcal{P}$-coloring of G with color set C and H is an induced subgraph of G, then the restriction ϕ of φ to $V(H)$ is a $\mathcal{P}$-coloring of H. Since φ is a $\mathcal{P}$-coloring of G, we have $G[\varphi^{-1}(c)] \in \mathcal{P}$ for all $c \in C$. Since $H[\phi^{-1}(c)]$ is an induced subgraph of $G[\varphi^{-1}(c)]$, it then follows from (P1) that $H[\phi^{-1}(c)] \in \mathcal{P}$ for all $c \in C$, that is, ϕ is a $\mathcal{P}$-coloring of H. Consequently, any induced subgraph H of G satisfies $\chi(H:\mathcal{P}) \le \chi(G:\mathcal{P})$ and $\chi_\ell(H:\mathcal{P}) \le \chi_\ell(G:\mathcal{P})$. In particular, for every vertex v of G, we have

$$\chi_\ell(G:\mathcal{P}) - 1 \le \chi_\ell(G-v:\mathcal{P}) \le \chi_\ell(G:\mathcal{P}).$$

If L is a list-assignment for G, we say that G is $(\mathcal{P},L)$-**vertex-critical** if $G-v$ is $(\mathcal{P},L)$-colorable for every vertex v of G, but G itself is not $(\mathcal{P},L)$-colorable.

Proposition 2.32 *Let $\mathcal{P}$ be a smooth graph property with $d(\mathcal{P}) = r$, let G be a nonempty graph, and let L be a list-assignment for G. If G is $(\mathcal{P},L)$-vertex–critical, then the following statements hold:*

(a) *$d_G(v) \ge r|L(v)|$ for all $v \in V(G)$.*
(b) *Let v be a vertex of G for which $d_G(v) = r|L(v)|$ and let φ be a $(\mathcal{P},L)$-coloring of $G-v$ with color set C. Then $|N_G(v)\cap\varphi^{-1}(c)| = r$ for all $c \in L(v)$ and $N_G(v) = \bigcup_{c\in L(v)}(N_G(v)\cap\varphi^{-1}(c))$.*

Proof Let v be a vertex of G. Since G is $(\mathcal{P},L)$-vertex-critical, there is a $(\mathcal{P},L)$-coloring φ of $G-v$. For a color $c \in L(v)$, let $N_c = N_G(v)\cap\varphi^{-1}(c)$. Since G is not $(\mathcal{P},L)$-colorable, for every color $c \in L(v)$ we have $G[\varphi^{-1}(c)\cup\{v\}] \notin \mathcal{P}$, but $G[\varphi^{-1}(c)] \in \mathcal{P}$. Then it follows from Proposition 2.31(e) that $|N_c| \ge r$. This gives $d_G(v) = \sum_{c\in L(v)}|N_c| \ge r|L(v)|$, where equality implies that $|N_c| = r$ for all $c \in L(v)$ and $N_G(v) = \bigcup_{c\in L(v)} N_c$. □

Let $\mathcal{P}$ be a smooth graph property with $d(\mathcal{P}) = r$, and let G be a $(\mathcal{P},L)$-vertex-critical graph. We denote by $V(G,\mathcal{P},L)$ the set of vertices v of G for which $d_G(v) = r|L(v)|$. A vertex of $V(G,\mathcal{P},L)$ is called a **low vertex** of G with respect to $(\mathcal{P},L)$, and $G[V(G,\mathcal{P},L)]$ is called the **low vertex subgraph** of G with respect to $(\mathcal{P},L)$.

The following result was obtained in 1995 by Borowiecki, Drgas-Burchardt, and Mihók [163]; it generalizes Gallai's theorem that the low vertex subgraph of any critical graph is a Gallai tree.

Theorem 2.33 *Let $\mathcal{P}$ be a smooth graph property with $d(\mathcal{P}) = r$, let G be a nonempty graph, and let L be a list-assignment for G. Suppose that G is $(\mathcal{P},L)$-vertex-critical and the low vertex subgraph $F = G[V(G,\mathcal{P},L)]$ is nonempty. If B is a block of F, then B is a brick, or $B \in \mathcal{F}(\mathcal{P})$ and B is r-regular, or $B \in \mathcal{P}$ and $\Delta(B) \le r$.*

Proof Let B be an arbitrary block of low vertex subgraph $F = G(V(G,\mathcal{P},L))$. Since G is vertex-$(\mathcal{P},L)$-critical, there is a $(\mathcal{P},L)$-coloring φ of $G - V(B)$ with a set C of p colors (possibly, $B = G$ and $\varphi = \varnothing$). By renaming the colors, we may assume $C = [1,p]$. For each color $c \in C$, the graph $G_c = G[\varphi^{-1}(c)]$ belongs to the property $\mathcal{P}$. Now we define a vector-function $f \in \mathcal{V}_p(B)$ as follows. For $v \in V(B)$ and $i \in [1,p]$, let $f_i(v) = \max\{0, r - d_{G_i+v}(v)\}$ if $i \in L(v)$ and $f_i(v) = 0$, otherwise.

We claim that B is not f-partitionable. Suppose, to the contrary, that B admits an f-partition $(G'_1, G'_2, \ldots, G'_p)$. For $i \in [1,p]$ let $\tilde{G}_i = G[V(G_i) \cup V(G'_i)]$. Clearly, $(\tilde{G}_1, \tilde{G}_2, \ldots, \tilde{G}_p)$ is a partition of G. Note that $v \in V(\tilde{G}_i)$ implies that $i \in L(v)$ (since $f_i(v) \geq 1$ for $v \in V(G'_i)$). If $\tilde{G}_i \in \mathcal{P}$ for all $i \in [1,p]$, it follows that G is $(\mathcal{P}, L)$-colorable, a contradiction. Consequently, there is an $i \in [1,p]$ such that $\tilde{G}_i \notin \mathcal{P}$ By Proposition 2.31(c), $\tilde{G}_i$ contains an induced subgraph H such that $H \in \mathcal{F}(\mathcal{P})$ and, thus, $\delta(H) \geq d(\mathcal{P}) = r$. Since G_i belongs $\mathcal{P}$ but H not, H contains a vertex of G'_i. Thus, the subgraph $H' = G'_i[V(H) \cap V(G'_i)]$ is nonempty. However, since G'_i is strictly f_i-degenerate, there is a vertex v in H' such that $d_{H'}(v) < f_i(v) = r - d_{G_i+v}(v)$ and thus $d_H(v) \leq d_{H'}(v) + d_{G_i+v}(v) < r$, a contradiction. Hence, B is not f-partitionable as claimed.

Since $d_G(v) = r|L(v)|$ for all $v \in V(B)$, we conclude that

$$\begin{aligned}\sum_{i=1}^{p} f_i(v) = \sum_{i \in L(v)} f_i(v) &\geq \sum_{i \in L(v)} (r - d_{G_i+v}(v)) \\ &= d_G(v) - \sum_{i \in L(v)} d_{G_i+v}(v) \geq d_B(v)\end{aligned}$$

Thus, by Theorem 2.26 and as B is a block, (B, f) is a hard pair that satisfies one of the three conditions (H1), (H2), or (H3). If (B, f) satisfies (H2) or (H3), then B is a brick. Hence it remains to consider the case that (B, f) satisfies (H1). Then, there is exactly one index $i \in [1,p]$ such that $f_i(v) = d_B(v)$ for all $v \in V(B)$ and $f_j(v) = 0$ for $j \neq i$. As a consequence, $d_{G_j+v}(v) \geq r$ for all $j \in L(v) \setminus \{i\}$ and thus, $d_B(v) \leq r$ for all $v \in V(B)$. If $B \in \mathcal{P}$, we have $\Delta(B) \leq r$ and there is nothing left to show. If $B \notin \mathcal{P}$, then by Proposition 2.31(c), B contains an induced subgraph B' from $\mathcal{F}(\mathcal{P})$. Since $d_B(v) \leq r$ for all $v \in V(B)$ and since $\delta(B') \geq d(\mathcal{P}) = r$, it must hold $B = B'$ and $d_B(v) = r$ for all $v \in V(B)$. Consequently, $B \in \mathcal{F}(\mathcal{P})$ and B is r-regular. This completes the proof. □

Let $\mathcal{P}$ be a smooth graph property with $d(\mathcal{P}) = r$, and let G be a graph. We say that G is $(\chi_\ell, \mathcal{P})$-**vertex-critical** if $\chi_\ell(H : \mathcal{P}) < \chi_\ell(G : \mathcal{P})$ for every proper induced subgraph H of G. Clearly, G is $(\chi_\ell, \mathcal{P})$-vertex-critical if and only if every vertex v of G satisfies $\chi_\ell(G - v : \mathcal{P}) = \chi_\ell(G : \mathcal{P}) - 1$.

Lemma 2.34 *Let $\mathcal{P}$ be a smooth graph property with $d(\mathcal{P}) = r \geq 1$. Then the following statements hold:*

(a) *Every graph G contains a $(\chi_\ell, \mathcal{P})$-vertex-critical induced subgraph H with $\chi_\ell(G : \mathcal{P}) = \chi_\ell(H : \mathcal{P})$.*
(b) *Let G be a $(\chi_\ell, \mathcal{P})$-vertex-critical graph with $\chi_\ell(G : \mathcal{P}) = k$. Then $\delta(G) \geq r(k-1)$. Furthermore, if $U = \{v \in V(G) \mid d_G(v) = r(k-1)\}$ is nonempty, then each block B of $G[U]$ is a brick, or $B \in \mathcal{F}(\mathcal{P})$ and B is r-regular, or $B \in \mathcal{P}$ and $\Delta(B) \leq r$.*
(c) *Every graph G satisfies $\chi_\ell(G : \mathcal{P}) \leq \frac{\Delta(G)}{r} + 1$.*

Proof Statement (a) follows from the fact that if H is an induced subgraph of G, then $\chi_\ell(H:\mathcal{P}) \le \chi_\ell(G:\mathcal{P})$. So among all induced subgraphs H of G with $\chi_\ell(H:\mathcal{P}) = \chi_\ell(G:\mathcal{P})$ we may choose one whose order is minimum. Then H is $(\chi_\ell,\mathcal{P})$-vertex-critical, since $\chi_\ell(H':\mathcal{P}) < \chi_\ell(H:\mathcal{P}) = \chi_\ell(G:\mathcal{P})$ for every proper induced subgraph H' of H.

For the proof of (b), assume that G is a $(\chi_\ell,\mathcal{P})$-vertex-critical graph such that $\chi_\ell(G:\mathcal{P}) = k$. Then there is a $(k-1)$-assignment L of G such that G is not $(\mathcal{P},L)$-colorable, but $G-v$ is $(\mathcal{P},L)$-colorable for every vertex v of G (since we have $\chi_\ell(G-v:\mathcal{P}) = k-1$). So G is $(\mathcal{P},L)$-vertex-critical and, by Proposition 2.32, we have $\delta(G) \ge r(k-1)$ and

$$U = V(G,\mathcal{P},L) = \{v \in V(G) \mid d_G(v) = r(k-1)\}.$$

By Theorem 2.33, it then follows that each block B of $G[U]$ is a brick, or $B \in \mathcal{F}(\mathcal{P})$ and B is r-regular, or $b \in \mathcal{P}$ and $\Delta(B) \le r$.

To prove (c), let G be an arbitrary graph. By (a), G contains a $(\chi_\ell,\mathcal{P})$-vertex-critical induced subgraph H with $\chi_\ell(G:\mathcal{P}) = \chi_\ell(H:\mathcal{P})$. If $\chi_\ell(G:\mathcal{P}) = k$, we deduce from (b) that $\Delta(G) \ge \Delta(H) \ge \delta(H) = r(k-1)$, which gives $\chi_\ell(G:\mathcal{P}) \le \frac{\Delta(G)}{r} + 1$. □

In order to establish a Brooks-type result for the $\mathcal{P}$-list-chromatic number, we shall restrict our attention to properties that are not only smooth, but also additive. A graph property $\mathcal{P}$ is called **additive** if $\mathcal{P}$ is closed under taking vertex disjoint unions. So if $\mathcal{P}$ is an additive graph property, then a nonempty graph belongs to $\mathcal{P}$ if and only if each of its components belong to $\mathcal{P}$. If $\mathcal{P}$ is also a smooth graph property, then any graph belonging to $\mathcal{F}(\mathcal{P})$ is connected and $d(\mathcal{P}) \ge 1$, since $K_0, K_1 \in \mathcal{P}$ (Proposition 2.31(a)).

Let $\mathcal{O}$ denote the class of all edgeless graphs. Clearly, $\mathcal{O}$ is a nontrivial, hereditary and additive graph property. Furthermore, if $\mathcal{P}$ is another such graph property, then Proposition 2.31(a) implies that $\mathcal{O} \subseteq \mathcal{P}$, and so any graph G satisfies $\chi_\ell(G:\mathcal{P}) \le \chi_\ell(G:\mathcal{O}) = \chi_\ell(G)$. The case $\mathcal{P} = \mathcal{O}$ of the following result immediately implies Brooks' theorem.

Theorem 2.35 *Let $\mathcal{P}$ be a nontrivial, hereditary and additive graph property with $d(\mathcal{P}) = r$ and let G be a connected graph. Then*

$$\chi_\ell(G:\mathcal{P}) \le \left\lceil \frac{\Delta(G)}{r} \right\rceil, \tag{2.5}$$

unless $G = K_{nr+1}$ for some integer $n \ge 0$, or G is r-regular and $G \in \mathcal{F}(\mathcal{P})$, or $\mathcal{P} = \mathcal{O}$ and G is an odd cycle.

Proof Let G be a connected graph. If $\Delta(G)$ is not divisible by r, then (2.5) is an immediate consequence of Lemma 2.34(c). So assume that $\Delta(G) = rn$ for some integer $n \ge 0$. Then Lemma 2.34(c) implies that $\chi_\ell(G:\mathcal{P}) \le n+1$. If $\chi_\ell(G:\mathcal{P}) = n$, we are done. Hence it remains to consider the case when $\chi_\ell(G:\mathcal{P}) = n+1$. By Lemma 2.34(a)(b), G contains a $(\chi_\ell,\mathcal{P})$-vertex-critical induced subgraph H with

$\chi_\ell(H:\mathcal{P}) = n+1$ and $\delta(H) \geq nr$. Since G is connected and $\Delta(H) \leq \Delta(G) = nr$, we deduce that $G = H$ and $\delta(G) = \Delta(G) = nr$, that is, G is $(\chi_\ell, \mathcal{P})$-vertex-critical and regular. Using again Lemma 2.34(b), it follows that each block B of G is a brick, or $B \in \mathcal{F}(\mathcal{P})$ and B is r-regular, or $B \in \mathcal{P}$ and $\Delta(B) \leq r$. Since G is regular of degree nr, this obviously implies that G itself is a block, unless $n = 1$. If $n = 1$, then G is r-regular and $\chi_\ell(G:\mathcal{P}) = 2$, which implies that $G \notin \mathcal{P}$. Since G is $(\chi_\ell, \mathcal{P})$-vertex-critical, $\chi_\ell(G-v:\mathcal{P}) = 1$ for every vertex v of G, and so $G - v \in \mathcal{P}$ for every vertex v of G. Consequently, $G \in \mathcal{F}(\mathcal{P})$ and we are done. Now assume that $n \neq 1$. Then G is a block and hence a brick. If G is a complete graph, then $G = K_{nr+1}$ and we are done. It remains to consider the case when G is an odd cycle. Since G is nr-regular and $n \neq 1$, this implies that $n = 2$ and $r = 1$. Then G is an odd cycle with $\chi_\ell(G:\mathcal{P}) = 3$. Since $\mathcal{O} \subseteq \mathcal{P}$, this implies that $K_2 \notin \mathcal{P}$ and so $K_2 \in \mathcal{F}(\mathcal{P})$. By Proposition 2.31(b), it then follows that $\mathcal{P} = \mathcal{O}$. This completes the proof. □

Erdős, Rubin, and Taylor [355] provided a degree version of Brooks' Theorem, see Corollary 1.10. To conclude this section, we present a related result for the generalized coloring.

Theorem 2.36 *Let $\mathcal{P}$ be a nontrivial, hereditary and additive graph property with $d(\mathcal{P}) = r$, and let G be a connected graph. Moreover, let L be a list-assignment for G such that $r|L(v)| \geq d_G(v)$ for all $v \in V(G)$ and G is not $(\mathcal{P}, L)$-colorable. If B is a block of G, then B is a brick, or $B \in \mathcal{F}(\mathcal{P})$ is r-regular, or $B \in \mathcal{P}$ and $\Delta(B) \leq r$.*

Proof Suppose that G is not $(\mathcal{P}, L)$-colorable. Then G contains a subgraph H which is $(\mathcal{P}, L)$-vertex-critical. By Proposition 2.32(a), it holds $d_H(v) \geq r|L(v)|$ for all $v \in V(H)$ and, thus, $d_G(v) = d_H(v) = r|L(v)|$ for all $v \in V(G)$. As G is connected, this implies that $G = H$, i.e., G is $(\mathcal{P}, L)$-vertex-critical. Moreover, it follows that $d_G(v) = r|L(v)|$ for all $v \in V(G)$ and so $V(G) = V(G, \mathcal{P}, L)$. Applying Theorem 2.33 then completes the proof. □

2.7 Exercises

2.1 Let G be a connected graph. Show that $\text{col}(G) = \Delta(G) + 1$ if and only if G is regular.

2.2 Let G be a nonempty graph. For a vertex u of G, let

$$N^{\leq}(u) = \{v \in N_G(u) \mid d_G(v) \leq d_G(u)\};$$

and define

$$\Delta_2(G) = \max_{u \in V(G)} \max_{v \in N^{\leq}(u)} d_G(v).$$

Show that $\text{col}(G) \leq \Delta_2(G)$. ($\Delta_2$ was introduced by Stacho [955])

2.3 Determine $\text{col}(K_{p,q})$ and $\chi_\ell(K_{n,n^n})$.

2.4 Let ρ be a graph parameter that satisfies the following two conditions:

- If $H \leqslant G$, then $\rho(H) \leq \rho(G)$.
- $\rho(G) \geq \delta(G)$ for all graphs G.

Show that $\text{col}(G) \leq \rho(G)+1$. (Szekeres and Wilf [985])

2.5 For a graph G with vertex set $V(G) = \{v_1, v_2, \dots, v_n\}$, let $\lambda(G)$ denote the largest eigenvalue of the adjacency matrix of G, that is, the matrix $A \in \mathbb{R}^{(n,n)}$ whose entry $A(i,j)$ is 1 if $v_iv_j \in E(G)$ and 0 otherwise. Show that if G is connected, then $\chi(G) \leq \lambda(G)+1$ with equality if and only if G is a complete graph or an odd cycle. (Wilf [1069])

2.6 Let G be a graph, and let H be an induced subgraph of G satisfying $\chi(H) \leq k$ for some integer $k \geq 3$. Suppose that G has a spanning forest F such that (1) $F[V(K)]$ is a tree for every component K of H, and (2) $d_G(v) \leq d_F(v)+k-2$ for every vertex v of $G-H$. Prove that for every coloring $\varphi \in \mathcal{CO}(H,k)$ there is a coloring $\phi \in \mathcal{CO}(G,k)$ such that $\phi|_K \sim \varphi|_K$ for every component K of H.

2.7 The Nordhaus-Gaddum Inequalities: Let G be a graph of order n with $n \geq 1$. Prove that

1. $2\sqrt{n} \leq \chi(G)+\chi(\overline{G}) \leq n+1$, and
2. $n \leq \chi(G)\chi(\overline{G}) \leq \frac{1}{4}(n+1)^2$.

(Hint: Use Proposition 1.6 and (2.1), and apply well known relations between the arithmetic mean and the geometric mean.) (Nordhaus and Gaddum [790])

2.8 For positive integers k and ℓ, let $u(k,\ell)$ be the maximum chromatic number of any graph which is the union of a k-critical graph and an ℓ-critical graph.

1. Show that $u(k,\ell) \leq (k-1)(\ell-1)+1$ if $k,\ell \geq 3$
2. Determine a lower bound for $u(k,\ell)$.

2.9 Show that for any graph G and any positive integer k, we have $\chi(G) \leq 2^k$ if and only if G is the union of k bipartite graphs.

2.10 Let G be a 5-regular graph which has a decomposition into a Hamilton cycle and pairwise vertex disjoint complete graphs on 4 vertices each. Show that G is the union of two bipartite graphs and so $\chi(G) = 4$. (Alon [43], see also Fleischner and Stiebitz [380])

2.11 The **rank number** $\text{rkn}(G)$ of a graph G is the least positive integer k for which there exists a mapping $\varphi : V(G) \to [1,k]$ such that any u-v path of G with $\varphi(u) = \varphi(v)$ contains a vertex w with $\varphi(w) > \varphi(u)$. Show that every graph G satisfies $\text{col}(G) \leq \text{rkn}(G)$. (Tuza and Voigt [1042])

2.12 The **Ramsey number** $R(k,\ell)$ is the smallest integer $n \in \mathbb{N}$ such that every graph G with $|G| \geq n$ satisfies $\omega(G) \geq k$ or $\alpha(G) \geq \ell$. To show that the Ramsey number is well defined, prove the following statements:

1. $R(k,\ell) = R(\ell,k)$.
2. $R(k,1) = 1$ and $R(k,2) = k$.
3. $R(k,\ell) \le R(k-1,\ell)+R(k,\ell-1)$ (for $k,\ell \ge 2$) (Hint: Use the pigeonhole principle and the fact that if v is a vertex of a graph G, then $d_G(v)+d_{\overline{G}}(v) = |G|-1$.)
4. $R(k,\ell) \le \binom{k+\ell-2}{2}$ and $R(k,k) \le 4^k$.

(Ramsey [847], Erdős and Szekeres [359])

2.13 Show that for $n \ge 1$ the class $\text{Forb}(K_{1,n})$ is χ-bounded and $f_n(x) = R(x,n)$ is a suitable bounding function. (Hint: Show that if $G \in \text{Forb}(K_{1,n}, K_k)$, then $\Delta(G) < R(k,n)$.) (Gyárfás [443])

2.14 Show that every bounding function f_n of the graph class $\text{Forb}(K_{1,n})$ satisfies $f_n(x) \ge \frac{R(x+1,n)-1}{n-1}$ for all fixed $n \ge 2$. (Gyárfás [443])

2.15 Show that every bounding function f_n of the graph class $\text{Forb}(P_n)$ satisfies $f_n(x) \ge \frac{R(x+1,\lceil n/2\rceil)-1}{\lceil n/2\rceil}$ for all fixed $n \ge 2$. (Hint: $\alpha(P_n) = \lceil n/2\rceil$ for $n \ge 2$.) (Gyárfás [443])

2.16 Show that $G \in \text{Forb}(P_3)$ if and only if each component of G is a complete graph.

2.17 Deduce from the Strong Perfect Graph Theorem that every graph in $\text{Forb}(P_4)$ is perfect. (Seinsche [924]). Furthermore, show without using this Theorem that every graph $G \in \text{Forb}(P_4)$ satisfies $\chi_{FF}(G) = \chi(G) = \omega(G)$. (Hint: Let $k = \chi_{FF}(G)$, let φ be the corresponding coloring, and let i be the smallest number such that there is a clique $\{v_i, v_{i+1}, \dots v_k\}$ of G with $\varphi(v_i) = i$ and $v_i \preceq v_{i+1} \preceq \dots \preceq v_k$; show that $i = 1$.) (Gyárfás and Lehel [446])

2.18 Show that if $\text{Forb}(T)$ is χ-bounded for all trees T, then $\text{Forb}(F)$ is χ-bounded for all forests F. (Hint: Use the fact that a forest can be extended to a tree)

2.19 Show that for every integer $n \ge 1$ there is an online tree $(T_n, \preceq_n)$ such that $\chi_{FF}(T_n, \preceq_n) = n$ and the largest vertex has color n. (Gyárfaás and Lehel [446]) (Hint: Construct $(T_n, \preceq_n)$ from $(T_1, \preceq_1), (T_2, \preceq_2), \dots, (T_{n-1}, \preceq_{n-1})$ by adding one vertex.)

2.20 Let $G_1, G_2, \dots, G_k$ be pairwise disjoint graphs. Let G denote the graph obtained from $G_1 \cup G_2 \cup \dots \cup G_k$ by adding for each subset A of the set $V(G_1) \cup V(G_2) \cup \dots \cup V(G_k)$ such that $|A \cap V(G_i)| = 1$ for all $i \in [1,k]$ a new vertex v_A and joining v_A to all vertices of A by an edge.

1. Show that if $\chi(G_i) = i$ and $\omega(G_i) = 2$ for $i \in [1,k]$, then $\chi(G) = k+1$ and $\omega(G) = 2$. (Zykov [1105])
2. Show that if G_i is i-critical for $i \in [1,k]$, then G is $(k+1)$-critical. (Schäuble [893])

2.21 Use the previous Exercise to show that for each positive integer k there exists a triangle-free k-chromatic graph.

2.22 The **shift graph** S_n ($n \geq 2$) is defined on ordered pairs (i,j) satisfying $1 \leq i < j \leq n$ as vertices and two pairs (i,j) and (k,ℓ) form an edge if $j = k$ or $\ell = i$. Show that S_n is a triangle-free graph and $\chi(S_n) \geq \log_2 n$. (Hint: Show that if φ is a k-coloring of S_n, then the n sets $A_i = \{\varphi(i,j) \mid i < j \leq n\}$ are pairwise distinct.) (Erdős and Hajnal [346])

2.23 Show that Forb(K_3) is not χ-bounded, but Forb$(\overline{K}_3)$ is χ-bounded. What is the best bounding function for Forb$(\overline{K}_3)$?

2.24 Let H_0 be an odd cycle and, for $s \geq 1$, let H_s be a complete graph of order $2s+3$. Show that if a graph G with $|CL_e(G)| = s$ contains H_s as a subgraph, then H_s is a block of G.

2.25 Let p and k be integers with $0 \leq p \leq k-1$ and $k \geq 2$. For a graph G, let

$$CL_{(p,k)}(G) = \{\ell \in CL(G) \mid \ell \equiv p \,(\text{mod } k)\}.$$

Show that every graph G with $|CL_{(1,k)}(G)| = r$ satisfies $\chi(G) \leq k(r+1)$. (Hint: As in the proof of Theorem 2.14 chose a DFS tree and arrange the levels into k groups L^p, where L^p is the union of the levels L_i with $i \equiv p \,(\text{mod } k)$.) (Diwan, Kenkre, and Vishwanathan [304])

2.26 Let $k \geq 2$. Show that every graph G with $|CL_{(2,k)}(G)| = s$ satisfies $\text{col}(G) \leq k(s+1)+1$. (Hint: As in the proof of Theorem 2.15 consider the sets $D_v^r(P) = \{\ell \in D_v(P) \mid \ell \equiv r \,(\text{mod } k)\}$ with $0 \leq r \leq k-1$.) (Diwan, Kenkre, and Vishwanathan [304])

2.27 Let G be a multigraph and let $f \in \mathcal{V}_p(G)$ be a vector function with $p \geq 1$. Then (G,f) is a hard pair if and only if for each block $B \in \mathfrak{B}(G)$ there is a unique function $f_B \in \mathcal{V}_p(B)$ such that (B, f_B) is a hard pair of type (H1), (H2), or (H3) and $f(v) = \sum_{B \in \mathfrak{B}_v(G)} f_B(v)$ for all $v \in V(G)$.

2.28 Let G be a nonempty graph and let $f \in \mathcal{V}_p(G)$ be a vector function with $p \geq 2$ such that

$$d_G(v) \leq f_1(v) + f_2(v) + \ldots + f_p(v) + p - 1$$

for all $v \in V(G)$. Show that there is a partition $(G_1, G_2, \ldots, G_p)$ of G such that $d_{G_i}(v) \leq f_i(v)$ for every vertex $v \in V(G_i)$ and for every $i \in [1,p]$. (Borodin and Kostochka [157])

2.29 Let G be a nonempty graph, let $f \in \mathcal{V}_p(G)$ be a vector-function, and let $h : V(G) \to \mathbb{N}$ be a function. Show that if $f_1(v) + f_2(v) + \cdots + f_p(v) \geq h(v)$ for all $v \in V(G)$ and G is strictly h-degenerate, then G is f-partionable. (Hint: Use induction on $|G|$.)

2.30 Let G be a nonempty graph and let $f \in \mathcal{V}_p(G)$ be a vector-function with $p \geq 2$ such that

$$f_1(v) + f_2(v) + \ldots + f_p(v) \geq d_G(v)$$

for all $v \in V(G)$. Show that if G is f-partitionable, then there is an f-partition $(G_1, G_2, \ldots, G_p)$ of G such that $d_{G_i}(v) \le f_i(v)$ for every vertex $v \in V(G_i)$ and for every $i \in [1, p]$. (Hint: Choose an f-partition $(G_1, G_2, \ldots, G_p)$ of G, for which the value

$$W = \sum_{i=1}^{p} \sum_{v \in V(G_i)} (d_{G_i}(v) - 2f_i(v))$$

is minimum.) (Borodin, Kostochka, and Toft [159])

2.31 Let G be a connected graph with maximum degree Δ such that $\Delta \ge 3$ and G is different from $K_{\Delta+1}$. Prove that if $f \in \mathcal{V}_p(G)$ is a vector function such that

$$f_1(v) + f_2(v) + \ldots + f_p(v) \ge d_G(v)$$

for all $v \in V(G)$, then either G has an f-partition $(G_1, G_2, \ldots, G_p)$ such that $d_{G_i}(v) \le f_i(v)$ for every vertex $v \in V(G_i)$ and for every $i \in [1, p]$, or G is a block and there is an index $j \in [1, p]$ such that each vertex v of G satisfies $f_j(v) = d_G(v)$ and $f_i(v) = 0$ for all $i \in [1, p] \setminus \{j\}$. (Borodin, Kostochka, and Toft [159])

2.32 Let G be a connected graph with maximum degree Δ such that $\Delta \ge 3$ and G is different from $K_{\Delta+1}$. Prove that if $k_1, k_2, \ldots, k_p$ are p positive integers such that $k_1 + k_2 + \cdots + k_p \ge \Delta$, then there is a partition $(G_1, G_2, \ldots, G_p)$ of G such that $\mathrm{col}(G_i) \le k_i$ and $\Delta(G_i) \le k_i$ for all $i \in [1, p]$. (Bollobás and Manvel [134], Borodin [147])

2.33 Let G be a connected graph with maximum degree Δ such that $3 \le \Delta \le st$ with $s \ge 2$ and G is different from $K_{\Delta+1}$. Prove that if L is a list-assignment of G such that $|L(v)| = s$ for every $v \in V(G)$, then G admits an L-coloring such that color each class induces a strictly t-degenerate subgraph of G having maximum degree not greater than t. (Borodin, Kostochka, and Toft [159])

2.34 For $k \ge 0$, let $\mathcal{DG}_k = \mathcal{G}(\mathrm{col} \le k+1)$ be the class of graphs whose coloring number is at most $k+1$; these graphs are called k-**degenerate**.

1. Show that $\mathcal{DG}_k$ is a smooth and additive graph property.
2. Show that $\mathcal{DG}_0 = \mathcal{O}$ and $\mathcal{DG}_1$ is the class of forests.
3. Deduce from Theorem 2.35 that if G is a connected graph, then

$$\chi_\ell(G : \mathcal{DG}_k) \le \left\lceil \frac{\Delta(G)}{k+1} \right\rceil,$$

unless $G = K_p$ with $p \equiv 1 \pmod{(k+1)}$, or G is $(k+1)$-regular, or $k = 0$ and G is an odd cycle. (Mitchem [744])

2.35 Show that a k-degenerate graph of order $n \ge k$ has at most $kn - \frac{k(k+1)}{2}$ edges.

2.8 Notes

The class $\mathcal{DG}_k$ of k-degenerate graphs was first introduced and investigated in 1970 by Lick and White [675]; similar concepts have been introduced under other names, both before and after. It is very natural to define the **degeneracy** of a graph as the smallest k such that it is k-degenerate. Even if this term is now commonly accepted, in particular in network theory, we do not know who used it first; Lick and White [675] do not use it at all. On the other hand, Proposition 1.5 implies that the degeneracy is just one less than the coloring number. An equally useful and closely related concept is the k**-core** of a graph G, usually defined as the maximal subgraph H of G with $\delta(H) \geq k$ (if no such subgraph exists, the k-core is defined to be the empty graph). The term k-core was introduced by several researchers, including Bollobás [128], Seidman [923], and possibly others. Seidman also defined the **core number** of a graph G as the largest k for which the k-core of G is nonempty. Again, Proposition 1.5 implies that the degeneracy and the core number are equal.

Theorem 2.1 may be considered as a strengthening of the trivial fact that every $(k-1)$-degenerate graph has chromatic number at most k. The proof of Theorem 2.1 follows Cranston and Rabern [261]; it is a refinement of the original argument of Stiebitz given in [970]. The Klein–Schönheim conjecture (Conjecture 2.5) was already mentioned in the book *Graph Coloring Problems* by Jensen and Toft [530, Problem 4.2], but no progress has been made since then. The special case $p = 2$ and $(d_1, d_2) = (1,2)$ suggests that a graph, which is the union of a forest and a 2-degenerate graph, has chromatic number $\chi \leq 5$. This problem was proposed first by M. Tarsi, see the paper [601]. The Klein–Schönheim conjecture is wide open and seems difficult to attack.

Gyárfás seminal 1978 paper [443] has motivated many graph theorist to search for χ-bounded graph classes (see the recent survey paper by Scott and Seymour [919]). Note that our definition of a χ-bounded graph class $\mathcal{P}$ deviates slightly from the definition proposed by Gyárfás in [443]. In our definition it is required that there is a function f such that the inequality $\chi(G) \leq f(\omega(G))$ holds for all graphs $G \in \mathcal{P}$; in Gyárfás' definition it is required that this inequality holds whenever G is an induced subgraph of a graph in $\mathcal{P}$. However, for hereditary graph properties, these two definitions are equivalent, and χ-boundedness has mostly been studied in the context of hereditary graph properties.

It is known that every hereditary graph property $\mathcal{P}$ can be defined in terms of forbidden induced subgraphs, that is, $\mathcal{P} = \text{Forb}(\mathcal{X})$ for an appropriate (possibly infinite) family of graphs $\mathcal{X}$ (see Appendix C.11). However, to find such a family for a given property $\mathcal{P}$ might be a difficult task. If the hereditary graph property $\mathcal{P}$ contains a nonempty graph, but not all graphs, then

$$\mathcal{F}(\mathcal{P}) = \{H \mid H \notin \mathcal{P}, \forall v \in V(H) : H - v \in \mathcal{P}\}$$

is the family of (vertex) minimal graphs not belonging to $\mathcal{P}$ and $\mathcal{P} = \text{Forb}(\mathcal{F}(\mathcal{P}))$ (see Proposition 2.31(c)). Moreover, if $\mathcal{P} = \text{Forb}(\mathcal{X})$, then every graph in $\mathcal{F}(\mathcal{P})$ is isomorphic to a graph in $\mathcal{X}$, and $\mathcal{F}(\mathcal{P})$ is equal to $\mathcal{X}$ (up to isomorphism) if and only if there are no two nonisomorphic graphs $H, H' \in \mathcal{X}$ such that H is an induced subgraph of H'.

Several theorems and conjectures state that, for certain families $\mathcal{X}$ of graphs, the graph property Forb($\mathcal{X}$) is χ-bounded. The most spectacular theorem in this context is the Strong Perfect Graph Theorem; and a beautiful but difficult conjecture is the Gyárfás-Sumner conjecture, saying that Forb(T) is χ-bounded for every tree T. Not much progress has been made since the conjecture was posed by Gyárfás [442] at the graph theory conference in Keszthely (Hungary) in 1973 and published in 1975 in the proceedings of the conference (the conjecture was independently posed in 1981 by Sumner [983]). Gyárfás [443] verified the conjecture for stars, paths, and brooms; a **broom** is a tree obtained from a star by subdividing one of its edges. Our proof of Theorem 2.10 follows the original proof from [443]. In 2013 Gravier, Hoàng, and Maffray [428, Corollary 1] proved that $f_n(x) = (n-2)^{x-1}$ is a χ-bounding function for Forb(P_n) with $n \geq 4$, thus improving Gyárfás' bounding function. Gyárfás, Szemerédi, and Tuza [447] showed that the class Forb(T, K_3) has bounded chromatic number for any radius two tree T. Kierstead and Penrice [579] strengthened this result by showing that Forb(T) is χ-bounded when T is a radius two tree. In 1997 Scott [912] obtained as a corollary of a more general result that Forb(T) is χ-bounded when T is a subdivision of a star. In 2004 Kierstead and Zhu [587] proved that Forb(T) is χ-bounded if T can be obtained from a rooted tree of hight 2 by subdividing once each edge incident with the root. It took more than a decade - until 2017 - before Chudnowsky, Scott, and Seymour [236] recognized two new types of trees T for which Forb(T) is χ-bounded, namely a tree obtained by adding one vertex to a subdivided star, or a tree obtained from a subdivided star and another star by adding a path joining their centers.

To attack the Gyárfás-Sumner conjecture one possibility is to show that certain subclasses of Forb(T) are χ-bounded. In 1996 Kierstaed and Rödl [582] proved that Forb($K_{n,n}, T$) is χ-bounded for any positive integer n and any tree T. For a graph H, let Forb$^*(H)$ denote the class of graphs such that no induced subgraph of G is isomorphic to a subdivision of H. For instance, Forb$^*(C_3)$ = Forb($\{C_n \mid n \geq 3\}$) is the class of forests. In 1997 Scott [912] proved that Forb$^*(T)$ is χ-bounded for any tree T. He also conjectured that the statement remains true with the tree T replaced by any other graph H. However, this was too optimistic and a bunch of counterexamples can be obtained using a result of Pawlik, Kozik, Krawczyk, Lasoń, Micek, Trotter, and Walczak [799]. On the other hand, it remains on open problem to decide when Scott's conjecture holds.

The class $\mathcal{PG}$ of perfect graphs is by definition a χ-bounded class with the identity function as its bounding function. The Strong Perfect Graph Theorem (SPGT) says that $\mathcal{F}(\mathcal{PG}) = \{C_n, \overline{C}_n \mid n \equiv 1 \pmod 2, n \geq 5\}$, that is, a graph is nonperfect if and only if it contains an odd hole or an odd antihole. A **hole** in a graph is an induced subgraph which is a cycle of length at least four, and a hole is **odd** if its length is odd. An **odd antihole** in a graph is an induced subgraph that is the complement of an odd hole. The SPGT raises the question for which subfamilies $\mathcal{X}$ of $\mathcal{F}(\mathcal{PG})$ the class Forb($\mathcal{X}$) remains χ-bounded. By Erdős' theorem (Theorem 2.9) such a family must be infinite. Gyárfás [443] conjectured that Forb($\mathcal{X}$) is χ-bounded provided that $\mathcal{X}$ is one of the following graph family:

- $\mathcal{H}^o = \{C_n \mid n \equiv 1 \,(\mathrm{mod}\, 2), n \geq 5\}$,
- $\mathcal{H}_p = \{C_n \mid n \geq p\}$ $(p \geq 3)$, or
- $\mathcal{H}_p^o = \{C_n \mid n \equiv 1 \,(\mathrm{mod}\, 2), n \geq p\}$ $(p \geq 3)$.

Since any shortest cycle of a graph has no chord, Forb($\mathcal{H}_3$) = Forb*(C_3) is the class of forests and Forb($\mathcal{H}_3^o$) is the class of bipartite graphs. The class Forb($\mathcal{H}_4$) is the class of chordal graphs, which are known to be perfect graphs (by Theorem 1.59). In a series of papers published by Scott and Seymour [913], [914], [915], [916], [917], [918], by Chudnovsky, Scott, and Seymour [233], [234], [235], [236], [237], and by Chudnovsky, Scott, Seymour, and Spirkl [238] the authors developed tools to solve several conjectures proposed by Gyárfás in [443].

Scott and Seymour [913] proved that Forb($\mathcal{H}^o$) is χ-bounded and $f(x) = 2^{2^{x+2}}$ is a suitable bounding function. Chudnovsky, Scott, and Seymour [233] [234] proved that Forb($\mathcal{H}_p$) is χ-bounded for all $p \geq 3$. Chudnovsky, Scott, Seymour, and Spirkl [238] proved that Forb($\mathcal{H}_p^o$) is χ-bounded for all $p \geq 3$. Clearly, the last result implies the two other results. Scott and Seymour [916] obtained the following result.

Theorem 2.37 (Scott and Seymour) *For all $n, p \geq 3$ there exists a c such that every K_n-free graph with chromatic number greater than c contains holes of every length modulo p .*

In 2023, Chudnovsky and Seymour [242] proved that any graph in Forb($\mathcal{H}_7^o$) whose girth is at least five is 3-colorable, thus solving a 2014 conjecture of Plummer and Zha [812]. For related problems and results we refer the interested reader to the paper by Chen [212] and to the paper by Wang and Wu [1060].

For a family $\mathcal{X}$ of graphs, let $\overline{\mathcal{X}} = \{\overline{G} \mid G \in \mathcal{X}\}$. Clearly, every graph G satisfies:

$$G \in \mathrm{Forb}(\overline{\mathcal{X}}) \Leftrightarrow \overline{G} \in \mathrm{Forb}(\mathcal{X}), \tag{2.6}$$

that is, if $\mathcal{P} = \mathrm{Forb}(\mathcal{X})$ then $\overline{\mathcal{P}} = \mathrm{Forb}(\overline{\mathcal{X}})$. For instance, as $\mathcal{PG} = \mathrm{Forb}(\mathcal{H}^o \cup \overline{\mathcal{H}^o})$, we obtain $\overline{\mathcal{PG}} = \mathcal{PG}$. If a hereditary graph property $\mathcal{P}$ is χ-bounded, then the hereditary graph property $\overline{\mathcal{P}}$ need not be χ-bounded (see Exercise 2.23). The following result due to Scott and Seymour [918]) provides a positive answer to a conjecture made by Gyárfás [443].

Theorem 2.38 (Scott and Seymour) *Let $c > 0$ be a constant. If $\mathcal{P}$ is a hereditary graph property and $\mathcal{P}$ is χ-bounded with $f(x) = x + c$ as a χ-bounding function, then $\overline{\mathcal{P}}$ is χ-bounded, too.*

Another well established graph property is the class $\mathcal{LG}$ of line graphs of graphs. Clearly, $\mathcal{LG}$ is a hereditary graph property and Vizing's theorem (Theorem 1.39) immediately implies that $\mathcal{LG}$ is χ-bounded with bounding function $f(x) = x + 1$. Beineke [90], [91] characterized line graphs of graphs in terms of forbidden induced subgraphs; he proved that $\mathcal{F}(\mathcal{LG})$ consists of nine graphs (up to isomorphism); six

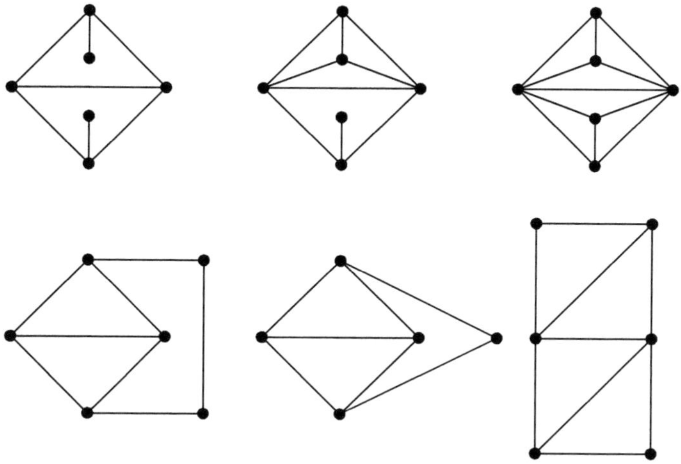

Fig. 2.3 Six members of the Beineke's family $\mathcal{F}(\mathcal{LG})$ of minimal nonline graphs.

of the nine graphs are shown in Figure 2.3, the three missing graphs are $K_{1,3}$ (the claw), $W_5 = K_1 \boxplus C_5$ (a wheel), and K_5^- (the complete graph on 5 vertices with one edge deleted). Translating Vizing's theorem using this, it becomes very natural to ask whether the statement remains true if the graph family $\mathcal{F}(\mathcal{LG})$ is replaced by a subfamily. That this is indeed the case was first shown by Choudum [220]. Javdekar [524] improved Choudum's result. A further improvement was obtained by Kierstead and Schmerl [583] and Kierstead [571]:

- *Every graph* $G \in \mathrm{Forb}(K_{1,3}, K_5^-)$ *satisfies* $\chi(G) \leq \omega(G) + 1$.

As proved by Kierstead and Schmerl [583], this statement is equivalent to the statement that $\chi'(H) \leq \Delta(H) + 1$ for every multigraph H with $\mu(H) \leq 2$ and without a 2-triangle. A proof of the latter statement was given by Kierstead [571]. This shows that interesting results on vertex colorings of graphs may be derived from improved results on the chromatic index of multigraphs, see also [572] and [584].

Let us say that a graph family $\mathcal{X}$ has the **Vizing property** VP_p if $\mathrm{Forb}(\mathcal{X})$ is χ-bounded with $f(x) = x + p$ as a suitable bounding function. By Theorem 2.38 it follows that if $\mathcal{X}$ has the Vizing property, then $\mathrm{Forb}(\overline{\mathcal{X}})$ is χ-bounded. Therefore, if a graph family $\mathcal{X} = \{H\}$ consisting of a single graph (up to isomorphism) has the Vizing property, then both graphs H and $\overline{H}$ are forests, which implies that H is an induced subgraph of the path P_4 on four vertices. For the Vizing property VP_1, this was proved by Randerath [848] in his dissertation (see also the survey by Randerath and Schiermeyer [850]). Furthermore, Randerath proved that if the family $\mathcal{X} = \{T, H\}$ consisting of two graphs T and H has the Vizing property VP_1, but neither the class $\{T\}$ nor the class $\{H\}$ has it, then one of the two, say T, has to be a tree not being an induced subgraph of P_4, and the other H has to belong to the set $\{K_3, K_4, K_4^-, K_5^-, K_3^+, K_4^*\}$, where K_3^+ is the graph obtained from K_3 by adding an additional vertex v and joining v to exactly one vertex of the K_3, and K_4^* is the graph obtained from K_4 by adding an additional vertex v and joining v to exactly two vertices of the K_4.

A class $\mathcal{P}$ of graphs has the **Erdős–Hajnal property**graph property if there is a constant $\varepsilon > 0$ such that every graph $G \in \mathcal{P}$ satisfies $\max\{\alpha(G), \omega(G)\} \geq |G|^{\varepsilon}$. For example, the class $\mathcal{PG}$ of perfect graphs has the Erdős–Hajnal property, since every graph $G \in \mathcal{PG}$ satisfies $|G| \leq \chi(G)\alpha(G) = \omega(G)\alpha(G)$ (by (1.2)) and hence $\max\{\alpha(G), \omega(G)\} \geq \sqrt{|G|}$. Following Chudnovsky [221], a single graph H is said to have the **Erdős–Hajnal property** if the class Forb(H) has the Erdős–Hajnal property. In the 1980s Erdős and Hajnal [349] [351] proposed the following conjecture.

Conjecture 2.39 (Erdős and Hajnal) *Every graph H has the Erdős–Hajnal property.*

The conjecture is wide open. Let $\mathcal{EH}$ denote the class of graphs having the Erdős–Hajnal property. Note that $\mathcal{EH}$ is hereditary graph property. Furthermore $\mathcal{EH}$ is closed under taking complements, i.e., $H \in \mathcal{EH}$ implies $\overline{H} \in \mathcal{EH}$. This follows from (2.6) and the fact that $\max\{\alpha(\overline{G}), \omega(\overline{G})\} = \max\{\alpha(G), \omega(G)\}$. If both H_1 and H_2 belong to $\mathcal{EH}$ and $v \in V(H_1)$, then the graph H obtained from the disjoint union of $H_1 - v$ and H_2 by adding all edges between $N_{H_1}(v)$ and H_2 belongs to $\mathcal{EH}$, too; this was proved by Alon, Pach, and Solymosi [56].

Clearly, Ramsey theory implies that $\mathcal{EH}$ contains all complete graphs. Since Forb(P_4) $\subseteq \mathcal{PG}$, we have $P_4 \in \mathcal{EH}$. On the other hand, it is not known whether $P_5 \in \mathcal{EH}$. That the bull (see Figure C.2) belongs to $\mathcal{EH}$ was proved by Chudnovsky and Safra [231]. That C_5 belongs to $\mathcal{EH}$ was proved by Chudnovsky, Scott, Seymour, and Spirkl [239].

Let $\mathcal{P}$ be a class of graphs such that, for a constant $c > 0$, we have $\chi(G) \leq \omega(G)^c$ for every graph $G \in \mathcal{P}$. Then (1.2) implies that every graph $G \in \mathcal{P}$ satisfies $|G| \leq \alpha(G)\omega(G)^c$, and hence $\mathcal{P}$ has the Erdős–Hajnal property. This observation and the conjecture of Erdős and Hajnal led Esperet [360] to propose the conjecture that every hereditary graph property $\mathcal{P}$ that is χ-bounded is **polynomially χ-bounded**, i.e., $\mathcal{P}$ has a polynomial χ-bounding function. However, this conjecture was disproved by Briański, Davies, and Walczak [168]. The existence of hereditary graph properties that are χ-bounded but not polynomially χ-bounded follows directly from a result in [168] which says that for any integers p and k with $k \geq p \geq 2$, there exists a graph G with $\chi(G) = k$ and $\omega(G) = p$ such that every induced graph H of G with $\omega(H) = q < p$ satisfies $\chi(H) \leq \binom{3q+1}{3}$. Although Esperet's conjecture has been disproved, it is still open whether the following strengthening of the Gyárfás–Sumner tree conjecture holds.

Conjecture 2.40 (The Polynomial Gyárfás–Sumner Tree Conjecture) *For every tree T the class* Forb(T) *is polynomially χ-bounded.*

The polynomial Gyárfás–Sumner tree conjecture holds for any tree of diameter at most 3, as proved by Scott, Seymour and Spirkl [920]. For additional information, see the paper by Chudnovsky, Cook, Davies, and Oum [222].

Let $\mathcal{M}$ be a family of sets. We say that a graph G can be represented as the **intersection graph** of $\mathcal{M}$ if there is a map S that assigns to each vertex v of G a set S_v such that $\mathcal{M} = (S_v)_{v \in V(G)}$ and $vw \in E(G)$ if and only if $S_v \cap S_w \neq \varnothing$. Every graph G can be represented as intersection graph of the family $(E_G(v))_{v \in V(G)}$. However, if we focus on special types of sets, then we get more interesting graph properties. A d**-dimensional box** is the cartesian product of d closed intervals in $\mathbb{R}$. Let $\mathcal{BG}_d$ denote the class of graphs that can be represented as intersection graphs of d-dimensional boxes. Then $\mathcal{BG}_1$ is known as the class of **interval graphs**. This class was introduced, independently, by György Hájos and Seymour Benzer. Hajós gave a talk at a conference (4. Östereichischer Mathematikerkonferenz), held September 17–22, 1956, in Vienna; in his talk he defined interval graphs as intersection graphs of intervals and discussed the recognition problem for this class (see [460] for a short abstract of his talk). Benzer [96] who was an biophysicists was concerned with the fine structures of genes, that is, the problem whether the sub-elements of genes are linked together in a linear order. The problem he was concerned with was to decide whether an overlap family of abstract sets can be represented as the overlap system of real intervals. Lekkerkerker and Boland [670] formalized Benzer's problem using graph theory; they introduced three classes of graphs which became later known as interval graphs, chordal graphs and AT-free graphs. A graph G is called **AT-free** if G contains no three vertices such that every two of them are joined by a path avoiding the neighborhood of the third. Lekkerkerker and Boland proved that a graph is an interval graph if and only if it chordal and AT-free. Furthermore, they characterized the family $\mathcal{F}(\mathcal{BG}_1)$ of minimal noninterval graphs; which consists of two special graphs and three infinite family of graphs including $\mathcal{H}^o$. They also proved that every connected chordal graph which is not a complete graph contains two nonadjacent simplicial vertices (see also Theorem 1.58). In 1948 Bielecki [115] proposed the problem (not using graph theory terminology) whether every K_3-free graph in $\mathcal{BG}_2$ has bounded chromatic number; he also noticed that every interval graph is perfect and presented an example of a 4-chromatic K_3-free graph in $\mathcal{BG}_2$ due to J. G. Mikusiński. That the maximum chromatic number of K_3-free graphs in $\mathcal{BG}_2$ is six was proved in 1960 by Asplund and Grünbaum [71]. They also proved that the class $\mathcal{BG}_2$ is χ-bounded with $f(x) = 8x^2$ as a suitable bounding function. As pointed out by Kostochka [617] the proof in [71] yields the bounding function $f(x) = 4x^2 - 3x$, which was improved by C. Hendler to $f(x) = 3x^2 - 2x - 1$ (see [621]); the best known lower bound is $f(x) \geq 3x$. A construction of Burling [184] shows that $\mathcal{BG}_3$ is not χ-bounded.

The study of the chromatic number of intersection graphs of geometric objects was initiated by Asplund and Grünbaum [71] almost half a century ago. Since then, this topic has received considerable attention, see the 2004 survey paper by Kostochka [621] or the 2012 paper by Fox and Pach [384]. As reported by Gyárfás [443] P. Erdős asked whether the intersection graphs of straight line segments in the plane, introduced in [323], forms a χ-bounded class. That the answer is negative was proved in 2014 by Pawlik, Kozik, Krawczyk, Lasoń, Micek, Trotter, and Walczak [799]; they constructed a family of triangle-free intersection graphs of line segments in the plane with unbounded chromatic number. This implies that $\mathrm{Forb}^*(H)$ is not

χ-bounded whenever H is obtained from a nonplanar graph by subdividing every edge at least once.

The **recognition problem** for a graph property $\mathcal{P}$ is the problem to decide whether a given graph belongs to $\mathcal{P}$; we are then interested in the complexity status of this problem. If the recognition problem for the property $\mathcal{P}$ belongs to a certain complexity class, we briefly say that $\mathcal{P}$ belongs to that class. For instance, the class $\mathcal{G}(\chi \leq k)$ of k-colorable graphs belongs to P if $k \leq 2$ and is NP-complete otherwise. If $\mathcal{P}$ is a hereditary graph property and $\mathcal{F}(\mathcal{P})$ is finite (up to isomorphism), then $\mathcal{P}$ belongs to P, and very often there is even an efficient **certifying algorithm** for $\mathcal{P}$, that is, a polynomial-time algorithm that does not only tell us whether an input graph belongs to $\mathcal{P}$ or not, but also provides a certificate which can be used to examine efficiently the correctness of the answer (see the survey of McConnell, Mehlhorn, Näher, and Schweitzer [727]). But even if $\mathcal{F}(\mathcal{P})$ is not finite, it might be possible that there is an efficient certifying algorithm. For instance, if $\mathcal{P}$ is the class of bipartite graphs, then there is a polynomial-time algorithm that either finds a bipartition or an odd cycle (see Section 1.1). The intersection of two hereditary graph properties remains a hereditary graph property, and for a graph property $\mathcal{P}$ we denote $\mathcal{P} \cap \mathcal{G}(\chi \leq k)$ by $\mathcal{P}(\chi \leq k)$. Note that the k**-coloring problem** for $\mathcal{P}$ and the recognition problem for $\mathcal{P}(\chi \leq k)$ are different. In the former case we want to decide whether a given graph $G \in \mathcal{P}$ has a k-coloring, while in the later case we want to decide whether a given graph G belongs to $\mathcal{P}$ and has a k-coloring. However, if the recognition problem for $\mathcal{P}$ is polynomial-time solvable, then these two problems are polynomial reducible to each other. For instance, the k-coloring problem, and even the coloring problem, for perfect graphs is in P as proved in 1984 by Grötschel, Lovász, and Schrijver [433]. That the recognition problem for the class $\text{Forb}(\mathcal{H}^o \cup \overline{\mathcal{H}^o})$ and hence for the class $\mathcal{PG}$ of perfect graphs is in P was proved much later by Chudnovsky, Cornuéjols, Liu, Seymour, and Vušković [223].

The **coloring problem** for a graph property $\mathcal{P}$ is the decision problem, whether, for a given graph $G \in \mathcal{P}$ and a given integer k, the graph G admits a k-coloring. For one forbidden graph H, Král', Kratochvíl, Tuza, and Woeginger [643] obtained a complete characterization of the complexity of the coloring problem for $\text{Forb}(H)$.

Theorem 2.41 (Král', Kratochvíl, Tuza, and Woeginger) *Given a graph H, the coloring problem for* $\text{Forb}(H)$ *is polynomial-time solvable if H is an induced subgraph of P_4 or $P_3 + K_1$, and* NP*-complete for any other graph H.*

To characterize the complexity of the k-coloring problem for various graph properties $\mathcal{P}$, respectively the recognition problem for $\mathcal{P}(\chi \leq k)$, is an ongoing process and it is almost impossible to be up to date, see e.g. [170, 171, 172, 643]. If we consider the list chromatic number instead of the chromatic number, the problem becomes even more difficult. A survey on the computational complexity of coloring and list coloring graphs with one or two forbidden subgraphs was published in 2017 by Golovach, Johnson, Paulusma, and Song [422]. For an integer $k \geq 1$ and a graph H, let $\mathcal{P}_k(H) = \text{Forb}(H)(\chi \leq k)$. If H contains a cycle, then $\mathcal{P}_k(H)$ is NP-complete. This follows from the result that the k-coloring problem for graphs with girth at least ℓ is NP-complete for any fixed integers $k \geq 3$ and $\ell \geq 3$. For $\ell = 3$, this was proved by

Král' et al. [643]. Kamiński and Lozin [551] extended this result to all $\ell \geq 3$ though in fact a stronger result had been obtained previously by Emden-Weinert, Hougardy, and Kreuter [326].

Theorem 2.42 (Emden-Weinert, Hougardy and Kreuter) *For all $k \geq 3$ and all $\ell \geq 3$, the decision problem whether a graph with girth at least ℓ and maximum degree at most $6k^{13}$ admits a k-coloring is* NP*-complete.*

Beineke's characterization of the minimal nonline graphs implies that for every forest H with $\Delta(H) \geq 3$, the class $\mathcal{LG}$ is a subclass of $\mathrm{Forb}(H)$. That the k-coloring problem for $\mathcal{LG}$ is NP-complete for all $k \geq 3$ was proved by Holyer [504] for $k = 3$ and by Leven and Galil [671] for all $k \geq 4$. Hence, for every forest H with $\Delta(H) \geq 3$ and for all $k \geq 3$, the property $\mathcal{P}_k(H)$ is NP-complete. Summarizing we obtain that if $\mathcal{P}_k(H)$ is in P (or, equivalently, the k-coloring problem for $\mathrm{Forb}(H)$ is in P), then $k \leq 2$ or H is a **linear forest**, that is, H is the disjoint union of paths. By a result of Huang [507] it follows that the k-coloring problem for $\mathrm{Forb}(H)$ is NP-complete if either $k = 4$ and $P_7 \leq H$, or $k \geq 5$ and $P_6 \leq H$. Recall that $G \leq H$ means that G is an induced subgraph of H. As P_4-free graphs are perfect (see Exercise 2.17), the k-coloring problem for P_n-free graphs is polynomial-time solvable if $k \leq 2$ or $n \leq 4$. The k-coloring problem for $\mathrm{Forb}(P_5)$ is polynomial-time solvable for every $k \geq 3$; for $k = 3$ this was proved by Sgall and Woeginger [1071] and for $k \geq 4$ by Hoàng, Kamiński, Sawada, and Shu [501]. That the 4-coloring problem for P_6-free graphs is in P was proved by Chudnovsky, Spirkl, and Zhong [243] [244]. Hence, for $k \geq 4$, the complexity of the k-coloring problem for $\mathrm{Forb}(P_n)$ is completely classified. The fact that the 3-coloring problem for P_7-free graphs can be solved by a poynomial-time algorithm was proved by Bonomo, Chudnovsky, Maceli, Schaudt, Stein, and Zhong [146], thereby strengthening an earlier result by Randerath and Schiermeyer [849] saying that 3-coloring for $\mathrm{Forb}(P_6)$ is in P. The complexity of the 3-coloring problem for P_n-free graphs with $n \geq 8$ seems to be unknown.

Recall that a graph G is critical if $\chi(G') < \chi(G)$ for every proper subgraph G' of G. Furthermore, G is called **vertex-critical** if every proper induced subgraph G' of G satisfies $\chi(G') < \chi(G)$. A vertex-critical k-chromatic graph is also called k**-vertex-critical**. Every critical graph is vertex-critical, and every vertex-critical graph contains a critical subgraph with the same chromatic number and each such subgraph has the same vertex set as the graph itself. It is easy to show that a graph G is $(k+1)$-vertex-critical if and only if $\chi(G) \geq k+1$ and $\chi(G-v) \leq k$ for every vertex v of G. This immediately implies that a graph G belongs $\mathcal{F}(P_k(H))$ if and only if $G = H$ or G is a $(k+1)$-vertex-critical H-free graph. If $\mathcal{F}(P_k(H))$ is finite (up to isomorphism), then it is likely that we can obtain an efficient certifying algorithm for the k-coloring problem for $\mathrm{Forb}(H)$. Such an algorithm would either return a k-coloring or a $(k+1)$-vertex-critical H-free graph. There are quite a few results regarding the number of vertex-critical H-free graphs, the reader may consult the paper by Chudnovsky, Goedgebeur, Schaudt, and Zhong [227], the paper by Cameron, Hoàng, and Sawada [188], and the paper by Cameron, Goedgebeuer, Huang, and Shi [189]. Here $P + P'$ means the disjoint union of P and P'.

Theorem 2.43 (Chudnovsky, Goedgebeur, Schaudt, and Zhong) *Let H be a graph. There are only finitely many 4-vertex-critical graphs in* $\mathrm{Forb}(H)$ *if and only if H is a subgraph of P_6, P_3+P_3, or $P_4+\overline{K}_k$ for some $k \in \mathbb{N}$.*

More results and references related to coloring problems for graph classes defined by forbidden induced subgraphs can be found in the following papers: [170], [201], [202], [228], [232], [419], and [453].

The Internet site *Information System on Graph Classes and their Inclusions* (ISGCI) provides a lot of information about certain graph classes. The ISGCI database contains information about 1600 graph classes including their relations and more than 1700 references.

As emphasized by Nešetřil [773] in his instructive survey about sparse graphs with high chromatic number, Erdős' theorem from 1959 that, for each fixed $\ell \geq 3$, the class of graphs with girth at least ℓ has unbounded chromatic number *is both a culmination of long development and the start of important consequent research and methods*. The basic case $\ell = 3$ was already considered in the late 1940s, independently, by Blanche Descartes and Alexander Zykov. Both presented a recursively defined sequence $(G_k)_{k=1}^{\infty}$ of triangle-free graphs with $\chi(G_k) \geq k$. The recursion given by Zykov [1105] (see Exercise 2.20) has infinite depth, while the recursion used by Descartes [280] has only depth one. Descartes' construction is a prototype of many subsequent constructions, proofs and variants. Here is a formal sketch. Given G_k with r vertices, we consider a set V with $n = k(r-1)+1$ vertices and for every r-subset Y of V we take an isomorphic copy G_Y of G. Then G_{k+1} is the graph obtained from the vertex set V and the union of the (disjoint) copies G_Y with $Y \in [V]^r$ by adding for each set Y a perfect matching between Y and G_Y. If there is a coloring of G_{k+1} with k colors, then one of the sets $Y \in [V]^r$ would be monochromatic (by the pigeonhole principle) and this would yield a coloring of G_Y and hence of G_k with $k-1$ colors. Hence, $\chi(G_k) \geq k$ implies $\chi(G_{k+1}) \geq k+1$. If we start the recursion with $G_1 = K_1$, then $G_2 = K_2$, $G_3 = C_9$ and it is easy to check that G_k remains triangle-free. In fact, $g(G_k) \geq 6$ as first observed by Kelly and Kelly [564]. A graph sequence with girth at least seven was first constructed by Nešetril [771]. The most notable feature of Descartes' construction is the fact that $H = (V, [V]^r)$ is the complete r-uniform hypergraph K_n^r (see Appendix C.14). As $n = k(r-1)+1$ the hypergraph K_n^r has chromatic number k and is even k-critical (see Section 7.1 for the coloring concept for hypergraphs). The main point, however, is the fact that Descartes' construction works well when the hypergraph is replaced by any other r-uniform hypergraph H with $\chi(H) \geq k$. We can even take a hypergraph H with $\chi(H) \geq k$ and $E(H) \subseteq [V(H)]^2 \cup [V(H)]^r$; then we apply the amalgamation procedure only to the hyperedges belonging to $[V(H)]^r$. Hence, the construction in Theorem 2.13 is an advanced version of Descartes' construction where the hypergraphs are chosen in a more sophisticated way. This leads to a graph sequence $(G_k)_{k=1}^{\infty}$ satisfying $\chi(G_k) \geq k$, $g(G_k) \geq \ell$ and $\mathrm{mad}(G_k) \leq 2(k-1)$ for any fixed $\ell \geq 3$. That the construction has asymptotically lowest average degree even in the broader class of triangle-free graphs follows from Corollary 5.26. The proof of Erdős' theorem given in Section 2.3 is due to Alon, Kostochka, Reiniger, West, and

Zhu [52]. The proof provides a direct construction of a graph with chromatic number at least k and girth at least ℓ by means of a $(k, k+1, 2\ell)$-graph. The advantage of this method is that it can be easily extended to construct an r-uniform hypergraph having chromatic number at least k and girth at least ℓ for any $r \geq 3$ and $k, \ell \in \mathbb{N}$ (see Exercise 8.5). The existence of such hypergraphs was first shown, using a probabilistic argument, by Erdős and Hajnal [347]. The first construction of such hypergraphs was obtained by Lovász [688]; his proof proceeds by double induction on r and ℓ (for a fixed k). Other constructions were provided by Nešetřil and Rödl [775] [777], see also Kostochka and Nešetřil [626]. The first construction of highly chromatic graphs without short cycles not using hypergraphs was given in the late 1980s by Lubotzky, Phillips, and Sarnak [704] and by Kříž [654]. Lubotzky et al. uses expanders and Ramanujan graphs to construct graphs having chromatic number at least k, girth at least ℓ and order at most $k^{4\ell}$. That such graphs exists was first proved by Erdős, however by probabilistic means. The constructions of Descartes and Zykov yield k-chromatic triangle-free graphs for all k, but the graphs obtained have exponential order in k, as in many other constructions. Erdős was the first to give a polynomial upper bound. A geometric construction by Erdős from 1958 shows the existence of triangle-free k-chromatic graphs of order $\leq c \cdot k^{50}$, as explained by Sachs [882]. Some results concerning bounds for the smallest k-chromatic graph of given girth can be found in the paper by Exoo and Goedgebeur [361]. Theorems 2.12 and 2.13 and their proofs are from [52].

A new interesting class of triangle-free graphs with unbounded chromatic number was recently discovered by Bonnet et al. [145]. Let $\mathcal{TC}$ be the smallest class of graphs that contain all graphs with at most two vertices and that is closed under duplicating a vertex (see Exercise 1.31) and under merging two disjoint graphs along an independent set of one or two vertices. Then it is easy to see that $\mathcal{TC}$ is a hereditary subclass of $\text{Forb}(K_3)$; and as proved in [145] this class has unbounded chromatic number.

The study of the cycle spectrum of a graph, that is, the distribution of its cycle lengths, is a well established area within graph theory. The research in this area focus on the question how certain graph properties, such as large minimum degree, large girth, large chromatic number, large connectivity, or nice expansion properties, affect the cycle spectrum and ensure the existence of cycles of particular lengths. The minimum number of different cycle length in a graph with given girth and minimum degree was first investigated by Erdős, Faudree, Rousseau, and Schelp [343]. There has been extensive research on cycle lengths in graphs with large minimum degree and the reader may find some interesting results about this subject in the recent paper by Liu and Ma [680]. As any graph with chromatic number at least k contains a 2-connected subgraph with minimum degree at least $k-1$, results about the cycle spectrum of graphs with given minimum degree lead to results relating the cycle spectrum to the chromatic number. Let $\text{ccl}_e(G)$ (respectively $\text{ccl}_o(G)$) denote the largest p such that G contains p cycles of consecutive even (respectively, odd) lengths. Furthermore, let $\text{ccl}(G)$ denote the largest p such that G contains p cycles of consecutive lengths. Liu and Ma [680] proved that every graph G satisfies

$$\chi(G) \leq 2\min\{\mathrm{ccl}_e(G), \mathrm{ccl}_o(G)\} + 3 \text{ and } \chi(G) \leq \mathrm{ccl}(G) + 4.$$

The first inequality is a strengthening of the result by Mihók and Schiermeyer (Corollary 2.16). Kostochka, Sudakov, and Verstraëte [637] proved that every triangle-free k-chromatic graph contains at least $\Omega(k^2 \log k)$ cycles of consecutive lengths. Our short proof of Gyárfás' result that $\chi(G) \leq 2|CL_o(G)| + 2$ is due to Diwan, Kernke, and Vishwanathan [304]. However, Gyárfás' proved a stronger result saying that K_{2k+2} is the only 2-connected graph with minimum degree $2k+1$ and $|CL_o(G)| = k \geq 1$. Our proof of Theorem 2.15 resembles the proof of Corollary 2.16 given by Mihók and Schiermeyer [737].

As emphasised by Tuza [1038] *the basic concept of depth-first search is not only an efficient tool in algorithmic graph theory but also it provides a powerful approach for proving theorems of a nonalgorithmic nature*, see also [1039]. Online coloring algorithms were introduced by Gyárfás and Lehel [446] to model a rectangle packing problem related to dynamical storage allocation. The Propositions 2.17, 2.18, Theorems 2.19, 2.21, and Corollary 2.20, 2.22, 2.23 as well as their proofs are mainly from the paper by Diwan, Kernke, and Vishwanathan [304].

The 1966 partition result of Lovász (Theorem 2.24) is one of the earliest results on partitioning a graph into a fixed number of induced subgraphs such that the ith part of the partition satisfies certain degree constraints. Theorem 2.24 provides a sufficient degree condition that guarantee the existence of a partition where the maximum degree of the ith part is bounded from above by a constant d_i. An extension of this result with pre-specified variable upper bounds for the degree of vertices in the ith part of the partition was presented by Borodin and Kostochka [157] (see Exercise 2.28). A counterpart of this result with lower bounds for the degrees was obtained by Stiebitz [971]. He proved that if G is a nonempty graph and $f \in \mathcal{V}_p(G)$ satisfies $d_G(v) \geq f_1(v) + f_2(v) + \cdots + f_p(v) + p - 1$ for all $v \in V(G)$, then there is a partition $(G_1, G_2, \ldots, G_p)$ of G such that for every $i \in [1, p]$ we have $G_i \neq \varnothing$ and $d_{G_i}(v) \geq f_i(v)$ for all $v \in V(G_i)$. A particular case of this theorem says that if G is a graph with $\delta(G) \geq s + t + 1$ $(s, t \geq 1)$, then G admits a partition (G_1, G_2) such that $\delta(G_1) \geq s$ and $\delta(G_2) \geq t$. This was conjectured by Thomassen [1003]. The complete graph K_{s+t+1} shows that the bound $s+t+1$ cannot be replaced by $s+t$. A weaker result was first proved by Thomassen [1001] and subsequently improved by R. Häggkvist, N. Alon and P. Hajnal [459] (with $2s+t-3$ instead of $s+t+1$ and $s, t \geq 3$).

Several interesting applications of Lovśz's partition result are discussed in the paper by Halldórsson and Lau [464]. Csiszár and Körner [265] used Lovász's argument to derive a continuous version of his partition result for edge weighted graphs; they used this result for proving coding theorems. However, the best known application of Lovász's result is the improvement of Brooks' theorem stated in Theorem 2.25. The problem of finding the best upper bound for the chromatic number χ of graphs in terms of the clique number ω and the maximum degree Δ was first proposed in 1968 by Vizing [1050]. Until now this problem has received a lot of attention and we shall discuss the matter further, particularly in Chapters 5 and 6.

Partition of graphs into degenerated subgraphs were considered and investigated by various researchers including Lick and White [675], Kronk and Mitchem [656], Mitchem [744], Borodin [147], and Bollobás and Manvel [134] (see Exercise 2.32). The k**-chromatic number** of a graph G is the least integer p for which G has a partition into p parts each of which is a k-degenerate subgraph of G. The k-chromatic number was introduced by Lick and White [675]. In 2000, Borodin, Kostochka, and Toft [159] introduced the concept of variable degeneracy of a graph extending that of k-degeneracy. This makes it possible to give a common generalization of the point partition numbers and the list chromatic number. The proofs of Propositions 2.27, 2.28, 2.29 and of Theorem 2.26 for graphs are due to Borodin, Kostochka, and Toft [159]. Schweser and Stiebitz [909] extended Theorem 2.26 to multigraphs and to hypergraph having multiple edges. That it is worthwhile to study the f-partition problem also for muligraphs is demonstrated in Chapter 3, Theorem 3.50. An extension of Theorem 2.26 to digraphs was obtained by Bang-Jensen, Schweser, and Stiebitz [80]; a DP-version of Theorem 2.26 for multigraphs was established by Kostochka, Schweser, and Stiebitz [628]. It seems very likely that one can deduce a polynomial time algorithm from the proof of Proposition 2.30, which, given a graph G and a vector-function $f \in \mathcal{V}_p(G)$ satisfying $f_1(v) + f_2(v) + \ldots + f_p(v) \geq d_G(v)$ for all $v \in V(G)$, finds an f-partition of G or shows that (G, f) is a hard pair. If we drop the assumption that $f_1(v) + f_2(v) + \ldots + f_p(v) \geq d_G(v)$ for all $v \in V(G)$, then the decision problem whether G admits an f-partition is NP-complete since it contains the NP-complete coloring problem. The complexity of the f-partition problem of graphs for constant vector-functions f and $p = 2$ was established in a series of papers [18], [137], [255], [1077], and [1078]. Let $s, t \geq 1$ be integers. A graph G is said to be (s,t)**-partitionable**, if there is a partition (G_1, G_2) of G such that G_1 is strictly s-degenarate and G_2 is strictly t-degenerate. As proved in [18], the decision problem whether a graph with maximum degree $\Delta \geq 3$ is (s,t)-partitionable is polynomial time solvable if $\Delta = 3$, or $s + t \geq \Delta$, or $(s,t) = (1,1)$; in all other cases the decision problem is NP-complete. Note that a graph G is (1,1)-partitionable if and only if G is bipartite.

In 2007, Matamala [720] proved that if G is a graph with maximum degree $\Delta \geq 3$ not containing a $K_{\Delta+1}$ as a subgraph and s, t are positive integers with $s + t \geq \Delta$, then there is a partition (G_1, G_2) of G such that G_1 is a maximum order strictly s-degenerate subgraph of G and G_2 is a strictly t-degenerate subgraph of G. This result improves earlier results obtained by Gerencsér [413], by Catlin [195], and by Catlin and Lai [196], see also Mitchem [746] and Exercise 1.7. In 2021, Schweser and Stiebitz [910] proved the following generalization of Matamala's result.

Theorem 2.44 *Let G be a connected multigraph and let $f \in \mathcal{V}_p(G)$ be a vector function of G with $p \geq 2$ such that $f_1(v) + f_2(v) + \cdots + f_p(v) \geq d_G(v)$ for all $v \in V(G)$ and (G, f) is not a hard pair. Then, there is a partition $(G_1, G_2, \ldots, G_p)$ of G such that G_1 is a maximum order strictly f_1-degenerate submultigraph of G, and for $i \in \{2, 3, \ldots, p-1\}$, the multigraph G_i is a maximum order strictly f_i-degenerate submultigraph of $G - (V(G_1) \cup V(G_2) \cup \cdots \cup V(G_{i-1}))$.*

The notion of generalized $\mathcal{P}$-coloring seems to have been independently discovered by several authors around 1968 (see [203], [204], [206], [251], [490], [675]) and again by several authors around 1985 (see [164], [169], [179], [469] [767]). The definition of the $\mathcal{P}$-chromatic number given in Section 2.6 is from 1968 and due to Hedetniemi [490], who studied, in particular, the $\mathcal{DG}_1$-chromatic number under the name **point aboricity**; he proved that every planar graph G satisfies $\chi(G : \mathcal{DG}_1) \leq 3$. Chartrand, Kronk, and Wall [206] also studied the point aboricity of a graph. The $\mathcal{DG}_k$-coloring problem, in general, was first considered by Lick and White [675]. For the parameter $\chi(G : \mathcal{DG}_k)$ Lick and White used the term **point partition number**, while Bolobás and Manvel [134] used the term k-chromatic number. In 1985 Harary [469] referred to generalized $\mathcal{P}$-colorings as **conditional colorings** and considered, in particular, the hereditary property Forb(H). Broere and Mynhardt [169], [767] also studied the Forb(H)-coloring problem. Brown and Corneil [179] studied the $\mathcal{P}$-coloring problem for general hereditary graph properties $\mathcal{P}$. A survey about the generalized coloring problem was published in 1989 by Albertson, Jamison, Hedetniemi, and Locke [38]; another survey was published in 1991 by Borowiecki and Mihók [164].

As emphasized by Jensen and Toft [530] *problems, results, and definitions in graph coloring theory have variations, extensions, and/or reformulations ad infinitum, it seems.* Cockayne [251] and, independently, Borowiecki and Mihok [164] developed a general framework that may be used in formulation graph coloring problems in a general setting. Let $\mathcal{P}_1, \mathcal{P}_2, \ldots, \mathcal{P}_k$ be graph properties. A **$(\mathcal{P}_1, \mathcal{P}_2, \ldots, \mathcal{P}_k)$-coloring** of a graph G is a mapping $\varphi : V(G) \to [1,k]$ such that $G[\varphi^{-1}(i)] \in \mathcal{P}_i$ for $i = 1,2,\ldots,k$. A graph G is called **$(\mathcal{P}_1, \mathcal{P}_2, \ldots, \mathcal{P}_k)$-colorable** if G has a **$(\mathcal{P}_1, \mathcal{P}_2, \ldots, \mathcal{P}_k)$-coloring**. The problem of recognizing $(\mathcal{P}_1, \mathcal{P}_2, \ldots, \mathcal{P}_k)$-colorable graphs is usually referred to as **Generalized Graph Coloring**. The **product** of graph properties $\mathcal{P}_1, \mathcal{P}_2, \ldots, \mathcal{P}_k$ is

$$\mathcal{P}_1 \circ \mathcal{P}_2 \circ \cdots \circ \mathcal{P}_k = \{G \mid G \text{ is } (\mathcal{P}_1, \mathcal{P}_2, \ldots, \mathcal{P}_k)\text{-colorable}\}.$$

The product of graph properties is a graph property, too. A graph property is **reducible** if it is the product of two other properties, otherwise it is **irreducible**. The most special case of the generalized graph coloring problem is the $\mathcal{P}$-coloring problem, with all properties identical. If $\mathcal{P}_1 = \mathcal{P}_2 = \cdots = \mathcal{P}_k = \mathcal{P}$, then $\mathcal{P}_1 \circ \mathcal{P}_2 \circ \cdots \circ \mathcal{P}_k = \mathcal{P}^k$. So $\mathcal{P}^k$ is the class of graphs having a $\mathcal{P}$-coloring with color set $C = [1,k]$. For an additive hereditary graph property $\mathcal{P}$, the $\mathcal{P}$-coloring problem behaves similar to the ordinary coloring problem. The results in Section 2.6 are from the paper by Borowiecki, Drgas-Burchardt, and Mihók [163]. The simple proof of Theorem 2.33 is due to Schweser [907]. Farrugia [364] proved that for any additive hereditary graph properties $\mathcal{P}$ and $\mathcal{P}'$, recognizing graphs in $\mathcal{P} \circ \mathcal{P}'$ is NP-hard with the only exception of bipartite graphs, that is, the case where $\mathcal{P} = \mathcal{P}' = \text{Forb}(K_2)$.

In 1956, Nordhaus and Gaddum [790] determined lower and upper bounds on the sum and the product, or equivalently, on the arithmetic mean and the geometric mean, of the chromatic number of a graph and its complement (see Exercise 2.7). Since then, relations of a similar type have been proposed for many other graph invariants. Aouchiche and Hanson [67] provide a survey of Nordhaus-Gaddum type relations. That every graph G satisfies $\chi(G)+\chi(\overline{G}) \leq |G|+1$ was proved already by Zykov [1105].

The following chain of five inequalties

$$\chi(G) \leq \chi_\ell(G) \leq \chi_{\text{DP}}(G) \leq \text{col}(G) \leq \text{mad}(G)+1 \leq \Delta(G)+1$$

involves six graph parameters. The first three of the graph parameters are defined in terms of colorings, whereas the last three are defined in terms of degrees. However, Alon [49] noticed a striking difference between χ and χ_ℓ (see also [45] and [530]). Alon proved that every graph G with average degree d satisfies

$$\chi_\ell(G) \geq (\tfrac{1}{2} - o(1)) \log_2 d, \tag{2.7}$$

where $o(1) \to 0$ as $d \to \infty$. Bernshteyn [104] proved that the DP-chromatic number of a graph grows much faster with the average degree than the list chromatic number, see Theorem 5.23. In 2023, Bernshteyn and Lee [112] introduced the **weak degeneracy** wdg(G) of a graph G as a new graph parameter and proved that we have $\chi_{\text{DP}}(G) \leq \text{wdg}(G)+1 \leq \text{col}(G)$ (see Exercise 3.32). They also showed that every planar graph G satisfies $\text{wde}(G) \leq 4$.

Chapter 3
Colorings and Orientations of Graphs

Colorings and orientations of graphs are related in different ways, but the deepness of these relations is not well understood. In this chapter we review some fundamental results.

3.1 The Gallai–Roy Theorem

A well-known result relating colorings to orientations is the Gallai–Roy theorem (Theorem 3.1). However, the first result relating colorings to orientations is due to Minty [741]. Minty's theorem (Theorem 3.2) published in 1962 says that a graph is k-colorable ($k \geq 2$) if and only if it has an orientation such that each cycle C has at least $|C|/k$ edges in each direction. Theorem 3.3 is a common extension of both the Gallai–Roy theorem and Minty's theorem – it was obtained in 1992 by Tuza [1039].

Given a graph G, an **orientation** of G is a simple digraph D such that $V(D) = V(G)$, $E(D) = E(G)$, and $\{v_D^+(e), v_D^-(e)\} = e$ for every edge $e \in E(G)$. Intuitively, an orientation of a graph is a digraph obtained from the graph by directing every edge from one of its ends to the other. If D is an orientation of G and $H \subseteq G$, then the subdigraph D' of D with $V(D') = V(H)$ and $E(D') = E(H)$ is denoted by $D|_H$. For a multidigraph D, let $\lambda(D)$ denote the maximum order of a directed path contained in D.

Theorem 3.1 *Let $k \geq 2$ be an integer. For any graph G the following statements are equivalent:*

(a) *G is k-colorable.*
(b) *G has an acyclic orientation D such that $\lambda(D) \leq k$.*
(c) *G has an orientation D such that any directed walk, where vertices but not edges may be repeated, has length at most $k - 1$.*
(d) *G has an orientation D such that $\lambda(D) \leq k$.*

M. Stiebitz et al., *Brooks' Theorem*, Springer Monographs in Mathematics,
https://doi.org/10.1007/978-3-031-50065-7_3

(e) *G has a spanning subgraph G' with an acyclic orientation D' such that $\lambda(D') \leq k$ and such that for any edge uv of G there is a directed path in D' between u and v, one way or the other.*

Proof For a given graph G and a given coloring φ of G with color set $C = [1, k]$, we associate an orientation $D = D_\varphi(G)$ by directing any edge uv of G from its end with the smaller color to its end with the larger color. Then it is easy to see that $\lambda(D) \leq k$. This shows that (a) implies (b). The implications from (b) to (c) and from (c) to (d) are trivial. To see that (d) implies (e), let D be an orientation of G such that $\lambda(D) \leq k$. Let E' be a minimal set of edges of G such that the orientation of $G' = G - E'$ induced by D is acyclic. This G' satisfies (e). That (e) implies (a) follows by letting $\varphi(v)$ be the order of a longest directed path in D' ending in the vertex $v \in V(G)$. This proves Theorem 3.1. □

The equivalence of (a) and (d) is the Gallai–Roy theorem, first obtained independently by Roy [879] and Gallai [400]. The equivalence between (a) and (b) had been noted earlier by Vitaver [1048] and Gupta [437], and possibly others. Finally, the equivalence between (a) and (e) was formulated in category-theoretical terms by Hasse [478]. The Gallai–Roy theorem is equivalent to the statement that each graph G satisfies

$$\chi(G) = \min\{\lambda(D) \mid D \text{ is an orientation of } G\}. \tag{3.1}$$

Let G be a graph, let D be an orientation of G, and let C be a cycle in G. Two edges $e, e' \in E(C)$ are said to be **C-equivalent** with respect to D, written $e \sim_C e'$, if either $e = e'$, or $e \neq e'$ and C contains a path P from $v_D^+(e)$ to $v_D^-(e')$ such that $v_D^-(e), v_D^+(e') \in V(P)$. Intuitively, $e \sim_C e'$ means that e and e' have the same direction in $D|_C$. It is easy to check that $\sim_C$ is an equivalence relation on $E(C)$, which induces a partition of $E(C)$ into at most two equivalence classes. Let F be one equivalence class with respect to $\sim_C$. Either F is the only equivalence class, i.e., $F = E(C)$, or $E(F) \setminus F$ is the other equivalence class. Define $r_D(C) = \min\{|F|, |C| - |F|\}$. Observe that $D|_C$ is a directed cycle if and only if $r_D(C) = 0$.

Theorem 3.2 (MINTY'S THEOREM) *Let $k \geq 2$ be an integer. A graph G is k-colorable if and only if it has an orientation D such that $r_D(C) \geq |C|/k$ for every cycle C of G.*

Suppose that a graph G has an orientation D satisfying $\lambda(D) \leq k$ for some integer k. Then D may contain directed cycles. However, if C is a cycle of G with $|C| \equiv 1 \pmod{k}$, then $r_D(C) \geq |C|/k$. Hence the following result of Tuza [1039] implies Theorem 3.1 as well as Theorem 3.2 and for $k = 2$ also König's theorem (Theorem C.7.

Theorem 3.3 (TUZA'S THEOREM) *Let $k \geq 2$ be an integer. A graph G is k-colorable if and only if it has an orientation D such that $r_D(C) \geq |C|/k$ for every cycle C of G having length $|C| \equiv 1 \pmod{k}$.*

Proof The "only if" implication is easy to see as we can take the orientation $D = D_\varphi(G)$, where φ is any coloring of G with color set $C = [1, k]$. To prove the "if" implication, we need only consider the case when G is connected (by (1.5)). So assume that D is an orientation of G such that $r_D(C) \geq |C|/k$ for every cycle C of G with $|C| \equiv 1 \pmod{k}$. To find a k-coloring of G, we shall use an appropriate spanning tree of G. First, choose a vertex r of G. Let T be an arbitrary spanning tree of G (rooted at r). An edge e of T is called *positive* if $v_D^-(e)$ belongs to the path $v_D^+(e)Tr$, otherwise e is called *negative*. Define a weight function $w_T : V(G) \longrightarrow \mathbb{N}_0$ as follows. Put $w_T(r) = 0$. If uv is an edge of T such that u belongs to the paths rTv, then put $w_T(v) = w_T(u) + 1$ if uv is positive (i.e., $v = v_D^+(uv)$), and put $w_T(v) = w_T(u) + 1 - k$ if uv is negative (i.e., $v = v_D^-(uv)$). Then the total weight $w(T)$ of T is defined by

$$w(T) = \sum_{v \in V(G)} w_T(v).$$

Among all spanning trees of G, let T be one with maximum total weight. Let $\varphi : V(G) \longrightarrow [1, k]$ be the map with $\varphi(v) \equiv w_T(v) \pmod{k}$ for all $v \in V(G)$. The proof is completed by showing that φ is a coloring of G. The proof is by contradiction. So suppose that there is an edge $e = uv$ of G such that $\varphi(u) = \varphi(v)$, implying that

$$w_T(u) \equiv w_T(v) \pmod{k} \tag{3.2}$$

Obviously, $e \in E(G) \setminus E(T)$ and hence $T + e$ contains a unique cycle C. For $w \in \{u, v\}$, let e_w denote the edge of C incident with w, but different from e; furthermore, let $T_w = T + e - e_w$. Clearly, T_u and T_v are spanning trees of G and the aim is to show that one of these trees has a larger total weight, which gives a contradiction. To this end, consider two cases.

Case 1: *$u \notin V(vTr)$ and $v \notin V(uTr)$.* By symmetry, we may assume that $w_T(u) \geq w_T(v)$. First suppose that $v_D^+(uv) = v$. Then uv is a positive edge of T_v and hence $w_{T_v}(w) > w_T(w)$ for for all vertices $w \in V(G)$ satisfying $v \in V(wTr)$. It then follows that $w(T_v) > w(T)$, a contradiction. Now suppose that $v_D^+(uv) = u$. If $w_T(u) > w_T(v)$, then (3.2) implies that $w_T(u) \geq w_T(v) + k$. The edge uv is a negative edge of T_v and, as before, it follows that $w(T_v) > w(T)$. If $w_T(u) = w_T(v)$, we can proceed as in the first case interchanging the roles of u and v. It then follows that $w(T_u) > w(T)$, a contradiction again.

Case 2: *$u \in V(vTr)$ or $v \in V(uTr)$.* By symmetry, we may assume that $u \in V(vTr)$. First suppose that $w_T(u) > w_T(v)$. Then (3.2) implies that $w_T(u) \geq w_T(v) + k$ and, no matter how uv is oriented, it follows that $w(T_v) > w(T)$, a contradiction. It remains to consider the case when $w_T(u) \leq w_T(v)$. Then (3.2) implies that $w_T(v) = w_T(u) + pk$ for an integer $p \geq 0$. Let q denote the number of negative edges of T belonging to the path $P = uTv$. Then $\ell(P) = (p+q)k$ and the cycle $C = uTv + uv$ has length $\ell(C) = (p+q)k + 1$. By hypothesis, $r_D(C) \geq p + q + 1/k$ and hence $r_D(C) \geq p + q + 1$. Since P has just q negative edges, it follows that $v_D^+(uv) = v$ and $p = 0$. This implies that $w_T(u) = w_T(v)$. Then uv is a positive edge of T_v and it follows that $w(T_v) > w(T)$, a contradiction again. □

If a graph G does not contain any cycle of length $1 \pmod{k}$ where $k \geq 2$, then Theorem 3.3 implies that $\chi(G) \leq k$ (for a short direct proof of this implication see

Exercise 3.4). In other words, any graph with chromatic number at least $k+1$ (≥ 3) has a cycle of length $1 \pmod{k}$. This generalizes König's theorem that a graph is bipartite if and only if it contains no odd cycle.

3.2 Kernel Perfect Digraphs

Given a multidigraph D, a **kernel** of D is defined as an independent set S of D such that every vertex of $D-S$ has an out-neighbor in D belonging to S. A multidigraph is said to be **kernel-perfect** if each of its induced subdigraphs has a kernel. The **underlying graph** of a multidigraph D is the graph G with $V(G)=V(D)$ and $E(G)=\{\{v^+(e),v^-(e)\} \mid e \in E(D)\}$. Observe that if D is a simple digraph, that is, D is a digraph without multiple edges, then D is (isomorphic to) an orientation of its underlying graph.

Lemma 3.4 (The Kernel Method) *Let D be a kernel-perfect digraph, let G be the underlying graph of D, and let $f,g: V(G) \to \mathbb{N}_0$ be two mappings such that*

$$f(v) \geq g(v) + \sum_{u \in N_D^+(v)} g(u) \tag{3.3}$$

whenever $g(v) \geq 1$. Then the weighted graph (G,g) is f-list-colorable.

Proof Let $V=V(D)$ and $W=\{v \in V \mid g(v) \geq 1\}$. We apply induction on $s=\sum_{v\in V} g(v)$. The base step $s=0$ is evident. So assume $s \geq 1$ and hence $W \neq \varnothing$. Let L be an f-assignment for G. Choose a color $c \in \bigcup_{v\in W} L(v)$. The vertices $v \in W$ with $c \in L(v)$ induces a nonempty subdigraph D' of D. As D is kernel-perfect, D' has a kernel S and $S \neq \varnothing$. Define a list-assignment L' of G by $L'(v)=L(v)\setminus\{c\}$ for $v \in V$; and let $f'(v)=|L'(v)|$ for $v \in V$. Define the function $g': V(G) \to \mathbb{N}_0$ by $g'(v)=g(v)-1$ if $v \in S$ and $g'(v)=g(v)$ otherwise. Then it is easy to check that

$$\sum_{v\in V} g'(v) < s \text{ and } f'(v) \geq g'(v) + \sum_{u \in N_D^+(v)} g'(u) \text{ whenever } g'(v) \geq 1.$$

Hence, the induction hypothesis implies that (G,g') is f'-list-colorable. So (G,g') has an L'-coloring φ', that is, there are sets $\varphi'(v) \subseteq L'(v)$ such that $|\varphi'(v)|=g'(v)$ and $\varphi'(v)\cap\varphi'(w)=\varnothing$ whenever $E_G(v,w) \neq \varnothing$. Define sets $\varphi(v)$ by setting $\varphi(v)=\varphi'(v)\cup\{c\}$ if $v \in S$ and $\varphi'(v)=\varphi(v)$ otherwise. Then $\varphi(v) \subseteq L(v)$ and $|\varphi(v)|=g(v)$ for all $v \in V$. Furthermore, $\varphi(v)\cap\varphi(w)=\varnothing$ whenever $E_G(v,w)\neq\varnothing$. Thus φ is an L-coloring of (G,g), and we are done. □

Corollary 3.5 (Bondy, Boppana, and Siegel) *Let D be a kernel-perfect digraph and let $f(v) \geq d_D^+(v)+1$ for every vertex $v \in V(D)$. Then the underlying graph G of D is f-list-colorable, and as a consequence $\chi_\ell(G) \leq \Delta^+(D)+1$.*

Recall that $\Delta^+(D)$ denotes the **maximum out-degree** of D. Corollary 3.5 was first used by Galvin [402] in 1995 to show that the list-edge-coloring conjecture holds for bipartite multigraphs. We prove an extension of this landmark result due to Borodin, Kostochka, and Woodall [160].

Theorem 3.6 (Borodin, Kostochka, and Woodall) *Let G be a bipartite multigraph and let $f(e) = \max\{d_G(u), d_G(v)\}$ for every edge $e \in E_G(u,v)$ with $u, v \in V(G)$. Then $L(G)$ is f-list-colorable, and as a consequence $\chi'_\ell(G) \le \Delta(G)$.*

Proof As the multigraph G is bipartite, G has a bipartition, say (X,Y). Let us say that two edges of G **meet** in X if they have a common end belonging to X, and correspondingly for Y.

For an arbitrary edge coloring of G, say $\varphi : E(G) \to \mathbb{N}$, we define a multidigraph $D = D_\varphi$ as follows. The vertex set is $V(D) = E(G)$. To define $E(D)$, consider two adjacent edges e and e' of G, say with $\varphi(e) < \varphi(e')$. If e and e' meet in X, let $a_X(e,e')$ be an edge directed from e' to e. If e and e' meet in Y, let $a_Y(e,e')$ be an edge directed from e to e'. The edge set $E(D)$ consists of all these edges. If e and e' are parallel edges, then both edges $a_X(e,e')$ and $a_Y(e,e')$ belong to D. Clearly, the underlying graph of D is the line graph of G. Our aim is to apply Corollary 3.5. To this end we prove the following two claims.

Claim 3.6.1 *The multidigraph $D = D_\varphi$ is kernel-perfect*

Proof: For the proof, it suffices to show that the induced subdigraph $D[F]$ has a kernel whenever $F \subseteq E(G)$. We apply induction on $|F|$. If $|F| = 0$, then the empty set is a kernel. For the induction step, let $|F| > 0$. Let S be the set of all edges $e \in F$ such that $\varphi(e) < \varphi(e')$ whenever $e' \neq e$ is an edge of F that meets e in X.

If S is an independent set of $D[F]$, then S is a kernel and we are done. Thus we assume that S is not independent in $D[F]$. Then there are two edges $e_1, e_2 \in S$ such that e_1 and e_2 meets in Y, say with $\varphi(e_1) < \varphi(e_2)$. Then $D[F \setminus \{e_1\}]$ has a kernel T by the induction hypothesis. If $e_2 \in T$, then the edge $a_Y(e_1,e_2)$ is directed from e_1 to e_2 and, therefore, T is a kernel of $D[F]$ and we are done. If $e_2 \notin T$, then there is an edge $e \in F \setminus \{e_1\}$ such that the multidigraph $D[F \setminus \{e_1\}]$ contains an edge directed from e_2 to e. By the choice of e_2, this implies that e and e_2 meets in Y and, therefore, $\varphi(e_2) < \varphi(e)$. Since the edges e, e_1, e_2 share the same vertex in Y and $\varphi(e_1) < \varphi(e_2) < \varphi(e)$, the edge $a_Y(e_1,e)$ is directed from e_1 to e and, therefore, T is a kernel of $D[F]$ and we are done, too. This proves the claim. △

For an edge coloring $\varphi : E(G) \to \mathbb{N}$ of G, let $d_\varphi(e)$ denote the out-degree of the vertex $e \in V(D_\varphi) = E(G)$ in D_φ.

Claim 3.6.2 *We can choose φ such that $d_\varphi(e) \le f(e) - 1$ for all $e \in E(G)$.*

Proof: We apply induction on $p = |E(G)| + |V(G)|$. If $|E(G)| = 0$, then we take the empty coloring. For the induction step, let $p > 0$. If G has an isolated vertex v, then the result follows from the induction hypothesis applied to $G - v$. Now assume that G has no isolated vertices. For a set $M \subseteq E$, let $V(M)$ denote the set of all vertices that are incident to some edge of M.

First, suppose that $|X| \le |Y|$. We claim that there is a nonempty matching M in G such that $N_G(V(M) \cap Y) \subseteq V(M) \cap X$. As G is bipartite and (X,Y) is a bipartition of G, $|N_G(Y)| \leq |X| \leq |Y|$. Hence there is a minimal nonempty set $Y' \subseteq Y$ such that $|N_G(Y')| \leq |Y'|$. Since G has no isolated vertices, $|N_G(Y')| > 0$. Furthermore, $|N_G(Y')| = |Y'|$ and $|N_G(Z)| > |Z|$ whenever $\varnothing \neq Z \subset Y'$. By Hall's theorem (Theorem C.11) it follows that there is a matching M of G such that $V(M) = Y' \cup N_G(Y')$. Clearly, M has the required property.

By the induction hypothesis, $H = G - M$ has an edge-coloring φ', say with color set $C = [1,k]$, such that $d_{\varphi'}(e) \leq f'(e) - 1$, where $f'(e) = \max\{d_H(u), d_H(v)\}$ for all edges $e \in E_H(u,v)$. Let φ be the edge-coloring of G with $\varphi(e) = \varphi'(e)$ for every edge $e \in E(H)$ and $\varphi(e) = k+1$ for every edge $e \in M$. Then $d_\varphi(e) \leq d_{\varphi'}(e) + 1$ for each edge $e \notin M$, and if $d_\varphi(e) = d_{\varphi'}(e) + 1$ then both ends of e are in $V(M)$, which implies $f(e) = f'(e) + 1$. If $e \in M$, say $e = uv$ with $u \in X$ and $v \in Y$, then $d_\varphi(e) = d_G(u) - 1 \leq \max\{d_G(u), d_G(v)\} - 1 = f(e) - 1$. Thus $d_\varphi(e) \leq f(e) - 1$ for each edge e of G, as required.

Now, suppose that $|X| > |Y|$. Then, as in the first case, there is a nonempty matching M in G such that $N_G(V(M) \cap X) \subseteq V(M) \cap Y$. By the induction hypothesis, $H = G - M$ has an edge-coloring φ', say with color set $C = [1,k]$, such that $d_{\varphi'}(e) \leq f'(e) - 1$, where $f'(e) = \max\{d_H(u), d_H(v)\}$ for all edges $e \in E_H(u,v)$. Let φ be the edge-coloring of G with $\varphi(e) = \varphi'(e) + 1$ for every edge $e \in E(H)$ and $\varphi(e) = 1$ for every edge $e \in M$. As before, it is easy to see that $d_\varphi(e) \leq f(e) - 1$ for each edge e of G, as required. This completes the proof of the claim. △

By Claim 3.6.2, there is an edge-coloring φ of G such that every edge e of the multidigraph D_φ satisfies $f(e) \geq d^+_{D_\varphi}(e) + 1$. By Claim 3.6.1, D_φ is kernel-perfect. Then Corollary 3.5 implies that the underlying graph of D_φ is f-list-colorable. Since this graph is the line graph of G, it follows that $L(G)$ is f-list-colorable. □

Theorem 3.6 together with (1.14) implies the following result due to Galvin [402].

Corollary 3.7 (Galvin) *Every bipartite multigraph G satisfies $\chi'_\ell(G) = \chi'(G) = \Delta(G)$.*

Borodin, Kostochka, and Woodall [160] used Theorem 3.6 to establish the following list analogue of Shannon's bound (Theorem 1.38).

Theorem 3.8 (Borodin, Kostochka, and Woodall) *let G be a multigraph, and let*

$$f(e) = \max\{d_G(u), d_G(v)\} + \lfloor \frac{1}{2} \min\{d_G(u), d_G(v)\} \rfloor$$

for every edge $e \in E_G(u,v)$ with $u, v \in V(G)$. Then $L(G)$ is f-list-colorable, and as a consequence $\chi'_\ell(G) \leq \lfloor \frac{3}{2}\Delta(G) \rfloor$.

Proof For the proof, let L be an arbitrary f-assignment of $L(G)$. Our aim is to show that G has an edge-coloring φ such that $\varphi(e) \in L(e)$ for all $e \in E(G)$.

Let $\{X,Y\}$ be a partition of $V(G)$ such that the set $E_G(X,Y)$ of all edges of G joining a vertex of X with a vertex of Y is as large as possible. Let H be the bipartite multigraph with $V(H) = X \cup Y$ and $E(H) = E_G(X,Y)$. Then it is easy to see that $d_H(v) \geq \frac{1}{2} d_G(v)$ for each vertex $v \in V(H)$. Otherwise move v to the other side of the partition. So if $G' = G - E(H)$ then $d_{G'}(v) \le \frac{1}{2} d_G(v)$ and $d_G(v) = d_{G'}(v) + d_H(v)$ for each vertex $v \in V(G)$. Hence, if $e \in E_G(u,v)$, then the assumption implies that

$$\begin{aligned} f(e) &\ge \max\{d_G(u) + \lfloor \frac{1}{2} d_G(v) \rfloor, d_G(v) + \lfloor \frac{1}{2} d_G(u) \rfloor\} \\ &\ge \max\{d_G(u) + d_{G'}(v), d_G(v) + d_{G'}(u)\} \\ &= \max\{d_H(u), d_H(v)\} + d_{G'}(u) + d_{G'}(v) \end{aligned}$$

Consequently, $|L(e)| = f(e) \ge d_{G'}(u) + d_{G'}(v) = d_{L(G')}(e) + 2$ for every edge $e \in E_{G'}(u,v)$, and so we can easily color the edges of G' from their lists. If now $e \in E_H(u,v)$ is an edge, then the number of different colors already used on edges adjacent to e in G is at most $d_{G'}(u) + d_{G'}(v)$, and so e still has a list of size at least

$$f(e) - d_{G'}(u) - d_{G'}(v) \ge \max\{d_H(u), d_H(v)\}.$$

Since the multigraph H is bipartite, the edges of H can be colored from their lists by Theorem 3.6. The union of these two edge colorings results in an edge coloring φ of G with $\varphi(e) \in L(e)$ for all $e \in E(G)$. □

3.3 Maximum Out-Degree

Given a multidigraph D, let G_D denote the underlying graph of D. Let $\mathcal{D}$ be a **digraph property**, that is, $\mathcal{D}$ is a class of multidigraphs closed under isomorphism. Then, for a graph G, define

$$\Delta^+_{\mathcal{D}}(G) = \min\{\Delta^+(D) \mid D \in \mathcal{D}, G = G_D\}$$

If $\mathcal{D}$ consists of all multidigraphs, then we write $\Delta^+_{\text{all}}(G)$ instead of $\Delta^+_{\mathcal{D}}(G)$, and if $\mathcal{D}$ consists of all acyclic multidigraphs, that is, those without directed cycles, then we write $\Delta^+_{\text{ac}}(G)$ instead of $\Delta^+_{\mathcal{D}}(G)$.

An acyclic multidigraph D has a kernel (see Exercise 3.11) and is therefore kernel-perfect. Hence, by Corollary 3.5, every graph G satisfies $\chi_\ell(G) \le \Delta^+_{\text{ac}}(G) + 1$. However, as pointed out by Finck and Sachs [372], every graph G also satisfies

$$\Delta^+_{\text{ac}}(G) = \text{col}(G) - 1, \tag{3.4}$$

so we only obtain the result from Proposition 1.3.

To see that $\text{col}(G) - 1 \le \Delta^+_{\text{ac}}(G)$, let H be a nonempty subgraph of G of maximum minimum degree. Then, by Proposition 1.5, $\text{col}(G) - 1 = \delta(H)$. In any acyclic multidigraph D with $G = G_D$ the vertices may be put in an order $v_1, v_2, \ldots, v_n$ in which every edge is directed from the vertex with the larger index to the vertex with the smaller index (see Exercise 3.11). If v_i is a vertex of H with the largest index, then $\text{col}(G) - 1 = \delta(H) \le d_H(v_i) \le d^+_D(v_i) \le \Delta^+_{\text{ac}}(G)$.

To see that $\mathrm{col}(G)-1 \geq \Delta^+_{\mathrm{ac}}(G)$, take a smallest last order $v_n, v_{n-1}, \ldots, v_1$ of G (see Section 1.3) and let $G_i = G[v_1, \ldots, v_i]$. Define an orientation D of G in which every edge is directed from the vertex with the larger index to the vertex with the smaller index. Clearly, D is acyclic and every vertex v_i satisfies $d^+_D(v_i) = d_{G_i}(v_i)$. Hence Proposition 1.5 implies that $\Delta^+_{\mathrm{ac}}(G) \leq \Delta^+(D) = \max_{1\leq i \leq n} d_{G_i}(v_i) = \mathrm{col}(G)-1$. This completes the proof of (3.4).

Clearly, any kernel of a multidigraph is a maximal independent set, but the converse statement is not true, and kernels may even fail to exists. For example, an odd directed cycle has no kernel. Chvátal [246] proved in 1973 that it is NP-complete to decide whether a multidigraph has a kernel, or not. On the other hand, Richardson [861] proved in 1953 the following positive result.

Theorem 3.9 (Richardson) *Every multidigraph having no odd directed cycle has a kernel and is therefore kernel-perfect.*

Obviously, every multidigraph whose underlying graph is bipartite contains no odd directed cycle. This simple observation together with Corollary 3.5 and Theorem 3.9 implies the following result about the list-chromatic number of bipartite graphs due to Alon and Tarsi [58].

Corollary 3.10 *Every bipartite graph G satisfies $\chi_\ell(G) \leq \Delta^+_{\mathrm{all}}(G)+1$.*

That every graph G satisfies

$$\Delta^+_{\mathrm{all}}(G) = \lceil \frac{1}{2}\mathrm{mad}(G) \rceil \tag{3.5}$$

seems to have been folklore at least from the second half of the 1980s. To prove (3.5) it suffices to show that if $G \neq \varnothing$ and $|E(G[X])| \leq d|X|$ for every $X \subseteq V(G)$, then G admits an orientation D with $\Delta^+(D) \leq d$. In the 1960s Hakimi [462] established a necessary and sufficient condition for the existence of an orientation of a graph G with prescribed out-degrees (Proposition 3.11). The inductive proof in [462] is rather long. An elegant proof from first principles was given by Frank and Gyarfás [387] (see also [385]); a proof of this result using Hall's theorem is included in the famous paper by Alon and Tarsi [58].

Proposition 3.11 *Let G be a nonempty graph and let $h: V(G) \to \mathbb{N}$ be a function. Then G has an orientation D such that $d^+_D(v) \leq h(v)$ for all $v \in V(G)$ if and only if $|E(G[X])| \leq \sum_{v\in X} h(v)$ for every set X with $\varnothing \neq X \subseteq V(G)$.*

Proof The "only if" part is evident. For the proof of the "if" part, we have to show that there is an appropriate orientation of G. Given an orientation D of G, let $\rho(D) = \sum(d^+_D(v) - h(v))$ where the sum is over all vertices v with $d^+_D(v) > h(v)$. Choose an orientation D of G such that $\rho(D)$ is minimum. If $\rho(D) = 0$, we are done. Otherwise, $\rho(D) > 0$ and there is a vertex v with $d^+_D(v) > h(v)$. Let X be the set of all vertices w such that there is a directed path in D from v to w. Then $v \in X$ and $w \in X$ implies $N^+_D(w) \subseteq X$. Consequently, X contains a vertex w with $d^+_D(w) \leq h(w)-1$, for otherwise $|E(G[X])| = \sum_{w\in X} d^+_D(w) > \sum_{v \in X} h(w)$, contradicting the hypothesis. Reorienting the edges of a directed path from v to w in D yields an orientation D' of G with $\rho(D') < \rho(D)$, a contradiction. □

Let G be a nonempty graph. A **fractional orientation** of G is a function f from the set $E(G) \times V(G)$ into the set of nonnegative rational numbers $\mathbb{Q}_+$ satisfying the following two conditions:

1. $f(uv,u) + f(uv,v) = 1$ for all $uv \in E(G)$, and
2. $f(uv,w) = 0$ for all $uv \in E(G)$ and all $w \in V(G) \setminus \{u,v\}$.

For a given fractional orientation f of G and a given vertex v of G, let

$$d^+_{G,f}(v) = \sum_{e \in E(G)} f(e,v)$$

be the **fractional out-degree** of v in G with respect to f. Obviously,

$$\sum_{v \in V(G)} d^+_{G,f}(v) = |E(G)|.$$

Furthermore, let $\Delta^+_f(G)$ denote the **maximum fractional out-degree** of G with respect to f, and let

$$\Delta^+_{\text{fr}}(G) = \inf_f \Delta^+_f(G),$$

where the infimum is taken over all fractional orientations f of G. For $G = \varnothing$ define $\Delta^+_{\text{fr}}(G) = 0$. The proof of the following result, due to Jensen and Toft [531], shows not only that the infimum is in fact a minimum, but also gives a good characterization of the graph parameter Δ^+_{fr} as well as a polynomial-time algorithm that computes this parameter for every graph G.

Theorem 3.12 (Jensen and Toft) *Every graph G satisfies*

$$\max_{\varnothing \neq H \subseteq G} \frac{|E(H)|}{|V(H)|} = \min_f \Delta^+_f(G),$$

where the minimum always exists and where the minimum is taken over all fractional orientations f of G.

Proof The statement is evident if $G = \varnothing$, so assume that $G \neq \varnothing$. For any nonempty subgraph H of G and any fractional orientation f of G:

$$\frac{|E(H)|}{|V(H)|} = \frac{1}{|V(H)|} \sum_{v \in V(H)} d^+_{H,f}(v) \leq \Delta^+_f(H) \leq \Delta^+_f(G).$$

Hence we need only to show that there is a nonempty subgraph H of G and a fractional orientation f such that $|E(H)|/|V(H)| = \Delta^+_f(G)$. To this end, we perform the following algorithm for each value of $q = 1, 2, \ldots, |G|$: Start off with any fractional orientation f, where all denominators are equal to q. A path $P = (v_0, v_1, \ldots, v_n)$ with $v_0 = v$ and $v_n = w$ is called an *augmenting path* w.r.t. f if $f(v_i v_{i+1}, v_i) > 0$ (and thus $\geq 1/q$) for all $i = 0, \ldots, n-1$. Modify f whenever there is such an augmenting path with $d^+_{G,f}(v) > d^+_{G,f}(w) + 1/q$ as follows: For $i \in [0, n-1]$, subtract $1/q$ from

$f(v_iv_{i+1},v_i)$ and add $1/q$ to $f(v_iv_{i+1},v_{i+1})$. The total sum over all pairs (v,w) of numerical differences

$$\sum |d^+_{G,f}(v) - d^+_{G,f}(w)|$$

decreases by at least $1/q$ ($\geq 1/|G|$), hence after at most $|E(G)|\cdot|G|^3$ modifications no further modifications as described above are possible (since the starting value of the sum is certainly at most $|E(G)|\cdot|G|^2$ and $q \leq |G|$).

Consider for a fixed q the final fractional orientation f. Let X be the set of vertices v of G with $d^+_{G,f}(v) = \Delta^+_f(G)$ and let H be the subgraph of G induced by X and the set Y of all vertices that can be reached from the set X by an augmenting paths, beginning the paths from X. Since no further modification of f is possible, the vertex set of H consists of the set X and a set $Y \subseteq V(G)\setminus X$ of vertices w with $d^+_{G,f}(w) = \Delta^+_f(G) - 1/q$.

Let H^* be a nonempty subgraph of G such that $|E(H^*)|/|V(H^*)| = \frac{1}{2}\mathrm{mad}(G)$. Then

$$\begin{aligned}\Delta^+_f(G) &\geq \frac{1}{2}\mathrm{mad}(G) = \frac{|E(H^*)|}{|V(H^*)|} \geq \frac{|E(H)|}{|V(H)|}\\ &= \frac{1}{|V(H)|}\sum_{v\in V(H)} d^+_{H,f}(v)) = \frac{1}{|V(H)|}\sum_{v\in V(H)} d^+_{G,f}(v)\\ &= \Delta^+_f(G) - \frac{1}{q}(\frac{|Y|}{|V(H)|}) > \Delta^+_f(G) - \frac{1}{q},\end{aligned}$$

where $\Delta^+_f(G)$ is a rational number with denominator q. For $q = |V(H^*)|$ it follows that

$$\Delta^+_f(G) = \frac{|E(H^*)|}{|V(H^*)|} = \frac{|E(H)|}{|V(H)|},$$

hence the optimal graph H^* can be found among the $|G|$ subgraphs H (there is a graph H for each value of q). This proves the theorem. □

Jensen and Toft [531] remarked that they did not try to obtain the most efficient algorithm, but rather a transparent proof of Theorem 3.12. The reader can find more information on efficient algorithms for computing the maximum average degree of graphs in the paper by Deligkas, Eiben, Gutin, Neary, and Yeo [275].

3.4 Normal Digraphs

A multidigraph D is said to be **normal** (or **clique-acyclic**) if for every clique X of D the subdigraph $D[X]$ has a kernel. Recall that a clique in a graph or digraph need not be maximal. Obviously, a multidigraph whose underlying graph is complete has a kernel if and only if it contains a vertex v such that there is an edge directed from every other vertex to v. In particular, an orientation of a complete graph is kernel-perfect if and only if it is acyclic.

Let G be a graph. If $\mathcal{D}$ is the class of all kernel-perfect multidigraphs D, then we write $\Delta^+_{\mathrm{ke}}(G)$ instead of $\Delta^+_{\mathcal{D}}(G)$. It follows from Corollary 3.5 that G satisfies $\chi_\ell(G) \leq \Delta^+_{\mathrm{ke}}(G)+1$.

By a **strict digraph** we mean a multidigraph without parallel edges. Clearly, if D' is a maximal strict subdigraph of a digraph D, then D has a kernel if and only if D' has kernel. Hence if $\mathcal{H}$ is the class of all kernel-perfect strict digraphs, then $\Delta^+_{\mathrm{ke}}(G) = \Delta^+_{\mathcal{H}}(G)$. Given a graph G, a strict digraph D such that $G_D = G$ is sometimes called a **super-orientation** of G. Several papers (see e.g. [23, 162, 402]) do not distinguish between orientations and super-orientations.

An edge of a multidigraph D directed from u to v is called **irreversible** if there is no edge in D directed from v to u. Hence in a super-orientation of a graph G we allow to replace an edge $e = uv$ of G by either an irreversible edge directed from u to v or from v to u, or by a pair of edges, one directed from u to v and the other one directed from v to u. By a **proper cycle** of a digraph D we mean a directed cycle of D such that all of its edges are irreversible in D.

Proposition 3.13 *A multidigraph D is normal if and only if D contains no proper cycle whose vertex set is a clique of D.*

Proof The "only if" part is evident: If D contains a proper cycle such that its vertex set X is a clique of D, then $D[X]$ has no kernel, a contradiction. For the proof of the "if" part let us consider a clique X of D. We prove by induction on $|X|$ that $D[X]$ has a kernel. This is evident if $|X| \leq 2$. For the induction step, let $|X| \geq 3$. From the induction hypothesis it then follows that $D[X \setminus \{u\}]$ has a kernel for each vertex $u \in X$, where such a kernel consists of a single vertex, say $v_u \in X$. If $D[X]$ has no kernel, then, since X is a clique in D, the multidigraph D contains for each vertex $u \in X$ an irreversible edge directed from v_u to u. This obviously yields a proper cycle in $D[X]$, a contradiction. Hence $D[X]$ has a kernel for each clique X of D and, therefore, D is normal. □

Evidently, every kernel-perfect multidigraph is normal, but a normal multidigraph need not be kernel-perfect. For instance, if C is an odd directed cycle with $|C| \geq 5$, then C is a normal digraph without containing any kernel.

A graph G is called **kernel-solvable** if every normal multidigraph whose underlying graph is G has a kernel. By the above observation about strict digraphs, a graph G is kernel-solvable if and only if every normal super-orientation of G has a kernel.

Let us first observe that kernel solvability is a hereditary graph property. Hence, every normal super-orientation of a kernel-solvable graph is kernel-perfect.

Proposition 3.14 *Let G be a kernel-solvable graph. Then every induced subgraph of G is kernel-solvable too, and as a consequence every normal multidigraph D with* $G_D = G$ *is kernel-perfect.*

Proof Let H be an induced subgraph of G, and let D' be a normal multidigraph such that $G_{D'} = H$. Let D'' be an acyclic orientation of $G - V(H)$, and let D be the multidigraph obtained from $D' \cup D''$ by adding, for each edge uv of G such that $u \in V(H)$ and $v \in V(G) \setminus V(H)$, an edge e to D with $v^+(e) = u$ and $v^-(e) = v$. Due to the above construction, $G_D = G$ and all directed cycles in D belong to D'. Thus, since D' is normal, D is normal, too (by Proposition 3.13). As G is kernel-solvable, D has a kernel S, and by definition of D, the set $S \cap V(D')$ is a kernel of D'. This proves that H is kernel-solvable. Since every induced subdigraph of a normal multidigraph is normal, too, it then follows that every normal multidigraph D with $G = G_D$ is kernel-perfect. □

Kernel-solvable graphs were introduced by Berge and Duchet [103], who conjectured that they are the same as perfect graphs. They also observed that odd cycles of length at least 5 and their complements are not kernel-solvable. Hence the strong perfect graph theorem (Theorem 1.60) and Proposition 3.14 imply one direction of their conjecture, namely that every kernel-solvable graph is perfect. The reverse direction of their conjecture was proved by Boros and Gurvich [162] in 1996. Hence we have the following result.

Theorem 3.15 (The Kernel Solvable Graph Theorem) *A graph G is perfect if and only if G is kernel-solvable.*

The proof of the "only if" part of Theorem 3.15 given in [162] is long and uses several results from cooperative game theory. A simpler proof, proposed by Aharoni and Holzman [23], is based on a fractional version of kernels and uses only a result of Scarf [892] about cores in cooperative games as well as a result of Lovász [691] (independently proved by Chvátal [248]) about fractional independent sets in perfect graphs.

Let V be an arbitrary set. Then $\mathbb{R}_+^V$ denotes the set of all functions from the set V into the set of nonnegative real numbers $\mathbb{R}_+$. For $X \subseteq V$ and $f \in \mathbb{R}_+^V$, define $f(X) = \sum_{v \in X} f(v)$. Furthermore, f is called the **characteristic function** of X, if $f(v) = 1$ if $v \in X$ and $f(v) = 0$ if $v \in V \setminus X$. Given a graph G with vertex set V, a function $f \in \mathbb{R}_+^V$ is called **fractionally independent** in G if $f(X) \leq 1$ for every clique X of G. Now let D be a multidigraph with vertex set V and let $G_D = G$. For a vertex $v \in V$, let $N_D^+[v]$ denote the **closed out-neighborhood** of v in D, that is the set of all vertices u of D such that $v = u$ or D contains an edge directed from v to u. Clearly, S is a kernel in D if S is an independent set of D and $N_D^+[v] \cap S \neq \varnothing$ for each vertex v of D. A function $f \in \mathbb{R}_+^V$ is called **fractionally independent** in D, if f is a fractionally independent in G. The function $f \in \mathbb{R}_+^V$ is called **fractionally absorbing** if $f(N_D^+[v]) \geq 1$ for all vertices $v \in V$, and **strongly absorbing** if for every vertex $v \in V$ the multidigraph D has a clique $X_v \subseteq N_D^+[v]$ for which $f(X_v) \geq 1$. Finally, f is called a **strong fractional kernel** of D if f is both fractionally independent and strongly absorbing.

The next result due to Aharoni and Holzman [23] says, in particular, that for super-orientations of perfect graphs kernels and strong fractional kernels coincides.

Theorem 3.16 *Let G be a perfect graph and let D be a multidigraph with $G = G_D$. Then D has a kernel if and only if D has a strong fractional kernel.*

Proof Obviously, the characteristic function of a kernel is a strong fractional kernel. This proves the "only if" implication. To prove the converse implication, let f be a strong fractional kernel of D. Then f is, in particular, a fractional independent set of D and so of G. As G is perfect, it then follows from a theorem of Lovász [691] (independently proved by Chvátal [248]), that there are independent sets of D, say $I_1, I_2, \ldots, I_p$ with $p \geq 1$, such that f is a convex combination of the characteristic functions of these independent sets. For $i \in [1,p]$ let f_i denote the characteristic function of I_i. Then there are positive real numbers $t_1, t_2, \ldots, t_p$ such that

$$f = t_1 f_1 + t_2 f_2 + \cdots t_p f_p \text{ and } t_1 + t_2 + \cdots + t_p = 1. \tag{3.6}$$

Now let $v \in V$ be an arbitrary vertex. As f is a strong fractional kernel, there is a clique $X_v \subseteq N_D^+[v]$ such that $f(X_v) \geq 1$ and so $f(X_v) = 1$. Clearly, $f_i(X_v) = |I_i \cap X_v| \leq 1$ for all $i \in [1,p]$. By (3.6), we then obtain

$$1 = f(X_v) = t_1 f_1(X_v) + t_2 f_2(X_v) + \cdots t_p f_p(X_v) \leq t_1 + t_2 + \cdots + t_p = 1,$$

which implies that $|X_v \cap I_i| = f_i(X_v) = 1$ for all $i \in [1,p]$. Hence $I_i \cap N_D^+[v] \neq \emptyset$ for all $i \in [1,p]$. Consequently, each independent set I_i is a kernel of D. Thus the proof of the theorem is complete. □

That every perfect graph is kernel-solvable follows from Theorem 3.16 and the next result from [23]. In this paper the reader can also find a proof of Scarf's theorem from [892] used in the proof of Theorem 3.17.

Theorem 3.17 *Every normal multidigraph D has a strong fractional kernel.*

Sketch of Proof: Let $U_1, U_2, \ldots, U_m$ be an enumeration of the maximal cliques of D. Construct a new digraph D' by adding to D for each clique U_i a new vertex u_i and for each vertex $u \in U_i$ an edge directed from u_i to u. Then $N_{D'}^+(u_i) = U_i$ and $N_{D'}^-(u_i) = \emptyset$ for all $i \in [1,m]$, and every vertex v of D satisfies $N_D^+(v) = N_{D'}^+(v)$. Furthermore, $V_i = U_i \cup \{u_i\}$ is a clique of D' for all $i \in [1,m]$. Since D is normal, D' is normal, too.

Since V_i is a clique of D' and D' is normal, there is a vertex order of V_i, say $v_0^i, v_1^i, \ldots, v_{m_i}^i$, such that, for $0 \leq h < k \leq m_i$, the multidigraph D' contains an edge directed from v_h^i to v_k^i. For a vertex $v \in V_i$, let $h_i(v)$ be the index k for which $v_k^i = v$. Clearly, $u_i \in Y_i$ and $h_i(u_i) = 0$.

Let $v_1, v_2, \ldots, v_n$ be a vertex order of D' such that $v_i = u_i$ for $i \in [1,m]$. Define an $m \times n$ matric $C = (c_{ij})$ as follows: For each i, j such that $v_j \notin Y_i$, define $c_{ij} = K_0$, where K_0 is some number larger than $|D'|$. If $v_j \in Y_i$ let $c_{ij} = h_i(v_j)$. Furthermore, let $B = (b_{ij})$ be the incidence matrix of the cliques $V_1, V_2, \ldots, V_m$ of D', that is, $b_{ij} = 1$ if $v_j \in Y_i$, and $b_{ij} = 0$ otherwise. Finally, let b be the all-one vector in $\mathbb{R}^m$.

Then B is an $m \times n$ matrix, where the first m columns form the $m \times m$ identity matrix. It is also not difficult to show that the set $\{x \in \mathbb{R}_+^n \mid Bx = b\}$ is bounded. The definition of the matrix C implies that $0 = c_{ii} \leq c_{ik} \leq c_{ij}$ whenever $i, j \in [1,m]$, $i \neq j$, and $k > m$.

By a theorem of Scarf [892] (see also [23]), it then follows that there is an index set $J \subseteq [1,n]$ and a vector $x \in \mathbb{R}_+^n$ such that

(a) $Bx = b$ and $x_j = 0$ whenever $j \notin J$, and
(b) for every $k \in [1,n]$ there exists an $i \in [1,m]$ such that $c_{ik} \leq c_{ij}$ for all $j \in J$.

Clearly, $V = \{v_{m+1}, v_{m+2}, \ldots, v_n\}$ is the vertex set of D. Let $f \in \mathbb{R}_+^V$ be the function defined by $f(v_j) = x_j$ for $j \in [m+1,n]$. To conclude the proof, it suffices to show that f is a strong fractional kernel of D, that is, f is fractionally independent and strongly absorbing.

To show that f is fractionally independent, let X be an arbitrary clique of D. Then $X \subseteq U_i$ for some $i \in [1,m]$. Since $Bx = b$ (by (a)), it follows that $x_i + f(U_i) = 1$ and so $f(X) \leq f(U_i) = 1 - x_i \leq 1$. This shows that f is fractionally independent.

To prove that f is strongly absorbing, let v_k be an arbitrary vertex of D. Then $k > m$ and there exists an $i \in [1,m]$ such that $c_{ik} \leq c_{ij}$ for all $j \in J$ (by (b)). Let $W_i = \{v_j \in V_i \mid j \in J\}$. We claim that $W_i \subseteq N_D^+[v_k]$. (Note that $N_D^+[v_k] = N_{D'}^+[v_k]$). Indeed, assume that $v_j \in W_i \setminus N_D^+[v_k]$. Since $v_j \in V_i$, it follows that $c_{ij} < K_0$. As $c_{ik} \leq c_{ij}$, this yields $c_{ik} < K_0$, which implies that $v_k \in V_i$. Since both v_j and v_k belong to the clique V_i of D', but $v_j \notin N_D^+[v_k]$, it follows that D contains an irreversible edge directed from v_j to v_k. Consequently, $h_i(v_j) < h_i(v_k)$, and hence $c_{ij} < c_{ik}$, giving a contradiction. Thus $W_i \subseteq N_D^+[v_k]$ and W_i is a clique of D. Since $W_i = \{v_j \in V_i \mid j \in J\}$, it follows from (a) that $f(W_i) = 1$. Hence f is strongly absorbing. Thus the proof is complete. □

3.5 The Graph Polynomial

All graphs, multigraphs, and multidigraphs considered in this section are labeled. To each vertex v we associate a variable x_v. If $v = v_i$, then we write x_i rather than x_{v_i}. In this section, n denotes a positive integer and $\{v_1, v_2, \ldots, v_n\}$ denotes a set of n distinct vertices.

With every multidigraph D on the vertex set $V(D) = \{v_1, v_2, \ldots, v_n\}$, we associate the polynomial

$$P_D = \prod_{e \in E(D)} (x_{v^+(e)} - x_{v^-(e)}). \tag{3.7}$$

Clearly, $P_D \in \mathbb{Z}[x_1, x_2, \ldots, x_n]$ is polynomial over the integer ring $\mathbb{Z}$ in n variables. As usual, the product over the empty set is the unit of the underlying ring.

If D and D' are two orientations of the same graph or multigraph G, then we have $P_D = \pm P_{D'}$. Hence the polynomial P_G of a graph or multigraph G can be defined as the polynomial P_D where D is some orientation of G, so P_G is unique up to sign. As a reference orientation of a (labeled) graph or multigraph G, with which other orientations will be compared, we usually take the **monotone orientation** of G, that is the orientation in which every edge is directed from the vertex with the smaller index to the vertex with the larger index. Hence for a graph G with $V(G) = \{v_1, \ldots, v_n\}$, we define

$$P_G = \prod_{v_i v_j \in E(G),\, i<j} (x_i - x_j). \tag{3.8}$$

If D is a multidigraph with $G_D = G$, then P_G is a divisor of P_D. Obviously, the graph polynomial P_G, in particular its zeros, encodes some information about the possible colorings of G. The basic observation is the following result, the proof of which is straightforward.

Observation 3.18 *Let G be a graph with vertex set $\{v_1, v_2, \dots, v_n\}$, and let L be a list-assignment of G, where $L(v)$ is a nonempty set of complex numbers for each vertex v of G. Moreover, let D be a multidigraph with $G_D = G$. Then G is not L-colorable if and only if P_G, respectively P_D, vanishes over the set $L(v_1) \times L(v_2) \times \cdots \times L(v_n)$.*

The classical concept of graph polynomials was studied already in the 19th century by Sylvester [984] and Petersen [803]. A result reminiscent of Hajós's theorem (see Theorem 4.3) was obtained by Kleitman and Lovász [699, 701].

Theorem 3.19 (Kleitman and Lovász) *Let G be a graph whose vertex set is $V = \{v_1, v_2, \dots, v_n\}$, and let k be an integer satisfying $2 \le k \le n$. Furthermore, let $\mathcal{I}$ be the ideal in $\mathbb{Z}[x_1, x_2, \dots, x_n]$ generated by all graph polynomials P_H of complete graphs H on k vertices of V. Then $\chi(G) \ge k$ if and only if P_G lies in $\mathcal{I}$.*

Let $\mathbb{F}$ be an arbitrary field, and let $\mathbb{F}'$ be a subring of $\mathbb{F}$. Every polynomial $P \in \mathbb{F}'[x_1, x_2, \dots, x_n]$ over $\mathbb{F}'$ in n variables $x_1, x_2, \dots, x_n$ can be written as a linear combination of monomials, that is,

$$P = \sum_{t_1, t_2, \dots, t_n} a_{t_1, t_2, \dots, t_n} x_1^{t_1} x_2^{t_2} \cdots x_n^{t_n}, \tag{3.9}$$

where the sum taken in $\mathbb{F}'$ is over all possible n-tuples $(t_1, t_2, \dots, t_n)$ of nonnegative integers. The coefficient $a_{t_1, t_2, \dots, t_n}$ belongs to the ring $\mathbb{F}'$ and is denoted by

$$\operatorname{coef}_{P, \mathbb{F}'}(t_1, t_2, \dots, t_n)$$

or briefly by $\operatorname{coef}_P(t_1, t_2, \dots, t_n)$ if it is clear that we refer to the ring $\mathbb{F}'$. For P to be a polynomial only a finite number of $\operatorname{coef}_P(t_1, t_2, \dots, t_n)$ are different from 0. If P is the zero polynomial, then we also write $P \equiv 0$. Recall that the degree $\deg(P)$ of P with $P \not\equiv 0$ is the maximum of $\sum_{i=1}^n t_i$ over all tuples $(t_1, t_2, \dots, t_n)$ for which $\operatorname{coef}_P(t_1, t_2, \dots, t_n) \neq 0$. If P is the zero polynomial, define $\deg(P) = -\infty$. The polynomial P may be considered as a polynomial in the variable x_i over the ring $\mathbb{F}[x_1, \dots, x_{i-1}, x_{i+1}, \dots, x_n]$ and we denote the degree of x_i in P by $\deg_{x_i}(P)$.

The next theorem due to Alon [48] provides some useful information about the zeros of multivariate polynomials; it shows, in particular, how Observation 3.18 can be exploited. First, we need the following simple lemma from [58].

Lemma 3.20 *Let $\mathbb{F}$ be a field, and let $P \in \mathbb{F}[x_1, x_2, \dots, x_n]$ be a polynomial. Suppose that, for $1 \le i \le n$, $\deg_{x_i}(P) \le t_i$ and L_i is a set of at least $t_i + 1$ distinct members of $\mathbb{F}$, where $t_i \in \mathbb{N}_0$. If $P(c_1, \dots, c_n) = 0$ for all tuples $(c_1, \dots, c_n) \in \prod_{i=1}^n L_i$, then $P \equiv 0$.*

Proof The proof is by induction on n. The base step $n = 1$ follows from the well known fact that a nonzero polynomial of degree at most t in one variable over a field can have at most t distinct zeros. For the induction step, let $n \geq 2$. We write P as a polynomial in x_n over $\mathbb{F}$, that is

$$P = \sum_{k=1}^{t_n} P_k \cdot x_n^k$$

with $P_k \in \mathbb{F}[x_1,x_2,\ldots,x_{n-1}]$ for $1 \leq k \leq t_n$. For each tuple $(c_1,c_2\ldots,c_{n-1}) \in \prod_{i=1}^{n-1} L_i$, the polynomial $P' \in \mathbb{F}[x_n]$ defined by

$$P' = \sum_{k=1}^{t_n} P_k(c_1,c_2,\ldots,c_{n-1}) \cdot x_n^k$$

has degree at most t_n and $P'(c_n) = 0$ for all $c_n \in L_n$, which yields $P' \equiv 0$. Consequently, $P_k(c_1,c_2,\ldots,c_{n-1}) = 0$ for all tuples $(c_1,c_2,\ldots,c_{n-1}) \in \prod_{i=1}^{n-1} L_i$. From the induction hypothesis it then follows that each polynomial P_k is the zero polynomial and, therefore, $P \equiv 0$, too. □

Theorem 3.21 (Combinatorial Nullstellensatz) *Let $\mathbb{F}$ be a field, let $P \in \mathbb{F}[x_1,x_2,\ldots,x_n]$ be a polynomial, and, for $i \in [1,n]$, let L_i be a nonempty subsets of $\mathbb{F}$ such that $|L_i| = d_i + 1$, and let $Q_i \in \mathbb{F}[x_i]$ be the polynomial defined by $Q_i = \prod_{c \in L_i}(x_i - c)$. Suppose that $P,Q_1,Q_2,\ldots,Q_n \in \mathbb{F}'[x_1,\ldots,x_n]$ for a subring $\mathbb{F}'$ of $\mathbb{F}$. Then there are polynomials $\overline{P},R_1,R_2,\ldots,R_n \in \mathbb{F}'[x_1,x_2,\ldots,x_n]$ such that*

$$P = \sum_{i=1}^{n} R_i Q_i + \overline{P}$$

and such that the following statements hold:

(a) $\deg(R_i) \leq \deg(P) - \deg(Q_i)$ *and* $\deg_{x_i}(\overline{P}) \leq d_i$ *for* $1 \leq i \leq n$.
(b) $\overline{P}(\underline{c}) = P(\underline{c})$ *for each tuple* $\underline{c} \in \prod_{i=1}^{n} L_i$.
(c) *If* $\overline{P}', R_1', R_2', \ldots, R_n' \in \mathbb{F}[x_1,x_2,\ldots,x_n]$ *such that* $P = \sum_{i=1}^{n} R_i' Q_i + \overline{P}'$ *and* $\deg_{x_i}(\overline{P}') \leq d_i$ *for* $1 \leq i \leq n$, *then* $\overline{P}' = \overline{P}$.
(d) $P(\underline{c}) = 0$ *for all tuples* $\underline{c} \in \prod_{i=1}^{n} L_i$ *if and only if* $\overline{P} \equiv 0$.
(e) *If* $\deg(P) = \sum_{i=1}^{n} d_i$, *then* $\mathrm{coef}_{\overline{P}}(d_1,d_2,\ldots,d_n) = \mathrm{coef}_P(d_1,d_2,\ldots,d_n)$.

Proof First, we claim that there exists polynomials $P_0 = P, P_1, P_2, \ldots, P_n$ and $R_1, R_2, \ldots, R_n$ in $\mathbb{F}'[x_1,x_2,\ldots,x_n]$ satisfying, for each $i \in [1,n]$, the following conditions:

$$P_{i-1} = R_i Q_i + P_i,$$

and, moreover,

$$\deg(P_i) \leq \deg(P), \deg(R_i Q_i) \leq \deg(P) \text{ and } \deg_{x_j}(P_i) \leq d_j \text{ for } 1 \leq j \leq i.$$

For the proof we apply induction on i. If $i=0$ then $P_0 = P$. For the induction step, let $i \geq 1$. The polynomial $Q_i = \prod_{c \in L_i}(x_i - c) \in \mathbb{F}'[x_i]$ is univariate and monic with degree $d_i + 1$ in x_i, that is,

$$Q_i = x_i^{d_i+1} + \alpha_{d_i} x_i^{d_i} + \cdots + \alpha_1 x_1 + \alpha_0$$

with $\alpha_{d_i}, \alpha_{d_i-1}, \ldots, \alpha_0 \in \mathbb{F}'$. Clearly, $Q_i \not\equiv 0$ and we can apply the well known division algorithm for polynomials, see for instance [508, page 159]. If P_{i-1} has degree at most d_i in x_i, then the division algorithm terminates with the co-divisor $R_i \equiv 0$ and the remainder $P_i = P_{i-1}$. Otherwise,

$$P_{i-1} = a_m x_i^m + a_{m-1} x_i^{m-1} + \cdots + a_1 x_i^1 + a_0$$

is a polynomial of degree $m \geq d_i + 1$ in x_i with

$$a_m, a_{m-1}, \ldots, a_0 \in \mathbb{F}'[x_1, \ldots, x_{i-1}, x_{i+1}, \ldots, x_n].$$

Then the first step of the division algorithm yields

$$P_{i-1} = a_m x_i^{m-d_i-1} Q_i + P'_{i-1},$$

where

$$P'_{i-1} = a_{m-1} x_i^{m-1} + \cdots + a_0 - a_m(\alpha_{d_i} x_i^{m-1} + \cdots + \alpha_0 x_i^{d_i})$$

is a polynomial from $\mathbb{F}'[x_1, \ldots, x_n]$ having degree at most $m-1$ in x_i and we can continue the division algorithm with P'_{i-1} and Q_i. Hence after a finite number of steps we obtain two polynomials $R_i, P_i \in \mathbb{F}'[x_1, \ldots, x_n]$, such that

$$P_{i-1} = R_i Q_i + P_i,$$

and such that the following degree conditions hold: $\deg_{x_i}(P_i) \leq d_i = \deg_{x_i}(Q_i) - 1$, $\deg(R_i Q_i) \leq \deg(P_{i-1})$, $\deg_{x_j}(P_i) \leq \deg_{x_j}(P_{i-1})$ for each $j \neq i$, and, $\deg(P_i) \leq \deg(P_{i-1})$. From the induction hypothesis we then conclude that

$$\deg(P_i) \leq \deg(P), \deg(R_i Q_i) \leq \deg(P) \text{ and } \deg_{x_j}(P_i) \leq d_j \text{ for } 1 \leq j \leq i.$$

This proves the claim. Hence, with $\overline{P} = P_n$ we obtain that

$$P = \sum_{i=1}^{n} R_i Q_i + \overline{P},$$

where $\deg(R_i Q_i) \leq \deg(P)$ and $\deg_{x_i}(P_n) \leq d_i$ for each $i \in [1,n]$. This implies, in particular, that $\deg(R_i) \leq \deg(P_i) - \deg(Q_i)$ for each $i \in [1,n]$. Thus (a) is proved.

For the proof of (b) and (c) assume that $\overline{P}', R'_1, R'_2, \ldots, R'_n$ are polynomials in $\mathbb{F}[x_1, x_2, \ldots, x_n]$ such that $P = \sum_{i=1}^{n} R'_i Q_i + \overline{P}'$ and $\deg_{x_i}(\overline{P}') \leq d_i$ for $1 \leq i \leq n$. If $\underline{c} = (c_1, c_2, \ldots, c_n) \in \prod_{i=1}^{n} L_i$, then $Q_i(c_i) = 0$ for each $i \in [1,n]$ and, therefore,

$$\overline{P}(c_1,c_2,\ldots,c_n) = P(c_1,c_2,\ldots,c_n) = \overline{P}'(c_1,c_2,\ldots,c_n).$$

Hence the polynomial $\tilde{P} = \overline{P} - \overline{P}'$ satisfies $\tilde{P}(\underline{c}) = 0$ for all tuples $\underline{c} \in \prod_{i=1}^n L_i$. Since $\deg_{x_i}(\tilde{P}) \le d_i$ for each $i \in [1,n]$, Lemma 3.20 then implies that $\tilde{P} \equiv 0$ and hence $\overline{P}' = \overline{P}$. This proves (b) and (c).

If $\overline{P} \equiv 0$, then (b) implies that $P(\underline{c}) = 0$ for all tuples $\underline{c} \in \prod_{i=1}^n L_i$. Conversely, if $P(\underline{c}) = 0$ for all tuples $\underline{c} \in \prod_{i=1}^n L_i$, then (b) implies that the same is true for the polynomial $\overline{P} \in \mathbb{F}[x_1,x_2,\ldots,x_n]$. Since $\deg_{x_i}(\overline{P}) \le d_i = |L_i| - 1$ for each $i \in [1,n]$, it then follows from Lemma 3.20 that $\overline{P} \equiv 0$. Hence (d) is proved.

To see (e), assume that $\deg(P) = \sum_{i=1}^n d_i$. Since $\deg(Q_i) = \deg_{x_i}(Q_i) = d_i + 1$ and $\deg(R_i Q_i) \le \deg(P)$ (by (1)), we easily conclude that $\mathrm{coef}_{R_i Q_i}(d_1,\ldots,d_n) = 0$ for $1 \le i \le n$. Since $P = \sum_{i=1}^n R_i Q_i + \overline{P}$, this implies that $\mathrm{coef}_{\overline{P}}(d_1,\ldots,d_n) = \mathrm{coef}_P(d_1,\ldots,d_n)$. This proves (e) and completes the proof of the theorem. □

Comment: The polynomial $\overline{P}$ is said to be the **remainder** of the polynomial P over $\mathbb{F}'[x_1,x_2,\ldots,x_n]$ with respect to the ideal generated by the polynomials $Q_1,Q_2,\ldots,Q_n \in \mathbb{F}'[x_1,x_2,\ldots,x_n]$.

Let D be a multidigraph with vertex set $V(D) = \{v_1,v_2,\ldots,v_n\}$. For every tuple $(d_1,d_2,\ldots,d_n)$ of nonnegative integers define

$$\mathrm{coef}_D(d_1,d_2,\ldots,d_n) = \mathrm{coef}_{P,\mathbb{Z}}(d_1,d_2,\ldots,d_n),$$

where $P = P_D$, and correspondingly for a graph or multigraph G. For an edge set $F \subseteq E(D)$, we introduce two new multidigraphs denoted by $D(F)$ and D/F, respectively. While the first multidigraph is obtained from D by deleting all edges from $E(D) \setminus F$, the second multidigraph is obtained from D by reversing the orientation of the edges in F. For the first multidigraph we obtain $V(D(F)) = V(D)$, $E(D(F)) = F$, and

$$v^{\pm}_{D(F)}(e) = v^{\pm}_D(e) \text{ for all edges } e \in F.$$

For the second multidigraph we obtain $V(D/F) = V(D)$, $E(D/F) = E(D)$,

$$v^{\pm}_{D/F}(e) = v^{\pm}_D(e) \text{ if } e \in E(D) \setminus F, \text{ and } v^{\pm}_{D/F}(e) = v^{\mp}_D(e) \text{ if } e \in F.$$

For every tuple $\underline{d} = (d_1,d_2,\ldots,d_n)$ of nonnegative integers, we define the following subsets of the power set $2^{E(D)}$:

$$\mathcal{E}^e_D(\underline{d}) = \{F \subseteq E(D) \,\big|\, |F| \equiv 0 \,(\mathrm{mod}\, 2),\ d^+_{D/F}(v_i) = d_i \text{ for } 1 \le i \le n\},$$

and

$$\mathcal{E}^o_D(\underline{d}) = \{F \subseteq E(D) \,\big|\, |F| \equiv 1 \,(\mathrm{mod}\, 2),\ d^+_{D/F}(v_i) = d_i \text{ for } 1 \le i \le n\}.$$

The next two results provide combinatorial interpretations for the coefficients of the polynomial P_D. Furthermore, the next result relates list colorability to coefficients in the polynomial P_D. Both results were obtained by Alon and Tarsi [58].

Lemma 3.22 (The Alon–Tarsi Polynomial Method) *Let D be a multidigraph with vertex set $V(D) = \{v_1, v_2, \dots, v_n\}$ and let $m = |E(D)|$. Then the polynomial $P_D \in \mathbb{Z}[x_1, x_2, \dots, x_n]$ is homogeneous of degree m and*

$$\mathrm{coef}_D(d_1, d_2, \dots, d_n) = |\mathcal{E}_D^e(d_1, d_2, \dots, d_n)| - |\mathcal{E}_D^o(d_1, d_2, \dots, d_n)|$$

for each tuple $(d_1, d_2, \dots, d_n)$ of nonnegative integers. Furthermore, if

$$\mathrm{coef}_D(d_1, d_2, \dots, d_n) \neq 0$$

for some tuple $(d_1, d_2, \dots, d_n)$ of nonnegative integers, then $d_1 + d_2 + \dots + d_n = m$ and the graph G_D is f-list-colorable, where $f(v_i) = d_i + 1$ for each $i \in [1, n]$.

Proof In expanding the product in (3.7), for every edge $e \in E(D)$, we can either choose $x_{v^+(e)}$ or $-x_{v^-(e)}$. This yields

$$\begin{aligned}
P_D &= \prod_{e \in E(D)} (x_{v^+(e)} - x_{v^-(e)}) \\
&= \sum_{F \subseteq E(D)} (-1)^{|F|} \prod_{e \in F} x_{v_D^-(e)} \prod_{e \in E(D) \setminus F} x_{v_D^+(e)} \\
&= \sum_{F \subseteq E(D)} (-1)^{|F|} \prod_{e \in E(D)} x_{v_{D/F}^+(e)} \\
&= \sum_{d_1, d_2, \dots, d_n \geq 0} (|\mathcal{E}_D^e(d_1, d_2, \dots, d_n)| - |\mathcal{E}_D^o(d_1, d_2, \dots, d_n)|) \prod_{i=1}^{n} x_i^{d_i}.
\end{aligned}$$

This shows that

$$\mathrm{coef}_D(d_1, d_2, \dots, d_n) = |\mathcal{E}_D^e(d_1, d_2, \dots, d_n)| - |\mathcal{E}_D^o(d_1, d_2, \dots, d_n)|.$$

Now assume that $\mathrm{coef}_D(d_1, d_2, \dots, d_n) \neq 0$ for some tuple $(d_1, d_2, \dots, d_n)$. Then, clearly, $d_1 + d_2 + \dots + d_n = |E(D)| = m$. Let L be an arbitrary list-assignment for $G = G_D$, where $|L(v_i)| = d_i + 1$ for each $i \in [1, n]$. By Theorem 3.21, statements (d) and (e), it then follows that $P_D(c_1, \dots, c_n) \neq 0$ for some tuple $(c_1, \dots, c_n) \in \prod_{i=1}^n L(v_i)$. Then (see Observation 3.18) G is L-colorable. This completes the proof. □

A multidigraph D is said to be **Eulerian** if the in-degree and out-degree are equal at each vertex of D, regardless of the number of components of D. With a multidigraph D we associate the following sets:

$$\mathcal{E}(D) = \{F \subseteq E(D) \,\big|\, D(F) \text{ is Eulerian}\},$$

$$\mathcal{E}^e(D) = \{F \in \mathcal{E}(D) \mid |F| \equiv 0 \,(\mathrm{mod}\ 2)\}$$

and

$$\mathcal{E}^o(D) = \{F \in \mathcal{E}(D) \mid |F| \equiv 1 \,(\mathrm{mod}\ 2)\}.$$

Furthermore, we introduce the following numbers:

$$\varepsilon(D) = |\mathcal{E}(D)|,\ \varepsilon_o(D) = |\mathcal{E}^o(D)| \text{ and } \varepsilon_e(D) = |\mathcal{E}^e(D)|.$$

Lemma 3.23 *Let D be a multidigraph with $V(D) = \{v_1, v_2, \ldots, v_n\}$ and let $d_D^+(v_i) = d_i$ for $i \in [1,n]$. Then $\mathcal{E}_D^e(d_1, d_2, \ldots, d_n) = \mathcal{E}^e(D)$, $\mathcal{E}_D^o(d_1, d_2, \ldots, d_n) = \mathcal{E}^o(D)$, and* $\mathrm{coef}_D(d_1, d_2, \ldots, d_n) = \varepsilon_e(D) - \varepsilon_o(D)$.

Proof For a set $F \subseteq E(D)$ and a vertex $v \in V(D)$, we obtain

$$d_D^+(v) = d_{D/F}^+(v) + (d_{D(F)}^-(v) - d_{D(F)}^+(v)).$$

Consequently, $\mathcal{E}_D^e(d_1, \ldots, d_n) = \mathcal{E}^e(D)$ and $\mathcal{E}_D^o(d_1, \ldots, d_n) = \mathcal{E}^o(D)$. Hence, by Lemma 3.22, $\mathrm{coef}_D(d_1, \ldots, d_n) = \varepsilon_e(D) - \varepsilon_o(D)$. □

Combining Lemma 3.22 with Lemma 3.23, we obtain the following result du to Alon and Tarsi [58].

Theorem 3.24 (Alon and Tarsi) *Let G be a graph, let D be a multidigraph with $G = G_D$, and let $f(v) = d_D^+(v) + 1$ for every vertex $v \in V(G)$. If $\varepsilon_e(D) \neq \varepsilon_o(D)$, then G is f-list-colorable, and as a consequence $\chi_\ell(G) \leq \Delta^+(D) + 1$.*

There are several classes $\mathcal{D}$ of multidigraphs satisfying $\varepsilon_e(D) \neq \varepsilon_o(D)$ for each $D \in \mathcal{D}$, including acyclic multidigraphs, multidigraphs without odd directed cycles and orientations of bipartite multigraphs. However, all these multidigraphs are also kernel-perfect. For a multidigraph D, in general, it seems to be a difficult task to decide whether $\varepsilon_e(D) \neq \varepsilon_o(D)$ holds, or not. As we shall see in the next section, this task becomes easier if the multidigraph D itself is Eulerian.

Let us conclude this section with a direct application of Lemma 3.22 and the Combinatorial Nullstellensatz due to Akbari, Mirrokni, and Sadjad [25].

Theorem 3.25 (Akbari, Mirrokni, and Sadjad) *Let G be a graph with vertex set $V = \{v_1, v_2, \ldots, v_n\}$ and m edges, let $(d_1, d_2, \ldots, d_n)$ be a sequence of nonnegative integers whose sum is equal to m, and let $f : V \to \mathbb{N}$ be a function where $f(v_i) = d_i + 1$ for each $i \in [1,n]$. If G is uniquely L-colorable for some f-list-assignment L of G, then G is f-list-colorable.*

Proof Suppose that L is an f-list-assignment for G such that G is uniquely L-colorable, i.e., there is only one L-coloring φ of G. Let $\underline{a} = (\varphi(v_1), \ldots, \varphi(v_n))$. If D is the monotone orientation of G, then each tuple $\underline{c} \in \prod_{i=1}^n L(v_i)$ satisfies $P_D(\underline{c}) = 0$ if and only if $\underline{c} \neq \underline{a}$. For $i \in [1,n]$, let $Q_i \in \mathbb{Z}[x_i]$ be the polynomial defined by $Q_i = \prod_{c \in L_i}(x_i - c)$. By Theorem 3.21, there exists the remainder $\overline{P}$ of P_D over $\mathbb{Z}[x_1, \ldots, x_n]$ with respect to the ideal generated by $Q_1, Q_2, \ldots, Q_n$. The remainder satisfies, in particular, that $\overline{P}(\underline{c}) = P_D(\underline{c})$ for each tuple $\underline{c} \in \prod_{i=1}^n L(v_i)$ and $\deg_{x_i}(\overline{P}) \leq d_i$ for each $i \in [1,n]$. Let $Q \in \mathbb{Z}[x_1, x_2, \ldots, x_n]$ be the polynomial defined by

$$Q = \prod_{i=1}^{n} \prod_{c \in L(v_i) \setminus \{a_i\}} (x_i - c).$$

Then the polynomial

$$Q' = \overline{P} - \frac{\overline{P}(\underline{a})}{Q(\underline{a})} Q$$

satisfies $Q'(\underline{c}) = 0$ for each tuple $\underline{c} \in \prod_{i=1}^{n} L(v_i)$ and $\deg_{x_i}(Q') \leq d_i$ for each $i \in [1,n]$. Hence Lemma 3.20 implies that $Q' \equiv 0$ and, therefore,

$$\overline{P} = \frac{\overline{P}(\underline{a})}{Q(\underline{a})} Q.$$

Consequently, $\mathrm{coef}_{\overline{P}}(d_1, d_2, \ldots, d_n) \neq 0$. Since $d_1 + d_2 + \cdots + d_n = m = \deg(P_D)$, it then follows from Theorem 3.21(e) that $\mathrm{coef}_D(d_1, d_2, \ldots, d_n) \neq 0$. Hence, by Lemma 3.22, G is f-list-colorable. □

3.6 Eulerian Digraphs

Let D be a multidigraph, let A, B be disjoint subsets of $E(D)$, and let U, W be disjoint subsets of $V(D)$. For a vertex v, define $E_D^+(v) = \{e \in E(D) \mid v_D^+(e) = v\}$ and $E_D^-(v) = \{e \in E(D) \mid v_D^-(e) = v\}$. Then let $\mathcal{E}(D, A, B, U, W)$ be the set of all edge sets $F \in \mathcal{E}(D)$ satisfying: $A \subseteq F$, $B \cap F = \emptyset$, $E_D^+(v) \subseteq F$ for each vertex $v \in U$, and $E_D^+(v) \cap F = \emptyset$ for each vertex $v \in W$. Eventually, let $\varepsilon(D, A, B, U, W) = |\mathcal{E}(D, A, B, U, W)|$. If $A = \{e_1, \ldots, e_p\}$, $B = \{f_1, \ldots, f_q\}$, $U = \{u_1. \ldots, u_s\}$, and $W = \{w_1, \ldots, w_t\}$, then instead of $\mathcal{E}(D, A, B, U, W)$ we also write

$$\mathcal{E}(D, e_1, \ldots, e_p, \overline{f}_1, \ldots, \overline{f}_q, u_1, \ldots, u_s, \overline{w}_1, \ldots, \overline{w}_t)$$

(regardless of the order of the edges and vertices), and correspondingly for the number $\varepsilon(D, A, B, U, W)$. The proof of the next result is straightforward.

Lemma 3.26 *Let D be an Eulerian multidigraph with $m \geq 1$ edges. Furthermore, define a mapping h by setting $h(F) = E(D) \setminus F$ for each $F \in \mathcal{E}(D)$. Then the following statements hold:*

(a) *The mapping h is a bijection from $\mathcal{E}(D)$ onto itself and $\varepsilon(D) \equiv 0 \pmod 2$.*
(b) *Let A, B be disjoint subsets of $E(D)$ and let U, W be disjoint subsets of $V(D)$. Then h induces a bijection from $\mathcal{E}(D, A, B, U, W)$ onto $\mathcal{E}(D, B, A, W, U)$, and $\varepsilon(D, A, B, U, W) = \varepsilon(D, B, A, W, U)$.*
(c) *If m is odd, then h induces a bijection from $\mathcal{E}^e(D)$ onto $\mathcal{E}^o(D)$, and as a consequence $\varepsilon_e(D) = \varepsilon_o(D)$.*
(d) *If m is even, then h induces a bijection from $\mathcal{E}^e(D)$ onto itself as well as a bijection from $\mathcal{E}^o(D)$ onto itself, and $\varepsilon_e(D) \equiv \varepsilon_o(D) \equiv 0 \pmod 2$*
(e) *If m is even and $\varepsilon(D) \equiv 2 \pmod 4$, then $\varepsilon_e(D) \neq \varepsilon_o(D)$.*

Combining Theorem 3.24 and Lemma 3.26(e), we obtain the following result about list colorings of graphs.

Corollary 3.27 *Let G be a graph, let D be a multidigraph with $G = G_D$, and let $f(v) = d_D^+(v) + 1$ for all $v \in V(D)$. Suppose that D is an Eulerian multidigraph with an even number of edges and with $\varepsilon(D) \equiv 2 \pmod 4$. Then G is f-list colorable, and as a consequence $\chi_\ell(G) \leq \Delta^+(D) + 1$.*

Let G be a multigraph. A coloring of G is a coloring of its underlying graph H, and so $\chi(G) = \chi(H)$ and $\chi_\ell(G) = \chi_\ell(H)$. An orientation D of G such that D is an Eulerian digraph is said to be an **Eulerian orientation** of G. It is well known that G has an Eulerian orientation if and only if its vertices all have even degree. Corollary 3.27 implies, in particular, the following result.

Corollary 3.28 *Let G be a multigraph with an even number of edges. If G has an Eulerian orientation D such that $\varepsilon(D) \equiv 2 \bmod 4$, then $\chi_\ell(G) \leq \frac{1}{2}\Delta(G) + 1$.*

Almost as soon as Theorem 3.24 became available, Fleischner and Stiebitz [379] used it to solve a problem raised by Du, Hsu, and Wang in [308], as well as a strengthening of it suggested by Paul Erdős.

Theorem 3.29 (The Cycle-Plus-Triangles Theorem) *Let n be a positive integer and let G be a 4-regular multigraph on $3n$ vertices. Suppose that G has a decomposition into a Hamilton cycle and n pairwise vertex disjoint triangles. Then $\chi(G) = \chi_\ell(G) = 3$.*

The proof of Theorem 3.29 is based on Corollary 3.28. Clearly, each Eulerian orientation of a **cycle-plus-triangles graph**, that is, a graph satisfying the assumption of Theorem 3.29, has maximum out-degree 2 as well as an even number of edges. Hence, it suffices to show that every cycle-plus-triangles graph has an Eulerian orientation D such that $\varepsilon(D) \equiv 2 \bmod 4$.

We denote the **symmetric difference** of two sets A and B by $A \oplus B$, that is $A \oplus B = (A \setminus B) \cup (B \setminus A) = (A \cup B) \setminus (A \cap B)$. Observe that every multidigraph D satisfies $D/\varnothing = D$ and $D/F/F' = D/(F \oplus F')$ for each sets $F, F' \subseteq E(D)$. For the proof of Theorem 3.29, we need the following result.

Lemma 3.30 *Let D be an Eulerian digraph and let C be a directed cycle in D with at least two edges. Furthermore, define a mapping h by setting $h(F) = F \oplus E(C)$ for each $F \in \mathcal{E}(D)$. Then the following statements hold:*

(a) *$D/E(C)$ is an Eulerian digraph, and $d_{D/E(C)}^+(v) = d_D^+(v)$ for each $v \in V(D)$.*
(b) *h is a bijection from $\mathcal{E}(D)$ onto $\mathcal{E}(D/E(C))$, and $\varepsilon(D) = \varepsilon(D/E(C))$.*

Proof Statement (a) is obviously true. To prove (b), let $F \in \mathcal{E}(D)$. Then $D(F)$ is an Eulerian digraph and we need to show that $D' = (D/E(C))(h(F))$ is an Eulerian digraph, too. Let v be a vertex of D'. If $v \notin V(C)$, then it follows from the definition of $h(F)$ that $d_{D'}^+(v) = d_{D(F)}^+(v) = d_{D(F)}^-(v) = d_{D'}^+(v)$. Now assume that $v \in V(C)$. Then there is exactly one edge $e^+ \in E_C^+(v)$ and exactly one edge $e^- \in E_C^-(v)$. Let $F' = F \cap \{e^+, e^-\}$. If $|F'| = 1$, then

$$d_{D'}^+(v) = d_{D(F)}^+(v) = d_{D(F)}^-(v) = d_{D'}^-(v).$$

If $|F'| = 0$, then

$$d^+_{D'}(v) = d^+_{D(F)}(v) + 1 = d^-_{D(F)}(v) + 1 = d^-_{D'}(v).$$

Eventually, if $|F'| = 2$, then

$$d^+_{D'}(v) = d^+_{D(F)}(v) - 1 = d^-_{D(F)}(v) - 1 = d^-_{D'}(v).$$

This shows that D' is an Eulerian digraph. Then it is easy to show that h is a bijection, which completes the proof of (b). □

Clearly, every cycle-plus-triangles graph G contains a triangle and has an even number of edges. Hence $\chi(G) \geq 3$ and Theorem 3.29 is an immediate consequence of Corollary 3.28 and the following result.

Theorem 3.31 *Let D be Eulerian digraph with at least two vertices. Suppose D has a decomposition into a directed Hamilton cycle and $n \geq 0$ pairwise vertex disjoint directed triangles. Then $\varepsilon(D) \equiv 2$ mod 4.*

Proof We apply induction on n. The base step $n = 0$ is evident, since then D is a directed cycle, which implies $\mathcal{E}(D) = \{\varnothing, E(D)\}$ and hence $\varepsilon(D) = 2$. For the induction step, let $n \geq 1$. Let D be an Eulerian digraph which has a decomposition into a directed Hamilton cycle, say C, and n pairwise vertex disjoint directed triangles. If e is an edge in a certain digraph directed from the vertex u to the vertex v, we briefly write $e = (u, v)$.

Let T be one directed triangle of the decomposition of D. Then T consists of three vertices v_1, v_2, v_3 and three edges e_1, e_2, e_2, where $e_1 = (v_3, v_2), e_2 = (v_1, v_3)$, and $e_3 = (v_2, v_1)$. Define

$$\begin{aligned}\varepsilon'(D) = \ &\varepsilon(D, \overline{e}_1, e_2, e_3) + \varepsilon(D, e_1, \overline{e}_2, \overline{e}_3) + \varepsilon(D, e_1, \overline{e}_2, e_3) + \\ &\varepsilon(D, \overline{e}_1, e_2, \overline{e}_3) + \varepsilon(D, e_1, e_2, \overline{e}_3) + \varepsilon(D, \overline{e}_1, \overline{e}_2, e_3).\end{aligned}$$

It then follows that

(a) $\varepsilon(D) = \varepsilon(D, e_1, e_2, e_3) + \varepsilon(D, \overline{e}_1, \overline{e}_2, \overline{e}_3) + \varepsilon'(D)$.

Let $D' = D - E(T)$. Since $E(T) \in \mathcal{E}(D)$, it is easy to see that $\varepsilon(D') = \varepsilon(D, \overline{e}_1, \overline{e}_2, \overline{e}_3) = \varepsilon(D, e_1, e_2, e_3)$. Then the induction hypothesis implies that $\varepsilon(D') \equiv 2$ mod 4 and hence $\varepsilon(D, \overline{e}_1, \overline{e}_2, \overline{e}_3) \equiv \varepsilon(D, e_1, e_2, e_3) \equiv 2$ mod 4. Together with (a) this implies that

(b) $\varepsilon(D) \equiv \varepsilon'(D) \pmod 4$.

To complete the proof, it suffices to show that $\varepsilon'(D) \equiv 2$ mod 4. To this end, we shall construct three new digraphs D'_1, D'_2, and D'_3 to which the induction hypothesis can be applied.

For $i = 1, 2, 3$, let v_i^+ be the successor of v_i on C, and let v_i^- be the predecessor of v_i on C, that is, there are edges $f_i, g_i \in E(C)$ such that $f_i = (v_i^-, v_i)$ and $g_i = (v_i, v_i^+)$. We may assume that C is directed from v_1 to v_2, from v_2 to v_3, and from v_3 to v_1

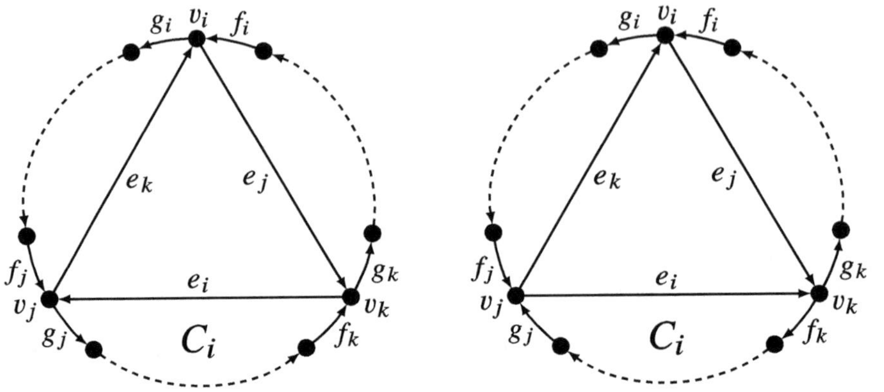

Fig. 3.1 The graph D (on the left side) and the graph $D_i = D/E(C_i)$ (on the right side).

(by Lemma 3.30). Then, for $i = 1,2,3$, there are unique directed cycles C_i such that $e_i \in E(C_i) \subseteq E(C) \cup \{e_i\}$. Clearly, the cycles C_1, C_2 and C_3 are edge disjoint and $E(C_1) \cup E(C_2) \cup E(C_3) = E(C) \cup E(T)$.

From now on, let (i, j, k) denote one of the triples $(1,2,3),(2,3,1)$, or $(3,1,2)$. Define $D_i = D/E(C_i)$, that is, D_i is the digraph obtained from D by reversing the orientation of C_i, see Figure 3.1. Now construct a digraph D_i' from D_i by splitting away the edges f_j, e_k, e_j and f_k. Let v_i', v_j', v_k' denote the three new vertices which arises by splitting each v_ℓ with $\ell \in \{i, j, k\}$ into two vertices of degree 2 in D_i'. The digraph D_i' arises from D_i by directing the edges f_j, e_k, e_j and f_k in D_i' so that $f_j = (v_j^-, v_j'), e_k = (v_j', v_i'), e_j = (v_i', v_k')$ and $f_k = (v_k', v_k^-)$, and by leaving all other incidences unaltered (see Figure 3.2).

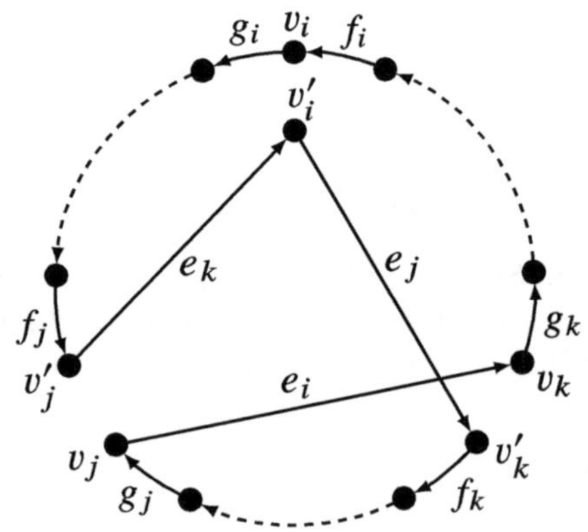

Fig. 3.2 The digraph D_i' obtained from D_i.

Obviously, D_i' is an Eulerian digraph which admits an edge disjoint decomposition into a Hamilton cycle and $n-1$ pairwise vertex disjoint directed triangles. It then follows from the induction hypothesis that

(c) $\varepsilon(D_1') \equiv \varepsilon(D_2') \equiv \varepsilon(D_3') \equiv 2 \pmod 4$,

which yields

(d) $\varepsilon(D_1') + \varepsilon(D_2') + \varepsilon(D_3') \equiv 2 \pmod 4$.

Claim 3.31.1 *Let* $(i,j,k) \in \{(1,2,3),(2,3,1),\ (3,1,2)\}$ *and* $\varepsilon_i = \varepsilon(D_i')$. *Then the following statements hold:*

(e) $\varepsilon_i = \varepsilon(D_i', e_i, e_j, e_k) + \varepsilon(D_i', \overline{e}_i, \overline{e}_j, \overline{e}_k) + \varepsilon(D_i', \overline{e}_i, e_j, e_k) + \varepsilon(D_i', e_i, \overline{e}_j, \overline{e}_k)$.
(f) $\varepsilon(D_i', e_i, e_j, e_k) = \varepsilon(D, \overline{e}_i, e_j, e_k)$.
(g) $\varepsilon(D_i', \overline{e}_i, \overline{e}_j, \overline{e}_k) = \varepsilon(D, e_i, \overline{e}_j, \overline{e}_k)$.
(h) $\varepsilon(D_i', e_i, \overline{e}_j, \overline{e}_k) = \varepsilon(D', \overline{v}_j, v_k) = \varepsilon(D', v_i, \overline{v}_j, \overline{v}_k) + \varepsilon(D', \overline{v}_i, \overline{v}_j, v_k)$.
(i) $\varepsilon(D_i', \overline{e}_i, e_j, e_k) = \varepsilon(D', v_j, \overline{v}_k) = \varepsilon(D', v_i, v_j, \overline{v}_k) + \varepsilon(D', \overline{v}_i, v_j, \overline{v}_k)$.

Proof: Statement (e) follows from the fact that if $F \in \mathcal{E}(D_i')$ then either both edges e_j and e_k are contained in F or both are not contained in F.

For the proof of (f), we show that there exists a bijection from $\mathcal{E} = \mathcal{E}(D, \overline{e}_i, e_j, e_k)$ onto $\mathcal{E}' = \mathcal{E}(D_i', e_i, e_j, e_k)$. It follows from the constructions of D_i and D_i' that

$$\mathcal{E}' = \mathcal{E}(D_i', g_j, e_i, g_i, f_j, e_k, e_j, f_k) = \mathcal{E}(D_i, g_j, e_i, g_i, f_j, e_k, e_j, f_k),$$

and

$$\mathcal{E} = \mathcal{E}(D, \overline{g}_j, \overline{e}_i, \overline{f}_k, f_j, e_k, e_j, g_k).$$

Now, define a mapping h by $h(F) = F \oplus E(C_i)$ for all $F \in \mathcal{E}(D)$. Then Lemma 3.30 implies that h is a bijection from $\mathcal{E}(D)$ onto $\mathcal{E}(D_i)$. From the definition of h it also follows that $h(F) \in \mathcal{E}'$ whenever $F \in \mathcal{E}$. Hence h induces a bijection from $\mathcal{E}$ onto $\mathcal{E}'$. This proves (f). Statement (g) is an immediate consequence of Lemma 3.26(b) and (f).

For the proof of (h), let $\mathcal{E}_1' = \mathcal{E}(D_i', e_i, \overline{e}_j, \overline{e}_k)$ and $\mathcal{E}_1 = \mathcal{E}(D, \overline{v}_j, v_k)$. Again from the constructions of D_i and D_i' it follows that

$$\mathcal{E}_1' = \mathcal{E}(D_i', g_j, e_i, g_k, \overline{f}_j, \overline{e}_j, \overline{e}_k, \overline{f}_k) = \mathcal{E}(D_i, g_j, e_i, g_k, \overline{f}_j, \overline{e}_j, \overline{e}_k, \overline{f}_k),$$

and

$$\mathcal{E}_1 = \mathcal{E}(D', \overline{f}_j, \overline{g}_j, f_k, g_k) = \mathcal{E}(D, \overline{e}_i, \overline{e}_j, \overline{e}_k, \overline{f}_j, \overline{g}_j, f_k, g_k).$$

In the same way as in the proof of (h) it now follows from Lemma 3.30 that there exist a bijection from $\mathcal{E}_1$ onto $\mathcal{E}_1'$, which implies that $\varepsilon(D_i', e_i, \overline{e}_j, \overline{e}_k) = \varepsilon(D', \overline{v}_j, v_k)$. Since v_i is a vertex of degree 2 in D', the second equation in (h) holds, too. This proves (h). Finally, using Lemma 3.26(b) and (h) we obtain (i), which completes the proof of the claim △

To finish the proof, let $\varepsilon = \varepsilon(D_1') + \varepsilon(D_2') + \varepsilon(D_3')$. It follows from Claim 3.31.1 that $\varepsilon = \varepsilon'(D) + \varepsilon'$, where

$$\begin{aligned}\varepsilon' = {} & \varepsilon(D',v_1,\overline{v}_2,v_3)+\varepsilon(D',\overline{v}_1,\overline{v}_2,v_3)+\varepsilon(D',v_1,v_2,\overline{v}_3)+\\ & \varepsilon(D',\overline{v}_1,v_2,\overline{v}_3)+\varepsilon(D',v_1,v_2,\overline{v}_3)+\varepsilon(D',v_1,\overline{v}_2,\overline{v}_3)+\\ & \varepsilon(D',\overline{v}_1,v_2,v_3)+\varepsilon(D',\overline{v}_1,\overline{v}_2,v_3)+\varepsilon(D',\overline{v}_1,v_2,v_3)+\\ & \varepsilon(D',\overline{v}_1,v_2,\overline{v}_3)+\varepsilon(D',v_1,\overline{v}_2,v_3)+\varepsilon(D',v_1,\overline{v}_2,\overline{v}_3).\end{aligned}$$

Using Lemma 3.26, we obtain that

$$\begin{aligned}\varepsilon' &= 2(\varepsilon(D',v_1,\overline{v}_2,v_3)+\varepsilon(D',\overline{v}_1,v_2,\overline{v}_3)+\varepsilon(D',v_1,v_2,\overline{v}_3)+\\ &\qquad \varepsilon(D',\overline{v}_1,\overline{v}_2,v_3)+\varepsilon(D',\overline{v}_1,v_2,v_3)+\varepsilon(D',v_1,\overline{v}_2,\overline{v}_3))\\ &= 4(\varepsilon(D',v_1,\overline{v}_2,v_3)+\varepsilon(D',v_1,v_2,\overline{v}_3)+\varepsilon(D',\overline{v}_1,v_2,v_3)).\end{aligned}$$

Consequently, $\varepsilon' \equiv 0 \pmod 4$ and hence $\varepsilon \equiv \varepsilon'(D) \pmod 4$. By (d), it then follows that $\varepsilon'(D) \equiv 2 \pmod 4$. Eventually, (b) implies that $\varepsilon(D) \equiv 2 \pmod 4$. □

Obviously, if G is a graph or multigraph and D is a reference orientation of G, then the function $F \in 2^{E(D)} \mapsto D/F$ is a bijection from the power set $2^{E(D)}$ onto the set of all orientations of G. For a graph or multigraph G, let $\mathrm{euo}(G)$ denote the number of Eulerian orientations of G.

Remark 3.32 *Let G be a graph or multigraph and let D be an Eulerian orientation of G. Then the function $F \in \mathcal{E}(D) \mapsto D/F$ is a bijection from $\mathcal{E}(D)$ onto the set of Eulerian orientations of G. As a consequence* $\mathrm{euo}(G) = \varepsilon(D)$ *and* $\varepsilon(D) = \varepsilon(D')$ *for each Eulerian orientation D' of G.*

Proof Evidently, the function is injective and D/F is an Eulerian orientation for each set $F \in \mathcal{E}(D)$. To see that the function is also surjective, let D' be an arbitrary Eulerian orientation of G. If F is the set of edges whose orientation is reversed when going from D to D', then $d_D^+(v) = d_{D/F}^+(v) + (d_{D(F)}^-(v) - d_{D(F)}^+(v))$ for each vertex v of D and, therefore, $F \in \mathcal{E}(D)$ and $D' = D/F$. □

3.7 Number of Colorings

As we have seen in Section 3.5 the graph polynomial encodes some information about the colorings of the graph. In this section we discuss the possibility to obtain information about the number of colorings of a graph from its polynomial.

Let n denote a positive integer, let G be a graph or multigraph of order n with vertex set $V(G) = \{v_1, v_2, \ldots, v_n\}$, and let $k \geq 2$ be an integer. As a reference orientation for G we use the monotone orientation D of G, i.e., $P_G = P_D$.

By definition, if G is not k-colorable and S is any set of k complex numbers, then the polynomial P_G vanishes for all values $(c_1, \ldots, c_n) \in S^n$, since for each such value of variables x_i there is some edge $v_i v_j$ of G with $x_i - x_j = 0$. Let $S = \{c \in \mathbb{C} \,|\, c^k - 1 = 0\}$ be the set of the kth roots of unity and let $Q \in \mathbb{C}[x]$ be the univariate polynomial defined by $Q = \prod_{c \in S}(x - c)$. Evidently, $Q = x^k - 1 \in \mathbb{Z}[x]$.

Let $\overline{P}_{G,k}$ denote the remainder of the polynomial P_G over $\mathbb{Z}[x_1,x_2,\ldots,x_n]$ with respect to the ideal generated by the polynomials $x_1^k-1,\ldots,x_n^k-1$. Then, see Theorem 3.21, $\overline{P}_{G,k}\in\mathbb{Z}[x_1,x_2,\ldots,x_n]$ is the unique polynomial such that $\deg_{x_i}(\overline{P}_{G,k})\le k-1$ for each $i\in[1,n]$ and such that there are polynomials $R_1,R_2,\ldots,R_n\in\mathbb{Z}[x_1,\ldots,x_n]$ satisfying $P_G=R_1(x_1^k-1)+\ldots R_n(x_n^k-1)+\overline{P}_{G,k}$.

Let $\mathbb{Z}_k=\{0,1,\ldots,k-1\}$ be the ground set of the integer ring modulo k, and let $C_k(G)$ the set of all colorings of G with color set $\mathbb{Z}_k$. For a polynomial $P\in\mathbb{Z}[x_1,x_2,\ldots,x_n]$, the 2-**norm** of P is defined by

$$||P||=\sqrt{\sum_{t_1,\ldots,t_n\ge 0}|\mathrm{coef}_P(t_1,t_2,\ldots,t_n)|^2}.$$

By Theorem 3.21, the remainder $\overline{P}_{G,k}$ is the zero polynomial if and only if G is not k-colorable. The following theorem due to Alon and Tarsi [59] provides a stronger result.

Theorem 3.33 *Every multigraph $G=(V,E)$ with $V=\{v_1,v_2,\ldots,v_n\}$ satisfies*

$$||\overline{P}_{G,k}||^2=\frac{4^{|E|}}{k^n}\sum_{\varphi\in C_k(G)}\prod_{e\in E}\sin^2\left[\frac{\pi(\varphi(v_D^+(e))-\varphi(v_D^-(e)))}{k}\right] \tag{3.10}$$

where $k\ge 2$ is an integer and D is the monotone orientation of G.

Proof Let $\overline{P}=\overline{P}_{G,k}$. By Theorem 3.21, $\deg_{x_i}(\overline{P})\le k-1$ for each $i\in[1,n]$ and, therefore,

$$\overline{P}=\sum_{(t_1,t_2,\ldots,t_n)\in\mathbb{Z}_k^n}\mathrm{coef}_{\overline{P}}(t_1,t_2,\ldots,t_n)\prod_{i=1}^n x_i^{t_i}.$$

Let $S=\{c\in\mathbb{C}\,|\,c^k-1=0\}$ and let $z=e^{2\pi i/k}$be a primitive kth root of unity. Then $S=\{z^0,z^1\ldots,z^{k-1}\}$ and, by Parseval's Formula, we obtain

$$||\overline{P}||^2=\frac{1}{k^n}\sum_{(t_1,t_2,\ldots,t_n)\in\mathbb{Z}_k^n}\left|\overline{P}(z^{t_1},z^{t_2},\ldots,z^{t_n})\right|^2. \tag{3.11}$$

By Theorem 3.21(2), it then follows that $\overline{P}(c_1,c_2,\ldots,c_n)=P_G(c_1,c_2,\ldots,c_n)$ for all tuples $(c_1,\ldots,c_n)\in S^n$. Let $\varphi:V\to\mathbb{Z}_k$ be an arbitrary map. If φ is not a proper coloring of G then

$$\overline{P}(z^{\varphi(v_1)},z^{\varphi(v_2)},\ldots,z^{\varphi(v_n)})=P_G(z^{\varphi(v_1)},z^{\varphi(v_2)},\ldots,z^{\varphi(v_n)})=0,$$

whereas if it is a proper coloring then

$$\begin{aligned}\left|\overline{P}(z^{\varphi(v_1)},\dots,z^{\varphi(v_n)})\right|^2 &= \left|P_G(z^{\varphi(v_1)},\dots,z^{\varphi(v_n)})\right|^2\\ &= \prod_{e\in E}\left|z^{\varphi(v_D^+(e))}-z^{\varphi(v_D^-(e))}\right|^2\\ &= 4^{|E|}\prod_{e\in E}\sin^2\left[\frac{\pi(\varphi(v_D^+(e))-\varphi(v_D^-(e)))}{k}\right]\end{aligned}$$

as $\left|e^{i\alpha}-e^{i\beta}\right|^2 = 4\sin^2\left(\frac{\alpha-\beta}{2}\right)$. Hence (3.11) implies (3.10). □

Formula (3.10) provides a lower bound for the number of k-colorings of a graph or multigraph G in terms of $\|\overline{P}_{G,k}\|^2$. For the special case $k = 3$ this is an equality. Since $\sin^2(2\pi/3) = \sin^2(\pi/3) = 3/4$, the number of 3-colorings of a graph or multigraph $G = (V,E)$ satisfies

$$\|\overline{P}_{G,3}\|^2 = 3^{|E|-|V|}|C_3(G)|. \tag{3.12}$$

Tarsi [993] found a different approach for computing the 2-norm of the remainder polynomial $\overline{P}_{G,k}$ of the graph or multigraph G, see Theorem 3.35. His approach is based on a combinatorial interpretation of the coefficients of the remainder polynomial of G, see the next lemma.

For every tuple $(d_1,d_2,\dots,d_n)\in\mathbb{Z}_k^n$, let

$$\mathrm{coef}_{G,k}(d_1,d_2,\dots,d_n) = \mathrm{coef}_{\overline{P}}(d_1,d_2,\dots,d_n)$$

be the coefficient of the remainder polynomial $\overline{P} = \overline{P}_{G,k}$. Note that all other coefficients of $\overline{P}$ are zero. For every tuple $\underline{d} = (d_1,d_2,\dots,d_n)$ of nonnegative integers, we define the following two subsets of the power set $2^{E(G)}$:

$$\mathcal{E}^e_{G,k}(\underline{d}) = \{F\subseteq E(G)\,\big|\,|F|\equiv 0\,(\mathrm{mod}\,2),\ d^+_{D/F}(v_i)\equiv d_i\,(\mathrm{mod}\,k),\ 1\le i\le n\},$$

and

$$\mathcal{E}^o_{G,k}(\underline{d}) = \{F\subseteq E(D)\,\big|\,|F|\equiv 1\,(\mathrm{mod}\,2),\ d^+_{D/F}(v_i)\equiv d_i\,(\mathrm{mod}\,k),\ 1\le i\le n\},$$

where D is the monotone orientation of G.

Lemma 3.34 *Every multigraph $G = (V,E)$ with $V = \{v_1,v_2,\dots,v_n\}$ satisfies*

$$\mathrm{coef}_{G,k}(d_1,d_2,\dots,d_n) = |\mathcal{E}^e_{G,k}(d_1,d_2,\dots,d_n)| - |\mathcal{E}^o_{G,k}(d_1,d_2,\dots,d_n)|$$

for each tuple $(d_1,d_2,\dots,d_n)\in\mathbb{Z}_k^n$, where $k\ge 2$ is an integer.

Proof Let $\overline{P} = \overline{P}_{G,k}$ and let D be the monotone orientation of G. The polynomial $\overline{P}$ is the remainder of the polynomial P_D over $\mathbb{Z}[x_1,x_2,\dots,x_n]$ with respect to the ideal generated by the polynomials $x_1^k-1, x_2^k-1,\dots,x_n^k-1$. Then, see the proof of Theorem 3.21, $\overline{P}$ can be obtained from P_D by using the division algorithm for polynomials, where we first compute the remainder P_1 of $P_0 = P$ with respect to $x_1^k - 1$, then the remainder P_2 of P_1 with respect to $x_2^k - 1$, and eventually, the remainder $\overline{P} = P_n$ of P_{n-1} with respect to $x_n^k - 1$. If in step i the polynomial $P_{i-1} = a_m x_i^m + a_{m-1}x_i^{m-1} + \cdots + a_1 x_i^1 + a_0$ has degree $m \geq k$ in x_i, where $a_m, \ldots, a_0 \in \mathbb{Z}[x_1, \ldots, x_{i-1}, x_{i+1}, \ldots, x_n]$, then the first step of the division algorithm yields $P_{i-1} = a_m x_i^{m-k}(x_i^k - 1) + P'_{i-1}$, where $P'_{i-1} = a_{m-1}x_i^{m-1} + \cdots + a_0 + a_m x_i^{m-k}$ is a polynomial of degree $\leq m-1$. This shows that each tuple $(d_1, d_2, \ldots, d_n) \in \mathbb{Z}_k^n$ satisfies

$$\mathrm{coef}_{\overline{P}}(d_1, d_2, \ldots, d_n) = \sum_{(t_1, t_2, \ldots, t_n)} \mathrm{coef}_D(t_1, \ldots, t_n),$$

where the sum is over all tuples $(t_1, t_2, \ldots, t_n)$ of nonnegative integers such that $t_i \equiv d_i \bmod k$ for each $i \in [1, n]$. Hence, the result follows from Lemma 3.22. □

To continue the discussion of Tarsi's result from [993], we need some further notation. For a multidigraph D' and an integer $k \geq 2$, define

$$\mathcal{E}_k(D') = \{F \subseteq E(D') \,|\, d^-_{D'(F)}(v) \equiv d^+_{D'(F)}(v) \,(\mathrm{mod}\, k), \forall v \in V(D')\},$$

$$\mathcal{E}_k^e(D') = \{F \in \mathcal{E}_k(D') \,|\, |F| \equiv 0 \,(\mathrm{mod}\, 2)\},$$

and finally

$$\mathcal{E}_k^o(D') = \{F \in \mathcal{E}_k(D) \,|\, |F| \equiv 1 \,(\mathrm{mod}\, 2)\}.$$

Given a multigraph G, let $\mathcal{F}_k(G)$ denote the set of all multidigraphs D' such that D' is an orientation of a spanning subgraph of G and every vertex $v \in V(G)$ satisfies $d^+_{D'}(v) \equiv d^-_{D'}(v) \,(\mathrm{mod}\, k)$. Furthermore, let $\mathcal{F}_k^{\neq 0}(G)$ denote the set of all multidigraphs $D' \in \mathcal{F}_k(G)$ such that $E(D') = E(G)$, i.e., D' is an orientation of G and $d^+_{D'}(v) \equiv d^-_{D'}(v) \,(\mathrm{mod}\, k)$ for every vertex $v \in V(G)$. Note that there is a one-to-one correspondence between the digraphs in $\mathcal{F}_k(G)$ and the $\mathbb{Z}_k$-flows, where the permitted flow values are $1, -1$ and 0. Here $\mathbb{Z}_k$ means the additive group of the residue ring modulo k. Eventually, let $\mathrm{NZF}_3(G)$ denote the set of all nowhere zero 3-flows of G. Then we have $|\mathrm{NZF}_3(G)| = |\mathcal{F}_3^{\neq 0}(G)|$ (see Appendix B.3).

Theorem 3.35 *Every multigraph $G = (V, E)$ with $V = \{v_1, v_2, \ldots, v_n\}$ satisfies*

$$\|\overline{P}_{G,k}\|^2 = (-1)^{|E|} \sum_{D' \in \mathcal{F}_k(G)} (-2)^{|E| - |E(D')|}, \tag{3.13}$$

where $k \geq 2$ is an integer.

Proof Let D be the monotone orientation of G. Let $F \in \mathcal{E}(\underline{d}) = \mathcal{E}^e_{G,k}(\underline{d}) \cup \mathcal{E}^e_{G,k}(\underline{d})$ for a tuple $\underline{d} = (d_1, d_2, \ldots, d_n)$ of $\mathbb{Z}_k^n$, and let $F' \subseteq E(D)$. Then every vertex v_i of G satisfy

$$d^+_{D/F}(v_i) \equiv d_i \,(\mathrm{mod}\, k)$$

as well as

$$d^+_{D/F}(v_i) = d^+_{D/F/F'}(v_i) + d^-_{(D/F)(F')}(v_i) - d^+_{(D/F)(F')}(v_i).$$

Since $D/F/F' = D/(F \oplus F')$, this implies that $F' \in \mathcal{E}_k(D/F)$ if and only if $F \oplus F' \in \mathcal{E}(\underline{d})$. From this it then follows that $|\mathcal{E}_k(D/F)| = |\mathcal{E}(\underline{d})|$. Let $\alpha \in \{e, o\}$ and let $\overline{e} = o$ and $\overline{o} = e$. A simple parity argument then shows that

$$\left|\mathcal{E}^{\alpha}_{G,k}(\underline{d})\right| = |\mathcal{E}^{\alpha}_k(D/F)|$$

if $|F|$ is even, and

$$\left|\mathcal{E}^{\alpha}_{G,k}(\underline{d})\right| = \left|\mathcal{E}^{\overline{\alpha}}_k(D/F)\right|$$

otherwise. By Lemma 3.34, it then follows that the coefficient $a_{\underline{d}} = \mathrm{coef}_{G,k}(\underline{d})$ satisfies

$$\begin{aligned} a_{\underline{d}}^2 &= a_{\underline{d}}\left(\left|\mathcal{E}^{e}_{G,k}(\underline{d})\right| - \left|\mathcal{E}^{o}_{G,k}(\underline{d})\right|\right) \\ &= \sum_{F \in \mathcal{E}^{e}_{G,k}(\underline{d})} \left(\left|\mathcal{E}^{e}_{G,k}(\underline{d})\right| - \left|\mathcal{E}^{o}_{G,k}(\underline{d})\right|\right) + \sum_{F \in \mathcal{E}^{0}_{G,k}(\underline{d})} \left(\left|\mathcal{E}^{o}_{G,k}(\underline{d})\right| - \left|\mathcal{E}^{e}_{G,k}(\underline{d})\right|\right) \\ &= \sum_{F \in \mathcal{E}(\underline{d})} \left(|\mathcal{E}^{e}_k(D/F)| - |\mathcal{E}^{o}_k(D/F)|\right) \end{aligned}$$

For the 2-norm of $\overline{P} = \overline{P}_{G,k}$, we then obtain that

$$\begin{aligned} ||\overline{P}||^2 &= \sum_{\underline{d} \in \mathbb{Z}_k^n} a_{\underline{d}}^2 \\ &= \sum_{F \subseteq E(D)} \left(|\mathcal{E}^{e}_k(D/F)| - |\mathcal{E}^{o}_k(D/F)|\right) \\ &= \sum_{F \subseteq E(D)} \sum_{F' \in \mathcal{E}_k(D/F)} (-1)^{|F'|}. \end{aligned}$$

Since $F' \in \mathcal{E}_k(D/F)$ if and only if $F' \subseteq E(D)$ and $(D/F)(F') = D(F')/(F \cap F') \in \mathcal{F}_k(G)$, we then obtain that

$$||\overline{P}||^2 = \sum_{D' \in \mathcal{F}_k(G)} (-1)^{|E(D')|} 2^{|E(D)| - |E(D')|}.$$

Obviously, this implies the equation (3.13). □

Corollary 3.36 *Every multigraph G satisfies $|C_3(G)| \equiv |\mathrm{NZF}_3(G)| \pmod 4$, provided that G has $m \geq 2$ edges.*

Proof For $0 \leq q \leq m$, let b_q denote the number of digraphs $D' \in \mathcal{F}_3(G)$ with $|E(D')| = q$. Evidently, $D' \in \mathcal{F}_3(G)$ if and only if $D'/E(D') \in \mathcal{F}_3(G)$. Hence b_q is even if $q \geq 1$ and $b_0 = 1$. Since $m \geq 2$, we then conclude from Theorem 3.35 that

$$||\overline{P}_{G,k}||^2 \equiv (-1)^m \sum_{q=0}^{m} b_q(-2)^{m-q} \equiv (-1)^m b_m \equiv b_m \pmod 4$$

Clearly, $|C_3(G)|$ is always divisible by 6 (permutations of the 3 colors). Hence, (3.12) implies $\|\overline{P}_{G,k}\|^2 \equiv |C_3(G)| \,(\mathrm{mod}\, 4)$. Since $b_m = |\mathcal{F}_3^{\neq 0}| = |\mathrm{NZF}_3(G)|$, the result follows. □

One can easily observe that if G is a 4-regular multigraph, then $D' \in \mathcal{F}_3^{\neq 0}$ if and only if D' is an Eulerian orientation of G. The following is then a direct consequence of Corollary 3.36

Corollary 3.37 *Every* 4*-regular multigraph* G *satisfies* $|C_3(G)| \equiv \mathrm{euo}(G) \,(\mathrm{mod}\, 4)$.

Corollary 3.38 *Every* 4*-regular multigraph* G *with* $|C_3(G)| \equiv 2 \,(\mathrm{mod}\, 4)$ *satisfies* $\chi_\ell(G) \leq 3$.

Proof Obviously, G has an Eulerian orientation D and $\varepsilon(D) = \mathrm{euo}(G)$ (see Remark 3.32). Since G has an even number of edges, the result follows from Corollary 3.28 and Corollary 3.37. □

Let G be a cycle-plus-triangles graph. Then it follows from Theorem 3.31 and Remark 3.32 that $\mathrm{euo}(G) \equiv 2 \,(\mathrm{mod}\, 4)$ and so $|C_3(G)| \equiv 2 \,(\mathrm{mod}\, 4)$ (by Corollary 3.37). The last congruence was proved independently by Sachs [884] by means of a direct induction.

3.8 Orientation vs. Brooks

The goal of this section is to prove Theorem 1.50. This theorem will be derived from the following result.

Theorem 3.39 *Let* (G, g) *be a connected weighted graph with* $g \geq 1$, *and let* f *be a vertex function of* G *such that*

$$f(v) \geq g(v) + \sum_{uv \in E(G)} g(u) - \min\{g(u) \mid uv \in E(G)\}$$

for every vertex $v \in V(G)$. *Then* (G, g) *is* f*-list colorable unless* G *is a Gallai tree.*

Recall that a Gallai tree is a connected graph all of whose blocks are complete graphs and/or odd cycles. The above theorem is an immediate consequence of the kernel method (Lemma 3.4) and the following result.

Lemma 3.40 *Let* G *be a connected graph. If* G *is not a Gallai tree, then* G *has a kernel-perfect super-orientation* D *such that* $d_D^+(v) \leq d_G(v) - 1$ *for all* $v \in V(G)$.

Proof Let G be a connected graph, but not a Gallai tree. Then Theorem 1.35 implies that G contains an even cycle C with at most one chord in G. If we contract C to a single vertex $w = w_C$, we obtain a connected graph G^*. Now, using a spanning tree of G^*, it is easy to construct an acyclic orientation of G^* such that each vertex of

G^* except w has in-degree at least one. Let D^* be the corresponding orientation of $G - E(G[V(C)])$. We extend D^* to an super-orientation D of G by orienting the edges of the cycle of C in an cyclic order and, moreover, by replacing the chord uv of C, if such a chord exists, by two directed edges, one directed from u to v and one directed from v to u. Clearly, $d_D^+(v) \leq d_G(v) - 1$ for every vertex v of G. To conclude the proof it suffices to show that D is kernel-perfect. To this end, let D' be an induced subdigraph of D. Then we show by induction on $|D'|$ that D' has a kernel. If $|D'| = 1$, this is evident. Now, assume that $|D'| \geq 2$. If D' has a vertex v with out-degree zero, then by induction, $D' - (N_{D'}^-(v) \cup \{v\})$ has a kernel S and $S \cup \{v\}$ is a kernel of D'. Otherwise, D' is an induced subdigraph of $D[V(C)]$ and we argue as follows. If C has no chord, then D' has no directed odd cycle and so D' has a kernel (by Theorem 3.9). Now, assume that C has a chord uv. If $D' = D[V(C)]$ then take as S the largest independent set. Since C is an even cycle with at most one chord, S is a kernel of D'. If D' is a proper induced subgraph of $D[V(C)]$, then let e_1, e_2 be the two edges of D such that $\{v_D^+(e_i), v_D^-(e_i)\} = \{u, v\}$ for $i \in \{1,2\}$. Then $D_1 = D' - e_1$ or $D_2 = D' - e_2$ does not contain a directed odd cycle. Hence Theorem 3.9 implies that either D_1 or D_2 has a kernel S. Clearly, S remains a kernel of the multidigraph D'. □

By Lemma 3.40, $G = K_4^-$ (= K_4 minus an edge) has a kernel-perfect super-orientation D such that $d_D^+(v) \leq d_G(v) - 1$ for all $v \in V(G)$ (see Figure 3.3), but there is no such orientation of G.

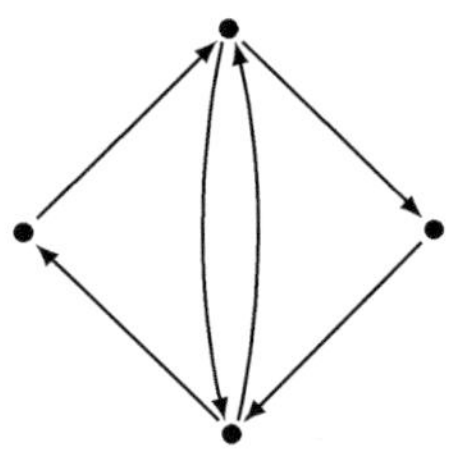

Fig. 3.3 A kernel-perfect super-orientation of K_4^-.

Lemma 3.41 *Let G be a connected graph. If G is not a Gallai tree, then G has an orientation D such that $d_D^+(v) \leq d_G(v) - 1$ for all $v \in V(G)$ and $\varepsilon_e(D) \neq \varepsilon_o(D)$.*

Proof We can argue similar as in the proof of Lemma 3.40. So let C be the even cycle of G with at most one chord in G, and let D be the corresponding kernel-perfect super-orientation of G as in the proof of Lemma 3.40 with the difference that the possible chord of C is replaced by only one arbitrarily directed edge in D. Then D is an orientation of D such that $d_D^+(v) \leq d_G(v) - 1$ for all $v \in V(G)$. As $D|_{G-E(H)}$ with $H = G[V(C)]$ is acyclic, every set in $\mathcal{E}(D)$ is a subset of $E(H)$. Clearly, the empty set and the edge set of C belong to $\mathcal{E}^e(D)$. There is at most one further set in $\mathcal{E}(D)$ depending on whether C has a chord or not. Consequently, $\varepsilon_e(D) \neq \varepsilon_o(D)$.□

Let G be a graph. If $\mathcal{D}$ is the class of all multidigraphs D such that $\varepsilon_e(D) \neq \varepsilon_o(D)$, then we write $\Delta^+_{\text{at}}(G)$ instead of $\Delta^+_{\mathcal{D}}(G)$. It follows from Theorem 3.24 that G satisfies $\chi_\ell(G) \leq \Delta^+_{\text{at}}(G) + 1$.

Corollary 3.42 *Let G be a connected graph. If G is neither a complete graph nor an odd cycle, then $\Delta^+_{\text{ke}}(G) \leq \Delta(G) - 1$ and $\Delta^+_{\text{at}}(G) \leq \Delta(G) - 1$.*

Proof If G is not a Gallai tree, then the result follows from Lemma 3.40 respectively from Lemma 3.41. If G is a Gallai tree, then G has at least two blocks, because G is neither a complete graph nor an odd cycle. Hence there is a a vertex v that $d_G(v) \leq \Delta(G) - 1$. Clearly, G has an acyclic orientation D such that each vertex of G except v has in-degree at least one and so $\Delta^+(D) \leq \Delta(G) - 1$. As D is kernel-perfect and $\varepsilon_e(D) \neq \varepsilon_o(D)$, we obtain $\max\{\Delta^+_{\text{ke}}(G), \Delta^+_{\text{at}}(G)\} \leq \Delta^+(D) \leq \Delta(G) - 1$. □

Based on Lemma 3.40 and Lemma 3.41, respectively, we can use the kernel method of Galvin as well as the polynomial method of Alon-Tarsi to show that any graph is degree-choosable if it is not a Gallai forest. However, while Galvin's kernel method also applies to list colorings of weighted graphs, the method of Alon and Tarsi does not (see Exercise 3.21).

Proposition 3.43 *Let (G, g) be a connected weighted graph with $g \geq 1$, let w be a vertex of G, and let f be a vertex function of G such that*

$$f(v) \geq g(v) + \sum_{uv \in E(G)} g(u) - \min\{g(u) \mid uv \in E(G)\} \tag{3.14}$$

for every vertex $v \in V(G) \setminus \{w\}$ and

$$f(w) \geq g(w) + \sum_{uw \in E(G)} g(u). \tag{3.15}$$

Then (G, g) is f-list colorable.

Proof Using a spanning tree of G, it is easy to construct an acyclic orientation D of G such that each vertex of G except w has in-degree at least one. Then it follows from (3.14) and (3.15) that

$$f(v) \geq g(v) + \sum_{u \in N^+_D(v)} g(u)$$

for all $v \in V(G)$. As D is kernel-perfect, (G, g) is f-list colorable (by Lemma 3.4). □

Combining Proposition 3.43 and Theorem 3.39, we are now ready to prove Theorem 1.50. For the readers convenience we will repeat this theorem.

Theorem 3.44 *Every connected weighted graph (G, g) with $g \geq 1$ satisfies*

$$\chi_\ell(G, g) \leq \Delta(G[g]) + 1 - \min_{u \in V(G)} g(u),$$

unless G is a complete graph or an odd cycle.

Proof Let $k = \Delta(G[g]) + 1 - \min_{u \in V(G)} g(u)$ and let L be an arbitrary k-assignment for G. Then every vertex v of G satisfies

$$|L(v)| = k \geq g(v) + \sum_{uv \in E(G)} g(u) - \min_{u \in V(G)} g(u) \tag{3.16}$$

We have to show that (G, g) has an L-coloring, provided that G is neither a complete graph nor an odd cycle. If G is not a Gallai tree, this follows immediately from Theorem 3.39 and (3.16).

It remains to consider the case that G is a Gallai tree. As G is neither a complete graph nor an odd cycle, this implies that G has at least two blocks. Hence G has an end-block, say B, and G has exactly one separating vertex, say w, contained in B. Let $w_1, w_2, \ldots, w_r$ be the neighbors of w in G belonging to B, and let $u_1, u_2, \ldots, u_s$ be the neighbors of w in G belonging to $G' = G - V(B)$. Note that $r, s \geq 1$. Now we can apply Proposition 3.43 to each component of G'. By (3.16) it then follows that (G', g) has an L-coloring φ'. Let L' be the list-assignment of B such that $L'(v) = L(v)$ if $v \in V(B - w)$ and $L'(w) = L(w) \setminus (\varphi'(u_1) \cup \cdots \cup \varphi'(u_s))$. Our aim is to show that (B, g) is L'-colorable. Clearly, this would imply that (G, g) has an L-coloring.

Let $m = \min_{u \in V(G)} g(u)$. Then $|L'(w)| = k - (g(u_1) + \cdots + g(u_s)) \geq g(w) + g(w_1) + \ldots + g(w_r) - m$ (by (3.16)) and, by deleting some colors from $L'(w)$ if necessary, we may assume that

$$|L'(w)| = g(w) + g(w_1) + \ldots + g(w_r) - m.$$

For the neighbor w_1 of w in B, we have $|L'(w_1)| = |L(w_1)| = |L(w)| = k \geq g(w) + g(w_1) + \ldots + g(w_r) + g(u_1) - m = |L'(w)| + g(u_1)$ and, therefore, $|L'(w_1) \setminus L'(w)| \geq g(u_1) \geq m$. To construct an L'-coloring of (B, g), we first color the vertex w_1 by a set A of $g(w_1)$ colors chosen from the list $L'(w_1)$. Since $|L'(w_1) \setminus L'(w)| \geq m$, this can be done in such a way that at most $g(w_1) - m$ colors of A are contained in $L'(w)$. Now, consider the list-assignment L_1 of the connected graph $B_1 = B - w_1$ such that $L_1(v) = L'(v) - A$ if $vw_1 \in E(B)$ and $L_1(v) = L'(v)$ otherwise. For a vertex $v \in V(B_1 - w)$, we have

$$|L_1(v)| \geq g(v) + \sum_{uv \in E(B_1)} g(u) - \min\{g(u) | uv \in E(B_1)\}.$$

Furthermore, we have

$$|L_1(w)| \geq |L'(w)| - (g(w_1) - m) = g(w) + g(w_2) + \cdots + g(w_r)$$

Since B_1 is connected, Proposition 3.43 then implies that there is an L_1 coloring of (B_1, g) and, therefore, also an L'-coloring of (B, g). This completes the proof. □

3.9 Exercises

3.1 Prove that the implication from (e) to (a) in Theorem 3.1 is not true if the word *acyclic* is left out in (a). (Hint: Let G be a K_5, and let G' be obtained from G by deleting the four edges of a Hamilton path.)

3.2 Let k be a positive integer. For a multidigraph D, let $\Delta^k(D)$ be the maximum out-degree of a vertex v such that D has a directed path of order k from a vertex u to v, if D has no such path let $\Delta^k(D) = 0$. Show that for any nonempty graph G and any acyclic orientation D of G, we have $\chi(G) \le \Delta^k(D) + k$. (Hint: Choose a vertex order $\ell = (v_1, v_2, \dots, v_n)$ in which all edges of D are oriented backwards. Let t be the number of colors used by the sequential coloring algorithm applied to this vertex order. Show that $\Delta^k(D) \ge t - k$) (Zaker [1089])

3.3 Let k be a positive integer and let G be a nonempty graph. A path $P = (v_1, v_2, \dots, v_p)$ is called **decreasing** if $d_G(v_i) \ge d_G(v_{i+1})$ for all $i \in [1, p-1]$; further, let $d(P)$ be the number of vertices $v \in N_G(v_p) \setminus V(P)$ satisfying $d_G(v) \le d_G(v_k)$. Define $\Delta^k(G) = 0$ if G has no decreasing path of order k, otherwise let

$$\Delta^k(G) = \max\{d(P) \mid P \text{ is a decreasing path of order } k\}.$$

1. Prove that $\chi(G) \le \Delta^k(G) + k$. (Hint: Show that $\Delta^k(D) \le \Delta^k(G)$ for an appropriate acyclic orientation D of G).
2. Construct a graph G such that $\Delta^2(G) + 2 < \mathrm{col}(G)$.

(Zaker [1089])

3.4 Let $k \ge 2$ be an integer and let G be a graph not containing any cycle of length $1 \pmod k$. Let T be a DFS tree rooted at r and let $\varphi : V(G) \to [0, k-1]$ be the mapping such that $\varphi(v) \equiv |rTv| \pmod k$. Show that φ is a coloring of G and so $\chi(G) \le k$. (Tuza [1039])

3.5 An orientation of a complete graph is called a **tournament**. Show that every tournament has a directed Hamilton path. (Rédei [853])

3.6 Let $P = (X, \prec)$ be a (strict) **partially ordered set**, that is, X is a set and $\prec$ is an irreflexive, asymmetric and transitive binary relation on X. Two distinct elements x and y of X are **comparable** if $x \prec y$ or $y \prec x$, and **incomparable** otherwise. A set of pairwise comparable elements of X is a **chain** of P, and a set of pairwise incomparable elements of X is an **antichain** of P. Prove that the maximum number of elements in a chain of P is equal to the minimum number of antichains into which X can be partitioned. (Hint: Apply the Gallai–Roy theorem to the simple digraph D whose vertex set is X and in which there is an edge directed from x to y if and only if $x \prec y$.) (Mirsky [742])

3.7 A graph G is a **comparability graph** if there is a partially ordered set $P = (X, \prec)$ such that $V(G) = X$ and two elements of X are joined by an edge in G if and only if they are comparable. Show that every comparability graph is perfect.

3.8 Dilworth's Theorem: Let $P = (X, \prec)$ be a partially ordered set. Prove that the maximum number of elements in an antichain of P is equal to the minimum number of chains into which X can be partitioned. (Dilworth [288])

3.9 A **transitive digraph** is a simple digraph D whose adjacency relation is transitive, that is, $(u,v) \in E(D)$ and $(v,w) \in E(D)$ implies $(u,w) \in E(D)$.

1. Show that there is a one-to-one correspondence between partially ordered sets and transitive digraphs.
2. Show that a graph G is a comparability graph if and only if it has an transitive orientation.

3.10 Suppose that G is a k-chromatic graph with $k \geq 1$, and φ is a coloring of G with color set $C = [1,k]$. Show that for any p distinct colors $c_1, c_2, \ldots, c_p$ there is a path $P = (v_1, v_2, \ldots, v_p)$ in G such that $\varphi(v_i) = c_i$ for all $i \in [1,p]$. (Hint: Use induction on p and consider the set X of all vertices v such that there is a path $P = (v_2, v_3, \ldots, v_p)$ with $v_2 = v$ and $\varphi(v_i) = c_i$ for all $i \in [2,p]$.) (Chang, Tong, Yan, and Yeh [199])

3.11 Prove the following statements for an arbitrary multidigraph D:

1. If D is acyclic, then there is a vertex $v \in V(D)$ such that $d_D^+(v) = 0$ and there is a vertex w such that $d_D^-(w) = 0$.
2. D is acyclic if and only if D has a vertex order $v_1, v_2, \ldots, v_n$ such that every edge of D is directed from the vertex with the larger index to the vertex with the smaller index.
3. D is acyclic if and only if every nonempty subdigraph D' of D has a vertex v with $d_{D'}^+(v) = 0$.
4. If D is acyclic, then D has a kernel. (Hint: Apply induction on $|D|$ and delete a vertex v with $d_D^+(v) = 0$ together with $N_D^-(v)$). (von Neumann and Morgenstern [779])
5. If D is acyclic, then D has a unique kernel.
6. If D is an orientation of a complete graph, then D is kernel perfect if and only if D is acyclic.

3.12 Let us consider the following simple two-person game. The game board is a digraph D and the vertex set of D are the positions of the game. The players, say 1 and 2, move a token each in turn. At the beginning of the game the token is placed on some vertex v_0 of D. Players alternate moves, the first move is made by player 1. A move consists in taking the token from the present position v and placing it on an out-neighbor of v in D. The first player unable to move loses the game and the other player wins. Show that if D is an acyclic digraph, then one of the two players has a winning strategy, which one depending on whether the vertex v_0 belongs to the unique kernel of D, or not.

3.13 Let $(N_1, N_2, \ldots, N_p)$ a sequence of positive integers and let V denote the set of all tuples $(n_1, n_2, \ldots, n_p)$ of integers satisfying $0 \leq n_i \leq N_i$ for all $i \in [1,p]$ Define a digraph D with vertex set V where two vertices $\underline{n} = (n_1, n_2, \ldots, n_p)$ and $\underline{n}' = (n_1', n_2', \ldots, n_p')$ are joined by an edge directed from $\underline{n}$ to $\underline{n}'$ if and only if there is exactly one index $i_0 \in [1, p]$ such that $n'_{i_0} < n_{i_0}$ and $n_i = n'_i$ for all $i \in [1, p] \setminus \{i_0\}$. Clearly, D is an acyclic digraph and has therefore a unique kernel S. To describe the kernel, we use the fact that every nonnegative integer n_i has a unique base 2 representation of the form

$$n_i = \sum_{k=0}^{m} n_{i,k} 2^k$$

for some m, where each $n_{i,k}$ is either zero or one. Prove that S consists of all tuples $(n_1, n_2, \ldots, n_p)$ of V such that $n_{1,k} + n_{2,k} + \cdots + n_{p,k} \equiv 0 \pmod 2$ for all $k \in [1, m]$.

3.14 The most famous take-away game is Nim, played as follows. There are three piles of chips containing N_1, N_2, and N_3 chips respectively. Players 1 and 2 alternate taking off any number of chips from a pile until there are no chips left. You may not remove chips from more than one pile in one turn, but from the pile you selected you may remove as many chips as desired, from one chip to the whole pile. The winner is the player who removes the last chip. Use the previous two exercises to show that one player has a winning strategy and to describe the winning strategy. Who has a winning strategy in case of $(N_1, N_2, N_3) = (9, 7, 5)$?

3.15 Let G be a bipartite multigraph and let φ be an edge coloring of G with color set $C = [1, n]$. Show that the multidigraph D_φ defined in the proof of Theorem 3.6 has maximum out-degree at most $n - 1$. Use this and the fact that D_φ is kernel perfect, to conclude that $\chi_\ell(L(G), p) \leq pn$ for every p. (Galvin [402])

3.16 Let G be a connected graph, and let p be a positive integer. Use Galvin's kernel method in order to prove the following statements:

1. $\chi_\ell(G, p) \leq p \cdot \mathrm{col}(G)$.
2. If G is a chordal graph, then $\chi_\ell(G, p) \leq p \cdot \chi(G)$.

(Tuza and Voigt [1042])

3.17 Let G be a connected chordal graph, and let p be a positive integer. For a vertex v of G, let $\omega(v)$ be the largest number of vertices in a clique of G containing v.

1. Show that G has an acyclic orientation D with $\Delta^+(D) = \omega(G) - 1$.
2. Show that (G, p) is f-choosable if $f(v) = p\omega(v)$ for all $v \in V(G)$.

(Hint: Use Theorem 1.58.) (Tuza and Voigt [1042])

3.18 Show that every planar bipartite graph G satisfies $\chi_\ell(G) \leq 3$. (Hint: Combine Corollary 3.10, (3.5), and Theorem C.18) (Alon and Tarsi [58])

3.19 Prove that Theorem 3.12 still holds if the definition of a fractional orientation uses the nonnegative real numbers $\mathbb{R}_+$ rather than the nonnegative rational numbers $\mathbb{Q}_+$.

3.20 **Stable Marriage Theorem**: Let G be a bipartite multigraph, let (X, Y) be a bipartition of G, and let $\preceq_v$ be a linear order on the edge set $E_G(v)$ for each

$v \in V(G)$. Show that G has a **stable matching**, i.e., a matching $M \subseteq E(G)$, such that for every edge $e \in E(G) \setminus M$, say $e \in E_G(v,w)$, there is an edge $e' \in M$ satisfying $e' \in E_G(v)$ and $e \preceq_v e'$, or $e' \in E_G(w)$ and $e \preceq_w e'$. (Hint: Use Exercise 1.25.2 and Theorem 3.15) (Gale and Shapley [396], and Maffray [713])

3.21 $G = K_4 - xy$ and let u, v be the two vertices of $G - x - y$. Let D be the orientation of G, where the edges of the cycle (x,u,y,v) are oriented in cyclic order an the edge uv is directed from v to u. Let g, f be the vertex functions with $g(x) = 1$, $g(u) = g(y) = g(v) = 2$, $f(x) = 3$, $f(u) = f(y) = 4$, and $f(v) = 5$. Show that D satisfies the Alon-Tarsi condition, that is, $\varepsilon_e(D) \neq \varepsilon_o(D)$, and f satisfies the inequality (3.3) for every vertex of G, but (G,g) is not f-list colorable.

3.22 Let G be a uniquely k-colorable graph with order $n \geq k$ and m edges. Show the following statements:

1. $m \geq (k-1)n + \binom{k}{2}$.
2. If $m = (k-1)n + \binom{k}{2}$, then $\chi(G) = \chi_\ell(G) = k$. (Hint: The vertex set of G has a unique partition into k independent sets, say $I_1, I_2, \ldots, I_k$. Choose a vertex $v_i \in I_i$ for $i \in [1,k]$ and let L be a list-assignment for G such that $L(u) = [1,i]$ for $u = v_i$ and $L(u) = [1,k]$ otherwise. Now apply Theorem 3.25) (Hefetz [492])

3.23 Let $P = P_G$ be the graph polynomial of a complete graph G with vertex set $V = \{v_1, v_2, \ldots, v_n\}$. Show that the coefficient of P have values $0, +1$ or -1, there are exactly $n!$ coefficient different from zero, and $\text{coef}_P(d_1, d_2, \ldots, d_n) \neq 0$ if and only if $d_1 + d_2 + \cdots + d_n = \binom{n}{2}$. (Hint: Express P as a determinant)

3.24 Prove that if a graph G has a decomposition into a Hamilton cycle C and pairwise disjoint triangles $T_1, T_2, \ldots, T_m$, then G has an Eulerian orientation D such that no triangle T_i is cyclically oriented, and so the set of all those vertices which are sinks of some triangle T_i with respect to D form an independent set of G. (Hint: Use Theorem 3.31 and the inclusion-exclusion principle.) (Fleischner and Stiebitz [380])

3.25 Let $k \geq 4$ be an integer and let G be a graph.

1. Show that if a graph G has a decomposition into an Hamilton cycle and m pairwise disjoint complete graphs each on at most k vertices, then $\chi(G) \leq k$. (Hint: Apply induction on k and use the cycle-plus-triangles theorem)
2. Show that the above statement is not true for $k = 3$.

(Fleischner and Stiebitz [380])

3.26 Suppose that a graph G is the disjoint union of two 2-factors each of which is the union of disjoint triangles. Prove that G is a perfect graph and so $\chi(G) = \omega(G) \leq 3$. (Hint: Show that G is the line graph of a cubic bipartite graph.) (A. Gyárfás, private communication 2015)

3.27 Variable Degeneracy: Let G be a graph and $f : V(G) \to \mathbb{N}_0$ be a vertex function. We say that G is **f-degenerate** if every nonempty subgraph H of G has a vertex v such that $d_H(v) \leq f(v)$. Show that the following statements hold:

1. G is f-degenerate if and only if G has an acyclic orientation D such that $d_D^+(v) \le f(v)$ for all $v \in V(G) = V(D)$.
2. For $k \ge 0$, G is k-degenerate if and only if G is f-degenerate for the constant function $f \equiv k$ (i.e. $f(v) = k$ for all $v \in V(G)$. (Hint: Use Proposition 1.5.)
3. If G is f-degenerate, then G is $(f+1)$-list-colorable. Here $f' = f+1$ is the vertex function of G with $f'(v) = f(v)+1$ for all $v \in V(G)$. (Hint: Use Exercise 3.11 and Corollary 3.5.)
4. If G is f-degenerate, then G is DP-$(f+1)$-colorable. (Hint: Use greedy coloring and the reduction introduced in Proposition 1.14.)

3.28 Let G be a graph and $f : V(G) \to \mathbb{N}_0$ be a vertex function. For a vertex v of G, let $f' = f/(G,v)$ denote the function $f' : V(G-v) \to \mathbb{Z}$ such that $f'(u) = f(u)-1$ if $u \in N_G(v)$ and $f'(u) = f(u)$ otherwise. We say, by a recursive definition, that G is f**-reducible** if either $G = \varnothing$ or there is a vertex $v \in V(G)$ such that $G' = G - v$ is f'-reducible for the function $f' = f/(G,v)$ and $f'(u) \ge 0$ for all $u \in V(G')$. Show that the following statements hold:

1. If $G = \varnothing$, then G is f-reducible.
2. If G is a K_n with $n \ge 1$, then G is f-reducible if and only if G has a vertex labelling $\ell = (v_1, v_2, \ldots, v_n)$ such that $f(v_i) \ge i-1$ for $i \in [1,n]$. (Hint: Use induction on n.)
3. G is f-reducible if and only if G is f-degenerate

3.29 Let G be a connected graph, let w be a vertex of G, and let $f : V(G) \to \mathbb{N}_0$ be a vertex function such that $f(v) \ge d_G(v)-1$ for all $v \in V(G) \setminus \{w\}$ and $f(w) \ge d_G(w)$. Show that G is f-degenerate. (Hint: Take the first part of the proof of Lemma 3.40.)

3.30 Show that if G is a connected graph, then every block of G is a complete graph or a cycle if and only if every induced subgraph of G without a separating vertex is regular. (Bernshteyn and Lee [112])

3.31 Let G be a graph and $f : V(G) \to \mathbb{N}_0$ be a vertex function. For two distinct vertices v, w of G, let $f' = f/(G,v,w)$ denote the function $f' : V(G-v) \to \mathbb{Z}$ such that $f'(u) = f(u)-1$ if $u \in N_G(v) \setminus \{w\}$ and $f'(u) = f(u)$ otherwise. We say, by a recursive definition, that G is **weakly** f**-reducible** if either $G = \varnothing$ or one of the following two conditions hold:

(a) there is a vertex $v \in V(G)$ such that $G' = G - v$ is weakly f'-reducible for the function $f' = f/(G,v)$ and $f'(u) \ge 0$ for all $u \in V(G')$, or
(b) there is a vertex $v \in V(G)$ and a vertex $w \in N_G(v)$ such that $f(v) > f(w)$ and $G' = G - v$ is weakly f'-reducible for the function $f' = f/(G,v,w)$ and $f'(u) \ge 0$ for all $u \in V(G')$.

Show that the following statements hold:

1. If G is f-reducible, then G is weakly f-reducible.
2. If G is weakly f-reducible and g is a vertex function of G such that $g(v) \ge f(v)$ for all $v \in V(G)$, then G is weakly g-reducible.

3. If G is weakly f-reducible, then G is DP-$(f+1)$-colorable. (Hint: Use greedy coloring, Proposition 1.14 and the following fact. If (X,H) is a cover of G, $vw \in E(G)$ and $|X_v| > |X_w|$, then there is a color $x \in X_v$ such that $N_H(x) \cap X_w = \varnothing$.)

(Bernshteyn and Lee [112])

3.32 Weak degeneracy: For a graph G, define the **weak degeneracy** $\mathrm{wdg}(G)$ of G as the least integer $k \geq 0$ such that G is weakly f-reducible for the constant function $f \equiv k$. Show that every graph G satisfies $\chi_{\mathrm{DP}}(G) \leq \mathrm{wdg}(G)+1 \leq \mathrm{col}(G)$. (Bernshteyn and Lee [112])

3.10 Notes

As mentioned in Section 3.1 the idea behind the Gallai–Roy theorem has repeatedly been rediscovered, and one may also call it the Gallai–Gupta–Hasse–Roy–Vitaver theorem. Let D be any multidigraph. The Gallai–Roy theorem says that the minimum number of disjoint independent sets of D whose union is $V(D)$ is at most the maximum number of vertices in a directed path of D. Conversely, the Gallai–Milgram theorem [401] says that the minimum number of disjoint directed paths whose vertex union is $V(D)$ is at most the maximum number of vertices in an independent set of D. Booth theorems solve special cases of the directed path partition conjecture proposed by Berge [100] in 1982, see also the paper by Aharoni, Ben–Arroyo Hartman, and Hoffmann [21] and the paper by Sebő [921]. The Gallai–Roy theorem implies that any orientation of an n-chromatic graph G contains a directed path of order n. The reader may find other results of this type in the papers by Bondy [140], by Havet and Thomassé [481], and by Addario-Berry, Havet, and Thomassé [19].

Based on Mintys theorem (Theorem 3.2), Youngs [1085] developed a technique for constructing a large family of 4-critical graphs which embed on the projective plane. A similar argument was already used by Gallai [397] for constructing a family of 4-regular 4-critical graphs that embed on the Klein bottle. Let us briefly describe this technique.

Let G be a graph. If $W = (v_0, v_1, \ldots, v_p)$ is a walk of G (see Appendix C.5 for the terminology), then we denote the inverse walk $W' = (v_p, v_{p-1}, \ldots, v_0)$ by $-W$. We call p the length of W, and write $|W| = p$. A walk is empty if its length $p = 0$. Given two walks W_1 and W_2 such that the last vertex of W_1 is the first vertex of W_2, we can form their concatenation (called sum), $W_1 + W_2$. A walk in which the first vertex is the same as the last vertex is called a **circuit**. We shall identify a circuit $W = (v_0, v_1, \ldots, v_p)$ with the shifted circuit $(v_1, v_2, \ldots, v_{p-1}, v_0, v_1)$ and so on, but distinguish W and its inverse walk $-W$. A circuit is trivial if it is a sum of walks of the form $W - W$. Thus a circuit of the form $U+V-V-U+W-W$ is equal to $(U+V)-(U+V)+(W-W)$ and hence trivial. Circuits are equivalent if they differ only by trivial circuits.

Given a nonempty walk $W = (v_0, v_1, \ldots, v_p)$ in a graph G and an orientation D of G, we define an integer sign function $\mathrm{Sg}_{D,W}$ as follows. For the edge $e = v_i v_{i+1}$ of W

with $i \in [0, p-1]$, let $\mathrm{Sg}_{D,W}(e) = 1$ if $v_D^+(e) = v_i$ (that is, the edge e is in D directed from v_i to v_{i+1}) and $\mathrm{Sg}_{D,W}(e) = -1$ otherwise. Then

$$S_D(W) = \sum_{i=0}^{p-1} \mathrm{Sg}_{D,W}(v_i v_{i+1})$$

is the **flow difference** of W with respect to D. If W is an empty walk, we put $S_D(W) = 0$. An important property of the flow difference is its linearity, that is if W is the sum of two walks, say $W = W_1 + W_2$, then $S_D(W_1 + W_2) = S_D(W_1) + S_D(W_2)$. Furthermore, $S_D(-W) = -S_D(W)$ for every walk W of G. So if two circuits W and W' are equivalent, then $S_D(W) = S_D(W')$. As pointed out by Youngs [1085], the following result is an equivalent formulation of Minty's theorem.

Theorem 3.45 (MINTY'S THEOREM) *Let G be a graph and let k be and integer with $k \geq 2$. Then G has a coloring with a set of k colors if and only if G has an orientation D such that each circuit C of G satisfies*

$$|S_D(C)| \leq \frac{k-2}{k}|C|. \tag{3.17}$$

Proof If G has an orientation D such that (3.17) holds, then it is easy to check that each cycle C of G satisfies $r_D(C) \geq |C|/k$. Hence Theorem 3.2 implies that $\chi(G) \leq k$. If G has a coloring φ with a set of k colors, say $[1, k]$, then let D be the orientation of G such that the edge $uv \in E(G)$ is directed from u to v provided that $\varphi(u) < \varphi(v)$. Let a be the number of edges e of W with $\mathrm{Sg}_{D,C}(e) = 1$, and let b be the number of edges e of W with $\mathrm{Sg}_{D,C}(e) = -1$. Then it is easy to check that $\min\{a, b\} \geq |C|/k$, and so $|S_D(C)| = |a - b| \leq ((k-2)/k)|C|$. □

Theorem 3.46 (GALLAI) *There is an infinite family of 4-regular 4-critical graphs all of which can be embedded on the Klein bottle.*

Sketch of Proof: Given integers p and q such that $p \geq 4$ is even and $q \geq 3$ is odd, we first construct a planar grid graph $H = H_{pq}$ with vertex set

$$V(H) = \{v_{ij} = (i, j) \mid i \in [0, p], j \in [0, q]\}$$

and edge set

$$E(H) = \{v_{ij} v_{k\ell} \mid |i - j| + |k - \ell| = 1\}$$

For $(i, j) \in [0, p-1] \times [0, q-1]$ the sequence

$$W_{ij} = (v_{ij}, v_{i,j+1}, v_{i+1,j+1}, v_{i+1,j}, v_{ij})$$

is a circuit in H with length $|W_{ij}| = 4$. Then the circuit

$$W = \sum_{i=0}^{p-1} \sum_{j=0}^{q-1} W_{ij}$$

is equivalent to the circuit

$$W \equiv (v_{00}, \ldots, v_{0q}, v_{1q}, \ldots, v_{pq}, v_{p,q-1}, \ldots, v_{p0}, v_{p-1,0}, \ldots, v_{00}).$$

Now let $G = G_{pq}$ be the graph obtained from H by identifying v_{i0} with v_{iq} for $i \in [0, p]$ and v_{0j} with $v_{p,q-j}$ for $j \in [0, q]$. Then G can be embedded on the Klein bottle and $|G| = pq$. Furthermore, G is connected and 4-regular. We claim that G is 4-critical. To prove this, it suffices to show that $G - e$ has a 3-coloring for every edge e of G, but G does not (see Proposition 1.29).

To show that $\chi(G) \geq 4$, assume that G has a coloring with color set $\{1, 2, 3\}$ and let D be the corresponding orientation of G that satisfies (3.17). For $(i, j) \in [0, p-1] \times [0, q-1]$, let C_{ij} be the circuit of G obtained from W_{ij} by the identification operation. Then $C = (v_{00}, v_{01}, \ldots, v_{0q})$ is a circuit in G as v_{0q} and v_{00} are identified in G, and the circuit

$$C' = \sum_{i=0}^{p-1} \sum_{j=0}^{q-1} C_{ij}$$

is in G equivalent to $C' \equiv 2C$. Clearly, $|C_{ij}| = 4$ and it easily follows from (3.17) that $S_D(C_{ij}) = 0$. Hence $S_D(C') = 0$ and so $S_D(C) = 0$, which is however impossible as $|C| = q$ is odd.

It remains to show that $G_e = G - e$ has a 3-coloring for every edge e of G. To this end, it suffices to find an independent set I_e of G_e such that $G_e - I_e$ contains no odd cycle and is therefore bipartite. By symmetry it suffices to consider the edges $e = v_{1,i}v_{2,i}$ ($i \in [0, q']$) and $e = v_{2,j}v_{2,j+1}$ ($j \in [0, q']$), where $q = 2q' + 1$ and $p = 2p'$. First, assume that $e = v_{1,i}v_{2,i}$ with $i \in [0, q']$. If $i = 2i'$, then $I_e = I_e^1 \cup I_e^2 \cup I_e^3$, where

- $I_e^1 = \{v_{2h-1,2q'}, v_{2h,2q'+1} \mid h \in [2, p']\}$,
- $I_e^2 = \{v_{2,2k-1}, v_{1,2k} \mid k \in [i'+1, q']\}$, and
- $I_e^3 = \{v_{2,2l}, v_{1,2l+1} \mid l \in [0, i'-1]\}$

has the property that $G_e - I_e$ is bipartite. If $i = 2i' - 1$, then $I_e = I_e^1 \cup I_e^4 \cup I_e^5$, where

- $I_e^4 = \{v_{1,2k}, v_{2,2k+1} \mid k \in [i', q']\}$, and
- $I_e^5 = \{v_{1,2l-1}, v_{2,2l} \mid l \in [1, i'-1]\}$

has the property that $G_e - I_e$ is bipartite. Now, assume that $e = v_{2,j}v_{2,j+1}$ with $j \in [0, q']$. If $j = 2j'$, then $I_e = \{v_{2,2j'+1}\} \cup I_e^6 \cup I_e^7 \cup I_e^8$, where

- $I_e^6 = \{v_{1,2h}, v_{2,2h+1} \mid h \in [j'+1, q']\}$,
- $I_e^7 = \{v_{2k-1,2q'}, v_{2k,2q'+1} \mid k \in [2, p']\}$, and
- $I_e^8 = \{v_{1,2l-1}, v_{2,2l} \mid l \in [1, j']\}$

has the property that $G_e - I_e$ is bipartite. If $j = 2j' - 1$, then $I_e = \{v_{00}, v_{11}\} \cup I_e^9 \cup I_e^{10} \cup I_e^{11}$, where

- $I_e^9 = \{v_{2,2h+1}, v_{3,2h} \mid h \in [j', q']\}$,
- $I_e^{10} = \{v_{2k,2q'+1}, v_{2k+1,2q'} \mid k \in [1, p'-1]\}$, and
- $I_e^{11} = \{v_{1,2l+1}, v_{2,2l} \mid l \in [1, j'-1]\}$

has the property that $G_e - I_e$ is bipartite. This shows that G is a 4-critical graph. □

The proof of the above theorem follows Gallai's argument from [397, Section (2.3)]; Gallai was the first to present an example of a 4-critical graph whose low vertex subgraph is empty. The smallest such graph G_{43} is shown in Figure 3.4. This graph is particularly interesting as it shows that the cycle-plus-triangles theorem (Theorem 3.29) cannot be relaxed to a cycles-plus-triangles theorem, even with just two cycles. As proved by Gallai, the graph $G = G_{pq}$ with $q = 2q' + 1$ and $p = 2q' \geq 4$ has odd girth $g_o(G) = 2q' + 1$, and so $|G| = 2q'(2q' + 1) < g_o(G)^2$. Hence for infinitely many values of n there are 4-critical graphs G having order n and odd girth $g_o(G) > \sqrt{n}$ (see also Theorem 8.38). To show that the Gallai graph $G = G_{pq}$ satisfies $\chi(G) > 3$ Gallai only used the trivial part of Minty's theorem; to show that $\chi(G - e) \leq 3$, Gallai constructed a 3-coloring, which of course gives an orientation that satisfies Minty's condition (3.17) with $k = 3$. As noticed by Gallai, the graph G_{pq} with $p \geq 4$ odd and $q \geq 3$ odd is 4-critical, too. Pretzel and Youngs [826, 827] developed some method which can be used to apply Minty's theorem without constructing an explicit coloring. Pretzel and Youngs [826] obtained, in particular, a characterization of functions defined on circuits which are flow differences. Combined with Minty's theorem this leads to the following result, see [1085].

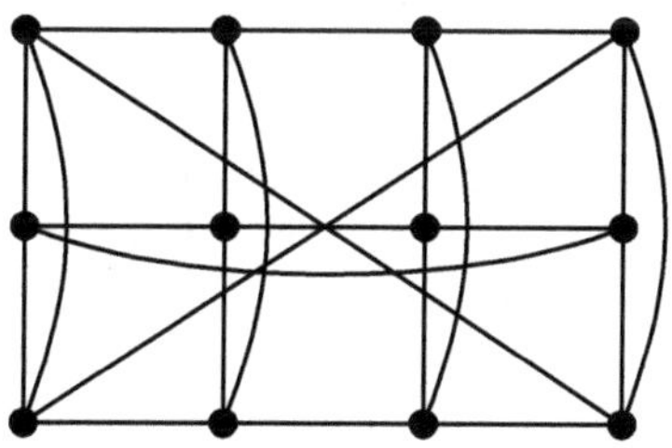

Fig. 3.4 Gallai's 4-critical graph G_{43}

Lemma 3.47 (Pretzel and Youngs) *Let G be a graph and let k be and integer with $k \geq 2$. Then G has a coloring with a set of k colors if and only if there exists an integer-valued function f defined on the circuits of G which satisfies, for all circuits C and D, the following conditions:*

$$1.\ |f(C)| \leq \frac{k-2}{k}|C|,\ 2.\ f(C) \equiv |C| \pmod 2,\ \text{and } 3.\ f(C+D) = f(C) + f(D).$$

Youngs [1085] presented a family of 4-critical projective planar graphs (see Section C.13 for the terminology). To show that a graph G of this family satisfies $\chi(G) > 3$ he used the trivial part of Minty's theorem; to show that $\chi(G - e) \leq 3$ he used Lemma 3.47.

Theorem 3.48 (Youngs) *Suppose that G is a nonbipartite graph that can be embedded in the projective plane such that each facial subgraph is a 4-cycle and each 4-cycle is a facial cycle. Then G is 4-critical.*

The family of projective plane graphs satisfying the hypothesis of Youngs' theorem includes many well-known graphs, in particular, the odd wheels and the generalized Mycielski graphs, defined in Section 8.4, including the Mycielski–Grötzsch graph shown in Figure 8.4. Some other interesting 4-critical graphs belonging to this family are presented by Youngs [1083] as well as by Nielsen and Toft [784].

In 1966, Gallai [400] (see also Toft [1028]) asked whether every k-critical graph admits an orientation having just one directed path of order $k-1$. Youngs [1084] proved that the answer in general is negative, but the answer is affirmative when $k \geq 5$ and the graph has maximum degree at most k.

The concept of a kernel in a digraph was introduced by von Neumann and Morgenstern [779] in the context of game theory. They proved that, in digraphs associated with certain combinatorial games, the existence of a kernel leads to a winning strategy for one of the players (see Exercise 3.12). They also proved that any (finite) acyclic digraph has a unique kernel (Exercise 3.11). Over the years, much attention has been paid to finding sufficient conditions for the existence of a kernel; the interested reader is referred to [162] for a detailed description. Early investigations in this direction were mostly motivated by application in game theory, but later the subject developed on its own. In the study of perfect graphs, kernels of digraphs are of particular interest, and much work has revolved around kernel-perfect digraphs. A lot of the theoretical interest in the field undoubtedly stems from the deep relation between kernel-perfect digraphs and orientations of perfect graphs, which finally led to the kernel solvable graph theorem (Theorem 3.15). Our proof of this theorem based on Scarf's lemma (see Theorem 3.16) is from the paper by Aharoni and Holzman [23]. The proof of Aharoni and Holzman was further simplified by Király and Pap [599] using Sperner's lemma

The kernel method (Lemma 3.4) is a simple, but very useful observation. Alon and Tarsi [58, Remark (2.4)] attributed this observation to Bondy, Boppana, and Siegel; it was stated for the special case when D has no odd directed cycle and $g \equiv 1$, but the generalization is obvious. The most interesting application to date of the kernel method is Galvin's proof [402] of the list edge coloring conjecture (LECC) for bipartite multigraphs. Galvin's proof is surprisingly simple and from first principle, without using earlier edge coloring results. Alternative versions of Galvin's proof have been given by Slivnik [949] as well as by Borodin, Kostochka, and Woodall [160]. Our proofs of Theorems 3.6 and 3.8 follow [160]. The advantage of the proof by Borodin, Kostochka, and Woodall [160] is that it not only leads to a list coloring result for bipartite multigraphs with variable list sizes, but also to a choosability analogue of the classical theorem of Shannon [934] that $\chi'(G) \leq \frac{3}{2}\Delta(G)$ for every multigraph G. However, the idea to associate with an edge coloring φ of a given bipartite multigraph G the digraph D_{φ} (see Claim 3.6.1) is the key also in Galvin's original proof. It is easy to see that if φ is an edge coloring of G with color set $C = [1,k]$, then D_{φ} has maximum out degree at most $k-1$. It is also easy to see that D_{φ} is a normal super-orientation of the line graph of G. As noted by Galvin in [402] the fact that a normal super-orientation of the line graph of a bipartite multigraph has a kernel was first proved for graphs by Gale and Shapley [396] and further extended to multigraphs by Maffray [713] (see also Exercise 3.20). Lloyd S.

Shapley received a Nobel Prize in Economic Science in 2012 for *the theory of stable allocations and the practice of market design*, where the concept of stable matchings was the basic idea. Maffray [713] verified the kernel solvable theorem for the class of line graphs of multigraphs. The kernel method was used by Plantholt and Tipnis [809] to prove the LECC for every multigraph whose underlying graph is a bipartite graph plus one edge; the proof was simplified by McDonald [730]. A multigraph is called **line perfect** if its line graph is perfect. Maffray [713] characterized line perfect multigraphs. That the LECC holds for line perfect multigraphs was proved by means of the kernel method by Peterson and Woodall [804]. Woodall [1073] extended this to multigraphs whose underlying graph is a cycle. An extension of Galvin's theorem was recently presented by Fleiner [378]. He proved that any graph G has an L-edge-coloring provided that $|L(e)| \geq \chi'(G)$ for all edges e of G and the edge-lists of an odd cycle never contain a common color.

There are various polynomials that are naturally associated with a graph (multigraph, or multidigraph), such as the graph polynomial introduced in this chapter, which is also referred to as the **adjacency polynomial**, the characteristic polynomial, the chromatic polynomial, and several others. It was the idea of Sylvester [984] in collaboration with Petersen [803] to associate with a polynomial of the form $P = (x_1 - x_2)^p (x_1 - x_3)^q \cdots (x_{n-1} - x_n)^r$ a multigraph G such that, in our terminology, $P = P_G$. One of the main problems of algebraic graph theory is to determine precisely how, or whether, properties of graphs are reflected in the algebraic properties of such polynomials. It is a trivial observation that the graph polynomial encodes information about its (proper) colorings. However, it was the merit of Alon and Tarsi [58] to develop the necessary algebraic and combinatorial tools in order to use this observation for proving upper bounds for the list chromatic number of graphs. One such tool is the Combinatorial Nullstellensatz (Theorem 3.21). The proof of this theorem follows Alon's paper [48]; in this paper the author discuss further applications of this result in combinatorial number theory, in graph theory, and in combinatorics. Another important ingredient of the Alon-Tarsi polynomial method is the connection between the coefficients of the graph polynomial and its orientations. The proofs of Lemma 3.22 and 3.23 are from [58]. The proof of Theorem 3.25 about uniquely list colorable graphs is due to Akbari, Mirrokni, and Sadjad [25].

The paper of Alon and Tarsi [58] has strongly influenced the research on list coloring of graphs. The first easy application of the Alon-Tarsi method, contained in [58], is the result that every planar bipartite graph is 3-list-colorable. However, this result can also be deduced by means of the kernel method. The next application of the Alon-Tarsi method is the cycle-plus-triangles-theorem; the results and proofs given in Section 3.6 are from the paper by Fleischner and Stiebitz [379]. Ellingham and Goddyn found a new interpretation for the coefficients of the graph polynomial and used the Alon-Tarsi approach to establish the LECC for the class of planar r-regular multigraphs having chromatic index r. Surprisingly, the LECC is unsolved for the class of complete graphs. Häggkvist and Jansen [451] proved by means of Lemma 3.22 that $\chi'_\ell(K_n) \leq n$ for all $n \geq 1$ thereby establishing the LECC for complete graphs of odd order. That the LECC is true for complete graphs of order $p+1$ where p is an odd prime number was proved by Schauz [896]. He proved, using the so-called Quantitative Combinatorial Nullstellensatz, that if $P = P_{L(K_{p+1})}$ then $\text{coef}_P(p-1, p-1, \ldots, p-1) \neq 0$. The parameter $\Delta^+_{\text{at}}(G)$ was introduced by Jensen and Toft [530, Problem 13.9] and as shown by Hefetz [492] $\chi(G) = \chi_\ell(G) = \Delta^+_{\text{at}}(G) + 1$ holds for certain families of graphs G. The parameter $\Delta^+_{\text{at}} + 1$ is usually called **Alon-Tarsi number**; it is not only an upper bound for the list chromatic number, but also for the **online list chromatic number** (also known as **paintability number**), as proved by Schauz [895]. Zhu [1102] proved that every planar graph G satisfies $\Delta^+_{\text{at}}(G) \leq 4$ and therefore has Alon-Tarsi number at most 5, thereby solving a problem posed by Hefetz [492]. Juvan, Mohar, and Thomas [543] proved by a more direct argument that the LECC holds for series-parallel graphs (but not for series-parallel multigraphs).

A graph G is called **chromatic-choosable** if $\chi(G) = \chi_\ell(G)$. Certain classes of graphs are known to be chromatic-choosable. Prowse and Woodall [828] proved by means of the Alon-Tarsi method that powers of cycles are chromatic-choosable. That every graph G with $|G| \leq 2\chi(G) + 1$ is chromatic choosable was conjectured by Ohba [792] and proved by Noel, Reed, and Wu [789].

Kostochka and Woodall [638] conjectured that the square of every graph is chromatic choosable. Recall that the square of a graph G is the graph $H = G^2$ with the same vertex set as G, where two vertices are adjacent in H if and only if their distance in G is at most 2. The conjecture became known as the List Square Coloring Conjecture. In 2015, Kim and Park [592] constructed infinitely many counterexamples to the List Square Coloring Conjecture. Furthermore, it is proved in [592] that the gap $\chi_\ell(G^2) - \chi(G^2)$ can be arbitrarily large.

The proof of Theorem 3.33 follows the paper by Alon and Tarsi [59]. The proofs of Lemma 3.34, Theorem 3.35, and Corollaries 3.36, 3.37, 3.38 are from Tarsi's paper [993]. A proof of Brooks' theorem based on the Alon-Tarsi method (see Lemma 3.41) was given by Hladký, Král', and Schauz [500]; they proved that $\Delta^+_{\text{at}}(G) \leq \Delta(G) - 1$ holds for any connected graph which is not an odd cycle or a complete graph (see Corollary 3.42). A proof of Brooks theorem based on the kernel method was given by Stiebitz, Tuza, and Voigt [979]. The proofs of Theorem 3.39, Lemma 3.40, Proposition 3.43, and Theorem 3.44 are all from [979].

Bernshteyn and Lee [112] defined the **weak degeneracy** $\text{wdg}(G)$ of a graph G as a new graph parameter (see Exercise 3.32) and introduced the concept of a weakly f-reducible graph (Bernshteyn and Lee use the term **weakly f-degenerate** instead of weakly f-reducible, see Exercise 3.31). One of the many interesting results obtained in [112] is the following strengthening of Corollary 1.19 (see also Lemmas 3.40 and 3.41 for similar results).

Theorem 3.49 (Bernshteyn and Lee) *Let G be a connected graph and let f be the vertex function of G with $f(v) = d_G(v) - 1$. Then G is not weakly f-reducible if and only if every block of G is a complete graph or a cycle.*

In this chapter we have discussed various results of the type that a given graph possess a certain coloring property if it admits an appropriate orientation. The most popular results of this type are the kernel method and the graph polynomial method.

On the other hand, the generalized coloring concept for graphs and multigraphs has a natural counterpart for multidigraphs. Let $\mathcal{D}$ be a digraph property and let D be a multidigraph. A **$\mathcal{D}$-coloring** of D with color set C is a mapping $\varphi : V(D) \to C$ such that $D[\varphi^{-1}(c)] \in \mathcal{D}$ for every color $c \in C$. The **$\mathcal{D}$-chromatic number** of D, denoted by $\vec{\chi}(D : \mathcal{D})$, is the least integer k for which there exists a $\mathcal{D}$-coloring of G with a set of k colors. If $L : V(D) \to 2^C$ is a list-assignment for D, then a **$(\mathcal{D}, L)$-coloring** of D is a $\mathcal{D}$-coloring φ of D such that $\varphi(v) \in L(v)$ for all $v \in V(D)$. The **$\mathcal{D}$-list-chromatic number** of D, denoted by $\vec{\chi}_\ell(D : \mathcal{D})$, is the smallest inter k such that D has a $(\mathcal{D}, L)$-coloring for every k-assignment L of D. Evidently, $\vec{\chi}(D : \mathcal{D}) \leq \vec{\chi}_\ell(D : \mathcal{D})$. Until now, this digraph coloring concept was only considered for the class $\mathcal{D}_{\text{ac}}$ of acyclic multidigraphs. In this case a $\mathcal{D}_{\text{ac}}$-coloring is also called an **acyclic coloring** and a $(\mathcal{D}_{\text{ac}}, L)$-coloring is also called an **acyclic L-coloring**; and we put $\vec{\chi}(D) = \vec{\chi}(D : \mathcal{D}_{\text{ac}})$ and $\vec{\chi}_\ell(D) = \vec{\chi}_\ell(D : \mathcal{D}_{\text{ac}})$. This particular coloring concept for digraphs was introduced by Neumann-Lara [780], he used the name **dichromatic number** for $\vec{\chi}(D)$. Note that if G is a graph, and $D = D^{\pm}(G)$ is the multidigraph obtained from G by replacing each edge with the pair of oppositely directed edges joining the same pair of vertices, then a mapping $\varphi : V(G) \to C$ is a coloring of G if and only if φ is an acyclic coloring of D. Consequently, every graph G satisfies

$$\vec{\chi}(D^{\pm}(G)) = \chi(G) \text{ and } \vec{\chi}_\ell(D^{\pm}(G)) = \chi_\ell(G). \tag{3.18}$$

Recall that a multidigraph is simple, if it has no multiple edges, i.e., it is an orientation of a graph. A multidigraph is strict if its has no parallel edges, i.e., it is a super-orientation of a graph. Clearly, if D is a multidigraph and D' is a maximal strict subdigraph of D, then $\vec{\chi}(D) = \vec{\chi}(D')$ and $\vec{\chi}_\ell(D) = \vec{\chi}_\ell(D')$. As observed by von Neumann-Lara [780], based on a sequential coloring argument it is easy to show that every multidigraph D satisfies

$$\vec{\chi}(D) \leq \vec{\chi}_\ell(D) \leq \min\{\Delta^+(D), \Delta^-(D)\} + 1. \tag{3.19}$$

Harutyunyan and Mohar [474] proved that if D is a connected multidigraph, then

$$\vec{\chi}_\ell(D) = \max\{\Delta^+(D), \Delta^-(D)\} + 1$$

if and only if D is a directed cycle of length ≥ 2, or $D = D^{\pm}(K_n)$ for some $n \geq 3$, or $D = D^{\pm}(C_n)$ for some odd $n \geq 3$. This generalization of Brooks' theorem is a simple consequence of the following result due to Harutyunyan and Mohar [474]. A multidigraph D is **connected** if its underlying graph G is connected; and a **block** of D is a an induced subgraph B of D such that $G[V(B)]$ is a block of G. The set of blocks of a multidigraph D is denoted by $\mathfrak{B}(D)$ and, for a vertex $v \in V(D)$, let $\mathfrak{B}_v(D) = \{B \in \mathfrak{B}(D) \mid v \in V(B)\}$. A **tournament** is an orientation of a complete graph.

Theorem 3.50 (Harutyunyan and Mohar) *Let D be a connected multidigraph, and let L be a list-assignment for D such that $|L(v)| \geq \max\{d_D^+(v), d_D^-(v)\}$ for all $v \in V(D)$. If D has no acyclic L-coloring, then the following statements hold:*

(a) *D is Eulerian and $|L(v)| = d_D^+(v) = d_D^-(v)$ for all $v \in V(D)$.*
(b) *If $B \in \mathfrak{B}(D)$, then B is a directed cycle of length ≥ 2, or $B = D^{\pm}(K_n)$ for some $n \geq 1$, or $B = D^{\pm}(C_n)$ for some odd $n \geq 3$.*
(c) *For each block B of D, there is a set C_B of $\Delta^+(B)$ colors, such that for every vertex v of D, the sets C_B with $B \in \mathfrak{B}_v(D)$ are pairwise disjoint and $L(v) = \bigcup_{B \in \mathfrak{B}_v(D)} C_B$.*

We shall deduce the above theorem from the following result about $\mathcal{DG}_1$-list colorings of multigraphs due to Schweser [906, Theorem 3.11], where $\mathcal{DG}_1$ is the class of forests.

Theorem 3.51 (SCHWESER) *Let G be a connected multigraph with $|G| \geq 2$, and let L be a list-assignment for G such that $2|L(v)| \geq d_G(v)$ for all $v \in V(G)$. Then G has no $(\mathcal{DG}_1, L)$-coloring if and only if the following two conditions hold:*

(a) *If $B \in \mathfrak{B}(G)$, then $B = tK_n$ with $n \geq 2$, $t \in \{1,2\}$ and $t(n-1) \equiv 0 \pmod 2$, or $B = 2C_n$ with $n \geq 3$ odd, or $B = C_n$ with $n \geq 2$.*
(b) *For $B \in \mathfrak{B}(G)$, there is a set C_B of $\Delta(B)/2$ colors, such that for every vertex $v \in V(G)$, the sets C_B with $B \in \mathfrak{B}_v(G)$ are pairwise disjoint and $L(v) = \bigcup_{B \in \mathfrak{B}_v(G)} C_B$.*

Proof We may assume that the color set $C = \bigcup_v L(v)$ satisfies $C = [1, p]$ with $p \geq 1$. Let $f \in \mathcal{V}_p(G)$ be the vector function with $f_i(v) = 2$ if $i \in L(v)$ and $f_i(v) = 0$ otherwise (for $v \in V(G)$ and $i \in [1, p]$). Then

$$(+) \quad f_1(v) + f_2(v) + \ldots + f_p(v) = 2|L(v)| \geq d_G(v) \text{ for all } v \in V(G).$$

Furthermore, G has a $(\mathcal{DG}_1, L)$-coloring if and only if G has an f-partition (see Section 2.5). Consequently, by (+) and Theorem 2.26, G has no $(\mathcal{DG}_1, L)$-coloring if and only if (G, f) is a hard pair.

If G and L satisfies (a) and (b), then it is easy to check that (G, f) is a hard pair (use Exercise 2.27) and hence G has no $(\mathcal{DG}_1, L)$-coloring. Now assume that G has no $(\mathcal{DG}_1, L)$-coloring. Then (G, f) is a hard pair and so $2|L(v)| = d_G(v)$ for all $v \in V(G)$ (by (+) and Proposition 2.27(a)). Furthermore, for each block $B \in \mathfrak{B}(G)$ there is a unique function $f_B \in \mathcal{V}_p(B)$ such that (B, f_B) is a hard pair of type (H1), (H2), or (H3) and $f(v) = \sum_{B \in \mathfrak{B}_v(G)} f_B(v)$ for all $v \in V(G)$ (see Exercise 2.27).

We claim that $f_B(v) \in \{0, 2\}^p$ whenever $B \in \mathfrak{B}(G)$ and $v \in V(B)$. The proof is by induction on the number of m of blocks of G. If $m = 1$, then G is a block and the claim is obvious. So assume that $m \geq 2$. Then G has an end-block, say B. Let v be the only separating vertex of G contained in B, let $G' = G - (V(B) \setminus \{v\})$, and let $f'(w) = f(w)$ for all $w \in V(G') \setminus \{v\}$ and $f'(v) = f(v) - f_B(v)$. Then (G', f') is a hard pair (see Exercise 2.27) satisfying $f'(w) \in \{0, 2\}^p$ for all $w \in V(G') \setminus \{v\}$ and $f'(v) \in \{0, 1, 2\}^p$. If $f'(v) \in \{0, 2\}^p$, then $f_B(v) \in \{0, 2\}^p$, too, and so we are done by induction. Otherwise one coordinate of $f'(v)$ equals 1, but then the same coordinate of $f_B(v)$ equals 1, too. Consequently, (B, f_B) is a mono-block (block of type (H1)) and $d_B(v) = 1$. Since B is an end-block of G, this implies that $B = K_2$, which is impossible. This proves the claim that $f_B(v) \in \{0,2\}^p$ whenever $B \in \mathfrak{B}(G)$ and $v \in V(G)$.

Let $B \in \mathfrak{B}(G)$ be an arbitrary block. Since $|G| \geq 2$, we have $|B| \geq 2$. Furthermore, (B, f_B) is a block of type (H1), (H2), or (H3). If (B, f_B) is of type (H1), then B is 2-regular and hence a cycle. If (B, f_B) is of type (H2), then $B = tK_n$ for some $t \geq 1$ and $n \geq 2$ and there are integers $n_1, n_2, \dots, n_p$ such that $n_1 + n_2 + \cdots + n_p = n - 1$ and $f(v) = (tn_1, tn_2, \dots, tn_p)$ for all $v \in V(B)$. Since $f_B(v) \in \{0,2\}^p$ for all $v \in V(B)$, this is only possible if $t \in \{1,2\}$ and $t(n-1) \equiv 0 \pmod 2$. If (B, f_B) is of type (H3), then $B = tC_n$ with $t \geq 1$ and $n \geq 3$ odd, and $f(v) = (t, t, 0, \dots, 0)$ (except for symmetry) for all $v \in V(B)$, which clearly implies that $t = 2$. This proves (a). Statement (b) is obvious. □

Proof of Theorem 3.50: Clearly, it suffices to prove the result for a connected strict digraph D. If $|D| = 1$, the result is obvious. So assume that $|D| \geq 2$. For $v \in V(D)$, let $a(v) = \max\{d_D^+(v), d_D^-(v)\}$ and let L be an a-assignment for which D has no acyclic L-coloring. Let G be the multigraph satisfying $V(G) = V(D)$, $E(G) = E(D)$, and $i_G(e) = \{v_D^+(e), v_D^-(e)\}$ for all $e \in E(G)$. So G is obtained from D just by ignoring the orientation of the edges. Consequently, if $X \subseteq V(G) = V(D)$, then $G[X] \in \mathfrak{B}(G)$ if and only if $D[X] \in \mathfrak{B}(D)$. Furthermore, $d_G(v) = d_D^+(v) + d_D^-(v) \leq 2a(v)$ for all $v \in V(G)$ and, since D is a strict digraph, $\mu(G) \leq 2$. Clearly, L is a list-assignment for G such that $2|L(v)| = 2a(v) \geq d_G(v)$ for all $v \in V(G)$. If G has an $(\mathcal{DG}_1, L)$-coloring, then every color class induces a forest in G and hence an acyclic subdigraph in D, and so D would have an acyclic L-coloring, which is impossible. Consequently, G has no $(\mathcal{DG}_1, L)$-coloring. From Theorem 3.51(a) it then follows that if $B \in \mathfrak{B}(G)$, then $B = tK_n$ with $n \geq 2$, $t \in \{1,2\}$ and $t(n-1) \equiv 0 \pmod 2$, or $B = 2C_n$ with $n \geq 3$ odd, or $B = C_n$ with $n \geq 2$. Moreover, Theorem 3.51(b) implies that every vertex v satisfies $2|L(v)| = d_G(v)$ and hence $d_G(v) = d_D^+(v) + d_D^-(v) = 2a(v)$. This implies, in particular, that D is Eulerian, i.e., $d_D^+(v) = d_D^-(v)$ for all $v \in V(D)$. Hence, $|L(v)| = d_D^+(v) = d_D^-(v)$ for all $v \in V(D)$, which proves (a). Since D is Eulerian, an easy induction on the number of blocks of D shows that each block of D is Eulerian, too (note that it is not possible that for all but one vertex of a digraph the in-degree equals the out-degree). Then Theorem 3.51(b) also implies statement (c). Now, let $B \in \mathfrak{B}(D)$ be an arbitrary block and let $B' = G[V(B)]$ be the corresponding block of G. If $B' = C_n$ with $n \geq 2$, then B is a directed cycle of length $n \geq 2$. If $B' = 2C_n$ with $n \geq 3$ odd, then $B = D^{\pm}(C_n)$. It remains to consider the case that $B' = tK_n$ with $n \geq 2$, $t \in \{1,2\}$, and $t(n-1) \equiv 0 \pmod 2$. If $t = 2$, then $B = D^{\pm}(K_n)$ and we are done. If $t = 1$, then $n \geq 3$ is odd and $n = 3$ implies again that B is a directed cycle. Otherwise, $B' = K_n$ with $n \geq 5$ odd, and so B is an Eulerian tournament of order n. The proof is completed by showing that this leads to an acyclic L-coloring of D.

So in what follows, assume that D has a block B which is an Eulerian tournament of odd order $n \geq 5$. Let C_B be the color set as described in statement (c). Then $|C_B| = (n-1)/2$ and $C_B \subseteq L(v)$ for all $v \in V(B)$. First, we claim that B admits an acyclic coloring with color set C_B. To this end, we chose a vertex $v \in V(B)$ and two distinct out-neigbors $u, w \in V(B)$ of v (which exist as $n \geq 5$ and $d_D^+(v) = (n-1)/2 \geq 2$). Then $B[\{u, v, w\}]$ is acyclic and so we can assign the three vertices u, v and w the same color c from C_B. Afterwards, we group the remaining vertices of B into $(n-3)/2$

pairs and assign each pair a unique color from the color set $C_B \setminus \{c\}$. Since B is a tournament, this leads to an acyclic coloring φ_B of B with color set C_B.

Next we want to show that we can extend φ_B to an acyclic L-coloring of D. For an arbitrary vertex $v \in V(B)$, let D_v be the component of $D - (V(B) \setminus \{v\})$ containing v, and let G_v be the component of $G - (V(B) \setminus \{v\})$ containing v. Note that $V(D_v) = V(G_v)$. Moreover, let L_v be the list-assignment of D_v (and hence of G_v) with $L_v(v) = L(v) \setminus C_B$ and $L_v(w) = L(w)$ for $w \in V(D_v) \setminus \{v\}$. If $v \in V(B)$ is a nonseparating vertex of D, then $G_v = K_1$ and $L_v(v) = \varnothing$, which implies that G_v has no $(\mathcal{DG}_1, L_v)$-coloring. If $v \in V(B)$ is a separating vertex of D, then $|G_v| \geq 2$ and (c) implies that $2|L_v(w)| \geq d_{G_v}(w)$ for all $w \in V(G_v)$ and (G_v, L_v) satisfy statements (a) and (b) of Theorem 3.51. Consequently, G_v has no $(\mathcal{DG}_1, L_v)$-coloring, too. Now, for $v \in V(B)$, let L'_v be the list-assignment of D_v (and G_v)) such that $L'_v(w) = L_v(w) = L(v)$ for all $v \in V(D_v) \setminus \{v\}$ and $L'_v(v) = L_v(v) \cup \{\varphi_B(v)\} = L(v) \setminus (C_B \setminus \{\varphi_B(v)\})$. If v is a nonseparating vertex of D, then $G_v = K_1$ and $L'_v(v) = \{\varphi_B(v)\}$, which implies that φ_v with $\varphi_v(v) = \varphi_B(v)$ is a $(\mathcal{DG}_1, L'_v)$-coloring of G_v and, therefore, an acyclic L'_v coloring of D_v. If v is a separating vertex of D, then $|G_v| \geq 2$, $2|L'_v(v)| > 2|L_v(v)| \geq d_{G_v}(v)$ and $2|L'_v(w)| \geq d_{G_v}(w)$ for all $w \in V(G_v) \setminus \{v\}$. Then Theorem 3.51 implies that G_v admits an $(\mathcal{DG}_1, L'_v)$-coloring φ_v, and as G_v has no $(\mathcal{DG}_1, L_v)$-coloring, we have $\varphi_v(v) = \varphi_B(v)$. Clearly, φ_v is an acyclic L_v-coloring of D_v. Since

$$D = B \cup \bigcup_{v \in V(B)} D_v$$

and since the digraphs D_v are pairwise disjoint, it follows that $\varphi = \bigcup_{v \in V(B)} \varphi_v$ is an acyclic L-coloring of D, which is impossible. This completes the proof of the theorem. □

In view of Theorem 3.50 and (3.19) it is natural to consider the decision problem whether, given multidigraph D and list-assignment L of D such that $|L(v)| \geq \min\{d^+_D(v), d^-_D(v)\}$, there exists an acyclic L-coloring of D. As proved by Harutyunyan and Mohar [474], this decision problem is surprisingly NP-complete, even restricted to planar digraphs with $\min\{d^+_D(v), d^-_D(v)\} \leq 3$ for all $v \in V(D)$. As proved in [126] the decision problem whether a given multidigraph D satisfies $\vec{\chi}(D) \leq 2$ is NP-complete, see also [503]. On the other hand, Neumann-Lara [781] proposed the following conjecture.

Conjecture 3.52 (Planar 2-Coloring Conjecture) *If D is an orientation of a planar graph, then $\vec{\chi}(D) \leq 2$.*

The above conjecture was independently proposed by Škrekovsky as mentioned in [126]. In 2016, Harutyunyan and Mohar [477] proved that planar digraphs of digirth at least five are acyclic 2-colorable; one year later Li and Mohar [674] proved the same conclusion for planar digraphs of digirth at least four. A first comprehensive study of the dichromatic number of digraphs on arbitrary surfaces was presented by Albouker, Havet, Knauer, and Rambaud [16] in 2022, see also the paper by Kostochka and Stiebitz [635].

After the introduction of the dichromatic number in the 1980s by Neumann-Lara, this coloring concept received only little attention and in the following years only few papers appeared on this subject. It was only when the concept was rediscovered at the beginning of the millennium by Mohar [748] and by Bokal, Fijavž, Juvan, Kayll, and Mohar [126] that the topic began to flourish and has since then attracted much more interest, see e.g. [13, 14, 16, 17, 22, 65, 77, 78, 80, 95, 126, 213, 423, 473, 474, 475, 476, 477, 503, 506, 563, 635, 674, 748, 750, 960, 962]. On the one hand, many interesting result on the chromatic number of graphs can be carried over to the dichromatic number of strict digraphs in a natural way; a typical example is Theorem 3.50. On the other hand, it follows from (3.18) that results about the dichromatic number of strict digraphs imply corresponding results about the chromatic number of graphs. In 2015, Chen, Ma, and Zang [213] continued the study of the relationship between colorings and cycle spectrum of graphs and digraphs. (see Section 2.4). In particular, they obtained the following result, thereby answering two questions proposed by Tuza [1041].

Theorem 3.53 (Chen, Ma, and Zang) *Let k and r be integers such that $k \geq 2$ and $k \geq r \geq 1$. Then the following statements hold:*

(a) *If a strict digraph D contains no directed cycle of length r modulo k, then $\vec{\chi}(D) \leq k$.*
(b) *If a graph G contains no cycle of length r modulo k, then $\chi(G) \leq k$ if $r \neq 2$ and $\chi(G) \leq k+1$ otherwise.*

Given a graph G, define

$$\vec{\chi}(G) = \max\{\vec{\chi}(D) \mid D \text{ is an orientation of } G\}$$

and

$$\vec{\chi}_\ell(G) = \max\{\vec{\chi}_\ell(D) \mid D \text{ is an orientation of } G\}.$$

Then the above conjecture claims that every planar graph G satisfies $\vec{\chi}(G) \leq 2$. Clearly, the point aboricity of a graph G is an upper bound for $\vec{\chi}(G)$, i.e. $\vec{\chi}(G) \leq \chi(G : \mathcal{DG}_1)$. Using the fact that planar graphs are 5-degenerate, it follows that every planar graph G satisfies $\vec{\chi}(G) \leq \vec{\chi}_\ell(G) \leq 3$ (use Exercise 2.29). However, Chartrand, Kronk, and Wall [206] constructed planar graphs G with $\chi(G : \mathcal{DG}_1) = 3$. While $\chi_\ell(K_n : \mathcal{DG}_1) = \lceil n/2 \rceil$, it was proved by Bensmail, Harutyunyan, and Le [95] that

$$\vec{\chi}_\ell(K_n) \leq \frac{n}{\log_2 n}(1 + o(n)).$$

An orientation of a graph is also referred to as an **oriented graph**. A directed cycle of length two is usually called a **digon**. So a digraph is an oriented graph if and only if it is a strict and digon-free digraph. It seems that the class of oriented graphs play a similar role with respect to the dichromatic number as triangle-free graphs do with respect the chromatic number. This is supported by the following conjecture proposed in the 1980s by Erdős [337].

Conjecture 3.54 (Digraph Conjecture) *Every graph G with maximum degree Δ satisfies $\vec{\chi}(G) \leq O(\frac{\Delta}{\log_2 \Delta})$.*

Several recent papers are devoted to this conjecture (e.g., [95, 423, 474, 962]). However, the conjecture is wide open.

Golowich [423] was the first to consider $\mathcal{D}$-coloring of digraphs for digraph properties $\mathcal{D}$ other than $\mathcal{D}_{\text{ac}}$. A digraph D is called **weakly k-degenerate** if every nonempty subdigraph D' of D has a vertex v satisfying

$$\min\{d^+_{D'}(v), d^+_{D'}(v)\} < k.$$

Let $\mathcal{D}_k$ denote the class of weakly k-degenerate digraphs. It is well known and easy to prove that $\mathcal{D}_{\text{ac}} = \mathcal{D}_1$. For a graph G, let

$$\vec{\chi}_k(G) = \max\{\vec{\chi}(D : \mathcal{D}_k) \mid D \text{ is an orientation of } G\}$$

Note that, for every graph G, we have $\vec{\chi}_1(G) = \vec{\chi}(G)$. Golowich proved that, for every $k \in \mathbb{N}$, every graph G satisfies

$$\vec{\chi}_k(G) \leq \left\lfloor \frac{\Delta(G) - \lfloor (\Delta(G)+1)/(4k+1) \rfloor}{2k} \right\rfloor.$$

For a digraph D, let $\Delta(D) = \max\{d^+_D(v) + d^-_D(v) \mid v \in V(D)\}$ be the **maximum total degree** of D, and let

$$\tilde{\Delta}(D) = \max\{\sqrt{d^+_D(v) \cdot d^-_D(v)} \mid v \in V(D)\}$$

be the **maximum geometric mean** of D. Note that if D is an orientation of a graph G, then $\Delta(D) = \Delta(G)$; and every digraph D satisfies $\tilde{\Delta}(D) \leq \Delta(D)/2$. For $k = 1$, the above inequality for $\vec{\chi}_k$ yields that every oriented graph D satisfies $\vec{\chi}(D) \leq \lfloor \frac{2}{5}(\Delta(D)+1) \rfloor + 1$. Using this bound, Golowich proved that every oriented graph D satisfies

$$\vec{\chi}(D) \leq \lfloor \sqrt{2/3}\, \tilde{\Delta}(D) + 7/5 \rfloor,$$

thus improving an earlier bound depending on $\tilde{\Delta}$ due to Harutyunyan and Mohar [475] (for a list version see [95]). Bang-Jensen, Schweser, and Stiebitz [80] established a Brooks type result for the $\mathcal{D}_k$-chromatic number and the $\mathcal{D}_k$-list-chromatic number. In particular, they proved that if D is a connected digraph and $m = \max_v\{d^+_D(v), d^-_D(v)\}$, then $\vec{\chi}_k(D) \leq \lceil m/k \rceil$, unless D is a bidirected odd cycle $D^{\pm}(C_{2p+1})$, a bidirected complete graph $D^{\pm}(K_n)$, or a digraph in which every vertex has in- and out-degree k. The notion of digraph degeneracy was introduced by Bokal et al. [126] in 2004; Bang-Jensen et al. [80] extended this notation to the case of variable degeneracy. Bang-Jensen, Bellitto, Schweser, and Stiebitz [77] extended DP-coloring of graphs to digraphs and obtained, in particular, a Brooks type result for the DP-chromatic number of digraphs.

Chapter 4
Properties of Critical Graphs

In this chapter we shall continue the study of critical graphs. Critical graphs were first introduced and investigated by G. A. Dirac in his doctoral thesis; he established the basic properties of critical graphs and published the results in the 1950s in a series of papers. The study of critical graphs was continued in the 1960s by G. Hajós, O. Ore, T. Gallai and others. It was believed (G. A. Dirac, private communication to the third author) that the investigation of structural properties of critical graphs might be of help in proving the four color theorem. However this was too optimistic. Unless $\mathsf{P} = \mathsf{NP}$, there is no good characterization of critical graphs with given chromatic number $\chi \geq 4$. .

4.1 Construction of Critical Graphs

Recall from Section 1.6 that a graph G is **critical** and k**-critical**, if G has chromatic number k, but every proper subgraph of G is $(k-1)$-colorable. So a graph G is critical if and only if all vertices and edges of G are critical. Furthermore, any graph G satisfies $\chi(G) \geq k$ if and only if G contains a k-critical subgraph (see Proposition 1.28(b)). For $k,n \in \mathbb{N}_0$, let $\mathrm{Crit}(k)$ denote the class of k-critical graphs and let

$$\mathrm{Crit}(k,n) = \{G \in \mathrm{Crit}(k) \mid |G| = n\}.$$

Since a graph G satisfies $\chi(G) = 0$ if and only if $G = \varnothing$, and $\chi(G) \leq 1$ if and only if $E(G) = \varnothing$, it follows from Proposition 1.28(b) that

$$\mathrm{Crit}(0) = \{\varnothing\}, \mathrm{Crit}(1) = \{K_1\} \text{ and } \mathrm{Crit}(2) = \{K_2\}.$$

König's characterization of bipartite graphs implies that the odd cycles are the only 3-critical graphs, that is,

$$\mathrm{Crit}(3) = \{C_n \mid n \equiv 1\,(\mathrm{mod}\,2), n \geq 3\}.$$

M. Stiebitz et al., *Brooks' Theorem*, Springer Monographs in Mathematics,
https://doi.org/10.1007/978-3-031-50065-7_4

For any fixed $k \geq 4$, a good characterization of the class $\mathrm{Crit}(k)$ is unlikely, and in what follows we shall mainly focus on the study of k-critical graphs with $k \geq 4$. A result due to V. Rödl and published by Toft [1027] says that there is a constant $c > 1$ such that for every $k \geq 4$ the number of nonisomorphic k-critical graphs of order n is at least c^{n^2} if n is large. To construct critical graphs, we first use two simple graph operations, one of which is very common in graph theory and the other of which was discovered by Hajós [461], perhaps as early as the 1940s.

Let G_1 and G_2 be two disjoint graphs. Let G be the graph obtained from the union $G_1 \cup G_2$ by adding all edges between G_1 and G_2, that is, $V(G) = V(G_1) \cup V(G_2)$ and $E(G) = E(G_1) \cup E(G_2) \cup \{uv \mid u \in V(G_1), v \in V(G_2)\}$. We call G the **Dirac join** of G_1 and G_2 and write $G = G_1 \boxplus G_2$. The Dirac join was first used by Dirac (see Gallai [397, (2.1) and (2.2)]).

Theorem 4.1 (Dirac Construction) *Let $G = G_1 \boxplus G_2$ be the Dirac join of two disjoint graphs G_1 and G_2. Then $\chi(G) = \chi(G_1) + \chi(G_2)$ and, moreover, G is critical if and only if both G_1 and G_2 are critical.*

Proof The statement is evident if G_1 or G_2 is empty. So assume that both graphs G_1 and G_2 are nonempty. Let $k_i = \chi(G_i)$ for $i \in \{1,2\}$. Note that $k_i \geq 1$ for $i \in \{1,2\}$. If φ_i is a coloring of G_i for $i \in \{1,2\}$, then $\varphi_1 \cup \varphi_2$ is a coloring of $G = G_1 \boxplus G_2$ if and only if $\mathrm{im}(\varphi_1) \cap \mathrm{im}(\varphi_2) = \varnothing$. This implies that $\chi(G) = k_1 + k_2 = \chi(G_1) + \chi(G_2)$.

Suppose that both G_1 and G_2 are critical. Then both G_1 and G_2 are connected implying that G is connected, too. Furthermore, $\chi(G) = k_1 + k_2 \geq 2$. Let e be an edge of G. If e belongs to G_1, then $G - e = (G_1 - e) \boxplus G_2$ and so

$$\chi(G - e) = \chi(G_1 - e) + \chi(G_2) \leq \chi(G_1) - 1 + \chi(G_2) = \chi(G) - 1.$$

If e belongs to G_2, then again $\chi(G - e) \leq \chi(G) - 1$. If $e = v_1v_2$ joins a vertex v_1 of G_1 with a vertex v_2 of G_2, then $\chi(G_i - v_i) \leq k_i - 1$ for $i \in \{1,2\}$. Hence G_1 can be colored with k_1 colors $1, 2, \ldots, k_1$ such that v_1 is the only vertex being colored with color 1, and G_2 can be colored with k_2 colors $1, k_1+1, k_1+2, \ldots, k_1+k_2-1$ such that v_2 is the only vertex being colored with color 1. Clearly this results in a coloring of $G - e$ with $k_1 + k_2 - 1$ colors. So $\chi(G - e) \leq \chi(G) - 1$ for all edges e of G, and hence G is critical (by Proposition 1.29).

Now suppose that G is critical. Let H be a proper subgraph of G_1. Then $G' = H \boxplus G_2$ is a proper subgraph of G and so $\chi(G') < \chi(G)$. Furthermore, $\chi(G') = \chi(H) + \chi(G_2)$ and $\chi(G) = \chi(G_1) + \chi(G_2)$. Consequently, $\chi(H) < \chi(G_1)$. This shows that G_1 is critical. By symmetry it follows that G_2 is critical, too. □

Let G_1 and G_2 be two disjoint graphs and, for $i \in \{1,2\}$, let $e_i = v_iw_i$ be an edge of G_i. Let G be the graph obtained from the union $G_1 \cup G_2$ by deleting the edges e_1 and e_2 from G_1 and G_2, respectively, identifying the vertices v_1 and v_2, and adding the new edge w_1w_2. We then say that G is the **Hajós join** of G_1 and G_2 and write $G = (G_1, v_1, w_1)\nabla(G_2, v_2, w_2)$ or briefly $G = G_1 \nabla G_2$. Figure 4.1 shows the Dirac join $C_5 \boxplus C_5$ and the Hajós join $K_4 \nabla K_4$.

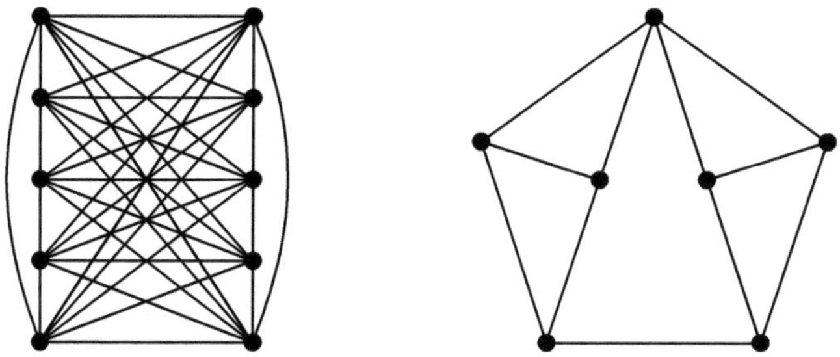

Fig. 4.1 The Dirac join $C_5 \boxplus C_5$ and the Hajós join $K_4 \nabla K_4$.

Theorem 4.2 (Hajós Construction) *Let G_1 and G_2 be disjoint graphs both with at least one edge, and let $G = G_1 \nabla G_2$ be the Hajós join of these two graphs. Then the following statements hold:*

(a) $\chi(G) \geq \min\{\chi(G_1), \chi(G_2)\}$.
(b) *If $\chi(G_1) = \chi(G_2) = k$ and $k \geq 3$, then $\chi(G) = k$.*
(c) *If both G_1 and G_2 are k-critical and $k \geq 3$, then G is k-critical*
(d) *If G is k-critical and $k \geq 4$, then both G_1 and G_2 are k-critical.*

Proof Suppose that $G = (G_1, v_1, w_1)\nabla(G_2, v_2, w_2)$, and let v^* denote the vertex of G obtained by identifying v_1 and v_2.

For the proof of (a), assume that $\chi(G) = k$. Then there is a k-coloring φ of G. Since w_1w_2 is an edge of G, we obtain that $\varphi(w_1) \neq \varphi(w_2)$ and so $\varphi(v^*) \neq \varphi(w_i)$ for some $i \in \{1,2\}$. Then φ yields a k-coloring of G_i, which proves (a).

For the proof of (b), assume that $\chi(G_1) = \chi(G_2) = k$ and $k \geq 3$. Since $v_iw_i \in E(G_i)$, there is a k-coloring φ_i of G_i such that $\varphi_i(v_i) \neq \varphi_i(w_i)$ with $i \in \{1,2\}$. As $k \geq 3$, we may permute the colors so that $\varphi_1(v_1) = \varphi_2(v_2)$ and $\varphi_1(w_1) \neq \varphi_2(w_2)$. Then $\varphi_1 \cup \varphi_2$ yields a k-coloring of G and hence $\chi(G) \leq k$. By (a) it follows that $\chi(G) \geq k$. Hence $\chi(G) = k$ and (b) is proved.

For the proof of (c), assume that both G_1 and G_2 are k-critical. To show that G is k-critical we apply Proposition 1.29. Both graphs G_1 and G_2 are 2-connected (by Corollary 1.31 and $k \geq 3$) and hence G is connected. By (b) it follows that $\chi(G) = k$. It remains to show that $\chi(G - e) \leq k - 1$ for every edge e of G. First suppose that $e = w_1w_2$. Since G_i is k-critical, there is a $(k-1)$-coloring φ_i of $G_i - v_iw_i$ for $i \in \{1,2\}$. As $\chi(G_i) = k$, we obtain that $\varphi_i(v_i) = \varphi_i(w_i)$. By permuting colors if necessary we may obtain that $\varphi_1(v_1) = \varphi_2(v_2)$. Then $\varphi_1 \cup \varphi_2$ yields a $(k-1)$-coloring of $G - e$. If $e \neq w_1w_2$, then e belongs to G_i for some $i \in \{1,2\}$, say $e \in E(G_1)$. Since G_1 is k-critical, $G_1 - e$ admits a $(k-1)$-coloring φ_1 such that $\varphi_1(v_1) \neq \varphi_1(w_1)$ (since v_1w_1 is an edge of $G_1 - e$). G_2 is k-critical, so there is a $(k-1)$-coloring φ_2 of $G - v_2w_2$ such that $\varphi_2(v_2) = \varphi_2(w_2)$. By permuting colors if necessary we may obtain that $\varphi_1(v_1) = \varphi_2(v_2)$. Then $\varphi_1(w_1) \neq \varphi_2(w_2)$ and hence $\varphi_1 \cup \varphi_2$ yields a $(k-1)$-coloring of $G - e$. Thus (c) is proved.

For the proof of (d), assume that G is k-critical. To show that both G_1 and G_2 are k-critical, we use Proposition 1.29. As $k \geq 3$ it follows from Corollary 1.31 that G is

2-connected and hence both G_1 and G_2 are connected. Next we claim that $\chi(G_1) \geq k$. Suppose this is false. Then there is a $(k-1)$-coloring φ_1 of G_1. As $v_1w_1 \in E(G_1)$, we obtain that $\varphi_1(v_1) \neq \varphi_1(w_1)$. As G is k-critical and $w_1w_2 \in E(G)$, there is a $(k-1)$-coloring φ of $G - w_1w_2$, which leads to a $(k-1)$-coloring φ_2 of $G_2 - v_2w_2$. As $k \geq 4$, we may permute the colors such that $\varphi_2(v_2) = \varphi_1(v_1)$ and $\varphi_2(w_2) \neq \varphi_1(w_1)$. But then $\varphi_1 \cup \varphi_2$ yields a $(k-1)$-coloring of G, which is impossible. This proves the claim that $\chi(G_1) \geq k$. Similarly, one can show that $\chi(G_2) \geq k$. It remains to show that $\chi(G_i - e) \leq k-1$ for every edge $e \in E(G_i)$. By symmetry it suffices to show this for $i = 1$. If $e = v_1w_1$ then the restriction of any $(k-1)$-coloring of $G - w_1w_2$ yields a $(k-1)$-coloring of $G_1 - e$. If $e \neq v_1w_1$, then there is a $(k-1)$-coloring φ of $G - e$. As $\chi(G_2) \geq k$ and $w_1w_2 \in E(G)$, we obtain that $\varphi(v^*) = \varphi(w_2) \neq \varphi(w_1)$. Consequently, φ induces a $(k-1)$-coloring of $G_1 - e$, and we are done. □

Note that statement (d) does not hold for $k = 3$ as demonstrated by the cycle C_7 being obtained as the Hajós join of two cycles C_4. In general, however, the advantage of the Hajós join is that it not only preserves the chromatic number, but also criticality. If $k \geq 3$, then

$$G \in \mathrm{Crit}(k,n) \Rightarrow G \boxplus K_1 \in \mathrm{Crit}(k+1,n+1) \text{ and } G\nabla K_k \in \mathrm{Crit}(k,n+k-1).$$

It easily follows that for every integer $k \geq 4$ there are k-critical graphs of order n if $n \geq k$ and $n \neq k+1$. Clearly, there is no k-critical graph with fewer than k vertices and, since no two independent vertices in a critical graph can have the same neighbourhood, it follows that there is no k-critical graph having order $k+1$ (see also Exercise 4.3). Hence, if $k \geq 4$, then

$$\mathrm{Crit}(k,k) = \{K_k\} \text{ and } \mathrm{Crit}(k,n) \neq \varnothing \text{ if and only if } n = k \text{ or } n \geq k+2.$$

There is another very common graph operation. Let G be a graph and let I be a nonempty independent set of G. Then we obtain a new graph H from $G - I$ by adding a new vertex $v = v_I$ and by joining v to a vertex u of $G - I$ if and only if u is adjacent in G to some vertex of I. We then write $H = G/(I \to v)$ or briefly $H = G/I$ and say that H is obtained from G by **identifying I to** v, or briefly by **identifying independent vertices**. Clearly, any k-coloring of G/I can be extended to a k-coloring of G by giving each vertex of I the same color as v_I. This shows that $\chi(G/I) \geq \chi(G)$.

The class of **Hajós-k-constructible graphs** is the smallest family of graphs that contains all complete graphs K_k and is closed under Hajós joins and identifying independent vertices. The following result is due to Hajós [461].

Theorem 4.3 (Hajós) *Let $k \geq 3$ be an integer. A graph has chromatic number at least k if and only if it contains a Hajós-k-constructible subgraph.*

Proof Clearly, every Hajós-k-constructible graph has chromatic number at least k. This follows from Theorem 4.2(a) and the fact that $\chi(G/I) \geq \chi(G)$ for every independent set I of G. This proves the "if" implication. The proof of the "only if" implication is by reductio ad absurdum. Suppose there is graph G with $\chi(G) \geq k$ without a Hajós-k-constructible subgraph. In addition, G may be assumed to be maximal in the sense that adding a new edge e to G results in a Hajós-k-constructible subgraph G_e of $G+e=(V(G),E(G)\cup\{e\})$ with $e\in E(G_e)$. If the relation *to be identical or nonadjacent* is an equivalence relation on $V(G)$, then the number of equivalence classes is at least k as $\chi(G)\geq k$. But then G contains the Hajós-k-constructible graph K_k. Consequently, G has three vertices u,v and w with $uv\notin E(G), vw\notin E(G)$ and $uw\in E(G)$. Then there are Hajós-k-constructible graphs G_{uv} and G_{vw}. Let G' be the graph obtained from the union $(G_{uv}-uv)\cup(G_{vw}-vw)$ by adding the edge uw. Then G' is a subgraph of G which can be obtained from disjoint copies of G_{uv} and G_{vw} in the following way. First we apply the Hajós join by removing the copies of the edges uv and vw, identifying the two copies of v and adding the copy of the edge uw. Then, for each vertex x belonging to both G_{uv} and G_{vw} we identify the two copies of x, thereby containing the Hajós-k-constructible subgraph G' of G. This contradiction completes the proof. □

The above proof of Hajós' theorem led Ore [794] to introduce a more restricted class of constructible graphs. Let G_1 and G_2 be two disjoint graphs and, for $i\in\{1,2\}$, let $e_i=v_iw_i$ be an edge of G_i. Moreover, let $\iota: A_1\rightarrow A_2$ be a bijection with $A_i\subseteq V(G_1-v_i)$ for $i\in\{1,2\}$ such that $\iota(w_1)\neq w_2$. Let G be the graph obtained from $(G_1,v_1,w_1)\nabla(G_2,v_2,w_2)$ by identifying v with $\iota(v)$ for each $v\in A_1$. We then say that G is the **Ore join** of G_1 and G_2 and write $G=(G_1,v_1,w_1,A_1)\nabla^{\circ}_{\iota}(G_2,v_2,w_2,A_2)$ or briefly $G=G_1\nabla^{\circ}G_2$.

The class of **Ore-k-constructible graphs** is the smallest family of graphs that contains all complete graphs K_k and is closed under Ore joins. So a graph G is Ore-k-constructible if and only if there is a sequence of graphs such that G is the last graph in the sequence and any graph of the sequence is either a K_k or it is the Ore join of two previous graphs. Clearly, each Ore-k-constructible graph is a Hajós-k-constructible graph. As noticed by Ore, the proof of Theorem 4.3 implies, in fact, the following result.

Theorem 4.4 (Ore) *Let $k\geq 3$ be an integer. A graph has chromatic number at least k if and only if it contains an Ore-k-constructible subgraph.*

The following result was proved by Urquhart [1045] and gives a positive answer to a conjecture of Hanson, Robinson and Toft [467] (see also Jensen and Toft [530, Problem 11.5]).

Theorem 4.5 (Urquhart) *For any graph G and any integer $k\geq 3$, the following conditions are equivalent:*

(a) *G is Ore-k-constructible.*
(b) *G is Hajós-k-constructible.*
(c) *The chromatic number of G satisfies $\chi(G)\geq k$.*

Proof Clearly, (a) implies (b) and (b) implies (c). So it suffices to show that (c) implies (a), that is, every graph with chromatic number $\chi\geq k$ is Ore-k-constructible. To this end we shall first show that several graphs are Ore-k-constructible. For an integer r with $0\leq r\leq k$, let $K_k[r]$ denote the graph obtained from K_k by adding a vertex and joining this vertex to r vertices of K_k. So $K_k[0]$ is the disjoint union of K_k and K_1. The disjoint union of K_k and K_2 is denoted by K_k+e, where e is the only edge of K_2. Let $\mathcal{O}_k$ denote the class of Ore-k-constructible graphs, and $\mathcal{O}_k^*$ the class of graphs of $\mathcal{O}_k$ containing a K_k. Note that $\mathcal{O}_k$ and $\mathcal{O}_k^*$ are graph properties, that is, both classes are closed under isomorphism.

Claim 4.5.1 *The graph $K_k[r]$ with $0 \leq r \leq k-2$ belongs to $\mathcal{O}_k^*$, and the graph K_k+e belongs to $\mathcal{O}_k^*$.*

Proof: Clearly, both graphs $K_k[r]$ and K_k+e contain a K_k, so it suffices to show that these graphs belong to $\mathcal{O}_k$. First we show that $K_k[k-2]$ belongs to $\mathcal{O}_k$. We take two disjoint copies of K_k, say G_1 and G_2. Then we chose an edge v_iw_i in G_i for $i \in \{1,2\}$, and a bijection ι from $A_1 = V(G_1) \setminus \{v_1,w_1\}$ onto $A_2 = V(G_2) \setminus \{v_2,w_2\}$. Then it is easy to check that $G = (G_1,v_1,w_1,A_1)\nabla_\iota^\circ(G_2,v_2,w_2,A_2)$ is a $K_k[k-2]$, and so $G \in \mathcal{O}_k$ as claimed. Next we claim that if $K_k[r]$ with $1 \leq r \leq k-2$ belongs to $\mathcal{O}_k$, then $K_k[r-1]$ belongs to $\mathcal{O}_k$. To this end, take two disjoint copies of $K_k[r]$, say G_1 and G_2. For $i \in \{1,2\}$ let v_i denote the vertex having degree r in G_i, let w_i be one of its neighbor in G_i, and let u_i be one of its nonneighbors in G_i (these vertices exists as $1 \leq r \leq k-2$). Now let ι be a bijection from $A_1 = V(G_1 - v_1)$ to $A_2 = V(G_2 - v_2)$ such that $\iota(w_1) = u_2$, $\iota(u_1) = w_2$ and ι maps $N_{G_1}(v_1) \setminus \{w_1\}$ onto $N_{G_2}(v_2) \setminus \{w_2\}$. Then it is easy to check that $G = (G_1,v_1,w_1,A_1)\nabla_\iota^\circ(G_2,v_2,w_2,A_2)$ is a $K_k[r-1]$, and so $K_k[r-1]$ belongs to $\mathcal{O}_k$ as claimed. It then follows by induction on r that $K_k[r]$ belongs to $\mathcal{O}_k$ whenever $0 \leq r \leq k-2$. To see that K_k+e belongs to $\mathcal{O}_k$, take two disjoint copies of $K_k[1]$, say G_1 and G_2. For $i \in \{1,2\}$ let w_i denote the vertex having degree 1 in G_i and let v_i be its only neighbor in G_i. Now let ι be a bijection from $A_1 = V(G_1) \setminus \{v_1,w_1\}$ onto $A_2 = V(G_2) \setminus \{v_2,w_2\}$. Then it is easy to check that $G = (G_1,v_1,w_1,A_1)\nabla_\iota^\circ(G_2,v_2,w_2,A_2)$ is a K_k+e, and so $G \in \mathcal{O}_k$ as claimed. This completes the proof of the claim. △

Claim 4.5.2 *Let G be a graph that belongs to $\mathcal{O}_k^*$. Then the graph G' obtained from G by adding an isolated vertex belongs to $\mathcal{O}_k^*$, too.*

Proof: Clearly, G' contains a K_k as G does, and hence it suffices to show that $G' \in \mathcal{O}_k$. Let G_2 be a copy of $K_k[0]$ that is disjoint from G. Note that G_2 belongs to $\mathcal{O}_k$ (by Claim 4.5.1). There is a vertex set $X_1 \subseteq V(G)$ such that $G[X_1]$ is a K_k. Let X_2 denote the vertex set of the K_k contained in G_2. For $i \in \{1,2\}$ let v_i, w_i and u_i be three vertices of X_i. Now let ι be a bijection from $A_1 = X_1 \setminus \{v_1\}$ onto $A_2 = X_2 \setminus \{v_2\}$ such that $\iota(w_1) = u_2$ and $\iota(u_1) = w_2$. Then it is easy to check that $(G,v_1,w_1,A_1)\nabla_\iota^\circ(G_2,v_2,w_2,A_2)$ is a copy of G', and so $G' \in \mathcal{O}_k$. △

Claim 4.5.3 *Let G be a graph that belongs to $\mathcal{O}_k^*$, and let e be an edge of $\overline{G}$. Then the graph $G+e$ obtained from G by adding e belongs to $\mathcal{O}_k$, too.*

Proof: Clearly, it suffice to show that $G+e \in \mathcal{O}_k$. There is a vertex set $X_1 \subseteq V(G)$ such that $G[X_1]$ is a K_k. Note that X_1 contains at most one end of e.

First assume that one end of e, say v_1, belongs to X_1. Let u_1 be the end of e distinct from v_1, let w_1 be a vertex of $X_1 \setminus \{v_1\}$, and let z_1 be a vertex of $X_1 \setminus \{v_1,w_1\}$. Let G_2

be a copy of $K_k[1]$, let u_2 be the vertex of degree 1 in G_2, let v_2 be the only neighbor of u_2 in G_2, let w_2 be a vertex of $G_2 - v_2 - u_2$, let z_2 be a vertex of $G_2 - \{u_2, v_2, w_2\}$, and let $X_2 = V(G_2 - u_2)$. Note that G_2 belongs to $\mathcal{O}_k$ (by Claim 4.5.1). Now let ι be a bijection from $A_1 = (X_1 \setminus \{v_1\}) \cup \{u_1\}$ onto $A_2 = (X_2 \setminus \{v_2\}) \cup \{u_2\}$ such that $\iota(u_1) = u_2$, $\iota(w_1) = z_2$, and $\iota(z_1) = w_2$. Then $(G, v_1, w_1, A_1)\nabla_\iota^\circ(G_2, v_2, w_2, A_2)$ is a copy of $G+e$, and so $G+e \in \mathcal{O}_k$.

Now assume that no end of e belongs to X_1 and let $e = x_1y_1$. Let G_2 be a copy of $K_k + e'$ that is disjoint from G, and let $e' = x_2y_2$. Note that G_2 belongs to $\mathcal{O}_k$ (by Claim 4.5.1). Let $X_2 \subseteq V(G_2)$ be the vertex set of the K_k contained in G_2. For $i \in \{1,2\}$, let u_i, v_i and w_i be three vertices of X_i. Now let ι be a bijection from $A_1 = X_1 \setminus \{v_1\} \cup \{x_1, y_1\}$ onto $A_2 = X_2 \setminus \{v_2\} \cup \{x_2, y_2\}$ such that $\iota(w_1) = u_2$, $\iota(u_1) = w_2$, $\iota(x_1) = x_2$, and $\iota(y_1) = y_2$. Then $(G, v_1, w_1, A_1)\nabla_\iota^\circ(G_2, v_2, w_2, A_2)$ is a copy of $G+e$, and so $G+e \in \mathcal{O}_k$. △

As K_k belongs to $\mathcal{O}_k^*$, it follows from Claim 4.5.2 and Claim 4.5.3 that any graph containing a K_k belongs to $\mathcal{O}_k^*$. To finish the proof we now show that any graph with $\chi \geq k$ belongs to $\mathcal{O}_k$. Suppose this is false and let G be a counterexample with minimum order and with a maximum number of edges with respect to this property. Then G does not contain a K_k. As $\chi(G) \geq k$, this implies that G is not complete. If e is an edge of $\overline{G}$, then $G+e$ belongs to $\mathcal{O}_k$. Now we continue as in the proof of Hajós' theorem. If the relation *to be identical or nonadjacent* is an equivalence relation on $V(G)$, then the number of equivalence classes is at least k as $\chi(G) \geq k$. But then G contains a K_k, a contradiction. Consequently, G has three vertices u, v and w with $uv \notin E(G), vw \notin E(G)$ and $uw \in E(G)$. Then both graphs $G+uv$ and $G+vw$ belong to $\mathcal{O}_k$, and G is the Ore join of two disjoint copies of these two graphs. Hence G belongs to $\mathcal{O}_k$. This contradiction completes the proof of the theorem. □

Following Pitassi and Urquhart [808], a **graph calculus** is a pair $(\mathcal{F}, \mathcal{R})$ consisting of a family $\mathcal{F}$ of graphs and a finite collections $\mathcal{R}$ of rules that allow us to derive new graphs. A **construction sequence** of a graph G in the calculus $(\mathcal{F}, \mathcal{R})$ is a sequence of graphs, ending with a copy of G, such that every graph is either a copy of a graph belonging to $\mathcal{F}$, or can be obtained from copies of previous graphs by one of the rules in $\mathcal{R}$; we then say that G is **constructible** in the calculus and denote by $\mathrm{Con}(\mathcal{F}, \mathcal{R})$ the class of graphs that are constructible in $(\mathcal{F}, \mathcal{R})$. Note that the class $\mathrm{Con}(\mathcal{F}, \mathcal{R})$ is a graph property, that is, a class of graphs closed under isomorphism. The size of a graph is the number of its edges. The **size** of a graph construction sequence is the sum of the sizes of all graphs occurring in the construction sequence. The **length** of a graph construction sequence in the calculus $(\mathcal{F}, \mathcal{R})$ is the the number of graphs occurring in the construction sequence, so it is the length of this sequence. The graph calculus $(\mathcal{F}, \mathcal{R})$ is **polynomially bounded** if there exists a polynomial $p = p(x)$ such that for every graph G in $\mathrm{Con}(\mathcal{F}, \mathcal{R})$ there is a construction sequence in $(\mathcal{F}, \mathcal{R})$ of size at most $p(|E(G)|)$. An alternative way of looking at a graph calculus is to regard a construction as a rooted tree, in particular, if we are interested in labelled graphs. A **construction tree** of a graph G in the calculus $(\mathcal{F}, \mathcal{R})$ is a rooted tree whose vertices are graphs, with root G, with leaves that are copies of graphs belonging to $\mathcal{F}$, and with each internal vertex being a graph obtained from graphs of its children by one of the rules in $\mathcal{R}$.

The standard **Hajós calculus** is a graph calculus $\mathcal{HC}_k = (\mathcal{F}_k, \mathcal{R})$ where $\mathcal{F}_k$ consists of K_k and $\mathcal{R}$ consists of the following three rules: the **addition rule** (add any number of vertices and edges), the **join rule** (form a Hajós-join of two disjoint graphs), and the **contraction rule** (identify two nonadjacent vertices in a graph). Let $\mathcal{G}(\chi \geq k)$ be the class of graphs G with $\chi(G) \geq k$. Then Hajós' theorem immediately implies that $\mathrm{Con}(\mathcal{HC}_k) = \mathcal{G}(\chi \geq k)$. Figure 4.2 shows a construction of the wheel $W_7 = K_1 \boxplus C_7$ in $\mathcal{HC}_4$. In the figure, we use a dashed arrow to indicate the edges that are used as inputs to a Hajós join, where the heads of the arrows are the vertices that will be identified. Furthermore, we use a vertex label to indicate the vertices as input to a contraction step, where vertices with the same labels are identified. The construction sequence $(K_4, K_4 \nabla K_4, W_5, W_5 \nabla K_4, W_7)$ of W_7 with respect to $\mathcal{HC}_4$ (see Figure 4.2) has length 5 and size 56, but the construction tree of W_7 has order 7.

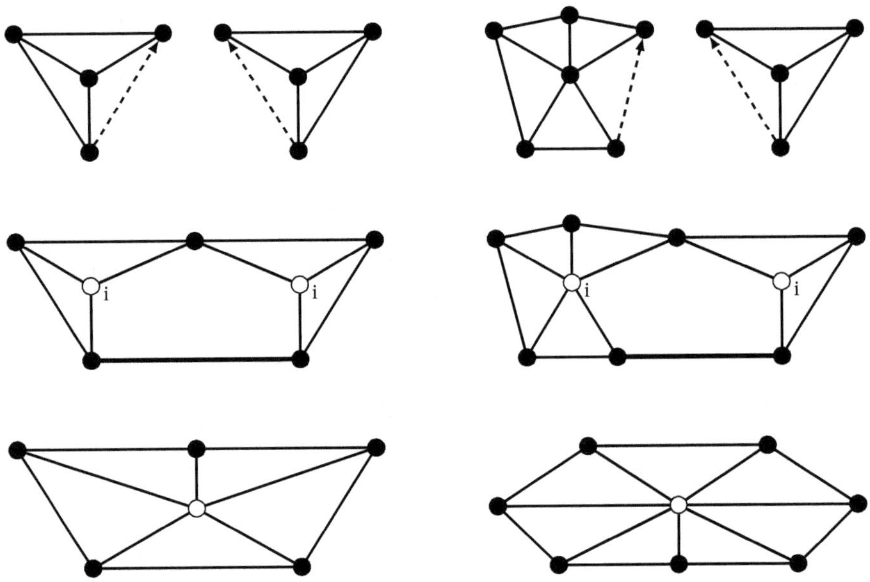

Fig. 4.2 A Hajós type construction of the wheel $W_7 = K_1 \boxplus C_7$.

In 1982 Lovász [699] wrote: *So far; Hajós' theorem was studied for its possible applications to planarity; its algorithmic complexity aspects are an unexplored territory.* The complexity of the Hajós construction was first investigated by Mansfield and Welsh [715], however, they used Ore's version of Hajós' theorem. So let $\mathcal{OC}_k = (\mathcal{F}_k, \mathcal{R})$ be the graph calculus where $\mathcal{F}_k$ consists of K_k and $\mathcal{R}$ consists of just one rule, namely the **Ore join rule** (form an Ore-join of two disjoint graphs). By the result of Urquhart (Theorem 4.5) $\mathrm{Con}(\mathcal{OC}_k) = \mathcal{G}(\chi \geq k)$. The construction sequence (K_4, W_5, W_7) of W_7 in $\mathcal{OC}_4$ has length 3 and size 30; the corresponding construction tree has order 5. Mansfield and Welsh were not aware of Urquhart's theorem, but of Ore's theorem, so they introduced the concept of a proof sequence. A **proof sequence** in $\mathcal{OC}_k$ for a graph G is a construction sequence in $\mathcal{OC}_k$ for a subgraph of

G. By Ore's theorem, any graph $G \in \mathcal{G}(\chi \geq k)$ has a proof sequence, and if G is a k-critical graph, then any proof sequence of G is a construction sequence of G. The k-**Hajós number** of a graph G, denoted by $\hbar_k(G)$, is the minimum length of a proof sequence of G in $\mathcal{O}C_k$ when $\chi(G) \geq k$ and 0 otherwise. So any graph G containing a K_k satisfies $\hbar_k(G) = 1$. Mansfield and Welsh [715] posed the problem of whether there exist graphs in $\mathcal{G}(\chi \geq 4)$ that require exponential-length proof sequences in $\mathcal{O}C_4$ with respect to the order of the input graph. Unless NP = coNP, there likely exist such graphs, but to date little progress has been made on this question. Hanson, Robinson, and Toft [467] proved that for all sufficiently large n there is a 4-critical graph G of order n such that $\hbar_4(G) \geq cn/\log n$ with $c > 0$. The proof is based on a counting argument and uses Rödl's result saying that there are exponential many 4-critical graphs of order n. Until now even a linear lower bound for the Hajós number $\hbar_4$ is unknown. On the other hand, Mansfield and Welsh [715] proved that any graph G of order n satisfies $\hbar_4(G) \leq 2^{n^2/3}$. Pitassi and Urquhart [808] proved that the Hajós calculus $\mathcal{H}C_4$ is polynomially bounded if and only if Extended Frege Systems, studied in logic, are polynomially bounded.

Hajós asked whether for every k-critical graph there is a construction in the Hajós calculus $\mathcal{H}C_k$ such that each intermediate graph of the construction is critical, too (see Jensen and Toft [530, Problem 11.7]). This is trivially true when $k = 3$ as $C_{n+2} = C_n \nabla K_3$. However, the answer is negative as shown by Hanson, Robinson and Toft [467] in the case of $k = 8$ (see Exercise 4.11), and by Jensen and Royle [528] in the case of $k \geq 4$, also in case of the Ore calculus. It is however still open if any k-critical graph may be obtained in the Hajós calculus with the Hajós join used only on k-critical graphs. A generalization and refinement of the Ore construction was provided by W. T. Tutte in 1992 (see Appendix B.1). The corresponding Tutte calculus $\mathcal{T}C_k$ was also described by Jensen and Royle [528].

4.2 Critical Graphs of Low Connectivity

As already mentioned in Section 1.6, any critical graph is connected and contains no separating vertex (see Corollary 1.31). Hence any k-critical graph with $k \geq 3$ is 2-connected. Dirac [298] as well as Gallai [397] provided a characterization of critical graphs having a 2-separator, that is, a separating vertex set of size 2. In particular, they proved the following two theorems. The first theorem shows how to decompose a critical graph having a 2-separator into two smaller critical graphs; the second theorem shows that every critical graph having a 2-separator can be obtained from two smaller critical graphs by a converse operation. In the 1970s Toft [1019] [1023] established counterparts of both results for critical hypergraphs. So we will not give separate proofs for the graph cases, but refer the reader to the proofs of the corresponding Theorems 7.19 and 7.20 in Section 7.5.

Theorem 4.6 (Dirac and Gallai) *Let G be a critical graph with chromatic number $k+1$ for an integer $k \geq 2$, and let $S \subseteq V(G)$ be a separating vertex set of G with*

$|S| \le 2$. Then S is an independent vertex set of G consisting of two vertices, say v and w, and $G - S$ has exactly two components H_1 and H_2. Moreover, if $G_i = G[V(H_i) \cup S]$ for $i = 1, 2$, we can adjust the notation so that there is a coloring $\varphi_1 \in \mathcal{CO}(G_1, k)$ with $\varphi_1(v) = \varphi_1(w)$. Then the following statements hold:

(a) *Every coloring $\varphi \in \mathcal{CO}(G_1, k)$ satisfies $\varphi(v) = \varphi(w)$ and every coloring $\varphi \in \mathcal{CO}(G_2, k)$ satisfies $\varphi(v) \neq \varphi(w)$.*
(b) *The graph $G_1' = G_1 + vw$ obtained from G_1 by adding the edge vw is critical and has chromatic number $k + 1$.*
(c) *The vertices v and w have no common neighbor in G_2 and the graph $G_2' = G_2/S$ obtained from G_2 by identifying v and w is critical and has chromatic number $k + 1$.*

Let G_1 and G_2 be two disjoint graphs. Let $e = v_1v_2$ be an edge of G_1, let w be a vertex of G_2 with $d_{G_2}(w) \ge 2$, and let $\{A_1, A_2\}$ be a 2-partition of $N_{G_2}(w)$. Note that $A_i \neq \varnothing$ for $i = 1, 2$. Let G be the graph obtained from $(G_1 - e) \cup (G_2 - w)$ by adding the edge set $E = \{v_iu \mid u \in A_i, i \in \{1, 2\}\}$. We then say that G is the **Gallai join** of G_1 and G_2 and write $G = (G_1, v_1, v_2) \blacktriangledown (G_2, w, A_1, A_2)$ or briefly $G = G_1 \blacktriangledown G_2$. Intuitively, G is obtained from $G_1 - e$ and G_2 by splitting the vertex w into the edge e. Note that $N_G(v_1) = (N_{G_1}(v_1) \setminus \{v_2\}) \cup A_1$ and $N_G(v_2) = (N_{G_1}(v_2) \setminus \{v_1\}) \cup A_2$. In particular, $S = \{v_1, v_2\}$ is a 2-separator of G. The graph $G_2' = G[V(G_2 - w) \cup \{v_1, v_2\}]$ is called the **split-up** of G_2 in G.

Theorem 4.7 (Dirac and Gallai) *Let $k \ge 2$ be an integer, let $G = G_1 \blacktriangledown G_2$ be the Gallai join of two disjoint $(k + 1)$-critical graphs G_1 and G_2, and let G_2' be the split-up of G_2 in G. If $\chi(G_2') \le k$, then G is $(k + 1)$-critical, too.*

Note that the Hajós join is a particular case of the Gallai join. If G is the Hajós join of two k-critical graphs with $k \ge 4$, then G is a k-critical graph that has not only a separating set of two vertices, but also a separating set consisting of one vertex and one edge. Theorem 4.2(c)(d) implies that the converse statement also holds (see also Theorem 7.21).

Proposition 4.8 *Let G be a k-critical graph graph with $k \ge 4$. If G has a separating set consisting of one edge and one vertex, then G is the Hajós join of two k-critical graphs.*

Let G be a k-critical graph with $k \ge 4$. Recall from Section 1.6 that G_H denotes the high vertex subgraph of G, that is, the subgraph of G induced by the vertices having degree at least k in G; and $G_L = G - V(G_H)$ denotes the low vertex subgraph. Clearly, each low vertex of G has degree $k - 1$ in G. A fundamental result of Gallai generalizing Brooks' theorem says that G_L is a Gallai forest, that is, G_L is empty or each block of G is a complete graph or an odd cycle (see Theorem 1.33). Furthermore, $G_H = \varnothing$ if and only if G is a complete graph or an odd cycle. A complete graph or an odd cycle is also referred to as a brick, respectively an r-brick if it is $(r - 1)$-regular. So if B is a r-brick, then $r \ge 1$ and $B = K_r$ or $r = 3$ and $B = C_{2q+1}$ for some $q \ge 1$. Following Gallai, an **ε_k-graph** with $k \ge 3$ is a Gallai tree T, such that each vertex of

T belongs to exactly one k-brick and to at most one 2-brick (see also Section 1.2). Gallai characterized k-critical graphs with $k \geq 4$ and $|G_H| \leq 1$ (see Exercises 4.5 and 4.6). In particular, he proved that if G is a k-critical graph with at most one high vertex v, then $G - v$ is a ε_{k-1}-graph. As a consequence, we obtain the following result. Recall that $W_n = K_1 \boxplus C_n$ with $n \geq 3$ is called a **wheel**. Note that $W_3 = K_4$. The wheel W_n is said to be **odd** or **even** depending on whether n is odd or even.

Proposition 4.9 *Let G be a k-critical graph graph with $k \geq 4$. If $|G_H| = 1$, then either G has a 2-separator or $k = 4$ and G is an odd wheel.*

Proof Let v be the only high vertex of G. As G is 2-connected, the low vertex subgraph $G_L = G - v$ is connected and hence a Gallai tree. Then G_L has an end-block which is a $(k-1)$-brick, say B. If B is the only block of G_L, then v is adjacent to all vertices of B. As $d_G(v) \geq k$, this implies that $k = 4$, $B = C_{2p+1}$ with $p \geq 2$, and hence G is an odd wheel. If B is not the only block of G_L, then G_L has a separating vertex u contained in B, and hence $\{u, v\}$ is a 2-separator of G. □

That a k-critical graph is $(k-1)$-edge-connected was proved by Dirac [293] in the 1950s (see Theorem 1.32). A decomposition result for k-critical graphs having a separating edge set of size $k-1$ was obtained by Toft [1019] and first by Gallai (oral communication to the third author). It is well known that a minimal separating edge set of a connected graph forms a cut (see also Appendix C.4).

Let G be an arbitrary graph. By a **cut** or **edge cut** of G we mean a triple (X, Y, F) such that X is a nonempty proper subset of $V(G)$, $Y = V(G) \setminus X$, and $F = \partial_G(X) = \partial_G(Y)$. If (X, Y, F) is an edge cut of G, then we denote by X_F (respectively Y_F) the set of vertices of X (respectively of Y) which are incident to some edge of F. An edge cut (X, Y, F) of G is nontrivial if $|X_F| \geq 2$ and $|Y_F| \geq 2$. The following decomposition result was proved independently by T. Gallai and B. Toft. Toft [1019] gave a complete characterization of the class of k critical graphs having a separating edge set of size $k-1$. The characterization, however, involves critical hypergraphs. For a proof of the next result the reader is referred to Theorem 7.22 and its proof, respectively to Remark 7.23.

Theorem 4.10 (Toft) *Let G be a k-critical graph with $k \geq 4$, and let $F \subseteq E(G)$ be a separating edge set of G with $|F| \leq k-1$. Then $|F| = k-1$ and there is an edge cut (X, Y, F) of G satisfying the following properties:*

(a) *Every coloring $\varphi \in \mathcal{CO}(G[X], k-1)$ satisfies $|\varphi(X_F)| = 1$, and every coloring $\varphi \in \mathcal{CO}(G[Y], k-1)$ satisfies $|\varphi(Y_F)| = k-1$.*
(b) *The graph G_1 obtained from $G[X \cup Y_F]$ by adding all missing edges between the vertices of Y_F, so that Y_F becomes a clique of G_1, is k-critical.*
(c) *The graph G_2 obtained from $G[Y]$ by adding a new vertex v and joining v to all vertices of Y_F is k-critical.*

Note that if $|X_F| = 1$ then $G_2 = G$, and if all edges between the vertices of Y_F are already in G then $G_1 = G$. Note also that $G[X]$ with X_F added as a hyperedge is

k-chromatic, and the theorem therefore is of the same type as Theorem 4.6. We shall expand on this in Chapter 7.

In what follows we are interested in connectivity parameters of graphs. Let G be a graph with at least two vertices. The **local connectivity** $\kappa_G(v,w)$ of distinct vertices v and w is the maximum number of internally vertex disjoint v-w paths of G. The **local edge connectivity** $\kappa'_G(v,w)$ of distinct vertices v and w is the maximum number of edge-disjoint v-w paths of G. The **maximum local connectivity** of G is

$$\bar{\kappa}(G) = \max\{\kappa_G(v,w) \mid v,w \in V(G), v \neq w\},$$

and the **maximum local edge connectivity** of G is

$$\bar{\kappa}'(G) = \max\{\kappa'_G(v,w) \mid v,w \in V(G), v \neq w\}.$$

For a graph G with $|G| \leq 1$, define $\bar{\kappa}(G) = \bar{\kappa}'(G) = 0$. From the definition it follows that $\bar{\kappa}(G) \leq \bar{\kappa}'(G)$ for every graph G. Mader [712] proved that every graph G satisfies $\delta(G) \leq \bar{\kappa}(G)$. Since $\bar{\kappa}$ is a monotone graph parameter in the sense that $H \subseteq G$ implies $\bar{\kappa}(H) \leq \bar{\kappa}(G)$, it follows from Maders result that every graph G satisfies $\operatorname{col}(G) \leq \bar{\kappa}(G)+1$. Consequently, every graph G satisfies

$$\chi(G) \leq \operatorname{col}(G) \leq \bar{\kappa}(G)+1 \leq \bar{\kappa}'(G)+1 \leq \Delta(G)+1. \tag{4.1}$$

The fact that $\chi(G) \leq \bar{\kappa}'(G)+1$ for every graph G is also an immediate consequence of Dirac's result about the edge connectivity of critical graphs (Theorem 1.32) and the critical graph method (Proposition 1.27(b)). Our goal is to characterize the class of graphs G for which $\chi(G) = \bar{\kappa}'(G)+1$. By Dirac's Theorem 1.32 the critical such graphs are those where any two different vertices are joined by $\chi - 1$ edge-disjoint paths, and not by more. For such a characterization we use the fact that if we have an optimal coloring of each block of a graph, then we can combine these colorings to an optimal coloring of the graph by permuting colors in the blocks if necessary (see also Proposition 1.30). Hence every nonempty graph G satisfies

$$\chi(G) = \max\{\chi(B) \mid B \in \mathfrak{B}(G)\}. \tag{4.2}$$

For $k \geq 1$ we define a subclass $\operatorname{Con}(k)$ of Hajós-k-constructible graphs as follows. For $1 \leq k \leq 3$ put $\operatorname{Con}(k) = \operatorname{Crit}(k)$. The class $\operatorname{Con}(4)$ is the smallest family of graphs that contains all odd wheels and is closed under taking Hajós joins. If $k \geq 5$, then $\operatorname{Con}(k)$ is the smallest family of graphs that contains all complete graphs of order k and is closed under taking Hajós join. Note that if $k \geq 3$, then $\operatorname{Con}(k)$ is the class of graphs constructible with respect to the restricted Hajós calculus $\mathcal{H}C'_k$ where the family of initial graphs consists of K_k when $k \neq 4$, respectively of the odd wheels W_n with $n \geq 3$ when $k = 4$, and the only rule is the join rule (form a Hajós join of two graphs). Then $\operatorname{Con}(k) = \operatorname{Con}(\mathcal{H}C'_k)$ and by Theorem 4.2(c) it follows that $\operatorname{Con}(k) \subseteq \operatorname{Crit}(k)$. The next theorem says that $\operatorname{Con}(k)$ consists of all k-critical graphs with $\bar{\kappa}' \leq k-1$.

Theorem 4.11 *Let $k \geq 1$ be an integer. A graph G is k-critical and $\bar{\kappa}'(G) \leq k-1$ if and only if G belongs to* $\mathrm{Con}(k)$.

Proof In what follows we denote by C_k the class of graphs G such that G is k-critical and $\bar{\kappa}'(G) \leq k-1$. Our aim is to show that the two classes C_k and $\mathrm{Con}(k)$ coincide for all $k \geq 1$. This is obvious when $1 \leq k \leq 3$, since in these cases $\mathrm{Con}(k) = \mathrm{Crit}(k) = C_k$. So assume that $k \geq 4$. The proof of the following claim is straightforward and left to the reader.

Claim 4.11.1 *The odd wheels belong to the class C_4 and the complete graphs of order k belong to the class C_k.*

Claim 4.11.2 *Let $G = G_1 \nabla G_2$ the Hajós join of two graphs G_1 and G_2. Then G belongs to the class C_k if and only if both G_1 and G_2 belong to the class C_k.*

Proof: Suppose that $G = (G_1, v_1, w_1)\nabla(G_2, v_2, w_2)$ and let v^* denote the vertex of G obtained by identifying v_1 and v_2. First suppose that $G_1, G_2 \in C_k$. As both G_1 and G_2 are k-critical, Theorem 4.2(c) implies that G is k-critical. So it suffices to prove that $\bar{\kappa}'(G) \leq k-1$. To this end let u and u' be distinct vertices of G and let $p = \bar{\kappa}'_G(u,u')$. Then there is a system $\mathcal{P}$ of p edge disjoint u-u' paths in G. If u and u' belong both to G_1, then only one path P of $\mathcal{P}$ may contain vertices not in G_1. In this case P contains the vertex v^* and the edge w_1w_2. If we replace in P the subpath v^*Pw_1 by the edge v_1w_1, we obtain a system of p edge disjoint u-u' paths in G_1, and hence $p \leq \bar{\kappa}'_{G_1}(u,u') \leq k-1$. If u and u' belong to G_2, a similar argument shows that $p \leq k-1$. It remains to consider the case that one vertex, say u, belongs to G_1 and the other vertex u' belongs to G_2. By symmetry we may assume that $u \neq v^*$. Again at most one path P of $\mathcal{P}$ uses the edge w_1w_2 and the remaining paths of $\mathcal{P}$ all uses the vertex $v^*(= v_1 = v_2)$. If we replace P by the path $uPw_1 + w_1v_1$, then we obtain p edge disjoint u-v_1 path in G_1, and hence $p \leq \bar{\kappa}'_{G_1}(u,v_1) \leq k-1$. This shows that $\bar{\kappa}'(G) \leq k-1$ and so $G \in C_k$.

Suppose conversely that $G \in C_k$. By Theorem 4.2(d), both graphs G_1 and G_1 are k-critical. So it suffices to show that $\bar{\kappa}'(G_i) \leq k-1$ for $i \in \{1,2\}$. By symmetry it suffices to show that $\bar{\kappa}'(G_1) \leq k-1$. To this end let u and u' be distinct vertices of G_1 and let $p = \bar{\kappa}'_G(u,u')$. Then there is a system $\mathcal{P}$ of p edge disjoint u-u' paths in G_1. At most one path of $\mathcal{P}$, say P, can contain the edge v_1w_1. Clearly, there is a v_2-w_2 path P' in G_2 not containing the edge v_2w_2. So if we replace the edge v_1w_1 of P by the path $P' + w_1w_2$, we get p edge disjoint u-u' paths of G, and hence $p \leq \bar{\kappa}'_G(u,u') \leq k-1$. This shows that $\bar{\kappa}'(G_1) \leq k-1$ and by symmetry $\bar{\kappa}'(G_2) \leq k-1$. Hence both G_1 and G_2 belong to C_k. △

As a consequence of Claim 4.11.1 and Claim 4.11.2 and the definition of the class $\mathrm{Con}(k)$ we obtain the following claim.

Claim 4.11.3 *The class* $\mathrm{Con}(k)$ *is a subclass of C_k.*

Claim 4.11.4 *Let G be a graph belonging to the class C_k. If G is 3-connected, then either $k = 4$ and G is an odd wheel, or $k \geq 5$ and G is a K_k.*

Proof: The proof is by contradiction, where we consider a counterexample G whose order $|G|$ is minimum. Then $G \in \mathcal{C}_k$ is a 3-connected graph, and either $k = 4$ and G is not an odd wheel, or $k \geq 5$ and G is not a K_k. First we claim that $|G_H| \geq 2$. If $G_H = \varnothing$, then Brooks' theorem implies that G is a K_k, a contradiction. If $|G_H| = 1$, then Proposition 4.9 implies that $k = 4$ and G is an odd wheel, a contradiction. This proves that $|G_H| \geq 2$ as claimed. Then let u and v be distinct high vertices of G. Since $G \in \mathcal{C}_k$, we obtain that $\bar{\kappa}'_G(u,v) \leq \bar{\kappa}'(G) \leq k-1$. Then Theorem 1.32 implies that $\bar{\kappa}'_G(u,v) = k-1$ and, therefore, G contains a separating edge set F of size $k-1$ which separates u and v. From Theorem 4.10 it then follows that there is an edge cut (X,Y,F) satisfying the three properties of that theorem. Since F separates u and v, we may assume that $u \in X$ and $v \in Y$. By Theorem4.10(a), $|Y_F| = k-1$ and hence each vertex of Y_F is incident to exactly one edge of F. Since Y contains the high vertex v, we conclude that $|Y_F| < |Y|$. Now we consider the graph G' obtained from $G[X \cup Y_F]$ by adding all missing edges between the vertices of Y_F, so that Y_F becomes a clique of G'. By Theorem 4.10(b), G' is k-critical. Clearly, every vertex of Y_F is a low vertex of G and every vertex of X has in G' the same degree as in G. Since X contains the high vertex u of G, this implies that $|X_F| < |X|$. Since G is 3-connected, we conclude that $|X_F| \geq 3$ and that G' is 3-connected.

Now we claim that $\bar{\kappa}'(G') \leq k-1$. To prove this, let x and y be distinct vertices of G'. If x or y is a low vertex of G', then $\bar{\kappa}'_{G'}(x,y) \leq k$ and there is nothing to prove. So assume that both x and y are high vertices of G'. Then both vertices x and y belong to X. Let $p = \bar{\kappa}'_{G'}(x,y)$ and let $\mathcal{P}$ be a system of p edge disjoint x-y paths in G'. We may choose $\mathcal{P}$ such that the number of edges in $\mathcal{P}$ is minimum. Let $\mathcal{P}_1$ be the paths in $\mathcal{P}$ which uses edges of F. Since $|Y_F| = k-1$ and each vertex of Y_F is incident with exactly one edge of F, this implies that each path P in $\mathcal{P}_1$ contains exactly two edges of F. Since $|X_F| < |X|$ and $|Y_F| < |Y|$, there are vertices $u' \in X \setminus X_F$ and $v' \in Y \setminus Y_F$. By Theorem 1.32 it follows that $\bar{\kappa}'_G(u',v') = k-1$ and, therefore, there are $k-1$ edge disjoint u'-v' paths in G. Since $|Y_F| = k-1$, for each vertex $z \in Y_F$, there is a v'-z path P_z in $G[Y]$ such that these paths are edge disjoint. Now let P be an arbitrary path in $\mathcal{P}_1$. Then P contains exactly two vertices of Y_F, say z and z', and we can replace the edge zz' of the path P by a z-z' path contained in $P_z \cup P_{z'}$. In this way we obtain a system of p edge disjoint x-y paths in G, which implies that $p \leq \bar{\kappa}'_G(x,y) \leq k-1$. This proves that $\bar{\kappa}'(G') \leq k-1$ as claimed. Consequently $G' \in \mathcal{C}_k$. Clearly, $|G'| < |G|$ and either $k = 4$ and G' is not an odd wheel, or $k \geq 5$ and G is not a K_k. This, however, is a contradiction to the choice of G. Thus the claim is proved. △

Claim 4.11.5 *Let G be a graph belonging to the class $\mathcal{C}_k$. If G has a 2-separator, then $G = G_1 \nabla G_2$ is the Hajós sum of two graphs G_1 and G_2, which both belong to $\mathcal{C}_k$.*

Proof: If G has a separating set consisting of one edge and one vertex, then Proposition 4.8 implies that G is the Hajoś join of two graphs G_1 and G_2. By Claim 4.11.2 it then follows that both G_1 and G_2 belong to $\mathcal{C}_k$ and we are done. It remains to consider the case that G does not contain a separating set consisting of one edge and one vertex. By assumption, G has a 2-separator, say $S = \{u,v\}$. Then

Theorem 4.6 implies that $G-S$ has exactly two components H_1 and H_2 such that the graphs $G_1 = G[V(H_1)\cup S]$ and $G_2 = G[V(H_2)\cup S]$ satisfy the three properties of that theorem. In particular, $G_1' = G_1 + uv$ is k-critical. By Theorem 1.32, it then follows that $\bar{\kappa}'_{G_1'}(u,v) \geq k-1$ implying that $\bar{\kappa}'_{G_1}(u,v) \geq k-2$. Since $G \in \mathcal{C}_k$, we then conclude that $\bar{\kappa}'_{G_2}(u,v) \leq 1$. Since G_2 is connected, this implies that G_2 has a bridge e. Since $H_2 \neq \varnothing$, we conclude that $\{u,e\}$ or $\{v,e\}$ is a separating set of G, a contradiction. △

As a consequence of Claim 4.11.4 and Claim 4.11.5, we conclude that the class $\mathcal{C}_k$ is a subclass of the class $\mathrm{Con}(k)$. Together with Claim 4.11.3 this yields that the two classes $\mathcal{C}_k$ and $\mathrm{Con}(k)$ coincide, as required. □

The next result follows immediately from Theorem 4.11 and is a counterpart of Brooks' theorem. In fact, Brooks' theorem may easily deduced from it. The result was proven by Aboulker, Brettell, Havet, Marx, and Trotignon [15] for $\bar{\kappa}'(G) = 3$ and by Stiebitz and Toft [977] for $\bar{\kappa}'(G) \geq 4$.

Theorem 4.12 *Let G be a graph with $\bar{\kappa}'(G) = k$ and $k \geq 3$. Then $\chi(G) \leq k+1$ and equality hods if and only if G has a block belonging to the class* $\mathrm{Con}(k+1)$.

Proof Let G be a graph with $\bar{\kappa}'(G) = k$ and $k \geq 3$. By (4.1) it follows that $\chi(G) \leq k+1$. If one block B of G belongs to $\mathrm{Con}(k+1)$, then $B \in \mathcal{C}_{k+1}$ (by Theorem 4.11) and so $\chi(G) = k+1$ (by (4.2)).

Assume conversely that $\chi(G) = k+1$. Then G contains a subgraph H which is $(k+1)$-critical (by Proposition 1.27(a)). Obviously, $\bar{\kappa}'(H) \leq \bar{\kappa}'(G) = k$, and so $H \in \mathcal{C}_{k+1}$. As H is critical, H is 2-connected (by Corollary 1.31). We claim that H is a block of G. For otherwise, H would be a proper subgraph of a block of G. This implies that there are distinct vertices u and v in H which are joined by a path P of G with $E(P)\cap E(H) = \varnothing$. Since $\bar{\kappa}'_H(u,v) \geq k$ (by Theorem 1.32), this implies that $\bar{\kappa}'_G(u,v) \geq k+1$, which is impossible. This proves the claim that H is a block of G. As $H \in \mathcal{C}_{k+1}$ we obtain that H belongs to $\mathrm{Con}(k+1)$ (by Theorem 4.11. This completes the proof of the theorem. □

4.3 Decomposable Critical Graphs

Following Gallai, a graph is called **decomposable** if it is the Dirac join of two nonempty disjoint subgraphs; otherwise the graph is called **indecomposable**. By Theorem 4.1 it follows that a decomposable critical graph is the Dirac join of its indecomposable critical subgraphs. So the indecomposable critical graphs are building elements of critical subgraphs. In 1963 Gallai [398] proved the following beautiful result about indecomposable critical graphs.

Theorem 4.13 (GALLAI) *Let $k \geq 2$ be an integer. Every indecomposable k-critical graph G satisfies $|G| \geq 2k-1$.*

Corollary 4.14 (GALLAI) *Let $k \geq 2$ be an integer. If G is a k-critical graph with $|G| \leq 2k-2$, then G is the Dirac join of two nonempty disjoint critical graphs.*

Before proving the theorem, we need to introduce some definitions. Let G be a graph. For a coloring $\varphi : V(G) \to C$ of G with color set C we shall use the following notation: $C_\varphi = \mathrm{im}(\varphi)$ is the **set of used colors**, $c(\varphi) = |C_\varphi|$ is the **number of used colors** and $\mathcal{X}_\varphi = \{\varphi^{-1}(c) \mid c \in C_\varphi\}$ is the set of nonempty color classes. Clearly, $c(\varphi) \geq \chi(G)$ and we call φ an **optimal coloring** of G if $c(\varphi) = \chi(G)$. If $G = \varnothing$, then $\mathcal{X}_\varphi = \varnothing$ else $\mathcal{X}_\varphi$ is a partition of the vertex set of G into independent sets of G. Furthermore, we denote by $H(\varphi)$ the hypergraph whose vertex set is $V(H(\varphi)) = V(G)$ and whose edge set is

$$E(H(\varphi)) = \{U \in \mathcal{X}_\varphi \mid |U| \geq 2\}.$$

The set of isolated vertices of $H(\varphi)$ is denoted by $I(\varphi)$. Note that a vertex v of G belongs to $I(\varphi)$ if and only if $\{v\}$ is a color class with respect to φ. The hypergraph $H(\varphi)$ has maximum degree $\Delta(H(\varphi)) \leq 1$ and

$$c(\varphi) = \mathrm{com}(H(\varphi)) = |I(\varphi)| + |E(H(\varphi))|.$$

A vertex set $X \subseteq V(G)$ is said to be **φ-closed** if X is the union of a set of color classes with respect to φ, that is, each color class $U \in \mathcal{X}_\varphi$ satisfies $U \subseteq X$ or $U \cap X = \varnothing$. For two colorings φ_1 and φ_2 of G, let $H(\varphi_1, \varphi_2) = H(\varphi_1) \cup H(\varphi_2)$. Obviously, this hypergraph has maximum degree at most 2.

In this section we shall identify a coloring φ of G with the set $\mathcal{X}_\varphi$ of all nonempty color classes. In particular, we suppress unused colors and do not distinguish between equivalent colorings. If φ is a coloring of G and $X \subseteq V(G)$, then we denote by $\varphi|_X$ the coloring φ' of $G[X]$ with $\mathcal{X}_{\varphi'} = \{U \cap X \mid U \in \mathcal{X}_\varphi \text{ and } U \cap X \neq \varnothing\}$. Furthermore, let $\overline{X} = V(G) \setminus X$. In what follows we shall use the following simple facts: X is φ-closed if and only if $\mathcal{X}_{\varphi'} \subseteq \mathcal{X}_\varphi$; if X is φ-closed, then $\overline{X}$ is φ-closed, $\varphi = \varphi|_X \cup \varphi|_{\overline{X}}$ and $c(\varphi) = c(\varphi|_X) + c(\varphi|_{\overline{X}})$.

The following three propositions lists some simple, but useful facts about optimal colorings. The first is an immediate consequence of Proposition 1.34(a)(c).

Proposition 4.15 *Let G be graph, let φ be an optimal coloring of G. Then the following statements hold:*

(a) *$I(\varphi)$ is a clique of G.*
(b) *If $X \subseteq V(G)$ is φ-closed, then $\overline{X}$ is φ-closed and, for $Y \in \{X, \overline{X}\}$, the restriction $\varphi|_Y$ is an optimal coloring of $G[Y]$ satisfying $\varphi = \varphi|_X \cup \varphi|_{\overline{X}}$ and $c(\varphi) = c(\varphi|_X) + c(\varphi|_{\overline{X}})$.*

Proposition 4.16 *For a vertex v of a graph G there exists an optimal coloring φ of G with $v \in I(\varphi)$ if and only if $\chi(G - v) < \chi(G)$.*

Proof If φ is an optimal coloring of G such that $v \in I(\varphi)$, then the set $\{v\}$ is φ-closed and it follows from Proposition 4.15(b) that $\varphi' = \varphi|_{V(G)\setminus\{v\}}$ is an optimal coloring of $G - v$ with $c(\varphi') = c(\varphi) - 1$ implying that $\chi(G - v) = \chi(G) - 1$. If conversely $\chi(G - v) < \chi(G)$, then $\chi(G - v) = \chi(G) - 1$ (by (1.13)). Then there is an optimal coloring φ' of $G - v$ with $c(\varphi') = \chi(G) - 1$. This coloring can be extended to a coloring φ of G by assigning to v an additional color, so that $c(\varphi) = \chi(G)$. Then φ is an optimal coloring of G and $v \in I(\varphi)$. □

Proposition 4.17 *Let φ_1 and φ_2 be two distinct optimal colorings of a graph G, and let $H = H(\varphi_1, \varphi_2)$. Then the following statements hold:*

(a) *If X is the vertex set of a component of the hypergraph H, then X is both φ_1-closed and φ_2-closed and $\varphi_3 = \varphi_1|_X \cup \varphi_2|_{\overline{X}}$ is an optimal coloring of G with $I(\varphi_3) = (I(\varphi_1) \cap X) \cup (I(\varphi_2) \cap \overline{X})$.*

(b) *If v_1v_2 is an edge of $\overline{G}$ such that $v_1 \in I(\varphi_1)$ and $v_2 \in I(\varphi_2)$, then v_1 and v_2 lie in the same component of the hypergraph H.*

Proof First we prove (a). Let U be any nonempty color class of φ_1 or φ_2. Then either $|U| = 1$ or U is an edge of H and, since X is the vertex set of a component of H, it follows that either $U \subseteq X$ or $U \cap X = \varnothing$. So X is both φ_1-closed and φ_2-closed. By Proposition 4.15(b) it follows that also $\overline{X}$ is both φ_1-closed and φ_2-closed. Furthermore, it follows that $\varphi_1' = \varphi_1|_X$ is an optimal coloring of $G[X]$ and $\varphi_2' = \varphi_2|_{\overline{X}}$ is an optimal coloring of $G[\overline{X}]$, where $\chi(G) = \chi(G[X]) + \chi(G[\overline{X}])$. From this we conclude that $\varphi_3 = \varphi_1' \cup \varphi_2'$ is an optimal coloring of G. Clearly $I(\varphi_3) = I(\varphi_1') \cup I(\varphi_2') = (I(\varphi_1) \cap X) \cup (I(\varphi_2) \cap \overline{X})$. This proves (a).

For the proof of (b), let X be the vertex set of the component of H containing v_1. Then it follows from (a) that $\varphi_3 = \varphi_1|_X \cup \varphi_2|_{\overline{X}}$ is an optimal coloring of G. If $v_2 \notin X$, then we conclude from (a) that $\{v_1, v_2\} \subseteq I(\varphi_3)$. But then Proposition 4.15(a) implies that v_1v_2 is an edge of G and not of $\overline{G}$ as assumed. This contradiction proves (b). □

A graph G is the Dirac join of two disjoint nonempty graphs if and only if $\overline{G}$ is disconnected. So a graph is indecomposable if and only if its complement is connected, and hence Theorem 4.13 is an immediate consequence of the following theorem.

Theorem 4.18 *Let G be a critical graph whose complement is connected. Then for every vertex v of G there is an optimal coloring φ of G such that $I(\varphi) = \{v\}$.*

Proof Let v be an arbitrary vertex of G. Since G is critical, $\chi(G - v) < \chi(G)$ and there exists an optimal coloring φ of G such that $v \in I(\varphi)$ (by Proposition 4.16). An optimal coloring φ of G is called a *v-extreme coloring* of G if $v \in I(\varphi)$ and $|I(\varphi)|$ is minimum subject to this condition. To complete the proof we must show that a v-extreme coloring φ of G satisfies $|I(\varphi)| = 1$. The proof is arranged in a series of three claims.

Claim 4.18.1 *Let v be a vertex of G, let φ_1 be a v-extreme coloring of G, and let φ_2 be any coloring of G. Then any component of the hypergraph $H = H(\varphi_1, \varphi_2)$ contains at most one vertex of $I(\varphi_1)$.*

Proof: Suppose some component H' of H contains at least two vertices of $I(\varphi_1)$. Then let

$$P = (v_0, e_0, v_1, e_1, \dots, v_{p-1}, e_{p-1}, v_p)$$

be a shortest hyperpath in H' such that v_0 and v_p are two distinct vertices of $I(\varphi_1)$. Clearly, such a hyperpath P exists and we may chose P so that v_p is distinct from v. By the minimality of P, the only vertices of P belonging to $I(\varphi_1)$ are v_0 and v_p. Clearly, the vertices v_i are distinct and the edges e_i are distinct. Also since the edges of $H(\varphi_i)$ are disjoint, the edges of P alternately lie in $H(\varphi_1)$ and $H(\varphi_2)$. As the edges e_0 and e_{p-1} both belong to $H(\varphi_2)$, the length p of P is odd, say $p = 2q+1$.

Let $X = \{v_0, v_1, \dots v_p\}$ be the vertex set of P. Then $\varphi_1|_X$ is a coloring of $G[X]$ with $c(\varphi_1|_X) = q+2$ and $\varphi_2|_X$ is a coloring of $G[X]$ with $c(\varphi_2|_X) = q+1$. So $\varphi_1|_X$ is no optimal coloring of $G[X]$. Since φ_1 is an optimal coloring of G, we then conclude from Proposition 4.15(b) that X is not φ_1-closed. Consequently, P contains an edge of $H(\varphi_1)$ of size greater than 2. Then there is a largest integer j such that v_j belongs to an edge of $H(\varphi_1)$ of size greater than 2, where $0 < j < p$. Let

$$P' = (v_j, e_j, v_{j+1}, \dots, v_{p-1}, e_{p-1}, v_p)$$

be the subhyperpath of P whose ends are v_j and v_p, and let Y be the vertex set of P', i.e., $Y = \{v_j, v_{j+1}, \dots, v_p\}$. By definition of v_j, the edge e_j belongs to $H(\varphi_2)$, so the length $p - j$ of P' is odd. For the restricted colorings we obtain that

$$c(\varphi_1|_Y) + c(\varphi_1|_{\overline{Y}}) = c(\varphi_1) + 1 \text{ and } I(\varphi_1|_{\overline{Y}}) = I(\varphi_1) - \{v_p\}.$$

The first equation follows from the fact that $U = e_{j-1}$ is a color class of φ_1 whose size is at least 3 and this is the only color class of φ_1 which is divided into two color classes, namely $U \cap Y = \{v_j\}$ and $U \cap \overline{Y} = U \setminus \{v_j\}$. The second equation follows from the fact that v_p is the only isolated vertex of $H(\varphi_1)$ belonging to Y. Furthermore, we conclude that $c(\varphi_1|_Y) = c(\varphi_2|_Y) + 1$ and $I(\varphi_2|_Y) = \emptyset$. Then $\varphi_3 = \varphi_2|_Y \cup \varphi_1|_{\overline{Y}}$ is a coloring of G satisfying

$$c(\varphi_3) = c(\varphi_2|_Y) + c(\varphi_1|_{\overline{Y}}) = c(\varphi_1|_Y) + c(\varphi_1|_{\overline{Y}}) - 1 = c(\varphi_1)$$

and

$$I(\varphi_3) = I(\varphi_2|_Y) \cup I(\varphi_1|_{\overline{Y}}) = I(\varphi_1) \setminus \{v_p\}.$$

This implies that φ_3 is an optimal coloring of G such that v belongs to $I(\varphi_3)$ and $|I(\varphi_3)| < I(\varphi_1)$. However, this contradicts the assumption that φ_1 is a v-extreme coloring of G. △

Claim 4.18.2 *If $v_1 v_2$ is an edge of $\overline{G}$ and φ_1 is a v_1-extreme coloring of G, then there exists a v_2-extreme coloring φ_2 such that $I(\varphi_1) \setminus \{v_1\} = I(\varphi_2) \setminus \{v_2\}$.*

Proof: There exists a v_2-extreme coloring φ_3 of G. By Proposition 4.17(b), the vertices v_1 and v_2 lie in the same component of the hypergraph $H(\varphi_1, \varphi_3)$. Let X be the vertex set of this component, and let

$$\varphi_2 = \varphi_3|_X \cup \varphi_1|_{\overline{X}}.$$

By Proposition 4.17(a), φ_2 is an optimal coloring of G. Using Claim 4.18.1, we conclude that $I(\varphi_1) \cap X = \{v_1\}$ and $I(\varphi_3) \cap X = \{v_2\}$, which gives $I(\varphi_1) \cap \overline{X} = I(\varphi_1) \setminus \{v_1\}$ and $I(\varphi_3) \cap \overline{X} = I(\varphi_3) \setminus \{v_1\}$. Using Proposition 4.17(a), we conclude that $v_2 \in I(\varphi_2)$ and $I(\varphi_1) \setminus \{v_1\} = I(\varphi_2) \setminus \{v_2\}$. So φ_2 is an optimal coloring of G such that $v_2 \in I(\varphi_2)$ and $|I(\varphi_2)| = |I(\varphi_1)|$. To show that φ_2 is a v_2-extreme coloring of G, it suffices to show that $|I(\varphi_2)| \leq |I(\varphi_3)|$. By symmetry, the coloring $\varphi_4 = \varphi_1|_X \cup \varphi_3|_{\overline{X}}$ is an optimal coloring of G such that $v_1 \in I(\varphi_4)$ and $|I(\varphi_4)| = |I(\varphi_3)|$. Since φ_1 is a v_1-extreme coloring of G, we then conclude that $|I(\varphi_3)| = |I(\varphi_4)| \geq |I(\varphi_1)| = |I(\varphi_2)|$ as required. △

Claim 4.18.3 *Let v be a vertex of G, and let φ be a v-extreme coloring of G. Then for every vertex v' of G there is a v'-extreme coloring φ' such that $I(\varphi) \setminus \{v\} = I(\varphi') \setminus \{v'\}$.*

Proof: The statement is evident if $v = v'$, otherwise, since $\overline{G}$ is connected, there is a path $P = (v_0, v_1, \ldots, v_p)$ in $\overline{G}$ with $p \geq 1$, $v_0 = v$ and $v_p = p$. Let $\varphi_0 = \varphi$. For $i \in [1,p]$, it then follows from Claim 4.18.2 by induction on i that there exists a v_i-extreme coloring φ_i of G such that $I(\varphi_{i-1}) \setminus \{v_{i-1}\} = I(\varphi_i) \setminus \{v_i\}$. This implies that $I(\varphi) \setminus \{v\} = I(\varphi_0) \setminus \{v_0\} = I(\varphi_p) \setminus \{v_p\}$, which proves the claim. △

To finish the proof of Theorem 4.18, let v be an arbitrary vertex of G. Suppose that there exists a v-extreme coloring φ of G such that $I(\varphi) \neq \{v\}$. Then there exists a vertex $v' \in I(\varphi) \setminus \{v\}$, and hence $|G| \geq 2$. By Claim 4.18.3, there is a v'-extreme coloring φ' such that $I(\varphi) \setminus \{v\} = I(\varphi') \setminus \{v'\}$, which is a contradiction since $v' \in I(\varphi) \setminus \{v\}$. This completes the proof of Theorem 4.18. □

A vertex of a graph G is called a **dominating vertex** of G if it is joined to every other vertex of G. So a vertex is a dominating vertex of G if and only if it is an isolated vertex of $\overline{G}$. If X is the set of dominating vertices of G, then $G[X]$ is a complete graph and $G = G[X] \boxplus H$, where $H = G - X$ is a graph without dominating vertices.

In what follows let G be an arbitrary k-critical graph of order $|G| \leq 2k-2$, where $k \geq 2$. Then

$$G = G_1 \boxplus G_2 \boxplus \cdots \boxplus G_t,$$

where $t = \mathrm{com}(\overline{G})$ and $\overline{G}_1, \overline{G}_2, \ldots, \overline{G}_t$ are the components of $\overline{G}$. From Corollary 4.14 we obtain that $t \geq 2$. For $i \in [1,t]$, let $k_i = \chi(G_i)$. By Theorem 4.1, we obtain that

(1) $k = k_1 + k_2 + \cdots + k_t$ and G_i is a k_i-critical graph for $i \in [1,t]$.

Since $\overline{G}_i$ is connected, Theorem 4.13 implies that

(2) $|G_i| \geq 2k_i - 1$ for $i \in [1,t]$.

Since $\mathrm{Crit}(1) = \{K_1\}$ and $\mathrm{Crit}(2) = \{K_2\}$, we conclude that either $k_i = 1$ and $G_i = K_1$, or $k_i \geq 3$ and $|G_i| \geq 5$. So if $k_i \geq 3$, then $|G_i| \geq 2k_i - 1 \geq \frac{5}{3}k_i$, where equality holds if and only if $k_i = 3$ and $G_i = C_5$.

Let s be the number of dominating vertices of G. Then $s = |\{i \in [1,t] \mid k_i = 1\}|$ and $s \geq 0$. We may assume that $k_i = 1$ for $i \in [1,s]$, which gives $k_i \geq 3$ for $i \in [s+1,t]$ and

(3) $G = K_1 \boxplus \cdots \boxplus K_1 \boxplus G_{s+1} \boxplus \cdots \boxplus G_t = K_s \boxplus G_{s+1} \boxplus \cdots \boxplus G_t$.

By (1), this yields $k = s + k_{s+1} + \cdots + k_t$. Then we obtain that

(4) $|G| = s + |G_{s+1}| + \cdots + |G_t| \geq s + \frac{5}{3}(k_{s+1} + \cdots k_t) = s + \frac{5}{3}(k - s)$,

which is equivalent to $s \geq \frac{3}{2}(\frac{5}{3}k - |G|)$. By the above, equality holds if and only if $G_i = C_5$ for all $i \in [s+1,t]$. Now assume that G has order $n \leq \frac{5}{3}k$. Then $s \geq 0$ and it follows from (3) and (4) that $G = K_s \boxplus G'$, where G' is an ℓ-critical graph with $\ell = k - s$ (and $0 \leq \ell \leq k$), G' has no dominating vertices, and $|G'| \geq \frac{5}{3}\ell$, where equality holds if and only if G' is the Dirac join of $\frac{1}{3}\ell$ disjoint cycles C_5. Summarizing we obtain the following result.

Corollary 4.19 *Let G be a k-critical graph of order $n \leq \frac{5}{3}k$, where $k \geq 3$, and let s be the number of dominating vertices of G. Then*

$$s \geq \tfrac{3}{2}(\tfrac{5}{3}k - n) \geq 0 \textit{ and } G = K_s \boxplus G'$$

is the Dirac join of a complete graph K_s and a graph G', where G' is an ℓ-critical graph with $\ell = k - s \geq 0$, G' has no dominating vertices, and $|G'| \geq \frac{5}{3}\ell$. Furthermore, $s = \frac{3}{2}(\frac{5}{3}k - n)$ if and only if G' is the Dirac join of of $\frac{1}{3}\ell$ disjoint cycles C_5.

Recall that $\text{Crit}(k,n)$ denotes the class of k-critical graphs of order n. Let $\text{Crit}^*(k,n)$ be the class consisting of all graphs in $\text{Crit}(k,n)$ with no dominating vertices. For a graph K and a graph property $\mathcal{P}$, define $K \boxplus \mathcal{P} = \{K \boxplus G \mid G \in \mathcal{P}\}$ if $\mathcal{P}$ is nonempty, and $K \boxplus \mathcal{P} = \varnothing$ otherwise.

Corollary 4.20 *Let $k \geq 4$ and $2 \leq p < \frac{2}{3}k$. Then*

$$\text{Crit}(k, k+p) = \bigcup_{\ell=\ell_p}^{\lfloor 3p/2 \rfloor} K_{k-\ell} \boxplus \text{Crit}^*(\ell, \ell+p),$$

where $\ell_p = 3$ if p is even, and $\ell_p = 4$ otherwise.

Proof Let $G \in \text{Crit}(k, k+p)$. Then $|G| < \frac{5}{3}k$ and $G = K_s \boxplus G'$ is the Dirac join of a complete graph K_s and an ℓ-critical graph with no dominating vertices, where $s = k - \ell \geq 1$ and $|G'| \geq \frac{5}{3}\ell$ (Corollary 4.19). Since G has order at least $k+2$, $G \neq K_k$, and so G' is nonempty. Since G' is an ℓ-critical graph with no dominating vertices, $\ell \geq 3$ and $|G'| \geq 5$. Furthermore, $|G'| = |G| - s = \ell + p \geq \frac{5}{3}\ell$. Consequently, we obtain $3 \leq \ell \leq \frac{3}{2}p$. Since no 3-critical graph has even order, we obtain $4 \leq \ell \leq \frac{3}{2}p$ if p is odd. □

4.4 Subgraphs of Critical Graphs

Our first goal in this section is to show that a graph is a subgraph of a k-critical graph if and only if the contraction of any edge results in a $(k-1)$-colorable graph.

To this end we shall first investigate the colorings of graphs that are direct products of complete graphs.

Let $p \in \mathbb{N}$ be fixed. For a set V we denote by V^p the p-fold cartesian product of V, that is, $V^p = \{(v_1, v_2, \ldots, v_p) \mid v_i \in V \text{ for all } i \in [1, p]\}$. Then $\pi_i : V^p \to V$ denotes the ith projection of V^p, that is, $\pi_i(v_1, v_2, \ldots, v_p) = v_i$. For an integer $n \geq 1$, let $K(n, p)$ denote the p-fold direct product of K_n, that is, $V(K(n, p)) = [1, n]^p$ and

$$E(K(n,p)) = \{vw \mid v, w \in [1,n]^p \text{ and } \pi_i(v) \neq \pi_i(w) \text{ for all } i \in [1,p]\}.$$

Note that $K(n, 1) = K_n$ and $K(n, p) = K(n, p-1) \times K_n$ if $p \geq 2$ (see Appendix C.7). Furthermore, it is easy to see that the projections $\pi_1, \pi_2, \ldots, \pi_p$ are colorings of $K(n, p)$ with color set $C = [1, n]$. As observed by Greenwell and Lovász [431] these are all colorings of $K(n, p)$ up to permutations.

Lemma 4.21 *Let $n \geq 3$ be an integer, and let φ be a coloring of $K(n, p)$ with color set $C = [1, n]$. Then there is an $i \in [1, p]$ such that φ is equivalent to π_i.*

Proof The proof is by induction on p. If $p = 1$, then the statement is obvious as $K(n, 1) = K_n$. So assume that $p \geq 2$, and let φ be a coloring of $K = K(n, p)$ with color set $C = [1, n]$. Then $V(K) = \{(v, k) \mid v \in [1, n]^{p-1} \text{ and } k \in [1, n]\}$ and $E(K) = \{(v, k)(w, \ell) \mid k \neq \ell \text{ and } vw \in E(K(n, p-1)\}$. We distinguish two cases.

Case 1: *There is an $v \in [1, n]^{p-1}$ such that the colors $\varphi(v, 1), \varphi(v, 2), \ldots, \varphi(v, n)$ are distinct.* We may assume that $\varphi(v, i) = i$ for $i \in [1, n]$. Let w be an arbitrary vertex of $K(n, p-1) - v$. Since $n \geq 3$, there is a vertex u of $K(n, p-1)$ such that u is adjacent to both v and w in $K(n, p-1)$. This implies that $\varphi(u, i) = i$ for $i \in [1, n]$. But then also $\varphi(w, i) = i$ for all $i \in [1, n]$. Since w was an arbitrarily vertex of $K(n, p-1) - v$ this means that $\varphi = \pi_p$.

Case 2: *For all $v \in [1, n]^{p-1}$ there is a color $c_v \in [1, n]$ and two distinct indices $i, j \in [1, n]$ such that $\varphi(v, i) = \varphi(v, j) = c_v$.* If vw is an edge of $K(n, p-1)$, then there are two distinct indices $i, j \in [1, n]$ such that $\varphi(v, i) = c_v$ and $\varphi(w, j) = c_w$. As (v, i) and (w, j) are adjacent in K, this implies that $c_v \neq c_w$. Consequently, the mapping φ' with $\varphi'(v) = c_v$ for all vertices v of $K(n, p-1)$ is a coloring of $K(n, p-1)$ with color set $C = [1, n]$. By the induction hypothesis, it follows that φ' is equivalent to π_k for some $k \in [1, p-1]$. Let now $w = (w_1, w_2, \ldots, w_p) = (w', w_p)$ be an arbitrary vertex of $K = K(n, p)$, and let i, j be the two distinct indices with $\varphi(w', i) = \varphi(w', j) = c_{w'}$. Clearly, $\varphi(w + (t, t, \ldots, t))$ for $t \in [0, n-1]$ are all different (where the addition is modulo n). For any $t \in [1, n-1]$ the vertex $w + (t, t, \ldots, t)$ is in K adjacent to (w', i) or (w', j), and hence $\varphi(w + (t, t, \ldots, t)) \neq c_{w'}$. But then $\varphi(w + (0, 0, \ldots, 0))$ is the remaining color $c_{w'}$, that is, $\varphi(w) = \varphi'(w')$. Thus φ is equivalent to π_k. This completes the proof. □

Let G be a graph and let $e = uv$ be an edge of G. Then G/e denotes the graph obtained from $G' = G - \{u, v\}$ by adding a new vertex v_e and joining v_e to all vertices w of G' such that $N_G(w) \cap \{u, v\} \neq \varnothing$. We then say that G/e is obtained from G by **contracting** the edge e. Clearly, $\chi(G/e) \leq \ell$ if and only if $G - e$ admits an ℓ-coloring in which the two ends of e receive the same color.

Theorem 4.22 (Greenwell and Lovász) *Let $k \geq 4$ be an integer. A graph H is a subgraph of a k-critical graph if and only if $\chi(H/e) \leq k-1$ for all edges $e \in E(H)$.*

Proof First suppose that H is a subgraph of a k-critical graph G. Let $e = uv$ be an arbitrary edge of H. Then $\chi(G-e) \leq k-1$, i.e., there is a coloring φ of $G-e$ with color set $C = [1, k-1]$. Since $\chi(G) = k$, we obtain $\varphi(u) = \varphi(v)$ and so φ induces a coloring of H/e implying $\chi(H/e) \leq k-1$.

Now suppose that $\chi(H/e) \leq k-1$ for all edges $e \in E(H)$. Our aim is to construct a k-critical graph G such that $H \subseteq G$. If H is edgeless, then we can take any k-critical graph G with $|G| \geq |H|$ and rename the vertices of G such that $H \subseteq G$. So assume that H has $p \geq 1$ edges. Let $e_1, e_2, \ldots, e_p$ be an enumeration of the edges of H, say $e_i = v_i w_i$ for $i \in [1, p]$. As $\chi(H/e_i) \leq k-1$, there is a coloring φ_i of $H - e_i$ with color set $C = [1, k-1]$ such that $\varphi_i(v_i) = \varphi_i(w_i)$ ($i \in [1, p]$). Now let K be the graph obtained from the disjoint union of H and the product graph $K(k-1, p)$ by joining a vertex u of H with a vertex v of $K(k-1, p)$ if and only if $\varphi_i(u) \neq \pi_i(v)$ for all $i \in [1, p]$. We claim that $\chi(K) \geq k$. Suppose this is false. Then there exists a coloring φ of K with color set $C = [1, k-1]$. As $K(k-1, p)$ is an induced subgraph of K, it follows from Lemma 4.21 that the restriction of φ to $K(k-1, p)$ is equivalent to the coloring π_i for some $i \in [1, p]$. By renaming the colors, we may assume that $\varphi(v) = \pi_i(v)$ for every vertex v of $K(k-1, p)$. Let u be an arbitrary vertex of H, and let $\ell \in C$ be an arbitrary color with $\ell \neq \varphi_i(u)$. Then there is a vertex $v = (v_1, v_2, \ldots, v_p)$ of $K(k-1, p)$ such that $\pi_i(v) = v_i = \ell$ and $\pi_j(v) = v_j \neq \varphi_j(u)$ for $j \in [1, p] \setminus \{i\}$. Consequently, uv is an edge of K and so $\varphi(u) \neq \varphi(v) = \pi_i(v) = \ell$, which leads to $\varphi(u) = \varphi_i(u)$. Hence $\varphi = \pi_i \cup \varphi_i$ and $\varphi(v_i) = \varphi(w_i)$ which is impossible as $e_i = v_i w_i$ is an edge of H. This shows that $\chi(K) \geq k$. Moreover, any edge e_i of H is a critical edge of K, as $\pi_i \cup \varphi_i$ is a $(k-1)$-coloring of $K - e_i$. Then $\chi(K) = k$ (by (1.13)). By Proposition 1.27, K contains a k-critical subgraph and any such k-critical subgraph G of K contains H. □

Let us now construct a family of forbidden subgraphs of critical graphs. The graph $W(\ell, d)$ is defined as the Dirac join $W(\ell, d) = C_\ell \boxplus K_d$ ($\ell \geq 3$ and $d \geq 0$) and it is called a d-**wheel**. Note that a 0-wheel is a cycle, and a 1-wheel is an ordinary **wheel**. Clearly, W is a d-wheel if and only if $W \boxplus K_1$ is a $(d+1)$-wheel. The d-wheel $W(\ell, d)$ is called **odd** or **even** depending on whether ℓ is odd or even. By the Dirac construction (see Theorem 4.1), an odd d-wheel is a $(d+3)$-critical graph. An even d-wheel $W = C_{2\ell} \boxplus K_d$ has chromatic number $d+2$; and if e is an edge of W belonging to the cycle $C_{2\ell}$, then $W/e = W(2\ell - 1, d)$ and so $\chi(W/e) = d+3$. Consequently, an even d-wheel cannot be contained in any $(d+3)$-critical graph as a subgraph, and we have obtained the following result from [967].

Lemma 4.23 (Stiebitz) *If G is a k-critical graph with $k \geq 3$ containing $W(\ell, k-3)$ as a subgraph, then $G = W(\ell, k-3)$ and $\ell \geq 3$ is an odd integer.*

For an arbitrary graph G and an integer $p \geq 1$, let $\mathcal{K}_p(G)$ denote the set of subgraphs K of G which are isomorphic to K_p, and let $\mathbb{k}_p(G) = |\mathcal{K}_p(G)|$. T. Gallai raised the problem, whether each graph $G \in \text{Crit}(k, n)$ satisfies $\mathbb{k}_{k-1}(G) \leq n$. This is evident when $n = k$ as K_k is the only k-critical graph of order k; and it is also true if $k = 3$ as the only 3-critical graphs are the odd cycles.

A powerful tool for counting the number of objects in certain graphs is to let vectors of some space correspond to the objects and to show that these vectors are linearly independent. Stiebitz [967] used a linear algebra approach to show that every 4-critical graph of order n has at most n triangles. Abbott and Zhou [11] adapted Stiebitz's argument to arbitrary critical graphs and settled Gallai's problem in the affirmative for all $k \geq 5$.

Theorem 4.24 (ABBOTT AND ZHOU) *If G is a k-critical graph of order n where $k \geq 4$, then $\Bbbk_{k-1}(G) \leq n$ with equality if and only if $n = k$ and $G = K_k$.*

In 2022, Gao and Ma [403] also used linear algebra to improve the above result of Abbott and Zhu, thus solving a problem posed by them.

Theorem 4.25 (GAO AND MA) *If G is a k-critical graph of order n where $n > k \geq 4$, then $\Bbbk_{k-1}(G) \leq n - k + 3$.*

Before we turn our attention to the proof of the above result, we need to make some preparations. In the following let G be an arbitrary graph. We denote by $\mathbb{Z}_2(G)$ the vector space over the residue ring modulo 2 consisting of all functions $f : V(G) \to \mathbb{Z}_2$. The **null vector** of $\mathbb{Z}_2(G)$ is denoted by $\mathbf{0}$, that is, $\mathbf{0}(v) = 0$ for all $v \in V(G)$. For a set $X \subseteq V(G)$, let $f_X \in \mathbb{Z}_2(G)$ denote the function with

$$f_X(v) = \begin{cases} 1 \text{ if } v \in X, \\ 0 \text{ if } v \in V(G) \setminus X \end{cases}$$

for all $v \in V(G)$. We call f_X the **characteristic function** of X in G. If $X = \{v\}$ for some vertex v of G, then we also write f_v for f_X. If $K \subseteq G$ is a subgraph of G, then we also write f_K for $f_{V(K)}$. If $Z \subseteq \mathbb{Z}_2(G)$, then we denote by span$(Z)$ the (linear) subspace of $\mathbb{Z}_2(G)$ spanned by the vector set Z. For two vectors $g, h \in \mathbb{Z}_2(G)$ let $\langle g, h \rangle$ denote the inner product, that is,

$$\langle g, h \rangle = \sum_{v \in V(G)} g(v)h(v).$$

For two distinct subsets X and Y of $V(G)$, we have

$$\langle f_X, f_Y \rangle = 0 \text{ in } \mathbb{Z}_2(G) \text{ if and only if } |X \cap Y| \equiv 0 \,(\mathrm{mod}\, 2). \tag{4.3}$$

Let $\mathbb{K}$ be a subspace of $\mathbb{Z}_2(G)$. Then we denote the dimension of $\mathbb{K}$ by $\dim(\mathbb{K})$. Furthermore, let $\mathbb{K}^{\perp}$ be the orthogonal complement of $\mathbb{K}$ in $\mathbb{Z}_2(G)$ with respect to the inner product. Then it is well known (for a proof see Lovász [698, Problem 5.31]) that we have

$$\dim(\mathbb{K}) + \dim(\mathbb{K}^{\perp}) = \dim(\mathbb{Z}_2(G)) = |G| \tag{4.4}$$

Let $p \geq 1$ be an integer. We denote by $\mathcal{K}_p$ the class of complete graphs of order p. Let $\mathcal{K} \subseteq \mathcal{K}_p$ be a finite set (i.e., $\mathcal{K}$ is a finite set of graphs each of which is a K_p). Then let

$$G[\mathcal{K}] = \bigcup_{K \in \mathcal{K}} K \text{ and } n(\mathcal{K}) = |G[\mathcal{K}]|;$$

we call $G[\mathcal{K}]$ the **graph** of $\mathcal{K}$ and $n(\mathcal{K})$ the **order** of $\mathcal{K}$. Moreover, let

$$F[\mathcal{K}] = \{f_K \in \mathbb{Z}_2(G[\mathcal{K}]) \mid K \in \mathcal{K}\} \text{ and } \mathbb{K}[\mathcal{K}] = \mathrm{span}(F[\mathcal{K}]).$$

For a vertex v of $G[\mathcal{K}]$, let

$$\mathcal{K}^v = \{K \in \mathcal{K} \mid v \in V(K)\} \text{ and } \mathcal{K}/v = \{K - v \mid K \in \mathcal{K}^v\}.$$

Furthermore, for two distinct vertices v, w of $G[\mathcal{K}]$, let

$$\mathcal{K}^{vw} = \mathcal{K}^v \cap \mathcal{K}^w.$$

Our goal is to show that if $G \in \mathrm{Crit}(k,n)$, then $\dim(\mathbb{K}_{k-1}(G)) \leq n-k+3$ unless G is a $(k-3)$-wheel. An important idea, going back to Abbott and Zhu, to achieve this goal, is to replace the class $\mathrm{Crit}(k)$ with a more appropriate class. For $k \geq 3$, we call a set $\mathcal{K} \subseteq \mathcal{K}_{k-1}$ a k**-cluster** if $\mathcal{K}$ is nonempty and finite, $G[\mathcal{K}]$ does not contain any $(k-3)$-wheel as a subgraph, and $\chi(G[\mathcal{K}]/e) \leq k-1$ for all edges $e \in E(G[\mathcal{K}])$. If $G \in \mathrm{Crit}(k)$ and G is not an odd $(k-3)$-wheel, then it follows from Lemma 4.23 and Theorem 4.22 that $\mathcal{K}_{k-1}(G)$ is empty or a k-cluster. Clearly, every nonempty subset of a k-cluster is a k-cluster, too.

Lemma 4.26 *Let $k \geq 3$ be an integer, let $\mathcal{K}$ be a nonempty finite subset of $\mathcal{K}_{k-1}$, and let $G = G[\mathcal{K}]$. Then the following statements hold:*

(a) *If $v \in V(G)$, then $|\mathcal{K}/v| = |\mathcal{K}^v| \geq 1$, $G[\mathcal{K}/v] = G[\mathcal{K}^v] - v$, and $\mathcal{K}/v \subseteq \mathcal{K}_{k-2}$.*
(b) $\sum_{u \in V(G)} |\mathcal{K}^u| = (k-1)|\mathcal{K}|$.
(c) *If $w \in V(G[\mathcal{K}/v])$, then $vw \in E(G)$ and $|(\mathcal{K}/v)^w| = |\mathcal{K}^{vw}| \geq 1$.*
(d) *$F[\mathcal{K}]$ is linearly independent in $\mathbb{Z}_2(G)$ if and only if for every nonempty subset $\mathcal{H}$ of $\mathcal{K}$ there is a vertex $v \in V(G)$ such that $|\mathcal{H}^v|$ is odd.*
(e) *If $\chi(G) \leq k-1$, then* $\dim(\mathbb{K}[\mathcal{K}]) \leq n(\mathcal{K}) - k + 2$.
(f) *If $\mathcal{K}$ is a 3-cluster, then G is a forest with $\delta(G) \geq 1$, $|\mathcal{K}| \leq n(\mathcal{K}) - 1$ and $|\mathcal{K}^v| = 1$ for some vertex v of G.*
(g) *If $\mathcal{K}$ is a k-cluster and $k \geq 4$, then $\mathcal{K}/v$ is a $(k-1)$-cluster for all $v \in V(G)$.*

Proof The proofs of statements (a), (b), (c) and (d) are straightforward and left to the reader. To prove (e) assume that $\chi(G) \leq k-1$ and let $\mathbb{K} = \mathbb{K}[\mathcal{K}]$. Then there is a coloring φ of G with color set $C = [1, k-1]$. For $c \in [1, k-2]$, let $X_c = \varphi^{-1}(c, k-1)$ and let $g_c \in \mathbb{Z}_2(G)$ be the characteristic function of X_c in G. If $K \in \mathcal{K}$, then $|V(K) \cap X_c| = 2$ for every color $c \in [1, k-2]$, which implies that g_c belongs to $\mathbb{K}^\perp$ for all $c \in [1, k-2]$ (see (4.3)). Since $\omega(G) \geq k-1$, we obtain that $\varphi^{-1}(c) \neq \varnothing$ for all $c \in [1, k-1]$. This implies that the vector set $\{g_c \mid c \in [1, k-2]\}$ is linearly independent in $\mathbb{Z}_2(G)$. Then we obtain that $\dim(\mathbb{K}^\perp) \geq k-2$. By (4.4), it follows that $\dim(\mathbb{K}) \leq n(\mathcal{K}) + k - 2$ (note that $n(\mathcal{K}) = |G|$). Thus (e) is proved. Statement (f) is obviously true, since the class of 0-wheels is the class of cycles. It remains to prove (g). So assume that $\mathcal{K}$ is a k-cluster with $k \geq 4$ and v is an arbitrary vertex of G. Let $H = G[\mathcal{K}^v]$ and $G' = G[\mathcal{K}/v]$. From (a) it follows that $\varnothing \neq \mathcal{K}/v \subseteq \mathcal{K}_{k-2}$ and $\mathcal{K}/v$ is finite. Furthermore, we obtain that $H \subseteq G$, $G' = H - v$ and $N_H(v) = V(G')$; which implies that $H = K_1 \boxplus G'$. Since $\mathcal{K}$ is a k-cluster, G does not contain a $(k-3)$-wheel, which implies that G' does not contain a $(k-4)$-wheel. Let $e = uw$ be an edge of G'. Since $\mathcal{K}$ is a k-cluster and $e \in E(G)$, there is a coloring φ of $G - e$ with color set $C = [1, k-1]$ such that $\varphi(u) = \varphi(w)$. By symmetry we may assume that $\varphi(v) = k - 1$. Since v is adjacent in G to all vertices of G', the restriction of φ to G' is a coloring of $G' - e$ with color set $C' = [1, k-2]$ and so $\chi(G'/e) \leq k-2$. This shows that $\mathcal{K}/v$ is a $(k-1)$-cluster. This completes the proof of the lemma. □

Lemma 4.27 *Let $k \geq 3$ be an integer, let $\mathcal{K}$ be a k-cluster, and let $G = G[\mathcal{K}]$. Then the following statements hold:*

(a) *For every edge $e = uv$ of G there is color class I with respect to a coloring of $G - e$ with color set $C = [1, k-1]$ such that $\sum_{w \in I} |\mathcal{K}^w| = |\mathcal{K}| + |\mathcal{K}^{uv}|$ and $\{u, v\} \subseteq I$.*
(b) *If $e = uv$ is an edge of G such that $|\mathcal{K}^{uv}| \leq k-3$, then there is a coloring φ of $G - e$ with color set $C = [1, k-1]$ such that $\varphi(u) = \varphi(v) = c$ and there is a color $c' \in C \setminus \{c\}$ such that $I = \varphi^{-1}(c')$ satisfies $|V(K) \cap I| = 1$ for all $K \in \mathcal{K}$ and hence $\sum_{w \in I} |\mathcal{K}^w| = |\mathcal{K}|$.*

Proof Let $e = uv$ be an arbitrary edge of G. Since G is the graph of the k-cluster $\mathcal{K}$, we have $\chi(G/e) \leq k-1$. Consequently, there is a coloring φ of $G - e$ with color set $C = [1, k-1]$ such that $\varphi(u) = \varphi(v) = c$ with $c \in C$. Let $I = \varphi^{-1}(c)$. then for $K \in \mathcal{K}$, we have $|V(K) \cap I| = 2$ if $K \in \mathcal{K}^{uv}$ and $|V(K) \cap I| = 1$ if $K \in \mathcal{K} \setminus \mathcal{K}^{uv}$. From this it follows that

$$\sum_{w \in I} |\mathcal{K}^w| = 2|\mathcal{K}^{uv}| + (|\mathcal{K}| - |\mathcal{K}^{uv}|) = |\mathcal{K}| + |\mathcal{K}^{uv}|.$$

Thus (a) is proved. For the color set $C' = C \setminus \{c\}$ we have $|C'| = k-2$. If $K \in \mathcal{K} \setminus \mathcal{K}^{uv}$, then each color of C' occurs in K. If $K \in \mathcal{K}^{uv}$, then exactly one color in C' does not occur in K. Hence, if $|\mathcal{K}^{uv}| \leq k-3$, then there is a color, say $c' \in C'$, that occurs in each complete graph $K \in \mathcal{K}$ exactly one time. Then, for $I = \varphi^{-1}(c')$ we have

$$\sum_{w \in I} |\mathcal{K}^w| = |\mathcal{K}|.$$

This proves (b). □

Lemma 4.28 *Let $k \geq 3$ be an integer, let $\mathcal{K}$ be a k-cluster, let $G = G[\mathcal{K}]$ and $\mathbb{K} = \mathbb{K}[\mathcal{K}]$. Then the following statements hold:*

(a) *If $k \geq 4$, then there is an edge $e = uv$ of G such that $|\mathcal{K}^{uv}| \leq k-3$.*
(b) *There is a vertex $w \in V(G)$ such that $|\mathcal{K}^w|$ is odd.*
(c) *The vector set $F[\mathcal{K}]$ is is linearly independent in the vector space $\mathbb{Z}_2(G)$, and as a consequence $|\mathcal{K}| = \dim(\mathbb{K})$.*
(d) $\dim(\mathbb{K}) \leq n(\mathcal{K}) - 1$.

Proof We prove (a), (b), (c) and (d) simultaneously by induction on k. Note that every nonempty subset of a k-cluster is a k-cluster, too. Consequently (b) implies (c) (by Lemma 4.26(d)). For $k = 3$, the statements (b), (c) and (d) follow from Lemma 4.26(e)(f). So assume that $k \geq 4$. Let v be an arbitrary vertex of $G = G[\mathcal{K}]$. Then $\mathcal{K}/v$ is a $(k-1)$-cluster (by Lemma 4.26(g)). Then $G' = G[\mathcal{K}/v]$ satisfies $G' = G[\mathcal{K}^v] - v$ (by Lemma 4.26(a)). Furthermore, we obtain that

$$\sum_{u \in V(G')} |\mathcal{K}^{vu}| = \sum_{u \in V(G')} |(\mathcal{K}/v)^u| = (k-2)|\mathcal{K}/v| \leq (k-2)(|G'|-1).$$

For the first equation we use Lemma 4.26(a), for the second equation we use Lemma 4.26(a) for the $(k-1)$-cluster $\mathcal{K}/v$, and for the last equation we use the induction hypothesis (note that $n(\mathcal{K}/v) = |G'|$). By the above inequality, there is a vertex $u \in V(G')$ such that $|\mathcal{K}^{vu}| \leq k-3$. By Lemma 4.26(c), we have $uv \in E(G)$. Thus (a) is proved.

For the proof of (b), we distinguish two cases. First, assume that $|\mathcal{K}|$ is even. Since $\mathcal{K}/v$ is a $(k-1)$-cluster, there is a vertex $u \in V(G')$ such that $|\mathcal{K}^{uv}| = |(\mathcal{K}/v)^u|$ is odd (by induction and Lemma 4.26(a)). From Lemma 4.27(a) it follows that there is an independent set I of $G - uv$ such that

$$\sum_{w \in I} |\mathcal{K}^w| = |\mathcal{K}| + |\mathcal{K}^{uv}|.$$

Since $|\mathcal{K}| + |\mathcal{K}^{uv}|$ is odd, it follows that $|\mathcal{K}^w|$ is odd for some vertex $w \in I \subseteq V(G)$. Now, assume that $|\mathcal{K}|$ is odd. By (a) there is an edge $e = uv$ in G such that $|\mathcal{K}^{uv}| \leq k-3$. Then it follows from Lemma 4.27 that there is an independent set I in $G - uv$ such that

$$\sum_{w \in I} |\mathcal{K}^w| = |\mathcal{K}|.$$

Since $|\mathcal{K}|$ is odd, it follows that $|\mathcal{K}^w|$ is odd for some vertex $w \in I \subseteq V(G)$. This proves (b). Statement (c) is an immediate consequence of (b).

For the proof of (d), it suffices to show that $\mathbb{K}' = \mathbb{K}^{\perp}$ contains a nonnull vector (by (4.4)). If k is odd, then each complete graph $K \in \mathcal{K}$ has even order and so $\langle f_K, \mathbf{1} \rangle = 0$, where $\mathbf{1}(v) = 1$ for all $v \in V(G)$. Hence $\mathbf{1} \in \mathbb{K}'$. If k is even, we obtain from (a) and Lemma 4.27(b) that there is a nonempty set $I \subseteq V(G)$ such that $|V(K) \cap I| = 1$ for all $K \in \mathcal{K}$. Then for $X = V(G) \setminus I$, we have $|V(K) \cap X| = k-2$ for all $K \in \mathcal{K}$. Since $k-2$ is even, it follows that the characteristic vector of X in G belongs to $\mathbb{K}'$ (by (4.3)). This completes the proof of (d). □

Theorem 4.29 *Let $k \geq 4$ be an integer, let $\mathcal{K}$ be a k-cluster and let $\mathbb{K} = \mathbb{K}[\mathcal{K}]$. Then* $\dim(\mathbb{K}) \leq n(\mathcal{K}) - k + 3$.

Proof Let $G = G[\mathcal{K}]$ be the graph of the k-cluster $\mathcal{K}$. If $\chi(G) \leq k-1$, then $\dim(\mathbb{K}) \leq n(\mathcal{K}) - k + 2$ (by Lemma 4.26(e)) and we are done. So we may assume that $\chi(G) \geq k$. Since $\mathcal{K}$ is a k-cluster, we have $\chi(G/e) \leq k-1$ for all $e \in E(G)$. Furthermore, $\mathcal{K}^v \neq \varnothing$ for all $v \in V(G)$. Consequently, $G \in \text{Crit}(k)$ and hence $\delta(G) \geq k-1$. Let $d = \dim(\mathbb{K})$. To show that $d \leq n(\mathcal{K}) - k + 3$, it suffices to show that there is a set of

$d+k-3$ vectors that are linearly independent in the vector space $\mathbb{Z}_2(G)$, since this implies that $d+k-3 \le \dim(\mathbb{Z}_2(G)) = |G| = n(\mathcal{K})$. To do this, we shall extend the basis $F[\mathcal{K}]$ of $\mathbb{K}$ to a larger independent set of $\mathbb{Z}_2(G)$. From Lemma 4.26(b) and Lemma 4.28 it follows that

$$\sum_{u\in V(G)} |\mathcal{K}^u| = (k-1)|\mathcal{K}| \le (k-1)(n(\mathcal{K})-1) = (k-1)(|G|-1).$$

This implies that there is a vertex $u \in V(G)$ such that $|\mathcal{K}^u| \le k-2$. We may choose u such that $|\mathcal{K}^u|$ is minimum. Since G contains no $(k-3)$-wheel, we have $\omega(G) = k-1$. Since $\delta(G) \ge k-1$, this implies that there are two distinct vertices $v, v' \in N_G(u)$ such that $vv' \notin E(G)$. Then $\mathcal{K}^{uv}$ and $\mathcal{K}^{uv'}$ are two nonempty disjoint sets, and we obtain that $|\mathcal{K}^{uv}| + |\mathcal{K}^{uv'}| \le |\mathcal{K}^u| \le k-2$. Consequently, we have

$$|\mathcal{K}^{uv}| \le k-3. \qquad (1)$$

We have $|\mathcal{K}^u| \le |\mathcal{K}^{v'}|$ (by the choice of u) and $\mathcal{K}^{uv} \ne \varnothing$. Let $K' \in \mathcal{K}^{uv}$. Since $vv' \notin E(G)$, we have $K' \notin \mathcal{K}^{v'}$. Consequently, there is a complete graph $H \in \mathcal{K}^{v'} \setminus \mathcal{K}^u$. Since $vv' \notin E(G)$, we obtain that $v \notin V(H)$. In summary, we have a complete graph $H \in \mathcal{K}$ satisfying

$$H \in \mathcal{K}^{v'} \text{ and } V(H) \cap \{u, v\} = \varnothing. \qquad (2)$$

Using (1) and Lemma 4.27(b), we obtain that there is a coloring φ of $G - uv$ with color set $C = [1, k-1]$ such that $\varphi(u) = \varphi(v) = c_1$ and there is a color $c_2 \in C \setminus \{c_1\}$ such that

$$I = \varphi^{-1}(c_2) \text{ satisfies } |V(K) \cap I| = 1 \text{ for all } K \in \mathcal{K}. \qquad (3)$$

From (2) it follows that $\varphi(V(H)) = C$. Let $C' = C \setminus \{c_1, c_2\}$ be the remaining color set; note that $|C'| = k-3$. Now let

$$X = F[\mathcal{K}] = \{f_K \in \mathbb{Z}_2(G) \mid K \in \mathcal{K}\}$$

and

$$Y = \{f_w \in \mathbb{Z}_2(G) \mid w \in V(H) \text{ and } \varphi(w) \in C'\}.$$

Then X and Y are disjoint vector sets in $\mathbb{Z}_2(G)$ satisfying $|X| = |\mathcal{K}| = \dim(\mathbb{K}) = d$ (by Lemma 4.28) and $|Y| = |C'| = k-3$. Then $Z = X \cup Y$ is a set of $d+k-3$ vectors in $\mathbb{Z}_2(G)$. Hence, if Z is linearly independent in $\mathbb{Z}_2(G)$, we are done. It remains to consider the case that Z is linearly dependent in $\mathbb{Z}_2(G)$. In this case we shall obtain a contradiction. There are subsets $\mathcal{H} \subseteq \mathcal{K}$ and $W \subseteq V(H) \cap \varphi^{-1}(C')$ such that

$$\sum_{K\in\mathcal{H}} f_K + \sum_{w\in W} f_w = \mathbf{0}, \qquad (4)$$

where the sum is the vector sum in $\mathbb{Z}_2(G)$. The vector set Y is obviously linearly independent in $\mathbb{Z}_2(G)$ (note that $B = \{f_w \in \mathbb{Z}_2(G) \mid w \in V(G)\}$ is the canonical basis of $\mathbb{Z}_2(G)$), implying that $\mathcal{H} \neq \varnothing$. Since the vector set $X = F[\mathcal{K}]$ is linearly independent in $\mathbb{Z}_2(G)$ (by Lemma 4.28(c)), we also obtain that $W \neq \varnothing$. Consequently, $\mathcal{H}$ is a k-cluster and $G' = G[\mathcal{H}]$ is a nonempty graph. For a vertex $w \in V(G) \setminus V(G')$, we have $\mathcal{H}^w = \varnothing$. As an immediate consequence of (4), we obtain that a vertex $w \in V(G)$ satisfies

$$|\mathcal{H}^w| \text{ is odd iff and only if } w \in W, \qquad (5)$$

which implies, in particular, that

$$|\mathcal{H}^w| \text{ is even whenever } \varphi(w) \in \{c_1, c_2\}. \qquad (6)$$

Furthermore, we have $\mathcal{H}^w \neq \varnothing$ for $w \in W$, which implies that $W \subseteq V(G')$. Since $H \in \mathcal{K}$ and $\varphi(V(H)) = C$, for every color $c \in C$ there exists a unique vertex $w(c)$ in H with $\varphi(w(c)) = c$. As a consequence of (3) we obtain that

$$\sum_{w \in \varphi^{-1}(c_2)} |\mathcal{H}^w| = |\mathcal{H}|,$$

which implies by (6) that $|\mathcal{H}|$ is even. Since $W \neq \varnothing$, there is a vertex $w \in W$. Then the color $c_3 = \varphi(w)$ belongs to $C' = C \setminus \{c_1, c_2\}$ and $w(c_3) \in V(G')$. The next crucial step is the proof of the following statement.

(a) If $w \in \varphi^{-1}(c_3)$, then $|\mathcal{H}^{ww(c_1)}|$ is odd if and only if $w = w(c_3)$

Proof of (a): Let $w \in \varphi^{-1}(c_3)$ be an arbitrary vertex. Note that we have $V(H) = \{w(c) \mid c \in C\}$. If $w \notin N_G(w(c_1))$, then $w \neq w(c_3)$ and $|\mathcal{H}^{ww(c_1)}| = 0$ is even as required. It remains to consider the case that $ww(c_1) \in E(G)$. Then there is a coloring ϕ of $G - ww(c_1)$ with color set $C = [1, k-1]$ such that $\phi(w) = \phi(w(c_1)) = c_0$. Now let $K \in \mathcal{H}$ be an arbitrary complete graph. If K contains the edge $ww(c_1)$, then we have $V(K) \cap \phi^{-1}(c_0) = \{w, w(c_1)\}$ otherwise we have $|V(K) \cap \phi^{-1}(c_0)| = 1$. This implies that

$$\sum_{v' \in \phi^{-1}(c_0)} |\mathcal{H}^{v'}| = |\mathcal{H}| + |\mathcal{H}^{ww(c_1)}|.$$

Since $|\mathcal{H}|$ is even, this gives

$$\sum_{v' \in \phi^{-1}(c_0)} |\mathcal{H}^{v'}| \equiv |\mathcal{H}^{ww(c_1)}| \pmod 2. \qquad (7)$$

Case 1: $w = w(c_3)$. Let $v' \in \phi^{-1}(c_0)$ be an arbitrary vertex. Then either $v' = w(c_3)$ or $v' \notin W$. In the former case, we have $|\mathcal{H}^{v'}|$ is odd (using (5) and $w(c_3) \in W$) and in the latter case we have $|\mathcal{H}^{v'}|$ is even (by (5)). From (7) it then follows that $|\mathcal{H}^{ww(c_1)}|$ is odd (note that $w(c_3) = w \in \phi^{-1}(c_0)$).

Case 2: $w \neq w(c_3)$. Let $v' \in \phi^{-1}(c_0)$ be an arbitrary vertex. Then either $v' \in \{w, w(c_1)\}$, or $v' \notin N_G(w(c_1))$ (since $\phi(w(c_1)) = c_0$). Consequently, $v' \notin W$ and $|\mathcal{H}^{v'}|$ is even (by (5)). From (7) it then follows that $|\mathcal{H}^{ww(c_1)}|$ is even.

This completes the proof of statement (a). ◇

By (a), $|\mathcal{H}^{w(c_3)w(c_1)}|$ is odd, and so $\mathcal{H}^{w(c_1)} \neq \emptyset$ is a k-cluster. Let $K \in \mathcal{H}^{w(c_1)}$ be an arbitrary complete graph. Since $\{u,v,w(c_1)\} \subseteq \varphi^{-1}(c_1)$ and $w(c_1) \notin \{u,v\}$, this implies that $V(K) \cap \{u,v\} = \emptyset$. Since K is a K_{k-1}, this implies that $|V(K) \cap \varphi^{-1}(c)| = 1$ for all colors $c \in C$. Consequently, we have

$$\sum_{w \in \varphi^{-1}(c_3)} |\mathcal{H}^{w(c_1)w}| = |\mathcal{H}^{w(c_1)}|.$$

By (a), it then follows that $|\mathcal{H}^{w(c_1)}|$ is odd. This contradiction to (6) completes the proof of Theorem 4.29. □

Proof of Theorem 4.25: Let G be a k-critical graph of order n with $k \geq 4$. If G contains a $(k-3)$-wheel, then Lemma 4.23 implies that G is an odd $(k-3)$-wheel, say $G = W(2\ell+1, k-3)$ with $\ell \geq 1$. If $\ell = 1$, then $n = k$, $G = K_n$ and $\Bbbk_{k-1}(G) = n$. If $\ell \geq 2$, then $n = 2\ell + k - 2$ and $\Bbbk_{k-1}(G) = 2\ell + 1 = n - k + 3 \leq n - 1$ as $k \geq 4$. It remains to consider the case when G contains no $(k-3)$-wheel. If $\Bbbk_{k-1}(G) = 0$ we are done. Otherwise $\mathcal{K} = \mathcal{K}_{k-1}(G)$ is a k-cluster and it follows from Lemma 4.28(c) and Theorem 4.29 that $|\mathcal{K}| = \dim(\mathbb{K}) \leq n(\mathcal{K}) + k - 3 \leq n - k + 3$. □

Clearly, every k-critical graph contains a $(k-1)$-critical subgraph and the smallest possible $(k-1)$-critical subgraph is K_{k-1}. Toft [1019] observed that the $(k-1)$-critical subgraphs together cover the whole k-critical graph. In particular, he proved the following simple but useful result.

Lemma 4.30 (Toft) *If e_1 and e_2 are two distinct edges of a k-critical graph G, then G has a $(k-1)$-critical subgraph G' containing e_1, but not e_2.*

Proof Since G is k-critical, there is a coloring φ of $G - e_1$ with a set of $k-1$ colors. Since $\chi(G) = k$, the two ends of e_1 receive the same color, say c. Since the two ends of e_2 get different colors, there is an end v of e_2 with $\varphi(v) \neq c$. Let $c' = \varphi(v)$. Then $G - \varphi^{-1}(c')$ has chromatic number $k-1$ (by Proposition 1.34(c)) and contains a $(k-1)$-critical subgraph G' (by Proposition 1.27(a)). Clearly, $e_2 \notin E(G')$, but G' contains e_1, since otherwise $G' \subseteq G - e_1 - \varphi^{-1}(c')$ and so $\chi(G') \leq k-2$, which is impossible. □

There are two natural problems concerning critical subgraphs of critical graphs. The first problem was raised in the 1970s by T. Gallai (oral communication). The question in the second problem was asked by J. Nešetřil and V. Rödl at the excursion of the *International Colloquium on Finite and Infinite Sets* in Keszthely, Hungary, in 1973. Not much progress has been made since the two problems became known, see Jensen and Toft [530, Problems 5.6 and 5.9] and Bang-Jensen, Reed, Schacht, Šámal, Toft, and Wagner [79, 5. Critical Graphs].

Problem 4.31 (Gallai) *Does every k-critical graph of order n contain n distinct $(k-1)$-critical subgraphs?*

Problem 4.32 (NEŠETŘIL AND RÖDL) *For $k \geq 4$, does a large k-critical graph contain a large $(k-1)$-critical subgraph?*

The answer to Gallai's question is obviously positive if $1 \leq k \leq 3$. Hare [471] proved that a 4-critical graph of order n has at least $(8n-29)/3$ odd cycles; a quadratic bound was established by Ma and Yang [709]. For $k \geq 5$, no linear bound is known (see [709] and Exercises 4.16 and 4.18). For $k = 4$, the question in Problem 4.32 was raised by Dirac [292] and a positive answer was given by Kelly and Kelly [564] and by Voss [1055] (see Corollary 4.50), but for $k \geq 5$ the problem is still open.

Problem 4.33 (TOFT) *Let P be a path of length 2 in a k-critical graph G with $k \geq 4$. Is there a $(k-1)$-critical subgraph of G containing P?*

The above problem is due to Toft [1024]. For $k = 4$ the answer is positive, that is, every path of length 2 in a 4-critical graph is contained in an odd cycle. This follows from a theorem of Dirac [302] and the fact that every 4-critical graph is 2-connected, and was also proved by Wessel [1065]. As observed by Toft [79], if the answer to the question is negative for some value $k \geq 5$, then based on Hajós joins it is possible to obtain arbitrarily large k-critical graphs where all $(k-1)$-critical subgraphs are of bounded order. Hence the answer to the above mentioned question of J. Nešetřil and V. Rödl would be negative, too.

The following result is due to Rödl [872]; its corollary provides an affirmative answer to a question raised by Erdős and Hajnal at the *International Colloquium on Finite and Infinite Sets* in Keszthely, Hungary, in 1973 (see also Erdős [335]).

Theorem 4.34 (RÖDL) *For given integers $k \geq 2$ and $n \geq 2$ there exists a number $a(k,n)$, such that if G is a graph with $\chi(G) \geq a(k,n)$, then G contains a complete graph of order n or a triangle-free k-chromatic graph.*

Proof The proof is by induction on $n \geq 2$. Obviously we can choose $a(k,2) = 2$ for every $k \geq 2$. Now suppose that $n \geq 3$ and $a(k,n-1)$ exists, that is, any graph G with $\chi(G) \geq a(k,n-1)$ contains a K_{n-1} or a triangle-free k-chromatic graph. Let $a(k,n) = (k-1)^{a(k,n-1)-1} + 1$ and let G be a graph with $\chi(G) \geq a(k,n)$. If G contains a K_n there is nothing to prove. So assume that G contains no K_n.

Let $\preceq$ be a linear order of $V(G)$. For a vertex v of G, let G^v denote the subgraph of G induced by the left neighborhood N^v of v, i.e.,

$$N^v = \{u \in V(G) \mid u \preceq v \text{ and } uv \in E(G)\}.$$

Since G contains no K_n, it follows that G^v contains no K_{n-1}. If $\chi(G^v) \geq a(k,n-1)$ for some $v \in V(G)$, then the induction hypothesis implies that G^v and hence G has a triangle-free k-chromatic subgraph. It remains to consider the case when $\chi(G^v) \leq a(k,n-1) - 1$ for every $v \in V(G)$. Let $p = a(k,n-1) - 1$ and let φ^v be a coloring of G^v with color set $C = [1,p]$. For a color $c \in C$, let G_c be the subgraph of G with

$$V(G_c) = V(G) \text{ and } E(G_c) = \{uv \in E(G) \mid \varphi^v(u) = c\}.$$

Clearly, each subgraph G_c is triangle-free. Furthermore, $G = G_1 \cup G_2 \cup \cdots \cup G_p$ and so $\chi(G) \leq \chi(G_1) \cdot \chi(G_2) \cdot \ldots \cdot \chi(G_p)$ (by (2.1)). As $\chi(G) \geq (k-1)^p + 1$, it follows that $\chi(G_c) \geq k$ for some $c \in C$ and so G contains a triangle-free k-chromatic graph. □

Corollary 4.35 *For every positive integer k there is an integer $f(k)$ such that any graph G with $\chi(G) \geq f(k)$ contains a triangle-free k-chromatic graph as a subgraph.*

Proof There is an integer $n(k)$ such that there exists a triangle-free k-chromatic graph of order $n(k)$ (see Exercises 2.20, 2.21 and 2.22). Now let $f(k) = a(k, n(k))$ where $a(k,n)$ is the function defined in Theorem 4.34. If $\chi(G) \geq a(k, n(k))$ then G contains a $K_{n(k)}$ and we find a triangle-free k-chromatic subgraph, or G contains a triangle-free k-chromatic subgraph. □

Problem 4.36 (Erdős and Hajnal) *Let k and ℓ be integers with $k \geq 3$ and $\ell \geq 4$. Does there exists an integer $f(k,\ell)$ such that every graph G with $\chi(G) \geq f(k,\ell)$ contains a subgraph H with chromatic number $\chi(H) \geq k$ and girth $g(H) \geq \ell$?*

4.5 Independent Sets and Degrees in Critical Graphs

In the 1960s Gallai suggested to investigate the independence number and the minimum degree of critical graphs and, in particular, the following two functions defined for $n \geq k \geq 4$ and $n \neq k+1$

$$\alpha_k(n) = \max\{\alpha(G) \mid G \in \mathrm{Crit}(k,n)\}$$

and

$$\delta_k(n) = \max\{\delta(G) \mid G \in \mathrm{Crit}(k,n)\}.$$

The following result due to Simonovits [939] has proved useful.

Lemma 4.37 (Simonovits) *Let G be a k-critical graph with $k \geq 4$, and let v be a vertex of G. Then there is a k-critical graph G' and an independent set I of low vertices in G' such that G can be obtained from G' by identifying I to v.*

Proof First, we construct a graph H out of $G - v$ by adding a new vertex v_A to $G - v$ for each set $A \in [N_G(v)]^{k-1}$ and joining v_A to all vertices of A. Note that v_A has degree $k-1$ in H. Let X be the set of all new vertices v_A. As $d_G(v) \geq k-1$, we have that $|X| \geq 1$. So X is a nonempty independent set in H, and G can be obtained from H by identifying X with v. In particular, $H - X = G - v$. Next we claim that $\chi(H) \geq k$. Suppose this is false and there is a coloring $\varphi \in \mathcal{CO}(H, k-1)$. Then φ induces a coloring of $G - v$ and, as $\chi(G) = k$, all $k-1$ colors occur in $N_G(v)$. Hence there is a set $A \in [N_G(v)]^{k-1}$ such that $|\varphi(A)| = k-1$, but then $\varphi(v_A) \in \varphi(A)$, which is impossible as v_A is joined to all vertices of A. This proves that $\chi(H) \geq k$. Consequently, H contains a k-critical graph G' as a subgraph. Then $I = V(G') \cap X$ is an independent set of G', and $G' - I$ is a subgraph of $G - v$. Hence $\chi(G' - I) \leq k-1$

which implies that $I \neq \varnothing$. As $\delta(G') \geq k-1$, we conclude that every vertex of I is a low vertex of G'. Note that $\chi(G'/I) \geq \chi(G') = k$. Next we claim that $G' - I = G - v$. For otherwise, $G' - I$ is a proper subgraph of $G - v$ and hence G'/I is a proper subgraph of G. As G is k-critical this implies that $\chi(G'/I) \leq k-1$, a contradiction. By the same argument, it follows that each vertex of $N_G(v)$ is adjacent to at least one vertex of I in G'. Consequently, $G'/I = G$, i.e., G can be obtained from G' by identifying I to v. □

The first construction of critical graphs having many independent vertices was presented by Brown and Moon [182]. Simonovits [939] used Lemma 4.37 with G 4-critical and many edges, to split all the vertices of a color class incident with many edges, to obtain another such construction. Also he obtained the first upper bound for the independence number. In 1973 Lovász [693] established the following bounds for $\alpha_k(n)$:

$$n - kn^{1/(k-2)} \leq \alpha_k(n) \leq n - \tfrac{k}{6} n^{1/(k-2)}. \tag{4.5}$$

While the lower bound holds for infinitely many values of n, the upper bounds hold whenever $n \geq k \geq 4$ and $n \neq k+1$. A proof of the lower bound for $k = 4$ is based on the construction mentioned above, and in general involves critical hypergraphs and methods discussed in Chapter 7. So we shall only give an outline here. Let G' be obtained from i-critical graphs G_i with $i \in [1, k-1]$ as in the construction of Zykov and Schäuble (see Exercise 2.20), where each of $G_3, G_4, \ldots, G_{k-1}$ has odd order $N \geq k+2$, except that we shall let G_2 be a 2-critical hypergraph, also on N vertices. Then G' is a k-critical hypergraph (see Exercise 7.12) with exactly one hyperedge of size N. Now add an odd $(k-3)$-wheel $W(N, k-3)$ and split a dominating vertex of the wheel into the hyperedge of G'. This results in a k-critical graph G (see Proposition 7.17) of order n such that $n \geq \alpha(G) \geq N^{k-2}$ and $n = 1 + (k-2)N + N^{k-2} + N + (k-2) \leq N^{k+2} + kN$, which leads to

$$n - \alpha_k(n) \leq n - \alpha(G) \leq n - N^{k+2} \leq kN \leq kn^{1/(k-2)}.$$

Lovász's proof of the upper bound in (4.5) is based on Simonovits' splitting lemma (Lemma 4.37). The proof is another interesting application of a linear algebra technique to solve a counting problem.

Theorem 4.38 (Lovász) *Every k-critical graph G with $k \geq 4$ has independence number $\alpha(G) \leq |G| - \frac{1}{6}k|G|^{1/(k-2)}$.*

Proof For an arbitrary graph G, let $\tau(G) = |G| - \alpha(G)$. Our aim is to show that if G is a k-critical graph, then $\tau(G) \geq c_k|G|^{1/(k-2)}$ with $c_k = \frac{1}{6}k$. To this end, let G be a k-critical graph, and let S be an maximum independent set of G. We may assume that each vertex of S is a low vertex of G. For otherwise, by repeated application of Lemma 4.37 to the high vertices of S, we obtain a k-critical graph G' and an independent set S' of G' consisting only of low vertices of G', such that $\tau(G') \leq \tau(G)$ and $|G| \leq |G'|$. Hence the theorem for G' implies the theorem for G. Let $T = V(G) \setminus S$, let $t = |T| (= \tau(G))$, let $s = |S|$, and, for each vertex $v \in S$, let $N_v = N_G(v)$. As G is critical and S is an independent set of G consisting only of low vertices of G, N_v is a subset of T having size $k-1$ for every $v \in S$, and $N_v \neq N_w$ for any two distinct vertices $v, w \in S$.

Now let S' be an arbitrary nonempty subset of S. Then we claim that there is a set $B \in [T]^{k-2}$ such that the number of vertices $v \in S'$ with $B \subseteq N_v$ is odd. To this end, let v' be a vertex of S'. As G is k-critical, there is a coloring $\varphi \in \mathcal{CO}(G - v', k-1)$. As $\chi(G) = k$, we obtain that $|\varphi(N_{v'})| = k-1$ and $|\varphi(N_v)| \leq k-2$ for all $v \in S' \setminus \{v'\}$. Hence, the total number of pairs (v, B), where $v \in S$ and $B \in [T]^{k-2}$ such that $\varphi(B) = [1, k-2]$ and $B \subseteq N_v$, is odd, since there is exactly 1 pair for $v = v'$, and 0 or 2 pairs for $v \in S \setminus \{v'\}$. Then at least one such set B must be contained in an odd number of sets N_v with $v \in S'$, as claimed.

Define the matrix $M : S \times [T]^{k-2} \to \mathbb{Z}_2$ by $M(v, B) = 1$ if $B \subseteq N_v$ and $M(v, B) = 0$ otherwise. By the above claim it follows that the matrix M has rank s over $\mathbb{Z}_2$. Consequently $s \leq \binom{t}{k-2}$ which leads to

$$|G| = s + t \leq \binom{t}{k-2} + t \leq \frac{2t^{k-2}}{(k-2)!} \leq \left(\frac{et}{(k-2)}\right)^{k-2}$$

where we use $m! \leq 2(m/e)^m$ by Stirling's formulae. Then we obtain

$$\tau(G) = t \geq \left(\frac{k-2}{3}\right)|G|^{1/(k-2)} \geq c_k |G|^{1/(k-2)}.$$

This completes the proof. □

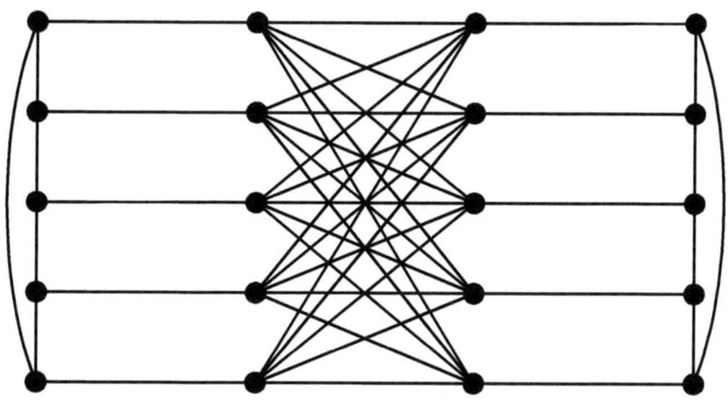

Fig. 4.3 The 4-critical Toft graph TG_5.

In 1949 P. Erdős visited England and met G. A. Dirac in London. Dirac told Erdős about the concept of critical graphs and presented the graph $C_{2p+1} \boxplus C_{2p+1}$ as an example of a 6-critical graph with many edges and with high minimum degree showing, in particular, that $\delta_6(2n) \geq n$ if $n \geq 3$ is odd. Then Erdős (see [338] [339]) suggested to investigate the order of magnitude of $\delta_4(n)$ and $\delta_5(n)$, respectively. In 1972 Simonovits [939] and Toft [1021] independently proved that $\delta_4(n) \geq c\sqrt[3]{n}$ for infinitely many values of n by construction. Toft [1020] presented his well known family of 4-critical graphs with many edges and used a splitting argument to obtain 4-critical graphs with high minimum degree.

For an odd integer $p \geq 3$, let G denote the graph whose vertex set consists of four disjoint sets A^1, A^2, A^3 and A^4 each of size p, say $A^i = \{a^i_1, a^i_2, \ldots, a^i_p\}$, such that the following conditions hold:

(T1) $G[A^3 \cup A^4]$ is a complete bipartite graph with parts A^3 and A^4.
(T2) For $i \in \{1,2\}$, the graph $G[A^i]$ is a cycle of the form $(a^i_1, a^i_2, \ldots, a^i_p, a^i_1)$ and $E_G(A^i, A^{i+2}) = \{a^i_j a^{i+2}_j \mid j \in [1,p]\}$ is a matching.
(T3) $E(G) = E_G(A^3, A^4) \cup \bigcup_{i=1}^{2} (E_G(A^i, A^i) \cup E_G(A^i, A^{i+2}))$

We then say that G is a **Toft graph** and write $G = TG_p(A^1, A^2, A^3, A^4)$ or briefly $G = TG_p$, see Figure 4.3. Toft [1020] proved that TG_p is a 4-critical graph. A short proof of this statement is given in Section 7.4 (see the discussion after Proposition 7.17); the proof uses the fact that TG_p can be obtained from a 4-critical hypergraph by Toft's reduction technique. Note that TG_p has order $n = 4p$ and size $m = p^2 + 4p$, and hence $m = \frac{1}{16}n^2 + n$. To this day, the constant $\frac{1}{16}$ has not been improved, nor has it been proved that this constant is best possible. In 2013, Pedgen [802] constructed dense triangle-free graphs in Crit(k,n) for all $k \geq 4$ and for infinitely many values of n. For $k \geq 6$, theses constructions have more that $(\frac{1}{4} - \varepsilon)n^2$ edges, which is asymptotically best possible by Turan's theorem. Other constructions of dense critical graphs are presented by Stiebitz [967], by Jensen [525], and by Năstase, Rödl, and Siggers [770].

Theorem 4.39 (Toft) *Let h be an integer with $h \geq 3$. Then for infinitely many values of n there exists a 4-critical graph G of order n containing an independent set I such that* (a) $d_G(v) \geq h$ *for all $v \in I$ and* (b) $n - |I| \leq 3\sqrt{(h-2)n}$.

Proof Let $h \geq 3$, let s be an even positive integer, let $p = (h-2)s + 1$, and let $n_i = i(h-2) + 1$ for $i \in [0,s]$. Clearly, p is odd and $p \geq 3$. Let G^0 be the Toft graph $TG_p(A^1, A^2, A^3, A^4)$ with $A^i = \{a^i_1, a^i_2, \ldots, a^i_p\}$ (satisfying (T1), (T2) and (T3)). For $1 \leq i \leq p$, let A^4_i be a set of s new vertices, say $A^4_i = \{a^4_i(1), a^4_i(2), \ldots, a^4_i(s)\}$, where the sets A^4_i are disjoint. Then

$$\tilde{A}^4 = A^4_1 \cup A^4_2 \cup \cdots \cup A^4_p$$

is a set of ps new vertices. Now let G be the unique graph obtained from $G^0 - A^4$ by adding the vertex set $\tilde{A}^4$ such that

$$N_G(a^4_i(\ell)) = \{a^3_j \mid n_{\ell-1} \leq j \leq n_\ell\} \cup \{a^2_i\}$$

for $i \in [1,p]$ and $\ell \in [1,s]$. Our aim is to show that G is a 4-critical graph. Note that $\tilde{A}^4$ is an independent set in G, and G^0 is obtained from G by simultaneously identifying the independent set A^4_i with the vertex a^4_i for $i \in [1,p]$. As $\tilde{A}^4$ is independent in G, this implies that $\chi(G) \leq \chi(G^0) = 4$.

Suppose that G has a coloring $\varphi \in CO(G,3)$. As $G[A^1] (= G^0[A^1])$ is an odd cycle, $|\varphi(A^1)| = 3$. As A^1 and A^3 are joined by a perfect matching in G, we obtain that $2 \le |\varphi(A^3)| \le 3$. Any two consecutive vertices in the sequence $a_1^3, a_2^3, \dots, a_p^3$ have in G a common neighbour in the set A_i^4 for each $i \in [1,p]$. If $|\varphi(A^3)| = 3$, this implies that $|\varphi(A_1^4) \cap \varphi(A_2^4)| \ge 2$, and hence $\varphi(a_1^2) = \varphi(a_2^2)$. However, this is impossible as $a_1^2 a_2^2$ is an edge of $G[A^2] (= G^0[A^2])$. Consequently $|\varphi(A^3)| = 2$, say $\varphi(A^3) = \{1,2\}$. But then $3 \in \varphi(A_i^4)$ for each $i \in [1,p]$. As $A_i^4 \subseteq N_G(a_i^2)$ for $i \in [1,p]$ and $G[A^2]$ is an odd cycle, this is impossible. Consequently, G has no 3-coloring and so $\chi(G) = 4$.

It remains to show that $G - e$ is 3-colorable for every edge e of G. So let e be an arbitrary edge of G. If e is not incident with $\tilde{A}^4$, then e belongs to G^0 and hence $G^0 - e$ has a 3-coloring φ, which obviously leads to a 3-coloring of $G - e$ by giving each vertex of A_i^4 the color $\varphi(a_i^4)$ for $i \in [1,p]$. It remains to consider the case when e is incident with some vertex of $\tilde{A}^4$, say $a_i^4(\ell)$ with $\ell \in [1,s]$ and $i \in [1,p]$. Then $e = a_i^4(\ell)w$ and $w \in \{a_j^3 \mid n_{\ell-1} \le j \le n_\ell\} \cup \{a_i^2\}$.

Case 1: *$e = a_i^4(\ell)a_j^3$ with $n_{\ell-1} \le j \le n_\ell$.* Then we obtain a coloring of $G - e$ with color set $C = \{1,2,3\}$ as follows. As $G[A^3 \cup A_i^4]$ is a tree with bipartition A^3 and A_i^4, it follows that $G[A^3 \cup A_i^4] - e$ has a 2-coloring with color set $\{1,2\}$ such that both colors occur in A^3. As each vertex $a \in A^1$ has precisely one neighbor in G belonging to A^3, a list $L(a)$ of two colors can be used for a and L is not a constant list-assignment. As $G[A^1]$ is an odd cycle, there is an L-coloring of $G[A^1]$ (by Theorem 1.7(b)), so we can color $G[A^1 \cup A^3 \cup A_i^4]$ with color set C. Now we give each vertex of $\tilde{A} \setminus A_i^4$ color 3. Eventually, we give a_i^2 color 3 and we use color 1 and 2 to color the path $G[A^2] - a_i^2$. The result is a 3-coloring of G.

Case 2: $e = a_i^4(\ell)a_i^2$. Then we obtain a coloring of $G - e$ with color set $C = \{1,2,3\}$ as follows. First we color $a_1^3, a_2^3, \dots a_{n_\ell - 1}^3$ with color 1 and the remaining vertices of A^3 with color 2. As in the former case we can then color the vertices of A^1. Now we give each vertex of $\tilde{A} \setminus A_i^4$ color 3. Then we color the vertices $a_i^4(1), a_i^4(2), \dots, a_i^4(\ell - 1)$ with color 2, vertex $a_i^4(j)$ with color 3 and the remaining vertices of A_i^4 get color 1. Eventually, a_i^2 is colored with color 3 and we use color 1 and 2 to color the path $G[A^2] - a_i^2$. The result is a 3-coloring of G.

Consequently, $\chi(G - e) \le 3$ for all edges e of G. As $\chi(G) = 4$ it follows that G is 4-critical (by Proposition 1.29). Then $I = \tilde{A}$ is an independent set of G such that $d_G(v) = h$ for all $v \in I$ and $|I| = ps$. For the order n of G we obtain that

$$n = 3p + ps = p\left(3 + \frac{p-1}{h-2}\right) > \frac{p^2}{h-2}$$

which implies that $n - |I| = 3p < 3\sqrt{(h-2)n}$. This complete the proof. □

Theorem 4.40 (Toft) *Let h be an even integer with $h \ge 4$. Then there is a 4-critical graph G of order $n = 2(h^2 - 4h + 5)(h-1)$ such that $\delta(G) \ge h$. As a consequence, $\delta_4(n) \ge \sqrt[3]{n/2}$ for infinitely many values of n.*

Proof Let h be an even integer with $h \ge 4$, let $s = h - 2$, let $p = (h-2)s + 1$, and let $n_i = i(h-2) + 1$ for $i \in [0,s]$. Clearly, s is even, p is odd, and $p \ge 3$. Let

G^0 be the Toft graph $TG_p(A^1, A^2, A^3, A^4)$ with $A^i = \{a^i_1, a^i_2, \ldots, a^i_p\}$ (satisfying (T1), (T2) and (T3)). For $1 \le i \le p$, let A^4_i be a set of s new vertices, say $A^4_i = \{a^4_i(1), a^4_i(2), \ldots, a^4_i(s)\}$, where the sets A^4_i are disjoint. Then

$$\tilde{A}^4 = A^4_1 \cup A^4_2 \cup \cdots \cup A^4_p$$

is a set of ps new vertices. Now let G^1 be the unique graph obtained from $G^0 - A^4$ by adding the vertex set $\tilde{A}^4$ such that

$$N_{G^1}(a^4_i(\ell)) = \{a^3_j \mid n_{\ell-1} \le j \le n_\ell\} \cup \{a^2_i\}$$

for $i \in [1,p]$ and $\ell \in [1,s]$. As proved in Theorem 4.39, G^1 is a 4-critical graph. As $s = h-2$, each vertex of $A^2 \cup \tilde{A}^4$ has degree at least h in G^1. For $k \in [1,s]$ let

$$A^4(k) = \{a^4_j(k) \mid j \in [1,p]\}.$$

Then $\tilde{A}^4 = A^4(1) \cup A^4(2) \cup \cdots \cup A^4(s)$ and

$$E_{G^1}(A^4(k), A^2) = \{a^4_j(k) a^2_j \mid j \in [1,p]\}$$

is a perfect matching. Furthermore, if $n_{k-1} < i < n_k$ and $k \in [2, s-1]$, or $k = 1$ and $1 \le i < n_1$, or $k = s$ and $n_{s-1} < i \le p$, then

$$N_{G^1}(a^3_i) = A^4(k) \cup \{a^1_i\}, \tag{4.6}$$

and if $i = n_k$ and $k \in [1, s-1]$, then

$$N_{G^1}(a^3_i) = A^4(k) \cup A^4(k+1) \cup \{a^1_i\}. \tag{4.7}$$

Next, we partition the index set $[1,p]$ into the sets $I = \{n_1, n_2, \ldots, n_{s-1}\}$ and $I' = [1,p] \setminus I$. Now we want to apply a splitting argument to the vertices in A^3 to construct a graph G in three steps. Clearly,

$$\tilde{A}^3 = \bigcup_{i \in I'} A^3_i$$

is a set of $(p-s+1)s$ new vertices. Now let G^2 be the unique graph obtained from $G^1 - \{a^3_i \mid i \in I'\}$ by adding the vertex set $\tilde{A}^3$ such that

$$N_{G^2}(a^3_i(\ell)) = \{a^4_j(k) \mid n_{\ell-1} \le j \le n_\ell\} \cup \{a^1_i\}$$

for $i \in I'$ and $\ell \in [1,s]$, where $N_{G^1}(a^3_i) = A^4(k) \cup \{a^1_i\}$ (see (4.6)). Our aim is to show that G^2 is a 4-critical graph. Note that $\tilde{A}^3$ is an independent set in G^2, and G^1 is obtained from G^2 by simultaneously identifying the independent set A^3_i with the vertex a^3_i for $i \in I'$. As $\tilde{A}^3$ is independent in G^2, this implies that $\chi(G^2) \le \chi(G^1) = 4$.

Suppose that G^2 has a coloring $\varphi \in \mathcal{CO}(G^2, 3)$. For $i \in I$, $N_{G^2}(a^3_i) = N_{G^1}(a^3_i)$. Hence it follows from (4.7) that $|\varphi(A^4(k))| \le 2$ for $k \in [1,2]$. Furthermore,

$G^2[A^2](=G^1[A^2])$ is an odd cycle, and $A^4(k)$ and A^2 are joined by a perfect matching in G^2. Consequently $|\varphi(A^2)|=3$ which leads to $|\varphi(A^4(k))|\le 2$, and hence to $|\varphi(A^4(k))|=2$. Let c_k be the third color not occurring in $\varphi(A^4(k))$. Let $i\in I'$ be an arbitrary index. Then there is exactly one $k\in[1,s]$ such that $N_{G^1}(a_i^3)=A^4(k)\cup\{a_i^1\}$. As any two consecutive vertices in the sequence $a_1^4(k),a_2^4(k),\dots,a_p^4(k)$ have a common neighbor in G^2 belonging to A_i^3, $c_k\in\varphi(A_i^3)$ and so $\varphi(a_i^1)\ne c_k$. But then we can give a_i^3 color c_k for each $i\in I'$ and obtain a 3-coloring of G^1, which is impossible Consequently, G^2 has no 3-coloring and so $\chi(G^2)=4$.

It remains to show that G^2-e is 3-colorable for every edge e of G^2. So let e be an arbitrary edge of G^2. If e is not incident with $\tilde{A}^3$, then e belongs to G^1 and hence G^1-e has a 3-coloring φ, which obviously leads to a 3-coloring of $G-e$ by giving each vertex of A_i^3 the color $\varphi(a_i^3)$ for $i\in I'$. It remains to consider the case when e is incident with some vertex of $\tilde{A}^3$, say $a_i^3(\ell)$ with $i\in I'$. Then $e=a_i^3(\ell)w$ and $w\in A^4(k)\cup\{a_i^2\}$, and we can argue as in Case 1 and Case 2 of Theorem 4.39. Consequently, G^2 is a 4-critical graph as claimed.

Eventually, we deal with the vertex set $A=\{a_i^3\mid i\in I\}$ of G^2. Then

$$\hat{A}^3=\bigcup_{i\in I}A_i^3$$

is a set of $s(s-1)$ new vertices. Now let G^3 be the unique graph obtained from G^2-A by adding the vertex set $\tilde{A}^3$ such that

$$N_{G^3}(a_i^3(\ell))=\{a_j^4(k),a_j^4(k+1)\mid n_{\ell-1}\le j\le n_\ell\}\cup\{a_i^1\} \tag{4.8}$$

for $i=n_k$ with $k\in[1,s-1]$ and $\ell\in[1,s]$. Our aim is to show that G^3 is a 4-chromatic graph. Note that $\hat{A}^3$ is an independent set in G^3, and G^2 is obtained from G^3 by simultaneously identifying the independent set A_i^3 with the vertex a_i^3 for $i\in I$ (see (4.7) and (4.8)). As $\hat{A}^3$ is independent in G^3, this implies that $\chi(G^3)\le\chi(G^2)=4$.

Suppose that G^3 has a coloring $\varphi\in CO(G^3,3)$. To arrive at a contradiction, we first claim that $|\varphi(A^4(k))|=2$ for $k\in[1,s]$. Note that $G^3[A^2]=G^0[A^2]$ is an odd cycle and $E_{G^3}(A^4(k),A^2)=E_{G^1}(A^4(k),A^2)$ is a perfect matching. This implies that $|\varphi(A^4(k))|\ge 2$. Suppose that $|\varphi(A^4(k))|=3$ for some $k\in[1,s]$. By (4.6) and (4.7) it follows that there is an edge $a_i^1a_{i+1}^1$ of the odd cycle $G^3[A^1]=(G^0[A^1])$ such that $A^4(k)\cup\{a_i^1\}\subseteq N_{G^1}(a_i^3)$ and $A^4(k)\cup\{a_{i+1}^1\}\subseteq N_{G^1}(a_{i+1}^3)$. Then $G^3[A^4(k)\cup A_i^3]$ and $G^3[A^4(k)\cup A_{i+1}^3]$ are isomorphic trees and it follows that the same two colors occur in $\varphi(A_i^3)$ and in $\varphi(A_{i+1}^3)$, but the $\varphi(a_i^1)=\varphi(a_{i+1}^1)$ which is impossible. This shows that $|\varphi(A^4(k))|=2$ for every $k\in[1,s]$ as claimed. Next, we claim that the color set $\varphi(A^4(k))$ consists of the same two colors for each $k\in[1,s]$. For otherwise, there is a $k\in[1,s-1]$ such that $\varphi(A^4(k))=\{1,2\}$ and $\varphi(A^4(k+1))=\{2,3\}$ (by symmetry). But then the colors 3 and 1 occur in $\varphi(A_i^3)$ with $i=n_k$ (see (4.7)), and so $\varphi(a_i^1)=2$ and $2\notin\varphi(A_i^3)$. By (4.8) this implies that $a_j^4(k)$ or $a_j^4(k+1)$ has color 2 for each $j\in[1,p]$. But then 2 does not occur in $\varphi(A^2)$, which is impossible. So, by symmetry, $\varphi(A^4(k))=\{1,2\}$ for each $k\in[1,s]$. Then color 3 occurs in each set

$\varphi(A_i^4)$ and so 3 does not occur in $\varphi(A^1)$ which is impossible, too. This that G^3 has no 3-coloring. Consequently, $\chi(G^3) = 4$. Then there is a 4-critical subgraph G of G^3.

If e is an edge of G^3 that is not incident with the new vertex set $\hat{A}^3$, then e belongs to G^2 and hence $G^2 - e$ has a 3-coloring, say φ. This coloring can be extended to a 3-coloring of $G^3 - e$ by giving each vertex of A_i^3 with $i \in I$ the color $\varphi(a_i^3)$. This implies that e belongs to G and $G^3 - \hat{A}^3 = G - \hat{A}^3$. Now let e be an edge of G^3 incident with $\hat{A}^3$. Then $e = a_i^3(\ell)w$ with $i = n_k$ for $k \in [1, s-1]$ and $w \in \{a_j^4(k), a_j^4(k+1) \mid n_{\ell-1} \leq j \leq n_\ell\} \cup \{a_i^1\}$ (see (4.8)). If $w = a_i^1$, then we can obtain a 3-coloring of $G^3 - e$ as follows. We identify each set A_k^4 with a_k^4 and each set A_k^3 with a_k^3 except for the set A_i^3. Let H denote the resulting graph. Then we can argue as in Case 2 of Theorem 4.39 to construct a 3-coloring of $H - e$ and hence of $G^3 - e$. Hence e belongs to G. If $w \in \{a_j^4(k), a_j^4(k+1)\}$ and $j \notin \{n_1, n_2, \ldots, n_{s-1}\}$, then a 3-coloring of $G^3 - e$ can be obtained as follows. We identify each set A_k^4 with a_k^4 and each set A_k^3 with a_k^3 except for the sets A_i^3 and A_j^4. Let H denote the resulting graph. Then we color vertex $w \in A_j^4$ with color 1 and the remaining vertices of A_j^4 with color 3, and vertex a_j^2 with color 2. Next we color each vertex a_k^4 with $k \neq j$ with color 2, and color the vertices of the path $H[A^2] - a_j^2 (= G^3[A^2] - a_j^2)$ with the colors 1 and 3. This coloring can be extended to a 3-coloring of $H - A^1 - e$ such that all vertices of A_i^3 have color 1, but not all vertices of $S = A^3 \setminus \{a_i^3\}$ have color 1. Note that if $v \in S$ then only two colors occur among the neighbors of v belonging to $(A^3 \setminus \{a_j^4\}) \cup A_j^4$. As $H[A^1](= G^3[A^1]$ is an odd cycle, we conclude that we can color the vertices of A^1 such that we obtain a 3-coloring of $H - e$ and hence of $G^3 - e$. Thus e belongs to G. It remains to consider the edges of the cycle $C(\ell) = (a_i(\ell), a_{n_\ell}^4(k), a_i(\ell+1), a_{n_\ell}^4(k+1), a_i(\ell))$ with $\ell \in [1, s-1]$. If e and e' are two adjacent edges of $C(\ell)$, then we can argue similar as in the former case to obtain a 3-coloring of $G^3 - e - e'$. Consequently, if E is the set of edges belonging to G^3 but not to G, then $E(\ell) = E \cap E(C(\ell))$ is a matching and $E = E(1) \cup E(2) \cdots \cup E(s-1)$. This implies, in particular, that $V(G) = V(G^3)$ and $\delta(G) \geq h$. For the order n of the 4-critical graph we obtain that

$$n = 2p + 2ps = 2p(s+1) = 2((h-2)^2 + 1)(h-1) \leq 2h^3,$$

which implies that $\delta(G) \geq h \geq \sqrt[3]{n/2}$. This completes the proof. □

Theorem 4.40 shows that $\delta_4(n) \geq \sqrt[3]{n/2}$ for infinitely many values of n. This may be the best order of of magnitude for the function $\delta_4(n)$. Simonovits [939] obtained a similar bound. He constructed for infinitely many values of n a 4-critical graph G on n vertices such that the edge connectivity of G satisfies $\kappa'(G) \geq \sqrt[3]{n}/6$.

For a given graph G and a vertex v of G, let G' be the graph obtained from $G - v$ by adding a set X of new independent vertices with $|X| \geq 2$ and by joining each vertex of $N_G(v)$ to exactly one vertex of X such that each vertex in X has a at least one neighbor in $N_G(v)$. This is possible if $d_G(v) \geq |X|$. We then say that G' is obtained from G by a **proper splitting** of v into X or briefly by **splitting a vertex**. Note that

X is an independent set of G', no two vertices of X have a common neighbor in G', and G is obtained from G' by identifying X to v. The splitting operation used in the proof of Toft's theorems is of a different type since it allows that the vertices of the new set X may have common neighbors.

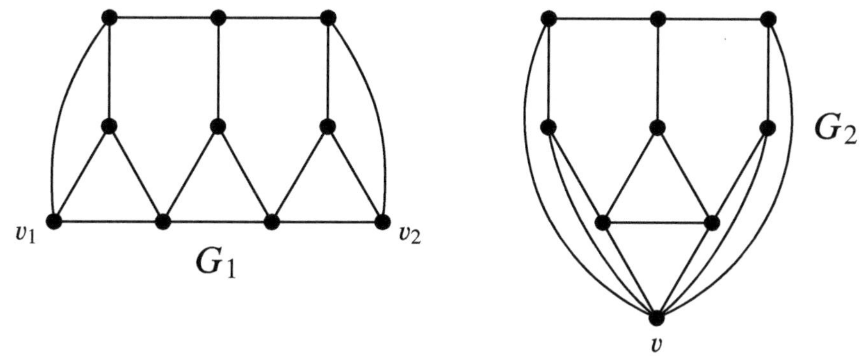

Fig. 4.4 Two 4-critical graphs.

Any graph obtained from a k-critical graph by splitting a vertex is either $(k-1)$-colorable or k-critical (see Exercise 4.22). An example showing that the later case can occur, even for $k = 4$, is shown in Figure 4.4; both graphs G_1 and G_2 are 4-critical and G_1 is obtained from G_2 by a proper splitting of v into the vertex set $\{v^1, v^2\}$. The graph G_1 is a Hajós join of the form $(K_4 \nabla K_4) \nabla K_4$ and hence 4-critical. That G_2 is 4-critical can easily be checked by hand. This example is mentioned in Toft's thesis [1019], attributed to T. Gallai. It is also contained in Lovśz's book [698]. On the other hand, a proper splitting of any vertex of the Toft graph TG_p results in a 3-colorable graph (see Exercise 4.23). This observation was used by V. Rödl (private communication to the third author) in order to obtain exponential many 4-critical graphs having order $n = 8p$.

Theorem 4.41 (RÖDL) *There is a constant $c > 0$ such that for every integer $n = 16p$ with $p \geq 5$ odd, the number of nonisomorphic 4-critical graphs on n vertices is at least c^{n^2}.*

Proof Let $p \geq 5$ be an odd integer and $n = 16p$. Let G^1 and G^2 be a decomposition of the Toft graph TG_p into two edge disjoint spanning subgraphs such that $\delta(G^i) \geq 1$ for $i \in \{1,2\}$. For $i \in \{1,2\}$ let H^i be a copy of G^i, where we replace each vertex v of TG_p by v^i. For each vertex v of TG_p, let K^v be a K_4^- where the missing edge is v^1v^2. We assume that H^1, H^2, and $K^v - v^1 - v^2$ with $v \in V(G)$ are disjoint. Now let G be the union of the graphs H^1, H^2, and K^v with $v \in V(G)$. Note that G has order $n = 16p$. We claim that G is 4-critical. Suppose that G has a 3-coloring φ. Then φ induces a 3-coloring of $K^v (= K_4 - v^1v^2)$ for each vertex v of TG_p. As K_4 is 4-chromatic, we conclude that $\varphi(v^1) = \varphi(v^2)$ for each vertex v of TG_p. Then φ induces a 3-coloring of TG_p, which is impossible. Hence $\chi(G) \geq 4$. Now let e be an arbitrary edge of

G. First assume that $e \in E(G_1) \cup E(G_2)$. If we delete the corresponding edge of TG_p the resulting graph has a 3-coloring, which yields a 3-coloring φ_i of $G^i - e$ for $i \in \{1,2\}$ such that $\varphi^1(v^1) = \varphi^2(v^2)$ for every vertex v of TG_p. Then $\varphi' = \varphi^1 \cup \varphi^2$ can be extended to a coloring of $K^v (= K_4 - v^1v^2)$ for all vertices v of TG_p. The result is a 3-coloring of $G - e$. Now assume that e is an edge of K^v for some vertex v of TG_p. As any proper splitting of v into a set of two vertices results in a 3-colorable graph (see Exercise 4.23), there is a 3-coloring φ^1 of G^1 and a 3-coloring φ^2 of G^2 such that $\varphi^1(v) \neq \varphi^2(v)$, but $\varphi^1(v') = \varphi^2(v')$ for every vertex $v' \neq v$ of TG_p. Then $\varphi^1 \cup \varphi^2$ can be easily extended to a 3-coloring of $G - e$. Consequently, G is 4-critical (By Proposition 1.29).

Let $TG_p = TG_p(A^1, A^2, A^3, A^4)$. If we delete a linear factor of $TG_p[A^3 \cup A^4] (= K_{p,p})$ then for any partition of the remaining $p^2 - p$ edges there is a feasible decomposition (G_1, G_2) of TG_p, and hence a 4-critical graph G on the same set of $n = 16p$ vertices. Hence the number of different such 4-critical graphs is at least $2^{(c'n^2)}$ with $c' > 0$. Some of these might be isomorphic, but each isomorphism class has at most $n!$ members. Hence for the number t_n of nonisomorphic 4-critical graphs on n vertices we obtain that

$$t_n \geq \frac{2^{(c'n^2)}}{n!} \geq \frac{2^{(c'n^2)}}{n^n} = \frac{c^{n^2}}{n^n} \geq \frac{c^{n^2}}{2^{n^2}} = \left(\frac{c}{2}\right)^{n^2}.$$

This completes the proof. □

4.6 Circumference in Critical Graphs

Recall that the circumference $c(G)$ of a graph G is the maximum length of a cycle in G. The study of the circumference of critical graphs was initiated by Dirac [292]. For integers n and k satisfying $k \geq 4$, $n \geq k$, and $n \neq k+1$, let

$$L_k(n) = \min\{c(G) \mid G \in \text{Crit}(k,n)\},$$

i.e., $L_k(n)$ is the length of a shortest possible longest cycle in k-critical graphs on n vertices. Dirac [292] proved that any graph G with $\delta(G) \geq |G|/2$ has a Hamilton cycle. Furthermore, he proved that if G is 2-connected, then $c(G) \geq \min\{|G|, 2\delta(G)\}$. As a k-critical graph is 2-connected and has minimum degree at least $k-1$, this implies that $L_k(n) \geq \min\{n, 2k-2\}$. However, Dirac conjectured that $L_k(n)$ tends to infinity as n tends to infinity, and that actually $L_k(n) \geq c\sqrt{n}$. The first conjecture was established by Kelly and Kelly [564]. They also proved that $L_4(n) = O(\log^2 n)$, thereby disproving Dirac's second conjecture for $k = 4$. Dirac [294] extended the upper bound of Kelly and Kelly to all $k \geq 4$, and an improved upper bound was given by Read [852]. In 1963 Gallai [397] obtained the best upper bound until now. Gallai not only improved the earlier bounds, but also significantly simplified the previous constructions.

Theorem 4.42 (Gallai) *For all $k \geq 4$ there are infinitely many integers n for which there exists a k-critical graph G on n vertices such that no cycle of G has length at least $\frac{2(k-1)}{\log(k-2)} \log n$ and no path of G has length greater than $\frac{4(k-1)}{\log(k-2)} \log n$.*

Proof Let $k \geq 4$ be a fixed integer. We construct a sequence $(H_p)_{p\geq 1}$ of Gallai trees as follows. Let H_1 be a K_{k-1}. For $p \geq 2$, let H_p be obtained from H_{p-1} by adding for every vertex v of H_{p-1}, having degree $k-2$ in H_{p-1}, a complete graph $K^v = K_{k-1}$ and joining this complete graph with v by precisely one edge, where the complete graphs K^v are disjoint. Then H_p is an $\mathcal{E}_{(k-1)}$-graph and, for $p \geq 2$, we obtain that

$$|H_p| = (k-1) + (k-1)^2 \sum_{i=0}^{p-2} (k-2)^i > (k-1)(k-2)^{p-1} - 2.$$

Let G_p be the graph obtained from H_p by adding an additional vertex w and joining w to a vertex v of H_p by an edge if and only if $d_{H_p}(v) = k-2$. Then, see Exercise 4.5, G_p is a k-critical graph and $|G_p| = |H_p| + 1 > (k-2)^p$. Now let C be a cycle of G_p. If w does not belong to C, then C is contained in some block of the low vertex subgraph H_p of G_p, and hence $|C| \leq k-1$. If w belongs to C, then $C - w$ is a path in the Gallai tree H_p and a simple induction on p shows that $|C| \leq (2p-1)(k-1)+1$. Hence any cycle in G_p satisfies $|C| < 2(k-1)p$. As $|G_p| > (k-2)^p$, this yields $|C| < (2(k-1)/\log(k-2)) \log |G_p|$. If P is a path of length ℓ in G_p, then $\ell \leq (4(k-1)/\log(k-2)) \log |G_p|$. This completes the proof.□

Corollary 4.43 (Gallai) *For all $k \geq 4$ there are infinitely many integers n such that*

$$L_k(n) < \frac{2(k-1)}{\log(k-2)} \log n.$$

The first nontrivial lower bound for the function $L_k(n)$ was obtained in 1954 by Kelly and Kelly [564], they proved that

$$L_k(n) \geq \sqrt{\frac{\log\log n}{\log\log\log n}}$$

for every fixed $k \geq 4$. The first improvement was obtained almost 50 years later by Alon, Krivelevich, and Seymour [53], they proved that

$$L_k(n) \geq 2\sqrt{\frac{\log(n-1)}{\log(k-2)}}$$

for all $k \geq 4$ and $n \geq k+2$. Alon, Krivelevich, and Seymour established a lower bound for a longest path in a critical graph. Then they used the fact that critical graphs are 2-connected and applied the following lemma due to Dirac [292] and Voss [1055].

Lemma 4.44 *If a 2-connected graph has a path of length ℓ, then it has a cycle of length at least $2\sqrt{\ell}$.*

For 3-connected graphs the bound in the above lemma can be substantially improved, as shown by Bondy and Locke [143].

Lemma 4.45 *If a 3-connected graph has a path of length ℓ, then it has a cycle of length at least $2\ell/5$.*

The next lemma can be used to show that a critical graph has a long path. The lemma is due to Shapira and Thomas [935], the case $X = \varnothing$ is implicit in [53].

Lemma 4.46 *Let G be a k-critical graph with $k \geq 4$, let $X \subseteq V(G)$ with $|X| = p \geq 0$, and let T be a DFS tree of $G - X$ rooted at r. Then for every integer $i \geq 1$, the number of vertices of T at distance i from r is at most $(k-1)^p$ if $i = 1$ and $(k-2)^{i-2}(k-1)^p$ otherwise.*

Proof Let G, X, T, r be as stated, let $X = \{v_1, v_2, \ldots, v_p\}$, and let V_i be the set of vertices $t \in V(T)$ having distance $i \geq 1$ from r in T. Let $t \in V_i$ be an arbitrary vertex. Then $rTt = (t_0 = r, t_1, \ldots, t_i = t)$ and $e_t = t_{i-1}t$ is an edge of T and hence of G. As G is k-critical, there is a coloring of $G - e_t$, say φ_t, with color set $C = [1, k-1]$ such that $\varphi_t(t_0) = 1$. If $i \geq 2$, then t_0t_1 is an edge of $G - e_t$ and we may choose the coloring φ_t such that $\varphi_t(t_1) = 2$. Define $S_t = (\varphi_t(t_2), \varphi_t(t_3), \ldots, \varphi_t(t_{i-1}), \varphi_t(v_1), \ldots, \varphi_t(v_p))$. We claim that $S_t \neq S_{t'}$ whenever $t, t' \in V_i$ are distinct. Suppose this is false, and let $t, t' \in V_i$ be distinct vertices such that $S_t = S_{t'}$. Then

$$rTt = (t_0 = r, t_1, \ldots, t_i = t) \text{ and } rTt' = (t'_0 = r, t'_1, \ldots, t'_i = t')$$

and there is a largest integer, say s, with $t_s = t'_s$. Note that $0 \leq s \leq i-1$ and

$$(+) \quad \varphi_t(v) = \varphi_{t'}(v) \text{ for all } v \in X \cup V(rTt_s).$$

Now let Y be the set of vertices u of T such that t_{s+1} belongs to the path rTu. Define a mapping $\phi : V(G) \to C$ by $\phi(v) = \varphi_t(v)$ for $v \in V(G) \setminus Y$ and $\phi(v) = \varphi_{t'}(v)$ for $v \in Y$. Since T is a DFS tree of $G - X$, it follows that every edge belonging to $E_G(Y, V(G) \setminus Y)$ has one end in $X \cup V(rTt_s)$. Then (+) implies that ϕ is a coloring of G with color set $C = [1, k-1]$, contrary to the k-criticality of G. This proves the claim that $S_t \neq S_{t'}$ whenever $t, t' \in V_i$ are distinct. The number of possible sequences S_t with $t \in V_i$ is at most $(k-1)^p$ if $i = 1$ and $(k-2)^{i-2}(k-1)^p$ if $i \geq 2$. This completes the proof. □

Lemma 4.47 (Alon, Krivelevich, and Seymour) *For every $k \geq 4$, every k-critical graph on n vertices has a path of length at least $\log n / \log(k-2)$.*

Proof Let G be a k-critical graph on n vertices, and let r be an arbitrary vertex of G. As G is 2-connected, G has a DFS tree T rooted at r and the height h of (T, r) satisfies $h \geq 2$. By Lemma 4.46 applied with $p = 0$, we obtain that

$$n = |T| \leq 1 + 1 + 1 + (k-2) + (k-2)^2 + \cdots + (k-2)^{h-2} \leq (k-2)^h,$$

which implies in turn that $h \geq \log n / \log(k-2)$, as required. □

Using Lemma 4.44 and Lemma 4.45, we obtain the following corollaries of Lemma 4.47.

Corollary 4.48 *For $k \geq 4$, every k-critical graph of order n has a cycle of length at least* $2\sqrt{\log n/\log(k-2)}$, *that is,* $L_k(n) \geq 2\sqrt{\log n/\log(k-2)}$.

Corollary 4.49 *For $k \geq 4$, every 3-connected k-critical graph of order n has a cycle of length at least* $2\log n/(5\log(k-2))$.

Corollary 4.50 *For $k \geq 4$, every k-critical graph of order n has an odd cycle of length at least* $\sqrt{\log n/\log(k-2)}$.

Corollary 4.50 follows from Corollary 4.48 and a result of Voss [1055, Theorem 2], which says that every 2-connected nonbipartite G satisfies $\mathrm{c}(G) \leq 2\mathrm{c_o}(G) - 2$.

4.7 Other Types of Color Critical Graphs

As emphasised by Toft [1019] in his thesis, *the idea behind the study of critical graphs is this: instead of examining the class of all graphs with given chromatic number k one studies those k-chromatic graphs that are minimal with respect to certain partial orderings in the class of all k-chromatic graphs.* There are three natural candidates for such an ordering, namely, 'to be a subgraph', 'to be an induced subgraph' and 'to be a minor'. This leads to three different types of criticality, called 'critical graphs', 'vertex-critical graphs' and 'contraction-critical graphs'. All tree classes were introduced by Dirac who noticed that the importance of the concept of 'criticality' lies in the fact that problems for k-chromatic graphs may often be reduced to problems for various types of critical k-chromatic graphs.

A graph is **vertex-critical**, respectively k**-vertex-critical**, if $\chi(G') < \chi(G) = k$ for every proper induced subgraph G' of G. While k-critical graphs are k-chromatic graphs that are minimal with respect to the relation 'to be a subgraph', k-vertex-critical graphs are k-chromatic graphs that are minimal with respect to the relation 'to be an induced subgraph'. So vertex critical graphs are of interest in studying hereditary graph properties (see Exercise 2.25). Clearly, every critical graph is vertex-critical, but not conversely. Examples of vertex-critical graphs that are not critical were given by Dirac. For $k \leq 3$, however, it is easy to show that a graph is k-critical if and only if it is k-vertex-critical. It follows from (1.13) that a graph G is k-vertex-critical if and only if $\chi(G - v) < k \leq \chi(G)$ for every vertex $v \in V(G)$. So a graph is vertex critical if and only if each of its vertices is critical. Clearly, a graph without isolated vertices is critical if and only if each of its edges is critical. Results about critical graphs can be often transformed into results about the larger class of vertex-critical graphs.

Proposition 4.51 *Let G be a k-vertex-critical graph. Then G contains a k-critical subgraph and any such subgraph has the same vertex set as G.*

Proof That G contains a k-critical subgraph follows from Proposition 1.27(a). Now let G' be such a subgraph. If a vertex v of G does not belong to G', then $G' \subseteq G - v$ and, since G is k-vertex-critical, we obtain that $\chi(G') \leq \chi(G-v) < k$, which is impossible. □

Clearly, any minimal imperfect graph G is vertex-critical with chromatic number $\omega(G)$ as every vertex v of G satisfies $\chi(G-v) = \omega(G) < \chi(G)$. Hence the odd holes C_{2p+1} $(p \geq 2)$ and the odd antiholes $\overline{C}_{2p+1}$ $(p \geq 2)$ are vertex-critical graphs. The reader may easily check that if $p \geq 3$ then $\overline{C}_{2p+1}$ is not critical (delete an edge whose ends have distance p in C_{2p+1}), but $\overline{C}_{2p+1}$ contains critical edges. The first example of a vertex-critical graph without any critical edge was presented by Brown [178]; the graph has chromatic number $\chi = 5$, vertex set is $\mathbb{Z}_{17}$ with a and b joined by an edge if and only if $a - b \in \{2,6,7,8,9,10,11,15\}$ (addition modulo 17). The first infinite family of vertex-critical graphs without any critical edge is due to Jensen [525]. He constructs for all integers k and p with $k \geq 5$ and $p \geq 1$, a k-vertex-critical graph that remains k-chromatic whenever a set of p edges, all incident with the same vertex, is deleted, thus providing an affirmative answer to a problem posed by Erdős [339] (see also Jensen and Toft [530, Problem 5.14]). Furthermore, Jensen constructed for every $k \geq 5$, a k-vertex-critical graph containing edge disjoint k-critical graphs, see also Martinsson and Steiner [718]. However, Dirac's original question whether there exists a 4-vertex-critical graph without critical edges remains unanswered.

The Strong Perfect Graph Theorem (SPGT) [230] is equivalent to the statement that the odd holes and the odd antiholes are the only minimal imperfect graphs. Hence the following theorem is a reformulation of the SPGT.

Theorem 4.52 (Chudnovsky, Robertson, Seymour, and Thomas) *Let $k \geq 3$ be an integer and let G be a graph. Then the following statements are equivalent:*

(a) *G is a k-vertex-critical such that any k'-vertex-critical induced subgraph of G is a $K_{k'}$ whenever $1 \leq k' < k$.*
(b) *G is an odd hole or an odd antihole.*

Problem 4.53 (Toft) *Let G be a graph such that both G and $\overline{G}$ are vertex-critical. Is it true that G is an odd hole or an odd antihole?*

If $G = G_1 \boxplus G_2$ is the Dirac join of two disjoint graphs G_1 and G_2, then G is vertex-critical if and only if both G_1 and G_2 are vertex-critical (one can similarly argue as in the proof of Theorem 4.1). Toft [1026] invented a modification of the Dirac join. Let G_1 and G_2 be two disjoint graphs and, for $i \in \{1,2\}$, let v_i be be a vertex of G_i. Let G be the graph obtained from the union $G_1 \cup G_2$ by identifying the vertices v_1 and v_2 and adding all edges between $G_1 - v_1$ and $G_2 - v_2$. We call G the **Toft join** of G_1 and G_2 and write $G = (G_1, v_1) \boxminus (G_2, v_2)$ or briefly $G = G_1 \boxminus G_2$. Figure 4.5 shows the Toft join $G = C_5 \boxminus C_5$. Before we prove our main result about the Toft join, we need a simple proposition.

Proposition 4.54 *Let $G = (G_1, v_1) \boxminus (G_2, v_2)$ be the Toft join of two nonempty disjoint graphs G_1 and G_2. Then*

$$\chi(G) = \min\{\chi(G_1 - v_1) + \chi(G_2), \chi(G_1) + \chi(G_2 - v_2)\}.$$

Proof Let v^* denote the vertex of G obtained by identifying the vertices v_1 and v_2. As $G_1 - v_1$ and $G_2 - v_2$ are vertex disjoint, we can color $G_1 - v_1$ and G_2, or G_1 and $G_2 - v_2$ separately, which implies that

$$\chi(G) \le \min\{\chi(G_1 - v_1) + \chi(G_2), \chi(G_1) + \chi(G_2 - v_2)\}.$$

Now let φ be a coloring of G with color set $C = [1, \chi(G)]$, and let $c = \varphi(v^*)$. As $G - v^* = (G_1 - v_1) \boxplus (G_2 - v_2)$, the color sets $C_1 = \varphi(V(G_1 - v^*))$ and $C_2 = \varphi(V(G_2 - v^*))$ are disjoint and c is not contained in both sets. We may assume that $c \notin C_1$. Then $|C_1| \ge \chi(G_1 - v_1)$, $|C_2 \cup \{c\}| \ge \chi(G_2)$, and so $\chi(G) = |C| \ge \chi(G_1 - v_1) + \chi(G_2)$, and hence $\chi(G) = \min\{\chi(G_1 - v_1) + \chi(G_2), \chi(G_1) + \chi(G_2 - v_2)\}$. □

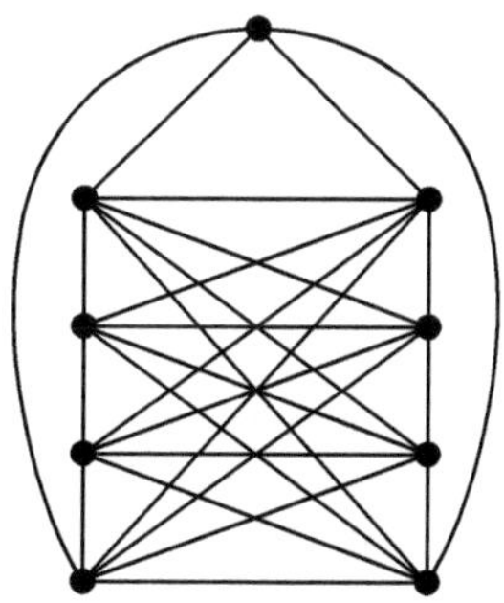

Fig. 4.5 The graph $C_5 \boxminus C_5$.

Theorem 4.55 (Toft) *Let $G = G_1 \boxminus G_2$ be the Toft join of two nonempty disjoint graphs G_1 and G_2. Then G is vertex-critical if and only if both G_1 and G_2 are vertex-critical. Furthermore, if G is vertex-critical, then $\chi(G) = \chi(G_1) + \chi(G_2) - 1$.*

Proof Let $G = (G_1, v_1) \boxminus (G_2, v_2)$ and let v^* denote the vertex of G obtained by identifying the vertices v_1 and v_2. Let $k = \chi(G)$ and, for $i \in \{1,2\}$, let $k_i = \chi(G_i)$, and $k_i^- = \chi(G_i - v_i)$. Then it follows from Proposition 4.54 that

$$k = \min\{k_1 + k_2^-, k_1^- + k_2\}. \tag{4.9}$$

Suppose first that G is vertex-critical. As $G - v^* = (G_1 - v_1) \boxplus (G_2 - v_2)$, we conclude that $k = \chi(G - v^*) + 1 = k_1^- + k_2^- + 1$ (by Theorem 4.1). By symmetry, it then follows from (1.13) and (4.9) that $k = k_1 + k_2^- \le k_1^- + k_2 \le k_1^- + k_2^- + 1 = k$. This implies that $k_1 = k_1^- + 1$ and $k_2 = k_2^- + 1$, and hence v^* is a critical vertex of both G_1 and G_2. Furthermore, $\chi(G) = k_1 + k_2 - 1$ (by (4.9)). Let v be a vertex of $G_i - v^*$. By symmetry we may assume that $i = 1$. Then $G - v = (G_1 - v, v_1) \boxminus (G_2, v_2)$ and Proposition 4.54 implies that

$$\chi(G-v) = \min\{\chi(G_1 - v - v_1) + \chi(G_2), \chi(G_1 - v) + \chi(G_2 - v_2)\}.$$

Since G is vertex-critical, we get $\chi(G-v) = k-1 = k_1+k_2-2 = \chi(G_1)+\chi(G_2)-2$, which implies that either $\chi(G_1 - v - v_1) = \chi(G_1) - 2$ or $\chi(G_1 - v) + \chi(G_2 - v_2) = \chi(G_1)+\chi(G_2)-2$. In both cases we obtain that $\chi(G_1 - v) = \chi(G_1) - 1$, that is, v is a critical vertex of G_1. We have then proved that both G_1 and G_2 are vertex-critical.

Suppose now that both G_1 and G_2 are vertex-critical. Then $k_i^- = k_i - 1$ ($i \in \{1,2\}$) and hence $\chi(G) = k = k_1 + k_2 - 1$ (by (4.9)). As $G - v^* = (G_1 - v_1) \boxplus (G_2 - v_2)$, we conclude that $\chi(G - v^*) = k_1^- + k_2^-$ (by Theorem 4.1), which leads to $\chi(G - v^*) = k_1 + k_2 - 2 = k - 1$. Hence v^* is a critical vertex of G. Now let $v \neq v^*$ be a vertex of G. By symmetry, we may assume that v belongs to $G_1 - v_1$. Then Proposition 4.54 implies that

$$\chi(G-v) = \min\{\chi(G_1 - v - v_1) + \chi(G_2), \chi(G_1 - v) + \chi(G_2 - v_2)\}.$$

As both G_1 and G_2 are vertex-critical, this gives

$$\chi(G-v) \leq \chi(G_1 - v) + \chi(G_2 - v_2) = k_1 + k_2 - 2 = k - 1.$$

Hence v is a critical vertex of G. Consequently, G is vertex-critical. □

As an immediate consequence of Theorem 4.55, Toft [1026] obtained that a graph G is the complement of a vertex-critical graph if and only if all blocks of G are complements of vertex-critical graphs (see Exercise 4.27).

Let $G_p = C_{2p+1} \boxminus C_{2p+1}$ be the Toft join of two odd cycles. Then G_p is a 5-vertex-critical graph on $n = 4p+1$ vertices (by Theorem 4.55); the graph G_2 is shown in Figure 4.5). The graph G_p has $m = 4(p^2+p)$ edges and for the number of K_4 in G we obtain that $\Bbbk_4(G) \geq cn^4$ in sharp contrast to Theorem 4.24. Clearly, G_p is not critical (see also Exercise 4.28), but it is difficult to describe the 5-critical factors of G_p.

The removal of an independent set of a graph decreases its chromatic number by at most one. However deleting two adjacent vertices of a graph may decrease its chromatic number by two. A graph G is called **double-critical** if G is connected and

$$\chi(G - v - w) \leq \chi(G) - 2$$

for every edge $vw \in E(G)$. So the complete graph K_k with $k \geq 1$ is a double-critical k-chromatic graph. It is easy to show that every double-critical graph is critical. So for $k \leq 3$, the complete graph K_k is the only double-critical graph with chromatic number $\chi = k$. Furthermore, it is easy to show (see Exercise 4.25) that K_4 is the is the only double-critical graph with chromatic number $\chi = 4$. That K_5 is the only double-critical 5-chromatic graph was proved, independently, by Mozhan [763] and Stiebitz [968], thereby solving the first nontrivial case of the following conjecture proposed in 1966 by Lovász (see Erdős [334]).

Conjecture 4.56 (THE DOUBLE-CRITICAL GRAPH CONJECTURE) *For every $k \geq 1$, the complete graph K_k is the only double-critical graph with chromatic number k.*

As discussed above, the conjecture is known to be true for $k \leq 5$ and unsolved for all $k \geq 6$. Results related to the conjecture are discussed in a recent paper by Kawarabayashi, Pedersen, and Toft [561].

Theorem 4.57 (MOZHAN / STIEBITZ) *The complete graph K_5 is the only double critical graph with chromatic number $\chi = 5$.*

Proof Let G be an arbitrary double-critical graph with chromatic number $\chi(G) = 5$. For a graph H and an edge $e = uv$ of H, let $T(e : H) = N_H(u) \cap N_H(v)$ be the common neighborhood of e in H. If H is an induced subgraph of G and v is a vertex of G, then $H + v$ denotes the induced subgraph of G with vertex set $V(H) \cup \{v\}$. We shall prove two claims.

Claim 4.57.1 *If $e = uv \in E(G)$ and $\varphi \in CO(G - u - v, 3)$, then $|\varphi(T(e : G))| = 3$ and hence $|T(e : G)| \geq 3$.*

Proof: The coloring φ of $G - u - v$ with color set $\{1,2,3\}$ can be easily extended to a coloring ϕ of G with color set $C = \{1,2,3,4,5\}$ by giving u color 4 and v color 5. Since $\chi(G) = 5$, for each color $c \in C$, there is a vertex of G, say v_c, such that $\phi(v_c) = c$ and $\phi(N_G(v_c)) = C \setminus \{c\}$ (by Proposition 1.34(a)). This implies that $\{v_1, v_2, v_3\} \subseteq T(e : G)$ and, therefore, $|\varphi(T(e : G))| = |\phi(T(e : G))| = 3$. △

Claim 4.57.2 *G contains a K_4.*

Proof: There is a longest sequence of distinct vertices of G, say $v_1, v_2, \ldots, v_r$, such that $G_i = G[\{v_1, v_2, \ldots, v_i\}]$ is a uniquely 3-colorable graph for $i \in [1,r]$. Since G contains a K_3 (by Claim 4.57.1), we have $r \geq 3$. Clearly, $G_r \neq G$. Let i_0 be the largest integer such that $e = v_{i_0}v_r$ is an edge of G_r. Then $T(e : G_r) \subseteq \{v_1, v_2, \ldots, v_{i_0-1}\} = V(G_{i_0})$. Since both G_{i_0-1} and G_r are uniquely 3-colorable, we get $|\varphi(T(e : G_r))| \leq 1$ for every 3-coloring φ of $G - v_{i_0} - v_r$. This implies that $|T(e : G) \setminus T(e : G_r)| \geq 2$ (by Claim 4.57.1). Let $X = V(G) \setminus \{v_1, v_2, \ldots, v_r\}$ and let $u, v \in X$ be any two distinct vertices of $T = T(e : G) \setminus T(e : G_r)$. There is no uniquely 3-colorable induced subgraph of G on $r+1$ vertices containing $v_1, v_2, \ldots, v_r$, hence, because of $u, v \in T$, both graphs $G_r + u$ and $G_r + v$ are 4-chromatic. Consequently, since G is double-critical and 5-chromatic, $X \setminus \{u\}$ as well as $X \setminus \{v\}$ are independent sets of G. But X cannot be an independent set of G, hence u and v are adjacent in G, and so $G[\{u, v, v_{i_0}, v_r\}]$ is a K_4. △

By Claim 4.57.2, there is a K_4 in G, say with vertex set X. Since G is double-critical and 5-chromatic, $Y = V(G) \setminus X$ is an independent set of G. Now let $v \in Y$ be an arbitrary vertex. Since G is connected, G has an edge e joining v with a vertex of X. By Claim 4.57.1, e is contained in at least three triangles of G, which implies that $N_G(v) = X$. Then $G[X \cup \{v\}]$ is a K_5 and it follows immediately that $|Y| = 1$, and so $G = K_5$. This completes the proof. □

The 'subgraph' relation and the 'induced subgraph' relation are two natural partial orderings for the class of (finite) graphs. However, there are two other such partial orderings that attracted a lot of attention: the 'minor' relation and the 'topological minor' relation (see Appendix C.8). In what follows we focus on the 'minor' relation.

Recall that a **minor** of a graph G is a graph H that can be obtained from a subgraph of G by a series of edge contractions. Every graph that is isomorphic to a minor of G is also called a minor of G. A minor of G that is not isomorphic to G itself is called a **proper minor** of G. If H is a minor of G, we write $H \preceq G$. From the definition it follows that every subgraph or induced subgraph of G is a minor of G. Furthermore, $H \preceq G$ is equivalent to the statement that there is a family $(X_v : v \in V(H))$ of pairwise disjoint subsets X_v of $V(G)$ such that $G[X_v]$ is connected for all $v \in V(H)$ and $E_G(X_v, X_w) \neq \varnothing$ whenever $vw \in E(H)$.

At the beginning of the 1980s, Neil Robertson and Paul Seymour started a research project with the aim to developed a theory of graph minors. It took more than 20 years (from [869] to [870]) to publish this seminal work, which culminated in a proof of a theorem usually cited as the graph minor theorem. This ultimate result of Robertson and Seymour says that the class of graphs are well-quasi ordered by the minor relation, or equivalently:

In any infinite sequence of graphs, there are two such graphs, differently placed in the sequence, where one is a minor of the other.

The statement above is sometimes referred to as Wagner's Conjecture. But it seems that in reality Wagner conjectured its negation ([1059, Problem 9, p. 61]), asking for an infinite set of graphs with no one being a minor of another.

A graph property $\mathcal{P}$ is called **$\preceq$-hereditary** or **minor-closed** if for every graph $G \in \mathcal{P}$, all minors of G are also in $\mathcal{P}$. Many interesting graph properties are minor-closed, including the class of planar graphs and, more generally, the class $\mathcal{G}(\hookrightarrow \mathbf{S})$ of graphs that can be imbedded in a fixed surface $\mathbf{S}$ (see Appendix C.13). The graph minor theorem is equivalent to the statement that for every minor-closed graph property the set of forbidden minors is finite (see also Appendix C.11).

Theorem 4.58 (Graph Minor Theorem of Robertson and Seymour) *For every minor-closed graph property $\mathcal{P}$ there is a finite set $\mathcal{X}$ of graphs such that* $\mathcal{P} = \mathrm{Forb}_{\preceq}(\mathcal{X})$.

The graph minor theorem is a far reaching extension of Kuratowski's theorem characterizing planar graphs in terms of forbidden substructures. For the sphere $\mathbf{S}_0$,

$$\mathcal{G}(\hookrightarrow \mathbf{S}_0) = \mathrm{Forb}_{\preceq_t}(K_5, K_{3,3}) = \mathrm{Forb}_{\preceq}(K_5, K_{3,3})$$

The first equality was proved by Kuratowski [661]. That the two classes $\mathcal{G}(\hookrightarrow \mathbf{S}_0)$ and $\mathrm{Forb}_{\preceq}(K_5, K_{3,3})$ are the same was proved by Wagner [1057]. A version of this result also appears in a paper of Harary and Tutte [470]. There it is explained that the theorems of Kuratowski and Wagner are (in a certain sense) dual to each other. Deleting an edge in a plane graph corresponds to contracting the corresponding edge in its dual plane graph, and vice versa. Note that every topological minor of a graph G is a minor of G, but not conversely.

A graph G is called **contraction-critical**, respectively **k-contraction-critical**, if $\chi(H) < \chi(G) = k$ for every proper minor H of G. While the chromatic number is monotone with respect to both the subgraph relation and the induced subgraph relation this is not the case with respect to the minor relation; any even cycle has K_3 as a minor. On the other hand, K_1 is a minor of any nonempty graph, and any graph with chromatic number at least k has a k-chromatic minor (and even a k-chromatic subgraph). Hence, every graph G has a minor that is contraction-critical and has the same chromatic number as G. Evidently, K_k is a k-contraction-critical graph and a famous conjecture due to Hadwiger [448] claims that the complete graphs are the only contraction-critical graphs.

Conjecture 4.59 (Hadwiger's Conjecture) *For every $k \geq 1$, the complete graph K_k is the only contraction critical graph with chromatic number k.*

The above formulation of Hadwiger's conjecture is due to Dirac [299] [302]. Hadwiger's original version states that any k-chromatic graph has K_k as a minor, or equivalently, any graph in $\mathrm{Forb}_{\preceq}(K_k)$ is $(k-1)$-colorable. To prove Hadwiger's conjecture it suffices to show that any k-contraction critical graph has K_k as a minor. As any subgraph of a graph G is a minor of G, it follows that any contraction-critical graph is critical. From this follows immediately that Hadwiger's conjecture is true when $k \leq 3$.

To prove Hadwiger's conjecture for $k = 4$, let G be a 4-contraction-critical graph. Then G is 4-critical and hence $\delta(G) \geq 3$. If G contains a K_3 as a subgraph, then $G - V(K_3)$ is connected (by Corollary 1.31), and hence K_4 is a minor of G. If G contains no K_3 as a subgraph, then let v be an arbitrary vertex of G. Contracting all edges incident with v into a vertex v^* results in a proper minor H of G. As G is 4-contraction-critical, H has a 3-coloring φ which yields a 3-coloring of G by giving each vertex in $N_G(v)$ the color $\varphi(v^*)$, and giving v a color different from $\varphi(v^*)$. This contradiction completes the proof.

That Conjecture 4.59 is true for $k \leq 4$ was noticed by Hadwiger [448]. For $k = 4$ he proved a slightly stronger statement. A different proof for $k = 4$ was given in 1951 by Dirac in his thesis.

Theorem 4.60 (Hadwiger) *Let G be a graph without K_4 as a minor. Then G is 3-colorable, in fact any 3-coloring of any induced cycle of G can be extended to a 3-coloring of G.*

Proof If G has no cycle, then G is a forest and hence 2-colorable. So assume that G has a cycle, and let C be an induced cycle of G with a given 3-coloring φ. Then we prove by induction on the order n of $G - V(C)$ that φ can be extended to a 3-coloring of G. If $n = 0$ there is nothing to prove. If $n = 1$, then the only vertex v of $G - V(C)$ has at most two neighbours belonging to C (for otherwise K_4 is a minor of G), and so φ can be extended to a 3-coloring of G. It remains to consider the case when $n \geq 2$. If $G - V(C)$ is disconnected, then the induction hypothesis implies that φ can be extended to of each component of $G - V(C)$, and hence to all of G. So assume that $G - V(C)$ is connected. Then at most two vertices of C, say v and w, have neighbors outside C, for otherwise K_4 is a minor of G. If at most one of v and w, say v, has a neighbor outside C, then $G - V(C)$ together with v induces a subgraph G' of G, which has a 3-coloring φ' by induction. By renaming the colors we may assume that $\varphi(v) = \varphi'(v)$ and hence $\varphi \cup \varphi'$ is a 3-coloring of G as required. Suppose finally that both v and w have neighbors in $G - V(C)$. Then there is a v-w path in $G - E(C)$ and P may be a shortest such path. Clearly, P has inner vertices and is an induced path of $G - E(C)$. The vertices v and w divide C into two v-w paths, say P_1 and P_2, where we may assume that P_1 has inner vertices. The coloring φ can be extended to a 3-coloring φ' of the cycle $C' = P_2 \cup P$. Then φ' can be extended to a 3-coloring of $G - (V(P_1) \setminus \{v, w\})$ (by induction). This results in a 3-coloring of G as required. □

Theorem 4.61 (Hadwiger / Dirac) *Any graph of minimum degree at least three contains a subdivision of* K_4.

Proof Let G be a graph with $\delta(G) \geq 3$ and let B be an end-block. Then $|B| \geq 4$, B is 2-connected, and B has at most one vertex, say u, with $d_B(u) < 3$. Let C be a longest cycle in B. For each vertex v of C, except u, there is a path P_v in B from v to a vertex v' of C all of whose edges and inner vertices are outside C. Furthermore, C contains a v-v' path Q_v that does not contain the vertex u. Select v and P_v such that Q_v is shortest possible. As C is a longest cycle in B, Q_v has a inner vertex, say v'. Then the union of C with P_v and $P_{v'}$ form a subdivision of K_4. □

By Kuratowski's theorem for planar graphs, Hadwiger's conjecture for $k = 5$, saying that any graph in $\text{Forb}_{\preceq}(K_5)$ is 4-colorable, obviously implies that any planar graph is 4-colorable. That the two statements are indeed equivalent follows from a deep result of Wagner [1057] characterizing the class $\text{Forb}_{\preceq}(K_5)$. The four color theorem saying that any planar graph is 4-colorable was proved by Appel and Haken [68] [69] in 1976 and in a more satisfactory way by Robertson, Sanders, Seymour, and Thomas [868]. That the case $k = 6$ of Hadwiger's conjecture is also equivalent to the four color theorem follows from a remarkable result obtained in 1993 by Robertson, Seymour, and Thomas [871].

Theorem 4.62 (Robertson, Seymour, and Thomas) *Every 6-contraction-critical graph* $G \neq K_6$ *has a vertex* v *such that* $G - v$ *is planar.*

So Hadwiger's conjecture is true when $k \leq 6$, however, all the remaining cases $k \geq 7$ are still unsolved. For $k = 7$, Kawarabayashi and Toft [562] proved that any 7-chromatic graphs has K_7 or $K_{4,4}$ as a minor. Jakobsen [518, 519] proved that every 9-chromatic graph has K_7 as a minor. Albar and Gonçalves [30] proved that every 11-chromatic graph has K_8 as a minor.

4.8 Exercises

4.1 Let G be a graph, and let $u, v \in V(G)$ be critical vertices of G. Show that if $N_G(u) \subseteq N_G(v)$, then $u = v$.

4.2 Show that a critical graph of order n cannot contain vertices of degree $n - 2$.

4.3 Let $k \geq 1$ be an integer. Prove the following statements:

1. If G is a graph of order $k+1$ with minimum degree at least $k-1$, then $G = K_{k+1}$ or there are is an edge vw of $\overline{G}$ such that $N_G(v) = N_G(w) = V(G) \setminus \{v,w\}$
2. Use (a) to show that no k-critical graph can have order $k+1$

4.4 Show that if $k \geq 4$, then $\text{Crit}(k,n) \neq \varnothing$ if and only if $n \geq k$ and $n \neq k+1$.

4.5 Use the Hajós join and Theorem 4.2(c) to show that if G is a graph obtained from an ε_k-graph T with $k \geq 3$ by adding a vertex v and joining v to each vertex u of T with $d_T(u) = k-1$, then G is a $(k+1)$-critical graph with $|G_H| \leq 1$. (Dirac [298] and Gallai [397])

4.6 Show that if G is a $(k+1)$-critical graph with $k \geq 3$ and v is the only high vertex of G, then $G - v$ is an ε_k-graph T and $N_G(v) = \{u \in V(T) \mid d_T(u) = k-1\}$. (Hint: Consider the list-assignment L of $T = G - v$ with $L(u) = [1, k-2]$ if $uv \in E(G)$ and $L(u) = [1, k-1]$ otherwise and use Theorem 1.8) (Dirac [298] and Gallai [397])

4.7 Let $k \geq 4$ be an integer, and let T be a tree with $\Delta(T) \leq k-1$. Furthermore, let $(H_t : t \in V(T))$ be a family of k-critical graphs, each containing the same vertex v^*, and otherwise pairwise disjoint. For every ordered pair (t,t') of adjacent vertices of T, let $v_{t,t'}$ be a vertex of H_t such that (1) $v_{t,t'}v^* \in E(H_t)$, and (2) if t' and t'' are distinct neighbors of t in T, then $v_{t,t'} \neq v_{t,t''}$. Now let G denote the graph obtained from the union of the graphs H_t with $t \in V(T)$ by, for every edge $tt' \in E(T)$, deleting the edges $v^*v_{t,t'}$ and $v^*v_{t',t}$ an adding the edge $v_{t,t'}v_{t',t}$. Show that the following statements hold:

1. G is k-critical and $|G| = \sum_{t \in V(T)}(|H_t| - 1)$.
2. For every path P in $G - v^*$ there is a path R in T such that $|P| \leq \sum_{t \in V(R)}(|H_t| - 1)$.

(Hint: Observe that G can be obtained from the graphs H_t by repeatedly taking Hajós joins.) (Shapira and Thomas [935])

4.8 Use the previous exercise to show that if $k \geq 4$ and $n \geq k+2$, then $L_k(n) \leq \frac{2(k-1)}{\log(k-1)} \log n + 2k$. (Shapira and Thomas [935])

4.9 Show that the wheel $W_n = K_1 \boxplus C_n$ with $n \geq 3$ odd has Hajós number $\hbar_4(W_n) \leq \lceil \log_2(n-1) \rceil$. (Mansfield and Welsch [715])

4.10 Show that for any graph G the following conditions are equivalent:

1. G is Hajós-2-constructible.
2. G is Ore-2-constructible.
3. G has exactly one edge.

(Urquhart [1045])

4.11 Let $G = C_5[K_3]$ be the Catlin graph (see Exercise 1.27). Prove that G is 8-critical and that in a construction in the Hajós calculus $\mathcal{H}C_8$ the final graph before G is not 8-critical. (Hanson, Robinson, and Toft [467])

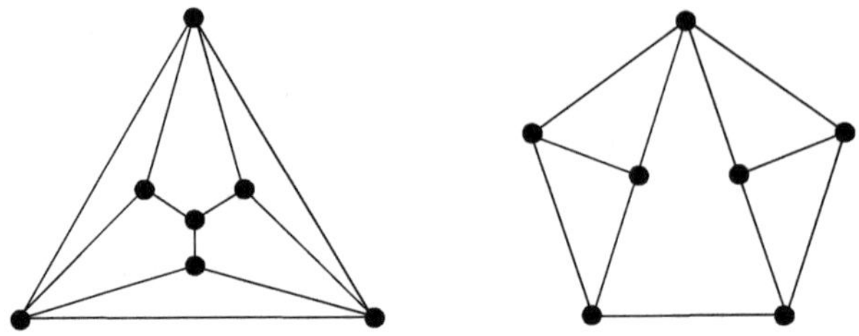

Fig. 4.6 The two graphs in $\mathrm{Crit}(4,7)$.

4.12 Deduce from Corollary 4.20 that $\mathrm{Crit}(k,k+2) = \langle K_{k-3} \boxplus C_5 \rangle$ when $k \geq 4$, and $\mathrm{Crit}(k,k+3) = K_1 \boxplus \mathrm{Crit}(k-1,k+2)$ and $\mathrm{Crit}(k,k+3) = K_{k-4} \boxplus \mathrm{Crit}(4,7)$ when $k \geq 5$.

4.13 Let H_1 and H_2 be the graphs shown in Figure 4.6. Prove that $\mathrm{Crit}(4,7) = \mathrm{Crit}^*(4,7) = \langle H_1, H_2 \rangle$, that is, H_1 and H_2 are the only 4-critical graphs of order 7 up to isomorphism. (Toft [1024])

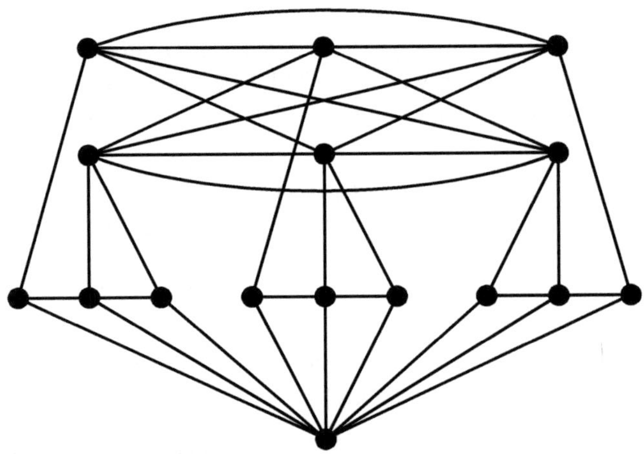

Fig. 4.7 The Toft graph with $k = 5$ and $|C^i| = |A_i| + 1 = 3$ for $i \in \{1,2,3\}$.

4.14 Let $k \geq 4$ be an integer. Let K be a complete $(k-2)$-partite graph with parts $A_1, A_2, \ldots, A_{k-2}$ each with at least two vertices, let $C^1, C^2, \ldots, C^{k-2}$ be odd cycles such that $|C^i| \geq |A_i|$ for all $i \in [1, k-2]$, and let $K' = K_{k-4}$. Let us assume that the graphs $K, K', C^1, C^2, \ldots, C^{k-2}$ are pairwise disjoint, and let G' be the union of all these graphs. We may now extend G' to a new graph G as follows: Join all vertices of K' to all vertices of $C^1 \cup C^2 \cup \cdots \cup C^{k-2}$. Then, for every $i \in [1, k-2]$, join each vertex of C^i to exactly one vertex of A_i such that each vertex of A_i is joined to at least one vertex of C^i. The graph obtained is G (see Figure 4.7). Show that G is

k-critical. Furthermore, show that for any graph H with $\chi(H) \leq k-2$, there is a k-critical graph of order n containing H, such that $n \leq 2|H|+c$ for some positive constant c (depending only on k). (Toft [1023])

4.15 Let G be a k-critical graph of order n that is not a $(k+3)$-wheel where $k \geq 4$. Proof that if G has an edge e that is contained in precisely one complete subgraph of G having order $k-1$, then $\mathbb{k}_{k-1}(G) \leq n-k+3$. (Hint: Consider the corresponding k-cluster of $G-e$, i.e. $\mathcal{K}_{k-1}(G-e)$ and use Lemma 4.26) (Su [982])

4.16 Let G be a k-critical graph of order n with $n \geq k \geq 4$. Show that the number of $(k-1)$-critical subgraphs of G is at least $\log_2 n$. (Hint: Let $\mathcal{K}$ denote the set of $(k-1)$-critical subgraphs, and for $e \in E(G)$ let $\mathcal{K}^e = \{H \in \mathcal{K} \mid e \in E(H)\}$. Use Lemma 4.30 to show that none of the sets $\mathcal{K}^e$ contains another one.) (Stiebitz [967])

4.17 Use Lemma 4.30 and Theorem 4.2(c) to show that for $k \geq 4$, the complete graph K_k is the only k-critical graph such that each $(k-1)$-critical subgraph is a K_{k-1}. (Stiebitz [967])

4.18 Let G be a k-critical graph of order n with $n > k \geq 4$, and let t be the number of $(k-1)$-critical subgraphs of G. Show that $\binom{t}{k-1} \geq n$ and hence $t > ((k-1)!n)^{1/(k-1)}$. (Hint: Let v be a vertex of G and let φ be a $(k-1)$-coloring of $G-v$ with color classes $(I_1, I_2, \ldots, I_{k-1})$. Then $G - I_i$ contains a $(k-1)$-critical subgraph H_i^v. Let $S^v = \{H_1^v, H_2^v, \ldots, H_{k-1}^v\}$. Show that the sets S^v are distinct.) (Abbott and Zhou [12])

4.19 Let G be a k-critical graph of order n with $n \geq k \geq 4$, let t be the number of $(k-1)$-critical subgraphs of G, and let s be the largest order of a $(k-1)$-critical subgraph of G. Show that $st \geq n$. (Abbott and Zhou [12])

4.20 Let G be a k-critical graph and let p be an integer satisfying $1 \leq p \leq k-2$. Show that the following statements hold:

1. By deleting at most p vertices from G, there exists for each vertex v of the remaining graph G', a $(k-p)$-critical subgraph of G' containing v.
2. By deleting at most p edges from G, there exists for each edge e of the remaining graph G', a $(k-p)$-critical subgraph of G' containing e, but not containing both ends of any deleted edges.

(Toft [1024])

4.21 Show that any graph obtained from a k-critical graph by splitting a vertex is either $(k-1)$-colorable or k-critical. (Lovász [698, Problem 9.20])

4.22 Show that any graph obtained from a k-critical graph by splitting a low vertex is $(k-1)$-colorable.

4.23 Show that any graph obtained from the Toft graph TG_p by splitting a vertex is 3-colorable.

4.24 Vertex-critical graph method: Let $\mathcal{P}$ be a hereditary graph property and let ρ be a graph parameter defined for $\mathcal{P}$ such that $\rho(H) \leq \rho(G)$ whenever $G \in \mathcal{P}$ and H is an induced subgraph of G. Show that the following statements hold:

(a) For every graph $G \in \mathcal{P}$ there exists a vertex-critical graph $H \in \mathcal{P}$ such that H is an induced subgraph of G and $\chi(H) = \chi(G)$.
(b) If $\chi(H) \leq \rho(H)$ for every vertex-critical graph $H \in \mathcal{P}$, then $\chi(G) \leq \rho(G)$ for every graph $G \in \mathcal{P}$.

4.25 Let k and n be integers satisfying $1 \leq k \leq n$. A graph G is called a **(k,n)-graph** if G is connected, $\chi(G) = n$, and $\chi(G - V(H)) = n - k$ for every k-chromatic induced subgraph H of G. Erdős and Lovász conjectured that if $2 \leq k \leq n-1$, then every (k,n)-graph contains a K_n (see Section 4.9)

1. Let G be a (k,n)-graph with $1 \leq k < n$, let X be a k-clique, and let φ be a coloring of $G - X$ with color set $C = [1, n-k]$. Show that for every color $c \in C$ there is a vertex v_c of $G - X$ such that $\varphi(v_c) = c$ and $X \cup \{v_c\}$ is a $(k+1)$-clique. (Hint: Use Proposition 1.34(a))
2. Show that every k-clique of a (k,n)-graph with $1 \leq k < n$ is contained in $n-k$ cliques of order $k+1$.
3. Show that every $(2,n)$-graph with $2 < n$ contains a K_3 as a subgraph.
4. Let G be a graph containing an odd cycle and let C be a shortest odd cycle. Show that if there exists a vertex v of $G - V(C)$ which is adjacent to at least three vertices of C in G, then C is a K_3 and $V(C) \cup \{v\}$ is a 4-clique of G.
5. Show that every $(3,n)$-graph with $3 < n$ contains a K_4 as a subgraph.
6. Show that if a (k,n)-graph with $2 \leq k \leq (n+1)/2$ contains a K_{n-k+1} as a subgraph, then G contains a K_n as a subgraph. (Hint: Use the fact that a (k,n)-graph is also an $(n-k+1,n)$-graph, consider an $(n-k+1)$-clique $X_0 = \{v_1, v_2, \ldots, v_{n-k+1}\}$ and let $u_1, u_2, \ldots, u_r$ be a largest sequence of distinct vertices of $G - X_0$ such that $X_i = \{u_1, u_2, \ldots, u_i, v_{i+1}, \ldots, v_{n-k+1}\}$ is an $(n-k+1)$-clique of G, and u_i is adjacent to all vertices of X_{i-1} (for $i \in [1,r]$).) (Stiebitz [969])
7. Show that if $(k,n) \in \{(2,4), (3,5), (3,6)\}$, then every (k,n)-graph contains a K_n as a subgraph.

4.26 Show that if G is not complete, then there is a partition (G_1, G_2) of G satisfying $\chi(G_1) + \chi(G_2) > \chi(G)$. (Hint: Consider a maximum clique in G.) (Lovász [698, Problem 9.2])

4.27 Show that a graph G is vertex-critical if and only if each block of the complement $\overline{G}$ is a complement of a vertex-critical graph. Moreover, show that if G is vertex-critical and $G_1, G_2, \ldots, G_p$ are the blocks of $\overline{G}$ and $q = \mathrm{com}(\overline{G})$, then

$$\chi(G) = \chi(G_1) + \chi(G_2) + \cdots + \chi(G_p) - (p - q).$$

(Hint: Use induction on p an apply Theorem 4.55.) (Toft [1026])

4.28 Suppose that $G = (G_1, v_1) \boxminus (G_2, v_2)$ is the Toft join of two disjoint vertex-critical graphs G_1 and G_2. For $i \in \{1,2\}$ let w_i be a vertex of $G_i - v_i$ such that $w_i v_i \notin E(G_i)$. Show that $w_1 w_2$ is not a critical edge of G. (Toft [1026])

4.29 Show that if G is a critical graph, then $\overline{G}$ has no separating vertex, that is, each connected component of $\overline{G}$ consists of precisely one block. (Toft [1026])

4.30 Show that if $\mathcal{P}$ is a minor closed graph property, then every graph $G \in \mathcal{P}$ has a contraction-critical minor H such that $\chi(H) = \chi(G)$.

4.31 Show that any k-contraction-critical graph G with $k \geq 1$ has minimum degree $\delta(G) \geq k$.

4.32 Show that Hadwiger's conjecture is equivalent to the statement that the largest number of colors needed to color any graph in a minor closed class is precisely the largest number of vertices in any complete graph belonging to the class. (Saaty and Kainen [880])

4.33 A **split graph** is a graph whose vertex set is the disjoint union of a clique and an independent set. Show that Hadwiger's conjecture is true for all graphs if and only if it is true for the squares of split graphs. (Chandran, Issac, Zhou [197])

4.9 Notes

Hajós' theorem (Theorem 4.3) was known in the 1940s, but its first appearance in print seems to have been in the book of Ringel [863] from 1959. In his talk given at the 25th anniversary celebration in 1992 at the Department of Combinatorics and Optimization, University of Waterloo, Canada, W. T. Tutte told that he learned Hajós' theorem from Hajós himself. That might have been at the first international meeting on graph theory at Dobogókő, Hungary, in October 1959. This meeting was organized by T. Gallai and had as participants C. Berge, G. Dirac, P. Erdős, G. Hajós, A. Renyi, H. Sachs, A. Stone, P. Turan, W. Tutte, and others. There are no proceedings from the meeting, but there is a short and interesting note about it by Tutte [1033]. In April 1960 H. Sachs organized the second international meeting on graph theory at the Martin Luther Universität, Halle Wittenberg, with C. Berge, G. Dirac, H. Grötzsch, G. Hajós, A. Kotzig, G. Ringel and others among the participants. In the extended abstracts of the Halle meeting, Hajós [461] presented a proof of his result as he was not satisfied with the (in his opinion) too complicated proof given by Ringel in his book [863]. As emphasised by Tutte in his talk from 1992, *Hajós theorem was a disappointment to its discover, it seemed to promise so much and yet achieved so little, and it had little publicity*. Tutte (see Appendix B.1) proved Hajós' theorem in a slightly stronger form than the original, so that it could be used to prove Brooks' theorem.

Over time, however, Hajós' theorem became popular among graph theorists. There are versions of Hajós theorem for the list chromatic number by Gravier [427] and Král [642], for the circular chromatic number by Zhu [1095, 1096], for the signed chromatic number by Kang [554], and for the chromatic number of edge weighted graphs by Mohar [749]. Hajós type theorems were obtained by Jaeger [515] for matroids representable over finite fields, by Nešetril [772] for graph homomorphisms, and by Jensen [526] for Grassmann homomorphisms, a homomorphism concept that provides a common generalization of Graph Colorings, Hypergraph Colorings and Nowhere-Zero Flows.

The Hajós join is a very useful tool for analyzing critical graphs, since it preserves not only the chromatic number but also criticality. That the Hajós join can be used also to obtain new critical graphs with respect to the chromatic index χ' from previously known ones, provided we add a degree condition, was proved by Jakobsen [520]. A Hajós construction for digraphs were invented by Hoshino and Kawarabayashi [506] in order to construct digraphs that are critical with respect to the dichromatic number $\vec{\chi}$. A Hajós type theorem for the dichromatic number was obtained by Bang-Jensen, Bellitto, Schweser, and Stiebitz [78].

The proof of Theorem 4.3 is Hajós' original proof from [461]. The proof of Theorem 4.5 follows [1045]. That the Hajós join can be used to construct 4-critical graphs with precisely one high vertex was first observed by Dirac [298]. Dirac [298] characterized the family of k-critical graphs with at most one high vertex, and in simpler terms also Gallai [397] did so using the Hajós join (see Exercise 4.6). Sachs and Stiebitz [885] [886] extended Gallai's result and provided a constructive characterization for the class of critical graphs with complete high vertex subgraphs. Gallai [397] generalized the Hajós join in several ways; see also the survey by Sachs and Stiebitz [887] on constructive methods for color-critical graphs.

As observed by Ore [794], Hajós' proof of Theorem 4.3 immediately leads to the concept of Ore-k-constructible graphs and to Theorem 4.4. Ore also noted that *for some of the applications it is essential to use the more specific form* of the construction, but he gave no example of such an application and we do not know any. Alon and Tarsi [59] presented a proof of Theorem 3.19 using Theorem 4.4, however they seem to overlook a small detail. The essential part in their argument is to show that if $G = (G_1, v_1, v_2, A_1)\nabla^{\circ}_{\iota}(G_2, v_1, v_3, A_2)$ and both graph polynomials P_{G_1} and P_{G_2} belong to a certain (ring) ideal, then P_G belongs to the ideal, too. We may assume that the vertices of G are $v_1, v_2, \ldots, v_n$ and that for G_i we use the labels of the vertices after the identification. Then $P_{G_1} = (x_{v_1} - x_{v_2})PA$ and $P_{G_2} = (x_{v_1} - x_{v_3})QA$, where A is the graph polynomial of the common part (after identification). If $P_G = (-1)^{\epsilon}(x_{v_2} - x_{v_3})PQA$, then $P_G = (-1)^{\epsilon'}(PP_{G_1} - QP_{G_2})$ and we are done. However, after the identification the term $x_{v_2} - x_{v_3}$ may occur in P_{G_1} or P_{G_2}, and then $P_G \neq (-1)^{\epsilon}(x_{v_2} - x_{v_3})PQA$. The proof of Theorem 3.19 given by Lovász [701] is purely algebraic.

For the graph constructions considered in this chapter various different names are used in the graph theory literature. For instance, the Hajós join is often called the Hajós sum and sometimes even the Dirac-Hajós construction (see [526]), and the Dirac join is briefly called just join as it is very commonly used also in other branches of graph theory.

The proofs of Theorems 4.11 and 4.12 are from the paper by Stiebitz and Toft [977]. Aboulker, Aubian, and Charbit [13] extended Theorem 4.12 to acyclic colorings of digraphs except for digraphs with $\vec{\chi} = 3$.

Let $\mathcal{L}_k$ denote the class of graphs G satisfying $\bar{\kappa}'(G) \leq k$. It is well known that membership in $\mathcal{L}_k$ can be tested in polynomial time. It is also easy to show that there is a polynomial-time algorithm that, given a graph $G \in \mathcal{L}_k$, decides whether G or one of its blocks belong to $\text{Con}(k+1)$. So it can be tested in polynomial time whether a graph $G \in \mathcal{L}_k$ satisfies $\chi(G) \leq k$. Moreover, the proof of Theorem 4.12

yields a polynomial-time algorithm that, given a graph $G \in \mathcal{L}_k$, finds a coloring of $CO(G,k)$ when such a coloring exists. This result provides an affirmative answer to a conjecture proposed by Aboulker et al. [15, Conjecture 1.8]. The case $k = 3$ was solved by Alboulker et al. [15], the case $k \geq 4$ was obtained in [977].

Theorem 4.63 *For fixed $k \geq 3$ there is a polynomial-time algorithm which, given a graph G with $\overline{\kappa}'(G) \leq k$, finds a k-coloring or a block belonging to* $\text{Con}(k+1)$.

Much attention has been paid to the problem of how to decompose a critical graph into smaller critical graphs. Clearly, to assemble and disassemble critical graphs are two sides of the same coin. Gallai's beautiful result about indecomposable critical graphs (Theorem 4.13) was originally formulated for vertex-critical graphs. That the two theorems are equivalent follows from Proposition 4.51.

Theorem 4.64 (Gallai) *If G is an indecomposable k-vertex-critical graph, then $|G| \geq 2k-1$.*

Let G be a graph. Clearly, G is decomposable if and only if its complement $\overline{G}$ is disconnected. For a vertex set $X \subseteq V(G)$, we obtain that X is an independent set (respectively, a clique) of G if and only if X is a clique (respectively, an independent set) of $\overline{G}$. Obviously, $H = \overline{G}$ if and only if $G = \overline{H}$. A k**-co-coloring** of G is a map $\varphi : V(G) \to [1,k]$ such that $\varphi^{-1}(c)$ is a clique of G for every $c \in [1,k]$. The **co-chromatic number** of G, denoted by $\overline{\chi}(G)$, is the least integer k for which G admits a k-co-coloring. A graph G is called (vertex) k**-co-critical** if $\overline{\chi}(G-v) < k \leq \overline{\chi}(G)$ for every vertex $v \in V(G)$. A mapping $\varphi : V(G) \to [1,k]$ is a k-co-coloring of G if and only if φ is a k-coloring of $\overline{G}$. Consequently, $\overline{\chi}(G) = \chi(\overline{G})$, and G is k-co-critical if and only if the complement $\overline{G}$ is k-vertex-critical. Therefore, Theorem 4.64 is equivalent to the following result.

Theorem 4.65 (Gallai) *If G is a connected k-co-critical graph, then $|G| \geq 2k-1$.*

There are three different proofs of Gallai's result in the graph theory literature. Gallai's original proof is an application of matching theory to co-critical graphs, so he first proved Theorem 4.65 and obtained Theorem 4.64 as a corollary. A second proof of Gallai's result was given by Molloy [753]. Molloy also applies matching theory to the complement of vertex-critical graphs; he uses Berge's version of Tutte's perfect matching theorem. A third proof was given by Stehlík [958]. He also prefers to work with co-critical graphs, but his proof is self-contained and uses no matching theory. Our proof of Theorem 4.13 is an adoption of Stehlík's proof, but deals directly with critical graphs instead of complements. Almost the same proof can be used to prove a corresponding decomposition result for hypergraphs that are critical with respect to the chromatic number; the hypergraph result was obtained by Stiebitz, Storch, and Toft [974]. Recently, Stehlík [960] proved a decomposition result for digraphs that are critical with respect to the dichromatic number (see Theorem 5.43); and Postel, Schweser, and Stiebitz [818] obtained a decomposition result for multigraphs that are critical with respect to the point partition number (see Theorem 5.41). Stehlík's proof for digraphs uses again matching theory. However, both results are immediate consequences of the decomposition result for hypergraphs, see Section 7.9 for a short proof of these two decomposition results. Our proof of Gallai's decomposition result, in particular its extension to hypergraphs, provides a general framework for proving decomposition results for graphs, multigraphs, multidigraphs, and hypergraphs that are critical with respect to various coloring parameters. A likewise general approach for the proof of Gallai-type decomposition results was presented by Sebő [922]. An interesting extension of Gallai's decomposition result to hypergraphs that are critical with respect to the transversal number was obtained by Stehlík [959].

The proofs of Lemma 4.21 and Theorem 4.22 follow Greenwell and Lóvasz [431]. The characterization of the possible subgraphs of k-critical graphs was obtained, independently, by Müller [764, 765]. He proved the following remarkable result.

Theorem 4.66 (MÜLLER) *Let r, k and ℓ be integers with $k, \ell \geq 3$ and $r \geq 1$. Let X be a finite set and let $\varphi_1, \varphi_2, \ldots, \varphi_r$ be r nonequivalent colorings with color set $[1,k]$ of an edgeless graph with vertex set X. Then there is a k-chromatic graph G such that $X \subseteq V(G)$, G has girth $g(G) \geq \ell$, and each coloring of $\mathcal{CO}(G,k)$ is equivalent to a coloring $\phi_i \in \mathcal{CO}(G,k)$ such that $\phi_i|_X = \varphi_i$ $(i \in [1,r])$.*

The proof of Theorem 4.22 gives an upper bound of order $O(k^{|H|^2})$ for the number of vertices in a k-critical graph containing a given graph H. Toft [1023] observed that for any $(k-2)$-colorable graph H there is a k-critical graph G such that $H \subseteq G$ and $|G| \leq 2|H| + c$ for a constant c depending only on k (see Exercise 4.14). Stiebitz [967] posed the following question.

Problem 4.67 (STIEBITZ) *If H is a subgraph of some k-critical graph ($k \geq 4$), is it then true that H is a subgraph of a k-critical graph G, where $|G| \leq c_k|H|$ for some constant depending only on k.*

The proof of Theorem 4.22 implies that a graph is a subgraph of some k-critical graph if and only if it is an induced subgraph of some k-critical graph. Another strengthening of this result was obtained by Sachs and Stiebitz [885, 886, 887]. They proved that if $d \geq 1$ is an integer, and G' is $(k-1)$-colorable graph such that the contraction of any edge of G' results in a $(k-1)$-colorable graph, then there is a k-critical graph G, such that the high vertex subgraph satisfies $G_H = G'$, the low vertex subgraph G_L is nonempty and has the same number of component as G_H, and $d_G(v) \geq d$ for every vertex $v \in V(G')$.

In 1982, Stiebitz [965] proved the following result dealing with connectivity properties of critical graphs, thus giving an affirmative answer to a question proposed by T. Gallai.

Theorem 4.68 (STIEBITZ) *If G is a k-critical graph different from K_k and X is a nonempty set of low vertices of G, then* $\mathrm{com}(G-X) \leq \mathrm{com}(G[X])$.

It seems unlikely, that Theorem 4.68 has a counterpart for the list chromatic number. On the other hand, Aboulker and Vermande [17] extended Theorem 4.68 to digraphs that are critical with respect to the dichromatic number.

Gallai's question whether every graph $G \in \mathrm{Crit}(k,n)$ satisfies $\Bbbk_{k-1}(G) \leq n$ became known with Stiebitz's paper [967] from 1987, which solved the case $k = 4$ using linear algebra. Köster [615] proved that every 4-critical planar graph G of order $n \geq 6$ has at most $n-1$ triangles, which is best possible as shown by the odd wheels. In 1995, Abbott and Zhu [11] made a decisive breakthrough by replacing the class $\mathrm{Crit}(k)$ by the class of k-cluster graphs. This allowed them to provide a positive solution to Gallai's question for all $k \geq 5$, see Lemma 4.28. Our proof of this lemma uses ideas from the paper by Abbott and Zhu [11]. In this paper from 1992, they also posed the question whether every graph $G \in \mathrm{Crit}(k,n)$ with $n \geq k+2$ satisfies $\Bbbk_{k-1}(G) \leq n-k+3$. Abbott and Zhu solved the case $k = 4$ by showing that every graph in $\mathrm{Crit}(4)$ is either a wheel or has at most $n-2$ triangles. The case $k \leq 7$ was solved by Su [982]; she proposed the following conjecture.

Conjecture 4.69 (SU) *Any k-critical graph G of order n with $n > k \geq 4$ has an edge that is contained in at most one complete subgraph K_{k-1} of G.*

It is noteworthy that Su's conjecture easily gives a positive answer to the question posed by Abott and Zhu (using Lemma 4.26(e), Lemma 4.28 and Exercise 4.15). Kézdy and Snevily [570] formulated the problem by Abbott and Zhu as a conjecture; they also proved that every graph $G \in \mathrm{Crit}(k,n)$ with $n > k \geq 4$ satisfies $\Bbbk_{k-1}(G) \leq n - 3k/5 + 2$. The conjecture was finally solved by Gao and Ma [403] in 2023, see Theorem 4.29. This result is slightly stronger than the result by Gao and Ma; however the proof is based on the same ideas as used by Gao and Ma. Let us conclude the discussion with a suggestion made by Gao and Ma.

Problem 4.70 (GAO AND MA) *Is it true that if $G \in \mathrm{Crit}(k,n)$ with $n > k \geq 4$ satisfies $\Bbbk_{k-1}(G) = n+k-3$ and $n-k+3$ is odd, then $G = W(n-k+3, k-3)$?*

The elegant proof of Theorem 4.34 is due to Rödl [872]. The question in Problem 4.36 is due to Erdős and Hajnal (see e.g. [335]). Rödl's result provides an affirmative answer to this question for $\ell = 4$; all other cases remain unsolved.

The theorems listed in Section 4.5 are in a certain sense negative; they show that the property of being critical is less restrictive than originally expected. For fixed $k \geq 4$ there are many k-critical graphs which significantly differ from one another with respect to a variety of graph properties and parameters. The proof of Lemma 4.37 is from Simonovits' paper [939]. The proof of Theorem 4.38 is from [693]. Lovász determined the order of magnitude for the function $\alpha_k(n)$ (see (4.5)). Simonovits [939] investigated the function $\alpha_k(n,h)$ defined as largest number m such that there exists a k-critical graph G on n vertices having an independent set I such that $|I| = m$ and $d_G(v) \geq h$ for all $v \in I$ ($4 \leq k \leq h+1 \leq n$), in particular, $\alpha_k(n) = \alpha(k,k-1)$. Simonovits proved by means of his splitting lemma that

$$\alpha_k(n,h) \leq \tfrac{1}{2} \sqrt[k-1]{(k-2)!nh} \text{ and } \alpha_4(n,h) \leq c_1(nh)^{2/5}.$$

He also constructed k-critical graphs with many independent vertices of high degree. In particular, he proved that if n is large enough and even, then $\alpha_4(n,h) \geq n - 20\sqrt{nh}$.

The corresponding bound by Toft [1021] is $\alpha_4(n,h) \geq n - 3\sqrt{(h-2)n}$ for infinitely many values of n.

The proofs of Theorems 4.39 and 4.40 are from [1021]. The proof of Theorem 4.41 is from [1027], the idea due to V. Rödl. Clearly, the result can be extended easily to k-critical graphs, using the Dirac join. The construction for 4-critical graphs was obtained by V. Rödl during a visit to Odense University in the spring of 1984. The critical graphs constructed in this way look very much random.

The proof of Theorem 4.42 about the circumference of critical graphs is from Gallai's paper [397]. Gallai used his characterization of k-critical graphs having at most one high vertex (see Exercises 4.5 and 4.6) to establish an upper bound for the function $L_k(n)$ for infinitely many values of n. The proof of Lemma 4.46 is due to Shapira and Thomas [935]. However, the special case $p = 0$ of this lemma is implicit in [53]. Alon, Krivelevich, and Seymour [53] used DFS trees to establish a lower bound for $L_k(n)$. In 2011, Shapira and Thomas [935] presented the following result saying that

$$\frac{\log n}{100 \log k} \leq L_k(n) \leq \frac{2(k-1)}{\log(k-2)} \log n + 2k.$$

whenever $k \geq 4$ and $n \geq k+2$. This result would solve the problem of determining the order of magnitude of the function $L_k(n)$ for every fixed $k \geq 4$; however the proof of the lower bound is problematic as we shall now explain. The upper bound follows by a minor modification of Gallai's construction presented in Theorem 4.42, see Exercise 4.7. So let us focus on the lower bound of $L_k(n)$. By Corollary 4.49, to prove the lower bound for $L_k(n)$ it suffices to analyse critical graphs that are not 3-connected. For such critical graphs one can apply the decomposition result by Gallai and Dirac, see Theorem 4.6. Shapira and Thomas use the concept of tree decomposition of graphs, to manage the decomposition of a critical graph that is not 3-connected.

Let G be an arbitrary graph. A **tree decomposition** of G is a pair $(T, \mathcal{W})$ where T is a tree and $\mathcal{W} = (W_t : t \in V(T))$ is a family of subsets of $V(G)$ such that the following two conditions hold:

- $V(G) = \bigcup_{t \in V(T)} W_t$ and every edge of G has both ends in some W_t, and
- if $t, t', t'' \in V(T)$ and t' belongs to the path tTt'', then $W_t \cap W_{t''} \subseteq W_{t'}$.

For $t \in V(T)$, the **torso** of $(G, T, \mathcal{W})$ at t is defined to be the graph with vertex set W_t in which two distinct vertices $u, v \in W_t$ are adjacent if and only if $uv \in E(G)$, or $u, v \in W_{t'}$ for some neighbor t' of t in T. The tree decomposition $(T, \mathcal{W})$ of G is said to be **standard** if $|W_t \cap W_{t'}| = 2$ for every edge $tt' \in E(T)$ and each torso of $(G, T, \mathcal{W})$ is either 3-connected or a cycle. The following result is well known, see Diestel [286, Exercise 12.20].

Lemma 4.71 *Every 2-connected graph has a standard tree-decomposition.*

By Theorem 4.6, if G is a critical graph and S is a separating vertex set of size at most 2, then S is an independent set of G with $|S| = 2$ and $G - S$ has exactly two components. This implies the following result from [935].

Lemma 4.72 *Let $k \geq 4$ be an integer. If G is a k-critical graph and $(T, \mathcal{W})$ is a standard tree-decomposition of G with $\mathcal{W} = (W_t : T \in V(T))$, then the following statements hold:*

(a) *If $tt' \in E(T)$, then $S = W_t \cap W_{t'}$ is an independent set of G with $|S| = 2$.*
(b) *If t, t' are distinct neighbors of t'' in T, then $W_t \cap W_{t''} \neq W_{t'} \cap W_{t''}$.*

Theorem 4.6 leads to the following construction. Let G be a k-critical graph with $k \geq 4$, and let $(T, \mathcal{W})$ be a standard tree-decomposition of G. Let $t \in V(T)$, and let $u, v \in W_t$ be distinct vertices. We call uv a **virtual edge** of W_t if $W_t \cap W_{t'} = \{u, v\}$ for some neighbor t' of t in T. By Lemma 4.72(b), the virtual edges of W_t are pairwise distinct, and they are not edges of G, but they are edges of the torso at t. Now let uv be a virtual edge of W_t, and let t' be the neighbor of t in T such that $W_t \cap W_{t'} = \{u, v\}$. Then $S = \{u, v\}$ is a separating set of G and there are two induced subgraphs G_1 and G_2 of G as in Theorem 4.6(a)(b)(c). Then W_t belongs to exactly one of the sets $V(G_1)$ and $V(G_2)$; if $W_t \subseteq V(G_1)$, then we say that the virtual edge uv is **additive**, otherwise we say that it is **contractive**. We now define the graph N_t as the graph obtained from $G[W_t]$ by adding the edge uv for every additive virtual edge uv of W_t, and identifying the vertices u and v for every contractive virtual edge uv of W_t. So N_t is obtained from the torso of $(G, T, \mathcal{W})$ at t by contracting all contractive virtual edges of W_t. The graph N_t is said to be the **nucleus** of $(G, T, \mathcal{W})$ at t. One crucial point in the proof is the claim that each nucleus N_t is a k-critical graph [935, Lemma 3.4(ii)] and so each torso at t is 3-connected [935, Lemma 3.6]. The problem, however, is with the first claim and the final part of its proof. To explain our concern take the following example. Let $G = K_4 \nabla K_4$ be the 4-critical graph obtained by the Hajós's join from two complete 4-graphs, that is, G is obtained from two complete 4-graphs on vertices a, b, c, d and a, x, y, z, respectively (with one vertex a in common), deleting edges ab and ax and adding bx. Thus G is one of the two 4-critical graphs on 7 vertices. A standard tree-decomposition $(T, \mathcal{W})$ of G has a tree T with three vertices t, t', t'' and two edges tt' and $t't''$, where $W_t = \{a, b, c, d\}$, $W_{t'} = \{a, b, x\}$, and $W_{t''} = \{a, x, y, z\}$. The torso at t' has only three vertices and is a K_3. The nucleus at t' is not 4-critical and the torso at t' is not 3-connected. This is in conflict with the statements in [935, Lemma 3.4(ii)] and [935, Lemma 3.6]. Also lager similar examples may be given. We may overlook something, maybe something obvious. And even if our concern is valid it may be possible to remedy. At present we consider the lower bound $\log n/(100 \log k)$ for $L_k(n)$ as a conjecture.

Proposition 4.54 and its proof is a special case of a more general statement obtained by Toft [1026, Lemma 1], the proof of Theorem 4.55 is also from [1026]. The short proof of Theorem 4.57 saying that K_5 is the only double-critical 5-chromatic graph is from [968]. The double critical graph conjecture (Conjecture 4.56) is an attractive case of a more general conjecture proposed by Erdős [334] at the graph theory conference in Tihany (Hungary) in 1966, and published in 1968 in the proceedings of the conference. Jensen and Toft [530, Problem 5.12] included this conjecture in their collection of graph coloring problems under the name Erdős-Lovász Tihany Conjecture.

Conjecture 4.73 (Erdős–Lovász Tihany Conjecture) *Let $s,t \geq 2$ be integers and let G be a graph with $\omega(G) < \chi(G) = s+t-1$. Then there exists a vertex partition (G_1, G_2) of G such that $\chi(G_1) \geq s$ and $\chi(G_2) \geq t$.*

The conjecture seems difficult to attack and might even be false. There are only a few pairs (s,t) with $s \leq t$ for which the conjecture is settled (in the affirmative), namely the pairs $(2,2), (2,3)$ (easy to prove), $(2,4)$ (Mozhan [763], Stiebitz [968]), $(3,3)$ (Brown and Jung [181]), and $(3,4), (3,5)$ (Stiebitz [969]). In 2008, Kostochka and Stiebitz [634] verified the conjecture for the class of line graphs of multigraphs. In 2009, Balogh, Kostochka, Prince, and Stiebitz [76] established the conjecture for quasi-line graphs and graphs with independence number $\alpha = 2$. Recently, Chudnovsky et al. [226] established the conjecture for the class of claw-free graphs, and Song [951] proved the conjecture for graphs with forbidden holes. Another relaxed version of the conjecture was obtained by Stiebitz [972]. He proved that if a graph G satisfies $\omega(G) < \chi(G) = s+t-1$ with $s,t \geq 2$, then there are two disjoint subgraphs G_1 and G_2 such that $\chi(G_1) \geq s$ and $\text{col}(G_2) \geq t$, or $\text{col}(G_1) \geq s$ and $\chi(G) \geq t$.

The proofs of Theorems 4.60 and 4.61 are from Toft's survey [1029] about Hadwiger's conjecture. Bollobás, Catlin, and Erdős [129] proved that Hadwiger's conjecture is true for almost every graph, moreover, they characterized Hadwiger's conjecture as one of the deepest unsolved problems in graph theory. Over the years, this conjecture has attracted much attention, and the number of publications related to Hadwiger's conjecture has increased immensely. Several survey papers help readers to keep track of the subject. The first summary published in 1996 by Toft [1029] contains a comprehensive survey of the history of the conjecture and the many partial results that have been obtained. The survey of Kawarabayashi [559] from 2015 focus also on the algorithmic aspects of Hadwiger's conjecture and on several variants of the conjecture. There is an excellent 2016 survey of Hadwiger's conjecture by Seymour [931], which discusses, in particular, the numerous weakenings and variations that have been obtained over the years. In 1984, Kostochka [619] and, independently, Thomason [999] proved that every graph in $\text{Forb}_{\preceq}(K_t)$ is $O(t\sqrt{\log t})$-colorable. In fact, both Kostochka and Thomason proved the stronger result that every graph in $\text{Forb}_{\preceq}(K_t)$ is $O(t\sqrt{\log t})$-degenerate, which improves earlier bounds obtained by Wagner [1058] and Mader [711]. A short, self-contained proof of this result has recently been given by Alon, Krivelevich, and Sudakov [55]. The advantage of this result is that it also gives a bound for the DP-chromatic number and, in particular, for the list chromatic number. On the other hand, it is known that Hadwiger's conjecture does not hold for the list chromatic number. As proved by Barát, Joret, and Wood [82], for large enough t, there a graphs in $\text{Forb}_{\preceq}(K_t)$ having list chromatic number larger than $4(t-3)/3$; this bound was improved by Steiner [964] to $(2-o(1))t$. The Kostochka–Thomason bound for the degeneracy of graphs in $\text{Forb}_{\preceq}(K_t)$ is tight as proved by Fernandez de la Vega [371] and by Kostochka [619]. It took more than three decades, before the Kostochka–Thomason bound for the chromatic number of graphs in $\text{Forb}_{\preceq}(K_t)$ was improved, see the paper by Postle [821]. The current state of the art is the upper bound $O(t \log\log t)$ for the chromatic number of graphs in $\text{Forb}_{\preceq}(K_t)$, obtained by Delcourt and Postle [274]; see also the paper by Steiner

[963]. The **Hadwiger number** of a graph G is defined as

$$\omega_{\preceq}(G) = \max\{n \mid K_n \preceq G\}.$$

So Hadwiger's conjecture is equivalent to the statement that every graph G satisfies $\chi(G) \leq \omega_{\preceq}(G)$. In 2004, Reed and Seymour [858] proved Hadwiger's conjecture for line graphs of multigraphs. That Hadwiger's conjecture holds for quasi-line graphs was proved in 2008 by Chudnovsky and Fradkin [224]. For claw-free graphs and for graphs with independence number $\alpha = 2$, Hadwiger's conjecture remains unsolved. However, Chudnovsky and Fradkin [225] proved that every claw-free graph in $\mathrm{Forb}_{\preceq}(K_t)$ is $O(t)$-colorable. Such a linear bound for graphs with $\alpha = 2$ is not known. On the other hand, this special case of Hadwiger's conjecture has attracted a lot of attention; see the papers [122], [165], [187], [190], [191], [241], [649], [801], [813] and [1091]. Since 2003 it has been shown, for a list of 33 graphs H, that every graph in $\mathrm{Forb}(\overline{K}_3, H)$ satisfies Hadwiger's conjecture, see Carter [191] for an overview.

Over the time, several enhancements of Hadwiger's conjecture were proposed in the literature, see e.g. the recent papers by Kriesell [650], Martinsson and Steiner [719] and Steiner [963]. An example from the early days is a conjecture, which became known under the name Hajós conjecture, saying that any k-chromatic graph contains a subdivision of K_k, i.e., $K_k \preceq_t G$. Hajós did not formulate the conjecture in writing, and he may have proposed it only for $k = 5$. Clearly, Hajós' conjecture implies Hadwiger's conjecture, and for $k \leq 4$ the two conjectures are indeed equivalent. However, Hajós' conjecture was disproved by Catlin [194] in 1979 for all k greater than 6 (see also Exercise 1.27). Whether Hajós conjecture holds for graphs G with $5 \leq \chi(G) \leq 6$ is still open. Erdős and Fajtlowicz [342] proved that Hajós' conjecture fails for almost all graphs. On the other hand, Kühn and Osthus [660] proved that all graphs of girth at least 186 satisfy Hajós' conjecture. Thomassen [1016] relates Hajós' conjecture to Ramsey theory, perfect graphs, and the maximum cut problem, thereby obtaining new classes of explicit counterexamples.

Let us conclude this chapter with an attractive formulation of Hadwiger's conjecture proposed by Saaty and Kainen [880] (see Exercise 4.32).

Metaconjecture 1 (Saaty and Kainen) *The largest number of colors needed to color any graph in a minor closed class $\mathcal{P}$ is precisely the largest number of vertices in a complete graph belonging to the class $\mathcal{P}$.*

In [880] the metaconjecture is formulated for the class of graphs closed under subdivision and its opposite operation, but in this formulation the statement is equivalent to Hajós' conjecture and thus false.

Chapter 5
Critical Graphs with few Edges

This chapter is concerned with the minimum number $\mathrm{ext}(k,n)$ of edges in k-critical graphs with n vertices. Brooks' theorem says that $2\mathrm{ext}(k,n) \geq (k-1)n+1$ for $n > k \geq 4$. That it is worthwhile to study critical graphs, and especially the function $\mathrm{ext}(k,n)$, was first emphasized by G. A. Dirac in his thesis, and subsequently by T. Gallai and O. Ore. In 2014, A. V. Kostochka and M. Yancey succeeded in determining the best linear approximation to the function $\mathrm{ext}(k,n)$ (see Sections 5.4 and 5.5); their result is a major breakthrough in the study of critical graphs. In particular, they proved that $\mathrm{ext}(4,n) = \lceil \frac{1}{3}(5n-2) \rceil$ (Theorem 5.10). As observed by Kostochka and Yancey this result immediately implies Grötzsch's theorem that any planar triangle-free graph is 3-colorable, thus yielding a short selfcontained proof of this fundamental result.

5.1 Preliminaries and Ore's Conjecture

One not surprising feature of critical graphs is the fact that their edge numbers increase with their order. It is an interesting task to investigate the extremal function $\mathrm{ext}(\cdot,\cdot)$ defined by

$$\mathrm{ext}(k,n) = \min\{|E(G)| \mid G \in \mathrm{Crit}(k,n)\}$$

and the corresponding set of extremal graphs defined by

$$\mathrm{Ext}(k,n) = \{G \in \mathrm{Crit}(k,n) \mid |E(G)| = \mathrm{ext}(k,n)\},$$

where k and n are positive integers. Over the years, much attention has been paid to this task. If $k \leq 3$, both $\mathrm{ext}(k,n)$ and $\mathrm{Ext}(k,n)$ are known, since in this case the only critical graphs are the complete graphs and the odd cycles; in particular, $\mathrm{Ext}(3,n) = \mathrm{Crit}(3,n) = \{C_n\}$ if n is odd and $n \geq 3$.

If $k \geq 4$, then there are k-critical graphs with n vertices if and only if $n \geq k$ and $n \neq k+1$ (see Exercise 4.3). Since every k-critical graph has minimum degree at least $k-1$, we obtain the trivial lower bound $2\ \mathrm{ext}(k,n) \geq (k-1)n$, and Brooks'

M. Stiebitz et al., *Brooks' Theorem*, Springer Monographs in Mathematics,
https://doi.org/10.1007/978-3-031-50065-7_5

theorem is equivalent to the fact that equality holds if and only if $n = k$. Evidently, $\mathrm{Ext}(k,k) = \mathrm{Crit}(k,k) = \{K_k\}$. In 1957 Dirac [298] proved that

$$2\,\mathrm{ext}(k,n) \geq (k-1)n + k - 3 \text{ for } n \geq k+2, \tag{5.1}$$

thus yielding the first improvement of Brooks' theorem. The first improvement (for n large enough) of Dirac's bound was established by Gallai [397] in 1963; he proved that

$$2\,\mathrm{ext}(k,n) \geq (k-1+\frac{k-3}{k^2-3})n \text{ for } n \geq k+2. \tag{5.2}$$

This result is a consequence of Gallai's characterization of the low vertex subgraph of a critical graph (Theorem 1.33), see Section 5.2 for a proof of Gallai's bound.

A proof of Dirac's bound seems to be more difficult. To prove this bound we can analyze the excess of a critical graph. For a graph G and an integer k, let $\mathrm{exc}_k(G) = 2|E(G)| - (k-1)|G|$ be the k-**excess** of G. Clearly,

$$\mathrm{exc}_k(G) = \sum_{v \in V(G)} ((d_G(v) - (k-1)),$$

and for any graph G with $\delta(G) \geq k-1$, we obtain that $\mathrm{exc}_k(G) \geq 0$, where equality holds if and only if G is k-regular. So if G is a k-critical graph with $k \geq 4$, then $\mathrm{exc}_k(G) \geq 0$, where equality holds if and only if $G = K_k$. Dirac's theorem is then equivalent to the statement that $\mathrm{exc}_k(G) \geq k-3$ for every k-critical graph $G \neq K_k$, where $k \geq 4$. On the one hand, Dirac's result implies Brooks' theorem. On the other hand, Dirac's bound is sharp.

To see that Dirac's bound is sharp, we introduce the following class of critical graphs. For an integer $k \geq 3$, let $\mathcal{DG}'(k)$ denote the family of all graphs G whose vertex set consists of four nonempty pairwise disjoint sets X_1, X_2, Y_1 and Y_2 with $|X_1| + |X_2| = |Y_1| + |Y_2| = k-1$ and $|X_2| + |Y_2| \leq k-1$ and an additional vertex v such that $X = X_1 \cup X_2$ and $Y = Y_1 \cup Y_2$ are cliques in G, $N_G(v) = X_1 \cup Y_1$, and a vertex $x \in X$ is joined to a vertex $y \in Y$ if and only if $x \in X_2$ and $y \in Y_2$. Furthermore, let $\mathcal{DG}(k)$ denote the family of all such graphs with $|X_2| = 1$ or $|Y_2| = 1$ (see Exercise 1.21). Obviously, any graph G of $\mathcal{DG}'(k)$ has order $|G| = 2k-1$ and independence number $\alpha(G) = 2$, which implies that $\chi(G) \geq k$. Furthermore, it is easy to check that the deletion of any edge of G results in a $(k-1)$-colorable graph, which implies that G is k-critical (by Proposition 1.29). If $|X_i| = p_i$ and $|Y_i| = q_i$, then $p_1 + p_2 = q_1 + q_2 = k-1$, $2 \leq p_2 + q_2 \leq k-1$, and for the k-excess of G we obtain that

$$\begin{aligned}\mathrm{exc}_k(G) &= \sum_{u \in X_2 \cup Y_2} (d_G(u) - (k-1)) + (d_G(v) - (k-1)) \\ &= p_2(q_2-1) + q_2(p_2-1) + (p_1 + q_1 - (k-1)) \\ &= p_2(q_2-2) + q_2(p_2-2) + k - 1 \geq k-3,\end{aligned}$$

where equality holds if and only if $p_2 = 1$ or $q_2 = 1$. So any graph $G \in \mathcal{DG}'(k)$ satisfies $\mathrm{exc}_k(G) \geq k-3$ and equality holds if and only if $G \in \mathcal{DG}(k)$. Note that a graph $G \in \mathcal{DG}'(k)$ belongs to $\mathcal{DG}(k)$ if and only if G has a separating vertex set of size 2; furthermore, any graph in $\mathcal{DG}(k)$ is a Gallai join and can be obtained from two disjoint complete graphs K_k by splitting a vertex of the first complete graph into an edge of the second complete graph (see Theorems 4.6, 4.7 and Exercise 1.21).

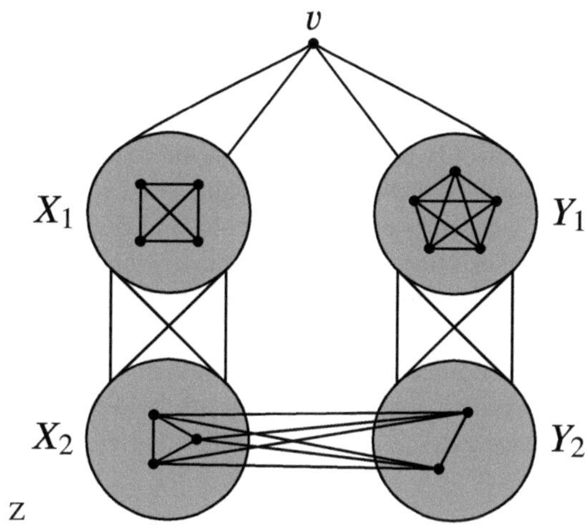

Fig. 5.1 The graph in $\mathcal{DG}'(k)$ with $k=8$, $|X_2|=3$ and $|Y_2|=2$

If G is a k-critical graph which is obtained from two disjoint k-critical graphs G_1 and G_2 by splitting a vertex of G_1 into an edge of G_2, then for the excess of G we obtain that $\text{exc}_k(G) = \text{exc}_k(G_1) + \text{exc}_k(G_2) + k - 3$, since $|G| = |G_1| + |G_2| - 1$ and $|E(G)| = |E(G_1)| + |E(G_2)| - 1$. This is in particular true if $G = G_1 \nabla G_2$ is a Hajós join of G_1 and G_2. So if we take a Hajós join of a graph belonging to $\text{Ext}(k,n)$ and a graph belonging to $\text{Ext}(k,n'+1)$, we obtain a k-critical graph of order $n+n'$ and with $\text{ext}(k,n) + \text{ext}(k,n'+1) - 1$ edges, and so

$$\text{ext}(k,n+n') \le \text{ext}(k,n) + \text{ext}(k,n'+1) - 1. \tag{5.3}$$

By Fekete's lemma (Lemma C.44), this implies that

$$\lim_{n\to\infty} \frac{\text{ext}(k,n)}{n}$$

exists and is finite for all $k \ge 4$.

We next establish an upper bound $\text{ore}(k,n)$ for $\text{ext}(k,n)$, where $k \ge 4$. Define $\text{ore}(k,k) = \binom{k}{2}$. If $n \ge k+2$ and $n \equiv p+1 \pmod{(k-1)}$, where $2 \le p \le k$, define

$$c_{k,p} = \text{ext}(k,k+p) - \tfrac{1}{2}(k - \frac{2}{k-1})(k+p) \tag{5.4}$$

and

$$\text{ore}(k,n) = \tfrac{1}{2}(k - \frac{2}{k-1})n + c_{k,p} = \tfrac{1}{2}(k - 1 + \frac{k-3}{k-1})n + c_{k,p}. \tag{5.5}$$

We claim that for all feasible values of n, there is a k-critical graph of order n with $\text{ore}(k,n)$ edges, implying that $\text{ext}(k,n) \le \text{ore}(k,n)$. This is obviously true if $n = k$ or $n = k + p$ with $2 \le p \le k$, since then $\text{ore}(k,n) = \text{ext}(k,n)$. Thus $\text{ore}(k,n)$ depends on the values of $\text{ext}(k,n)$ for small n. If G is a k-critical graph with $n \ge k+2$ vertices and $\text{ore}(k,n)$ edges, then a Hajós join of G and K_k results in a k-critical graph (by Theorem 4.2(c)) with $n' = n + k - 1$ vertices and $m = |E(G)| + \binom{k}{2} - 1$ edges. Since $n' \equiv n \,(\text{mod } (k-1))$ and $|E(G)| = \text{ore}(k,n)$, an easy calculation shows that $m' = \text{ore}(k,n) + \binom{k}{2} - 1 = \text{ore}(k, n+k-1)$, which proves the claim. We call $\text{ore}(k,n)$ the **Ore-function**. Note that the Ore-function is not linear in n, since the additive constant $c_{k,p}$ depends not only on k, but also on the residue class $p+1$ of n (mod $(k-1)$). In Section 5.3 we shall establish the values $\text{ext}(k, k+p)$ for $2 \le p \le k-1$. That $\text{ext}(k, 2k) = k^2 - 3$ was proved by Kostochka and Stiebitz [630]. This enables us to give an explicit description of the constant $c_{k,p}$ and so $\text{ore}(k,n)$ is a known function (it was not known to Ore since he did not know the value of $\text{ext}(k,2k)$ not even for $k = 4$).

Ore [794] proposed the conjecture that a Hajós join of an extremal graph belonging to $\text{Ext}(k,n)$ and a K_k results in an extremal graph belonging to $\text{Ext}(k, n+k-1)$. This suggests that we have $\text{ext}(k,n) = \text{ore}(k,n)$ for all integers n and k which satisfy $n \ge k$, $n \ne k+1$, and $k \ge 4$.

5.2 List-Critical Graphs with Few Edges

In this section we prove Gallai's bound (5.2) for the number of edges in critical graphs, moreover, we extend this bound to the class of list-critical graphs.

A list-assignment L of a graph G may be considered also as a list-assignment for every subgraph of G. So we denote the restriction of L to a subgraph H of G also by L and say that L is a list-assignment for H. If G is L-colorable, then each of its subgraphs is L-colorable, too. In this sense, L-colorability is a monotone graph property.

Let G be a graph, and let L be a list-assignment of G. The **L-core** of G is the unique maximal subgraph H of G satisfying $d_H(v) \ge |L(v)|$ for all $v \in V(H)$. Clearly, H is an induced subgraph of G (possibly empty). If L is a k-assignment, the corresponding term is **k-core**. Note that, for $k \ge 1$, the k-core of a graph G is empty if and only if G is $(k-1)$-degenerate.

Proposition 5.1 *Let G be a graph and let L be a list-assignment for G. Then G is L-colorable if and only if the L-core of G is L-colorable.*

Proof The "only if" part is evident. For the proof of the "if" part suppose that G is not L-colorable. Then there is a smallest subgraph H of G which is not L-colorable. Then H is nonempty and $H - v$ is L-colorable for every vertex v of H. Since H is not L-colorable, we conclude that $d_H(v) \ge |L(v)|$ for all $v \in V(H)$. Otherwise we could extend an L-coloring of $H - v$ to an L-coloring of H, a contradiction. Hence

H is a subgraph of the L-core of G and, therefore, the L-core is not L-colorable, a contradiction. □

A graph G is called **L-critical** or **list-critical**, where L is a list-assignment of G, if every proper subgraph of G is L-colorable, but G itself is not L-colorable. Furthermore, a graph G is called **k-list-critical** if there is a $(k-1)$-list-assignment L for G for which G is L-critical.

Note that any list-critical graph is connected. Furthermore, a connected graph is list-critical if and only if it is not list-colorable, but every edge deleted subgraph is list-colorable. From Proposition 1.29 we conclude that a graph G is k-critical ($k \geq 1$) if and only if G is L-critical for the constant list-assignment L with $L(v) = [1, k-1]$ for all $v \in V(G)$. So any k-critical graph is k-list-critical. The following example shows that the converse statement is not true. This example also shows that a k-list-critical graph may contain another k-list-critical graph as a proper subgraph. For k-critical graphs, however, this does not occur.

Remark 5.2 *Let G be the graph obtained from two disjoint copies K and K' of the complete graph K_k of order $k \geq 3$ by adding an edge joining a vertex $u \in V(K)$ with a vertex $u' \in V(K')$. Furthermore, let L be the $(k-1)$-assignment for G defined by*

$$L(v) = \begin{cases} \{1, 2, \ldots, k-1\} & \text{if } v \in V(G) \setminus \{u, u'\}, \\ \{2, 3, \ldots, k\} & \text{if } v \in \{u, u'\}. \end{cases}$$

Then G is L-critical and hence k-list-critical, but not k-critical. Furthermore, $\chi_\ell(G) = \chi_\ell(K_k) = k$ and K_k is k-list-critical.

Proof Since K and K' are complete graphs both of order k, in any L-coloring of $K \cup K'$, both vertices u and u' receive color k. Hence G is not L-colorable. If e is an edge of G, then the L-core of $G - e$ is either K or K' or $K \cup K'$. From Proposition 5.1 it then follows that $G - e$ is L-colorable. Consequently, G is L-critical. Since $\Delta(G) \leq k$, Theorem 1.11 implies that $\chi_\ell(G) \leq k$ and hence $\chi_\ell(G) = k$. Obviously, $\chi_\ell(K_k) = k$ and K_k is k-list-critical. □

Let G be a graph and let L be a list-assignment for G. Then G is not L-colorable if and only if G contains an L-critical subgraph. Hence, for $k \geq 1$, $\chi_\ell(G) \geq k$ if and only if G contains a k-list-critical subgraph. Obviously, for $k \in \{1, 2\}$, the complete graph K_k is the only k-list critical graph. By a **minimal k-list-critical** graph we mean a k-list-critical graph that does not contain any k-list-critical graph as a proper subgraph.

Proposition 5.3 *For $k \geq 1$, a graph G is minimal k-list-critical if and only if $\chi_\ell(H) < \chi_\ell(G) = k$ for every proper subgraph H of G.*

Proof First, assume that G is minimal k-list-critical. Then every subgraph of G, except G itself, is $(k-1)$-list-colorable. Hence $\chi_\ell(G) \geq k > \chi_\ell(H)$ for every proper subgraph H of G and we only need to show that $\chi_\ell(G) = k$. To this end, let L be an arbitrary k-assignment for G. Then let $u \in V(G)$ and $c \in L(u)$. Deleting color c from each list $L(v)$ with $v \in N_G(u)$ results in a list-assignment L' of $G' = G - u$ such that $|L'(v)| \geq k-1$ for all $v \in V(G')$. Since $\chi_\ell(G') \leq k-1$, G' has an L'-coloring which can be extended to an L-coloring of G by assigning color c to u. This proves that $\chi_\ell(G) = k$.

Now, assume that $\chi_\ell(H) < \chi_\ell(G) = k$ for every proper subgraph H of G. Then there is a $(k-1)$-assignment L for which G is not L-colorable, but every proper subgraph of G is L-colorable. Since every proper subgraph of G is $(k-1)$-list-colorable, this implies that G is minimal k-list critical. □

The following theorem was obtained in 1997 by Thomassen [1012, Theorem 4.2] and, independently, by Kostochka, Stiebitz, and Wirth [636, Theorem 2].

Theorem 5.4 *Let G be a graph and let L be a list-assignment of G such that G is L-critical. Let $U = \{u \in V(G) \mid d_G(u) > |L(v)|\}$ and let $W = V(G) \setminus U$. Then $G[W]$ is a Gallai forest (possibly empty) and $d_G(w) = |L(w)|$ for all $w \in W$.*

Proof Let G' be a component of $G[W]$ and let $H = G - V(G')$. Since G is L-critical, there is an L-coloring φ of H. For a vertex v of G' let $N_v = N_G(v) \cap V(H)$ and let $L'(v) = L(v) \setminus \varphi(N_v)$. Then L' is a list-assignment of G' such that $|L'(v)| \geq |L(v)| - |\varphi(N_v)| \geq d_G(v) - |N_v| = d_{G'}(v)$ for all $v \in V(G')$. If G' has an L'-coloring φ', then $\varphi \cup \varphi'$ would be an L-coloring of G, which is impossible. So G' has no L'-coloring and hence (G', L') is an uncolorable pair. From Theorem 1.7 it then follows that G' is a Gallai tree and $|L'(v)| = d_{G'}(v)$ for all $v \in V(G')$, which implies that $|L(v)| = d_{G'}(v) + |N_v| = d_G(v)$ for all $v \in V(G')$. This completes the proof. □

Proposition 5.5 *Let G be a k-list-critical graph, let $U = \{u \in V(G) \mid d_G(u) \geq k\}$ and let $W = V(G) \setminus U$. Then $G[W]$ is a Gallai forest and $d_G(w) = k-1$ for all $w \in W$. If $U = \varnothing$, then $G = K_k$, or $k = 3$ and G is an odd cycle. Furthermore, if $G[W]$ contains a K_k, then $G = K_k$.*

Proof By assumption, there is a $(k-1)$-assignment L of G such that G is L-critical. Then Theorem 5.4 implies that $G[W]$ is a Gallai forest and $d_G(w) = k-1$ for all $w \in W$. Now assume that $U = \varnothing$. Then $G = G[W]$ and, since G is L-critical, G is connected. Therefore, G is a $(k$-$1)$-regular Gallai tree. Since every block of a Gallai tree is regular, this implies that G itself is a block. Hence G is a brick, that is, $G = K_k$, or $k = 3$ and G is and odd cycle. If $G[W]$ contains a K_k, then $d_G(v) = k-1$ for all $v \in V(K_k)$. Hence K_k is a component of G. Since G is connected, this implies that $G = K_k$. □

Let G be a k-list-critical graph with $k \geq 4$. Then $2|E(G)| \geq (k-1)|G|$ and equality holds if and only if G is a K_k. This follows from Proposition 5.5 and is, in fact, equivalent to Theorem 1.11, the choosability version of Brooks's theorem. That the excess $2|E(G)| - (k-1)|V|$ tends to infinity if $|G|$ tends to infinity is a simple consequence of Proposition 5.5 and the following two results due to Gallai [397].

Proposition 5.6 *Let T be a Gallai tree different from K_k and with maximum degree $\Delta(T) \leq k-1$, where $k \geq 4$. Then*

$$(k-2+\frac{2}{k-1})|T| - 2|E(T)| \geq 2.$$

Proof Throughout the proof, let $\mathcal{T}$ denote the set of Gallai trees distinct from the complete graph K_k and with maximum degree at most $k-1$. Furthermore, let

$$r = (k - 2 + \frac{2}{k-1})$$

and, for $T \in \mathcal{T}$, let

$$\sigma(T) = r|V(T)| - 2|E(T)| = \sum_{v \in V(T)} (r - d_T(v)).$$

Let $T \in \mathcal{T}$ be a Gallai tree. We prove by induction on the number of blocks of T that $\sigma(T) \geq 2$. If $T = B$ consists of one block B, then either B is a complete graph K_b with $1 \leq b \leq k-1$ and we obtain that

$$\sigma(B) = b(r - b + 1) \begin{cases} \geq r \text{ if } 1 \leq b \leq k-2, \\ = 2 \text{ if } b = k-1, \end{cases}$$

or B is an odd cycle of order at least five and we obtain that $\sigma(B) \geq (r-2)5 \geq r$. Now suppose that T has at least two blocks. We consider two cases.

Case 1: *T has an end-block B distinct from K_{k-1}.* Let $T' = T - (V(B) \setminus \{v\})$ where v is the only separating vertex of T contained in B. Since T' and B have only vertex v in common and $T = T' \cup B$, we obtain that $\sigma(T) = \sigma(T') + \sigma(B) - r$. Since B belongs to $\mathcal{T}$ and is not a K_{k-1}, we conclude that $\sigma(B) \geq r$ and hence $\sigma(T) \geq \sigma(T')$. From the induction hypothesis it then follows that $\sigma(T) \geq 2$.

Case 2: *Every end-block of T is a K_{k-1}.* Choose an arbitrary end-block B of T and let v be the only separating vertex of T contained in B. Then B is a K_{k-1} and, since T has at least two end-blocks and $\Delta(T) \leq k-1$, the vertex v is contained in a block B' where B' is a K_2. Then $T' = T - V(B)$ belongs to $\mathcal{T}$ and we obtain that $\sigma(T) = \sigma(T') + \sigma(B) - 2$. Since B is K_{k-1}, this yields $\sigma(B) = 2$ and hence $\sigma(T) = \sigma(T')$. From the induction hypothesis it then follows that $\sigma(T) \geq 2$.

This completes the proof of the induction step and hence the proof of the proposition. □

Proposition 5.7 *Let $k \geq 3$ be an integer and let G be a k-list-critical graph with $G \neq K_k$. Let $U = \{u \in V(G) \mid d_G(u) \geq k\}$ and $W = \{w \in V \mid d_G(w) = k-1\}$. Furthermore, let c be a constant satisfying $0 \leq c \leq k - \frac{2}{k-1}$, and define*

$$\sigma = (k - 2 + \frac{2}{k-1})|W| - 2|E(G[W])|,$$

and

$$\varrho_c = 2|E(G[U])| + (k - c - \frac{2}{k-1}) \sum_{u \in U} (d_G(u) - k).$$

Suppose that $\sigma + \varrho_c \geq c|U|$. Then $2|E(G)| \geq g_k(|G|, c)$, where

$$g_k(n, c) = (k - 1 + \frac{k-3}{(k-c)(k-1) + k - 3})n.$$

Proof For $X \subseteq V(G)$, let $e(X) = |E(G[X])|$. By Proposition 5.5, $V(G) = U \cup W$ and $|G| = |U| + |W|$. On one hand, since $\sigma + \varrho_c \geq c|U|$ and $d_G(w) = k-1$ for all $w \in W$, we obtain

$$\begin{aligned}
2|E(G)| &= 2e(U) + 2(k-1)|W| - 2e(W) \\
&= 2e(U) + \sigma + (k - \frac{2}{k-1})|W| \\
&\geq 2e(U) + c|U| - \varrho_c + (k - \frac{2}{k-1})|W| \\
&\geq c|U| - (k - c - \frac{2}{k-1}) \sum_{u \in U} (d_G(u) - k) + (k - \frac{2}{k-1})|W| \\
&= c|G| - (k - c - \frac{2}{k-1}) \sum_{u \in U} (d_G(u) - k) + (k - c - \frac{2}{k-1})|W|.
\end{aligned}$$

On the other hand,

$$2|E(G)| = k|G| - |W| + \sum_{u \in U} (d_G(u) - k).$$

Adding the first inequality to the second equality multiplied with $(k - c - \frac{2}{k-1})$ leads to

$$2|E(G)|(1 + k - c - \frac{2}{k-1}) \geq (c + k(k - c - \frac{2}{k-1}))|G|.$$

Since $k - c - \frac{2}{k-1} \geq 0$, the last inequality is equivalent to

$$2|E(G)| \geq (k - 1 + \frac{k-3}{(k-c)(k-1) + k - 3})|V| = g_k(|V|, c).$$

This completes the proof. □

Theorem 5.8 *Suppose that G is a k-list-critical graph distinct from K_k where $k \geq 4$. Then*

$$2|E(G)| \geq (k - 1 + \frac{k-3}{(k^2-3)})|G| = g_k(|G|, 0).$$

Proof Let $U = \{u \in V(G) \mid d_G(u) \geq k\}$ and $W = \{w \in V(G) \mid d_G(w) = k-1\}$. If $W = \varnothing$, then $2|E(G)| \geq k|G| \geq g_k(|G|, 0)$ and we are done. Otherwise, by Proposition 5.5, $G[W]$ is a nonempty Gallai forest not containing K_k as a subgraph. Let

$$\sigma = (k - 2 + \frac{2}{k-1})|W| - 2|E(G[W])|$$

and

$$\varrho_0 = 2|E(G[U])| + (k - \frac{2}{k-1}) \sum_{u \in U} (d_G(u) - k).$$

Since $\Delta(G[W]) \leq k-1$, it follows from Proposition 5.6 that $\sigma \geq 2$. Since $\varrho_0 \geq 0$, Proposition 5.7 implies that $2|E(G)| \geq g_k(|G|, 0)$. □

5.3 Extremal Graphs whose Order is Close to χ

Theorem 5.9 (GALLAI) *Let k and p be integers satisfying $2 \le p \le k-1$. Then*

$$2\text{ext}(k, k+p) = (k-1)(k+p) + p(k-p) - 2$$

and $\text{Ext}(k, k+p) = K_{k-p-1} \boxplus \mathcal{DG}(p+1)$.

Proof For $2 \le p \le k-1$, let $\mathcal{G}(k,p) = K_{k-p-1} \boxplus \mathcal{DG}(p+1)$ and

$$e(k,p) = (k-1)(k+p) + p(k-p) - 2.$$

Obviously, we have $\mathcal{G}(k,p) \subseteq \text{Crit}(k, k+p)$ and an easy calculation shows that every graph G in $\mathcal{G}(k,p)$ satisfies $2|E(G)| = e(k,p)$. Hence for the proof it suffices to show that $\text{Ext}(k, k+p) \subseteq \mathcal{G}(k,p)$. To do this, we use induction on k. The statement is evident for $k=3$, since then $p=2$ and C_5 is the only 3-critical graph of order 5, which gives $\mathcal{G}(3,2) = \mathcal{DG}(3) = \{C_5\} = \text{Crit}(3,5) = \text{Ext}(3,5)$.

Now, assume that $k \ge 4$ and $G \in \text{Ext}(k,n)$ with $n = k+p$ and $2 \le p \le k-1$. First, consider the case that G has a dominating vertex, that is, $G = K_1 \boxplus G'$. Then it follows from Theorem 4.1 that $G' \in \text{Crit}(k', n')$ with $k' = k-1$ and $n' = n-1$. Note that $n' = k' + p$. Furthermore, we have $G' \in \text{Ext}(k', n')$. For otherwise, there would be a graph $H' \in \text{Crit}(k', n')$ with $|E(H')| < |E(G')|$. Then $H = K_1 \boxplus H'$ would belong to $\text{Crit}(k,n)$ (by Theorem 4.1) with $|E(H)| < |E(G)|$, a contradiction to the choice of G. If $p \le k-2$, then $n \le 2k-2$ and $n' \le 2k'-1$. Hence, the induction hypothesis implies that $G' \in \mathcal{G}(k', p) = K_{k-p-2} \boxplus \mathcal{DG}(p+1)$. This implies that $G = K_1 \boxplus G'$ belongs to $\mathcal{G}(k,p) = K_{k-p-1} \boxplus \mathcal{DG}(p+1)$ as required. Otherwise, we have $p = k-1$ and $n = 2k-1$. Then we obtain that

$$2|E(G)| \ge (k-1)n + (n-k) = 2k^2 - 2k = e(k, k-1) + 2,$$

which is impossible. It remains to consider the case that G has no dominating vertex. We distinguish two cases.

Case 1: $p \le k-2$. Then we have $n \le 2k-2$ and it follows from Theorem 4.13 that G is decomposable. Hence $t = \text{com}(\overline{G})$ satisfies $t \ge 2$ and we have

$$G = G_1 \boxplus G_2 \boxplus \cdots \boxplus G_t,$$

where $\overline{G_1}, \overline{G_2}, \ldots, \overline{G_t}$ are the components of $\overline{G}$. For $i \in [1,t]$, let $k_i = \chi(G_i)$ and $n_i = |G_i|$. By Theorem 4.1, we obtain that

$$k = k_1 + k_2 + \cdots + k_t \text{ and } G_i \in \text{Crit}(k_i, n_i) \text{ for } i \in [1,t].$$

Since $\overline{G_i}$ is connected, Theorem 4.18 (respectively, Theorem 4.13) implies that $|G_i| = n_i \ge 2k_i - 1$ for $i \in [1,t]$. Since $\text{Crit}(1) = \{K_1\}$, $\text{Crit}(2) = \{K_2\}$ and G has no dominating vertex, it follows that $k_i \ge 3$ for all $i \in [1,t]$. Let

$$S = \{i \in [1,t] \mid n_i = 2k_i - 1\}$$

and let $T = [1,t] \setminus S$. Note that if $i \in T$, then $n_i \geq 2k_i$. If $|S| \leq 1$, then we obtain that

$$n = \sum_{i=1}^{t} n_i \geq \Big(\sum_{i=1}^{t} 2k_i\Big) - 1 \geq 2k - 1,$$

which is impossible. Consequently, we have $|S| \geq 2$. By symmetry, we may assume that $\{1,2\} \subseteq S$. Hence $n_i = 2k_i - 1$ and $k_i \geq 3$ for $i \in \{1,2\}$. Since $G \in \mathrm{Ext}(k,n)$, we obtain (as before) that $G_i \in \mathrm{Ext}(k_i, n_i)$ for $i \in \{1,2\}$. The induction hypothesis then implies that $G_i \in \mathcal{G}(k_i, k_i - 1)$ and $2|E(G_i)| = e(k_i, k_i - 1) = 2k_i^2 - 2k_i - 2$ for $i \in \{1,2\}$. Furthermore, we obtain that $G' = G_1 \boxplus G_2$ satisfies $G \in \mathrm{Crit}(k', n')$ with $k' = k_1 + k_2$ and $n' = n_1 + n_2$. Since $G \in \mathrm{Ext}(k,n)$, it follows that $G' \in \mathrm{Ext}(k', n')$. On the one hand, we have that

$$2|E(G')| = 2|E(G_1) + 2|E(G_2)| + 2n_1n_2 = 2k_1^2 + 2k_2^2 - 6(k_1 + k_2) + 8k_1k_2 - 2.$$

On the other hand, for $p' = n' - k'$ we have $p' = k_1 + k_2 - 2$, which gives

$$\begin{aligned} e(k', p') &= (k_1 + k_2 - 1)(2k_1 + 2k_2 - 2) + (k_1 + k_2 - 2)2 - 2 \\ &= 2k_1^2 + 2k_2^2 + 4k_1k_2 - 2(k_1 + k_2) - 4. \end{aligned}$$

Since $k_1 \geq 3$ and $k_2 \geq 3$, this leads to

$$2|E(G')| - e(k', p') = 4k_1k_2 - 4(k_1 + k_2) + 2 \geq 2,$$

which is impossible. This settles Case 1.

Case 2: $p = k - 1$. Then $n = 2k - 1$ and our aim is to show that $G \in \mathcal{G}(k, k-1) = \mathcal{DG}(k)$. Since $G \in \mathrm{Crit}(k)$ and $k \geq 4$, G is 2-connected (by Corollary 1.31). If G has a separating vertex set $U = \{u, v\}$, then it follows from Theorem 4.6 that U is an independent set of G and there are two graphs G_1 and G_2 such that we have: $G = G_1 \cup G_2$, $V(G_1 \cap G_2) = U$, both graphs $G_1' = G_1 + uv$ and $G_2' = G_2/U$ are k-critical, and $N_G(u) \cap N_G(v) \cap V(G_2) = \varnothing$. Then $|G_i'| \geq k$ for $i \in \{1,2\}$ and $|G_1'| + |G_2'| = |G| + 1 = 2k$, which implies that $|G_i'| = k$ and so $G_i' = K_k$ for $i \in \{1,2\}$. Consequently, $G \in \mathcal{DG}(k)$ as required.

It remains to consider the case when G is 3-connected. Since $G \in \mathrm{Ext}(k,n)$ and $n = 2k - 1$, we obtain that $G \neq K_k$ and $\omega(G) \leq k - 1$. Furthermore, G has at least one high vertex. If G has no low vertices, then $2|E(G)| \geq kn > (k-1)n + (k-3) = e(k,p)$, which is impossible. So we may assume that G has low vertices. Then there are edges uv of G such that u is a low vertex and v is a high vertex of G; we call such an edge an *(l,h)-edge*.

First we claim that the low vertex subgraph G_L of G has clique number $\omega(G_L) \leq k - 2$. Suppose this is false and let X be a clique of size $k - 1$ in G_L. Then each vertex v of X has exactly one neighbor z_v in $I = N_G(X) \setminus X$, which implies that $|I| \leq k - 1$. The graph $G' = G - X$ has order $|G'| = |G| - |X| = k$, which implies that I is a separating vertex set of G. Since G is 3-connected, $|I| \geq 3$ and so $|G' - I| \leq k - 3$. Since G is critical, $\chi(G') \leq k - 1$. As $|G' - I| \leq k - 3$, this implies that there is a coloring φ of G' with color set $C = [1, k-1]$ such that $|\varphi(I)| \geq 2$. Now let L be the list assignment of the complete graph $G[X] = K_{k-1}$ defined by $L(v) = C \setminus \varphi(z_v)$ for all $v \in X$. Then $|L(v)| = k-2 = d_{G[X]}(v)$ for all $v \in X$. Since L is not the constant list assignment, it follows from Theorem 1.7(b) that $G[X]$ has an L-coloring φ'. Then $\varphi \cup \varphi'$ is a coloring of G with color set C, and so $\chi(G) \leq |C| = k-1$, which is impossible. This proves the claim that $\omega(G_L) \leq k-2$.

Next we claim that G has an (l,h)-edge uv such that $N_G(u) \setminus \{v\}$ is not a clique of G. Suppose this is false. Clearly, G contains an (l,h)-edge uv. Then $X = N_G(u) \setminus \{v\}$ is a clique of G and so is $X' = X \cup \{u\}$. Since $|X'| = k-1$ and $\omega(G_L) \leq k-2$, it then follows that X' contains a high vertex v'. Clearly, $v' \notin \{v, u\}$ and so uv' is an (l,h)-edge, which implies that $N_G(u) \setminus \{v'\}$ is a clique of G. Since $\omega(G) \leq k-1$, we then conclude that $U = \{v, v'\}$ is an independent set in G. This implies that if $w \in N_G(u) \setminus U$, then $N_G(u) \setminus \{w\}$ is not a clique and so w is a low vertex of G. Therefore, for $Y = N_G(u) \cup \{u\}$ we obtain that $G[Y] = K_k - vv'$ and each vertex of $Y \setminus U$ is a low vertex. This implies that U is a separating set, which is impossible. This prove the claim that there is an (l,h) edge uv of G such that $X = N_G(u) \setminus \{v\}$ is not a clique of G.

Since G is k-critical, $\chi(G-u) \leq k-1$. This implies that there is a coloring φ of G with a set of k colors such that $\varphi^{-1}(c) = \{u\}$ for some color $c \in C$. Clearly, color $c' = \varphi(v)$ is different from c. Let I be a maximal independent set of G containing the color class $\varphi^{-1}(c')$. Since G has no dominating vertex, $|I| \geq 2$. Clearly, $\chi(G) \leq \chi(G-I)+1$ and, since G is k-critical, we then conclude that $\chi(G-I) = k-1$. Then $G-I$ contains a $(k-1)$-critical subgraph H (by Proposition 1.27(a)). Since $\chi(G - \varphi^{-1}(c,c')) = k-2$ (by Proposition 1.34(c)), we conclude that u belongs to H. Since $\delta(H) \geq k-2$ (by Theorem 1.32), and since $d_G(u) = k-1$ and $v \notin V(H)$, we conclude that $X = N_G(u) \setminus \{v\}$ belongs to H. Since X is not a clique, this implies that $H \neq K_{k-1}$ and so $|H| \geq k+1$. The set $Y = V(G) \setminus V(H)$ contains I and so $|Y| \geq |I| \geq 2$. Since $|H| = |G| - |Y| \leq 2k-3$, we get $|H| = (k-1)+p'$ with $2 \leq p' \leq k-2$. Subtracting the excess

$$\mathrm{exc}_{k-1}(H) = 2|E(H)| - (k-2)|H| = \sum_{x \in V(H)} (d_H(x) - (k-2))$$

of H from the excess

$$\mathrm{exc}_k(G) = 2|E(G)| - (k-1)n = \sum_{x \in V(G)} (d_G(x) - (k-1))$$

of G leads to

$$\mathrm{exc}_k(G) - \mathrm{exc}_{k-1}(H) = \sum_{x \in V(H)} (d_G(x) - d_H(x) - 1) + \sum_{x \in Y} (d_G(x) - (k-1)).$$

Denote the first sum by $a(H)$ and the second sum by $b(Y)$, so that

$$\mathrm{exc}_k(G) = \mathrm{exc}_{k-1}(H) + a(H) + b(Y).$$

By the induction hypothesis, we obtain that

$$\mathrm{exc}_{k-1}(H) \geq e(k-1, p') - (k-2)|H| = p'(k-1-p') - 2,$$

where equality holds if and only if $H \in K_{k-2-p'} \boxplus \mathcal{DG}(p'+1)$. Since the high vertex v of G belongs to Y, we obtain $b(Y) \geq 1$. Since the set Y contains the maximal independent set I of G, we have $d_G(x) \geq d_H(x)+1$ for all $x \in V(H)$ and so $a(H) \geq 0$. This gives $a(H)+b(Y) \geq 1$. First assume that $a(H)+b(Y) \geq 2$. This leads to

$$\mathrm{exc}_k(G) \geq \mathrm{exc}_{k-1}(H) + 2 \geq p'(k-1-p') \geq k-2,$$

since $2 \leq p' \leq k-2$ and $k \geq 4$. Then $2|E(G)| \geq (k-1)n + (k-2) = e(k,p)+1$, which is impossible. So we are done in this case. It remains to consider the case when $a(H)+b(Y) = 1$, which gives $a(H) = 0$ and $b(Y) = 1$. Then we obtain that

$$\mathrm{exc}_k(G) = \mathrm{exc}_{k-1}(H) + 1 \geq p'(k-1-p') - 1.$$

Since $2 \leq p' \leq k-2$, this gives $\mathrm{exc}_k(G) \geq k-2$, unless $p' = k-2$ and $\mathrm{exc}_k(G) = k-3$. From $p' = k-2$ it follows that $|H| = 2k-3$ and $|Y| = 2$. This implies that $Y = I$ and so $I = \{v, v'\}$, where $v \neq v'$, $d_G(v) \geq k$ and $d_G(v') \geq k-1$. Since I is an independent set of G and $|H| = 2k-3$, we then conclude that v and v' have in G a common neighbor x belonging to H. This gives $d_G(x) \geq d_H(x)+2$ and so $a(H) \geq 1$, a contradiction. So we are also done if $a(H)+b(Y) = 1$. Thus the proof is complete. □

5.4 A Bound for 4-Critical Graphs

For 4-critical graphs it is known that $\mathrm{ext}(4,4) = 6$, $\mathrm{ext}(4,7) = 11$ and $\mathrm{ext}(4,8) = 13$ (see Figure 5.2). By (5.3), we obtain that $\mathrm{ext}(4, n+3) \leq \mathrm{ext}(4,n)+5$. From this we conclude that $\mathrm{ext}(4,n) \leq \lceil \frac{1}{3}(5n-2) \rceil$ and Ore's conjecture suggests that equality holds. That this is indeed the case was proved in 2014 by Kostochka and Yancey [639].

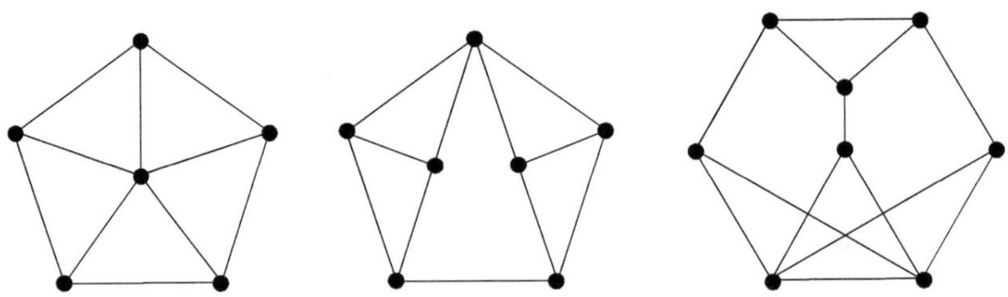

Fig. 5.2 The graphs in $\mathrm{Ext}(4,n)$ for $n = 6, 7$ and 8.

Theorem 5.10 (Kostochka and Yancey) *Every 4-critical graph of order n has at least $\frac{1}{3}(5n-2)$ edges, and as a consequence* $\text{ext}(4,n) = \lceil \frac{1}{3}(5n-2) \rceil$.

The proof is based on the so-called potential method introduced by Montassier, Ossona de Mendez, Raspaud, and Zhu in [758]. Let G be an arbitrary graph. We denote by $e(G)$ the number of edges of G. For vertex sets $U, W \subseteq V(G)$, let $e_G(U,W) = |E_G(U,W)|$ and $e_G(U) = |E_G(U,U)|$. For a vertex set $X \subseteq V(G)$, we define the **potential** of X in G by

$$\rho_G(X) = 5|X| - 3e_G(X)$$

and set $\rho(G) = \rho_G(V(G))$. For the complete graph K_n we have $\rho(K_n) = 5n - 3\binom{n}{2}$, which in particular leads to

$$\rho(K_1) = 5, \rho(K_2) = 7, \rho(K_3) = 6, \text{ and } \rho(K_4) = 2. \tag{5.6}$$

The potential function is monotone in the following sense.

$$\text{If } F \subseteq E(G) \text{ and } G' = G - F, \text{ then } \rho(G') = \rho(G) + 3|F| \geq \rho(G). \tag{5.7}$$

Let G be a 4-critical graph, let $R \subset V(G)$, and let $\varphi \in CO(G[R],3)$ be a coloring of $G[R]$ with color set $\{1,2,3\}$. Let $X = \{v_1, v_2, v_3\}$ be a set of three new vertices, and let G' be the graph obtained from $G - R$ by adding the vertex set X such that $G'[X] = K_3$, $E(G') = E(G'[X]) \cup E(G-R) \cup F$, and

$$F = \{uv_i \mid u \in V(G) \setminus R, i \in \{1,2,3\}, N_G(u) \cap \varphi^{-1}(i) \neq \varnothing\}.$$

So G' is obtained from G by deleting all edges from $G[R]$, contracting the color class $\varphi^{-1}(i)$ to the vertex v_i for $i \in \{1,2,3\}$, and adding all possible edges between the vertices of X. In what follows we write $G' = G/(R, \varphi, X)$.

Proposition 5.11 *Let G be a 4-critical graph, let $R \subset V(G)$, let $\varphi \in CO(G[R],3)$ be a coloring, and let $G' = G/(R,\varphi,X)$ with $X = \{v_1, v_2, v_3\}$. Then $\chi(G') \geq 4$. Furthermore, if $W \subseteq V(G')$ and $U = (W \setminus X) \cup R$, then*

$$\rho_G(U) = \rho_{G'}(W) - \rho_{G'}(W \cap X) + \rho_G(R) + 3s, \tag{5.8}$$

with $s = e_{G'}(W \setminus X, W \cap X) - e_G(W \setminus X, R) \leq 0$.

Proof First suppose, to the contrary, that $\chi(G') \leq 3$. Then there is a coloring $\phi \in CO(G',3)$ and, as $G'[X] = K_3$, we may assume that $\phi(v_i) = i$ for $i \in \{1,2,3\}$. By construction of G', this implies that the map $\tilde{\varphi} = \phi|_{V(G')\setminus X} \cup \varphi$ belongs to $CO(G,3)$, which is impossible. This proves the claim that $\chi(G') \geq 4$.

Now let $W \subseteq V(G')$ and $U = (W \setminus X) \cup R$. Then $|U| = |W| + |R| - |W \cap X|$ and $e_G(U) = e_G(R) + e_G(W \setminus X) + e_G(W \setminus X, R)$. As $G' - X = G - R$, we obtain that $e_G(W \setminus X) = e_{G'}(W \setminus X)$. Consequently, for

$$s' = \rho_G(U) - \rho_G(R) - \rho_{G'}(W) + \rho_{G'}(W \cap X)$$

we obtain that

$$\begin{aligned} s' &= 3(-e_G(U)+e_G(R)+e_{G'}(W)-e_{G'}(W\cap X)) \\ &= 3(-e_{G'}(W\setminus X)-e_G(W\setminus X,R)+e_{G'}(W)-e_{G'}(W\cap X)) \\ &= 3(e_{G'}(W\setminus X,W\cap X)-e_G(W\setminus X,R)). \end{aligned}$$

By construction of G', it follows that $e_{G'}(W\setminus X,W\cap X)\le e_G(W\setminus X,R)$ and so $s=e_{G'}(W\setminus X,W\cap X)-e_G(W\setminus X,R)\le 0$ and $s'=3s$. This completes the proof of the proposition. □

For the proof of Theorem 5.10 it suffices to show that every graph $G\in \mathrm{Crit}(4)$ satisfies $3e(G)\ge 5|G|-2$, which is equivalent to $\rho(G)=5|G|-3e(G)\le 2$. Hence Theorem 5.10 is a consequence of the following result.

Theorem 5.12 (Kostochka and Yancey) *Every graph $G\in \mathrm{Crit}(4)$ has $\rho(G)\le 2$.*

Proof The proof is by reductio ad absurdum. So we choose a counterexample G whose order is minimum. Then $G\in \mathrm{Crit}(4)$ and $\rho(G)\ge 3$. Moreover, $\rho(\tilde G)\le 2$ for every graph $\tilde G\in \mathrm{Crit}(4)$ with $|\tilde G|<|G|$. To derive at a contradiction we shall prove several claims and then use discharging. As an immediate consequence we obtain the following statement.

Claim 5.12.1 *Let G' be a graph with $|G'|<|G|$ such that $\rho_{G'}(X)\ge 3$ for every set $X\subseteq V(G')$ with $|X|\ge 4$. Then $\chi(G')\le 3$.*

Proof: If $\chi(G')\ge 4$, then G' contains a 4-critical subgraph $\tilde G$. Then $|\tilde G|\ge 4$ and hence $\rho(\tilde G)\ge \rho_{G'}(V(\tilde G))\ge 3$ (by (5.7) and the assumption of the claim), contradicting the choice of G. △

Claim 5.12.2 *If $R\subseteq V(G)$ and $2\le |R|<|G|$, then $\rho_G(R)\ge 6$.*

Proof: Let $\mathcal{R}$ denote the set of subsets R of $V(G)$ such that $2\le |R|<|G|$. Let $R\in\mathcal{R}$ be a set such that $m=\rho_G(R)$ is minimum. If $m\ge 6$ then there is nothing to prove. So assume that $m\le 5$. Then, as $|R|\ge 2$, we conclude from (5.6) and (5.7) that $|R|\ge 4$. Since $|R|<|G|$ and G is 4-critical, there is a coloring $\varphi\in \mathcal{CO}(G[R],3)$. Consequently, the graph $G'=G/(R,\varphi,X)$ satisfies $\chi(G')\ge 4$ (by Proposition 5.11). Then G' contains a 4-critical subgraph $\tilde G$. Let $W=V(\tilde G)$. Since $|R|\ge 4$, we obtain that $|\tilde G|<|G|$, which implies that $\rho(\tilde G)\le 2$ (by the choice of G). Consequently, $\rho_{G'}(W)\le \rho(\tilde G)\le 2$ (by (5.7)). Since $G'-X=G-R$ and $\chi(G-R)\le 3$, we obtain that $W\cap X\ne\varnothing$ and so $1\le |W\cap X|\le 3$. From (5.6) and (5.7) it then follows that $\rho_{G'}(W\cap X)\ge 5$. For the set $U=(W\setminus X)\cup R$ we then obtain that

$$\rho_G(U)\le \rho_{G'}(W)-\rho_{G'}(W\cap X)+\rho_G(R)\le \rho_{G'}(W)-5+m\le m-3$$

(by Proposition 5.11 and (5.8)). Since $|U|>|R|\ge 2$ and $\rho_G(U)<m=\rho_G(R)$, we obtain that $U=V(G)$. This implies that $\rho(G)=\rho_G(U)\le m-3\le 2$, which is impossible. This proves the claim. △

Claim 5.12.3 *Let $v \in V(G)$ and let G' be a graph obtained from a subgraph of $G - v$ by adding at most one edge. Then $\chi(G') \le 3$.*

Proof: Clearly, $|G'| < |G|$. Let $X \subseteq V(G')$ be an arbitrary set with $|X| \ge 4$. Since $|X| < |G|$, it follows from Claim 5.12.2 that $\rho_G(X) \ge 6$. Hence $\rho_{G'}(X) \ge \rho_G(X) - 3 \ge 3$ (by (5.7)). From Claim 5.12.1 it then follows that $\chi(G') \le 3$. △

Claim 5.12.4 *If $R \subseteq V(G)$ and $2 \le |R| < |G|$, then $\rho_G(R) \ge 6$ and equality implies that $G[R] = K_3$.*

Proof: Suppose this is false. Then there is a set $R \subseteq V(G)$ such that $2 \le |R| < |G|$ and $\rho_G(R) \le 6$, but $G[R]$ is not a K_3. Using (5.6) and (5.7), we obtain that $|R| \ge 4$. Since G is 4-critical, G is 2-connected and, as $4 \le |R| < |G|$, there are two vertices $u, v \in R$ both having neighbors in G belonging to $V(G) \setminus R$. By Claim 5.12.3, $\chi(G[R] + uv) \le 3$ and, therefore, there is a coloring $\varphi \in \mathcal{CO}(G[R], 3)$ such that $\varphi(u) \ne \varphi(v)$. Then the graph $G' = G/(R, \varphi, X)$ satisfies $\chi(G') \ge 4$ (by Proposition 5.11). Hence G' contains a 4-critical subgraph $\tilde{G}$. Let $W = V(\tilde{G})$ and let $U = (W \setminus X) \cup R$. Since $|R| \ge 4$, we obtain that $|\tilde{G}| < |G|$, which implies that $\rho(\tilde{G}) \le 2$ (by the choice of G). Consequently, $\rho_{G'}(W) \le \rho(\tilde{G}) \le 2$ (by (5.7)). Since $G' - X = G - R$ and $\chi(G - R) \le 3$, we obtain that $W \cap X \ne \varnothing$ and so $1 \le |W \cap X| \le 3$. If $|W \cap X| \ge 2$, then $\rho_{G'}(W \cap X) \ge 6$ (by (5.6) and (5.7)). Using Proposition 5.11 and (5.8), we then obtain that

$$\rho_G(U) \le \rho_{G'}(W) - \rho_{G'}(W \cap X) + \rho_G(R) \le \rho_{G'}(W) - 6 + 6 = \rho_{G'}(W) \le 2.$$

Since $|U| \ge 2$, Claim 5.12.2 implies that $U = V(G)$ and hence $\rho(G) = \rho_G(U) \le 2$, a contradiction. If $|W \cap X| = 1$, then $\rho_{G'}(W \cap X) = \rho(K_1) = 5$ and

$$s = 3(e_{G'}(W \setminus X, W \cap X) - e_D(W \setminus X, R)) \le -3$$

as $\varphi(u) \ne \varphi(v)$ and both vertices u and v have neighbors in G belonging to $G - R$. Again, using Proposition 5.11 and (5.8), we obtain that

$$\rho_G(U) = \rho_{G'}(W) - \rho_{G'}(W \cap X) + \rho_G(R) + s \le 0.$$

Since $|U| \ge 2$, Claim 5.12.2 implies that $U = V(G)$ and hence $\rho(G) = \rho_G(U) \le 0$, a contradiction. This proves the claim. △

Claim 5.12.5 *G does not contain a K_4^- (a complete graph on 4 vertices with one edge deleted) as a subgraph.*

Proof: Clearly, G is not a K_4, as K_4 is a 4-critical graph with $\rho(K_4) = 2$. Hence $|G| \ge 5$. If $G[R] = K_4^-$, then $\rho_G(R) = 5|R| - 3e_G(R) = 5$, which is impossibly by Claim 5.12.4. This proves the claim. △

Claim 5.12.6 *Each triangle of G contains at most one low vertex of G.*

Proof: Suppose, to the contrary, that G has a triangle K, say with vertices v_1, v_2, v_3, such that both v_1 and v_2 are low vertices of G. Then, for $i \in \{1,2\}$, there is a vertex u_i such that $N_G(v_i) = (V(K) \setminus \{v_i\}) \cup \{u_i\}$. By Claim 5.12.5, $u_1 \neq u_2$. By Claim 5.12.3, $G' = G - v_1 - v_2 + u_1u_2$ satisfies $\chi(G') \leq 3$. Hence there is a coloring $\varphi' \in CO(G - v_1 - v_2, 3)$ such that $\varphi'(u_1) \neq \varphi'(u_2)$. Then φ' can be easily extended to a 3-coloring of G, which is impossible. △

Claim 5.12.7 *Let uv be an edge of the low vertex subgraph G_L. Then both u and v are contained in a triangle.*

Proof: Suppose that u is not in a K_3 and let $N_G(u) = \{v, x, y\}$. Then $xy \notin E(G)$. Let G' be the graph obtained from $G - u - v$ by identifying x and y to a new vertex w. Since a 3-coloring of G' can be extended to a 3-coloring of G, we conclude that $\chi(G') \geq 4$. Hence G' contains a 4-critical subgraph $\tilde{G}$. Since $\tilde{G}$ has fewer vertices than G, this implies that that $W = V(\tilde{G})$ satisfies $\rho_{G'}(W) \leq \rho_{\tilde{G}}(W) = \rho(\tilde{G}) \leq 2$. Since $\tilde{G}$ is not a subgraph of G, the vertex w belongs to W. Then the set $U = (W \setminus \{w\}) \cup \{x, y, u\}$ satisfies $|U| = |W| + 2 < |G|$ and $e_G(U) \geq e_{G'}(W) + 2$, which leads to $\rho_G(U) \leq \rho_{G'}(W) + 10 - 6 \leq 6$. By Claim 5.12.4, this implies that $G[U] = K_3$, a contradiction. Hence the claim is proved. △

From the last two claims it follows that $\Delta(G_L) \leq 1$, that is, any low vertex of G is adjacent to at most one other low vertex of G. A discharging argument may then be used to arrive at a contradiction. The initial charge on each vertex v is the degree $d_G(v)$. Then each vertex of degree at least 4 sends a charge of $\frac{1}{6}$ to each of its neighbors of degree 3. Since $\Delta(G_L) \leq 1$, each vertex of degree 3 gets a new charge of at least $3 + 2\frac{1}{6} = \frac{10}{3}$, and a vertex of degree $d \geq 4$ gets a new charge of at least $d - \frac{1}{6}d = \frac{5}{6}d \geq \frac{10}{3}$. The total amount of charge is $2e(G)$, and hence $2e(G)| \geq \frac{10}{3}|G|$, that is, $\rho(G) = 5|G| - 3e(G)| \leq 0$, contradicting the initial assumption that $\rho(G) \geq 3$. This completes the proof of the theorem. □

As pointed out by Kostochka and Yancey [639] Theorem 5.10 leads to a very short proof of Grötzsch's famous theorem [434] that any planar triangle-free graph is 3-colourable.

Theorem 5.13 (GRÖTZSCH'S THEOREM) *Every triangle-free planar graph G is 3-colorable.*

Proof If this is false, then there is a plane triangle-free graph G with $\chi(G) \geq 4$. Let $n = |V(G)|$, $m = |E(G)|$ and $\ell = |F(G)|$. We may choose G such that $n + m$ is minimum. Then $G \in \text{Crit}(4, n)$. If a face of G is bounded by a 4-cycle $C = (u, v, w, x, u)$, then, for $(a, b) \in \{(u, w), (v, x)\}$, $ab \notin E(G)$ and the graph G_{ab} obtained from G by identifying a and b satisfies $\chi(G_{ab}) \geq 4$. Then it is easy to see that G_{uw} or G_{vx} is triangle-free (see Lemma 9.15), which results in a smaller counterexample. Otherwise each face of G is bounded by a p-cycle with $p \geq 5$. Then $5\ell \leq 2m$ (by (C.5)) and $n - m + \ell = 2$ (by Euler's formula, see C.13), which leads to $m \leq (5n - 10)/3$. However, by Theorem 5.10, $m \geq (5n - 2)/3$, a contradiction. □

5.5 A Bound for k-Critical Graphs

As we have seen in Section 5.1 (see (5.4) and (5.5)), the Ore-function

$$\text{ore}(k,n) = \tfrac{1}{2}(k - \frac{2}{k-1})n + c_{k,p}$$

is an upper bound for $\text{ext}(k,n)$. Using (5.4) and Theorem 5.9, a simple computation shows that

$$c_{k,p} \geq c_{k,k-1} = -\frac{k(k-3)}{2(k-1)}$$

for $2 \leq p \leq k$. For $p = k$ we can use the result by Kostochka and Stiebitz [630] saying that $\text{ext}(k,2k) = k^2 - 3$. We now define the **Kostochka-Yancey** function as

$$\text{koy}(k,n) = \tfrac{1}{2}(k - \frac{2}{k-1})n + c_{k,k-1} = \frac{(k+1)(k-2)n - k(k-3)}{2(k-1)}.$$

In 2014 Kostochka and Yancey [640] proved the following remarkable result, saying that

$$\text{koy}(k,n) \leq \text{ext}(k,n) \leq \text{ore}(k,n) \tag{5.9}$$

whenever $n \geq k \geq 4$ and $n \neq k+1$. If $n \equiv 1 \pmod{(k-1)}$, then $\text{koy}(k,n) = \text{ore}(k,n)$, and so (5.9) implies that $\text{ext}(k,n) = \text{ore}(k,n)$. This solves Ore's conjecture (see the end of Section 5.1) for infinitely many values of n. Since the limit of $\text{ext}(k,n)/n$ exists for any fixed $k \geq 4$ (see Section 5.1), we conclude from (5.9) that

$$\lim_{n\to\infty} \frac{\text{ext}(k,n)}{n} = \frac{1}{2}(k - \frac{2}{k-1}). \tag{5.10}$$

Note that $\text{koy}(4,n) = (5n-2)/3$ and so Theorem 5.10 implies (5.9) for $k = 4$. The proof of (5.9) for $k \geq 5$ is similar to the proof of Theorem 5.10. In particular, we use again the potential method. However, the proof for $k \geq 5$ is much more sophisticated and needs more preparation.

The next lemma is an extension of the simple fact that any super-orientation of a bipartite graph results in a kernel perfect digraph. The two succeeding lemmas, which analyse the structure of k-critical graphs, are needed in the proof of (5.9). All three lemmas are due to Kostochka and Yancey [640].

Lemma 5.14 *Let G be a graph and let I be an independent set of G. Let D be any digraph obtained from G by replacing each edge of $G - I$ by a pair of oppositely directed edges and by orienting each edge of $E_G(I, V(G) \setminus I)$ arbitrarily. Then D is a kernel perfect digraph.*

Proof Clearly, it suffices to show that D has a kernel. The proof is by induction on the order of D. The statement is evident if $I = \varnothing$ as in this case $D = D^{\pm}(G)$, which implies that any maximal independent set of D is a kernel of D. So assume that $I \neq \varnothing$. If every vertex of $D - I$ has an out-neighbor in D belonging to I, then I is a kernel of D and we are done. Otherwise, some vertex v of $D - I$ satisfies $N_D^+(v) \cap I = \varnothing$. By induction, $D' = D - v - N_D^-(v)$ has a kernel S. Then $S \cup \{v\}$ is a kernel of D. □

Lemma 5.15 *Let G be a k-critical graph with $k \geq 4$, and let A and B be two disjoint subsets of $V(G)$ such that at least one of A and B is an independent set of G, $d_G(a) = k - 1$ for every $a \in A$, and $d_G(b) = k$ for every $b \in B$. Then the following statements hold:*

(a) *The bipartite graph H with $V(H) = A \cup B$ and $E(H) = E_G(A, B)$ has $\delta(H) \leq 2$.*
(b) *In G, some vertex of A has at most one neighbor in B, or some vertex of B has at most three neighbors in A.*

Proof Both statements are evident if $A \cup B = \varnothing$. Otherwise, as G is k-critical, there is a coloring φ of $G - (A \cup B)$ with color set $C = [1, k-1]$. Let L be the list-assignment for $G' = G[A \cup B]$ such that $L(v) = C \setminus \varphi(N_G(v) \setminus (A \cup B))$ for all $v \in V(G')$. By the degree assumptions, we then obtain that (1) $|L(a)| \geq d_{G'}(a)$ for all $a \in A$, and (2) $|L(b)| \geq d_{G'}(b) - 1$ for all $b \in B$.

To prove (a), assume that $\delta(H) \geq 3$. Let H' be the graph obtained from H by splitting each vertex $b \in B$ into $\lceil d_H(b)/3 \rceil$ vertices of degree at most 3. Since H is bipartite, H' is bipartite with parts A and B', where B' is the set of vertices obtained from B by the splitting operation. Then $d_{H'}(a) \geq 3$ for all $a \in A$ and $d_{H'}(b) \leq 3$ for all $b \in B$. Then using Hall's theorem (see Theorem C.11), we conclude that H' has a matching M such that $A \subseteq V(M)$, that is, each vertex of A is covered by M. We now construct a super-orientation D of G' as follows. Replace each edge of $G[A]$ or $G[B]$ by two oppositely directed edges. By assumption, one of the two graphs $G[A]$ or $G[B]$ is edgeless. Now orient each edge e of H towards A if e corresponds to an edge in M, and towards B otherwise. By Lemma 5.14, D is a kernel perfect digraph. By (1) it follows that $|L(a)| \geq d_{G'}(a) = d_D^+(a) + 1$ for all $a \in A$, and by (2) it follows that each vertex $b \in B$ satisfies

$$d_D^+(b) \leq d_{G'}(b) - \left\lfloor \tfrac{2}{3} d_H(b) \right\rfloor \leq (|L(b)| + 1) - 2 = |L(b)| - 1.$$

Then Corollary 3.5 implies that G' is L-colorable. But if φ' is an L-coloring of G, then $\varphi \cup \varphi'$ is a $(k-1)$-coloring of G, which is impossible. This proves (a).

To prove (b), assume that in G each vertex $a \in A$ has at least two neighbors in B and each vertex $b \in B$ has at least four neighbors in A. To arrive at a contradiction, we argue similar as in the proof of (a). Let H' be the graph obtained from H by splitting each vertex $b \in B$ into $\lceil d_H(b)/2 \rceil$ vertices of degree at most 2. Since H is bipartite, H' is bipartite with parts A and B', where B' is the set of vertices obtained from B by the splitting operation. Then $d_{H'}(a) \geq 2$ for all $a \in A$ and $d_{H'}(b) \leq 2$ for all $b \in B$. Again, using Hall's theorem (see Theorem C.11), we conclude that H' has a matching M such that $A \subseteq V(M)$. Now we construct a super-orientation D of G' as in the proof of (a). Clearly, D is a kernel perfect digraph. By (1) it follows that $|L(a)| \geq d_{G'}(a) = d_D^+(a) + 1$ for all $a \in A$, and by (2) it follows that each vertex $b \in B$ satisfies

$$d_D^+(b) \leq d_{G'}(b) - \left\lfloor \tfrac{1}{2} d_H(b) \right\rfloor \leq (|L(b)| + 1) - 2 = |L(b)| - 1,$$

as $d_H(b) \geq 4$. Then Corollary 3.5 implies that G' is L-colorable, which leads to a $(k-1)$-coloring of G, a contradiction. This complete the proof of the lemma. □

Lemma 5.16 *Let G be a k-critical graph with $k \geq 4$, and let A and B be two disjoint subsets of $V(G)$ such that at least one of A and B is an independent set of G, $d_G(a) = k-1$ for every $a \in A$, and $d_G(b) = k$ for every $b \in B$. If $|A|+|B| \geq 3$, then* (a) $e_G(A,B) \leq 2(|A|+|B|)-4$ *and* (b) $e_G(A,B) \leq |A|+3|B|-3$.

Proof Let H be the bipartite graph with $V(H) = A \cup B$ and $E(H) = E_G(A,B)$. Then $e(H) = e_G(A,B)$. First we prove (a) by induction on $m = |A|+|B|$. If $m = 3$, then $e(H) \leq 2$ and we are done. Suppose that $m \geq 4$. By Lemma 5.15(a), H has a vertex v with $d_H(v) \leq 2$. Then the induction hypothesis implies that $e(H-v) \leq 2(m-1)-4$, which leads to $e(H) \leq 2+e(H-v) \leq 2m-4$, as claimed. Now we prove (b) by induction on $m = |A|+|B|$. Let us first consider the base case $m = 3$. If $|A| = 3$, then $e(H) = 0 = |A|+3|B|-3$. Otherwise, $|B| \geq 1$, which gives $|A|+3|B| \geq 5$ and hence $e(H) \leq 2 \leq |A|+3|B|-3$. Suppose that $m \geq 4$. By Lemma 5.15(b), there is a vertex $a \in A$ with $d_H(a) \leq 1$ or a vertex $b \in B$ with $d_H(b) \leq 3$. Applying the induction hypothesi to $H-a$ or $H-b$ gives the desired bound for $e(H)$. □

To continue the proof of (5.9), let $k \geq 5$ be a fixed integer. Given a graph G and a vertex set $X \subseteq V(G)$, we define the k-**potential** of X in G by

$$\hat{\rho}_G(X) = (k-2)(k+1)|X| - 2(k-1)e_G(X) \tag{5.11}$$

and set $\hat{\rho}(G) = \hat{\rho}_G(V(G))$. First we list some basic properties of the potential function.

Proposition 5.17 *The following statements hold for the k-potential function $\hat{\rho}$ with $k \geq 5$ and for a graph G:*

(a) $\hat{\rho}(K_1) = (k-2)(k+1), \hat{\rho}(K_2) = 2(k^2-2k-1), \hat{\rho}(K_{k-1}) = 2(k-2)(k-1)$, *and* $\hat{\rho}(K_k) = k(k-3)$.
(b) *If $F \subseteq E(G)$ and $G' = G-F$, then $\hat{\rho}(G') = \hat{\rho}(G)+2(k-1)|F| \geq \hat{\rho}(G)$.*
(c) *If $R \subseteq V(G)$ such that $|R| \geq 2$ and $\hat{\rho}_G(R) \leq (k-2)(k+1)$, then $|R| \geq k$.*
(d) *If $R \subseteq V(G)$ such that $|R| \geq 2$, $\hat{\rho}_G(R) \leq 2(k-2)(k-1)$ and $G[R] \neq K_{k-1}$, then $|R| \geq k$.*
(e) *If $X \subseteq V(G)$ such that $1 \leq |X| \leq k-1$, then $\hat{\rho}_G(X) \geq (k-2)(k+1)$.*
(f) *If $X \subseteq V(G)$ such that $2 \leq |X| \leq k-1$, then $\hat{\rho}_G(X) \geq 2(k-2)(k-1)$.*

Proof Clearly, $\hat{\rho}(K_n) = (k-2)(k+1)n - 2(k-1)\binom{n}{2}$ and a simple calculation yields (a). Statement (b) is evident. If $R \subseteq V(G)$ and $|R| = n$, then (b) implies that $\hat{\rho}_G(R) \geq \hat{\rho}(K_n)$. For fixed k, the function $f(n) = \hat{\rho}(K_n)$ is concave in n, which combined with (b) implies (c), (d), (e) and (f). □

Let G be a k-critical graph, let $R \subset V(G)$, and let $\varphi \in CO(G[R], k-1)$ be a coloring of $G[R]$ with color set $[1, k-1]$. Let $X = \{v_1, v_2, \ldots, v_{k-1}\}$ be a set of $k-1$ new vertices, and let G' be the graph obtained from $G-R$ by adding the vertex set X such that $G'[X] = K_{k-1}$, $E(G') = E(G'[X]) \cup E(G-R) \cup F$, and

$$F = \{uv_i \mid u \in V(G) \setminus R, i \in [1, k-1], N_G(u) \cap \varphi^{-1}(i) \neq \varnothing\}.$$

So G' is obtained from G by deleting all edges from $G[R]$, contracting the color class $\varphi^{-1}(i)$ to the vertex v_i for $i \in [1, k-1]$, and adding all possible edges between the vertices of X. In what follows we write $G' = G/(R, \varphi, X)$. The proof of the following proposition is similar to the proof of Proposition 5.11 and it is left to the reader.

Proposition 5.18 *Let G be a k-critical graph with $k \geq 5$, let $R \subset V(G)$, let $\varphi \in \mathcal{CO}(G[R], k-1)$ be a coloring, and let $G' = G/(R, \varphi, X)$ with $X = \{v_1, v_2, \ldots, v_{k-1}\}$. Then $\chi(G') \geq k$. Furthermore, if $W \subseteq V(G')$ and $U = (W \setminus X) \cup R$, then*

$$\hat{\rho}_G(U) = \hat{\rho}_{G'}(W) - \hat{\rho}_{G'}(W \cap X) + \hat{\rho}_G(R) + 2(k-1)s, \tag{5.12}$$

with $s = e_{G'}(W \setminus X, W \cap X) - e_G(W \setminus X, R) \leq 0$.

The next proposition describe the construction of an auxiliary graph used in the proof of our main result. Let S be a set and let $g : S \to \mathbb{N}$ be a weight function for S. If $X \subseteq S$, we set $g(X) = \sum_{v \in X} g(v)$.

Proposition 5.19 *Let S be a set with $|S| \geq 2$, let $g : S \to \mathbb{N}$ be a weight function with $g(S) \geq k-1$ ($k \geq 5$), and let i be an integer with $1 \leq i \leq (k-1)/2$. Then there exists a graph H with $V(H) = S$ and $e(H) \leq i$ such that every independent set I of H with $|I| \geq 2$ satisfies $g(S \setminus I) \geq i$.*

Proof Assume that $|S| = p$ and let $S = \{v_1, v_2, \ldots, v_p\}$ such that $g(v_1) \geq g(v_2) \geq \cdots \geq g(v_p) \geq 1$. Let $S' = S \setminus \{v_1\}$. Clearly, $g(S') \geq p-1$. First assume that $g(S') \leq i$. Since $2i \leq k-1$, this implies that $g(v_1) \geq i$. Now let $E(H) = \{v_1 v_j \mid 2 \leq j \leq p\}$, which leads to $e(H) = p-1 \leq i$. Then any independent set I of H with $|I| \geq 2$ satisfies $v_1 \notin I$ and hence $g(S \setminus I) \geq g(v_1) \geq i$, as claimed. Now assume that $g(S') \geq i+1$. Let j be the largest index for which $g(v_j) + g(v_{j+1}) + \cdots + g(v_p) \geq i$ and let $d = i - (g(v_{j+1}) + g(v_{j+2}) + \cdots + g(v_p))$. Then $j \geq 2$ and $d \geq 1$. Since $g(S) \geq k-1 \geq 2i$, we get that $g(v_1) + g(v_2) + \cdots + g(v_j) \geq i + d$. Furthermore, by the choice of j and the ordering of the vertices, $d \leq g(v_j) \leq g(v_i)$. Consequently, there is a multigraph $\tilde{H}$ with vertex set S and i edges, such that $d_{\tilde{H}}(v) \leq g(v)$ for all $v \in S$. We may first join v_1 and v_j by d edges and then $\{v_1, v_2, \ldots v_j\}$ and $\{v_{j+1}, v_{j+2}, \ldots, v_p\}$ by $i-d$ edges. Let H be the underlying simple graph of $\tilde{H}$. Then $e(H) \leq i$ and if I is an independent set of H, then I is an independent set of $\tilde{H}$ and we obtain that

$$g(S \setminus I) = \sum_{v \in S \setminus I} g(v) \geq \sum_{v \in S \setminus I} d_{\tilde{H}}(v) \geq \tfrac{1}{2} \sum_{v \in S} d_{\tilde{H}}(v) = e(\tilde{H}) = i,$$

as claimed. This proves the proposition. □

For a vertex v of a graph G, we denote by $N_G[v]$ the **closed neighborhood** of v in G, that is, $N_G[v] = N_G(v) \cup \{v\}$. An unordered pair uv of distinct vertices of G is called a **pair of twins** if $N_G[u] = N_G[v]$. Note that if uv is a pair of twins of G, then $uv \in E(G)$ and $d_G(u) = d_G(v)$. Let $TW(G)$ denote the set of edges of G which form a pair of twins, and let $tw(G) = |TW(G)|$.

If the graph G has order n, then $e(G) \geq \text{koy}(k,n) = \frac{(k+1)(k-2)n-k(k-3)}{2(k-1)}$ is equivalent to $\hat{\rho}(G) = (k-2)(k+1)n - 2(k-1)e(G) \leq k(k-3)$. Hence (5.9) is an immediate consequence of Theorem 5.10 and the following result.

Theorem 5.20 (Kostochka and Yancey) *Every k-critical graph with $k \geq 5$ satisfies $\hat{\rho}(G) \leq k(k-3)$.*

Proof The proof is by reductio ad absurdum. So let G be a counterexample, that is, G is a k-critical graph with $\hat{\rho}(G) > k(k-3)$. Among all such graphs G we choose one for which (1) $|G|$ is minimum, (2) $e(G)$ is minimum subject to (1), and (3) $tw(G)$ is maximum subject to (1) and (2). Clearly, $G \neq K_k$ and so $|G| \geq k+1$. Furthermore, G does not contain K_k as a subgraph, i.e. $\omega(G) \leq k-1$.

Claim 5.20.1 *Let G' be a graph with $|G'| < |G|$ such that $\hat{\rho}_{G'}(X) > k(k-3)$ for every set $X \subseteq V(G')$ with $|X| \geq k$. Then $\chi(G') \leq k-1$.*

Proof: If $\chi(G') \geq k$, then G' contains a k-critical subgraph $\tilde{G}$. Then $|\tilde{G}| \geq k$ and hence $\hat{\rho}(\tilde{G}) \geq \hat{\rho}_{G'}(V(\tilde{G})) > k(k-3)$ (by Proposition 5.17(b) and the assumption of the claim), contradicting the choice of G. △

Claim 5.20.2 *If $R \subseteq V(G)$ and $2 \leq |R| < |G|$, then $\hat{\rho}_G(R) > (k-2)(k+1)$.*

Proof: Let $\mathcal{R}$ denote the set of subsets R of $V(G)$ such that $2 \leq |R| < |G|$. Let $R \in \mathcal{R}$ be a set such that $m = \hat{\rho}_G(R)$ is minimum. If $m > (k-2)(k+1)$ then there is nothing to prove. So assume that $m \leq (k-2)(k+1)$. Since $|R| \geq 2$, we conclude from Proposition 5.17(c) that $|R| \geq k$. Since $|R| < |G|$ and G is k-critical, there is a coloring $\varphi \in \mathcal{CO}(G[R], k-1)$. Consequently, the graph $G' = G/(R, \varphi, X)$ satisfies $\chi(G') \geq k$ (by Proposition 5.18). Then G' contains a k-critical subgraph $\tilde{G}$. Let $W = V(\tilde{G})$. Since $|R| \geq k$, we obtain that $|\tilde{G}| < |G|$, which implies that $\hat{\rho}(\tilde{G}) \leq k(k-3)$ (by the choice of G). Consequently, $\hat{\rho}_{G'}(W) \leq \hat{\rho}(\tilde{G}) \leq k(k-3)$ (by Proposition 5.17(b)). Since $G' - X = G - R$ and $\chi(G-R) \leq k-1$, we conclude that $W \cap X \neq \varnothing$ and so $1 \leq |W \cap X| \leq k-1$. From Proposition 5.17(e) it then follows that $\hat{\rho}_{G'}(W \cap X) \geq (k-2)(k+1)$. For the set $U = (W \setminus X) \cup R$ it then follows from Proposition 5.18 and (5.12) that

$$\hat{\rho}_G(U) \leq \hat{\rho}_{G'}(W) - (k-2)(k+1) + m \leq m - 2k + 2 < m.$$

Since $|U| \geq |R| \geq 2$ and $\hat{\rho}_G(U) < m = \hat{\rho}_G(R)$, we obtain that $U = V(G)$. This implies that $\hat{\rho}(G) = \hat{\rho}_G(U) \leq m - 2k + 2 = k(k-3)$, which is impossible. This proves the claim. △

Claim 5.20.3 *If $R \subseteq V(D)$ and $2 \leq |R| < |G|$, then $\hat{\rho}_G(R) \geq 2(k-2)(k-1)$ and equality implies that $G[R] = K_{k-1}$.*

Proof: Let $\mathcal{R}$ denote the set of subsets R of $V(G)$ such that $2 \le |R| < |G|$. Let $R \in \mathcal{R}$ be a set such that $m = \hat{\rho}_G(R)$ is minimum. Suppose $m \le 2(k-2)(k-1)$ and $G[R] \ne K_{k-1}$. By Proposition 5.17(d), this implies that $|R| \ge k$. Furthermore, there is an integer i such that

$$(+)\ 1 + k(k-3) + 2i(k-1) \le \hat{\rho}_G(R) \le k(k-3) + 2(i+1)(k-1).$$

By Claim 5.20.2, we get $i \ge 1$. Since $k \ge 5$,

$$1 + k(k-3) + \frac{k-1}{2} 2(k-1) > 2(k-2)(k-1) \ge \hat{\rho}_G(R),$$

which implies together with (+) that $i \le (k-1)/2$.

For $v \in R$, let $g(v) = |N_G(v) \setminus R|$. Let $S = \{v \in R \mid g(v) \ge 1\}$. Since G is a k-critical graph, G is 2-connected (by Corollary 1.31) as well as $(k-1)$-edge-connected (by Theorem 1.32). This implies that $|S| \ge 2$ and $g(S) = g(R) = |E_G(R, V(G) \setminus R)| \ge k-1$. From Proposition 5.19 it then follows that we can add to $G[S]$ a set E_0 of at most i edges such that every independent set I of $G[S] + E_0$ with $|I| \ge 2$ satisfies $g(S \setminus I) \ge i$.

Let $H = G[R] + E_0$ be the graph with $V(H) = R$ and $E(H) = E(G[R]) \cup E_0$. Clearly, $|H| < |G|$. Let $X \subseteq V(H) = R$ be an arbitrary set with $|X| \ge k$. Since $X \in \mathcal{R}$, we obtain from (+) that $\hat{\rho}_G(X) \ge \hat{\rho}_G(R) \ge 1 + k(k-3) + 2i(k-1)$. Since $|E_0| \le i$, this leads to $\hat{\rho}_H(X) \ge \hat{\rho}_G(X) - 2(k-1)i > k(k-3)$ (by Proposition 5.17(b)). From Claim 5.20.1 it then follows that $\chi(H) \le k-1$. Consequently, there is a coloring $\varphi \in \mathcal{CO}(H, k-1)$. Since $\varphi \in \mathcal{CO}(G[R], k-1)$, the graph $G' = G/(R, \varphi, X)$ satisfies $\chi(G') \ge k$ (by Proposition 5.18). Then G' contains a k-critical subgraph $\tilde{G}$. Let $W = V(\tilde{G})$. Since $|R| \ge k$, we obtain that $|\tilde{G}| < |G|$, which implies that $\hat{\rho}(\tilde{G}) \le k(k-3)$ (by the choice of G). Consequently, $\hat{\rho}_{G'}(W) \le \hat{\rho}(\tilde{G}) \le k(k-3)$ (by Proposition 5.17(b)). Since $G' - X = G - R$ and $\chi(G - R) \le k-1$, we conclude that $W \cap X \ne \varnothing$ and so $1 \le |W \cap X| \le k-1$. We distinguish two cases.

Case 1: $|W \cap X| \ge 2$. Then Proposition 5.17(f) implies that $\hat{\rho}_{G'}(W \cap X) \ge 2(k-2)(k-1)$. Since $\hat{\rho}_G(R) = m \le 2(k-2)(k-1)$, Proposition 5.18 then leads to

$$\hat{\rho}_G(U) \le \hat{\rho}_{G'}(W) - \hat{\rho}_{G'}(W \cap X) + \hat{\rho}_G(R) \le \hat{\rho}_{G'}(W) \le k(k-3).$$

Since $|U| \ge 2$ and $(k-2)(k+1) > k(k-3)$, Claim 5.20.2 implies that $U = V(G)$ and hence $\hat{\rho}(G) = \hat{\rho}_G(U) \le k(k-3)$, a contradiction.

Case 2: $|W \cap X| = 1$. By symmetry, we may assume that $W \cap X = \{v_1\}$. Recall from the construction of $G' = G/(R, \varphi, X)$ that v_1 is the vertex obtained from G by contraction the color class $\varphi^{-1}(1) = \{u \in R \mid \varphi(u) = 1\}$.

Since $|W \cap X| = 1$, we obtain $\hat{\rho}_{G'}(W \cap X) = \hat{\rho}(K_1) = (k-2)(k+1)$ and so

$$(*)\ \hat{\rho}_{G'}(W) - \hat{\rho}_{G'}(W \cap X) \le k(k-3) - (k-2)(k+1) = -2k + 2.$$

By Proposition 5.18, this leads to

$$\hat{\rho}_G(U) \leq \hat{\rho}_{G'}(W) - \hat{\rho}_{G'}(W \cap X) + \hat{\rho}_G(R) \leq \hat{\rho}_G(R) - 2k + 2 < \hat{\rho}_G(R).$$

Since $|U| \geq 2$, the minimality of $m = \hat{\rho}_G(R)$ implies that $U = V(G)$ and so $W = (V(G') \setminus X) + \{v_1\}$.

Let $I = \{u \in S \mid \varphi(u) = 1\} = \varphi^{-1}(1) \cap S$. Clearly, I is an independent set of H as well as of $G[R]$. First assume that $|I| = 1$. Then for $U' = (W \setminus \{v_1\}) \cup I$, we obtain that $\hat{\rho}_G(U') = \hat{\rho}_{G'}(W) \leq k(k-3)$. Since $|U'| \geq 2$ and $(k-2)(k+1) > k(k-3)$, Claim 5.20.2 implies that $U = V(G)$ and hence $\hat{\rho}(G) = \hat{\rho}_G(U) \leq k(k-3)$, a contradiction. Now assume that $|I| \geq 2$. Since I is an independent set of H, we obtain $g(S \setminus I) \geq i$ and so $s = e_{G'}(W \setminus X, W \cap X) - e_G(W \setminus X, R)$ satisfies $s \leq -i$. By Proposition 5.18 and (5.12), we then obtain

$$\hat{\rho}_G(U) \leq \hat{\rho}_{G'}(W) - \hat{\rho}_{G'}(W \cap X) + \hat{\rho}_G(R) - 2(k-1)i.$$

Using (+) and (*), this results in

$$\hat{\rho}_G(U) \leq -2k + 2 + k(k-3) + 2(i+1)(k-1) - 2(k-1)i = k(k-3).$$

Since $|U| \geq 2$ and $(k-2)(k+1) > k(k-3)$, Claim 5.20.2 implies that $U = V(G)$ and hence $\hat{\rho}(G) = \hat{\rho}_G(U) \leq k(k-3)$, a contradiction. This completes the proof of the claim. △

Claim 5.20.4 *Let X be a $(k-1)$-clique of G and let $u \in V(G) \setminus X$. Then $|N_G(u) \cap X| < k/2$. As a consequence, G does not contain a K_k^- (a K_k with one edge deleted) as a subgraph.*

Proof: Let $R = X \cup \{u\}$ and $d = |N_G(u) \cap X|$. Then $|R| = k$ and so $2 \leq |R| < |G|$. As $|R| = |X| + 1$ and $e_G(R) = e_G(X) + d$, we get $\hat{\rho}_G(R) = \hat{\rho}_G(X) + (k-2)(k+1) - 2(k-1)d$. If $d \geq k/2$, this leads to $\hat{\rho}_G(R) < \hat{\rho}_G(X) = \hat{\rho}(K_{k-1}) = 2(k-2)(k-1)$ (by Proposition 5.17(a)), a contradiction to Claim 5.20.3. Hence $d < k/2$ and we are done. △

Recall that $TW(G)$ denotes the set of edges uv of G such that $N_G[u] = N_G[v]$, and $tw(G) = |TW(G)|$. For a vertex v of G, let $tw(v)$ denote the number of edges of $TW(G)$ that are incident with v. A *cluster* of G is a maximal set $T \subseteq V(G)$ such that each vertex v of T is a low vertex of G (that is, $d_G(v) = k-1$) and each pair uv of distinct vertices of T belongs to $TW(G)$. If T is a cluster of G, then $N_G[v]$ is the same set S for each vertex $v \in T$, and we call S the *closure* of T in G: Note that if S is the closure of a cluster T of G, then $|S| = k$, $G[S] \neq K_k$, and every $(k-1)$-clique X of G with $X \subseteq S$ satisfies $T \subseteq X$.

Claim 5.20.5 *If v_1 and v_2 are two low vertices of G contained in the same $(k-1)$-clique X of G, then $v_1v_2 \in TW(G)$.*

Proof: Clearly, for $i \in \{1,2\}$, there is a vertex u_i such that $N_G[v_i] = X \cup \{u_i\}$. If $u_1 = u_2$, then $v_1v_2 \in TW(G)$ as claimed. So suppose that $u_1 \neq u_2$, and let $G' = G - v_1 - v_2 + u_1u_2$. Let $X \subseteq V(G')$ be an arbitrary set with $|X| \geq k$. Since $|X| \leq$

$|G'| < |G|$, it follows from Claim 5.20.3 that $\hat{\rho}_G(X) \geq 2(k-2)(k-1)$, which leads (by Proposition 5.17(b)) to

$$\hat{\rho}_{G'}(X) \geq \hat{\rho}_G(X) - 2(k-1) \geq 2(k-1)(k-3) > k(k-3).$$

Consequently, $\chi(G') \leq k-1$ (by Claim 5.20.1) and so there exists a coloring $\varphi' \in \mathcal{CO}(G - v_1 - v_2, k-1)$ such that $\varphi(u_1) \neq \varphi(u_2)$. Then φ' can be easily extended to a $(k-1)$-coloring of G, which is impossible. △

Claim 5.20.6 *Let T be a cluster of G. Then $|T| \leq k-3$ and T is contained in at most one $(k-1)$-clique of G. Furthermore, if T is contained in a $(k-1)$-clique of G, then $|T| \leq (k-1)/2$.*

Proof: Let S be the closure of the cluster T. If $|T| \geq k-2$ or T is contained in at least two $(k-1)$-cliques of G, then $G[S]$ contains a K_k^-, a contradiction to Claim 5.20.4. Now suppose that there is a $(k-1)$-clique X of G with $T \subseteq X$. Then $X \subseteq S$ and there is a vertex $u \in S \setminus T$ such that $S = X \cup \{u\}$ and $T \subseteq N_G(u)$. Then Claim 5.20.4 implies that $|T| \leq |N_G(u) \cap T| < k/2$, which gives $|T| \leq (k-1)/2$ as claimed. △

Claim 5.20.7 *Let uv be an edge of the low vertex subgraph G_L of G, such that $uv \notin TW(G)$ and $tw(u) \geq tw(v)$. Then $N_G[u] \setminus \{v\}$ is a clique of G and $tw(v) = 0$.*

Proof: Let G' be the graph obtained from $G - v$ by adding a new vertex, say u', and joining u' to all vertices of $N_G[u] \setminus \{v\}$. Then $|G'| = |G|$, $e(G') = e(G)$, and $N_{G'}[u] = N_{G'}[u']$. By the construction of G', it follows that if e is an edge of $TW(G)$ not incident with v, then e belongs to $TW(G')$. Furthermore, if $uw \in TW(G)$, then $w \neq v$ and both edges uw and $u'w$ belong to $TW(G')$. Since $tw(u) \geq tw(v)$ and $uu' \in TW(G')$, this implies that $tw(G') > tw(G)$.

If there is a coloring $\varphi' \in \mathcal{CO}(G', k-1)$, then we can extend the coloring $\varphi'|_{V(G'-u-v)}$ to a coloring $\varphi \in \mathcal{CO}(G, k-1)$ as follows. Since $N = N_G(v) \setminus \{u\}$ is a set of $k-2$ vertices, we first set $\varphi(v) = c$ for some color $c \in [1, k-1] \setminus \varphi'(N)$. Then we set $\varphi(u) = c'$ with $c' \in \{\varphi'(u), \varphi'(u')\} \setminus \{c\}$. Since $\chi(G) = k$, this implies that $\chi(G') \geq k$. Then G' contains k-critical subgraph $\tilde{G}$. Since $|G'| = |G|$, $e(G') = e(G)$, and $tw(G') > tw(G)$, the choice of G implies that $\tilde{G}$ is no counterexample, that is, $\hat{\rho}(\tilde{G}) \leq k(k-3)$. Clearly, $\tilde{G}$ is not a subgraph of $G' - u' = G - v$, and so u' belongs to $\tilde{G}$. Then u' is a low vertex of $\tilde{G}$ and hence $N_{G'}[u'] \subseteq V(\tilde{G})$. Let $W = V(\tilde{G})$ and $U = W \setminus \{u'\}$. Then $|U| \geq k-1$, $|W| = |U| + 1$, and $e_{G'}(W) = e_G(U) + k - 1$. Consequently, $\hat{\rho}_{G'}(W) = \hat{\rho}_G(U) + (k-2)(k+1) - 2(k-1)(k-1)$ and $\hat{\rho}_G(W) \leq \hat{\rho}(\tilde{G}) \leq k(k-3)$. This leads to

$$\hat{\rho}_G(U) \leq k(k-3) - (k-2)(k+1) + 2(k-1)^2 = 2(k-2)(k-1).$$

Hence, $G[U] = K_{k-1}$ (by Claim 5.20.3). Since $N_{G'}[u] = N_{G'}[u'] \subseteq W$, we obtain that $U = N_G(u) \setminus \{v\}$, and so $N_G[u] \setminus \{v\}$ is a clique of G as claimed. Since $uv \notin TW(G)$, we obtain that $N_G[u] \neq N_G[v]$, which implies that $tw(v) = 0$. △

Claim 5.20.8 *If T is a cluster of G, then $tw(v) = |T| - 1$ for all $v \in T$. Furthermore, each low vertex of G belongs to exactly one cluster of G.*

Proof: Let T be an arbitrary cluster of G, and let S be the closure of T in G. Then T is a clique and $tw(v) \geq |T| - 1$ for all $v \in T$. Suppose there is a $v \in T$ and $tw(v) \geq |T|$. Then there is a low vertex $u \in S \setminus T$ and $uv \in TW(G)$. By the maximality of T, there is also a vertex $w \in T$ such that $wu \notin TW(G)$. But then $tw(w) \geq 1$ and $tw(u) \geq 1$, a contradiction to Claim5.20.7. This shows that $tw(v) = |T| - 1$ for all $v \in T$. As a consequence we obtain that each low vertex of G belongs to exactly one cluster of the graph G. △

Claim 5.20.9 *If $k = 5$, then the following statements hold:*

(a) *Each low vertex v of G satisfies $tw(v) = 0$.*
(b) *Each 4-clique of G contains at most one low vertex of G.*
(c) *If uv is an edge of the low vertex subgraph G_L of G, then each of u and v is contained in a 4-clique of G.*

Proof: To prove (a), suppose this is false. By Claim 5.20.6 this implies that there is a cluster T of G with $|T| = 2$, say $T = \{u, v\}$. Let S be the closure of T in G. Then $|S| = k = 5$ and, as $\omega(G) \leq k - 1$, there are two vertices in $S \setminus T$, say x and y, such that $xy \notin E(G)$. Let G' be the graph obtained from $G - u - v$ by identifying x and y to a new vertex w. Since a 4-coloring of G' can be easily extended to a 4-coloring of G, we obtain that $\chi(G') \geq 5$ as $\chi(G) = 5$. Hence G' contains a 5-critical subgraph $\tilde{G}$. Since $\tilde{G}$ has fewer vertices than G, this implies that the set $W = V(\tilde{G})$ satisfies $\hat{\rho}_{G'}(W) \leq \hat{\rho}_{\tilde{G}}(W) = \hat{\rho}(\tilde{G}) \leq k(k-3)$. Let $U = (W \setminus \{w\}) \cup \{x, y, u, v\}$. Then $e_G(U) \geq e_{G'}(W) + 5$ and $|U| = |W| + 3$. Since $(k-2)(k+1) = 18$ and $2(k-1) = 8$, this leads to $\hat{\rho}_G(U) \leq \hat{\rho}_{G'}(W) + 54 - 40 \leq 24 = 2(k-2)(k-1)$. Since $xy \notin E(G)$, $G[U] \neq K_{k-1} = K_4$. Hence Claim 5.20.3 implies that $U = V(G)$. But then $e_G(U) \geq e_{G'}(W) + 7$, and so $\hat{\rho}_G(U) \leq 24 - 16 = 8 < k(k-3)$, a contradiction. This proves (a).

Statement (b) follows from (a) and Claim 5.20.5; and statement (c) follows from (a) and Claim 5.20.7. △

Claim 5.20.10 $k \geq 6$.

Proof: Suppose this is false, that is, $k = 5$. Let A be the set of low vertices that are not contained in a 4-clique of G, and let B be the set of vertices of degree 5 in G that are not contained in a 4-clique of G. Set $a = |A|$ and $b = |B|$. By Claim 5.20.9(c), A is an independent set of G. We claim that $e_G(A, B) \leq a + 3b$. If if $a + b \leq 2$, this is evident. If $a + b \geq 3$, this follows from Lemma 5.16(b). A discharging argument may then be used to arrive at a contradiction. The initial charge on each vertex v is $d_G(v)$. Then each vertex of $V(G) \setminus B$ having degree at least 5 sends a charge of $\frac{1}{6}$ to each of its neighbors. First we claim that (a) each vertex of $v \in V(G) \setminus (A \cup B)$ has a new charge of at least $\frac{9}{2}$. If $d_G(v) = 4$, then v is contained in a 4-clique X of G, and each vertex of $X \setminus \{v\}$ has degree at least 5 and does not belong to B (by Claim 5.20.9(b)). Hence v get a new charge of at least $4 + 3\frac{1}{6} = \frac{9}{2}$. If $d_G(v) = 5$, then at least two neighbors of v have degree at least 5 and do not belong to B (by Claim 5.20.9(b)). So the new charge of v is at least $5 - 3\frac{1}{6} = \frac{9}{2}$. If $d_G(v) = d \geq 6$, then v has a new charge of at least $d - \frac{1}{6}d = \frac{5}{6}d \geq \frac{9}{2}$. This proves (a). Next we claim that (b) the sum of the new charges on the vertices in $A \cup B$ is at least $\frac{9}{2}(a+b)$. Let $m = e_G(A, V(G) \setminus (A \cup B))$. Since A is an independent set of G, $m + e_G(A, B) = 4a$. Since $e_G(A, B) \leq a + 3b$, this gives $m \geq 3(a-b)$. By Claim 5.20.9(c), each neighbor of a vertex in A has degree at least 5 in G. So the sum of the new charges of the vertices in $A \cup B$ is at least

$$4a + 5b + \tfrac{1}{6}m \geq 4a + 5b + \tfrac{1}{2}(a-b) = \tfrac{9}{2}(a+b).$$

Since the total amount of charge is $2e(G)$, it follows from (a) and (b) that $2e(G) \geq \frac{9}{2}|G|$, which is equivalent to

$$\hat{\rho}(G) = (k-2)(k+1)|G| - 2(k-1)e(G) = 18|G| - 8e(G) \leq 0,$$

contradicting the assumption that $\hat{\rho}(G) > k(k-3) = 10$. This proves the claim. △

Claim 5.20.11 *Let T be a cluster of G such that $t = |T| \geq 2$, and let S be the closure of T in G. Then the following statements hold:*

(a) *If S contains no $(k-1)$-clique of G, then $d_G(v) \geq k - 1 + t$ for every $v \in S \setminus T$.*
(b) *If S contains a $(k-1)$-clique X of G, then $T \subseteq X$ and $d_G(v) \geq k - 1 + t$ for every $v \in X \setminus T$.*

Proof: Note that $|S| = k$ and $N_G[u] = S$ for every vertex $u \in T$. Since $\omega(G) \leq k-1$, every $(k-1)$-clique X of G contained in S satisfies $T \subseteq X$. Hence S contains at most one $(k-1)$-clique of G (by Claim 5.20.6). Since $|T| \geq 2$, we obtain that $tw(u) \geq 2$ for every $u \in T$. Hence, if $S \setminus T$ contains a low vertex v of G, then it follows from Claim 5.20.6 that $X = S \setminus \{v\}$ is a $(k-1)$-clique of G and $d_G(u) \geq k$ for every $u \in S \setminus (T \cup \{v\}$. So if (a) or (b) is false, then there is a vertex $v \in S \setminus T$ such that (1) $k \leq d_G(v) \leq k - 2 + t$ and (2) if S contains a $(k-1)$-clique X of G, then $v \in X$. Since S contains at most one $(k-1)$-clique of G, this implies that (3) $S \setminus \{v\}$ is not a $(k-1)$-clique. Furthermore, (4) $|N_G(v) \setminus T| \leq k-2$. Now let G' be the graph obtained from $G - v$ by adding a new vertex, say v', and joining v' to each vertex of $S \setminus \{v\}$. So each vertex $u \in T \cup \{v'\}$ satisfies $N_{G'}[u] = (S \setminus \{v\}) \cup \{v'\}$. Furthermore, $|G'| = |G|$ and $e(G') < e(G)$ as $d_{G'}(v') = k-1$ and $d_G(v) \geq k$. Next we claim that $\chi(G') \geq k$. Suppose this is false, that is, there is a coloring $\varphi' \in CO(G', k-1)$. Clearly, $H = G - (T \cup \{v\})$ satisfies $H = G' - (T \cup \{v'\})$, and so $\varphi = \varphi'|_{V(H)}$ is a coloring of H. Then we can extend φ to a $(k-1)$ coloring of G as follows. By (4), there is a color $c \in [1, k-1] \setminus \varphi(N_G(v) \setminus T)$ and we give color c to v. Then we can color the vertices of T by using colors from the color set $\varphi'(T \cup \{v'\}) \setminus \{c\}$. This results in a $(k-1)$-coloring of G, which is impossible. Hence $\chi(G') \geq k$ as claimed. Then G' contains a k-critical subgraph $\tilde{G}$. Since $|G'| = |G|$ and $e(G') < e(G)$, the choice of G implies that $\hat{\rho}(\tilde{G}) \leq k(k-3)$. Let $W = V(\tilde{G})$. Then we obtain that $\hat{\rho}_{G'}(W) \leq \hat{\rho}(\tilde{G}) \leq k(k-3)$ Since $G' - v' = G - v$ and $d_{G'}(v') = k-1$, we conclude that v' belongs to W and $S \setminus \{v\} \subseteq W$. Let $U = W \setminus \{v'\}$. Then $|W| = |U| + 1$ and $e_{G'}(W) \leq e_G(U) + (k-1)$, which leads to

$$\hat{\rho}_G(U) \le \hat{\rho}_{G'}(W) - (k-2)(k+1) + 2(k-1)(k-1).$$

Since $\hat{\rho}_{G'}(W) \le k(k-3)$, we obtain that $\hat{\rho}_G(U) \le 2(k-2)(k-1)$. Since $|U| < |W| \le |G'| = |G|$, this implies that $G[U] = K_{k-1}$ (by Claim 5.20.3). But then $U = S \setminus \{v\}$, a contradiction to (3). This proves the claim. △

Claim 5.20.12 *If X is a $(k-1)$-clique of G with a unique low vertex of G, then X contains at least $(k-1)/2$ vertices having degree at least $k+1$ in G.*

Proof: Let v be the unique low vertex of G contained in X. Then there is a unique vertex u of G such that $N_G[v] = X \cup \{u\}$. Let Y be the set of vertices $y \in X$ such that $d_G(y) \ge k+1$. Suppose, to the contrary, that $|Y| < (k-1)/2$. Since $|N_G(u) \cap X| < k/2$ (by Claim 5.20.4), this implies that there is a vertex $w \in X \setminus (Y \cup \{v\})$ such that $uw \notin E(G)$. Clearly, $d_G(w) = k$ and there are two distinct vertices, say w_1 and w_2, such that $N_G(w) \setminus X = \{w_1, w_2\}$. Let $G' = G - v + uw_1 + uw_2$. First suppose that there is a coloring $\varphi' \in \mathcal{CO}(G', k-1)$. Then $\varphi' \in \mathcal{CO}(G - v, k-1)$ such that $\varphi'(u) \ne \varphi'(w_i)$ for $i \in \{1,2\}$, and we can construct a $(k-1)$-coloring of G as follows. If $C = \varphi'((X \setminus \{v\}) \cup \{u\})$ satisfies $|C| \le k-2$, then we color v with a color $c \in [1, k-1] \setminus C$. If $|C| = k-1$, then we recolor w by $\varphi'(u)$ and color v by $\varphi'(w)$. In both cases this results in a $(k-1)$-coloring of G, a contradiction. Hence it remains to consider the case when $\chi(G') \ge k$. Then G' contains a k-critical subgraph $\tilde{G}$. Let $W = V(\tilde{G})$. Since $|G'| < |G|$, we obtain that $\hat{\rho}_{G'}(W) \le \hat{\rho}(\tilde{G}) \le k(k-3)$. If $W = V(G')$, then for $U = V(G)$ we have $|U| = |W| + 1$ and $e_G(U) = e_{G'}(W) + (k-3)$, which gives

$$\hat{\rho}(G) = \hat{\rho}_G(U) = \hat{\rho}_{G'}(W) + (k-2)(k+1) - 2(k-1)(k-3) < \hat{\rho}_{G'}(W)$$

as $k \ge 6$ (by Claim 5.20.10). But then $\hat{\rho}(G) < k(k-3)$, which is impossible. If $W \ne V(G')$, then $|W| < |G|$ and $e_G(W) \ge e_{G'}(W) - 2$ which leads to

$$\hat{\rho}_G(W) \le \hat{\rho}_{G'}(W) + 2(k-1)2 \le k(k-3) + 4(k-1) < 2(k-2)(k-1)$$

as $k \ge 6$, a contradiction to Claim 5.20.3. Thus the proof is complete. △

Claim 5.20.13 *If $k = 6$ and T is a cluster contained in a 5-clique X of G, then $|T| = 1$.*

Proof: By Claim 5.20.6, $|T| \le 2$. Suppose, to the contrary, that $|T| = 2$, say $T = \{v_1, v_2\}$. Let u be the unique vertex satisfying $N_G[v_i] = X \cup \{u\}$ ($i \in \{1,2\}$), and let $X \setminus T = \{u_1, u_2, u_3\}$. Since $\omega(G) \le 5$, there is vertex in $X \setminus T$, say u_1, such that $uu_1 \notin E(G)$. Let G' be the graph obtained from $G - v_1 - v_2$ by identifying u and u_1 to a new vertex w. If G' has a 5-coloring, then there is a coloring $\varphi \in \mathcal{CO}(G - v_1 - v_2, 5)$ such that $\varphi(u) = \varphi(u_1)$. So only three colors occur in the neighborhood of T and we can easily extend φ to a 5-coloring of G, which is impossible. Hence $\chi(G') \ge 6$ and G' contains a 6-critical subgraph $\tilde{G}$. Clearly, $\tilde{G}$ contains the new vertex w. Let $W = V(\tilde{G})$ and let $p = |W \cap \{u_2, u_3\}|$. Since $|\tilde{G}| < |G|$, we obtain that $\hat{\rho}_{G'}(W) \le k(k-3) = 18$. We distinguish three cases.

Case 1: $p = 0$. Then $U = W \setminus \{w\} \cup (X \cup \{u\})$ satisfies $|U| = |W| + 5$ and $e_G(U) \geq e_{G'}(W) + 12$, which leads to

$$\hat{\rho}_G(U) \leq \hat{\rho}_{G'}(W) + 28 \cdot 5 - 10 \cdot 12 \leq 38.$$

By Claim 5.20.6, this yields that $U = V(G)$. But then $e_G(U) \geq e_{G'}(W) + 12$ as we did not count $e_G(\{u_2u_3\}, V(G) \setminus X)$. So we obtain that $\hat{\rho}(G) = \hat{\rho}_G(U) \leq 23 - 2 \cdot 10 = 18$, which is impossible.

Case 2: $p = 1$. Then $U = (W \setminus \{w\}) \cup T \cup \{u, u_1\}$ satisfies $|U| = |W| + 3$ and $e_G(U) \geq e_{G'}(W) + 7$, which leads to

$$\hat{\rho}_G(U) \leq \hat{\rho}_{G'}(W) + 28 \cdot 3 - 10 \cdot 7 \leq 32.$$

Since $|U| < |G|$, this is a contradiction to Claim 5.20.3.

Case 3: $p = 3$. Then $U = (W \setminus \{w\}) \cup T \cup \{u, u_1\}$ satisfies $|U| = |W| + 3$ and $e_G(U) \geq e_{G'}(W) + 9$, which leads to

$$\hat{\rho}_G(U) \leq \hat{\rho}_{G'}(W) + 28 \cdot 3 - 10 \cdot 9 \leq 12,$$

a contradiction to Claim 5.20.3. △

For the rest of the proof, let A be the set of low vertices of G, and let $B = V(G) \setminus A$ be the set of high vertices of G. Furthermore, let

$$A_0 = \{v \in A \mid N_G(v) \subseteq B\} \text{ and } A_1 = A \setminus A_0;$$

and let

$$B_0 = \{v \in B \mid d_G(v) = k\} \text{ and } B_1 = B \setminus B_0.$$

Clearly, A_0, A_1, B_0, B_1 are pairwise disjoint sets whose union is $V(G)$, each vertex in B_1 has degree at least $k+1$ in G, and A_0 is an independent set of G. To simplify the notation, let $d(v) = d_G(v)$ for $v \in V(G)$, and let

$$q = k - \tfrac{2}{k-1}$$

To arrive at a final contradiction, we may now use discharging. The initial charge of each vertex v of G is $d_G(v)$. Then the total charge assigned is $2e(G)$. We now redistribute the charge according to the following two rules:

1. Each vertex $v \in B_1$ keeps charge q for itself and distributes the rest $d(v) - q$ equally among its neighbors in G belonging to A.
2. Each vertex $v \in A_1$ that belongs to a cluster T of G such that $t = |T| \geq 2$ and T is contained in a $(k-1)$-clique of G gives charge $\frac{k-3}{t(k-1)}$ to each vertex $u \in A \setminus T$, provided that $uv \in E(G)$ and no $(k-1)$-clique of G contains u.

Denote the resulting charge on each vertex $v \in V(G)$ by $M(v)$. For $X \subseteq V(G)$, let

$$M(X) = \sum_{v \in X} M(v).$$

Our discharging rules moves charge around, but the sum over all charges remains unchanged. Hence $M(V(G)) = 2e(G)$ and our aim is to show that $M(V(G)) \geq q|G|$. To this end, we establish two claims.

Claim 5.20.14 $M(A_0 \cup B_0) \geq (|A_0| + |B_0|)q$.

Proof: Set $a = |A_0|$ and $b = |B_0|$. Note that A_0 is an independent set of G. Then $e_G(A_0, B_0) \leq 2(a+b)$. This is evident if $a+b \leq 2$ and follows from Lemma 5.16(a) for $a+b \geq 3$. Let $m = e_G(A_0, V(G) \setminus (A_0 \cup B_0))$. Since A_0 is an independent set of G, $m + e_G(A, B) = (k-1)a$, which leads to (+) $m \geq (k-1)a - 2(a+b)$. Let $uv \in E(G)$ be an edge with $u \in A_0$ and $v \in V(G) \setminus (A_0 \cup B_0))$. Then $v \in B_1$ and so v transfers charge at least $\gamma = \frac{d(v)-q}{d(v)}$ to u (by rule 1). Since $q > 0$ and $d(v) \geq k+1$, this gives

$$\gamma = \frac{d(v)-q}{d(v)} \geq \frac{k+1-q}{k+1} = \frac{1+2/(k-1)}{k+1} = \frac{1}{k-1}.$$

Since the charges of the vertices in B_0 remains unchanged, we obtain that

$$M(A_0 \cup B_0)) \geq (k-1)a + kb + \frac{m}{k-1}.$$

Using (+), this leads to

$$M(A_0 \cup B_0)) \geq ka + kb - \frac{2(a+b)}{k-1} = (a+b)q.$$

as claimed. △

Claim 5.20.15 $M(v) \geq q$ *for every vertex* $v \in A_1 \cup B_1$.

Proof: Obviously, $M(v) = q$ if $v \in B_1$. So assume that $v \in A_1$. Then v belongs to a unique cluster T of G (by Claim 5.20.8). Let S be the closure of T in G and let $t = |T|$. We distinguish three main cases.

Case 1: $t \geq 2$ *and* T *is contained in a* $(k-1)$*-clique* X *of* G. Then $k \geq 7$ (by Claims 5.20.10 and 5.20.13) and $t \leq (k-1)/2$ (by Claim 5.20.6). Clearly, $X \subseteq S$ and Claim 5.20.11(b) implies that $d(w) \geq k-1+t \geq k+1$ for all $w \in X \setminus T$. Then $|X \setminus T| = k-1-t$ and every vertex $w \in X \setminus T$ belongs to B_1 and has at most $d(w) - (k-2-t)$ neighbors in A, and so w transfers charge at least $\gamma = \frac{d(w)-q}{d(w)-k+2+t}$ to v (by rule 1). The function $\gamma = \gamma(d(w))$ is increasing, and so

$$\gamma \geq \frac{k-1+t-q}{2t+1} = \frac{t-1+2/(k-1)}{2t+1}.$$

Furthermore, v has at most one neighbor in $A \setminus T$. According to rule 2, we therefore obtain that

$$M(v) \geq k-1+(k-1-t)\frac{t-1+2/(k-1)}{2t+1} - \frac{k-3}{t(k-1)} = f(t).$$

Let

$$g(t) = (k-1-t)((t-1)(k-1)+2) - (2t+1)(k-3)(1+\frac{1}{t}).$$

and

$$h(t) = (k-1-t)((t-1)(k-1)+2) - (2t+1)(k-3)(3/2).$$

Then $f(t) \geq q$ is equivalent to $f(t)(2t+1) \geq q(2t+1)$ and this is equivalent to $g(t) \geq 0$. For $t \geq 2$ we have $g(t) \geq h(t)$. The function $h(t)$ is quadratic in t with a negative coefficient at t^2, and so $h(t) \geq \min\{h(2), h((k-1)/2)\}$ for all real numbers t with $2 \leq t \leq (k-1)/2$. Since $k \geq 7$, we have

$$h(2) = (k-3)(k-\frac{13}{2}) \geq 0$$

and

$$\begin{aligned} 4 \cdot h((k-1)/2)) &= (k-1)((k-3)(k-1)+4) - 6k(k-3) \\ &= k^1 3 - 11k^2 + 29k - 7 \\ &= (k-7)(k^2 - 4k + 1) \geq 0. \end{aligned}$$

As a consequence, we have $M(v) \geq q$ as claimed.

Case 2: *$t \geq 2$ and T is not contained in a $(k-1)$-clique of G.* Then $t \leq k-3$ (by Claim 5.20.6). Clearly, S contains no $(k-1)$-clique and Claim 5.20.11(b) implies that $d(w) \geq k-1+t \geq k+1$ for all $w \in S \setminus T$. Then $|S \setminus T| = k-t$ and every vertex $w \in S \setminus T$ belongs to B_1, and so w transfers charge at least $\gamma = \frac{d(w)-q}{d(w)}$ to v (by rule 1). Since v has at no neighbor in $A \setminus T$, we obtain that that

$$M(v) \geq k-1+(k-t)\frac{t-1+2/(k-1)}{k-1+t} - \frac{k-3}{t(k-1)} = f(t).$$

Let

$$\begin{aligned} g(t) &= (k-t)(t-1+2/(k-1)) - \frac{k-3}{k-1}(k-1-t) \\ &= t(k-t) - 2(1-2/(k-1))(k-1) \\ &= t(k-t) - 2(k-3). \end{aligned}$$

Then $M(v) \geq q$ is equivalent to $g(t) \geq 0$. Since $2 \leq t \leq k-3$ and $g(2) = 2$ and $g(k-3) = k-3$, we conclude that $g(t) \geq 0$, and so $M(v) \geq q$ as claimed. This settles Case 2.

Case 3: $t = 1$. Then $tw(v) = 0$ (by Claim 5.20.8). First assume that v is not contained in a $(k-1)$-clique of G. Since $v \in A_1$, vertex v has a neighbor $u \in A$. Clearly, $uv \notin TW(G)$ and $tw(u) \geq 0 = tw(v)$. Then Claim 5.20.7 implies that $N_G[u] \setminus \{v\}$ is a $(k-1)$-clique of G and $tw(u) \geq 2$. Hence u belongs to a cluster T' of G with $|T'| \geq 2$ and S is the closure of T', too. Then each vertex of T' sends a charge to v according to rule 2, and we have $M(v) = k-1+\frac{k-3}{k-1} = k - \frac{2}{k-1} = q$ as claimed. Now assume that v belongs to a $(k-1)$-clique X of G. Then $X \subseteq S$ and v is the only low vertex of G

contained in X (by Claim 5.20.5 and $tw(v) = 0$). Consequently, there is a set $W \subset X$ such that $|W| \geq (k-1)/2$ and $d_G(w) \geq k+1$ for all $w \in W$ (by Claim 5.20.12). If $w \in W$, then w has at most $d(w) - (k-3)$ neighbors in A, and so w transfers charge at least $\gamma = \frac{d(w)-q}{d(w)-k+3}$ to v (by rule 1). The function $\gamma = \gamma(d(w))$ is increasing, and so

$$\gamma \geq \frac{k+1-q}{4} = \frac{1+2/(k-1)}{4}.$$

Since $k \geq 6$, we then have

$$M(v) \geq k-1+\frac{k-1}{2}\left(\frac{1+2/(k-1)}{4}\right) = k+\frac{k-7}{8} \geq k-\frac{2}{k-1} = q$$

as claimed. This completes the proof of Claim 5.20.15 △

Combining the last two claims, we have

$$2e(G) = M(V(G)) \geq q|G| = \left(k-\frac{2}{k-1}\right)|G| = \left(\frac{(k-2)(k+1)}{k-1}\right)|G|,$$

which is equivalent to $\hat{\rho}(G) = (k-2)(k+1)|G| - 2(k-1)e(G) \leq 0$, contradicting our initial assumption $\hat{\rho}(G) > k(k-3)$. Thus Theorem 5.20 is proved. □

Corollary 5.21 (Kostochka and Yancey) $\mathrm{ext}(k,n) = \mathrm{ore}(k,n)$ *for any* k, n *with* $k \geq 4$ *and* $n \equiv 1 \pmod{k-1}$.

5.6 Triangle-Free Graphs

The problem of finding the best upper bound for the chromatic number of a triangle-free graph in terms of its maximum degree has a long history and was posed by Vizing [1050] half a century ago. Let $\mathrm{tri}(\Delta)$ denote the maximum chromatic number of a triangle-free graph with maximum degree Δ. In the late 1970s Borodin and Kostochka [157], Catlin [193] and Lawrence [668] independently proved that $\mathrm{tri}(\Delta) \leq \frac{3}{4}(\Delta+2)$ as a special case of a more general Brooks type bound for the chromatic number of a graph in terms of its maximum degree and clique number (see Theorem 2.25). Some years later Kostochka [618] improved this bound to $\mathrm{tri}(\Delta) \leq \frac{2}{3}(\Delta+3)$, which is still the best bound for small values of Δ. Better asymptotic bounds were achieved in the mid 1990s using probabilistic arguments. In 1996, Johansson [534] proved that $\mathrm{tri}(\Delta) \leq (9+o(1))\frac{\Delta}{\ln \Delta}$, but his manuscript was never published. In 1995, Kim [589] proved that the upper bound holds with a leading constant of $1+o(1)$, but only for graphs of girth at least five. Here and in the rest of the section, $o(1)$ stands for a positive function of Δ that converges to 0 when Δ tends to infinity. Lower bounds for $\mathrm{tri}(\Delta)$ were first established by Bollobás [127] and, independently, by Kostochka and Mazurova [623]. Both proved by using probabilistic arguments that $\mathrm{tri}(\Delta) = \Omega(\frac{\Delta}{\ln \Delta})$. Johansson proved that his bound not only holds for the chromatic number, but indeed for the list chromatic number. The leading constant $9+o(1)$ in Johansson's bound for the list chromatic number was subsequently improved first to $4+o(1)$ by Pettie and Su [805] and then to $1+o(1)$ by Molloy [754]. Bernshteyn [106] not only succeeded in providing a shorter proof of Molloy's result, but also to extend it to the DP-chromatic number. While the proof by Molloy is based on the entropy compression method developed by Moser and Tardos [761], Bernshteyn's proof uses a decomposition result due to Aharoni and Haxell, the lopsided local lemma and Chernoff bounds for negatively correlated random variables; the reader will find all three results in Appendix C.15.

The DP-coloring concept was introduced in Section 1.4 for multigraphs, but in this section we only deal with graphs. For the convenience of the reader we shall recall the main notation from Section 1.4. Let G be a graph. A **cover** of G is a pair (X,H) consisting of a graph H and a map $X: V(G) \to 2^{V(H)}$ such that following conditions hold:

(C1) The sets $X(v)$ with $v \in V(G)$ are pairwise disjoint independent sets of H and their union is $V(H)$.
(C2) If $u,v \in V(G)$ are distinct vertices, then the edge set $E_H(X(u),X(v))$ is a matching in H (possibly empty) if $uv \in E(G)$ and empty if $uv \notin E(G)$.

Let (X,H) be an arbitrary cover of G. For $U \subseteq V(G)$, let

$$X(U) = \bigcup_{u \in U} X(u) \text{ and } H(U) = H[X(U)].$$

Note that $(X|_U, H(U))$ is a cover of $G[U]$ and we also denote this cover by $(X(U), H(U))$. We say that (X,H) is a k**-cover** if $|X(v)| = k$ for all $v \in V(G)$. A vertex set $T \subseteq V(H)$ is called a **transversal** of (X,H) if $|T \cap X_v| = 1$ for each vertex $v \in V(G)$. An **independent transversal** of (X,H) is a transversal of (X,H) which is an independent set of H. An independent transversal of (X,H) is also called an (X,H)**-coloring** of G, and the vertices of H are referred to as **colors**. The **DP-chromatic number** $\chi_{\rm DP}(G)$ of G is the least integer k such that G has a (X,H)-coloring for every k-cover (X,H) of G.

The proof of Bernshteyn's theorem uses the concept of a partial coloring. Let (X,H) be a cover of G. A set $T \subseteq V(G)$ is a **partial transversal** of (X,H) if $|T \cap X_v| \leq 1$ for all $v \in V(G)$. For a partial transversal T of (X,H), define the **domain** of T in G by $\mathrm{dom}(T) = \{v \in V(G) \mid |T \cap X_v| = 1\}$. Furthermore, let

$$G_T = G - \mathrm{dom}(T), H_T = H - X(\mathrm{dom}(T)) - N_H(T),$$

and let $X_T: V(G_T) \to 2^{V(H)}$ be the map with $X_T(v) = X(v) \setminus N_H(T)$ for all $v \in V(G_T)$. Note that (X_T, H_T) is a cover of G_T. An **independent partial transversal** of (X,H) is a partial transversal of (X,H) which is an independent set of H. An independent partial transversal of (X,H) is also called a **partial** (X,H)**-coloring** of G. The crucial observation is the obvious fact that if T is a partial (X,H)-coloring of G and T' is a (X_T,H_T)-coloring of G_T, then $T \cup T'$ is an (X,H)-coloring of G.

Theorem 5.22 (Bernshteyn) *Every triangle-free graph G with maximum degree Δ satisfies*

$$\chi_{\mathrm{DP}}(G) \leq (1+o(1))\frac{\Delta}{\ln \Delta}.$$

Proof Fix an ε with $0 < \varepsilon < 1$, a triangle-free graph G whose maximum degree Δ is sufficiently large, and a k-cover (X,H) of G with $k = (1+\varepsilon)\Delta/\ln \Delta$. Furthermore, let $\ell = \Delta^{\varepsilon/2}$. For the proof of the theorem, it suffices to show that G has an (X,H)-coloring.

A partial (X,H)-coloring T of G is said to be *extendable* if $|X_T(v)| \geq \ell$ for all $v \in V(G_T)$ and $d_{H_T}(x) \leq \ell/2$ for all $x \in V(H_T)$. If G has an extendable partial (X,H)-coloring T, then it follows from a result due to Aharoni and Haxell (see Theorem C.38 or Theorem 6.10) that G_T has an (X_T,H_T)-coloring T', and hence $T \cup T'$ is an (X,H)-coloring of G and we are done. So it suffices to show that G has an extendable partial (X,H)-coloring.

To prove the existence of an extendable partial coloring, we shall use the lopsided Lovász Local Lemma (see Lemma C.39) and the Chernoff bounds for negatively correlated random variables (see Lemma C.40). The proof is divided into two claims.

First we need some more notation. For a vertex v of G, let $N(v) = N_G(v)$ be the neighborhood of v in G and $N[v] = N(v) \cup \{v\}$ be the closed neighborhood of v in G, and for $d \in \mathbb{N}_0$, let $N^d[v] = \{w \in V(G) \mid \mathrm{dist}_G(v,w) \leq d\}$. So $N^0[v] = \{v\}$ and $N^1[v] = N[v]$. As usual, for $U \subseteq V(G)$, let $\overline{U} = V(G) \setminus U$.

Let $u \in V(G)$ be a vertex, and let S be a partial independent transversal of the cover $(X(\overline{N[u]}), H(\overline{N[u]}))$ of $G[\overline{N[u]}]$. If T' is partial $(X_S(N(u)), H_S(N(u)))$-coloring of $G[N(u)]$, then $T = S \cup T'$ is a partial (X,H)-coloring of G.

Claim 5.22.1 *Let $u \in V(G)$ be a vertex, and let S be an independent partial transversal of the cover $(X(\overline{N[u]}), H(\overline{N[u]}))$ of $G[\overline{N[u]}]$. If $\mathbf{T}'$ is a uniformly random partial $(X_S(N(u)), H_S(N(u)))$-coloring of $G[N(u)]$, then the random set $\mathbf{T} = S \cup \mathbf{T}'$ is a partial (X,H)-coloring of G such that*

(a) $\Pr(|X_{\mathbf{T}}(u)| < \ell) \leq \frac{1}{8}\Delta^{-3}$; *and*
(b) $\Pr($*there is* $x \in X_{\mathbf{T}}(u)$ *with* $d_{H_{\mathbf{T}}}(x) > \ell/2) \leq \frac{1}{8}\Delta^{-3}$ □

Proof: Let u, S, and $\mathbf{T}'$ satisfy the hypothesis of the claim. Since G is triangle-free, $G[N(u)]$ is edgeless and so $E_H(X(v), X(w)) = \varnothing$ whenever $v, w \in N(u)$ (by (C2)). Hence the random partial coloring $\mathbf{T}'$ can be constructed by the following random process. Let $\bullet$ be a special element that does not belong to $V(H)$. For each vertex $v \in N(u)$, choose a color $x_v \in X_S(v) \cup \{\bullet\}$ independently and uniform at random. Let $\mathbf{T}' = \{x_v \mid v \in N(u) \text{ and } x_v \neq \bullet\}$. Then $\mathbf{T}'$ is a uniformly random partial $(X_S(N(u)), H_S(N(u)))$-coloring of $G[N(u)]$ and $\mathbf{T} = S \cup \mathbf{T}'$ is a partial (X,H)-coloring of G.

For a vertex v of G_S, let $k_v = |X_S(v)|$. Let $x \in X(u)$ be an arbitrary color. Define $\tilde{N}(x) = \{v \in N(u) \mid N_H(x) \cap X_S(v) \neq \varnothing\}$. For $v \in \tilde{N}(x)$, we have $N_H(x) \cap X_S(v) = \{x'\}$ (by (C2)) and $\Pr(x' \in \mathbf{T}') = 1/(k_v+1)$. Consequently,

$$\Pr(x \in X_{\mathbf{T}}(u)) = \Pr(\mathbf{T}' \cap N_H(x) = \varnothing) = \prod_{v \in \tilde{N}(x)} \left(1 - \frac{1}{k_v + 1}\right).$$

Since $k_v \geq 1$ for all $v \in \tilde{N}(x)$ and $\exp(-1/a) \leq 1 - 1/(a+1) \leq \exp(-1/(a+1))$ for all $a > 0$, this implies that

$$\exp\left(-\sum_{v \in \tilde{N}(x)} \frac{1}{k_v}\right) \leq \Pr(x \in X_{\mathbf{T}}(u)) \leq \exp\left(-\sum_{v \in \tilde{N}(x)} \frac{1}{k_v + 1}\right) \tag{5.13}$$

For the expectation of the random variable $|X_{\mathbf{T}}(u)|$, we then obtain that

$$\mathbb{E}(|X_{\mathbf{T}}(u)|) = \sum_{x \in X(u)} \Pr(x \in X_{\mathbf{T}}(u)) \geq \sum_{x \in X(u)} \exp\left(-\sum_{v \in \tilde{N}(x)} \frac{1}{k_v}\right).$$

Since $k_v = |X_S(v)|$, it follows from (C2) that

$$\sum_{x \in X(u)} \sum_{v \in \tilde{N}(x)} \frac{1}{k_v} \leq \sum_{v \in N(u): X_S(v) \neq \varnothing} \sum_{y \in X_S(v)} \frac{1}{k_v} \leq d_G(u) \leq \Delta.$$

Since $|X(u)| = k$, $k = (1+\varepsilon)\Delta/\ln\Delta$, and $\ell = \Delta^{\varepsilon/2}$, the above inequality and the convexity of the exponential function leads to

$$\sum_{x \in X(u)} \exp\left(-\sum_{v \in \tilde{N}(x)} \frac{1}{k_v}\right) \geq k \exp(-\Delta/k) = \frac{(1+\varepsilon)\Delta}{\ln\Delta} \Delta^{-1/(1+\varepsilon)} \geq 2\ell,$$

provided that Δ is large enough. In summary, we thus obtain that

$$\mathbb{E}(|X_{\mathbf{T}}(u)|) \geq 2\ell. \tag{5.14}$$

For a proof of (a), let $p_0 = \Pr(|X_{\mathbf{T}}(u)| < \ell)$. By (5.14), we obtain

$$p_0 \leq \Pr(|X_{\mathbf{T}}(u)| < \tfrac{1}{2}\mathbb{E}(|X_{\mathbf{T}}(u)|)).$$

Obviously, the random variable $|X_{\mathbf{T}}(u)|$ is the sum of the indicator random variables of the events $A_x = \{x \in X_{\mathbf{T}}(u)\}$ for $x \in X(u)$. Now we claim that the indicator random variables of the events $B_x = \{x \notin X_{\mathbf{T}}(u)\}$ for $x \in X(u)$ are negatively correlated (see (C.14)). To prove the claim, it suffices to show that whenever $x \in X(u)$ and $Y \subseteq X(u) \setminus \{x\}$ we have

$$\Pr(B_x \mid Y \cap X_{\mathbf{T}}(u) = \varnothing) \leq \Pr(B_x),$$

which is equivalently to

$$\Pr(Y \cap X_{\mathbf{T}}(u) = \varnothing \mid A_x) \geq \Pr(Y \cap X_{\mathbf{T}}(u) = \varnothing).$$

This can in turn be rewritten as

$$\Pr(\mathbf{T}' \cap N_H(y) \neq \varnothing \forall y \in Y | \mathbf{T}' \cap N_H(x) = \varnothing) \geq \Pr(\mathbf{T}' \cap N_H(y) \neq \varnothing \forall y \in Y),$$

which holds, since the sets $N_H(x)$ and $N_H(Y)$ are disjoint (by (C2)). Hence the indicator variables of the event B_x for $x \in X(u)$ are negatively correlated as claimed. Then we can apply the Chernoff bound, namely Lemma C.40(b) with $t = \frac{1}{2}\mathbb{E}(|X_{\mathbf{T}}(u)|)$, which leads to

$$p_0 \leq \Pr(|X_{\mathbf{T}}(u)| < \tfrac{1}{2}\mathbb{E}(|X_{\mathbf{T}}(u)|)) \leq \exp(-\tfrac{1}{8}\mathbb{E}(|X_{\mathbf{T}}(u)|)).$$

Since $-\frac{1}{8}\mathbb{E}(|X_{\mathbf{T}}(u)|) \leq -\frac{1}{4}\ell$ (by (5.14)) and $\ell = \Delta^{\varepsilon/2}$, we obtain that

$$p_0 \leq \exp(-\tfrac{1}{4}\ell) = \exp(-\tfrac{1}{4}\Delta^{\varepsilon/2}) < \tfrac{1}{8}\Delta^{-3},$$

provided that Δ is large enough (see also Exercise 5.18). Thus (a) is proved.

For a proof of (b), let

$$p_1 = \Pr(\text{there is } x \in X(u) \text{ such that } x \in X_{\mathbf{T}}(u) \text{ and } d_{H_{\mathbf{T}}}(x) > \ell/2),$$

and, for $x \in X(u)$, let

$$p_x = \Pr(x \in X_{\mathbf{T}}(u) \text{ and } d_{H_{\mathbf{T}}}(x) > \ell/2).$$

Then $p_1 \leq \sum_{x \in X(u)} p_x$ and $|X(u)| = k < \Delta/8$ when Δ is large enough. Hence, to prove that $p_1 \leq \frac{1}{8}\Delta^{-3}$, it suffices to show that $p_x \leq \Delta^{-4}$ for all $x \in X(u)$. The proof is by reductio ad absurdum. So assume that $p_x \geq \Delta^{-4}$ for some $x \in X(u)$. Using the second inequality in (5.13), we obtain that

$$\Delta^{-4} \leq p_x \leq \Pr(x \in X_{\mathbf{T}}(u)) \leq \exp\left(-\sum_{v \in \tilde{N}(x)} \frac{1}{k_v + 1}\right),$$

which leads to

$$\sum_{v \in \tilde{N}(x)} \frac{1}{k_v + 1} \leq 4 \ln \Delta.$$

Since $k_v = |X_S(v)|$, we then conclude that

$$\mathbb{E}(d_{H_{\mathbf{T}}}(x)) = \sum_{v \in \tilde{N}(x)} \Pr(v \notin \operatorname{dom}(\mathbf{T}')) = \sum_{v \in \tilde{N}(x)} \frac{1}{k_v + 1} \leq 4 \ln \Delta \leq \ell/4,$$

provided that Δ is large enough. The events $\{v \notin \operatorname{dom}(\mathbf{T}')\}$ for $v \in \tilde{N}(x)$ are mutually independent, and so the standard Chernoff bound for independent $\{0, 1\}$-valued random variables (see Lemma C.41 with $\vartheta\mathbb{E}(d_{H_{\mathbf{T}}}(x)) = \ell/4$ and $\vartheta \geq 1$) leads to

$$\begin{aligned} p_x &\leq \Pr(d_{H_{\mathbf{T}}}(x) > \ell/2) \leq \Pr(d_{H_{\mathbf{T}}}(x) > \mathbb{E}(d_{H_{\mathbf{T}}}(x)) + \ell/4)) \\ &\leq \exp(-\tfrac{1}{12}\ell) = \exp(-\tfrac{1}{12}\Delta^{\varepsilon/2}) \leq \Delta^{-4}, \end{aligned}$$

where the last inequality holds if Δ is large enough. This contradiction proves (b). So the proof of the claim is complete. △

Claim 5.22.2 *G has an extendable partial (X,H)-coloring.* □

Proof: Choose an partial (X,H)-coloring $\mathbf{T}$ of G uniformly at random. As emphasized by Bernshteyn, the following immediate observation is crucial in the proof:

(1) *Fix a set $U \subseteq V(G)$ and a partial independent transversal S of the cover $(X(\overline{U}), H(\overline{U}))$ of $G[\overline{U}]$. Then the random variable $\mathbf{T} \cap X(U)$, conditioned to the event $\{\mathbf{T} \cap X(\overline{U}) = S\}$, is uniformly distributed over the partial $(X_S(U), H_S(U))$-colorings of $G[U]$.*

For each vertex $u \in V(G)$, the random set $\mathbf{T}$ is said to be u**-bad** if $u \notin \mathrm{dom}(\mathbf{T})$, but $|X_{\mathbf{T}}(u)| < \ell$ or there is a color $x \in X_{\mathbf{T}}(u)$ such that $d_{H_{\mathbf{T}}}(x) > \ell/2$. Clearly, if $\mathbf{T}$ is not u-bad for all $u \in V(G)$, then $\mathbf{T}$ is an extendable partial (X,H)-coloring of G, and the proof is finished. Let A_u denote the event that $\mathbf{T}$ is u-bad. Our aim is to show that

$$\Pr\left(\bigcap_{u\in V(G)} \overline{A}_u\right) > 0.$$

To this end, we apply Lemma C.39 with $p = \frac{1}{4}\Delta^{-3}$, $d = \Delta^3$, and $N_u = N^3[u]$ for all $u \in V(G)$. Then $|N_u| \le \Delta^3 = d$ for all $u \in V(G)$. Hence, it only remains to show that if $u \in V(G)$, then for all $M \subseteq V(G) \setminus N_u$,

$$\Pr\left(A_u \mid \bigcap_{v\in M} \overline{A}_v\right) \le \tfrac{1}{4}\Delta^{-3} = p.$$

The outcome of the event A_v is determined by the set $\mathbf{T} \cap X(N^2[v])$. If $v \notin N_u$, then $\mathrm{dist}_G(u,v) \ge 4$, hence $N^2[v] \subseteq \overline{N(u)}$. Hence, the outcome of the event A_v with $v \notin N_u$ is determined by the set $\mathbf{T} \cap X(\overline{N(u)})$. Therefore, it sufficed to show that for every partial independent transversal S of the cover $(X(\overline{N(u)}), H(\overline{N(u)}))$ of $G[\overline{N(u)}]$,

$$\Pr(A_u \mid \mathbf{T} \cap X(\overline{N(u)}) = S) \le \tfrac{1}{4}\Delta^{-3}.$$

To show this, let $u \in V(G)$ be a vertex, and let S be a independent partial transversal of the cover $(X(\overline{N(u)}), H(\overline{N(u)}))$ of $G[\overline{N(u)}]$. We may assume that $u \notin \mathrm{dom}(S)$, otherwise the event A_v is incompatible with $\{\mathbf{T} \cap X(\overline{N(u)}) = S\}$. Thus $S \subseteq X(\overline{N[u]})$. Then (1) implies that the random variable $\mathbf{T}' = \mathbf{T} \cap X(N(u))$, conditioned to the random event $\{\mathbf{T} \cap X(\overline{N(u)}) = S\}$, is uniformly distributed over the partial $(X_S(N(u)), H_S(N(u)))$-colorings of $G[\overline{N(u)}]$. Then it follows from Claim 5.22.1 that

$$\Pr(A_u \mid \mathbf{T} \cap X(\overline{N(u)}) = S) \le \tfrac{1}{8}\Delta^{-3} + \tfrac{1}{8}\Delta^{-3} = \tfrac{1}{4}\Delta^{-3}.$$

This proves the claim. △

Hence, the proof of Theorem 5.22 is complete. □

That the bound in Theorem 5.22 is rather tight follows from another result by Bernshteyn [104]; in this paper he proved a weaker upper bound for the DP-chromatic number of triangle-free graphs and the following result.

Theorem 5.23 (Bernshteyn) *Every graph G with average degree $d > 2e$ satisfies*

$$\chi_{\mathrm{DP}}(G) \geq \frac{d/2}{\ln(d/2)}.$$

Proof Let G be a graph with average degree $d > 2e$, where $n = |G|$ and $m = |E(G)|$. Let k be an positive integer with

$$k \leq \frac{d/2}{\ln d/2}. \quad (+)$$

Our aim is to show that G has a k-cover without an independent transversal. To this end, fix a sequence $(X_v)_{v \in V(G)}$ of pairwise disjoint sets, each of size k. Randomly construct a graph H with vertex set $V(H) = \bigcup_{v \in V(H)} X_v$ as follows. For each edge $e = uv$ of G choose a perfect matching M_e of the complete bipartite graph with bipartition (X_u, X_v) independently and uniform at random. Then $E(H)$ is the union of the perfect matchings M_e with $e \in E(G)$. Clearly, (X, H) is a k-cover of G. Let I be any transversal of (X, H), and let p_I be the probability that I is an independent set of H. Since for every edge $e \in E(G)$ the probability that a specific edge of the corresponding complete bipartite graph belongs to M_e is $1/k$, we obtain that

$$p_I = (1 - \frac{1}{k})^m \leq e^{-m/k}$$

Consequently, if p is the probability that (X, H) has an independent transversal, then we obtain that

$$p \leq k^n \cdot e^{-m/k} = e^{n \ln k - m/k} < 1.$$

To see that the last inequality holds, observe that $d/2 > e$ and hence (+) leads to $\ln k < \ln(d/2)$ and so to

$$k \ln k < \frac{d}{2} = \frac{m}{n},$$

which is equivalent to $n \ln k - m/k < 0$. Since $p < 1$, G has a k-cover without an independent transversal. Hence $\chi_{\mathrm{DP}}(G) > k$. This proves the theorem. □

Corollary 5.24 (Bernshteyn) *Every Δ-regular triangle-free graph satisfies*

$$(\tfrac{1}{2} - o(1))\frac{\Delta}{\ln \Delta} \leq \chi_{\mathrm{DP}}(G) \leq (1 + o(1))\frac{\Delta}{\ln \Delta}.$$

In the remaining part of this section we shall discuss some interesting applications of Theorem 5.22. However, we only need the bound for the list chromatic number, which was already obtained by Molloy [754].

As we have seen in the previous section, for any fixed $k \geq 4$, there are infinitely many k-critical graphs whose average degree is less than k. That this is not the case for the class of triangle-free k-critical graphs, provided that k is large, follows from Theorem 5.22 and the following observation due to Kostochka and Stiebitz [631].

Lemma 5.25 *Let $\mathcal{P}$ be a hereditary graph property, and let f be a function such that $f(k) = o(k)$ and $\chi_\ell(G) \leq f(\Delta)$ for every graph $G \in \mathcal{P}$ with maximum degree at most Δ. Then every k-list-critical graph $G \in \mathcal{P}$ on n vertices has at least $(k - o(k))n$ edges.*

Proof We may assume that f is a continues and monotonically increasing function and $f(0) \geq 1$. Then there is a function g such that $g(k)f(g(k)) = k^2$ for every integer $k \geq 1$. Furthermore, it is easy to check that this leads to

$$(1) \quad (k - f(g(k)))(1 - \frac{k}{g(k)}) \geq k - 2f(g(k))$$

for every integer $k \geq 1$. Since $f(k) = o(k)$, it follows that $f(g(k)) = o(k)$. Now, let G be a k-list-critical member of $\mathcal{P}$ such that $n = |G|$ and $m = |E(G)|$. Our aim is to show that if k is sufficiently large, then $m \geq (k - 2f(g(k))) = (k - o(k)n$ as desired.

Since G is k-list-critical, there is a $(k-1)$-list-assignment L of G such that G is L-critical. For $v \in V(G)$ and $U \subseteq V(G)$, let $d(v : U) = |N_G(v) \cap U|$. Moreover, let $X = \{v \in V(G) \mid d_G(v) \geq g(k)\}$ and let $Y = V(G) \setminus X$. We consider two cases.

Case 1: *There exists a nonempty subset Y' of Y such that every vertex $v \in Y'$ satisfies $d(v : V(G) \setminus Y') \leq k - 1 - \lfloor f(g(k)) \rfloor$.* Since G is L-critical, $G - Y'$ has an L-colouring φ. For the graph $G' = G[Y']$, define a list-assignment L' by $L'(v) = L(v) - \varphi(N_G(v) \cap (V(G) \setminus Y'))$ for each $v \in Y'$. Then for all $v \in Y'$ we have

$$|L'(v)| \geq k - 1 - d(v : V(G) \setminus Y') \geq \lfloor f(g(k)) \rfloor.$$

Since $Y' \subseteq Y$, for all $v \in Y'$ we have

$$d_{G'}(v) \leq d_G(v) < g(k).$$

Since G belongs to $\mathcal{P}$, it follows from the hypothesis of the theorem that G' has an L'-coloring φ', but then $\varphi \cup \varphi'$ is an L-coloring of G, which is impossible.

Case 2: *For every nonempty subset Y' of Y there is a vertex $v \in Y'$ such that $d(v : V(G) \setminus Y') \geq k - \lfloor f(g(k)) \rfloor \geq k - f(g(k))$.* This implies, in particular, that there is a strict linear order $\prec$ on $V(G)$ such that:

(2) all vertices in X precede all vertices in Y, and
(3) every vertex of Y is adjacent to at least $k - f(g(k))$ preceding vertices.

Let $p(k) = \frac{k}{g(k)}$. Since

$$\frac{k^2}{g(k)} = f(g(k)) = o(k),$$

it follows that $1 - p(k) \geq p(k)$ provided that k is sufficiently large. Now, we use a discharging argument. The initial charge is 1 for every edge and 0 for every vertex implying that the sum is m. Then we transfer from every edge uv of G with $u \prec v$ a charge of $p(k)$ to u and a charge of $1 - p(k)$ to v. We may assume that $1 - p(k) \geq p(k)$. Since $d_G(v) \geq g(k)$ for all $v \in X$, the new charge of every vertex in X is at least $g(k)p(k) \geq k$. By (1) and (3), the new charge of every vertex in Y is at least $(k - f(g(k)))(1 - p(k)) \geq k - 2f(g(k))$. The new charge of every edge is zero. Therefore, $m \geq (k - 2f(g(k)))n$ as claimed. □

Corollary 5.26 (Kostochka and Stiebitz) *Let G be a triangle-free graph on n vertices. If G is k-list-critical, then G has at least $(k - o(k))n$ edges and, in particular, the average degree of G is at least $2k - o(k)$.*

Constructions of triangle-free k-critical graphs whose average degree is at most $2(k-2)$ have been obtained by various authors, in particular, by Abbott, Hare, and Zhou [3], [4] and by Kostochka and Nešetřil [626].

Evidently, every graph G of order n satisfies $\alpha(G)\chi(G) \geq n$ (see (1.2)). If G has maximum degree Δ, then a simple greedy coloring argument gives $\chi(G) \leq \Delta + 1$ and so $\alpha(G) \geq \frac{n}{\Delta+1}$. In 1980, Ajtai, Komlós, and Szemerédi [24] proved that, for the class of triangle-free graphs, this trivial lower bound can be improved by a logarithmic factor, namely $\alpha(G) \geq c\, n(\ln \Delta/\Delta)$, where $c = \frac{1}{100}$ is an absolute constant and $\Delta \geq \Delta_0$. Shortly afterwards Shearer [936] not only simplified the proof, but also improved the constant factor to $c = (1 - o(1))$. Shearer's result is another simple consequence of Theorem 5.22. Note that $\frac{1}{1+\varepsilon} \geq 1 - \varepsilon$ when $0 < \varepsilon < 1$ and that the bound in Theorem 5.22 also applies for the class of triangle-free graphs with maximum degree at most Δ.

Corollary 5.27 (Ajtai, Komlós, Szemerédi / Shearer) *Let $\Delta \geq 1$, and let G is a triangle-free graph on n vertices and with maximum degree at most Δ. Then $\alpha(G) \geq (1 - o(1))\, n\, \frac{\ln \Delta}{\Delta}$.*

It is noteworthy that the inequality in Corollary 5.27 was originally proved for the average degree in place of the maximum degree; this is a slightly stronger statement. On the other hand, it is easy to see that a graph of average degree d contains an induced subgraph H with $|H| \geq \frac{1}{2}|G|$ and $\Delta(G) \leq 2d$ (see Exercise 5.17.2). Hence the above result implies that a triangle-free graph having order n and average degree $d \geq 1$, satisfies $\alpha(G) \geq \frac{1}{2}(1 - o_d(1))n\frac{\ln d}{d}$, which is of course only half of what Shearer obtained. Here $o_d(1)$ means a positive function of d that tends to zero if d tends to infinity. Alon [47] obtained in an elegant way a bound with a decent absolute constant (see Exercise 5.18).

The **Ramsey number** $R(s,t)$ with $r, s \in \mathbb{N}$ is the smallest integer $n \in \mathbb{N}$ such that every graph G with $|G| \geq n$ satisfies $\omega(G) \geq s$ or $\alpha(G) \geq t$. Some basic properties of the Ramsey numbers are listed in Exercise 2.12. In particular, we have $R(s,t) = R(t,s)$, $R(1,t) = 1$ and $R(2,t) = t$. The first result about the function $R(3,t)$ was obtained in 1935 by Erdős and Szekeres, who proved that $R(3,t) \leq \binom{t+1}{2} = \Theta(t^2)$ (see Exercise 2.12). Half a century later, after many intermediate steps, it finally turned out that $R(3,t) = \Theta(t^2/\ln t)$. The lower bound was proved in 1995 by Kim [590] based on a sophisticated probabilistic construction and thirty pages of computation; he obtained $R(3,t) \geq \frac{1}{162}(1-o_t(1))\frac{t^2}{\ln t}$. The upper bound was proved by Ajtai, Komlós, and Szemerédi [24]; the constant term was later improved by Shearer. The tigth upper bound for $R(3,t)$ is also a simple consequence of Theorem 5.22.

Corollary 5.28 (Ajtai, Komlós, Szemerédi / Shearer) *The off-diagonal Ramsey number satisfies* $R(3,t) \leq (1+o_t(1))\frac{t^2}{\ln t}$.

Proof Fix an ε with $0 < \varepsilon < 1$. Our aim is to show that $R(3,t) \leq (1+2\varepsilon)\frac{t^2}{\ln t}$ provided that t is large. To this end, let G be a triangle-free graph on n vertices with $n = \lfloor (1+2\varepsilon)\frac{t^2}{\ln t} \rfloor$. It suffice to show that $\alpha(G) \geq t$. If the triangle-free graph G has a vertex of degree at least t, then the neighborhood of this vertex is an independent set and we are done. Otherwise, $\Delta(G) < t$ and we can apply Theorem 5.22. There is a Δ_ε such that $\chi(G) \leq (1+\varepsilon)\frac{t}{\ln t}$, provided that $t \geq \Delta_\varepsilon$. Since $n \leq \alpha(G)\chi(G)$, this leads to

$$\alpha(G) \geq \frac{n}{\chi(G)} \geq \frac{\lfloor (1+2\varepsilon)\frac{t^2}{\ln t} \rfloor}{(1+\varepsilon)\frac{t}{\ln t}} \geq t,$$

where the last inequality holds if t is large, say $t \geq \frac{1}{\varepsilon}$. □

Clearly, we can replace the asymptotic constant $(1+o_t(1)$ by an absolute positive constant c such that $R(3,t) \leq c(t^2/\ln t)$ for all $t \geq 2$. However, this constant might be very large. On the other hand, this implies the following corollary.

Corollary 5.29 (Ajtai, Komlós, and Szemerédi) *There is a constant $c > 0$, such that every triangle-free graph on $n \geq 2$ vertices satisfies $\alpha(G) \geq c\sqrt{n \ln n}$.*

Recall from Section 1.12 that we can combine the maximal independent set algorithm with the greedy strategy to find a (proper) coloring; so we color as many vertices as possible with color 1, then as many as possible with color 2, and we continue like that until every vertex is colored. However, the problem with this coloring algorithm is to estimate an upper bound for the number of used colors. One possible way to find such an upper bound was pointed out by Jensen and Toft [530, Problem 7.3].

Lemma 5.30 *Let $\mathcal{P}$ be a hereditary graph property, and let f be a positive, nondecreasing, continues function defined for all real numbers $x \geq 2$. If $\alpha(G) \geq f(|G|)$ for all members G in $\mathcal{P}$ with $|G| \geq 2$, then*

$$\chi(G) \leq 2 + \int_2^{|G|} \frac{1}{f(x)}dx$$

holds for every $G \in \mathcal{P}$ with $|G| \geq 2$.

Proof The proof is by induction on the order n of G. If G is bipartite, the result trivially holds. Clearly, G has an independent set I of size $\alpha(G)$, and hence $|I| \geq f(n)$.

Furthermore, for $G' = G - I$ we have $|G'| = n - \alpha(G) \geq 2$ (otherwise G would be bipartite). Clearly, $G' \in \mathcal{P}$ and $\chi(G) \leq \chi(G-I)+1$. Hence, by applying the induction hypothesis to G', we obtain

$$\begin{aligned}
\int_2^n \frac{1}{f(x)}dx &= \int_2^{n-\alpha(G)} \frac{1}{f(x)}dx + \int_{n-\alpha(G)}^n \frac{1}{f(x)}dx \\
&\geq \chi(G-I) - 2 + \int_{n-\alpha(G)}^n \frac{1}{f(x)}dx \\
&\geq \chi(G-I) - 2 + \int_{n-\alpha(G)}^n \frac{1}{f(n)}dx \\
&= \chi(G-I) - 2 + \alpha(G)\frac{1}{f(n)} \\
&\geq \chi(G-I) - 2 + 1 \\
&\geq \chi(G) - 2.
\end{aligned}$$

This proves the result. □

By using elementary analysis, we obtain that

$$2 + \int_2^n \frac{1}{c\sqrt{x \ln x}}dx \leq c_2 \frac{\sqrt{n}}{\sqrt{\ln n}},$$

which implies the following result, by combining Corollary 5.29 with Lemma 5.30.

Corollary 5.31 (Jensen and Toft) *Every triangle-free graph G on n vertices satisfies*

$$\chi(G) \leq c\frac{\sqrt{n}}{\sqrt{\ln n}},$$

where c is an absolute positive constant and $n \geq 2$.

By using the upper bound for the chromatic number of a triangle-free graph in terms of its maximum degree derived from Theorem 5.22, we eventually obtained another upper bound that depends only on the order of the graph. As pointed out by Gimbel and Thomassen [416] and, independently, by Nilli [788], combining these two bounds leads to a bound in terms of the size of a graph.

Corollary 5.32 (Gimbel and Thomassen / Nilli) *Every triangle-free graph G on m edges satisfies*

$$\chi(G) \leq c\frac{m^{1/3}}{(\ln m)^{2/3}},$$

where c is an absolute positive constant and $m \geq 2$.

Proof Let $\Delta = m^{1/3}(\ln m)^{1/3}$ and let $U = \{v \in V(G) \mid d_G(v) \leq \Delta\}$. Then $\chi(G) \leq \chi(G[U]) + \chi(G-U)$. Since $\Delta(G[U]) \leq \Delta$, Theorem 5.22 implies that

$$\chi(G[U]) \le c_1 \frac{\Delta}{\ln \Delta} \le c_1 \frac{m^{1/3}}{(\ln m)^{2/3}},$$

where c_1 is an absolute positive constant. Let $G' = G - U$ and $n = |G'|$. Clearly, $\delta(G') \ge \Delta$ and so $2m \ge \Delta n$, which yields

$$n \le 2\frac{m}{\Delta} = 2\frac{m^{2/3}}{(\ln m)^{1/3}}.$$

Then Corollary 5.31 implies that

$$\chi(G') \le c_2 \frac{\sqrt{n}}{\sqrt{\ln n}} \le c_3 \frac{m^{1/3}}{(\ln m)^{2/3}}$$

where c_3 is an absolute positive constant. This completes the proof. □

That the bounds in the last three corollaries all have the right order of magnitude follows from the probabilistic constructions due to Kim [590], these constructions being related to the Ramsey number $R(3,t)$.

5.7 Exercises

5.1 Let G be a k-critical graph with $k \ge 4$, let w be a low vertex of G, and let $u_1, u_2 \in N_G(w)$ be two distinct neighbors such that $u_1u_2 \notin E(G)$. Let $G^* = G/(\{u_1,u_2\} \to u^*)$ be the graph obtained from G by identifying $\{u_1,u_2\}$ to u^*. Prove that the following statements hold:

1. G^* contains a k-critical subgraph and any such subgraph contains u^*, but not w.
2. Let H be a k-critical subgraph of G^*, let $U = (V(H) \setminus \{u^*\}) \cup \{u_1,u_2\}$, and $W = V(G^*) \setminus V(H) = V(G) \setminus U$. Then

$$\mathrm{exc}_k(G) \ge \mathrm{exc}_k(H) + \sum_{v \in W} (d_G(v) - (k-1)) + E_G(U,W) - (k-1),$$

and, moreover, $\mathrm{exc}_k(G) \ge \mathrm{exc}_k(H)$. (Hint: Use Theorem 1.32)

(Kostochka [622])

5.2 A **k-clique-cluster** of a graph G is a triple (X, u_1, u_2) consisting of a clique X of G with $|X| = k-1$ and two distinct vertices $u_1, u_2 \in V(G)$ such that $X \subseteq N_G(u_1) \cup N_G(u_2)$ and $u_1u_2 \notin E(G)$. Suppose that G is a k-critical graph with $k \ge 4$ such that (1) $G \ne K_k$ and $\mathrm{exc}_k(G) \le k-4$, and (2) $|G|$ is minimum subject to (1). Prove the following statements:

1. G contains a k-clique-cluster and any such k-clique-cluster (X, u_1, u_2) satisfies that X contains a low vertex of G and $X \setminus N_G(u_i) \ne \varnothing$ for $i \in \{1,2\}$.

2. Let (X,u_1,u_2) be a k-clique-cluster such that $m = |N_G(u_1) \cap X|$ is maximum, let $u_1' = u_1$ if $N_G(u_1) \cap X$ contains a low vertex of G and let $u_1' = u_2$ otherwise, let $w \in N_G(u_1') \cap X$ be a low vertex of G, and let $u_2' \in X \setminus N_G(u_1')$ be a vertex of smallest degree. Then there is a k-clique-cluster (X', u_1', u_2') such that $M' \cap (M \setminus \{u_2\} = \varnothing$. Furthermore, $\mathrm{exc}_k(G) \geq (k-2-m)(k-2-m') + m' - 1$, where $m' = |N_G(u_1') \cap X|$.

(Hint: Use Exercise 5.1) (Kostochka [622])

5.3 Use the first two exercises to show that if G is a k-critical graph with $k \geq 4$, then $G = K_k$ or $\mathrm{exc}_k(G) \geq k-3$. (Dirac [298], see also Kostochka [622, Theorem 3])

5.4 The Dirac bound: Use Theorem 5.9 and Theorem 5.20 to show that, for $k \geq 4$, every k-critical graph $G \neq K_k$ satisfies $2|E(G)| \geq (k-1)|G| + (k-3)$, where equality holds if and only if $G \in \mathcal{DG}(k)$. (Dirac [303])

5.5 A **bad list-assignment** of a graph G is a list-assignment L of G such that G has no L-coloring. A graph G is said to be **strong k-critical** if G is k-critical and if every bad $(k-1)$-assignment L of G is constant, i.e., there is a set C of $k-1$ colors such that $L(v) = C$ for all $v \in V(G)$. Prove the following statements:

1. Every strong k-critical graph with $k \geq 1$ is minimal k-list-critical and has list chromatic number k.
2. The complete graph K_k is strong k-critical.
3. The odd cycle is strong 3-critical.
4. Every graph in $\mathcal{DG}(k)$ is strong k-critical.
5. If G is strong k-critical, that $G \boxplus K_p$ is strong $(k+p)$-critical.

(Stiebitz, Tuza, and Voigt [978])

5.6 Let G be a K_n with vertex set V and let L be a list-assignment of G. Use Exercise 1.13 to show that G is L-critical if and only if $|\bigcup_{v \in V} L(v)| = n-1$, and for every two distinct vertices $u, u' \in V$, there is a color $c \in L(u) \cap L(u')$ such that the set system $(L(v) \setminus \{c\} : v \in V \setminus \{u.u'\})$ satisfies Hall's condition. (Vizing [1051], see also [978])

5.7 Let k, n be integers with $5 \leq k \leq n$, let G be a K_n with vertex set $V = [1,n]$, and let $\pi \in S_{n-2}$ be a cyclic permutation of the set $[1, n-2]$ such that $\pi(1) = 2, \pi(2) = 3, \ldots, \pi(n-3) = n-2$ and $\pi(n-2) = 1$. For $i \in V$, let

$$S_i = \{\pi^{i-1}(j) \mid j \in [1, k-2]\}$$

and $L(i) = S_i \cup \{n-1\}$. Show that G is L-critical and hence k-list-critical. (Stiebitz, Tuza, and Voigt [978])

5.8 Let G be a graph and let (X,H) be a cover of G. We say that G is **(X,H)-critical** if G has no (X,H)-coloring, but for any vertex $u \in V(G)$, there is an independent set I of $H - X_u$ such that $|I \cap X_v| = 1$ for all $v \in V(G) \setminus \{u\}$. A graph all of whose blocks are complete graphs or cycles is called a **DP-forest** and a **DP-tree** if it is connected.

Use Corollary 1.19 and a reduction similar as in the proof of Theorem 5.4 to show the following result: Let (X,H) be a cover of G, let $U = \{u \in V(G) \mid |X_u| > d_G(u)\}$, and let $W = V(G) \setminus U$. If G is (X,H)-critical, then $G[W]$ is a DP-forest and $|X_w| = d_G(w)$ for all $w \in W$. (Bernshteyn and Kostochka [108])

5.9 A graph G is called k**-cover-critical** if there is a cover (X,H) of G such that $|X_v| \geq k-1$ for all $v \in V(G)$ and G is (X,H)-critical. Show that if G is a k-cover-critical graph with $k \geq 4$ and $G \neq K_k$, then

$$2|E(G)| \geq \left(k-1+\frac{k-3}{(k^2-3)}\right)|G|.$$

(Bernshteyn, Kostochka, and Pron [110])

5.10 Show that every k-list-critical graph is k-cover-critical.

5.11 By modifying the proof of Lemma 5.25 show that the following statement hold: Let $\mathcal{P}$ be a hereditary graph property, and let f be a function such that $f(k) = o(k)$ and $\chi_{\rm DP}(G) \leq f(\Delta)$ for every graph $G \in \mathcal{P}$ with maximum degree at most Δ. Then every k-cover-critical graph $G \in \mathcal{P}$ on n vertices has at least $(k - o(k))n$ edges.

5.12 Show that $\mathrm{ext}(5,n) = \mathrm{koy}(5,n)$ if $n \equiv 1 \pmod 4$, or $n \equiv 2 \pmod 4$ and $n \geq 10$ (Kostochka any Yancey [640])

5.13 Use Euler's formulae (see Theorem C.18) and the Kostochka–Yancey bound to show that any triangle-free graph embedded on the torus or the projective plane is 4-colorable. (Kronk and White [657])

5.14 Show that every graph of girth at least five embeddable in the projective plane is 3-colorable (Thomassen [1009]).

5.15 Show that if G is a graph embedded on the torus, then $\chi(G) \leq \mathrm{col}(G) \leq 7$ and $\chi(G) = 7$ implies $K_7 \subseteq G$ (Use Theorems C.18, C.21 and the Kostochka–Yancey bound for 6-critical graphs.) (Peter Ungar [unpublished], Dirac [297])

5.16 Dirac's map color theorem: Let $\mathbf{S}$ be a surface, and let G be a graph embeddable in $\mathbf{S}$. If the Euler characteristic $\varepsilon(\mathbf{S}) \leq -2$, then $\chi(G) \leq H(\mathbf{S})$, where equality holds if and only if G contains a complete subgraph on $H(\mathbf{S})$ vertices. Here $H(\mathbf{S})$ is the Heawood number of the surface, see (C.8). (Hint: Dirac used Theorems C.18, C.21 and his bound for the number of edges in critical graphs) (Dirac [297])

5.17 Maximum degree vs. average degree: Let G be a graph on n vertices and with average degree d. Show that the following statements hold:

1. If $V_c = \{v \in V(G) \mid d_G(v) > cd\}$ for $c \in \mathbb{R}$ and $c \geq 1$, then $|V_c| \leq \frac{1}{c}n$.
2. G has an induced subgraph H such that $|H| \geq \frac{1}{2}n$ and $\Delta(H) \leq 2d$.
3. The function $f(\Delta) = \frac{\ln \Delta}{\Delta}$ is monotonically decreasing for $\Delta \geq e$.

5.18 Fix an ε with $0 < \varepsilon < 1$. Let f be the function with $f(x) = \exp(\frac{1}{4}x^{\varepsilon})/x^3$. Use the derivate of f and L'Hôspital's rule to show that f is monotonically increasing when $x > (\frac{12}{\varepsilon})^{1/\varepsilon}$ and $\lim_{x\to\infty} f(x) = \infty$. Furthermore show that there is a Δ_ε such that $\exp(-\frac{1}{4}\Delta^{\varepsilon/2}) < \frac{1}{8}\Delta^{-3}$ for all $\Delta \geq \Delta_\varepsilon$.

5.19 Alon's bound for α for triangle-free graphs: Let G be a triangle-free graph on n vertices and with maximum degree $\Delta \geq 1$. Our aim is to how that

$$\alpha(G) \geq n\frac{\log_2 \Delta}{8\Delta}. \tag{5.15}$$

First use the trivial bound $\alpha(G) \geq n/(\Delta+1)$ to show that (5.15) holds if $\Delta < 16$. Hence in what follows we assume $\Delta \geq 16$. Let $\mathbf{I}$ be a random independent set of G, chosen uniformly among all independent sets of G. For a vertex v of G, define the random variable $X_v = \Delta|\mathbf{I} \cap \{v\}| + |\mathbf{I} \cap N_G(v)|$, and let $X = \sum_{v\in V(G)} X_v$. Show that the following statements hold:

1. Let $v \in V(G)$, let S be an independent set of $H = G - (N_G(v) \cup \{v\})$, let $W = \{w \in N_G(v) \mid \mathbf{I} \cup \{w\}$ is independent in $G\}$, and let $x = |W|$. Then

$$\mathbb{E}(X_v \mid \mathbf{I} \cap V(H) = S) = \frac{\Delta}{2^x+1} + \frac{\sum_{k=0}^{x} k\binom{x}{k}}{2^x+1} = \frac{\Delta + x2^{x-1}}{2^x+1}$$

   and

$$\mathbb{E}(X_v \mid \mathbf{I} \cap V(H) = S) \geq \frac{\log_2 \Delta}{4}.$$

   (Hint: Show otherwise $4\Delta - \log_2 \Delta < 2^x(\log_2 \Delta - 2x) < \sqrt{\Delta}\log_2 \Delta$)
2. $\mathbb{E}(X_v) \geq \frac{\log_2 \Delta}{4}$ for all $v \in V(G)$ and $\mathbb{E}(X) \geq n\frac{\log_2 \Delta}{4}$
3. $X \leq 2\Delta|\mathbf{I}|$ and $\mathbb{E}(|\mathbf{I}|) \geq n\frac{\log_2 \Delta}{8\Delta}$

(Alon [47], see also the third edition of [57])

5.8 Notes

In his thesis from 1951 Dirac noticed that Brooks' theorem is equivalent to the statement that, for any $k \geq 4$, $2\text{ext}(k,n) \geq (k-1)n$, where equality holds only for $n = k$ and with $\text{Ext}(k,k) = \{K_k\}$. This motivated Dirac [298] to investigate the function $\text{ext}(k,n)$ further, and in 1957 he published the first improvement of Brooks' theorem by showing that $2\text{ext}(k,n) \geq (k-1)n + k - 3$ if $n \geq k+2$ and $k \geq 4$. Dirac's proof in [298] was rather long and complicated. Shorter proofs were given by Kronk and Mitchem [655] as well as by Weinstein [1063]. Weinstein uses induction on k and the fact that deleting an independent set of a k-critical graph results in a graph that contains a $(k-1)$-critical graph. A very short proof of Dirac's bound was given by Kostochka [622]; he uses induction on the order and the fact that identifying two nonadjacent vertices of a k-critical graph results in a graph that contains a smaller

k-critical graph, see also Exercises 5.1, 5.2 and 5.3. In 1974 Dirac [303] extended his result by characterizing the extremal graphs, see Exercise 5.4. A selfcontained proof of this extended result by means of Kempe-changes was given by Mitchem [745]; a short proof based on the contraction method (see Exercise. 5.1) was given by Deuber, Kostochka, and Sachs [281].

In 1963 Gallai published two influential papers [397] and [398] that were both devoted to the structure of critical graphs. On the one hand, he was the first to obtain, for any fixed $k \geq 4$, a lower bound for the function $\text{ext}(k,n)$ where the excess tends to infinity with n, see (5.2). On the other hand, he used his decomposition result (Theorem 4.13) to establish the value of $\text{ext}(k,n)$, including a description of the extremal class $\text{Ext}(k,n)$, provided that $k \geq 4$ and $k+2 \leq n \leq 2k-1$. The proof of Theorem 5.9 is a refinement of Gallai's original proof in [398]. Furthermore, Gallai [397, Section 4] determined the minimum number of edges possible in a k-critical graph having exactly one high vertex; this led him to propose the suggestion that - in our terminology - $\text{ext}(k,n) = \text{koy}(k,n)$ provided that $k \geq 4$ and $n \equiv 1 \pmod{k-1}$. This is a special case of Ore's conjecture [794] that $\text{ext}(k,n) = \text{ore}(k,n)$.

In the past 30 years several authors presented improvements of both the Dirac bound and the Gallai bound for the extremal function $\text{ext}(k,n)$, see e.g. [365, 629, 630, 631, 632, 633, 651, 652]. For instance, Krivelevich [651] uses both Theorem 4.68 as well as Proposition 5.7 with $c = 2$ in addition to Gallai's arguments to show that $\text{ext}(k,n) \geq g_k(n,2)$ holds for $n \geq k+2$ and $k \geq 4$. For a detailed discussion of the many partial results dealing with the minimum number of edges in critical graphs and hypergraphs, the reader is refereed to the instructive survey by Kostochka [622]. However, only the method change by Kostochka and Yancey [639, 640] made it possible to achieve a qualitative leap in the improvement of the lower bounds for $\text{ext}(k,n)$. Their impressively short proof that $\text{ext}(4,n) = \lceil \frac{1}{3}(5n-2) \rceil$ gives not only an elegant proof, but also a better understanding of Grötzsch's famous theorem that every triangle free planar graph is 3-colorable. The reason for this is simply that triangle free planar graphs are not dense enough to contain 4-critical subgraphs. The proof of Theorem 5.10 is from [639], and the proof of Theorem 5.20 is from [640]. That lower bounds for the function $\text{ext}(k,n)$ have interesting applications in graph coloring theory, Ramsey theory, and the theory of random graphs was pointed out by Krivelevich [651] and later by Kostochka and Yancey [639, 640]. However, already Dirac [290, 295, 296, 297] used his bound for $\text{ext}(k,n)$ to extend Heawood's map color theorem, which says that the chromatic number of a graph embedded on a given surface is bounded from above by the Heawood number of the surface, see Exercises 5.15 and 5.16. Further applications to map colorings are discussed below and in Chapter 9.

In what follows we shall discuss another application of the Kostochka–Yancey bound from [640]. Let G be a graph. The **Ore-degree** of an edge $uv \in E(G)$ is defined by $\theta_G(uv) = d_G(u) + d_G(v)$, and the **Ore-degree** of G is $\theta(G) = \max_{uv \in E(G)} \theta_G(uv)$. Clearly, the Ore degree is related to a basic theorem of Ore [793] that if a graph G has Ore-degree $\theta(G) \leq |G| - 2$, then the complement of G has an Hamilton cycle. Kierstead and Kostochka [575, 578] observed that any graph G satisfies

$$\chi(G) \leq \lfloor \theta(G)/2 \rfloor + 1. \tag{5.16}$$

By Proposition 1.27, it suffices to prove (5.16) only when G is critical, but then $\chi(G) - 1 \leq \delta(G) \leq \lfloor \theta(G)/2 \rfloor$ and so we are done. Obviously, equality holds in (5.16) for complete graphs and odd cycles since $\theta(K_n) = 2n - 2$ and $\theta(C_n) = 4$. However, for small odd θ there are other critical graphs for which (5.16) holds with equality.

Let $\mathcal{OG}_4$ denote the class of graphs defined recursively as follows. The Hajós join $K_4 \nabla K_4$ belongs to $\mathcal{OG}_4$. If G belongs to $\mathcal{OG}_4$ and if $d_G(w) = 4$, then the Gallai join $(K_4, v_1, v_2) \blacktriangledown (G, w, A_1, A_2)$ with $|A_1| = |A_2| = 2$ (see Section 4.2) belongs to $\mathcal{OG}_4$, too. By using Theorem 4.7, it is easy to see that every graph $G \in \mathcal{OG}_4$ is 4-critical and satisfies $\theta(G) = 7$. Furthermore, let $O_5 = (K_5, v_1, v_2) \blacktriangledown (K_5, w, A_1, A_2)$ be the Gallai join of two graphs K_5 again with $|A_1| = |A_2| = 2$. Then O_5 is 5-critical, $|O_5| = 9$, and $\theta(O_5) = 9$. Note that O_5 belongs to the class $\text{Ext}(5,9) = \mathcal{DG}(5)$.

Theorem 5.33 *If G is a graph satisfying $\chi(G) = \lfloor \theta(G)/2 \rfloor + 1 \geq 5$, then either $\chi(G) = \omega(G)$, or $\chi(G) = 5$ and $O_5 \subseteq G$.*

For graphs with $\chi(G) \geq 7$, the above theorem was proved by Kierstead and Kostochka [575]; the case $\chi(G) = 6$ was solved by Rabern [836], and the final case $\chi(G) = 5$ was obtained by Kostochka, Rabern, and Stiebitz [627]. For a proof of Theorem 5.33, we may consider a critical subgraph G' of G such that $\chi(G') = \chi(G)$. If $k = \chi(G')$, then $k = \lfloor \theta(G')/2 \rfloor + 1$ and hence $\delta(G') \geq k - 1 = \lfloor \theta(G')/2 \rfloor$. This clearly implies that $\Delta(G') \leq k + 1$ and the vertices of degree $k + 1$ form an independent set. Hence Theorem 5.33 is a consequence of the following result.

Theorem 5.34 *If G is a k-critical graph with $k \geq 5$ such that $\Delta(G) \leq k + 1$ and the high vertex subgraph G_H is edgeless, then either $G = K_k$ or $k = 5$ and $G = O_5$.*

Proof The proof is by reductio ad absurdum. So suppose that G is a k-critical graph with $k \geq 5$ such that $\Delta(G) \leq k + 1$ and the high vertex subgraph G_H is edgeless, but G is neither a complete graph nor O_5. Let $n = |G|$ and $p = |G_H|$. As $G \neq K_k$, we have $n \geq k + 2$. Since $\text{Ext}(k, k + 2) = \langle K_{k-3} \boxplus C_5 \rangle$ (see Corollary 4.20 or Exercise 4.12) and $\Delta(K_{k-3} \boxplus C_5) = k + 2$, we obtain that $n \geq k + 3$. For the number $e(G)$ of edges of G we have $2e(G) = (k - 1)n + p$. By (5.9), this yields

$$(k-1)n + p \geq 2\lceil \text{koy}(k, n) \rceil \geq (k - \frac{2}{k-1})n + \frac{k(k-3)}{k-1},$$

which leads to

$$(1)\ p \geq \left\lceil \frac{(k-3)n - k(k-3)}{k-1} \right\rceil.$$

If $B = V(G_H)$ and $A = V(G) \setminus B$, then $e_G(A, B) = pk$. By Lemma 5.16(b), this leads to $pk \leq |A| + 3|B| - 3 = 2p + n - 3$ and hence to

$$(2)\ p \leq \left\lfloor \frac{n-3}{k-2} \right\rfloor.$$

Since $n \geq k+3$ and $k \geq 5$, it follows from (1) that

$$(3)\ p \geq \left\lceil \frac{(k-3)(k+3)-k(k-3)}{k-1} \right\rceil = \left\lceil \frac{(3(k-3)}{k-1} \right\rceil \geq 2.$$

If $n \leq 2(k-1)$, then (2) implies that $p \leq \lceil (2k-5)/(k-2) \rceil \leq 1$. Consequently, we obtain that $n \geq 2k-1$. Combining (1) and (2), we get

$$\frac{(k-3)n-k(k-3)}{k-1} \leq \frac{n-3}{k-2}$$

and resolving with respect to n yields

$$n \leq \frac{k(k-2)(k-3)-3(k-1)}{(k-1)(k-5)+2}.$$

If $k \geq 6$, the above inequality leads to $n < 2k-1$, which is impossible. If $k = 5$, then $n \geq 9$ and so $n = 9$ and $p = 2$ (by (2) and (3)). Then $2e(G) = (k-1)n + p = 38 = 2\mathrm{ext}(5,9)$ and hence $G \in \mathcal{DG}(5)$ (by Theorem 5.9), that is, G is obtained from two K_5's by splitting a vertex of one K_5 into an edge of the other K_5. Since $\Delta(G) \leq k = 5$, this implies that $G = O_5$. This completes the proof of the theorem. □

The above proof is due to Kostochka and Yancey [640]; only when $k = 5$ they use Theorem 4.68 instead of Theorem 5.9. Since then G_H consists of two independent vertices, Theorem 4.68 implies that the two high vertices form a separating set of the 5-critical graph G. Since $|G| = 9$ this also leads to $G = O_5$ (one may argue directly or use Theorem 4.6). In any case, the above proof is much shorter than the original one, but it uses deep results about the extremal function $\mathrm{ext}(k,n)$. A far reaching extension of Lemma 5.16 was obtained by Kierstead and Rabern [580]; they applied the new lemma to several coloring problems in a novel manner. In particular they established an Ore-degree version of Brooks' theorem for online list-coloring provided that $\theta \geq 10$. Theorem 5.34 was generalized by Kostochka, Rabern, and Stiebitz [627] and further by Rabern [838] to the following result. Note that this result does not supply much information about 4-critical graphs. In particular, to characterize the class of 4-critical graphs G with $\omega(G_H) \leq 1$ is an unsolved problem.

Theorem 5.35 *If G is a k-critical graph with $k \geq \Delta(G)+1-p \geq 4$ for some $p \in \mathbb{N}_0$ such that $\omega(G_H) \leq \frac{k+1}{p+1} - 2$, then either $G = K_k$ or $k = 5$ and $G = O_5$.*

The results of Kostochka and Yancey are landmarks in the theory of critical graphs. The Kostochka-Yancey bound $\mathrm{koy}(k,n)$ for $\mathrm{ext}(k,n)$ is better than Dirac's bound, but only if $n \geq 2k$. However, for $n \leq 2k-1$ we know the values of $\mathrm{ext}(k,n)$ by Gallai's exact theorem (Theorem 5.9). Our proof follows Gallai's original proof from [398] with some simplifications in Case 1. On the one hand, Gallai's result combined with the Kostochka–Yancey bound implies Dirac's bound (see Exercise 5.4). On the other hand, using Dirac's bound would simplify Gallai's proof; Case 2 is a direct consequence of the result in Exercise 5.4. Picasarri-Arrieta and Stiebitz [806] extended Gallai's exact theorem to the dichromatic number. In the late 1990s, Kostochka and

Stiebitz presented a bound in [629] that combines Gallai's and Dirac's bound, and in [630] they proved that if $n \geq k+2$ and $n \neq 2k-1$, then $2\mathrm{ext}(k,n) \geq (k-1)n+2(k-3)$ where equality holds if $n \in \{k+2, 2k, 3k-2\}$. Again, the Kostochka–Yancey bound is better than this bound, but only if $n \geq 3k-1$. As pointed out by Kostochka and Yancey, we have

$$0 \leq \mathrm{ext}(k,n) - \lceil \mathrm{koy}(k,n) \rceil \leq \frac{k(k-1)}{8} - 1.$$

For fixed $k \geq 4$, the Ore function $\mathrm{ore}(k,n)$ is the union of $k-1$ linear functions, which all have the same slope and differ only in the additive constant, and the Kostochka–Yancey function $\mathrm{koy}(k,n)$ ist the smallest of these linear functions. This may be one of the reasons why a further improvement of the Kostochka–Yancey bound will be difficult to achieve. Note that $\mathrm{koy}(k,n)$ is an integer if and only if $n \equiv 1 \pmod{k-1}$, and in case of $n \equiv 1 \pmod{k-1}$ we have $\mathrm{ext}(k,n) = \mathrm{koy}(k,n) = \mathrm{ore}(k,n)$. Shortly after the publication of their fundamental results, Kostochka and Yancey [641] succeeded in characterizing the class $\mathrm{Ext}(k,n)$ of extremal graphs provided that $n \equiv 1 \pmod{k-1}$. They proved that the union $\mathcal{EX}_k$ of all these classes is the smallest class of graphs that contains K_k and that is closed with respect to Gallai joins of the form $G \blacktriangledown K_k$ (so $G \in \mathcal{EX}_k$ implies $G \blacktriangledown K_k \in \mathcal{EX}_k$). Furthermore, they proved that if G is a k-critical graph, but $G \notin \mathcal{EX}_k$, then

$$2|E(G)| \geq \frac{(k+1)(k-2)|G| - C_k}{(k-1)}$$

where $C_4 = 2$, $C_5 = 4$, and $C_k = k^2 - 5k + 2$ for $k \geq 6$. As pointed out by Kostochka and Yancey [641] the constant C_k is best possible. As a consequence of this improved bound, Kostochka and Yancey [641] obtained that

$$\mathrm{ext}(5,n) = \begin{cases} \frac{9n-5}{4} & \text{if } n \equiv 1 \pmod 4 \\ \lceil \frac{9n-3}{4} \rceil & \text{otherwise,} \end{cases}$$

where $n \geq 5$ and $n \neq 6$. Note that all extremal graphs in $\mathcal{EX}_k$ except K_k have a separating set of size two and therefore split into smaller k-critical graphs. However, in general there are extremal graphs that are 3-connected; an example of such a graph belonging to $\mathrm{Ext}(4,8)$ is shown in Figure 5.2.

The results from Section 5.6 about the number of edges of k-critical triangle-free graphs (see Corollary 5.26) are meaningless if k is small. Moore and Smith-Roberge [759] proved that every triangle-free graph $G \in \mathrm{Crit}(4,n)$ satisfies

$$|E(G)| \geq \frac{5n+2}{3}.$$

That this bound holds if $G \in \mathrm{Crit}(4,n)$ has girth at least five was proved in 2017 by Liu and Postle [681]. They also conjectured that every triangle-free graph $G \in \mathrm{Crit}(4,n)$ satisfies $e(G) \geq (5n+5)/3$; note that equality holds if G is the Mycielsky–Grötzsch graph.

The bound by Liu and Postle has a nontrivial corollary about graphs embeddable on the torus or the Klein bottle. If G is such a graph with girth at least five and order n, then $e(G) \le \frac{5}{3}n$ (see Theorem C.18). The coloring result for the torus is due to Thomassen [1009], and the coloring result for the Klein bottle is due to Thomas and Walls [996]

Corollary 5.36 *Every graph of girth at least five embeddable on the torus or Klein bottle is* 3*-colorable.*

In 2017 Postle [820] proposed the following conjecture, which he verified for $k = 5$ and $\varepsilon_5 = \frac{1}{84}$. A proof of the conjecture for $k = 6$ was published in 2018 by Gao and Postle [404] on arXive, and a proof of the conjecture for large k was obtained in 2014 by Larsen [667] as a main part of his Ph.D thesis, see also [425].

Conjecture 5.37 (Postle) *For every $k \ge 5$ there is an $\varepsilon_k > 0$ such that if G is a k-critical K_{k-2}-free graph, then $|E(G)| \ge (\frac{k}{2} - \frac{1}{k-1} + \varepsilon_k)|G| - \frac{k(k-3)}{2(k-1)}$.*

The proof of Gallai's bound for $\text{ext}(k,n)$ in (5.2) relies on the fact that the low vertex subgraph of a critical graph is a Gallai forest (Theorem 1.33) and uses then a bound for the average degree of Gallai trees, see Proposition 5.6. Gallai's proposition was extended by Mihók and Škrekovsky [738] to a larger class of graphs.

Lemma 5.38 (Mihók and Škrekovsky) *Let $p \ge 1$ be an integer. Let T be a nonempty graph such that $\Delta(T) \le p$ and $\Delta(B) < p$ for all blocks $B \in \mathfrak{B}(T)$. Then*

$$\left(p - 1 + \frac{2}{p}\right)|T| - 2|E(T)| \ge 2.$$

This may be one of the reasons why it is often easy to obtain a Gallai type bound for the number of edges of graphs that are critical with respect to various coloring parameters or coloring properties, in particular for list-critical graphs (see Theorem 5.8) and for cover-critical graphs (see Exercise 5.9). Given a graph parameter ξ, a graph G is called **ξ-critical** or **critical with respect to ξ** if $\xi(H) < \xi(G)$ for every proper subgraph H of G. Gallai type bounds for critical graphs (or even digraphs) with respect to various kinds of graph parameters ξ were established by several authors, see e.g. [17, 77, 261, 628, 738, 818, 860, 907, 908, 947].

The term list-critical graph was introduced by Thomassen [1012]. About the same time Kostochka, Stiebitz, and Wirth [636] investigated χ_ℓ-critical graphs. However, these two concepts are closely related to each other by Proposition 5.3, which was first stated in [978]. That Gallai's bound also holds for χ_ℓ-critical graphs with $\chi_\ell \ge 4$ was first pointed out in [636]. Our short proof of Theorem 5.4 is based on the characterization of uncolorable pairs given in Theorem 1.7. The original proof due to Thomassen [1012] combines a slightly weaker version of Theorem 1.35 with the shifting argument applied by Gallai in his proof of Theorem 1.33, dating back to Brooks' own proof of his result. Therefore we would like to present Thomassen's proof here.

First, we discuss the shifting argument. Let G be a graph, let L be a list-assignment of G such that G is L-critical, let $W = \{v \in V(G) \mid d_G(v) \le |L(v)|\}$, and let $v \in W$

be an arbitrary vertex. Since G is L-critical, there is an L-coloring φ of $G - v$. Since G is not L-colorable, we have $\varphi(N_G(v)) = L(v)$. Since $v \in W$, it follows that $|\varphi(N_G(v))| = d_G(v)$, that is, for each color $c \in L(v)$ there is a unique vertex $w_c \in N_G(v)$ such that $\varphi(w_c) = c$. Consequently, we have $|L(v)| = d_G(v)$. If $w = w_c$ is a neighbor of v in G, we can uncolor w and color v with c. This results in an L-coloring φ' of $G - w$. We then say that φ' is obtained from φ by **shifting** the color from w to v. If $w \in W$, we can repeat the shifting argument with another neighbor of w in G.

Theorem 5.39 (Thomassen) *Let G be a graph, let L be a list-assignment of G such that G is L-critical, and let $W = \{v \in V(G) \mid d_G(v) \leq |L(v)|\}$. Then $G[W]$ is a Gallai forest (possibly empty) and $d_G(w) = |L(w)|$ for all $w \in W$.*

Proof It suffices to prove that $G[W]$ is a Gallai forest. Suppose this is false. Then it follows from Theorem 1.35 that $G[W]$ contains an even cycle, say $C = (v_0, v_1, \ldots, v_p, v_0)$ such that no chord of C in G is incident with v_0. Let $k = d_G(v_0)$ and $U = N_G(v_0) \setminus \{v_p, v_1\}$. Since G is L-critical, there is an L-coloring φ of $G - v_0$. Since $v \in W$, the list $L(v)$ is a set of k-colors and, by symmetry, we may assume that $L(v) = [1,k]$, $\varphi(v_1) = 1$, $\varphi(v_p) = 2$ and $\varphi(U) = [3,k]$. Starting with $\varphi_0 = \varphi$, we obtain for $i \in [1,p]$, a coloring φ_i from φ_{i-1} by shifting the color from v_i to v_{i-1}. Finally, we obtain a coloring φ' from φ_p by shifting the color from v_0 to v_p. Since $V(C) \subseteq W$, it follows that φ' is an L-coloring of $G - v_0$, where $\varphi'(v_p) = 1$, $\varphi'(v_1) = \varphi(v_2)$ and $\varphi'(U) = [3,k]$ (since $U \cap V(C) = \varnothing$). Since G has no L-coloring, this implies that $\varphi'(v_1) = \varphi(v_2) = 2$. Repeating this argument, we obtain that φ is a (proper) 2-coloring of the path $P = C - v_0$ such that the two ends of P have distinct colors. However, this is impossible, since $|P|$ is odd. Thus the theorem is proved. □

That Theorem 5.39 easily implies Theorem 1.7 was pointed out by Thomassen [1012]. So the two theorems can easily be derived from each other.

Let $\mathrm{Crit}_\ell(k)$ denote the class of χ_ℓ-critical graphs G with $\chi_\ell(G) = k$, let $\mathrm{Crit}_\ell(k,n) = \{G \in \mathrm{Crit}_\ell(k) \mid |G| = n\}$, and let

$$\mathrm{ext}_\ell(k,n) = \min\{|E(G)| \mid G \in \mathrm{Crit}_\ell(k,n)\}.$$

Recall from Proposition 5.7 that

$$g_k(n,c) = \left(k - 1 + \frac{k-3}{(k-c)(k-1)+k-3}\right)n.$$

Gallai's bound says that $2\mathrm{ext}_\ell(k,n) \geq g_k(n,0)$ provided that $k \geq 4$ and $n > k$. Over the years several improvements of Gallai's bound were obtained, in particular, by Kostochka and Stiebitz [633], by Rabern [840] and [841], by Cranston and Rabern [262] and by Kierstead and Rabern [581]. Let $\alpha_k = \frac{1}{2} - \frac{1}{(k-1)(k-2)}$. In [633] it is proved that $2\mathrm{ext}_\ell(k,n) \geq g_k(n,c)$ with $c = (k-5)\alpha_k$ and for $n > k \geq 6$, this was improved in [581] to $2\mathrm{ext}_\ell(k,n) \geq g_k(n,c)$ with $c = (k-3)\alpha_k$ and for $n > k \geq 7$ and to $2\mathrm{ext}_\ell(k,n) \geq g_k(n,c)$ with $c = (k-4)\alpha_k$ and for $n > k$ and $k \in \{5,6\}$. The bounds by Kierstead and Rabern [581] were further improved for $k \geq 7$ by Cranston and

Rabern [262] and for $k \in \{4,5,6\}$ by Rabern [840], who proved that $2\text{ext}_\ell(k,n) \geq (k-1+(k-3)/(k^2-2k+2))n$, provided that $k \geq 4$ and $n > k$. One primary tool used by Rabern is a lemma from [580] that generalizes the kernel technique described in Section 5.5. The **maximum independent cover number** of a graph G, denoted by $\text{mic}(G)$ is the maximum of $\sum_{v \in I} d_G(v)$ over all independent sets I of G. Then every graph $G \in \text{Crit}_\ell(k,n)$ satisfies

$$2|E(G)| \geq (k-2)n + \text{mic}(G) + 1$$

(see [580] and [840]). The bounds for $\text{ext}_\ell(k,n)$ obtained in [581] and in [840] also holds for graphs that are critical with respect to the online list chromatic number and to the Alon-Tarsi number. Kostochka and Stiebitz [632] succeeded to prove a Dirac type bound for $\text{ext}_\ell(k,n)$ namely

$$2\ \text{ext}_\ell(k,n) \geq (k-1)n + k - 3 \text{ for } n \geq k+2 \geq 6. \tag{5.17}$$

The above bound does not hold for k-list-critical graphs. If G is the graph obtained from two disjoint K_k's by adding one edge, then G is k-list-critical (see Remark 5.2), but $2|E(G)| = (k-1)n + 2$. However, the bounds in [632] and [633] were originally established for k-list-critical graphs not containing K_k as a subgraph. In contrast to Dirac (see Exercise 5.4), Kostochka and Stiebitz could not characterize the χ_ℓ-critical graphs that satisfy Dirac's bound with equality. However, investigating χ_{DP}-critical graphs and k-cover-critical graphs (see Exercises 5.8 and 5.9 for the definition). Bernshteyn and Kostochka [108] could answer the question.

Theorem 5.40 (Bernshteyn and Kostochka) *For $k \geq 4$, every k-cover-critical graph G with $\omega(G) < k$ satisfies $2|E(G)| \geq (k-1)|G| + (k-3)$, where equality holds if and only if $G \in \mathcal{DG}(k)$.*

Dirac [296] used his bound for $\text{ext}(k,n)$ to verify Hadwiger's conjecture for graphs that can be embedded in a surface and whose chromatic number is close to the Heawood number of that surface. Stiebitz and Toft [975] extended the results obtained by Dirac and proved that if G is a k-contraction critical graph, but not a K_k, then the number of vertices having degree at least $k+1$ in G is at least $k-4$ and, therefore, $2|E(G)| \geq k|G| + k - 4$.

Let us briefly discuss another coloring parameter. Let $t \in \mathbb{N}$. For a graph or multigraph G, define $\chi_t(G) = \chi(G : \mathcal{DG}_{t-1})$ (see Section 2.6). So $\chi_t(G)$ is the least integer p for which G has a partition into p parts each of which is strictly t-degenerate (i.e., $(t-1)$-degenerate). In particular, we have $\chi_1 = \chi$. There are only few papers dealing with χ_t-critical graphs for $t \geq 2$. The papers by Kronk and Mitchen [656], by Bollobás and Harary [131], by Mihok [735], and by Škrekovski [947] are all devoted to the structure of χ_2-critical graphs. The two papers by Thomason [997] [998] deal with χ_t-critical graphs for $t \geq 2$, and the papers by Schweser [907] and by Schweser and Stiebitz [909] contain some results about χ_t-critical multigraphs for $t \geq 2$. Škrekovski [947] established a Gallai type bound as well as a Dirac type bound for the minimum number of edges in χ_2-critical graphs; a Gallai type bound for χ_t-critical graphs in general was proved by Kostochka, Schweser, and Stiebitz [628] as a special case of a more general result. If G is a χ_t-critical graph of order n such that $\chi_t(G) = k+1 \geq 4$ and $n > kt+1$, then

$$2|E(G)| \geq \left(kt + \frac{kt-2}{(kt+1)^2-3}\right)n + \frac{2kt}{(kt+1)^2-3}.$$

The following two results from [818] indicate that it makes sense to look at χ_t-critical multigraphs.

Theorem 5.41 (Postel, Schweser, and Stiebitz) *Let G be a χ_t-critical multigraph of order n and with $\chi_t(G) = k$ for $k \geq 2$. If $n \leq 2k-2$, then G is obtained from the disjoint union of two nonempty submultigraphs of G, say G_1 and G_2, by joining each vertex of G_1 to each vertex of G_2 by exactly t parallel edges.*

Theorem 5.42 (Postel, Schweser, and Stiebitz) *Let n and k be integers with $n = k+p$ and $1 \leq p \leq k-1$. If $\mathrm{ext}_t(k,n)$ is the minimum number of edges in a χ_t-critical multigraph G satisfying $\chi_t(G) = k$ and $|G| = n$, then $\mathrm{ext}_t(k,n) = t\binom{n}{2} - \frac{t}{2}(2p+1)p$, provided that t is even.*

Recall from Section 3.10 that the dichromatic number of a multidigraph D is denoted by $\vec{\chi}(D)$. So D is $\vec{\chi}$-critical if $\vec{\chi}(D') < \vec{\chi}(D)$ for every subdigraph D' of D. Clearly, every $\vec{\chi}$-critical digraph is strict, but may contain digons. By $\mathrm{extd}(k,n)$ we denote the minimum number of edges in a $\vec{\chi}$-critical digraph D such that $\vec{\chi}(D) = k$ and $|D| = n$; and by $\mathrm{exto}(k,n)$ we denote the minimum number of edges in a $\vec{\chi}$-critical oriented graph D such that $\vec{\chi}(D) = k$ and $|D| = n$. Aboulker, Bellitto, Havet, and Rambaud [14] used the characterization of the low vertex subdigraph, due to Bang-Jensen, Belitto, Schweser, and Stiebitz [78], to establish a Gallai bound of the form

$$\mathrm{exto}(k,n) \geq \left(k-1+\frac{2k-5}{4(2k-3)}\right)n + \frac{3}{4(2k-3)} \quad \text{for } n > k \geq,4$$

thus improving the lower bound of [78]. Using the potential methods, Aboulker et al. proved that $\mathrm{exto}(3,n) \geq (7n+2)/3$ and Havet et al. proved in [480] that $\mathrm{exto}(4,n) \geq (10/3+1/51)n-1$. Stehlík [960] extended Gallai's decomposition result for critical graphs to digraphs, thereby answering a question that was raised in an earlier version of [78].

Theorem 5.43 (Stehlík) *Let D be a $\vec{\chi}$-critical digraph of order n and with $\vec{\chi}(D) = k$ for $k \geq 3$. If $n \leq 2k-2$, then D is obtained from the disjoint union of two nonempty subdigraphs of D, say D_1 and D_2, by joining each vertex of D_1 to each vertex of D_2 by two opposite directed edges.*

It is easy to see that if D is a $\vec{\chi}$-critical digraph with dichromatic number $k \geq 1$, then each vertex $v \in V(D)$ satisfies $\min\{d_D^+(v), d_D^-(v)\} \geq k-1$. Mohar [750] proved a Brooks type result for the dichromatic number which is equivalent to the following immediate consequence of Theorem 3.50.

Theorem 5.44 (MOHAR) *If D is a $\vec{\chi}$-critical digraph with $\vec{\chi}(D) = k \geq 2$ such that each vertex $v \in V(D)$ satisfies $d_D^+(v) = d_D^-(v) = k-1$, then* (1) $k = 2$ *and D is a directed cycle, or* (2) $k = 3$ *and $D = D^{\pm}(C_{2p+1})$ with $p \geq 1$, or* (3) $k \geq 4$ *and $D = D^{\pm}(K_k)$.*

As a consequence of the above result, for $k \geq 2$, we have $\text{extd}(k,n) \geq (k-1)n$, and equality holds if and only if $k = 2$ and $n \geq 3$, or $k = 3$ and $n \geq 3$ odd, or $n = k \geq 4$. Furthermore, it follows from (3.18) that $\text{extd}(k,n) \leq 2\text{ext}(k,n)$ for $n \geq k \geq 4$ and $n \neq k+1$. The set $\text{Crit}(k,k+1)$ is empty, but the corresponding set for digraphs is not. This set consists of exactly one member (by Theorem 5.43), namely the digraph that is obtained from $D^{\pm}(K_{k+1})$ by deleting the edges of a directed cycle of length three. Hence $\text{extd}(k,k+1) = \binom{k}{2} - 3$. Kostochka and Stiebitz [635] were able to apply the potential method to critical digraphs with dichromatic number 4 and to determine the function $\text{extd}(4,n)$ with a small error term by showing that for $n \geq 4$ we have

$$\text{extd}(4,n) \geq \frac{10n-4}{3}.$$

Consequently, for $n \geq 4$ and $n \neq 5$, we have $\text{extd}(4,n) = 2\text{ext}(4,n)$ if $n \equiv 1 \pmod 3$ or $n \equiv 2 \pmod 3$; otherwise we have $2\text{ext}(4,n) - 1 \leq \text{extd}(4,n) \leq 2\text{ext}(4,n)$. This motivated Kostochka and Stiebitz [635] to propose the conjecture that $\text{extd}(k,n) = 2\text{ext}(k,n)$ provided that $n \geq k \geq 4$ and $n \neq k+1$, see also the paper by Havet, Picasarri-Arrieta, and Rambaud [480]). By (5.10) this implies the following conjecture.

Conjecture 5.45 (KOSTOCHKA AND STIEBITZ) *If $k \geq 4$ is a fixed integer, then*

$$\lim_{n\to\infty} \frac{\text{extd}(k,n)}{n} = k - \frac{2}{k-1}.$$

That the limit indeed exists follows from Fekete's lemma (Lemma C.44) and the fact that there is a Hajós type construction for digraphs that preserves criticality (see [78, Theorem 2]), which implies, similar to (5.3), that

$$\text{extd}(k,n+n') \leq \text{extd}(k,n) + \text{extd}(k,n'+1) - 1$$

for $n,n' \geq k \geq 4$. Aboulker and Vermande [17] established a Gallai type bound for $\text{extd}(k,n)$ and a Dirac type bound of the form $\text{extd}(k,n) \geq (k-1)n + (k-3)$ if $n > k \geq 4$.

If G is a critical graph, then $D^{\pm}(G)$ is a critical digraph. On the other hand, it is a nontrivial task to construct critical digraphs without digons. For $k \in \mathbb{N}$, let $N^o(k)$ denote the least integer n for which there exists a k-critical digraph that is an oriented graph and has order n. Since every k-critical digraph satisfies $\min\{d_D^+(v), d_D^-(v)\} \geq k-1$, we have $N^o(k) \geq 2k-1$. The directed triangle shows that $N^o(2) = 3$. From Theorem 5.43 it follows that $N^o(k) \geq 2k$ when $k \geq 3$. By a result of Neumann-Lara [782], we have $N^o(3) = 7$, $N^o(4) = 11$ and $17 \leq N^o(5) \leq 19$. In a recent paper by Bellitto, Bousquet, Kabela, and Pierron [93] it was shown, based on a computer search, that $N^o(5) = 19$, thus disproving a conjecture of Neumann-Lara.

The study of the class $\mathcal{G}(\omega \le 2)$ of triangle-free graphs dates back to the pioneering work of Blanche Descartes from 1947 when she proved that triangle-free graphs may have arbitrarily high chromatic numbers. Contrary to that, Herbert Grötzsch proved in 1958 that every graph in the class $\mathcal{G}(\omega \le 2)$ embeddable in the plane has chromatic number at most 3. These two outstanding publications [280] and [434] mark the beginning of a long history of studying the class $\mathcal{G}(\omega \le 2)$ and related classes. Today, the AMS database lists more than 400 publications that mention triangle-free graphs already in their title. The theory of triangle-free graphs uses a broad array of tools and results from other fields like linear algebra, probability theory, geometry, information theory, and algorithm theory. The study of the chromatic number and the independence number of triangle-free graphs has become a classic topic in graph theory; this topic is closely related to quantitative Ramsey theory. The fact that this field of research has become so popular and has remained so until today is undoubtedly due to the great interest and the many trend-setting contributions of Paul Erdős on this topic. Not only did he develop a new method (see [327, 328, 329]) - the *probabilistic method* - with the help of which he achieved surprising new results, but he also presented a large number of problems and conjectures which contributed significantly to the continuing popularity of this field. In 1947, Erdős [327] proved that there exists a graph on $2^{k/2}$ vertices such that $\omega(G) < k$ and $\alpha(G) < k$, and hence $R(k,k) > 2^{k/2}$. The proof is not based on a deterministic construction, but on a 12-line counting argument. In the two papers [328] and [329], Erdős used advanced probabilistic arguments to show – among other results – the existence of graphs with arbitrary high girth and chromatic number, and the existence of a triangle-free graph on n vertices and independence number at most $c\, n^{1/2} \ln n$ for every $n \ge n_0$, which implies that $R(3,t) = \Omega(t^2/\ln t)$. The asymptotic behaviour of the function $R(3,t)$ remained Erdős' favorite problem for many years, see the survey paper by Graham and Nešetřil [426]. In 1980, Ajtai, Komlós, and Szemerédi [24] proved that $R(3,t) = O(t^2/\log t)$. A simpler proof of a slight generalization was presented by Shearer [936]. Finally, in 1995 Kim [590] proved this to be best possible. So far $R(3,t)$ is the only nontrivial infinite set for which the asymptotic behaviour of Ramsey numbers is known. In his paper [330] from 1962, Erdős defined the function

$$\mathrm{erd}(m,k,n) = \max\{\chi(G) \mid |G| \le n, \forall H \subseteq G : (|H| \le m \Rightarrow \chi(H) \le k)\},$$

in particular,

$$\mathrm{erd}(3,2,n) = \max\{\chi(G) \mid G \in \mathcal{G}(\omega \le 2), |G| \le n\}.$$

Erdős' result in [329] shows that $\mathrm{erd}(3,2,n) = \Omega(n^{1/2}/\ln n)$. He also established a lower bound for the function $\mathrm{erd}(m,3,n)$. In 1995, Kim proved that $\mathrm{erd}(3,2,n) \ge \frac{1}{9}\sqrt{n/\ln n}$ for $n \ge n_0$. By Corollary 5.31 this yields

$$\mathrm{erd}(3,2,n) = \Theta(\frac{\sqrt{n}}{\sqrt{\ln n}})$$

So for the function $\text{erd}(3,2,n)$ the order of magnitude is known, but it would be of interest to have an asymptotic formula. The upper bound can be obtained easily from Corollary 5.27 due to Ajtai, Komlós, and Szemerédi [24], as pointed out by Jensen and Toft [530]; that this can be done was already noted by Erdős and Hajnal [350], but without giving details. The paper of Ajtai, Komlós, and Szemerédi [24] from 1980 is a landmark in the further development of the probabilistic method, as it is the origin of the *semi-random method*. This method is often called *Rödl nibble method*, as it was used by Rödl [873] in an ingenious manner to solve a conjecture of Erdős and Hanani [352] about perfect covers in uniform hypergraphs.

For the function $\text{tri}(\Delta)$ the order of magnitude is also known, namely

$$\text{tri}(\Delta) = \Theta\left(\frac{\Delta}{\ln \Delta}\right). \tag{5.18}$$

The lower bound in (5.18) follows from random constructions due to Bollobás [127] and Kostochka and Mazurova [623]. It is notable that the lower bound also applies if we replace the class $\mathcal{G}(\omega \leq 2)$ by the subclass $\mathcal{G}(\text{girth} \geq \ell)$ for any fixed $\ell \geq 3$. A proof of the upper bound in (5.18) by means of the semi-random method were obtained by Johansson [534], by Molloy and Reed [756], and by Jamall [522], and possibly others. In 1995, Kim [589] proved such a bound, however only for the class $\mathcal{G}(\text{girth} \geq 5)$. The first proof for the class $\mathcal{G}(\omega \leq 2)$ was obtained in 1996 by Johannsen with the leading constant $c = (9 + o(1))$, but we are not aware of any copy of Johannsen's manuscript; a first available proof, however with the larger constant $c = (160 + o(1))$, was given in 2001 by Molloy and Read [756]. Jamall [522] improved the constant to $c = (67 + o(1))$. In 2010 Moser and Tardos [761] achieved another landmark in the further development of the probabilistic method. They developed a general framework – the *entropy compression method* – that can be used to provide efficient probabilistic algorithms for finding the combinatorial objects whose existence are guaranteed by Lovász Local Lemma. The entropy compression method was subsequently used by Jamall [523], by Petti and Su [805], and by Molloy [754] to improve the leading constant for the upper bound in (5.18) eventually to $c = (1 + o(1))$. As pointed out by Molloy [754] the best known constant for the lower bound in (5.18) is $c = \frac{1}{2}$ and comes from random Δ-regular graphs (see Frieze and Luzcak [391]). It should be highlighted that the upper bounds obtained in [522, 523, 534, 535, 589, 754, 756, 805] apply to the list chromatic number and not only to the chromatic number. That these bounds, in particular Molloy's bound with the leading constant $c = (1 + o(1))$, also apply to the DP-chromatic number was proved by Bernshteyn [106]. Bernshteyn cleverly realized that the use of the entropy compression method in Molloy's proof can be replaced by the lopsided Lovász Local Lemma, which leads to a substantial simplification of the proof. The short proof of Theorem 5.22 is from Bernshteyn's paper [106], and the short proof of Theorem 5.23 is from Bernshteyn's paper [104]. Obtaining an improved constant or an asymptotic formula would be a major achievement.

In 2020, Rosenfeld [877] proposed a new counting technique that aims to be applied to the same problems as Lovász Local Lemma or the entropy compression method. Bernshteyn, Brazelton, Cao, and Kang [107] used Rosenfeld's technique to give an alternative proof of Theorem 5.22. A different proof of Molloy's result using Rosenfeld's counting argument was given by Hurley and Pirot [509] and simplified by Martinsson [717]. To demonstrate Rosenfeld's technique, we will prove the following result. The proof is due to Dvořák [309] and is similar to the proof of Martinsson; however the bound is slightly weaker than the bound in [717].

Theorem 5.46 (Martinsson / Dvorák) *Let G be a triangle-free graph having order n and maximum degree at most Δ, where $n \geq 1$ and $\Delta \geq 10^{10}$. Let $k = \lceil 4\Delta/\log\Delta \rceil$, and let $\ell = \Delta^c/\log\Delta$ with $c = 1 - \log 2$. Then $|CO(G,k)| \geq \ell^n$ and so $\chi(G) \leq k$.*

Proof Note that $\log x = \log_{10} x$, and so $c = 0.698..$ We claim that for every vertex v of G, we have

$$|CO(G,k)| \geq |CO(G-v,k)| \cdot \ell. \tag{5.19}$$

Note that $|CO(K_1,k)| = k \geq \ell$ and $|CO(\varnothing,k)| = 1$. Hence it follows from (5.19) by induction on n that $|CO(G,k)| \geq \ell^n$. We proof (5.19) by induction on n. If $n = 1$, the claim is evident. So assume that $n \geq 2$, and let v be an arbitrary fixed vertex of G. Then the induction hypothesis implies that

$$|CO(G-v,k)| \geq |CO(G-v-u,k)| \cdot \ell \text{ for all } u \in V(G-v). \tag{5.20}$$

For a partial coloring ϕ of G with color set $S = [1,k]$, that is, $\phi \in CO(G',k)$ where G' is an induced subgraph of G, let $L_\phi(u) = [1,k] \setminus \phi(N_G(u))$ be the set of colors that are not used among the neighbours of u in G, whether $\phi(u)$ is defined or not. For an induced subgraph H of G', let $\phi|_H$ denote the restriction of ϕ to $V(H)$. If $\varphi \in CO(G-v,k)$, then each coloring in $CO(G,k)$ can be obtained from φ by coloring v with a color from $L_\varphi(v)$. Hence we obtain that

$$|CO(G,k)| = \sum_{\varphi \in CO(G-v,k)} |L_\varphi(v)|. \tag{5.21}$$

We now consider the probability space $(\Omega, \Pr)$ with $\Omega = CO(G-v,k)$ and with uniform distribution $\Pr(\varphi) = m$ for all $\varphi \in \Omega$. Then for $\varphi \in \Omega$, $X = |L_\varphi(v)|$ is a random variable. Our aim is to show that

$$\mathbb{E}(|L_\varphi(v)|) \geq \ell \text{ for all } \varphi \in \Omega. \tag{5.22}$$

Note that (5.22) combined with (5.21) implies (5.19), by the linearity of the expectation and because $|CO(G,k)|$ is constant. For the proof of (5.22), let $\varphi \in \Omega$ be any coloring (chosen uniformly at random from $\Omega = CO(G-v,k)$).

Let G_0 be the subgraph obtained from G by deleting the vertex set $N_G(v) \cup \{v\}$. Since G is triangle-free, we obtain that the following statements hold:

(a) $N_G(v)$ is an independent set of G.
(b) If ϕ and ϕ' are two partial k-colorings of $G-v$ such that $\phi|_{G_0} = \phi'|_{G_0}$, then $L_\phi(u) = L_{\phi'}(u)$ for all $u \in N_G(v)$.

Let $t = \ell/\log\Delta$; note that $t \geq 2$. For a vertex $u \in N_G(v)$, let

$$M_u = \{\varphi' \in \mathcal{CO}(G-v,k) \mid |L_{\varphi'}(u)| \leq t\}.$$

Using (b) and (5.20), we obtain that

$$\Pr(|L_\varphi(u)| \leq t) = \frac{|M_u|}{|\mathcal{CO}(G-v,k)|} \leq \frac{t|\mathcal{CO}(G-v-u)|}{|\mathcal{CO}(G-v,k)|} \leq \frac{t}{\ell} \tag{5.23}$$

for all $u \in N_G(v)$. For a partial k-coloring ϕ of $G-v$, let

$$W_\phi = \{u \in N_G(v) \mid |L_\phi(u)| \leq t\}.$$

From (5.23) it then follows that

$$\mathbb{E}(|W_\varphi|) = \sum_{u \in N_G(v)} \Pr(|L_\varphi(u)| \leq t) \leq \frac{t}{\ell}\Delta \leq \frac{k}{4}.$$

By Markov's inequality, this implies that

$$\Pr\left(|W_\varphi| > \frac{k}{2}\right) \leq \frac{1}{2} \text{ and hence } \Pr\left(|W_\varphi| \leq \frac{k}{2}\right) \geq \frac{1}{2}.$$

If X is a nonnegative random discrete variable and A is an event, then it is known that

$$\mathbb{E}(X) = \Pr(A)\cdot\mathbb{E}(X|A) + \Pr(\overline{A})\cdot\mathbb{E}(X|\overline{A}) \geq \Pr(A)\cdot\mathbb{E}(X|A).$$

Using this with $A = (|W_\varphi| \leq \frac{k}{2})$, we obtain that

$$\mathbb{E}(|L_\varphi(v)|) \geq \Pr(A)\cdot\mathbb{E}(|L_\varphi(v)||A) \geq \frac{1}{2}\mathbb{E}(|L_\varphi(v)| \mid |W_\varphi| \leq \frac{k}{2})).$$

Hence, to finish the proof of (5.22) it suffices to show that

$$\mathbb{E}(|L_\varphi(v)| \mid (|W_\varphi| \leq \frac{k}{2})) \geq 2\ell \text{ for all } \varphi \in \Omega. \tag{5.24}$$

Let $\phi_0 \in \mathcal{CO}(G_0,k)$ be an arbitrary coloring such that $|W_{\phi_0}| \leq k/2$. Then every coloring $\varphi \in \Omega$ with $\varphi|_{G_0} = \phi_0$ satisfies $L_\varphi(u) = L_{\phi_0}(u)$ for all $u \in N_G(v)$ (by (b)) and hence $W_\varphi = W_{\phi_0}$. Consequently, in order to prove (5.22) it suffices to show that for all colorings $\varphi \in \Omega$, we have

$$\mathbb{E}(|L_\varphi(v)| \mid (\varphi|_{G_0} = \phi_0)) \geq 2\ell.$$

Any coloring $\varphi \in \Omega$ such that $\varphi|_{G_0} = \phi_0$ can be constructed from ϕ_0 by coloring each neighbor $u \in N_G(v)$ by an arbitrary color in $L_{\phi_0}(u)$ (by (a)). As $\varphi \in \Omega$ is uniformly distributed, any such extension of ϕ_0 is equally likely. Let $U = N_G(v) \setminus W_{\phi_0}$, let $G_1 = G - v - U$, and let $\phi_1 \in \mathcal{CO}(G_1,k)$ be any coloring such that $\phi_1|_{G_0} = \phi_0$. Then $W_{\phi_1} = W_{\phi_0}$ and so $|W_{\phi_1}| \leq k/2$. To finish the proof, it suffices to show that for all colorings $\varphi \in \Omega$, we have

$$\mathbb{E}(|L_\varphi(v)| \mid (\varphi|_{G_1} = \phi_1)) \geq 2\ell.$$

So let $\varphi \in \Omega$ be any coloring such that $\varphi|_{G_1} = \phi_1$. Let $T = L_{\phi_1}(v)$ be the set of colors of $S = [1, k]$ that are not used by ϕ_1 among the neighbors of v in G. Since $|W_{\phi_1}| \leq k/2$, we have $|T| \geq k/2$. For $u \in U = N_G(v) \setminus W_{\phi_1}$, let $L(u) = L_{\phi_1}(u)$. Note that $|L(u)| > t$ for all $u \in U$. The coloring φ can be obtained from ϕ_1 by coloring each vertex $u \in U$ by a color from the set $L(u)$ (independently and uniformly at random). Then we only use colors from the set T. For $c \in T$, let Y_c be the binary random variable with $Y_c = 1$ if $c \in L_\varphi(v)$. Then, for $E = \mathbb{E}(|L_\varphi(v)| \mid (\varphi|_{G_1} = \phi_1))$ we obtain that

$$E = \sum_{c \in T} \mathbb{E}(Y_c \mid (\varphi|_{G_1} = \phi_1)) = \sum_{c \in T} \Pr((Y_c = 1) \mid (\varphi|_{G_1} = \phi_1))$$

Since we color each vertex $u \in U$ with a color from $L(u)$ independently and with equal probability, we obtain that

$$E = \sum_{c \in T} \prod_{u \in U : c \in L(u)} (1 - 1/|L(u)|) \geq |T| \left(\prod_{c \in T} \prod_{u \in U : c \in L(u)} (1 - 1/|L(u)|) \right)^{1/|T|},$$

where we use the inequality for the arithmetic mean and the geometric mean. By rearranging the double product, we obtain that

$$E \geq |T| \left(\prod_{u \in U} \prod_{c \in L(u) \cap T} (1 - 1/|L(u)|) \right)^{1/|T|} \geq |T| \left(\prod_{u \in U} (1 - 1/|L(u)|)^{|L(u)|} \right)^{1/|T|}$$

Now we use that $|L(u)| \geq t$ for all $u \in U$, $|U| \leq d_G(v) \leq \Delta$, $t \geq 2$ and $|T| \geq k/2$. Then we obtain that

$$\begin{aligned} E &\geq |T| \left(\prod_{u \in U} (1 - 1/t)^t \right)^{1/|T|} \geq |T| (1 - 1/t)^{t\Delta/|A|} \\ &\geq |T| 4^{-\Delta/|A|} \geq \tfrac{k}{2} 4^{-2\Delta/k} \geq \frac{k}{k 4^{(\log \Delta)/2}} \\ &\geq 2 \frac{\Delta^{1 - \log 2}}{\log \Delta} = 2\ell. \end{aligned}$$

This completes the proof of Theorem 5.46. □

Martinsson [717] used almost the same argument, but with $k = \lceil (1 + \varepsilon)\Delta / \ln \Delta \rceil$, $\ell = (\ln \Delta)^2$ and $1 \ll t \ll \ell / \ln \Delta$. The result obtained by Hurley and Pirot [509], also based on Rosenfeld's technique, is quite a bit stronger since it applies to list colorings and to graphs containing a small number of edges in the neighborhood of each vertex, see the discussion below. Wanless and Wood [1061] used Rosenfeld's technique to obtain an upper bound for the list chromatic number of an r-uniform hypergraph depending on its maximum degree (see Theorem 7.40). In this paper, the reader can find further references to papers showing that exponentially many colorings exists.

The following conjecture about the chromatic number of triangle-free graphs is a special case of a more general conjecture proposed by Bruce Reed in his seminal paper [854], see Conjecture 6.3.

Conjecture 5.47 (Reed) *If G is a triangle-free graph, then $\chi(G) \leq \frac{\Delta(G)}{2} + 2$.*

So Reed's conjecture says that $\mathrm{tri}(\Delta) \leq \Delta/2+2$. When Δ is large, this follows from (5.18). Evidently, $\mathrm{tri}(\Delta) \leq \Delta+1$ and Brooks' theorem implies that equality holds if and only if $\Delta \in \{0,1,2\}$. So for $\Delta \geq 3$ we have $\mathrm{tri}(\Delta) \leq \Delta$. The Petersen graph shows that $\mathrm{tri}(3) = 3$. Gallai [397] constructed 4-regular 4-critical graphs that are triangle-free (see Theorem 3.46) and so $\mathrm{tri}(4) = 4$. It seems to be unknown whether $\mathrm{tri}(5) \leq 4$ and $\mathrm{tri}(6) \leq 5$. The first improved linear upper bound, namely $\mathrm{tri}(\Delta) \leq \frac{3}{4}(\Delta+2)$ was obtained independently, by Borodin and Kostochka [157], Catlin [193] and Lawrence [668]. The proof (see Theorem 2.25) follows by combining Lovaśz's decomposition result (Theorem 2.24) with Brooks' theorem. However, to apply Brooks' theorem for $\Delta = 2$ we have to exclude all odd cycles and not only triangles. Hence the bound applies for the larger class of K_4-free graphs. A better linear bound, namely $\mathrm{tri}(\Delta) \leq \frac{2}{3}(\Delta+3)$ was established by Kostochka [618]. An alternative proof of this bound was obtained in 2012 by Rabern [837] as a consequence of the following decomposition result, which is similar to Lovász's decomposition result.

Lemma 5.48 (Rabern) *Let $p \in \mathbb{N}$, let $d_1, d_2, \ldots, d_p \in \mathbb{N}_0$, and let G be a triangle-free graph. If*

$$\Delta(G) \leq d_1 + d_2 + \cdots d_p + p - 2,$$

then G has a partition $(G_1, G_2, \ldots, G_p)$ such that $\Delta(G_i) \leq d_i$ and each d_i-regular component of G_i is a K_{d_i+1} whenever $i \in [1,p]$.

Proof If H is an induced subgraph of G and $v \in V(G)$, then let $H+v = G[V(H) \cup \{v\}]$, and for $v \in V(H)$ let $d(v:H) = d_H(v)$. If $\mathfrak{G} = (G_1, G_2, \ldots, G_p)$ is a partition of G, let $G_i(\mathfrak{G}) = G_i$ for $i \in [1,p]$. For $p = 1$, the theorem is evident. So assume that $p \geq 2$. A partition $\mathfrak{G}$ of G is *optimal* if its weight

$$W(\mathfrak{G}) = \sum_{i=1}^{p} (d_i|G_i(\mathfrak{G})| - e(G_i(\mathfrak{G}))$$

is maximum. Now let $\mathfrak{G}$ be an arbitrary optimal partition of G. Let $v \in V(G_i(\mathfrak{G}))$ be a vertex with $d(v:G_i(\mathfrak{G})) \geq d_i$. Since $d_G(v) \leq \Delta(G) \leq d_1+d_2+\cdots d_p+p-2$, there is an index $j \in [1,p] \setminus \{i\}$, such that $d(v:G_j(\mathfrak{G})+v) \leq d_j$ and we then say that v is a *movable vertex* and a *j-movable vertex* of $\mathfrak{G}$. Now let $\mathfrak{G}'$ be the partition obtained from $\mathfrak{G}$ by moving v to $G_j(\mathfrak{G})$, that is, $G_i(\mathfrak{G}') = G_i(\mathfrak{G}) - v$, $G_j(\mathfrak{G}') = G_j(\mathfrak{G}) + v$, and $G_k(\mathfrak{G}') = G_k(\mathfrak{G})$ for all $k \in [1,p] \setminus \{i,j\}$. In what follows we write $\mathfrak{G}' = \mathfrak{G}(v,j)$. Then

$$W(\mathfrak{G}') - W(\mathfrak{G}) = (d(v:G_i(\mathfrak{G})) - d_i) + (d_j - d(v:G_j(\mathfrak{G})+v)) \geq 0,$$

which implies that $d(v:G_i(\mathfrak{G}))=d_i$ and $d(v:G_j(\mathfrak{G}'))=d_j$. Consequently, $\mathfrak{G}'$ is an optimal partition, too. Hence if $\mathfrak{G}$ is an optimal partition of G, then $\Delta(G_i(\mathfrak{G}))\leq d_i$ for all $i\in[1,n]$ and, moreover, every vertex $v\in G_i(\mathfrak{G})$ with $d(v:G_i(\mathfrak{G}))=d_i$ is a movable vertex of $\mathfrak{G}$.

Let $\mathfrak{G}$ be an optimal partition of G. A *bad i-component* of $\mathfrak{G}$ is a component H of $G_i(\mathfrak{G})$ such that H is d_i-regular, but not a K_{d_i+1}. Note that this implies $d_i\geq 2$. Let $c(\mathfrak{G})$ be the number of components of the $G_i(\mathfrak{G})$'s summed over $i\in[1,p]$, and let $b(\mathfrak{G})$ be the number of bad i-components of $\mathfrak{G}$ summed over $i\in[1,p]$. We call $\mathfrak{G}$ a *good partition* of G if (1) $\mathfrak{G}$ is an optimal partition of G, (2) $b(\mathfrak{G})$ is minimum subject to (1), and (3) $c(\mathfrak{G})$ is minimum subject to (2).

Now let $\mathfrak{G}^1$ be a good partition of G. If $b(\mathfrak{G}^1)=0$, we are done. Otherwise, there is a bad i_1-component H_1 of $\mathfrak{G}^1$ for some index $i_1\in[1,p]$, and there is a nonseparating vertex v_1 of H_1. Now there is an index $i_2\neq i_1$ such that v_1 is i_2-moveable. Then $\mathfrak{G}^2=\mathfrak{G}^1(v_1,i_2)$ is an optimal partition, and hence $b(\mathfrak{G}^2)\geq b(\mathfrak{G}^1)$. Consequently, $G_{i_2}(\mathfrak{G}^2)=G_{i_2}(\mathfrak{G}^1)+v_1$ has a bad i_2-component H_2 containing v_1, which leads to $b(\mathfrak{G}^2)=b(\mathfrak{G}^1)$. Then $c(\mathfrak{G}^2)\geq c(\mathfrak{G}^1)$ and, therefore, $H_2'=H_2-v_1$ is a component of G_{i_2} and so $c(\mathfrak{G}^2)=c(\mathfrak{G}^1)$. Thus $\mathfrak{G}^2$ is a good partition of G. Since H_2 is not a complete graph, H_2 has a nonseparating vertex v_2 such that $v_2v_1\notin E(G)$. Continuing in this way we get a sequence $(\mathfrak{G}^k,H_k,v_k,i_k)$ with $k\in\mathbb{N}$ such that $\mathfrak{G}^k$ is a good partition of G, H_k is a i_k-bad component of $\mathfrak{G}^k$, v_k is a nonseparating vertex of H_k, $v_kv_{k+1}\notin E(G)$, $\mathfrak{G}^{k+1}=\mathfrak{G}^k(v_k,i_{k+1})$, $i_{k+1}\neq i_k$, and $v_k\in H_{k+1}$.

Since G is finite, we have to reuse an earlier destroyed bad component, that is, there is a smallest index k such that $H_{k+1}-v_k=H_\ell-v_\ell$ for some index $\ell<k$. The graph H_ℓ is an i_ℓ-bad component of $\mathfrak{G}^\ell$, which implies, in particular, that H_ℓ is a d_{i_ℓ}-regular graph and $d_{i_\ell}\geq 2$. The graph $H_\ell'=H_\ell-v_\ell$ is a component of $G_{i_\ell}(\mathfrak{G}^k)$ and $H_{k+1}=H_\ell'+v_k$ is a d_{i_ℓ}-regular component of $G_{i_\ell}(\mathfrak{G}^k)+v_k$. Consequently, v_k and v_ℓ have at least d_{i_ℓ} common neighbors in G. Since G is triangle-free, this implies that v_k is in G not adjacent to v_ℓ. Let $X=V(G_{i_\ell}(\mathfrak{G}^\ell))\cap\{v_i\mid i\in[\ell,k-1]\}$. The main observation is that the vertices of X belong to different components of $G_{i_\ell}(\mathfrak{G}^\ell)$ and form an independent set of G, each vertex $v\in X$ satisfies $d(v:G_{i_\ell}(\mathfrak{G}^\ell))=d_i$, and $N_G(v_k)\cap V(G_{i_\ell}(\mathfrak{G}^\ell))\subseteq(V(H_\ell)\cup X)\setminus\{v_\ell\}$. Consequently, each vertex of X is a movable vertex of $\mathfrak{G}^\ell$, and we can even move the vertices of $X\setminus\{v_\ell\}$ simultaneously out of $G_{i_\ell}(\mathfrak{G}^\ell)$. This results in an optimal partition $\mathfrak{G}$ of G such that H_ℓ is an i_ℓ-bad component of $G_{i_\ell}(\mathfrak{G})$. Since $\mathfrak{G}$ is an optimal partition of G and $v_k\in G_{i_k}(\mathfrak{G})$ with $i_k\neq i_\ell$, we have $d(v_k:G_{i_k}(\mathfrak{G}))=d_{i_k}$, which implies that v_k is a movable vertex of $\mathfrak{G}$. Furthermore, v_k has d_{i_ℓ} neighbors in $G_{i_\ell}(\mathfrak{G})$ all belonging to H_ℓ. Hence v_k is an i_ℓ-movable vertex of the partition $\mathfrak{G}$. Since $\mathfrak{G}$ is an optimal partition of G, $\mathfrak{G}'=\mathfrak{G}(v_k,i_\ell)$ is an optimal partition of G, too, but $G_{i_\ell}(\mathfrak{G}')$ has a vertex of degree $d_{i_\ell}+1$ (take $w\in N_G(v_k)\cap V(H_\ell)$), which is impossible. □

The above proof essentially follows Rabern's reasoning. He formulated his result for graphs in general, but we were then unable to show that v_ℓ and v_k are not adjacent in the general case. So we added the condition that G is triangle-free. It implies the stronger conclusion that a component of G_i can be d_i-regular only for d_i equal to 0 or 1. However, the above lemma suffices to obtain Kostocha's result that $\text{tri}(\Delta) \le \frac{2}{3}(\Delta+3)$: put $p = \lceil(\Delta+2)/3\rceil$, $d_i = 2$ for $i \in [1, p-1]$, and $d_p = 0$ if $\Delta \equiv 2 \pmod 3$ else $d_p = 2$.

It was first observed by Alon, Krivelevich, and Sudakov [54] that graphs with few triangles have nearly the same behaviour with respect to χ as triangle-free graphs. For a graph G and a vertex v of G, let $d^t_G(v)$ be the number of triangles of G incident with v, or equivalently, the number of edges of $G[N_G(v)]$. Let $\Delta^t(G) = \max_{v \in V(G)} d^t_G(v)$ be the **maximum triangle degree** of G. Note that if G is connected and has maximum degree Δ, then $\Delta^t(G) \le \binom{\Delta}{2}$ and equality holds if and only if $G = K_{\Delta+1}$. Define

$$\text{tri}(\Delta, \Delta^t) = \max\{\chi(G) \mid \Delta(G) = \Delta, \Delta^t(G) \le \Delta^t\}.$$

Then Alon, Krivelevich, and Sudakov [54] proved that

$$\text{tri}(\Delta, \Delta^t) = O\Big(\frac{\Delta}{\ln(\Delta^2/\Delta^t)}\Big),$$

where $1 \le \Delta^t \le \Delta^2/e$. That a similar result holds for the list chromatic number was proved by Vu [1056]. Whether this result can also be extended to the DP-chromatic number seems to be unknown. The reader will find more results of this kind as well as relevant references in the recent paper by Davies, Kang, Pirot, and Sereni [271] and, moreover, in the paper by Harris [472]. Another way to extend Theorem 5.22 is to prove a local version, similar to what Erdős, Rubin, and Taylor [355] did with Brooks' theorem. The reader will find results of this kind in the paper by Bonamy, Kelly, Nelson, and Postle [138] and in the paper by Davies, de Joannis de Verclos, Kang, and Pirot [270]; see also Section 6.9.

At the end of this section we want to point out another colouring problem for triangle-free graphs, which has attracted some attention in recent years. For a real number c, let

$$\text{adr}(c) = \sup\{\chi(G) \mid G \in \mathcal{G}(\omega \le 2), \delta(G) \ge c|G|\}.$$

By Turan's theorem, $\text{adr}(\frac{1}{2}) = 2$, and Andrásfai [63] proved in 1964 that $\text{adr}(c) = 2$, provided that $\frac{2}{5} < c \le \frac{1}{2}$. The cycle C_5 shows that $\text{adr}(\frac{2}{5}) \ge 3$, and Jin proved that $\text{adr}(c) = 3$ if $\frac{10}{29} < c \le \frac{2}{5}$. Using Kneser graphs (see Section 8.2), Hajnal (see [356]) constructed graphs showing that $\text{adr}(c) = \infty$ if $0 < c < \frac{1}{3}$. In 2002, Thomassen [1013] proved that $\text{adr}(c) < \infty$ as long as $\frac{1}{3} < c < \frac{10}{29}$, and shortly after Brandt and Thomassé proved in an unpublished paper [167] that $\text{adr}(c) = 4$ for this interval. We have not found any information about the value of $\text{adr}(\frac{1}{3})$. Further results about the structure of triangle-free graphs with large minimum degree are contained in references [208] and [705].

Chapter 6
Bounding χ by Δ and ω

Brooks' theorem may be stated as an upper bound on the chromatic number χ of a graph as a function of its maximum degree Δ and its clique number ω, so the question of a best possible such bound arises quite naturally. An answer to this question might be too difficult, but there are some interesting suggestions which we will address in this chapter.

6.1 Conjectures by Borodin, Kostochka and Reed

We define the function $\mathrm{bro}(\cdot,\cdot)$ by

$$\mathrm{bro}(\Delta,\omega) = \max\{\chi(G) \mid \Delta(G) = \Delta \text{ and } \omega(G) = \omega\},$$

where $2 \leq \omega \leq \Delta + 1$. Trivially, we have $\mathrm{bro}(1,2) = 2$ and $\mathrm{bro}(2,2) = \mathrm{bro}(2,3) = 3$. Brooks' theorem then says that $\mathrm{bro}(\Delta,\omega) \leq \max\{\Delta,\omega\}$ provided that $\Delta \geq 3$. By (2.2), we have

$$\mathrm{bro}(\Delta,\omega) \leq \Delta + 1 - \left\lfloor \frac{\Delta+1}{\omega+1} \right\rfloor \text{ for } 3 \leq \omega \leq \Delta. \tag{6.1}$$

From Theorem 5.22 and Lemma 5.48 we know that

$$\mathrm{bro}(\Delta,2) = O\left(\frac{\Delta}{\ln \Delta}\right)$$

and

$$\mathrm{bro}(\Delta,2) \leq \tfrac{2}{3}(\Delta+3) \text{ if } \Delta \geq 3.$$

As observed by Borodin and Kostochka [157] for $3 \leq \omega \leq \frac{1}{2}(\Delta - 1)$ we obtain that $\mathrm{bro}(\Delta,\omega) \leq \Delta - 1$ (by (6.1)). This observation led Borodin and Kostochka in 1977 to propose the following conjecture; for equivalent formulations see Proposition 6.15.

Conjecture 6.1 (Borodin and Kostochka) *Every graph G with $\chi(G) > \omega(G)$ and $\Delta(G) \geq 9$ satisfies $\chi(G) \leq \Delta(G) - 1$.*

M. Stiebitz et al., *Brooks' Theorem*, Springer Monographs in Mathematics,
https://doi.org/10.1007/978-3-031-50065-7_6

So the Borodin-Kostochka conjecture says that $\text{bro}(\Delta,\omega) \leq \max\{\Delta-1,\omega\}$ when $\Delta \geq 9$. Let us first show that the assumption $\Delta \geq 9$ is necessary. To see this consider the Catlin graph $G = C_5[K_3]$, see Figure 6.1. Then $\chi(G) = 8$ (by Corollary 1.49), $\Delta(G) = 8$ and $\omega(G) = 6$. Various other examples of graphs satisfying $\chi = \Delta \leq 8$ and $\omega \leq \Delta - 1$ can be found in the paper by Cranston and Rabern [258]. On the other hand, in 1999 Reed [855] verified the Kostochka-Borodin conjecture for graphs with sufficiently large maximum degree. The proof combines probabilistic arguments, in particular the Lovász Local Lemma, with decomposition results for graphs.

Theorem 6.2 (Reed) *Every graph G with $\chi(G) > \omega(G)$ and $\Delta(G) \geq 10^{14}$ satisfies $\chi(G) \leq \Delta(G) - 1$.*

In 1998 Reed [854] published an influential paper dealing with Brooks' theorem. In particular, he proposed an upper bound for the function $\text{bro}(\Delta,\omega)$ that is a convex combination of the two trivial bounds, namely $\omega \leq \text{bro}(\Delta,\omega) \leq \Delta + 1$.

Conjecture 6.3 (Reed) *Every graph G satisfies $\chi(G) \leq \left\lceil \frac{\Delta(G)+1+\omega(G)}{2} \right\rceil$.*

Conjecture 6.4 (Reed) *Every graph G satisfies $\chi(G) \leq \frac{2}{3}(\Delta(G)+1) + \frac{1}{3}\omega(G)$, unless $\Delta(G) \leq 2$.*

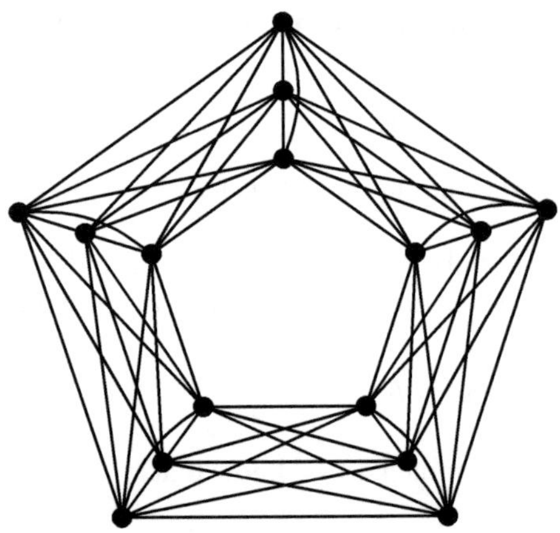

Fig. 6.1 The Catlin graph $C_5[K_3]$.

Gallai [398] constructed an infinite family of 4-regular triangle-free graphs all belonging to Crit(4), see also Theorem 3.46. This family shows that the ceiling function in the first conjecture of Reed is necessary. Another family of graphs showing this was found by Kostochka, namely the Catlin graphs $G = C_5[K_n]$ with $n \geq 1$ satisfying $\omega(G) = 2n$, $\Delta(G) = 3n - 1$ and $\chi(G) = \lceil (5n)/2 \rceil$.

The following two results, obtained in 1998 by Reed [854], support Conjecture 6.3, see also [855, Theorems 1 and 2].

Theorem 6.5 (Reed) *There exists an $\varepsilon > 0$ such that every graph G satisfies*

$$\chi(G) \leq \lceil (1-\varepsilon)(\Delta(G)+1)+\varepsilon\omega(G)\rceil .$$

Theorem 6.6 (Reed) *For every positive real number b, there exists a number Δ_b such that every graph G with $\Delta(G) \geq \Delta_b$ and $\omega(G) \leq \Delta(G)+1-2b$ satisfies $\chi(G) \leq \Delta(G)+1-b$.*

The original proofs of these two theorems in [854] are quite long and complicated; the proofs combine probabilistic coloring procedures with decomposition results for graphs. A simpler proof of Theorem 6.5 was found by King and Reed [596]; we shall discuss the proof in Section 6.3.

Proposition 6.7 (Reed) *Suppose that for $0 < \varepsilon \leq 1$ there is a Δ_ε such that every graph G with $\Delta(G) \geq \Delta_\varepsilon$ satisfies $\chi(G) \leq \lceil (1-\varepsilon)(\Delta(G)+1)+\varepsilon\omega(G)\rceil$. Then $\varepsilon \leq \frac{1}{2}$.*

Proof Let $G = C_5[K_n]$ with $n \geq 1$. Then $\Delta(G) = 3n-1$, $\omega(G) = 2n$ and $\chi(G) = \lceil (5n)/2\rceil$. So for sufficiently large n, we have $\Delta(G) \geq \Delta_\varepsilon$ and hence $\lceil 5n/2\rceil \leq \lceil (1-\varepsilon)3n+\varepsilon 2n\rceil = \lceil n(3-\varepsilon)\rceil$. This clearly implies that $\varepsilon \leq \frac{1}{2}$. □

6.2 Hitting Sets and Independent Transversals

Let G be a graph. A **maximum clique** of G is a clique of G having cardinality $\omega(G)$. We denote by $C(G)$ the set of maximum cliques of G. We are interested in the existence of an independent set I of G such that $I \cap C \neq \varnothing$ for every $C \in C(G)$; in this case we say that I is **hitting** every maximum clique of G. Note that if I is an independent set of G hitting all maximum cliques, then every independent set I' of G with $I \subseteq I'$ is hitting all maximum cliques of G, too.

Proposition 6.8 *Let G be a counterexample to Conjecture 6.3 having minimum order. Then G has no independent set hitting all maximum cliques of G.*

Proof Suppose this is false. Then there is a maximal independent set I of G hitting every maximum clique of G. This implies that $\omega(G-I) = \omega(G)-1$. Since I is maximal independent in G, for every vertex v of G, either v or a neighbor of v belongs to I, which leads to $\Delta(G-I) \leq \Delta(G)-1$. A coloring of $G-I$ can be extended to a coloring of G by assigning a new color to the vertices of I. Hence $\chi(G) \leq \chi(G-I)+1$. By the choice of G, $G-I$ satisfies Conjecture 6.3, and so we obtain that

$$\begin{aligned}\chi(G) &\leq \chi(G-I)+1 \leq \left\lceil \frac{\omega(G-I)+\Delta(G-I)+1}{2}\right\rceil +1\\ &\leq \left\lceil \frac{\omega(G)-1+\Delta(G)-1+1}{2}\right\rceil +1 = \left\lceil \frac{\omega(G)+\Delta(G)+1}{2}\right\rceil ,\end{aligned}$$

which is impossible. □

Let G be a graph, let $\mathcal{X}$ be a partition of the vertex set of G, and let $T \subseteq V(G)$. We call T a **transversal** of $\mathcal{X}$ in G if $|T \cap X| = 1$ for every $X \in \mathcal{X}$; and we call T a **partial transversal** of $\mathcal{X}$ in G if $|T \cap X| \leq 1$ for every $X \in \mathcal{X}$. An **independent transversal** (respectively, **independent partial transversal**) of $\mathcal{X}$ in G is a transversal (respectively, partial transversal) of $\mathcal{X}$ in G that is also an independent set of G. A **totally dominating set** of G is a set $D \subseteq V(G)$ such that $N_G(v) \cap D \neq \varnothing$ for every $v \in V(G)$.

In 1980, Kostochka [617] proved that every graph with $\omega > \Delta - \sqrt{\Delta} + \frac{3}{2}$ has an independent set hitting every maximum clique. This bound was improved by Rabern [834] to $\omega \geq \frac{3}{4}(\Delta+1)$, and by King [594] to $\omega > \frac{2}{3}(\Delta+1)$. To show the existence of a hitting set, King proved the following result, which extends an earlier result of Haxell and Aharoni (see Theorem 6.10).

Theorem 6.9 (King) *For a positive integer k, let G be a graph and let $\mathcal{X}$ be a partition of the vertex set of G such that, for every $X \in \mathcal{X}$ and every $v \in X$, v has at most* $\min\{k, |X| - k\}$ *neighbors outside X in G. Then, for every $v \in V(G)$, there is an independent transversal of $\mathcal{X}$ in G that contains v.*

Proof The proof is by contradiction. So let $(G, \mathcal{X}, k)$ be a counterexample for which $|V(G)| + |E(G)|$ is minimum. This implies that every set $X \in \mathcal{X}$ is an independent set of G with $|X| > k$. Furthermore, there is a vertex $v^* \in V(G)$ such that there is no independent transversal of $\mathcal{X}$ in G containing v^*. Let X^* be the set of $\mathcal{X}$ containing v^* and let $\mathcal{X}^* = \mathcal{X} \setminus \{X^*\}$. By our choice, there is an independent transversal of $\mathcal{X}^*$ in $G - X^*$.

Let $\mathcal{X}' \subseteq \mathcal{X}^*$. Then we denote by $G(\mathcal{X}')$ the subgraph of G induced by the set $U = \bigcup \mathcal{X}' \cup \{v^*\}$. We call $(\mathcal{X}', R, S)$ a *good triple* if the following conditions hold:

1. $R \cap S = \varnothing$ and $D = R \cup S$ is a totally dominating set in $G(\mathcal{X}')$.
2. S is an independent partial transversal of $\mathcal{X}'$ in $G(\mathcal{X}') - v^*$
3. R is an independent set in $G(\mathcal{X}')$ with $v^* \in R$ and $|R| \leq |S|$.

Case 1: *There exists a good triple* $(\mathcal{X}', R, S)$. Then we have $1 \leq |R| \leq |S| \leq |\mathcal{X}'|$ (by 2. and 3.). Since $D = R \cup S$ is a totally dominating set in $G' = G(\mathcal{X}')$, we obtain that

$$\sum_{v \in D} d_G(v) \geq \sum_{v \in D} d_{G'}(v) \geq |G'| = 1 + \sum_{X \in \mathcal{X}'} |X|. \tag{6.2}$$

If a vertex $v \in V(G)$ belongs to the set $X \in \mathcal{X}$, then X is an independent set of G and so $d_G(v) \leq \min\{k, |X| - k\}$. Furthermore, $|X| > k$ for all $X \in \mathcal{X}$. Since S is an independent partial transversal of $\mathcal{X}' \subseteq \mathcal{X}^*$, we then obtain that

$$\sum_{v \in S} d_G(v) \leq \sum_{X \in \mathcal{X}'} (|X| - k). \tag{6.3}$$

For the set R we obtain that

$$\sum_{v \in R} d_G(v) \leq k|R| \leq k|\mathcal{X}'|. \tag{6.4}$$

Since D is the disjoint union of R and S, we obtain from (6.3) and (6.4) that

$$\sum_{v \in D} d_G(v) \leq k|\mathcal{X}'| + \sum_{X \in \mathcal{X}'} (|X| - k) = \sum_{X \in \mathcal{X}'} |X|,$$

a contradiction to (6.2).

Case 2: *There exists no good triple.* Let $G^* = G(\mathcal{X}^*) = G - (X^* \setminus \{v^*\})$. For a set $S \subseteq V(G)$, let $\mathcal{X}(S) = \{X \in \mathcal{X} \mid X \cap S \neq \varnothing\}$. In order to get a contradiction, we will construct an infinite sequence of growing independent partial transversals of $\mathcal{X}^*$ in the graph G^*.

There exists an independent transversal of $\mathcal{X}^*$ in $G^* - v^*$. Now, let T_1 be an independent transversal of $\mathcal{X}^*$ in $G^* - v^*$ such that $S'_1 = T_1 \cap N_G(v^*)$ has minimum size. Note that S'_1 is nonempty, for otherwise $T_1 \cup \{v^*\}$ would be an independent transversal of $\mathcal{X}$ in G, which is impossible. Moreover, let $v_1 = v^*$, let $S_1 = S'_1$ and let $R_1 = \{v_1\}$. Starting from (R_1, S_1, T_1), we recursively define a sequence $(R_j, S_j, T_j)_{j \geq 1}$. Let $i \geq 2$ and suppose we have $\{(R_j, S_j, T_j) \mid j \in [1, i-1]\}$ such that the following holds for all $j \in [1, i-1]$:

(a) R_j is an independent set in G^* consisting of distinct vertices $v_1, v_2, \ldots, v_j$. If $j > 1$, then v_j is a vertex belonging to $G(\mathcal{X}(S_{j-1})) - (R_{j-1} \cup S_{j-1})$ with $N_G(v_j) \cap (R_{j-1} \cup S_{j-1}) = \varnothing$.

(b) T_j is an independent transversal of $\mathcal{X}^*$ in $G^* - v^*$ satisfying $T_j \cap N_G(v_\ell) = S'_\ell$ for all $\ell \in [1, j]$. Subject to this, T_j is chosen such that $S'_j = T_j \cap N_G(v_j)$ has minimum size.

(c) $S_j = S'_1 \cup S'_2 \cup \cdots \cup S'_j$ and $S'_j \neq \varnothing$.

Note that (a) and (b) imply that the sets $S'_1, S'_2, \ldots, S'_{i-1}$ are pairwise disjoint. In order to obtain (R_i, S_i, T_i), let v_i be a vertex of $G(\mathcal{X}(S_{i-1})) - (R_{i-1} \cup S_{i-1})$ with $N_G(v_i) \cap (R_{i-1} \cup S_{i-1}) = \varnothing$. If such a vertex does not exist, let $D = R_{i-1} \cup S_{i-1}$. Then D is a totally dominating set in $G(\mathcal{X}(S_{i-1}))$ (by (a) and (b)). Moreover, it follows from (a), (b) and (c) that R_{i-1} and S_{i-1} are disjoint, that $S_{i-1} \subseteq T_{i-1}$ is an independent partial transversal of $\mathcal{X}(S_{i-1})$ in $G(\mathcal{X}(S_{i-1}))$, and that R_{i-1} is an independent set in G containing v^* with $|R_{i-1}| \leq |S_{i-1}|$. Hence $(\mathcal{X}(S_{i-1}), R_{i-1}, S_{i-1})$ is a good triple, contradicting the assumption of Case 2. Thus, v_i exists. Now choose such a vertex v_i so that there is an independent transversal T_i of $\mathcal{X}^*$ in $G^* - v^*$ such that $T_i \cap N_G(v_\ell) = S'_\ell$ for all $\ell \in [1, i-1]$ and $S'_i = T_i \cap N_G(v_i)$ has minimum size. Note that T_i exists since T_{i-1} would be a possible candidate for T_i. We claim that S'_i is nonempty. For otherwise, v_i has no neighbor in T_i. There is a unique set $X \in \mathcal{X}$ such that $v_i \in X$, and there unique vertex w_i belonging to $T_i \cap X$. Since $i \geq 2$, v_i is a vertex of $G(\mathcal{X}(S_{i-1})) - (R_{i-1} \cup S_{i-1})$ and so there is a unique index $j \in [1, i-1]$ such that $w_i \in S'_j$ (by (a)). Then $T'_j = (T_j \setminus \{w_i\}) \cup \{v_i\}$ is an independent transversal of $\mathcal{X}^*$ in $G^* - (R_{j-1} \cup S_{j-1})$. For $\ell \in [1, j-1]$, we have $T'_j \cap N_G(v_\ell) = T_j \cap N_G(v_\ell) = S'_\ell$ and $T'_j \cap N_G(v_j) = (T_j \cap N_G(v_j)) \setminus \{w_i\}$, a contradiction to (b).

Thus, S_i' is nonempty and by setting $R_i = R_{i-1} \cup \{v_i\}$ and $S_i = S_{i-1} \cup S_i'$ we obtain a new sequence $(R_j, S_j, T_j)_{1 \le j \le i}$ satisfying (a), (b) and (c).

This construction gives us an infinite sequence $S_1 \subset S_2 \subset \ldots$ of partial independent transversals of $\mathcal{X}$ in G, which is impossible since G is finite. This proves the theorem. □

As an immediate consequence of Theorem 6.9 we obtain a classical result due to Aharoni and Haxell, see [482] for the history of this result.

Theorem 6.10 (Aharoni and Haxell) *For a positive integer k, let G be a graph with $\Delta(G) \le k$, and let $\mathcal{X}$ be a partition of the vertex set of G such that $|X| \ge 2k$ for every $X \in \mathcal{X}$. Then there is an independent transversal of $\mathcal{X}$ in G.*

For the proof of King's result that every graph with $\omega > \frac{2}{3}(\Delta + 1)$ has an independent set hitting every maximum clique, we need two further auxiliary results. The first result is due to Hajnal [456], while the second result combines the results of Kostochka [617] and of Christofides, Edwards, and King [264].

Lemma 6.11 (Hajnal) *Let G be a graph and let $\mathcal{C} \subseteq \mathcal{C}(G)$ be a nonempty set of maximum cliques in G. Then,*

$$\left|\bigcap \mathcal{C}\right| + \left|\bigcup \mathcal{C}\right| \ge 2\omega(G).$$

Proof The proof is by induction on $|\mathcal{C}|$. The statement is evident if $|\mathcal{C}| = 1$. Now assume that $|\mathcal{C}| \ge 2$. Choose a clique $C \in \mathcal{C}$ and let $\mathcal{C}' = \mathcal{C} \setminus \{C\}$. By the induction hypothesis, we obtain that

$$\left|\bigcap \mathcal{C}'\right| + \left|\bigcup \mathcal{C}'\right| \ge 2\omega(G). \tag{6.5}$$

The set $X = (C \cap \bigcup \mathcal{C}') \cup \bigcap \mathcal{C}'$ is a clique in G, and so we obtain that

$$\begin{aligned}
\omega(G) \ge |X| &= |C \cap \bigcup \mathcal{C}'| + |\bigcap \mathcal{C}'| - |(C \cap \bigcup \mathcal{C}') \cap \bigcap \mathcal{C}'| \\
&= |C \cap \bigcup \mathcal{C}'| + |\bigcap \mathcal{C}'| - |\bigcap \mathcal{C}| \\
&= |C| + |\bigcup \mathcal{C}'| - |\bigcup \mathcal{C}| + |\bigcap \mathcal{C}'| - |\bigcap \mathcal{C}| \\
&\ge \omega(G) + 2\omega(G) - (|\bigcup \mathcal{C}| + |\bigcap \mathcal{C}|),
\end{aligned}$$

where the last inequality follows from (6.5) and since $|C| = \omega(G)$. As a consequence, we obtain that

$$\left|\bigcap \mathcal{C}\right| + \left|\bigcup \mathcal{C}\right| \ge 2\omega(G).$$

This proves the lemma. □

For a set $\mathcal{C} \subseteq \mathcal{C}(G)$ of maximum cliques of G, let $\mathrm{IG}(\mathcal{C})$ denote the **intersection graph** of $\mathcal{C}$, that is, $V(\mathrm{IG}(\mathcal{C})) = \mathcal{C}$ and two two distinct members of $\mathcal{C}$ are adjacent in IG(C) if and only if they have a nonempty intersection. The following Lemma was proved by Kostochka [617] for $\omega > \frac{2}{3}(\Delta+1)$ and by Christofides, Edwards, and King [264] for $\omega = \frac{2}{3}(\Delta+1)$.

Lemma 6.12 *Let G be a graph with $\omega(G) \geq \frac{2}{3}(\Delta(G)+1)$, let $C \subseteq \mathcal{C}(G)$ such that $IG(C)$ is connected, let $X = \bigcap C$ and $Y = \bigcup C$. Then the following statements hold:*

(a) *If $X \neq \varnothing$, then $|Y| \leq \Delta(G)+1 \leq \frac{3}{2}\omega(G)$ and $|X| \geq \frac{1}{2}\omega(G) \geq \frac{1}{3}(\Delta(G)+1)$.*

(b) *If $X = \varnothing$, then $\omega(G) = \frac{2}{3}(\Delta(G)+1)$ and, moreover, Y has a partition $\mathcal{Q} = \{Q_i \mid i \in [1,n]\}$ into $n \geq 4$ cliques of G each of cardinality $\frac{1}{2}\omega(G)$ such that $C \setminus \{Q_1 \cup Q_n\} = \{Q_i \cup Q_{i+1} \mid i \in [1,n-1]\}$.*

Proof For $C' \subseteq C$, let $X(C') = \bigcap C'$ and $Y(C') = \bigcup C'$. Obviously, $X(C') \subseteq Y(C')$. Furthermore, $X = X(C)$ and $Y = Y(C)$. For the proof of (a) assume that X is nonempty. Then there is a vertex $v \in X$ and $N_G(v)$ contains $Y \setminus \{v\}$, which leads to $|Y| \leq \Delta(G)+1 \leq \frac{3}{2}\omega(G)$. By Lemma 6.11, we then obtain that

$$|X| \geq 2\omega(G) - |Y| \geq \tfrac{1}{2}\omega(G) \geq \tfrac{1}{3}(\Delta(G)+1).$$

This proves (a). The proof of (b) is by induction on $|C|$. So assume that $X(C) = \varnothing$. Since IG(C) is connected, this implies that $|C| \geq 3$. Furthermore, there is a nonseparating vertex C in IG(C). Let $C' = C \setminus \{C\}$. Clearly, IG(C') is connected and C has in IG(C) a neighbor belonging to C'. We distinguish two cases.

Case 1: $X(C') = \varnothing$. Then the induction hypothesis implies that $\omega(G) = \frac{2}{3}(\Delta(G)+1)$, and $Y' = Y(C')$ has a partition $\mathcal{Q} = \{Q_i \mid i \in [1,n]\}$ into $n \geq 4$ cliques of G each of cardinality $\frac{1}{2}\omega(G)$ such that

$$C' \setminus \{Q_1 \cup Q_n\} = \{Q_i \cup Q_{i+1} \mid i \in [1,n-1]\}.$$

Then $H' = P_n[K_p]$ with $p = \frac{1}{2}\omega(G)$ is a spanning subgraph of $G[Y']$. Let $R = Q_2 \cup Q_3 \cup \cdots \cup Q_{n-1}$. Then every vertex $v \in R$ has in H' degree $r = 3p - 1 = \frac{3}{2}\omega(G) - 1 = \Delta(G)$ implying that

$$G[Y'] - E_G(Q_1, Q_n) = H' = P_n[K_p] \text{ and } N_G(v) = N_{H'}(v) \text{ for all } v \in R. \quad (6.6)$$

There is a clique $B \in C'$ that is in IG(C) adjacent to C. Then, by (a), we obtain that $|B \cap C| \geq \frac{1}{2}\omega(G) = p$. By (6.6), this implies that $C \cap Y(C') \in \{Q_1, Q_n, Q_1 \cup Q_n\}$. It is easy to check that in all three cases we are done.

Case 2: $X(C') \neq \varnothing$. Let $p = \frac{1}{2}\omega(G)$. From (a) we obtain that $|X(C')| \geq p$ and $|Y(C')| \leq \Delta(G)+1 \leq 3p$. There is a clique $B \in C'$ such that B is adjacent to C in IG(C). Then, by (a), $|B \cup C| \leq \Delta(G)+1 \leq 3p$ and $|B \cap C| \geq p$. Consequently, $|C \setminus B| = |C \cup B| - |B| \leq 3p - 2p = p$. This implies that

$$|Y(C)| = |Y(C') \cup (C \setminus B)| \leq |Y(C')| + |C \setminus B| \leq 3p + p = 4p.$$

Since $X(C) = \varnothing$, it follows from Lemma 6.11 that $|Y(C)| = |X(C)| + |Y(C)| \geq 2\omega(G) = 4p$. Therefore, we obtain that $|Y(C)| = 4p$, $|Y(C')| = 3p$, $|C \setminus B| = p$,

$Y(C') \cap (C \setminus B) = \varnothing$, $|C \cup B| = 3p$ and $\Delta(G)+1 = 3p = \frac{3}{2}\omega(G)$. Furthermore, $X(C) = X(C') \cap C = \varnothing$, which leads to $|X(C')| = |B \cap C| = p$ and B is the disjoint union of $X(C')$ and $B \cap C$. Since $|\mathcal{C}| \geq 3$, there is a clique $A \in \mathcal{C}'$ with $A \neq B$. Clearly, $A \subseteq Y(C')$ and so $A \cap C = \varnothing$. For otherwise, there is a vertex $v \in A \cap (C \cap B)$ and so $d_G(v) \geq |C \setminus B| + |X(C')| + |B \cap C| - 1 + 1 = 3p = \Delta(G)+1$, which is impossible. Since $X(C') \subseteq A$, we obtain that $A \cap B = X(C')$. We claim that $\mathcal{C} = \{A, B, C\}$. For otherwise, there is a $D \in \mathcal{C}' \setminus \{A, B\}$ and $X(C') \subseteq D$, which leads to a vertex $v \in X(C')$ such that $d_G(v) \geq 3p = \Delta(G)+1$, which is impossible. Let $Q_1 = A \setminus B$, $Q_2 = B \setminus C = X(C')$, $Q_3 = B \cap C = B \setminus A$ and $Q_4 = C \setminus B$. Then it is easy to check that $\mathcal{Q} = \{Q_1, Q_2, Q_3, Q_4\}$ is a partition of $Y(\mathcal{C})$, Q_i is a p-clique for $i \in [1,3]$ and $\mathcal{C} = \{Q_i \cup Q_{i+1} \mid i \in [1,3]\}$. So we are done. □

Theorem 6.13 *Let G be a connected graph with $\omega(G) \geq \frac{2}{3}(\Delta(G)+1)$. Then G has an independent set I such that $\omega(G-I) = \omega(G)-1$, unless $\omega(G) = \frac{2}{3}(\Delta(G)+1)$ and $G = C_n[K_p]$ with $n \geq 5$ odd and $p = \frac{1}{2}\omega(G)$.*

Proof The proof is by induction on the order of G. So let G be a connected graph with $\omega(G) \geq \frac{2}{3}(\Delta(G)+1)$. If $|G| \leq 2$, the statement is evident. So assume that $|G| \geq 3$. Then $\Delta(G) \geq 2$ and $\frac{1}{3}(\Delta(G)+1) \geq 1$. First, we claim that every vertex of G is contained in a maximum clique of G. For otherwise, let v be a vertex of G that is not contained in a maximum clique of G. Then $G' = G - v$ satisfies $\omega(G') = \omega(G) \geq \frac{2}{3}(\Delta(G)+1) \geq \frac{2}{3}(\Delta(G')+1)$. Hence, the induction hypothesis implies that G' has an independent set I such that I is hitting every maximum clique of G', or $\omega(G') = \frac{2}{3}(\Delta(G')+1)$ and $G' = C_{2n+1}[K_p]$ with $n \geq 2$ and $p = \frac{1}{2}\omega(G')$. In the former case, I is an independent set of G hitting every maximum clique of G. In the later case $\omega(G') = \omega(G) = \frac{2}{3}(\Delta(G)+1) = \frac{2}{3}(\Delta(G')+1)$ and so G' is regular of degree r with $r = \Delta(G') = \Delta(G)$. This implies that v is an isolated vertex of G. Since G is connected, this is a contradiction. This proves the claim that every vertex of G is contained in a maximum clique of G.

For a component $\mathcal{H}$ of the intersection graph $\mathrm{IG}(\mathcal{C}(G))$, let $X_{\mathcal{H}} = \bigcap V(\mathcal{H})$ and let $Y_{\mathcal{H}} = \bigcup V(\mathcal{H})$. We distinguish two cases.

Case 1: *For every component $\mathcal{H}$ of $IG(\mathcal{C}(G))$ we have $X_{\mathcal{H}} \neq \varnothing$.* By Lemma 6.11, we obtain that

$$|X_{\mathcal{H}}| + |Y_{\mathcal{H}}| \geq 2\omega(G) \geq \tfrac{4}{3}(\Delta(G)+1) \tag{6.7}$$

for every component $\mathcal{H}$ of $\mathrm{IG}(\mathcal{C}(G))$. Let

$$\mathcal{X} = \{X_{\mathcal{H}} \mid \mathcal{H} \text{ is a component of } \mathrm{IG}(\mathcal{C}(G))\}$$

and let G' be the subgraph of G induced by $X = \bigcup \mathcal{X}$. Clearly, $\mathcal{X}$ is a partition of the vertex set of G', and, by Lemma 6.12(a), we have $|X_{\mathcal{H}}| \geq \frac{1}{3}(\Delta(G)+1) \geq 1$ for every component $\mathcal{H}$ of $\mathrm{IG}(\mathcal{C}(G))$. Let $\mathcal{H}$ be an arbitrary component of $\mathrm{IG}(\mathcal{C}(G))$, let $v \in X_{\mathcal{H}}$ and let d_v be the number of neighbors of v in G' outside $X_{\mathcal{H}}$. Then there is a maximum clique C of G with $v \in C \subseteq Y_{\mathcal{H}}$, and so $d_v \leq \Delta(G) - (\omega(G)-1) \leq \frac{1}{3}(\Delta(G)+1)$. Furthermore, v is joined in G to all vertices of $Y_{\mathcal{H}} \setminus \{v\}$ implying that $d_v \leq \Delta(G) - (|Y_{\mathcal{H}}|-1) \leq |X_{\mathcal{H}}| - \frac{1}{3}(\Delta(G)+1)$, where the last inequality follows from

(6.7). So there is a positive integer k such that, for every $X \in \mathcal{X}$, every vertex of X has at most $\min\{k, |X|-k\}$ neighbors outside X in G'. Then Theorem 6.9 implies that there is an independent set I of G' such that $|I \cap X_{\mathcal{H}}| = 1$ for every component $\mathcal{H}$ of $\mathrm{IG}(\mathcal{C}(G))$. Clearly, I is an independent set of G. If C is a maximum clique of G, then C belongs to a component $\mathcal{H}$ of $\mathrm{IG}(\mathcal{C}(G))$ and hence $X_{\mathcal{H}} \subseteq C$ implying that $I \cap C \neq \varnothing$. Consequently, we obtain that $\omega(G-I) = \omega(G)-1$.

Case 2: *$X_{\mathcal{H}} = \varnothing$ for some component $\mathcal{H}$ of $IG(\mathcal{C}(G))$.* Let $\mathcal{C} = V(\mathcal{H})$ and $Y = Y_{\mathcal{H}}$. Then it follows from Lemma 6.12(b) that $\omega(G) = \frac{2}{3}(\Delta(G)+1)$ and Y has a partition $\mathcal{Q} = \{Q_i \mid i \in [1,n]\}$ into $n \geq 4$ cliques of G each of cardinality $p = \frac{1}{2}\omega(G)$ such that $\mathcal{C} \setminus \{Q_1 \cup Q_n\} = \{Q_i \cup Q_{i+1} \mid i \in [1, n-1]\}$.

First, assume that $Q_1 \cup Q_n \in \mathcal{C}$. Then $H = C_n[K_p]$ is a spanning subgraph of $G[Y]$, and H is regular of degree $r = 3p-1 = \frac{3}{2}\omega(G) = \Delta(G)$. Since G is connected, this implies that $G = H = C_n[K_p]$. So if n is odd, we are done. If n is even, then it is easy to see that G has an an independent set I hitting every maximum clique of G, and we are done, too.

It remains to consider the case that $Q' = Q_1 \cup Q_n \notin \mathcal{C}$, and so Q' is not a clique in G. Consequently, $\mathcal{C} = \{Q_i \cup Q_{i+1} \mid i \in [1, n-1]\}$ and $H' = P_n[K_p]$ is a spanning subgraph of $G[Y]$. Let $R = Q_2 \cup Q_3 \cup \cdots \cup Q_{n-1}$. Then every vertex $v \in R$ has in H' degree $r = 3p-1 = \frac{3}{2}\omega(G)-1 = \Delta(G)$ implying that

$$G[Y] - E_G(Q_1, Q_n) = H' = P_n[K_p] \text{ and } N_G(v) = N_{H'}(v) \text{ for all } v \in R. \tag{6.8}$$

Now, let G' be the graph obtained from $G-R$ by adding edges between Q_1 and Q_2 so that $Q' = Q_1 \cup Q_n$ is a clique of G'. Note that $|Q'| = 2p = \omega(G)$, $\Delta(G') \leq \Delta(G)$ and $\omega(G) \leq \omega(G')$.

We claim that $\omega(G) = \omega(G')$. Suppose this is false. Then G' has a clique C with $|C| \geq \omega(G)+1 = 2p+1$. Hence, there is a vertex $v \in C \setminus Q'$. Then v belongs to a maximum clique C' of G. Clearly, C' does not belong to $\mathcal{H}$ and so $C' \cap Q' = \varnothing$. Since $d_G(v) \leq \Delta(G) = 3p-1$, this implies that $|C \cap Q'| \leq p$, and so $|C \setminus Q'| \geq p+1$. Since $\omega(G'-Q') \leq \omega(G)$, we obtain that $C \cap Q' \neq \varnothing$. Hence there is a vertex $u \in C \cap Q'$ and, since u belongs to a maximum clique in $H' = P_n[K_p]$, we obtain that $d_G(u) \geq (2p-1) + |C' \setminus Q'| \geq (2p-1) + (p+1) = 3p = \Delta(G)+1$, which is impossible. This proves the claim that $\omega(G') = \omega(G) = 2p$. Hence $\omega(G') = \omega(G) \geq \frac{2}{3}(\Delta(G)+1) \geq \frac{2}{3}(\Delta(G')+1)$ and $\mathcal{C}(G') = (\mathcal{C}(G) \setminus \mathcal{C}) \cup \{Q'\}$. By the induction hypothesis, there is an independent set I' of G' hitting every maximum clique of G'. Then $I' \cap Q' = \{v\}$ and, by symmetry we may assume that $v \in Q_1$. If n is even, choose a vertex from every set Q_i with odd $i \in [3, n-1]$, and let I^0 be the set of all these vertices. Then it follows from (6.8) that $I = I' \cup I^0$ is an independent set of G, and it is easy to check that I is hitting every maximum clique of G. If n is odd, choose a vertex from every set Q_i with even $i \in [2, n-1]$, and let I^1 be the set of all these vertices. Then it follows from (6.8) that $I = (I' \setminus \{v\}) \cup I^1$ is an independent set of G, and it is easy to check that I is hitting every maximum clique of G. Thus the proof of Theorem 6.13 is complete. □

Theorem 6.13 was proved by King [594] for $\omega > \frac{2}{3}(\Delta+1)$ and by Christofides, Edwards, and King [264] for $\omega = \frac{2}{3}(\Delta+1)$. As an immediate consequence of Theorem 6.13, we obtain the following result due to Cranston and Rabern [259]. Note that if $\omega \geq \Delta - 4$ and $\Delta \geq 14$, then $\omega \geq \frac{2}{3}(\Delta+1)$ and equality holds and only if $\Delta = 14$ and $\omega = 10$.

Corollary 6.14 (Cranston and Rabern) *Every connected graph satisfying $\Delta(G) \geq 14$ and $\omega(G) \geq \Delta(G) - 4$ has an independent set hitting every maximum clique of G, unless $G = C_{2n+1}[K_5]$ with $n \geq 2$.*

Proposition 6.15 *The following statements are equivalent:*

(a) *Every graph G with $\chi(G) > \omega(G)$ and $\Delta(G) \geq 9$ satisfies $\chi(G) \leq \Delta(G) - 1$.*
(b) *Every graph G with $\chi(G) \geq \Delta(G) \geq 9$ satisfies $\omega(G) \geq \Delta(G)$.*
(c) *For every $k \geq 9$, K_k is the only k-critical graph with maximum degree $\leq k$.*
(d) *K_9 is the only 9-critical graph with maximum degree ≤ 9.*
(e) *Every graph G with $\chi(G) \geq \Delta(G) = 9$ satisfies $\omega(G) \geq 9$.*

Proof To show that (a) implies (b), suppose that (b) is false. Then there is a graph G, such that $\chi(G) \geq \Delta(G) \geq 9$, but $\omega(G) \leq \Delta(G) - 1$. Then $\chi(G) > \omega(G)$ and so $\chi(G) \leq \Delta(G) - 1$ (by (a)), a contradiction. To show that (b) implies (a), suppose that (a) is false. Then there is a graph G such that $\chi(G) > \omega(G)$ and $\chi(G) \geq \Delta(G) \geq 9$. This implies $\chi(G) > \omega(G) \geq \Delta(G) \geq 9$ (by (b)), and hence $\chi(G) = \omega(G)$ (by Brooks' theorem), a contradiction.

To see that (b) implies (c), let $G \in \mathrm{Crit}(k)$ be a graph with $\Delta(G) \leq k$ and $k \geq 9$. Then $\Delta(G) \geq \delta(G) \geq k - 1$ and $\chi(G) = k$. If $\Delta(G) = k$, then $\omega(G) \geq \Delta(G) = k$ (by (b)), and so $G = K_k$, which is impossible. If $\Delta(G) = k - 1$, then Brooks' theorem implies that $G = K_k$, and so (c) holds. Thus (b) implies (c). Evidently, (c) implies (d). To see that (d) implies (e), let G be a graph with $k = \chi(G) \geq \Delta(G) = 9$. Clearly, $k \leq \Delta(G) + 1 = 10$. If $k = 10$, then Brooks' theorem implies that $\omega(G) \geq k$, and so (e) holds. If $k = 9$, then G contains a subgraph $G' \in \mathrm{Crit}(9)$ (by Proposition 1.27) and $\Delta(G') \leq \Delta(G) = 9$. Hence (d) implies that $G' = K_9$, and so $\omega(G) \geq \omega(G') = 9$, i.e. (e) holds. This shows that (d) implies (e).

To see that (e) implies (b), assume that (b) is false and let G be a counterexample whose order is minimum. This implies that $\chi(G) \geq \Delta(G) \geq 9$ and $\omega(G) \leq \Delta(G) - 1$. By (e), we have $\Delta(G) \geq 10$. If $\omega(G) \leq \Delta(G) - 2$, then let I be a maximal independent set of G; otherwise let I be a maximal independent set of G such that $\omega(G - I) = \omega(G) - 1$ (such a set exist by Theorem 6.13, since $\omega(G) = \Delta(G) - 1 \geq 9$ implies $\omega(G) > \frac{2}{3}(\Delta(G) + 1)$). For $G' = G - I$, we then obtain that $\omega(G') \leq \Delta(G) - 2$, $\Delta(G') \leq \Delta(G) - 1$ and $\Delta(G') + 1 \geq \chi(G') \geq \chi(G) - 1 \geq \Delta(G) - 1 \geq 9$. Consequently, $\Delta(G') \geq 8$ and Brooks' theorem implies that

$$\chi(G') \leq \max\{\omega(G'), \Delta(G')\} \leq \max\{\Delta(G) - 2, \Delta(G) - 1\} = \Delta(G) - 1.$$

Then we obtain that $\chi(G') = \Delta(G) - 1 = \Delta(G') \geq 9$ and $\omega(G') \leq \Delta(G) - 2 = \Delta(G') - 1$, a contradiction to the choice of G. □

Proposition 6.15 was proved by Kostochka [617] and, independently, by Catlin [192, Corollary 6.7] in his thesis submitted 1976 to the Ohio State University. The proposition shows that it is sufficient to prove the Borodin-Kostochka conjecture (Conjecture 6.1) for the base case $\Delta = 9$, i.e., to show that every graph G with $\chi(G) = \Delta(G) = 9$ contains a K_9 as a subgraph. Cranston and Rabern [257] proved that the Borodin-Kostochka conjecture is equivalent to the following conjecture.

Conjecture 6.16 (Cranston and Rabern) *Every graph G satisfying $\chi(G) = \Delta(G) = 9$ contains $K_3 \boxplus \overline{K}_6$ as a subgraph.*

6.3 A Weakening of Reed's Conjecture

In this section we present a short proof of Theorem 6.5 due to King and Reed [596]. Their proof uses the simple fact that a minimal counterexample to Theorem 6.5 has no independent set hitting all maximum cliques; hence a minimal counterexample satisfies the inequality $\omega \leq \frac{2}{3}(\Delta+1)$ (by Theorem 6.13). The proof then distinguishes between two cases to arrive at a contradiction. If the number of edges in the neighborhood of a vertex is close to $\binom{\Delta}{2}$, we can apply Theorem 6.17 due to King and Reed [596, Theorem 6]. Otherwise, the neighborhood of each vertex contains much less than $\binom{\Delta}{2}$ edges, and we can apply Theorem 6.18 due to Molloy and Reed [756, Theorem 10.5].

Recall from Section 5.8 that, for a graph G and a vertex v of G, we denote by $d_G^t(v)$ the number of triangles of G incident with v, or equivalently, the number of edges of the graph $G[N_G(v)]$.

Theorem 6.17 (King and Reed) *Let a and ε be constants such that $0 < a$ and $0 < \varepsilon < \frac{1}{6} - 2\sqrt{a}$. Let G be a graph with $\omega(G) \leq \frac{2}{3}(\Delta(G)+1)$. If G has a vertex v such that $d_G^t(v) > (1-a)\binom{\Delta(G)}{2}$, then $\chi(G) \leq \max\{\chi(G-v), (1-\varepsilon)(\Delta(G)+1)\}$.*

Theorem 6.18 (Molloy and Reed) *There is a $\Delta_0 \geq 1$ such that for any graph G with maximum degree $\Delta > \Delta_0$ and $B > \Delta(\log\Delta)^3$, if $d_G^t(v) \leq \binom{\Delta}{2} - B$ for all $v \in V(G)$, then $\chi(G) \leq (\Delta+1) - \frac{B}{e^6\Delta}$.*

Corollary 6.19 *There is a $\Delta_0 \geq 1$ such that for any graph G with maximum degree $\Delta > \Delta_0$ and $a > 2(\log\Delta)^3/(\Delta-1)$, if $d_G^t(v) \leq (1-a)\binom{\Delta}{2}$ for all $v \in V(G)$, then*

$$\chi(G) \leq (\Delta+1) - \frac{a(\Delta-1)}{2e^6} \leq (1 - \frac{a}{2e^6})(\Delta+1) + \frac{a}{2e^6}\omega,$$

where $\omega = \omega(G)$.

Proof Apply Theorem 6.18 with $B = a\binom{\Delta}{2}$, and note that $\omega \geq 2$. □

Before we discuss the proofs of Theorems 6.17 and 6.18, we shall show how these results can be used together with Theorem 6.13 to prove Theorem 6.5, i.e. the following result.

Theorem 6.20 *There is an $\varepsilon > 0$ such that every graph G satisfies $\chi(G) \leq \lceil(1-\varepsilon)(\Delta(G)+1)+\varepsilon\omega(G)\rceil$.*

Proof We choose Δ_0 from the statement of Corollary 6.19 such that, for $a = \frac{1}{160}$, we have $2(\log\Delta)^3/(\Delta-1) < a$ for all $\Delta > \Delta_0$. Let $\varepsilon = \min\{\frac{1}{\Delta_0}, \frac{a}{2e^6}\}$. Then it is easy to check that $0 < \varepsilon \leq \frac{a}{2e^6} < \frac{1}{6} - 2\sqrt{a}$. Hence, we can apply Theorem 6.17 and Corollary 6.19. Our aim is to show that every graph G satisfies

$$\chi(G) \leq \lceil(1-\varepsilon)(\Delta(G)+1)+\varepsilon\omega(G)\rceil.$$

To this end, we shall analyze a possible counterexample, say G, whose order is minimum. This implies that G is vertex critical, and so G is connected. Let $\Delta = \Delta(G)$ and $\omega = \omega(G)$. Then, by assumption, we have

$$\chi(G) > \lceil(1-\varepsilon)(\Delta+1)+\varepsilon\omega\rceil. \tag{6.9}$$

Then we have $\omega \leq \Delta$. For otherwise, $\omega = \Delta+1$ and (6.9) yields $\chi(G) > \Delta+1$ which is impossible. Hence, $2 \leq \omega \leq \Delta$. Furthermore, we have $\Delta > \Delta_0$. For otherwise, we have $\Delta \leq \Delta_0$ and hence $\varepsilon\Delta \leq \varepsilon\Delta_0 \leq 1$, and so (6.9) yields

$$\chi(G) > \lceil\Delta+(1-\varepsilon\Delta)+\varepsilon(\omega-1)\rceil \geq \lceil\Delta+\varepsilon\rceil = \Delta+1,$$

which is impossible. Next we claim that $\omega \leq \frac{2}{3}(\Delta+1)$. For otherwise, Theorem 6.13 implies that G has an independent set I hitting every maximum clique of G; we may choose I such that it is a maximal independent set of G. Then, for $G' = G-I$, we have $\Delta(G') \leq \Delta-1$, $\omega(G') = \omega-1$ and $\chi(G) \leq \chi(G')+1$. By the minimality of G, this yields

$$\begin{aligned}\chi(G) &\leq \chi(G')+1\\ &\leq \lceil(1-\varepsilon)(\Delta(G')+1)+\varepsilon\omega(G')\rceil+1\\ &\leq \lceil(1-\varepsilon)\Delta+\varepsilon(\omega-1)+1\rceil\\ &= \lceil(1-\varepsilon)(\Delta+1)+\varepsilon\omega\rceil,\end{aligned}$$

a contradiction to (6.9). This proves the claim that $\omega \leq \frac{2}{3}(\Delta+1)$. Since $\omega \leq \Delta$ and $0 < \varepsilon \leq \frac{a}{2e^6} < 1$, we obtain from (6.9) that

$$\chi(G) > (1-\tfrac{a}{2e^6})(\Delta+1)+\tfrac{a}{2e^6}\omega.$$

Then Corollary 6.19 implies that G has a vertex v such that $d^t_G(v) > (1-a)\binom{\Delta}{2}$. But then Theorem 6.17 implies that $\chi(G) \leq (1-\varepsilon)(\Delta+1)$, a contradiction to (6.9). This completes the proof. □

Proof of Theorem 6.17: Let a and ε as in the statement of Theorem 6.17. This implies that $a < 1/144$. Let G be a graph with maximum degree Δ and clique number ω, where $\omega \leq \frac{2}{3}(\Delta+1)$. Let v be a vertex of G such that $d^t_G(v) > (1-a)\binom{\Delta}{2}$. Then $\Delta \geq 2$.

Now put $k = \lfloor(1-\varepsilon)(\Delta+1)\rfloor$ and $k' = \max\{\chi(G-v), k\}$. For the proof of the theorem it suffices to show that G has a coloring with color set $S = [1, k']$. We may assume that $d_G(v) = \Delta$. For otherwise, we can add to G a set of $\Delta - d_G(v)$ new vertices that are only joined to v.

Let $N = N_G(v) \cup \{v\}$ be the closed neighborhood of v in G. Since $d^t_G(v) > (1-a)\binom{\Delta}{2}$ and $d_G(v) = \Delta$, we obtain that

$$\begin{aligned}\Delta^2 \geq \sum_{u \in N \setminus \{v\}} d_G(u) &= 2d^t_G(v) + |E_G(N, V(G) \setminus N)| + \Delta \\ &> |E_G(N, V(G) \setminus N)| + (1-a)\Delta(\Delta-1) + \Delta,\end{aligned}$$

which yields

$$(1)\ \ |E_G(N, V(G) \setminus N)| < a\Delta(\Delta-1) < a\Delta^2.$$

To construct an appropriate coloring of G, we decompose N into three disjoint sets U_1, U_2 and U_3 as follows:

- $U_1 = \{u \in N \mid |N_G(u) \setminus N| > \frac{1}{2}(\Delta+1)\}$
- $U_2 = \{u \in N \setminus U_1 \mid |N_G(u) \setminus (N \setminus U_1)| > \sqrt{a}(\Delta+1)\}$
- $U_3 = N \setminus (U_1 \cup U_2)$

From (1) it follows that $|U_1| < 2a\Delta$. Since $0 < \frac{1}{6} - 2\sqrt{a}$, this implies that

$$(2)\ \ |U_1| < \tfrac{1}{6}\sqrt{a}(\Delta+1).$$

Since $N_G(v) = N \setminus \{v\}$, we have $v \notin U_1$ and, because of (2), $v \notin U_2$. Consequently, $v \in U_3$. Every vertex in U_1 has more neighbors outside N than in N, which implies that $|E_G(U_1, U_2 \cup U_3)| < |E_G(U_1, V(G) \setminus N)|$. Using (1), it then follows that we have $|E_G(V(G) \setminus (U_2 \cup U_3), U_2 \cup U_3)| < a\Delta^2$ and so

$$(3)\ \ |U_2| < \sqrt{a}(\Delta+1).$$

Clearly, we have $|U_1| + |U_2| + |U_3| = \Delta+1$. Let $b_1 = |U_1|/(\Delta+1)$ and $b_2 = |U_2|/(\Delta+1)$. Then $|U_3| = (1 - b_1 - b_2)(\Delta+1)$ and it follows from (2) and (3) that

$$(4)\ \ b_1 < \tfrac{1}{6}\sqrt{a} \text{ and } b_2 < \sqrt{a}.$$

Since $v \in U_3$ and $k' \geq \chi(G-v)$, $G' = G - (U_2 \cup U_3)$ has chromatic number at most k', so we can fix a coloring of G' with color set $S = [1, k']$. We now use a greedy strategy to extend this coloring to a coloring of $G - U_3$ with color set S. Then, in each step, every vertex in U_2 has at most $p = |U_1| + |U_2| - 1 + \frac{1}{2}(\Delta+1)$ colored neighbors and $p = (b_1 + b_2 + \frac{1}{2})(\Delta+1) - 1$. Using (4) and $k' \geq k = \lfloor(1-\varepsilon)(\Delta+1)\rfloor$, it then follows that $k' - p > (1 - \varepsilon - \frac{7}{6}\sqrt{a} - \frac{1}{2})(\Delta+1) > 0$; note that $0 < \varepsilon < \frac{1}{6} - 2\sqrt{a}$. Hence, for each vertex in U_2, there is a free color in S, thereby we obtain a coloring φ of $G - U_3$ with color set S.

It remains to show that the coloring φ of $G - U_3$ can be extended to a coloring of G without using new colors. To this end, let $H = G[U_3]$ and let L be a list-assignment for H such that, for every $u \in U_3$, we have $L(u) = S \setminus \varphi(N_G(u) \setminus U_3)$. If H has an L-coloring φ', then $\varphi \cup \varphi'$ is a coloring of G with color set S and we are done. For $u \in U_3$, we have $|N_G(u) \setminus U_3| \leq \sqrt{a}(\Delta+1) + |U_2| < 2\sqrt{a}(\Delta+1)$ (by (3)). This yields $|L(u)| \geq k' - |N_G(u) \setminus U_3| > (1-\varepsilon)(\Delta+1) - 1 - 2\sqrt{a}(\Delta+1)$. Since $0 < \varepsilon < \frac{1}{6} - 2\sqrt{a}$, we obtain that $|L(u)| > \frac{5}{6}(\Delta+1) - 1$ for every $u \in U_3$. So H has an L-coloring provided that $\chi_\ell(H) \leq \frac{5}{6}(\Delta+1)$.

To bound the list chromatic number of $H = G[U_3]$, we use a result of Erdős, Rubin, and Taylor [355], see Exercise 6.16. Let M be a maximum matching of the complement of H, and let X be the set of vertices of H that are in $\overline{H}$ not incident with edges of M. Let $p = |M|$ and $q = |X|$. Clearly, X is an independent set of $\overline{H}$ and hence a clique of H, which yields $q \leq \omega \leq \frac{2}{3}(\Delta+1)$. For $n = |H|$ we obtain that $n = 2p + q \leq \Delta + 1$. We claim that $p + q \leq \frac{5}{6}(\Delta+1)$. If $p \leq \frac{1}{6}(\Delta+1)$, then $p + q \leq \frac{1}{6}(\Delta+1) + \frac{2}{3}(\Delta+1) = \frac{5}{6}(\Delta+1)$ and we are done. If $p \geq \frac{1}{6}(\Delta+1)$, then $p + q = n - p \leq (\Delta+1) - \frac{1}{6}(\Delta+1) = \frac{5}{6}(\Delta+1)$ and we are done, too. We can extend H to a graph H' that is obtained from K_n by deleting a matching of size p. Then, see Exercise 6.16, we have $\chi_\ell(H) \leq \chi_\ell(H') = n - p = p + q \leq \frac{5}{6}(\Delta+1)$. This complete the proof. □

Theorem 6.18 deals with coloring graphs with sparse neighborhoods. The next proposition is folklore and shows that it suffices to prove Theorem 6.18 for regular graphs.

Proposition 6.21 *Let G be a graph with maximum degree Δ such that $d^t_G(v) \leq D$ for all $v \in V(G)$. Then there exists a graph H such that H is regular of degree Δ, $G \subseteq H$, and $d^t_H(v) \leq D$ for all $v \in V(G')$.*

Proof The graph H can be obtained from G by iterating the following process. As long as G is not regular, take two copies of G and add edges between corresponding vertices of degree less than Δ. □

Let G be a graph. A **partial coloring** of G is a coloring of an induced subgraph of G. Let k be an integer and let x be a real number with $0 < x \leq k$. A **partial** (k,x)**-coloring of** G is a partial coloring of G with color set $S = [1,k]$ in which every vertex has at least x repeated colors in his neighborhood.

Proposition 6.22 *Let G be a Δ-regular graph, let k be an integer, and let x be a real number such that $0 < x \leq k$ and $x + k \leq \Delta$. If G has a partial (k,x)-coloring, than $\chi(G) \leq (\Delta+1) - x$.*

Proof Fix a partial (k,x)-coloring of G. First we extend each of the k color classes such that if a vertex is uncolored, all k colors appears on its neighborhood; we can do this greedily one color at the time. Then, deleting all colored vertices from G results in a graph G' with $\Delta(G') \leq \Delta - k - x$ and hence with $\chi(G') \leq (\Delta+1) - k - x$. Clearly, this yields $\chi(G) \leq \chi(G') + k \leq (\Delta+1) - x$. □

Theorem 6.23 (Molloy and Reed) *Let G be a Δ-regular graph and let $B > \Delta(\log \Delta)^3$. Assume that Δ is sufficiently large and $d^t_G(v) \leq \binom{\Delta}{2} - B$ for all $v \in V(G)$. Then $\chi(G) \leq (\Delta+1) - \frac{B}{e^6\Delta}$.*

Proof Let G, Δ and B as in the statement of the theorem. Let $k = \lfloor \Delta/2 \rfloor$ and $x = \frac{B}{e^6 \Delta}$. Our aim is to show that $\chi(G) \leq (\Delta+1) - x$. To this end, it suffices to show that G has a partial (k,x)-coloring (by Proposition 6.22). To construct a partial coloring of G, we fix a vertex order $v_1, v_2, \ldots, v_n$ of G, where $n = |G|$, and apply the following **naive coloring procedure** consisting of two steps:

1. (Coloring step) To each vertex v_i of G assign a color ω_i chosen uniformly at random from $S = [1,k]$. This results in a random improper coloring $\omega \in S^n$ of G; the corresponding probability space is the product space $(\Omega, \Pr)$ with $\Omega = S^n$ and $\Pr(\omega) = 1/k^n$.
2. (Correction step) Whenever two adjacent vertices of G receive the same color in the coloring step, both become uncolored. This results in a partial coloring $\varphi = \varphi(\omega)$ of G with color set S. If a vertex v belongs to the domain of φ (i.e., $\varphi(v)$ is defined), we say that v *retains* it color.

We want to show that with positive probability there is a partial (k,x)-coloring of G. To this end, we consider, for every vertex v of G, the random variable $X_v : \Omega \to \mathbb{R}$, defined as the number of colors $c \in S$ such that at least two nonadjacent neighbors of v receive color c in the coloring step, and every neighbor of v that receives color c in the coloring step retains its color. Furthermore, let A_v be the event that $X_v < x$. To prove that a partial (k,x)-coloring of G exists, it suffices to show that with positive probability none of the events A_v occurs. To this end, we will use the symmetric version of the Lovász Local Lemma (see Corollary C.36). To bound the probability of the event A_v, we introduce, for every vertex v of G, the following two random variables over Ω:

- Let Y_v be the number of colors $c \in S$ such that there are two nonadjacent neighbors of v that receive color c in the coloring step.
- Let Z_v be the number of colors $c \in S$ such that there are at least two nonadjacent neighbors of v that receives color c in the coloring step; and there is a neighbor of v that receives color c in the coloring step, but does not retain its color.

Note that $X_v = Y_v - Z_v$. For a vertex v of G, let $B_v = \binom{\Delta}{2} - d_G^t(v)$. Since G is Δ-regular, B_v is the number of nonedges in neighborhood graph $G[N_G(v)]$. By assumption, we have $B_v \geq B$. We shall prove in three separate claims that, for every vertex v of G, the following statements hold:

(a) $\mathbb{E}(X_v) \geq \frac{1.99 B_v}{e^6 \Delta} \geq \frac{1.99 B}{e^6 \Delta} = 1.99\, x$.

(b) $\mathbb{E}(X_v) \leq \mathbb{E}(Y_v) \leq \frac{3B_v}{\Delta} \leq 2\, e^6 \mathbb{E}(X_v)$.

(c) $\Pr\left(|X_v - \mathbb{E}(X_v)| > \log \Delta \sqrt{\mathbb{E}(X_v)}\right) < \frac{1}{4\Delta^5}$.

Before proving these statements, let us show that this implies that G has a partial (k,x)-coloring. Since $B \geq \Delta(\log \Delta)^3$ and Δ is sufficiently large, we obtain from (a) and (b) that

$$\mathbb{E}(X_v) - \log \Delta \sqrt{\mathbb{E}(X_v)} \geq \frac{1.99\, B_v}{e^6 \Delta} - \log \Delta \sqrt{\frac{3B_v}{\Delta}} > \frac{B_v}{e^6 \Delta} \geq \frac{B}{e^6 \Delta} = x.$$

Consequently, for every vertex v of G, we obtain from (c) that

$$\begin{aligned}\Pr(A_v) &= \Pr(X_v < x) \le \Pr\left(X_v < \mathbb{E}(X_v) - \log\Delta\sqrt{\mathbb{E}(X_v)}\right)\\ &\le \Pr\left(|X_v - \mathbb{E}(X_v)| > \log\Delta\sqrt{\mathbb{E}(X_v)}\right) < \tfrac{1}{4\Delta^5}\end{aligned}$$

Each event A_v is mutually independent of all the other events A_u, unless $\mathrm{dist}_G(u,v) \le 4$. So we can apply Corollary C.36 with $p = 1/(4\Delta^5)$ and $d = \Delta^4$. Then $ep(d+1) \le 1$ and so

$$\Pr\left(\bigcap_{v\in V(G)} \overline{A}_v\right) > 0$$

This implies that G has a partial (k,x)-coloring and so $\chi(G) \le (\Delta+1) - x$, which proves the theorem. So it remains to prove statements (a), (b) and (c).

Claim 6.23.1 $\mathbb{E}(X_v) \ge \frac{1.99B_v}{e^6\Delta}$. □

Proof: Let X_v' denote the number of colors $c \in S = [1,k]$ such that exactly two non-adjacent neighbors of v receive color c in the coloring step and both retains there color. Clearly, $X_v \ge X_v'$ and so $\mathbb{E}(X_v) \ge \mathbb{E}(X_v')$. Let u,w be two nonadjacent neighbors of v, and let $c \in S$ be a color. Let $p = p(u,w,c)$ be the probability that u and v are the only neighbors of v that receive color c in the coloring step and both retain their color. This happens if no vertex in $N = (N_G(u) \cup N_G(v) \cup N_G(w)) \setminus \{u,w\}$ receive color c in the coloring step. Since G is Δ-regular and $k = \lfloor \Delta/2 \rfloor$, $|N| \le 3\Delta - 3 \le 6k$. Consequently,

$$p = (\tfrac{1}{k})^2(1-\tfrac{1}{k})^{|N|} \ge (\tfrac{1}{k})^2(1-\tfrac{1}{k})^{6k}.$$

Since $B_v = \binom{\Delta}{2} - d_G^t(v)$, we have B_v choices for the vertex pair u,w; and we have k choices for c. Consequently, we obtain that

$$\mathbb{E}(X_v') \ge kB_v(\tfrac{1}{k})^2(1-\tfrac{1}{k})^{6k} = \tfrac{B_v}{k}(1-\tfrac{1}{k})^{6k} \ge \tfrac{2B_v}{\Delta}(1-\tfrac{1}{k})^{6k} \ge \tfrac{1.99B_v}{e^6\Delta}.$$

This proves the claim. △

Claim 6.23.2 $\mathbb{E}(X_v) \le \mathbb{E}(Y_v) \le \frac{3B_v}{\Delta} \le 2\,e^6\mathbb{E}(X_v)$. □

Proof: Clearly, $X_v \le Y_v$ and so $\mathbb{E}(X_v) \le \mathbb{E}(Y_v)$. The probability that a (fixed) color being assigned to at least two nonadjacent neighbors of v in the coloring step is at most $B_v(1/k)^2$. Consequently, we obtain that

$$\mathbb{E}(Y_v) \le kB_v(\tfrac{1}{k})^2 = \tfrac{B_v}{k} \le \tfrac{3B_v}{\Delta} \le 2\,e^6\mathbb{E}(X_v),$$

where the last inequality follows from Claim 6.23.1. This proves the claim. △

Claim 6.23.3 $\Pr\left(|X_v - \mathbb{E}(X_v)| > \log\Delta\sqrt{\mathbb{E}(X_v)}\right) < \frac{1}{4\Delta^5}$ □

Proof: Let $t = \frac{1}{2}\log\Delta\sqrt{\mathbb{E}(X_v)}$. We have $X_v = Y_v - Z_v$ and so $\mathbb{E}(X_v) = \mathbb{E}(Y_v) - \mathbb{E}(Z_v)$. This implies that $|X_v - \mathbb{E}(X_v)| \le |Y_v - \mathbb{E}(Y_v)| + |Z_v - \mathbb{E}(Z_v)|$. From this we obtain that

(1) $\Pr(|X_v - \mathbb{E}(X_v)| > 2t) \le \Pr(|Y_v - \mathbb{E}(Y_v)| > t) + \Pr(|Z_v - \mathbb{E}(Z_v)| > t)$.

Let $p_v = \Pr(|Y_v - \mathbb{E}(Y_v)| > t)$ and $p'_v = \Pr(|Z_v - \mathbb{E}(Z_v)| > t)$. To establish an upper bound for p_v and p'_v, we shall use Talagrand's inequality (see Theorem C.42). First, we have to show that if $X \in \{Y_v, Z_v\}$, then X is (r,ℓ)-certifiable (see Appendix C.15 for the definition). It is not difficult to check that for $X = Y_v$ we can take $(\ell, r) = (2,2)$ and for $X = Z_v$ we can take $(\ell, r) = (2,3)$. Note that $Z_v \geq s$ can be verified by a set of at most $3s$ vertices (for each color counted by Z_v we need two nonadjacent neighbors of v and a third vertex that is adjacent to one of the first two).

Let us first consider Y_v, and let $(\ell, r) = (2,2)$. By Claim 6.23.2, we have $\mathbb{E}(X_v) \geq \frac{1}{2e^6}\mathbb{E}(Y_v)$. For Δ sufficiently large, we then obtain that

$$t = \tfrac{1}{2}\log\Delta\sqrt{\mathbb{E}(X_v)} > 96\ell\sqrt{r\mathbb{E}(Y_v)} + 128r\ell^2.$$

Then it follows from Theorem C.42 that

$$p_v \leq 4\exp\left(-\frac{t^2}{8\ell^2 r(4\mathbb{E}(Y_v)+t)}\right) = 4\exp\left(-\frac{t^2}{64(4\mathbb{E}(Y_v)+t)}\right).$$

For Z_v we can use a similar argument with $(\ell, r) = (2,3)$. Since $\mathbb{E}(Y_v) \geq \mathbb{E}(Z_v)$, it then follows from Theorem C.42 that

$$p'_v \leq 4\exp\left(-\frac{t^2}{8\ell^2 r(4\mathbb{E}(Z_v)+t)}\right) \leq 4\exp\left(-\frac{t^2}{96(4\mathbb{E}(Y_v)+t)}\right).$$

By (1), we then obtain that

$$\begin{aligned}
\Pr\left(|X_v - \mathbb{E}(X_v)| > \log\Delta\sqrt{\mathbb{E}(X_v)}\right) &= \Pr\left(|X_v - \mathbb{E}(X_v)| > 2t\right) \\
&\leq p_v + p'_v \leq 2p'_v \\
&\leq 8\exp\left(-\frac{t^2}{96(4\mathbb{E}(Y_v)+t)}\right) \\
&< \frac{1}{4\Delta^5},
\end{aligned}$$

which proves the claim. To see that the last inequality holds, first observe that this inequality is equivalent to

$$480\ln(2\Delta) < \frac{t^2}{4\mathbb{E}(Y_v)+t} = M$$

For $t = \frac{1}{2}\log\Delta\sqrt{\mathbb{E}(X_v)}$, we have $t \geq \frac{1}{2\sqrt{2e^6}}(\log\Delta)\sqrt{\mathbb{E}(Y_v)}$ (by Claim 6.23.2), which implies that

$$\begin{aligned}
M &\geq \frac{\frac{1}{8e^6}(\log\Delta)^2\mathbb{E}(Y_v)}{4\mathbb{E}(Y_v) + \frac{1}{2\sqrt{2e^6}}(\log\Delta)\sqrt{\mathbb{E}(Y_v)}} \\
&= \frac{(\log\Delta)^2}{32e^6 + 2\sqrt{2}e^3(\log\Delta)(1/\sqrt{\mathbb{E}(Y_v)})}
\end{aligned}$$

Since $B_v \geq B > \Delta(\log\Delta)^3$, it follows from Claim 6.23.1 and Claim 6.23.2 that

$$\mathbb{E}(Y_v) \geq \mathbb{E}(X_v) \geq \frac{B}{e^6\Delta} \geq \frac{(\log\Delta)^3}{e^6},$$

which implies that

$$\begin{aligned} M &\geq \frac{(\log\Delta)^2}{32e^6 + 2\sqrt{2}e^6(\log\Delta)(1/\sqrt{(\log\Delta)^3})} \\ &\geq \frac{(\log\Delta)^2}{32e^6 + 2\sqrt{2}e^6} = M'. \end{aligned}$$

For Δ arbitrarily large, we then obtain that $M \geq M' > 480\ln(2\Delta)$. This completes the proof of the claim. △

Hence, Theorem 6.23 is proved. □

6.4 Independent Sets

Every graph G satisfies $|G| \leq \chi(G)\alpha(G)$ (see (1.2)). So upper bounds for the chromatic number leads to lower bounds for the independence number, but not conversely. The following bound for the independence number was obtained by Fajtlowicz [362]; it is almost a consequence of Reed's conjecture.

Theorem 6.24 (Fajtlowicz) *Every graph G of order n satisfies*

$$\alpha(G) \geq \tfrac{2n}{\Delta(G)+1+\omega(G)}.$$

Proof It suffices to show that $\alpha(G)(\Delta(G)+1+\omega(G)) \geq 2n$. This is evident if G is a complete graph. So assume that G is not complete. Then $\alpha(G) \geq 2$. Let I be an independent set of G with $|I| = \alpha(G)$. For $i \in [0, |I|]$, define

$$a_i(I) = |\{v \in V(G) \setminus I \mid |N_G(v) \cap I| = i\}|.$$

Since $\alpha(G) = |I|$, we obtain that $a_0(I) = 0$ implying that

$$n - |I| = |V(G) \setminus I| = \sum_{i=1}^{|I|} a_i(I).$$

Furthermore, we obtain that

$$\sum_{v \in I} d_G(v) = |E_G(I, V(G) \setminus I)| = \sum_{i=1}^{|I|} i a_i(I).$$

Multiplying the first equation with 2 and subtracting the second equation then yields

$$2n-2|I|-\sum_{v\in I}d_G(v)=\sum_{i=1}^{|I|}(2-i)a_i(I)\leq a_1(I). \tag{6.10}$$

For a vertex $v\in I$, let X_v the set of vertices in $V(G)\setminus I$ that are adjacent to v in G, but to no other vertex of I. We claim that X_v is a clique for every vertex $v\in I$. For otherwise, there are vertices $u,w\in X_v$ such that $u\neq w$ and $uw\notin E(G)$, implying that $I'=(I\setminus\{v\})\cup\{u,w\}$ is an independent set of G with $|I'|=|I|+1=\alpha(G)+1$, which is impossible. This proves the claim that, for every $v\in I$, the set X_v is a clique of G, and hence $X_v\cup\{v\}$ is a clique of G, too. Consequently, $|X_v|\leq\omega(G)-1$ for all $v\in I$. This implies that $a_1(I)\leq|I|(\omega(G)-1)=\alpha(G)(\omega(G)-1)$. Combined with (6.10), we obtain that

$$\begin{aligned}2n&\leq 2|I|+\sum_{v\in I}d_G(v)+a_1(I)\\&\leq 2\alpha(G)+\Delta(G)\alpha(G)+\alpha(G)(\omega(G)-1)\\&=\alpha(G)(\Delta(G)+1+\omega(G)).\end{aligned}$$

Thus the proof is complete. □

Let G be a graph and let I be a nonempty independent set of G. Then we obtain a new graph H from $G-I$ by adding a new vertex $v(I)$ and by joining $v(I)$ to a vertex u of $G-I$ if and only if $N_G(u)\cap I\neq\varnothing$. We then write $H=G/(I\to v(I))$ or briefly $H=G/I$ and say that H is obtained from G by **identifying I to $v(I)$**, or briefly by **identifying independent vertices**. Clearly, any k-coloring of G/I can be extended to a k-coloring of G by giving each vertex of I the same color as v_I. This shows that $\chi(G/I)\geq\chi(G)$. If $I_1,I_2,\ldots,I_p$ are disjoint nonempty independent sets of G, let $G/(I_1,I_2,\ldots,I_p)=G/I_1/I_2/\cdots/I_p$. The map $\phi: V(G)\to G/(I_1,I_2,\ldots,I_p)$ with $\phi(v)=v$ if $v\notin I_1\cup I_2\cup\cdots\cup I_p$ and $\phi(v)=v(I_i)$ if $v\in I_i$ is an homomorphism from G to $H=G/(I_1,I_2,\ldots,I_p)$ (see Exercise 8.2). We obtain that $\chi(G)\leq\chi(H)$ and $\omega(G)\leq\omega(H)$. Furthermore, it is easy to show that $\omega(H)\leq\omega(G)+p$. The following results are due to Rabern [831, 832]

Theorem 6.25 (Rabern) *Let G be a graph, let $p\in\mathbb{N}$, and let $I_1,I_2,\ldots,I_p$ be disjoint nonempty independent sets of G. Then*

$$\chi(G)\leq\tfrac{1}{2}\Big(\omega(G)+|G|-\sum_{i=1}^{p}|I_i|+2p-1\Big)$$

Sketch of Proof: For the proof we distinguish two cases depending on whether the graph $K=G/(I_1,I_2,\ldots,I_p)$ is complete or not.

Case 1: *K is a complete graph.* Let $H=G/(I_1,I_2,\ldots,I_{p-1})$. Then I_p is an independent set of H and $K=H/I_p$. This implies that $X=V(H)\setminus I_p$ is a clique of H and every vertex $v\in X$ satisfies $N_H(v)\cap I_p\neq\varnothing$. In $\overline{H}$, the set X is independent, I_p is a clique and, for every vertex $v\in X$ there is a vertex $u\in I_p$ with $vu\notin E(\overline{H})$. This implies that $\chi(\overline{H})=|I_p|$. By Proposition 1.6, this leads to

$$(1)\ \chi(H) \leq |H| + 1 - \chi(\overline{H}) = |H| + 1 - |I_p|.$$

On the other hand, a coloring of the clique X of H can be extended to a coloring of H without using a new color, unless there is a vertex $v \in I_p$ such that $N_H(v) = X$. So we obtain that

$$(2)\ \chi(H) = \omega(H) \leq \omega(G) + p - 1.$$

Furthermore, we obtain that

$$|G| - |H| = \sum_{i=1}^{p-1} (|I_i| - 1) = \sum_{i=1}^{p} |I_i| - |I_p| - p + 1,$$

which is equivalent to

$$(3)\ |H| - |I_p| = |G| - \sum_{i=1}^{p} |I_i| + p - 1.$$

Using (1), (2) and (3), we obtain that

$$\begin{aligned} 2\chi(G) \leq 2\chi(H) &\leq \omega(G) + p - 1 + |H| + 1 - |I_p| \\ &= \omega(G) + p - 1 + 1 + |G| - \sum_{i=1}^{p} |I_i| + p - 1 \\ &= \omega(G) + |G| - \sum_{i=1}^{p} |I_i| + 2p - 1. \end{aligned}$$

This settles the first case.

Case 2: *K is not a complete graph.* Then K is homomorphic to a complete graph K_k with $k = \chi(K)$ (see (8.1)). Furthermore, K is homomorphic to a graph H such that there is an independent set I of H satisfying $|I| = 2$ and $K_k = H/I$. So we have $K \to H \to K_k$ and, therefore, $\chi(K) = \chi(H) = k$ (by Proposition 8.1). The set $X = V(H) \setminus I$ is a clique of H and every vertex $v \in X$ satisfies $N_H(v) \cap I_p \neq \varnothing$. As in the former case, this leads to $\chi(H) = \omega(H)$, and hence to

$$(4)\ k = \chi(H) = |H| - 1 = \omega(H) \leq \omega(G) + p + |K| - |H|.$$

Furthermore, we obtain that

$$|G| - |K| = \sum_{i=1}^{p} |I_i| - p.$$

Combining this with (4), we obtain that

$$\begin{aligned}2\chi(G) &\le 2\chi(K) = 2\chi(H)\\ &\le |H| - 1 + \omega(G) + p + |K| - |H|\\ &= \omega(G) + |G| - \sum_{i=1}^{p} |I_i| + 2p - 1.\end{aligned}$$

This settles the second case. Hence the theorem is proved. □

Corollary 6.26 (Rabern) *Every graph G satisfies*

$$\chi(G) \le \tfrac{1}{2}(\omega(G) + |G| - \alpha(G) + 1) \text{ and } \chi(G) \le \tfrac{1}{2}(\omega(G) + |G|).$$

Theorem 6.27 (Rabern) *Every graph G with* $\alpha(G) \le 2$ *satisfies*

$$\chi(G) \le \tfrac{1}{2}(\omega(G) + \Delta(G) + 2).$$

Proof Suppose this is false, and let G be a counterexample whose order n is minimum. Then G is vertex critical and $n \ge 1$. Let v be a vertex of G with $d_G(v) = \delta(G)$. Since $\alpha(G) \le 2$, $V(G) \setminus (N_G(v) \cup \{v\})$ is a clique of G, and so $\omega(G) \ge n - (\delta(G)+1)$. This leads to $\Delta(G) + 1 \ge \delta(G) + 1 \ge n - \omega(G)$, and hence to

$$\chi(G) > \tfrac{1}{2}(\omega(G) + \Delta(G) + 2) \ge \tfrac{n+1}{2}.$$

Since G is vertex critical, this implies that $G = H \boxplus H'$ for two nonempty graphs H and H' (see Theorem 4.13 respectively Theorem 4.64). The minimality of G then implies that $\chi(H) \le \frac{1}{2}(\omega(H) + \Delta(H) + 2)$. By Corollary 6.26, we have $\chi(H') \le \frac{1}{2}(\omega(H') + |H'|)$. Combining both inequalities, we then obtain that

$$\chi(G) = \chi(H \boxplus H') = \chi(H) + \chi(H') \le \tfrac{1}{2}(\omega(H) + \Delta(H) + 2 + \omega(H') + |H'|),$$

where the second equation follows from Theorem 4.1. Clearly, we have $\omega(G) = \omega(H \boxplus H') = \omega(H) + \omega(H')$ and $\Delta(H) + |H'| \le \Delta(H \boxplus H') = \Delta(G)$. Consequently, we obtain that $\chi(G) \le \frac{1}{2}(\omega(G) + \Delta(G) + 2)$, a contradiction. □

Lemma 6.28 (Rabern) *Every graph G satisfies* $2\chi(G) \le \omega(G) + \frac{|G| + \Delta(G) + 1}{2}$.

Proof Let $I_1, I_2, \ldots, I_p$ be disjoint independent sets of G such that (1) $|I_j| \ge 3$ for $j \in [1, p]$ and (2) $G' = G - \bigcup_{j=1}^{m} I_j$ has minimum order subject to (1). We claim that (3) $\chi(G) \le \frac{1}{2}(\omega(G) + \Delta(G) + p + 2)$. First, assume that $G' = \varnothing$. Then we have $\chi(G) \le p$, which yields $\chi(G) \le \frac{1}{2}(\chi(G) + p) \le \frac{1}{2}(\omega(G) + \Delta(G) + p + 2)$ and we are done. Now, assume that $G' \ne \varnothing$. By (2) we obtain that each vertex of G' has a neighbor in G in each independent set I_j, and so $\Delta(G') \le \Delta(G) - p$. Furthermore, (2) implies that $\alpha(G') \le 2$. From Theorem 6.27 it then follows that

$$\begin{aligned}\chi(G) &\le p + \chi(G') \le p + \tfrac{1}{2}(\omega(G') + \Delta(G') + 2)\\ &\le p + \tfrac{1}{2}(\omega(G) + \Delta(G) - p + 2) = \tfrac{1}{2}(\omega(G) + \Delta(G) + p + 2).\end{aligned}$$

This completes the proof of (3). From Theorem 6.25 and (1) it follows that $\chi(G) \le \frac{1}{2}(\omega(G)+|G|-p-1)$. Together with (3), this gives $2\chi(G) \le \omega(G)+\frac{|G|+\Delta(G)+1}{2}$. □

Corollary 6.29 (Rabern) *Every graph G whose complement is disconnected satisfies $\chi(G) \le \frac{1}{2}(\omega(G)+\Delta(G)+1)$.*

Proof By assumption, we have $G = G_1 \boxplus G_2$, where both graphs G_1 and G_2 are nonempty. Hence, $\Delta(G) = \max\{\Delta(G_1)+|G_2|, \Delta(G_2)+|G_1|\}$. Using Theorem 4.1 and Lemma 6.28, we then obtain that

$$\begin{aligned}\chi(G) &= \chi(G_1)+\chi(G_2)\\ &\le \tfrac{1}{2}\Big(\omega(G_1)+\omega(G_2)+\frac{|G_1|+\Delta(G_1)+|G_2|+\Delta(G_2)+2}{2}\Big)\\ &\le \tfrac{1}{2}\Big(\omega(G)+\frac{2\Delta(G)+2}{2}\Big) = \tfrac{1}{2}(\omega(G)+\Delta(G)+1).\end{aligned}$$

Hence the proof is complete. □

Corollary 6.30 (Rabern) *Every graph G with $\chi(G) > \lceil |G|/2 \rceil$ satisfies $\chi(G) \le \frac{1}{2}(\omega(G)+\Delta(G)+1)$.*

Proof Obviously, G contains a vertex-critical induced subgraph G' with $\chi(G') = \chi(G)$ (see Exercise 4.24). From Theorem 4.64 it then follows that the complement of G' is disconnected. By Corollary 6.29, we then obtain that

$$\chi(G) = \chi(G') \le \tfrac{1}{2}(\omega(G')+\Delta(G')+1) \le \tfrac{1}{2}(\omega(G)+\Delta(G)+1)$$

Hence the proof is complete. □

6.5 Mozhan Partitions

In 1983, Mozhan [762] developed a method that can be used to show that a graph whose chromatic number is close to its maximum degree has a large clique number. The idea behind this method is to consider partitions of a graph into a fixed number of parts where each part has a prescribed chromatic number and is a subgraph induced by the disjoint union of color classes in some fixed optimal coloring of the graph. Then Brooks' theorem is used to to show that several parts of such a partition have small cliques that can be combined into a large clique.

Mozhan's method was further developed by Cranston and Rabern [259]. They proved that every graph with $\chi \ge \Delta \ge 13$ satisfies $\omega \ge \Delta - 3$. From Corollary 6.14 it follows that it is sufficient to confirm the base case, i.e. to show that every graph with $\chi = \Delta = 13$ contains a K_{10}. Our aim in this section is to prove the result of Cranston and Rabern (Corollary 6.46

In the following, let G be a given graph and $p \in \mathbb{N}$. If H is an induced subgraph of G and $v \in V(G)$, then let $H+v = G[V(H) \cup \{v\}]$, further let $d(v:H) = d_{H+v}(v)$, and let $e(H) = |E(H)|$. Recall from Section 2.5 that a **partition** and **p-partition** of a graph is a sequence of p induced subgraphs such that each vertex of the graph belongs to exactly one graph of the sequence. If $\mathfrak{G}$ is a p-partition of G, we denote by $G_i(\mathfrak{G})$, for $i \in [1,p]$, the **ith part** of $\mathfrak{G}$, such that $\mathfrak{G} = (G_1(\mathfrak{G}), G_2(\mathfrak{G}), \ldots, G_p(\mathfrak{G}))$.

Let $\mathfrak{G}$ be a p-partition of G. Then we can combine optimal colorings of the parts, using disjoint color sets, to a coloring of G, showing that

$$\chi(G) \leq \sum_{i=1}^{p} \chi(G_i(\mathfrak{G})). \tag{6.11}$$

We call $\mathfrak{G}$ an **exact p-partition** of G if equality holds in (6.11). Let φ be an optimal coloring of G, that is, a coloring of G with color set $C = [1, \chi(G)]$, and let $(C_1, C_2, \ldots, C_p)$ be a partition of C. Then we denote by

$$\mathfrak{G} = \mathfrak{G}(G, \varphi, C_1, C_2, \ldots, C_p)$$

the p-partition of G such that, for $i \in [1,p]$, we have

$$G_i(\mathfrak{G}) = G\big[\bigcup_{c \in C_i} \varphi^{-1}(c)\big].$$

Then it follows from Proposition 1.34(c) that $\mathfrak{G}$ is an exact p-partition of G such that $\chi(G_i(\mathfrak{G})) = |C_i|$. As a consequence we obtain the following simple result.

Proposition 6.31 *Let G be a graph and let $(k_1, k_2, \ldots, k_p) \in \mathbb{N}^p$ be a sequence such that $\chi(G) = k_1 + k_2 + \cdots + k_p$. Then there exists an exact p-partition $\mathfrak{G}$ of G such that $\chi(G_i(\mathfrak{G})) = k_i$ for $i \in [1,p]$.*

Let $\mathfrak{G}$ be a p-partition of G. A **component** of $\mathfrak{G}$ is a component of some part of $\mathfrak{G}$. Let v be a vertex of G. Then we denote by $i(v, \mathfrak{G})$ the unique index $i \in [1,p]$ such that v belongs to $G_i(\mathfrak{G})$, and we denote by $K_v(\mathfrak{G})$ the unique component of $\mathfrak{G}$ containing v. We say that v is a **critical vertex** of $\mathfrak{G}$ if v is a critical vertex of $G_i(\mathfrak{G})$ with $i = i(v, \mathfrak{G})$, that is, v belongs to $G_i(\mathfrak{G})$ and $\chi(G_i(\mathfrak{G}) - v) = \chi(G_i(\mathfrak{G})) - 1$. Let $i = i(v, \mathfrak{G})$ and $j \in [1,p] \setminus \{i\}$. Then there exists a p-partition $\mathfrak{G}'$ such that

$$G_\ell(\mathfrak{G}') = \begin{cases} G_\ell(\mathfrak{G}) & \text{if } \ell \notin \{i,j\}, \\ G_\ell(\mathfrak{G}) - v & \text{if } \ell = i, \\ G_\ell(\mathfrak{G}) + v & \text{if } \ell = j, \end{cases}$$

for all $\ell \in [1,p]$. We say that $\mathfrak{G}'$ is obtained from $\mathfrak{G}$ by **moving vertex** v from the ith part to the jth part, and we call $(\mathfrak{G}, v, j, \mathfrak{G}')$ a **move**. Clearly, if $(\mathfrak{G}, v, j, \mathfrak{G}')$ is a move and $i = i(v, \mathfrak{G})$, then $j = i(v, \mathfrak{G}')$ and $(\mathfrak{G}', v, i, \mathfrak{G})$ is a move, too. We call $(\mathfrak{G}', v, i, \mathfrak{G})$ the **reverse move** of $(\mathfrak{G}, v, j, \mathfrak{G}')$ and denote it by $(\mathfrak{G}, v, j, \mathfrak{G}')^{-1}$.

Let G be a graph and let $(k_1, k_2, \ldots, k_p) \in \mathbb{N}^p$. A **$(k_1, k_2, \ldots, k_p)$-M-partition** of G is a pair $(\mathfrak{G}, v)$ satisfying the following conditions:

(M1) $\mathfrak{G}$ is an exact p-partition of G,

(M2) v is a critical vertex of $\mathfrak{G}$,
(M3) $\chi(G_i(\mathfrak{G})) = k_i$ for $i \in [1,p] \setminus \{i(v,\mathfrak{G})\}$ and $\chi(G_i(\mathfrak{G})) = k_i + 1$ for $i = i(v,\mathfrak{G})$,

The letter M in the above definition stands for the name Mozhan; we use the name M-partition in order to distinguish this type of partition from the f-partitions considered in Section 2.5.

If $(\mathfrak{G}, v)$ is a $(k_1, k_2, \dots, k_p)$-M-partition of G, then $\chi(G) = 1 + k_1 + k_2 + \cdots + k_p$ (by (M1) and (M3)) and v is a critical vertex of G (by (M1), (M2) and (6.11)). The next proposition shows that these two conditions are sufficient for the existence of such a partition of G.

Proposition 6.32 *Let G be a graph, let v be a critical vertex of G, let $(k_1, k_2, \dots, k_p)$ be a sequence of positive integers such that $\chi(G) = 1 + k_1 + k_2 + \cdots + k_p$ and let $i_0 \in [1,p]$. Then G has a $(k_1, k_2, \dots, k_p)$-M-partition $(\mathfrak{G}, v)$ with $i(v,\mathfrak{G}) = i_0$. Furthermore, every such partition $(\mathfrak{G}, v)$ of G satisfies the following statements:*

(a) *$\chi(K_v(\mathfrak{G})) = k_{i_0} + 1$ and $\chi(K) = k_{i_0}$ for every component $K \neq K_v(\mathfrak{G})$ of $G_{i_0}(\mathfrak{G})$.*
(b) *v is a critical vertex of $K_v(\mathfrak{G})$ and $d(v : K_v(\mathfrak{G})) = d(v : G_{i_0}(\mathfrak{G})) \geq k_{i_0}$.*
(c) *If $j \in [1,p] \setminus \{i_0\}$, then $\chi(G_j(\mathfrak{G}) + v) = \chi(G_j(\mathfrak{G})) + 1 = k_i + 1$ and there is a component K of $G_j(\mathfrak{G})$ such that $d(v : K) \geq k_j$.*

Proof Since v is a critical vertex of G, we get $\chi(G - v) = \chi(G) - 1 = k_1 + k_2 + \cdots + k_p$ (by (1.13)). From Proposition 6.31 it follows that there is an exact partition $\mathfrak{G}'$ of $G' = G - v$ such that $\chi(G_i(\mathfrak{G}')) = k_i$. Let $\mathfrak{G}$ be the partition obtained from $\mathfrak{G}'$ by adding v to the i_0th part, that is, $G_i(\mathfrak{G}) = G_i(\mathfrak{G}')$ if $i \neq i_0$ and $G_{i_0}(\mathfrak{G}) = G_{i_0}(\mathfrak{G}') + v$. Then we have $\chi(G_{i_0}(\mathfrak{G})) = k_{i_0} + 1$. For otherwise, we have $\chi(G_{i_0}(\mathfrak{G})) = k_{i_0}$ (by (1.13)), which yields that

$$\sum_{i=1}^{p} \chi(G_i(\mathfrak{G})) = \sum_{i=1}^{p} k_i = \chi(G) - 1,$$

a contradiction to (6.11). Consequently, $(\mathfrak{G}, v)$ is a $(k_1, k_2, \dots, k_p)$-M-partition of G with $i(v,\mathfrak{G}) = i_0$.

Now let $(\mathfrak{G}, v)$ be an arbitrary $(k_1, k_2, \dots, k_p)$-M-partition of G with $i(v,\mathfrak{G}) = i_0$. Statement (a) follows from (M2), (M3) and (1.5). Statement (b) is an immediate consequence of (a) and (M2). For the proof of (c), let $j \in [1,p] \setminus \{i_0\}$ and let $G_j = G_j(\mathfrak{G})$. First, we claim that $\chi(G_j + v) = \chi(G_j) + 1 = k_j + 1$. For otherwise, we have $\chi(G_j + v) = \chi(G_j) = k_j$ (by (1.13)). For the partition $\mathfrak{G}'$ obtained from $\mathfrak{G}$ by moving v from the i_0th part to the jth part, we then obtain that

$$\sum_{i=1}^{p} \chi(G_i(\mathfrak{G}')) = \sum_{i=1}^{p} k_i < \chi(G),$$

a contradiction to (6.11). This proves the claim that $\chi(G_j + v) = \chi(G_j) + 1 = k_j + 1$, which immediately implies that $d(v : G_j) \geq k_j$. Suppose that for every component K of G_j we have $d(v : K) \leq k_j - 1$. Since $\chi(G_j) = k_j$, this implies that, for every component K of G_j, there exists a k_j-coloring φ_K of $K + v$. By permuting colors, we may assume that $\varphi_K(v) = c$ for all such components K, which yields a k_j-coloring of $G_j + v$, a contradiction. Hence there is a component K of G_j such that $d(v : K) \geq k_j$. This completes the proof of the proposition. □

Let G be a graph and let $(\mathfrak{G}, v)$ be a $(k_1, k_2, \ldots, k_p)$-M-partition of G with $i(v, \mathfrak{G}) = i_0$. Then we define the **weight** of $(\mathfrak{G}, v)$ by

$$w(\mathfrak{G}, v) = \sum_{i=1}^{p} e(G_i(\mathfrak{G})) - d(v : G_{i_0}(\mathfrak{G})).$$

The following proposition is an immediate consequence of Proposition 6.32 and the definition of the weight.

Proposition 6.33 *Let G be a graph, let $(\mathfrak{G}, v)$ be a $(k_1, k_2, \ldots, k_p)$-M-partition of G, and let $(\mathfrak{G}, v, j, \mathfrak{G}')$ be a move. Then $\mathfrak{G}'$ is a $(k_1, k_2, \ldots, k_p)$-M-partition of G and $w(\mathfrak{G}, v) = w(\mathfrak{G}', v)$.*

Proposition 6.34 *Let K be a $(k+1)$-chromatic connected graph with $k \geq 3$, and let there be a critical vertex of K with degree k. If $\Delta(K) \geq k+1$, then K has a critical vertex u with $d_K(u) \geq k+1$.*

Proof Let V be the set of critical vertices of K whose degree is k. By assumption, we have $V \neq \varnothing$ and $V \neq V(K)$. Since K is connected, this implies that there is an edge uv in K such that $v \in V$ and $u \notin V$. Since $v \in V$, there is a k-coloring φ of $K - v$. Since $d_K(v) = k$ and $\chi(K) = k+1$, this implies that among the k neighbors of v all k colors occur exactly once. Hence if we color v with $\varphi(u)$ and uncolor u, we obtain a k-coloring of $K - u$. So u is a critical vertex of K. Since $u \notin V$, we have $d_K(u) \geq k+1$. □

A $(k_1, k_2, \ldots, k_p)$-M-partition $(\mathfrak{G}, v)$ of G with $i(v, \mathfrak{G}) = i_0$ is said to be **optimal** if it has minimum weight $w(\mathfrak{G}, v)$ among all such partitions and, subject to this, minimizes the value $d(v : \mathfrak{G}) = d(v : G_{i_0}(\mathfrak{G})) - k_{i_0}$.

Lemma 6.35 (MOZHAN'S LEMMA) *Let G be a graph and let $(\mathfrak{G}, v)$ be an optimal $(k_1, k_2, \ldots, k_p)$-M-partition of G with $i(v, \mathfrak{G}) = i_0$. If $\Delta(G) \leq \chi(G) + p - 2$ and $k_i \geq 3$ for $i \in [1, p]$, then $K_v(\mathfrak{G})$ is a complete graph of order $k_{i_0} + 1$.*

Proof By Proposition 6.32(b), we have $d(v : \mathfrak{G}) \geq 0$. First, we claim that $d(v : \mathfrak{G}) = 0$. For otherwise, $d(v : \mathfrak{G}) \geq 1$. If $j \in [1, p] \setminus \{i_0\}$, then there is a move $(\mathfrak{G}, v, j, \mathfrak{G}')$ and $(\mathfrak{G}', v)$ is a $(k_1, k_2, \ldots, k_p)$-M-partition of G with $w(\mathfrak{G}, v) = w(\mathfrak{G}', v)$ (by Proposition 6.33), which implies that $d(v : \mathfrak{G}') \geq 1$. Consequently, for $i \in [1, p]$, we have $d(v : G_i(\mathfrak{G})) \geq k_i + 1$ (by Proposition 6.32). This implies that

$$\Delta(G) \geq d_G(v) = \sum_{i=1}^{p} d(v : G_i(\mathfrak{G})) \geq \sum_{i=1}^{p} (k_i + 1) = \chi(G) + p - 1,$$

which is impossible. So we have $d(v : \mathfrak{G}) = 0$ and hence, for the component $K = K_v(\mathfrak{G})$, we have $d(v : K) = k_{i_0}$. By Proposition 6.32(a), we have $\chi(K) = k_{i_0} + 1$ and

$\chi(G_{i_0}(\mathfrak{G}) - v) \le k_{i_0}$. Hence v is a critical vertex of K. To complete the proof, it suffices to show that K is a complete graph. Since $k_{i_0} \ge 3$, we can use Brooks' theorem. Hence it suffices to show that $\Delta(K) \le k_{i_0}$. Suppose this is false. Then, by Proposition 6.34, K contains a critical vertex u with $d(u:K) \ge k_{i_0}+1$. Then $(\mathfrak{G}, u)$ is a $(k_1, k_2, \ldots, k_p)$-M-partition of G with $w(\mathfrak{G}, u) < w(\mathfrak{G}, v)$, a contradiction. □

Proposition 6.36 *Let $G \in \mathrm{Crit}(k)$ and $(k_1, k_2, \ldots, k_p) \in \mathbb{N}^p$. Suppose that $\Delta(G) \le k+p-2$, $k = 1+k_1+k_2+\cdots+k_p$, and $k_i \ge 3$ for $i \in [1,p]$. Then G has an optimal $(k_1, k_2, \ldots, k_p)$-M-partition. Furthermore, if $(\mathfrak{G}, v)$ is an optimal $(k_1, k_2, \ldots, k_p)$-M-partition of G with $i(v, \mathfrak{G}) = i_0$, then the following statements hold:*

(a) *If K is a component of $G_{i_0}(\mathfrak{G})$, then $\chi(K) = k_{i_0}+1$ if $K = K_v(\mathfrak{G})$, otherwise $\chi(K) \le k_{i_0}$.*
(b) *$K_v(\mathfrak{G}) = K_{k_{i_0}+1}$ and, moreover, $(\mathfrak{G}, u)$ is an optimal $(k_1, k_2, \ldots, k_p)$-M-partition of G if and only if u belongs to $K_v(\mathfrak{G})$.*
(c) *If $j \in [1,p] \setminus \{i_0\}$, then $\chi(G_j(\mathfrak{G})+v) = \chi(G_j(\mathfrak{G}))+1 = k_j+1$ and there is a component K of $G_j(\mathfrak{G})$ such that $d(v:K) \ge k_j$.*
(d) *Let $j \in [1,p] \setminus \{i_0\}$ and let $(\mathfrak{G}, v, j, \mathfrak{G}')$ be a move. Then $(\mathfrak{G}', v)$ is a $(k_1, k_2, \ldots, k_p)$-M-partition and this partition is optimal if $d(v:G_j(\mathfrak{G})) = k_j$.*
(e) *Let $j \in [1,p] \setminus \{i_0\}$ and let $(\mathfrak{G}, v, j, \mathfrak{G}')$ be a move. If $d(v:G_j(\mathfrak{G})) = k_j$, then $K_v(\mathfrak{G}') = K_{k_j+1}$ and $(\mathfrak{G}', u)$ is an optimal $(k_1, k_2, \ldots, k_p)$-M-partition of G for every vertex u belonging to $K_v(\mathfrak{G}')$.*
(f) *If $\Delta(G) \le \chi(G)+p-4$, then here are at least two indices $j \in [1,p] \setminus \{i_0\}$ such that $d(v:G_j(\mathfrak{G})) = k_j$.*

Proof From the assumption about G and Proposition 6.32 it follows that G has a $(k_1, k_2, \ldots, k_p)$-M-partition and hence an optimal $(k_1, k_2, \ldots, k_p)$-M-partition. Now let $(\mathfrak{G}, v)$ be an optimal $(k_1, k_2, \ldots, k_p)$-M-partition of G with $i(\mathfrak{G}, v) = i_0$. Then statements (a), (b), (c) and (d) are consequences of Proposition 6.32, Proposition 6.33, and Lemma 6.35. Statement (e) is a consequence of (d) and (b). If (f) is false, then we obtain that $\Delta(G) \ge d_G(v) \ge k_1+k_2+\cdots k_p+(p-2) = \chi(G)+p-3$ (by (c) and Lemma 6.35), contradicting the assumption. □

Let $G \in \mathrm{Crit}(k)$ and $(k_1, k_2, \ldots, k_p) \in \mathbb{N}^p$ such that $k = 1+k_1+k_2+\cdots+k_p$ and $k_i \ge 3$ for $i \in [1,p]$. Let $(\mathfrak{G}, v)$ be a $(k_1, k_2, \ldots, k_p)$-M-partition of G with $i(v, \mathfrak{G}) = i_0$. Then define

$$I(v, \mathfrak{G}) = \{j \in [1,p] \setminus \{i_0\} \mid d(v:G_j(\mathfrak{G})) = k_j\}.$$

If $(\mathfrak{G}, v)$ is an optimal $(k_1, k_2, \ldots, k_p)$-M-partition of G and $\Delta(G) \le k+p-2$, then it follows from Proposition 6.36 that $K_v(\mathfrak{G})$ is a $K_{k_{i_0}+1}$ and, moreover, if $j \in I(v, \mathfrak{G})$ then there is a unique component $K_j(v, \mathfrak{G})$ of $G_j(\mathfrak{G})$ which is a K_{k_j} and satisfies $N_G(v) \cap V(G_j(\mathfrak{G})) = V(K_j(v, \mathfrak{G}))$. Our aim is to show that the complete graphs $K_v(\mathfrak{G})$ and $K_j(v, \mathfrak{G})$ with $j \in I(v, \mathfrak{G})$ can be combined to a larger complete graph. For a component K of $\mathfrak{G}$, we denote by $i(K, \mathfrak{G})$ the unique index $i \in [1,p]$ such that K is a component of $G_i(\mathfrak{G})$. We call K a **full component** of $(\mathfrak{G}, v)$ if K is a component of $\mathfrak{G}$ and $K = K_{k_i}$ for $i = i(K, \mathfrak{G})$. The component $K_v(\mathfrak{G})$ is called the **active component** of $(\mathfrak{G}, v)$.

First, we need an auxiliary coloring result and some more notation. Let G be a graph. If H and H' are two disjoint subgraphs of G such that $H \boxplus H'$ is also a subgraph of G, then we say that H and H' are **complete** in G, or H is **complete** to H' in G. If H consists of a single vertex v, we also say that v is **complete** to H' in G. Two disjoint subsets of $V(G)$ are **complete** in G if the associated induced subgraphs of G are complete in G. Let φ be a partial coloring of G with color set S, let v be a vertex of G, and let $c \in S$. If exactly one neighbor u of v in G belongs to the color class $\varphi^{-1}(c)$, we call c a **singleton color** for v under φ and u a **singleton neighbor** of v under φ. If at least two neighbors of v in G belong to the color class $\varphi^{-1}(c)$, we call c a **multiple color** for v under φ. The following two results are due to Haxell and McDonald [483].

Proposition 6.37 *Let G be a graph with chromatic number $k = \Delta(G) - t$, let v be a critical vertex of G, and let φ be a coloring of $G - v$ with a set S of $k - 1$ colors. Then the following statements hold:*

(a) *The number of multiple colors of v under φ is at most $t + 1$.*
(b) *If u is a singleton neighbor of v under φ, then at most $t + 1$ colors are multiple for u under φ.*
(c) *If u is a singleton neighbor of v under φ and $w \notin N_G(v)$ is a singleton neighbor of u under φ, then at most $t + 2$ colors are multiple for w under φ.*

Proof To prove (a), let m denote the number of multiple colors for v and let s denote the number of singleton colors of v. Since φ is a coloring of $G - v$ using $k - 1$ colors and G is k-chromatic, we have $m + s = k - 1 = \Delta(G) - t - 1$. On the other hand, $2m + s \leq \Delta(G)$, which yields $m \leq \Delta(G) - (m + s) \leq \Delta(G) - (\Delta(G) - t - 1) = t + 1$. This proves (a). To prove (b), assume that u is a singleton neighbor of v. Then coloring v with $\varphi(u)$ and uncolor u results in a coloring φ' of $G - u$ with color set S. Hence (b) follows from (a) applied to φ'. To prove (c), assume that u is a singleton neighbor of v and $w \notin N_G(v)$ is a singleton neighbor of u. As before, coloring v with $\varphi(u)$ and uncolor u results in a coloring φ' of $G - u$ with color set S. By (b), at most $t + 1$ colors are multiple for w under φ'. Taking the color $\varphi(u) = \varphi'(v)$ into account, at most $t + 2$ colors are multiple for w under φ. □

Proposition 6.38 *Let G be a graph with chromatic number $k = \Delta(G) - t$, let v be a critical vertex of G, and let K be a complete subgraph of $G - v$. If v has at least $2t + 4$ neighbors in K, then v has at most $t + 1$ nonneighbors in K.*

Proof Suppose this is false. Then K contains a set B of $t + 2$ vertices such that no vertex of B is a neighbor of v in G. By assumption on G, there is a coloring φ of $G - v$ with a set S of $k - 1$ colors, and $C = N_G(v) \cap V(K)$ has at least $2t + 4$ vertices. So $\varphi(C)$ is a set of at least $2t + 4$ colors. Since the number of multiple colors of v under φ is at most $t + 1$ (by Proposition 6.37(a)), C contains a set A of $t + 3$ singleton neighbors of v under φ. Let H denote the complete bipartite subgraph of G with parts A and B; note that H is a $K_{t+3,t+2}$ and a subgraph of K. If $e = uw$ is an edge of H, we orient e from u to w if $\varphi(u)$ is a singleton color for w under φ. This results in a strict digraph D; so D has no parallel edges, but may contain digons. Let u be an arbitrary vertex of A. Then u is a singleton neighbor of v, which implies that at most $t+1$ colors are multiple for u (by Proposition 6.37(b)). Since $|\varphi(B)| = |B| = t+2$, it then follows that $d_D^-(u) \geq 1$. Since the underlying graph of D is a bipartite graph with parts A and B, this implies that the set $B' = \{w \in B \mid d_D^+(w) \geq 1\}$ is nonempty. So in $D' = D - (B \setminus B')$ each vertex of A has positive in-degree. Now let w be an arbitrary vertex of B'. Then there is a vertex $u \in A$ such that the edge uw is directed from w to u, that is, w is a singleton neighbor of u under φ. Since u is a singleton neighbor of v and $w \notin N_G(v)$, at most $t+2$ colors are multiple for w (by Proposition 6.37(c)). Since $|\varphi(A)| = |A| = t+3$, there is a vertex $u' \in A$ such that the edge $u'w$ is directed from u' to w. Hence each vertex of D' has positive in-degree. Consequently, D' has a directed cycle, possibly a digon. So there is a sequence $v_0, v_1, \ldots, v_n$ of distinct vertices of $A \cup B$ with $v_0 \in A$ and n odd such that v_i is a singleton neighbor of v_{i+1} (where the indices are taken modulo n). Then v_n belongs to B. Furthermore, we can shift the color $\varphi(v_i)$ to v_{i+1} and then we can color v with color $\varphi(v_0)$. This results in a coloring of G with a set of $k-1$ colors. This is a contradiction to $\chi(G) = k$ proves the proposition. □

Proposition 6.39 *Let $G \in \text{Crit}(k)$ and let $(k_1, k_2, \ldots, k_p) \in \mathbb{N}^p$ such that $k = \Delta(G) = 1 + k_1 + k_2 + \cdots + k_p$ and $k_i \geq 3$ for $i \in [1, p]$. If $(\mathfrak{G}, v)$ is a $(k_1, k_2, \ldots, k_p)$-M-partition of G with $i(v, \mathfrak{G}) = i_0$, then the following statements hold:*

(a) *If $j \in [1, p]$, then $d(v : G_j(\mathfrak{G})) = k_j$ if $j \in I(v, \mathfrak{G}) \cup \{i_0\}$ else $d(v : G_j(\mathfrak{G})) = k_j + 1$. Furthermore, $d(v : G_j(\mathfrak{G})) > k_j$ for at most one index $j \in [1, p]$.*
(b) *If a component K of $\mathfrak{G}$ satisfies $d(v : K) \geq 2$ and $i = i(K, \mathfrak{G})$, then $d(v : K) \geq k_i$.*

Proof For the proof of (a), let $d_i = d(v : G_i(\mathfrak{G})) - k_i$ for $i \in [1, p]$. By Proposition 6.32, we have $d_i \geq 0$ for $i \in [1, p]$. Since $k = \Delta(G) = 1 + k_1 + k_2 + \cdots + k_p$, we obtain that

$$0 \leq \sum_{i=1}^{p} d_i = d_G(v) - \sum_{i=1}^{p} k_i = d_G(v) - \Delta(G) + 1 \leq 1,$$

which implies that $d_i = 0$ for every index $i \in [1, p]$ except for at most one index i where we can have $d_i = 1$. This proves (a). For the proof of (b), let K be a component of $\mathfrak{G}$ such that $d(v : K) \geq 2$ and $i = i(\mathfrak{G}, K)$. Suppose, on the contrary, that $d(v : K) \leq k_i - 1$. Then it follows from Proposition 6.32 that $i \neq i_0$ and there is a component K' in $G_i(\mathfrak{G})$ such that $d(v : K') \geq k_i$. Clearly, K' and K are disjoint and so $d(v : G_i(\mathfrak{G})) \geq k_i + 2$, a contradiction to (a). This proves (b). □

Proposition 6.40 *Let $G \in \text{Crit}(k)$ and let $(k_1, k_2, \ldots, k_p) \in \mathbb{N}^p$ such that $k = \Delta(G) = 1 + k_1 + k_2 + \cdots + k_p$ and $k_i \geq 3$ for $i \in [1, p]$. If $(\mathfrak{G}, v)$ is an optimal $(k_1, k_2, \ldots, k_p)$-M-partition of G with $i(v, \mathfrak{G}) = i_0$, then the following statements hold:*

(a) *Let K be a component of $\mathfrak{G}$ whose index $i = i(K, \mathfrak{G})$ belongs to $I(v, \mathfrak{G})$. If $d(v : K) \geq 1$, then K is a full component of $(\mathfrak{G}, v)$ and $N_G(v) \cap V(G_i(\mathfrak{G})) = V(K)$. Furthermore, if $(\mathfrak{G}, v, i, \mathfrak{G}')$ is a move, then $(\mathfrak{G}', v)$ is an optimal $(k_1, k_2, \ldots, k_p)$-M-partition of G and $K_v(\mathfrak{G}') = K + v$ is a K_{k_i+1}.*

(b) *Let K be a full component of $(\mathfrak{G}, v)$. If $K_v(\mathfrak{G}) - v$ is complete to K in G, then $K_v(\mathfrak{G})$ is complete to K in G.*

(c) *Let K and K' be distinct full components of $(\mathfrak{G}, v)$ that are both complete to the active component of $(\mathfrak{G}, v)$ in G. Then K and K' are complete in G.*

Proof We first prove (a). By the assumption on K, we have $d(v : G_i(\mathfrak{G})) = k_i$ and $p \geq 2$. From Proposition 6.36(e) it then follows that $(\mathfrak{G}', v)$ is an optimal $(k_1, k_2, \ldots, k_p)$-M-partition of G and $K_v(\mathfrak{G}') = K_{k_i+1}$. Then $K_v(\mathfrak{G}') - v = K$ and so statement (a) holds.

For the proof of (b), suppose this is false. For $i = i(K, \mathfrak{G})$ we have $i \neq i_0$ and there is a vertex u in K such that $vu \notin E(G)$. Let w be a vertex of $K_v(\mathfrak{G}) - v$. Then $(\mathfrak{G}, w)$ is an optimal $(k_1, k_2, \ldots, k_p)$-M-partition of G and $K_w(\mathfrak{G}) = K_v(\mathfrak{G})$. We may now move w into the ith part, i.e., we consider the move $(\mathfrak{G}, w, i, \mathfrak{G}')$. By Proposition 6.36(d), $(\mathfrak{G}', w)$ is a $(k_1, k_2, \ldots, k_p)$-M-partition of G. By assumption, $K = K_{k_i}$ and w is completely joined to K in G, and $k_i \leq d(w : G_i(\mathfrak{G})) \leq k_i + 1$ (by Proposition 6.39(a)). Since w is a critical vertex of $\mathfrak{G}'$, this implies that u is a critical vertex of $\mathfrak{G}'$ and so $(\mathfrak{G}', u)$ is $(k_1, k_2, \ldots, k_p)$-M-partition of G. Furthermore, $K' = K_v(\mathfrak{G}) - w$ is a component of $\mathfrak{G}'$ with $i(K', \mathfrak{G}') = i_0$ and $d(u : K') = |K' - v| = k_{i_0} - 1 \geq 2$, a contradiction to Proposition 6.39(b). This proves (b).

To prove (c), suppose this is false. Let $i = i(K, \mathfrak{G})$ and $i' = i(K', \mathfrak{G})$. Since $K \neq K'$ and both are complete to $K_v(\mathfrak{G})$, it follows from Proposition 6.39(a) that $i \neq i'$, and $d(v : K) = k_i$ or $d(v : K') = k_{i'}$. By symmetry, we may assume that $d(v : K) = k_i$. Since K and K' are not complete in G, there are vertices $u \in V(K)$ and $u' \in V(K')$ such that $uu' \notin E(G)$. Since $k_j \geq 3$ for all $j \in [1, p]$ and $|K_v(\mathfrak{G})| \geq 4$, we obtain from Proposition 6.38 (with $t = 0$) that u is completely joined to $K' - u'$ and so $d(u : K') = k_{i'} - 1$. Now consider the move $(\mathfrak{G}, v, i, \mathfrak{G}')$. Then it follows from Proposition 6.36(e) that $(\mathfrak{G}', v)$ is an optimal $(k_1, k_2, \ldots, k_p)$-M-partition of G with $K_v(\mathfrak{G}') = K + v$ implying that $(\mathfrak{G}', u)$ is an optimal $(k_1, k_2, \ldots, k_p)$-M-partition of G, too. Clearly, K' is a component of $\mathfrak{G}'$ with $i(K', \mathfrak{G}') = i'$ and $d(u : K') = k_{i'} - 1 \geq 2$, a contradiction to Proposition 6.39(b). This proves (c). □

Our aim is to show that a graph G with $\chi(G) = \Delta(G) = 13$ contains a K_{10}. We may assume that G is critical. Then there is an optimal $(3,3,3,3)$-M-partition $(\mathfrak{G}, v)$ of G and $|I(v, \mathfrak{G})| \geq 2$. The active component $K_v(\mathfrak{G})$ of $(\mathfrak{G}, v)$ is a K_4 and each full component is a K_3. By Proposition 6.40(c), we are done if two distinct full components of $(\mathfrak{G}, v)$ are completely joined to the active component of $(\mathfrak{G}, v)$. If this is not the case, we may consider a move $(\mathfrak{G}, v, i, \mathfrak{G}')$ with $i \in I(v, \mathfrak{G})$; such a move is called a **good move**. Then $(\mathfrak{G}', v)$ is another optimal $(3,3,3,3)$-M-partition of G. Finally, to be successful, we need a sequence of good moves with certain special restrictions. One restriction is that $i \in I(\mathfrak{G}, v)$, that is, we move v to a part in which all neighbors of v belong to a full component K, then $K + v$ is the active component of the new partition and we may move any vertex of $K + v$. The second condition is that the full component K is not completely joined to the former active component. A third condition is that we never move a vertex twice.

In the following, let $G \in \mathrm{Crit}(k)$ be a graph and let $(k_1, k_2, \ldots, k_p) \in \mathbb{N}^p$ such that $k = \Delta(G) = 1 + k_1 + k_2 + \cdots + k_p$ and $k_i \geq 3$ for $i \in [1, p]$. In this case we say that

$(G,k_1,k_2,\ldots,k_p)$ satisfies the condition (GA). A **good move sequence** of length q for G is a sequence $\mathcal{S} = ((\mathfrak{G}^\ell, v_\ell, i_\ell, \mathfrak{G}^{\ell+1})_{\ell=1}^q)$ satisfying the following conditions:

(S1) $(\mathfrak{G}^1, v_1)$ is an optimal $(k_1,k_2,\ldots,k_p)$-M-partition of G.
(S2) For $\ell \in [1,q]$, $(\mathfrak{G}^\ell, v_\ell, i_\ell, \mathfrak{G}^{\ell+1})$ is a good move, i.e. $i_\ell \in I(v_\ell, \mathfrak{G}^\ell)$.

A good move sequence $\mathcal{S} = ((\mathfrak{G}^\ell, v_\ell, i_\ell, \mathfrak{G}^{\ell+1})_{\ell=1}^q)$ for G is said to be **nonrepetitive** if $v_\ell \neq v_{\ell'}$ whenever ℓ and ℓ' are distinct indices from $[1,q]$.

Let $\mathcal{S} = ((\mathfrak{G}^\ell, v_\ell, i_\ell, \mathfrak{G}^{\ell+1})_{\ell=1}^q)$ be a good move sequence for G. In this case, we always use the convention that $v_{q+1} = v_q$. Then the pairs

$$(\mathfrak{G}^1, v_1), (\mathfrak{G}^2, v_2), \ldots, (\mathfrak{G}^q, v_q), \text{ and } (\mathfrak{G}^{q+1}, v_{q+1})$$

are all optimal $(k_1,k_2,\ldots,k_p)$-M-partitions of G (by Proposition 6.40(a), (S1) and (S2)). Furthermore, the subsequence $\mathcal{S}(i,j) = ((\mathfrak{G}^\ell, v_\ell, i_\ell, \mathfrak{G}^{\ell+1})_{\ell=i}^j)$ with $1 \le i \le j \le q$ is a good move sequence for G; and the reverse sequence

$$\mathcal{S}^{-1} = ((\mathfrak{G}^q, v_q, i_q, \mathfrak{G}^{q+1})^{-1}, (\mathfrak{G}^{q-1}, v_{q-1}, i_{q-1}, \mathfrak{G}^q)^{-1}, \ldots, (\mathfrak{G}^1, v_1, i_1, \mathfrak{G}^2)^{-1})$$

is a good move sequence for G. Furthermore, $K_{v_\ell}(\mathfrak{G}^\ell)$ is the active component of $(\mathfrak{G}^\ell, v_\ell)$ for $\ell \in [1,q+1]$, and $K_{v_\ell}(\mathfrak{G}^\ell) = K_{v_{\ell-1}}(\mathfrak{G}^\ell)$ for $\ell \in [2,q]$. A **club** of $\mathcal{S}$ is a sequence $\mathbf{K} = (K^1, K^2, \ldots, K^{q+1})$ such that K^1 is a component of $(\mathfrak{G}^1, v_1)$ and, for $\ell \in [1,q]$ we have

$$K^{\ell+1} = \begin{cases} K^\ell - v_\ell & \text{if } K^\ell \text{ is the active component of } (\mathfrak{G}^\ell, v_\ell), \\ K^\ell + v_\ell & \text{if } K^\ell + v \text{ is the active component of } (\mathfrak{G}^{\ell+1}, v_{\ell+1}), \\ K^\ell & \text{otherwise.} \end{cases}$$

Let $V(\mathcal{S}) = \{v_\ell \mid \ell \in [1,q]\}$; a vertex $v \in V(G)$ is called a **traveler** of $\mathcal{S}$ if $v \in V(\mathcal{S})$ and otherwise it is called a **nontraveler** of $\mathcal{S}$.

Let $\mathbf{K} = (K^1, K^2, \ldots, K^{q+1})$ be an arbitrary club of $\mathcal{S}$. Then K^ℓ is a component of $\mathfrak{G}^\ell$ for $\ell \in [1,q+1]$; and there is an index $i \in [1,p]$ such that $i(K^\ell, \mathfrak{G}^\ell) = i$ for all $\ell \in [1,q+1]$, we call i the **index** of $\mathbf{K}$ and write $i = i(\mathbf{K}, \mathcal{S})$. Furthermore, let

$$V(\mathbf{K}) = V(K^1) \cup V(K^2) \cup \cdots \cup V(K^{q+1})$$

be the vertex set of the club $\mathbf{K}$. Note that we have three types of components; the active components, the full components, and the remaining components (of $(\mathfrak{G}^\ell, v_\ell)$ with $\ell \in [1,q+1]$). If a vertex $v \in V(\mathbf{K})$ is a nontraveler of $\mathcal{S}$, then $v \in V(K^\ell)$ for all $\ell \in [1,q+1]$. If every vertex in $V(\mathbf{K})$ is a nontraveler of $\mathcal{S}$, then the sequence $\mathbf{K}$ is constant, so it consists of a component K of $\mathfrak{G}^1$ which remains unchanged during the moving process. If one component of $\mathbf{K}$ is an active or full component, then all components of $\mathbf{K}$ are active or full components; in this case we say that $\mathbf{K}$ is a **complete club**. If $\mathbf{K}$ is a complete club with $i(\mathbf{K}, \mathcal{S}) = i$ and $K = K^\ell$ with $\ell \in [1,q+1]$, then K is a K_{k_i} if K is a full component of $(\mathfrak{G}^\ell, v_\ell)$, or K is a K_{k_i+1} if K is the active component of $(\mathfrak{G}^\ell, v_\ell)$. Furthermore, let

$$I_a(\mathbf{K},\mathcal{S}) = \{\ell \in [1,q+1] \mid K^\ell \text{ is an active component of } (\mathfrak{G}^\ell, v_\ell)\}.$$

For a complete club $\mathbf{K}$ of $\mathcal{S}$, we define the **spread** of $\mathbf{K}$ by

$$SP(\mathbf{K},\mathcal{S})) = \{i_\ell \mid \ell \in I_a(\mathbf{K},\mathcal{S})\}.$$

Furthermore, let

$$sp(\mathcal{S}) = \max\{|SP(\mathbf{K},\mathcal{S})| \mid \mathbf{K} \text{ is a complete club of } \mathcal{S}\}$$

be the **spread** of the move sequence $\mathcal{S}$.

For a component K of $\mathfrak{G}^\ell$ with $\ell \in [1,q+1]$ there exists a unique club $\mathbf{K} = (K^1,K^2,\ldots,K^{q+1})$ of $\mathcal{S}$ such that $K^\ell = K$; we then say that $\mathbf{K}$ is the **club containing** K. If $v \in V(G)$ is a nontraveler of $\mathcal{S}$, then there is a unique club $\mathbf{K}$ of $\mathcal{S}$ such that $v \in V(\mathbf{K})$ and hence $v \in V(K^\ell)$ for all $\ell \in [1,q+1]$. If $\mathbf{K} = (K^1,K^2,\ldots,K^{q+1})$ is a club of $\mathcal{S}$, then

$$\mathbf{K}(i,j) = (K^i,K^{i+1},\ldots K^j,K^{j+1})$$

is a club of $\mathcal{S}(i,j)$ for $1 \le i \le j \le q$. Furthermore, $\mathbf{K}^{-1} = (K^{q+1},K^q,\ldots,K^2,K^1)$ is a club of the reverse move sequence $\mathcal{S}^{-1}$.

Proposition 6.41 *Suppose $(G,k_1,k_2,\ldots,k_p)$ satisfies the condition (GA). Let $\mathcal{S} = ((\mathfrak{G}^\ell,v_\ell,i_\ell,\mathfrak{G}^{\ell+1})_{\ell=1}^q)$ be a good move sequence for G, and, moreover, let $\mathbf{K} = (K^1,K^2,\ldots,K^{q+1})$ and $\mathbf{L} = (L^1,L^2,\ldots,L^{q+1})$ be two complete clubs of $\mathcal{S}$. Then the following statements hold:*

(a) *If K^ℓ and L^ℓ are complete in G for some $\ell \in [1,q+1]$, then K^ℓ and L^ℓ are complete in G for all $\ell \in [1,q+1]$.*
(b) *Let $\ell \in I_a(\mathbf{K},\mathcal{S})$ and $v \in V(K^\ell)$. If $d(v:L^\ell) \ge 2$, then v is complete to L^ℓ in G.*
(c) *Let $\ell \in I_a(\mathbf{K},\mathcal{S})$, $v \in V(K^\ell)$ and $u \in V(L^\ell)$. If $d(v:L^\ell) \ge 2$ and $d(u:K^\ell - v) \ge 2$, then u is complete to K^ℓ in G.*

Proof For the proof of (a), suppose this is false. We may choose a counterexample $(\mathcal{S},\mathbf{K},\mathbf{L})$ such that the length q of $\mathcal{S}$ is minimum. Since each subsequence $\mathcal{S}(i,j)$ is a good move sequence for G, too, we obtain that $q = 1$. Hence $\mathcal{S} = (\mathfrak{G}^1,v_1,i_1,\mathfrak{G}^2)$. Since the reverse sequence is a good move sequence for G, too, we may assume that K^1 and L^1 are not complete in G, but K^2 and L^2 are complete in G. By symmetry, this implies that $K^1 = K_{v_1}(\mathfrak{G}^1)$ is the active component of $(\mathfrak{G}^1,v_1)$ and L^1 is a full component of $(\mathfrak{G}^1,v_1)$, Since K^2 and L^2 are complete in G, we have that $K^2 = K^1 - v_1$ and $L^2 = L^1$. This yield a contradiction to Proposition 6.40(b).

For the proofs of (b) and (c) assume that $\ell \in I_a(\mathbf{K},\mathcal{S})$. Then $K^\ell = K_{v_\ell}(\mathfrak{G}^\ell)$ is the active component of $(\mathfrak{G}^\ell,v_\ell)$; and L^ℓ is a full component of $\mathfrak{G}^\ell$ and so $L^\ell = K_{k_i}$ with $i = i(L^\ell,\mathfrak{G}^\ell)$. Furthermore, $(\mathfrak{G}^\ell,v_\ell)$ is an optimal $(k_1,k_2,\ldots,k_p)$-M-partition of G. By assumption, we have $G \in \text{Crit}(k)$ and $\Delta(G) = k$. Hence, statement (b) is an immediate consequence of Proposition 6.39(b). Furthermore, we have $d(v:L^\ell) \le k_i + 1$ (by Proposition 6.39(a)). Clearly, $(\mathfrak{G}^\ell,v)$ is an optimal $(k_1,k_2,\ldots,k_p)$-M-partition of G whose active component remains K^ℓ (by Proposition 6.36(b)). Now we may consider the move $(\mathfrak{G}^\ell, v, i, \mathfrak{G}')$. Then $(\mathfrak{G}', v)$ is a $(k_1, k_2, \ldots, k_p)$-M-partition of G (by Proposition 6.36(d)). Since v is completely joined to $L^\ell = K_{k_i}$ and has at most one neighbor outside L^ℓ in $G_i(\mathfrak{G}')$, we obtain that $\chi(G_i(\mathfrak{G}') - u) \le k_i$ and so $(\mathfrak{G}', u)$ is a $(k_1, k_2, \ldots, k_p)$-M-partition of G. By assumption on (c), $K' = K^\ell - v$ is a full component of $\mathfrak{G}'$ such that $d(u : K') \ge 2$. Hence Proposition 6.39(b) implies that u is complete to K' in G and hence also complete to K^ℓ, since $uv \in E(G)$ (by (b)). This proves the proposition. □

In what follows, let $\mathcal{S} = ((\mathfrak{G}^\ell, v_\ell, i_\ell, \mathfrak{G}^{\ell+1})_{\ell=1}^q)$ be a nonrepetitive good move sequence of G; and let $v = v_\ell$ be a traveler of $\mathcal{S}$ with $\ell \in [1, q]$. Then there is exactly one club $\mathbf{K} = (K^1, K^2, \ldots, K^{q+1})$ of $\mathcal{S}$ such that $\ell \in I_a(\mathbf{K}, \mathcal{S})$. Then we have $v_\ell \in V(K^h)$ for $h \in [1, \ell]$, $v_\ell \notin V(K^h)$ for $h \in [\ell+1, q+1]$, K^ℓ is the active component of $(\mathfrak{G}^\ell, v_\ell)$, $K^{\ell+1} = K^\ell - v_\ell$ is a full component, and, moreover, either $\ell = 1$ or $v_{\ell-1} \in V(K^h)$ for $h \in [\ell-1, q+1]$. Furthermore, there is a unique club $\mathbf{L} = (L^1, L^2, \ldots, L^{q+1})$ such that $L^{\ell+1}$ is the active component of $(\mathfrak{G}^{\ell+1}, v_{\ell+1})$ implying that $\ell+1 \in I_a(\mathbf{L}, \mathcal{S})$ and $v_\ell \in L^h$ for $h \in [\ell+1, q+1]$. Then we say that v_ℓ **travels** from the club $\mathbf{K}$ to the club $\mathbf{L}$, and $\mathbf{K}$ **sends** v_ℓ to $\mathbf{L}$. If v_ℓ travels from $\mathbf{K}$ to $\mathbf{L}$, then $i(\mathbf{L}, \mathcal{S}) = i_\ell \in SP(\mathbf{K}, \mathcal{S})$.

Proposition 6.42 *Suppose $(G, k_1, k_2, \ldots, k_p)$ satisfies the condition (GA). Let $\mathcal{S} = ((\mathfrak{G}^\ell, v_\ell, i_\ell, \mathfrak{G}^{\ell+1})_{\ell=1}^q)$ be a nonrepetitive good move sequence for G, and, moreover, let $\mathbf{K} = (K^1, K^2, \ldots K^{q+1})$ be a complete clubs of $\mathcal{S}$. If there are two distinct indices ℓ and ℓ' in $I_a(\mathbf{K}, \mathcal{S})$ such that $i_\ell = i_{\ell'} = i$, then there is a club $\mathbf{L}$ of $\mathcal{S}$ such that both v_ℓ and $v_{\ell'}$ travels from $\mathbf{K}$ to $\mathbf{L}$.*

Proof We may assume that $\ell < \ell'$. This implies that $v_\ell \in K^h$ for $h \in [1, \ell]$, $v_\ell \ne v_{\ell'}$, and $v_{\ell'} \in K^h$ for $h \in [1, \ell']$. Since the components of $\mathbf{K}$ are full components or active components, we obtain that $v_{\ell'}$ is adjacent in G to v_ℓ. Since $\ell \in I_a(\mathbf{K}, \mathcal{S})$, there is a unique club $\mathbf{L} = (L^1, L^2, \ldots, L^{q+1})$ of $\mathcal{S}$ such that v_ℓ travels from $\mathbf{K}$ to $\mathbf{L}$. Then $v_\ell \in L^h$ for $h \in [\ell+1, q+1]$. Note that $i(\mathbf{L}, \mathcal{S}) = i$. Since $\ell' > \ell$ and $\ell' \in I_a(\mathbf{K}, \mathcal{S})$, we obtain that $K^{\ell'}$ is the active component of $(\mathfrak{G}^{\ell'}, v_{\ell'})$. Furthermore, $L^{\ell'}$ is a full component of $\mathfrak{G}^{\ell'}$ satisfying $i(L^{\ell'}, \mathfrak{G}^{\ell'}) = i = i_{\ell'} \in I(\mathfrak{G}^{\ell'}, v_{\ell'})$ and $d(v_{\ell'} : L^{\ell'}) \ge 1$, since $v_\ell \in V(L^{\ell'})$. From Proposition 6.40(a) it then follows that $L^{\ell'+1} = L^{\ell'} + v_{\ell'}$ is the active component of $(\mathfrak{G}^{\ell'+1}, v_{\ell'+1})$. This implies that $v_{\ell'}$ travels from $\mathbf{K}$ to $\mathbf{L}$. □

A **feasible move sequence** of length q for G is a good move sequence $\mathcal{S} = ((\mathfrak{G}^\ell, v_\ell, i_\ell, \mathfrak{G}^{\ell+1})_{\ell=1}^q)$ of length q for G such that the following two conditions hold:

(S3) $\mathcal{S}$ is nonrepetitive, and
(S4) for $\ell \in [1, q]$, $K_{v_\ell}(\mathfrak{G}^\ell)$ is not complete to $K_{v_\ell}(\mathfrak{G}^{\ell+1}) - v_\ell$ in G.

Let $\mathcal{S} = ((\mathfrak{G}^\ell, v_\ell, i_\ell, \mathfrak{G}^{\ell+1})_{\ell=1}^q)$ be a feasible move sequence for G with $sp(\mathcal{S}) \le 2$. From Proposition 6.42 it then follows that each club of $\mathcal{S}$ sends vertices to at most two other clubs. Furthermore, if a club $\mathbf{K}$ sends v_ℓ to a club $\mathbf{L}$, then it follows from (S4) that K^ℓ is not complete in G to L^ℓ. Clearly, every subsequence $\mathcal{S}' = \mathcal{S}(i, j)$ of $\mathcal{S}$ is a feasible move sequence for G with $sp(\mathcal{S}') \le 2$. The next lemma is crucial for a successful use of Mozhan partitions.

Lemma 6.43 (Cranston and Rabern) *Suppose $(G, k_1, k_2, \ldots, k_p)$ satisfies (GA). Let $\mathcal{S} = ((\mathfrak{G}^\ell, v_\ell, i_\ell, \mathfrak{G}^{\ell+1})_{\ell=1}^q)$ be a feasible move sequence for G such that $sp(\mathcal{S}) \leq 2$. Then the active component of $(\mathfrak{G}^{q+1}, v_{q+1})$ contains a nontraveler of $\mathcal{S}$.*

Proof Suppose this is false. Let $\mathbf{K}$ be the club of $\mathcal{S}$ containing the active component of $(\mathfrak{G}^{q+1}, v_{q+1})$, that is $K^{q+1} = K_{v_{q+1}}(\mathfrak{G}^{q+1})$. By symmetry, we may assume that $i(\mathbf{K}, \mathcal{S}) = 1$ and so K^{q+1} is a K_{k_1+1} and $k_1 \geq 3$. Since every vertex of K^{q+1} is a traveler of $\mathcal{S}$, this implies that $\mathbf{K}$ sends at least three vertices to other clubs of $\mathcal{S}$. Since $sp(\mathcal{S}) \leq 2$, it then follows from Proposition 6.42 that at least two vertices of $\mathbf{K}$ are send to the same club $\mathbf{L}$. We may replace $\mathcal{S}$, $\mathbf{K}$ and $\mathbf{L}$ by the corresponding shortest subsequences such that $\mathbf{K}$ sends two vertices to $\mathbf{L}$ and is again active. So we may assume that the following conditions hold for $(\mathcal{S}, \mathbf{K}, \mathbf{L}, \mathbf{L}')$:

(1) $\mathcal{S} = ((\mathfrak{G}^\ell, v_\ell, i_\ell, \mathfrak{G}^{\ell+1})_{\ell=1}^q)$ is a feasible move sequence for G such that $sp(\mathcal{S}) \leq 2$.
(2) $\mathbf{K} = (K^1, K^2, \ldots, K^{q+1})$ is the complete club of $\mathcal{S}$ such that K^{q+1} is the active component of $(\mathfrak{G}^{q+1}, v_{q+1})$, and $i(\mathbf{K}, \mathcal{S}) = 1$. Furthermore, $|I_a(\mathbf{K}, \mathcal{S})| \leq 3$ and there are two indices $a = 1$ and b in $I_a(\mathbf{K}, \mathcal{S})$ such that $i_a = i_b = 2$, that is, $\mathbf{K}$ sends v_1 and v_b to a complete club $\mathbf{L} = (L^1, L^2, \ldots, L^{q+1})$ of $\mathcal{S}$ with $i(\mathbf{L}, \mathcal{S}) = 2$.
(3) There is at most one index d in $I_a(\mathbf{K}, \mathcal{S})$ such that $a = 1 < d < b$ and $i_d = 3$; that is, $\mathbf{K}$ sends v_d to a club $\mathbf{L}'$ with $i(\mathbf{L}', \mathcal{S}) = 3$.

Furthermore, we obtain that

(4) $|I_a(\mathbf{L}, \mathcal{S})| \leq 3$,

for otherwise we find a shorter subsequence of $\mathcal{S}$ such that a club sends two vertices to another club and is again active. Our aim is to show that K^{q+1} and L^{q+1} are completely joined in G. To this end, we shall prove several claims. The first claim is an immediate consequence of (1), (2), (3) and (4).

Claim 6.43.1 *The following statements hold for* $\mathbf{K}$ *and* $\mathbf{L}$*:*

(a) *If $\ell \in \{1, b\}$, then $v_\ell \in K^i$ for $i \in [1, \ell]$ and $v_\ell \in L^j$ for $j \in [\ell+1, q+1]$.*
(b) *There is a nontraveler u of $\mathcal{S}$ contained in* $\mathbf{K}$*, that is, $u \in K^i$ for every $i \in [1, q+1]$.*
(c) *If d exists, then $v_d \in K^i$ for $i \in [1, d]$. If d does not exists, then there is a nontraveler $u' \neq u$ of $\mathcal{S}$ contained in* $\mathbf{K}$.
(d) $\{2, b+1\} \subseteq I_a(\mathbf{L}, \mathcal{S})$.

Claim 6.43.2 *v_b is complete to L^d in G.*

Proof: By Claim 6.43.1, we obtain that L^{b+1} contains v_1, v_b and v_{b+1}, which implies that $v_b \in K^d$ has two neighbors in L^d namely v_1 and v_{b+1}. Then Proposition 6.41(b) with $\ell = d$ implies that v_b is complete to L^d in G. △

Claim 6.43.3 *v_1 is complete to K^d in G.*

Proof: By Claim 6.43.2, v_b is complete to L^d in G. By Claim 6.43.1, we have $v_1, v_d, v_b, u \in K^1$ and so v_1 is complete to $A = \{v_d, v_b, u\}$ in G. Since $v_1 \in L^d$ and $A \subseteq K^d$, we obtain from Proposition 6.41(c) with $\ell = d$ that v_1 is complete to K^d in G, as claimed. △

Claim 6.43.4 *v_1 is complete to K^b in G.*

Proof: Since **K** sends v_b to **L** (by (2)), we have that v_b is complete to L^b in G. By (2) and (3), we obtain that $|K^d \cap K^b| \geq 3$. By Claim 6.43.3, v_1 is complete to K^d in G, which implies that $v_1 \in L^b$ has at least three neighbors in K^b. By Claim 6.43.2, we obtain that v_b has at least two neighbors in L^b. Then it follows from Proposition 6.41(c) with $\ell = b$ that v_1 is complete to K^b in G, as claimed. △

Claim 6.43.5 *K^{q+1} is complete to L^{q+1} in G.*

Proof: Using (1), (2), (3) and (4), we obtain that $1 < d < b < q$, $K^{b+1} = K^b - v_b$, $L^{b+1} = L^b + v_b$, $K^{q+1} = K^{b+1} + v_q$, and $v_q \in V(\mathbf{K})$. By Claim 6.43.3, we then obtain that both v_1 and v_b are complete to K^{b+1} in G. Since both vertices v_1 and v_b belong to L^{q+1}, we then obtain from Proposition 6.41(b) with $\ell = q+1$ that K^{b+1} is complete to L^{q+1} in G. Since $|K^{b+1}| \geq 3$, it then follows from Proposition 6.41(c) with $\ell = q+1$ that L^{q+1} is complete to K^{q+1} in G. △

Since K^{q+1} and L^{q+1} are complete in G (by Claim 6.43.5), it follows from Proposition 6.41(a) that K^1 and L^1 are completely joined in G. Since **K** sends v_1 to **L** (by (2)), this is a contradiction to (1). This contradiction completes the proof of Lemma 6.43. □

Theorem 6.44 (Cranston and Rabern) *Let G be a critical graph with $\chi(G) = \Delta(G)$. Then $\omega(G) \geq \Delta(G) - 3$ if $\Delta(G) \equiv 1 \pmod 3$ and $\omega(G) \geq \Delta(G) - 4$ otherwise.*

Proof The statement is evident if $\Delta(G) \leq 6$. So assume that G is a critical graph with $\chi(G) = \Delta(G) \geq 7$. Let $p = \lfloor (\Delta(G) - 1)/3 \rfloor$. Then $p \geq 2$ and there is a sequence $(k_1, k_2, \ldots, k_p) \in \mathbb{N}^p$ such that $k_\ell \in \{3, 4\}$ for $\ell \in [1, p]$ and $\Delta(G) = 1 + k_1 + k_2 + \cdots + k_p$. Note that $k_\ell = 3$ for all $\ell \in [1, p]$, provided that $\Delta(G) \equiv 1 \pmod 3$. By Proposition 6.36, G has an optimal $(k_1, k_2, \ldots, k_p)$-M-partition $(\mathfrak{G}, v)$, say with $i_0 = i(\mathfrak{G}, v)$, and the active component $K_v(\mathfrak{G})$ is a $K_{k_{i_0}+1}$, yielding $\omega(G) \geq k_{i_0} + 1$. So we may assume that $\Delta(G) \geq 10$. By Proposition 6.39(a) it then follows that $|I(v, \mathfrak{G})| \geq p - 2 \geq 1$ and, for every $i \in I(v, \mathfrak{G})$ we have a full component K of $(\mathfrak{G}, v)$ with $i(K, \mathfrak{G}) = i$. We distinguish two cases.

Case 1: *For every $i \in I(\mathfrak{G}, v)$ there is a full component K of $(\mathfrak{G}, v)$ with $i(K, \mathfrak{G}) = i$ such that K is complete to $K_v(\mathfrak{G})$ in G.* Then it follows from Proposition 6.40(a) that $\omega(G) \geq \Delta(G) - k_j$ where $j \in [1, p]$ is the only index such that $j \notin I(\mathfrak{G}, v) \cup \{i_0\}$. This clearly implies that $\omega(G) \geq \Delta(G) - 3$ if $\Delta(G) \equiv 1 \pmod 3$ and $\omega(G) \geq \Delta(G) - 4$ otherwise.

Case 2: *There is an index $i \in I(\mathfrak{G}, v)$ such that no full component K of $(\mathfrak{G}, v)$ with $i(K, \mathfrak{G}) = i$ is complete to $K_v(\mathfrak{G})$ in G.* Now let $\mathcal{S} = ((\mathfrak{G}^\ell, v_\ell, i_\ell, \mathfrak{G}^{\ell+1})_{\ell=1}^q)$ be a feasible move sequence for G whose length q is maximum and such that

$(\mathfrak{G}^1, v_1) = (\mathfrak{G}, v)$. By the assumption of the case, there is a good move $(\mathfrak{G}, v, i, \mathfrak{G}')$, which yields that $q \geq 1$. Let K be the active component of $(\mathfrak{G}^{q+1}, v_{q+1}) = (\mathfrak{G}^{q+1}, v_q)$. By Lemma 6.43, K contains a nontraveler of $\mathcal{S}$, say w. Then $K = K_w(\mathfrak{G}^{q+1})$ and we may replace v_{q+1} by w. Let $\mathbf{K} = (K^1, K^2, \dots, K^{q+1})$ be the club of $\mathcal{S}$ containing K, that is, $K^{q+1} = K$. Since $\mathcal{S}$ is a feasible move sequence, $\mathbf{K}$ sends vertices to at most two other clubs of $\mathcal{S}$. Suppose, there is a club $\mathbf{L}$ such that $\mathbf{K}$ send v_ℓ to $\mathbf{L}$. This implies that $\ell < q - 1$ and K^ℓ is not complete to L^ℓ in G (by (S4)). From Proposition 6.41 it then follows that K^{q+1} and L^{q+1} are not complete in G. If $i' = i(\mathbf{L}, \mathcal{S})$ belongs to $I' = I(\mathfrak{G}^{q+1}, v_{q+1})$, then there is a good move $(\mathfrak{G}^{q+1}, v_{q+1} = w, i', \mathfrak{G}')$, contradicting the maximality condition for $\mathcal{S}$. Consequently, $\mathbf{K}$ send vertices to at most one club of $\mathcal{S}$ and $i(\mathbf{L}, \mathcal{S})$ does not belong to I' for such a club $\mathbf{L}$ of $\mathcal{S}$. By the maximality condition for $\mathcal{S}$, we then obtain that for every $i \in I'$ there is a full component K of $(\mathfrak{G}^{q+1}, v_{q+1})$ with $i(K, \mathfrak{G}^{q+1}) = i$ that is complete to the active component of $(\mathfrak{G}^{q+1}, v_{q+1})$ in G. Then as in the first case we obtain that $\omega(G) \geq \Delta(G) - 3$ if $\Delta(G) \equiv 1 \pmod 3$ and $\omega(G) \geq \Delta(G) - 4$ otherwise. □

Corollary 6.45 (Cranston and Rabern) *If G is a graph with $\chi(G) \geq \Delta(G)$, then $\omega(G) \geq \Delta(G) - 3$ if $\Delta(G) \equiv 1 \pmod 3$ and $\omega(G) \geq \Delta(G) - 4$ otherwise.*

Proof Clearly, G contains a critical graph G' with $\chi(G') = \chi(G)$ as a subgraph. Then we obtain that $\Delta(G') \leq \Delta(G) \leq \chi(G) = \chi(G') \leq \Delta(G') + 1$. If $\chi(G') = \Delta(G') + 1$, then Brooks' theorem implies that $\omega(G) \geq \omega(G') = \Delta(G') + 1 \geq \Delta(G)$ and we are done. Otherwise, we have $\chi(G') = \Delta(G') = \Delta(G)$ and the result follows from Theorem 6.44 □

Corollary 6.46 (Cranston and Rabern) *Every graph G with $\chi(G) \geq \Delta(G) \geq 13$ satisfies $\omega(G) \geq \Delta(G) - 3$.*

Proof Suppose this is false, and let G be a counterexample whose order is minimum. Hence, $\chi(G) \geq \Delta(G) \geq 13$ and $\omega(G) \leq \Delta(G) - 4$. Then $\chi(G) = \Delta(G)$ (by Brooks' theorem) and $\Delta(G) \geq 14$ (by Corollary 6.45). Furthermore, G is vertex critical and hence connected. If $\omega(G) \leq \Delta(G) - 5$, then let I be a maximal independent set of G; otherwise let I be a maximal independent set of G such that $\omega(G - I)) = \omega(G) - 1$ (such a set exists by Corollary 6.14, note that $G' = C_{2n+1}(K_5)$ satisfies $\chi(G') \neq \Delta(G')$ (by Theorem 1.48)). For $G' = G - I$, we then obtain that $\omega(G') \leq \Delta(G) - 5$, $\Delta(G') \leq \Delta(G) - 1$, and $\chi(G') \geq \chi(G) - 1 = \Delta(G) - 1 \geq 13$. By Brooks' theorem, we have $\chi(G') \leq \max\{\omega(G'), \Delta(G')\} \leq \Delta(G) - 1$, which implies that $\chi(G') = \Delta(G) - 1 = \Delta(G') \geq 13$ and $\omega(G') \leq \Delta(G) - 5 \leq \Delta(G') - 4$, a contradiction to the choice of G.□

6.6 Graphs with χ Close to Δ

In 1972, Karp [556] proved that graph k-colorability is an NP-complete decision problem whenever $k \geq 3$. Garrey and Johnson [407] proved that graph 3-colorability remains NP-complete even when restricted to planar graphs with maximum degree 4;

see Theorem 7.38 for further complexity results. In 1998, Emden-Weinert, Hougardy, and Kreuter [326] proved the following complexity result.

Theorem 6.47 (Emden-Weinert, Hougardy, and Kreuter) *For any fixed integer $k \geq 3$, graph k-colorability is* NP*-complete even when restricted to graphs with maximum degree at most $k + \lceil\sqrt{k}\rceil - 1$.*

Sketch of Proof: We use reducibility from graph k-colorability. So let G be an arbitrary graph with m edges. If G has a vertex v with $d_G(v) > k + \lceil\sqrt{k}\rceil - 1$, then we construct a new graph G' as follows. First, we add a set I of $\lceil\sqrt{k}\rceil$ new vertices to G and replace for k edges in $E_G(v)$ their end v by a vertex of I, so that every vertex of I has degree at least 1 and at most $\lceil\sqrt{k}\rceil$. Then we add a complete graph K on $k-1$ new vertices and join each vertex of K to all vertices of $I \cup \{v\}$ by an edge. The resulting graph is G'. It is easy to see that G is k-colorable if and only if G' is k-colorable. Furthermore, each new vertex u has degree at most $k + \lceil\sqrt{k}\rceil - 1$ and $d_{G'}(v) = d_G(v) - 1$. Hence, by iterating this construction $O(m)$ times we obtain a graph G'' with $\Delta(G'') \leq k + \lceil\sqrt{k}\rceil - 1$ such that G is k-colorable if and only if G'' is k-colorable. □

As an immediate consequence of the above complexity result we obtain that (if $\mathsf{NP} \neq \mathsf{P}$) there are arbitrarily large graphs in $\mathrm{Crit}(k+1)$ with maximum degree at most $k + \lceil\sqrt{k}\rceil - 1$ whenever $k \geq 3$. Based on the construction used in the above reduction, we can obtain an explicit family of such graphs. For $\Delta \geq 2$, define

$$h_\Delta = \max\{h \in \mathbb{N}_0 \mid (h+2)(h+1) \leq \Delta\}; \tag{6.12}$$

hence, we have $\sqrt{\Delta} - 3 < h_\Delta < \sqrt{\Delta} - 1$. The following result was obtained by Molloy and Reed [757].

Theorem 6.48 (Molloy and Reed) *Let $\Delta \geq 3$ and let $2 \leq k \leq \Delta - 1 - h_\Delta$. Then* $\mathrm{Crit}(k+1)$ *contains arbitrarily large graphs with maximum degree at most Δ.*

Proof The statement is evident if $k = 2$, since $\mathrm{Crit}(3)$ is the family of odd cycles. So we may assume that $k \geq 3$. Let $\mathrm{Crit}_\Delta(k+1) = \{G \in \mathrm{Crit}(k+1) \mid \Delta(G) \leq \Delta\}$. Since $h_\Delta \geq 0$, we have $\Delta \geq k+1$ and so $K_{k+1} \in \mathrm{Crit}_\Delta(k+1)$.

Now let $G \in \mathrm{Crit}_\Delta(k+1)$ be an arbitrary graph. We shall construct a new graph $G' \in \mathrm{Crit}_\Delta(k+1)$ with $|G'| > |G|$. This implies that $\mathrm{Crit}_\Delta(k+1)$ contains arbitrarily large graphs. Let v be an arbitrary vertex of G. Then $k \leq d_G(v) \leq \Delta$. Let $s = \min\{\Delta - k + 2, d_G(v)\}$ and $t = \min\{\Delta - k + 1, k - 1\}$. The graph G' is constructed in two steps. First we replace v by a set I of s new vertices and we replace in each edge $e \in E_G(v)$ the end v by a vertex of I such that each vertex $u \in I$ has degree d with $1 \leq d \leq t$. Intuitively, we split the vertex v into the independent set I. Since $s \leq d_G(v)$, this is possible if $d_G(v) \leq st$. This is obviously true if $s = d_G(v)$, since $t \geq 1$. So assume that $s = \Delta - k + 2$. If $t = \Delta - k + 1$, then we obtain that

$$st = (\Delta - k + 2)(\Delta - k + 1) \geq (h_\Delta + 3)(h_\Delta + 2) > \Delta \geq d_G(v),$$

and we are done. Otherwise, $t = k - 1 < \Delta - k + 1$, which yields that $k < \frac{1}{2}\Delta + 1$ and $s = \Delta - k + 2 > \frac{1}{2}\Delta$. Since $k \geq 3$, this implies that

$$st > \tfrac{1}{2}\Delta(k-1) \geq \Delta \geq d_G(v),$$

and we are done, too. Note that $|I| = s \geq 2$. To complete the construction of G', we add a complete graph K on $k-1$ new vertices and join each vertex of K to all vertices of I. The resulting graph is G'. A vertex $u \in V(K)$ satisfies $d_{G'}(u) = k-2+s \leq \Delta$, and a vertex $u \in I$ satisfies $d_{G'}(u) \leq k-1+t \leq \Delta$. Consequently, we have $\Delta(G') \leq \Delta$. To complete the proof, it suffices to show that $G' \in \text{Crit}(k+1)$. First, we claim that $\chi(G') \geq k+1$. For otherwise, G' has a coloring with a set of k colors. Then all vertices in I receive the same color, say c. If we color v with c, we obtain a coloring of G with a set of k colors, contradicting $G \in \text{Crit}(k+1)$. Let e' be an arbitrary edge of G'. We claim that $G' - e'$ has coloring with k colors. Since G' is connected, this implies that $G' \in \text{Crit}(k+1)$ (by Proposition 1.29). For the edge $e' \in E(G')$ we have two possibilities. First, assume that e' is incident with a vertex of $G - v$. Then let e be the corresponding edge of G, that is, $e = e'$ or $e \in E_G(v)$. Since $G \in \text{Crit}(k+1)$, $G - e$ has a coloring with a set S of k colors. To get a coloring of $G' - e'$ with color set S, we color all vertices of I with the color of v and then we can color the vertices of K with the remaining $k-1$ colors. Now, assume that e' is incident with a vertex of K. Then it is easy to see that we can color $G'' = G'[V(K) \cup I] - e'$ by a set S of k colors such that exactly one vertex $w \in I$ has color c and all other vertices of I have color $c' \neq c$. Note that $|I| = s \geq 2$. Then w has d neighbors in $G - v$ with $1 \leq d \leq t \leq k-1$. Let w' be one such neighbor, Since $G \in \text{Crit}(k+1)$, there is a coloring of $G - vw'$ with color set S, and we may assume that v and w' have color c'. Since w has only $k-1$ neighbors in $G - v$, by permuting colors if necessary, this leads to a coloring of $G' - V(K)$ with color set S such that w has color c and all other vertices of I have color c'. Combining this with the coloring of G'', we obtain a coloring of $G' - e'$ with color set S. □

Theorem 6.49 (Molloy and Reed) *Let $\Delta \geq 3$ and $k = \Delta - h_\Delta$ such that $k \geq 2$ and $(h_\Delta+1)(h_\Delta+2) = \Delta$. Then* Crit$(k+1)$ *contains arbitrarily large graphs with maximum degree at most Δ.*

Proof We can use the same construction as proving Theorem 6.48. If $s = \Delta - k + 2$ and $t = \Delta - k + 1$, then $st = (\Delta-k+2)(\Delta-k+1) = (h_\Delta+2)(h_\Delta+1) = \Delta \geq d_G(v)$, as required. □

The construction used in Theorem 6.48 is a special case of a general construction introduced by Toft [1019, 1023] for the class of hypergraphs (see Theorem 7.13). We can also use this construction to decompose critical graphs into smaller critical graphs. Let G be a graph, let $k \geq 2$ be an integer, and let X, I be disjoint subsets of $V(G)$. We say that (X, I) is a k-**reducer** of G if $G[X]$ is a K_{k-1} and $N_G(u) \cup \{u\} = X \cup I$ for all $u \in X$. Note that if (X, I) is a k-reducer of G, then $G[X] \boxplus G[I] \subseteq G$, and I is an independent set, or G contains a K_{k+1}. If (X, I) is a k-reducer of G, then let G' denote the graph obtained from $G - X - I$ by adding a new vertex $v = v(I)$ and joining v to all vertices of $G - X - I$ that have in G a neighbor belonging to I. In this case we write $G' = G/(X, I)$ and call G' a k-**reduction** of G with respect to (X, I).

Theorem 6.50 *Let $k \geq 2$, let $G \in \mathrm{Crit}(k+1)$ be a graph with maximum degree at most Δ, let (X, I) be a k-reducer of G, let $U = V(G - X - I)$, and let $G' = G/(X, I)$. If G is not a K_{k+1}, then the following statements hold:*

(a) *I is an independent set with $2 \leq |I| \leq \Delta - k + 2$.*
(b) *$|E_G(I, U)| \geq k$ and $1 \leq |E_G(u, U)| \leq \Delta - k + 1$ for all $u \in I$.*
(c) *Every vertex w of U has at most one neighbor in G belonging to I.*
(d) *$G' \in \mathrm{Crit}(k+1)$.*
(e) *$k \leq \Delta - h_\Delta$ and, if $k = \Delta - h_\Delta$, then $\Delta - h_\Delta \leq d_{G'}(v(I)) \leq (h_\Delta + 1)(h_\Delta + 2)$ and $\Delta(G') \leq \Delta$.*
(f) *If $k = \Delta - h_\Delta$ and $(h_\Delta + 1)(h_\Delta + 2) \leq \Delta - 1$, then $|I| = h_\Delta + 2$, $d_G(v) = \Delta$ for all $v \in X$, and $\sum_{v \in I}(\Delta - d_G(v)) = \Delta - 1 - d_{G'}(v(I)) \leq h_\Delta - 1$.*

Proof Since $G \in \mathrm{Crit}(k+1)$, we have $k \leq \delta(G) \leq \Delta(G) \leq \Delta$. Since $G \neq K_{k+1}$, we obtain that $\omega(G) \leq k$. Since (X, I) is a k-reducer of G, we have that $G[X] = K_{k-1}$, $G[X] \boxplus G[I] \subseteq G$, and $E_G(X, V(G) \setminus (X \cup I)) = \varnothing$. Hence, I is an independent set and $U = V(G - X - I)$ is nonempty. Since G has no separating vertex (by Corollary 1.31), we have $|I| \geq 2$. If $u \in X$, then we have $d_G(u) = k - 2 + |I| \leq \Delta$, which yields $|I| \leq \Delta - k + 2$. Thus (a) is proved. Since G is k-edge connected (by Theorem 1.32), for $m = E_G(I, U)$ we get that $m = E_G(X \cup I, V(G) \setminus (X \cup I)) \geq k$. If $u \in I$, then $k \leq d_G(u) \leq \Delta$. Since $|E_G(u, X)| = k - 1$, this implies that $1 \leq |E_G(u, U)| \leq \Delta - k + 1$. This proves (b). For the proof of (c), assume that a vertex w of U has two neighbors in G belonging to I, say u and u'. Since $G \in \mathrm{Crit}(k+1)$, there is a coloring φ of $G - wu$ with a set S of k colors. Then we have $\varphi(u) = \varphi(w)$. Since $G[X] \boxplus G[I] = K_{k-1} \boxplus G[I] \subseteq G - uw$, we obtain that $\varphi(u') = \varphi(u)$ and so $\varphi(u') = \varphi(w)$, which is impossible. This proves statement (c).

For the proof of (d), we use Proposition 1.29. Let $v(I)$ denote the new vertex of $G' = G/(X, I)$. If G' has a coloring with a set S of k colors, then we can obtain a coloring of G with color set S by giving the color of $v(I)$ to all vertices of I and by using the remaining $k - 1$ colors to color $G[X] = K_{k-1}$. Since $G \in \mathrm{Crit}(k+1)$, this shows that $\chi(G') \geq k + 1$. Now let e' be an arbitrary edge of G'. Let e be the corresponding edge of G, that is, $e = e'$ if $e' \notin E_{G'}(v(I))$ else $e' \in E_{G'}(v(I), w)$ with $w \in V(G' - v(I))$ and $e \in E_G(u, w)$ for a unique vertex $u \in I$ (by (c)). Since $G \in \mathrm{Crit}(k+1)$, there is a coloring of $G - e$ with a set S of k colors, where all vertices of I receive the same color $c \in S$. If we color $v(I)$ with c, then it is easy to check that this yields a coloring of $G' - e'$ with color set S. This proves that $G' \in \mathrm{Crit}(k+1)$.

For the proof of (e) and (f), consider the new vertex $v(I)$ of $G' = G/(X, I)$. From (a), (b) and (c) we obtain that

$$d_{G'}(v(I)) = \sum_{u \in I} |E_G(u, U)| \leq (\Delta - k + 2)(\Delta - k + 1).$$

Suppose that $k \geq \Delta + 1 - h_\Delta$. Since $(h_\Delta + 1)(h_\Delta + 2) \leq \Delta$ and $G' \in \mathrm{Crit}(k+1)$, we then obtain that

$$k \leq d_{G'}(v(I)) \leq (h_\Delta + 1)h_\Delta \leq \Delta - 2h_\Delta - 2 \leq k - 1,$$

which is impossible. This proves that $k \le \Delta - h_\Delta$. If $k = \Delta - h_\Delta$, then we obtain that

$$\Delta - h_\Delta = k \le d_{G'}(v(I)) \le (\Delta - k + 2)(\Delta - k + 1) = (h_\Delta + 2)(h_\Delta + 1) \le \Delta,$$

which implies that $\Delta(G') \le \Delta$. Thus (e) is proved. Now assume that $k = \Delta - h_\Delta$ and $(h_\Delta + 1)(h_\Delta + 2) \le \Delta - 1$. By (a), we have $|I| \le \Delta - k + 2 = h_\Delta + 2$. If $|I| \le h_\Delta + 1$, then we obtain that

$$d_{G'}(v(I)) \le |I|(\Delta - k + 1) = (h_\Delta + 1)^2 \le \Delta - h_\Delta - 1 \le k - 1,$$

which is impossible. Consequently, we have $|I| = \Delta - k + 2 = h_\Delta + 2$. For $v \in X$, we then obtain that $d_G(v) = k - 2 + |I| = \Delta$. Furthermore, using (c) we obtain that

$$\begin{aligned}
\sum_{v \in I} (\Delta - d_G(v)) &= \Delta|I| - \sum_{v \in I} d_G(v) \\
&= \Delta|I| - (k-1)|I| - d_{G'}(v(I)) \\
&= |I|(\Delta - k + 1) - d_{G'}(v(I)) \\
&= (h_\Delta + 2)(h_\Delta + 1) - d_{G'}(v(I)) \\
&\le \Delta - 1 - d_{G'}(v(I)) \le h_\Delta - 1.
\end{aligned}$$

Note that the last inequality holds, since $d_{G'}(v(I)) \ge \Delta - h_\Delta$ (by (e)). This proves (f). Hence, the proof of the theorem is complete. □

Let $G \in \mathrm{Crit}(k+1)$ be a graph with $G \ne K_{k+1}$ and $k \ge 2$, and let (X, I) be a k-reducer of G. Then we can decompose G into one critical graph and one critical hypergraph. By Theorem 6.50, $G/(X, I) \in \mathrm{Crit}(k+1)$. Furthermore, I is an independent set of G and $G[X] \boxplus G[I]$ is an induced subgraph of G, and the hypergraph H obtained from this graph by adding a hyperedge with vertex set I is $(k+1)$-critical, because it is the Dirac join of $K_{k-1} \in \mathrm{Crit}(k-1)$ and a 2-critical hypergraph $H[I]$ (see Theorem 7.10).

The following theorem, proved by Molloy and Reed [757] in 2014, completes the characterization of those pairs (Δ, k) for which there are arbitrarily large graphs in $\mathrm{Crit}(k+1)$ having maximum degree at most Δ.

Theorem 6.51 (Molloy and Reed) *There is a constant Δ_0 such that for any integers $\Delta \ge \Delta_0$ and $k \ge \Delta - h_\Delta$, if a graph G satisfies $\Delta(G) \le \Delta$ and $\chi(G) = k+1$, and either $k \ge \Delta + 1 - h_\Delta$, or G has no k-reducer, then there is a vertex $v \in G$ such that $G' = G[N_G(v) \cup \{v\}]$ satisfies $\chi(G') = k+1$.*

Molloy and Reed's proof of the above result uses techniques similar to those used in Section 6.3 to prove Theorem 6.20. The proof of Theorem 6.51 combines structural results with probabilistic arguments, but it is much more elaborate than the proof in Section 6.3. A proof of the above theorem is beyond the scope of the book. The size of the constant Δ_0 results from the many inequalities which are used in the proof. On the one hand, this leads to a large constant Δ_0. On the other hand, Molloy and Reed conjectured that the constant Δ_0 in Theorem 6.51 can be small.

Corollary 6.52 (MALLOY AND REED) *For integers k and Δ with $2 \le k \le \Delta$, let* $\text{Crit}_\Delta(k+1) = \{G \in \text{Crit}(k+1) \mid \Delta(G) \le \Delta\}$. *If $\Delta \ge \Delta_0$, then the following statements hold:*

(a) *If $k \le \Delta - 1 - h_\Delta$, then* $\text{Crit}_\Delta(k+1)$ *contains graphs of arbitrarily large order.*
(b) *If $k = \Delta - h_\Delta$, then every graph $G \in \text{Crit}_\Delta(k+1)$ with $|G| \ge \Delta + 2$ contains a k-reducer (X, I) and $G/(X, I) \in \text{Crit}_\Delta(k+1)$.*
(c) *If $k \ge \Delta + 1 - h_\Delta$, then every graph in* $\text{Crit}_\Delta(k+1)$ *has order at most $\Delta + 1$.*

Proof Statement (a) is an immediate consequence of Theorem 6.48. For the proof of (b) and (c), let $G \in \text{Crit}_\Delta(k+1)$ and assume that $k \ge \Delta - h_\Delta$. If G has a vertex v such that $G' = G[N_G(v) \cup \{c\}]$ has chromatic number $k+1$, then $G = G'$ (since $G \in \text{Crit}(k+1)$) and so $|G| = |G'| = d_G(v) + 1 \le \Delta(G) + 1 \le \Delta + 1$. Otherwise, it follows from Theorem 6.51 that $k = \Delta - h_\Delta$ and G has a k-reducer (X, I). Then Theorem 6.50 implies that $G/(X, I) \in \text{Crit}_\Delta(k+1)$. □

As observed by Molloy and Reed, in case (b) of the above theorem, we have two possibilities. If $k = \Delta - h_\Delta$ and $(h_\Delta + 1)(h_\Delta + 2) = \Delta$, then $\text{Crit}_\Delta(k+1)$ contains graph of arbitrarily large order (by Theorem 6.49). It remains the case when $k = \Delta - h_\Delta$ and $(h_\Delta + 1)(h_\Delta + 2) \le \Delta - 1$ holds. Then, as proved by Molloy and Reed, every graph in $\text{Crit}_\Delta(k+1)$ has order at most $(\Delta + 1)^2 h_\Delta$ (see Exercise 6.22).

Farzad, Molloy, and Reed [365] observed that Corollary 6.52(c) combined with Corollary 4.20 can be used to characterize graphs with maximum degree at most Δ and $\chi \le \Delta - h$ by a set of finite forbidden subgraphs, provided that Δ is large enough and $h < h_\Delta$.

Let Δ and h be integers such that $0 \le h < h_\Delta$. Then let $\mathcal{S}_{\Delta,h}$ denote the set of nonisomorphic graphs H in $\text{Crit}(\Delta + 1 - h)$ with $|H| \le \Delta + 1$, that is, a graph belongs to $\text{Crit}(\Delta + 1 - h)$ and has order at most $\Delta + 1$ if and only if the graph is isomorph to a graph in $\mathcal{S}_{\Delta,h}$. Clearly, $\mathcal{S}_{\Delta,h}$ is finite.

Recall from Chapter 4 that $\text{Crit}(k, n)$ is the class of k-critical graphs of order n; and $\text{Crit}^*(k, n)$ is the class consisting of all graphs in $\text{Crit}(k, n)$ that have no dominating vertex. A vertex of a graph G is dominating if it is adjacent to all the remaining vertices of G. For a graph K and a graph property $\mathcal{G}$, define $K \boxplus \mathcal{G} = \{K \boxplus G \mid G \in \mathcal{G}\}$ if $\mathcal{G}$ is nonempty and $K \boxplus \mathcal{G} = \varnothing$ otherwise. If $\mathcal{X}$ is a family of graphs, then $\mathcal{G} = \langle \mathcal{X} \rangle$ denotes the graph property that a graph G belongs to $\mathcal{G}$ if and only if G is isomorphic to a graph in $\mathcal{X}$.

Theorem 6.53 (FARZAD, MALLOY, AND REED) *Let $\Delta \ge \Delta_0$ and $0 \le h < h_\Delta$. Then every graph G with $\Delta(G) \le \Delta$ satisfies*

(T_h) $\chi(G) \ge \Delta - h + 1$ *if and only if G contains a subgraph isomorphic to a graph in $\mathcal{S}_{\Delta,h}$.*

Furthermore, $\mathcal{S}_{\Delta,0} = \{K_{\Delta+1}\}$, $\mathcal{S}_{\Delta,1} = \{K_\Delta\}$, $\mathcal{S}_{\Delta,2} = \{K_{\Delta-1}, K_{\Delta-4} \boxplus C_5\}$, *and* $\mathcal{S}_{\Delta,3} = \{K_{\Delta-2}, K_{\Delta-5} \boxplus C_5, K_{\Delta-6} \boxplus H_1, K_{\Delta-6} \boxplus H_2\}$, *where H_1 and H_2 are the nonisomorphic graphs in* $\text{Crit}(4, 7)$ *shown in Figure 4.6.*

Proof Let G be a graph with $\Delta(G) \le \Delta$, where $\Delta \ge \Delta_0$, and let $0 \le h < h_\Delta$. Then $k = \Delta - h$ satisfies $k \ge \Delta - h_\Delta + 1$. By Proposition 1.28, we have $\chi(G) \ge k+1 = \Delta - h + 1$ if and only if G has a subgraph H with $H \in \text{Crit}(k+1)$. Since $\Delta(H) \le \Delta(G) \le \Delta$, Corollary 6.52(c) implies that $|H| \le \Delta + 1$. Since $\{H \in \text{Crit}(\Delta - h + 1) \mid |H| \le \Delta + 1\} = \langle \mathcal{S}_{\Delta,h} \rangle$, we have that $\chi(G) \ge \Delta - h + 1$ if and only if G has a subgraph that is isomorphic to a graph in $\mathcal{S}_{\Delta,h}$. Thus (T_h) holds.

For an integer $k \ge 4$, we have $\text{Crit}(k,k) = \langle K_k \rangle$ and $\text{Crit}(k, k+1) = \varnothing$. Define the class $\mathcal{G}_{\Delta,h}$ by

$$\mathcal{G}_{\Delta,h} = \text{Crit}(\Delta - h + 1, \Delta - h + 1) \cup \bigcup_{s=3}^{h+1} \text{Crit}(\Delta - h + 1, \Delta - h + s).$$

Then we have $\mathcal{G}_{\Delta,h} = \langle \mathcal{S}_{\Delta,h} \rangle$. In particular, we have $\mathcal{G}_{\Delta,0} = \text{Crit}(\Delta+1, \Delta+1) = \langle \mathcal{S}_{\Delta,0} \rangle$ and $\mathcal{G}_{\Delta,1} = \text{Crit}(\Delta,\Delta) = \langle \mathcal{S}_{\Delta,1} \rangle$. Consequently, we obtain that $\mathcal{S}_{\Delta,0} = \{K_{\Delta+1}\}$ and $\mathcal{S}_{\Delta,1} = \{K_\Delta\}$. Since $0 \le h < h_\Delta$, we get that $\Delta \ge (h_\Delta + 1)(h_\Delta + 2) \ge (h+2)(h+3) > \frac{5}{2}h - 1$. Then $\Delta + 1 < \frac{5}{3}(\Delta - h + 1)$ and we can use Corollary 4.20 with $k = \Delta - h + 1$ and $2 \le p = s - 1 \le h$, which implies that

$$\text{Crit}(k, k+p) = \bigcup_{\ell=\ell_p}^{\lfloor 3p/2 \rfloor} K_{k-\ell} \boxplus \text{Crit}^*(\ell, \ell + p),$$

where $\ell_p = 3$ if p is even, and $\ell_p = 4$ otherwise.

Case 1: $h = 2$. Then $k = \Delta - 1$, $p = s - 1 = 2$, and $\ell_2 = 3$. Hence, we obtain that

$$\mathcal{G}_{\Delta,2} = \langle \mathcal{S}_{\Delta,2} \rangle = \text{Crit}(\Delta - 1, \Delta - 1) \cup \text{Crit}(\Delta - 1, \Delta + 1),$$

and

$$\text{Crit}(\Delta - 1, \Delta + 1) = K_{\Delta-4} \boxplus \text{Crit}^*(3,5).$$

Since $\text{Crit}^*(3,5) = \langle C_5 \rangle$, we then obtain that $\mathcal{S}_{\Delta,2} = \{K_{\Delta-1}, K_{\Delta-4} \boxplus C_5\}$.

Case 2: $h = 3$. Then $k = \Delta - 2$, $2 \le p = s - 1 \le 3$, $\ell_2 = 3$ and $\ell_3 = 4$. Hence, we obtain that

$$\mathcal{G}_{\Delta,3} = \text{Crit}(\Delta - 2, \Delta - 2) \cup \text{Crit}(\Delta - 2, \Delta) \cup \text{Crit}(\Delta - 2, \Delta + 1).$$

Furthermore, we have

$$\text{Crit}(\Delta - 2, \Delta) = K_{\Delta-5} \boxplus \text{Crit}^*(3,5)$$

and

$$\text{Crit}(\Delta - 2, \Delta + 1) = K_{\Delta-6} \boxplus \text{Crit}^*(4,7).$$

As proved by Toft [1024] (see also Exercise 4.13), there are two nonisomorphic graphs in $\text{Crit}^*(4,7)$, denoted by H_1 and H_2 and shown in Figure 4.6. Summarizing, we obtain that $\mathcal{S}_{\Delta,3} = \{K_{\Delta-2}, K_{\Delta-5} \boxplus C_5, K_{\Delta-6} \boxplus H_1, K_{\Delta-6} \boxplus H_2\}$. This completes the proof of the theorem. □

Let Δ and h be given integers with $0 \le h < h_\Delta$. Then $\Delta \ge (h+2)(h+3) \ge 6$. If $\Delta \ge \Delta_0$ (the constant from Theorem 6.51), then it follows from Theorem 6.53 that a graph G with $\Delta(G) \le \Delta$ satisfies $\chi(G) \le \Delta - h$ if and only if $G \in \text{Forb}_{\subseteq}(\mathcal{S}_{\Delta,h})$. Since $\mathcal{S}_{\Delta,h}$ is finite, there exists a polynomial time algorithm to decide whether a graph with maximum degree at most Δ has a $(\Delta - h)$-coloring. To construct such an algorithm, however, we need an explicit list of graphs belonging to $\mathcal{S}_{\Delta,h}$. The proof of Theorem 6.53 shows us how such a list can be obtained in principle, provided we have a list of small critical graphs. Theorem 6.53 provides the list $\mathcal{S}_{\Delta,h}$ for $h \in [0,3]$. To describe all graphs in $\mathcal{S}_{\Delta,4}$ a complete list of nonisomorphic graphs in $\text{Crit}^*(4,8) \cup \text{Crit}^*(5,9) \cup \text{Crit}^*(6,10)$ is needed. Toft [1024] proved that $\text{Crit}^*(4,8)$ contains 4 nonisomorphic graphs; Jensen and Royle [527] showed that $\text{Crit}^*(5,9)$ has 16 nonisomorphic graphs. From Corollary 4.19 it easily follows that $\text{Crit}^*(6,10) = \langle C_5 \boxplus C_5 \rangle$. This implies that $\mathcal{S}_{\Delta,4}$ consists of 26 graphs (see Exercise 6.24). As discussed in [366], $\mathcal{S}_{\Delta,5}$ has 420 graphs. To show this, a complete list of nonisomorphic graphs in $\bigcup_{\ell=4}^{7} \text{Crit}^*(\ell, \ell+5)$ is needed. The nonisomorphic graphs in $\text{Crit}(\ell, \ell+5)$ for $\ell \in [4,7]$ were computed by Gordon Royle and published on the internet (see the corresponding reference in [366]); however, the internet page seems to be no longer available.

Note that, for the two graphs H_1 and H_2 in $\text{Crit}^*(4,7)$, we have $\omega(H_1) = \omega(H_2) = 3$. Consequently, each graph in $\mathcal{S}_{\Delta,h}$ for $h \in [0,3]$ contains a $K_{\Delta-h}$ (by Theorem 6.53). Hence, Theorem 6.53 implies the following result.

Corollary 6.54 (Farzad, Malloy, and Reed) *Every graph G with $\chi(G) \ge \Delta(G) - 2$ satisfies $\omega(G) \ge \chi(G) - 1$, provided that $\Delta(G) \ge \max\{\Delta_0, 3\}$.*

Farzad, Molloy, and Reed [366] noticed that the threshold $\Delta - 2$ in Corollary 6.54 is best possible. To see this, let $G = K_{\Delta-9} \boxplus C_5 \boxplus C_5$ with $\Delta \ge 9$. Then $\chi(G) = \Delta - 3$ and $\omega(G) = \Delta - 5 = \chi(G) - 2$.

6.7 Graphs with Bounded Clique Number

In 1988, Alon, Krivelevich, and Sudakov [54] proposed the following conjecture, dealing with the maximum possible chromatic numbers of graphs with forbidden subgraphs.

Conjecture 6.55 (Alon, Krivelevich, and Sudakov) *For every fixed graph H, there is a constant $c(H)$ such that every graph $G \in \text{Forb}_{\subseteq}(H)$ of maximum degree Δ satisfies $\chi(G) \le (c(H) + o(1))\Delta/\ln\Delta$, where $o(1)$ tends to zero if Δ tends to infinity.*

The conjecture is evident if $|H| \le 2$. So in what follows we assume that $|H| \ge 3$. The conjecture was confirmed for $H = K_3$, even if we replace the chromatic number by the DP-chromatic number, see Theorem 5.22. On the one hand, it suffices to establish the conjecture for complete graphs. This follows from the fact that if $|H| = r$, then $\text{Forb}_{\subseteq}(H) \subseteq \text{Forb}_{\subseteq}(K_r) = \text{Forb}(K_r)$. On the other hand, K_3 is the only complete graph for which the conjecture is confirmed. Johannson [535] proved that, for any fixed $r \geq 4$, every graph in $\mathrm{Forb}(K_r)$ with maximum degree $\Delta \geq 3$ has list chromatic number $O(\Delta \log_2 \log_2 \Delta / \log_2 \Delta)$. The paper [535] seems to be available on the internet, but has never been published. Molloy [754] proved a similar result and Bernshteyn [106] extended Molloy's result to the DP-chromatic number. Before we formulate this result, we need two auxiliary results about independent sets in K_r-free graphs.

For a graph G, let $\mathcal{I}(G)$ denote the set of all independent sets of G (including the empty set). The following two lemmas are due to Shearer [937] (see also Molloy [754, Section 4.1]).

Lemma 6.56 *Let $H \in \mathrm{Forb}(K_r)$ be a graph with $n = |H| \geq 1$ and $r \geq 2$. Then $\alpha(H) \geq n^{1/(r-1)} - 1$.*

Proof The proof is by induction on $r \geq 2$. Let H be a K_r-free graph of order $n \geq 1$. If $r = 2$, then $\alpha(H) = n$ and we are done. So assume that $r \geq 3$, and let $d = n^{(r-2)/(r-1)}$. If $\Delta(H) < d$, then $n \leq \alpha(H)\chi(H) \leq \alpha(H)(d+1)$, which yields $\alpha(H) \geq n/(d+1) \geq n^{1/(r-1)} - 1$. Otherwise, there is a vertex v in H with $d_H(v) \geq d$. Since $H' = H[N_H(v)]$ is K_{r-1}-free, the induction hypothesis implies that $\alpha(H) \geq \alpha(H') \geq d^{1/(r-2)} - 1 = n^{1/(r-1)} - 1$. □

Lemma 6.57 *Let $H \in \mathrm{Forb}(K_r)$ be a graph with $n = |H| \geq 2$ and $r \geq 4$, let $m = |\mathcal{I}(H)|$ and let $p = \lfloor \log_2 m / (2r \log_2 \log_2 m) \rfloor$. Then*

$$\sum_{i=0}^{p} \binom{n}{i} < \tfrac{1}{2} m, \tag{6.13}$$

and as a consequence at least half of the sets in $\mathcal{I}(H)$ have at least $p+1$ elements.

Proof Since $n \geq 2$, we have $m \geq 3$ and hence $p \geq 0$. Furthermore, (6.13) holds if $p = 0$. So assume that $p \geq 1$. Then, for $x = \log_2 m$, we have $x \geq 2$ and $r \leq \log_2 m / (2 \log_2 \log_2 m)$. Since $m = |\mathcal{I}(H)| \geq 2^{\alpha(H)}$, it follows from Lemma 6.56 that $x = \log_2 m \geq \alpha(H) \geq n^{1/(r-1)} - 1$, which leads to $n \leq (x+1)^{r-1}$. Since $x \geq 2$ and $r \geq 4$, for $h = (x+1)^{r-1}$, we obtain that $h \geq 27$ and $n \leq h < \frac{1}{4} x^{2r}$. An induction on $\ell \geq 0$ shows that

$$\sum_{i=0}^{\ell} \binom{q}{i} \leq 2q^{\ell},$$

provided that $q \geq 2$, where we use that $\binom{q}{\ell} \leq (q^{\ell})/(\ell!)$. Since $2rp \leq x/\log_2 x$, we obtain that $x^{2rp} \leq 2^x$. Hence, if M denotes the left hand side of (6.13), then we obtain that

$$M \leq 2n^p \leq 2h^p < \tfrac{1}{2} x^{2rp} \leq \tfrac{1}{2} 2^x = \tfrac{1}{2} m.$$

This proves the lemma. □

Theorem 6.58 (Bernshteyn) *There is an absolute constant C such that, for any $r \geq 4$, every K_r-free graph G with maximum degree $\Delta \geq 3$ satisfies $\chi_{\mathrm{DP}}(G) \leq C r \Delta \log_2 \log_2 \Delta / \log_2 \Delta$*

Note that the bound in Theorem 6.58 holds for any $r \geq 4$, but it is evident unless $r \leq \log_2 \Delta / C \log_2 \log_2 \Delta$; it also implies a bound for graphs of $\mathrm{Forb}_{\subseteq}(H)$ in general. The factor $\log_2 \log_2 \Delta$ in the bound is due the fact that Bernshteyn's proof uses Lemma 6.56; this is the only place where the assumption that G is K_r-free is used. The proof of Theorem 6.58 is very much similar to the proof of Theorem 5.22 and we shall only give a brief outline of the proof.

Sketch of the Proof of Theorem 6.58: Let $r \geq 4$ be a fixed integer, let $G \in \mathrm{Forb}(K_r)$ be a graph with maximum degree $\Delta \geq \Delta_1$ (where Δ_1 can be large, but is independent of r), let (X,H) be an arbitrary k-cover with $k = 2\lceil 100r\Delta \log_2 \log_2 \Delta / \log_2 \Delta \rceil \geq 200r\Delta \log_2 \log_2 \Delta / \log_2 \Delta$, and let $\ell = \Delta^{9/10}$. Since $G \in \mathrm{Forb}(K_r)$, it follows from (C1) and (C2) that $H \in \mathrm{Forb}(K_r)$, too. For the proof of the theorem, it suffices to show that G has an (X,H)-coloring.

As in the proof of Theorem 5.22 we want to show that G has a partial (X,H)-coloring which can be extended to an (X,H)-coloring of G. To this end, we use the fact that every graph G' satisfies $\chi_{\mathrm{DP}}(G') \leq \Delta(G') + 1$. A partial (X,H)-coloring T of G is said to be *extendable* if $|X_T(v)| \geq \ell$ for all $v \in V(G_T)$ and $\Delta(G_T) < \ell$. If G has an extendable partial (X,H)-coloring T, then G_T has an (X_T, H_T)-coloring T^*, and hence $T \cup T^*$ is an (X,H)-coloring of G and we are done. We use the same notation as in the proof of Theorem 5.22.

Claim 6.58.1 *Let $u \in V(G)$ be a vertex, and let S be an independent partial transversal of the cover $(X(\overline{N[u]}), H(\overline{N[u]}))$ of $G[\overline{N[u]}]$. If $\mathbf{T}'$ is a uniformly random partial $(X_S(N(u)), H_S(N(u)))$-coloring of $G[N(u)]$, then the random set $\mathbf{T} = S \cup \mathbf{T}'$ is a partial (X,H)-coloring of G such that*

(a) $\Pr(|X_{\mathbf{T}}(u)| < \ell) \leq \frac{1}{8}\Delta^{-3}$; *and*

(b) $\Pr(d_{G_{\mathbf{T}}}(u) \geq \ell$ *and* $|X_{\mathbf{T}}(v)| \geq \ell$ *for all* $v \in N_{G_{\mathbf{T}}}(u)) \leq \frac{1}{8}\Delta^{-3}$.

Proof: We denote by Ω the set of all independent partial transversals of the cover $(X_S(N(u)), H_S(N(u)))$ of $G[N(u)]$. Before we describe a random process for constructing $\mathbf{T}'$, we need some more notation. For a color $x \in X(u)$, let $L(x) = X_S(N(u)) \cap N_H(x)$. It follows from (C2) that $|X_S(v) \cap L(x)| \leq 1$ for all $v \in N(u)$, and the sets $L(x)$ with $x \in X(u)$ are disjoint. Let $x \in X(u)$ be a color, and let $Q \in \Omega$ with $Q \cap L(x) = \varnothing$. Then let $L(x,Q)$ be the set of colors $y \in L(x)$ such that there is a color $y' \in Q$ with $yy' \in E(H)$ or $\{y,y'\} \subseteq X_S(v)$ for some $v \in N(u)$. Furthermore, let $F(x,Q)$ be the graph $H[L(x) \setminus L(x,Q)]$. Note that if Q' is an independent set of $F(x,Q)$, then $Q \cap Q' = \varnothing$ and $Q \cup Q'$ belongs to Ω. As noticed by Bernshteyn, if $\mathbf{T}'$ is an uniform random member of Ω, then the following statement holds:

(1) *Fix $x \in X(u)$ and $Q \in \Omega$ with $Q \cap L(x) = \varnothing$. Then the random variable $\mathbf{T}' \cap L(x)$, conditioned to the event $\{\mathbf{T}' \setminus L(x) = Q\}$, is uniformly distributed over the independent sets in $F(x,Q)$.*

A uniform random member $\mathbf{T}'$ of Ω can be constructed by the the following randomized procedure. Let $\mathbf{T}_0$ be a uniform random member of Ω, and let $(x_1, x_2, \ldots, x_k)$ be an arbitrary ordering of $X(u)$. Note that $X_S(u) = X(u)$ has k colors. Repeat the following step for $i \in [1,k]$:

- Let $\mathbf{F}_i = F(x_i, \mathbf{T}_{i-1} \setminus L(x_i))$, let $\mathbf{Q}_i$ be a uniform random independent set in $\mathbf{F}_i$, and let $\mathbf{T}_i = (\mathbf{T}_{i-1} \setminus L(x_i)) \cup \mathbf{Q}_i$.
- Finally set $\mathbf{T}' = \mathbf{T}_k$.

It follows from (1), by induction on i, that $\mathbf{T}_i$ is a uniform random member of Ω; in particular, we have that $\mathbf{T}'$ is a uniform random member of Ω (see also [754, Lemma 15]).

Let us consider step i of the above procedure. If the values $\mathbf{T}_0, \mathbf{Q}_1, \dots, \mathbf{Q}_{i-1}$ are fixed, then $\mathbf{F}_i$ is fully determined and we choose $\mathbf{Q}_i$ uniformly in $\mathcal{I}(\mathbf{F}_i)$. Let $m_i = |\mathcal{I}(\mathbf{F}_i)|$. We call $\mathbf{F}_i$ *large* if $m_i > \Delta^{1/20} > 2$, otherwise $\mathbf{F}_i$ is called *small*. If $\mathbf{F}_i$ is large, i.e. $m_i > \Delta^{1/20}$, then

$$\frac{\log_2 m_i}{2r \log_2 \log_2 m_i} > \frac{\log_2 \Delta}{40r \log_2 \log_2 \Delta} (= c(\Delta, r))$$

and it follows from Lemma 6.57 that the corresponding conditional probability of the event $(|\mathbf{Q}_i| > c(\Delta, r))$ is at least $1/2$. Note that $\mathbf{F}_i$ is K_r-free and $r \geq 4$. If $\mathbf{F}_i$ is small, i.e. $m_i \leq \Delta^{1/20}$, then the corresponding conditional probability of the event $(\mathbf{Q}_i = \varnothing)$ is precisely $1/m_i$ and $1/m_i \geq \Delta^{-1/20}$.

Define integers $\mathbf{s}_i, \overline{\mathbf{s}}_i$ recursively by $\mathbf{s}_0 = \overline{\mathbf{s}}_0 = 0$, and if $\mathbf{F}_i$ is small, then let $\mathbf{s}_i = \mathbf{s}_{i-1} + 1$ and $\overline{\mathbf{s}}_i = \overline{\mathbf{s}}_{i-1}$, while if $\mathbf{F}_i$ is large, then let $\mathbf{s}_i = \mathbf{s}_{i-1}$ and $\overline{\mathbf{s}}_i = \overline{\mathbf{s}}_{i-1} + 1$. Note that $\mathbf{s}_i = |\{j \in [1, i] \mid \mathbf{F}_j \text{ is small}\}|$, $\overline{\mathbf{s}}_i = |\{j \in [1, i] \mid \mathbf{F}_j \text{ is large}\}|$, and $\mathbf{s}_i + \overline{\mathbf{s}}_i = i$ for all $i \in [1, k]$.

Let $(Z_1, Z_2, \dots, Z_k)$ and $(Z_1', Z_2', \dots, Z_k')$ be two sequences of $\{0,1\}$-valued random variables, that are mutually independent such that $\Pr(Z_i = 1) = 1/2$ and $\Pr(Z_i' = 1) = \Delta^{-1/20}$ for $i \in [1, k]$. Then we can couple the distribution of these sequences with the randomized procedure described above such that the following holds for $i \in [1, k]$:

(2) if $\mathbf{F}_i$ is small and $Z'_{\mathbf{s}_i} = 1$, then $\mathbf{Q}_i = \varnothing$, while
(3) if $\mathbf{F}_i$ is large and $Z_{\overline{\mathbf{s}}_i} = 1$, then $|\mathbf{Q}_i| > c(\Delta, r)$.

If in step i of the randomized procedure $\mathbf{F}_i$ is small, i.e. $m_i = |\mathcal{I}(\mathbf{F}_i)| \leq \Delta^{1/20}$, then $\mathbf{s}_i = \mathbf{s}_{i-1} + 1$ and we consider $Z'_{\mathbf{s}_i}$. If $Z'_{\mathbf{s}_i} = 1$, then we choose $\mathbf{Q}_i = \varnothing$; and if $Z'_{\mathbf{s}_i} = 0$, then, for $Q \in \mathcal{I}(\mathbf{F}_i)$, we choose $\mathbf{Q}_i = Q$ with probability $p(Q)$, where $p(\varnothing) = 1/m_i - \Delta^{-1/20}$ and $p(Q) = 1/m_i$ for $Q \neq \varnothing$. When $\mathbf{F}_i$ is large, we consider $Z_{\overline{\mathbf{s}}_i}$ and proceed similarly.

Note that k is even, and define

$$Z = \sum_{i=1}^{k/2} Z_i \text{ and } Z' = \sum_{i=1}^{k/2} Z_i'.$$

Furthermore, define the events $A = (Z < k/5)$ and $B = (Z' < \ell)$.

Proof of (a): First, we claim that if the outcome of the procedure yields $|X_T(u)| < \ell$, then A or B holds. To see this, suppose to the contrary, that both $\overline{A} = (X \geq k/5)$ and $\overline{B} = (Z' \geq \ell)$ hold. Since $\mathbf{s}_k + \overline{\mathbf{s}}_k = k$, we have $\mathbf{s}_k \geq k/2$ or $\overline{\mathbf{s}}_k \geq k/2$. First, assume that $\mathbf{s}_k \geq k/2$. Then $(Z' \geq \ell)$ and (2) imply that there are at least ℓ indices $i \in [1,k]$ such that $\mathbf{F}_i$ is small and $\mathbf{Q}_i = \varnothing$. But, for any such index i, we have $x_i \in X_{\mathbf{T}}(u)$, and so $|X_{\mathbf{T}}(u)| \geq \ell$, a contradiction. Now, assume that $\bar{\mathbf{s}}_k \geq k/2$. Then $(Z \geq k/5)$ and (3) imply that there are at least $k/5$ indices $i \in [1,k]$ such that $\mathbf{F}_i$ is large and $|\mathbf{Q}_i| > c(\Delta,r)$. Since the independent sets $\mathbf{Q}_i$ with $i \in [1,k]$ are disjoint subsets of $\mathbf{T}'$, this implies that

$$|\mathbf{T}'| > \frac{k}{5}c(\Delta,r) = \frac{k}{5} \cdot \frac{\log_2 \Delta}{40r \log_2 \log_2 \Delta} \geq \Delta.$$

Since $\mathbf{T}'$ is an independent partial transversal of the cover $(X_S(N(u)), H_S(N(u)))$ of $G[N(u)]$, we have $|\mathbf{T}'| \leq d(u) \leq \Delta$, a contradiction. This proves the claim that $|X_T(u)| < \ell$ implies A or B. Hence, we have

$$\Pr(|X_{\mathbf{T}}(u)| < \ell) \leq \Pr(Z < k/5) + \Pr(Z' < \ell).$$

Using Chernoff's bound for independent random variables (see Lemma C.40(b)), we show that $\Pr(Z < k/5) \leq \Delta^{-3}/16$ and $\Pr(Z' < \ell) \leq \Delta^{-3}/16$ for $\Delta \geq \Delta_1$, which gives $\Pr(|X_{\mathbf{T}}(u)| < \ell) \leq \Delta^{-3}/8$ as required.

For $Y \in \{Z, Z'\}$, Chernoff's bound from Lemma C.40(b) with $t = b\mathbb{E}(Y)$ and $0 < b < 1$ implies that

$$\Pr(Y < (1-b)\mathbb{E}(Y)) \leq \exp(-b^2\mathbb{E}(Y)/2),$$

where $(1-b)\mathbb{E}(Z) = k/5$ and $(1-b)\mathbb{E}(Z') = \ell$. Then it suffices to show that $\exp(-b^2\mathbb{E}(Y)/2) \leq \Delta^{-3}/16$, which is equivalent to (+) $2\ln(16\Delta^3) \leq b^2\mathbb{E}(Y)$. Note that $r \geq 4$ and $k \geq 200r\Delta \log_2 \log_2 \Delta/\log_2 \Delta$.

First, consider the case $Y = Z$. Then we get $\mathbb{E}(Z) = k/2$ and, moreover, $k/5 = (1-b)\mathbb{E}(Z) = (1-b)(k/2)$, which yields $b = 3/5$. Then we obtain that

$$\begin{aligned} b^2\mathbb{E}(Z) = (9/50)k &\geq 36r\Delta \log_2 \log_2 \Delta/\log_2 \Delta \\ &\geq 146\Delta \log_2 \log_2 \Delta/\log_2 \Delta \\ &\geq 2\ln(16\Delta^3), \end{aligned}$$

where the last inequality holds provided that Δ is large (independent of r). Hence, for $\Delta \geq \Delta_1$, (+) holds and so $\Pr(Z < k/5) \leq \Delta^{-3}/16$.

Now, consider the case $Y = Z'$. Then $\mathbb{E}(Z') = (k/2)\Delta^{-1/20}$ and $(1-b)\mathbb{E}(Z') = \ell = \Delta^{9/10}$, which yields $(1-b)k = 2\Delta^{19/20}$ and hence

$$b \geq \left(1 - \frac{\log_2 \Delta}{400\Delta^{1/20} \log_2 \log_2 \Delta}\right).$$

Note that $0 < b < 1$ provided that Δ is large (independent of r). Then we obtain that

$$\begin{aligned} b^2\mathbb{E}(Z') &= b^2(k/2)\Delta^{-1/20} \\ &\geq \left(1 - \frac{\log_2 \Delta}{400\Delta^{1/20}\log_2\log_2 \Delta}\right)^2 400\Delta^{19/20}\log_2\log_2 \Delta/\log_2 \Delta \\ &\geq 2\ln(16\Delta^3), \end{aligned}$$

where the last inequality holds provided that Δ is large (but independent of r). Hence, for $\Delta \geq \Delta_1$, (+) holds and so $\Pr(Z' < \ell) \leq \Delta^{-3}/16$. This completes the proof of statement (a). ◇

Proof of (b): Let $p = \Pr(d_{G_{\mathbf{T}}}(u) \geq \ell \text{ and } |X_{\mathbf{T}}(v)| \geq \ell \text{ for all } v \in N_{G_{\mathbf{T}}}(u))$, and let $h = \lceil \ell \rceil = \lceil \Delta^{9/10} \rceil$. We may assume that $d(u) \geq h$, for otherwise $p = 0$ and we are done. Let U be an arbitrary h-subset of $N(u)$, and let

$$p_U = \Pr(U \subseteq \mathrm{dom}(\mathbf{T}') \text{ and } |X_T(v)| \geq \ell \text{ for all } v \in U).$$

We claim that $p_U \leq 1/h!$. To prove the claim, consider an arbitrary partial independent transversal Q of $(X_S(N(u)), H_S(N(u)))$ with $Q \cap X(U) = \varnothing$. We either have $|X_S(v) \setminus N_H(Q)| < \ell$ for some vertex $v \in U$, or else, there are at least $h!$ ways to greedily choose a h-set $U' \subseteq X_S(U) \setminus N_H(Q)$ such that $Q \cup U'$ is a partial independent transversal of $(X_S(U), H_S(U))$. Consequently, we obtain that

$$\Pr(U \subseteq \mathrm{dom}(\mathbf{T}') \text{ and } |X_T(v)| \geq \ell \text{ for all } v \in U \mid \mathbf{T}' \setminus X(U) = Q) \leq 1/h!,$$

which implies the claim that $p_U \leq 1/h!$. As a consequence, we obtain that $p \leq \binom{\Delta}{h}/h!$. To see that this yields $p \leq \Delta^{-3}/8$, we use the following standard inequalities for binomial coefficients (see Stirling's formulae). If $1 \leq h \leq n$, then

$$\frac{n^h}{h^h} \leq \binom{n}{h} \leq \frac{n^h}{h!} < \left(\frac{ne}{h}\right)^h.$$

Since $h = \lceil \ell \rceil = \lceil \Delta^{9/10} \rceil$, we also obtain that

$$\left(\frac{e\Delta}{h}\right)^h 8\Delta^3 < \left(\frac{h}{e}\right)^h,$$

provided that Δ is large (independent of r). Combining these inequalities, we obtain that

$$\frac{1}{h!} < \frac{1}{\left(\frac{h}{e}\right)^h} < \frac{1}{\left(\frac{e\Delta}{h}\right)^h 8\Delta^3} \leq \frac{1}{\binom{\Delta}{h} 8\Delta^3},$$

which yields $p \leq \binom{\Delta}{h}/h! \leq \Delta^{-3}/8$. Thus (b) is proved. ◇

This completes the proof of Claim 6.58.1 △

Claim 6.58.2 *G has an extendable partial (X,H)-coloring, and as a consequence G has a (X,H)-coloring.*

Proof: To show that Claim 6.58.2 follows from Claim 6.58.1, we can repeat the argument showing that Claim 5.22.2 follows from Claim 5.22.1 (see Theorem 5.22); we just need to adjust the definition of u-bad accordingly. △

Hence, the proof of Theorem 6.58 is complete. □

Corollary 6.59 (Bernshteyn) *There is a constant Δ_1 such that for every $r \geq 4$ every graph $G \in \mathrm{Forb}(K_r)$ with maximum degree $\Delta \geq \Delta_1$ satisfies $\chi_{\mathrm{DP}}(G) \leq 200r\Delta\log_2\log_2\Delta/\log_2\Delta$.*

The assumption $G \in \mathrm{Forb}(K_r)$ in Theorem 6.58 is used in Bernshteyn's proof only insofar as it implies that $H \in \mathrm{Forb}(K_r)$ whenever (X,H) is a cover of G. A cover (X,H) of a graph G is said to be K_r-free if $H \in \mathrm{Forb}(K_r)$, regardless of whether G contains a K_r or not. For a graph G and an integer $r \geq 3$, define the ***r*-DP-chromatic number** $\chi^r_{\mathrm{DP}}(G)$ as the least k such that G is (X,H)-colorable for every K_r-free k-cover (X,H). Note that if $G \in \mathrm{Forb}(K_r)$, then every cover of G is K_r-free and $\chi_{\mathrm{DP}}(G) = \chi^r_{\mathrm{DP}}(G)$. Bernshteyn's proof of Theorem 6.58 then implies that, for every $r \geq 4$, every graph G with maximum degree Δ satisfies

$$\chi^r_{\mathrm{DP}}(G) \leq 200r\Delta\frac{\log_2\log_2\Delta}{\log_2\Delta}, \tag{6.14}$$

provided that Δ is large. The question arises whether the factor $200r$ in the above bound can be improved. The subgraphs $\mathbf{F}_i$ of H, constructed in Bernshteyn's proof, have a common neighbor, so these subgraphs belong not only to $\mathrm{Forb}(K_r)$, but also $\mathrm{Forb}(K_{r-1})$; and we can replace r by $r-1$, provided that $r \geq 5$ (note that Lemma 6.57 holds for $r \geq 4$). What happens, if we set $k = 100r\Delta\log_2\log_2\Delta/\log_2\Delta$. Then we need to adjust the definition of being a *small graph*. So we may change the threshold $\Delta^{1/20}$ to $\Delta^{1/10}$, which still gives $(k/5)c(\Delta,r) \geq \Delta$ as required in the proof. But then, for $Y = Z'$, we obtain that $(1-b)k = 2\Delta$ implying that $(1-b)$ tends to infinity if Δ tends to infinity. Using Chernoff's bound, however, requires $0 < b \leq 1$.

Przybylo [829] was able to lower the factor $200r$ in Bernshteyn's bound. Using an extension of the approach proposed by Bernshteyn and a strengthening of Lemma 6.57, Przybylo provided an elegant proof of the following result.

Theorem 6.60 (Przybylo) *For every $r \geq 4$, there is a constant Δ_r such that every graph G with maximum degree $\Delta \geq \Delta_r$ satisfies*

$$\chi^r_{\mathrm{DP}}(G) \leq 2(r-2)\Delta\frac{\log_2\log_2\Delta}{\log_2\Delta}.$$

A bound for $\chi^3_{\mathrm{DP}}(G)$ cannot be derived directly from Bernshteyn's proof of Theorem 5.22. Bernshteyn's proof uses the simple fact that if (X,H) is a cover of a graph $G \in \mathrm{Forb}(K_3)$, then $H(N_G(u))$ has no edges for every vertex $u \in V(G)$; this need not be the case if we only assume that $H \in \mathrm{Forb}(K_3)$, but allow triangles in G. However, Przybylo [829] could adopt Bernshteyn's argument to prove the following result.

Theorem 6.61 (Przybylo) *For any* $\varepsilon \in \mathbb{R}(0, \frac{1}{3})$, *there is a* Δ_ε *such that every graph with maximum degree* $\Delta \geq \Delta_\varepsilon$ *satisfies*

$$\chi^3_{\text{DP}}(G) \leq \lfloor (1+4\varepsilon)\Delta/\ln\Delta \rfloor.$$

6.8 Exercises

6.1 Show that Conjecture 6.3 holds for graphs G satisfying $\Delta(G)-1 \leq \omega(G) \leq \Delta(G)+1$.

6.2 For a graph G, let $\alpha'(G)$ be the matching number of G, and let $\text{odd}(G)$ be the number of components of G having odd order. Show that every graph G with $\alpha(G)=2$ satisfies $\chi(G) = |G| - \alpha'(\overline{G})$ and

$$\chi(G) = \min\{\tfrac{1}{2}(|G| - \text{odd}(\overline{G}-P) + |P|) \mid P \subset V(G)\}.$$

(Hint: Use the Tutte-Berge Formula, see Theorem C.12).

6.3 Show that Conjecture 6.3 holds for every graph G with $\Delta(G) = |G|-1$. (Hint: Choose a maximum matching in the complement $\overline{G}$ and extend it to a coloring of G.) (Reed [854])

6.4 Show that Conjecture 6.3 implies that every graph G with $\chi(G) = \Delta(G) - t$ satisfies $\omega(G) \geq \Delta(G) - 2t - 2$.

6.5 Show that if G is a graph and L is a list-assignment for G such that for every vertex v and every color $c \in L(v)$ the number of neighbors w of v in G for which $c \in L(w)$ is at most $\min\{k, |L(v)| - k\}$, then G has an L-coloring. (Hint: Use Theorem 6.9 and the DP-coloring concept.)

6.6 The List Coloring Constant: Let G be a graph, let L be a list-assignment for G, and let $k \in \mathbb{N}$. For a pair $(v,c) \in V(G) \times L(v)$, let $N_L(v:c) = \{u \in N_G(v) \mid c \in L(u)\}$. Prove that G is L-colorable if $|L(v)| = 2k$ and $|N_L(v:c)| \leq k$ for each $v \in V(G)$ and $c \in L(v)$. (Haxell [482], see also Reed and Sudakov [859])

6.7 A graph G is **breakable** if every induced subgraph H of G has an independent set hitting all maximum cliques of H. Show that a graph G is breakable if and only if G is perfect.

6.8 Show that a triangle-free graph G has an independent set hitting all maximum cliques of G if and only if G is bipartite.

6.9 Show that every P_4-free graph G has an independent set hitting all maximal cliques of G. (Kathrick and Maffray [557])

6.10 Show that Conjecture 6.3 holds for graphs G with $\alpha(G) \leq 2$. (Rabern [832])

6.11 Show that every graph G with $\chi(G) > \frac{1}{2}(|G|+3-\alpha(G))$ satisfies $\chi(G) \leq \frac{1}{2}(\omega(G)+\Delta(G)+2.)$ (Hint: Consider $H = G - I$, where I is a maximum independent set and use Corollary 6.30) (Rabern [832])

6.12 A graph G is called a (t_0,t_1)**-graph** if $\chi(G) \geq \Delta(G) \geq t_0$ and $\omega(G) \leq \Delta(G) - t_1$. Use Brooks' theorem, in order to show that the following statements hold for $t_0 \geq 4$ and $t_1 \geq 1$:

1. If G is a (t_0,t_1)-graph, then $\chi(G) = \Delta(G)$.
2. If there exists a (t_0,t_1)-graph, then there is a vertex critical (t_0,t_1)-graph.

(Cranston and Rabern [259] and [257])

6.13 A graph H is said to be d_1**-choosable** if H is L-colorable for every list-assignment L of H satisfying $|L(v)| \geq d_H(v) - 1$ for all $v \in V(H)$. Show that no vertex critical graph G with $\chi(G) = \Delta(G)$ contains an induced subgraph that is d_1-choosable. (Cranston and Rabern [259])

6.14 Small Pot Lemma: Show that if G is a graph and $f : V(G) \to [0, |G|-1]$ is a vertex function, then G is f-list-colorable if and only if G is L-colorable for every list-assignment L of G such that $|L(v)| = f(v)$ for all $v \in V(G)$ and $\left|\bigcup_{v \in V(G)} L(v)\right| < |G|$. (Hint: Use Hall's theorem, see also Exercise 1.13) (Kierstead [573], Cranston and Rabern [259])

6.15 Use the previous exercise to show that both graphs $K_4 \boxplus C_4$ and $K_4 \boxplus \overline{C_4}$ are d_1-choosable. (Cranston and Rabern [259])

6.16 Show that if a graph G is obtained from K_n by deleting a matching of size p, then $\chi_\ell(G) \leq n - p$. (Hint: Use induction on n and if $n = 2p$ use Exercise 6.14).) (Erdős, Rubin, and Taylor [355]).

6.17 Let G be a graph with maximum degree Δ, let v be a vertex of G, and let φ be a coloring of $G - v$ with $\Delta - t - 1$ colors. Let A be a set of $t+3$ singleton neighbors of v under φ, and let B be a set set of $t+2$ nonneighbors of v in G. Show that $\chi(G) \leq \Delta - t - 1$, provided that $|\varphi(B)| = |B|$ and, moreover, A and B are complete in G. (Haxell and MacDonald [483])

6.18 Show that for every constant $\Delta \geq 4$ and every integer k with $3 \leq k \leq \Delta - h_\Delta - 1$, it is NP-hard to decide whether graphs of maximum degree Δ are k-colorable. (Emden-Weinert, Hougardy, Kreuter [326], Molloy and Reed [757])

6.19 Show that, for $3 \leq k \leq 6$, the class Crit(k) contains arbitrarily large graphs with maximum degree at most k. (Hint: Use Theorem 6.48 and Theorem 6.49).

6.20 Show that if $4 \leq k \leq 8$, then there are graphs in Crit(k) with maximum degree k. (Beutelspacher und Hering [114], Haxell and Naserasr [485])

6.21 Show that if $G \in$ Crit(8), $\Delta(G) \leq 8$ and $|G| \geq 10$, then G contains no 7-reducer. (Hint: Use Theorem 6.50)

6.22 Let $\Delta \geq 2$ and $k = \Delta - h_\Delta$. Suppose that $(h_\Delta + 1)(h_\Delta + 2) \leq \Delta - 1$ and every graph $G \in \text{Crit}_\Delta(k+1)$ with $|G| \geq \Delta + 2$ has a k-reducer. Show that every graph $G \in \text{Crit}_\Delta(k+1)$ satisfies $|G| \leq (\Delta+1)^2 h_\Delta$. (Hint: Use Theorem 6.50) (Molloy and Reed [757, Corollary 7])

6.23 Deduce from Corollary 6.52 that if $\Delta \geq \max\{\Delta_0, 9\}$, then $\text{Crit}_\Delta(\Delta) = \langle K_\Delta \rangle$. Furthermore show that every graph G satisfies

$$\chi(G) \leq \max\{\Delta(G) - 1, \omega(G)\}$$

provided that $\Delta(G) \geq \max\{\Delta_0, 9\}$

6.24 Let $\Delta \geq \Delta_0$ and $4 < h_\Delta$. Furthermore, let $\mathcal{A}$, $\mathcal{B}$ and C be sets of nonisomorphic graphs such that $\text{Crit}^*(4,7) = \langle \mathcal{A} \rangle$, $\text{Crit}^*(4,8) = \langle \mathcal{B} \rangle$ and $\text{Crit}^*(5,9) = \langle C \rangle$. Show that a graph G belongs to $\mathcal{S}_{\Delta,4}$ if and only if $G = K_{\Delta-3}$, $G = K_{\Delta-6} \boxplus C_5$, $G = K_{\Delta-7} \boxplus H$ with $H \in \mathcal{A}$, $G = K_{\Delta-6} \boxplus C_7$, $G = K_{\Delta-7} \boxplus H$ with $H \in \mathcal{B}$, $G = K_{\Delta-8} \boxplus H$ with $H \in C$, or $G = K_{\Delta-9} \boxplus C_5 \boxplus C_5$. (Hint: Use the proof of Theorem 6.53.) (Farzad, Molloy, and Reed [366])

6.9 Notes

The Borodin–Kostochka conjecture (in short: BKC) proposes an interesting extension of Brooks' theorem, saying that $\text{bro}(\Delta, \omega) \leq \max\{\Delta - 1, \omega\}$, when $\Delta \geq 9$. Although the conjecture is still unsolved, it has led to several interesting papers, see e.g. [114] [157] [256]) [257] [258] [259] [285] [440] [485] [584] [617] [762] [833] [835] [855]; our list is far from being complete. The two strongest results to date were obtained by Reed [855] and by Cranston and Rabern [259]. Reed verified BKC provided Δ is sufficiently large (see Theorem 6.2). Cranston and Rabern proved that $\text{bro}(\Delta, \omega) \leq \max\{\Delta - 1, \omega + 3\}$ when $\Delta \geq 13$ (see Corollary 6.46), thereby improving earlier bounds obtained by Borodin and Kostochka [157], by Kostochka [617] and by Mozhan [762] (see [259] for details). Since BKC is difficult for graphs in general, it is reasonable to search for hereditary graph properties for which BKC can be proved. So BKC holds for line graphs of multigraphs (Rabern [835]), for claw-free graphs (Cranston and Rabern [258]), for the graph class $\text{Forb}(C_5, P_4)$ (Gupta and Pradhan [440]), and for the graph class $\text{Forb}(P_5, K_1 \boxplus P_4)$ (Cranston, Lafayette and Rabern [256]); for other graph classes see [114], [285], [584], and [833]. For an integer $k \geq 3$, let $\mathcal{BK}(k)$ denote the class of graphs $G \in \text{Crit}(k)$ with $\Delta(G) \leq k$. Then an attractive equivalent formulation of BKC say that $\mathcal{BK}(k) = \langle K_k \rangle$ whenever $k \geq 9$. An interesting observation, made independently by Kostochka [617] and by Catlin [192], says that for solving BKC it suffices to establish the base case, that is, BKC is equivalent to the statement that $\mathcal{BK}(9) = \langle K_9 \rangle$ (see Proposition 6.15). A graph property $\mathcal{P}$ is **finite** if there exists a finite set $\mathcal{X}$ of graphs such that $\mathcal{P} = \langle \mathcal{X} \rangle$, that is, $G \in \mathcal{P}$ if and only if G is isomorphic to a graph belonging to $\mathcal{X}$; otherwise $\mathcal{P}$ is **infinite**. The only known graphs belonging to $\mathcal{BK}(8)$ are K_8 and $C_5[K_3]$. That

$C_5[K_3]$ is 8-critical follows from Theorem 1.48 (see also Exercise 1.28); note that $C_5[K_3] = \mathrm{L}(3C_5)$. Haxell and Naserasr [485] proved that $\mathcal{BK}(8) = \langle K_8, C_5[K_3]\rangle$ implies $\mathcal{BK}(9) = \langle K_9\rangle$ and hence BKC. They also proposed the following problem.

Problem 6.62 (HAXELL AND NASERASR) *Is the graph family $\mathcal{BK}(7)$ finite?*

As observed by Haxell and Naserass if every graph $G \in \mathcal{BK}(7)$ with $|G| \geq 9$ has a 6-reducer, then $\mathcal{BK}(7)$ is finite. Benedict and Chinn [94] observed that for $3 \leq k \leq 7$, there are graphs G with $\chi(G) = \Delta(G) = k$ and $\omega(G) = k-1$. Catlin [192] constructed infinitely many graphs G satisfying $\chi(G) = \Delta(G) = 6$ and $\omega(G) = 5$; he used the splitting operation described in the proof of Theorem 6.48. For $3 \leq k \leq 6$, the graph family $\mathcal{BK}(k)$ is infinite (see Exercise 6.19). The study of the class $\mathcal{BK}(k)$ was initiated by Beutelspacher and Hering [114]; they proved the following result.

Theorem 6.63 (BEUTELSPACHER AND HERING) *If $G \in \mathrm{Crit}(k)$ has maximum degree k, then $|G| \geq 2k-1$, $k \geq 4$, and $|G| = 2k-1$ implies that $4 \leq k \leq 8$.*

Sketch of Proof: Let $G \in \mathrm{Crit}(k)$ with $\Delta(G) = k$. Clearly, this implies that $k \geq 4$. To see that $|G| \geq 2k-1$, suppose this is false. Then it follows from Corollary 4.14 that G is the Dirac join of two nonempty graphs, say $G = G_0 \boxplus G_1$. For $i \in \{0,1\}$, let $k_i = \chi(G_i)$. Then $G_i \in \mathrm{Crit}(k_i)$ and $k_0 + k_1 = k$ (by Theorem 4.1). Furthermore, we have $k_i - 1 \leq \delta(G_i) \leq \Delta(G_i)$ and $\Delta(G_i) + |G_{1-i}| \leq \Delta(G) = k$. Consequently, we obtain that

$$k-2+|G| = (k_0-1)+|G_1|+(k_1-1)+|G_0| \leq 2\Delta(G) = 2k,$$

which yields $|G| \leq k+2$. This implies that $G = K_k$ or $G = K_{k-3} \boxplus C_5$ (see Corollary 4.20 and Exercise 4.12). But then $\Delta(G) \neq k$, a contradiction. This shows that $|G| \geq 2k-1$.

Now assume that $|G| = 2k-1$. It remains to show that $k \leq 8$. Suppose this is false and choose G and k smallest possible. Note that $\omega(G) \leq k-1$. If $\omega(G) \leq \Delta(G)-2 = k-2$, then let I be a maximal independent set of G; otherwise let I be a maximal independent set of G such that $\omega(G-I) = \omega(G)-1$ (such a set exist by Theorem 6.13, since $\omega(G) = \Delta(G)-1 = k-1 \geq 8$ implies $\omega(G) > \frac{2}{3}(\Delta(G)+1)$). For the graph $G' = G-I$, we then obtain that $\omega(G') \leq \Delta(G)-2 = k-2$, $\Delta(G') \leq \Delta(G)-1 = k-1$ and $\chi(G') \geq \chi(G)-1 \geq \Delta(G)-1 = k-1 \geq 8$. Consequently, $\Delta(G') \geq 3$ and Brooks' theorem implies that $\chi(G') \leq \max\{\omega(G'), \Delta(G')\} \leq \Delta(G)-1 = k-1$. Then we obtain that $\chi(G') = \Delta(G)-1 = k-1 = \Delta(G')$ and $\omega(G') \leq \Delta(G')-1$. There is a subgraph H of G such that $H \in \mathrm{Crit}(k-1)$. Then $\Delta(H) \leq \Delta(G) = k-1$ and $\omega(H) \leq \omega(G) \leq k-2$. By Brooks' theorem, this implies that $\Delta(H) = k-1$. Then $|H| \geq 2(k-1)-1 = 2k-3 = |G|-2$. Since $|I| \geq 2$, this implies that $|I| = 2$ and $|H| = 2k-3$. This shows that $k = 9$; for otherwise we can replace the counterexample (G,k) by the smaller counterexample $(H, k-1)$. Furthermore, $G \in \mathrm{Crit}(9)$, $\Delta(G) = 9$, $|G| = 17$, $H \in \mathrm{Crit}(8)$, $\Delta(H) = 8$ and $|H| = 15$. Up to isomorphism there is only on such graph H, namely $H = C_5[K_3]$. For the proof it suffices to show that H is regular of degree $r = 8$ (see [114]). Since $\Delta(G) = 9$, this implies that every vertex of

H has only one neighbor in G belonging to I. But then for one vertex in I, say v, we have $d_G(v) \leq 7$. Since $G \in \mathrm{Crit}(9)$, this is impossible. This completes the proof. □

Beutelspacher and Hering [114] used a computer search to give a complete list of all graphs $G \in \mathrm{Crit}(k)$ with maximum degree k and order $2k-1$; exhibiting 13 such graphs.

As mentioned by Cranston and Rabern [258], Borodin and Kostochka are convinced that their conjecture also holds for the list chromatic number; whether BKC holds for the DP-chromatic number is unknown.

Conjecture 6.64 (Borodin and Kostochka) *Every graph G satisfies $\chi_\ell(G) \leq \max\{\Delta(G)-1, \omega(G)\}$, provided that $\Delta(G) \geq 9$.*

Reed's conjecture (in short: RC), saying that $\mathrm{bro}(\Delta,\omega) \leq \lceil (\Delta+1+\omega)/2 \rceil$, has attracted a lot of attention since it was published by Reed [854] in 1998. For a graph G and a vertex $v \in V(G)$, let $\omega_G(v)$ denote the largest p such that G has a p-clique containing v. In 2009, King [593] proposed the following local version of RC.

Conjecture 6.65 (King) *Every graph G satisfies*

$$\chi(G) \leq \max_{v \in V(G)} \left\lceil \frac{d_G(v)+1+\omega_G(v)}{2} \right\rceil.$$

It is natural to wonder if RC and its local version are true for the list-chromatic number and even for the DP-chromatic number. Kelly [565] (see also Kelly and Postle [566]) conjectured that this is the case for the list-chromatic number. The following conjecture would be a strengthening of these conjectures.

Conjecture 6.66 (The Local DP Conjecture) *Let G be a graph, and let (X,H) be a cover of G such that $|X(v)| \geq \lceil (d_G(v)+1+\omega_G(v))/2 \rceil$ for all $v \in V(G)$. Then G is (X,H)-colorable.*

In his seminal paper, Reed [854] not only published his conjecture, but also provided initial results justifying his conjecture; in particular, he proved that RC holds provided that $\omega(G)$ and $\Delta(G)+1$ are sufficiently close to each other.

Theorem 6.67 (Reed) *Let $c = 1/70000000$. There is a constant Δ' such that every graph G satisfies $\chi(G) \leq (\Delta(G)+1+\omega(G))/2$, provided that $\Delta(G) \geq \Delta'$ and $\omega(G) \geq (1-c)(\Delta(G)+1)$.*

Corollary 6.68 (Reed) *There is a constant Δ_0 and an $\varepsilon \in \mathbb{R}(0,\frac{1}{2})$ such that every graph G with $\Delta(G) \geq \Delta_0$ satisfies $\chi(G) \leq \lceil (1-\varepsilon)(\Delta(G)+1)+\varepsilon\omega(G) \rceil$.*

Proof Let c and Δ' as in Theorem 6.67, and put $\Delta_0 = \Delta'$ and $\varepsilon = c/2$. Suppose that G is a graph with $\Delta(G) \geq \Delta_0 = \Delta'$. If $\omega(G) \geq (1-c)(\Delta(G)+1)$, then it follows from Theorem 6.67 that

$$\chi(G) \leq \tfrac{\Delta(G)+1}{2} + \tfrac{\omega(G)}{2} \leq (1-\varepsilon)(\Delta(G)+1)+\varepsilon\omega(G),$$

where the second inequality holds because $\omega(G) \le \Delta(G)+1$ and $0 < \varepsilon < \frac{1}{2}$. If $\omega(G) < (1-c)(\Delta(G)+1)$, let G' be the graph obtained from G by adding a complete graph of order $p = \lceil (1-c)(\Delta(G)+1) \rceil$. Then $\Delta(G') = \Delta(G)$ and $\omega(G') = p$, and so Theorem 6.67 yields

$$\begin{aligned}\chi(G) \le \chi(G') &\le \tfrac{\Delta(G')+1}{2} + \tfrac{\omega(G')}{2}\\ &= \tfrac{\Delta(G)+1}{2} + \tfrac{\lceil (1-c)(\Delta(G)+1)\rceil}{2}\\ &\le \lceil (1-\tfrac{c}{2})(\Delta(G)+1)\rceil \le \lceil (1-\varepsilon)(\Delta(G)+1)+\varepsilon\omega(G)\rceil.\end{aligned}$$

This completes the proof. □

Corollary 6.69 (Reed) *There is an $\varepsilon' \in \mathbb{R}(0, \frac{1}{2})$ such that every graph G satisfies $\chi(G) \le \lceil (1-\varepsilon')(\Delta(G)+1)+\varepsilon'\omega(G)\rceil$.*

Proof Let Δ_0 and ε as in Corollary 6.68, and put $\varepsilon' = \min\{\frac{1}{\Delta_0}, \varepsilon\}$. Let G be an arbitrary graph. If $\Delta(G) \ge \Delta_0$, then the result follows from Corollary 6.68 because $0 < \varepsilon' \le \varepsilon$. If $\Delta(G) < \Delta_0$, then the result follows because $\varepsilon'\Delta_0 \le 1$ (see the proof of Theorem 6.20). □

We refer to the inequality in Corollary 6.68 as an *epsilon version* of RC. So the first epsilon version of RC was obtained by Reed from Theorem 6.67 with $\varepsilon = 1/140000000$; the proof of this theorem, however, is long and sophisticated. An improved epsilon version of RC with a shorter proof was obtained by King and Reed [596] with $\varepsilon = 1/(320e^6)$; see Theorem 6.20 in Section 6.3. Our proof follows King and Reed [596], however, in the proof of Claim 6.23.3 we use a different version of Talagrand's inequality. King and Reed used a version of Talagrand's inequality mentioned in the book of Molloy and Reeed [756, Chapter 10]; however, as explained by Kelly and Postle [566], this version is flawed. So we use a version of Talagrand's inequality proved by Kelly and Postle [566], see Theorem C.42. Molloy and Reed [757] later published a corrected version of their Talagrand inequality, which can also be used to prove Claim 6.23.3. Bonamy, Perrett, and Postle [139] proved that for large Δ, $\varepsilon = 1/26$ suffices for an epsilon version of RC. Their proof proceeds in a similar fashion as the proof by King and Reed. Edwards and King [322] proposed a further strengthening of RC that involves a bound supplied by the neighborhood of two adjacent vertices. Kelly and Postle [566, 567] proved the following result supporting Conjecture 6.66.

Theorem 6.70 (Kelly and Postle) *Let $\varepsilon = 1/330$. If G is a graph whose maximum degree is sufficiently large and L is a list-assignment for G such that for all $v \in V(G)$, $|L(v)| \ge \omega_G(v) + \log_{10}\Delta(G)$ and*

$$|L(v)| \ge (1-\varepsilon)(d_G(v)+1)+\varepsilon\omega_G(v),$$

then G is L-colorable.

Further support for RC comes from the fact that a fractional relaxation holds, as proved by Molloy and Reed [756, Theorem 21.7]. That such a fractional relaxation also holds for the local version of RC was noted by C. McDirmid, for a proof see King [593].

Theorem 6.71 (McDiarmid) *Every graph G satisfies*

$$\chi^*(G) \leq \max_{v \in V(G)} \frac{d_G(v)+1+\omega_G(v)}{2}.$$

Note that the rounding in RC is not needed in the above theorem. As proved in Section 8.2, every graph G satisfies $\alpha(G)\chi^*(G) \geq |G|$ (see Theorem 8.6). Hence, Theorem 6.71 implies that every graph of order n has an independent set of size at least $(2n)/(\Delta(G)+1+\omega(G))$, a result that was obtained much earlier by Fajtlowicz [362] (see Theorem 6.24).

There are several attempts to prove RC directly, but only for restricted classes of graphs, in particular for hereditary graph properties. Rabern [831, 832] proved RC for several classes of graphs, including graphs with $\alpha \leq 2$ and graphs whose complements are disconnected; see the results mentioned in Section 6.4. That the local version of RC holds for graphs with $\alpha \leq 2$ was proved by King [593], see also [322]. Further results on RC with respect to the independence number can be found in the paper by Kohl and Schiermeyer [606]. It is known that RC holds for line graphs of multigraphs (King, Reed, and Vetta [597]), for quasi-line graphs (King and Reed [595]), and for claw-free graphs with $\alpha \leq 3$ (King [593]). Recall that a graph is a **quasi-line graph** if the neighborhood of any vertex induces the complement of a bipartite graph. Chudnovsky, King, Plumettaz, and Seymour [229, Theorem 4] proved that the local version of RC holds for quasi-line graphs.

Theorem 6.72 (Chudnovsky, King, Plumettaz, Seymour) *Every quasi-line graph G satisfies*

$$\chi(G) \leq \max_{v \in V(G)} \left\lceil \frac{d_G(v)+1+\omega_G(v)}{2} \right\rceil.$$

The following result is a strengthening of a result obtained by Rabern [834, Theorem D], see also Karthick and Maffray [557, Theorem 3.9], and King, Reed, and Vetta [597].

Theorem 6.73 *Let $\mathcal{P}$ be a hereditary graph property such that every graph $G \in \mathcal{P}$ satisfies $\chi(G) \leq \lceil 5\omega(G)/4 \rceil$. Then $\chi(G) \leq \lceil (\Delta(G)+1+\omega(G))/2 \rceil$ for every graph $G \in \mathcal{P}$.*

Proof The proof is by induction on the order of $G \in \mathcal{P}$. The statement is evident if $G = \varnothing$. So suppose that $G \neq \varnothing$. First assume that $\omega(G) > \frac{2}{3}(\Delta(G)+1)$. Then it follows from Theorem 6.13 that there is an independent set I of G with $\omega(G-I) = \omega(G)-1$. We may extend I to a maximal independent set, so that $\Delta(G-I) \leq \Delta(G)-1$. Clearly, we also have $\chi(G) \leq \chi(G-I)+1$. Since $G-I \in \mathcal{P}$ and $I \neq \varnothing$, it then follows from the induction hypothesis that

$$\begin{aligned}\chi(G) \le \chi(G-I)+1 &\le \left\lceil \frac{\Delta(G-I)+1+\omega(G-I)}{2} \right\rceil + 1 \\ &\le \left\lceil \frac{(\Delta(G)-1)+1+(\omega(G)-1)}{2} \right\rceil + 1 \\ &= \left\lceil \frac{\Delta(G)+1+\omega(G)}{2} \right\rceil,\end{aligned}$$

and we are done. Now assume that $\omega(G) \le \frac{2}{3}(\Delta(G)+1)$. Since $G \in \mathcal{P}$, we then obtain that

$$\begin{aligned}\chi(G) &\le \left\lceil \frac{5\omega(G)}{4} \right\rceil = \left\lceil \frac{3\omega(G)}{4} + \frac{\omega(G)}{2} \right\rceil \\ &\le \left\lceil \frac{\Delta(G)+1+\omega(G)}{2} \right\rceil,\end{aligned}$$

and we are also done. □

Corollary 6.74 *Let $\mathcal{P}$ be a hereditary graph property such that every graph $G \in \mathcal{P}$ satisfies $\chi(G) \le \max\{\lceil 5\omega(G)/4 \rceil, \lceil(\Delta(G)+1+\omega(G))/2\rceil\}$. Then every graph $G \in \mathcal{P}$ satisfies $\chi(G) \le \lceil(\Delta(G)+1+\omega(G))/2\rceil$.*

Proof The proof is by induction on the order of $G \in \mathcal{P}$. If $\omega(G) > \frac{2}{3}(\Delta(G)+1)$, then there is an independent set hitting all maximum cliques of G and we can argue as in the proof of Theorem 6.73. If $\omega(G) \le \frac{2}{3}(\Delta(G)+1)$ and, moreover, $\max\{\lceil 5\omega(G)/4 \rceil, \lceil(\Delta(G)+1+\omega(G))/2\rceil\} = \lceil 5\omega(G)/4 \rceil$, then we can argue as in the proof of Theorem 6.73; otherwise there is nothing to prove. □

Theorem 6.75 (King, Reed, and Vetta) *Let $\mathcal{P}$ be the class of line graphs of multigraphs. Then every graph $G \in \mathcal{P}$ satisfies $\chi(G) \le \lceil(\Delta(G)+1+\omega(G))/2\rceil$.*

Sketch of Proof: Clearly, $\mathcal{P}$ is a hereditary graph property. Let $G \in \mathcal{P}$ be an arbitrary graph, we may assume that $G \ne \varnothing$. Then there is a multigraph H such that $G = \mathrm{L}(H)$. By a result of Anderson [61] (see also [973, Chapter 6]), we obtain that

$$\chi(G) \le \max\{\left\lfloor \frac{7}{6}\Delta(H) + \frac{4}{6} \right\rfloor, \lceil \chi^*(G) \rceil\}.$$

Now we use Theorem 6.71 and the fact that $\Delta(H) \le \omega(G)$. Then we obtain that

$$\begin{aligned}\chi(G) &\le \max\{\left\lfloor \frac{7}{6}\omega(G) + \frac{4}{6} \right\rfloor, \left\lceil \frac{\Delta(G)+1+\omega(G)}{2} \right\rceil\} \\ &\le \max\{\left\lceil \frac{5}{4}\omega(G) \right\rceil, \left\lceil \frac{\Delta(G)+1+\omega(G)}{2} \right\rceil\}.\end{aligned}$$

From Corollary 6.74 it then follows that $\chi(G) \le \lceil(\Delta(G)+1+\omega(G))/2\rceil$. □

As a consequence of Theorem 6.73, we obtain that RC holds for all graphs belonging to $\mathrm{Forb}(P_4)$ because all these graphs are perfect. On the other hand, it is not known whether RC holds for all graphs belonging to Forb(P_5). Partial results related to this problem can be found in [70], [228], [383], [557], and [899]. For instance, Chudnovsky, Karthick, Maceli, and Maffray [228] proved that every graph G in $\mathcal{P} = \text{Forb}(P_5, K_1 \boxplus P_4)$ has chromatic number at most $\lceil 5\omega(G)/4 \rceil$, and so every such graph satisfies RC.

Given a finite set system $\mathcal{F}$, a **hitting set** for $\mathcal{F}$ is a subset S of $\bigcup \mathcal{F}$ that has a nonempty intersection with each set in $\mathcal{F}$. The determination of a hitting set of minimum cardinality for a finite set system was one of the first combinatorial optimization problem to be identified as NP-hard by Karp [556]; the corresponding decision problem is NP-complete. Recall that for a graph G, we denote by $C(G)$ the set of maximum cliques of G, that is, the set of cliques of G having cardinality $\omega(G)$. The hitting set problem for $C(G)$ was first investigated by Chang, Kloks, and Lee [200]. They considered the problem for planar graphs and presented fixed parameter and approximation results, and also for other classes of graphs. Let S be a hitting set for $C(G)$. For the clique number, we then have that $\omega(G-S) \leq \omega(G) - 1$, for the chromatic number, however, we have only $\chi(G) \leq \chi(G-S) + \chi(G[S])$. To investigate coloring problems, it would be of much interest to find an independent hitting set of $C(G)$, that is, an independent set of G hitting all maximum cliques of G. However, this is not always possible as the Catlin graph $C_5[K_3]$ shows. That independent hitting sets of $C(G)$ are useful to solve coloring problems was first noted by Borodin and Kostochka [157] and, independently, by Catlin [192]; it is the key for proving Proposition 6.15 showing that BKC can be reduced to the base case. It is easy to see that every perfect graph G has an independent set hitting all maximum cliques of G (see Exercise 6.7); finding such a set efficiently is the key to coloring perfect graphs in polynomial time (see Reed [856]). The **Independent Maximum Clique Cover Problem** (IMCCP for short) is the algorithmic problem of deciding whether or not a given graph G has an independent hitting set for $C(G)$. The complexity status of IMCCP is open. On the one hand, it is very unlikely that IMCCP belongs to NP, because the determination of the clique number of a graph is NP-hard. On the other hand, it is easy to see that IMCCP is NP-hard as noted by T. Fischer and F. Hörsch (oral communication to the first author).

Theorem 6.76 (Fischer and Hörsch) *IMCCP is* NP*-hard even for graphs belonging to the class* Forb(K_4).

Proof The reduction is from 3-SAT. Let I be an instance of 3-SAT with n variables $x_1, x_2, \ldots, x_n$ and m clauses, where $m, n \in \mathbb{N}$. We now construct a graph $G(I)$ as follows. For a clause $C = \ell_{C,1} \vee \ell_{C,2} \vee \ell_{C,3}$ of I, let $T(C) = (\ell_{C,1}, \ell_{C,2}, \ell_{C,3}, \ell_{C,1})$ be a triangle, whose vertices are labeled by the literals of C, where $\ell_{C,i} \in \{x, \overline{x} \mid x \in \{x_1, x_2, \ldots, x_n\}\}$. We may assume that the three variables in each clause are different. From the disjoint union off all these triangles $T(C)$ we form a graph $G(I)$ by adding edges according to the following rule: we join a literal $\ell_{C,i}$ with a literal $\ell_{C',j}$ by an edge if and only if C and C' are distinct clauses of I, and $\ell_{C,i} = \overline{\ell_{C',j}}$. Then $G(I)$ belongs to Forb(K_4), and if T is a triangle of $G(I)$, then $T = T(C)$ for some clause C. Furthermore, it is easy to show that I is satisfiable if and only if $G(I)$ has an independent set hitting all maximum cliques of $G(I)$. □

Kostochka [617] proved that a graph G has an independent hitting set for $C(G)$, provided that ω is close to $\Delta+1$. Rabern [834] improved Kostochka results by using a result of Haxell, see Theorem 6.10. King [594] improved Haxell's theorem, see Theorem 6.9, and used this result to give an improvement of Rabern's result. Our proof of Theorem 6.9 follows King [594]. The proof of Lemma 6.11 is due to Hajnal [456]; the proof of Lemma 6.12 combines ideas of Kostochka [617] and of Christofides, Edwards, and King [264]. The proof of Theorem 6.13 combines ideas of King [594] and of Christofides, Edwards, and King [264]; this result is best possible. Corollary 6.14 is a strengthening of a result obtained by Cranston and Rabern [259].

As explained in Section 6.5 the method of Mozhan partitions was introduced in 1983 by Mozhan [762] and further developed by Rabern [836], by Cranston and Rabern [258] [259], by Haxell and McDonald [483], and possibly by others. Mozhan partitions have become important to make significant progress on the BKC. Recall that BKC states that $\chi \geq \Delta \geq 9$ implies $\omega \geq \Delta$. The first result supporting this conjecture was obtained by Borodin and Kostochka [157], who proved that $\omega \geq \lfloor(\Delta+1)/2\rfloor$. The first improvement was obtained by Mozhan [762], who proved that $\omega \geq \lfloor(2\Delta+1)/3\rfloor$ when $\Delta \geq 10$. The next substantial improvement was achieved by Kostochka [617], who proved that $\omega \geq \Delta-28$. Rabern in his thesis proved that $\omega \geq \Delta-3$ when $\Delta \geq 31$; the thesis, however, is not easy available. The best result to date is due to Cranston and Rabern [259], who prove that $\omega \geq \Delta-3$ when $\Delta \geq 13$ (see Corollary 6.46). The results in Section 6.5 and their proofs mainly follow the paper of Cranston and Rabern [259], our exposition is perhaps somewhat more detailed. Mozhan's Lemma (see Lemma 6.35) and its proof, however, goes back to Mozhan [762]. A given graph has many Mozhan partitions of a certain type, so a search technique is required for exploring the set of all these partitions to find one that is particulary favorable. To this end, the concept of move sequences is used. The move sequences used by Cranston and Rabern have the property that no vertex is moved twice. In a 2023 paper by Haxell and McDonald [483], a different type of move sequences is used, so-called depth-first move sequences. This approach leads to a simpler proof of the result by Cranston and Rabern when $\Delta \geq 16$ and to some results related to RC.

Theorem 6.77 (Haxell and McDonald) *Let $t \in \mathbb{N}_0$. Every graph with $\chi(G) = \Delta(G)-t$ and $\Delta(G) \geq 6t^2+20t+16$ satisfies $\omega(G) \geq \Delta(G)-2t^2-6t-3$.*

As already noted by Cranston and Rabern, the method of Mozhan partitions leads to a polynomial time algorithm that, for a given graph G and $t \in \mathbb{N}_0$ with $\Delta(G) \geq 6t^2+20t+16$, finds either a coloring of G with $\Delta(G)-t-1$ colors or a complete graph of G with $\Delta(G)-2t^2-6t-3$ vertices.

In 1998, Maffray and Preissmann [714] proved that, for each $k \geq 3$, graph k-colorability remains NP-complete among graphs with maximum degree at most $2k-2$. Emden-Weinert, Hougardy, and Kreuter [326] improved this bound to $k+\lceil\sqrt{k}\rceil-1$ (see Theorem 6.47). For the reduction to k-colorability, they use a splitting operation first used by Catlin in his thesis [192, page 43] to construct infinitely many

6-chromatic graphs with maximum degree 6 and clique number 5. This splitting technique was further developed and extended to hypergraphs by Toft [1019] [1023], see Section 7.4. As observed by Molloy and Reed [757, Theorem 42], the splitting operation can be used to prove the following strengthening of Theorem 6.47; see also the proof of Theorem 6.48 (h_Δ is defined in (6.12)).

Theorem 6.78 *For every constant $\Delta \geq 3$ and every k with $3 \leq k \leq \Delta - h_\Delta - 1$, it is* NP*-hard to decide whether graphs of maximum degree Δ are k-colorable.*

On the other hand, Molloy and Reed [755] proved a first counterpart of Theorem 6.78 which states that, for fixed $\Delta \geq 3$ and k with $f(\Delta) \leq k \leq \Delta + 1$, the decision problem whether a given graph of maximum degree Δ is k-colorable belongs to the complexity class P. Note that this is trivial for $k = \Delta + 1$ and an immediate consequence of Brooks' theorem for $k = \Delta$. In 2014, Molloy and Reed [757] succeeded to show that the above positive result holds for $f(\Delta) = \Delta - h_\Delta$, provided that Δ is large enough.

Theorem 6.79 (Molloy and Reed) *There exists a constant Δ_0 such that for any integers $\Delta \geq \Delta_0$ and $k \geq \Delta - h_\Delta$, the problem to decide whether graphs of maximum degree Δ are k-colorable belongs to* P.

Theorem 6.79 is an immediate consequence of Theorem 6.51 which is the main result in [757]; the proof of this result is about 50 pages long; too long to present here. The use of probabilistic methods in the proof of this theorem leads to a large value of Δ_0, which is, however, not specified. As a main conclusion of Theorem 6.51 we obtain (see Corollary 6.52 and the remark afterwards) that if $\Delta \geq \Delta_0$ and $k \geq \Delta - h_\Delta$, then $\mathrm{Crit}_\Delta(k+1)$ is finite, unless $k = \Delta - h_\Delta$ and $(h_\Delta + 1)(h_\Delta + 2) = \Delta - 1$. In this case, however, every graph $G \in \mathrm{Crit}(k+1)$ with $|G| \geq \Delta + 2$ has a k-reducer and can be reduced to a smaller graph belonging to $\mathrm{Crit}_\Delta(k+1)$ (by Theorem 6.50). The proof of Theorem 6.50 uses standard arguments and follows Molloy and Reed [757]. Our proof of Theorem 6.53 resembles the original proof of Farzad, Molloy and Reed [366]. Molloy and Reed, see the remark after Theorem 2 in [757], made the following conjecture.

Conjecture 6.80 (Molloy and Reed) *Let $\Delta \geq 3$ and let $k \geq \Delta + 1 - h_\Delta$. Then $\mathrm{Crit}_\Delta(k+1)$ is finite.*

Note that this conjecture holds for $3 \leq \Delta \leq 11$ by Brooks' theorem. On the other hand, Theorem 6.51 does not hold for $\Delta = 8$. Then $h_\Delta = 1$ and we can take $k = \Delta - h_\Delta = 7$ and $G = C_5[K_3]$. Then $\chi(G) = k + 1 = 8$, $\Delta(G) = \Delta = 8$, but G has neither a k-reducer, nor a vertex v such that $G' = G[N_G(v) \cup \{v\}]$ satisfies $\chi(G') = k + 1 = 8$, see also Exercise 6.21.

Lemma 6.56 appears in Molloy [754]; the proof is straightforward. Our proof of Lemma 6.57 follows Molloy [754]. Improvements of Lemma 6.57 are given by Przybylo [829] and by Bonamy, Kelly, Nelson, and Postle [138]. Our proof of Theorem 6.58 and Corollary 6.59, follows Bernshteyn [106], but our presentation is more detailed. Bernshteyn's proof is similar to Molloy's proof, who obtained

Bernshteyn's bound for the list chromatic number in [754]. However, the advantage of Bernshteyn's argument is that it gives a stronger result, see (6.14).

The following result is from the paper by Bonamy, Kelly, Nelson, and Postle [138]. They used similar ideas as Bernshteyn [106] and provided local versions for DP-colorings of graphs. For a graph G and $v \in V(G)$, let $\chi_G(v) = \chi(G[N_G(v) \cup \{v\}]$; note that if $G \in \text{Forb}(K_3)$ and $\delta(G) \geq 1$, then $\chi_G(v) = 2$ for all $v \in V(G)$.

Theorem 6.81 (Bonamy, Kelly, Nelson, Postle) *There is a constant Δ_0 such that for all $\Delta \geq \Delta_0$ the following statements hold: Let G be a graph with maximum degree at most Δ, and let (X, H) be a cover of G. For every vertex $v \in V(G)$, let*

$$f(v) = 72 \cdot \min\{\sqrt{\frac{\ln \omega_G(v)}{\ln d_G(v)}}, \frac{\omega_G(v) \ln \ln d_G(v)}{\ln d_G(v)}, \frac{\log_2 \chi_G(v)}{\ln d_G v)}\}.$$

If for each $v \in V(G)$,

$$|X(v)| \geq f(v) \cdot d_G(v)$$

and $d_G(v) \geq (\ln \Delta)^2$, then G is (X, H)-colorable.

The condition $\delta(G) \geq (\ln \Delta)^2$ in the above theorem is not very restrictive. Clearly, we may assume that G is connected and $\delta(G) \geq 1$. If the minimum degree of G is to small, we can duplicate G, and we add edges between each vertex of minimum degree and its duplicate (see also Proposition 6.21). Then the minimum degree of the resulting graph G' is increased by one and, for every $v \in V(G)$, we have $\omega_{G'}(v) = \omega_G(v)$ and $\chi_{G'}(v) = \chi_G(v)$ (because $\delta(G) \geq 1$). Hence as a consequence of Theorem 6.81, we obtain the following result.

Corollary 6.82 (Bonamy, Kelly, Nelson, Postle) *There is a constant Δ_0 such that for all $\Delta \geq \Delta_0$ and $r \geq 2$ the following statements holds: If $G \in \text{Forb}(K_{r+1})$ has maximum degree at most Δ and $\Delta \geq \Delta_0$, then*

$$\chi_{\text{DP}}(G) \leq \Delta \sqrt{\frac{\ln r}{\ln \Delta}},$$

and as a consequence

$$\chi_{\text{DP}}(G) \leq \tfrac{1}{c} \Delta,$$

provided that $c > 1$ and $\Delta \geq r^{(72c)^2}$

Combining Lemma 5.25 with the result about DP-coloring of K_r-free graphs we obtain that for every $k \geq 4$, if $G \in \text{Forb}(K_r)$ is a k-list-critical graph of order n, then G has at least $(k - o(k))n$ many edges and hence an average degree of at least $2k - o(k)$; this was observed by Kostochka and Stiebitz [631].

The only class of graphs, for which Conjecture 6.55 is known to be true, is the class of **nearly bipartite graphs**, that is, the class of subgraphs of $K_{1,t,t}$ with $t \geq 1$. Alon, Krivelevich, and Sudakov [54] verified Conjecture 6.55 for every subgraph H of $K_{1,t,t}$ with $c(H) = O(t)$. This bound was improved to $c(H) \leq t$ by Davies,

Kang, Pirot, and Sereni [271]. It was further improved to $c(H) \leq 4$ by Anderson, Bernshteyn, and Dhawan [62].

Inspired by Turan's theorem and other classical results, Andrásfai [63] asked about structural properties of H-free graphs caused by a high minimum degree compared to the order of the graph. Obviously, the coloring properties are of particular interest. For a graph H, let

$$\mathrm{adr}_\chi(H) = \inf\{c \in \mathbb{R}[0,1] \mid \exists k : \forall G \in \mathrm{Forb}(H) : \delta(G) \geq c|G| \Rightarrow \chi(G) \leq k\}$$

In 1974, Andrasfai, Erdős, and Sós [64] proved that for $r \geq 3$, every graph $G \in \mathrm{Forb}(K_r)$ with $\delta(G) > \frac{3r-7}{3r-4}|G|$ satisfies $\chi(G) \leq r-1$. This leads to an upper bound for $\mathrm{adr}_\chi(K_r)$. That $\mathrm{adr}_\chi(K_3) \geq \frac{1}{3}$ was proved by A. Hajnal (see the paper by Erdős and Simonovits [356]); Hajnal proved that for $0 < c < \frac{1}{3}$ graphs $G \in \mathrm{Forb}(K_3)$ with $\delta(G) \geq c|G|$ can have arbitrarily large chromatic number. Combined with a result of Thomassen [1013], this yields $\mathrm{adr}_\chi(K_3) = \frac{1}{3}$; see also the discussion at the end of Section 5.8. Nikoforov [786] as well as Goddard and Lyle [417] extended these results from K_3 to K_r and proved that for every $r \geq 3$,

$$\mathrm{adr}_\chi(K_r) = \frac{2r-5}{2r-3},$$

and, moreover, $\chi(G) \leq r+1$ for every graph $G \in \mathrm{Forb}(K_r)$ with $\delta(G) > \frac{2r-5}{2r-3}|G|$. For further results related to Andrásfai's problem, we refer the reader to the 2014 paper by Lyle [708] and to the 2020 paper by Oberkampf and Schacht [791].

Chapter 7
Coloring of Hypergraphs

Hypergraphs are discrete structures that generalize graphs in a very natural way. While in a graph every edge is incident with exactly two vertices, in a hypergraph an edge may be incident with more than two vertices. So hypergraphs model more general types of relations than graphs, and hypergraphs have proved also to be of major interest in applications to real-world problems. Most concepts and problems for graphs can be extended to hypergraphs in natural ways, and sometimes this is indispensable and leads to a better understanding of the original concept and problem. Clearly, since graphs are special cases of hypergraphs, problems for hypergraphs are at least as hard as its specialized versions to the graph case. Coloring of hypergraphs were first defined and investigated by Erdős and Hajnal in the 1960s.

7.1 Coloring Preliminaries

A **hypergraph** H is a pair of sets, $V(H)$ and $E(H)$, where $V(H)$ is finite and $E(H)$ is a subset of $2^{V(H)}$ such that $|e| \geq 2$ for all $e \in E(H)$. The set $V(H)$ is the **vertex set** of H and its elements are the **vertices** of H. The set $E(H)$ is the **edge set** of H and its elements are the **edges** of H. An edge e with $|e| \geq 3$ is called a **hyperedge**, and an edge e with $|e| = 2$ is called an **ordinary edge**. If all edges of H have the same size q, then H is said to be q**-uniform**. So a graph is a 2-uniform hypergraph, that is, a hypergraph in which each edge is ordinary. Thus graphs are special hypergraphs and we adopt the notation for graphs, see Appendix C.14. In particular, we denote the **order** of H by $|H|$, that is, the number of vertices of H. If $|H| = 0$, then H is said to be **empty** and we write $H = \varnothing$. A hypergraph is called **nontrivial** if it has at least two vertices. For a vertex v of H, let $E_H(v) = \{e \in E(H) \mid v \in e\}$ be the set of edges of H that are **incident** with v. Then $d_H(v) = |E_H(v)|$ is the **degree** of v in H.

As for graphs, there are a few basic operations to obtain one hypergraph from another. Let H be a hypergraph, let $v \in V(H)$ and $e \in E(H)$. Then the vertex deleted subhypergraph is

$$H - v = (V(H) \setminus \{v\}, E(H) \setminus E_H(v))$$

M. Stiebitz et al., *Brooks' Theorem*, Springer Monographs in Mathematics,
https://doi.org/10.1007/978-3-031-50065-7_7

and the edge deleted subhypergraph is

$$H - e = (V(H), E(H) \setminus \{e\}).$$

Furthermore, $H \div v$ denotes the hypergraph obtained from $H - v$ by adding all edges belonging to the set $E' = \{e \setminus \{v\} \mid e \in E_H(v), |e| \geq 3\}$, that is,

$$H \div v = (V(H) \setminus \{v\}, (E(H) \setminus E_H(v)) \cup E').$$

We call $\div$ the **shrinking operation**. It is notable that $H \div v$ need not be a subhypergraph of H, but if H is a graph, then $E' = \varnothing$ and so $H \div v = H - v$. Furthermore, if u, v are distinct vertices of H, then $(H \div u) \div v = (H \div v) \div u$.

A **coloring** of a hypergraph H with **color set** C is a mapping $\varphi : V(H) \to C$ that assigns to each vertex $v \in V(H)$ a color $\varphi(v) \in C$ such that $|\varphi(e)| \geq 2$ for every edge e of H. In particular, if $e = uv \in E(H)$ is an ordinary edge of H, then $\varphi(u) \neq \varphi(v)$. Hence this coloring concept for hypergraphs restricted to graphs coincides with the ordinary coloring concept for graphs. For hypergraphs, we use the same terminology as for graphs. A **k-coloring** of H is a coloring of H with a set of k colors, and H is called **k-colorable** if it has a k-coloring. We denote by $\mathcal{CO}(H,k)$ the set of all colorings of H with color set $[1,k]$. The least integer k for which a hypergraph H has a k-coloring is called the **chromatic number** of H, written $\chi(H)$.

Every coloring of a hypergraph H induces a coloring with the same color set of each of its subhypergraphs. Consequently, the chromatic number is a monotone hypergraph parameter, that is,

$$H' \subseteq H \Rightarrow \chi(H') \leq \chi(H). \tag{7.1}$$

Here $H' \subseteq H$ means that H' is a subhypergraph of H, that is, $V(H') \subseteq V(H)$ and $E(H') \subseteq E(H)$. The components of a hypergraph can be colored independently, so if $H \neq \varnothing$, then

$$\chi(H) = \max\{\chi(H') \mid H' \text{ is a component of } H\}. \tag{7.2}$$

By definition, a mapping $\varphi : V(H) \to C$ is a coloring of H with color set C if and only if, for every color $c \in C$, the preimage

$$\varphi^{-1}(c) = \{v \in V(H) \mid \varphi(v) = c\}$$

is an **independent set** of H, that is, no edge of H is a subset of $\varphi^{-1}(c)$. These preimages of colors are also referred to as **color classes**. Thus there is a one-to-one correspondence between colorings of H with a set of k colors and sequences $(I_1, I_2, \ldots, I_k)$ of disjoint independent sets of H whose union is $V(H)$. The maximum cardinality of an independent set of H is the **independence number** of H, denoted by $\alpha(H)$. Thus any color class has at most $\alpha(H)$ vertices, which implies that any coloring of H with a set of k colors satisfies $|H| \leq k\alpha(H)$, and so $|H| \leq \chi(H)\alpha(H)$. Evidently, $\chi(H) \leq |H|$ and

$$\chi(H) = |H| \Leftrightarrow \alpha(H) = 1 \Leftrightarrow V(H) \text{ is a clique of } H. \tag{7.3}$$

Recall that a vertex set $X \subseteq V(H)$ is a **clique** of H if $[X]^2 \subseteq E(H)$, and the **clique number** $\omega(H)$ is the largest cardinality of a clique of H. Consequently, we obtain that $\omega(H) \leq \chi(H)$.

The chromatic number $\chi(H)$ of a hypergraph H is the least integer k for which $V(H)$ can be partitioned into k independent sets. So $\chi(H) \leq 1$ if and only if H is edgeless, and $\chi(H) \leq 2$ if and only if H is bipartite. To decide whether a graph is bipartite, we can use König's theorem, which says that a graph is bipartite if and only if it contains no odd cycle as a subgraph. However, a good characterization for the class of bipartite hypergraphs is unknown. As proved by Lovśz [692] the class of bipartite hypergraphs forms an NP-complete hypergraph property (see Exercise 7.6).

The list coloring concept for graphs can easily be extended to hypergraphs, we only need to replace the word graph by hypergraph in the second paragraph of Section 1.3. In particular, the **list chromatic number** $\chi_\ell(H)$ of a hypergraph H is the least integer k for which H is k**-list-colorable**, that is, for every k**-assignment** L of H there exists a coloring φ of H such that $\varphi(v) \in L(v)$ for all $v \in V(H)$. Clearly, every hypergraph H satisfies $\omega(H) \leq \chi(H) \leq \chi_\ell(H)$.

7.2 Brooks' Theorem for Hypergraphs

A hypergraph H is **degree-choosable** if H has an L-coloring for every list-assignment L of H satisfying $|L(v)| \geq d_H(v)$ for all $v \in V(H)$. It follows from Theorem 1.7 and Theorem 1.8 that a graph G is not degree choosable if and only if each block of G is a brick, that is, a complete graph or an odd cycle. In the hypergraph case, one more type of forbidden blocks has to be taken into account.

A hypergraph H is said to be a **complete** q**-uniform hypergraph** if $E(H) = [V(H)]^q$. If H is a complete q-uniform hypergraph of order n, we briefly say that H is a K_n^q and we write $H = K_n^q$. So K_n^2 is a complete graph of order n and K_n^n is a hypergraph consisting of one edge of size n and its vertex set. A hypergraph H is called a **hyperbrick** if $H = K_n^2$ with $n \geq 1$, or $H = C_n$ with $n \geq 3$ odd, or $H = K_n^n$ with $n \geq 3$. In particular, every brick is a hyperbrick.

Let H be a connected hypergraph. A vertex v of H is called a **separating vertex** of H if H is the union of two nontrivial subhypergraphs having only vertex v in common. A vertex v is a separating vertex of H if and only if $H \div v$ is disconnected (see Appendix C.14, Proposition C.34).

A **block** of a hypergraph H is a maximal connected subhypergraph of H that has no separating vertex. The set of all blocks of H is denoted by $\mathfrak{B}(H)$, and for a vertex $v \in V(H)$, let $\mathfrak{B}_v(H) = \{B \in \mathfrak{B}(H) \mid v \in V(B)\}$. Note that if a connected hypergraph H is the union of two nontrivial subhypergraphs H_1 and H_2 having exactly one vertex in common, then $\mathfrak{B}(H) = \mathfrak{B}(H_1) \cup \mathfrak{B}(H_2)$ and $\mathfrak{B}(H_1) \cap \mathfrak{B}(H_2) = \varnothing$. Clearly, $\mathfrak{B}(\varnothing) = \varnothing$. It is easy to see that two distinct blocks of a hypergraph H can have at most one vertex in common; and a vertex of H is a separating vertex of H if and only if it belongs to more than one block of H.

Let H be a connected hypergraph. An edge e of H is called a **strong bridge** of H if $H - e$ has precisely $|e|$ components, that is, the vertices of e belong to different components of the edge deleted subhypergraph $H - e$. If e is a strong bridge of H, then the hypergraph $K(e) = (e, \{e\})$ is a block of H. The following result is due to Kostochka, Stiebitz and Wirth [636].

Theorem 7.1 (Kostochka, Stiebitz, and Wirth) *A connected hypergraph H is not degree-choosable if and only if each block of H is a hyperbrick.*

For the proof of Theorem 7.1, we shall investigate pairs of the form (H, L) consisting of a connected hypergraph H and a list-assignment L of H such that $|L(v)| \geq d_H(v)$ for all $v \in V(H)$, but H has no L-coloring. We then say that (H, L) is an **uncolorable hyperpair**.

A hypergraph all of whose blocks are hyperbricks is called a **Gallai hypertree** when it is connected and a **Gallai hyperforest** in general. If B is a hyperbrick, then B is regular and we define $s(B) = k$ if B is $(k-1)$-regular. For a Gallai hypertree H, let $\Lambda(H)$ denote the set of all mappings λ that assign to each block $B \in \mathfrak{B}(H)$ a set $\lambda(B)$ of $s(B) - 1$ colors such that $\lambda(B) \cap \lambda(B') = \varnothing$ whenever $B, B' \in \mathfrak{B}(H)$ are distinct blocks having a common vertex. Note that two blocks of H are disjoint or have exactly one vertex in common. For a mapping $\lambda \in \Lambda(H)$, let L_λ denote the list-assignment of H with

$$L_\lambda(v) = \bigcup_{B \in \mathfrak{B}_v(H)} \lambda(B)$$

for all $v \in V(H)$. Note that $|L_\lambda(v)| = d_H(v)$ for all $v \in V(H)$. Theorem 7.1 is an immediate consequence of the following result.

Theorem 7.2 *Let H be a connected hypergraph and let L be a list-assignment for H. Then (H, L) is an uncolorable hyperpair if and only if H is a Gallai hypertree and $L = L_\lambda$ for some $\lambda \in \Lambda(H)$.*

Proof First assume that H is a Gallai hypertree and $L = L_\lambda$ for some $\lambda \in \Lambda(H)$. Then $|L(v)| = d_H(v)$ for all $v \in V(H)$. Our aim is to show that H is not L-colorable, and hence (H, L) is an uncolorable hyperpair. The proof is by induction on the number of blocks of H. If H has only one block, then H is a hyperbrick and there is a color set C of $s(H) - 1$ colors such that $L(v) = C$ for all $v \in V(H)$. Since H is a hyperbrick, $\chi(H) = s(H)$ and hence H is not L-colorable. Now assume that H has at least two blocks. Then H is the union of two nontrivial induced subhypergraphs, say H_1 and H_2, which have exactly one vertex, say v, in common. Since H is connected, $\mathfrak{B}(H)$ is the disjoint union of $\mathfrak{B}(H_1)$ and $\mathfrak{B}(H_2)$, and hence both H_1 and H_2 are Gallai hypertrees. For $i = 1, 2$, let λ_i be the restriction of λ to $\mathfrak{B}(H_i)$ and let $L_i = L_{\lambda_i}$. Then $\lambda_i \in \Lambda(H_i)$ and the induction hypothesis implies that (H_i, L_i) is an uncolorable hyperpair. Clearly, $L = L_1 \cup L_2$ and $L_1(v) \cap L_2(v) = \varnothing$. Since H_i has no L_i-coloring for $i \in \{1, 2\}$, we then conclude that H has no L-coloring. This proves the "if" implication.

To prove the "only if" implication, assume that (H, L) is an uncolorable hyperpair. Let $h(H)$ be the sum of $|e|$ over all edges $e \in E(H)$. Note that $h(H) \geq 2|E(H)|$ and equality holds if and only if H has no hyperedge. We prove by induction on $h(H)$ that H is a Gallai hypertree and $L = L_\lambda$ for some $\lambda \in \Lambda(H)$. If $h(H) = 2|E(H)|$, then H is a connected graph and the statement follows from Corollary 1.9.

Now assume that $h(H) > 2|E(H)|$, which implies that H has a hyperedge. First we claim that every hyperedge of H is a strong bridge of H. For otherwise, there is a hyperedge e of H and a vertex $v \in e$, such that the hypergraph H' obtained from H by replacing the hyperedge e by the edge $e' = e \setminus \{v\}$ is connected. Since (H, L) is an uncolorable pair, (H', L) is an uncolorable pair, too. Since $h(H') < h(H)$, the induction hypothesis implies that H' is a Gallai hypertree and $L = L_\lambda$ for some $\lambda \in \Lambda(H')$. But then $|L(v)| = d_{H'}(v) < d_H(v)$, a contradiction. This proves that claim that every hyperedge e of H is a strong bridge of H, and hence $K(e)$ is a block of H. If H has exactly one block, then $H = K(e) = (e, \{e\})$, where e is the only edge of H. Then $|L(v)| \geq 1$ for all $v \in e$ and since H is not L-colorable, it follows that $L(v) = \{c\}$ for all $v \in e$, that is, H is a Gallai hypertree and $L = L_\lambda$ for some $\lambda \in \Lambda(H)$. Now assume that H has at least two blocks. Then H is the union of two nontrivial induced subhypergraphs, say H_1 and H_2, which have exactly one vertex, say v, in common, and $\mathfrak{B}(H)$ is the disjoint union of $\mathfrak{B}(H_1)$ and $\mathfrak{B}(H_2)$. For $i \in \{1,2\}$, let C_i be the set of colors $c \in L(v)$ such that no L-coloring φ_i of H_i satisfies $\varphi_i(v) = c$. If $L(v) \setminus (C_1 \cup C_2)$ contains a color c, then there is an L-coloring φ_i of H_i with $\varphi_i(v) = c$ (for $i \in \{1,2\}$), and so $\varphi = \varphi_1 \cup \varphi_2$ is an L-coloring of H. This contradiction shows that $L(v) = C_1 \cup C_2$.

For the hypergraph H_i (with $i \in \{1,2\}$) define the list-assignment L_i by $L_i(w) = L(w)$ if $w \in V(H_i) \setminus \{v\}$ and $L_i(v) = C_i$. By definition of C_i, H_i is not L_i-colorable. Since (H, L) is an uncolorable pair, $|L_i(w)| = |L(w)| \geq d_H(w) = d_{H_i}(w)$ for all $w \in V(H_i) \setminus \{v\}$. Since $L(v) = C_1 \cup C_2$, we obtain that

$$|C_1| + |C_2| \geq |C_1 \cup C_2| = |L(v)| \geq d_H(v) = d_{H_1}(v) + d_{H_2}(v).$$

We claim that $|C_i| = d_{H_i}(v)$ for $i \in \{1,2\}$. Otherwise, we have $|C_i| > d_{H_i}(v)$ for some $i \in \{1,2\}$, which implies that (H_i, L_i) is an uncolorable pair. Since $h(H_i) < h(H)$, the induction hypothesis then implies that H_i is a Gallai hypertree and $L_i = L_{\lambda_i}$ for some map $\lambda_i \in \Lambda(H_i)$. But the $|L_i(v)| = |C_i| = d_{H_i}(v)$, a contradiction. This proves that claim that $|C_i| = |L_i(v)| = d_{H_i}(v)$ for $i \in \{1,2\}$. Then $C_1 \cap C_2 = \varnothing$, and (H_i, L_i) is an uncolorable hyperpair for $i \in \{1,2\}$. By the induction hypothesis, H_i is a Gallai hypertree and $L_i = L_{\lambda_i}$ for some map $\lambda_i \in \Lambda(H_i)$. Then H is a Gallai hypertree and, since $C_1 \cap C_2 = \varnothing$, the mapping $\lambda = \lambda_1 \cup \lambda_2$ belongs to $\Lambda(H)$ and $L = L_\lambda$. This completes the proof. □

As a further consequence of Theorem 7.1, we obtain the following extension of Brooks' theorem to hypergraphs.

Theorem 7.3 (Kostochka, Stiebitz, and Wirth) *Let H be a connected hypergraph. Then $\chi_\ell(H) \leq \Delta(H) + 1$ and equality holds if and only if H is a hyperbrick.*

Proof By Theorem 7.2, every hyperbrick H satisfies $\chi_\ell(H) = \Delta(H)+1$. Now suppose that H is a connected hypergraph with $\chi_\ell(H) \geq k+1$ and $k = \Delta(H)$. Then there exists a k-assignment L of H such that H has no L-coloring. But then (H, L) is an uncolorable hyperpair, and it then follows from Theorem 7.2 that H is k-regular and every block of H is a hyperbrick. Since every hyperbrick is also regular, we conclude that H is a hyperbrick. □

7.3 Properties of Critical Hypergraphs

In studying the coloring problem for hypergraphs, critical hypergraphs play an important role. A hypergraph H is called **critical** or k**-critical** if $\chi(H') < \chi(H) = k$ for every proper subhypergraph H' of H. So k-critical hypergraphs are minimal k-chromatic hypergraphs with respect to the relation $\subseteq$ (to be a subhypergraph). To see why critical hypergraphs form a useful concept, let us consider a **hypergraph property** $\mathcal{H}$, that is, a class of hypergraphs closed under isomorphism. Suppose that $\mathcal{H}$ is **monotone** in the sense that $H' \subseteq H \in \mathcal{H}$ implies $H' \in \mathcal{H}$. Furthermore, consider a **hypergraph parameter** ρ defined for $\mathcal{H}$, that is, a mapping that assigns to each hypergraph of $\mathcal{H}$ a real number such that $\rho(H') = \rho(H)$ whenever H' and H are isomorphic hypergraphs of $\mathcal{H}$. If we want to bound the chromatic number for the hypergraphs of $\mathcal{H}$ from above by the parameter ρ, then we can apply the critical hypergraph method, provided that ρ is **monotone**, that is, $H' \subseteq H \in \mathcal{H}$ implies $\rho(H') \leq \rho(H)$. The proof of the following proposition is exactly the same as the simple proof for graphs (see Proposition 1.27).

Proposition 7.4 (The Critical Hypergraph Method) *Let $\mathcal{H}$ be a monotone hypergraph property and let ρ be a monotone hypergraph parameter defined for $\mathcal{H}$. Then the following statements hold:*

(a) *For every hypergraph $H \in \mathcal{H}$ there exists a critical hypergraph $H' \in \mathcal{H}$ such that $H' \subseteq H$ and $\chi(H') = \chi(H)$.*
(b) *If $\chi(H') \leq \rho(H')$ for every critical hypergraph $H' \in \mathcal{H}$, then $\chi(H) \leq \rho(H)$ for every hypergraph $H \in \mathcal{H}$.*

The chromatic number is a monotone hypergraph parameter and it is easy to check that if we delete a vertex or an edge from a hypergraph, then the chromatic number decreases by at most one. So if H is a hypergraph and $t \in V(H) \cup E(H)$, then

$$\chi(H) - 1 \leq \chi(H - \{t\}) \leq \chi(H). \tag{7.4}$$

As a consequence we obtain the following well known result saying that the class of $(k-1)$-colorable hypergraphs can be characterized in terms of forbidden k-critical subhypergraphs.

Proposition 7.5 *Let H be a hypergraph and let $k \in \mathbb{N}$. Then $\chi(H) \leq k-1$ if and only if there is no k-critical hypergraph H' with $H' \subseteq H$.*

Proof If H contains a k-critical hypergraph H' as a subgraph, then $\chi(H) \geq \chi(H') = k$. Conversely, if $\chi(H) \geq k$, then it follows from (7.4) that there is a subhypergraph G of H with $\chi(G) = k$. By Lemma 7.4(a), G and hence H contains a k-critical subhypergraph H'. □

Some basic properties of critical hypergraphs are listed in the following proposition. A hypergraph H is called **simple** if no edge of H is contained in another edge of H.

Proposition 7.6 *Let H be a k-critical hypergraph, where $k \geq 1$. Then the following statements hold:*

(a) *$|H| \geq k$ and equality holds if and only if $H = K_k^2$.*
(b) *If $v \in V(H)$ and $\varphi \in \mathcal{CO}(H - v, k-1)$, then for every color $c \in [1, k-1]$ there is an edge $e_c \in E_H(v)$ such that $\varphi(e_c \setminus \{v\}) = \{c\}$. As a consequence, v is contained in $k-1$ edges having pairwise only v in common.*
(c) *H is connected and $\delta(H) \geq k-1$.*
(d) *If $e \in E(H)$ and $\varphi \in \mathcal{CO}(H - e, k-1)$, then $|\varphi(e)| = 1$. As a consequence, H is a simple hypergraph.*

Proof Statement (a) follows from (7.3) and the fact that K_k^2 is k-critical. Statement (b) is evident. If such an edge e_c does not exists, we can extend φ to a $(k-1)$-coloring of G by assigning color c to v, which is impossible. Statement (c) is a consequence of (7.2) and (b). Statement (d) is obvious. □

Let H be a nonempty hypergraph. On the one hand, it seems very natural to define the degree of a vertex v in H as the number of edges incident with v. On the other hand, Proposition 7.6(b) give rise to another definition of vertex degree. Let $d_H^*(v)$ denote the maximum number of edges of H having only vertex v in common, we call this number the **star degree** of v in H. As usual, $\delta^*(H) = \min_{v \in V(H)} d_H^*(v)$ is the **minimum star degree** of H and $\Delta^*(H) = \max_{v \in V(H)} d_H^*(v)$ is the **maximum star degree** of H. The star degree is sometimes also called the **linear degree**. A hypergraph is called **linear** if any two edges of H have at most one vertex in common. Every vertex v of H satisfies $d_H^*(v) \leq d_H(v)$ and equality holds if H is linear. If H is critical, then Proposition 7.6(b) implies that $\chi(H) - 1 \leq \delta^*(H)$, which yields $\chi(H) \leq \Delta^*(H) + 1$. Since Δ^* is a monotone hypergraph parameter, it follows from Proposition 7.4 that every nonempty hypergraph H satisfies $\chi(H) \leq \Delta^*(H) + 1$. In 1975 Jones [541] proposed the question whether there exists a Brooks' type result for the maximum star degree, that is, whether there exists a good characterization of the class of connected hypergraphs H satisfying $\chi(H) = \Delta^*(H) + 1$. The question is still unsolved.

If we want to check whether a given hypergraph is critical, it suffices to investigate all edge deleted subhypergraphs. This follows from the monotonicity of the chromatic number, see (7.1), and the fact that if a hypergraph contains an edge and has minimum degree $\delta(H) \geq 1$, then any proper subhypergraph is a subhypergraph of an edge deleted subhypergraph.

Proposition 7.7 *Let H be a hypergraph with $\delta(H) \geq 1$ and let $k \geq 2$ be an integer. Then H is k-critical if and only if $\chi(H-e) < k \leq \chi(H)$ for every edge $e \in E(H)$.*

For integers $k,n \in \mathbb{N}_0$, let $\mathrm{Crih}(k)$ denote the class of k-critical hypergraphs and let

$$\mathrm{Crih}(k,n) = \{H \in \mathrm{Crih}(k) \mid |H| = n\}.$$

Since a hypergraph H satisfies $\chi(H) = 0$ if and only if $H = \varnothing$, and $\chi(H) \leq 1$ if and only if $E(H) = \varnothing$, it follows from Proposition 7.5 that

$$\mathrm{Crih}(0) = \{\varnothing\}, \mathrm{Crih}(1) = \{K_1^2\} \text{ and } \mathrm{Crih}(2) = \{K_n^n \mid n \geq 2\}.$$

While König's characterization of bipartite graphs implies that the odd cycles are the only 3-critical graphs, investigations of the class of 3-critical hypergraphs have received significant attention in the literature. For any fixed $k \geq 3$, a good characterization of the class $\mathrm{Crih}(k)$ is unlikely.

A hypergraph is **vertex-critical** or **k-vertex-critical** if $\chi(H') < \chi(H) = k$ for every proper induced subhypergraph H' of H. While k-critical hypergraphs are k-chromatic hypergraphs that are minimal with respect to the relation 'to be a subhypergraph', k-vertex-critical hypergraphs are k-chromatic hypergraphs that are minimal with respect to the relation 'to be an induced subhypergraph'. Clearly, every critical hypergraph is vertex-critical, but not conversely. Examples of vertex-critical graphs that are not critical were given by Dirac. For $k \leq 3$, however, a graph is k-critical if and only if it is k-vertex-critical. It follows from (7.4) that a hypergraph H is k-vertex-critical if and only if $\chi(H-v) < k \leq \chi(H)$ for every vertex $v \in V(H)$. Results about critical hypergraphs can be often transformed into results about the larger class of vertex-critical hypergraphs.

Proposition 7.8 *Every k-vertex-critical hypergraph H contains a k-critical subhypergraph and any such subhypergraph has the same vertex set as H.*

Proof That H contains a k-critical subhypergraph follows from Proposition 7.4(a). Now let H' be such a subhypergraph. If a vertex v of H does not belong to H', then $H' \subseteq H - v$ and, since H is k-vertex-critical, we obtain that $\chi(H') \leq \chi(H-v) < k$, which is impossible. □

Let H be an arbitrary hypergraph. If L is a list-assignment of H such that every subhypergraph of H, except H, has an L-coloring, then H is said to be **L-critical**. The hypergraph H is called **k-list-critical** if H is L-critical for some $(k-1)$-assignment L of H. Notice that every k-critical hypergraph is L-critical for the constant list-assignment $L \equiv [1, k-1]$ and so H is a k-list-critical hypergraph. Furthermore, it is easy to show that $\chi_\ell(H') < \chi_\ell(H) = k$ for every proper subhypergraph H' of H if and only if H is k-list-critical and no proper subhypergraph of H is (see Proposition 5.3). Recall that if H is a hypergraph and $X \subseteq V(H)$, then $H[X]$ is the subhypergraph of H induced by X, that is, $V(H[X]) = X$ and $E(H[X]) = \{e \in E(H) \mid e \subseteq X\}$.

Theorem 7.9 *Let H be hypergraph and let L be a list-assignment for H such that H is L-critical. Furthermore, let $U = \{v \in V(H) \mid d_H(v) > |L(v)|\}$ and $W = V(H) \setminus U$. Then the following statements hold:*

(a) *$H[W]$ is a Gallai hyperforest (possibly empty) and $d_H(v) = |L(v)|$ for all $v \in W$.*

(b) *If L is a $(k-1)$-assignment with $k \geq 1$, then $U \neq \varnothing$, or H is a hyperbrick. Furthermore, if $H[W]$ contains a K_k, then $H = K_k$.*

Proof For the proof of (a), let F be a component of $H[W]$. Then $H' = H - V(F)$ is a proper subhypergraph of H. Since H is L-critical, there is an L-coloring φ of H'. For each edge $e \in E(H)$ with $e \setminus V(F) \neq \varnothing$, choose a vertex $v(e) \in e \setminus V(F)$. For the connected hypergraph F, define a list-assignment L' by

$$L'(u) = L(u) \setminus \{\varphi(v(e)) \mid e \in E_H(u), e \setminus V(F) \neq \varnothing\}$$

for each vertex $u \in V(F)$. For $u \in V(F)$, we have $|L(u)| \geq d_H(u)$ and hence $|L'(u)| \geq d_F(u)$. If φ' is an L'-coloring of F, then $\varphi' \cup \varphi$ is an L-coloring of H. Therefore, F is not L'-colorable and so (F, L') is an uncolorable hyperpair. By Theorem 7.1 it then follows that F is a Gallai hypertree and $|L'(u)| = d_F(u)$ for all $u \in V(F)$ implying that $|L(u)| = d_H(u)$ for all $u \in V(F)$. This proves (a).

To prove (b), assume that $|L(v)| = k - 1$ for all $v \in V(H)$. Then we have $U = \{v \in V(H) \mid d_H(v) \geq k\}$ and $W = \{v \in V(H) \mid d_H(v) = k - 1\}$ (by (a)). Suppose that $U = \varnothing$. Then $H = H[W]$ is connected and hence a Gallai hypertree (by (a)) that is regular. Since each block of a Gallai hypertree is regular, too, this implies that H is a block and hence a hyperbrick. Now suppose that $H[W]$ contains a K_k. Since H is connected, $H[W]$ is a Gallai hyperforest and each vertex of W has degree $k-1$ in H, this implies that $H = H[W] = K_k$. This proves (b). □

Let H be an L-critical hypergraph for a list-assignment L of H. Theorem 7.9(a) characterizes the structure of the subhypergraph induced by the set W of vertices having low degree in H with respect to L, that is, $W = \{v \in V(H) \mid d_H(v) = |L(v)|\}$. However, we may also consider the hypergraph $H\langle W\rangle$ obtained by **shrinking** H to W, that is, $V(H\langle W\rangle) = W$ and

$$E(H\langle W\rangle) = \{e \cap W \mid e \in E(H) \text{ and } |e \cap W| \geq 2\}.$$

Note that $H\langle W\rangle$ need not be a subhypergraph of H, but it contains $H[W]$ as a subhypergraph. As proved in [633], $H\langle W\rangle$ is also a Gallai hyperforest (see also Exercise 7.14)). For $U \subseteq V(H)$ we define $H \div U = H\langle V(H) \setminus U\rangle$, in accordance with the earlier definition of $\div$.

7.4 Constructions of Critical Hypergraphs

In Section 4.1 we discussed the two most important constructions for critical graphs, namely the Dirac join and the Hajós join. Both constructions can also be applied to hypergraphs. For critical hypergraphs there is a specific construction, the so-called enlarging operation. A further construction is the splitting operation that generalizes the Gallai join for graphs.

Let H_1 and H_2 be two vertex disjoint hypergraphs, and let H be the hypergraph obtained from the union $H_1 \cup H_2$ by adding all ordinary edges between H_1 and H_2, that is, $V(H) = V(H_1) \cup V(H_2)$ and

$$E(H) = E(H_1) \cup E(H_2) \cup \{uv \mid u \in V(H_1), v \in V(H_2)\}.$$

We call H the **Dirac join** of H_1 and H_2 and write $H = H_1 \boxplus H_2$, see Figure 7.1. The proof of the next theorem is the same as the proof of Theorem 4.1.

Theorem 7.10 (DIRAC CONSTRUCTION) *Let $H = H_1 \boxplus H_2$ be the Dirac join of two disjoint hypergraphs H_1 and H_2. Then $\chi(H) = \chi(H_1) + \chi(H_2)$ and, moreover, H is critical if and only if both H_1 and H_2 are critical.*

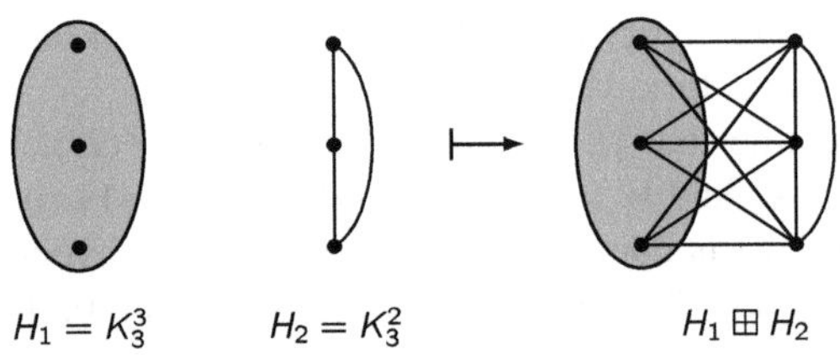

Fig. 7.1 The Dirac join $K_3^3 \boxplus K_3^2$ belongs to Crih(5).

Let H_1 and H_2 be two vertex disjoint hypergraphs. For $i \in \{1,2\}$, let v_i be a vertex of H_i, and let $e_i \in E_{H_i}(v_i)$ be an edge. Let H be the hypergraph obtained from $H_1 - e_1$ and $H_2 - e_2$ by identifying the vertices v_1 and v_2 to a new vertex v^* and by adding the new edge $e^* \in E(H)$ either with $e^* = (e_1 \cup e_2) \setminus \{v_1, v_2\}$ or with $e^* = ((e_1 \cup e_2 \setminus \{v_1, v_2\}) \cup \{v^*\}$. We call H the **Hajós join** of H_1 and H_2 and write $H = (H_1, v_1, e_1)\nabla(H_2, v_2, e_2)$ or, briefly, $H = H_1 \nabla H_2$. Figure 7.2 shows the two possible Hajós joins of type $K_4 \nabla K_4$.

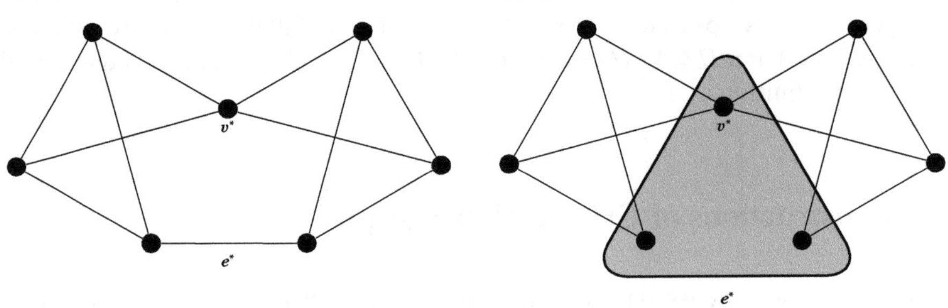

Fig. 7.2 The two possible Hajós joins of two graphs K_4.

Theorem 7.11 (Hajós Construction) *Let H_1 and H_2 be disjoint hypergraphs both with at least one edge, and let $H = H_1 \nabla H_2$ be the Hajós join of these two hypergraphs. Then the following statements hold:*

(a) $\chi(H) \geq \min\{\chi(H_1), \chi(H_2)\}$.
(b) *If $\chi(H_1) = \chi(H_2) = k$ and $k \geq 3$, then $\chi(H) = k$.*
(c) *If both H_1 and H_2 are k-critical and $k \geq 3$, then H is k-critical.*
(d) *If H is k-critical and $k \geq 4$, then both H_1 and H_2 are k-critical.*

Proof Suppose that $H = (H_1, v_1, e_1)\nabla(H_2, v_2, e_2)$. Let v^* denote the vertex of H obtained by identifying v_1 and v_2, and let e^* denote the new edge. To simplify the proof, we may assume that $v^* = v_1 = v_2$.

For the proof of (a), assume that $\chi(H) = k$. Then there is a k-coloring φ of H. Since $e^* \in E(H)$, we have $|\varphi(e^*)| \geq 2$. This implies that $|\varphi(e_1)| \geq 2$ or $|\varphi(e_2)| \geq 2$ and hence φ yields a k-coloring of H_1 or H_2, which proves (a).

For the proof of (b), assume that $\chi(H_1) = \chi(H_2) = k$ and $k \geq 3$. Then, for $i \in \{1,2\}$, there is a k-coloring φ_i of H_i. Since $v_i \in e_i \in E(H_i)$, there is a vertex $w_i \in e_i \setminus \{v_i\}$ such that $\varphi_i(v_i) \neq \varphi_i(w_i)$. As $k \geq 3$, we may permute the colors so that $\varphi_1(v_1) = \varphi_2(v_2)$ and $\varphi_1(w_1) \neq \varphi_2(w_2)$. Then $\varphi_1 \cup \varphi_2$ is a k-coloring of H and so $\chi(H) \leq k$. By (a) it follows that $\chi(H) \geq k$. Hence $\chi(H) = k$ and (b) is proved.

For the proof of (c), assume that both H_1 and H_2 are k-critical. To show that H is k-critical we apply Proposition 7.7. Clearly, we have $\delta(H_i) \geq k-1 \geq 2$ for $i \in \{1,2\}$, and hence $\delta(H) \geq 1$. By (b) it follows that $\chi(H) = k$. It remains to show that $\chi(H-e) \leq k-1$ for every edge $e \in E(H)$. First suppose that $e = e^*$. Since H_i is k-critical, there is a $(k-1)$-coloring φ_i of $H_i - e_i$ for $i \in \{1,2\}$. As $\chi(H_i) = k$, we obtain that $|\varphi_i(e_i)| = 1$. By permuting colors if necessary we may obtain that $\varphi_1(e_1) = \varphi_2(e_2)$. Then $\varphi_1 \cup \varphi_2$ is a $(k-1)$-coloring of $H - e^*$. If $e \neq e^*$, then e belongs to $H_i - e_i$ for some $i \in \{1,2\}$, say $e \in E(H_1) \setminus \{e_1\}$. Since H_1 is k-critical, $H_1 - e$ admits a $(k-1)$-coloring φ_1 such that $|\varphi_1(e_1)| \geq 2$ (since e_1 is an edge of $H_1 - e$). H_2 being k-critical, so there is a $(k-1)$-coloring φ_2 of $H - e_2$ such that $|\varphi_2(e_2)| = 1$. By permuting colors if necessary we may obtain that $\varphi_1(v_1) = \varphi_2(v_2)$. Then $\varphi = \varphi_1 \cup \varphi_2$ satisfies $|\varphi(e^*)| \geq 2$, and so φ is a $(k-1)$-coloring of $H - e$. Thus (c) is proved.

For the proof of (d), assume that H is k-critical and $k \geq 4$. To show that both H_1 and H_2 are k-critical, we use Proposition 7.7. We have $\delta(H) \geq k-1 \geq 3$ and hence $\delta(H_i) \geq 1$ for $i \in \{1,2\}$. Next we claim that $\chi(H_1) \geq k$. Suppose this is false. Then there is a $(k-1)$-coloring φ_1 of H_1. As $e_1 \in E(H_1)$, we obtain that $|\varphi_1(e_1)| \geq 2$, say $\varphi_1(v_1) = \alpha$ and $\varphi(u) = \beta \neq \alpha$ for a vertex $u \in e_1 \setminus \{v_1\}$. Since H is k-critical and $e^* \in E(H)$, there is a $(k-1)$-coloring φ of $H - e^*$, which leads to a $(k-1)$-coloring φ_2 of $H_2 - e_2$. As $k \geq 4$, we may permute the colors such that $\varphi_2(v_2) = \alpha$ and $\varphi_2(u') \neq \beta$ for at least one vertex $u' \in e_2 \setminus \{v_2\}$. But then $\varphi = \varphi_1 \cup \varphi_2$ is a $(k-1)$-coloring of H, which is impossible. This proves the claim that $\chi(H_1) \geq k$. Similarly, one can show that $\chi(H_2) \geq k$. It remains to show that $\chi(H_i - e) \leq k-1$ for every edge $e \in E(H_i)$. By symmetry it suffices to show this for $i = 1$. If $e = e_1$, then there is a $(k-1)$-coloring φ of $H - e^*$, which yields a $(k-1)$-coloring of $H_1 - e_1$. If $e \neq e_1$, then there is a $(k-1)$-coloring φ of $H - e$. Since e^* belongs to $H - e$, we have $|\varphi(e^*)| \geq 2$. Since $\chi(H_2) \geq k$, we have $|\varphi(e_2)| = 1$, which leads to $|\varphi(e_1)| \geq 2$. Consequently, φ induces a $(k-1)$-coloring of $H_1 - e$. This proves (d). □

A 7-cycle is the Hajós join of two 4-cycles. This shows that the condition $k \geq 4$ in (d) is needed.

The next construction consists in replacing an edge of a hypergraph by a larger edge. Let H be a hypergraph, let e' be an edge of H, and let e^+ be a set with $e' \subseteq e^+ \subseteq V(H)$. Let $H' = (H \setminus e') + e^+$ be the hypergraph obtained from H by replacing the edge e' by e^+. We then write $H' = H(e' \prec e^+)$ and say that H' is obtained from H by **enlarging** the edge e' to e^+. Note that if H is a simple hypergraph and $H' = H(e' \prec e^+)$, then either $e^+ = e'$ and $H' = H$, or $e^+ \notin E(H)$. The next theorem says under which condition an edge of a critical hypergraph can be enlarged so that the enlarged hypergraph remains critical and has the same chromatic number.

Theorem 7.12 (Enlarging Operation) *Let H be a k-critical hypergraph with $k \geq 3$, and let $e' \in E(H)$ be an edge of H. Suppose that e^+ is a vertex set of H such that $e' \subseteq e^+$ and $|\varphi(e^+)| = 1$ for any coloring $\varphi \in \mathcal{CO}(H - e', k-1)$. Then the hypergraph $H' = H(e' \prec e^+)$ is k-critical.*

Proof Obviously we have $\delta(H') \geq 1$ and $\chi(H') \geq k$ (because of the assumption about e^+). Now let $e \in E(H')$. If $e = e^+$, then $H' - e = H - e'$ and so $\chi(H' - e) = \chi(H - e') \leq k - 1$. If $e \neq e^+$, then $e \in E(H) \setminus \{e'\}$, and there is a $(k-1)$-coloring of $H - e$, which is also a $(k-1)$-coloring of $H' - e$. This proves that $\chi(H' - e) \leq k - 1$ for all $e \in E(H')$. Then it follows from Proposition 7.7 that H' is k-critical. □

An important operation for analyzing critical hypergraphs is the following splitting operation. Let H_1 and H_2 be two disjoint hypergraphs, let $\tilde{e} \in E(H_1)$ and $\tilde{v} \in V(H_2)$. Furthermore, let $s : E_{H_2}(\tilde{v}) \to 2^{\tilde{e}}$ be a mapping such that $s(e) \neq \varnothing$ for all $e \in E_{H_2}(\tilde{v})$ and

$$\bigcup_{e \in E_{H_2}(\tilde{v})} s(e) = \tilde{e}.$$

Define a hypergraph H whose vertex set is $V(H) = V(H_1) \cup V(H_2 - \tilde{v})$ and whose edge set is

$$E(H) = (E(H_1) \setminus \{\tilde{e}\}) \cup (E(H_2) \setminus E_{H_2}(v)) \cup \{(e \setminus \{\tilde{v}\}) \cup s(e) \mid e \in E_{H_2}(\tilde{v})\}.$$

We then say that the hypergraph H is obtained from H_1 and H_2 by **splitting** the vertex $\tilde{v}$ into the edge $\tilde{e}$, and we write $H = S(H_1, \tilde{e}, H_2, \tilde{v}, s)$. If $|s(e)| = 1$ for all $e \in E_{H_2}(\tilde{v})$, then s is said to be a **simple splitting**, see Figure 7.5 for an example.

Theorem 7.13 (Splitting Operation) *Let H_1 and H_2 be two disjoint hypergraphs that are both $(k+1)$-critical with $k \geq 2$, let $\tilde{e} \in E(H_1)$ be an arbitrary edge, and let $\tilde{v} \in V(H_2)$ be an arbitrary vertex. Furthermore, let $H = S(H_1, \tilde{e}, H_2, \tilde{v}, s)$ and let $H_2' = H[(V(H_2) \setminus \{\tilde{v}\}) \cup \tilde{e}]$. Suppose that for every coloring $\varphi \in \mathcal{CO}(H[\tilde{e}], k)$ with $|\varphi(\tilde{e})| \geq 2$ there is a coloring $\varphi' \in \mathcal{CO}(H_2', k)$ such that $\varphi'|_{\tilde{e}} = \varphi$. Then, H is a $(k+1)$-critical hypergraph.*

Proof Since $H_i \in \text{Crih}(k+1)$, $\delta(H_i) \geq k$ and H_i is a simple hypergraph for $i \in \{1,2\}$. Hence $\delta(H) \geq k$ and $\tilde{e}$ is an independent set of H. To show that H is $(k+1)$-critical, we use Proposition 7.7.

First we claim that $\chi(H) \geq k+1$. Otherwise, there is a coloring $\varphi \in \mathcal{CO}(H,k)$. If $\varphi(\tilde{e}) = \{c\}$ for some color $c \in [1,k]$, then the map φ' with $\varphi'(v) = \varphi(v)$ for all $v \in V(H_2 - \tilde{v})$ and $\varphi(\tilde{v}) = c$ is a k-coloring of H_2, which is impossible. If $|\varphi(\tilde{e})| \geq 2$, then the restriction φ' of φ to $V(H_1)$ is a k-coloring of H_1, which is impossible. This proves the claim that $\chi(H) \geq k+1$.

Now let e' be an arbitrary edge of H. Our aim is to show that $H - e'$ is k-colorable. First assume that e' is an edge of $H_1 - \tilde{e}$. Since H_1 is $(k+1)$-critical, there is a coloring $\varphi_1 \in \mathcal{CO}(H_1 - e', k)$. Since $\tilde{e}$ is an edge of $H_1 - e'$, we have $|\varphi_1(\tilde{e})| \geq 2$. Then, by the assumption of the theorem, there is a coloring $\varphi' \in \mathcal{CO}(H_2', k)$ such that $\varphi'|_{\tilde{e}} = \varphi_1$. Then $\varphi' \cup \varphi_1$ is a k-coloring of $H - e'$. It remains to consider the case that e' is an edge of H_2'. This implies that e' is an edge of H_2 or $e' = (e - \{\tilde{v}\}) \cup s(e)$ for some edge $e \in E_{H_2}(v^*)$. Since H_2 is $(k+1)$-critical, in both cases we conclude that there is a coloring $\varphi_2 \in \mathcal{CO}(H_2' - e', k)$ such that $|\varphi_2(\tilde{e})| = 1$. Since H_1 is $(k+1)$-critical, there is a coloring $\varphi_1 \in \mathcal{CO}(H_1 - \tilde{e}, k)$. Clearly, $|\varphi_1(\tilde{e})| = 1$ and by permuting colors if necessary we may assume that $\varphi_1(\tilde{e}) = \varphi_2(\tilde{e})$. Then $\varphi_1 \cup \varphi_2$ is a k-coloring of $H - e'$.

Summarizing, we have $\chi(H) \geq k+1$ and $\chi(H - e') \leq k$ for all $e' \in E(H)$. Hence H is $(k+1)$-critical. □

Let $H_1, H_2, \tilde{e}, \tilde{v}, H$ and H_2' be as in Theorem 7.13. Since H_1 is critical, H_1 is a simple hypergraph (by Proposition 7.6(d)). Consequently, $\tilde{e}$ is an independent set of both hypergraphs H and H_2', and so $H[\tilde{e}] = H_2'[\tilde{e}]$. We then say that H_2' is obtained from H_2 by **splitting $\tilde{v}$ into the independent set** $\tilde{e}$, and write $H_2' = S(H_2, \tilde{v}, \tilde{e}, s)$.

Let $H \in \text{Crih}(k+1)$ with $k \geq 2$, and let v be a vertex of H. We call v a **universal vertex** of H, if for every hypergraph $H' = S(H, v, X, s)$, where X is an independent set of H', and every coloring $\varphi' \in \mathcal{CO}(H'[X], k)$ with $|\varphi'(X)| \geq 2$ there is a coloring $\varphi \in \mathcal{CO}(H', k)$ with $\varphi|_X = \varphi'$.

Theorem 7.13 then says that if H_1 and H_2 are disjoint $(k+1)$-critical hypergraphs, and $\tilde{v}$ is a universal vertex of H_2, then any hypergraph H obtained from H_1 and H_2 by splitting $\tilde{v}$ into an edge $\tilde{e}$ of H_2 is a $(k+1)$-critical hypergraph, too. However, a good characterization of universal vertices in critical hypergraphs or graphs seems not available. To obtain some partial results about universal vertices in critical hypergraphs, the following lemma is useful. This lemma is a consequence of the fact that complements of bipartite graphs are perfect.

Lemma 7.14 *Let G be a graph and let $k \geq 1$ be an integer. Suppose that (U, W, F') is an edge cut of G such that $|F'| \leq k$ and both sets U and W are cliques of G with $|U| = |W| = k$. If $\chi(G) \geq k+1$, then $|F'| = k$ and $F' \subseteq E_G(v)$ for some vertex v of G*

Proof As G is the complement of a bipartite graph, G is perfect and hence $\omega(G) = \chi(G) \geq k+1$. Consequently, G contains a clique Z with $|Z| = k+1$. Let $s = |U \cap Z|$ and $t = |W \cap Z|$. Clearly, $s + t = k+1$ and, since $|U| = |W| = k$, we have $s, t \geq 1$. Let $E' = \{uw \mid u \in U \cap Z, w \in W \cap Z\}$. Clearly, $E' \subseteq F'$ and, since Z is a clique,

$|E'| = st = s(k+1-s)$. The function $g(s) = t(k+1-s)$ is strictly concave in the interval $I = \{x \in \mathbb{R} \mid 1 \leq x \leq k\}$ as $g''(s) = -2$. Since $g(1) = g(k) = k$, this implies that $g(s) > k$ for $1 < s < k$. Since $g(s) = |E'| \leq |F'| \leq k$, this implies that $s = 1$ or $s = k$. In both cases, we obtain that $E' = F' \subseteq E_G(v)$ for a vertex v of G and $|E'| = |F'| = k$. □

Let $H \in \text{Crih}(k+1)$ with $k \geq 0$. A vertex v of H is called a **low vertex** if $d_H(v) = k$, otherwise v is called a **high vertex** of H.

Proposition 7.15 *Each low vertex of a hypergraph $H \in \text{Crih}(k+1)$ with $k \geq 2$ is a universal vertex of H.*

Proof Let $H \in \text{Crih}(k+1)$ and let v be a low vertex of H. Let $H' = S(H, v, X, s)$, where X is an independent set of H', and let $\varphi' \in CO(H'[X], k)$ such that $|\varphi'(X)| \geq 2$. Since $H \in \text{Crih}(k+1)$ and v is a low vertex of H, it follows from Proposition 7.6(b) that if $\varphi \in CO(H - v, k)$ then for every color $c \in [1, k]$ there is an edge $e_c \in E_H(v)$ such that $\varphi(e_c \setminus \{v\}) = \{c\}$. Then $E_H(v) = \{e_c \mid c \in [1, k]\}$. In H' the edge e_c is replaced by the edge $(e_c \setminus \{v\}) \cup s(e_c)$ where $s(e_c) \subseteq X$. Since $|\varphi'(X)| \geq 2$, we conclude from Lemma 7.14 that we can permute the colors in φ such that $\varphi \cup \varphi' \in CO(H', k)$. This shows that v is a universal vertex of H. □

Let $KC_{n,p} = K_n \boxplus C_{2p+1}$ be the Dirac sum of a complete graph and an odd cycle, where $n \geq 1$ and $p \geq 1$. Since $K_n \in \text{Crit}(n)$ and $C_{2p+1} \in \text{Crit}(3)$, Theorem 7.10 implies that $KC_{n,p} \in \text{Crit}(n+3)$. Recall that $KC_{1,p}$ is an odd wheel; and $KC_{n,p}$ is an odd n-wheel, see Figure 7.3.

Proposition 7.16 *The graph $G = KC_{n,p}$ with $n \geq 0$ and $p \geq 1$ is $(n+3)$-critical and the only bad $(n+2)$-assignment for G is the constant list-assignment.*

Proof Clearly, G is $(n+3)$-critical. Now let L be a bad $(n+2)$-assignment for G. We prove by induction on n that L is the constant list-assignment, that is, $L(v) = C$ for all $v \in V(G)$ where C is a set of $n+2$-colors. If $n = 0$, then G is an odd cycle and the statement is evident (see also Corollary 1.9). Now let $n \geq 1$. Then G has a vertex v that is joined to every other vertex of G by an edge and $G - v = KC_{n-1,p}$. Then for an arbitrary color c from the set $C = L(v)$, let L_c be the list-assignment of $G - v$ defined by $L_c(u) = L(u) \setminus \{c\}$ for all vertices u of $G - v$. Clearly, $|L_c(u)| \geq n+1$ for all vertices u of $G - v$, and L_c is a bad list-assignment of $G - v$, since an L_c-coloring of $G - v$ can be extended to an L-coloring of G by assigning color c to v. By the induction hypothesis, $L_c(u) = C_c$ for all vertices $u \in V(G - v)$, where C_c is a set of $n+1$-colors. Since c was an arbitrary color from the set $C = L(v)$, we then conclude that $L(u) = C$ for all $u \in V(G)$. □

Proposition 7.17 *Each vertex of the critical graph $KC_{n,p}$ is universal $(n, p \geq 1)$.*

Proof Let $H = KC_{n,p}$ and let v be an arbitrary vertex of H. Our aim is to show that v is a universal vertex of H. If v is a low vertex of H, this follows from Proposition 7.15. So assume that v is a high vertex of H. Then v is a vertex of the complete graph K_n of H, $p \geq 2$ and $H - v = KC_{n-1,p}$. Note that H is $(n+3)$-critical, so let $k = n+2$. Let $H' = S(H,v,X,s)$, where X is an independent set of H', and let $\varphi' \in CO(H'[X],k)$ with $|\varphi'(X)| \geq 2$. If $e = vu$ is an edge of H incident with v, then in H' the edge e is replaced by the edge $e' = \{u\} \cup s(e)$, where $s(e) \subseteq X$. Then for each edges $e \in E_H(v)$ we can chose a vertex $v(e) \in s(e)$ such that $C = \{\varphi(v(e)) \mid e \in E_H(v)\}$ is a set of at least two colors. Let L be the list-assignment of $H - v = KC_{n-1,p}$ where $L(u) = [1,k] \setminus \{\varphi'(v(e))\}$ for $e = uv$. Then L is a $(n+1)$-assignment of $H - v = KC_{n-1,p}$, but not a constant list-assignment. Then it follows from Proposition 7.16 that there is a L-coloring φ of $H - v$. Then $\varphi \cup \varphi' \in CO(H',k)$ and we are done. □

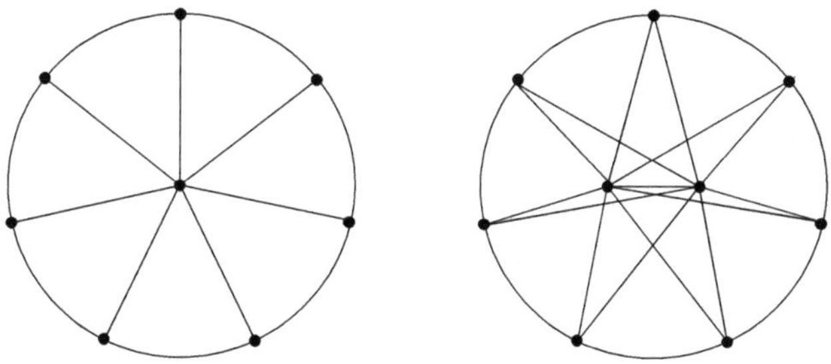

Fig. 7.3 The odd wheel $KC_{1,3}$ and the odd 2-wheel $KC_{2,3}$.

The graphs from the family $KC_{n,p}$ enable us to construct from any $(k+1)$-critical hypergraph a $(k+1)$-critical graph, by repeated applying the splitting operation with copies of $KC_{k-2,p} \in \text{Crit}(k+1)$, provided that $k \geq 3$. Note that if $H = S(H_1,\tilde{e},H_2,\tilde{v},s)$ and s is a simple splitting, then $d_{H_2}(\tilde{v}) \geq |\tilde{e}|$. One popular family of critical graphs obtained from critical hypergraphs was presented by Toft [1019]. Since $K_n^n \in \text{Crih}(2)$ for all $n \geq 2$, $H_n = K_n^n \boxplus K_n^n \in \text{Crih}(4)$ for all $n \geq 2$ (by Theorem 7.10). If we start with the 4-critical hypergraph H_{2p+1} with $p \geq 2$ having the two hyperedges e_1 and e_2 and apply the splitting operation with two copies of the odd wheels $KC_{1,p}$ and its unique high vertex v, that is, we first construct $H' = S(H_{2p+1},e_1,KC_{1,p},v,s)$ with a simple splitting s and then $TG_{2p+1} = S(H',e_2,KC_{1,p},v,s')$ with a simple splitting s', then the resulting graph TG_{2p+1} is a 4-critical graph of order $n = 8p+4$ and with $m = (2p+1)^2 + 8p + 4 = \frac{1}{16}n^2 + n$ edges, i.e., G_{2p+1} has many edges. The constant $\frac{1}{16}$ has not been improved. The Toft graph TG_5 is shown in Figure 4.3.

In Section 7.7 we want to apply the reduction technique with another family of critical graphs. Let $k \geq 4$ be a given integer. For $d \geq 1$, let $TK_{k,d}$ be the graph obtained from d vertex disjoint copies of K_{k-1}, say $K^1, K^2, \ldots, K^d$, and an additional vertex v, first by adding $d-1$ edges $u_1w_2, u_2w_3, \ldots, u_{d-1}w_d$, where $u_i \in V(K^i)$ and $w_{i+1} \in V(K_{i+1})$ for $i \in [1,d-1]$ and $u_i \neq w_i$ for $i \in [2,d-1]$, and then by joining v to the remaining vertices of degree $k-2$. Figure 7.4 shows the graph $TK_{5,3}$. Note that $TK_{k,1} = K_k$ and $TK_{k,d+1} = TK_{k,d} \nabla K_k$. By Theorem 7.11, this implies that

$TK_{k,d} \in \text{Crit}(k)$. It is notable, that $TK_{k,d}$ is a k-critical graph with at most one high vertex, namely v, see Exercises 4.5 and 4.6.

Proposition 7.18 *Each vertex of the k-critical graph $TK_{k,d}$ with $k \geq 4$ and $d \geq 1$ is universal.*

Proof We use the same notation as in the above definition. By Proposition 7.15 it suffices to show that v is universal, since all other vertices are low vertices of the k-critical graph $H = TK_{k,d}$. Let $H' = S(H, v, X, s)$, where X is an independent set of H', and let $\varphi' \in CO(H'[X], k)$ with $|\varphi'(X)| \geq 2$. If $e = vu$ is an edge of H incident with v, then in H' the edge e is replaced by the edge $e' = \{u\} \cup s(e)$, where $s(e) \subseteq X$. Then for each edges $e \in E_H(v)$ we can chose a vertex $v(e) \in s(e)$ such that $C = \{\varphi(v(e)) \mid e \in E_H(v)\}$ is a set of at least two colors. Let L be the list-assignment of $G = H - v$ defined as follows. If $u \in V(G)$ and $e = vu \in E(H)$, let $L(u) = [1, k-1] \setminus \{\varphi'(v(e))\}$, otherwise let $L(u) = [1, k-1]$. Then $G = H - v$ is a Gallai tree and L is a list-assignment of G such that $|L(u)| = d_G(u)$ for all $u \in V(G)$. If there is an L-coloring φ of G, then $\varphi \cup \varphi' \in CO(H', k)$ and we are done. Otherwise, (G, L) is a uncolorable pair and so $L = L_\lambda$ for some $\lambda \in \Lambda(G)$ (by Theorem 7.2). Note that the blocks of G are the copies $K^1, K^2, \ldots, K^d$ of K_{k-1} and the copies of K_2 belonging to the edges $u_1w_2, u_2w_3, \ldots, u_{d-1}w_d$. But then it easily follows that $L(u) = [1, k-1] \setminus \{c\}$ for all vertices u of G with $d_G(u) = k-2$ (and hence with $uv \in E(H)$), where c is one color. This implies that $|C| = 1$, a contradiction. □

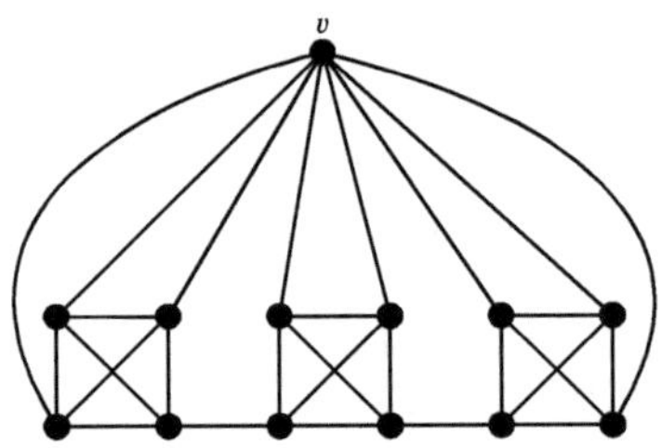

Fig. 7.4 The graph $TK_{5,3}$.

7.5 Critical Hypergraphs with Low Connectivity

In the 1960s, G. A. Dirac and T. Gallai studied connectivity properties of critical graphs. We discussed their results in Section 4.2, but without giving any proofs. The results of Dirac and Gallai were later generalized to hypergraphs by B. Toft, and we will present these results together with proofs in this section. First, we deal with the vertex connectivity of critical hypergraphs. The basic observation is that any nonempty hypergraph H satisfies

$$\chi(H) = \max\{\chi(B) \mid B \in \mathfrak{B}(H)\}.$$

This is due to the fact that if we have optimal colorings of the blocks of H, then by permuting the colors in the blocks, we can obtain an optimal coloring of H without using any additional color. An immediate consequence of this observation is that a critical hypergraph has no separating vertices. This raises the problem of characterizing the critical hypergraphs that have a separating vertex set consisting of two vertices. For graphs this was done by Dirac [302] and Gallai [397], and it was extended to hypergraphs by Toft [1023].

Theorem 7.19 (Toft) *Let H be a $(k+1)$-critical hypergraph for an integer $k \geq 2$, and let $S \subseteq V(H)$ be a separating vertex set of H satisfying $|S| \leq 2$. Then S is an independent set of H consisting of two vertices, say v and w, and $H \div S$ has exactly two components F_1 and F_2. Moreover, if $H_i = H[V(F_i) \cup S]$ for $i \in \{1,2\}$, we can adjust the notation so that so that there is a coloring $\varphi_1 \in \mathcal{CO}(H_1,k)$ with $\varphi_1(v) = \varphi_1(w)$. Then, the following statements hold:*

(a) *Each coloring $\varphi \in \mathcal{CO}(H_1,k)$ satisfies $\varphi(v) = \varphi(w)$ and each coloring $\varphi \in \mathcal{CO}(H_2,k)$ satisfies $\varphi(v) \neq \varphi(w)$.*
(b) *The hypergraph $H_1' = H_1 + vw$ obtained from H_1 by adding the edge vw is $(k+1)$-critical.*
(c) *The hypergraph H_2' obtained from H_2 by identifying v and w is $(k+1)$-critical.*

Proof Since $H \in \text{Crih}(k+1)$ and $k \geq 2$, the separating set S consist of exactly two vertices, say $S = \{v,w\}$. Since H is connected, H is the union of two nontrivial induced subhypergraphs, say H_1 and H_2, with $V(H_1) \cap V(H_2) = S$. In particular, H_i is a proper subhypergraph of H, and hence there is a coloring $\varphi_i \in \mathcal{CO}(H_i,k)$ ($i \in \{1,2\}$). Then, for one coloring, say φ_1, we have $\varphi_1(v) = \varphi_1(w)$, and for the other coloring we have $\varphi_2(v) \neq \varphi_2(w)$. For otherwise, we could permute the colors such that $\varphi_1 \cup \varphi_2 \in \mathcal{CO}(H,k)$, which is impossible. Consequently, S is an independent set of H. A further consequence is that each coloring $\varphi \in \mathcal{CO}(H_1,k)$ satisfies $\varphi(v) = \varphi(w)$, and each coloring $\varphi \in \mathcal{CO}(H_2,k)$ satisfies $\varphi(v) \neq \varphi(w)$. This proves (a).

For the proof of (b), let $H_1' = H_1 + vw$. Clearly, $\delta(H_1') \geq 1$. So in order to show that H_1' is $(k+1)$-critical, we can use Proposition 7.7. From (a) it follows that $\chi(H_1') \geq k+1$. Now let e be an arbitrary edge of H_1'. Our aim is to show that $H_1' - e$ is k-colorable. This is evident if $e = vw$. Otherwise, $e \in E(H_1)$ and, since H is $(k+1)$-critical, there is a coloring $\varphi \in \mathcal{CO}(H-e,k)$. Then φ induces a k-coloring of H_2, and it follows from (a) that $\varphi(v) \neq \varphi(w)$, which implies that φ induces a k-coloring of $H_1' - e$. This proves that $\chi(H_1' - e) \leq k$ for all $e \in E(H_1')$. Hence H_1' is $(k+1)$-critical and (b) is proved.

For the proof of (c), let H_2' be the hypergraph obtained from H_2 by identifying v and w to a new vertex v^*. Clearly $\delta(H_2') \geq 2$ and we apply Proposition 7.7 to show that H_2' is $(k+1)$-critical. Again, (a) implies that $\chi(H_2') \geq k+1$. Now, let e' be an arbitrary edge of H_2'. and let e be the corresponding edge of H_2. Then there is a coloring $\varphi \in \mathcal{CO}(H-e,k)$ and, by (a), we have $\varphi(v) = \varphi(w)$. Consequently, φ induces a k-coloring of $H_2' - e'$. Hence H_2' is $(k+1)$-critical and (c) is proved.

For the hypergraph H we obviously have

$$H \div S = (H_1 \div S) \cup (H_2 \div S) = (H_1' \div S) \cup (H_2' \div v).$$

Since $H_1' \in \mathrm{Crih}(k+1)$ and S is not an independent set of H_1', it follows that $H_1' \div S$ is connected. Since $H_2' \in \mathrm{Crih}(k+1)$, it follows that $(H_2' \div v)$ is connected, too. This implies that $H \div S$ has exactly two components, say F_1 and F_2, and $H_i = H[V(F_i) \cup S]$ for $i \in \{1,2\}$. This completes the proof of the theorem. □

The above theorem says that if a hypergraph $H \in \mathrm{Crih}(k+1)$ has a separating vertex set of size two, then H can be decomposed into two hypergraphs $H_1', H_2' \in \mathrm{Crih}(k+1)$, provided that $k \geq 2$. The next result due to Toft [1023] shows that every critical hypergraph having a separating vertex set of size two can be obtained in this way.

Theorem 7.20 (Toft) *Let H_1 and H_2 be two disjoint $(k+1)$-critical hypergraphs with $k \geq 2$, let $\tilde{e} = vw$ be an ordinary edge of H_1, and let $\tilde{v}$ be a vertex of H_2. Let $H = S(H_1, \tilde{e}, H_2, \tilde{v}, s)$ and let $H_2' = H[(V(H_2) \setminus \{\tilde{v}\}) \cup \tilde{e}]$. If $\chi(H_2') \leq k$, then H is a $(k+1)$-critical hypergraph, and $\{v, w\}$ is a separating vertex set of G of size two.*

Proof For the proof we apply Theorem 7.13. We have $\tilde{e} = vw$ with $v, w \in V(H_1)$. Now let $\varphi \in CO(H[\tilde{e}], k)$ with $|\varphi(\tilde{e})| \geq 2$, that is, with $\varphi(v) \neq \varphi(w)$. Since $\chi(H_2') \leq k$, there is a coloring $\varphi' \in CO(H_2', k)$. Since $\chi(H_2) = k+1$, it follows that $\varphi'(v) \neq \varphi'(w)$. Hence by permuting colors if necessary, we may assume that $\varphi(v) = \varphi'(v)$ and $\varphi(w) = \varphi'(w)$. Then Theorem 7.13 implies that $H \in \mathrm{Crih}(k+1)$. □

Let $H \in \mathrm{Crih}(k+1)$ with $k \geq 2$. Then $H - e$ is connected whenever $e \in E(H)$. If $H = H_1 \nabla H_2$ and $k \geq 3$, then $H_1, H_2 \in \mathrm{Crih}(k+1)$ (by Theorem 7.11(d)), and, moreover, H has a edge e^* and a vertex v^* such that v^* is a separating vertex of $H - e^*$. That the converse statement is also true is folklore.

Theorem 7.21 *Let H be a $(k+1)$-critical hypergraph with $k \geq 3$. If H has an edge e^* and a vertex v^* such that v^* is a separating vertex of $H - e^*$, then H is a Hajós join of two hypergraphs.*

Proof By assumption and since $H - e^*$ is connected, $H - e^*$ is the union of two nontrivial induced subhypergraphs, say H_1 and H_2, that have only vertex v^* in common. Since $H \in \mathrm{Crih}(k+1)$, H is a block and $e^* \cap V(H_i) \neq \varnothing$ for $i \in \{1,2\}$. For $i \in \{1,2\}$, let $e_i = (e^* \cap V(H_i)) \cup \{v^*\}$. If both edges e_1 and e_2 are not in H, then H is the Hajós join of (two disjoint copies of) $H_1 + e_1$ and $H_2 + e_2$, and we are done. For otherwise, we may assume that $e_1 \in E(H)$. Since $H \in \mathrm{Crih}(k+1)$, there is a coloring $\varphi \in CO(H - e^*, k)$ and $\varphi(e^*) = \{c\}$ for some color $c \in [1,k]$. Since e_1 is an edge of $H - e^*$, this implies that $\varphi(v^*) \neq c$. Since $k \geq 3$, there is a color $c' \in [1,k] \setminus \{c, \varphi(v^*)\}$. By coloring all vertices of H_2 having color c with c' and vice versa, we obtain a k-coloring of H, which is impossible. This completes the proof. □

The concept of critical graphs was introduced by Dirac in the early 1950s and he obtained the first results on the structure of such graphs. Already in his first investigations he was concerned with connectivity properties of critical graphs, and in his 1953 paper [293] he proved that every graph $G \in \mathrm{Crit}(k+1)$ has edge connectivity

$\kappa'(G) \geq k$ provided that $k \geq 1$. A characterization of graphs $G \in \text{Crit}(k+1)$ with $\kappa'(G) = k$ by means of Lemma 7.14 was obtained much later by Gallai [unpublished] and, independently, by Toft [1019]. Toft also extended this result to hypergraphs, see [1023].

Recall from Appendix C.14 that an **edge cut** of a hypergraph H is a triple (X,Y,F) such that $\varnothing \neq X \subset V(H)$, $Y = V(H) \setminus X$, and $F = \partial_H(X) = \partial_H(Y)$. We then denote by X_F (respectively Y_F) the set of vertices of X (respectively of Y) that are incident to some edge of F. An edge cut (X,Y,F) of H is nontrivial if $|X_F| \geq 2$ and $|Y_F| \geq 2$.

Theorem 7.22 (Toft) *Let H be a $(k+1)$-critical hypergraph with $k \geq 2$, and let $F \subseteq E(H)$ be a separating edge set of H with $|F| \leq k$. Then $|F| = k$ and there is an edge cut (X,Y,F) of H satisfying the following properties:*

(a) *Every coloring $\varphi \in \mathcal{CO}(H[X],k)$ satisfies $|\varphi(X_F)| = 1$, and every coloring $\varphi \in \mathcal{CO}(H[Y],k)$ satisfies $|\varphi(Y_F)| = k$ and for every color $c \in [1,k]$ there is an edge $e \in F$ such that $\varphi(e \cap Y) = \{c\}$.*
(b) *Each vertex of Y_F is incident to exactly one edge of F.*
(c) *If $|X_F| \geq 2$, then the hypergraph H_1 obtained from $H[X]$ by adding the edge with vertex set X_F is $(k+1)$-critical.*
(d) *The hypergraph H_2 obtained from $H[Y]$ by adding a new vertex v^* and adding for each edge $e \in F$ the new edge $(e - X) \cup \{v^*\}$ is $(k+1)$-critical.*

Proof We may assume that F is a minimal separating edge set of H. Then, by Proposition C.33, $F = \partial_H(X)$ for a set X with $\varnothing \neq X \subset V(H)$. Hence (X,Y,F) with $Y = V(H) \setminus X$ is an edge cut of H. Since $H \in \text{Crih}(k+1)$, for $Z \in \{X,Y\}$, there is a coloring $\varphi_Z \in \mathcal{CO}(H[Z],k)$. We now construct an auxiliary graph G as follows. The vertex set of G is the disjoint union of two sets both having cardinality k, say $U = \{u_i \mid i \in [1,k]\}$ and $W = \{w_i \mid i \in [1,k]\}$. The edges set of G is $E(G) = [U]^2 \cup [W]^2 \cup F'$, where F' is the set of edges $u_i w_j$ for which there is an edge $e \in F$, such that $\varphi(e \cap X) = \{i\}$ and $\varphi(e \cap Y) = \{j\}$. Then both sets U and W are cliques of G, and (U,W,F') is an edge cut of G with $|F'| \leq |F| \leq k$. If there exists a coloring $\varphi' \in \mathcal{CO}(G,k)$, then we may assume that $\varphi(u_i) = i$ for $i \in [1,k]$, and $\varphi(w_j) = \pi(j)$ for $j \in [1,k]$, where $\pi \in \text{Sym}([1,k])$. But then $\varphi'_Y = \pi \circ \varphi_Y \in \mathcal{CO}(H[Y],k)$ and $\varphi_X \cup \varphi'_Y \in \mathcal{CO}(H,k)$, which is impossible. Consequently, $\chi(G) \geq k+1$ and Lemma 7.14 implies that $|F'| = k$ and $F' \subseteq E_G(v)$ for a vertex $v \in U \cup W$. By symmetry, we may assume that $v \in U$. This implies that $|\varphi_X(X_F)| = 1$ and X_F is an independent set of H. Furthermore, it follows that $\varphi_Y(Y_F) = [1,k]$ and for every color $c \in [1,k]$ there is an edge $e \in F$ such that $\varphi_Y(e \cap Y) = \{c\}$. Consequently, $|F| = k$. If $\varphi \in \mathcal{CO}(G[X],k)$, then we can apply the same argument to the colorings φ and φ_Y, which yields $|\varphi(X_F)| = 1$. If $\varphi \in \mathcal{CO}(H[Y],k)$, then we can apply the argument to the colorings φ_X and φ, which yields that for every color $c \in [1,k]$ there is an edge $e \in F$ such that $\varphi(e \cap Y) = \{c\}$. This proves (a) and (b).

For the proof of (c) suppose that $|X_F| \geq 2$, and let H_1 be the hypergraph obtained from $H[X]$ by adding the edge $e^* = X_F$. Clearly, $\delta(H_1) \geq 1$ and by (a) it follows that $\chi(H_1) \geq k+1$. Now let e be an arbitrary edge of H_1. We claim that $\chi(H_1 - e) \leq k$. If $e = e^*$, this is obvious as $H_1 - e^* = H[X]$. Otherwise, e is an edge of $H[X]$ and, since

$H \in \text{Crih}(k+1)$, there is a coloring $\varphi' \in CO(H-e,k)$. Then $\varphi'|_Y \in CO(H[Y],k)$ which implies by (a) that $|\varphi'(X_F)| \geq 2$. Consequently, $\varphi'|_X \in CO(H_1-e,k)$. This proves the claim that $\chi(H_1-e) \leq k$ for all edges $e \in E(H_1)$. Then Proposition 7.7 implies that $H_1 \in \text{Crih}(k+1)$. This proves (c).

In order to prove (d) let H_2 be the hypergraph obtained from $H[Y]$ by adding a new vertex v^* and adding for each edge $e \in F$ the new edge $(e-X) \cup \{v^*\}$. Clearly, $\delta(H_2) \geq 1$ and (a) implies that $\chi(H_2) \geq k+1$. Now let e' be an arbitrary edge of H_2, and let e be the corresponding edge of H, i.e., $e' = e \in E(H[Y])$, or $e \in F$ and $e' = (e-X) \cup \{v^*\}$. Since $H \in \text{Crih}(k+1)$, there is a coloring $\varphi \in CO(H-e,k)$. Then $\varphi|_X \in CO(H[X],k)$ and (a) implies that $\varphi(X_F) = \{c\}$ for some color $c \in [1,k]$. Then the map $\varphi' : V(H_2) \to [1,k]$ with $\varphi'(v) = \varphi(v)$ if $v \in V(H_2 - v^*)$ and $\varphi'(v^*) = c$ belongs to $CO(H_2-e',k)$. This shows that $\chi(H_2-e') \leq k$ for all edges $e' \in E(H_2)$. Then Proposition 7.7 implies that $H_2 \in \text{Crih}(k+1)$. This proves (d). □

Remark 7.23 Note that Theorem 4.10(a)(c) follows immediately from Theorem 7.22 (a)(d). To see that Theorem 4.10(b) holds, assume that the $(k+1)$-critical hypergraph H in Theorem 7.22 has only ordinary edges. Then the hypergraph H_1 obtained from $H[X]$ by adding the edge whose vertex set is X_F, has also only ordinary edges except possibly the new edge $\tilde{e} = X_F$. Now let H_1' be the hypergraph obtained from $H[X \cup Y_F]$ by adding all missing edges between the vertices of Y_F, so that Y_F becomes a clique of H_1'. Note that $|Y_F| = k$, and hence H_1' is obtained from H_1 and K_{k+1} by splitting a vertex of K_{k+1} into the edge $\tilde{e}$. Since both H_1 and K_{k+1} belong to $\text{Crih}(k+1)$, it follows from Theorem 7.13 and Proposition 7.15 that $H_1' \in \text{Crih}(k+1)$. Clearly, H_1' has only ordinary edges. This shows that Theorem 4.10(b) also follows from Theorem 7.22.

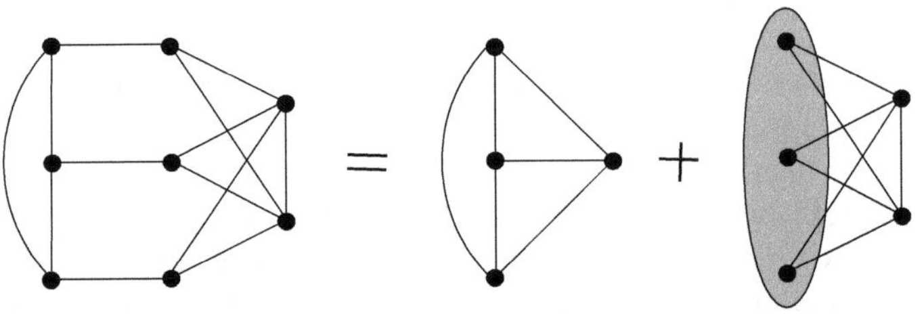

Fig. 7.5 An illustration to Theorems 7.22 and 7.24 for 4-critical hypergraphs.

The following result is the counterpart to Theorem 7.22. The result is an immediate consequence of Theorem 7.13 and Proposition 7.15, see also Figure 7.5.

Theorem 7.24 (Toft) *Let H_1 and H_2 be two disjoint $(k+1)$-critical hypergraphs with $k \geq 2$, let $\tilde{e} \in E(H_1)$, and let $\tilde{v} \in V(H_2)$ be a low vertex of H_2. Then the hypergraph $H = S(H_1, \tilde{e_1}, H_2, \tilde{v}, s)$ is $(k+1)$-critical, too, and $F = \partial_H(V(H_1))$ is a separating edge set of size k.*

7.6 Decomposable Critical Hypergraphs

In the early 1960s Gallai published his spectacular theorem that any graph in $\mathrm{Crit}(k,n)$ is decomposable unless $n \geq 2k-1$, see Section 4.3. In this section we shall prove that Gallai's decomposition result can be extended to hypergraphs. A hypergraph is called **decomposable** if it is the Dirac join of two disjoint nonempty hypergraphs, otherwise the hypergraph is called **indecomposable**. The following extension of Gallai's decomposition result is due to Stiebitz, Storch, and Toft [974].

Theorem 7.25 (Stiebitz, Storch, and Toft) *Each hypergraph in* $\mathrm{Crih}(k,n)$ *with* $n \leq 2k-2$ *and* $k \geq 2$ *is decomposable.*

The proof of this result is almost identical to the proof of the graph result, we just have to adapt the concept of complementary graphs to hypergraphs. To make the comparison with the graph proof easier for the reader, we denote the critical hypergraph by G. Let G be a hypergraph. The **relative complement** of G, denoted by $R(G)$ is the graph defined by

$$V(R(G)) = V(G) \text{ and } E(R(G)) = [V(G)]^2 \setminus E(G).$$

If G is a graph, then $R(G)$ is the usual complement $\overline{G}$. Furthermore, any hypergraph G satisfies $R(R(R(G))) = R(G)$. If G is a simple hypergraph, then G is decomposable if and only if $R(G)$ is disconnected. The only difference in the proof of the hypergraph result is that in the graph proof we replace the complement by the relative complement. For the convenience of the reader, we repeat the notation from Section 4.3.

Let G be a hypergraph. For a coloring φ of G with color set C, let $C_\varphi = \mathrm{im}(\varphi)$ be the **set of used colors**, let $c(\varphi) = |C_\varphi|$ be the **number of used colors**, and let $\mathcal{X}_\varphi = \{\varphi^{-1}(c) \mid c \in C_\varphi\}$ be the set of nonempty color classes. Clearly, $c(\varphi) \geq \chi(G)$ and we call φ an **optimal coloring** of G if $c(\varphi) = \chi(G)$. If $G = \varnothing$, then $\mathcal{X}_\varphi = \varnothing$, else $\mathcal{X}_\varphi$ is a partition of the vertex set of G into independent sets of G. Furthermore, we denote by $H(\varphi)$ the hypergraph whose vertex set is $V(H(\varphi)) = V(G)$ and whose edge set is

$$E(H(\varphi)) = \{U \in \mathcal{X}_\varphi \mid |U| \geq 2\}.$$

The set of isolated vertices of $H(\varphi)$ is denoted by $I(\varphi)$. Note that a vertex v of G belongs to $I(\varphi)$ if and only if $\{v\}$ is a color class with respect to φ. The hypergraph $H(\varphi)$ has maximum degree $\Delta(H(\varphi)) \leq 1$ and

$$c(\varphi) = \mathrm{com}(H(\varphi)) = |I(\varphi)| + |E(H(\varphi))|.$$

A vertex set $X \subseteq V(G)$ is said to be **φ-closed** if X is the union of a set of color classes with respect to φ, that is, each color class $U \in \mathcal{X}_\varphi$ satisfies $U \subseteq X$ or $U \cap X = \varnothing$. For two colorings φ_1 and φ_2 of G, let $H(\varphi_1, \varphi_2) = H(\varphi_1) \cup H(\varphi_2)$. Obviously, this hypergraph has maximum degree at most 2.

In this section we shall identify a coloring φ of G with the set $\mathcal{X}_\varphi$ of all nonempty color classes. In particular, we suppress unused colors and do not distinguish between equivalent colorings. If φ is a coloring of G and $X \subseteq V(G)$, then we denote by $\varphi|_X$ the coloring φ' of $G[X]$ with $\mathcal{X}_{\varphi'} = \{U \cap X \mid U \in \mathcal{X}_\varphi \text{ and } U \cap X \neq \varnothing\}$. Furthermore, let $\overline{X} = V(G) \setminus X$. In what follows we shall use the following simple facts: X is φ-closed if and only if $\mathcal{X}_{\varphi'} \subseteq \mathcal{X}_\varphi$; if X is φ-closed, then $\overline{X}$ is φ-closed, $\varphi = \varphi|_X \cup \varphi|_{\overline{X}}$ and $c(\varphi) = c(\varphi|_X) + c(\varphi|_{\overline{X}})$.

The following three propositions lists some simple, but useful facts about optimal colorings.

Proposition 7.26 *Let G be hypergraph, let φ be an optimal coloring of G. Then the following statements hold:*

(a) *$I(\varphi)$ is a clique of G.*
(b) *If $X \subseteq V(G)$ is φ-closed, then $\overline{X}$ is φ-closed and, for $Y \in \{X, \overline{X}\}$, the restriction $\varphi|_Y$ is an optimal coloring of $G[Y]$ satisfying $\varphi = \varphi|_X \cup \varphi|_{\overline{X}}$ and $c(\varphi) = c(\varphi|_X) + c(\varphi|_{\overline{X}})$.*

Proof Suppose (a) is false. Then there are two distinct vertices $u, v \in I(\varphi)$ such that $U = \{u, v\}$ is an independent set of G. Consequently, we can combine the color classes $\{u\}, \{v\} \in \mathcal{X}_\varphi$ to one color class U, which yields a coloring φ' of G with $c(\varphi') = c(\varphi) - 1$, which is impossible. Thus (a) is proved. Statement (b) is obvious.□

Proposition 7.27 *For a vertex v of a graph G there exists an optimal coloring φ of G with $v \in I(\varphi)$ if and only if $\chi(G - v) < \chi(G)$.*

Proof The proof is the same as for Proposition 4.16. □

Proposition 7.28 *Let φ_1 and φ_2 be two distinct optimal colorings of a hypergraph G, and let $H = H(\varphi_1, \varphi_2)$. Then the following statements hold:*

(a) *If X is the vertex set of a component of the hypergraph H, then X is both φ_1-closed and φ_2-closed and $\varphi_3 = \varphi_1|_X \cup \varphi_2|_{\overline{X}}$ is an optimal coloring of G with $I(\varphi_3) = (I(\varphi_1) \cap X) \cup (I(\varphi_2) \cap \overline{X})$.*
(b) *If v_1v_2 is an edge of $R(G)$ such that $v_1 \in I(\varphi_1)$ and $v_2 \in I(\varphi_2)$, then v_1 and v_2 lie in the same component of the hypergraph H.*

Proof The proof is the same as for Proposition 4.17. □

If a hypergraph G has an optimal coloring φ such that $|I(\varphi)| = 1$, then we have $|G| \geq 2c(\varphi) - 1$. Hence, Theorem 7.25 is an immediate consequence of the following result.

Theorem 7.29 *Let G be a critical hypergraph whose relative complement is connected. Then for every vertex v of G there is an optimal coloring φ of G such that $I(\varphi) = \{v\}$.*

Proof Let v be an arbitrary vertex of G. Since G is critical, $\chi(G-v) < \chi(G)$ and there exists an optimal coloring φ of G such that $v \in I(\varphi)$ (by Proposition 7.27). An optimal coloring φ of G is called a *v-extreme coloring* of G if $v \in I(\varphi)$ and $|I(\varphi)|$ is minimum subject to this condition. To complete the proof we must show that a v-extreme coloring φ of G satisfies $|I(\varphi)| = 1$. The proof is arranged in a series of three claims.

Claim 7.29.1 *Let v be a vertex of G, let φ_1 be a v-extreme coloring of G, and let φ_2 be any coloring of G. Then any component of the hypergraph $H = H(\varphi_1, \varphi_2)$ contains at most one vertex of $I(\varphi_1)$.*

Proof: The proof is the same as for Claim 4.18.1. △

Claim 7.29.2 *If v_1v_2 is an edge of $R(G)$ and φ_1 is a v_1-extreme coloring of G, then there exists a v_2-extreme coloring φ_2 such that $I(\varphi_1) \setminus \{v_1\} = I(\varphi_2) \setminus \{v_2\}$.*

Proof: The proof is the same as for Claim 4.18.2. △

Claim 7.29.3 *Let v be a vertex of G, and let φ be a v-extreme coloring of G. Then for every vertex v' of G there is a v'-extreme coloring φ' such that $I(\varphi) \setminus \{v\} = I(\varphi') \setminus \{v'\}$.*

Proof: The proof is the same as for Claim 4.18.3, we only have to replace the complement $\overline{G}$ by the relative complement $R(G)$. △

To conclude the proof of Theorem 7.29, let v be an arbitrary vertex of G. Suppose, to the contrary, that there exists a v-extreme coloring φ of G such that $I(\varphi) \neq \{v\}$. Then there exists a vertex $v' \in I(\varphi) \setminus \{v\}$ and, by Claim 7.29.3, there is a v'-extreme coloring φ' such that $I(\varphi) \setminus \{v\} = I(\varphi') \setminus \{v'\}$, which is a contradiction since $v' \in I(\varphi) \setminus \{v\}$. Thus the proof of the theorem is complete. □

The advantage of the decomposition result for hypergraphs is that it can be used to derive decomposition results for graphs, multigraphs and digraphs which are critical with respect to various coloring parameters. We have discussed two such results in Section 5.8, see Theorems 5.41 and 5.43. Stehlik [960] proved the following decomposition result for digraphs that are critical with respect to the dichromatic number.

Theorem 7.30 (Stehlík) *Let D be a $\vec{\chi}$-critical digraph of order n and with $\vec{\chi}(D) = k$ for $k \geq 3$. If $n \leq 2k-2$, then D is obtained from the disjoint union of two nonempty subdigraphs of D, say D_1 and D_2, by joining each vertex of D_1 to each vertex of D_2 by two opposite directed edges.*

Proof Construct a hypergraph H as follows. The vertex set is $V(H) = V(D)$, and a subset $e \subseteq V(D)$ is an edge of H if and only if $D[e]$ is a directed cycle. Clearly, H is a simple hypergraph. Furthermore, if $\varphi : V(D) \to C$ is a map, then φ is an acyclic coloring of D if and only if φ is a coloring of H. Consequently $\chi(H) = \vec{\chi}(D)$. Since D is $\vec{\chi}$-critical and $\vec{\chi}(D) = k$, it follows that H is k-vertex-critical. By Proposition 7.8,

H has a subhypergraph H' such that H' is k-critical and $V(H') = V(H)$. Hence $|H'| = |D| \leq 2k-2$. Then Theorem 7.25 implies that H' is the Dirac join of two nonempty hypergraphs. Since H is a simple hypergraph and $V(H) = V(H')$, the same holds for H. Clearly, if e is an ordinary edge of H, then $D[e]$ is a digon. Hence D is obtained from the disjoint union of two nonempty subdigraphs of D, say D_1 and D_2, by joining each vertex of D_1 to each vertex of D_2 by two opposite directed edges. □

Let H be a simple hypergraph. A **dominating vertex** of H is a vertex that is joined to all other vertices of H by an ordinary edge. A **dominating edge** of H is an edge $e \in E(H)$ such that $vw \in E(H)$ for all $v \in e$ and all $w \in V(H) \setminus e$. Let X be the set of dominating vertices of H with $|X| = p$, let $e_1, e_2, \ldots, e_q$ be the dominating hyperedges of H, where $|e_i| = n_i \geq 3$, and let $Y = V(H) \setminus (X \cup e_1 \cup e_2 \cup \cdots \cup e_q)$. Then $H[X]$ is a complete graph, $H[e_i] = K_{n_i}^{n_i}$ and $n_i \geq 3$ for $i = 1, 2, \ldots, q$, and

$$H = K_p \boxplus K_{n_1}^{n_1} \boxplus \cdots \boxplus K_{n_t}^{n_t} \boxplus H[Y],$$

where the remaining hypergraph $H[Y]$ has neither dominating vertices nor dominating edges. We shall apply Theorem 7.25 to prove the following result.

Theorem 7.31 (Stiebitz, Storch, and Toft) *Let H be a k-critical hypergraph, let p be the number of dominating vertices of H, and let q be the number of dominating hyperedges of H. Then the following statements hold:*

(a) $0 \leq p \leq k$ *and there exists a hypergraph* $H' \in \mathrm{Crih}(k-p)$ *such that* $H = K_p \boxplus H'$, *H' has no dominating vertices, and* $|H'| \geq \frac{3}{2}(k-p)$. *Furthermore,* $p \geq 3k - 2|H|$ *and equality holds if and only if H' is the Dirac sum of* $\frac{1}{2}(k-p)$ *disjoint hypergraphs* K_3^3.

(b) $0 \leq p + 2q \leq k$ *and there exists hypergraph* $H_1 \in \mathrm{Crih}(2q)$ *and a hypergraph* $H_2 \in \mathrm{Crih}(k-p-2q)$ *such that* $H = K_p \boxplus H_1 \boxplus H_2$, *$H_1$ is the Dirac sum of q hypergraphs all belonging to* $\mathrm{Crih}(2,3)$, *H_2 has no dominating vertices and no dominating edges, and* $|H_2| \geq \frac{5}{3}(k-p-2q)$. *Furthermore,* $2p + q \geq 5k - 3|H|$ *and equality holds if and only if H_1 is the Dirac sum of q disjoint hypergraphs* K_3^3 *and* $\frac{1}{3}(k-p-2q)$ *disjoint hypergraphs belonging to* $\mathrm{Crih}(3,5)$.

Proof In what follows, let $H \in \mathrm{Crih}(k)$ be an arbitrary hypergraph. Then H is a simple hypergraph and

$$H = H_1 \boxplus H_2 \boxplus \cdots \boxplus H_t,$$

where $R(H_1), R(H_2), \ldots, R(H_t)$ are the components of $R(H)$. For $i \in [1,t]$, let $k_i = \chi(H_i)$ and $n_i = |H_i|$. From Theorem 7.10 it follows that

(i) $k = k_1 + k_2 + \cdots + k_t$ and $H_i \in \mathrm{Crih}(k_i)$ for $i \in [1,t]$.

Since $R(H_i)$ is connected, Theorem 7.25 implies that

(ii) $|H_i| \geq 2k_i - 1$ for $i \in [1,t]$.

Since $\mathrm{Crih}(1) = \{K_1\}$ and $\mathrm{Crih}(2) = \{K_n^n \mid n \geq 2\}$, we then obtain that either $k_i = 1$ and $H_i = K_1$, or $k_i = 2$ and $H_i = K_{n_i}^{n_i}$ with $n_i \geq 3$, or $k_i \geq 3$ and $|H_i| \geq 5$. For a subset X of $[1,t]$, let

$$H_X = \sum_{i\in X} H_i \text{ (Dirac sum) and } k_X = \sum_{i\in X} k_i,$$

where $H_\varnothing = \varnothing$ and $k_\varnothing = 0$. By Theorem 7.10, $H_X \in \text{Crih}(k_X)$. Let $P = \{i \in [1,t] \mid k_i = 1\}$, $Q = \{i \in [1,t] \mid k_i = 2\}$, $R = [1,t] \setminus (P \cup Q)$, $p = |P|$, $q = |Q|$, and $r = |R|$. The sets P, Q and R are pairwise disjoint and their union is $[1,t]$. Hence we obtain that

(iii) $H = H_P \boxplus H_Q \boxplus H_R$, where $H_P = K_p$ and $H_Q = \sum_{i\in Q} K_{n_i}^{n_i}$.

Note that p is the number of dominating vertices of H and q is the number of dominating hyperedges of H. First we want to establish a lower bound for p. So let $\overline{P} = [1,t] \setminus P$. Then $\overline{P} = R \cup Q$ and $H = H_P \boxplus H_{\overline{P}}$. For $i \in \overline{P}$, we have that $k_i \geq 3$ and so, by (ii), $|H_i| \geq 2k_i - 1 \geq \frac{3}{2}k_i$, where equality holds if and only if $H_i = K_3^3$. For the order of H, it follows from (i) that

$$|H| = p + \sum_{i\in\overline{P}} |H_i| \geq p + \tfrac{3}{2}\sum_{i\in\overline{S}} k_i = p + \tfrac{3}{2}(k-p),$$

which is equivalent to $p \geq 3k - 2|H|$. Clearly, $p = 3k - 2|H|$ if and only if $H_{\overline{P}}$ is the Dirac sum of $\frac{1}{2}(k-p)$ disjoint hypergraphs K_3^3. This proves (a).

For $i \in R$ we have that $k_i \geq 3$ and so, by (ii), $|H_i| \geq 2k_i - 1 \geq \frac{5}{3}k_i$, where equality holds if and only if $H_i \in \text{Crit}(3,5)$. For the order of H we then obtain that

$$|H| = p + \sum_{i\in Q} |H_i| + \sum_{i\in R} |H_i| \geq p + 3q + \tfrac{5}{3}\sum_{i\in R} k_i = p + 3q + \tfrac{5}{3}k_R.$$

Since $k = k_P + k_Q + k_R = p + 2q + k_R$ (by (i)), it follows that $|H| \geq p + 3q + \frac{5}{3}(k - p - 2q)$, which is equivalent to $2p + q \geq 5k - 3|H|$. Clearly, $2p + q = 5k - 3|H|$ if and only if $H_i = K_3^3$ for all $i \in Q$ and $H_i \in \text{Crih}(3,5)$ for all $i \in R$. Thus, (b) is proved. □

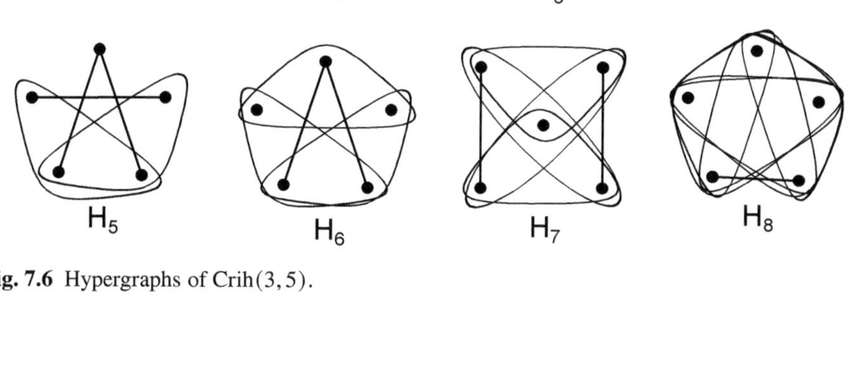

Fig. 7.6 Hypergraphs of Crih$(3,5)$.

The first interesting classes of critical hypergraphs not covered by the decomposition theorem are the classes $\mathrm{Crih}(3,5)$, $\mathrm{Crih}(3,6)$, and $\mathrm{Crih}(4,7)$. Based on a computer search, Storch (unpublished) found that the class $\mathrm{Crih}(3,5)$ consists of 9 graphs (up to isomorphism) and the class $\mathrm{Crih}(6,3)$ consists of 64 graphs (up to isomorphism). A proof by simple case analysis, showing that $\mathrm{Crih}(3,5)$ consists of nine hypergraphs up to isomorphism, is given in [974, Proposition 7.2]. Figure 7.6 shows eight of the nine hyhergraphs from $\mathrm{Crih}(3,5)$; the missing hypergraph is $H_9 = K_5^3$. Let $\mathrm{Crih}^*(k,n)$ denote the hypergraphs from $\mathrm{Crih}(k,n)$ that have no dominating vertices and no dominating edges. Since $\mathrm{Crih}(2,5) = \{K_5^5\}$, we obtain that $\mathrm{Crih}(3,6) = \mathrm{Crih}^*(3,6) \cup \{K_1 \boxplus K_5^5\}$. Lists of small critical graphs were determined by Toft [1024] and by Jensen and Royle [527].

7.7 Critical Hypergraphs with Few Edges

As for graphs, it is an interesting task to investigate the extremal function $\mathrm{exth}(\cdot,\cdot)$ defined by

$$\mathrm{exth}(k,n) = \min\{|E(H)| \mid H \in \mathrm{Crih}(k,n)\}$$

and the corresponding class of extremal hypergraphs defined by

$$\mathrm{Exth}(k,n) = \{H \in \mathrm{Crih}(k,n) \mid |E(H)| = \mathrm{exth}(k,n)\},$$

where k and n are positive integers.

Recall that $\mathrm{Crih}(1) = \{K_1\}$ and $\mathrm{Crih}(2) = \{K_n^n \mid n \geq 2\}$, and so $\mathrm{exth}(2,n) = 1$ for all $n \geq 2$. If $k \geq 3$, then $\mathrm{Crih}(k,n) \neq \varnothing$ if and only if $n \geq k$. In particular, we have $\mathrm{Crih}(k,k) = \{K_k\}$ (see Proposition 7.6(a)) and $\mathrm{Crih}(k,k+1) = \{K_{k-2} \boxplus K_3^3\}$ (see Exercise 7.15)). Consequently, for $k \geq 3$ we have

$$\mathrm{exth}(k,k) = \binom{k}{2} \text{ and } \mathrm{exth}(k,k+1) = \binom{k+1}{2} - 2. \tag{7.5}$$

While $\mathrm{Crit}(3)$ is the class of odd cycles, a good characterization of the class $\mathrm{Crih}(3)$ is unknown and very unlikely. Clearly, $K_1 \boxplus K_p^p$ belongs to $\mathrm{Crih}(3,p+1)$ for $p \geq 2$, which implies that $\mathrm{exth}(3,n) \leq n$ for all $n \geq 2$. That $\mathrm{exth}(3,n) = n$ was proved by several authors including Lovász [689, 692], Woodall [1072], Seymour [925], and possibly others. Seymour's proof is particularly simple, based on elementary linear algebra.

Theorem 7.32 *If H is a 3-critical hypergraph, then $|E(H)| \geq |V(H)|$ implying that* $\mathrm{exth}(3,n) = n$ *for all* $n \geq 3$.

Proof Let $H = (V,E)$ be a 3-critical hypergraph with m edges and n vertices. Suppose, to the contrary, that $m < n$. Let $\mathbf{I} : E \times V \to \mathbb{R}$ be the edge-vertex incidence matrix of H, that is, $\mathbf{I}(e,v) = 1$ if $v \in e$ and $\mathbf{I}(e,v) = 0$ if $v \notin e$. Since $m < n$, the rang of $\mathbf{I}$ is smaller that n. This implies that the linear equation $\mathbf{Ix} = \mathbf{0}$ has a nontrivial solution $\mathbf{x}$, that is, there is a nonnull vector $\mathbf{x} : V \to \mathbb{R}$ such that (+) $\sum_{v \in e} \mathbf{x}(v) = 0$ for all edges e of H. Then V is the disjoint union of the sets $X_1 = \{v \in V \mid \mathbf{x}(v) > 0\}$, $X_2 = \{v \in V \mid \mathbf{x}(v) < 0\}$, and $Y = \{v \in V \mid \mathbf{x}(v) = 0\}$. Clearly, Y is a proper subset of V and so the induced subhypergraph $H[Y] = (Y, E(H) \cap 2^Y)$ has a 2-coloring, say with color classes Y_1 and Y_2. But then (+) implies that H is a bipartite graph with parts $X_1 \cup Y_1$ and $X_2 \cup Y_2$, contradicting our assumption that H is 3-critical. □

Determining lower bounds for the number of edges of critical hypergraphs is obviously more difficult than for graphs. A k-critical hypergraph has minimum degree at least $k-1$. In the case of graphs, this leads to the trivial lower bound $\text{ext}(k,n) \geq \frac{1}{2}(k-1)n$. Even to show that this also holds for hypergraphs, that is, $\text{exth}(k,n) \geq \frac{1}{2}(k-1)n$, seems to be nontrivial. However, the situation improved after Kostochka and Yancey presented their new bound for $\text{ext}(k,n)$ for $k \geq 4$. Recall that

$$\text{koy}(k,n) = \tfrac{1}{2}(k - \frac{2}{k-1})n + c_{k,k-1} = \frac{(k+1)(k-2)n - k(k-3)}{2(k-1)}.$$

Then it is easy to check that the Kostochka–Yancey function satisfies the following recursion.

$$\text{koy}(k,k) = \binom{k}{2} \text{ and } \text{koy}(k,n+k-1) = \text{koy}(k,n) + \binom{k}{2} - 1.$$

As observed during a discussion between A. V. Kostochka and authors of this book the Kostochka–Yancey bound also holds for critical hypergraphs. This result was published in the survey [976] about Brooks' theorem.

Theorem 7.33 *Every k-critical hypergraph on n vertices with $n \geq k \geq 4$ has at least* $\text{koy}(k,n)$ *edges. As a consequence* $\text{exth}(k,n) \geq \text{koy}(k,n)$ *for $n \geq k \geq 4$.*

Proof Let $H \in \text{Crih}(k,n)$ be a hypergraph with m edges. To show that $m \geq \text{koy}(k,n)$, we use induction on the number h of hyperedges in H. If $h = 0$, then the result follows from the Kostochka–Yancey bound (5.9). Now assume that $h \geq 1$. Then H has a hyperedge, say e. To get rid of the hyper edge e we apply the splitting argument introduced in Section 7.4. To this end, consider the k-critical graph $G = TK_{k,d}$ with $d \geq 2$ defined in Section 7.4. Then G has exactly one high vertex v and we may choose $d \geq 2$ such that $d_G(v) \geq |e|$. Now, let $H' = S(H,e,G,v,s)$ be the hypergraph obtained from H and G by splitting the vertex v into the hyperedge e, where s is a simple splitting. By Theorem 7.13 and Proposition 7.18, $H' \in \text{Crih}(k)$. Since s is a simple splitting, H' has $h-1$ hyperedges. For $n' = |H'|$ and $m' = |E(H')|$ we obtain that $n' = n + d(k-1)$ and

$$m' = |E(H)| + |E(G)| - 1 = m + d\left(\frac{(k-2)(k+1)}{2}\right).$$

On the other hand, since H' has $h-1$ hyperedges, the induction hypothesis implies

$$\begin{aligned} m' &\geq \mathrm{koy}(k,n') = \mathrm{koy}(k,n) + \frac{1}{2}(k - \frac{2}{k-1})(d(k-1)) \\ &= \mathrm{koy}(k,n) + d\left(\frac{(k-2)(k+1)}{2}\right), \end{aligned}$$

which yields $m \geq \mathrm{koy}(k,n)$ as claimed. □

Since $\mathrm{Crit}(k,n) \subseteq \mathrm{Crih}(k,n)$, we obtain that $\mathrm{ext}(k,n) \geq \mathrm{exth}(k,n)$, which implies, by Theorem 7.33 and (5.9), that

$$\mathrm{koy}(k,n) \leq \mathrm{exth}(k,n) \leq \mathrm{ext}(k,n) \leq \mathrm{ore}(k,n),$$

provided that $n \geq k \geq 4$ and $n \neq k+1$. In particular, if $n \equiv 1 \pmod{k-1}$, then we have

$$\mathrm{koy}(k,n) = \mathrm{exth}(k,n) = \mathrm{ext}(k,n) = \mathrm{ore}(k,n).$$

If Ore's conjecture that $\mathrm{ext}(k,n) = \mathrm{ore}(k,n)$ for $n \geq k \geq 4$ and $n \neq k+1$ is true, we can repeat the proof of Theorem 7.33 with the Ore-function $\mathrm{ore}(k,n)$ instead of the Kostochka–Yancey function $\mathrm{koy}(k,n)$ to show that $\mathrm{exth}(k,n) = \mathrm{ore}(k,n)$. So Ore's conjecture implies the following conjecture.

Conjecture 7.34 (Ore's Conjecture for Hypergraphs) *If* $n \geq k \geq 4$ *and* $n \neq k+1$, *then* $\mathrm{exth}(k,n) = \mathrm{ext}(k,n)$.

As an immediate consequence of Theorem 7.32 and Theorem 7.33, we obtain another lower bound for $\mathrm{exth}(k,n)$. It is notable that we are not able to prove this seemingly trivial lower bound without using the Kostochka–Yancey bound. The second corollary supports Ore's conjecture for hypergraphs. That the limit of $\mathrm{exth}(k,n)/n$ for $n \to \infty$ exists and is finite follows from Fekete's lemma, since the function $\mathrm{exth}(k,n)$ is subadditive in n (use the Hajós join and Theorem 7.11). The value of the limit follows from Theorem 7.32 and Theorem 7.33 (see also (5.10)).

Corollary 7.35 *If* $n \geq k \geq 3$, *then* $\mathrm{exth}(k,n) \geq \frac{1}{2}(k-1)n$.

Corollary 7.36 *If* $k \geq 3$, *then*

$$\lim_{n\to\infty} \frac{\mathrm{exth}(k,n)}{n} = \lim_{n\to\infty} \frac{\mathrm{ext}(k,n)}{n} = \frac{1}{2}(k - \frac{2}{k-1})$$

Gallai used his decomposition result to determine the values of the extremal function $\mathrm{ext}(k,n)$ for $k+2 \leq n \leq 2k-1$ as well as the description of the corresponding extremal classes $\mathrm{Ext}(k,n)$. From the decomposition result for hypergraphs the values of the extremal function $\mathrm{exth}(k,n)$ for $k \leq n \leq 2k-1$ can be derived likewise. However, a full description of the corresponding extremal classes $\mathrm{Exth}(k,n)$ seems not possible since we were not able to characterize the class $\mathrm{Exth}(k,2k-1)$ when $k \geq 4$. So we only obtain the following result due to Stiebitz, Storch, and Toft [974].

Theorem 7.37 (Stiebitz, Storch, and Toft) *Let* $n = k+p$ *be an integer such that* $1 \leq p \leq k-1$. *Then*

$$\text{exth}(k,n) = \binom{n}{2} - (p^2+1) = \frac{1}{2}((k-1)n + p(k-p) - 2)$$

and $\text{Exth}(k,k+p) = \{K_{k-p-1} \boxplus H \mid H \in \text{Exth}(p+1,2p+1)\}$.

The proof of the above theorem uses induction over k and is similar to the proof of Theorem 5.9, but it is somewhat easier, because we do not characterize the class $\text{Exth}(p+1,2p+1)$. We shall give a short outline of the proof. Let $H \in \text{Crih}(k,n)$, where $n = k+p$ and $1 \le p \le k-1$. The number of edges of a hypergraph H' is denoted by $e(H')$. Furthermore, let

$$e(k,p) = \frac{1}{2}((k-1)(k+p) + p(k-p) - 2).$$

Our aim is to show that $e(H) \ge e(k,p)$ and that equality holds if and only if $H = K_{k-p-1} \boxplus H'$ with $H' \in \text{Exth}(p+1,2p+1)$. The proof is by induction on k. The statement is evident if $k=2$, since then we have $p=1$, $H = K_3^3$ and $e(H) = e(2,3) = 1$. So assume that $k \ge 3$. If $p = 1$, then $H = K_{k-2}^2 \boxplus K_3^3$ (by Theorem 7.31, see also Exercise 7.15)) and $2e(H) = 2e(k,1) = k^2 - 4 + k$, so we are done. If $p = k-1$, then $n = 2k-1$ and, by using Theorem 7.32 if $k=3$ and Theorem 7.33 if $k \ge 4$, we obtain that $e(H) \ge \text{exth}(k,n) \ge \text{koy}(k,2k-1) = k^2 - k - 1 = e(k,k-1)$, which yields the desired result. Note that $\text{koy}(3,n) = n$. So it remains to consider the case when $2 \le p \le k-2$. By Theorem 7.25 it follows that H is decomposable. Then we distinguish three cases.

The first case is when H has a dominating vertex. Then Then $H = K_1 \boxplus H'$ with $H' \in \text{Crih}(k-1,n')$, $n' = k+p-1$ and $e(H) = e(H') + n'$. From the induction hypothesis it follows that

$$e(H) = e(H') + k + p - 1 \ge e(k-1,p) + k + p - 1 = e(k,p).$$

Consequently, $e(H) = e(k,p)$ if and only if $e(H') = e(k-1,p)$, which is equivalent to $H' = K_{k-p-2} \boxplus H''$ with $H'' \in \text{Exth}(p+1,2p+1)$ and hence to $H = K_{k-p-1} \boxplus H''$. So we are done.

The second case is when H has a dominating edge, but no dominating vertex. Then $H = K_q^q \boxplus H'$, where $q \ge 3$ and $H' \in \text{Crih}(k-2,n')$ with $n' = k+p-q$. Since $k \ge 4$ and H' has no dominating vertex, $n' \ge k$ and so $q \le p$. Then $n' = k+p-q \le 2k-5$ and the induction hypothesis implies that $e(H') \ge e(k-2,p+2-q)$. Since $e(H) = e(H') + qn' + 1$ and $3 \le q \le p$, this leads to

$$\begin{aligned} 2e(H) &\ge 2e(k-2,p+2-q) + 2q(k+p-q) + 2 \\ &= 2e(k,p) + (q-2)(4p-3q+3) > 2e(k,p). \end{aligned}$$

So we are done.

It remains to consider the case that H has neither a dominating edge nor a dominating vertex. Since $H \in \text{Crih}(k)$, H is a simple hypergraph, and hence

$$H = H_1 \boxplus H_2 \boxplus \cdots \boxplus H_t,$$

where $R(H_1), R(H_2), \ldots, R(H_t)$ are the components of $R(H)$ and $t \geq 2$ (by Theorem 7.25). For $i \in [1,t]$, let $k_i = \chi(H_i)$, $n_i = |H_i|$, $m_i = e(H_i)$ and $\overline{m}_i = \binom{n_i}{2} - m_i$. By Theorem 7.10, we obtain that

$$k = k_1 + k_2 + \cdots + k_t \text{ and } H_i \in \text{Crih}(k_i) \text{ for } i \in [1,t].$$

By the assumption of the case, $k_i \geq 3$ for $i \in [1,t]$. Our aim is to show that $e(H) \geq e(k,p) + 1$. For that we may argue as in the first case of the proof of Theorem 5.9. Here we need Corollary 7.35, saying the $2m_i \geq (k_i - 1)n_i$. For the details of the remaining part of the proof, we refer the reader to [974].

7.8 Exercises

7.1 Every graph is the intersection graph of a hypergraph.

7.2 Prove that if a hypergraph H satisfies $|e \cap e'| \neq 1$ for any two edges e, e' of H, then $\chi(H) \leq 2$. (Hint: Use Proposition 7.6(b)) (Erdős and Lovász [354])

7.3 Prove that if $\chi(H) \geq 4$, then H contains two disjoint edges.

7.4 Given a graph G, let H denote the hypergraph with $V(H) = V(G)$ and $E(H) = \{V(C) \mid C$ is an induced odd cyle of $G\}$. Show that $\chi(G) \leq 4$ if and only if $\chi(H) \leq 2$. (Woodall [1072], Berge [101])

7.5 Uniform hypergraph with large chromatic number and large girth: Let k, ℓ and r be integers with $k \geq 2, \ell \geq 3$ and $r \geq 2$. Construct an r-uniform hypergraph H such that $\chi(H) \geq k+1$ and $g(H) \geq \ell$. (Hint: Start with an $(k, (r-1)k+1, 2\ell)$-graph G and proceed as in the proof of Theorem 2.9 in Section 2.3.) (Alon, Kostochka, Reiniger, West, and Zhu [52])

7.6 Let G be a hypergraph, and let $k \geq 3$ be an integer. Construct a hypergraph H as follows. First, take k copies of G and two additional vertices v, w. Next, replace each edge e of the k copies of G with the edge $e \cup \{v\}$. Then, for each vertex u of G, add an edge consisting of w and the k copies of u. Finally, add the ordinary edge vw. The resulting hypergraph is H. Show that $\chi(H) \leq 2$ if and only if $\chi(G) \leq k$. (Hint: A 2-coloring of H encodes a k-coloring of G and vice versa, because of the k copies of $V(G)$ in H) (Lovász [692])

7.7 Let H be a hypergraph, and let E' be the hyperedges of H. Construct a graph G as follows. For each edge $e \in E'$, let $W_e = KC_{1,p}$ be an odd wheel with $p = \lfloor |e|/2 \rfloor$, and let v_e be a vertex of degree $2p+1$ in W_e. Assume that the wheels W_e with $e \in E'$ are disjoint. For each edge $e \in E'$, split the vertex v_e of W_e into the edge e; the resulting graph is G. Use Proposition 7.17 to show that $\chi(H) \leq 3$ if and only if $\chi(G) \leq 3$. (Toft [1023])

7.8 For a graph G, let $\text{bip}(G) = \min\{|F| \mid F \subseteq E(G), \chi(G-F) \leq 2\}$. Show that there is a constant c such that for all $p \geq 2$ the Toft graph $G = TG_{2p+1}$ satisfies $\text{bip}(G) \leq c|G|$.

7.9 Let H be the hypergraph whose vertex set is the union of five disjoint sets X_0, X_1, X_2, X_3 and X_4 each having $q \geq 2$ elements, where the edge set $E(H)$ consists of all ordinary edges uv with $u \in X_i$ and $v \in X_{i+1}$ for $i \in [0,4]$, and the hyperedge $X_i \cup X_{i+2}$ for $i \in [0,4]$ (where the indices are taken modulo 5). Show that $H \in \text{Crih}(4)$. Use the splitting argument from Exercise 7.7 to show that there is a positive constant c such that for infinitely many values of n there is a graph $G \in \text{Crit}(4,n)$ satisfying $\text{bip}(G) \geq cn^2$. (Stiebitz [967])

7.10 Let H be the 3-uniform 3-regular hypergraph associated with the Fano plane (see Figure 7.7), that is, the seven vertices of H are the points of the Fano plane and the seven edges of H are the lines of the Fano plane. Show that $H \in \text{Crih}(3,7)$ and $H \in \text{Exth}(3,n)$.

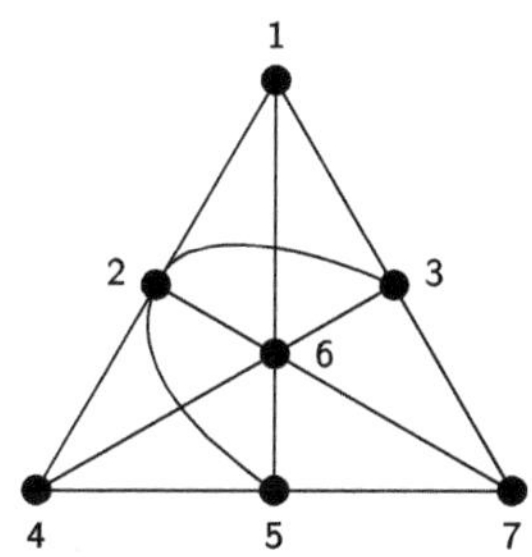

Fig. 7.7 The Fano plane.

7.11 Let $q \geq 2$ be an integer. For the complete q-uniform hypergraph K_n^q determine $\Delta(K_n^q)$ and $\Delta^*(K_n^q)$. Furthermore show that $\chi(K_n^q) = \lceil n/(q-1) \rceil$ and that $K_n^q \in \text{Crih}(k+1,n)$ provided that $n = k(q-1)+1$ and $k \geq 0$.

7.12 The Zykov-Schäuble-Lovász construction: Let $H_1, H_2, \ldots, H_k$ be pairwise disjoint hypergraphs. Furthermore, let H denote the hypergraph obtained from the union $H_1 \cup H_2 \cup \cdots \cup H_k$ by adding, for each subset A of $V(H_1) \cup V(H_2) \cup \cdots \cup V(H_k)$ such that $|A \cap V(H_i)| = 1$ for all $i \in [1,k]$, a new vertex v_A and joining v_A to all vertices of A by an ordinary edge.

1. Show that if H_i is i-critical for $i \in [1,k]$, then H is $(k+1)$-critical.
2. Show that if the hypergraphs H_i are all triangle-free then so is H.

(Lovász [693], see also Sachs and Stiebitz [887])

7.13 Let T be an arbitrary rooted tree on n vertices such that the root has degree at least two and the distance between the leaves and the root have all the same parity. Let H be the hypergraph obtained from T by adding the edge consisting of the leaves of T, see Figure 7.8. Show that $H \in \text{Crih}(3,n)$ and $H \in \text{Exth}(3,n)$.

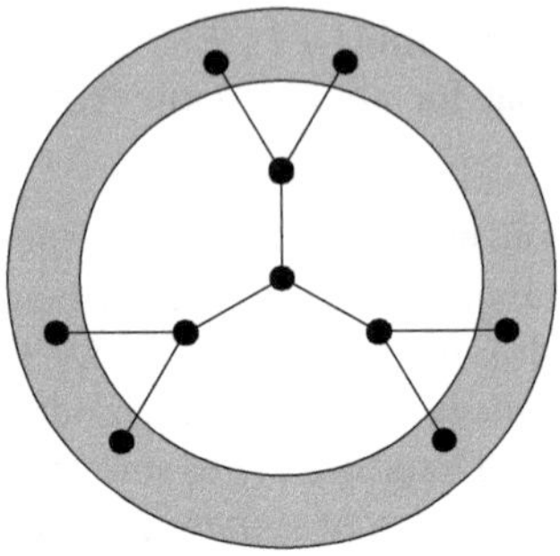

Fig. 7.8 An extremal hypergraph belonging to $\text{Exth}(3,10)$.

7.14 Let H be hypergraph and let L be a list-assignment of H such that H is L-critical. Furthermore, let $W \subseteq \{v \in V(H) \mid d_H(v) = |L(v)|\}$ and $E' = \{e \in E(H) \mid |e \cap W| \geq 2 \text{ and } e \setminus W \neq \varnothing\}$. Show that the following statements hold:

1. $H\langle W\rangle$ is a Gallai forest.
2. If $v \in W$, then every two distinct edges $e, e' \in E_H(v)$ have only v in common.
3. If $e, e' \in E'$ are distinct then $e \cap W \neq e' \cap W$.
4. If $e \in E'$, then $e \cap W$ is a strong bridge of $H\langle W\rangle$

(Hint: Use Theorem 7.2 and a similar reduction as in the proof of Theorem 7.9(a)) (Kostochka and Stiebitz [633])

7.15 For a hypergraph K and a hypergraph property $\mathcal{H}$, define

$$K \boxplus \mathcal{H} = \{K \boxplus H \mid H \in \mathcal{H}\}$$

if $\mathcal{H} \neq \varnothing$, and $K \boxplus \mathcal{H} = \varnothing$ otherwise. Use Theorem 7.31 to show that the following statements hold:

1. $\text{Crih}(k, k+1) = \{K_{k-2} \boxplus K_3^3\}$ for $k \geq 3$.
2. $\text{Crih}(4,6) = (K_1 \boxplus \text{Crih}(3,5)) \cup (K_3^3 \boxplus \text{Crih}(2,3))$.
3. $\text{Crih}(k, k+2) = \{K_{k-4} \boxplus K_3^3 \boxplus K_3^3\} \cup (K_{k-3} \boxplus \text{Crih}(3,5))$ for $k \geq 5$.

7.16 Show that Theorem 5.41 is an immediate consequence of Theorem 7.25.

7.17 Let $H \in \text{Crih}(k+1)$ with $k \geq 2$. If H has exactly one high vertex, then either H has a separating vertex set if size 2, or $k = 2$ and $H = K_1 \boxplus K_n^n$ for some $n \geq 1$, or H is an odd wheel. (Hint: Use Exercise 7.14) (Schweser, Stiebitz, and Toft [911])

7.18 Let H be a nontrivial hypergraph. The **local edge connectivity** $\kappa'_H(v,w)$ of distinct vertices $v, w \in V(H)$ is the maximum number of edge disjoint v-w hyperpaths. Then

$$\kappa'_H(v,w) = \min\{|\partial_H(X)| \mid X \subseteq V(H), v \in X, w \notin X\}$$

(see [386, Theorem 2.5.28] and [598]). The **maximum local edge connectivity** of H is

$$\bar{\kappa}'(H) = \max\{\kappa'_H(v,w) \mid v,w \in V(H), v \neq w\}.$$

Show that every hypergraph H satisfies $\chi(H) \leq \bar{\kappa}'(H)+1$. (Hint: Use Proposition 7.4 and Theorem 7.22) (Toft [1019])

7.19 Let $k \geq 4$ be an integer, and let $C_k = \{H \in \text{Crih}(k) \mid \bar{\kappa}'(H) \leq k-1\}$. Show that the following statements hold:

1. The odd wheels belong to the class C_4 and the complete graphs K_k belong to the class C_k.
2. If $H = H_1 \nabla H_2$ is the Hajós join of two hypergraphs H_1 and H_2, then $H \in C_k$ if and only if $H_1, H_2 \in C_k$.
3. If $H \in C_k$ does not admit a separating vertex set of size at most 2, then either $k = 4$ and H is an odd wheel, or $k \geq 5$ and H is a K_k.
4. If $H \in C_k$ has aseparating vertex set of size 2, then $H = H_1 \nabla H_2$ is the Hajós join of two hypergraphs $H_1, H_2 \in C_k$.

(Hint: Use the results from Section 7.5 and adopt the proof of Theorem 4.11) (Schweser, Stiebitz, and Toft [911])

7.20 For $k \geq 4$ we define a class $\text{Conh}(k)$ of hypergraphs as follows. The class $\text{Conh}(4)$ is the smallest family of hypergraphs that contains all odd wheels and is closed under taking Hajós joins. If $k \geq 5$, then $\text{Conh}(k)$ is the smallest family of hypergraphs that contains all complete graphs of order k and is closed under taking Hajós join. Sow that a hypergraph H belongs to the class C_k if and only if H belongs to the class $\text{Conh}(k)$.

7.21 Prove that if H is a hypergraph with $\bar{\kappa}'(H) = k$ and $k \geq 3$, then $\chi(H) \leq k+1$ and equality holds if and only if H has a block belonging to the class $\text{Conh}(k+1)$. (Schweser, Stiebitz, and Toft [911])

7.22 Show that if an r-uniform hyprgraph H has less than k^{r-1} edges, then $\chi(H) \leq k$. (Hint: Color the vertices uniformly at random with k colors and calculate the expected number of monochromatic edges) (Erdős [331, 332], see also Kostochka [622])

7.23 Let H be a hypergraph in which every edge contains at least r vertices and intersects at most d other edges. If $e(d+1) \leq 2^{r-1}$, then $\chi(H) \leq 2$. (Hint: Color the vertices uniformly at random with two colors and use the Lovász Local Lemma) (Erdős and Lovász [354])

7.24 Use Lovász Local Lemma to show that if H be an r-uniform hypergraph with $\Delta(H) \leq k^{r-1}/(4r)$, then $\chi(H) \leq k$. (Erdős and Lovász [354])

7.25 Use the previous exercise to show that every r-uniform r-regular hypergraph is bipartite, provided that $r \geq 9$.

7.9 Notes

Hypergraphs in general behave similar to graphs with respect to coloring properties. That a connected hypergraph is degree-choosable, unless each of its block is a hyperbrick, that is, a brick or a connected hypergraph consisting of one edge, was proved by Kostochka, Stiebitz and Wirth [636]. The original proof in [636] uses a reduction similar to that used to prove Theorem 1.7. Let H be a hypergraph, let L be a list-assignment of H, let $v \in V(H)$ and $c \in L(v)$. Then let $H' = H \div v$ and let L' be the list-assignment of H' with $L'(u) = L(u) \setminus \{c\}$ if uv is an ordinary edge of H and $L'(u) = L(u)$ otherwise. If H' is L'-colorable, then H is L-colorable. The characterization of the lists in an uncolorable hyperpair described in Theorem 7.2 was first given by Kostochka and Stiebitz [633]. Brooks' theorem for the chromatic number of hypergraphs was obtained in 1975 by Jones [541]; its extension to the list chromatic number (see Theorem 7.3) is due to Kostochka, Stiebitz, and Wirth [636]. Their paper also contains a characterization of the low vertex subhypergraphs of critical and list-critical hypergraphs (see Theorem 7.9(a)). That the Dirac join and the Hajós join can be also applied to hypergraphs was first noticed by Toft [1019, 1023]; he also extended the Gallai join for graphs to hypergraphs (see Theorem 7.13). Propositions 7.15, 7.17, and 7.18 are due to Toft [1019, 1023]. The results in Section 7.5 about connectivity properties of critical hypergraphs are mainly due to Toft [1019]; some of the proofs are from [911]. An immediate consequence of Theorem 7.22 is the fact that every hypergraph H satisfies $\chi(H) \leq \bar{\kappa}'(H) + 1$, where $\bar{\kappa}'(H)$ is the maximum local edge connectivity of H (see Exercise 7.18). This raises the question to characterize those hypergraphs H, for which $\chi(H) = \bar{\kappa}'(H) + 1$ holds. The corresponding class of graphs can be characterized by means of the Hajós join (see Theorem 4.12). This characterization can easily be extended to hypergraphs (see Exercise 7.21), provided that $\bar{\kappa}'(H) \geq 3$. Note that if H is a connected hypergraph with $\bar{\kappa}'(H) \in \{0, 1\}$, then $\chi(H) = \bar{\kappa}'(H) + 1$ if and only if $\bar{\kappa}'(H) = 0$ and $H = K_1$, or $\bar{\kappa}'(H) = 1$ and each block of G consists of just one edge (so it is a K_q^q with $q \geq 2$). If G is a connected graph with $\bar{\kappa}'(G) = 2$, then $\chi(G) = 3$ if and only if at least one block of G is an odd cycle. This follows from the fact that Crit(3) is the family of odd cycles. One cannot hope for such a simple characterization of the class Crih(3), since complexity theory tells us that the problem of deciding if a given hypergraph is 2-colorable is NP-hard. We are not even able to characterize the class of hypergraphs $H \in$ Crih(3) with $\bar{\kappa}'(H) \leq 2$. Therefore, we have not been able to give a characterization for the class of connected hypergraphs H that satisfies $\bar{\kappa}'(H) \leq 2$ and $\chi(H) = 3$.

Theorem 7.38 (The Reducibility Theorem) *Let $k \geq 3$ be fixed. The following types of decision problems are equivalent in the sense that if there is a polynomial time algorithm to solve one of them, then there are polynomial time algorithms for all of them.*

(a) *Hypergraph k-colorability.*
(b) *Hypergraph 3-colorability.*
(c) *Hypergraph 2-colorability.*
(d) *2-colorability of 3-uniform hypergraphs.*

(e) *Graph k-colorability.*
(f) *Graph* 3*-colorability.*
(g) 3*-colorability for planar graphs.*
(h) 3*-colorability for graphs of maximum degree* 4.
(i) 3*-colorability for planar graphs of maximum degree* 4.

Sketch of Proof: Problem (c), that is, the decision problem whether a given hypergraph is 2-colorable, is *reducible* to (b). This follows from the fact that if H is an instance of (c), then we can construct an instance $K_1 \boxplus H$ of (b) in polynomial time, and $\chi(H) \leq 2$ if and only if $\chi(K_1 \boxplus H) \leq 3$ (by Theorem 7.10). Hence a polynomial time algorithm solving (b) leads to a polynomial time algorithm solving (c). By a similar argument it follows that each of the remaining problems listed in the theorem are reducible to (a). That (a) is reducible to (c) follows from the construction described in Exercise 7.6. Since the relation *to be reducible* is transitive, the remaining problems are reducible to (c). Trivially, problems (f), (g), (h) and (i) are reducible to (e), problems (g), (h) and (i) are reducible to (f), and problem (i) is reducible to (g) and (h). That hypergraph 3-colorability is reducible to graph 3-colorability follows from the construction described in Exercise 7.7. Hence problem (b) is reducible to (f). So wee need only to prove that problem (c) is reducible to (d), and problem (f) is reducible to (i). Then any problem listed in the theorem is reducible to any other problem and we are done.

First we want to show that hypergraph 2-colorability is reducible to 3-uniform hypergraph 2-colorability. For an integer p, let WH_p be the hypergraph with $2p+4$ vertices, say $v_1, v_2, \ldots, v_{2p+1}, u_1, u_2, u_3$, and the edges $\{u_1, u_2, u_3\}$, $\{u_i, v_j, v_{j+1}\}$ for $i \in \{1,2,3\}$ and $j \in [1,2p]$, and $\{u_i, v_1, v_{2p+1}\}$ for $i \in \{1,2,3\}$. Evidently, WH_p is 3-uniform and $WH_p \div \{u_1, u_2, u_3\} = C_{2p+1}$. First we claim that $\chi(WH_p) \geq 3$. For otherwise, there is a 2-coloring and both colors occur in the edge $\{u_1, u_2, u_3\}$ and two consecutive vertices of the odd cycle $WH_p \div \{u_1, u_2, u_3\}$, say v_1 and v_2, have the same color. But then one of the three edges $\{u_i, v_1, v_2\}$ ($i \in \{1,2,3\}$) is monochromatic, a contradiction. Hence $\chi(WH_p) \geq 3$ as claimed. Furthermore, it is easy to check that $\chi(WH_p - e) \leq 2$ for every edge e of WH_p. By Propsition 7.7, it then follows that WH_p is a 3-critical hypergraph. Next, we claim that u_1 is a universal vertex of WH_p. To this end, let $H' = S(WH_p, u_1, X, s)$ be the hypergraph obtained from WH_p by splitting u_1 into the independent set X, and let φ' be a 2-coloring of $H'[X]$ where both colors occur in X. Then it is an easy exercise to show that φ' can be extended to a 2-coloring of H'. Consequently, u_1 is a universal vertex of WH_p having degree $2p+2$. Now let H be an arbitrary hypergraph. For each edge e of H, let W_e be a hypergraph WH_p with $p = \lceil (|e|-1)/2 \rceil$ and let u_e be a universal vertex of W_e having degree $2p+2$. We may assume that the hypergraphs W_e with $e \in E(H)$ are disjoint. Now let H' be the hypergraph obtained from H by splitting, for each edge $e \in H$, the vertex u_e of W_e into e (using a simple splitting). Then the resulting hypergraph H' is 3-uniform, and from Theorem 7.13 it follows that $\chi(H) \leq 2$ if and only if $\chi(H') \leq 2$. This shows that problem (c) is reducible to (d).

It remains to show that graph 3-colorability is reducible to graph 3-colorability for planar graphs having maximum degree 4. The reduction is done in two steps.

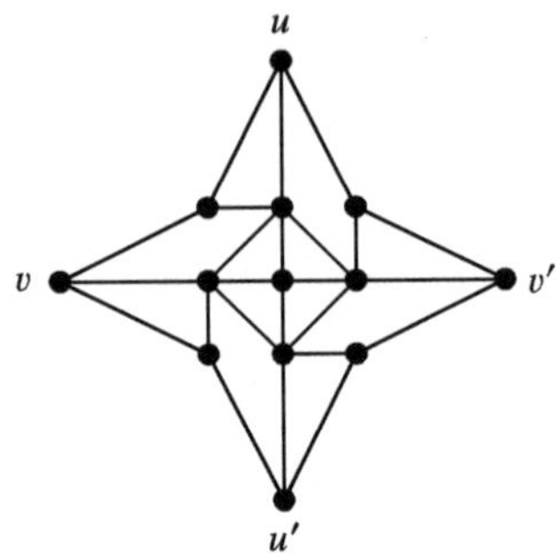

Fig. 7.9 A crossover $CO = CO(u,u',v,v')$.

In the first step, we show that graph 3-colorability is reducible to planar graph 3-colorability, that is, for a given graph G a planar graph G' can be constructed (in polynomial time) such that $\chi(G) \leq 3$ if and only if $\chi(G') \leq 3$. To this end, consider a drawing of G in the plane, also denoted by G, so that each vertex v is represented by a point, also denoted by v, and each edge uv is represented by a piecewise linear curve between the endpoints of the edge, but without double points; this curve is called the (u,v)-line. Crossings are allowed, but two edges (respectively, the corresponding lines) have at most one crossing point, which is an inner point of both edges or a common endpoint. This can be done in polynomial time. To get rid of the crossings, we use the plane graph CO pictured in Figure 7.9, we call CO a crossover with outlets u,u',v,v' as labelled and write $CO = CO(u,u',v,v')$. The two important coloring properties of a crossover $CO = CO(u,u',v,v')$ are the following:

- $\chi(CO) \leq 3$ and if $\varphi \in CO(CO,3)$, then $\varphi(u) = \varphi(u')$ and $\varphi(v) = \varphi(v')$.
- For any color pair (c,c') with $c,c' \in \{1,2,3\}$, there is a $\varphi \in CO(CO,3)$ such that $\varphi(u) = \varphi(u') = c$ and $\varphi(v) = \varphi(v') = c'$.

To construct the plane graph G' from the drawing of G, we proceed as follows:

- To each (u,v)-line of G that is crossed by other lines of G, add new vertices (called extra-points), one between each endpoint u and v of the line and the nearest crossing to it and one between each pair of adjacent crossing-points of the line.
- Replace each crossing by a copy of $CO(u,u',v,v')$, identify u and u' with the nearest two extra-points on one of the lines involved, and identify v and v' with the nearest two extra-points on the other line.
- For each (u,v)-line L of G that is crossed by other lines of G, choose one endpoint, say u, and coalesce it with the nearest extra-point w on L (i.e., we contract the (u,w)-line segment of the L to a single point).

The resulting graph G' is a plane graph. Furthermore, by using the above mentioned coloring properties of a crossover it is an easy exercise to show that $\chi(G) \leq 3$ if and only if $\chi(G') \leq 3$. This completes the proof that graph 3-colorability is reducible to planar graph 3-colorability.

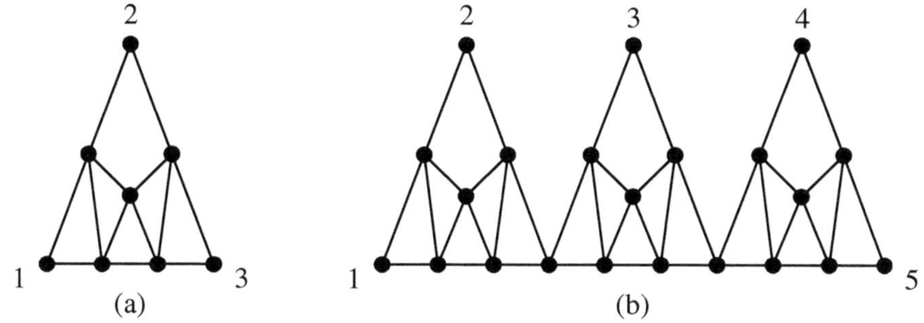

Fig. 7.10 The high degree graph HD_3 and HD_5.

In a second step, we show that planar graph 3-colorability is reducible to planar graph 3-colorability for graphs with $\Delta \leq 4$, that is, for a given planar graph G a planar graph G' with $\Delta(G) \leq 4$ can be constructed, so that $\chi(G) \leq 3$ if and only if $\chi(G') \leq 3$. To get rid of high degree vertices, we use the plane graph $HD_d = HD_d(1,2,\ldots,d)$, where $1,2,\ldots,d$ are the vertices of degree 2 in a clockwise order, see Figure 7.10. We may assume that G is a plane graph (an embedding of G in the plane can be constructed in polynomial time). Let $v_1, v_2, \ldots, v_p$ be the vertices of G having degree at least 5 in G. We construct a sequence $G_0, G_1, G_2, \ldots, G_p$ of plane graphs as follows. We start with $G_0 = G$. Now let $i \in [1,p]$. Then let $d = d_G(v_i) = d_{G_{i-1}}(v)$ and v_iu_1, v_iu_2, ..., v_iu_d be the d edges incident with v_i in G_{i-1} taken in clockwise order. The plane graph G_i is obtained from $G_{i-1} - v_i$ by adding a copy of $HD_d = HD_d(1,2,\ldots,d)$ and by replacing the edge v_iu_j by an edge joining u_j and with the vertex j of HD_d. This can be done so that the resulting graph G_i is a plane graph. Then $G' = G_p$ is a plane graph with $\Delta(G') \leq 4$. Furthermore, it is an easy exercise to show that, for $i \in [1,p]$, we have $\chi(G_{i-1}) \leq 3$ if and only if $\chi(G_i) \leq 3$. This follows from the fact that $\chi(HD_d) = 3$ and every 3-coloring of HD_d assigns the same color to all the vertices having degree 2 in HD_d. Consequently, $\chi(G) \leq 3$ if and only if $\chi(G') \leq 3$. This completes the proof of the second reduction and hence of Theorem 7.38 □

The classical textbook about computational complexity is due to Garey and Johnson [407]. Our outline of the proof that graph 3-colorability is reducible to graph 3-colorability for planar graphs having maximum degree 4 is due to Garey, Johnson, and Stockmeyer [408] and the reader may consult [408] or the book [407, Theorems 4.1 and 4.2] for a more detailed proof. In [408], the authors also show that 3-Satisfiability (3-SAT) is reducible to graph 3-colorability. It is known that 3-SAT is one of the basic NP-complete problems, see [407]. Consequently, all decision problems listed in Theorem 7.38 are NP-complete. By Brooks' theorem planar graph 3-colorability for graphs with $\Delta \leq 3$ belongs to the complexity class P. That hypergraph k-colorability is reducible to hypergraph 2-colorability was proved by Lovász [692]. That hypergraph 3-colorability is reducible to graph 3-colorability follows from Toft's general reduction method discussed in Section 7.4, see also [1023]. This method has proved to be a useful tool and we have shown several interesting applications. The proof that hypergraph 2-colorability is reducible to 3-uniform hypergraph 2-colorability is from Toft's paper [1025]; the reduction is a special case of a more general theorem in that paper.

Theorem 7.39 (TOFT) *Let $k \geq 3$, and let H_1 and H_2 be two disjoint hypergraphs. Let $E(H_1)$ be the disjoint union of E_1 and E_2, where $E_1 \neq \varnothing$. Let $v \in V(H_2)$ and let $E_1' = \{e_1 \cup (e_2 \setminus \{v\}) \mid e_1 \in E_1, e_2 \in E_{H_2}(v)\}$. Furthermore, let H be the hypergraph whose vertex set is $V(H) = V(H_1) \cup V(H_2 - v)$, and whose edge is $E(H) = E_1' \cup E_2 \cup E(H_2 - v)$. If any two of the three hypergraphs H_1, H_2 and H belong to the class* $\mathrm{Crih}(k)$, *then they all belong to the class* $\mathrm{Crih}(k)$.

If we apply Toft's construction to the hypergraphs $H_1 = C_{2p+1}$ and $H_2 = K_1 \boxplus K_3^3$, where $E_2 = \varnothing$ and v is the vertex of degree 3 in H_2, then the resulting hypergraph H is the hypergraph WH_p introduced in the proof of Theorem 7.38 to show that the 2-colorability problem of hypergraphs can be reduced to the 2-colorability problem of 3-uniform hypergraphs. Obviously, $H_1, H_2 \in \mathrm{Crih}(3)$ and thus Toft's theorem implies that $WH_p \in \mathrm{Crih}(3)$.

The decomposition result for hypergraphs (see Theorem 7.25) and its proof is due to Stiebitz, Storch, and Toft [974]. That critical hypergraphs can be used to analyze graphs, multigraphs, and digraphs that are critical with respect to various coloring parameters seems to be folklore; early examples of such applications can be found in the paper by Miller [740] and in the paper by Brown and Corneil [180].

All known results related to the function $\mathrm{exth}(k,n)$ are presented in Section 7.7. Note that Theorems 7.33 and 7.37 lead to a Dirac-type bound, namely

$$2\mathrm{exth}(k,n) \geq (k-1)n + k - 3 \text{ for } n \geq k \geq 4.$$

However, we are not able to characterize the k-critical hypergraphs that satisfy this bound with equality. The corresponding function for k-list-critical hypergraphs does not seem to have been investigated so far; at least no results seem known. However, instead of investigating the number of edges of a hypergraph, one may also investigate the degree sum of a hypergraph H, that is, the parameter

$$d(H) = \sum_{v \in V(H)} d_H(v).$$

Clearly, if H is a r-uniform hypergraph, then $d(H) = r|E(H)|$. Kostochka and Stiebitz [632] proved that if H is a k-list-critical hypergraph with $k \geq 4$ not containing K_k, then $d(H) \geq (k-1)|H| + k - 3$. That Theorem 7.9 and its extension (see Exercise 7.14) can be used to prove a Gallai-type bound for the degree sum was pointed out by Kostochka and Stiebitz [633]; this paper also provides an improvement of the Gallai bound for the degree sum of k-list-critical hypergraphs.

Two subclasses of hypergraphs that are popular among graph theorists are uniform hypergraphs and linear hypergraphs. Recall that a hypergraph H is **linear** if any two distinct edges of H have at most one vertex in common. Let $\mathcal{LH}$ denote the class of linear hypergraph. For an integer $r \geq 2$, let $\mathcal{UH}(r)$ denote the class of r-uniform hypergraphs, which are also refereed to as r-**graphs**. Furthermore, let $\mathcal{WH}(r)$ denote the class of hypergraphs H such that every edge of H has at least r vertices. Clearly, $\mathcal{UH}(r) \subseteq \mathcal{WH}(r)$ and $\mathcal{WH}(3)$ is the class of hypergraphs having no ordinary edges.

For the class of uniform hypergraphs, Wanless and Wood [1061] established an upper bound for the list chromatic number in terms of its maximum degree, thus improving such a bound for the chromatic number obtained earlier by Erdős and Lovász [354]. While the proof of Erdős and Lovász uses the Lovász Local Lemma, the proof of Wanless and Wood uses Rosenfeld's counting technique (see also the proof of Theorem 5.46) and needs no probabilistic argument. The bound by Wanless and Wood is a simple consequence of a more general result about coloring of hypergraphs with forbidden coloring patterns. First, we need some notation.

For a set X, let $\mathbb{Z}^X$ be the set of mappings from X to $\mathbb{Z}$. Let H be a hypergraph. Then we also use $\mathbb{Z}^H$ instead of $\mathbb{Z}^{V(H)}$. So $\mathbb{Z}^H$ is the set of improper colorings of H with the color set $\mathbb{Z}$. Let $\mathcal{F}$ be a map that assigns to every edge e of H a set $\mathcal{F}(e) \subseteq \mathbb{Z}^e$; we call $\mathcal{F}$ a **coloring pattern** for H (Wanless and Wood call $(H,\mathcal{F})$ an **instance**). A map $\phi \in \mathbb{Z}^H$ is called an $\mathcal{F}$**-bad** coloring of H if there is an edge $e \in E(H)$ such that $\phi|_e \in \mathcal{F}(e)$; otherwise ϕ is called an $\mathcal{F}$**-good** coloring of H. Let L be a list-assignment for H with color set $\mathbb{Z}$. Then we denote by $\mathcal{CO}(H,\mathcal{F},L)$ the set of all $\mathcal{F}$-good colorings ϕ of H such that $\phi(v) \in L(v)$ for all $v \in V(G)$. Let $\mathcal{F}^c(H)$ denote the coloring pattern for H, where $\mathcal{F}^c(e)$ consists of all constant maps in $\mathbb{Z}^e$ whenever $e \in E(H)$. Then, we have

$$\mathcal{CO}(H,\mathcal{F}^c(H),L) \neq \varnothing \text{ if and only if } H \text{ is } L\text{-colorable.}$$

Let $(H,\mathcal{F})$ be an arbitrary instance. If $e \in E(H)$, then we say that a set $S \subseteq e$ **determines** $\mathcal{F}(e)$ if any two maps in $\mathcal{F}(e)$ that agree on S are identical. Let $D(e)$ be the sets $S \subseteq e$ that determines $\mathcal{F}(e)$; note that $e \in D(e)$. We say that $(H,\mathcal{F})$ is a **good instance** if for every edge $e \in E(H)$ and every vertex $v \in e$, there is a set $S \in D(e)$ with $S \subseteq e \setminus \{v\}$. Assume that $(H,\mathcal{F})$ is a good instance. Then, for a pair (v,e) with $e \in E_H(v)$, let

$$W(e,v) = |e| - 1 - \min\{|S| \mid S \in D(e), S \subseteq e \setminus \{v\}\}$$

be the **weight** of (e,v). For a vertex $v \in V(H)$ and $p \in \mathbb{N}_0$, define

$$\mathrm{Ex}_p(v) = |\{e \in E_H(v) \mid W(v,e) = p\}|$$

The following theorem is the main result in [1061]. However, it has several interesting implications, as shown in [1061].

Theorem 7.40 (Wanless and Wood) *Let $(H,\mathcal{F})$ be a good instance. Let $k \in \mathbb{N}$ and let $b \in \mathbb{R}$ with $b \geq 1$ such that every vertex v of H satisfies*

$$k \geq b + \sum_{p \in \mathbb{N}_0} b^{-p} \cdot \mathrm{Ex}_p(v).$$

Then $|\mathcal{CO}(H,\mathcal{F},L)| \geq b^{|H|}$ *for every k-assignment L of H.*

Let us consider the interesting case, when H is an r-uniform hypergraph with maximum degree Δ where $r \geq 2$. If $\mathcal{F} = \mathcal{F}^c(H)$, then $(H,\mathcal{F})$ is a good instance and $W(e,v) = r-2$ whenever $e \in E_H(v)$, implying that for every $v \in V(H)$, we have $\mathrm{Ex}_{r-2}(v) = d_H(v) \leq \Delta$ and $\mathrm{Ex}_p(v) = 0$ for $p \neq r-2$. If $r \geq 3$, then we can use Theorem 7.40 with $b = ((r-2)\Delta)^{1/(r-1)}$ and

$$k = \left\lceil \left(\frac{r-1}{r-2}\right)((r-2)\Delta)^{1/(r-1)} \right\rceil.$$

Thus Theorem 7.40 implies that $\chi_\ell(H) \leq k$ and H has at least $b^{|H|}$ list colorings for every k-assignment. If $r = 2$, we can use Theorem 7.40 with $b \geq 1$ and $k = \Delta + b$. For $b = 1$, we obtain that $\chi_\ell(H) \leq \Delta + 1$, and for $b \geq 2$ we obtain that H has at least $b^{|H|}$ list colorings for every $(\Delta + b)$-assignment of H. However, it is easy to prove these well-known facts by means of a greedy algorithm.

Let H be a hypergraph. A graph G is said to be a **substitute** of H if $V(G) = V(H)$ and every edge of H contains an edge of G. It is obvious that if G is a substitute of H, then any coloring of G is a coloring of H, and therefore $\chi(H) \leq \chi(G)$ and $\chi_\ell(H) \leq \chi_\ell(G)$. Gravin and Karpov [429] proved that if $H \in \mathcal{W}(r)$, then there is a substitute G of H, such that $\Delta(G) \leq \lceil 2\Delta(H)/r \rceil$. As a consequence, they obtained the following result, which also holds for the list-chromatic number.

Theorem 7.41 (Gravin and Karpov) *Every hypergraph* $H \in \mathcal{WH}(r)$ *with* $r \geq 3$ *satisfies* $\chi(H) \leq \left\lceil \frac{2\Delta(H)}{r} \right\rceil$.

The proof of the following remarkable result of Frieze and Mubayi [392] is based on the semi-random method (see Section 5.8). This result shows that hypergraphs belonging to the class $\mathcal{UH}(r) \cap \mathcal{LH}$, where $r \geq 3$ is fixed, behave similarly to triangle-free graphs.

Theorem 7.42 (Frieze and Mubayi) *For every integer* $r \geq 3$, *there is a constant c depending on r, such that every linear r-uniform hypergraph H with maximum degree* Δ *satisfies*

$$\chi(H) \leq c\left(\frac{\Delta}{\ln \Delta}\right)^{\frac{1}{r-1}}.$$

Let $k \geq 3$ and $r \geq 2$ be integers. Let $\mathrm{Crih}(k,r,n) = \mathrm{Crih}(k,n) \cap \mathcal{UH}(r)$ and $\mathrm{Crih}^*(k,r,n) = \mathrm{Crih}(k,r,n) \cap \mathcal{LH}$. Note that $\mathrm{Crih}(2) = \{K_r^r \mid r \geq 2\}$. Using the Dirac join, it is easy to show that $\mathrm{Crih}(k,n) \neq \varnothing$ if and only if $n \geq k$. For the two restricted classes of critical hypergraphs such a characterization for the existence problem is not so easy to obtain. As shown in [1] and [1025] if $r \geq 3$, then we have $\mathrm{Crih}(k,r,n) \neq \varnothing$ if and only if $n \geq (r-1)(k-1)+1$. Abott and Liu [5] proved that there is a least integer $L(k,r)$ such that for all $n \geq L(r,k)$ we have $\mathrm{Crih}^*(r,k,n) \neq \varnothing$. The proof uses Ramsey's theorem and gives no direct calculation for the value $L(r,k)$. Note that $\mathrm{Crih}(k,2,n) = \mathrm{Crih}^*(k,2,n) = \mathrm{Crit}(k,n)$ and so $L(k,2) = k+2$. Only one more value seems to be known, namely $L(3,3) = 9$. The study of the minimum number of edges of uniform and linear critical hypergraphs has also attracted some interest. Let

$$\text{exth}(k,r,n) = \min\{|E(H)| \mid H \in \text{Crih}(k,r,n)\}$$

and

$$\text{exth}^*(k,r,n) = \min\{|E(H)| \mid H \in \text{Crih}^*(k,r,n)\}.$$

We obviously have $\text{exth}(k,2,n) = \text{exth}^*(k,2,n) = \text{ext}(k,n)$, and if $r \geq 3$, then

$$\text{exth}(k,n) \leq \text{exth}(k,r,n) \leq \text{exth}^*(k,r,n).$$

If $r \geq 3$ is fixed, then the first interesting value is $k = 3$. By Theorem 7.32, we have $\text{exth}(3,n) = n$ for all $n \geq 3$. Examples of extremal hypergraphs from $\text{Exth}(3,n)$ are $K_1 \boxplus K_{n-1}^{n-1}$ and the hypergraphs obtained from a rooted tree, where the root has degree at least 2 and the distance of the root to the leaves have the same parity, by adding an edge containing all leaves of the tree (see Exercise 7.13). However, none of these hypergraphs are uniform. Liu gave constructions which shows that, as n tends to infinity, $\text{exth}(3,r,n) = n + O(1)$. Burstein [186] proved that for all n, except a finite number of values, $\text{exth}(3,r,n) = n$ for any $r \geq 3$. In his thesis from 1976, Liu [679] showed that for $k,r \geq 3$, the limits

$$a(k,r) = \lim_{n\to\infty} \frac{\text{exth}(k,r,n)}{n} \text{ and } a^*(k,r) = \lim_{n\to\infty} \frac{\text{exth}^*(k,r,n)}{n}$$

exists and are finite. Thus $\text{exth}(k,r,n)$ and $\text{exth}^*(k,r,n)$ grow in essential linear fashion in n. In his proof, Liu uses Fekete's lemma and the fact that both functions are subadditive in n. In the graph case $r = 2$ the subadditivity can be shown by means of the Hajós join. However, the Hajós join leaves only the linearity invariant, but not the uniformity. Therefore, Liu developed a construction that is similar to the Hajós join. Let $H \in \text{Crih}^*(k,r,p)$ with $V(H) = \{v_1, v_2, \dots, v_p\}$ and, for $i \in [1,p]$, let $H_i \in \text{Crih}^*(k+1,r,n_i)$, let $e_i \in E(H_i)$ and $u_i \in e_i$. Assume that the hypergraphs $H, H_1, H_2, \dots, H_p$ are disjoint. Let H^* be the hypergraph obtained from the union

$$H \cup \bigcup_{i=1}^{p} (H_i - e_i)$$

by identifying v_i with u_i, and then by adding a new vertex v^* and the new edge $e_i^* = (e_i \setminus \{u_i\}) \cup \{v^*\}$ for $i \in [1,p]$. The resulting hypergraph H^* is obviously linear and r-uniform. Furthermore, $\chi(H^*) \geq k+1$. For otherwise, there is a coloring $\varphi \in \mathcal{CO}(H^*,k)$ such that $\varphi(v^*) = k$. Then φ induces a coloring of $H_i - e_i$ and so $|\varphi(e_i)| = 1$ as $H_i \in \text{Crih}(k+1)$. This implies that $\varphi(v_i) \neq k$ for $i \in [1,p]$. But then φ induces a coloring of H with color set $[1,k-1]$, which is impossible as $H \in \text{Crit}(k)$. This shows that $\chi(H^*) \geq k+1$. It is also easy to check that $\chi(H^* - e) \leq k$ for every edge e of H^*. Then Proposition 7.7 implies that $H^* \in \text{Crih}(k+1)$ and hence $H^* \in \text{Crih}^*(k+1,r,n)$ with $n = n_1 + n_2 + \cdots + n_p + 1$. As a consequence, we obtain the recursion

$$\text{exth}^*(k+1,r,n) \le \sum_{i=1}^{p} \text{exth}^*(k+1,r,n_i) + \text{exth}^*(k,r,p),$$

where $\text{exth}^*(2,r,p) = 1$. Clearly, the same recursion holds for the other function $\text{exth}(k+1,r,n)$ for $k \ge 2$.

By the results about the class Crih(3) due to Liu and Burstein we have $a(3,r) = 1$ for all $r \ge 3$. Both the construction of Liu and Burstein mentioned above yield linear hypergraphs in the case $r = 3$, so that $a^*(3,3) = 1$. Abott and Hare [2] constructed 3-critical hypergraphs showing that $a^*(3,r) = 1$ for all $r \ge 3$. Further constructions of critical hypergraphs with few edges and no short cycles were developed by Abott, Hare, and Zhou [3, 4]; improved constructions were later given by Kostochka and Nešetřil [626].

Theorem 7.43 (Kostochka and Nešetřil) *For each $r \ge 3, k \ge 3, g \ge 3$ and $\varepsilon > 0$, there exists an r-uniform $(k+1)$-critical hypergraph H with girth at least g such that $|E(H)| < (k-1+\varepsilon)|H|$.*

That the upper bound in the above theorem has the right order of magnitude follows from a result by Kostochka and Stiebitz [631], the asymptotic part was later improved by Bernshteyn [105].

Theorem 7.44 (Kostochka and Stiebitz / Bernshteyn) *Let $k \ge 3$ be an integer. Every $(k+1)$-list-critical hypergraph H without ordinary edges satisfies $|E(H)| \ge (k-o(k))|H|$.*

While Kostochka and Stiebitz use the Lovász Local Lemma and get $o(k) = 3k^{2/3}$, Bernshteyn uses the Local Cut Lemma and gets $o(k) = 4\sqrt{k}$. However, Bernshteyn established his improvement only for critical hypergraphs and not for list-critical hypergraphs.

All known r-uniform $(k+1)$-critical hypergraphs whose average degree is small have many vertices, unless $r = 2$. This raises the question to determine the minimum number of edges of such hypergraphs, no matter what their order is. Given $r, k \ge 2$, define

$$m(r,k) = \min\{|E(H)| \mid H \in \mathcal{UH}(r), \chi(H) > k\}.$$

Note that the corresponding extremal hypergraphs belong to the hypergraph class $\text{Crih}(k+1) \cap \mathcal{UH}(r)$. The problem of determining or estimating $m(r,k)$ was raised in 1961 by Erdős and Hajnal [345]. Their paper is devoted to set theory and uses the corresponding terminology. As emphasised by Erdős and Hajnal the investigation of the function $m(r,k)$ was motivated by a result of Miller [739] about family of sets. Miller introduced the following notation. A family $\mathcal{F}$ of sets possess **property B** if there is a subset X of the ground set of $\mathcal{F}$ such that every set of $\mathcal{F}$ contains an element of X as well as an element of its complement (with respect to the ground set). Using graph theory terminology, we may consider the hypergraph $H = H(\mathcal{F})$ whose vertex set is the ground set of $\mathcal{F}$ and whose edges are the members of $\mathcal{F}$; then $\mathcal{F}$ has property B if and only if H is 2-colorable, i.e. bipartite. However the letter B in Miller's definition does not derive from the hypergraph property *to be bipartite*, but from the name Felix Bernstein, who in his paper [113] from 1908 first stated a necessary condition for a family of sets to possess property B.

Erdős and Hajnal [345] observed that a certain complete uniform hypergraph provides an upper bound for the function $m(r,k)$, namely

$$m(r,k) \le e(K^r_{k(r-1)+1}) = \binom{k(r-1)+1}{r},$$

and that in the graph case $r = 2$ equality applies (see Proposition 1.34(b)). They also conjectured that equality holds when r is large enough, but this was disproved by Alon [42]. Based on simple probabilistic arguments Erdős [331, 332] proved that

$$2^{r-1} \le m(r,2) \le r^2 2^r.$$

It seems that the upper bound has not yet been improved. But the lower bound has been improved several times in the meantime, especially by Beck [87], Spencer [954], and Radhakrishnan and Srinivasan [844]. The last authors proved that for every c with $0 < c < 1/2$, there is an r_0 such that $m(r,2) \ge c2^r\sqrt{r/\ln r}$ for all $r > r_0$. Finding bounds for $m(r,2)$ for small values of r has also attracted a lot of attention; the reader will find a survey of such results in the paper [721] from 2015 and in the paper [20] from 2019. In particular it is known that $m(2,2) = 3$, $m(3,2) = 7$ (see Exercise 7.10), and $m(4,2) = 23$ (see Toft [1025], Seymour [926] and Östergård [795]). That there are close relations between list coloring of bipartite graphs and 2-coloring of hypergraphs was first pointed out by Erdős, Rubin, and Taylor [355]. Denote by $N(k,r)$ the minimum number of vertices in a k-colorable graph with list chromatic number greater than r. For $k = 2$ we have the following result.

Lemma 7.45 *For every $r \ge 2$, $m(r,2) \le N(2,r) \le 2m(r,2)$.*

Proof There exists an r-uniform hypergraph H which is not 2-colorable and has $m = m(2,r)$ edges. We claim that $K = K_{m,m}$ is not r-list colorable. To see this, view the vertices of H as colors and let $A = \{a_e \mid e \in E(H)\}$ and $B = \{b_e \mid e \in E(H)\}$ be the two partite sets of K. Since $|e| = r$ for every edge $e \in E(H)$, the function L with $L(a_e) = L(b_e) = e$ defines an uncolorable r-assignment for K. Suppose on the contrary that there exists an L-coloring φ of K. Then $\varphi(a) \ne \varphi(b)$ for every $a \in A, b \in B$ and $\{\varphi(a_e), \varphi(b_e)\} \subseteq e$ for every $e \in E(H)$. Therefore, (X_1, X_2) with $X_1 = \{\varphi(a_e) \mid e \in E(H)\}$ and $X_2 = X - X_1$ is a partition of X into two independent sets of H, contradicting $\chi(H) \ge 3$. This proves that $N(2,r) \le 2m(r,2)$.

Now, consider a bipartite graph $G = (V,E)$ with $|V| < m(r,2)$ and with the two partite sets V_1 and V_2. Let L be an arbitrary r-assignment for G. Let H be the hypergraph with $V(H) = \bigcup_{v \in V} L(v)$ and $E(H) = \{L(v) \mid v \in V\}$. Since the hypergraph H is r-uniform and $|E(H)| \le |V| < m(r,2)$, there is a 2-coloring φ of H, say with color classes (X_1, X_2). Then $L(v) \cap X_i \ne \emptyset$ for every $v \in V$ and for $i = 1,2$. For $v \in V_i$ with $i \in [1,2]$, let $\varphi(v)$ be a vertex of $L(v) \cap X_i$ in H. Clearly, this defines an L-coloring of G. This proves $N(2,r) \le m(r,2)$. □

Erdős, Rubin, and Taylor [355] proved that $\chi_\ell(K_{n,n}) = \log_2 n + o(\log_2 n)$. Alon [44] proved that if $K(r,t)$ is the complete r-partite graph with t vertices in each of its r parts, then there are two constants c_1 and c_2 such that

$$c_1 \cdot r \ln t \le \chi_\ell(K(r,t)) \le c_2 \cdot r \ln t$$

for all $r,t \ge 2$.

As pointed out by Kostochka [622], the probabilistic argument applied by Erdős to establish bounds for the function $m(r,2)$ can be easily extended to show that

$$k^{r-1} \le m(r,k) \le 20r^2k^r \ln r$$

for every $r,k \ge 2$. For the lower bound see Exercise 7.22. For the upper bound it suffices to show that there is a hypergraph $H \in \mathcal{UH}(r)$ such that $|E(H)| \le m$ for $m = \lfloor 20r^2k^r \ln r \rfloor$ and $\chi(H) > k$. If $m \ge \binom{kr+1}{r}$, we can take $H = K^r_{kr+1}$. Otherwise we use a probabilistic construction. Let V be a set of kr^2 labelled vertices, and let H be the hypergraph with vertex set V, where a set $e \in [V]^r$ belongs to $E(H)$ with probability

$$p = \tfrac{1}{2}m\binom{kr^2}{r}^{-1}$$

independently over all r-subsets of V. Then $\Pr(|E(H)| > m) < 1/2$. The difficult part is to show that the probability that there is some independent set $I \subseteq V$ with $|I| = r^2$ is at most $1/4$. Then with positive probability H has at most m edges and independence number $\alpha(H) < r^2$, and so $\chi(H) > |H|/\alpha(H) = k$.

The best known lower bound for $m(r,k)$ seems to be that of Akolzin and Shabanov [26] who showed that $m(r,k) > c(r/\ln r)k^r$ for all $r,k \ge 2$. Cherkashin and Petrov [216] proved that, for any fixed $r \ge 2$, the sequence $m(r,k)/k^r$ has a finite limit when $k \to \infty$, thereby solving a conjecture due to Alon [42].

No other coloring problem for hypergraphs has attracted so much interest as the investigation of the function $m(r,k)$ and related functions. This is confirmed by the large number of related publications, including the papers [1, 6, 8, 20, 26, 42, 87, 186, 215, 216, 279, 284, 331, 332, 345, 347, 354, 410, 539, 620, 622, 624, 679, 721, 795, 811, 844, 845, 926, 932, 933, 954, 1023], although our list is certainly not complete. Needless to say, that there is both a linear version (consider linear r-uniform hypergraphs) and a list version (replace χ by χ_ℓ) of the the function $m(r,k)$.

A simple probabilistic argument shows that every r-regular r-uniform hypergraph with $r \ge 9$ is bipartite (see Exercise 7.25). That the same statement holds with 9 replaced by 8 was proved by Alon and Bregman [50] by applying a factor theorem to the bipartite graph G associated to a hypergraph H by defining $V(G) = V(H) \cup E(H)$ and $E(G) = \{ve \mid v \in V(H), e \in E(H), v \in e\}$. Thomassen [1005] proved that every r-regular r-uniform hypergraph with $r \ge 4$ is bipartite as conjectured in [50]. The hypergraph H associated to the Fano plane (see Exercise 7.10) is 3-regular and 3-uniform, but has $\chi(H) = 3$. Henning and Yeo [497] proved that if H is a 4-regular 4-uniform hypergraph, then there exists a vertex $v \in V(H)$ such that $H - v$ has a 2-coloring in which each edge has vertices of both colors. Several authors have established sufficient conditions for the existence of a 2-coloring in uniform hypergraphs, e.g. Alon and Tarsi [58], Henning and Yeo [496], Krul and Thoma [658], Seymour [925], Thomassen [1002], Radhakrishnan and Srinivasan [844], and Vishwanathan [1047].

On the one hand it seems difficult to construct hypergraphs belonging to the class $\text{Crih}^*(k,r,n)$, on the other hand the number of such hypergraphs grows exponentially in n (when $k \geq 3$ and $r \geq 3$ are fixed) as proved by Abbott, Liu, and Toft [7]. Their bound for the number of nonisomorphic hypergraphs in $\text{Crih}^*(k,r,n)$ was improved by Rödl and Siggers [874] and even extended to **(r,ℓ)-graphs**, that is, to the class of r-uniform hypergraphs such that no ℓ-subset of $V(H)$ occurs in more than one edge of H. In particular, the class of $(r,2)$-graphs coincides with the class of linear r-uniform hypergraphs. Let $NI(k,r,\ell,n)$ denote the number of nonisomorphic k-critical (r,ℓ)-graphs on n vertices. Then, as proved by Rödl and Siggers [874] for all $k \geq 3$ and $2 \leq \ell < r$, there exists $d = d(k,r,\ell) > 1$ and $n_0 = n_0(k,r,\ell)$ such that $NI(k,r,\ell,n) > d^{n^\ell}$ for all $n > n_0$; for the graph case $r = 2$ see Theorem 4.41.

In their paper [874], Rödl and Siggers continued the investigation on k-critical hypergraphs having many edges, which goes back to a question from Erdős to Dirac (see Section 4.5, the paragraph after Theorem 4.38 and its proof). In [1025], Toft proved that the maximum number of edges of a hypergraph in $\text{Crit}(k,r,n)$ is of order less that n^r for $k = 3$ and is of order n^r for $k > 3$. Evidently, every hypergraph $H \in \text{Crit}(k,r,n)$ satisfies $|E(H)| \leq \binom{n}{r}$. Toft constructed hypergraphs belonging to $\text{Crit}(3,r,n)$ with at least $c_r n^{r-1}$ edges, and quotes a result of Erdős saying that any such hypergraph has at most $c'_r n^{r-1+1/r}$ edges. Lovász used a refined argument from linear algebra to show that every hypergraph belonging to the class $\text{Crit}(3,r,n)$ has at most $\binom{n}{r-1}$ edges. In fact, Lovász [696] proved the following stronger result.

Theorem 7.46 (Lovász) *Let $H \in \text{Crit}(3,r,n)$ be a hypergraph with m edges, and let*

$$\mathbf{I} : [V(H)]^{r-1} \times E(H) \to \mathbb{R}$$

be the incidence matrix defined by $\mathbf{I}(A,e) = 1$ if $A \subseteq e$ and $\mathbf{I}(A,e) = 0$ otherwise. Then the rank of $\mathbf{I}$ is at least m.

Proof The proof is by reductio ad absurdum. So suppose that the rank of $\mathbf{I}$ is less than m. Then the linear equation $\mathbf{Ix} = \mathbf{0}$ has a nontrivial solution $\mathbf{x}$, that is, there is a nonnull vector $\mathbf{x} : E(H) \to \mathbb{R}$ such that

$$(+) \qquad \sum_{e:A\subseteq e} \mathbf{x}(e) = 0 \text{ for all sets } A \in [V(H)]^{r-1}.$$

Then there is an edge $e' \in E(H)$ such that $\mathbf{x}(e') \neq 0$. Since $H \in \text{Crit}(3)$, the hypergraph $H - e'$ is bipartite, that is, the vertex set of H is the disjoint union of two sets, say X and Y, such that both sets are independent sets of $H - e'$. By symmetry, we may assume that e' is not contained in Y, and so $H[Y]$ is edgeless. Then e' is the only edge of H contained in X. For $p \in [0,r]$, let

$$E_p = \{e \in E(H) \mid |e \cap X| = p\} \text{ and let } a_p = \sum_{e\in E_p} \mathbf{x}(e).$$

Since $H[Y]$ is edgeless, $E_0 = \varnothing$ and $a_0 = 0$. Furthermore, $E_r = \{e'\}$ and, therefore, $a_r = \mathbf{x}(e') \neq 0$. Now let

$$\mathcal{A}_p = \{A \in [V(H)]^{r-1} \mid |A \cap X| = p\}.$$

Let $p \in [0, r-1]$. If an edge $e \in E(H)$ contains a subset from $\mathcal{A}_p$, then $e \in E_p \cup E_{p+1}$. Using (+), we then conclude that

$$0 = \sum_{A \in \mathcal{A}_p} \sum_{e : A \subseteq e} \mathbf{x}(e) = (r-p) \sum_{e \in E_p} \mathbf{x}(e) + (p+1) \sum_{e \in E_{p+1}} \mathbf{x}(e),$$

which leads to the recurrence $(r-p)a_p + (p+1)a_{p+1} = 0$. Since $a_0 = 0$, this implies that $a_r = 0$, too. This contradiction proves the theorem. □

Rödl and Siggers obtained the following result about critical (r,ℓ)-graphs with many edges, thereby extending the results of Toft [1025] about critical uniform hypergraphs with many edges. As remarked by Rödl and Siggers an (r,ℓ)-graph on n vertices can have at most $\binom{n}{\ell}/\binom{r}{\ell} = \Theta(n^\ell)$ edges.

Theorem 7.47 (Rödl and Siggers) *For all $k \geq 3$ and $2 \leq \ell < r$, there exists a positive constant $c = c(k,r,\ell)$ and an positive integer $n_0 = n_0(k,r,\ell)$ such that for all $n \geq n_0$, there exists a k-critical (r,ℓ)-graph on n vertices, with at least cn^ℓ edges.*

The coloring of hypergraphs began in 1966 with the publication of the influential paper [347] by Erdős and Hajnal. In this paper they introduced the notions of coloring and the chromatic number of a hypergraph, and obtained the first important results. In particular, they defined the colouring number $\text{col}(H)$ of a hypergraph H; this became an important tool in studying the chromatic number of a hypergraph. Note that the main focus in [347] are infinite graphs and hypergraphs. They also continued the study of hypergraphs having property B, started five years earlier with their paper [345]. The graph concept of degeneracy also applies to hypergraphs. A hypergraph H is **strict k-degenerate** if every nonempty subhypergraph H' of H has a vertex v with $d_{H'}(v) < k$. Then $\text{col}(H)$ is the smallest integer k such that H is strict k-degenerate. As for graphs, every hypergraph H satisfies $\omega(H) \leq \chi(H) \leq \chi_\ell(H) \leq \chi_{\text{DP}}(H) \leq \text{col}(H) \leq \Delta(H) + 1$. That the results about f-partitionable multigraphs, discussed in Section 2.5, can be extended to hypergraphs having multiple edges was established in 2020 by Schweser and Stiebitz [909].

The DP-coloring concept for graphs (see Section 1.4) has a natural counterpart for hypergraphs. In [109], Bernshteyn and Kostochka investigated the minimum number of edges in an r-uniform hypergraph with $\chi_{\text{DP}}(H) > k$, that is, a DP-version of the function $m(r,k)$. Potapov [825] provides some results about hypergraphs H with $\chi_{\text{DP}}(H) > 2$. Schweser [905] characterizes DP-degree colorable hypergraphs (having multiple edges), thereby obtaining a Brooks-type result for the DP-chromatic number of hypergraphs. Schweser proved that if H is a connected hypergraph, then $\chi_{\text{DP}}(H) \leq \Delta(H) + 1$ and equality holds if and only if H is a hyperbrick.

Chapter 8
Homomorphisms and Colorings

Homomorphisms and colorings of graphs are closely related. On the one hand, colorings are just homomorphisms into complete graphs. On the other hand, homomorphisms give rise to various generalizations of ordinary coloring concepts, like fractional colorings, circular colorings and odd colorings.

8.1 Homomorphism Universal Graph Properties

Given graphs G and H, a **homomorphism** from G to H is a map $\phi : V(G) \to V(H)$ for which

$$vw \in E(G) \Longrightarrow \phi(v)\phi(w) \in E(H)$$

for all $v, w \in V(G)$. If there exists a homomorphism from G to H, we say that G is **homomorphic** to H and write $G \to H$. If there exists no homomorphism from G to H we write $G \nrightarrow H$. The homomorphism relation $\to$ is reflexive and transitive, but not symmetric. For example, K_3 is homomorphic to K_4, but not conversely. The homomorphism relation also satisfies

$$G \subseteq H \Longrightarrow G \to H$$

and

$$G \cong H \Longrightarrow G \to H \text{ and } H \to G.$$

The converse of the last implication is not true. For example, K_2 and C_4 are not isomorphic, but $K_2 \to C_4$ and $C_4 \to K_2$. There is a bijective homomorphism from C_4 to K_4, but no isomorphism from C_4 to K_4. A homomorphism ϕ from G to H is an isomorphism from G to H if and only if ϕ is bijective and vw is an edge of G whenever $\phi(v)\phi(w)$ is an edge of H (see also Exercise 8.4).

Let G be a graph. If ϕ is a homomorphism from G to a graph H, then $\phi^{-1}(v)$ is an independent set for all $v \in V(H)$. Hence any homomorphism from G to a graph H is a coloring of G with color set $V(H)$. This implies that G satisfies

M. Stiebitz et al., *Brooks' Theorem*, Springer Monographs in Mathematics,
https://doi.org/10.1007/978-3-031-50065-7_8

$$\chi(G) \leq k \Longleftrightarrow G \to K_k. \tag{8.1}$$

Conversely, any homomorphism from a complete graph K to G is injective and hence

$$\omega(G) \geq k \Longleftrightarrow K_k \to G. \tag{8.2}$$

A homomorphism from a cycle C to G need not be injective, so there is no relation between the existence of such homomorphisms and the girth of G. On the other hand, G satisfies

$$g_o(G) \leq 2p+1 \Longleftrightarrow C_{2p+1} \to G. \tag{8.3}$$

Let us first prove the reverse implication. So let ϕ be a homomorphism from a cycle $C = (v_1, v_2, \ldots, v_{2p+1}, v_1)$ to G. Then $C' = (\phi(v_1), \phi(v_2), \ldots, \phi(v_{2p+1}), \phi(v_1))$ is a closed walk in G with $\ell(C') = 2p+1$. Proposition C.4 then implies that $g_o(G) \leq 2p+1$. Now assume that $g_o(G) \leq 2p+1$. Then G contains an odd cycle $C' = (v_1, v_2, \ldots, v_{2q+1}, v_1)$ with $q \leq p$. If $C = (1, 2, \ldots, 2p+1, 1)$ is an odd cycle of length $2p+1$, then the mapping $\phi : V(C) \to V(C')$ with

$$\phi(i) = \begin{cases} v_i & \text{if } i \leq 2q+1, \\ v_{2q} & \text{if } i > 2q+1 \text{ and } i \equiv 0 \pmod 2, \\ v_{2q+1} & \text{if } i > 2q+1 \text{ and } i \equiv 1 \pmod 2) \end{cases}$$

is a homomorphism from C to C' and hence from C to G.

Proposition 8.1 *For every two graphs G and H,*

$$G \to H \Longrightarrow \chi(G) \leq \chi(H), \omega(G) \leq \omega(H) \text{ and } g_o(G) \geq g_o(H).$$

Proof For the proof we use the fact that the homomorphism relation is transitive. Suppose that $G \to H$. By (8.1), we obtain that

$$\chi(H) = k \Rightarrow H \to K_k \Rightarrow G \to K_k \Rightarrow \chi(G) \leq k.$$

For the clique number, we obtain from (8.2) that

$$\omega(G) = k \Rightarrow K_k \to G \Rightarrow K_k \to H \Rightarrow \omega(H) \geq k.$$

If $g_o(G) = \infty$, then $g_o(H) \leq g_o(G)$. Otherwise, (8.3) implies that

$$g_o(G) = 2p+1 \Rightarrow C_{2p+1} \to G \Rightarrow C_{2p+1} \to H \Rightarrow g_o(H) \leq 2p+1.$$

This proves the proposition. □

Proposition 8.2 *If G is a critical graph, then any homomorphism from G to G is an automorphism of G.*

Proof Let $\phi : V(G) \to V(G)$ be a homomorphism from G to G. Let H be an arbitrary subgraph of G, such that ϕ is also a homomorphism from G to H. By Proposition 8.1, it follows that $\chi(G) \leq \chi(H)$, which gives $H = G$, since G is critical.

So $\text{im}(\phi) = V(G)$ and hence ϕ is a bijective map from $V(G)$ to $V(G)$. Furthermore, $\phi(v)\phi(w)$ is an edge of G if and only if vw is an edge of G. This implies that ϕ is an automorphism. □

Many interesting graphs, such as cycles and complete graphs, are transitive. To decide whether a given graph is homomorphic to a transitive graph, the following result of Albertson and Collins [33] may be of use. As a simple example, no graph G with $\alpha(G) = 2$ and $|G| \geq 6$ is homomorphic to the Petersen graph.

Proposition 8.3 *Let G and H be two graphs with $|G| \geq 1$. If H is transitive, then*

$$G \to H \Longrightarrow \frac{|G|}{\alpha(G)} \leq \frac{|H|}{\alpha(H)}.$$

Proof Let $\mathcal{I}$ denote the set of all independent sets of H with cardinality $\alpha(H)$. If $\varphi \in \text{Aut}(H)$, then $v \in I \in \mathcal{I}$ implies that $\varphi(v) \in \varphi(I) \in \mathcal{I}$. Since H is transitive, it follows that the number of independent sets of $\mathcal{I}$ containing a vertex of H is the same, say m, for any vertex of H. This implies that

$$(1)\ \alpha(H)|\mathcal{I}| = m|H|.$$

Now suppose that $G \to H$, and let $\phi : V(G) \to V(H)$ be a homomorphism from G to H. For every independent set $I \in \mathcal{I}$, the inverse image $\phi^{-1}(I)$ is an independent set of G, which leads to

$$(2)\ \sum_{I \in \mathcal{I}} |\phi^{-1}(I)| \leq \alpha(G)|\mathcal{I}|.$$

Every vertex v of G belongs to exactly m sets of the form $\phi^{-1}(I)$ with $I \in \mathcal{I}$, which gives

$$(3)\ \sum_{I \in \mathcal{I}} |\phi^{-1}(I)| = m|G|.$$

Combining (1), (2) and (3) we obtain

$$\frac{|G|}{\alpha(G)} \leq \frac{|\mathcal{I}|}{m} = \frac{|H|}{\alpha(H)}.$$

This proves the proposition. □

Let $\mathcal{P}$ be a graph property. A graph H is called **homomorphism universal** with respect to $\mathcal{P}$ if $\mathcal{P} = \{G \mid G \to H\}$; in this case $\mathcal{P}$ is said to be a **homomorphism universal graph property**. Clearly, any homomorphism universal graph property is a monotone graph property. The class of k-colorable graphs is homomorphism universal for any fixed k, and K_k is a homomorphism universal graph.

8.2 Fractional Colorings and Kneser Graphs

Let G be an arbitrary nonempty graph. Then $\mathcal{I}(G)$ denotes the set of all independents sets of G, and $\mathcal{I}(G, v)$ denotes the set of all independent sets of G that contain the vertex v. Any k-coloring of G can be viewed as an partition of its vertex set into at most k independent sets; so the chromatic number of G can be defined as the optimal value of a binary linear program. The linear relaxation of this program leads to a new graph invariant.

A weight function $\mathbf{x} : \mathcal{I}(G) \to \mathbb{R}[0, 1] = \{r \in \mathbb{R} \mid 0 \leq r \leq 1\}$ is called a **fractional coloring** of G if every vertex v of G satisfies

$$\sum_{I \in \mathcal{I}(G,v)} \mathbf{x}(I) = 1.$$

Let $\mathcal{FC}(G)$ denote the set of all fractional colorings of G. For a fractional coloring $\mathbf{x} \in \mathcal{FC}(G)$, we call

$$\sum_{I \in \mathcal{I}(G)} \mathbf{x}(I)$$

the **value** of $\mathbf{x}$. The minimum value over all fractional colorings of G is the **fractional chromatic number** of G, denoted by $\chi^*(G)$, that is,

$$\chi^*(G) = \min\{ \sum_{I \in \mathcal{I}(G)} \mathbf{x}(I) \mid \mathbf{x} \in \mathcal{FC}(G)\}.$$

The minimum exists, since this is a feasible LP-problem bounded from below. Any subset of an independent set is itself an independent set, and the chromatic number is the smallest integer k for which its vertex set can be covered by k independent sets. The linear relaxation of this formulation leads to the set $\mathcal{FC}'(G)$ of all weight functions $\mathbf{x} : \mathcal{I}(G) \to \mathbb{R}_{\geq 0}$ such that each vertex v of G satisfies

$$\sum_{I \in \mathcal{I}(G,v)} \mathbf{x}(I) \geq 1.$$

Proposition 8.4 *Every nonempty graph G satisfies*

$$\chi^*(G) = \min\{ \sum_{I \in \mathcal{I}(G)} \mathbf{x}(I) \mid \mathbf{x} \in \mathcal{FC}'(G)\}. \tag{8.4}$$

Proof Let $\mathcal{FC}^o$ be the set of weight functions $\mathbf{x} \in \mathcal{FC}'(G)$ having minimum value $\sum_{I \in \mathcal{I}(G)} \mathbf{x}(I)$. Since $\mathcal{FC}(G) \subseteq \mathcal{FC}'(G)$, it suffices to show that there exists a weight function in $\mathcal{FC}^o \cap \mathcal{FC}(G)$. To this end, we chose a weight function $\mathbf{x} \in \mathcal{FC}^o$ for which

$$p(\mathbf{x}) = |\{v \in V(G) \mid \sum_{I \in \mathcal{I}(G,v)} \mathbf{x}(I) = 1\}|$$

is maximum. If $p(\mathbf{x}) = |G|$, then $\mathbf{x} \in \mathcal{FC}(G)$ and we are done. Otherwise, there is a vertex $v \in V(G)$ for which

$$\sum_{I \in \mathcal{I}(G,v)} \mathbf{x}(I) = b > 1.$$

Let $I_1, I_2, \ldots, I_t$ be an enumeration of the independent sets in $\mathcal{I}(G,v)$. Then there are nonnegative real numbers $a_i \leq \mathbf{x}(I_i)$ such that $\sum_{i=1}^{t} a_i = b - 1$. Define a weight function $\mathbf{x}'$ by

$$\mathbf{x}'(I) = \begin{cases} \mathbf{x}(I) - a_i & \text{if } I = I_i, \\ \mathbf{x}(I) + a_i & \text{if } I = I_i \setminus \{v\} \text{ and } I \neq \varnothing, \\ \mathbf{x}(I) & \text{otherwise.} \end{cases}$$

Then $\mathbf{x}'(I) \geq 0$ for all $I \in \mathcal{I}(G)$. Furthermore, we obtain that for the vertex v we have

$$\sum_{I \in \mathcal{I}(G,v)} \mathbf{x}'(I) = 1,$$

and for a vertex $u \in V(G) \setminus \{v\}$ we have

$$\sum_{I \in \mathcal{I}(G,u)} \mathbf{x}'(I) = \sum_{I \in \mathcal{I}(G,u)} \mathbf{x}(I).$$

Hence, $\mathbf{x}' \in \mathcal{FC}'(G)$ and $\sum_{I \in \mathcal{I}(G)} \mathbf{x}'(I) \leq \sum_{I \in \mathcal{I}(G)} \mathbf{x}(I)$ implying that $\mathbf{x}' \in \mathcal{FC}^o$. Furthermore, $p(\mathbf{x}') > p(\mathbf{x})$, contradicting the existence of v. □

If a nonempty graph G has chromatic number k, then there is a partition $\mathcal{X}$ of $V(G)$ into k independent set. The weight function $\mathbf{x}$ with $\mathbf{x}(I) = 1$ if $I \in \mathcal{X}$ and $\mathbf{x}(I) = 0$ if $I \in \mathcal{I}(G) \setminus \mathcal{X}$ then defines a fractional coloring of G and we have

$$\chi^*(G) \leq \sum_{I \in \mathcal{I}(G)} \mathbf{x}(I) = k = \chi(G).$$

For the empty graph G, put $\chi^*(G) = 0$. Then any graph G satisfies $\chi^*(G) \leq \chi(G)$.

For a nonempty graph G, let $\mathcal{FK}(G)$ denote the set of all weight function $\mathbf{y} : V(G) \to \mathbb{R}_{\geq 0}$ such that each independent set $I \in \mathcal{I}(G)$ satisfies

$$\sum_{v \in I} \mathbf{y}(v) \leq 1;$$

and define

$$\omega^*(G) = \max\{ \sum_{v \in V(G)} \mathbf{y}(v) \mid \mathbf{y} \in \mathcal{FK}(G)\}. \tag{8.5}$$

If X is a clique of G, then the weight function $\mathbf{y}$ with $\mathbf{y}(v) = 1$ if $v \in X$, and $\mathbf{y}(v) = 0$ otherwise, belongs to $\mathcal{FK}(G)$ and we have $|X| = \sum_{v \in V(G)} \mathbf{y}(v) \leq \omega^*(G)$. This shows that the linear program (8.5) is feasible. Since this linear program is also bounded from above, it follows that $\omega^*(G)$ exists and $\omega(G) \leq \omega^*(G)$. A weight function $\mathbf{y} \in \mathcal{FK}(G)$ is called a **fractional clique** of G, and $\omega^*(G)$ is the **fractional clique**

number of G. For the empty graph G, put $\omega^*(G) = 0$. The two programs (8.4) and (8.5) are obviously dual. The strong duality theorem of linear programming then implies that $\omega^*(G) = \chi^*(G)$. Summarizing, we obtain the following result, which is mentioned in the book of Godsil and Roy [418] and in the book of Hell and Nešetřil [495].

Theorem 8.5 *Every graph G satisfies* $\omega(G) \le \omega^*(G) = \chi^*(G) \le \chi(G)$.

The next theorem, also mentioned in [418] and [495], enables us to determine the fractional chromatic number for various types of graphs.

Theorem 8.6 *Every nonempty graph G satisfies*

$$\chi^*(G) \ge \frac{|G|}{\alpha(G)}$$

where equality holds if G is transitive.

Proof The weight function $\mathbf{y}$ with $\mathbf{y}(v) = \frac{1}{\alpha(G)}$ for all $v \in V(G)$ is a fractional clique of G, implying that

$$\frac{|G|}{\alpha(G)} \le \sum_{v \in V(G)} \mathbf{y}(v) \le \omega^*(G) = \chi^*(G)$$

(by Theorem 8.5). So assume that G is transitive. Now let $\mathbf{y}$ be a fractional clique of G whose value is $\omega^*(G)$, i.e., $\sum_{v \in V(G)} \mathbf{y}(v) = \omega^*(G)$. Since G is nonempty, $\omega^*(G) \ge 1$ and $\mathbf{y}$ is nonzero. If $\gamma \in \mathrm{Aut}(G)$, then $\mathbf{y}^\gamma = \mathbf{y} \circ \gamma$ is a fractional clique of G, since $\gamma(I)$ is an independent set for each independent set I of G. It then follows that the weight function $\mathbf{y}'$ defined by

$$\mathbf{y}' = \frac{1}{|\mathrm{Aut}(G)|} \sum_{\gamma \in \mathrm{Aut}(G)} \mathbf{y}^\gamma$$

is also a fractional clique of G. Clearly, $\mathbf{y}^\gamma$ has the same value as $\mathbf{y}$, and so $\mathbf{y}'$ has the same value as $\mathbf{y}$, too. Since $\mathrm{Aut}(G)$ acts transitively on $V(G)$, it then follows that the weight function $\mathbf{y}'$ is constant, say $\mathbf{y}'(v) = c$ for all $v \in V(G)$. Since $\mathbf{y}'$ is a fractional clique of G, this implies that $c \le \frac{1}{\alpha(G)}$, which gives

$$\chi^*(G) = \omega^*(G) = \sum_{v \in V(G)} \mathbf{y}'(v) \le \sum_{v \in V(G)} \frac{1}{\alpha(G)} \le \frac{|G|}{\alpha(G)}$$

and the result follows. □

Corollary 8.7 *For* $n \ge 1$, $\chi^*(K_n) = n$, $\chi^*(C_{2n}) = 2$ *and* $\chi^*(C_{2n+1}) = 2 + \frac{1}{n}$.

In the following we will discuss a different but equivalent way of defining the fractional chromatic number. The fractional chromatic number of a nonempty graph G gives some information about the chromatic number of the composition $G[K_p]$

with $p \geq 1$. Recall from Section 1.9 that $V(G[K_p]) = V(G) \times [1, p]$ and two vertices (v, i) and (w, j) are adjacent in $G[K_p]$ if and only if either (1) $vw \in E(G)$, or (2) $v = w$ and $1 \leq i < j \leq p$. If $\mathbf{y}$ is a fractional clique of G, then the weight function $\mathbf{y}'$ of $G[K_p]$ with $\mathbf{y}'(v, i) = \mathbf{y}(v)$ for all $v \in V(G)$ and $i \in [1, p]$ is a fractional clique of $G[K_p]$ and $\sum_{(v,i)} \mathbf{y}'(v, i) = p \sum_v \mathbf{y}(v)$. Consequently, we obtain $p\omega^*(G) \leq \omega^*(G[K_p])$.

Theorem 8.8 *Let G be a nonempty graph and let $\chi^*(G) = r$. Then r is rational and $\chi(G[K_p]) \geq pr$ for every positive integer p, where equality holds if and only if pr is an integer and there exists a family of pr independents sets in G containing each vertex exactly p times. Furthermore, there are infinitely many positive integers p such that $\chi(G[K_p]) = pr$.*

Proof That r is rational follows from the fact that the fractional chromatic number is the optimal value of a linear program with integer coefficients. By Theorem 8.5, $\chi^*(G) = r$ implies

$$\chi(G[K_p]) \geq \chi^*(G[K_p]) = \omega^*(G[K_p]) \geq p\omega^*(G) = p\chi^*(G) = pr.$$

Clearly, $\chi(G[K_p]) = pr$ if and only if pr is an integer and the vertex set of $G[K_p]$ can be partitioned into pr independent set. This is equivalent to the statement that there exists a family of pr independent sets in G using each vertex exactly p times, because any independent set of $G[K_p]$ contains at most one vertex of each vertical clique $X_v = \{v\} \times [1, p]$ with $v \in V(G)$.

Since the fractional chromatic number is the optimal value of a linear program with integer coefficients, there is an optimal fractional coloring $\mathbf{x}$ of G such that $\mathbf{x}(I)$ is rational for every independent set I of G, where $\chi^*(G) = r = \sum_I \mathbf{x}(I)$. Let s denote the least common multiple of the denominators of all the values $\mathbf{x}(I)$. Then s is a positive integer and $k_I = s\mathbf{x}(I)$ is a nonnegative integer for all independent sets I of G. Since $\mathbf{x}$ is an optimal fractional coloring of G, we deduce that

$$\forall v: \sum_{I \in \mathcal{I}(G,v)} k_I = s \sum_{I \in \mathcal{I}(G,v)} \mathbf{x}(I) = s \text{ and } \sum_{I \in \mathcal{I}(G)} k_I = s \sum_{I \in \mathcal{I}(G)} \mathbf{x}(I) = sr.$$

This means that there is a family of sr independent sets in G using each vertex of G exactly s times. Clearly, this is equivalent to $\chi(G[K_s]) = sr$. If $p = ks$ for a positive integer k, then there is a family of $pr = ksr$ independent sets in G using each vertex exactly $p = ks$ times, and so $\chi(G[K_p]) = pr$. This completes the proof. □

Combining Theorem 8.8 and Proposition 1.45(a)(d), we obtain the following characterization of the fractional chromatic number of an arbitrary graph G:

$$\chi^*(G) = \lim_{p \to \infty} \frac{\chi(G[K_p])}{p} = \min_{p \in \mathbb{N}} \frac{\chi(G[K_p])}{p} = \min_{p \in \mathbb{N}} \frac{\chi(G, p)}{p}. \tag{8.6}$$

Next, we show that colorings of weighted graphs are just homomorphisms into suitable target graphs. Let n and p be integers with $n \geq 2p - 1$ and $p \geq 1$. The **Kneser graph** $KG(n, p)$ is the graph with

$$V(KG(n,p)) = \{A \mid A \subseteq [1,n], |A| = p\}$$
$$E(KG(n,p)) = \{AB \mid A \cap B = \varnothing\}.$$

The Kneser graph $KG(2p-1,p)$ is edgeless and the Kneser graph $KG(2p,p)$ is 1-regular. Furthermore, $KG(n,1) = K_n$ and $KG(5,2)$ is known as the **Petersen graph** (see Figure 8.1).

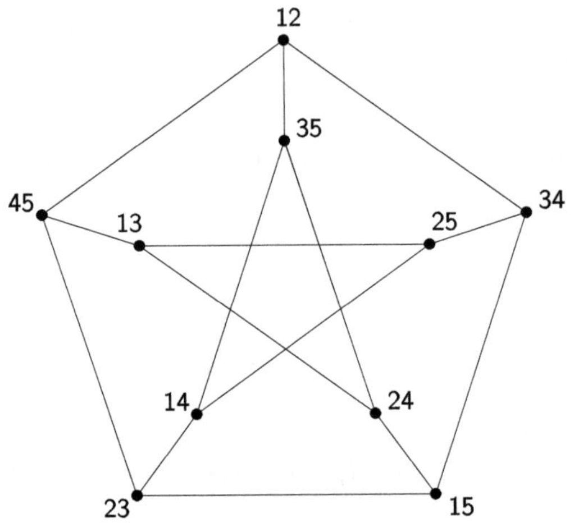

Fig. 8.1 The Petersen graph as the Kneser graph $KG(5,2)$.

Let G be a nonempty graph, and let n,p be integers with $n \geq 2p-1$ and $p \geq 1$. A homomorphism from G to the Kneser graph $KG(n,p)$ is a map that assigns to each vertex of G a p-element subset of $[1,n]$ such that the sets assigned to adjacent vertices are disjoint. But such a map is nothing else than an n-coloring of the weighted graph (G,p), implying that

$$\chi(G,p) \leq n \Longleftrightarrow G \to KG(n,p). \tag{8.7}$$

Theorem 8.9 *Every nonempty graph G satisfies*

$$\chi^*(G) = \min\{\tfrac{n}{p} \mid 1 \leq 2p-1 \leq n, G \to KG(n,p)\}.$$

Proof Because of (8.6) and (8.7), it suffices to show that the set

$$S(G) = \{(n,p) \mid \chi(G,p) = n, \chi^*(G) = \frac{n}{p}, p \geq 1\}$$

contains a pair (n,p) with $n \geq 2p-1$. If G is edgeless, $\chi(G,p) = p$ for all $p \geq 1$ and so $(1,1)$ belongs to $S(G)$. If G has at least one edge, then any pair $(n,p) \in S(G)$ satisfies $n = \chi(G,p) \geq 2p$. Observe that $S(G) \neq \varnothing$, by (8.6). □

Let n and p be integers with $n \geq 2p$ and $p \geq 1$. If $\varphi \in \mathrm{Sym}([1,n])$, then the mapping $A \mapsto \varphi(A)$ defines an automorphism of the Kneser graph $KG(n,p)$. Consequently, $KG(n,p)$ is transitive and $\mathrm{Aut}(KG(n,p))$ contains a subgroup isomorphic to $\mathrm{Sym}([1,n])$. Godsil and Royle [418, Corollary 7.8.2] proved that if $n > 2p$, then the automorphism group of $KG(n,p)$ is indeed isomorphic to the symmetric group $\mathrm{Sym}([1,n])$. Since $KG(n,p)$ is transitive, it follows from Theorem 8.6 that

$$\chi^*(KG(n,p)) \cdot \alpha(KG(n,p)) = |KG(n,p)| = \binom{n}{p}. \tag{8.8}$$

We shall determine the fractional chromatic number of Kneser graphs in the next section (Corollary 8.25). From Theorem 8.9 we deduce that $\chi^*(KG(n,p)) \leq \frac{n}{p}$. A fractional coloring of $KG(n,p)$ can be obtained as follows. Obviously, the set $\mathcal{J}_i = \{A \in V(KG(n,p)) \mid i \in A\}$ forms an independent set in $KG(n,p)$ (for $i \in [1,n]$), and each vertex of $KG(n,p)$ lies in exactly p of these sets. Then the weight function with value $\frac{1}{p}$ on each set $\mathcal{J}_i$, and zero elsewhere, is a fractional coloring of $KG(n,p)$ with value $\frac{n}{p}$, and so $\chi^*(KG(n,p) \leq \frac{n}{p}$.

As a simple consequence of Theorem 8.6, we obtain the following monotonicity property of the fractional chromatic number.

Corollary 8.10 *For every two graphs G and H,*

$$G \to H \Longrightarrow \chi^*(G) \leq \chi^*(H).$$

Proof If $|G| = 0$, this is evident. So assume that $|G| \geq 1$. Since $G \to H$, we have $|H| \geq 1$. From Theorem 8.9 it follows that $\chi^*(H) = \frac{n}{p}$ with $H \to KG(n,p)$ and $n \geq 2p-1 \geq 1$. Since $G \to H$, we deduce that $G \to KG(n,p)$. Using again Theorem 8.9, we obtain that $\chi^*(G) \leq \frac{n}{p} = \chi^*(H)$, as required. □

The name *Kneser graphs* is derived from a conjecture proposed by Kneser [605] in 1955. A graph-theoretic formulation of Kneser's conjecture (due to Lovász) is that $\chi(KG(n,p)) = n - 2p + 2$. It is easy to check that the map

$$\varphi : V(KG(n,p)) \to [1, n-2p+2]$$

with $\varphi(A) = \min\{\min A, n-2p+2\}$ for all $A \in V(K(n,p))$ is a coloring of $KG(n,p)$, and so $\chi(KG(n,p)) \leq n - 2p + 2$. The difficult part of the conjecture is to show that $\chi(KG(n,p)) \geq n - 2p + 2$. That this is indeed the case was first proved by Lovász [697] in 1978.

Theorem 8.11 (The Lovász–Kneser Theorem) $\chi(KG(n,p)) = n - 2p + 2$ *whenever* $n \geq 2p - 1$ *and* $p \geq 1$.

Lovász derived Theorem 8.11 from a general result, which bounds the chromatic number of a graph from below in terms of the connectivity of a topological space associated with the graph (Theorem 8.12). We shall apply this result in Section 8.4 to generalized Mycielski graphs. We need some concepts from algebraic topology, see [120] or [953] for more details.

A(n abstract) **simplicial complex** K is a hereditary set system, that is, $\sigma \subseteq \tau \in \mathsf{K}$ implies $\sigma \in \mathsf{K}$. The elements of K are called (abstract) **simplices** of K (including the empty set). We call $V(\mathsf{K}) = \{x \mid \{x\} \in \mathsf{K}\}$ the **vertex set** of K, so the elements of $V(\mathsf{K})$ are the **vertices** of K. If $V(\mathsf{K})$ is a finite set, then K is said to be **finite**. In this book we only consider finite simplicial complexes.

Let K be a simplicial complex. Then there is an integer $d \geq 1$ and an injective map $i : V(\mathsf{K}) \to \mathbb{R}^d$ such that the point set $i(\sigma \cup \tau)$ is affinely independent in $\mathbb{R}^d$ for all $\sigma, \tau \in \mathsf{K}$. Hence, for every (abstract) simplex $\sigma \in \mathsf{K}$, the convex hull $||\sigma|| = \text{conv}(i(\sigma))$ is a (geometric) simplex in $\mathbb{R}^d$; and $||\sigma \cap \tau|| = ||\sigma|| \cap ||\tau||$ for $\sigma, \tau \in \mathsf{K}$. For $x \in V(\mathsf{K})$, we put $||x|| = i(x)$. Furthermore, $||\mathsf{K}|| = \bigcup_{\sigma \in \mathsf{K}} ||\sigma||$ is a topological subspace of the Euclidean space $\mathbb{R}^d$. This topological space associated with K is unique up to homeomorphism (for a precise definition of this space we refer the reader to Matoušek's book [722]). We call $||\mathsf{K}||$ the **body** of K (induced by the map i). Note that if $\mathsf{K} = \{\varnothing\}$, then $V(\mathsf{K}) = \varnothing$ and $||\mathsf{K}|| = \text{conv}(\varnothing) = \varnothing$.

For $d \geq 0$, let $\mathbf{S}^{d-1} = \{\mathbf{x} \in \mathbb{R}^d \mid ||\mathbf{x}|| = 1\}$ and $\mathbf{B}^d = \{\mathbf{x} \in \mathbb{R}^d \mid ||\mathbf{x}|| \leq 1\}$ denote the ordinary **$(d-1)$-sphere** and the **d-ball**, respectively; both are topological subspaces of the Euclidean space $\mathbb{R}^d$. Here $||\mathbf{x}||$ is the Euclidean norm of the vector $\mathbf{x} \in \mathbb{R}^d$. Note that $\mathbf{S}^{-1} = \varnothing$, $\mathbf{S}^0 = \{-1, 1\}$, $\mathbf{B}^0 = \mathbb{R}^0$ consists of a single point and $\mathbf{B}^1 = \mathbb{R}[-1, 1]$ is an interval. Given an integer $k \geq -1$, a topological space $\mathbf{T}$ is called **k-connected** if for for every integer $d \in [-1, k]$ each continuous map $f : \mathbf{S}^d \longrightarrow \mathbf{T}$ has a continuous extension over $\mathbf{B}^{d+1}$ (or, equivalently, each continuous map $f : \mathbf{S}^d \longrightarrow \mathbf{T}$ is homotopic equivalent to a constant map). Note that $\mathbf{T}$ is (-1)-connected if and only if $\mathbf{T} \neq \varnothing$, and $\mathbf{T}$ is 0-connected if and only if it is (arcwise) connected. Clearly, any k-connected space is also ℓ-connected for every $\ell \in [-1, k]$. The sphere $\mathbf{S}^d$ with $d \geq 0$ is a $(d-1)$-connected topological space (but not d-connected), and so is any topological space homeomorphic, or homotopy equivalent, to $\mathbf{S}^d$ (see [722, Theorem 4.3.2]).

For a nonempty graph G, let $\mathsf{N}(G)$ be the finite simplicial complex, called the **neighborhood complex** of G, whose vertices are the nonisolated vertices of G and whose simplices are those sets of vertices which have a common neighbor in G, i.e. $\sigma \in \mathsf{N}(G)$ if and only if $\sigma \subseteq N_G(v)$ for some $v \in V(G)$.

In his pioneering work of 1978, Lovász [697] established a lower bound for the chromatic number of a graph in terms of the connectivity of the topological space related to its neighborhood complex (see Theorem 8.12). He then used the result to prove Kneser's conjecture (see Theorem 8.11) by showing that $||\mathsf{N}(KG(n, p))||$ is an $(n-2p-1)$-connected topological space.

Theorem 8.12 (Lovász) *Let G be a nonempty graph and $k \geq -1$ be an integer. If $||N(G))||$ is a k-connected topological space, then $\chi(G) \geq k+3$.*

Let G be a nonempty graph and let N be the neighborhood complex of G. If G is edgeless, then $||\mathsf{N}|| = \varnothing$ and so $||\mathsf{N}||$ is k-connected for no k. If G is an even cycle, then $||\mathsf{N}||$ is homeomorphic to the union of two disjoint spheres $\mathbf{S}^1$, and so $||\mathsf{N}||$ is k-connected only for $k = -1$. If G is an odd cycle, then $||\mathsf{N}||$ is homeomorphic to $\mathbf{S}^1$, and so $||\mathsf{N}||$ is 0-connected. If $G = K_n$ with $n \geq 2$, then N consists of all subsets σ of $V(K_n) = [1, n]$ with $|\sigma| \leq n-1$, and so $||\mathsf{N}||$ is homeomorphic to $\mathbf{S}^{n-2}$, which implies that $||\mathsf{N}||$ is $(n-3)$-connected.

Lovász's proof of Kneser's conjecture is one of the earliest applications of topological methods in combinatorics. Over the years, several different proofs and enhancements of the Lovász–Kneser theorem have been given; we refer the reader to the excellent book by Matoušek [722]. All proofs of the Lovász–Kneser theorem listed in [722] use topology, inspired by the Borsuk–Ulam theorem. In 2004, Matoušek [723] gave a purely combinatorial proof of the Lovász–Kneser theorem. He realized that Kneser's conjecture can be directly derived from (a particular version of) Tucker's lemma. Tucker's lemma (see [722, Theorem 2.3.1]) is a statement about labeling the vertices of triangulations of $\mathbf{B}^n$; it resembles Sperner's lemma and can be used to prove the Borsuk–Ulam theorem. Using a proof of Tucker's lemma due to Freund and Todd [390], Matousek obtained a self-contained proof of the Lovász–Kneser theorem.

Now we will state the octahedral Tucker lemma, an adjusted version of Tucker's original lemma that suffices for most applications. The lemma deals with vectors of signs and was first stated explicitly by Ziegler [1104, Lemma 4.1]. For a vector $\mathbf{x} \in \{+,-,0\}^n$, let $\mathbf{x}^+ = \{i \in [1,n] \mid x_i = +\}$ and $\mathbf{x}^- = \{i \in [1,n] \mid x_i = -\}$. Given $\mathbf{x}, \mathbf{y} \in \{+,-,0\}^n$, define $\mathbf{x} \preceq \mathbf{y}$ if $\mathbf{x}^+ \subseteq \mathbf{y}^+$ and $\mathbf{x}^- \subseteq \mathbf{y}^-$. As pointed out in the survey by De Loera, Goaoc, Meunier, and Mustafa [276] there is an interpretation of $(\{+,-,0\}^n, \preceq)$ as a simplicial complex.

Lemma 8.13 (The Octahedral–Tucker Lemma) *Let λ be a map from the set $\{+,-,0\}^n \setminus \{\mathbf{0}\}$ into the set $[-m,m] \setminus \{0\}$ such that $\lambda(-\mathbf{x}) = -\lambda(\mathbf{x})$ for all $\mathbf{x}$. If $\lambda(\mathbf{x}) + \lambda(\mathbf{y}) \neq 0$ whenever $\mathbf{x} \preceq \mathbf{y}$, then $n \leq m$.*

Proof of Theorem 8.11: Let n and p be integers with $n \geq 2p-1$ and $p \geq 1$, let $G = KG(n,p)$, and let $k = \chi(G)$. Then G has a coloring φ with color set $[1,k]$. Our aim is to show that $k \geq n-2p+2$. To this end, we apply Lemma 8.13 and define an appropriate map $\lambda : \{+,-,0\}^n \setminus \{\mathbf{0}\} \to \{\pm 1, \pm 2, \ldots, \pm m\}$ with $m = 2p-2+k$ as follows.

Choose $\mathbf{x} \in \{+,-,0\}^n \setminus \{\mathbf{0}\}$. Let $E^+_{\mathbf{x}} = \{A \in V(G) \mid A \subseteq \mathbf{x}^+\}$, let $E^-_{\mathbf{x}} = \{A \in V(G) \mid A \subseteq \mathbf{x}^-\}$, and let $s(\mathbf{x})$ be the sign of the first nonzero component of $\mathbf{x}$. First assume that $|\mathbf{x}^+| \leq p-1$ and $|\mathbf{x}^-| \leq p-1$. Then $E^+_{\mathbf{x}} = E^-_{\mathbf{x}} = \varnothing$ and we define $\lambda(\mathbf{x}) = s(x)(|\mathbf{x}^+| + |\mathbf{x}^-|)$. Since $\mathbf{x} \neq \mathbf{0}$, we have $\lambda(x) \in \{\pm 1, \pm 2, \ldots, \pm(2p-2)\}$. Clearly, $s(-\mathbf{x}) = -s(\mathbf{x})$ and so $\lambda(-\mathbf{x}) = -\lambda(\mathbf{x})$. Now assume that $|\mathbf{x}^+| \geq p$ or $|\mathbf{x}^-| \geq p$. Then $E_{\mathbf{x}} = E^+_{\mathbf{x}} \cup E^-_{\mathbf{x}}$ is nonempty and $c = \min\{\varphi(A) \mid A \in E_{\mathbf{x}}\}$ belongs to the set $[1,k]$. Note that if $A \in E^+_{\mathbf{x}}$ and $B \in E^-_{\mathbf{x}}$, then $A \cap B = \varnothing$ and so $\varphi(A) \neq \varphi(B)$. Consequently, $E_{\mathbf{x},c} = \{A \in E_{\mathbf{x}} \mid \varphi(A) = c\}$ is either a subset of $E^+_{\mathbf{x}}$ or a subset of $E^-_{\mathbf{x}}$, and we define

$$\lambda(\mathbf{x}) = \begin{cases} +(2p-2+c) \text{ if } E_{\mathbf{x},c} \subseteq E^+_{\mathbf{x}}, \\ -(2p-2+c) \text{ if } E_{\mathbf{x},c} \subseteq E^-_{\mathbf{x}}. \end{cases}$$

Hence $\lambda(\mathbf{x}) \in \{\pm(2p-1), \pm 2p, \ldots, \pm(2p-2+k)\}$ and $\lambda(-\mathbf{x}) = -\lambda(\mathbf{x})$. Furthermore, it is easy to see that if both $\mathbf{x}$ and $\mathbf{y}$ belong to $\{+,-,0\}^n \setminus \{\mathbf{0}\}$ and $\mathbf{x} \preceq \mathbf{y}$, then $\lambda(\mathbf{x}) + \lambda(\mathbf{y}) \neq 0$. Then Lemma 8.13 implies that $n \leq m = 2p-2+k$ and so $\chi(G) = k \geq n-2p+2$. □

Kneser graphs have several interesting properties. In particular, they can be used to construct graphs whose chromatic number and odd girth are arbitrarily large.

Proposition 8.14 *Given positive integers m and k with $m \geq 3$, let $p = k(m-2)$ and $n = (2k+1)(m-2)$. Then $\chi(KG(n,p)) = m$ and $g_o(KG(n,p)) \geq 2k+1$.*

Proof That $G = KG(n,p)$ has chromatic number m follows from Theorem 8.11. Let C be a shortest odd cycle in G. Since $\chi(G) = m \geq 3$, $|C| = g_o(G) = 2\ell+1$ for some integer $\ell \geq 1$. Since $C \subseteq G$, we have $C \to G$. Since the Kneser graph G is transitive, we deduce from Proposition 8.3 and (8.8) that

$$\frac{2\ell+1}{\ell} = \frac{|C|}{\alpha(C)} \leq \frac{|G|}{\alpha(G)} = \chi^*(G) \leq \frac{n}{p} = \frac{2k+1}{k},$$

which yields $\ell \geq k$, and so $g_o(G) \geq 2k+1$. □

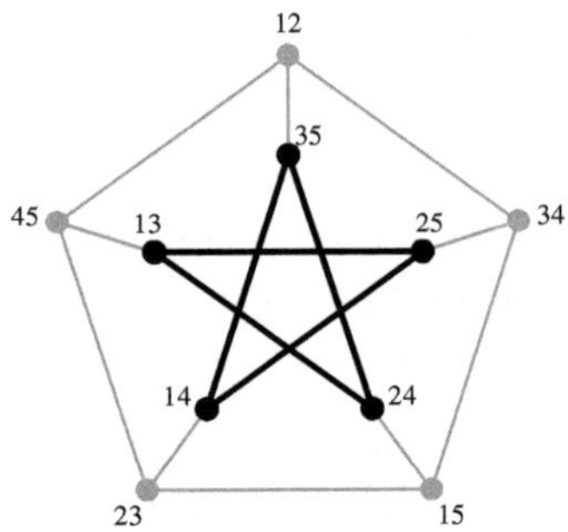

Fig. 8.2 The Schrijver graph $SG(5,2) = C_5$.

Schrijver [902] proved that one can remove a large number of vertices from the Kneser graph $KG(n,p)$ without reducing its chromatic number. Let $SG(n,p)$ denote the graph obtained from the Kneser graph $KG(n,p)$ with $n \geq 2p \geq 2$ by deleting all vertices $A \in [1,n]^p$ that contain either a pair $\{i,i+1\}$ with $i \in [1,n]$, or $\{1,n\}$. The graph $SG(n,p)$ is called the **Schrijver graph** and the **stable Kneser graph**. Viewing $[1,n]$ as the vertex set of a cycle $C_n = (1,2,\ldots,n,1)$, let

$$\mathcal{I}_p(C_n) = \{I \in \mathcal{I}(C_n) \mid |I| = p\}.$$

Then $SG(n,p)$ is the subgraph of $KG(n,p)$ induced by the vertex set $\mathcal{I}_p(C_n)$. For instance, $SG(2p,p)$ is a K_2 and $SG(5,2)$ is a C_5 (see Figure 8.2). By adapting Bárány's short proof [81] of the Lovász–Kneser theorem, Schrijver obtained the following result.

Theorem 8.15 (Schrijver) *For $n \geq 2p \geq 2$, the graph $SG(n,p)$ is a vertex critical induced subgraph of $KG(n,p)$ having chromatic number $n-2p+2$.*

8.3 Circular Colorings and Circular Graphs

In this section we study another generalization of the chromatic number, the so-called circular chromatic number. Let n and d be integers with $1 \le d \le n$. Given a graph G, a **circular** (n,d)**-coloring** of G is a map $\varphi : V(G) \to [0, n-1]$ such that $d \le |\varphi(v) - \varphi(w)| \le n-d$ whenever vw is an edge of G. If G is nonempty, then the **circular chromatic number** of G, written as $\chi^\circ(G)$, is defined by putting

$$\chi^\circ(G) = \inf\{\tfrac{n}{d} \mid G \text{ has a cicular } (n,d)\text{-coloring}\}.$$

For the empty graph G, let $\chi^\circ(G) = 0$. That every graph G satisfies $\chi^\circ(G) \le \chi(G)$ follows from the fact that a circular $(n,1)$-coloring of G is an n-coloring of G.

The circular chromatic number is indeed a 'circular' version of the ordinary chromatic number; this follows from the following characterization of circular colorings due to Zhu [1092], [1094]. Let $t \ge 1$ be a real number, and let I_t° be a circle of length t in the Euclidean plane $\mathbb{R}^2$. An I_t°**-coloring** of a graph G is a mapping φ which assigns to each vertex v of G an unit length open arc of I_t°, so that $\varphi(v) \cap \varphi(w) = \varnothing$ for every edge vw of G. For any nonempty graph G, we have

$$\chi^\circ(G) = \inf\{t \in \mathbb{R} \mid G \text{ has an } I_t^\circ\text{-coloring}, t \ge 1\}.$$

Using the following proposition and the fact that the rational numbers are dense in $\mathbb{R}$, we easily conclude that the two definitions of the circular chromatic number are indeed equivalent.

Proposition 8.16 *Let G be a graph, let n and d be integers with $1 \le d \le n$, and let $t = \frac{n}{d}$. Then the the following statements are equivalent:*

(a) *G has an I_t°-coloring.*
(b) *There is a map $\varphi : V(G) \to I_t^\circ$ such that $d_{I_t^\circ}(\varphi(v), \varphi(w)) \ge 1$ for every edge vw of G, where $d_{I_t^\circ}(x,y)$ denotes the circular distance of $x, y \in I_t^\circ$.*
(c) *There is a map $\varphi : V(G) \to I_t$ with $I_t = \mathbb{R}[0,t]$ such that $1 \le |\varphi(v) - \varphi(w)| \le t-1$ for every edge vw of G.*
(d) *G has a circular (n,d)-coloring.*

Proof For an open unit length arc a of I_t°, let $x(a)$ be the counterclockwise endpoint of a. If a and a' are two such arcs of I_t°, then $a \cap a' = \varnothing$ if and only if $d_{I_t^\circ}(x(a), x(b)) \ge 1$. From this it follows that (a) and (b) are equivalent.

We may view I_t° as obtained from the real interval $I_t = \mathbb{R}[0,t]$ by identifying the two ends. Thus there is a bijection between I_t° and I_t. Then two pints of I_t° have circular distance at least 1 if and only if the corresponding vertices x, y of I_t satisfy $1 \le |x-y| \le t-1$. This implies that (b) and (c) are equivalent.

It remains to show that (c) and (d) are equivalent. First assume that $\varphi : V(G) \to I_t$ satisfies the hypothesis of (c). Let $\phi : V(G) \to [0, n-1]$ be the mapping defined as $\phi(v) = \lfloor d\,\varphi(v) \rfloor$. If vw is an edge of G, then $1 \le |\varphi(v) - \varphi(w)| \le t-1$, which implies $d \le |\phi(v) - \phi(w)| \le d(t-1) = n-d$. So ϕ is a circular (n,d)-coloring of G. Now assume that ϕ is an (n,d)-coloring of G. Let $\varphi : V(G) \to I_t$ be the mapping defined as

$\varphi(v) = \frac{\phi(v)}{d}$. If vw is an edge of G, then we have $d \le |\phi(v) - \phi(w)| \le n - d = d(t-1)$, which implies that $1 \le |\varphi(v) - \varphi(w)| \le t - 1$, and so φ satisfies the conclusion of statement (c). □

It is noteworthy that we obtain a definition of the usual chromatic number if we replace the circle I_t° by the interval (line segment) I_t of length t and the circular arcs also by open subintervals of unit length of I_t in Zhu's definition of the circular chromatic number.

Let n and d be integers with $1 \le d \le n$. A circular (n,d)-coloring can be regarded as a homomorphism into a suitable target graph. The **circular graph** $CG(n,d)$ is the graph with vertex set $[0, n-1]$ and edge set $\{ij \mid d \le |i - j| \le n - d\}$. Note that $CG(n,d)$ is edgeless unless $n \ge 2d$, that $CG(n,1)$ is isomorphic to the complete graph K_n, that $CG(2k,k)$ is 1-regular, and that $CG(2k+1,k)$ is isomorphic to the odd cycle C_{2k+1} (see Figure 8.3). Clearly, if G is a nonempty graph, then G has a circular (n,d)-coloring if and only if $G \to CG(n,d)$, and so

$$\chi^\circ(G) = \inf\{\tfrac{n}{d} \mid G \to CG(n,d), 1 \le d \le n\} \tag{8.9}$$

For an integer $n \ge 2$, let $(\mathbb{Z}_n, +, \cdot)$ denote the ring of integers modulo n. We identify a congruence class in $\mathbb{Z}_n$ with its unique representative, so $\mathbb{Z}_n = [0, n-1]$. Then the circular graph $C = CG(n,d)$ has vertex set $\mathbb{Z}_n$, and two vertices v and w are adjacent in C if and only if $v - w = \pm p$ (in $\mathbb{Z}_n$) for some integer $p \in [d, n-d]$.

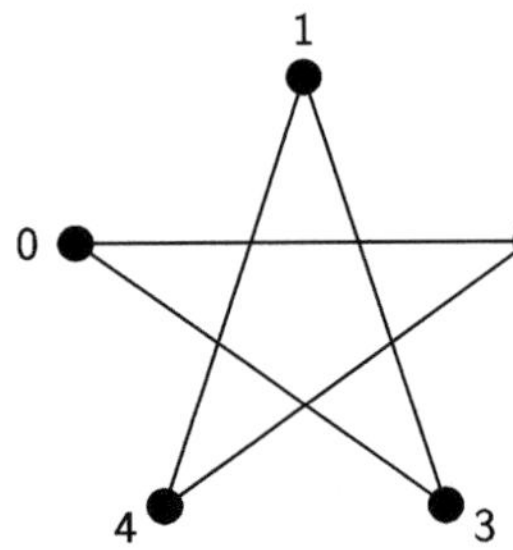

Fig. 8.3 The circular graph $CG(5,2)$.

Bondy and Hell [142] established some basic properties of circular graphs (see the next five propositions). In particular, they proved that the infimum in the definition of the circular chromatic number is actually a minimum (see Theorem 8.22).

Proposition 8.17 *Let n and d be integers with $n \ge 2d$ and $d \ge 1$. Then $CG(n,d)$ is a transitive graph with n vertices, $\alpha(CG(n,d)) = d$, and $\chi^*(CG(n,d)) = \frac{n}{d}$*

Proof Let $C = CG(n,d)$ and let $u, u' \in V(C) = \mathbb{Z}_n$. Then the map $\varphi : \mathbb{Z}_n \to \mathbb{Z}_n$ defined by $\varphi(v) = v + (u' - u)$ for all $v \in \mathbb{Z}_n$ is an automorphism of C (since $\varphi(v) - \varphi(w) = v - w$) and, moreover, $\varphi(u) = u'$. This shows that C is transitive.

That $\alpha(C) \ge d$ follows from the fact that $[0, d-1]$ is an independent set in C. Let I be an independent set of C, and let v be an element of I. If $S = \mathbb{Z}_n \setminus N_C(v)$, then

$I \subseteq S$ and $S \subseteq \{v-i, v+(d-i) \mid i \in [0, d-1]\}$. Since the vertices $v-i$ and $v+(d-i)$ are adjacent in C, we deduce that $|I| \leq d$. This shows that $\alpha(C) = d$. Since C is transitive, we deduce from Theorem 8.6 that

$$\chi^*(C) = \frac{|C|}{\alpha(C)} = \frac{n}{d}.$$

This completes the proof of the lemma. □

Proposition 8.18 *Let n and d be integers with $n \geq 2d-1$ and $d \geq 1$. Then $CG(n,d)$ is isomorphic to an induced subgraph of $KG(n,d)$.*

Proof The statement is evident if $n = 2d-1$, since then $CG(n,d)$ and $KG(n,d)$ are both edgeless and $|CG(n,d)| \leq |KG(n,d)|$. So assume that $n \geq 2d$. Let $C = CG(n,d)$ and $K = KG(n,d)$. For $v \in \mathbb{Z}_n$, let $I_v = \{v, v+1, \ldots, v+d-1\}$ be an interval of $\mathbb{Z}_n$. Note that $|I_v| = d$, and so I_v is a vertex of K. Let v and w be two vertices of C, so $v, w \in \mathbb{Z}_n$. Then $vw \in E(C)$ if and only if $v - w = \pm p$ (in $\mathbb{Z}_n$) for some integer $p \in [d, n-d]$. This implies that $vw \in E(G)$ if and only if $I_v \cap I_w = \varnothing$. It follows that the map $\varphi : V(C) \to V(K)$ with $\varphi(v) = I_v$ for all $v \in V(C)$ is an isomorphism from C to the subgraph of K induced by the vertices $I_0, I_1, \ldots, I_{n-1}$. This proves the lemma. □

Proposition 8.19 *Let n, d and h be positive integers with $n \geq 2d$. Then $CG(n,d) \to CG(hn, hd)$ and $CG(hn, hd) \to CG(n,d)$.*

Proof There is a unique (group) homomorphism φ of $(\mathbb{Z}_n, +)$ to $(\mathbb{Z}_{hn}, +)$ with $\varphi(1) = h$. Then it is easy to check that φ is a homomorphism from $CG(n,d)$ to $CG(hn, hd)$. Conversely, let $\phi : \mathbb{Z}_{hn} \to \mathbb{Z}_n$ be the mapping with $\phi(v) = u$ if and only if $v \in \{\varphi(u), \varphi(u)+1, \ldots, \varphi(u)+h-1\}$. Then it is also easy to check that ϕ is a homomorphism from $CG(hn, hd)$ to $CG(n,d)$. □

Proposition 8.20 *Let n, n', d and d' be positive integers such that $n \geq 2d$. Then $CG(n,d) \to CG(n', d')$ if and only if $\frac{n}{d} \leq \frac{n'}{d'}$.*

Proof First assume that $CG(n,d) \to CG(n',d')$. Since $n \geq 2d$, the circular graph $CG(n,d)$ has edges and so has its homomorphic image $CG(n',d')$, which gives $n' \geq 2d'$. By Proposition 8.17 and Corollary 8.10, we then deduce that

$$\frac{n}{d} = \chi^*(CG(n,d)) \leq \chi^*(CG(n',d')) = \frac{n'}{d'}$$

Next assume that $\frac{n}{d} \leq \frac{n'}{d'}$. By Proposition 8.19, we have $CG(n,d) \to CG(n'n, n'd)$ as well as $CG(nn', nd') \to CG(n', d')$. But $n'd \geq nd'$, so $CG(nn', n'd) \subseteq CG(nn', nd')$, which gives $CG(nn', n'd) \to CG(nn', nd')$. Since the homomorphism relation is transitive, we deduce that $CG(n,d) \to CG(n', d')$. □

Proposition 8.21 *Let n and d be relatively prime integers such that $n \geq 2d$ and $d \geq 1$. Then there are integers n' and d' such that $1 \leq d' \leq n' < n$, $\frac{n'}{d'} < \frac{n}{d}$ and $CG(n,d) - v \to CG(n', d')$ for any $v \in V(CG(n,d))$.*

Proof Since n and d are relatively prime, there exists an integer $a \in \mathbb{Z}_n$ with $ad = 1$ (in $\mathbb{Z}_n$). Then there is an integer $b \in \mathbb{Z}$ such that $ad - 1 = bn$ (in $\mathbb{Z}$). Clearly, we have $1 \le a \le n-1$ and $0 \le b \le d-1$. Define $n' = n - a$ and $d' = d - b$. Then $1 \le d' \le n' < n$. Furthermore, $ad - 1 = bn$ implies that

$$\frac{n'}{d'} = \frac{n-a}{d-b} = \frac{n}{d} - \frac{1}{d(d-b)} < \frac{n}{d}.$$

It remains to show that $CG(n,d) - v \to CG(n',d')$ for any $v \in V(CG(n,d))$. By Proposition 8.17, $CG(n,d)$ is transitive, so it suffices to consider the case $v = d$. In the integer ring $\mathbb{Z}_n$, let $S = \{d, 2d, \ldots, ad\}$ and $I_v = \{v, v+1, \ldots, v+d-1\}$ (for $v \in \mathbb{Z}_n$). Since $ad = 1$ (in $\mathbb{Z}_n$), we have $|S| = a$ and $1 \in S$. Furthermore, it is easy to show that each interval I_v, $v \neq 1$, contains exactly b elements of S, while I_1 contains exactly $b+1$ elements. If $v, w \in \mathbb{Z}_n$, then vw is an edge of $CG(n,d)$ if and only if $I_v \cap I_w = \emptyset$ (see the proof of Proposition 8.18). We shall now construct a homomorphism φ of $CG(n,d) - d$ to $CG(n',d')$. For each vertex v of $CG(n,d)$ different from d, let

$$\varphi(v) = v - |\{u \in S \mid u \le v \text{ (in } \mathbb{N})\}|.$$

If vw is an edge of $CG(n,d) - d$ with $v < w$, then $d \le w - v \le n - d$. Taking into account the above remarks, we then deduce that

$$d' = (d-b) \le \varphi(w) - \varphi(v) \le (n-a) - (d-b) = (n' - d'),$$

which implies that $\varphi(v)\varphi(w)$ is an edge of $CG(n',d')$. This completes the proof. □

Theorem 8.22 (Bondy and Hell) *Every nonempty graph G satisfies*

$$\chi^\circ(G) = \min\{\frac{n}{d} \mid G \to CG(n,d), 1 \le 2d-1 \le n \le |G|\}.$$

Proof The statement is evident if G is edgeless, since then $G \to CG(1,1) = K_1$ and $\chi^\circ(G) = 1$. So assume that G is not edgeless. If $G \to CG(n,d)$, then $n \ge 2d$ and $d \ge 1$. By Proposition 8.19, we may assume that n and d are relatively prime. If $n > |G|$, then $G \to CG(n,d) - v$ for some vertex v of $CG(n,d)$, and so it follows from Proposition 8.21 that $G \to CG(n',d')$, where n' and d' are integers such that

$$1 \le d' \le n' < n \text{ and } \frac{n'}{d'} < \frac{n}{d}.$$

Since G has edges, $G \to C(n',d')$ implies that $n' \ge 2d'$. By (8.9), this implies that

$$\chi^\circ(G) = \inf\{\tfrac{n}{d} \mid G \to CG(n,d), 1 \le 2d-1 \le n \le |G|\}.$$

Since the set $\{\tfrac{n}{d} \mid G \to CG(n,d), 1 \le 2d-1 \le n \le |G|\}$ is finite, the infimum can be replaced by a minimum. □

Corollary 8.23 *For every two graphs G and H,*

$$G \to H \Longrightarrow \chi^\circ(G) \le \chi^\circ(H).$$

Proof If $|G| = 0$, this is evident. So assume that $|G| \geq 1$. Since $G \to H$, we have $|H| \geq 1$. From Theorem 8.22 it follows that $\chi^\circ(H) = \frac{n}{d}$ with $H \to CG(n,d)$. Since $G \to H$, we deduce that $G \to CG(n,d)$. Using Theorem 8.22 again, we get that $\chi^\circ(G) \leq \frac{n}{d} = \chi^\circ(H)$, as needed. □

Corollary 8.24 *Every graph G satisfies* $\chi(G) - 1 < \chi^\circ(G) \leq \chi(G)$ *and, moreover,* $\chi^*(G) \leq \chi^\circ(G)$.

Proof If $|G| = 0$, this is evident. So assume that $|G| \geq 1$ and $\chi(G) = k'$. Then $G \to K_{k'} = CG(k',1)$ (by (8.1)), and so $\chi^\circ(G) \leq k'/1 = k'$ (by Theorem 8.22). Furthermore, Theorem 8.22 implies that $\chi^\circ(G) = \frac{n}{d}$, where $1 \leq 2d-1 \leq n \leq |G|$ and $G \to CG(n,d)$. By Proposition 8.18, $CG(n,d) \to KG(n,d)$, and so $G \to KG(n,d)$, which gives $\chi^*(G) \leq \chi^*(KG(n,d)) \leq \frac{n}{d} = \chi^\circ(G)$ (by Corollary 8.10 and Theorem 8.9). It remains to show that $\chi(G) - 1 < \chi^\circ(G)$. If $n \leq 2d-1$, then $CG(n,d)$ is edgeless, and so is G, since $G \to CG(n,d)$. Consequently, $\chi(G) = 1$, which gives $\chi(G) - 1 = 0 < \chi^\circ(G)$, as required. If $n \geq 2d$, then we argue as follows. Suppose, to the contrary, that $\chi^\circ(G) \leq \chi(G) - 1$, and so $2 \leq \frac{n}{d} \leq k' - 1$. Then it follows from Proposition 8.20 that $G \to CG(k'-1,1) = K_{k'-1}$, which gives $\chi(G) \leq k'-1$ (by (8.1)). This contradiction shows that $\chi(G) - 1 < \chi^\circ(G)$, as required. □

Corollary 8.25 $\chi^*(KG(n,p)) = \frac{n}{p}$ *if* $n \geq 2p$ *and* $p \geq 1$.

Proof By Theorem 8.9, we have $\chi^*(KG(n,p)) \leq \frac{n}{p}$. By Proposition 8.18, we have $CG(n,p) \to KG(n,p)$. Using Proposition 8.17 and Corollary 8.10, we obtain that $\frac{n}{p} = \chi^*(CG(n,p)) \leq \chi^*(KG(n,p))$. Thus the proof is complete. □

Combining Corollary 8.25 and (8.8), we obtain the following result, known as (a part of) the Erdős–Ko–Rado theorem [353].

Corollary 8.26 $\alpha(KG(n,p)) = \binom{n-1}{p-1}$ *if* $n \geq 2p$ *and* $p \geq 1$.

Let $n, d \in \mathbb{N}$ with $n \geq 2d$ and $d \geq 1$. From Theorem 8.22 and Proposition 8.20 we deduce that $\chi^\circ(CG(n,d)) = \frac{n}{d}$. Furthermore, we obtain that every graph G satisfies

$$\chi^\circ(G) \leq \frac{n}{d} \Longleftrightarrow G \to CG(n,d).$$

Since $CG(2p+1,p) = C_{2p+1}$ ($p \geq 1$), this implies that

$$\chi^\circ(G) \leq 2 + \frac{1}{p} \Longleftrightarrow G \to C_{2p+1}. \tag{8.10}$$

Another consequence of Theorem 8.22 is that if $\chi^\circ(G) = r$, then r is a rational number, where $r \in \{0,1\}$ or $r \geq 2$. Clearly, for a nonempty graph G, we have $\chi^\circ(G) \geq 1$, where equality holds if and only if G is edgeless. If G has at least one edge, then $\chi^\circ(G) \geq 2$ with equality if and only if G is bipartite. For the complete graph $K_n = CG(n,1)$ and the odd cycle $C_{2n+1} = CG(2n+1,n)$ with $n \geq 1$, we obtain

$$\chi^\circ(K_n) = n \text{ and } \chi^\circ(C_{2n+1}) = 2 + \frac{1}{n}.$$

The family of Kneser graphs shows that the difference between the fractional chromatic number and the chromatic number can be arbitrarily large (by Theorem 8.11 and Corollary 8.25). On the other hand, Corollary 8.24 implies that the chromatic number of a graph is obtained by taking the ceiling of its circular chromatic number. So the circular chromatic number is a refinement of the chromatic number of a graph, and it gives more information about the graph.

8.4 Odd Girth and Mycielski Graphs

While Kneser graphs and circular graphs are target graphs for homomorphism universal graph properties, Mycielski graphs are a class of graphs recursively defined by means of the following graph transformation.

Given a graph G and an integer $r \geq 1$, one can transform G into a new graph $\mathrm{M}_r(G)$ defined by

$$\begin{aligned} V(\mathrm{M}_r(G)) &= V(G) \times [1,r] \cup \{z\} \\ E(\mathrm{M}_r(G)) &= \{(u,1)(v,1) \mid uv \in E(G)\} \cup \{(u,i)(v,j) \mid uv \in E(G), \\ &\qquad |i-j| = 1, i,j \in [1,r]\} \cup \{(u,r)z \mid u \in V(G)\}. \end{aligned}$$

Furthermore, for positive integers $r_1, r_2, \ldots, r_k$, we define the graph $\mathrm{M}_{r_1,r_2,\ldots,r_k}(G)$ recursively by

$$\mathrm{M}_{r_1,r_2,\ldots,r_k}(G) = \mathrm{M}_{r_1}(\mathrm{M}_{r_2,\ldots,r_k}(G))$$

when $k \geq 2$. In case of $r_1 = r_2 = \cdots = r_k = r$ we also write

$$\mathrm{M}_r^k(G) = \mathrm{M}_{r_1,r_2\ldots,r_k}(G).$$

Note that $\mathrm{M}_1^k(G) = G \boxplus K_k$. The graph transformation M_2 was invented in 1955 by Mycielski [766], so we call $\mathrm{M}_r(G)$ a (generalized) **Mycielski graph** of G. Mycielski proved that $\mathrm{M}_2^k(K_2)$ is a triangle-free graph with chromatic number $k+2$. The first graph in this sequence is $\mathrm{M}_2(K_2) = C_5$ and the second graph $\mathrm{M}_2^2(K_2) = \mathrm{M}_2(C_5)$ is also known as Grötzsch's graph (see Figure 8.4).

Let $H = \mathrm{M}_r(G)$ be a Mycielski graph of G. The set $X_i = V(G) \times \{i\}$ is called the ***i*th level** of H and z is called the **top vertex** of H. Note that $H[X_1] \cong G$, $H[X_i]$ is edgeless, $H[X_i \cup X_{i+1}] \cong G \times K_2$ (for $i \in [2, r-1]$), and $N_H(z) = X_r$. Here $G \times K_2$ is the direct product of G and K_2 (see Appendix C.7 or Section 8.6). Furthermore, $H - E(H[X_1])$ is bipartite, and so is $H - X_1$.

An advantage of the Mycielski transformation M_r is that it preserves the clique number, except in the cases where $r = 1$ or $\omega \leq 1$. On the other hand, $\Delta(G) \geq 2$ implies that $g(\mathrm{M}_r(G)) \leq 4$. For the odd girth, however, we have the following result.

Proposition 8.27 *Let G be a graph with at least one edge and let $r \geq 1$. Then $g_o(M_r(G)) = \min\{g_o(G), 2r+1\}$.*

Proof Let $H = \mathrm{M}_r(G)$, let X_i be the ith level of H, and let z be the top vertex of H. Since G has an edge, H contains $C_{2r+1} = \mathrm{M}_r(K_2)$ and, therefore, $g_o(H) \le 2r+1$. Since $G \cong H[X_1]$, we also have $g_o(H) \le g_o(G)$.

Now consider an odd cycle C of H. First, suppose that $z \in V(C)$. Since the graph $H \setminus E(H[X_1])$ is bipartite, this implies that $|V(C) \cap X_i| \ge 2$ for every $i \in [1,r]$. Thus we have $\ell(C) \ge 2r+1$. Next, suppose that $z \notin V(C)$. Let C' be the projection of C into X_1, that is, we replace each vertex (u,i) of C by the vertex $(u,1)$. Then C' is an odd closed walk in $H[X_1]$. Since $H[X_1] \cong G$, we deduce that $\ell(C) = \ell(C') \ge g_o(G)$ (see Proposition C.4). This completes the proof. □

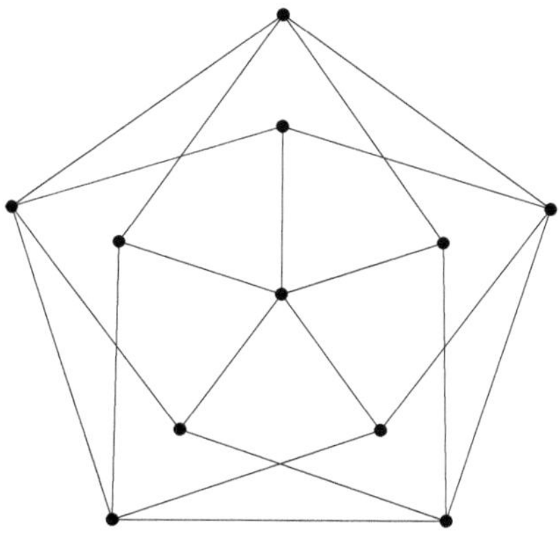

Fig. 8.4 The Mycielski–Grötzsch graph $M_2(C_5)$.

Unlike the clique number, the chromatic number does not behave uniformly with respect to the Mycielski transformation, except when $r \in \{1,2\}$, in which case the chromatic number always increases by one.

Proposition 8.28 *Let G be a nonempty graph and let $r \ge 1$. Then*

$$\chi(G) \le \chi(M_r(G)) \le \chi(G)+1 \tag{8.11}$$

and $\chi(M_r(G)-z) = \chi(G)$, where z is the top vertex of $M_r(G)$. Furthermore, if $r \in \{1,2\}$, then $\chi(M_r(G)) = \chi(G)+1$.

Proof It suffices to show that $\chi(\mathrm{M}_r(G)) = \chi(G)+1$, provided that $r \in \{1,2\}$. For $r = 1$, this follows from Theorem 4.1, since $\mathrm{M}_1(G) = G \boxplus K_1$. So let $H = \mathrm{M}_2(G)$ and suppose, to the contrary, that $\chi(H) = \chi(G) = k$. Then there exists a coloring φ of H with a set C of k colors, and $V(H) = X_1 \cup X_2 \cup \{z\}$, where $X_i = \{(v,i) \mid v \in V(G)\}$. Since the top vertex z is completely joined to X_2, the color $c = \varphi(z)$ does not belong to the set $C' = \varphi(X_2)$. Since $G[X_1] \cong G$ and $\chi(G) = k$, it follows from Proposition 1.34 that there is a vertex $u = (v,1)$ in X_1 such that $\varphi(u) = c$ and all other colors occur in the neighborhood $Y = N_H(u) \cap X_1$, that is, $\varphi(Y) = C \setminus \{c\}$. But then $Z = Y \cup \{z\}$ belongs to the neighborhood of $u' = (v,2)$ in H, and $\varphi(u') \in C = \varphi(Z)$, which is impossible. □

Proposition 8.29 *Let G be a k-critical graph with $k \geq 2$ and let $r \geq 1$. Then $M_r(G)$ is $(k+1)$-critical, provided that $\chi(M_r(G)) = k+1$.*

Proof Let $H = \mathrm{M}_r(G)$, let X_i be the ith level of H, and let z be the top vertex of H. As G is k-critical and $k \geq 2$, G is connected and has at least one edge. This implies that H is connected, too. Thus it suffices to show that $H - e$ has a k-coloring for all edges e of H. We distinguish three cases.

Case 1: $e \in E(H[X_1])$. Since $G \cong H[X_1]$ is k-critical, there is a $(k-1)$-coloring of $H[X_1] - e$ with color set $C = [1, k-1]$. This coloring can be extended to a k-coloring of $H - e$ by alternately coloring the remaining vertices in $X_2, X_3, \ldots, X_r, \{z\}$ with the colors k and 1.

Case 2: $e = (w,i)(w',i+1)$ *with* $1 \leq i \leq r+1$. Then $e' = ww'$ is an edge of G. Again, as G is k-critical, there is a $(k-1)$-coloring φ of $G - e'$ with color set $C = [1, k-1]$. Since G is not $(k-1)$-colorable, $\varphi(w) = \varphi(w')$. Now we define a map $\phi : X \to [1,k]$, where $X = \bigcup_{j=1}^{i+1} X_j$. If $1 \leq j \leq i$, then put

$$\phi((v,j))) = \begin{cases} \varphi(v) & \text{if } v \neq w', \\ k & \text{if } v = w'; \end{cases}$$

furthermore, let $\phi(v, i+1) = \varphi(v)$. Then it is easy to check that ϕ is a k-coloring of $H[X] - e$, which can be extended to a k-coloring of $H - e$.

Case 3: $e = (u,r)z$. Since G is k-critical, there is a k-coloring φ of G with color set $C = [1,k]$ such that $\varphi(v) = k$ only for $v = u$. Then color each vertex (v,i) of H with the color $\varphi(v)$, and color the vertex z with k. This results in a k-coloring of $H - e$. □

Combining Proposition 8.28 and Proposition 8.29 we obtain the following result of Schäuble [893].

Theorem 8.30 *If G is a graph with at least one edge, then $\chi(M_2(G)) = \chi(G) + 1$. Furthermore, $M_2(G)$ is critical, provided that G is critical.*

Note that $\chi(\mathrm{M}_r(G)) = \chi(G) + 1$ does not hold in general when $r \geq 3$. For every integer $k \geq 4$, let $F_k = K_{k-4} \boxplus C_7^2$. Then, as proved by Tardif [987], $\chi(\mathrm{M}_3(F_k)) = \chi(F_k) = k$. The graph C_7^2 is the complement of a 7-cycle $C = (v_1, v_2, v_3, v_4, v_5, v_6, v_7, v_1)$. Since $\alpha(\overline{C}) \geq 2$, it follows from (1.2) that $\chi(\overline{C}) \geq 4$. So it suffices to show that $H = \mathrm{M}_3(\overline{C})$ has a 4-coloring, which gives that $\chi(\mathrm{M}_3(\overline{C})) = \chi(\overline{C}) = 4$. For a map $\varphi : V(H) \to [1,4]$, let

$$\vec{v}_i = (\varphi(v_1, i), \varphi(v_2, i), \ldots, \varphi(v_7, i))$$

for $i \in \{1,2,3\}$. If we choose φ such that

$$\vec{v}_1 = (1,1,4,3,3,2,2),$$
$$\vec{v}_2 = (1,4,4,4,3,2,2),$$
$$\vec{v}_3 = (1,1,4,3,3,3,1),$$

and $\varphi(z) = 2$ for the top vertex z of H, then it is easy to check that φ is a 4-coloring of H.

On the other hand, it is easy to see that $\mathrm{M}_r(K_k) = k+1$ for all $r \geq 1$. This follows from the fact that K_k is uniquely k-colorable, which implies that $\mathrm{M}_r(K_k) - z$ is again uniquely k-colorable (where z is the top vertex), and so in any k-coloring of $\mathrm{M}_r(K_k) - z$ all k colors occur in each of the r levels of $\mathrm{M}_r(K_k)$. Thus $\mathrm{M}_r(K_k)$ is a $(k+1)$-critical graph (by Proposition 8.29), which can be made bipartite by the deletion of only $\binom{k}{2}$ edges. This was observed by Rödl and Tuza [875], who also proved that the bound is best possible (Theorem 8.31).

For positive integers n and p, let $tc(n, p)$ denote the minimum number of edges in any graph having n vertices and independence number at most p. This number is closely related to the Turan number $t(n, p)$ (see Appendix C.9); obviously, we have $tc(n, p) + t(n, p) = \binom{n}{2}$.

Theorem 8.31 (Rödl and Tuza) *Let G be a $(k+1)$-chromatic graph, and let $F \subseteq E(G)$ such that $\chi(G - F) \leq p$, where $2 \leq p \leq k$. Then $|F| \geq tc(k+1, p)$. If G is in addition critical and $|G| > 2(tc(k, p-1) - 1)$, then $|F| \geq tc(k, p-1)$.*

Proof Let φ be a coloring of $G - F$ with a set C of p colors. For a color $c \in C$, let $k_c = \chi(G[\varphi^{-1}(c)])$. On the one hand, we have $\chi(G) = k+1 \leq \sum_{c \in C} k_c$. On the other hand, $G[\varphi^{-1}(c)]$ has at least $\binom{k_c}{2}$ edges (by Proposition 1.34(b)). Since all these edges belong to F, we deduce that $|F| \geq s$ with $s = \sum_{c \in C} \binom{k_c}{2}$. A simple calculation (or Turan's theorem (Theorem C.8)) implies that s is minimum if $|k_c - k_{c'}| \leq 1$ whenever c and c' are different colors of C. In this case, we obtain that $s = tc(k+1, p)$.

It remains to consider the case that G is critical and $|G| > 2(tc(k, p-1) - 1)$. Let $X = \{v \in V(G) \mid E_G(v) \cap F \neq \varnothing\}$, and let $Y = V(G) \setminus Y$. If $|X| > 2(tc(k, p-1) - 1)$, then $|F| \geq tc(k, p-1)$ and we are done. So assume that $|X| \leq 2(tc(k, p-1) - 1)$, which implies that $G' = G[X]$ is a proper subgraph of G. Since G is $(k+1)$-critical, G' is k-colorable, and so there exists a coloring φ of G' with a set C of k colors. Let H be an auxiliary graph with vertex set C, where two colors c and c' are adjacent if and only if $E_G(\varphi^{-1}(c), \varphi^{-1}(c')) \cap F \neq \varnothing$. We claim that $\alpha(H) \leq p-1$. To this end, suppose that $\alpha(H) \geq p$. Then there exists a set $C' \subset C$ of p colors which form an independent set in H. This implies that the subgraph of G induced by $X' = Y \cup \bigcup_{c \in C'} \varphi^{-1}(c)$ contains no edge of F, and so $\chi(G[X']) \leq \chi(G - F) \leq p$. Clearly, if $Z = \bigcup_{c \in C \setminus C'} \varphi^{-1}(c)$ then

$$\chi(G - X') \leq \chi(G[Z]) \leq |C \setminus C'| = k - p,$$

and so $\chi(G) \leq \chi(G[X]) + \chi(G - X) \leq k$, which contradicts $\chi(G) = k+1$. This contradiction proves the claim that $\alpha(H) \leq p-1$. It then follows that $|F| \geq |E(H)| \geq tc(k, p-1)$, as required. □

Note that $tc(k, 1) = \binom{k}{2}$, so at least $\binom{k}{2}$ edges need to be removed from a $(k+1)$-critical graph with at least $k^2 - k - 1$ vertices to make it bipartite.

For an integer $k \geq 1$, let

$$\mathcal{MG}(k+2) = \{\mathrm{M}_{r_1, r_2, \ldots, r_k}(K_2) \mid r_1, r_2, \ldots, r_k \in \mathbb{N}\}.$$

Then $\mathcal{MG}(3)$ is the class consisting of all odd cycles. If $k \geq 3$, then the class $\mathcal{MG}(k)$ contains the Mycielski graphs $\mathrm{M}_1^{k-2}(K_2) = K_k$ and $\mathrm{M}_2^{k-2}(K_2)$. If we define $\mathcal{MG}(2) = \{G \in \mathcal{G} \mid G = K_2\}$, then $\mathcal{MG}(k+1) = \{\mathrm{M}_r(G) \mid G \in \mathcal{MG}(k), r \in \mathbb{N}\}$ for $k \geq 2$. Our goal is to prove the following generalization of Mycielski's result obtained by Stiebitz [966] in 1985.

Theorem 8.32 *For any integer $k \geq 2$, any graph belonging to $\mathcal{MG}(k)$ is k-critical.*

Using induction on $k \geq 2$, we deduce from (8.11) that $\chi(G) \leq k$ for every graph of $\mathcal{MG}(k)$. So the difficult part of Theorem 8.32 is to show that $\chi(G) \geq k$ for every graph $G \in \mathcal{MG}(k)$, which then implies that any graph of $\mathcal{MG}(k)$ is k-critical (by Proposition 8.29). To show that every graph of $\mathcal{MG}(k)$ has chromatic number at least k, we shall use Lovász' topological lower bound for the chromatic number of a graph in terms of the connectivity of its neighborhood complex. Thus, Theorem 8.32 is a consequence of Theorem 8.12 and the following result.

Theorem 8.33 (Stiebitz) *For any graph $G \in \mathcal{MG}(k)$, where $k \geq 2$, the topological space $\|\mathsf{N}(G)\|$ is $(k-3)$-connected.*

The proof of the above theorem uses two well known topological constructions. Let K be a finite simplicial complex. The **cone** of K, denoted by $\mathrm{cone}(\mathsf{K})$, is the simplicial complex defined by

$$\mathrm{cone}(\mathsf{K}) = \mathsf{K} \cup \{\sigma \cup \{a\} \mid \sigma \in \mathsf{K}\},$$

where a is a new vertex, so $V(\mathrm{cone}(\mathsf{K})) = V(\mathsf{K}) \cup \{a\}$. The **suspension** of K, denoted by $\mathrm{susp}(\mathsf{K})$, is the simplicial complex defined by

$$\mathrm{susp}(\mathsf{K}) = \mathsf{K} \cup \{\sigma \cup \{x\} \mid \sigma \in \mathsf{K}, x \in \{a,b\}\},$$

where a and b are two distinct new vertices, so $V(\mathrm{susp}(\mathsf{K})) = V(\mathsf{K}) \cup \{a,b\}$. Both the cone and the suspension are also well known constructions for topological spaces, and we have that $\|\mathrm{cone}(\mathsf{K})\|$ is homotopy equivalent to $\mathrm{cone}(\|\mathsf{K}\|)$ and $\|\mathrm{susp}(\mathsf{K})\|$ is homotopy equivalent to $\mathrm{susp}(\|\mathsf{K}\|)$ (see the book of Matoušek [722, Section 4.2]).

Theorem 8.34 *For any nonempty graph G and any integer $r \geq 1$, the space $\|\mathsf{N}(M_r(G))\|$ is homotopy equivalent to the suspension* $\mathrm{susp}(\|(\mathsf{N}(G))\|)$.

Sketch of Proof: Let us write $\mathsf{N} = \mathsf{N}(G)$ and $\mathsf{K} = \mathsf{N}(\mathrm{M}_r(G))$. To characterize the neighborhood complex K of $\mathrm{M}_r(G)$, we define a bijective mapping p_i from $V(G)$ to a set Y_i, where the sets $Y_1, Y_2, \ldots, Y_r$ are pairwise disjoint. If $r = 2s$, then put $p_i(u) = (u, r-2i+2)$ and $p_{s+i}(u) = (u, 2i-1)$ for $i \in [1,s]$ and $u \in V(G)$. If $r = 2s+1$, then put $p_i(u) = (u, r-2i+2)$ for $i \in [1, s+1]$ and $p_{s+i}(u) = (u, 2i)$ for $i \in [1,s]$, where $u \in V(G)$. Note that $Y_1, Y_2, \ldots, Y_r$ is a rearrangement of the r levels $X_1, X_2, \ldots, X_r$ of $\mathrm{M}_r(G)$. For example, if $r=4$, then $(Y_1, Y_2, Y_3, Y_4) = (X_4, X_2, X_1, X_3)$. For $i \in [1, r-1]$, put

$$\mathsf{K}_i = \{\sigma \mid \sigma \subseteq p_i(\tau) \cup p_{i+1}(\tau), \tau \in \mathsf{N}\},$$

furthermore, put

$$\mathsf{K}_0 = \{\sigma \mid \sigma \subseteq Y_1\},$$

and

$$\mathsf{K}_r = \{\sigma \mid \sigma \subseteq p_r(\tau) \cup \{z\}, \tau \in \mathsf{N}\},$$

where z is the top vertex of $M_r(G)$. Then it is easy to check that

$$\mathsf{K} = \mathsf{K}_0 \cup \mathsf{K}_1 \cup \cdots \cup \mathsf{K}_r.$$

Note that the simplicial complex $\mathsf{L}_i = \{\sigma \mid \sigma \subseteq p_i(\tau), \tau \in \mathsf{K}\}$ with $i \in [1, r]$ is an isomorphic copy of N, and p_i is a simplicial map from N to L_i. The subcomplex K_0 is a full simplex, and so $\|\mathsf{K}_0\|$ is a contractible subspace of $\|\mathsf{K}\|$. The subcomplex K_r is isomorphic to cone(N). If G is edgeless, then $\mathsf{N} = \varnothing$ and K is the disjoint union of K_0 and K_r, which are booth full simplices. Thus $\|\mathsf{K}\|$ is homotopy equivalent to $\mathbf{S}^0$ and hence to susp($\|\mathsf{N}\|$) as required. So assume that G has at least one edge. Then, for $1 \le i \le r-1$, the space $\|\mathsf{K}_i\|$ is homotopy equivalent to the space $\|\mathsf{N}\| \times I$, where I is the unit interval in $\mathbb{R}$ (see Matoušek [722, Theorem 5.9.6]). Let $\mathsf{K}' = \mathsf{K}_1 \cup \mathsf{K}_2 \cup \cdots \cup \mathsf{K}_{r-1}$. If $r \ge 2$, then $\|\mathsf{K}'\|$ is homotopy equivalent to the space $\|\mathsf{N}\| \times I$. Furthermore, $\mathsf{K} = \mathsf{K}_0 \cup \mathsf{K}' \cup \mathsf{K}_r$, where $\mathsf{K}' = \varnothing$ when $r = 1$. Therefore, when $r = 1$ the space $\|\mathsf{K}\|$ is obtained from the cone $\|\text{cone}(\mathsf{N})\|$ by identifying the base of the cone; and when $r \ge 2$ the space $\|\mathsf{K}\|$ is homotopy equivalent to a space which is the quotient with respect to $\|\mathsf{K}\|$ of a full simplex $\|\mathsf{K}_0\|$, a homotopy $\|\mathsf{N}\| \times I$ and a cone $\|\text{cone}(\mathsf{N})\|$. This implies that the space $\|\mathsf{K}\|$ is homotopy equivalent to the suspension susp($\|(\mathsf{N})\|$). □

A more formal proof of Theorem 8.34 can be found in Matoušek's book [722, Theorem 5.9.6]. A strengthening of this result was obtained by Csorba in his thesis [266, Theorem 2.23].

Proof of Theorem 8.33: We show by induction on $k \ge 2$ that if $G \in \mathcal{MG}(k)$, then the space $\|\mathsf{N}(G)\|$ is homotopy equivalent to the sphere $\mathbf{S}^{k-2}$. which implies that $\|\mathsf{N}(G)\|$ is $(k-3)$-connected since $\mathbf{S}^{k-2}$ is so. For $k = 2$ this is evident, since then $G = K_2$ and $\|\mathsf{N}(K_2)\|$ is homeomorphic to $\mathbf{S}^0$. So let $G \in \mathcal{MG}(k+1)$, where $k \ge 2$. Then $G = \mathrm{M}_r(G')$ for some graph $G' \in \mathcal{MG}(k)$ and for some integer $r \ge 1$. Then the induction hypothesis implies that $\|\mathsf{N}(G')\|$ is homotopy equivalent to $\mathbf{S}^{k-2}$. By Theorem 8.34, $\|\mathsf{N}(G)\|$ is homotopy equivalent to the suspension susp$\|(\mathsf{N}(G'))\|$. Hence, $\|\mathsf{N}(G)\|$ is homotopy equivalent to susp($\mathbf{S}^{k-2}$), and so to $\mathbf{S}^{k-1}$, as susp($\mathbf{S}^{k-2}$) is homeomorphic to $\mathbf{S}^{k-1}$ (see [722, Section 4.2]). □

Theorem 8.35 *For every integer $k \ge 3$ there are infinitely many k-critical graphs G with $g_o(G) \ge 2^{(k-3)/(k-2)}|G|^{1/(k-2)}$.*

Proof Given integers $k \ge 3$ and $r \ge 1$, let $G = \mathrm{M}_r^{k-2}(K_2)$. Then $G \in \mathcal{MG}(k)$, which implies that G is k-critical (by Theorem 8.32). From Proposition 8.27 we deduce that $g_o(G) = 2r+1$. Let $n = n(r,k)$ the order of G. We claim that $n \le (1/2^{k-3})(2r+1)^{k-2}$, which implies that

$$g_o(G) = (2r+1) \ge 2^{(k-3)/(k-2)}|G|^{1/(k-2)}.$$

To prove the claim, we apply induction on $k \geq 3$. For $k = 3$, we have $n(r,3) = |\mathrm{M}_r(K_2)| = 2r+1$ as required. Using the induction hypothesis, we deduce that

$$n(r,k+1) = rn(r,k)+1 \leq r\frac{1}{2^{k-3}}(2r+1)^{k-2}+1 \leq \frac{1}{2^{k-2}}(2r+1)^{k-1}$$

as required. This completes the proof. □

In 1973 Erdős [336] conjectured that for every $k \geq 4$, there is a constant c_k such that every graph G of chromatic number $k+2$ has odd girth at most $c_k|G|^{1/k}$. Ten years later, Erdős' conjecture was settled in the affirmative by Kierstead, Szemerédi, and Trotter [585]; they actually proved a stronger result (see Theorem 8.37).

Theorem 8.36 (Kierstead, Szemerédi, and Trotter) *Every graph G with $\chi(G) = k$ and $k \geq 4$ satisfies $g_o(G) \leq (4k-8)\lceil |G|^{1/(k-2)}\rceil + 1$.*

If G is a graph and $X \subseteq V(G)$ is a vertex set, then the **radius** of X in G, denoted by $\mathrm{rad}_G(X)$, is

$$\mathrm{rad}_G(X) = \min_{v\in X}\max_{w\in X}\mathrm{dist}_G(v,w),$$

where $\mathrm{dist}_G(v,w)$ is the distance of v and w in G (see Appendix C.5). If H is a subgraph of G, then we write $\mathrm{rad}_G(H)$ instead of $\mathrm{rad}_G(V(H))$. If H contains an odd cycle C, then $|C| \leq 2\mathrm{rad}_G(H)+1$. So the special case $p = 2$ of the following result implies Theorem 8.36.

Theorem 8.37 *Any graph G with $\chi(G) \geq k(p-1)+2$, where $k,p \in \mathbb{N}$, contains a subgraph H with $\chi(H) > p$ and* $\mathrm{rad}_G(H) \leq 2k\lceil |G|^{1/k}\rceil$.

Proof Let G be a graph on n vertices and let $a = \lceil n^{1/k}\rceil$. Suppose that for any subgraph H of G, if $\mathrm{rad}_G(H) \leq 2ka$, then $\chi(H) \leq p$. We shall prove the following claim, which gives the required contradiction in the case $h = k$, $H = G$, and $|I| = 1$, since if $\chi(G) \geq k(p-1)+2$, there are not enough vertices in $G-I$ for the required vertex set F.

Claim *Let $h \in [0,k]$ and let $H \subseteq G$ with $\chi(H) \geq h(p-1)+2$. Then for any independent set I of H there exists a set $F \subseteq V(H)\setminus I$ such that $|F| \geq a^h$ and* $\mathrm{rad}_G(F) \leq 2ha$.

Proof: The proof is by induction on h. To prove the basis step, suppose that $h = 0$. Let $H \subseteq G$ with $\chi(H) \geq 2$, and let I be an independent set of H. Then H contains an edge, and so $H-I$ contains a vertex set F with $|F|$=1 and $\mathrm{rad}_G(F) = 0$. Thus we are done.

To prove the induction step, suppose that $1 \leq h+1 \leq k$. Let $H \subseteq G$ with $\chi(H) \geq (h+1)(p-1)+2$, and let I be an independent set in H. Clearly, H contains an induced subgraph with chromatic number p. Among all such subgraphs, let H' be one whose order is maximum. Then the vertex set of H' has a partition $I_1, I_2, \ldots, I_p$ into p independent sets of H. Clearly, the subgraph $H_1 = H-(I_1\cup I_2\cup\cdots\cup I_{p-1})$ satisfies $\chi(H_1) \geq h(p-1)+2$. Then the induction hypothesis implies that $H_1 - I_p$ contains a vertex set F_0 satisfying $|F_0| \geq a^h$ and $\mathrm{rad}_G(F_0) \leq 2ha$. Note that $H_1 - I_p = H - V(H')$ and so $F_0 \cap V(H') = \varnothing$. For an positive integer i, let

$$F_i = \{v \in V(H') \mid \min_{w \in F_0} \mathrm{dist}_G(v,w) = i\}.$$

Let $F = \bigcup_{i=0}^{2a} F_i$. Then, we have $\mathrm{rad}_G(F) \le \mathrm{rad}_G(F_0) + 2a \le 2ha + 2a = 2(h+1)a$, which yields $\mathrm{rad}_G(F-I) \le 2(h+1)a$. To complete the proof, it suffices to show that $|F-I| \ge a^{h+1}$. To this end, we claim that $|(F_j \cup F_{j+1}) \setminus I| \ge a^h$ for every $j \in [1, 2a-1]$. If not, then the order of the subgraph K' of H induced by the vertex set $(V(H') \setminus ((F_j \cup F_{j+1}) \setminus I)) \cup F_0$ is greater than the order of H'. Hence we obtain a contradiction by showing that $\chi(K') \le p$. Let K_0 be the subgraph of H induced by $F_0 \cup F_1 \cdots \cup F_{j-1} \cup (F_j \cap I)$, and let $K_1 = K' - V(K_0)$. Since I is an independent set of H, there is no edge between K_0 and K_1. Since $\mathrm{rad}_G(K_0) \le 2ka$, we have $\chi(K_0) \le p$. Since $K_1 \subseteq H'$, we also have $\chi(K_1) \le p$. Thus $\chi(K) \le p$. This contradiction proves the claim that $|(F_j \cup F_{j+1}) - I| \ge a^h$ for every $j \in [1, 2a-1]$. It then follows that $|F - I| \ge \frac{1}{2}(2a)a^h = a^{h+1}$ as required. This proves the induction step. △

Thus the proof of the theorem is finished. □

By Theorem 8.36, any 4-chromatic graph on n vertices contains an odd cycle of length at most $\lceil 8\sqrt{n}\rceil + 1$. Jiang [532] improved this bound to $2\sqrt{n}+3$. On the other hand, there are infinitely many values of n for which there exists a 4-critical graph on n vertices containing no odd cycle of length less than $\sqrt{2n}$ (Theorem 8.35).

Theorem 8.38 (Jiang) *Every 4-chromatic graph G of order n satisfies $g_o(G) < 2\sqrt{n}+3$.*

Proof Let G be a 4-chromatic graph on n vertices. Assuming that $g_o(G) = 2p+3$ with $p \ge 0$, we prove that $n > p^2$, from which the result follows. To this end, we consider pairs of disjoint independent sets in G. Among all such pairs, let (X,Y) be one for which $|X \cup Y|$ is maximum. Our goal is to show that both sets X and Y have at least $\binom{p}{2}$ elements. Since G is 4-chromatic, $G - X$ contains an odd cycle. Let C be a shortest odd cycle of $G - X$. Then C is an induced cycle of $G - X$ and, by our assumption, $|C| \ge 2p+3$. Let v be an arbitrary vertex of C. For $i \in \mathbb{N}_0$, let $N_i = \{w \in V(G) \mid \mathrm{dist}_G(v,w) = i\}$. Since $g_0(G) = 2p+3$, N_i is an independent set of G whenever $i \in [0,p]$. For $i \in [1,p]$, let

$$S_i = \Big(\bigcup_{j=0}^{i-1} N_j\Big) \cup (N_i \cap Y),\ T_i = \bigcup_{j=i+1}^{\infty} (N_j \cap (X \cup Y)) \text{ and } W_i = S_i \cup T_i.$$

Clearly, $G[S_i]$ is a bipartite graph having parts X_i and Y_i with $N_i \cap Y \subseteq Y_i$. Furthermore, $G[T_i]$ is a bipartite graph with parts $T_i \cap X$ and $T_i \cap Y$. We then deduce that $G[W_i]$ is a bipartite graph with parts $X_i \cup (T_i \cap X)$ and $Y_i \cup (T_i \cap Y)$. By the choice of (X,Y), it follows that $|X \cup Y| \ge |W_i|$, which yields $|(X \cup Y) \setminus W_i| \ge |W_i \setminus (X \cup Y)|$. Since C is an induced cycle of $G - X$ and $v \in V(C)$, C contains a path P on $2i-1$ vertices with $V(P) \subseteq N_0 \cup N_1 \cup \cdots \cup N_{i-1}$. Since Y is an independent set of G, we have $|Y \cap V(P)| \le i$, which implies that $|W_i \setminus (X \cup Y)| \ge |V(P) \setminus (X \cup Y)| \ge i-1$. Obviously, $(X \cup Y) \setminus W_i = N_i \cap X$. Thus we deduce that $|N_i \cap X| = |(X \cup Y) \setminus W_i| \ge |W_i \setminus (X \cup Y)| \ge i-1$. Hence, we obtain that

$$|X| \ge \sum_{i=1}^{p} |N_i \cap X| \ge \sum_{i=1}^{p} (i-1) = \binom{p}{2}.$$

By symmetry, we also have $|Y| \ge \binom{p}{2}$. Since C is a cycle of length $2p+1$ of $G - X$, at least $p+1$ vertices of C do not lie in $X \cup Y$. Then we have

$$n \ge |X| + |Y| + p + 1 \ge 2\binom{p}{2} + p + 1 > p^2$$

as required. □

8.5 Odd Colorings and Gyárfás Graphs

Let G be an arbitrary graph, let $X \subseteq V(G)$, and let $p \in \mathbb{N}_0$. For a vertex v of G, the **distance** of v to X in G is $\mathrm{dist}_G(v, X) = \min\{\mathrm{dist}_G(v, w) \mid w \in X\}$. The p**th distance class** of X in G, denoted by $N_G^p(X)$, is defined by

$$N_G^p(X) = \{v \in V(G) \mid \mathrm{dist}_G(v, X) = p\}.$$

Note that a vertex v belongs to $N_G^p(X)$ if and only if G contains some path of length p from v to a vertex in X, but no shorter such path. In particular, we have $N_G^0(X) = X$ and, for an independent set X of G, we have $N_G^1(X) = N_G(X)$.

Let k and ℓ be integers with $k \ge 1$ and $\ell \ge 0$. An **odd** (k, ℓ)**-coloring** of G is a mapping $\varphi : V(G) \to [1, k]$ such that the pth distance class of the color class $\varphi^{-1}(c)$ is an independent set in G whenever $c \in [1, k]$ and $p \in [0, \ell]$. We denote by $\mathcal{OC}(k, \ell)$ the set of all graphs that have an odd (k, ℓ)-coloring. Note that an odd $(k, 0)$-coloring is simply a k-coloring, and any odd (k, ℓ)-coloring is a particular k-coloring with color set $[1, k]$. Furthermore, we have

$$\mathcal{OC}(k, \ell + 1) \subseteq \mathcal{OC}(k, \ell) \subseteq \mathcal{OC}(k + 1, \ell).$$

The name odd coloring has been chosen because of the following characterization of an odd (k, ℓ)-coloring.

Theorem 8.39 *Let k and ℓ be integers with $k \ge 1$ and $\ell \ge 0$. For a graph G and a mapping $\varphi : V(G) \to [1, k]$, the following statements are equivalent:*

(a) *φ is an odd (k, ℓ)-coloring of G.*
(b) *If P is an odd v-w walk in G such that $\varphi(v) = \varphi(w)$, then $\ell(P) > 2\ell + 1$.*

Proof Let $\varphi : V(G) \to [1, k]$ be a map and let $I_c = \varphi^{-1}(c)$ for every $c \in [1, k]$. First suppose that φ is an odd (k, ℓ)-coloring of G. Let P be an v-w walk of G, where $\varphi(v) = \varphi(w) = c$ and $\ell(P) \le 2\ell + 1$. Then $N_G^p(I_c)$ is an independent set of G for all $p \in [0, \ell]$. This implies that $X = \bigcup_{p=0}^{\ell} N_G^p(I_c)$ induces a bipartite subgraph of G

with parts U and W such that $I_c = N_G^0(I_c)$ is contained in one of the two parts. Since $\ell(P) \leq 2\ell+1$ and $u, v \in I_c$, this implies that $V(P) \subseteq X$, and so P is an even walk, a contradiction.

Next, suppose that φ is not an odd (k,ℓ)-coloring of G. Then there exists a color $c \in [1,k]$ and an integer $p \in [1,k]$ such that $N_G^p(I_c)$ is not an independent set of G. Thus, two vertices of $N_G^p(I_c)$, say v_1 and v_2, are adjacent in G. Then there is a vertex $w_i \in I_c = N_G^0(I_c)$ and w_i-v_i path P_i of length p in G ($i \in \{1,2\}$). Then $P = w_1 P_1 v_1 v_2 P_2 w_2$ is an odd w_1-w_2 walk of G satisfying $\varphi(w_1) = \varphi(w_2)$ and $\ell(P) = 2p+1 \leq 2\ell+1$. This completes the proof. □

Corollary 8.40 *Let k and ℓ be integers with $k \geq 1$ and $\ell \geq 0$. If $G \in OC(k,\ell)$, then $g_o(G) > 2\ell+1$.*

Let G be a graph, and let $p \geq 1$ be an integer. The p**th power** G^p of G is the graph with $V(G^p) = V(G)$ and

$$E(G^p) = \{vw \mid \text{there is a } v\text{-}w \text{ path } P \text{ in } G \text{ with } \ell(P) \in [1,p]\}.$$

Furthermore, the p**th partial power** $G^{<p>}$ of G is the graph with $V(G^{<p>}) = V(G)$ and

$$E(G^{<p>}) = \{vw \mid \text{there is an odd } v\text{-}w \text{ path } P \text{ in } G \text{ with } \ell(P) \in [1,p]\}.$$

Evidently, $G^1 = G^{<1>} = G$ and, for all $p \geq 1$, we have $G \subseteq G^{<p>} \subseteq G^p \subseteq G^{p+1}$.

Theorem 8.41 *Let k and ℓ be integers with $k \geq 1$ and $\ell \geq 0$. For a graph G with $g_o(G) > 2\ell+1$ and for a mapping $\varphi : V(G) \to [1,k]$, the following statements are equivalent:*

(a) *φ is an odd (k,ℓ)-coloring of G.*
(b) *φ is a k-coloring of $G^{<2\ell+1>}$*

Proof That (a) implies (b) follows immediately from Theorem 8.39. To prove that (b) implies (a), suppose that φ is k-coloring of $G^{<2\ell+1>}$, but not an odd (k,ℓ)-coloring of G. It then follows from Theorem 8.39 that there exist an odd v-w walk P of G such that $\varphi(v) = \varphi(w)$ and $\ell(P) \leq 2\ell+1$. Since $g_o(G) > 2\ell+1$, we deduce that $v \neq w$ and there is an odd v-w path in G with length at most $2\ell+1$ (see Proposition C.5). Consequently, $vw \in E(G^{<2\ell+1>})$ and, therefore, φ is no k-coloring of $G^{<2\ell+1>}$, a contradiction. □

Obviously $K_k \in OC(k,0)$, $K_1 \in OC(1,\ell)$ and $K_2 \in OC(2,\ell)$. It is also easy to verify that the odd cycle $C_{6\ell+3}$ belongs to $OC(3,\ell)$. However, it is not clear a priori whether each class $OC(k,\ell)$ contains a k-chromatic graph. For $\ell = 1$, this problem was addressed by N. J. A. Harvey and U. S. R. Murty. In 2004, Gyárfás, Jensen, and Stiebitz [445] proved the following result.

Theorem 8.42 (Gyárfás, Jensen, and Stiebitz) *For any $k \geq 1$ and any $\ell \geq 0$, the class $OC(k,\ell)$ contains a k-chromatic graph.*

Proof This is obvious if $k = 1$ or $\ell = 0$. So assume that $k \geq 2$ and $\ell \geq 1$. We claim that $\mathcal{OC}(k,\ell) \cap \mathcal{MG}(k) \neq \emptyset$, which implies that $\mathcal{OC}(k,\ell)$ contains a k-chromatic graph (by Theorem 8.32). To this end, we apply induction on k (where ℓ is fixed). For $k = 2$ the claim is evident, since $K_2 \in \mathcal{OC}(2,\ell) \cap \mathcal{MG}(2)$. To prove the induction step, assume that $k \geq 3$ and $G \in \mathcal{OC}(k-1,\ell) \cap \mathcal{MG}(k-1)$. Let $H = \mathrm{M}_{3\ell+1}(G)$. Clearly, $H \in \mathcal{MG}(k)$, so it suffices to show that $H \in \mathcal{OC}(k,\ell)$. Since $G \in \mathcal{OC}(k-1,\ell)$, there exists an odd $(k-1,\ell)$-coloring φ of G. Define a map $\varphi' : V(H) \to [1,k]$ as follows. Partition the set $[1,3\ell+1]$ in to sets A and B, where $A = \{\ell+2, \ell+4, \ldots, 3\ell-2, 3\ell\}$ and $B = [1,3\ell+1] \setminus A$, so $B = [1,\ell+1] \cup \{\ell+3, \ell+5, \ldots, 3\ell-1, 3\ell+1\}$. If v is a vertex of G and $i \in [1,3\ell+1]$, then let

$$\varphi'(v,i) = \begin{cases} \varphi(v) & \text{if } i \in B, \\ k & \text{if } i \in A; \end{cases}$$

furthermore, let $\varphi'(z) = k$, where z is the top vertex of H. What remains is to show that φ' is an odd (k,ℓ)-coloring of H. For $i \in [1,3\ell+1]$, let $X_i = V(G) \times \{i\}$ be the ith level of H. Since $H[X_1] \cong G$ and $\varphi'(v,1) = \varphi(v)$ for all $v \in V(G)$, we deduce that the restriction of φ' to X_1 is an odd $(k-1,\ell)$-coloring of $H[X_1]$. To show that φ' is an odd (k,ℓ)-coloring of H, we shall use Theorem 8.39. So let P be an x-y walk in H with $\varphi'(x) = \varphi'(y) = c$. It suffices to show that $\ell(P)$ is even or $\ell(P) \geq 2\ell+2$. First, assume that $c = k$. Then x as well as y belong to $\bigcup_{i \in A} X_i \cup \{z\}$. If P contains a vertex of X_1, this implies that $\ell(P) \geq 2\ell+2$. Otherwise, $\ell(P)$ is even, since $H - X_1$ is a bipartite graph where x, y belong to the same part. Now, assume that $c \in [1,k-1]$ and $\ell(P) \leq 2\ell+1$. Then both x and y belong to $\bigcup_{i \in B} X_i$. If P contains the top vertex z, then it follows that the vertices of P belong to $\bigcup_{i \in [\ell+1,3\ell+1]} X_i \cup \{z\}$ implying that $\ell(P)$ is even. Otherwise, projecting P down to X_1 (i.e., replacing each vertex (v,i) of P by its projection $(v,1)$) results in an x'-y' walk P' of $H[X_1]$ such that $\varphi'(x') = \varphi'(y') = c$ and $\ell(P') = \ell(P) \leq 2\ell+1$. Using the induction hypothesis and Theorem 8.39, it follows that $\ell(P) = \ell(P')$ is even. This completes the proof. □

Clearly, K_k is a homomorphism universal graph in $\mathcal{OC}(k,0)$ as well as the smallest k-chromatic graph in $\mathcal{CO}(k,0)$. For an integer $k \geq 2$, let $GY(k)$ denote the **Gyárfás graph** defined by

$$\begin{aligned} V(GY(k)) &= \{(i,A) \mid i \in [1,k], i \notin A \subseteq [1,k], A \neq \emptyset\} \\ E(GY(k)) &= \{(i,A)(j,B) \mid j \in A, i \in B, A \cap B = \emptyset\}. \end{aligned}$$

The first two graphs in this sequence are $GY(2) = K_2$ and $GY(3) = C_9$. The graph $GY(4)$ is shown in Figure 8.5 (the vertex $(1,\{2,3\})$ is marked by "1,23"). Then it is easy to check that the map $\varphi : V(GY(k)) \to [1,k]$ with $\varphi(i,A) = i$ is an odd $(k,1)$-coloring of $GY(k)$, and so $GY(k) \in \mathcal{OC}(k,1)$. One of the main results in [445] says that the graph $GY(k)$ is k-critical and homomorphism universal in $\mathcal{OC}(k,1)$. That the mapping $\varphi : V(GY(k)) \to [1,k]$ with $\varphi(i,A) = i$ is an odd $(k,1)$-coloring of $GY(k)$ is easy to check. It is also easy to see that $G \in \mathcal{OC}(k,1)$ if and only if $G \to GY(k)$. However, to show that $\chi(GY(k)) \geq k$ is a nontrivial task.

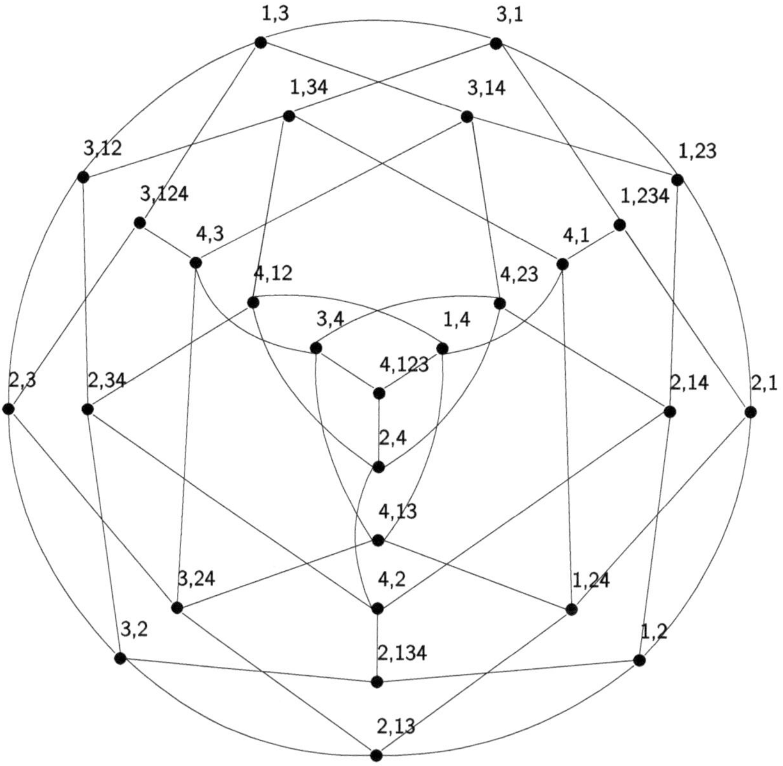

Fig. 8.5 The Gyárfás graph $GY(4)$.

Let $k \geq 2$ and $\ell \geq 0$ be integers. Our aim is to construct a graph $GY(k,\ell)$ which is k-critical and homomorphism universal in $OC(k,\ell)$. We denote by $A(k,\ell)$ the set of all mappings $a : [1,k] \longrightarrow [0,\ell+1]$ such that

$$|a^{-1}(0)| = 1 \text{ and } |a^{-1}(1)| \geq 1. \tag{8.12}$$

Then, the graph $GY(k,\ell)$ is defined as the graph whose vertices are the mappings of $A(k,\ell)$ and whose edges are those tuples aa' that satisfy, for every $i \in [1,k]$, the condition

$$|a(i) - a'(i)| = \begin{cases} 0 \text{ if } a(i) = a'(i) = \ell+1, \\ 1 \text{ otherwise.} \end{cases} \tag{8.13}$$

If a is a mapping of $A(k,\ell)$, then (8.12) implies that there is exactly one element $i \in [1,k]$ such that $a(i) = 0$. We call this element the **index** of a and write $i = i_a$. Clearly, the number of vertices of $GY(k,\ell)$ satisfies

$$|A(k,\ell)| = k((\ell+1)^{k-1} - \ell^{k-1}), \tag{8.14}$$

and, moreover, we have $GY(2,\ell) = K_2$ for all $\ell \geq 1$ and $GY(k,0) = K_k$ for all $k \geq 3$. If $\ell = 1$, then (8.12) and (8.13) implies immediately that the mapping $\varphi : V(GY(k,1)) \to V(GY(k))$ defined by $\varphi(a) = (i_a, a^{-1}(1))$ for all $a \in A(k,1)$ is an isomorphism from $GY(k,1)$ onto $GY(k)$. Hence $GY(k,1) \cong GY(k)$. As we shall see below $GY(3,\ell) = C_{6\ell+3}$ for all integers $\ell \geq 1$. The graph $GY(4,2)$ is shown in Figure 8.6 (the vertex a is marked by the 4-tuple $a(1)a(2)a(3)a(4)$).

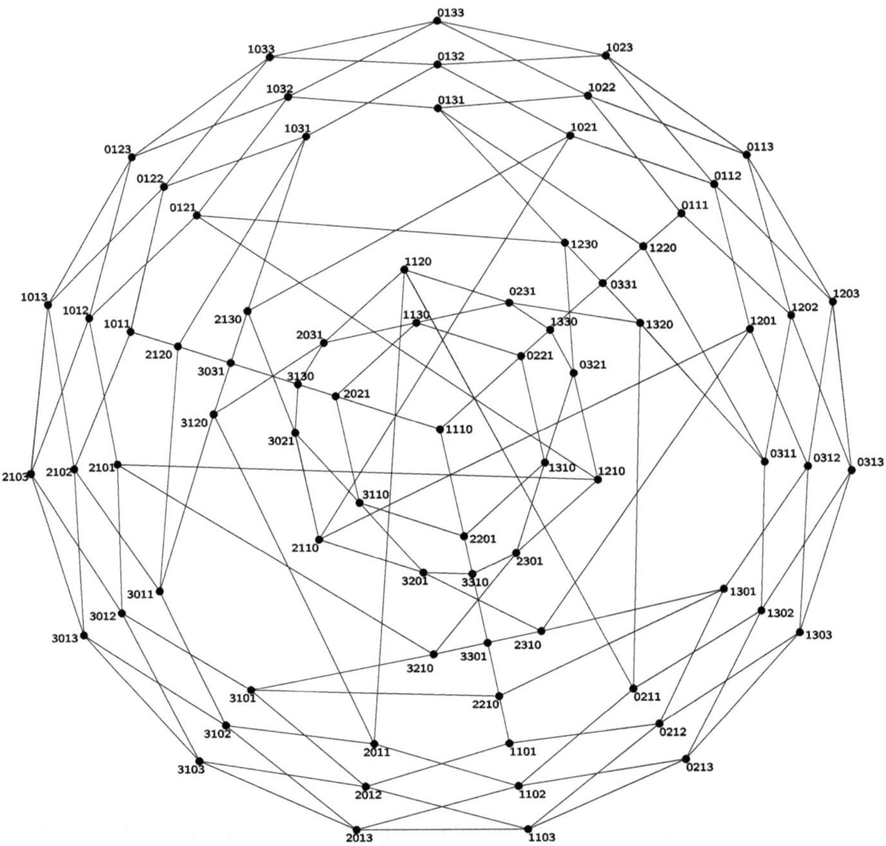

Fig. 8.6 The Gyárfás–Simonyi–Tardos graph $GY(4,2)$.

Before we prove that $GY(k,\ell)$ is homomorphism universal in $OC(k,\ell)$ (see Corollary 8.46), we shall establish some basic properties of the this graph family.

Lemma 8.43 *Let $k \geq 2$ and $\ell \geq 0$ be integers. Then, for the graph $G = GY(k,\ell)$, the following statements hold:*

(a) *Let $aa' \in E(G)$ be an edge of G. Then $a(i_{a'}) = a'(i_a) = 1$ and $i_a \neq i_{a'}$.*
(b) *Let $P = (a_0, \ldots, a_p)$ be a walk in G of length $p \leq \ell + 1$ and let $i = i_{a_0}$ be the index of a_0. Then $a_p(i) \leq p$ and $a_p(i) \equiv p \pmod 2$.*

(c) *Let $a \in V(G)$ such that $|a^{-1}(1)| = r$. Then $1 \le r \le k-1$ and $d_G(a) = r2^{k-1-r}$. Furthermore, $\delta(G) = k-1$.*
(d) *Let $a \in V(G)$. If $k \ge 4$, then $d_G(a) = k-1$ if and only if $|a^{-1}(1)| = k-1$.*

Proof Statement (a) is an immediate consequence of (8.12) and (8.13). Statement (b) follows easily from (8.13) by induction on $p \ge 1$. Statement (c) is also an immediate consequence of (8.12) and (8.13). Let $d(r) = r2^{k-1-r}$ where $k \ge 4$. Then $d(1) = d(2) = 2^{k-2} > k-1 = d(k-1)$ and, for $2 \le r \le k-1$, the function d is monotone decreasing. This proves (d). □

Lemma 8.44 *Let $k \ge 2$ and $\ell \ge 0$ be integers. For a permutation $\pi \in S_k$, let $\phi_\pi : A(k,\ell) \longrightarrow A(k,\ell)$ be the mapping with $\phi_\pi(a) = a \circ \pi^{-1}$ for all $a \in A(k,\ell)$. Then the following statements hold:*

(a) *$\phi_\pi \in \mathrm{Aut}(GY(k,\ell))$ for all $\pi \in S_k$.*
(b) *Let $\pi_1, \pi_2 \in S_k$ and $a \in A(k,\ell)$. If $a \circ \pi_1^{-1} = a \circ \pi_2^{-1}$, then $\pi_1(i_a) = \pi_2(i_a)$.*
(c) *The mapping $\pi \longmapsto \phi_\pi$ is an injective homomorphism from the symmetric group S_k into the group $\mathrm{Aut}(GY(k,\ell))$.*
(d) *$\Gamma' = \{\phi_\pi \mid \pi \in S_k\}$ is a subgroup of $\mathrm{Aut}(GY(k,\ell))$ and $|\Gamma'| = k!$.*

Proof For the proof of (a), consider a permutation $\pi \in S_k$. A map $a : [1,k] \rightarrow [0,\ell+1]$ can be considered as a k-tuple $a = (a(1), a(2), \ldots, a(k))$ with elements from the set $[0,\ell+1]$. Then the k-tuple $a \circ \pi^{-1} = (a(\pi^{-1}(1)), \ldots, a(\pi^{-1}(k))$ consists of the same elements as a and, clearly, $a \in A(k,\ell)$ if and only if $a \circ \pi^{-1} \in A(k,\ell)$. Hence, ϕ_π is a mapping from $A(k,\ell)$ in $A(k,\ell)$ and it is easy to check that this mapping is bijective. Furthermore, aa' is an edge of $GY(k,\ell)$ if and only if $(a \circ \pi^{-1})(a' \circ \pi^{-1})$ is an edge of $GY(k,\ell)$. This proves (a).

From the hypothesis in (b) it follows that $a = (a \circ \pi_2^{-1}) \circ \pi_1$. This implies that $0 = a(i_a) = a(\pi_2^{-1}(\pi_1(i_a)))$. Consequently, by (8.12), $i_a = \pi_2^{-1}(\pi_1(i_a))$ and, therefore, $\pi_1(i_a) = \pi_2(i_a)$. This proves (b).

By (a), it follows that $\pi \longmapsto \phi_\pi$ defines a mapping from S_k to $\mathrm{Aut}(GY(k,\ell))$. Consider two permutations π_1 and π_2 of S_k. Then

$$\begin{aligned}\phi_{\pi_1 \circ \pi_2}(a) &= a \circ (\pi_1 \circ \pi_2)^{-1} \\ &= (a \circ \pi_2^{-1}) \circ \pi_1^{-1} \\ &= \phi_{\pi_1^{-1}}(a \circ \pi_2^{-1}) \\ &= \phi_{\pi_1}(\phi_{\pi_2}(a)) \\ &= (\phi_{\pi_1} \circ \phi_{\pi_2})(a)\end{aligned}$$

for all $a \in A(k,\ell)$. Consequently $\phi_{\pi_1 \circ \pi_2} = \phi_{\pi_1} \circ \phi_{\pi_2}$. This proves that the mapping $\pi \longmapsto \phi_\pi$ is a homomorphism from the group S_k into the group $\mathrm{Aut}(GY(k,\ell))$. Now, suppose that $\phi_{\pi_1} = \phi_{\pi_2}$. Then, for all $a \in A(k,\ell)$, we have $a \circ \pi_1^{-1} = a \circ \pi_2^{-1}$ and $\pi_1(i_a) = \pi_2(i_a)$ (by (b)). Since $\{i_a \mid a \in A(k,\ell)\} = [1,k]$, this implies that $\pi_1 = \pi_2$. Hence, the mapping $\pi \longmapsto \phi_\pi$ is injective. This proves (c).

Clearly, statement (d) follows immediately from (c). This completes the proof of Lemma 8.44 □

Theorem 8.45 *Let $k \geq 2$ and $\ell \geq 0$ be integers, and let G be an arbitrary graph. Furthermore, let $f : A(k,\ell) \to [1,k]$ be the mapping satisfying $f(a) = i_a$ for all $a \in A(k,\ell)$. Then the following statements hold:*

(a) *f is an odd (k,ℓ)-coloring of $GY(k,\ell)$.*
(b) *If $\phi : V(G) \to V(GY(k,\ell))$ is a homomorphism from G to $GY(k,\ell)$, then $g = f \circ \phi : V(G) \to [1,k]$ is an odd (k,ℓ)-coloring of G.*
(c) *If $g : V(G) \to [1,k]$ is an odd (k,ℓ)-coloring of G, then there exists a homomorphism $\phi : V(G) \to V(GY(k,\ell))$ from G to $GY(k,\ell)$ such that $g = f \circ \phi$.*

Proof In order to prove (a), suppose that it is false. Then, by Theorem 8.39, there is odd walk $P = (a_0, a_1, \ldots, a_p, a_{p+1}, \ldots, a_{2p+1})$ in $GY(k,\ell)$ such that $p \leq \ell$ and $f(a_0) = f(a_{2p+1}) = i$ for some color $i \in [1,k]$. Then $i = i_{a_0} = i_{a_{2p+1}}$ and, therefore, Lemma 8.43 implies $a_p(i) \leq p$ and $a_p(i) \equiv p \,(\text{mod } 2)$ as well as $a_{p+1}(i) \leq p$ and $a_{p+1}(i) \equiv p \,(\text{mod } 2)$. Since $a_p a_{p+1}$ is an edge of $GY(k,\ell)$, this gives a contradiction to (8.13). This proves (a).

For the proof of (b), assume that ϕ is a homomorphism from G to $GY(k,\ell)$. Then $g = f \circ \phi$ is a mapping from $V(G)$ to $[1,k]$. Suppose, to the contrary, that g is not an odd (k,ℓ)-coloring of G. Then, by Theorem 8.39, there is an odd walk $P = (v_0, v_1, \ldots, v_{2p+1})$ in G such that $p \leq \ell$ and $g(v_0) = g(v_{2p+1})$. Since ϕ is a homomorphism from G to $GY(k,\ell)$, it then follows that $P' = (\phi(v_0), \phi(v_1), \ldots, \phi(v_{2p+1}))$ is an odd walk in $GY(k,\ell)$, where

$$f(\phi(v_0)) = g(v_0) = g(v_{2p+1}) = f(\phi(v_{2p+2}))$$

and $\ell(P') = \ell(P) \leq 2\ell + 1$, contradicting Theorem 8.39 (by (a)). This proves (b).

For the proof of (c), assume that $g : V(G) \to [1,k]$ is an odd (k,ℓ)-coloring of G. First, for every vertex u of G with $d_G(u) \geq 1$, we define a map $a_u : [1,k] \to [0, \ell+1]$ by

$$a_u(i) = \min\{\ell + 1, p \mid \text{there is an } u\text{-}u' \text{ walk of length } p \text{ in } G \text{ with } g(u') = i\}$$

for all $i \in [1,k]$. Then, clearly, $a_u^{-1}(0) = \{g(u)\}$ and $\{g(v) | uv \in E(G)\} \subseteq a_u^{-1}(1)$. Since $d_G(u) \geq 1$, this implies that $a_u \in A(k,\ell)$. Next, for every vertex u of G with $d_G(u) = 0$, let a_u be a mapping in $A(k,\ell)$ satisfying $a_u^{-1}(0) = \{g(u)\}$. Then, the assignment ϕ with $\phi(u) = a_u$ for every $u \in V(G)$ defines a mapping from $V(G)$ to $V(GY(k,\ell))$. Since $f(a_u) = i_{a_u} = g(u)$ for all $u \in V(G)$, we have $g = f \circ \phi$. The proof is completed by showing that ϕ is a homomorphism from G to $GY(k,\ell)$.

Let uv be an arbitrary edge of G. We need to show that $a_u a_v$ is an edge of $GY(k,\ell)$. To this end, let $i \in [1,k]$ be an arbitrary color. Then $a_u(i) = p_u$ and $a_v(i) = p_v$, where $p_u, p_v \in [0, \ell+1]$. By symmetry, we can assume that $p_u \leq p_v$. If $p_u = \ell + 1$, then $p_v = \ell + 1$ and $|p_u - p_v| = 0$. Otherwise, $p_u \leq \ell$ and we argue as follows. Since $uv \in E(G)$, we deduce that $p_v \leq p_u + 1$. Now, suppose that $p_v = p_u = p$. Then there is an u-u' walk of length p in G such that $g(u') = i$ as well as a v-v' walk of length p in G such that $g(v') = i$. Since $uv \in E(G)$, this implies that there is an odd u'-v' walk P in G such that $\ell(P) = 2p + 1$ and $g(u') = g(v') = i$. Since g is an odd (k,ℓ)-coloring of G and $p \leq \ell$, this yields a contradiction to Theorem 8.39. Hence, $p_u \neq p_v$ and, therefore, $|p_u - p_v| = 1$. This shows that $a_u a_v$ is an edge of $GY(k,\ell)$ and, therefore, ϕ is a homomorphism from G to $GY(k,\ell)$. Thus (c) is proved. □

To complete the graph family $GY(k,\ell)$, we put $GY(1,\ell) = K_1$ for all $\ell \geq 0$. Theorem 8.45 implies the following result.

Corollary 8.46 *Let $k \geq 1$ and $\ell \geq 0$ be integers. Then the following statements hold:*

(a) $OC(k,\ell) = \{G \mid G \to GY(k,\ell)\}$.
(b) $GY(k,\ell) \in OC(k,\ell)$ *and* $\chi(GY(k,\ell)) = k$.

Proof For $k \geq 2$ and $\ell \geq 0$, statement (a) follows immediately from Theorem 8.45. Furthermore, we have $OC(1,\ell) = \{G \mid G \to K_1\}$. This proves (a). Consequently, $GY(k,\ell) \in OC(k,\ell)$ and, therefore, $\chi(G(k,\ell)) \leq k$. By Theorem 8.42, $OC(k,\ell)$ contains a k-chromatic graph G. Then $G \to GY(k,\ell)$ (by (a)), which implies that $k = \chi(G) \leq \chi(GY(k,\ell))$ (by Proposition 8.1). This proves (b). □

Theorem 8.47 *The graph $GY(k,\ell)$ is k-critical for every $k \geq 1$ and every $\ell \geq 0$.*

Proof By Corollary 8.46, $\chi(GY(k,\ell)) = k$ and, therefore, we only need to show that every proper subgraph of $GY(k,\ell)$ is colorable with less than k colors. We prove this by induction on $k \geq 1$. The bottom cases are trivial with $GY(1,\ell) = K_1$, $GY(2,\ell) = K_2$ and $GY(k,0) = K_k$.

Now assume that $k \geq 3$ and $\ell \geq 1$. Let G denote the graph $GY(k,\ell)$. Then, by Lemma 8.43, $\delta(G) = k - 1 \geq 2$. Hence, to show that $\chi(H) < k$ for every proper subgraph H of G, it is sufficient to show that $G - e$ has a (k-1)-coloring for every edge e of G. To this end, consider an arbitrary edge $a_1 a_2$ of G. By symmetry and since $k \geq 3$, we can assume that $i_{a_1} = 1$ and $i_{a_2} = 2$. Then, by (8.12), we have $a_1(k) \neq 0$ as well as $a_2(k) \neq 0$. Furthermore, by Lemma 8.43, $a_1(2) = 1$ and $a_2(1) = 1$. This implies, in particular, that both mappings $a_1|_{[1,k-1]}$ and $a_2|_{[1,k-1]}$ belong to $A(k-1,\ell)$.

For a mapping $a' \in A(k-1,\ell)$ and an integer $p \in [1,\ell+1]$, let $a = (a',p)$ denote the mapping from $A(k,\ell)$ defined by

$$a(i) = \begin{cases} p & \text{if } i = k, \\ a'(i) & \text{if } i \in [1,k-1]. \end{cases}$$

Furthermore, let $A_p = \{a = (a',p) \mid a' \in A(k-1,\ell)\}$ and $A = A_1 \cup A_2 \cup \cdots \cup A_{\ell+1}$. Then, by (8.12), $V(G) = A(k,\ell) = A \cup A_0 \cup A_{-1}$ where

$$A_0 = \{a \in A(k,\ell) \mid a(k) = 0\}$$

and

$$A_{-1} = \{a \in A(k,\ell) \mid a(k) = 1, |a^{-1}(1)| = 1\}.$$

Note that

$$A_1 = \{a \in A(k,\ell) \mid a(k) = 1, |a^{-1}(1)| \geq 2\}$$

and, for $p \in [2,\ell+1]$,

$$A_p = \{a \in A(k,\ell) \mid a(k) = p\}.$$

If e is an edge in G, then it follows from (8.13), that e joins two vertices of $A_{\ell+1}$ or a vertex of A_p with a vertex of A_{p-1} for some $p \in [0, \ell+1]$. Furthermore, for two vertices $a = (a', p)$ and $\hat{a} = (\hat{a}', q)$ of A, we deduce from (8.13), that $a\hat{a}$ is an edge of $G = GY(k,\ell)$ if and only if $a'\hat{a}'$ is an edge of $G(k-1,\ell)$ and $p = q = \ell+1$ or $|p-q| = 1$. This implies, in particular, that $G' = G[A_{\ell+1}]$ is isomorphic to $GY(k-1,\ell)$.

Now consider the edge a_1a_2 of $G = G(k,\ell)$. Then both a_1 and a_2 belong to A, and, therefore, we have $a_1 = (a_1', p_1)$ and $a_2 = (a_2', p_2)$, where $a_1'a_2'$ is an edge of $GY(k-1,\ell)$ and $p_1 = p_2 = \ell+1$ or $|p_1 - p_2| = 1$. By induction, there is a coloring φ of $GY(k-1,\ell) - a_1'a_2'$ with color set $S' = [1, k-2]$. Since $\chi(GY(k-1,\ell)) = k-1$, this implies that $\varphi(a_1') = \varphi(a_2')$.

If $p_1 = p_2 = \ell+1$, then it is easy to check that the mapping $\phi : A(k,\ell) \to [1, k-1]$ defined by

$$\phi(a) = \begin{cases} \varphi(a') & \text{if } a = (a', \ell+1) \in A_{\ell+1}, \\ k-1 & \text{if } a \in A_\ell \cup A_{\ell-2} \cup \cdots, \\ 1 & \text{if } a \in A_{\ell-1} \cup A_{\ell-3} \cup \cdots, \end{cases}$$

is a coloring of $G - a_1a_2$ with color set $S = [1, k-1]$. If $|p_1 - p_2| = 1$ we construct a coloring ϕ of $G - a_1a_2$ with color set S as follows. By symmetry, we have $1 \le p_1 \le \ell$ and $p_2 = p_1 + 1$. If $a = (a', p) \in A_{\ell+1} \cup \cdots \cup A_{p_1}$, then let

$$\phi(a) = \begin{cases} k-1 & \text{if } a' = a_1' \text{ and } p \in [p_1+1, \ell+1], \\ \varphi(a') & \text{otherwise.} \end{cases}$$

Note that φ uses only colors from the set $[1, k-2]$ and, therefore, in A_{p_1} only colors from the set $[1, k-2]$ occur. For the remaining vertices $a \in A_{p_1-1} \cup \cdots \cup A_{-1}$ we define

$$\phi(a) = \begin{cases} k-1 & \text{if } a \in A_{p_1-1} \cup A_{p_1-3} \cup \cdots, \\ 1 & \text{if } a \in A_{p_1-2} \cup A_{p_1-4} \cup \cdots. \end{cases}$$

This results in a (k-1)-coloring of $G - a_1a_2$. This proves Theorem 8.47. □

The following result is an immediate consequence of Theorem 8.47 and Proposition 8.2. Note that (8.14) implies that $|GY(3,\ell)| = 3((\ell+1)^2 - \ell^2) = 6\ell + 3$.

Corollary 8.48 *Let $k \ge 1$ and $\ell \ge 0$ be integers. Then the following statements hold:*

(a) *$GY(3,\ell) = C_{6\ell+3}$.*
(b) *$G(k,\ell)$ is the smallest k-chromatic graph in $OC(k,\ell)$.*
(c) *Any homomorphism from $GY(k,\ell)$ to $GY(k,\ell)$ belongs to* $\mathrm{Aut}(GY(k,\ell))$.

Next we want to show that the automorphism group of $GY(k,\ell)$ is isomorphic to the symmetric group S_k, provided that $k \ge 4$. That $\mathrm{Aut}(G(k,\ell))$ contains a subgroup isomorphic to S_k follows from Lemma 8.44. Furthermore, we know that $GY(k,\ell)$ contains exactly k vertices of degree $k-1$ provided that $k \ge 4$. Our aim is to show that the only automorphism of $GY(k,\ell)$ that fix all vertices of degree $k-1$ in $G(k,\ell)$ is the identity map. For the proof of this statement, we shall study the distance function in $GY(k,\ell)$.

Let $a \in A(k,\ell)$ be a vertex of $GY(k,\ell)$, where $k \geq 3$ and $\ell \geq 1$. Moreover, let $i \in [1,k]$. By (8.12), the set $\{a(j) \mid a(j) \not\equiv a(i) \,(\text{mod } 2),\ j \in [1,k]\}$ is not empty and we define

$$\mathrm{MU}^i(a) = \max\{a(j) \mid a(j) \not\equiv a(i) \,(\text{mod } 2),\ j \in [1,k]\}.$$

Furthermore, we define

$$\mathrm{me}^i(a) = \min\{a(j) \mid a(j) \equiv a(i) \,(\text{mod } 2),\ j \in [1,k] \setminus \{i\}\},$$

provided that the underlying set is nonempty; otherwise, we set $\mathrm{me}^i(a) = \infty$. For given integers $k \geq 2$ and $\ell \geq 0$, let

$$B(k,\ell) = \{b \in A(k,\ell) \mid |b^{-1}(1)| = k-1\}.$$

Then, by (8.12), $|B(k,\ell)| = k$ and, for every $i \in [1,k]$, there is exactly one vertex $b \in B(k,\ell)$ such that $b(i) = 0$. By Lemma 8.43(c), every vertex of $B(k,\ell)$ has degree $k-1$ in $GY(k,\ell)$.

Lemma 8.49 *Let $k \geq 3$ and $\ell \geq 1$ be integers. Furthermore, let $G = GY(k,\ell)$ and let $b \in B(k,\ell)$ be a vertex such that $b(i) = 0$ for some $i \in [1,k]$. Then $N_G^p(b) = N^p$ for all $p \in [0, 3\ell+1]$, where*

$$N^p = \{a \in A(k,\ell) \mid a(i) \in [0,1],\ \mathrm{MU}^i(a) = p+1,\ \mathrm{me}^i(a) = \infty\}$$

if $p \in [0,\ell]$,

$$N^p = \{a \in A(k,\ell) \mid a(i) \in [0,1],\ \mathrm{me}^i(a) = 2\ell+2-p\}$$

if $p \in [\ell+1, 2\ell+1]$, and,

$$N^p = \{a \in A(k,\ell) \mid a(i) = p - 2\ell\}$$

if $p \in [2\ell+2, 3\ell+1]$.

Sketch of Proof: Let a be a vertex of $GY(k,\ell)$. Then $a(i) \not\equiv \mathrm{MU}^i(a) \,(\text{mod } 2)$. Furthermore, if $\mathrm{me}^i(a) = \infty$, then $a(j) \not\equiv a(i) \,(\text{mod } 2)$ for every $j \in [1,k] \setminus \{i\}$. This implies that

$$N^0 = \{a \mid a(i) = 0,\ a(j) = 1,\ j \neq i\} = \{b\} = N_G^0(b).$$

Now, consider an arbitrary vertex $a \in N^p$ with $p \in [0, 3\ell+1]$. We claim that $N_G(a) \subseteq N^{p-1} \cup N^{p+1}$ where $N^{-1} = \varnothing$ and $N^{3\ell+2} = N^{3\ell+1}$. Furthermore, we claim that a has at least one neighbor in N^{p-1}, provided that $p \geq 1$. The proofs of both claims are straightforward and left to the reader; a detailed proof can be found in Baum [84]. From these two claims it then follows that $N_G^p(b) = N^p$ for all $p \in [0, 3\ell+1]$. □

Corollary 8.50 *Let $k \geq 3$ and $\ell \geq 1$ be integers. If a_1 and a_2 are two distinct vertices of $G = GY(k,\ell)$, then there is a vertex $b \in B(k,\ell)$ such that* $\mathrm{dist}_G(a_1,b) \neq \mathrm{dist}_G(a_2,b)$.

Proof Since $a_1 \neq a_2$, there is an $i \in [1,k]$ such that $a_1(i) \neq a_2(i)$. Now, consider the vertex $b \in B(k,\ell)$ with $b(i) = 0$. For a vertex $a \in A(k,\ell)$, Lemma 8.49 implies that

$$\mathrm{dist}_G(a,b) \in \begin{cases} \{0,2,\ldots,2\ell\} & \text{if } a(i) = 0, \\ \{1,3,\ldots,2\ell+1\} & \text{if } a(i) = 1, \\ \{2\ell + a(i)\} & \text{if } a(i) \geq 2. \end{cases}$$

Consequently, $\mathrm{dist}_G(a_1,b) \neq \mathrm{dist}_G(a_2,b)$. □

Theorem 8.51 *Let $k \geq 4$ and $\ell \geq 1$ be integers. For a permutation $\pi \in S_k$, denote by $\phi_\pi : A(k,\ell) \to A(k,\ell)$ the map with $\phi_\pi(a) = a \circ \pi^{-1}$ for all $a \in A(k,\ell)$. Then* $\mathrm{Aut}(GY(k,\ell)) = \{\phi_\pi \mid \pi \in S_k\}$.

Proof Let $\Gamma = \mathrm{Aut}(GY(k,\ell))$ and $\Gamma' = \{\phi_\pi \mid \pi \in S_k\}$. By Lemma 8.44, Γ' is a subgroup of Γ that is isomorphic to S_k, which gives $|\Gamma| \geq |\Gamma'| = k!$. Therefore, to prove that $\Gamma = \Gamma'$, it is sufficient to show that $|\Gamma| = k!$.

Since $k \geq 4$, Lemma 8.43 implies that for the vertex set $B = B(k,\ell)$ we have $B = \{a \in A(k,\ell) \mid d_{GY(k,\ell)}(a) = k-1\}$ and $|B| = k$. Consequently, $\phi(B) = B$ for all $\phi \in \Gamma$ (by Proposition C.9). Then $\Gamma|_B = \{\phi|_B \mid \phi \in \Gamma\}$ is a subgroup of Γ with $|\Gamma|_B| \leq k!$. We claim that $|\Gamma|_B| = k!$. To prove this claim, consider two permutations $\pi_1, \pi_2 \in S_k$ such that $\phi_{\pi_1}|_B = \phi_{\pi_2}|_B$. Then $b \circ \pi_1^{-1} = b \circ \pi_2^{-1}$ for every $b \in B$, and so $\pi_1(i_b) = \pi_2(i_b)$. Since $\{i_b \mid b \in B\} = [1,k]$, this implies that $\pi_1 = \pi_2$. This proves the claim that $|\Gamma|_B| = k!$.

Combining Corollary 8.50 and Proposition C.9(b), we obtain that the subgroup

$$\Gamma_B = \{\phi \in \Gamma \mid \phi(b) = b \text{ for all } b \in B\}$$

consists only of the identity map, which gives $|\Gamma_B| = 1$. Since $|\Gamma|_B|$ is equal to the index $|\Gamma/\Gamma_B|$ of Γ_B in Γ, Lagrange's Theorem (Proposition C.10) implies that $|\Gamma| = |\Gamma|_B||\Gamma_B| = k!$. □

Let G be a graph. Two mappings (colorings) $f, g : V(G) \to [1,k]$ are called **equivalent** if there is a permutation $\pi \in S_k$ such that $g(v) = \pi(f(v))$ for every vertex v of G. The graph G is called **uniquely odd** (k,ℓ)**-colorable** if $G \in OC(k,\ell)$ and if every two odd (k,ℓ)-colorings of G are equivalent.

Theorem 8.52 *Let $k \geq 1$ and $\ell \geq 0$ be integers with $k \neq 3$. Then $GY(k,\ell)$ is uniquely odd (k,ℓ)-colorable.*

Proof Since $GY(1,\ell) = K_1$, $GY(2,\ell) = K_2$ and $GY(k,0) = K_k$, we need only to consider the case $k \geq 4$ and $\ell \geq 1$. Let $f : A(k,\ell) \to [1,k]$ be the mapping with $f(a) = i_a$. Then, by Theorem 8.45, f is an odd (k,ℓ)-coloring of $GY(k,\ell)$. Clearly, it is sufficient to show that if $g : A(k,\ell) \to [1,k]$ is an arbitrary odd (k,ℓ)-coloring of $GY(k,\ell)$, then g is equivalent to f. From Theorem 8.45 it follows that there is a homomorphism ϕ from $GY(k,\ell)$ to $GY(k,\ell)$ such that $g = f \circ \phi$. By Corollary 8.48,

$\phi \in \mathrm{Aut}(G(k,\ell))$. Then Theorem 8.51 implies that there is a permutation $\pi \in S_k$ such that $\phi(a) = a \circ \pi^{-1}$ for all $a \in A(k,\ell)$. Then, for every vertex $a \in A(k,\ell)$, we have $g(a) = f(\phi(a)) = f(a \circ \pi^{-1}) = i_{a\circ\pi^{-1}} = \pi(i_a) = \pi(f(a))$. □

Theorem 8.53 *Let $k \geq 3$ and $\ell \geq 0$ be integers. Then*

$$g_o(GY(k,\ell)) = 2\ell + 1 + 2\left\lceil \frac{2\ell+1}{k-2} \right\rceil.$$

Proof Since $GY(k,0) = K_k$, the statement is evident for $\ell = 0$. So assume that $\ell \geq 1$. First, we prove that $g_o(G(k,\ell)) \geq 2\ell + 1 + 2\lceil (2\ell+1)/(k-2) \rceil$. Suppose this is false. Then there is an odd closed walk $P = (a_0, a_1, \ldots, a_{t-1}, a_t)$ in $GY(k,\ell)$ with $t = \ell(P) = 2\ell - 1 + 2\lceil (2\ell+1)/(k-2) \rceil$. By (8.12), $|a_p^{-1}(0)| = 1$ for all $p \in [0,t]$. Consequently, there is an $i \in [1,k]$ for which

$$m = |\{p \in [1,t] \mid a_p(i) = 0\}| \geq \lceil t/k \rceil.$$

Now, we claim that there is a $p \in [0, t-1]$ for which $a_p(i) = a_{p+1}(i) = \ell + 1$. For otherwise, we deduce from (8.13), that

$$0 = a_0(i) - a_t(i) \equiv \sum_{p=0}^{t-1} (a_p(i) - a_{p+1}(i)) \pmod 2 \equiv t \pmod 2 = 1.$$

This contradiction proves the claim. If $a_p(i) = 0$ for some $p \in [0,t]$, then Lemma 8.43 implies that $a_{p\pm1}(i) = 1$ and $a_{p\pm q}(i) \neq \ell + 1$ for $q \in [0,\ell]$, where all indices are taken modulo t. This, clearly, implies that $t = \ell(P) \geq 2\ell + 1 + 2m$. Since $k \geq 3$, we then conclude that

$$\ell(P) \geq 2\ell + 1 + 2\left\lceil \frac{2\ell - 1 + 2\lceil (2\ell+1)/(k-2) \rceil}{k} \right\rceil \geq 2\ell + 1 + 2\left\lceil \frac{2\ell+1}{k-2} \right\rceil,$$

giving a contradiction.

Next, we prove that $g_o(GY(k,\ell)) \leq t$ where $t = 2\ell + 1 + 2\lceil (2\ell+1)/(k-2) \rceil$. For the proof of this inequality it is sufficient to construct an odd closed walk P in $GY(k,\ell)$ whose length is t. Let $m = \lceil (2\ell+1)/(k-2) \rceil$ and

$$\mathbf{v} = (v_0, v_1, \ldots, v_{t-1}) = (\underbrace{0,1,0,\ldots,1,0,1}_{\text{length } 2m}, 2, 3, \ldots, \ell, \ell+1, \ell+1, \ell, \ldots, 3, 2, 1).$$

For an integer $p \in [0, t-1]$, let $a_p : [1,k] \to [0, \ell+1]$ be the mapping defined by

$$a_p(i) = \begin{cases} 2 & \text{if } p + 2(i-1)m > 2t - 2 \text{ and } v_{p+2(i-1)m} = 0, \\ v_{p+2(i-1)m} & \text{otherwise,} \end{cases}$$

for all $i \in [1,k]$, where all indices of v are taken modulo t. Since t is odd and

$$\begin{aligned}2m-2+2(k-1)m &= 4m-2+2(k-2)m\\ &= 4m-2+2(k-2)\left\lceil\frac{2\ell+1}{k-2}\right\rceil\\ &\geq 4m-2+2(2\ell+1)\\ &= 4m+4\ell\\ &= 2t-2\end{aligned}$$

we deduce that all mappings $a_0, a_1, \ldots, a_{t-1}$ belong to $A(k,\ell)$ and, moreover, that $P = (a_0, a_1, \ldots a_{t-1}, a_0)$ is an odd closed walk of length t in $GY(k,\ell)$. □

Combining Theorem 8.53, Proposition 8.1 and Corollary 8.46, we immediately obtain the following result.

Corollary 8.54 *Let $k \geq 3$ and $\ell \geq 0$ be integers. If $G \in OC(k,\ell)$, then*

$$g_o(G) \geq 2\ell+1+2\left\lceil\frac{2\ell+1}{k-2}\right\rceil.$$

We define the **odd chromatic number** of a graph G, written as $\chi^{\mathrm{od}}(G)$, by

$$\chi^{\mathrm{od}}(G) = \inf\{\tfrac{k}{2\ell+1} \mid G \in OC(k,\ell)\}.$$

By Corollary 8.46, this is equivalent to

$$\chi^{\mathrm{od}}(G) = \inf\{\tfrac{k}{2\ell+1} \mid G \to GY(k,\ell)\}.$$

Then we have the following monotonicity property of the odd chromatic number.

Proposition 8.55 *Let G and H be two graphs. If $G \to H$, then $\chi^{\mathrm{od}}(G) \leq \chi^{\mathrm{od}}(H)$.*

Proof From $G \to H$ it follows that

$$\{\tfrac{k}{2\ell+1} \mid H \to GY(k,\ell)\} \subseteq \{\tfrac{k}{2\ell+1} \mid G \to GY(k,\ell)\},$$

which yields $\chi^{\mathrm{od}}(G) \leq \chi^{\mathrm{od}}(H)$. □

Theorem 8.56 *Let G be an arbitrary graph. If G is bipartite, then $\chi^{\mathrm{od}}(G) = 0$. Otherwise, G satisfies*

$$\chi^{\mathrm{od}}(G) = \min\{\tfrac{k}{2\ell+1} \mid G \to GY(k,\ell), k \leq |G| \text{ and } 2\ell+3 \leq g_o(G)\}$$

and

$$0 < \frac{\chi(G)}{g_o(G)-2} \leq \chi^{\mathrm{od}}(G) \leq \chi(G).$$

Proof If G is bipartite, then $G \to GY(2,\ell) = K_2$ for all $\ell \geq 0$, and so $\chi^{\mathrm{od}}(G) = 0$. Now assume that G is not bipartite. Then $g_o(G)$ is finite and $\chi(G) \geq 3$. If $G \to GY(k,\ell)$, then Proposition 8.1 and Corollary 8.46 implies that

$$\chi(G) \leq \chi(GY(k,\ell)) = k$$

and, moreover, Corollary 8.40 imply that $g_o(G) \geq 2\ell+3$. Consequently, we have

$$\frac{\chi(G)}{g_o(G)-2} \leq \frac{k}{2\ell+1}$$

whenever $G \to G(k,\ell)$. Hence

$$\frac{\chi(G)}{g_o(G)-2} \leq \chi^{\mathrm{od}}(G).$$

On the other hand, if $\chi(G)=k'$, then $G \to K_{k'} = GY(k',0)$ and, therefore, $\chi^{\mathrm{od}}(G) \leq k' = \chi(G)$. Furthermore, if $G \in OC(k,\ell)$ for an integer $k \geq |G| = n$, then we have $G \in OC(n,\ell)$ and $n/(2\ell+1) \leq k/(2\ell+1)$. Hence $\chi^{\mathrm{od}}(G)$ is the infimum over the set $S = \{k/(2\ell+1) \mid G \to GY(k,\ell),\, k \leq |G|,\, 2\ell+3 \leq g_o(G)\}$. Since the set S is finite, the infimum is in fact a minimum. This completes the proof of the theorem. □

Corollary 8.57 *The following statements hold:*

(a) *If G is a graph containing a triangle, then $\chi^{\mathrm{od}}(G) = \chi(G)$.*
(b) *$\chi^{\mathrm{od}}(K_1) = \chi^{\mathrm{od}}(K_2) = 0$ and $\chi^{\mathrm{od}}(K_n) = n$ for all $n \geq 3$.*
(c) *If $k \geq 3$ and $\ell \geq 0$ are integers, then*

$$\frac{k}{2\ell+1+2\lceil\frac{2\ell+1}{k-2}\rceil} \leq \chi^{\mathrm{od}}(GY(k,\ell)) \leq \frac{k}{2\ell+1}.$$

(d) *If $k \geq 2\ell+3$ and $\ell \geq 0$, then $\chi^{\mathrm{od}}(GY(k,\ell)) = \frac{k}{2\ell+1}$.*

8.6 The Rise and Fall of the Product Conjecture

Given two graphs G and H, their **direct product** $G \times H$ is the graph whose vertex set is $V(G)\times V(H)$ and in which two vertices (v,w) and (v',w') are adjacent if and only if $vv' \in E(G)$ and $ww' \in E(H)$. This product is most natural in the category of graphs, and it is also called the **categorial product**.

It is easy to check that the direct product is commutative in the sense that, for all graphs G and H, we have $G\times H \cong H \times G$; it is also associative. Furthermore, $G\times H$ is the empty graph if and only if G or H is the empty graph.

Let G and H be two graphs. We then denote by $\mathrm{Hom}(G,H)$ the set of homomorphisms from G to H. For each vertex v of G, the set $X_v = \{v\}\times V(H)$ is an independent set of $G\times H$; and so the map $p_G : V(G\times H) \to V(G)$ with $p_G(v,w) = v$ is a homomorphism from $G\times H$ to G. Similarly, the map $p_H : V(G\times H) \to V(H)$ with $p_H(v,w) = w$ is a homomorphism from $G\times H$ to H. The two maps p_G and p_H are called **projections** of $G\times H$ to G and H, respectively, and we have

$$G\times H \to G \text{ and } G\times H \to H \tag{8.15}$$

Furthermore, for any graph K we have

$$K \to G \text{ and } K \to H \Longleftrightarrow K \to G \times H \tag{8.16}$$

If $\phi_1 \in \mathrm{Hom}(K,G)$ and $\phi_2 \in \mathrm{Hom}(K,H)$, then the map ϕ defined by

$$\phi(u) = (\phi_1(u), \phi_2(u))$$

for all $u \in V(K)$ is the unique homomorphism from K to $G \times H$ satisfying $\phi_1 = p_G \circ \phi$ and $\phi_2 = p_H \circ \phi$ (see Exercise 8.19).

Both the clique number and the odd girth of a given graph can be characterized by homomorphisms from complete graphs and odd cycles, respectively, into the graph (see (8.2) and (8.3)). Using Proposition 8.1, (8.15) and (8.16) we easily get that

$$\omega(G \times H) = \min\{\omega(G), \omega(H)\} \text{ and } g_o(G \times H) = \max\{g_o(G), g_o(H)\}$$

(see Exercise 8.20). For the chromatic number of $G \times H$, however, we only obtain the following upper bound

$$\chi(G \times H) \leq \min\{\chi(G), \chi(H)\}. \tag{8.17}$$

This is an immediate consequence of (8.15) and Proposition 8.1. Observation (8.17) led Hedetniemi [489] to propose the following conjecture, also known as the Lovász–Hedetniemi product conjecture.

Conjecture 8.58 (PRODUCT CONJECTURE I) $\chi(G \times H) = \min\{\chi(G), \chi(H)\}$ *for all graphs G and H.*

Let us first discuss a parameterized version of Hedetniemi's conjecture. A graph K is called **multiplicative** if $G \times H \to K$ implies $G \to K$ or $H \to K$ for all graphs G and H. The class of multiplicative graphs was first introduced by Häggkvist, Hell, Miller, and Neumann-Lara [450]; they proved that all cycles are multiplicative. However, we are more interested in deciding which complete graphs are multiplicative. The following proposition is an immediate consequence of (8.1) and (8.17) (see also Exercise 8.16).

Proposition 8.59 *For every integer* $n \in \mathbb{N}_0$, *the following statements are equivalent:*

(a) K_n *is multiplicative.*
(b) *For all graphs G and H,* $\chi(G \times H) \leq n$ *implies* $\min\{\chi(G), \chi(H)\} \leq n$.
(c) *For all graphs G and H,* $\min\{\chi(G), \chi(H)\} = n+1$ *implies* $\chi(G \times H) = \min\{\chi(G), \chi(H)\}$.

Hedetniemi's product conjecture is obviously true for graphs G and H satisfying $\min\{\chi(G), \chi(H)\} = 0$. Moreover, Proposition 8.59 implies that the product conjecture is equivalent to the following conjecture.

Conjecture 8.60 (PRODUCT CONJECTURE II) *For all* $n \in \mathbb{N}_0$, *the complete graph* K_n *is multiplicative.*

Since Hedetniemi proposed this conjecture in 1966, little progress was made until it was refuted by Shitov [938] in 2019. Hedetniemi himself [489] confirmed that K_0, K_1 and K_2 are multiplicative (see Theorem 8.65). A major breakthrough was achieved in 1985 when El-Zahar and Sauer [325] succeeded to show that K_3 is multiplicative (see Theorem 8.68) implying that $\chi(G\times H) = 4$, provided that G and H are graphs satisfying $\min\{\chi(G), \chi(H)\} = 4$. Unfortunately, their method does not apply to any complete graph K_n with $n \geq 4$. Before we turn to the proofs of these theorems, we introduce a further concept for graphs.

Let G and K be graphs. The **map graph** K^G is the pseudograph whose vertices are the maps from $V(G)$ to $V(K)$, where two such maps f and f' are adjacent in K^G if and only if $vv' \in E(G)$ implies $f(v)f'(v') \in E(K)$ for all vertices $v, v' \in V(G)$. Note that a vertex f of K^G is a loop (i.e. $\{f\} \in E(K^G)$) if and only if $f \in \mathrm{Hom}(G,K)$. So $|\mathrm{Hom}(G,K)|$ is equal to the number of loops of K^G; and K^G is an ordinary graph if and only if $G \nrightarrow K$.

The definitions of homomorphisms and isomorphisms can be easily extended to pseudographs (loops must be mapped to loops, and an edge can be mapped to a loop). Also the definition of the direct product can be easily extended to pseudographs. Note that a vertex (v,w) of $G\times H$ forms a loop in $G\times H$ if and only if v forms a loop in G and w forms a loop in H.

The **looped path** P_n^o is a path of order n where one end of the path forms a loop; so it is a pseudograph with exactly one loop. In particular, P_1^o is the pseudograph having one vertex and one loop. Note that if K is any graph, then $K^\varnothing = P_1^o$ and every pseudograph is homomorphic to P_1^o. The direct product $G = P_3 \times P_4^o$ is depicted Figure 8.7. If w is the only vertex of P_4^o having degree one, then the vertex set $Y_w = \{(v,w) \in V(G) \mid v \in V(P_3)\}$ is independent in G and identifying Y_w to a new vertex results in the Mycielski graph $\mathrm{M}_3(P_3) = (P_3 \times P_4^o)/Y_w$.

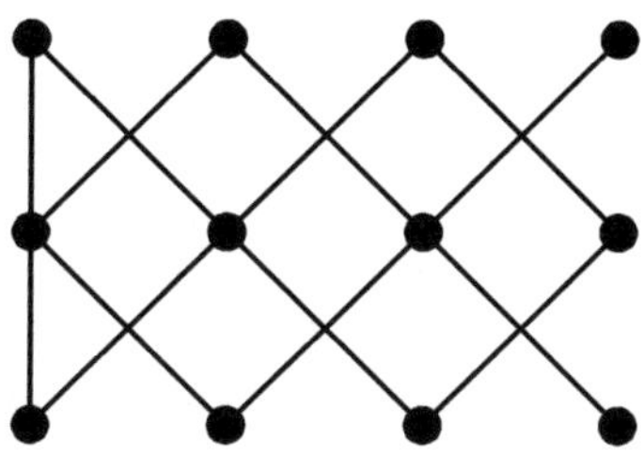

Fig. 8.7 The direct product $G = P_3 \times P_4^o$.

Theorem 8.61 *Let G, H and K be graphs. Then the following statements hold:*

(a) *If $G \to H$, then $K^H \to K^G$.*
(b) $K^{G\times H} \cong (K^G)^H$.
(c) $|\mathrm{Hom}(G\times H, K)| = |\mathrm{Hom}(H, K^G)|$.
(d) *If G has at least one edge, then the constant mappings from $V(G)$ to $V(K)$ induce a subgraph of K^G isomorphic to K.*

Proof In order to prove (a), assume that φ is a homomorphism from G to H. Define a map ϕ from $V(K^H)$ to $V(K^G)$ by putting $\phi(f) = f \circ \varphi$ for all $f \in V(K^H)$. To show that ϕ is a homomorphism from K^H to K^G, let ff' be an edge of K^H. The proof is completed by showing that $g = f \circ \varphi$ and $g' = f' \circ \varphi$ are adjacent in K^G. To show this, let vv' be an arbitrary edge of G. Since φ is a homomorphism, $\varphi(v)\varphi(v')$ is an edge of H. Since ff' is an edge of K^H, $g(v) = f(\varphi(v))$ is adjacent to $g'(v') = f'(\varphi(v'))$ in K. Consequently, gg' is an edge of K^G as required. Thus (a) is proved.

To prove (b), we first construct a bijection Φ between $K^{G\times H}$ and $(K^G)^H$, and then we shall show that Φ is an isomorphism. Consider a vertex f of $K^{G\times H}$, that is, f is a map from $V(G \times H)$ to $V(K)$. Then, for every vertex w of H, let f_w denote the map $v \mapsto f(v,w)$ from $V(G)$ to $V(K)$, so f_w is a vertex of K^G. Now let $\Phi(f)$ be the map $w \mapsto f_w$. Clearly, this map belongs to $(K^G)^H$. This defines a map Φ from $K^{G\times H}$ to $(K^G)^H$, and it is easy to check that this map is injective. It is also easy to see that both graphs $K^{G\times H}$ and $(K^G)^H$ have the same order. This implies that Φ is a bijection between these two graphs. It remains to show that Φ is also an isomorphism. So let f and f' be two vertices of $K^{G\times H}$. Let $g = \Phi(f)$ and $g' = \Phi(f')$ be the corresponding vertices of $(K^G)^H$

First assume that ff' is an edge of $K^{G\times H}$. To see that gg' is an edge of $(K^G)^H$, let ww' be an edge of H. Then $g(w) = f_w$ and $g'(w') = f'_{w'}$. Now, we must show that $f_w f'_{w'}$ is an edge of K^G. This follows from the fact that if vv' is an edge of G, then $f_w(v) = f(v,w)$ and $f'_{w'}(v') = f'(v',w')$ are adjacent in K, since (v,w) is adjacent to (v',w') in $G \times H$ and ff' is an edge of $K^{G\times H}$. So gg' is an edge of $(K^G)^H$ as required.

Now assume that gg' is an edge of $(K^G)^H$. To see that ff' is an edge of $K^{G\times H}$, let (v,w) and (v',w') be two adjacent vertices of $G \times H$. Then ww' is an edge of H, which implies that $g(w) = f_w$ and $g'(w') = f'_{w'}$ are adjacent in K^G. Furthermore vv' is an edge of G, which implies that $f_w(v) = f(v,w)$ and $f'_{w'}(v') = f'(v',w')$ are adjacent in K. So ff' is an edge of $K^{G\times H}$ as required.

Statement (c) follows from (b) and the fact that isomorphic pseudographs have the same number of loops, which are precisely the homomorphisms.

It remains to prove (d). For a vertex u of K, let f_u be the constant map $v \mapsto u$ from $V(G)$ to $V(K)$. The proof is completed by showing that uu' is an edge of K if and only if $f_u f_{u'}$ is an edge of K^G. First, assume that uu' is an edge of K. This obviously implies that $f_u f_{u'}$ is an edge of K^G, since $f_u(v) f_{u'}(v') = uu'$ is an edge of of K whenever vv' is an edge of G. Assume conversely that $f_u f_{u'}$ is an edge of K^G. By hypothesis, G has an edge vv', and so $f_u(v) = u$ and $f_{u'}(v') = u'$ are adjacent in K as required. □

For a graph G and an integer $n \in \mathbb{N}_0$, the map graph K_n^G is also called the ***n*-coloring graph** of G. The vertex set of K_n^G consists of all improper colorings of G with color set $V(K_n)$; and two such improper colorings f and f' are adjacent in K_n^G if and only if $vv' \in E(G)$ implies $f(v) \neq f'(v')$ for all $v, v' \in V(G)$. If $\chi(G) > n$, then $G \nrightarrow K_n$ (by 8.1), which implies that the pseudograph K_n^G has no loops and is therefore a graph. Note that $K_0^G = K_0$ unless G is the empty graph. The next two results are due to El-Zahar and Sauer [325]; they used these results to simplify the study of the product conjecture by means of the map graph.

Theorem 8.62 (El-Zahar and Sauer) *Let G be a connected graph such that $\chi(G) > n$, where n is a positive integer. Then the graph K_n^G contains a unique complete subgraph of order n, namely the subgraph induced by the constant maps.*

Proof Since $\chi(G) > n$ and $n \geq 1$, G has an edge. Then it follows from Theorem 8.61(d) that the constant maps from $V(G)$ to $V(K_n)$ induces a complete subgraph of K_n^G having order n.

Now, let $f_1, f_2, \ldots, f_n$ denote the vertices of a complete subgraph in K_n^G having order n. If $i, j \in [1,n]$ and $i \neq j$, then $f_i f_j \in E(K_n^G)$ which implies that (1) $f_i(u) \neq f_j(v)$ whenever $uv \in E(G)$.

First assume that there is a vertex $v \in V(G)$ such that the values of the sequence $f_1(v), f_2(v), \ldots, f_n(v)$ are all distinct, that is, $V(K_n) = \{f_1(v), f_2(v), \ldots, f_n(v)\}$. By (1) it then follows that if $vu \in E(G)$, then $f_i(u) = f_i(v)$ for $i \in [1,n]$. Since G is connected, this implies that $f_i(u) = f_i(v)$ for all $u, v \in V(G)$ and $i \in [1,n]$. Consequently, $f_1, f_2, \ldots, f_n$ are the constant maps from $V(G)$ to $V(K_n)$.

It remains to consider the case that, for all vertices $v \in V(G)$, the sequence $f_1(v), f_2(v), \ldots, f_n(v)$ contains at least one repetition. For a vertex $v \in V(G)$, let $\phi(v)$ denote a vertex of K_n that appears twice in the sequence $f_1(v), f_2(v), \ldots, f_n(v)$. Then ϕ is a map from $V(G)$ to $V(K_n)$ and it follows from (1) that every edge $uv \in E(G)$ satisfies $\phi(u) \neq \phi(v)$. Therefore, ϕ is a homomorphism from G to K_n and so $\chi(G) \leq n$, which is impossible. □

Theorem 8.63 (El-Zahar and Sauer) *Let G be a graph such that $\chi(G) > n$ with $n \in \mathbb{N}_0$. Then the graph K_n^G is n-colorable if and only if $\chi(G \times H) > n$ for all graphs H satisfying $\chi(H) > n$.*

Proof First assume that K_n^G is n-colorable. We must show that $\chi(G \times H) > n$ for all graphs H satisfying $\chi(H) > n$. Since $\chi(H) > n$, it follows from Proposition 8.1, that there is no homomorphism from H to K_n^G. Using Theorem 8.61(c), we obtain that

$$0 = |\mathrm{Hom}(H, K_n^G)| = |\mathrm{Hom}(G \times H, K_n)|,$$

which implies that $\chi(G \times H) > n$. Assume conversely, that $\chi(G \times H) > n$ for all graphs H satisfying $\chi(H) > n$. By Theorem 8.61(c),

$$|\mathrm{Hom}(G \times K_n^G, K_n)| = |\mathrm{Hom}(K_n^G, K_n^G)| > 0,$$

which implies that $\chi(G \times K_n^G) \leq n$, and so $\chi(K_n^G) \leq n$. □

Corollary 8.64 *For an integer $n \in \mathbb{N}_0$, the following statements are equivalent:*

(a) *K_n is multiplicative.*
(b) *The map graph K_n^G is n-colorable for every graph G with $\chi(G) > n$.*
(c) *The map graph K_n^G is n-colorable for every graph $G \in \mathrm{Crit}(n+1)$.*

Proof To show that (a) implies (b), assume that K_n is multiplicative. Let G be a graph with $\chi(G) > n$. Since K_n is multiplicative, Proposition 8.59(b) implies that

$\chi(G\times H) > n$ for all graphs H satisfying $\chi(H) > n$. Then, by Theorem 8.63, K_n^G is n-colorable and we are done. To show that (b) implies (a), assume that, for every graph G with $\chi(G) > n$, the map graph K_n^G is n-colorable. Then Theorem 8.63 implies that $\chi(G\times H) > n$ for all graphs H satisfying $\chi(H) > n$. From Proposition 8.59(b) it then follows that K_n is multiplicative. Obviously, (b) implies (c). To see the converse implication, let G be a graph with $\chi(G) > n$. Then G contains a subgraph G' such that G' is critical and $\chi(G') = n+1$ (by Proposition 1.28(b)). By (b), we obtain that $\chi(K_n^{G'}) \le n$. Since $G' \subseteq G$, we have $G' \to G$ which implies that $K_n^G \to K_n^{G'}$ (by Theorem 8.61) and so $\chi(K_n^G) \le \chi(K_n^{G'}) \le n$. This proves that (c) implies (b). □

Recall from Section 4.1 that Crit(1) = $\{K_1\}$, Crit(2) = $\{K_2\}$, and Crit(3) = $\{C_n \mid n \equiv 1 \,(\mathrm{mod}\ 2), n \ge 3\}$. We have $K_0^{K_1} = K_0$ and $K_1^{K_2} = K_1$ implying that $\chi(K_0^{K_1}) = 0$ and $\chi(K_1^{K_2}) = 1$. If C is an odd cycle, then K_2^C has exactly one edge (by Theorem 8.62) and so $\chi(K_2^C) = 2$. By Corollary 8.64 this yields the following result of Hedetniemi [489].

Theorem 8.65 (Hedetniemi) *K_0, K_1 and K_2 are multiplicative.*

Next, we want to prove Hedetniemi's conjecture for 4-chromatic graphs by showing that K_3 is multiplicative. To this end, we use Corollary 8.64 and analyse the class of 3-coloring graphs of 4-chromatic graphs. This enables us to reduce arguments to odd cycles.

Let C be a cycle or the disjoint union of cycles, and let f be a vertex of K_3^C. A vertex v of C is **fixed** by f if the two neighbors of v in C get different values with respect to f, i.e. $|f(N_C(v))| = 2$. Let $v(C:f)$ denote the number of vertices of C that are fixed by f.

Lemma 8.66 (El-Zahar and Sauer) *Let C be a cycle, and let f be a vertex of K_3^C. Then the following statements hold:*

(a) *If $v(C:f)$ is odd, then C contains a path P such that $|f(V(P))| = |P| = 3$.*
(b) *If $f \in \mathrm{Hom}(C, K_3)$, then $v(C:f) \equiv |C| \,(\mathrm{mod}\ 2)$.*
(c) *If g is a vertex of K_3^C and $fg \in E(K_3^C)$, then $v(C:f) \equiv v(C:g) \,(\mathrm{mod}\ 2)$.*

Proof Let $C_n = (v_1, v_2, \ldots, v_n, v_1)$ be a cycle on n vertices. We will compute the indices modulo n. Let f be a vertex of the map graph $K_3^{C_n}$.

For the proof of (a), assume that $v(C_n:f)$ is odd. Then f is not a constant map. Let P be a maximal subpath of C such that f restricted to $V(P)$ is a constant map. Then $1 \le |P| < |C|$ and we can partition C into such paths. If $|P| \ge 2$, then the only vertices of P that are fixed by f are the two ends of P. If P consists of one vertex, say v_i, then v_i is fixed by f if and only if $f(v_{i-1}), f(v_i)$ and $f(v_{i+1})$ are all distinct. From this statement (a) easily follows.

For the proof of (b) assume that $f \in \mathrm{Hom}(C_n, K_3)$, i.e., f is a (proper) coloring of C_n with color set $V(K_3) = \{1,2,3\}$. Then, we prove by induction of n that $v(C_n:f) \equiv n \,(\mathrm{mod}\ 2)$. Clearly, $v(C_3:f) = 3$. Furthermore, $v(C_4:f)$ is zero or two depending on whether f uses two or three colors. So we are done if $n \in \{3,4\}$. Now assume that $n \ge 5$. If each vertex of C_n is fixed by f, the statement is evident. So assume that one vertex, say v_n, is not fixed by f. By symmetry, we may assume that $f(v_{n-1}) = f(v_1) = 1$ and $f(v_n) = 2$. If we delete v_n and identify v_{n-1} and v_1, we obtain a cycle C_{n-2} and $v_i \mapsto f(v_i)$ remains a (proper) coloring of C_{n-2}. If $f(v_2) = f(v_{n-2}) = 3$, then the number of fixed vertices decreases by two; in all other cases the number of fixed vertices remains the same. Hence $v(C_n : f) \equiv v(C_{n-2} : f)\ (\text{mod}\ 2)$. The induction hypothesis implies that $v(C_{n-2} : f) \equiv n-2\ (\text{mod}\ 2)$. Consequently, we get $v(C_n : f) \equiv n\ (\text{mod}\ 2)$. This proves statement (b).

For the proof of (c) assume that fg is an edge of $K_3^{C_n}$. Let $H = C_n \times K_2$, where $V(K_2) = \{a, b\}$. Since f and g are adjacent in $K_3^{C_n}$, the map $h : V(H) \to V(K_2)$ with $h(v_i, a) = f(v_i)$ and $h(v_i, b) = g(v_i)$ for $i \in [1, n]$ is a homomorphism from H to K_2. The graph H consists of one cycle of order $2n$ if n is odd and two disjoint cycles of order n if n is even. From (b) we then deduce that the number of vertices of H that are fixed by h is even. On the other hand, it is easy to check that the sum of the number of fixed vertices by f and g is equal to the number of fixed vertices by h. Consequently, $v(C_n : f)$ and $v(C_n : g)$ have the same parity. This proves (c). □

Let G be a graph and let $C \subseteq G$ be a cycle. For a vertex f of K_3^G, let f^C denote the restriction of f to $V(C)$. Then $f^C \in K_3^C$. Furthermore, if fg is an edge of K_3^G, then $f^C g^C$ is an edge of K_3^C. To simplify the notation we use the letter f also for the restricted map f^C.

Lemma 8.67 (El-Zahar and Sauer) *Let G be a graph with $\chi(G) \geq 4$, and let f be a nonisolated vertex of K_3^G. Then there is an odd cycle C of G such that $v(C : f)$ is even.*

Proof By assumption, there is a vertex g of K_3^G such that $gf \in E(K_3^G)$. Then (+) $g(u) \neq f(v)$ for every edge uv of G. Let X be the set of vertices $v \in V(G)$ such that there exists a vertex $u \in N_G(v)$ with $f(u) = f(v)$.

First, we claim that $G[X]$ contains an odd cycle. Otherwise, $G[X]$ is a bipartite graph, say with parts X_1 and X_2. Let h be a vertex of K_3^G defined as

$$h(v) = f(v) \text{ for } v \in V(G) \setminus X_1 \text{ and } h(v) = g(v) \text{ for } v \in X_1.$$

Let uv be an arbitrary edge of G. If $uv \in E(G[X])$, then, by symmetry, u belongs to X_1 and v belongs to X_2 which leads to $h(u) = g(u) \neq f(v) = h(v)$ (by (+)). If $uv \notin E(G[X])$, then, by symmetry, $v \notin X = X_1 \cup X_2$ implying that $f(v) \neq f(u)$ and so $h(v) \neq h(u)$. Consequently, $h \in \text{Hom}(G, K_3)$ and so $\chi(G) \leq 3$, a contradiction. This proves the claim that $G[X]$ contains an odd cycle, say C.

Next, we claim that $v(C : f)$ is even. Suppose this is false. Then it follows from Lemma 8.66(a) that C contains a path $P = (u, v, w)$ of three vertices such that $f(u), f(v)$ and $f(w)$ are all distinct. Since uv and vw are edges of G, we obtain that $g(v) \notin \{f(u), f(w)\}$ (by (+)) implying that $g(v) = f(v)$. Since $v \in X$, there is a vertex $v' \in N_G(v)$ such that $f(v) = f(v')$. Since vv' is an edge of G, we have $f(v') \neq g(v)$ (by (+)). However, this is impossible. Thus $v(C : f)$ is even and the proof of the lemma is complete. □

The next theorem is due to El-Zahar and Sauer [325]; this result together with Theorem 8.65 implies that the product conjecture is true for graphs with chromatic number at most 4.

Theorem 8.68 (El-Zahar and Sauer) *K_3 is multiplicative.*

Sketch of Proof: The proof is by contradiction. So we assume that K_3 is not multiplicative. Then there are two graphs G and H such that $G \times H \to K_3$, but $G \nrightarrow K_3$ and $H \nrightarrow K_3$. Consequently, $\chi(G) \geq 4$ and $\chi(H) \geq 4$, but $\chi(G \times H) \leq 3$. Clearly, we may assume that both G and H are connected. There is a homomorphism $\varphi \in \text{Hom}(G \times H, K_3)$ where $V(K_3) = \{1,2,3\}$. For a vertex $v \in V(G)$, let f^v be the vertex of K_3^H with

$$f^v(w) = \varphi(v,w) \text{ for all } w \in V(H);$$

and for a vertex $w \in V(H)$, let f_w be the vertex of K_3^G with

$$f_w(v) = \varphi(v,w) \text{ for all } v \in V(G).$$

Claim 8.68.1 *$f^v f^{v'} \in E(K_3^H)$ if $vv' \in E(G)$, and $f_w f_{w'} \in E(K_3^G)$ if $ww' \in E(H)$.*

Proof: Let $vv' \in E(G)$ and $ww' \in E(H)$ be arbitrary edges. Then $(v,w)(v',w')$ is an edge of $G \times H$ and so $\varphi(v,w) \neq \varphi(v',w')$. This implies that $f^v(w) \neq f^{v'}(w')$ and $f_w(v) \neq f_{v'}(w')$. From this the claim easily follows. △

Claim 8.68.2 *There is an odd cycle $C \subseteq G$ such that $\upsilon(C : f_w)$ is even for all $w \in V(H)$; and there is an odd cycle $C' \subseteq H$ such that $\upsilon(C' : f^v)$ is even for all $v \in V(G)$.*

Proof: Fix a vertex $w' \in V(H)$. Since G is a connected graph with $\chi(G) \geq 4$, the vertex w' is not isolated in G and so $f_{w'}$ is a nonisolated vertex of K_3^G (by Claim 8.68.1). From Lemma 8.67 it then follows that there is an odd cycle $C \subseteq G$ such that $\upsilon(C : f_{w'})$ is even. Since H is connected, it follows from Claim 8.68.1 that the maps f_w with $w \in V(H)$ belong all to the same component of K_3^G. From Lemma 8.66(c) it then follows that $\upsilon(C : f_w)$ is even for all $w \in V(H)$. The proof that there is an odd cycle $C' \subseteq H$ such that $\upsilon(C' : f^v)$ is even for all $v \in V(G)$ can be done analogously. △

Then $C = (v_1, v_2, \ldots, v_n, v_1)$ and $C' = (w_1, w_2, \ldots, w_m, w_1)$ where both n and m are odd and $n, m \geq 3$. Then $D = C \times C'$ is a subgraph of $G \times H$ and for the vertex set of D we have $V(D) = \{(v_i, w_j) \mid i \in [1,n], j \in [1,m]\}$. In what follows we will compute the indices of v_i modulo n and the indices of w_j modulo m. Let

$$A = mn - \sum_{j=1}^{m} \upsilon(C : f_{w_j}) - \sum_{i=1}^{n} \upsilon(C' : f^{v_i}).$$

From Claim 8.68.2 it then follows that A is odd. To arrive at a contradiction, we shall show that A is even. To this end, we use the fact that $\varphi \in \text{Hom}(G \times H, K_3)$ induces a coloring of D with color set $V(K_3) = \{1,2,3\}$. For $(i,j) \in [1,n] \times [1,m]$,

$$C(i,j) = ((v_i, w_{j-1}), (v_{i-1}, w_j), (v_i, w_{j+1}), (v_{i+1}, w_j), (v_i, w_{j-1}))$$

is a quadrilateral (i.e. a 4-cycle) of D centered by the vertex (v_i, w_j) (see Figure 8.8). Note that these are all quadrilaterals of D, and all these cycles are induced cycles. If Q is a quadrilateral of D, then the coloring φ of D uses two or three colors for the vertices of Q. Let B be the number of quadrilaterals Q of D such that φ uses two colors for the vertices of Q.

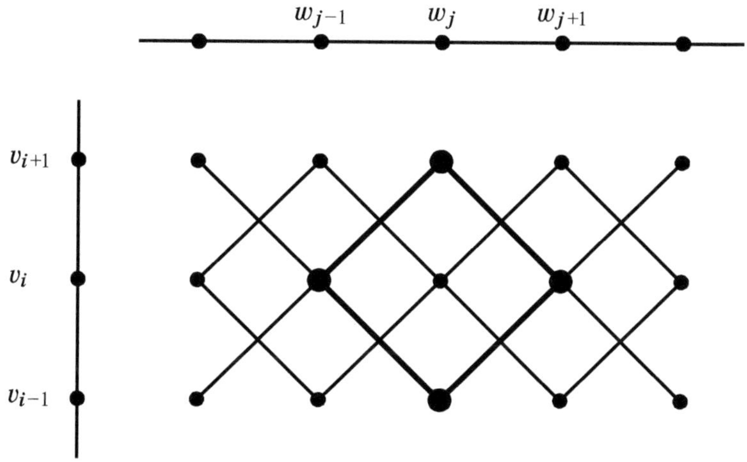

Fig. 8.8 The 4-cycle $C(i,j)$ of $D = C \times C'$.

Claim 8.68.3 $A = B$.

Proof: Since the coloring φ of D uses only three colors, there are exactly three different ways to color the quadrilateral $C(i,j)$, namely

(1) $\varphi(v_{i-1}, w_j) \neq \varphi(v_{i+1}, w_j)$,
(2) $\varphi(v_i, w_{j-1}) \neq \varphi(v_i, w_{j+1})$,
(3) $\varphi(v_{i-1}, w_j) = \varphi(v_{i+1}, w_j)$ and $\varphi(v_i, w_{j-1}) = \varphi(v_i, w_{j+1})$.

Note that the three possibilities are mutually exclusive. Furthermore it is easy to check that (1) holds if and only if v_i is fixed by f_{w_j}; and (2) holds if and only if w_j is fixed by f^{v_i}. Since D has nm quadrilaterals, A counts the number of quadrilaterals $C(i,j)$ for which (3) holds. Since (3) holds if and only if φ uses two colors for $C(i,j)$, we obtain that $B = A$. △

Define an auxiliary graph C_D as follows. The vertex set of C_D are the quadrilaterals of D. A quadrilateral $C(i,j)$ is adjacent in C_D to $C(i+1, j+1)$ and to $C(i-1, j-1)$. Here we compute the first index modulo n and the second index modulo m. Then it is easy to check that C_D is the disjoint union of d cycles, where d is the greatest common divisor of the pair (n,m). For instance, if $(n,m) = (3,3)$, then $d = 3$ and the cycles of C_D are

$$(C(1,1),C(2,2),C(3,3),C(1,1))$$
$$(C(1,2),C(2,3),C(3,1),C(1,2))$$
$$(C(1,3),C(2,1),C(3,2),C(1,3)).$$

Let $C = (Q_1, Q_2, \ldots, Q_p, Q_1)$ be a cycle of C_D. Then there are distinct edges $e_1, e_2, \ldots, e_p$ of D such that $e_k = u_k u'_k$ and $Q_k = (u_k, u'_k, u'_{k+1}, u_{k+1}, u_k)$ for $k \in [1, p]$, where the indices are computed modulo p. So $\{e_k, e_{k+1}\}$ is a perfect matching of the quadrilateral Q_k. Direct the edges of D according to the coloring φ such that the arrows (directions) goes from colors 1 to 2, 2 to 3, and 3 to 1. Color the edge $e_k = u_k u'_k$ red if the edge is directed from u_k to u'_k, otherwise color the edge blue. Then it is easy to see that Q_k uses two colors with respect to φ if and only if the edges e_k and e_{k+1} have different colors. From this it easily follows that C has an even number of quadrilaterals that uses two colors with respect to φ. Consequently, we obtain that B is even. On the other hand, A is odd and $B = A$ (by Claim 8.68.3). This contradiction completes the proof of Theorem 8.68. □

Combining Theorems 8.65 and 8.68 with Propositions 8.59 implies the following positive result concerning the product conjecture.

Corollary 8.69 *For any two graphs G and H, $\chi(G \times H) = \min\{\chi(G), \chi(H)\}$, provided that* $\min\{\chi(G), \chi(H)\} \le 4$.

On the one hand, Corollary 8.64 can be used to verify that K_n is multiplicative as it was successfully done by El-Zharar and Sauer for $n = 3$. On the other hand, Corollary 8.64 may be also used to refute the product conjecture. To do this, it suffices to find a pair (n, G), consisting of an integer n and a graph G, for which both inequalities $\chi(G) > n$ and $\chi(K_n^G) > n$ hold; we call such a pair (n, G) a **negative pair**. Note that $\chi(G \times K_n^G) \le n$ for all $n \in \mathbb{N}$ (by Theorem 8.61(c), see also Exercise 8.23). So if (n, G) is a negative pair, then (G, K_n^G) is a counterexample for the product conjecture.

That negative pairs do indeed exist was first proved in 2019 by Yaroslav Shitov. His construction [938] is surprisingly simple and elegant, but it is based on nontrivial results about the fractional chromatic number. Let us give a brief summary of Shitov's argument. Erdős' classical probabilistic argument from [328] combined with Theorem 8.6 shows the existence of graphs with arbitrarily large girth and fractional chromatic number; so there is a graph F satisfying

$$g(F) \ge 6 \text{ and } \chi^*(F) > \tfrac{31}{10}.$$

The crucial point in Shitov's construction is the statement that if $p \ge p_0$ and $n = \lceil 3.1p \rceil$, then

$$\chi(K_n^{F[K_p]}) > n.$$

The one page proof of this inequality is by reductio ad absurdum and uses only the assumption that $g(F) \ge 6$. Using Theorem 8.8, we obtain that

$$\chi(F[K_p]) \ge p\chi^*(G) > \lceil 3.1p \rceil,$$

where the last inequality holds for infinitely many values of p. So for a suitable large p and $n = \lceil 3.1p \rceil$, $(n, F[K_p])$ is a negative pair implying that K_n is not multiplicative. Shitov [938] made no comments about the value of n. On the other hand, Shitov's argument can be adapted to show the existence of a smallest integer n_0 such that no complete graph K_n is multiplicative if $n \geq n_0$. As shown by Zhu [1103], $n_0 \leq 128$. Tardif [988] was able to improve this bound to $n_0 \leq 13$, and Wrochna [1076] even to $n_0 \leq 5$. So the only remaining case is whether the complete graph K_4 is multiplicative or not.

Theorem 8.70 (Shitov / Tardiv / Wrochna) *For $n \geq 5$, K_n is not multiplicative.*

The main difficulty in finding a negative pair (n, G) is to identify among all graphs G with $\chi(G) > n$ a suitable candidate G for which it can be shown that the corresponding map graph K_n^G admits no n-coloring. Shitov's argument shows that if we take a graph F whose girth and fractional chromatic number are large, then $G = F[K_p]$ with large p is a suitable candidate for a negative pair. Tardif [988] constructed negative pairs which are the lexicographic product of the graph $GY(n,2)$ (see Section 8.5) and K_4; he showed that $(n+5, GY(n,2)[K_4])$ is a negative pair, unless $n \leq 8$, and $(13, GY(9,2)[\overline{C_7}])$ is a negative pair, too. Wrochna [1076] constructed negative pairs of the form $(n, GY(dk,6))$ based on the following result.

Theorem 8.71 (Wrochna) *Let $k, n, d \in \mathbb{N}$ such that $dk > n \geq d+1$ and $n \geq 2k+1$. Then $\chi(K_n^{GY(dk,6)}) > n$.*

Proof Since $dk > d+1$ and $dk > n \geq 2k+1$, we obtain that $k \geq 2$, $d \geq 3$, and $n \geq 5$. Let $G = GY(dk,6)$ and let $G' = K_n^G$. From Corollary 8.46 it follows that that $\chi(G) = dk$. The vertex set of G' consists of all maps from $V(G)$ to $V(K_n) = [1,n]$. If f and g are two such maps, then fg is an edge of G' if and only if for every edge $vw \in E(G)$ we have $f(v) \neq g(w)$. Since $\chi(G) = dk > n$, G' has no loops.

We have to show that $\chi(G') > n$ holds. The proof is by reductio ad absurdum. So suppose that $\chi(G') \leq n$, which implies that there is a homomorphism, say ϕ, from G' to K_n, where $V(K_n) = [1,n]$. By Theorem 8.61(d), the constant maps $f_c \equiv c$ with $c \in [1,n]$ induces a K_n in G'. Hence, by permuting colors if necessary, we may assume that $\phi(f_c) = c$ for all $c \in [1,n]$. If f is a vertex of G' and $c \in [1,n] \setminus \mathrm{im}(f)$, then ff_c is an edge of G' and so $\phi(f) \neq \phi(f_c) = c$. Consequently,

$$\phi(f) \in \mathrm{im}(f) \text{ for all } f \in V(G') \tag{8.18}$$

By Corollary 8.46, $G = GY(dk,6)$ has an odd $(dk,6)$-coloring φ using the color set $[1,d] \times [1,k]$. Let $\alpha : [1,d] \times [1,k] \to [1,d]$ be the projection map, that is, $\alpha(i,j) = i$. For a vertex set $U \subseteq V(G)$ and an integer $p \in \mathbb{N}_0$, let $W^p(U)$ denote the set of vertices $v \in V(G)$ for which there exists an u-v walk of length p in G with $u \in U$. Note that $W^0(U) = U$.

Claim 8.71.1 *If I is a color class of G with respect to the coloring φ and $p \in [0,6]$, then $W^p(I)$ is an independent set of G.*

Proof: Suppose this is false. Then there is a color class $I = \varphi^{-1}(c)$ with $c \in [1,d]\times[1,k]$ such that $W^p(I)$ contains an edge v_1v_2 of G. Then, for $i \in \{1,2\}$, there is a vertex $u_i \in I$ and an u_i-v_i walk W_i of G having length $\ell(W_i) = p$. Consequently, $W = u_1W_1v_1v_2W_2^{-1}u_2$ is an odd u_1-u_2 walk in G with $\ell(W) = 2p+1$. Since $G = GY(dk,6)$ and $dk \ge 3$, Theorem 8.53 implies that $g_o(G) > 13$. Hence, $\ell(W) \le 13 < g_o(G)$ and, by Proposition C.4, $u_1 \ne u_2$. On the other hand, $\varphi(u_1) = \varphi(u_2) = c$, which leads to a contradiction to Theorem 8.39 as φ is an odd $(dk,6)$-coloring of G. This proves the claim. △

In what follows, let f^o be the vertex of G' such that $f^o(v) = \alpha(\varphi(v))$ for $v \in V(G)$, and let $i_0 = \phi(f^o)$. Since $\chi(G) = dk$, $\mathrm{im}(\varphi) = [1,d]\times[1,k]$. Hence $i_0 \in \mathrm{im}(f^o) = [1,d]$ (by (8.18)) and $U = \varphi^{-1}(\alpha^{-1}(i_0))$ is the union of k color classes of G with respect to φ. Clearly, $f^o(v) = i_0$ if and only if $v \in U$.

Claim 8.71.2 *If $p \in [0,6]$, then $W^p(U) \ne V(G)$.*

Proof: The set U is the union of k color classes with respect to ϕ, and $W^p(I) \in \mathcal{I}(G)$ for each color class (by Claim 8.71.1). Consequently, $W^p(U)$ is the union of k independent sets of G. Since $\chi(G) = dk > n \ge 2k+1$, we obtain that $W^p(U) \ne V(G)$ as claimed. △

Let $\tau \in S_n$ be the transposition $\tau = (1,i_0)$, i.e., $\tau(1) = i_0$, $\tau(i_0) = 1$ and $\tau(i) = i$ for $i \in [1,n]\setminus\{1,i_0\}$. Note that if $i_0 = 1$, then $\tau(i) = i$ for all $i \in [1,n]$. For $i,j \in [1,n]$ and $p \in [1,5]$, let $h_{i,j}^{(p)}$ be the vertex of G' such that, for every vertex $v \in V(G)$, we have

$$h_{i,j}^{(p)}(v) = \begin{cases} \tau(i) & \text{if } v \notin W^p(U) \\ \tau(j) & \text{if } v \in W^p(U). \end{cases}$$

Claim 8.71.3 *Let $i,j \in [1,n]$ and $p \in [1,5]$. For the vertex $h_{i,j}^{(p)}$ of G' the following statements hold:*

(a) $\phi(h_{i,j}^{(p)}) \in \mathrm{im}(\phi(h_{i,j}^{(p)})) = \{\tau(i),\tau(j)\}$.
(b) $h_{i,j}^{(p+1)}h_{i',j'}^{(p)}$ *is an edge of G', provided that $p \le 4$, $i \ne i'$, $j \ne j'$ and $j \ne i'$.*
(c) $\phi(h_{1,j}^{(1)}) = \tau(j)$ *for $j \in [1,n]\setminus[1,d]$.*
(d) $\phi(h_{n,j}^{(2)}) = \tau(j)$ *for $j \in [2,n-1]$.*
(e) $\phi(h_{i,j}^{(3)}) = \tau(j)$ *for $i \in [2,n-1]$ and $j \in [1,n-1]\setminus\{i\}$.*
(f) $\phi(h_{i,j}^{(4)}) = \tau(j)$ *for $i \in [1,n-1]$ and $j \in [1,n]\setminus\{i\}$.*
(g) $\phi(h_{i,j}^{(5)}) = \tau(j)$ *for $i,j \in [1,n]$ and $i \ne j$.*

Proof: Statement (a) follows from (8.18) and Claim 8.71.2. To prove (b), suppose this is false. Then there is an edge $vv' \in E(G)$ such that

$$h_{i,j}^{(p+1)}(v) = h_{i',j'}^{(p)}(v') = c,$$

where $c \in \{\tau(i),\tau(j)\}\cap\{\tau(i'),\tau(j')\}$. Since τ is a bijection, this is only possible if $c = \tau(i) = \tau(j')$ (by the assumptions on (i,j,i',j')). But then $v \notin W^{p+1}(U)$ and $v' \in W^p(U)$, which is impossible.

For the proof of (c), let $h = h^{(1)}_{1,j}$ and $j \in [1,n] \setminus [1,d]$. Then $j > i_o$ and so $\tau(j) = j \notin [1,d] = \mathrm{im}(f^o)$. We claim that $f^o h \in E(G')$. Suppose this is false. Then there is an edge $vv' \in E(G)$ such that $f^o(v) = h(v')$. Since $f^o(v) \neq \tau(j)$ and $\mathrm{im}(h) = \{\tau(1), \tau(j)\}$ (by (a)), this implies that $f^o(v) = h(v') = \tau(1) = i_0$. But then $v \in U$ and $v' \notin W^1(U)$, which is impossible. This proves that claim that $f^o h \in E(G')$. Hence $\phi(h) \neq \phi(f^o) = i_0 = \tau(1)$. Since $\phi(h) \in \{\tau(1), \tau(j)\}$ (by (a)), this implies that $\phi(h) = \tau(j)$. Thus (c) is proved.

To prove (d), suppose that $j \in [2, n-1]$. Then $h = h^{(2)}_{n,j}$ is adjacent to $h' = h^{(1)}_{1,n}$ (by (b)) implying that $\phi(h) \neq \phi(h') = \tau(n)$ (by (c)), which leads to $\phi(h) = \tau(j)$ (by (a)). Thus (d) is proved.

For the proof of (e), suppose that $i \in [2, n-1]$ and $j \in [1, n-1] \setminus \{i\}$. Then $h = h^{(3)}_{i,j}$ is adjacent to $h^{(2)}_{n,i}$ in G (by (b)) implying that $\phi(h) \neq \phi(h^{(2)}_{n,i}) = \tau(i)$ (by (d)), and so $\phi(h) = \tau(j)$ (by (a)). This proves (e).

For the proof of (f), suppose that $i \in [1, n-1]$ and $j \in [1,n] \setminus \{i\}$. Since $n \geq 5$, there exists an $i' \in [2, n-1] \setminus \{i, j\}$. Then $h = h^{(4)}_{i,j}$ is adjacent to $h' = h^{(3)}_{i',i}$ (by (b)) and, therefore, $\phi(h) \neq \phi(h') = \tau(i)$ (by (e)). Consequently, $\phi(h) = \tau(j)$ (by (a)). This proves (f).

For the proof of (g), let $i, j \in [1,n]$ and $i \neq j$. Since $n \geq 5$, there exists an $i' \in [1, n-1] \setminus \{i, j\}$. Then (b) implies that $h = h^{(5)}_{i,j}$ is adjacent to $h' = h^{(4)}_{i',i}$ and, therefore, $\phi(h) \neq \phi(h') = \tau(i)$ (by (f)). Consequently, $\phi(h) = \tau(j)$ (by (a)). This proves (g). △

For a vertex $v \in W^6(U)$, let k_v be an integer such that $k_v \in [1,k]$ and $v \in W^6(\varphi^{-1}(i_0, k_v))$. Since $U = \varphi^{-1}(\alpha^{-1}(i_0))$, such an integer k_v does indeed exist. Let $I = [1,n] \setminus \tau^{-1}([1,k])$. Since $\tau \in S_n$, $|I| = n - k$. For $j \in I$, let g_j be the vertex of the map graph G' such that, for $v \in V(G)$, we have

$$g_j(v) = \begin{cases} \tau(j) & \text{if } v \notin W^6(U) \\ k_v & \text{if } v \in W^6(U). \end{cases}$$

By Claim 8.71.2, $W^6(U) \neq V(G)$ implying that

$$\mathrm{im}(g_j) = \{k_v \mid v \in W^6(U)\} \cup \{\tau(j)\}$$

for $j \in I$. Since $\tau \in S_n$ and $k_v \in [1,k]$, this implies that $g_j \neq g_{j'}$ provided that $j, j' \in I$ and $j \neq j'$.

Claim 8.71.4 *Let $j \in I$. Then the following statements hold:*

(a) *There is an integer $i \in [1,n] \setminus \{j\}$ such that g_j is adjacent to $h^{(5)}_{i,j}$.*
(b) $\phi(g_i) \in [1,k]$.

Proof: Since $n \geq 2k+1$, we obtain that $|I| = n - k \geq k + 1 \geq 2$ and so there is an $i \in I \setminus \{j\}$. We claim that g_j is adjacent to $h = h^{(5)}_{i,j}$. Suppose this is false. Then there is an edge $vv' \in E(G)$ such that $g_j(v) = h(v') = c$ and $c \in \{\tau(j), k_v\} \cap \{\tau(i), \tau(j)\}$. Since $i, j \in I$ and $i \neq j$, we obtain that $\{\tau(i), \tau(j)\} \cap [1,k] = \varnothing$ and $\tau(i) \neq \tau(j)$. Since $k_v \in [1,k]$, the only possible value for c is $c = \tau(j)$ and so $v \notin W^6(U)$ and

$v' \in W^5(U)$, which is impossible. This proves the claim that g_j is adjacent to h. Then $\phi(g_j) \neq \phi(h) = \tau(j)$ (by Claim 8.71.3(g)). Since $\phi(g_j) \in \mathrm{im}(g_j) \subseteq [1,k] \cup \{\tau(j)\}$, it then follows that $\phi(g_j) \in [1,k]$. This proves the claim. △

To complete the proof, let $X = \{g_j \mid j \in I\}$. Then $X \subseteq V(G')$ and $|X| = |I| = n-k \geq k+1$. From Claim 8.71.4(b) it then follows that there are $j, j' \in I$ with $j \neq j'$ such that $\phi(g_j) = \phi(g_{j'})$. This implies that $g_j g_{j'} \notin E(G')$. Consequently, there is an edge $vv' \in E(G)$ such that $g_j(v) = g_{j'}(v') = c$ where $c \in \{\tau(j), k_v\} \cap \{\tau(j'), k_{v'}\}$. Since $\tau(j) \neq \tau(j')$ and $\{\tau(j), \tau(j'\} \cap [1,k] = \varnothing$, we then conclude that $k_v = k_{v'}$. This implies that $v, v' \in W^6(\varphi^{-1}(i_0, k_v))$, a contradiction to Claim 8.71.1. This completes the proof of Theorem 8.71 □

Proof of Theorem 8.70: Let $n \in \mathbb{N}$ such that $n \geq 5$, then there is an integer $d \geq 3$ such that $d+1 \leq n < 2d$. Using Corollary 8.46 and Theorem 8.71, we obtain that $G = GY(2d,6)$ satisfies $\chi(G) > n$ and $\chi(K_n^G) > n$. Hence, (n,G) is a negative pair and so K_n is not multiplicative. □

Taking $n = 5$ and $d = 3$ it follows that $G = GY(6,6)$ and $G' = K_5^G$ is a counterexample to the product conjecture, since $\chi(G) \geq 6$ and $\chi(G') \geq 6$, but $\chi(G \times G') \leq 5$. By (8.14), we have $|G| = 6(7^5 - 6^5) = 54186$. The order of the map graph G' is $|G'| = 5^{|G|}$. However, the proof of Theorem 8.71 shows that we can replace G' by an induced subgraph H' of G' whose vertex set consists of the constant maps f_c with $c \in [1,5]$, the special map f^o, the maps $h_{i,j}^{(p)}$ with $i, j \in [1,5]$ and $p \in [1,5]$, and the maps g_j with $j \in I$. This leads to $|H'| = 5 + 1 + 5^3 + 3 = 134$. As pointed out by Wrochna [1076, Section 4], based on a more careful analysis of the proof of Claim 8.71.3, one can replace H' by an induced subgraph H with $|H| = 30$.

The remaining open case of the product conjecture, whether K_4 is multiplicative, was solved by Tardif [989] in 2023. He showed that K_4 is not multiplicative either. Tardif used similar ideas as Wrochna and proved, in particular, that if $G = GY(8,6)$, then $\chi(G) = 8$ and $\chi(K_4^G) > 4$ implying that $(4, GY(8,6))$ is a negative pair. Hence, the product of two 5-chromatic graphs can be 4. Summarizing we obtain the following result.

Theorem 8.72 *For $n \in \mathbb{N}_0$, K_n is multiplicative if and only if $n \leq 3$.*

8.7 Topological Lower Bounds for χ

Lovász's 1978 proof of the Kneser conjecture is a masterpiece of imagination and opened up a new view on the applications of algebraic topology in combinatorics especially chromatic graph theory. Thereby Lovász [697] became the founder of a new branch of graph theory which is mainly concerned with the study of topological lower bounds for the chromatic number of graphs. More general, topological combinatorics solves combinatorial problems by applying topological tools; see also the paper [277] by Mark de Longueville from 2004. Matousek's 2003 textbook [722] provides a comprehensive overview of the state of research of using the Borsuk–Ulam theorem up to that time; it remains an important source to this day. Since the publication of Matousek's monograph, however, an enormous number of important new contributions have appeared on the subject. In 2013 de Longueville [278] published an undergraduate textbook on topological combinatorics. In this section, we will give an overview of some results obtained mainly in the last two decades, leaving out most of the proofs.

In the previous sections we introduced, for any fixed integer $n \geq 0$, four families of graphs: the family of Kneser graphs $\mathcal{KF}(n) = \{KG(n+2p, p) \mid p \in \mathbb{N}\}$, the family of Shrijver graphs $\mathcal{SF}(n) = \{SG(n+2p, p) \mid p \in \mathbb{N}\}$, the family of generalized Mycielski graphs $\mathcal{MF}(n) = \mathcal{MG}(n+2)$ (see Section 8.4), and the family of Gyárfás–Simony–Tardos graphs $\mathcal{GF}(n) = \{GY(n+2, \ell) \mid \ell \in \mathbb{N}_0\}$. All graphs belonging to these four families have chromatic number $n+2$, and the difficulty in proving this equality in each case is to show that $n+2$ is indeed a lower bound for χ. All these graph families are suitable for proving the existence of graphs with chromatic number $n+2$ and arbitrarily large odd girth. As we shall see all four graph families are of the same topological type, and there are various homomorphism relations between these families.

Let $\mathcal{P}$ and $\mathcal{P}'$ be two graph families. We say that $\mathcal{P}'$ is **hom-bounded from below** by $\mathcal{P}$ if for every graph $H \in \mathcal{P}'$ there exists a graph $G \in \mathcal{P}$ such that $G \to H$; in this case we write $\mathcal{P} \to \mathcal{P}'$. Clearly, if every graph in $\mathcal{P}$ has chromatic number at least k and $\mathcal{P}'$ is hom-bounded from below by $\mathcal{P}$, then every graph in $\mathcal{P}'$ has chromatic number at least k, too. Marshall, Naserasr, and Nešetřil [716] investigated a similar concept, namely $\mathcal{P}$ is **hom-bounded from above** by $\mathcal{P}'$, that is, every graph in $\mathcal{P}$ is homeomorphic to some graph in $\mathcal{P}'$. We say that $\mathcal{P}$ is **hom-equivalent** to $\mathcal{P}'$, written $\mathcal{P} \leftrightarrow \mathcal{P}'$, if $\mathcal{P} \to \mathcal{P}'$ and $\mathcal{P}' \to \mathcal{P}$. It is straightforward to show that $\to$ is a quasi-order for graph families, and $\leftrightarrow$ is an equivalence relation for graph families.

The following 1933 result is commonly referred to as the Borsuk–Ulam theorem, since it was conjectured by Stanislav Ulam and proved by Karol Borsuk; here $n \in \mathbb{N}_0$ is an arbitrary fixed integer.

(B1) *For every continuous mapping $f : \mathbf{S}^n \to \mathbb{R}^n$ there is a point $\mathbf{x} \in \mathbf{S}^n$ such that $f(\mathbf{x}) = f(-\mathbf{x})$.*

For relevant information about the history and the importance of the Borsuk–Ulam theorem we refer the reader to Matousek's book. In this book one can also find all references left out here and all concepts from topology not defined here. Note that all topological spaces considered in this section are subspaces of the Euclidean space $(\mathbb{R}^d, \|\cdot\|)$ for some $d \geq 0$. The Borsuk–Ulam theorem has many different proofs, various equivalent formulations and countless applications in mathematics, computer science, and theoretical physic. We are interested in the following two equivalent formulations of the Borsuk–Ulam theorem.

(B2) *If $\mathbf{S}^n$ is the union of $n+1$ subsets each of which is closed, then at least one subset contains a pair of antipodal points.*

(B3) *If $\mathbf{S}^n$ is the union of $n+1$ subsets, then at least one subset has diameter 2.*

That (B1) and (B2) are equivalent was already mentioned in Borsuk's paper from 1933. A proof of (B2) was obtained by Lyusternik and Shnirel'man in 1930. That (B2) and (B3) are equivalent is evident. Two points $\mathbf{x},\mathbf{y} \in \mathbb{R}^{n+1}$ are **antipodal** if $\mathbf{y} = -\mathbf{x}$. Since the Euclidean norm can be computed by the Euclidean scalar product, it is easy to see that if $\mathbf{x},\mathbf{y} \in \mathbb{R}^{n+1}$, then $||\mathbf{x}-\mathbf{y}||^2 + ||\mathbf{x}+\mathbf{y}||^2 = 2(||\mathbf{x}||^2 + ||\mathbf{y}||^2)$. As an immediate consequence we obtain that

$$||\mathbf{x}-\mathbf{y}||^2 + ||\mathbf{x}+\mathbf{y}||^2 = 4 \text{ provided that } \mathbf{x},\mathbf{y} \in \mathbf{S}^n. \tag{8.19}$$

So two points $\mathbf{x},\mathbf{y} \in \mathbf{S}^n$ are antipodal if and only if $||\mathbf{x}-\mathbf{y}|| = 2$. Recall that if $A \subseteq \mathbb{R}^{n+1}$, then the **diameter** of A is defined by

$$\operatorname{diam}(A) = \sup\{||\mathbf{x}-\mathbf{y}|| \mid \mathbf{x},\mathbf{y} \in A\}.$$

Any set $A \subseteq \mathbf{S}^n$, satisfies $\operatorname{diam}(A) \le \operatorname{diam}(\mathbf{S}^n) = 2$; and if A is is closed set, then $\operatorname{diam}(A) = 2$ if and only if A contains a pair of antipodal points. Consequently, we can reformulate (B3) as a coloring result about graphs associated with the unit n-sphere, however these are infinite graphs. The **Borsuk graph** $BG(n,\alpha)$ is the infinite graph whose vertex set is $\mathbf{S}^n$ and in which two vertices $\mathbf{x}$ and $\mathbf{y}$ are adjacent if and only if $||\mathbf{x}-\mathbf{y}|| > \alpha$, where $0 < \alpha < 2$ is a real number. Then, for every fixed $\alpha_0 \in \mathbb{R}$ with $0 \le \alpha_0 < 2$, the statement

(B4) $\chi(BG(n,\alpha)) \ge n+2$ *whenever* $\alpha_0 < \alpha < 2$

is equivalent to (B3) (since $\alpha = 2-\epsilon$ for some $\epsilon > 0$) and hence to the Borsuk–Ulam theorem. This was first observed by Erdős and Hajnal [348] in 1967. The Borsuk graph $BG'(n,\alpha)$ is the infinite graph whose vertex set is $\mathbf{S}^n$ and in which two vertices $\mathbf{x}$ and $\mathbf{y}$ are adjacent if and only if $||\mathbf{x}-\mathbf{y}|| \ge \alpha$; these graphs were first studied in a paper by Lovász [700]. Clearly, if $0 < \alpha' < \alpha < 2$, then

$$BG(n,\alpha) \subseteq BG'(n,\alpha) \subseteq BG(n,\alpha').$$

That the lower bound for the chromatic number of Borsuk graphs can be attained if α is large enough was observed by Erdős and Hajnal as well as by Lovász; so there is an α_n with $0 < \alpha_n < 2$ such that

$$\chi(BG(n,\alpha)) = \chi(BG'(n,\alpha)) = n+2 \text{ for } \alpha_n < \alpha < 2.$$

A coloring of $BG'(n,\alpha)$ can be obtained as follows. Let σ be a regular $(n+1)$-simplex inscribed in the sphere $\mathbf{S}^n$. Then, there is a set $V \subseteq \mathbf{S}^n$ of $n+2$ affinely independent points such that σ is the convex hull of V. Let S be the boundary of σ. Then there is a mapping $\varphi : S \to V$ such that every point $\mathbf{x} \in S$ belongs to the convex hull of $V \setminus \{\varphi(x)\}$; note that this convex hull is a facet of σ. The mapping $\mathbf{x} \mapsto \mathbf{x}/||\mathbf{x}||$ is a homeomorphism between S and $\mathbf{S}^n$; and we put $\phi(\mathbf{x}/||\mathbf{x}||) = \varphi(\mathbf{x})$ which defines a map $\phi : \mathbf{S}^n \to V$. Let τ' be the image of any facet of σ on the n-sphere. If α is larger than the diameter of τ', then ϕ is a (proper) coloring of $BG'(n,\alpha)$; we call ϕ a **regular coloring** of $BG'(n,\alpha)$. As remarked by Simonyi and Tardos [941,

Corollary 15] we can choose $\alpha_n = 2\sqrt{1-1/(n+3)}$, see also the paper by Kahle and Martinez-Figueroa [545].

The famous de Bruijn–Erdős compactness theorem, proved in [272], says that if every finite subgraph of an (infinite) graph is k-colorable ($k \in \mathbb{N}_0$ fixed), then so is the graph itself. Lovász [700] constructed a finite subgraph $LB(n,\alpha)$ of $BG'(n,\alpha)$ such that $\chi(LB(n,\alpha)) = \chi(BG'(n,\alpha)) = n+2$, provided that $\alpha_n \leq \alpha'_n < \alpha < 2$. We call $LB(n,\alpha)$ a **Lovász–Borsuk graph**; the value α'_n is introduced to ensures the existence of a certain convex polytope inscribed in the n-sphere that is used in the construction of the Lovász–Borsuk graph. Lovász's proof that $\chi(LB(n,\alpha)) \geq n+2$ is the second application of his topological lower bound for the chromatic number (Theorem 8.12). The Lovász–Borsuk graph $LB(n,\alpha)$ is a finite subgraph of a distance graph. Given a topological subspace $\mathbf{T}$ of an Euclidean space and a set $D \subseteq \mathbb{R}$, the **distance graph** $DG(\mathbf{T}, D)$ is the infinite graph whose vertex set is $\mathbf{T}$ and in which two vertices $\mathbf{x}$ and $\mathbf{y}$ are adjacent if and only if $\|\mathbf{x}-\mathbf{y}\| \in D$. Instead of $DG(\mathbf{T},\{a\})$ we also write $DG(\mathbf{T},a)$. Distance graphs have attracted a lot of attention over the years and the determination of the chromatic number of certain distance graphs is a challenging problem (see e.g. [530, Section 9]). The Lovász–Borsuk graph $LB(n,\alpha)$ is a finite subgraph of the distance graph $DG(\mathbf{S}^n,\alpha)$ and shows that $\chi(DG(\mathbf{S}^n,\alpha)) = n+2$ for $\alpha'_n < \alpha < 2$, thereby solving a problem proposed by Erdős and Graham at the Hungarian Combinatorial Conference held in Eger, July 1981. The most popular distance graph, however, is the **plane unit distance graph** $DG(\mathbb{R}^2,1)$. The problem of determining the chromatic number of this graph is known as the **Hadwiger–Nelson problem**; see [530, Problem 9.1] for the history of this problem. That $DG(\mathbb{R}^2,1)$ contains a 4-chromatic graph was observed by Nelson in 1950; that the 4-critical graph $K_4 \nabla K_4$ is the smallest such graph was proved by Leo and William Moser [760] in 1961, so this graph is refereed to as the **Moser spindle**. A fellow student of Nelson, J. Isabell, provided a 7-coloring of the plane unit distance graph. Hence,

$$4 \leq \chi(DG(\mathbb{R}^2,1)) \leq 7.$$

Since the popularization of the problem by Martin Gardner in his *Mathematical Games* columns from 1960 [405], no improvement has been provided on either bound except recently when De Grey [273] discovered a non-4-colorable plane unit distance graph of order $n = 1581$. In 1992, Fisher and Ullman first investigated the fractional chromatic number of the plane. The Moser spindle shows that $\chi^*(DG(\mathbb{R}^2,1)) \geq 7/2$; they also found a coloring that proved the upper bound $\chi^*(DG(\mathbb{R}^2,1)) \leq 8\sqrt{3}/\pi \approx 4.4106$. The best known bounds are $76/21 \leq \chi^*(DG(\mathbb{R}^2,1)) \leq 4.3599$. The upper bound is due to Hochberg and O'Donnell [502]; the lower bound is due to Cranston and Rabern [260]. Whether the graph discovered by De Grey leads to a further improvement of the lower bound seems unknown.

In 1994, Johnson [540] posed the problem of deciding whether the list chromatic number of the plane unit distance graph $DG(\mathbb{R}^2,1)$ is finite or not. As pointed out by Jensen and Toft [531], the answer follows immediately from Alon's result that if one of χ_ℓ, col and mad tend to infinity, then they all do (see (2.7)). Hence, the subgraph of $DG(\mathbb{R}^2,1)$ induced by the the union of two open disks of radius ϵ with

$0 < \epsilon < 1/2$, whose centers are of distance 1 from each other, is infinite. Schmerl [901] proved that $DG(\mathbb{R}^2, 1)$ and $DG(\mathbb{R}^3, 1)$ have countable list chromatic number, but $DG(\mathbb{R}^4, 1)$ has uncountable list chromatic number; for further information about the list chromatic number of Euclidean distance graphs the reader is referred to the 2011 paper by Komjáth [607]. It is notable, that the compactness theorem of de Bruijn and Erdős [272] holds for the list chromatic number, but not for the coloring number. The result about the list chromatic number was proved by Johnson [540], while the result about the coloring number was proved by Erdős and Hajnal [347].

Another interesting feature of the Borsuk graph $BG(n,\alpha)$, observed by Erdős and Hajnal, is the fact that its odd girth tends to infinity if α tends to 2. To see this consider a path $P = (\mathbf{x},\mathbf{y},\mathbf{z})$ in $BG(n,\alpha)$. Then $||\mathbf{x}-\mathbf{y}|| > \alpha$ and $||\mathbf{y}-\mathbf{z}|| > \alpha$. To obtain an upper bound for the distance between $\mathbf{x}$ and $\mathbf{z}$, we first apply the triangle inequality with $-\mathbf{y}$ and then (8.19); this leads to

$$\begin{aligned} ||\mathbf{x}-\mathbf{z}|| &\leq ||\mathbf{x}+\mathbf{y}|| + ||\mathbf{y}+\mathbf{z}|| \\ &= \sqrt{4-||\mathbf{x}-\mathbf{y}||^2} + \sqrt{4-||\mathbf{y}-\mathbf{z}||^2} \\ &< 2\sqrt{4-\alpha^2}. \end{aligned}$$

Now, assume that $C = (\mathbf{x}_0, \mathbf{x}_1, \ldots, \mathbf{x}_{2p}, \mathbf{x}_0)$ is a shortest odd cycle in $BG(n,\alpha)$. Then, using the triangle inequality and the above result, we obtain that

$$||\mathbf{x}_0-\mathbf{x}_{2p}|| \leq ||\mathbf{x}_0-\mathbf{x}_2|| + ||\mathbf{x}_2-\mathbf{x}_4|| + \cdots + ||\mathbf{x}_{2p-2}-\mathbf{x}_{2p}|| < 2p\sqrt{4-\alpha^2}.$$

Furthermore, $||\mathbf{x}_0-\mathbf{x}_{2p}|| > \alpha$ and, by (8.19), $||\mathbf{x}_0+\mathbf{x}_{2p}|| < \sqrt{4-\alpha^2}$. This implies that

$$2 = ||2\mathbf{x}_0|| \leq ||\mathbf{x}_0-\mathbf{x}_{2p}|| + ||\mathbf{x}_{2p}+\mathbf{x}_0|| < (2p+1)\sqrt{4-\alpha^2}.$$

Consequently, since the odd girth satisfies $g_o(BG(n,\alpha)) = 2p+1$, we obtain that

$$g_o(BG(n,\alpha)) > \frac{2}{\sqrt{4-\alpha^2}}.$$

A regular coloring ϕ of the the Borsuk graph $BG(n,\alpha)$ with $\alpha > \alpha_n$ has a further remarkable property as observed by Simonyi and Tardos [943]. For every $\ell \in \mathbb{N}_0$, there exists an $\alpha_{n,\ell}$ with $\alpha_n < \alpha_{n,\ell} < 2$ such that if $\alpha_{n,\ell} < \alpha < 2$, then every odd $\mathbf{x}$-$\mathbf{y}$ walk P in $BG(n,\alpha)$ with $\phi(\mathbf{x}) = \phi(\mathbf{y})$ satisfies $\ell(P) > 2\ell+1$. Then, see Theorem 8.39, ϕ is an odd $(n+2,\ell)$-coloring. Consequently, $BG(n,\alpha) \in OC(n+2,\ell)$ and $BG(n,\alpha) \to GY(n+2,\ell)$ (see Corollary 8.46). Note that the results in Section 8.5 are proved for graphs, but not for infinite graphs; so one should be a bit careful. On the other hand there is a generalization of the de Bruijn–Erdős theorem saying that if H is a (finite) graph, then an infinite graph admits an homomorphism to H if and only if all of its finite subgraphs do (see [452]). In another paper by Simons, Tardif, and Wehlau [940, Lemma 2] it is proved that for every Borsuk graph $BG'(n,\alpha)$ with $1 < \alpha < 2$ there exists a Mycielski graph $G \in \mathcal{MG}(n+2) = \mathcal{MF}(n)$ such that $G \to BG'(n,\alpha)$; note that $BG'(0,\alpha) = K_2$. The short elementary proof of this statement is based on induction by $n \geq 0$. So we have three more graph families, the family of Borsuk graphs $\mathcal{BF}(n) = \{BG(n,\alpha) \mid \alpha \in \mathbb{R}, 0 < \alpha < 2\}$, respectively $\mathcal{BF}'(n) = \{BG'(n,\alpha) \mid \alpha \in \mathbb{R}, 0 < \alpha < 2\}$, and the family of Lovász-Borsuk graphs $\mathcal{LF}(n) = \{LB(n,\alpha) \mid \alpha \in \mathbb{R}, \alpha'_n < \alpha < 2\}$. Note that $\mathcal{BF}(n) \leftrightarrow \mathcal{BF}'(n)$. From the results mentioned above we then obtain that the following relations hold:

$$\mathcal{MF}(n) \to \mathcal{BF}(n), \mathcal{LF}(n) \to \mathcal{BF}(n) \text{ and } \mathcal{BF}(n) \to \mathcal{GF}(n). \tag{8.20}$$

It is easy to see that the Mycielski construction preserves graph homomorphisms, that is for every two graphs G and H and every integer $r \geq 1$ we have

$$G \to H \text{ implies } \mathrm{M}_r(G) \to \mathrm{M}_r(H). \tag{8.21}$$

Kaiser and Stehlík [549, Lemma 6.2] proved that $\mathrm{M}_p(SG(m,p)) \to SG(m+1,p)$ if $m \geq 2p \geq 2$; note that $SG(2p,p) = K_2$ and $SG(2p+1,p) = C_{2p+1}$. Recall that $\mathcal{MF}(0) = \mathcal{MG}(2) = \{K_2\}$ and $\mathcal{MF}(n+1) = \{\mathrm{M}_p(G) \mid G \in \mathcal{MF}(n), p \in \mathbb{N}\}$. Hence, by a simple induction on $n \geq 0$, we obtain that

$$\mathcal{MF}(n) \to \mathcal{SF}(n). \tag{8.22}$$

As an immediate consequence of (8.20) and (8.22) we obtain that the graph families $\mathcal{KF}(n), \mathcal{SF}(n), \mathcal{GF}(n)$ and $\mathcal{BF}(n)$ are hom-bounded from below by the Mycielski family $\mathcal{MF}(n)$ for all $n \geq 0$. That $\mathcal{MF}(n) \to \mathcal{GF}(n)$ follows from (8.20) but also from the proof of Theorem 8.42.

That the statements (B2) and (B4) are equivalent is obvious. However, that the Borsuk–Ulam theorem implies Lovász's topological lower bound for the chromatic number of (finite) graphs (see Theorem 8.12) is less obvious. Since chromatic number and graph homomorphisms are closely related, it is natural to consider the graph category `GRA` whose objects are (finite) graphs and whose morphisms are graph homomorphisms. It was the innovative idea of Lovász to transform a coloring problem into a topological problem using the topological category `TOP` whose objects are the topological spaces (subspaces of Euclidean spaces) and whose morphisms are the continuous maps. To construct an appropriate functor from `GRA` to `TOP`, a third category is used. The complex category `COM` has as objects the (finite) simplicial complexes and as morphisms the simplicial maps.

Let us first introduce some more notation. Let $\mathbf{T}$ and $\mathbf{T}'$ be two topological spaces. If there exists a continuous map from $\mathbf{T}$ to $\mathbf{T}'$, we write $\mathbf{T} \to \mathbf{T}'$. Furthermore, we write $\mathbf{T} \cong \mathbf{T}'$ if there exists a homeomorphism from $\mathbf{T}$ to $\mathbf{T}'$, that is a bijective continuous map whose inverse map is continuous, too; and we write $\mathbf{T} \simeq \mathbf{T}'$ if the two spaces are homotopic equivalent (or have the same homotopy type). Let $\mathrm{conn}(\mathbf{T})$ denote the **connectivity** of $\mathbf{T}$, that is, the largest integer $n \in \mathbb{N}_0$ such that $\mathbf{T}$ is n-connected. The Borsuk–Ulam theorem is equivalent to the statement that $\mathrm{conn}(\mathbf{S}^n) = n-1$ for all $n \in \mathbb{N}_0$. Furthermore, we have

$$\mathbf{T} \cong \mathbf{T}' \Longrightarrow \mathbf{T} \simeq \mathbf{T}' \Longrightarrow \mathrm{conn}(\mathbf{T}) = \mathrm{conn}(\mathbf{T}').$$

Let K and K' be two simplicial complexes. We denote by $\mathrm{sd}(\mathsf{K})$ the **barycentric subdivision** of K. The vertices of $\mathrm{sd}(\mathsf{K})$ are the nonempty simplices of K, and the simplices of $\mathrm{sd}(\mathsf{K})$ are chains of simplices of K ordered by inclusion. A **simplicial map** (or **homomorphism**) from K to K' is a map $f : V(\mathsf{K}) \to V(\mathsf{K}')$ such that $f(\sigma) \in \mathsf{K}'$ for every simplex $\sigma \in \mathsf{K}$. Let $\mathrm{Hom}(\mathsf{K},\mathsf{K}')$ denote the set of simplicial maps from K to K'. If there exists a simplicial map from K to K', then we write $\mathsf{K} \to \mathsf{K}'$. Recall from Section 8.2 that $||\mathsf{K}||$ denotes the **body** (or geometric realization) of K; this is an object of the category $\mathtt{TOP}$. Then it is known that $||\mathrm{sd}(\mathsf{K}))|| \cong ||\mathsf{K}||$. Furthermore, it is known that if $f \in \mathrm{Hom}(\mathsf{K},\mathsf{K}')$, then f can be extended to a continuous map $||f||$ from $||\mathsf{K}||$ to $||\mathsf{K}'||$; extended means that every vertex $x \in V(\mathsf{K})$ satisfies $||f||(||x||) = ||f(x)||$. Then it can be shown that $||\cdot||$ is indeed a functor from $\mathtt{COM}$ to $\mathtt{TOP}$.

Let G and H be two nonempty graphs. For a set $X \subseteq V(G)$, let

$$CN_G(X) = \{v \in V(G) \mid uv \in E(G) \text{ for all } u \in X\}$$

denote the **common neighborhood** of X in G; so $CN_G(\emptyset) = V(G)$. Then $\mathsf{N}(G) = \{A \subseteq V(G) \mid CN_G(A) \neq \emptyset\}$ is the neighbourhood complex of G; and we call the body $\mathbf{T}_N(G) = ||\mathsf{N}(G)|| \cong ||\mathrm{sd}(\mathsf{N}(G))||$ the **N-space** of G. It is easy to show that $\mathbf{T}_N(K_n) \simeq \mathbf{S}^{n-2}$ for $n \geq 1$. Now assume that $f \in \mathrm{Hom}(G,H)$. Then $A \in \mathsf{N}(G)$ implies $f(A) \in \mathsf{N}(G)$. Then it follows that $f \in \mathrm{Hom}(\mathrm{sd}(\mathsf{N}(G)),\mathrm{sd}(\mathsf{N}(H)))$, and so $||f||$ is a continuous map from $\mathbf{T}_N(G)$ to $\mathbf{T}_N(H)$. Consequently, $\mathbf{T}_N$ is a functor from $\mathtt{GRA}$ to $\mathtt{TOP}$. If $\chi(G) = n$, then $G \to K_n$ and $n \geq 1$, which leads to $\mathbf{T}_N(G) \to \mathbf{T}(K_n) = \mathbf{S}^{n-2}$. This implies that $\mathrm{conn}(\mathbf{T}_N(G)) \leq \mathrm{conn}(\mathbf{S}^{n-2}) = n-3$; however, the inequality is not at all obvious and needs a proof (see [697]). Summarizing, we obtain Lovász' bound $\chi(G) \geq \mathrm{conn}(\mathbf{T}_N(G)) + 3$.

If the N-space of a graph is homotopy equivalent to $\mathbf{S}^n$, then the chromatic number of this graph is at least $n+2$. It is known that for $n \geq 0$ we have

$$\mathbf{T}_N(G) \simeq \mathbf{S}^n \text{ for } G \in \mathcal{LF}(n) \cup \mathcal{MF}(n) \cup \mathcal{SF}(n). \tag{8.23}$$

For the Lovász–Borsuk graphs this was proved by Lovász [700], for the generalized Mycielski graphs this was first proved by Stiebitz [966] (see also [887] and [445]), the result for the Shrijver graphs was conjectured by Lovász and Shrijver and proved by Björner and de Longueville [121].

Lovász [697] proved that every Kneser graph $KG(n+2p,p) \in \mathcal{KF}(n)$ has connectivity $\mathrm{conn}(\mathbf{T}_N(KG(n+2p,p))) \geq n-1$ and hence $\chi(KG(n+2p,p)) \geq n+2$. Bárány [81] observed that statement (B2) remains true for $n+1$ open sets; which gives another equivalent version (B4) of the Borsuk–Ulam theorem. Combining (B4) with a result due to Gale, Bárány found a direct proof showing that $\chi(KG(n+2p,p)) \geq n+2$. In 2002, Greene [430] discovered that (B2) remains true if each of the $n+1$ sets is open or closed; Greene used this observation to give a book proof of the Lovász–Kneser theorem. In 1986, Alon, Frankl, and Lovász [51] established Erdős' generalization of Kneser's conjecture for hypergraphs (see also [1104] [942]); for this purpose, they introduced another simplicial graph complex, which is more suitable than the neighborhood complex. For an overview about various graph complexes yielding lower bounds for the chromatic number, we refer to the survey by Matousek and Ziegler [724] as well as to the paper by Babson and Kozlov [74]. We shall discuss three of these complexes.

We start with some further notation from topology. A $\mathbb{Z}_2$**-space** is a pair $(\mathbf{T},\nu)$ consisting of a topological space $\mathbf{T}$ and a continuous mapping $\nu:\mathbf{T}\to\mathbf{T}$ satisfying $\nu^2=\mathrm{id}_{\mathbf{T}}$; the mapping ν is called an **involution**. We call the $\mathbb{Z}_2$-space $(\mathbf{T},\nu)$ **free** if $\nu(\mathbf{x})\neq\mathbf{x}$ for every $\mathbf{x}\in\mathbf{T}$. The standard involution for the sphere $\mathbf{S}^n$ is the central reflection $\mathbf{x}\mapsto-\mathbf{x}$ and we denote the corresponding $\mathbb{Z}_2$-space by $(\mathbf{S}^n,-)$ or briefly by $\mathbf{S}^n$. Let $(\mathbf{T},\nu)$ and $(\mathbf{T}',\nu')$ be two $\mathbb{Z}_2$-spaces. If there exists a continuous mapping $f:\mathbf{T}\to\mathbf{T}'$ such that $f\circ\nu=\nu'\circ f$, we call f a $\mathbb{Z}_2$**-map** and write $(\mathbf{T},\nu)\stackrel{2}{\to}(\mathbf{T}',\nu')$. Clearly, the relation $\stackrel{2}{\to}$ defines a quasi order. If we have $(\mathbf{T},\nu)\stackrel{2}{\to}(\mathbf{T}',\nu')$ and $(\mathbf{T}',\nu')\stackrel{2}{\to}(\mathbf{T},\nu)$, then we say that the $\mathbb{Z}_2$-spaces $(\mathbf{T},\nu)$ and $(\mathbf{T}',\nu')$ are $\mathbb{Z}_2$**-equivalent** and write $(\mathbf{T},\nu)\stackrel{2}{\leftrightarrow}(\mathbf{T}',\nu')$. For an $\mathbb{Z}_2$-space $(\mathbf{T},\nu)$ we consider the following two topological parameters (see [722], [724]): the $\mathbb{Z}_2$**-index** defined by

$$\mathrm{ind}(\mathbf{T},\nu)=\min\{n\in\mathbb{N}_0\mid(\mathbf{T},\nu)\stackrel{2}{\to}(\mathbf{S}^n,-)\},$$

and the $\mathbb{Z}_2$**-coindex** defined by

$$\mathrm{coind}(\mathbf{T},\nu)=\max\{n\in\mathbb{N}_0\mid(\mathbf{S}^n,-)\stackrel{2}{\to}(\mathbf{T},\nu)\}.$$

To simplify the notation we write $\mathbf{T}$ instead of $(\mathbf{T},\nu)$, in particular when it is clear to which involution ν we refer. The $\mathbb{Z}_2$-spaces $\mathbf{S}^n$ always refers to the standard involution space, i.e. to $(\mathbf{S}^n,-)$.

If $\mathbf{T}$ is a $\mathbb{Z}_2$-space with an involution ν, then the product space $\mathbf{T}'=\mathbf{T}\times\mathbb{R}[-1,1]$ is a $\mathbb{Z}_2$-space with the involution $(\mathbf{x},t)\mapsto(\nu(\mathbf{x}),-t)$ implying that the suspension $\mathrm{susp}(\mathbf{T})=\mathbf{T}'/(\mathbf{T}\times\{-1\},\mathbf{T}\times\{-1\})$ is a $\mathbb{Z}_2$-space, too (see [722]). If two $\mathbb{Z}_2$-spaces $\mathbf{T}$ and $\mathbf{T}'$ are $\mathbb{Z}_2$**-homeomorphic**, then we write $\mathbf{T}\stackrel{2}{\cong}\mathbf{T}'$; and if they are $\mathbb{Z}_2$**-homotopic equivalent** then we write $\mathbf{T}\stackrel{2}{\simeq}\mathbf{T}'$ (see [724] for a definition). The following lemma lists some well known basic properties of $\mathbb{Z}_2$-spaces; statement (a) as well as (b) are evident, and both statements (c) and (d) are equivalent to the Borsuk–Ulam theorem (see also [941]).

Lemma 8.73 *Let* $\mathbf{T}$ *and* $\mathbf{T}'$ *be two* $\mathbb{Z}_2$*-spaces. Then the following statements hold:*

(a) $\mathbf{T}\stackrel{2}{\to}\mathbf{T}'$ *implies* $\mathrm{ind}(\mathbf{T})\le\mathrm{ind}(\mathbf{T}')$, $\mathrm{coind}(\mathbf{T})\le\mathrm{coind}(\mathbf{T}')$ *and* $\mathrm{susp}(\mathbf{T})\stackrel{2}{\to}\mathrm{susp}(\mathbf{T}')$.
(b) $\mathbf{T}\stackrel{2}{\cong}\mathbf{T}'$ *implies* $\mathbf{T}\stackrel{2}{\simeq}\mathbf{T}'$ *and this implies* $\mathbf{T}\stackrel{2}{\leftrightarrow}\mathbf{T}'$.
(c) $\mathrm{coind}(\mathbf{T})\le\mathrm{ind}(\mathbf{T})$.
(d) $\mathrm{coind}(\mathbf{S}^n)=\mathrm{ind}(\mathbf{S}^n)=n$ *for all* $n\in\mathbb{N}_0$.
(e) $\mathrm{susp}(\mathbf{S}^n)\stackrel{2}{\simeq}\mathbf{S}^{n+1}$ *for all* $n\in\mathbb{N}_0$.
(f) $\mathrm{ind}(\mathrm{susp}(\mathbf{T}))\le\mathrm{ind}(\mathbf{T})+1$ *and* $\mathrm{coind}(\mathrm{susp}(\mathbf{T}))\ge\mathrm{coind}(\mathbf{T})+1$.

A proof of the following result can be found in Matousek's book (see the proof of Proposition 5.3.2, statement (iv)). The construction of the required $\mathbb{Z}_2$-map from

$\mathbf{S}^n$ to $\mathbf{T}$ is done step by step, starting with $\mathbf{S}^0$, and resembles the construction used by Lovász in his proof of Theorem 8.12. The second part of the statement follows from Lemma 8.73(a)(e).

Lemma 8.74 *Let* $\mathbf{T}$ *be a* $\mathbb{Z}_2$*-space such that* $\text{conn}(\mathbf{T}) = n-1$ *with* $n \in \mathbb{N}$. *Then* $\mathbf{S}^n \xrightarrow{2} \mathbf{T}$ *and as a consequence* $\text{coind}(\mathbf{T}) \geq n$.

The N-space $\mathbf{T}_N(G)$ of a graph has no canonical involution; the simplicial map $CN_G : \text{sd}(\mathsf{N}(G)) \to \text{sd}(\mathsf{N}(G))$ leads only to a continuous map $\nu : \mathbf{T}_N(G) \to \mathbf{T}_N(G)$ with $\nu^3 = \nu$, but for most graphs G we have $\nu^2 \neq \text{id}$. In what follows we introduce two simplicial complexes for a nonempty graph G (see Matousek and Ziegler [724]), namely

$$\mathsf{B}(G) = \{A \uplus B \mid A, B \subseteq V(G), A \subseteq CN_G(B) \neq \varnothing, B \subseteq CN_G(A) \neq \varnothing\}$$

and

$$\mathsf{B}_0(G) = \{A \uplus B \mid A, B \subseteq V(G), A \cap B = \varnothing, A \subseteq CN_G(B), B \subseteq CN_G(A)\}$$

(here, $A \uplus B = A \times \{1\} \cup B \times \{2\}$). These complexes belong to a series of similar complexes that are usually refereed to as **box-complexes**. Note that $A \uplus \varnothing \in \mathsf{B}(G)$ if and only if $CN_G(A) \neq \varnothing$; and $A \uplus \varnothing \in \mathsf{B}_0(G)$ if and only if $A \subseteq V(G)$. Then we define two topological spaces for the graph G:

$$\mathbf{T}_B(G) = \|\mathsf{B}(G)\| \text{ and } \mathbf{T}_B^o(G) = \|\mathsf{B}_0(G)\|.$$

The map $A \uplus B \mapsto B \uplus A$ yields a simplicial map from the set $\text{Hom}(\mathsf{B}(G), \mathsf{B}(G))$, respectively from $\text{Hom}(\mathsf{B}_0(G), \mathsf{B}_0(G))$; the corresponding continuous mapping from $\mathbf{T}_B(G)$ to $\mathbf{T}_B(G)$, respectively from $\mathbf{T}_B^o(G)$ to $\mathbf{T}_B^o(G)$, is an involution. Hence, both spaces $\mathbf{T}_B(G)$ and $\mathbf{T}_B^o(G)$ are free $\mathbb{Z}_2$-spaces. Lovász introduced a polyhedral complex $\mathsf{Hom}(H,G)$ defined for any two graphs H and G; the vertices of this complex are the homomorphisms in $\text{Hom}(H,G)$. The geometric realization of the complex $\mathsf{Hom}(K_2,G)$ leads to a free $\mathbb{Z}_2$-space denoted by $\mathbf{T}_H(G)$; for some more details see Babson and Kozlov [74]. As proved in [74], every graph G satisfies $\mathbf{T}_H(G) \simeq \mathbf{T}_N(G)$; note that this is not a $\mathbb{Z}_2$-relation. The three $\mathbb{Z}_2$-spaces $\mathbf{T}_B(G)$, $\mathbf{T}_B^o(G)$ and $\mathbf{T}_H(G)$ are closely related to each other, as every nonempty graph G satisfies

$$\mathbf{T}_B(G) \overset{2}{\leftrightarrow} \mathbf{T}_H(G), \mathbf{T}_B^o(G) \overset{2}{\simeq} \text{susp}(\mathbf{T}_B(G)), \mathbf{T}_B^o(G) \overset{2}{\leftrightarrow} \text{susp}(\mathbf{T}_H(G)). \tag{8.24}$$

The first result was proved by Csorba [266], the second result is due to Matousek and Ziegler [724], and the third result is a consequence of the first two relations and Lemma 8.73(b). Furthermore, it is well know that for every $n \in \mathbb{N}$, we have

$$\mathbf{T}_N(K_n) \simeq \mathbf{S}^{n-2}, \mathbf{T}_B(K_n) \overset{2}{\simeq} \mathbf{T}_H(K_n) \overset{2}{\simeq} \mathbf{S}^{n-2} \text{ and } \mathbf{T}_B^o(K^n) \overset{2}{\simeq} \mathbf{S}^{n-1} \tag{8.25}$$

The most important property, however, is the functoriality of the operators $\mathbf{T}_B, \mathbf{T}_B^o$ and $\mathbf{T}_H$, that is, if $\mathbf{T}$ is one of these operators, then

$$G \to H \text{ implies } \mathbf{T}(G) \xrightarrow{2} \mathbf{T}(H) \tag{8.26}$$

(see [722], [724] and [74]). Combining these results, we obtain the following well known sequence of inequalities for a nonempty graph G, namely

$$\begin{aligned}\chi(G) &\geq \operatorname{ind}(\mathbf{T}_H(G))+2 \geq \operatorname{ind}(\mathbf{T}_B^o(G))+1 \geq \\ &\geq \operatorname{coind}(\mathbf{T}_B^o(G))+1 \geq \operatorname{coind}(\mathbf{T}_H(G))+2 \geq \\ &\geq \operatorname{conn}(\mathbf{T}_H(G))+3 = \operatorname{conn}(\mathbf{T}_N(G))+3.\end{aligned} \tag{8.27}$$

Proof Since the two spaces $\mathbf{T}_N(G)$ and $\mathbf{T}_H(G)$ are homotopic equivalent, we get $\operatorname{conn}(\mathbf{T}_N(G)) = \operatorname{conn}(\mathbf{T}_H(G))$. By Lemma 8.74, we have $\operatorname{conn}(\mathbf{T}_H(G)) \leq \operatorname{coind}(\mathbf{T}_H(G)) - 1$. Assume that $\chi(G) = n$. Then $n \geq 1$ and $G \to K_n$. This implies that $\mathbf{T}_H(G) \xrightarrow{2} \mathbf{T}_H(K_n) \overset{2}{\simeq} \mathbf{S}^{n-2}$ (by (8.26) and (8.25)). Using (8.24) and Lemma 8.73, we obtain that

$$\begin{aligned}
\operatorname{coind}(\mathbf{T}_H(G))+2 &\leq \operatorname{coind}(\operatorname{susp}(\mathbf{T}_H(G)))+1 && \text{(by L 8.73(f))}\\
&= \operatorname{coind}(\mathbf{T}_B^o(G))+1 && \text{(by L 8.73(a), (8.24))}\\
&\leq \operatorname{ind}(\mathbf{T}_B^o(G))+1 && \text{(by L 8.73(c))}\\
&= \operatorname{ind}(\operatorname{susp}(\mathbf{T}_H(G)))+1 && \text{(by L 8.73(a), (8.24))}\\
&\leq \operatorname{ind}(\mathbf{T}_H(G))+2 && \text{(by L 8.73(f))}\\
&\leq \operatorname{ind}(\mathbf{T}_H(K_n))+2 && \text{(by L 8.73(a))}\\
&= \operatorname{ind}(\mathbf{S}^{n-2})+2 && \text{(by 8.73(a)(b))}\\
&= n = \chi(G). && \text{(by L 8.73(d))}
\end{aligned}$$

This complete the proof. □

The first inequality in (8.27) is a reformulation of Lovász' original bound, see the paper by Alon, Frankl, and Lovász [51]. For a discussion of the strength of the inequalities in (8.27), see the paper by Simonyi, Tardos, and Vrećica [944]. Each of the first four inequalities in (8.27) can be strict. A graph G without 4-cycles satisfies $\operatorname{ind}(\mathbf{T}_H(G)) \leq 1$ (see Matousek and Ziegler [724]); hence there are graphs G for which the difference $\chi(G) - (\operatorname{ind}(\mathbf{T}_H(G)) + 2)$ is arbitrarily large. The same holds for the difference $\operatorname{ind}(\mathbf{T}_B^o(G)) - \operatorname{coind}(\mathbf{T}_B^0(G))$.

For an integer $n \in \mathbb{N}_0$, define

$$\mathcal{CT}(n) = \{G \in \mathcal{G} \mid \operatorname{coind}(\mathbf{T}_H(G)) \geq n\}.$$

By Lemma 8.74, every graph G with $\operatorname{conn}(\mathbf{T}_N(G)) \geq n-1$ belongs to the graph family $\mathcal{CT}(n)$. This implies, in particular, that

$$\mathcal{CT}(n) \supseteq \mathcal{KF}(n) \cup \mathcal{SF}(n) \cup \mathcal{MF}(n) \cup \mathcal{LF}(n).$$

Suppose that G and G' are two nonempty graphs such that $G \to G'$. Then we obtain that $\mathbf{T}_H(G) \xrightarrow{2} \mathbf{T}_H(G')$ (by (8.26)) and so $\operatorname{coind}(\mathbf{T}_H(G)) \leq \operatorname{coind}(\mathbf{T}_H(G'))$

(by Lemma 8.73(a)). Consequently, if $G \in \mathcal{CT}(n)$, then $G' \in \mathcal{CT}(n)$, too. From (8.20) it then follows that

$$\mathcal{CT}(n) \supseteq \mathcal{GF}(n),$$

and as an immediate consequence of (8.27) we obtain the following result.

Corollary 8.75 $\chi(G) = \mathrm{coind}(\mathbf{T}_H(G)) = n+2$ *for all graphs G belonging to the family $\mathcal{KF}(n) \cup \mathcal{SF}(n) \cup \mathcal{MF}(n) \cup \mathcal{LF}(n) \cup \mathcal{GF}(n)$, where $n \in \mathbb{N}_0$.*

The following result due to Simonyi and Tardos [941, Lemma 4.4] provides a combinatorial characterization for the graph family $\mathcal{CT}(n)$.

Lemma 8.76 (Simonyi and Tardos) *For $n \in \mathbb{N}_0$, a graph G belongs to $\mathcal{CT}(n)$ if and only if $BG'(n,\alpha) \to G$ for some $\alpha \in \mathbb{R}(0,2)$.*

Sketch of Proof: If $BG(n,\alpha) \to G$, then there is an α with $\alpha < \alpha' < 2$ such that $LG(n,\alpha') \subseteq BG(n,\alpha)$. Consequently, we have $LG(n,\alpha') \to G$ which gives $n \le \mathrm{coind}(\mathbf{T}_H(LG(n,\alpha'))) \le \mathrm{coind}(\mathbf{T}_H(G))$ and hence $G \in \mathcal{CT}(n)$. If $G \in \mathcal{CT}(n)$, then there is a $\mathbb{Z}_2$-map $f : \mathbf{S}^n \to \mathbf{T}_H(G)$. Using f one can construct a homomorphism from a Borsuk graph $BG'(n,\alpha)$ to G for some α close enough to 2. □

Combining Corollary 8.75, Lemma 8.76, (8.20), (8.22), and the fact that the graph families $\mathcal{BF}(n)$ and $\mathcal{BF}'(n)$ are hom-equivalent, we obtain the following two results. The second result is due to Simons, Tardif, and Wehlau [940].

Corollary 8.77 *For any $n \in \mathbb{N}_0$,*

$$\mathcal{CT}(n) \leftrightarrow \mathcal{BF}'(n) \leftrightarrow \mathcal{BF}(n) \leftrightarrow \mathcal{LF}(n) \leftrightarrow \mathcal{MF}(n).$$

Corollary 8.78 *For any graph G,*

$$\mathrm{coind}(\mathbf{T}_H(G)) = \max\{n \in \mathbb{N}_0 \mid \mathcal{MF}(n) \to \{G\}\}.$$

8.8 Exercises

8.1 Show that the homomorphism relation for graphs is reflexive and transitive.

8.2 Let G be a graph, let $I \in \mathcal{I}(G)$, and let $H = G(I \to v_I)$ be the graph obtained from G by identifying the independent set I to a new vertex v_I. Show that $G \to H$ and the map $\phi : V(G) \to V(H)$, defined by $\phi(v) = v$ if $v \notin I$ and $\phi(v) = v_I$ if $v \in I$, is an homomorphism from G to H.

8.3 Show that for every graph G, the following statements hold:

1. $\mathrm{Hom}(G, K_0) \neq \varnothing$ if and only if $G = \varnothing$.
2. $\mathrm{Hom}(G, K_1) \neq \varnothing$ if and only if $E(G) = \varnothing$.
3. $\mathrm{Hom}(G, K_2) \neq \varnothing$ if and only if G is bipartite.

4. If H is bipartite, then $\mathrm{Hom}(G,H) \neq \varnothing$ if and only if G is bipartite.
5. $|\mathrm{Hom}(K_0,G)| = 1$.

8.4 Show that if ϕ is a homomorphism from G to H, then ϕ is an isomorphism from G to H if and only if ϕ is bijective and ϕ^{-1} is an homomorphism from H to G.

8.5 Show that if ϕ is an **endomorphism** of a graph G, i.e., ϕ is a homomorphism from G to G, then the following statements are equivalent:

1. ϕ is an **automorphism** of G, i.e. an isomorphism from G to G.
2. ϕ is bijective.
3. ϕ is surjective
4. ϕ is injective

8.6 Show that for graphs G and H there is an isomorphism from G to H if and only if there is a surjective homomorphism from G to H and a surjective homomorphism from H to G.

8.7 Show that if G and H are two nonempty graphs such that $|G|\alpha(H) \leq |H|\alpha(G)$, and each vertex of H is contained in the same number of maximum independent sets of H, then G is not homomorphic to H. (G. Sabbidussi, private communication)

8.8 If G and H are graphs such that $G \to H$ and $H \to G$, then we say that G and H are **hom-equivalent** and write $G \leftrightarrow H$. Show that $\leftrightarrow$ is an equivalence relation for graphs.

8.9 Show that $H \leftrightarrow K_2$ if and only if H is bipartite and $E(H) \neq \varnothing$.

8.10 For a graph G, let $\mathrm{Aut}(G)$ be the set of automorphisms of G. Show that $\mathrm{Aut}(G)$ is a subgroup of $\mathrm{Sym}(V(G))$ with respect to the composition of maps; this group is called the **automorphism group** of G.

8.11 A graph is called a **core** if every endomorphism of this graph is an automorphism. Show that the following statements hold:

1. If G and H are two cores such that $G \leftrightarrow H$, then $G \cong H$.
2. For every graph G there is a unique subgraph H of G such that $G \leftrightarrow H$ and H is a core. We call H the **core** of G.
3. The core H of a graph G is a **retract** of G, i.e., there exists a homomorphism ϕ from G to H such that $\phi|_{V(H)} = \mathrm{id}_{V(H)}$.
4. The core of a graph G is an induced subgraph of G.
5. If two graphs G and H are hom-equivalent, then their cores are isomorphic.

8.12 Show that every vertex critical graph is a core.

8.13 Show that if ϕ is an endomorphism of a graph G and I is an independent set of G, then $\phi^{-1}(I)$ is an independent set of G and $|\phi^{-1}(I)| \geq |I|$.

8.14 Show that every Kneser graph is a core. (Hint: Let ϕ be an endomorphism of the Kneser graph $KG(n,p)$. Consider the independent sets $\mathcal{I}_i = \{A \in V(KG(n,p)) \mid i \in A\}$ (for $i \in [1,n]$) and the sets $\phi^{-1}(\mathcal{I}_i)$; use Corolarry 8.26). (see e.g. Hahn and Tardif [452])

8.15 Dol'nikov's Theorem: To any hypergraph H we associate the **generalized Kneser graph** $G = KG(H)$ with $V(G) = E(H)$ and $E(G) = \{ef \mid e,f \in E(H), e \cap f = \emptyset\}$. The **colorability defect** of a hypergraph H is defined by $\mathrm{cd}(H) = \min\{|X| \mid X \subseteq V(H), \chi(H-X) \le 2\}$. Show the following statements:

1. Every graph is isomorphic to a generalized Kneser graph.
2. For $n \ge 2p-1 \ge 1$, we have $KG(n,p) = KG(K_n^p)$ and $\mathrm{cd}(K_n^p) = n-2p+2$.
3. Every hypergraph H satisfies $\chi(KG(H)) \ge \mathrm{cd}(H)$. (Hint: Use the octahedral Tucker lemma) (Dol'nikov [307] , see also [276])

8.16 Show that if G, G', H and H' are graphs, then $G' \subseteq G$ and $H' \subseteq H$ implies $G' \times H' \subseteq G \times H$.

8.17 Show that if G and H are two connected bipartite graphs, then $G \times H$ has two components, unless $G = H = K_1$. (Weichsel [1062])

8.18 Show that if G is a connected graph with $|G| \ge 2$, then $G \times K_2$ is connected if and only if G is not bipartite.

8.19 Show that, for all graphs K, G and H, if $\phi_1 \in \mathrm{Hom}(K,G)$ and $\phi_2 \in \mathrm{Hom}(K,H)$, then the map ϕ defined by $\phi(u) = (\phi_1(u), \phi_2(u))$ for all $u \in V(K)$ is the unique homomorphism from K to $G \times H$ satisfying $\phi_1 = p_G \circ \phi$ and $\phi_2 = p_H \circ \phi$.

8.20 Show that for two graphs G and H we have $\omega(G \times H) = \min\{\omega(G), \omega(H)\}$ and $g_o(G \times H) = \max\{g_o(G), g_o(H)\}$.

8.21 Let K and K' be hom-equivalent graphs. Show that K is multiplicative if and only if K' is multiplicative.

8.22 Show that each even cycle is multiplicative.

8.23 Deduce from Theorem 8.61(c) that if G and H are two graphs and $n \in \mathbb{N}$, then $\chi(G \times H) \le n$ if and only if $H \to K_n^G$. As a consequence, $\chi(G \times K_n^G) \le n$.

8.24 Let G be a graph with $\chi(G) = n \ge 3$ and let $r \in \mathbb{N}$. Show that $\chi(M_r(G)) = n$ if and only if K_n^G contains a path of length at most r between a (proper) n-coloring of G and a constant map. (Tardif [987])

8.25 Deduce from Lemma 8.66 that if C is an odd cycle, then K_3^C contains no path between a (proper) 3-coloring of G and a constant map. By the previous exercise this implies that $\chi(M_r(C)) = 4$ for every odd cycle C and every $r \in \mathbb{N}$.

8.26 Let $n \ge 4$ be an even integer, and let G be the complement of the odd cycle C_{2n-1}. Show that $\chi(G) = n$ and K_n^G contains a path between a (proper) n-coloring of G and a constant map.

8.9 Notes

Homomorphism as a mathematical concept and its various specifications, endomorphism, isomorphism, automorphism and others, were first used in abstract algebra in order to compare algebraic structures, such as groups, rings and vector spaces. One of the early publications in which the term homomorphism is used is due to Ernst Ritter; he [867, footnote page 22] refers to a lecture by his teacher Felix Klein. A homomorphism is an operation-preserving map between two algebraic structures of the same type, or more precisely between their ground sets; and an isomorphism is usually a homomorphism such that the inverse map exists and is a homomorphism, too. While isomorphisms induce an equivalence relation on the structures under consideration, a homomorphism in general leads only to a **quasi-order** (i.e., a binary relation that is reflexive and transitive). In the 20th century the concept of homomorphism has been generalized, under the name morphism, to many other mathematical structures. This eventually led to the mathematical discipline of category theory, which occupy a central position in both contemporary mathematics and theoretical computer science.

Graph homomorphisms is an important tool in the study of graph problems, especially coloring problems. While the first paper on this subject was published in 1961 by Sabidussi [881], the first monograph on graph homomorphisms by Hell and Nešetřil [495] appeared over forty years later. Gert Sabidussi (1929–2022) maintained a lifelong leading role in algebraic graph theory and during the last years of his life in Vienna he worked on a manuscript *Soft Graph Theory*, containing a personal basic account of the subject. We recommend the intensive survey paper on graph homomorphisms by Hahn and Tardif [452] from 1979 and the appealing textbook on algebraic graph theory by Godsil and Royle [418] from 2001, which contains two special chapters, one on homomorphisms and one on Kneser graphs. One of the early results on graph homomorphisms was obtained in 1967 by Lovász [687]; this result states that a graph G can be characterized by counting homomorphisms from all graphs to G.

Theorem 8.79 (Lovász) *Two graphs G and G' are isomorphic if and only if $|\mathrm{Hom}(H,G)| = |\mathrm{Hom}(H,G')|$ for all graphs H.*

A short proof of Lovász' result is contained in [418, Lemma 6.5.1]. This result has some far-reaching consequences; in particular, it is one of the starting points for the application of methods from functional analysis to graph theory, which eventually led to the theory of graph limits (see the book by Lovász [702] from 2012).

Section 8.1 lists some basic properties of graph homomorphisms, most are folklore. Proposition 8.3 is often referred to as the *no-homomorphism lemma*, since it provides restrictions for the existence of homomorphisms between two graphs, when the target graph is transitive. Our proof of this proposition is standard and taken from [452]. As remarked earlier, a (proper) n-coloring of a graph may be considered as a homomorphism of this graph into the complete graph K_n. Considering this, it is natural to call a homomorphism of a graph G to a graph H a (proper) H-**coloring** of G. Then a graph G is called **H-colorable** if there exists an H-coloring of G, i.e. $\mathrm{Hom}(G,H) \neq \varnothing$. Given a fixed graph H, the **H-coloring problem** is the decision problem whether a given input graph G admits an H-coloring. The complexity of the H-coloring problem was investigated by several authors and finally solved by Hell and Nešetřil [494]; in their paper the reader will also find references to many papers related to the H-coloring problem. The H-coloring problem is in P if H is bipartite, this follows from Exercise 8.3; however, the H-coloring problem is NP-complete for any nonbipartite graph H.

Let H be a fixed graph. A graph G is called **H-critical** if all proper subgraphs of G are H-colorable, but G itself is not H-colorable. Let $\mathrm{Crit}(H)$ denote the class of H-critical graphs, and define

$$\mathrm{ext}(H,n) = \min\{|E(G)| \mid G \in \mathrm{Crit}(H) \text{ and } |G| = n\}.$$

This concept generalizes the notion of χ-critical graphs since $\mathrm{Crit}(k) = \mathrm{Crit}(K_{k-1})$ and $\mathrm{ext}(k,n) = \mathrm{ext}(K_{k-1},n)$. Evidently, the critical graph methods (Proposition 1.27) has a counterpart for H-critical graphs. According to Exercise 8.8 two graphs K and K' are hom-equivalent (written $K \leftrightarrow K'$) if and only if booth $K \to K'$ and $K' \to K$ holds. It is straightforward to show that $\mathrm{Crit}(H) = \mathrm{Crit}(H')$, provided that $H' \leftrightarrow H$. Consequently, if H is bipartite and $E(H) \neq \varnothing$, then $\mathrm{Crit}(H) = \mathrm{Crit}(K_2) = \mathrm{Crit}(3) = \{C_n \mid n \text{ is odd and } n \geq 3\}$ (see also Exercise 8.9). Another consequence is that we may restrict our attention to the case that the fixed graph H is a core (see Exercise 8.11 for the definition of core).

Assume that $H = K_p \boxplus H'$, $\delta(H') \geq 1$, $p \geq 0$ and $G \in \mathrm{Crit}(H)$. If v is an arbitrary vertex of G, then there is a coloring $\varphi \in \mathrm{Hom}(G-v,H)$. If $A = \varphi(N_G(v))$ is the set of colors (vertices of H) that occur in the neighborhood of v, then $|A| \geq p+2$ and A contains all colors belonging to K_p. For otherwise, as $\delta(H') \geq 1$, there is a color $c \in V(H)$ such that $A \subseteq N_H(c)$ and assigning color c to v yields a H-coloring of G, which is impossible. Hence, $\delta(G) \geq p+2$. By adapting Gallai's recoloring procedure (see also the proof of Theorem 5.39) it can be shown that if $X = \{v \in V(G) \mid d_G(v) = p+2\}$, then $G[X]$ is a Gallai forest (possibly empty). This result is due to Negussie [785]; he used this result to give a positive answer to a problem raised by Nešetril and Negussie [774]. Another consequence is that if $\Delta(G) \leq p+2$, then either $p = 0$ and G is an odd cycle, or $p \geq 1$ and $G = K_{p+3}$. In the later case a Gallai type bound for the number of edges of G can be obtained (see the proof of Theorem 5.8) implying that

$$2\mathrm{ext}(H,n) \geq (p+2+\frac{p}{(p+3)^2-3})n \text{ for } n \geq p+4.$$

In studying H-critical graphs the first interesting cases for H are the odd cycles. Recall from (8.10) that if G is a graph and $p \geq 1$, then $\chi^{\circ}(G) \leq 2+\frac{1}{p}$ if and only if $G \to C_{2p+1}$. Grötzsch's theorem (see Theorem 5.13) says that every planar graph of girth at least 4 is C_3-colorable; a short proof of this result can be given by using the Kostochka–Yancey bound $\mathrm{ext}(C_3,n) = \mathrm{ext}(4,n) \geq \frac{1}{3}(5n-2)$ (see Theorem 5.10 and the proof of Theorem 5.13). Hence, the following conjecture may be considered as a possible generalization of Grötzsch's theorem.

Conjecture 8.80 (Jaeger) *Every planar graph of girth at least $4p$ is C_{2p+1}-colorable.*

For the motivation and history of this conjecture, we refer the interested reader to the papers [156], [317], [703], [778] and [823]. Using duality of planar graphs, an equivalent formulation of the above conjecture is the claim that every planar $4p$-edge-connected graph G admits a **modulo $(2p+1)$-orientation**, that is, an orientation D of G such that $d_D^+(v) \equiv d_D^-(v) \pmod{2p+1}$. In 1984, Jaeger [516] conjectured that this holds not only for planar graphs, but for all graphs; however, this was refuted in 2018 by Han, Li, Wu, and Zhang [466].

For $p \geq 1$, let $\mathrm{gpl}(p)$ be the least integer $g \geq 3$ for which every planar graph with girth at least g is C_{2p+1}-colorable. Then Conjecture 8.80 claims that $\mathrm{gpl}(p) \leq 4p$. DeVoss [282] constructed planar graphs showing that $\mathrm{gpl}(p) \geq 4p$. The case $p = 1$ is the only case where the conjecture is confirmed. Lovász, Thomassen, Wu, and Zhang [703] proved that $\mathrm{gpl}(p) \leq 6p$ for all $p \geq 2$; this seems to be the best known general bound. However, $\mathrm{gpl}(2) \leq 10$ as proved by Dvořák and Postle [317], and $\mathrm{gpl}(3) \leq 16$ as proved by Postle and Smith-Roberge [823]. Upper bounds for $\mathrm{gpl}(p)$ can be obtained by means of density bounds for C_{2p+1}-critical graphs. However, as explained by Postle and Smith-Roberge [823, Section 5] it is not possible to prove Conjecture 8.80 using only density arguments.

Every graph in $\mathrm{Crit}(C_{2p+1})$ is connected and even 2-connected, and has minimum degree at least 2 hence $e(G) \geq |G|$. By (8.10), $\mathrm{Crit}(C_{2p+1})$ contains all odd cycles of order at most $2p-1$, implying that if $G \in \mathrm{Crit}(C_{2p+1})$, then $e(G) = |G|$ if and only if G is an odd cycle with $|G| \leq 2p-1$. So let

$$\mathrm{Crit}^*(C_{2p+1}) = \mathrm{Crit}(C_{2p+1}) \setminus \{C_{2q+1} \mid 1 \leq q \leq p-1\}.$$

Dvořák and Postle [317] proved that every graph $G \in \mathrm{Crit}^*(C_5)$ of order n satisfies $e(G) \geq (5n-2)/4$; the proof uses the potential method (see Section 5.4) and is tough. Now assume that there is a planar graph of girth at least 10 that is not C_5-colorable, then there is also such a graph G that is C_5-critical. If G has n vertices and m edges, then $m \leq \frac{5}{4}(n-2)$ (by Euler's Formulae, see Theorem C.18) and $m \geq (5n-2)/4$ (by the Dvořák–Postle bound), a contradiction. This proves that $\mathrm{gpl}(2) \leq 10$. Postle and Smith-Roberge [823] proved that

$$e(G) \geq \left(1 + \frac{1}{4p}\right)|G| + \frac{1}{3p} \quad \text{for all } G \in \mathrm{Crit}^*(C_{2p+1}),$$

and

$$e(G) \geq \frac{17|G|-2}{15} \quad \text{for all } G \in \mathrm{Crit}^*(C_7).$$

While the proof of the first inequality uses a simple discharging argument and some structural properties of C_{2p+1}-critical graphs, the proof of the second inequality is much harder and uses the potential method (see Section 5.4). Postle and Smith-Roberge asked whether the inequality

$$\mathrm{ext}(C_{2p+1},n) \geq \frac{p(2p+3)n-(p+1)(2p-1)}{2p^2+2p-1}$$

holds for any fixed $p \geq 3$; they also showed that if this is true, then equality holds for infinitely many values of n. That the above inequality holds for $p = 2$ was conjectured by Dvořák and Postle [317].

Fractional graph theory deals with real valued (fractional) analogues of integral graph parameters and concepts. Fractional graph theory explores the close relation between combinatorial optimization and linear programming; and it uses, in particular, linear programming relaxations for analyzing difficult combinatorial problems. The first booklet on this subject [99], due to Claude Berge, appeared in 1978. Since then, fractional graph theory has become an important method in combinatorics and combinatorial optimization. The first monograph, providing a rational approach to the theory of graphs, was published by Scheinerman and Ullmann [898] in 1997. Another very comprehensive source is the three volume set about combinatorial optimization by Schrijver [903] from 2003; this in-depth work contains results about most fractional graph parameters.

The definition of the fractional chromatic number $\chi^*(G)$ as an optimal value of a linear program, which is a relaxation of the integer program whose value equals the chromatic number $\chi(G)$, is very common in the graph theory literature; most fractional graph parameters associated to integral graph parameters are defined in this way. However, the parameter χ^* first appears in the paper by Hilton, Rado, and Scott [499] from 1973 (see Section 1.12); for the definition of $\chi^*(G)$ they use Fekete's lemma and (8.6) (see Proposition 1.45). That the two definitions are indeed equivalent seems to be first mentioned in a paper by Chvátal, Garey, and Johnson [249]; they refer to an unpublished technical memorandum of Gilbert from 1972. While no term for χ^* was introduced in [499], the parameter χ^* was called the **ultimate chromatic number** in [249] and **multichromatic number** in [250]; however, the term **fractional chromatic number** for χ^* is now more common and also more adequate.

The first part of Section 8.2 up to the Lovász-Kneser-theorem lists some well-known simple facts about the fractional chromatic number, with standard proofs (we refer to [418, Chapter 7] and, for Theorem 8.8, to [495]). An immediate consequence of the definition is that χ^* is a lower bound for χ. On the other hand, as proved by Johnson [538] and, independently, by Lovász [695], every graph G satisfies

$$\chi(G) \leq (1+\ln\alpha(G))\chi^*(G).$$

A short proof of this inequality, based on maximal independent set coloring (see the discussion in Section 1.12), can be found in [903, Theorem 64.13].

That the fractional chromatic number can be characterized by homomorphisms into the Kneser graphs (see (8.7) and Theorem 8.9) is folklore; (8.7) was first mentioned in the paper by Stahl [956] from 1976. He investigated homomorphisms between Kneser graphs and proposed the following conjecture (our formulation is due to Poljak and Roberts [814], see also Tardif and Zhu [990]).

Conjecture 8.81 (Stahl) *$KG(n,p) \to KG(m,q)$ if and only if $m \geq an-2b$, where $q = ap-b$, $a \geq 1$ and $0 \leq b < p$.*

For $q=1$, the conjecture says that $KG(n,p) \to KG(m,1) = K_m$ if and only if $m \geq n-2p+2$, or equivalently, $\chi(KG(n,p)) = n-2p+2$. So Stahl's conjecture is a strengthening of Kneser's conjecture and only partial results are known, see e.g. [956], [957], [990] and [814]. Note that Stahl's conjecture proposes an exact value for the chromatic number of the weighted Kneser graph $(KG(n,p),q)$; that the proposed value is indeed an upper bound was confirmed by Stahl [956] himself. Csorba and Oszttényi [266] showed that Lovász's topological lower bound cannot be used to prove Stahl's conjecture. Another result by Stahl [956, Corollary], supporting his conjecture, says that $\chi(KG(n,p),kp) = kn$ for all $k \geq 1$. Using (8.6) and the fact that $\chi(G,p) = \chi(G[K_p])$, this implies that $\chi^*(KG(n,p)) = \frac{n}{p}$. Our proof of this result (see Corollary 8.25) uses a different approach. As an immediate consequence, we obtain that $\chi^*(KG(3p,p)) = 3$ for all $p \geq 1$. On the other hand, the Lovász–Kneser–theorem implies that $\chi(KG(3p,p)) = p+2$. This is bad news for the possibility to approximate χ by χ^*. Other bad news concern the computational complexity of the fractional chromatic number. One possibility to compute $\chi^*(G)$ is to use the definition and to compute the optimal value of the associated linear program. A landmark result in computational complexity says that linear programs can be solved in polynomial time. However polynomial refers to the size of the problem; and the linear program for $\chi^*(G)$ has exponential many variables, one for each independent set of G. In fact, for every fixed real number $t > 2$, the decision problem whether a given graph G satisfies $\chi^*(G) \leq t$ is NP-complete, as proved by Grötschel, Lovász, and Schrijver [432]. It should be noted that for other fractional graph parameters the situation is often better. The fractional chromatic index $\chi'^*(G)$ of a multigraph G is usually defined by a linear program and $\chi'^*(G) = \chi^*(\mathrm{L}(G))$. (see [898] or [903]). Furthermore, the Goldberg–Seymour Conjecture (see Section 1.8) implies that

$$\lceil \chi'^*(G) \rceil \leq \chi'(G) \leq \lceil \chi'^*(G) \rceil + 1$$

for every multigraph G (see [973]). The first paper about the fractional chromatic number by Hilton, Rado, and Scott [499] predates the proof of the four color theorem and proposed a fractional version of the four color problem. However, the main result in this paper only says that $\chi^*(G) < 5$ for every planar graph G. Even today it would be of interest to find a proof of the fractional four color theorem that is simpler than the present proof of the four color theorem. The only result in this direction was obtained by Cranston and Rabern [263]; they proved that every planar graph G admits a $KG(9,2)$-coloring and so $\chi^*(G) \leq 9/2$ holds.

The Kneser graph $KG_k = KG(2k+1,k)$ with $k \geq 1$ has chromatic number $\chi(KG_k) = 3$ (by Theorem 8.11) and odd girth $g_0(KG_k) \geq 2k+1$ (by Proposition 8.14). It is easy to show that we have $g_0(KG_k) = 2k+1$. Consequently, every KG_k-colorable graph G satisfies $\chi(G) \leq 3$ and $g_0(G) \geq 2k+1$ (by Proposition 8.1).

Let G_k be the graph obtained from a K_4 with vertex set $V = \{u,v,w,x\}$ by replacing the edge uv with a path P_1 of length $2k-1$ and replacing the edge wx with a path P_2 of length $2k-1$, where P_1 and P_2 are disjoint. Then G_k is a planar graph with $g_o(G_k) = 2k+1$ and, moreover, $\chi(G_k) \le 3$ unless $k = 1$. Note that $G_1 = K_4$, $KG_1 = K_3$ and KG_2 is the Petersen graph. On the one hand, Klostermeyer and Zhang [604] proved that G_k has no KG_k-coloring. On the other hand, they proved that every planar graph with odd girth at least $10k-7$, where $k \ge 2$, has a KG_k-coloring. Pirnazar and Ullman [807] proved that every planar graph with girth at least $8k-4$, where $k \ge 1$, has a KG_k-coloring. The proof is done by induction on k and for the base case $k = 1$ the theorem of Grötzsch is used. Chen and Raspaud [210] proved that every triangle-free graph G with $\mathrm{mad}(G) < 5/2$ is homomorphic to the Petersen graph $KG_2 = KG(5,2)$. They also proposed the following conjecture.

Conjecture 8.82 (Chen and Raspaud) *Every graph G with $g_o(G) \ge 2k+1$ and $\mathrm{mad}(G) < 2+\frac{1}{k}$, where $k \ge 1$, satisfies $G \to KG(2k+1,k)$.*

If the conjecture is true, the graph G_k constructed by Klostermeyer and Zhang for every $k \ge 1$ shows that the bound on $\mathrm{mad}(G)$ is best possible. Note that $g_o(G_k) = 2k+1$, $G_k \nrightarrow KG(2k+1,k)$ (as proved in [604]), and it is easy to show that we have $\mathrm{mad}(G_k) = 2+\frac{1}{k}$.

Gimbel, Kündgen, Li, and Thomassen [414] studied fractional coloring methods with applications to degenerate graphs and graphs on surfaces; they proposed several conjectures. A far reaching generalization of the four color theorem is Hadwiger's conjecture saying that $\chi(G) \le \omega_{\preceq}(G)$ for every graph G (see Section 4.9). The conjecture is wide open and the best fractional result, obtained by Reed and Seymour [857], says that $\chi^*(G) \le 2\omega_{\preceq}(G)$ for every graph G. Let us mention another very popular coloring conjecture, known as the Erdős–Faber–Lovász conjecture (see [530, Problem 9.11] for the history of the conjecture). The conjecture says that if a graph G is the union of n complete graphs of order n no two of which share more than one vertex, then $\chi(G) = n$. In 1992, Kahn and Seymour [546] proved that the conjecture holds for all n when χ is replaced by χ^*, see also [898]. In 2021, Kang, Kelly, Kühn, Methuku, and Osthus [552] proved that the conjecture holds if n is sufficiently large. A fractional version of the total coloring conjecture was proved by Kilakos and Reed [588], that is, every graph G satisfies $\chi^*(\mathrm{T}(G)) \le \Delta(G)+2$.

There seems to be no canonical way for defining the fractional list chromatic number respectively the fractional DP-chromatic number by means of linear programming relaxations. However, one can adapt the original definition of $\chi^*(G)$ from [499] that is based on the chromatic number of weighted graphs (G,p) with $p \in \mathbb{N}$, see Theorem 8.9 and (8.6). So

$$\chi^*(G) = \inf\{\frac{n}{p} \mid \chi(G,p) \le n\}.$$

Note that we can replace the infimum by minimum and $\le$ by $=$. The list chromatic number of weighted graphs was defined in Section 1.9. Then the **fractional list chromatic number** of a graph G is defined by

$$\chi_\ell^*(G) = \inf\{\frac{n}{p} \mid \chi_\ell(G,p) \le n\}.$$

This definition is due to Alon, Tuza, and Voigt [60]; they use the term **choice ratio** for χ_ℓ^*. The main results in [60] say that the infimum is indeed a minimum and that every graph G satisfies $\chi_\ell^*(G) = \chi^*(G)$. The DP-chromatic number of weighted graphs was defined in Section 1.12; and the **fractional DP-chromatic number** of a graph is defined by

$$\chi_{\mathrm{DP}}^*(G) = \inf\{\frac{n}{p} \mid \chi_{\mathrm{DP}}(G,p) \le n\}.$$

This definition is due to Bernshteyn, Kostochka, and Zhu [111] (see also Kaul and Mudrock [558]). As proved in [111], unlike the fractional chromatic number and fractional list chromatic number, the infimum in the definition of the fractional DP-chromatic number is not always a minimum. A remarkable result in [111] says that every graph G with maximum average degree $d \ge 4$, satisfies $\chi_{\mathrm{DP}}^*(G) \ge d/(2\ln d)$. Consequently, the fractional DP-chromatic number of bipartite graphs can be arbitrarily large; on the other hand any bipartite graph G with $E(G) \ne \varnothing$ satisfies $\chi(G) = \chi^*(G) = \chi_\ell^*(G) = 2$. As proved in [558], for odd cycles all three fractional parameters χ^*, χ_ℓ^* and χ_{DP}^* coincide.

The circular chromatic number χ° of a graph was introduced in a short paper by Vince [1046] in 1988, using the name **star chromatic number**. That the name for χ° was later changed into **circular chromatic number** is not only justified by the equivalent definition of χ° in terms of homomorphisms into circular target graphs $CG(n,d)$, but also by the characterization of the parameter obtained by Zhu [1092] and [1094] (see Proposition 8.16). Section 8.3 discusses only some basic facts about the circular chromatic number. The proofs of Propositions 8.17, 8.18, 8.19, 8.20 and 8.21 are from the original paper by Bondy and Hell [142] (see also [452] and [495]). The main conclusion of these propositions is the fact that the infimum in the definition of the circular chromatic number is always attained and, hence, a rational number. A result equivalent to this was first proved by Vince [1046], using a different language and methods of continuous mathematics. Another important conclusion, as shown in [1046], says that $\chi - 1 < \chi^\circ \le \chi + 1$. Thus, the circular chromatic number is both a fractional coloring parameter and a refinement of the chromatic number. This is certainly one of the reasons why the circular chromatic number has attracted so much interest since its introduction by Vince. The relation between the circular chromatic number and the chromatic number of a graph has been one of the main subjects of research in this topic. Xuding Zhu published two review articles [1094] and [1097] on the circular chromatic number in 2001 and 2006. In his first survey he proposed 28 problems related to the circular chromatic number some of which have since been solved.

By a **graph construction** we usually mean an algorithm C whose input is a sequence of p graphs, say $(G_1, G_2, \ldots, G_p)$, and whose output is another graph $G = \mathrm{C}(G_1, G_2, \ldots, G_p)$. Very often we require that if $(G_1, G_2, \ldots, G_p)$ and $(G_1', G_2', \ldots, G_p')$ satisfies that $G_i \cong G_i'$ for $i \in [1,p]$, then $\mathrm{C}(G_1, G_2, \ldots, G_p) \cong \mathrm{C}(G_1', G_2', \ldots, G_p')$. In this case we say that C is a **p-ary abstract graph construc-**

tion. If C is a binary graph construction, we write $G_1 \mathrm{C} G_2$ for $\mathrm{C}(G_1, G_2)$. In general, we restrict our attention to those constructions that can be executed in polynomial time. Graph constructions are an important tool in graph theory, and they are a dime a dozen.

The Mycielski construction M_2 is a unary abstract graph construction introduced by Mycielski [766] in 1955 in order to construct triangle-free graphs of arbitrarily high chromatic number. The construction preserves the property of being triangle-free and increases the chromatic number by one; that it also preserve criticality was proved by Schäuble [893]. Hence, for any $k \in \mathbb{N}$, $\mathrm{M}_2^k(K_2) \in \mathrm{Crit}(k+2)$ is a triangle-free graph of order $n = 3 \cdot 2^k - 1$. That $\mathrm{M}_2^2(K_2) = \mathrm{M}_2(C_5)$ is the unique triangle-free 4-chromatic graph with at most eleven vertices was proved by Chvátal [247]. The generalized Mycielski construction $\mathrm{M}_r : \mathcal{G} \to \mathcal{G}$ with $r \in \mathbb{N}$ was introduced by Stiebitz [966] in 1985. The graph $\mathrm{M}_r(G)$ is refereed to as the r-**Mycielskian** of G, and it is also known as the **cone** over G (see Tardif [987]). Recall that P_n^o denotes the path on n vertices with a loop at one end; let $v(P_n^o)$ be the only vertex of degree one in P_n^o. Then $\mathrm{M}_r(G)$ is obtained from $G \times P_{r+1}^o$ by identifying the independent set $I = V(G) \times \{v(P_{r+1}^o)\}$ to a single vertex. While the chromatic number is always increased by one when the construction M_1 or M_2 is applied, this need not be the case when the construction M_r with $r \geq 3$ is applied. For $r \geq 3$, we only obtain that every graph G satisfies $\chi(G) \leq \chi(\mathrm{M}_r(G)) \leq \chi(G)+1$. This leads to a classification problem similar to the classification problem for graphs with respect to the chromatic index. However we do not know any result about the complexity of this classification problem. The most interesting question, however, is to find a characterization of those graphs for which any n-fold application of the generalized Mycielski construction increases the chromatic number by n. So let $\mathcal{PM}$ denote the class of graphs G such that

$$\chi(\mathrm{M}_{r_1,r_2,\ldots,r_n}(G)) = \chi(G)+n$$

for all $n \in \mathbb{N}$ and all $r_1, r_2, \ldots, r_n \in \mathbb{N}$. The main result in [966] shows that $K_2 \in \mathcal{PM}$. The proof of this result is the third application of Lovász's topological lower bound, see Theorem 8.32. This result is an immediate consequence of Theorem 8.34 saying that every graph G satisfies $\mathbf{T}_N(\mathrm{M}_r(G)) \simeq \mathrm{susp}(\mathbf{T}_N(G))$ for all $r \in \mathbb{N}$. Our proof of Theorem 8.34 is from [966]. The main point is that the neighborhood complex $\mathsf{N}(\mathrm{M}_r(G))$ can be obtained from the neighborhood complex $\mathsf{N}(G)$ by a Mycielski-type construction for simplicial complexes that is a combination of known constructions for complexes. As a consequence, we obtain that every graph $G \in \mathcal{MF}(n)$ satisfies $\mathbf{T}_N(G) \simeq \mathbf{S}^n$ and hence $\chi(G) = n+2$. In 2005, Csorba [266] (see also [267]) generalized Theorem 8.34 by showing that every graph G satisfies $\mathbf{T}_H(\mathrm{M}_r(G)) \overset{2}{\simeq} \mathrm{susp}(\mathbf{T}_H(G))$; this also shows that every graph $G \in \mathcal{MF}(n)$ satisfies $\chi(G) = n+2$. The results about the class $\mathcal{MF}(n)$ were first published in a survey paper by Sachs and Stiebitz [887] in 1989; however, both Theorems 8.32 and 8.34 were stated without proofs; the proofs were first published in a paper by Gyárfás, Jensen, and Stiebitz [445] in 2004. In [887] the main aim was to show that $\mathcal{MF}(n)$ is one of the many families of graphs with fixed chromatic number $n+2$, but with arbitrarily large odd girth; see Theorem 8.35. So define

$\text{odd}(k,n) = \max\{g_o(G) \mid \chi(G) = k, |G| = n\}$. For $k \geq 4$ fixed, there are constants c_k and c'_k such that

$$c_k n^{1/(k-2)} \leq \text{odd}(k,n) \leq c'_k n^{1/(k-2)} + 1,$$

where the upper bound holds for all $n \geq k$ and the lower bound holds for infinitely many value of n. The generalized Mycielski graphs seems to provide the best values for the constant c_k (see Theorem 8.35). For example, the Mycielski graphs give the value $c_4 = \sqrt{2}$ while the Shrijver graphs only gives the value $c_4 = 1$. The best known value for c'_k for general k was established by Kierstead, Szemerédi, and Trotter [585]; our proof of Theorem 8.37 follows their original one from [585]. The value c'_4, however, was first improved by Nilli [787] to $c'_4 = \sqrt{8}$ and then by Jiang [532] to $c'_4 = 2$. Our proof of Theorem 8.38 is from [532]. The following conjecture might be too optimistic (see also Ngoc and Tuza [783]); however the case $r = 1$ holds trivially and the case $r = 2$ was settled by Chvátal [247].

Conjecture 8.83 (Ngoc and Tuza) *For any $r \in \mathbb{N}$, if a graph $G \in \text{Crit}(4,n)$ satisfies $g_o(G) = 2r+1$, then $n \geq 2r^2+r+1$ and equality holds if and only if $G = M_r(C_{2r+1})$.*

As the r-Mycielskian of K_2 is an odd cycle of order $2r+1$, every graph in $\mathcal{MG}(3)$ is 3-critical. However, proving that every r-Mycielskian of an odd cycle is not 3-colorable and hence a 4-critical graph is much more challenging. Our proof first shows that the N-space of $G_{r,s} = \text{M}_r(C_{2s+1})$ is homotopic equivalent to the sphere $\mathbf{S}^2$ and then uses Lovász' topological bound.

Payan [800] investigated the chromatic number of cube-like graphs, i.e. binary Cayley graphs. Given an group $(\Gamma,+)$ and a subset $\Omega \subset \Gamma \setminus \{0\}$, the **Cayley graph** $CG(\Gamma,\Omega)$ is the graph with vertex set $V(CG(\Gamma,\Omega)) = \Gamma$ and edge set $E(CG(\Gamma,\Omega)) = \{\{a,b\} \mid a-b \in \Omega\}$. If $x+x = 0$ for all $x \in \Gamma$, we say that $CG(\Gamma,\Omega)$ is a **binary Cayley graph**. Payan proved that if G is a binary Cayley graph with $g_o(G) = 2k+1$, then $G_{k,k} = \text{M}_k(C_{2k+1})$ is a subgraph of G; a strengthening of this result was obtained by Beaudou, Naserasr, and Tardif [86]. Furthermore, Payan gave a combinatorial proof showing that $\chi(G_{r,s}) = 4$. A similar proof was given by Ngoc and Tuza [783]. A third proof was given by Tardif [987] using a result of El-Zahar and Sauer [325] (see Exercise 8.25). A fourth proof is due to Youngs [1085]; he observed that $G_{r,s}$ is isomorphic to a quadrangulation of the projective plane and so $G_{r,s}$ is 4-critical because of his Theorem 3.48. Figure 8.9 shows two graphs satisfying the hypothesis of Youngs' theorem. The graph on the left side in Figure 8.9 is the Mycielski graph $\text{M}_2(K_3)$. The graph on the right side is up to isomorphism the unique 4-critical subgraph of the triangle free 4-regular 4-chromatic graph of order $n = 12$ reported by Chvátal [245], contained in it twice. Youngs' theorem was generalized by Kaiser and Stehlík [547] in 2015.

Theorem 8.84 (Kaiser and Stehlík) *If G is a nonbipartite quadrangulation of the n-dimensional projective space, then $\chi(G) \geq n+2$.*

For $n \in \mathbb{N}_0$, let $\mathcal{PS}(n)$ denote the class of nonbipartite graphs that are isomorphic to a quadrangulation of the n-dimensional projective space. Then $\mathcal{PS}(0) = \varnothing$ and $\mathcal{PS}(1)$ is the class of odd cycles. The proof of Theorem 8.84 is based on the

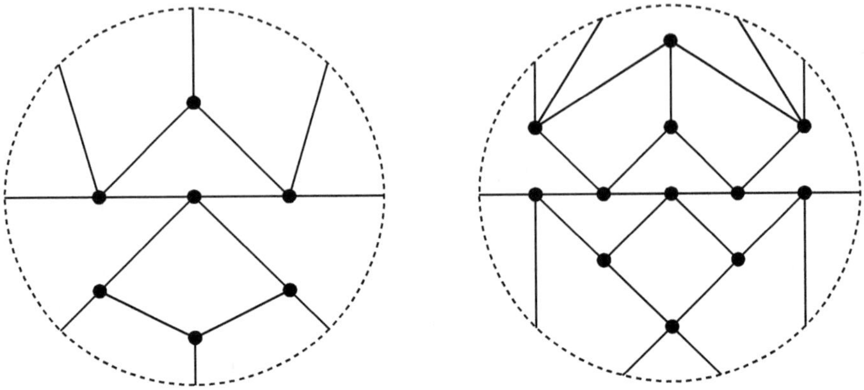

Fig. 8.9 Two 4-critical projective plane graphs

topological bounds listed in (8.27) by showing that every graph $G \in \mathcal{PS}(n)$ satisfies $\mathrm{coind}(\mathbf{T}_B(G)) \geq n$. By (8.24) and Lemma 8.73(a), we have $\mathrm{coind}(\mathbf{T}_B(G)) = \mathrm{coind}(\mathbf{T}_H(G))$. Hence, for $n \in \mathbb{N}$, $\mathcal{PS}(n) \subseteq \mathcal{CT}(n)$, implying that $\chi(G) \geq n+2$ for all $G \in \mathcal{PS}(n)$. Youngs' theorem says that $\chi(G) = 4$ for every graph $G \in \mathcal{PS}(2)$. Contrary to that, Kaiser and Stehlík have shown that for fixed $n \geq 3$ the chromatic number of graphs in $\mathcal{PS}(n)$ is unbounded. Furthermore, Kaiser and Stehlík have shown that $\mathcal{MF}(n) \subseteq \mathcal{PS}(n)$, thereby giving an alternative proof showing that $K_2 \in \mathcal{PM}$. Since the family of Shrijver graphs $\mathcal{SF}(n)$ is hom-bounded from below by $\mathcal{MF}(n)$ as proved in [547], this also yield a new proof of the Lovász–Kneser theorem. The Shrijver graph $SG(n+2p, p)$ is a vertex critical subgraph of the Kneser graph $KG(n+2p, p)$ having the same chromatic number $n+2$. In general, it is not known whether $KG(n+2p, p)$ contains a $(n+2)$-chromatic subgraph whose order is smaller than the order of the corresponding Shrijver graph. On the other hand, $SG(n+2p, p)$ is not critical, unless $n \in \{0,1\}$ or $p = 1$. Critical subgraphs of Shrijver graphs with the same chromatic number were exhibited by Kaiser and Stehlík in [549] and [550].

The family $\mathcal{PM}$ has the following two properties. If $G \in \mathcal{PM}$, then $\mathcal{PM}$ contains every graph $\mathrm{M}_r(G)$ for $r \in \mathbb{N}$ (by definition of $\mathcal{PM}$) and every graph G' satisfying $G \to G'$ and $\chi(G) = \chi(G')$ (by (8.21)). By Corollary 8.78, this implies that every graph $G \in \mathcal{CT}(n)$ with $\chi(G) = n+2$ belongs to $\mathcal{PM}$. From Corollary 8.75 we then obtain that $\mathcal{PM}$ contains most of the classes introduced in this chapter. The results in Section 8.7 also show that the two families $\mathcal{LF}(n)$ and $\mathcal{MF}(n)$ with $n \in \mathbb{N}_0$ have a special place in the homomorphism hierarchy of the graph classes discussed in this chapter. In particular, we obtain the following result.

Lemma 8.85 *For any integer $n \in \mathbb{N}_0$, the following statements are equivalent:*

(a) *There is no $\mathbb{Z}_2$-map from $\mathbf{S}^n$ to $\mathbf{S}^{n-1}$.*
(b) *If $\mathbf{S}^n$ is the union of $n+1$ subsets each of which is either open or closed, then at least one subset contains a pair of antipodal points.*
(c) *Every graph $G \in \mathcal{BF}(n)$ satisfies $\chi(G) \geq n+2$.*

(d) *Every graph* $G \in \mathcal{LF}(n)$ *satisfies* $\chi(G) \geq n+2$.
(e) *Every graph* $G \in \mathcal{MF}(n)$ *satisfies* $\chi(G) \geq n+2$.

Statement (a) is another very popular version of the Borsuk–Ulam theorem that was already mentioned by Borsuk in 1933. That (b) is equivalent to the Borsuk–Ulam theorem was proved by Greene [430]. That (a) and (c) are equivalent was observed by Erdős and Hajnál. That (c), (d), and (e) are equivalent follows from Corollary 8.77. That Theorem 8.32 is equivalent to the Borsuk–Ulam theorem was first proved by Kaiser and Stehlík [548]; they showed that (a) and (e) are equivalent. A key observation in [548] is that for every $n \in \mathbb{N}_0$ and every $\epsilon > 0$, there exists a graph $G \in \mathcal{MF}(n)$ such that $G \subseteq BG(n, \sqrt{4-\epsilon^2})$. Statement (b) was used by Greene to give a book proof of the Lovász–Kneser theorem. As pointed out by Simonyi and Tardos [941, Proposition 9], this proof can be easily extended to show that every graph $G \in \mathcal{KF}(n)$ satisfies $\mathbf{S}^n \xrightarrow{2} \mathbf{T}_H(G)$ and hence $G \in \mathcal{CT}(n)$.

Generalized Mycielski graphs have the property that the deletion of relatively few edges leads to bipartite graphs and hence to large independent sets. This observation is related to Theorem 8.31; our proof of this theorem goes back to Rödl and Tuza [875]. Năstase, Rödl, and Siggers [770] proved that for any integers $k \geq 3$ and $\ell \geq 5$ there exists a positive constant $c = c(k,\ell)$ such that for all n' there exists a graph $G \in \mathrm{Crit}(k+1,n)$ with $n > n'$ and $g_o(G) \geq \ell$, which can be made $(k-1)$-colorable by the deletion of at least cn^2 edges, only.

Immediately after Erdős was introduced to the concept of critical graphs by Dirac in the late 1940s, he proposed to study the maximum number of edges in k-critical graphs of order n, see Erdős [338] and [339]. So define the density function $\mathrm{mde}(\cdot,\cdot)$ by

$$\mathrm{mde}(k,n) = \max\{e(G) \mid G \in \mathrm{Crit}(k,n)\}.$$

Dirac observed that $G = C_{2p+1} \boxplus C_{2p+2} \in \mathrm{Crit}(6,n)$ with $n = 4p+2$ and $e(G) = \frac{1}{4}n^2 + n$; and there is, for every $k \geq 6$, a constant c_k such that $\mathrm{mde}(k,n) > c_k n^2$ for infinitely many values of n. In particular, we have $c_6 = \frac{1}{4}$. That such a constant also exists for $k \in \{4,5\}$, was proved by Toft [1020] with $c_4 = \frac{1}{16}$ and $c_5 = \frac{4}{31}$. Since then, no progress has been made. To this day it is not known whether the constants $1/16$, $4/31$ and $1/4$ are best possible. The first nontrivial upper bound for $\mathrm{mde}(k,n)$ using the Turan number was obtained by Stiebitz [967, Theorem 2.2]; he proved that if $k \geq 4$ and n is large enough, then $\mathrm{mde}(k,n) \leq t(n,k-2)$. More than 35 years passed before Luo, Ma, and Yang [707] improved this bound to $\mathrm{mde}(k,n) \leq t(n,k-2) - c_k n^2$. Surprisingly, at least for $k \geq 6$, the situation becomes simpler if we additionally require constraints on the odd girth. So let

$$\mathrm{mde}(k,n,\ell) = \max\{e(G) \mid G \in \mathrm{Crit}(k,n) \text{ and } g_o(G) > \ell\},$$

where $\ell \geq 3$ is odd. Then, for $k \geq 4$ and $\ell \geq 3$ odd, there is a "best" constant $oc(k,\ell)$ such that $\mathrm{mde}(k,n,\ell) \geq oc(k,\ell)n^2$ holds for infinitely many values of n, that is, $oc(k,\ell)$ is the supremum over all constants c such that $\mathrm{mde}(k,n,\ell) \geq cn^2$ holds for infinitely many values of n. Let us first consider the case $\ell = 3$, that is, the class of triangle-free graphs. The Toft graph TG_p with $p \geq 5$ odd is triangle-free and no better construction seems to be known, implying that $oc(4,3) \geq \frac{1}{16}$. For $k = 5$ Gyárfás (see Pedgen [802]) extended Toft's construction to obtain a triangle-free graph belonging to Crit(5). First we take four disjoint copies T_1, T_2, T_3 and T_4 of the Toft graph TG_p (see Section 4.5), then we take four disjoint independent sets I_1, I_2, I_3 and I_4 each of cardinality $4p$. Now we add a perfect matching between $V(T_i)$ and I_i for $i \in [1,4]$, then we join I_i and I_{i+1} by all possible edges, for $i \in [1,3]$, and, finally, we join an additional vertex to all vertices of $I_1 \cup I_4$. The resulting graph G belongs to Crit(5) and is triangle-free. If $p = 5$, then G has order $n = 161$ and size $e(G) = (\frac{13}{256} + o(1))n^2$. So Gyárfás' graphs show that $oc(5,3) \geq 13/256$. Pedgen improved this bound to $oc(5,3) \geq 4/31$; this is the same density shown by Toft [1020] for 5-critical graphs without any girth requirement. Pedgen extended Toft's construction for $k \geq 6$ in order to show that $oc(k,3) \geq \frac{1}{4}$ which implies that $oc(k,3) = \frac{1}{4}$ by Turan's theorem. To handle the case $\ell \geq 5$, Pedgen [802] used the result about generalized Mycielski graphs, in particular the two Theorems 8.32 and 8.35. He then proved that, for $\ell \geq 5$ odd, we have $oc(4,\ell) \geq 1/(\ell+1)^2$, $oc(5,\ell) \geq 1/2(\ell+1)$, and $oc(k,\ell) \geq 1/4$ for all $k \geq 6$. By a famous result of Erdős and Stone [358], we then obtain that $oc(k,\ell) = 1/4$ for all $k \geq 6$. Pedgen asked whether it is true that $oc(4,\ell)$ tends to zero if ℓ tends to infinity.

For a graph G, let

$$\mathrm{ipn}(G) = \max\{|H| - 2\alpha(H) + 2 \mid H \text{ is an induced subgraph of } G\}.$$

Note that $\mathrm{ipn}(G) \geq 1$, and if $\mathrm{ipn}(G) \leq 2$ then G is bipartite. A deep result obtained by Folkman [381], conjectured by Erdős and Hajnal at the graph theory conference in Tihany (Hungary) in 1966, says that every graph G satisfies $\chi(G) \leq \mathrm{ipn}(G)$. The proof by Folkman from 1970 is long and difficult; a slightly shorter proof was given by Bonamy, Charbit, Defrain, Joret, Lagoutte, and Limouzy [136] in 2020.

The graph class $OC(k,\ell)$ from Section 8.5 was introduced by Gyárfás, Jensen, and Stiebitz [445]. Gyárfás observed that the graph $GY(k) = GY(k,1)$ is homomorphisms universal with respect to the graph property $OC(k,1)$. He proved from scratch that $GY(4)$ has chromatic number 4, Jensen observed that $GY(4)$ isomorphic to a quadrangulation of the projective plane (see Figure 3 in [445]) and hence 4-chromatic by Youngs' theorem, and Stiebitz observed that $\mathrm{M}_4(C_9) \to GY(4)$ which implies $\chi(GY(4)) = 4$ by Theorem 8.32. This theorem was the right key to show that $OC(k,\ell)$ contains a k-chromatic graph whenever $k \geq 1$ and $\ell \geq 0$, see Theorem 8.42; our proof of this theorem is from [445]. A homomorphism universal graph G_k^ℓ with respect to the class $OC(k,\ell)$, proposed by one of the referees of the paper [445], was presented by the authors, but without giving a proof. The graph G_k^ℓ, however, is not minimal in $OC(k,\ell)$, so it is not a core. The remaining results of Section 8.5 and its proofs are from the 2005 paper by Baum and Stiebitz [85] (see also [84]). However, this paper was published only as a preprint at the University of Odense, but the preprint server was later closed. The reason why the paper [85] was never submitted to a journal is due to the fact that the main results in [85] were obtained at the same time independently by Simonyi and Tardos [941]. They investigated a slightly different, but equivalent concept. An s-**wide coloring** of a graph G with t colors (for short, an (s,t)-**wide coloring**) is a coloring of G with color set $[1,t]$ such that the two ends of any walk of length $2s-1$ receive different colors. From Theorem 8.41 it then follows that any (s,t)-wide coloring of G is an odd $(t,s-1)$-coloring of G, and vice versa. Simonyi and Tardos introduce a graph $W(s,t)$ such that a graph G admits an (s,t)-wide coloring if and only if $G \to W(s,t)$. Then they show that $W(s,t)$ is a t-critical graph. For the proof that $\chi(W(s,t)) \geq t$, they first show that $W(s,t)$ is the homomorphic image of a t-chromatic Shrijver graph. It is easy to show that the two graphs $W(s,t)$ and $GY(t,s-1)$ are isomorph.

However, the main topic of the paper by Simonyi and Tardos [941] is not the wide coloring concept, but the local coloring concept and a conjecture by Johnson, Holroyd, and Stahl [536], saying that Kneser graphs have equal circular and ordinary chromatic number. The local chromatic number of a graph was introduced in [344] as the minimum number of colors that must appear in the closed neighborhood of a vertex no matter how many colors are used. Let G be a nonempty graph, and let m,r be integers such that $1 \leq r \leq m$. A **local** (m,r)**-coloring** of G is a coloring φ of G with a set of m colors such that $|\varphi(N_G[v])| \leq r$ for every vertex $v \in V(G)$. Recall that $N_G[v] = N_G(v) \cup \{v\}$ is the closed neighborhood of v in G. Note that if φ is an optimal coloring of G, that is, a coloring of G with $\chi(G)$ colors, then $|\varphi(N_G[v])| = \chi(G)$ for some vertex v of G (see Proposition 1.34(a)). Hence if G admits a local (m,r)-coloring, then $m > \chi(G)$, unless $r = m$. In particular, G has a local (m,m)-coloring with $m = \chi(G)$. In [344] a graph $GU(m,r)$ is defined by

$$\begin{aligned} V(GU(m,r)) &= \{(i,A) \mid i \in [1,m], A \subseteq [1,m], |A| = r-1, i \notin A\} \\ E(GU(m,r)) &= \{(i,A)(j,B) \mid i \in B, j \in A\}. \end{aligned}$$

Then it is easy to check that the map $(i,A) \mapsto i$ is a local (m,r)-coloring of $GU(m,r)$. As proved in [611, Theorem 3], for every graph G we obtain that

$$G \text{ admits a local } (m,r)\text{-coloring if and only if } G \to GU(m,r).$$

The **local chromatic number** of G is defined by

$$\Psi(G) = \min\{r \mid G \to GU(m,r), r \leq m \leq |G|\};$$

for the empty graph we set $\Psi(G) = 0$. Then every graph G satisfies $\Psi(G) \leq \chi(G)$. On the other hand, it was shown by Körner, Pilotto, and Simonyi [611, Theorem 3] that $\Psi(G) \geq \chi^*(G)$. Hence, every graph G satisfies

$$\chi^*(G) \leq \Psi(G) \leq \chi(G).$$

Furthermore, as proved in [611], we have $\chi^*(GU(m,r)) = \Psi(GU(m,r)) = r$ when $m \geq r \geq 2$. For a bipartite graph G with at least one edge we have $\chi^*(G) = \Psi(G) = \chi(G) = 2$. On the other hand, as proved in [344], the class of graphs G with $\Psi(G) = 3$ has an unbounded chromatic number. As mentioned in [941], it has been observed several times that $\Psi(KG(n,p)) \geq n-3p+3$ holds, since the neighborhood of any vertex of $KG(n,p)$ induces a $KG(n-p,p)$. On the other hand, we have $\chi^*(KG(n,p)) = n/p$. Hence $\chi^*(KG(4p,p)) = 4$ and $\Psi(KG(4p,p)) \geq p+3$; this shows that the gap between the fractional chromatic number and the local chromatic number can be arbitrarily large.

Simonyi and Tardos [941] used the topological lower bounds listed in (8.27) to establish a general lower bound for the local chromatic number. By (8.27), every nonempty graph G satisfies

$$\chi(G) \geq \text{coind}(\mathbf{T}_B^o(G))+1 \geq \text{coind}(\mathbf{T}_H(G))+2.$$

Using a result of Fan [363] from 1952, generalizing the Borsuk–Ulam theorem, Simonyi and Tardos proved the following result which became known as the zig-zag theorem.

Theorem 8.86 (THE ZIG–ZAG THEOREM) *Let G be a graph such that* $\text{coind}(\mathbf{T}_B^o(G)) \geq n-1$ *for an integer $n \geq 2$, and let φ be a coloring of G with an arbitrary number of colors that are linearly ordered. Then G contains a complete bipartite graph $K_{s,t}$ with $(s,t) = (\lceil \frac{n}{2} \rceil, \lfloor \frac{n}{2} \rfloor)$ such that φ assigns distinct colors to all n vertices of this subgraph and these colors appear alternating on the two parts of the bipartite subgraph with respect to their order.*

Recall from (8.27) that every graph $G \in \mathcal{CT}(n-2)$ with $n \geq 2$ satisfies

$$\chi(G) \geq \text{coind}(\mathbf{T}_B^o(G))+1 \geq \text{coind}(\mathbf{T}_H(G))+2 \geq n.$$

The zig-zag theorem can be used to establish a lower bound for both Ψ and χ°.

Theorem 8.87 (SIMONYI AND TARDOS) *Let G be a graph and let n be an integer with $n \geq 2$. Then the following statements hold:*

(a) *If* $\text{coind}(\mathbf{T}_B^o(G)) \geq n-1$, *then* $\Psi(G) \geq \lceil \frac{n}{2} \rceil + 1$.
(b) *If* $\text{coind}(\mathbf{T}_B^o(G)) = n-1$ *and n is even, then* $\chi^\circ(G) \geq n$.
(c) *If $G \in \mathcal{OC}(n,3)$, then* $\Psi(G) \leq \lfloor \frac{n}{2} \rfloor + 2$.

Sketch of Proof: Statement (a) is an immediate consequence of the zig-zag theorem, since $\frac{n}{2} \geq 1$. For the proof of (b) assume that $n = \text{coind}(\mathbf{T}_B^o(G))+1$ is an even number and let φ be a circular (m,d)-coloring of G. Since n is even, the zig-zag theorem implies that G contains a cycle $C = (v_1, v_2, \ldots, v_n, v_1)$ such that $\varphi(v_1) < \varphi(v_2) < \cdots < \varphi(v_n)$. Then we have $\varphi(v_{i+1}) - \varphi(v_i) \geq d$ for $i \in [1, n-1]$ which leads to $\varphi(v_n) - \varphi(v_1) \geq (n-1)d$. On the other hand we have $\varphi(v_n) - \varphi(v_1) \leq m-d$. The last two inequalities give $m/d \geq n$ and hence $\chi^\circ(G) \geq n$ as claimed. For the proof of (c) assume that $G \in \mathcal{OC}(n,3)$. Then there exists an odd $(n,3)$-coloring φ of G. A vertex v of G is called *bad* if $|\varphi(N_G(v))| > n/2$. Let c be a color not used by φ, and let $\phi : V(G) \to \text{im}(\varphi) \cup \{c\}$ be the map defined by $\phi(v) = c$ if v has a bad neighbor and $\phi(v) = \varphi(v)$ otherwise. Since φ is an odd $(n,3)$-coloring of G, it follows that $\phi^{-1}(c)$ is an independent set of G and so ϕ is a coloring of G. If v is a bad vertex, then $\phi(N_G(v)) = \{c\}$ otherwise we have $|\phi(N_G(v))| \le |\varphi(N_G(v))| + 1 \le n/2 + 1$. Consequently, $\Psi(G) \le n/2 + 2$ as claimed. □

Corollary 8.88 (Simonyi and Tardos) *Let $G \in \mathcal{CT}(n-2)$ be a graph with $n \ge 2$ such that $\chi(G) = n$. Then the following statements hold:*

(a) *If n is even, then $\chi^\circ(G) = \chi(G) = n$.*
(b) *If n is odd and $G \in OC(n,3)$, then $\Psi(G) = \lfloor \frac{n}{2} \rfloor + 2$*

Combining Corollary 8.75 and Corollary 8.88, we obtain that every graph $G \in \mathcal{KF}(n) \cup \mathcal{SF}(n) \cup \mathcal{MF}(n) \cup \mathcal{LF}(n) \cup \mathcal{GF}(n)$ with $n \ge 0$ satisfies (a) $\chi^\circ(G) = \chi(G)$, provided that n is even, and (b) $\Psi(G) = \lfloor \chi(G)/2 \rfloor + 1$, provided that n is odd and $G \in OC(n+2,3)$. From the proof of Theorem 8.42 we obtain the following results: $K_2 \in OC(2,\ell) \cap \mathcal{MG}(2)$ for all $\ell \ge 0$, and if $G \in OC(k,\ell) \cap \mathcal{MG}(k)$, then $\mathrm{M}_{3\ell+1}(G) \in OC(k+1,\ell) \cap \mathcal{MG}(k+1)$. Hence, the generalized Mycielski graph $\mathrm{M}_{r_1,r_2,\dots,r_k}(K_2)$ with $r_i \ge 7$ for all $i \in [1,k]$ belongs to $OC(k+2,3) \cap \mathcal{MG}(k+2)$ and satisfies $\Psi = \lfloor \chi/2 \rfloor + 1$ if k is odd. The local chromatic number of the Mycielski graphs in $\mathcal{MG}(k)$ for even k was studied by Simonyi, Tardos, and Vrećica [944] in 2009. In 2001, Tardif [987] proved that every graph G with at least one edge satisfies

$$\chi^*(\mathrm{M}_r(G)) = \chi^*(G) + \frac{1}{\sum_{k=0}^{r-1} (\chi^*(G) - 1)^k};$$

the formula for $r = 2$ was obtained in 1995 by by Larsen, Propp, and Ullman [666]. In the paper by Lin, Wu, Lam, and Gu [678], the behaviour of various graph parameters with respect to the Mycielski construction M_r is studied; the paper by Collins, Heenehan, and McDonald [253] deals with the immersion number of r-Mycielskians of graphs.

The results of Simonyi and Tardos [941] imply that the circular chromatic number of the Kneser graph $KG(n,p)$ is equal to its chromatic number, unless n is odd. That this holds for all n was first proved by Chen [211], thereby solving the Johnson–Holroyd–Stahl conjecture. A shorter proof of this result was given by Chang, Liu, and Zhu [198]. Booth proofs use Fan's lemma from [363]; a further simplified proof was obtained by Liu and Zhu [682]. Alishahi, Hajiabolhassan, and Meunier [41] established an extension of the zig-zag theorem and used it to show that if H is a nonempty hypergraph such that $\chi(KG(H)) = \mathrm{cd}(H)$, then $\chi(KG(H)) = \chi^\circ(KG(H))$ (see Exercise 8.15).

Graph products are very special abstract binary graph constructions that occur naturally in graph theory as important tools in combinatorial constructions. The definitive reference on product graphs is the textbook by Imrich and Klavžar [512] from 2000. Following a proposal of Nešetril and Rödl [776], a **graph product** is an abstract binary graph construction $*$ such that the vertex set of the product $G * H$ is the Cartesian product $V(G) \times V(H)$ of the vertex set of the two factors G and H, and adjacency in the product $G * H$ depends on the adjacency properties of the projections of pairs of vertices into the factors. This leads to 256 different graph products. The most commonly used graph products are the Cartesian product $G\square H$, the lexicographic product $G[H]$, the direct product $G \times H$ (see Appendix C.7), and the **strong product** $G \boxtimes H = G\square H \cup G \times H$. As discussed in [512, Appendix C], these four products are the most relevant graph products. The four graph products occur in the graph theory literature under various names. The lexicographic product of G and H is often called the **composition**, and instead of $G[H]$ also $G \circ H$ is used. For the product $G \times H$ introduced in Section 8.6 many names are used in the graph theory literature, including **direct product**, **categorial product**, **tensor product**, **Kronecker product**, **weak product**, and possibly others. In linear algebra, especially for matrices, the term tensor product or Kronecker product is used synonymously. The adjacency matrix of the direct product of graphs is the tensor product of the adjacency matrices of its two constituents.

Given a graph product $*$ and a graph parameter $\rho : \mathcal{G} \to \mathbb{N}_0$, we say that $*$ is **ρ-conform** if there is a function $f : \mathbb{N}_0^2 \to \mathbb{N}_0$ such that for every graphs $G, H \in \mathcal{G}$ we have

$$\rho(G * H) = f(\rho(G), \rho(H)).$$

If f is such a function for the chromatic number, then either $f(p,q) = 1$, $f(p,q) = p$, $f(p,q) = q$, or $f(p,q) = \max\{p,q\}$, or $*$ is the direct product and $f(p,q) = \min\{p,q\}$. This result is due to Puš [830]. It is not difficult to show that for any graphs G and H, we have

$$\chi(G\square H) = \max\{\chi(G), \chi(H)\},$$

and

$$\chi(G \boxtimes H) \leq \chi(G[H]) \leq \chi(G)\chi(H).$$

So the Cartesian product $\square$ is χ-conform. For the composition we have $\chi(G[H]) = \chi(G[K_n])$ if $\chi(H) = n$ and $\chi(C_{2n+1}[K_p]) = 2p + \lceil \frac{p}{n} \rceil$. (see Section 1.9). This shows that the composition is not χ-conform.

Hedetniemi's product conjecture claimed the direct product to be χ-conform with the function $f(p,q) = \min\{p,q\}$. A coloring of the direct product of two graphs can be obtained by using a coloring of one of its two constituents. This follows from the fact that if $\phi \in \mathrm{Hom}(G, K_n)$ is a coloring of G with n colors, then $\phi \circ p_G \in \mathrm{Hom}(G \times H, K_n)$ is a coloring of $G \times H$ with n colors. However, Shitov [938] showed, in a mere three page paper, that there are better ways to color certain direct products of graphs than many mathematicians had thought possible. Thus, Hedetniemi's conjecture turned out to be false in general.

The map graph K_n^G was introduced by El-Zahar and Sauer [325] and extended to K^G by Häggkvist, Hell, Miller, and Neumann-Lara [450]. Theorem 8.61 lists some basic properties of the map graph, the proof of this result is straightforward and influenced by Godsil and Royle's treatment in [418, Section 6.4]. Theorem 8.62 and its proof is from [325]. Theorem 8.68 is the strongest positive result concerning Hedetniemi's conjecture; our proof (including Lemmas 8.66 and 8.67) is essentially the original proof from El-Zahar and Sauer with some re-arrangements. While K_n is multiplicative if $n \leq 3$, this is not the case when $n \geq 5$. This is an direct consequence of Theorem 8.71; the proof of this result is essentially from Wrochna's paper [1076]. However, our proof is somewhat longer than Wrochna's original proof. This is due to the fact that Wrochna uses the wide coloring concept while we use the odd coloring concept. In the former case, the first two claims in our proof do not need to be proved. That K_4 is not multiplicative was proved by Tardif [989]. There are also some positive results related to the product conjecture. In 1976, Burr, Erdős, and Lovász [185] proved that if $\chi(G \times H) = n+1$ if $\chi(G) = \chi(H) = n+1$ and each vertex of G belongs to an n-clique. Häggkvist et al. [450] proved that all cycles are multiplicative. The surveys by Zhu [1093] and by Sauer [891] are excellent sources of information on the product conjecture; these articles survey methods and partial results, and discuss problems for graphs and digraphs related to the product conjecture. The **Poljak–Rödl function** pro : $\mathbb{N}_0 \to \mathbb{N}_0$ introduced in [815] is defined by

$$\text{pro}(n) = \min\{\chi(G \times H) \mid \chi(G) = \chi(H) = n\}.$$

Clearly, Hedetniemi's conjecture is equivalent to the statement that $\text{pro}(n) = n$ for all n. So this is false and from Theorem 8.72 it follows that $\text{pro}(n) = n$ for $n \le 4$ and $\text{pro}(n) \le n-1$ for all $n \ge 5$. On the other hand, it is unknown whether the Poljak–Rödl function is bounded by a constant. As proved by Poljak and Rödl, if the function is bounded, then $\text{pro}(n) \le 9$ for all $n \in \mathbb{N}_0$ (see [815], [891] and [1093]). Based on Shitov's counterexamples for the product conjecture, Tardif and Zhu [991] proved the following results:

1. $\lim_{n\to\infty}(n - \text{pro}(n)) = \infty$, and
2. $\limsup_{n\to\infty} \text{pro}(n)/n \le 1/2$, provided that Stahl's conjecture holds.

It was proved by He and Widgerson [486] that there is a constant $\epsilon > 10^{-9}$ such that $\text{pro}(n) \le (1-\epsilon)n$ for all sufficiently large n. Tardif [989] proved that $\text{pro}(n) \le \lceil n/2 \rceil + 3$. Zhu [1100] proved that

$$\chi^*(G \times H) = \min\{\chi^*(G), \chi^*(H)\}$$

for any graphs G and H, hence the direct product is χ^*-conform, and in [1092] he conjectured that the direct product is also χ°-conform.

For the graph category GRA, the following operation $\Lambda_k, \Gamma_k, \Omega_k$, parameterized by an odd integer k, are of particular interest. The first operation, the **graph k-subdivision** $\Lambda_k(G)$ is obtained by replacing every edge by a path on k edges (this is sometimes denoted by $G^{1/k}$). The k**th power** $\Gamma_k(G)$ is the graph on the same vertex set $V(G)$, with two vertices are joined by an edge if they were joined by a walk of length k in G; this is sometimes denoted G^k (note that this is not the same as the classical power joining vertices at distance at most k). Each of these operation Π is a functor in the (thin) category GRA, that is, $G \to H$ implies $\Pi(G) \to \Pi(H)$; moreover, Γ_k is a right adjoin to Λ_k, and Ω_k is a right adjoin to Γ_k. Hence $\Gamma_k(G) \to K$ if and only if $G \to \Omega_k(K)$. Then Theorem 8.41 implies that $\Omega_{2\ell+1}(K_n)$ is a graph having an odd (n,ℓ)-coloring. For more information about the importance of these operations in the study of graph homomorphisms and the product conjecture, we refer to the papers [382], [454], [455], [494] and [1075].

Chapter 9
Coloring Graphs on Surfaces

The coloring of maps or, equivalently, the coloring of graphs on surfaces is one of the most popular topics in graph coloring theory. This chapter focuses on such map coloring results that can be derived from the density results for critical graphs discussed in Chapter 5.

9.1 Map Color Theorems for Arbitrary Surfaces

Recall from the first chapter and (1.8) that the clique number ω, the chromatic number χ, the list chromatic number χ_ℓ, the DP-chromatic number χ_{DP} and the coloring number col satisfy the following sequence of inequalities

$$\omega(G) \leq \chi(G) \leq \chi_\ell(G) \leq \chi_{\text{DP}}(G) \leq \text{col}(G) \tag{9.1}$$

for every graph G. In this chapter we consider graphs embedded on an arbitrary surface $\mathbf{S}$; see Appendix C.13 for the terminology. We denote by $\mathcal{G}(\hookrightarrow \mathbf{S})$ the class of graphs that can be embedded on the surface $\mathbf{S}$. Recall that every surface is homomorphic to precisely one of the following surfaces:

- the orientable surface $\mathbf{S}_g$ homeomorphic to the sphere with g handles ($g \geq 0$), and
- the nonorientable surface $\mathbf{N}_h$ homeomorphic to the sphere with h crosscaps ($h \geq 0$).

The **Euler characteristic** $\varepsilon(\mathbf{S})$ of the surface $\mathbf{S}$ is $2-2g$ if $\mathbf{S} = \mathbf{S}_g$ and $2-h$ if $\mathbf{S} = \mathbf{N}_h$. The derived invariant $\text{eg}(\mathbf{S}) = 2 - \varepsilon(\mathbf{S})$ is the **Euler genus** of $\mathbf{S}$. In 1890 Percy John Heawood published his celebrated paper *Map color theorems*. In this paper he established an upper bound for the number of colors required for maps on higher surfaces. For a surface $\mathbf{S}$, we call

M. Stiebitz et al., *Brooks' Theorem*, Springer Monographs in Mathematics,
https://doi.org/10.1007/978-3-031-50065-7_9

$$H(\mathbf{S}) = \left\lfloor \frac{7+\sqrt{49-24\,\varepsilon(\mathbf{S})}}{2} \right\rfloor = \left\lfloor \frac{7+\sqrt{24\,\mathrm{eg}(\mathbf{S})+1}}{2} \right\rfloor \tag{9.2}$$

the **Heawood number** of $\mathbf{S}$; see the table below.

$\mathbf{S} =$	$\mathbf{S}_0$	$\mathbf{N}_1$	$\mathbf{S}_1, \mathbf{N}_2$	$\mathbf{N}_3$	$\mathbf{S}_2, \mathbf{N}_4$	$\mathbf{N}_5$	$\mathbf{S}_3, \mathbf{N}_6$	$\mathbf{N}_7$	......	$\mathbf{N}_{11}$
$\varepsilon(\mathbf{S}) =$	2	1	0	-1	-2	-3	-4	-5	......	-9
$H(\mathbf{S}) =$	4	6	7	7	8	9	9	10	......	11

Table 9.1 The Heawood function $H(\mathbf{S})$ for a surface $\mathbf{S}$

One of the main results obtained by Heawood [487] says that the Heawood number of a surface $\mathbf{S}$ is an upper bound for the coloring number and hence for the chromatic number for graphs embedded on $\mathbf{S}$, unless $\mathbf{S}$ is the sphere $\mathbf{S}_0$. For a short proof of this result, see Theorem C.21. Note that $\mathcal{G}(\hookrightarrow \mathbf{S})$ is a monotone graph property.

Theorem 9.1 (HEAWOOD'S MAP COLOR THEOREM) *Let* $\mathbf{S}$ *be a surface with* $\varepsilon(\mathbf{S}) \leq 1$. *Then every graph* $G \in \mathcal{G}(\hookrightarrow \mathbf{S})$ *satisfies* $\mathrm{col}(G) \leq H(\mathbf{S})$.

For a surface $\mathbf{S}$ and a graph parameter ρ, define

$$\rho(\mathbf{S}) = \max\{\rho(G) \mid G \in \mathcal{G}(\hookrightarrow \mathbf{S})\}.$$

Heawood's result implies that for each surface $\mathbf{S}$, except for the sphere $\mathbf{S}_0$, we have

$$\omega(\mathbf{S}) \leq \chi(\mathbf{S}) \leq \chi_\ell(\mathbf{S}) \leq \chi_{\mathrm{DP}}(\mathbf{S}) \leq \mathrm{col}(\mathbf{S}) \leq H(\mathbf{S}).$$

Heawood also conjectured that we always have $\chi(\mathbf{S}) = H(\mathbf{S})$. He verified the conjecture for the torus $\mathbf{S}_1$ by showing that K_7 can be embedded on the torus. However, almost a century passed before Ringel and Youngs [865], [866] generally confirmed the conjecture. They proved that, for each surface $\mathbf{S}$, except for the Klein bottle $\mathbf{N}_2$, the complete graph of order $H(\mathbf{S})$ can be embedded on $\mathbf{S}$ and so $\omega(\mathbf{S}) = H(\mathbf{S})$. This shows that if $\mathbf{S}$ is neither the sphere nor the Klein bottle, then $\chi(\mathbf{S}) = H(\mathbf{S})$. For the Klein bottle $\mathbf{N}_2$, we have $H(\mathbf{N}_2) = 7$, but K_7 cannot be embedded on $\mathbf{N}_2$; for a proof see [752, Theorem 4.4.6]. For the sphere $\mathbf{S}_0$, we have $H(\mathbf{S}_0) = 4$, $\omega(\mathbf{S}_0) = 4$ and $\mathrm{col}(\mathbf{S}_0) = 6$. In 1852, Francis Guthrie asked his brother Frederic Guthrie (see [441]) whether it is true that $\chi(\mathbf{S}_0) \leq 4$. This problem became known as the four color problem and has influenced the development of graph theory for more than a century before Appel, Haken, and Koch [68], [69] provided the first computer aided proof of this result. In 1997, Robertson, Sanders, Seymour, and Thomas [868] provided a new proof that uses essentially the same approach as Appel, Haken, and Koch, but the proof is shorter and clearer and avoids problematic details of the original proof. Thus the four color problem has become the four color theorem. Recall from Appendix C.13 that a graph is **planar** if it can be embedded in the plane; furthermore, the two classes $\mathcal{G}(\hookrightarrow \mathbb{R}^2)$ and $\mathcal{G}(\hookrightarrow \mathbf{S}_0)$ are the same. A graph embedded in the plane is called a **plane graph**.

Four Color Theorem *Every planar graph is* 4*-colorable.*

There are several books and survey papers about the four color theorem, among them Wilson [1070].

In the 1950s Dirac became interested in Heawood's map color theorem; see his papers [290], [295], [296] and [297]. In particular, he introduced the concept of critical graphs and proved the first results on the density of critical graphs. Based on these density results, he [290], [297] proved the following extension of Heawood's theorem for the chromatic number. The result was extended to the list chromatic number by Böhme, Mohar and Stiebitz [125]. However, Dirac's proof from [297] can be easily extended to the DP-chromatic number, by using the Dirac-type bound for $\chi_{\rm DP}$-critical graphs established by Bernshteyn and Kostochka [108]; see Theorem 5.40.

Theorem 9.2 (Dirac's Map Color Theorem) *Let* $\mathbf{S}$ *be a surface with* $\varepsilon(\mathbf{S}) \leq 0$ *and* $\varepsilon(\mathbf{S}) \neq -1$. *If* $G \in \mathcal{G}(\hookrightarrow \mathbf{S})$ *then* $\chi_{\rm DP}(G) < H(\mathbf{S})$ *unless* G *contains a complete graph on* $H(\mathbf{S})$ *vertices as a subgraph.*

Proof Let $k = H(\mathbf{S})$, $\varepsilon = \varepsilon(\mathbf{S})$ and let $G \in \mathcal{G}(\hookrightarrow \mathbf{S})$ be a graph not containing K_k as a subgraph. For the proof it suffices to show that $\chi_{\rm DP}(G) < k$. Suppose this is false and let G be a smallest counterexample. Then G is $\chi_{\rm DP}$-critical and $\chi_{\rm DP}(G) = k$. This implies that G is k-cover-critical (see Exercises 5.8 and 5.9). Let $n = |G|$ and $m = |E(G)|$. Since K_k is not contained in G, it follows from the bound by Bernstein and Kostochka (see Theorem 5.40) that $n \geq k+1$ and

$$2m \geq (k-1)n + k - 3,$$

where equality holds if and only if $G \in \mathcal{DG}(k)$. First, we show that this implies $n \geq k+2$. For otherwise, we have $|G| = k+1$ and $2m \geq (k-1)(k+1) + k - 2 = k^2 + k - 3$. To arrive at a contradiction, let $W = \{v \in V(G) \mid d_G(v) = k-1\}$. Then $\delta(G) \geq k-1$ and $G[W]$ is a DP-forest, that is, every block of $G[W]$ is a complete graph or a cycle (see Exercise 5.8). Consequently, G is obtained from K_{k+1} by deleting edges of some matching M with $1 \leq |M| \leq 2$. If $|M| = 1$, then G contains K_k, a contradiction. If $|M| = 2$, then $2m = (k+1)k - 4 < k^2 + k - 3$, a contradiction, too. Hence, as claimed $n \geq k+2$. On the other hand, by Corollary C.19, we obtain that

$$m \leq 3(n - \varepsilon).$$

Note that $n \geq k+2$ and $k \geq 7$. Combining the two inequalities for m, we obtain that

$$(k-7)n \leq 3 - k - 6\varepsilon,$$

and hence

$$(k-7)(k+2) \leq 3 - k - 6\varepsilon. \tag{9.3}$$

It is easy to check that this inequality leads to a contradiction for $\varepsilon = 0$ and when $-9 \leq \varepsilon \leq -2$, so we may assume that $\varepsilon \leq -10$. By (9.3), $k^2 - 4k - 17 + 6\varepsilon \leq 0$, which leads to

$$k \leq 2 + \sqrt{21 - 6\varepsilon}.$$

On the other hand,

$$k = H(\mathbf{S}) = \left\lfloor \frac{7 + \sqrt{49 - 24\,\varepsilon}}{2} \right\rfloor > \frac{5 + \sqrt{49 - 24\,\varepsilon}}{2}$$

which yields

$$1 + \sqrt{49 - 24\,\varepsilon} < \sqrt{84 - 24\varepsilon}.$$

This implies that $2\sqrt{49 - 24\,\varepsilon} < 34$ and, therefore, $\varepsilon > -10$, a contradiction. □

The three excluded cases of Dirac's map color theorem for the chromatic number were treated by Albertson and Hutchinson [34] in 1979. For the sphere it is known that there are infinitely many 4-critical graphs that can be embedded on $\mathbf{S}_0$. For the surfaces $\mathbf{N}_3$ with $\varepsilon(\mathbf{N}_3) = -1$ and $H(\mathbf{N}_3) = 7$, Dirac's density bound is not strong enough to show that K_7 is the only 7-critical graph that can be embedded on $\mathbf{N}_3$. To show that this is indeed the case, Albertson and Hutchinson analyzed embeddings of 7-chromatic graphs on $\mathbf{N}_3$. On the one hand, Dirac's bound saying that $2\text{ext}(k,n) \geq (k-1)n + (k-3)$ is tight. On the other hand, we can use an extension of Dirac's bound obtained by Kostochka and Stiebitz [630] in order to give a short proof of the map color result for $\mathbf{N}_3$. Let us first state the density result from [630]. For the definition of the graph family $\mathcal{DG}'(k)$ see Section 5.1; note that every graph in $\mathcal{DG}'(k)$ has order $n = 2k - 1$ and $\mathcal{DG}(k) \subseteq \mathcal{DG}'(k)$.

Theorem 9.3 (Kostochka and Stiebitz) *Let $G \in \text{Crit}(k,n)$ with $k \geq 4$. Then $2|E(G)| \geq (k-1)n + 2(k-3)$ unless $G = K_k$ or $G \in \mathcal{DG}'(k)$.*

Theorem 9.4 (Albertson and Hutchinson) *Every graph $G \in \mathcal{G}(\hookrightarrow \mathbf{N}_3)$ satisfies $\chi(G) < H(\mathbf{N}_3) = 7$ unless G contains K_7 as a subgraph.*

Proof Let $G \in \text{Crit}(7)$ be a graph that can be embedded on $\mathbf{N}_3$, let $n = |G|$ and $m = |E(G)|$. For the proof it suffices to show that $G = K_7$. The proof is by reductio ad absurdum. So we assume that $G \neq K_7$. Since $G \in \mathcal{G}(\hookrightarrow \mathbf{N}_3)$ is a connected graph, it follows from Corollary C.19 that

$$2m \leq 6(n - \varepsilon(\mathbf{N}_3)) = 6n + 6.$$

On the other hand, by Theorem 9.3, we obtain that

$$2m \geq 6n + 8,$$

unless $G \in \mathcal{DG}'(7)$. So if $G \notin \mathcal{DG}'(7)$ we are done. It remains to consider the case when $G \in \mathcal{DG}'(7)$. Then $V(G)$ consists of four nonempty pairwise disjoint sets X_1, X_2, Y_1 and Y_2 with $|X_1| + |X_2| = |Y_1| + |Y_2| = 6$ and $|X_2| + |Y_2| \leq 6$ and an additional vertex v such that $X = X_1 \cup X_2$ and $Y = Y_1 \cup Y_2$ are cliques in G, $N_G(v) = X_1 \cup Y_1$, and a vertex $u \in X$ is joined to a vertex $w \in Y$ if and only if $u \in X_2$ and $w \in Y_2$. Then, see the discussion in Section 5.1, it is easy to check that $2m \geq 6n+4$. Consequently, $G \in \mathcal{G}(\hookrightarrow \mathbf{N}_3)$ is a connected graph with $n = 13$ and $3n+2 \leq m \leq 3n+3$. Furthermore, we obtain that every embedding of G in $\mathbf{N}_3$ is a 2-cell embedding, and each face of such an embedding is bounded by a 3-cycle except for at most one face which is bounded by a 4-cycle (see Remark C.23). Since the subgraph $G[N_G(v)]$ of G consists of two components and $\delta(G) \geq 6$, this is impossible. This contradiction completes the proof. □

Theorem 9.4 was extended to the list chromatic number by Král' and Škrekovsky [646] in 2006 with a long and complicated proof. The one page proof that K_6 is the only 6-critical graph embedded on the projective plane $\mathbf{N}_1$ given by Albertson and Hutchinson uses induction on the order. The choosability version of this result was obtained by Böhme, Mohar, and Stiebitz [125] in 1999. Summarizing, we have the following result.

Theorem 9.5 (Dirac's Map Color Theorem) *Let* $\mathbf{S}$ *be a surface with* $\varepsilon(\mathbf{S}) \leq 1$. *If* $G \in \mathcal{G}(\hookrightarrow \mathbf{S})$, *then* $\chi_\ell(G) < H(\mathbf{S})$ *unless* G *contains a complete graph on* $H(\mathbf{S})$ *vertices as a subgraph.*

Whether Theorem 9.2 holds for $\mathbf{N}_3$ is unknown, that it holds for the projective plane $\mathbf{N}_1$ is shown in Section 9.3. For the Klein bottle $\mathbf{N}_2$, Theorem 9.2 says that every graph $G \in \mathcal{G}(\hookrightarrow \mathbf{N}_2)$ satisfies $\chi_{\mathrm{DP}}(G) < H(\mathbf{N}_2) = 7$ unless K_7 is a subgraph of G. Since K_7 cannot be embedded on the Klein bottle, this implies that $\chi_{\mathrm{DP}}(\mathbf{N}_2) \leq 6$; for the chromatic number this was proved by Franklin [388]. That equality holds follows from the fact that $K_3 \boxplus C_5 \in \mathcal{G}(\hookrightarrow \mathbf{N}_2)$ (see [34, Figure 1]) and $\chi(K_3 \boxplus C_5) = \mathrm{col}(K_3 \boxplus C_5) = 6$.

Corollary 9.6 *Every graph* $G \in \mathcal{G}(\hookrightarrow \mathbf{N}_2)$ *satisfies* $\chi_{\mathrm{DP}}(G) \leq 6$.

Dirac [295], [297] used his density results for critical graphs to show that the number of k-critical (nonisomorphic) graphs that can be embedded on a surface $\mathbf{S}$ is finite, provided that $k \geq 8$ and $\varepsilon(\mathbf{S}) \leq -2$. From the Gallai bound it follows that this result can be extended to 7-critical graphs. For a surface $\mathbf{S}$ and an integer $k \geq 3$, let

$$\mathrm{Crit}(\mathbf{S}, k) = \mathrm{Crit}(k) \cap \mathcal{G}(\hookrightarrow \mathbf{S})$$

and let

$$N(\mathbf{S}, k) = \max\{|G| \mid G \in \mathrm{Crit}(\mathbf{S}, k)\},$$

provided that $\mathrm{Crit}(\mathbf{S}, k) \neq \varnothing$. Note that $\mathrm{Crit}(\mathbf{S}, 3) = \mathrm{Crit}(3)$ and hence $N(\mathbf{S}, 3) = \infty$ for every surface $\mathbf{S}$. Furthermore, $\mathrm{Crit}(\mathbf{S}, k) = \varnothing$ if and only if $k > H(\mathbf{S})$ or $\mathbf{S} = \mathbf{N}_2$ and $k = 7$. By Dirac's map color theorem, we have $N(\mathbf{S}, H(\mathbf{S})) = H(\mathbf{S})$ for every surface $\mathbf{S} \neq \mathbf{N}_2$ with $\varepsilon(\mathbf{S}) \leq 1$. Note that $H(\mathbf{S}) \geq 7$, unless $1 \leq \varepsilon(\mathbf{S}) \leq 2$.

Proposition 9.7 *Let* $\mathbf{S}$ *be a surface satisfying* $\varepsilon(\mathbf{S}) \leq 0$ *and let* $7 \leq k \leq H(\mathbf{S})$. *Then*

$$N(\mathbf{S}, k) \leq \frac{k(k-3) - 6(k-1)\varepsilon(\mathbf{S})}{k^2 - 7k + 4}.$$

Proof By combining the Kostochka–Yancey bound for $\mathrm{ext}(k,n)$ (see (5.9)) with the bound in Corollary C.19 we obtain that if $G \in \mathrm{Crit}(k,n) \cap \mathcal{G}(\hookrightarrow \mathbf{S})$, then

$$\frac{(k+1)(k-2)n - k(k-3)}{2(k-1)} \leq 3(n - \varepsilon(\mathbf{S})),$$

which leads to the required bound. □

The Kostochka–Yancey bound for $\mathrm{ext}(k,n)$ is best possible, but only if $n \geq 2k-1$. For $n \leq 2k-2$, we can use the bound in Theorem 9.3 and even the exact values established by Gallai (see Theorem 5.9). Let us consider the case when $\mathbf{S} = \mathbf{N}_{17}$ and $k = H(\mathbf{S}) - 1 = 12$, note that $\varepsilon(\mathbf{S}) = -15$. The Kostochka–Yancey bound yields $N(\mathbf{S},k) \leq 17 < 2k-1 = 23$. Then we get $N(\mathbf{S},k) \leq \lfloor(-2k+6-6\varepsilon(\mathbf{S}))/(k-7)\rfloor = 14$ (using Theorem 9.3). Since we have $k = 12$, $\mathrm{Crit}(k,k+1) = \varnothing$ and $\mathrm{Crit}(k,k+2) = \langle K_{k-3} \boxplus C_5\rangle$ (see Exercise 4.12), we obtain that every graph in $\mathrm{Crit}(\mathbf{N}_{17}, 12)$ is isomorph to K_{12} or to $K_9 \boxplus C_5$. Whether $G = K_9 \boxplus C_5$ can be embedded in $\mathbf{N}_{17}$ is not clear; note that $|G| = 14$ and $|E(G)| = 86$ and a graph of order $n = 14$ embedded in $\mathbf{N}_{17}$ can have $m = 3(n - \varepsilon(\mathbf{N}_{17})) = 3(14+15)) = 87$ edges. Škrekovski [946] proved that, for any surface $\mathbf{S}$ with $\varepsilon(\mathbf{S}) \leq -8$, $N(\mathbf{S}, H(\mathbf{S}) - 1) \leq H(\mathbf{S}) + 1$.

We say that a graph property $\mathcal{P}$ is **finite** if there exists a finite set $\mathcal{X}$ of graphs such that a graph G belongs to $\mathcal{P}$ if and only if G is isomorphic to a graph in $\mathcal{X}$; we then write $\mathcal{P} = \langle \mathcal{X} \rangle$. Hence $\mathrm{Crit}(\mathbf{S},k)$ is finite for every surface $\mathbf{S}$ and for all $k \geq 7$. Thomassen [1012] was able to extend this results to 6-critical graphs. However, the proof of this remarkable result requires a more sophisticated technique and not just results on the density of 6-critical graphs.

Theorem 9.8 (Thomassen) *For each surface* $\mathbf{S}$*, the graph property* $\mathrm{Crit}(\mathbf{S},6)$ *is finite.*

Theorem 9.8 is best possible as it does not extend to k-critical graphs with $3 \leq k \leq 5$. Indeed, Thomassen [1012, Section 3], using a construction of Fisk [376], proved that for every surface $\mathbf{S}$, other than the sphere, $\mathrm{Crit}(\mathbf{S},5)$ is not finite.

Corollary 9.9 *For each surface* $\mathbf{S}$ *and each integer* $k \geq 6$*, there exists a polynomial time algorithms to decide for a given graph* $G \in \mathcal{G}(\hookrightarrow \mathbf{S})$ *if it admits a coloring with* $k-1$ *colors.*

9.2 Map Color Theorems for the Plane

In his 1890 paper [487], Heawood pointed out an error in Kempe's proof [568] of the four color theorem. On the other hand, Heawood [487] proved a five color theorem for planar graphs. While Kempe's proof used his recoloring method, Heawood gave an alternative proof that avoids any recoloring. Our proof is a dual version of Heawood's proof.

Theorem 9.10 (FIVE COLOR THEOREM) *Every graph* $G \in \mathcal{G}(\hookrightarrow \mathbf{S}_0)$ *has chromatic number* $\chi(G) \leq 5$.

Proof The proof is by induction on the order of G. If $|G| \leq 4$, we are done, since $\chi(G) \leq |G|$. So assume that $|G| \geq 5$. Since $\mathrm{col}(\mathbf{S}_0) \leq 6$, there is a vertex v of G with $d_G(v) \leq 5$. If $d_G(v) \leq 4$, then $G - v$ is planar and, by the induction hypothesis, there is a coloring φ of $G - v$ with a set S of five colors. Hence, there is a color $c \in S \setminus \varphi(N_G(v))$ and we may color v with c to get a coloring of G with color set S. If $d_G(v) = 5$, then there are two distinct vertices $u, w \in N_G(v)$ such that $uw \notin E(G)$, since K_5 is not planar. Let G' be the graph obtained from $G - v$ by identifying u and w to a new vertex. Then G' is planar and, by the induction hypothesis, there is a coloring of G' with a set S of five colors. This results in a coloring φ of $G - v$ with color set S such that $\varphi(u) = \varphi(w)$. Then there is a color $c \in S \setminus \varphi(N_G(v))$ and, if we color v with c, we obtain a coloring of G with color set S. This completes the proof. □

The four color theorem says that $\chi(\mathbf{S}_0) = 4$. In 1979, Erdős, Rubin, and Taylor [355] conjectured that $\chi_\ell(\mathbf{S}_0) = 5$. That this is indeed the case was proved in two papers in the early 1990s. On the one hand, Voigt [1052] presented an example of a plane graph showing that $\chi_\ell(\mathbf{S}_0) \geq 5$; on the other hand, Thomassen [1008] proved that $\chi_\ell(\mathbf{S}_0) \leq 5$. Thomassen's short and tricky proof can be applied to the DP-chromatic number without modification.

Theorem 9.11 (THOMASSEN) *Every graph* $G \in \mathcal{G}(\hookrightarrow \mathbf{S}_0)$ *has DP-chromatic number* $\chi_{\mathrm{DP}}(G) \leq 5$.

First, let us recall the DP-coloring concept, see Section 1.4, or Section 5.6. Let G be an arbitrary graph. Let (X,H) be a **cover** of G, that is, (C1) H is a graph and $X : V(G) \to 2^{V(H)}$ is a map such that $V(H)$ is the disjoint union of the independent sets $X(v)$ of H with $v \in V(G)$, and (C2) the edge set $E_H(X(u), X(v))$ is a matching in H (possibly empty) if $uv \in E(G)$ and empty if $uv \notin E(G)$. A **transversal** of (X,H) is a set $T \subseteq V(H)$ such that $|T \cap X(v)| = 1$ for all $v \in V(H)$. An **independent transversal** of (X,H) is a transversal of (X,H) which is an independent set of H. Clearly, a set T is a transversal of (X,H) if and only if T is the image of a mapping $\varphi : V(G) \to V(H)$ such that $\varphi(v) \in X(v)$ for all $v \in V(G)$. An (X,H)**-coloring** of G is an independent transversal of G, or, equivalently, a mapping $\varphi : V(G) \to V(H)$ such that $\varphi(v) \in X(v)$ for all $v \in V(G)$ and $T = \{\varphi(v) \mid v \in V(G)\}$ is an independent set of H. Let G' be an induced subgraph of G, let U be the union of the sets $X(v)$ with $v \in V(G')$, let $H' = H[U]$ and $X' = X|_{V(G')}$. Then (X',H') is a cover of G' and we write $(X',H') = (X,H)/G'$. To simplify the notion, we call an (X',H')-coloring of G' also an (X,H)-coloring of G'. So an (X,H)**-coloring** of G' is a mapping $\varphi' : V(G') \to V(H)$ such $\varphi'(v) \in X(v)$ for all $v \in V(G')$ and $T' = \{\varphi'(v) \mid v \in V(G')\}$ is an independent set of H. We call (X,H) a k**-cover** of G if $|X(v)| = k$ for all $v \in V(G)$. Then the **DP-chromatic number** $\chi_{\mathrm{DP}}(G)$ of G is the least integer k such that G admits an (X,H)-coloring for every k-cover (X,H) of G.

Recall that a graph G is χ_{DP}**-critical** if $\chi_{\mathrm{DP}}(H) < \chi_{\mathrm{DP}}(G)$ for every proper subgraph H of G. Let $\mathrm{Crit}_{\mathrm{DP}}(k)$ denote the class of χ_{DP}-critical graphs G with $\chi_{\mathrm{DP}}(G) =$

k; furthermore, let $\mathrm{Crit}_{\mathrm{DP}}(k,n) = \{G \in \mathrm{Crit}_{\mathrm{DP}}(k) \mid |G| = n\}$. We have $\chi_{\mathrm{DP}}(G) \le \mathrm{col}(G)$ for every graph G, and $\chi_{\mathrm{DP}}(C) = 3$ for every cycle C. This implies that $\mathrm{Crit}_{\mathrm{DP}}(k) = \langle K_k \rangle$ for $k \le 2$, $\mathrm{Crit}_{\mathrm{DP}}(3) = \langle C_n : n \ge 3 \rangle$, $\mathrm{Crit}_{\mathrm{DP}}(k,k) = \langle K_k \rangle$ for $k \ge 0$, and $\mathrm{Crit}_{\mathrm{DP}}(k,n) = \varnothing$ for $n < k$. If $G \in \mathrm{Crit}_{\mathrm{DP}}(k)$, then there is a $(k-1)$-cover (X,H) such that G has no (X,H)-coloring, but for every $v \in V(G)$ there is an independent set I of H such that $|I \cap X_u| = 1$ for all $u \in V(G-v)$. Hence every graph in $\mathrm{Crit}_{\mathrm{DP}}(k)$ is k-cover-critical (see Exercise 5.9); the converse statement is not true. For $n \ge k \ge 3$, let

$$\mathrm{ext}_{\mathrm{DP}}(k,n) = \min\{|E(G)| \mid G \in \mathrm{Crit}_{\mathrm{DP}}(k,n)\}.$$

For the function $\mathrm{ext}_{\mathrm{DP}}(\cdot,\cdot)$ only two bounds are known. The Gallai-type bound by Bernshteyn, Kostochka, and Pron [110] saying that

$$2\mathrm{ext}_{\mathrm{DP}}(k,n) \ge \Big(k - 1 + \frac{k-3}{(k^2-3)}\Big)n \text{ for } n > k \ge 4; \tag{9.4}$$

and the Dirac-type bound established by Bernshteyn and Kostochka [108] saying that

$$2\mathrm{ext}_{\mathrm{DP}}(k,n) \ge (k-1)n + (k-3) \text{ for } n > k \ge 4,$$

where equality holds only for $n = 2k-1$ (see Theorem 5.40).

Clearly, the DP-chromatic number is monotone, that is, if H and G are graphs, then $H \subseteq G$ implies $\chi_{\mathrm{DP}}(H) \le \chi_{\mathrm{DP}}(G)$. Hence in order to prove that $\chi_{\mathrm{DP}}(G) \le 5$ for every graph $G \in \mathcal{G}(\hookrightarrow \mathbf{S}_0)$, it suffices to prove this for maximal planar graphs, and hence for plane triangulations (see Corollary C.28).

A **plane near-triangulation** is a plane graph G such that all inner faces are bounded by a triangle and the outer face is bounded by a cycle C; this cycle C is called the **outer cycle** of G, and the vertices of $V(G) \setminus V(C)$ are the vertices **inside** C, or the vertices belonging to the **interior** of C. Theorem 9.11 is a consequence of the following result due to Thomassen. A strengthening of Theorem 9.11 using weak degeneracy was obtained by Bernshteyn and Lee [112].

Theorem 9.12 (THOMASSEN) *Let G be a plane near-triangulation, let C be the outer cycle of G, and let $uu' \in E(C)$ be an edge. Furthermore, let (X,H) be a cover of G such that $|X(v)| \ge 3$ for all $v \in V(C)$ else $|X(v)| \ge 5$, and let $x,x' \in V(H)$ be two colors such that $x \in X(u)$, $x' \in X(u')$ and $xx' \notin E(H)$. Then there exists an (X,H)-coloring φ of G such that $\varphi(u) = x$ and $\varphi(u') = x'$.*

Proof The proof is by induction on the order n of G. Clearly, $n \ge 3$ and if $n = 3$, then $G = C = K_3$ and the statement is evident. So assume that $n \ge 4$. Let $w \in N_C(u) \setminus \{u'\}$ be the second neighbor of u in C. First assume that there is a chord of C in G incident with w, say ww'. Then ww' lies on two unique cycles $C_1, C_2 \subseteq C + ww'$ with $uu' \in E(C_1) \setminus E(C_2)$. For $i \in \{1,2\}$, let G_i denote the plane subgraph of G consisting of C_i and the part of G inside C_i. Clearly, G_i is a plane near-triangulation whose outer cycle is C_i, and $|G_i| < n$. By induction, G_1 has an (X,H)-coloring φ_1 such that $\varphi_1(u) = x$ and $\varphi_1(u') = x'$. Let $y = \varphi_1(w)$ and $y' = \varphi_1(w')$. Then $yy' \notin E(H)$, and again, by induction, there is an (X,H)-coloring φ_2 of G_2 such that $\varphi_2(w) = y$ and

$\varphi_2(w') = y'$. Consequently, $\varphi = \varphi_1 \cup \varphi_2$ is an (X,H)-coloring of G with $\varphi(u) = x$ and $\varphi(u') = x'$, and we are done.

It remains to consider the case that no chord of C in G is incident with w. Let $w' \in N_C(w) \setminus \{u\}$ be the second neighbor of w in C. Since G is a near-triangulation whose outer cycle is C, the neighbors of w in G belong to a path P whose ends are w' and u, and whose inner vertices lie inside C. Then $C' = (C-w) \cup P$ is a cycle in G, and $G' = G - w$ is a plane near-triangulation whose outer cycle is C'. Let $H' = H - X(w)$. Since $|X(w)| \geq 3$ and $x \in X(u)$, there are two colors $y, y' \in X(w) \setminus N_H(x)$ (by (C2)). Let $X' : V(G') \to 2^{V(H')}$ be the map with $X'(v) = X(v) \setminus (N_H(y) \cup N_H(y'))$ if v is an inner vertex of P else $X'(v) = X(v)$. If v is an inner vertex of P, then $|X'(v)| \geq |X(v)| - 2 \geq 3$ (by (C2)). Hence, (X', H') is a cover of the plane near-triangulation G' whose outer cycle is C', where $|X'(v)| \geq 3$ for all $v \in V(C')$ and $|X'(v)| = |X(v)| \geq 5$ for all $v \in V(G') \setminus V(C')$. By the induction hypothesis, there is an (X', H')-coloring φ' of G' such that $\varphi'(u) = x$ and $\varphi'(u') = x'$. If $v \in N_G(w) \setminus \{w'\}$, then the color $\varphi'(v)$ is neither adjacent to y nor to y' in H, and one of the colors y and y' is not adjacent to the color $\varphi'(w')$ in H. By assigning that color to w, we get an (X,H)-coloring φ of G such that $\varphi(u) = x$ and $\varphi(u') = x'$. Thus the proof is complete. □

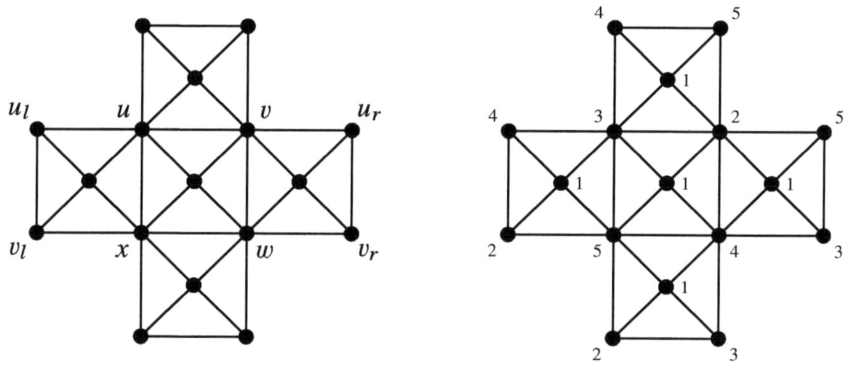

Fig. 9.1 A plane near-triangulation T (on the left) and the labelling λ_1 (on the right).

In 1993, Voigt [1052] constructed the first example of a plane graph that is not 4-list-colorable. This construction is not only extremely clever, but also the model for many later designs. On the other hand, Voigt's graph has 238 vertices which caused the search for graphs of smaller order. In 1996, Mirzakhani [743] presented a non-4-choosable plane graph of order 63, which seems to be the best result so far. The construction of the Mirzakhani graph MG together with an uncolorable 4-list-assignment is carried out as follows. First, let T be a plane near-triangulation of order 17 shown on the left hand side of Figure 9.1. We call $u_r v_r$ the right edge, $u_l v_l$ the left edge of T, and we denote by C the outer cycle of T. Next, we construct, for $i \in [1,4]$, a pair (G_i, λ_i) consisting of a copy G_i of T and a vertex labelling $\lambda_i : V(G_i) \to \Lambda$ with $\Lambda = [1,5]$. We assume that the copies G_1, G_2, G_3 and G_4 are disjoint. If v is a vertex of T, then we denote by v^i its copy in G_i; furthermore, we denote by C^i the copy of the outer cycle C of T in G_i. Let $\pi = (4,5)(1,2,3)$ and $\pi' = (1,2)(3,4,5)$ be two permutations of the symmetric group S_5, using cycle representation. The vertex labelling λ_1 of G_1 is shown on the right hand side of Figure 9.1. Then, for a vertex v of T, we define $\lambda_2(v^2) = \pi(\lambda_1(v^1))$, $\lambda_3(v^3) = \pi(\lambda_2(v^2))$, and $\lambda_4(v^4) = \pi'(\lambda_3(v^3))$. For the right edge and the left edge of T we then obtain that

$$\lambda_i(u_r^i) = \lambda_{i+1}(u_l^{i+1}) \text{ and } \lambda_i(v_r^i) = \lambda_{i+1}(v_l^{i+1}) \text{ for } i \in \{1,2,3\} \tag{9.5}$$

For $i \in [1,4]$, let L_i be the list-assignment of G_i defined by $L_i(v^i) = \Lambda \setminus \lambda_i(v^i)$. Clearly, L_i is a 4-list-assignment of G_i with color set $\Lambda = [1,5]$. Furthermore, by (9.5), we have

$$L_i(u_r^i) = L_{i+1}(u_l^{i+1}) \text{ and } L_i(v_r^i) = L_{i+1}(v_l^{i+1}) \text{ for } i \in \{1,2,3\} \tag{9.6}$$

The key property of the list-assignment L_i of G_i is the following.

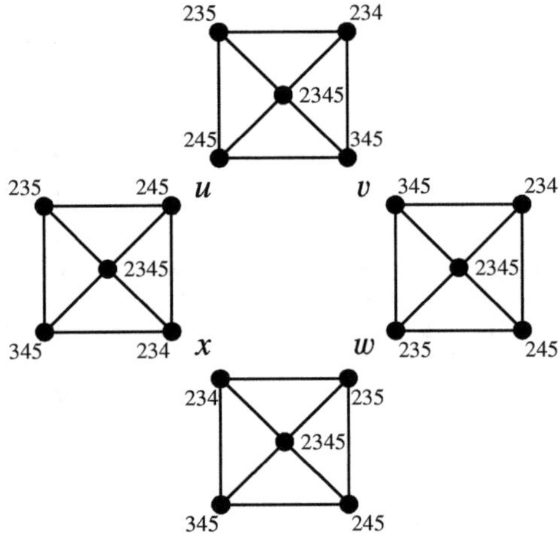

Fig. 9.2 The list-assignment L' of G_1.

Proposition 9.13 (Mirzakhani) *In any L_i-coloring of G_i ($i \in [1,4]$) at least one vertex of its outer cycle C^i has color i.*

Sketch of Proof: Note that if v is a vertex belonging to the interior of the outer cycle C of the plane near-triangulation T, then $\lambda_i(v^i) = i$, and so $L_i(v^i) = \Lambda \setminus \{i\}$ for $i \in [1,4]$. Hence, it suffice to prove the proposition for $i = 1$. Let L' be the list-assignment of G_1 such that $L'(v^1) = L_1(v^1) \setminus \{1\} = \Lambda \setminus \{\lambda_1(v^1), 1\}$ if v belongs to the outer cycle C else $L'(v^1) = L_1(v^1)$. For the proof it suffices to show that G_1 has no L'-coloring. To simplify the notation, we shall not distinguish between a vertex v of T and its copy v^1 in G_1. Let u^* denote the common neighbor of the vertices u, v, w and x in G_1. Note that $L'(u) = \{2,4,5\}$, $L'(v) = \{3,4,5\}$, $L'(w) = \{2,3,5\}$

and $L'(x) = \{2,3,4\}$ (see Figure 9.2). Let φ be an arbitrary L'-coloring of $G_1 - u^*$. Then it is easy to check that $\varphi(u) = 5$ forces $\varphi(v) = 3$ which, in turn, forces $\varphi(w) = 2$ and $\varphi(x) = 4$. In fact, any one of these equations forces the three others. Arguing counterclockwise, the same holds for the four equations $\varphi(v) = 4$, $\varphi(u) = 2$, $\varphi(x) = 3$ and $\varphi(w) = 5$. The only remaining choice is then $\varphi(u) = 4$, $\varphi(v) = 5$, $\varphi(w) = 3$ and $\varphi(x) = 2$. Summarizing we obtain that in any L'-coloring φ of $G_1 - u^*$ all four colors 2, 3, 4 and 5 occurs in the neighborhood of the vertex u^*. Since $L'(u^*) = \Lambda \setminus \{1\}$, this implies that there is no choice of a color for the vertex u^*. Hence G_1 is not L'-colorable. This completes the proof. □

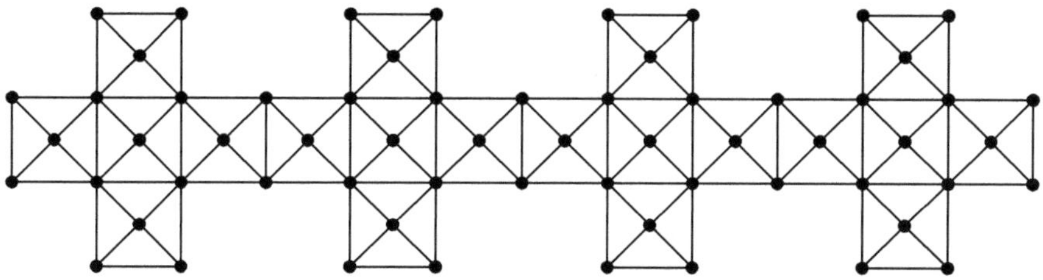

Fig. 9.3 The Mirzakhani graph $MG - v^*$ embedded in the plane.

The **Mirzakhani graph** MG' consists of the four copies G_1, G_2, G_3 and G_4 of T and an additional vertex v^* that is joined to all vertices v^i whenever v belongs to the outer cycle C of T and $i \in [1,4]$. Let L' be the list-assignment of MG' such that $L'(v^i) = L_i(v^i)$ for all $v \in V(T)$ and $L'(v^*) = \Lambda \setminus \{5\} = [1,4]$. Clearly, MG' is a planar graph, L' is a 4-list-assignment of MG' with color set Λ, and, by Proposition 9.13, MG' is not L'-colorable. Hence $\chi_\ell(MG') \geq 5$; note that MG' has order $n = 69$. The **Mirzakhani graph** MG is obtained from MG' by identifying, for $i \in \{1,2,3\}$, the vertex u_r^i with u_l^{i+1} and the vertex v_r^i with v_l^{i+1}. Clearly, MG is a planar graph of order $n = 63$ which is isomorphic to a plane triangulation; an embedding of the graph $MG - v^*$ in the plane is shown in Figure 9.3. By (9.6), the 4-list-assignment L' of MG' yields a 4-list-assignment L of MG such that MG has no L-coloring. Hence $\chi_\ell(MG) \geq 5$. Another interesting feature of the Mirzakhani graph MG is that $\chi(MG) = 3$. To see this, observe that $MG - v^*$ is isomorphic to a plane near-triangulation G' (see Figure 9.3) whose outer cycle C' has even length, $N_{MG}(v^*) = V(C')$, and the set of vertices belonging to the interior of C' form an independent set of MG. So the Mirzakhani graph MG is a planar 3-chromatic graph which is not 4-list-colorable, disproving a conjecture of Jensen suggesting that every planar 3-chromatic graph is 4-list-colorable. On the other hand, Alon and Tarsi [58] proved that every planar bipartite graph is 3-list-colorable (see Exercise 5.18).

Certainly, one of the most popular map color theorems is Grötzsch's three color theorem of 1959, which states that every triangle-free planar graph is 3-colorable (see Theorem 5.13). While the original proof by Grötzsch [434] was rather complicated, Kostochka and Yancey gave a much simpler proof using their density result for 4-critical graphs (see Theorem 5.10).

A planar triangle-free graph G of order $n \geq 4$ has at most $2n-4$ edges (by Theorem C.18), implying that $\chi_\ell(G) \leq \mathrm{col}(G) \leq 4$. In 1995, Voigt [1053] constructed planar triangle-free graphs that are not 3-list-colorable; thereby providing a negative answer to a problem raised by Kratochvíl and Tuza [647]. On the other hand, Thomassen [1010] proved the following result, which can be used to give an alternative proof of Grötzsch's three color theorem.

Theorem 9.14 (THOMASSEN) *Every planar graph of girth at least* 5 *is 3-list-colorable.*

In the following, we will discuss another extension of Grötzsch's three color result known as the Grünbaum–Aksionov theorem. Grünbaum [435] published the result in 1963, but his proof was incorrect; the first correct proof was given by Aksionov [27] in 1974, see also Borodin [151], and Borodin, Kostochka, Lidický, and Yancey [158]. First, we need a technical lemma.

Lemma 9.15 *Let G be a plane graph and let $C = (v_0, v_1, v_2, v_3, v_0)$ be a facial* 4*-cycle of G such that $v_0v_2, v_1v_3 \notin E(G)$. For $i \in \{0,1\}$, let G_i be the graph obtained from G by identifying v_i and v_{i+2}. If both G_0 and G_1 have more triangles than G, then, for some vertex $u \in V(G)$ and some index $i \in [1,4]$, there is a triangle in G with vertex set $\{v_i, v_{i+1}, u\}$. Moreover, G contains vertices v and w not in C such that (v_{i+1}, u, v, v_{i-1}) and (v_i, u, w, v_{i+2}) are paths in G (where indices are modulo* 4*).*

Proof Since G_0 has more triangles than G, there must be a path (v_0, u, w, v_2) in G where $u, w \notin V(C)$. Since G_1 has more triangles that G, there must be a path (v_1, u', v, v_3) in G where $u', v \notin V(C)$. Since G is a plane graph, the sets $\{u, w\}$ and $\{u', v\}$ are not disjoint. By symmetry, we may assume that $u = u'$. Then G contains a triangle with vertex set $\{v_0, v_1, u\}$ and paths (v_1, u, v, v_3) and (v_0, u, w, v_2). Note that $v = w$ is possible, see Figure 9.4. □

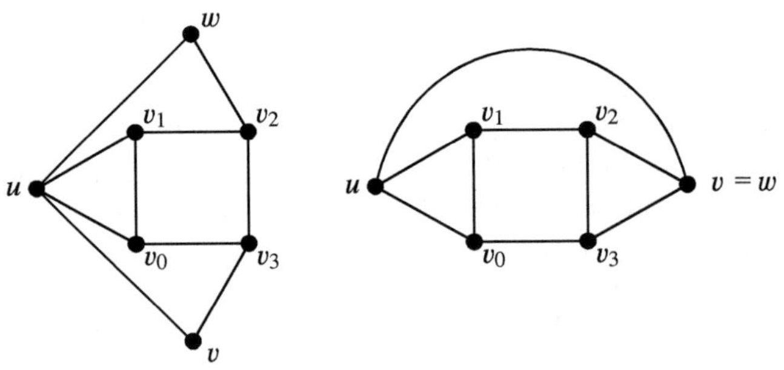

Fig. 9.4 Illustration to Lemma 9.15.

For a graph G and two distinct vertices u and v of G, we denote by $G + uv$ the graph obtained from G by adding the edge uv, where $G + uv = G$ if $uv \in E(G)$. If H is a plane graph, then a face $f \in F(H)$ with $d_H(f) = k$ is called a ***k*-face** of H.

The following extension of Grötzsch's theorem was obtained by Aksionov [28] and, independently, by Jensen and Thomassen [529]. Note that our proof does not use Grötzsch's result.

Theorem 9.16 *Let G be a plane triangle-free graph, and let u, v be two distinct vertices of G. Then $G + uv$ is 3-colorable.*

Proof Suppose this is false, and let G be a counterexample with minimum order. Then G is a plane triangle-free graph and there are two distinct vertices u, v in G such that $\chi(G+uv) \geq 4$. Clearly, $G+uv$ contains a subgraph G' such that $G' \in \text{Crit}(4)$. By the minimality of G, we have $n = |G'| = |G|$. Furthermore, $H = G' - uv$ is a plane triangle-free graph of order n. Let $m = |E(H)|$ and $\ell = |F(H)|$. Note that $|E(G')| \leq m+1$, where equality holds if and only if $uv \in E(G')$. Since $G' \in \text{Crit}(4)$, G' is 3-edge-connected (by Theorem 1.32) and, hence, H is 2-edge-connected. Since H is a plane triangle-free graph, this implies that $d_H(f) \geq 4$ for all $f \in F(H)$. Furthermore, if f is a 4-face of H, then f is bounded by a 4-cycle, say $C = (v_0, v_1, v_2, v_3, v_0)$; in this case we write $f = v_0v_1v_2v_3$.

Claim 9.16.1 *H has at least two 4-faces.*

Proof: Suppose H has at most one 4-face. Then by (C.5), we obtain that

$$2m = \sum_{f \in F(H)} d_H(f) \geq 5(\ell - 1) + 4 = 5\ell - 1.$$

Euler's formula for H, see (C.9), yields $n - m + \ell = 2$. Consequently, we obtain that $10 \leq 5n - 3m + 1$, i.e., $m \leq (5n-9)/3$. Since $G' \in \text{Crit}(4, n)$ and $|E(G')| \leq m+1$, Theorem 5.10 implies that $m + 1 \geq |E(G')| \geq (5n-2)/3$ and, hence, $m \geq (5n-5)/3$, a contradiction. △

Claim 9.16.2 *$uv \in E(G')$.*

Proof: Suppose this is false. Then $G' = H$ is a plane triangle-free graph and, by Claim 9.16.1, G' has a 4-face, say $f = v_0v_1v_2v_3$, where $v_0v_2, v_1v_3 \notin E(G')$. Let G_i be the graph obtained from G' by identifying v_i and v_{i+2} where $i \in \{0, 1\}$. Clearly, both G_0 and G_1 are planar graphs. Since G' is triangle-free, Lemma 9.15 implies that G_0 or G_1 is triangle-free, say G_0 is triangle free (by symmetry). By the minimality of G, this implies that $\chi(G_0) \leq 3$, which yields $\chi(G') \leq 3$, a contradiction. △

Claim 9.16.3 *H has a 4-face $f = v_0v_1v_2v_3$ such that uv is neither v_0v_2 nor v_1v_3*

Proof: Suppose this is false. Then it follows from Claim 9.16.1 and Claim 9.16.2 that H has two distinct 4 faces, say f and f', such that $f = uv_1vv_2$ and $f' = uw_1vw_2$. Since H is triangle-free, we have $E(H) \cap \{v_1v_2, w_1w_2\} = \varnothing$. Since $G' \in \text{Crit}(4)$, we have $\delta(G') \geq 3$. If $v' \in \{v_1, v_2\} \cap \{w_1, w_2\}$, then v' has degree two in H and hence also in G', a contradiction. This shows that $\{v_1, v_2\} \cap \{w_1, w_2\} = \varnothing$. Let G^* be the graph obtained from H by identifying v_1 and v_2 into a new vertex v^*. Note that G^*

is planar. We claim that G^* is triangle-free. Suppose this is false. Since H is a plane triangle-free graph, this implies that there exists a path $P = (v_1, u_1, u_2, v_2)$ in H where $u_1, u_2 \notin \{v_1, v_2, u, v\}$. On the one hand, this implies that $\{u_1, u_2\} = \{w_1, w_2\}$. On the other hand, $w_1 w_2 \notin E(H)$, a contradiction. This proves the claim that G^* is a planar triangle-free graph. By the minimality of G, the graph $G^* + uv$ has a 3-coloring φ. Then φ can be extended to a 3-coloring of G', by assigning color $\varphi(v^*)$ to v_1 and v_2, a contradiction. △

By Claim 9.16.3, H has a 4-face $f = v_0 v_1 v_2 v_3$ and uv is neither $v_0 v_2$ nor $v_1 v_3$. Since H is a plane triangle-free graph, this implies that $v_0 v_2, v_1 v_3 \notin E(H)$. Let G_i be the graph obtained from H by identifying v_i and v_{i+2} to a new vertex v_i^* where $i \in \{0, 1\}$. From Lemma 9.15 we then obtain that G_0 or G_1 is triangle-free. By symmetry we may assume that G_0 is triangle-free. By the minimality of G, the graph $G_0 + uv$ has a 3-coloring φ. Then φ can be extended to a 3-coloring of G', by assigning color $\varphi(v_0^*)$ to v_0 and v_2, a contradiction. This completes the proof of the theorem. □

Theorem 9.17 *Let G be a graph, let $v \in V(G)$ be a vertex with $d_G(v) \le 4$. If $G - v$ is a plane triangle-free graph, then G is 3-colorable.*

Proof For the proof, let G be a smallest counterexample, that is, (1) $G' = G - v$ is a plane triangle-free graph for some vertex v of G with $d_G(v) \le 4$, (2) $\chi(G) \ge 4$ and (3) $|V(G)| + |E(G)|$ is minimum subject to (1) and (2). By Grötzsch's theorem, $\chi(G') \le 3$. Hence (3) implies that $G \in \mathrm{Crit}(4)$. Let $n = |G|, m = |E(G)|$ and $\ell = |F(G')|$. Since $G \in \mathrm{Crit}(4)$, G is 2-connected and, hence, G' is connected. Furthermore, $|G'| \ge 3$ and $\delta(G') \ge 2$. Let $d = d_G(v)$.

Case 1: *$d_{G'}(f) \ge 5$ for all $f \in F(G')$.* Then, by (C.5), $5\ell \le 2|E(G')| = 2(m - d)$. Euler's formula for G' yields $(n-1) - (m-d) + \ell = 2$. Consequently, we obtain that $10 \le 5(n-1) - 3(m-d) \le 5n - 3m + 7$, i.e., $m \le (5n-3)/3$. Since $G \in \mathrm{Crit}(4, n)$, Theorem 5.10 implies that $m \ge (5n-2)/3$, a contradiction.

Case 2: *$d_{G'}(f) \le 4$ for a face $f \in F(G')$.* Since G' is triangle-free and $\delta(G') \ge 2$, this implies that f is bounded by a 4-cycle, say $C = (v_0, v_1, v_2, v_3, v_0)$, where $v_0 v_2, v_1 v_3 \notin E(G')$. Let G_i be the graph obtained from G' by identifying v_i and v_{i+2} to a new vertex v_i^* where $i \in \{0, 1\}$. From Lemma 9.15 we then obtain that G_0 or G_1 is triangle free. By symmetry we may assume that G_0 is triangle free. Then G_0 has a 3-coloring (by the choice of G), which gives a 3-coloring of G, a contradiction. □

Theorem 9.18 *Let G be a plane triangle-free graph, and let C be a facial cycle of G with $|C| \le 5$. Then each 3-coloring of C can be extended to a 3-coloring of G.*

Proof Let φ be a 3-coloring of C. Since G is triangle-free, $4 \le |C| \le 5$ and C has no chord in G. First consider the case that C is a 4-cycle, say $C = (v_0, v_1, v_2, v_3, v_0)$. If $|\varphi(V(C))| = 2$, then let G' be the graph obtained from G by adding a vertex v and joining v to all vertices of C. By Theorem 9.17, G' has a 3-coloring φ'. Then $\varphi'(v) \notin \varphi(V(C))$ and, by permuting colors if necessary, we may assume that $\varphi'|_{V(C)} = \varphi$. Hence φ can be extended to a 3-coloring of G. If $|\varphi(V(C))| = 3$, then, by symmetry, we may assume that $\varphi(v_0) = \varphi(v_2)$ and $\varphi(v_1) \ne \varphi(v_3)$. By Theorem 9.16,

$G+v_1v_3$ has a 3-coloring φ', and, by permuting colors if necessary, we may assume that $\varphi'|_{V(C)}=\varphi$. Hence φ can be extended to a 3-coloring of G.

It remains to consider the case that $|C|=5$, say $C=(v_0,v_1,v_2,v_3,v_4,v_0)$. Then $|\varphi(V(C))|=3$, and, by symmetry, we may assume that $\varphi(v_0)=\varphi(v_2)$ and $\varphi(v_1)=\varphi(v_3)$. Let G' be the graph obtained from G by adding a vertex v and joining v to all vertices of $C-v_4$. By Theorem 9.17, G' has a 3-coloring φ'. By permuting colors if necessary, we may assume that $\varphi'|_{V(C)}=\varphi$. Hence φ can be extended to a 3-coloring of G. □

For a plane graph with just one triangle Theorem 9.18 is not true. Let $C=(v_0,v_1,v_2,v_3,v_4,v_0)$ be a 5-cycle and let u be a vertex joined to v_0,v_2 and v_3. Then the 3-coloring φ of C with $(\varphi(v_0),\varphi(v_1),\varphi(v_2),\varphi(v_3),\varphi(v_4))=(1,2,3,2,3)$ cannot be extended to u. This example is due to Tibor Gallai, presented as a counter-example to a statement by Grünbaum [435].

Theorem 9.19 (GRÜNBAUM–AKSIONOV) *Let G be a planar graph containing at most three triangles. Then G is 3-colorable.*

Proof Suppose this is false. Then, let G be a smallest counterexample, that is, (1) G is a plane graph with at most three triangles, (2) $\chi(G)\geq 4$, and (3) $s(G)=|V(G)|+|E(G)|$ is minimum subject to (1) and (2). Clearly, this implies that $G\in \text{Crit}(4)$. Consequently, $\delta(G)\geq 3$, G is 2-connected and each face of G is bounded by a cycle. Let $n=|G|, m=|E(G)$ and $\ell=|F(G)|$.

Claim 9.19.1 *Let C be a nonfacial cycle of G such that $|C|\leq 5$. Furthermore, let G_0, respectively G_1, be the subgraph of G consisting of C and the part of G inside of C, respectively of C and the part of G outside of C. Then $|C|\geq 4$ and both graphs G_0 and G_1 contain triangles.*

Proof: Clearly, G_i is a plane graph and C is a facial cycle of G_i; furthermore, $s(G_i)<s(G)$. By (3), this implies that, for $i\in\{0,1\}$, there is a 3-coloring φ_i of G_i. Since $G\in\text{Crit}(4)$, no separating set of G is a clique (by Corollary1.31), implying that $4\leq |C|\leq 5$. If G_i is triangle-free for some $i\in\{0,1\}$, then it follows from Theorem 9.18, there is a 3-coloring φ' of G_i such that $\varphi'(v)=\varphi_{1-i}(v)$ for all $v\in V(C)$. Then $\varphi=\varphi'\cup\varphi_{1-i}$ is a 3-coloring of G, a contradiction to (2). Hence the claim is proved. △

Claim 9.19.2 *G has a facial 4-cycle.*

Proof: Otherwise, using (C.5) and (1), we obtain that $5\ell-6\leq 2m$. By Euler's formula, we have $n-m+\ell=2$. Consequently, $10\leq 5n-3m+6$, i.e. $m\leq (5n-4)/3$. Since $G\in\text{Crit}(4,n)$, Theorem 5.10 implies that $m\geq (5n-2)/3$, a contradiction. △

Claim 9.19.3 *No facial 4-cycle of G has a chord in G.*

Proof: Otherwise, there is a 4-face $f=v_0v_1v_2v_3$ of G such that $v_0v_2\in E(G)$. Then it follows from Claim 9.19.1 that (v_0,v_1,v_2,v_0) and (v_0,v_2,v_3,v_0) are both facial 3-cycles of G. Hence, $G=C+v_0v_2$ is a K_4 minus an edge implying that $\chi(G)\leq 3$, a contradiction to (2). △

Let $C = (v_0, v_1, v_2, v_3, v_0)$ be an arbitrary facial 4-cycle of G. Then, by Claim 9.19.3, neither v_0v_2 nor v_1v_3 is an edge of G. For $i \in \{0,1\}$, let G_i be the graph obtained from G by identifying v_i and v_{i+2} to a new vertex. Then G_i is a planar graph (by (1)), $\chi(G_i) \geq 4$ (by (2)), and $s(G_i) < s(G)$. By (3), this implies that both G_0 and G_1 have more triangles than G. From Lemma 9.15 it then follows that we may adjust the numbering in C such that both $P = (v_0, u, w, v_2)$ and $P' = (v_1, u, v, v_3)$ are paths in G, where u, v, w are not in C.

Claim 9.19.4 *The two paths P and P' satisfy $v \neq w$.*

Proof: Suppose that $v = w$. Then $f_u = v_0v_1u$ and $f_v = v_2v_3v$ are both 3-faces of G (by Claim 9.19.1), and $f = v_0v_1v_2v_3$ is a 4-face of G. Furthermore, $C_1 = (u, v_1, v_2, v, u)$ and $C_2 = (u, v_0, v_3, v, u)$ are both 4-cycles of G. If neither C_1 nor C_2 is a facial cycle of G, then it follows from Claim 9.19.1 that G has at least four triangles. This contradiction to (1) shows that C_1 or C_2 is a facial 4-cycle. By symmetry, we may assume that C_1 is a facial 4-cycle. Note that neither v_0v_2 nor v_1v_3 is an edge edge of G. For $i \in \{0,1\}$, let G_i be the graph obtained from G by identifying v_i and v_{i+2} to a new vertex v_i^*. Then $\chi(G_i) \geq 4$ (by (2)) and, hence, G_i contains a 4-critical subgraph G_i'. Clearly, both G_0' and G_1' are planar graphs, and both have more triangles than G (by (3)). Note that $d_{G_0}(v_1) = 2$ implying that $v_1 \notin V(G_0')$ and, hence, the triangle (v_1, v_0^*, u, v_1) is not contained in G_0'. Thus there is another triangle that is in G_0 but not in G. Since G is a plane graph, this implies that there is a vertex w_1 in $G - v_3$ such that $w_1v_0, w_1v \in E(G)$. Using a similar argument for G_1, we obtain that there is a vertex w_2 in $G - v_0$ such that $w_2v_3, w_2u \in E(G)$. Since G is a plane graph, this implies that $w_1 = w_2$. But then G has more than three triangles, a contradiction to (1). This proves the claim that $v \neq w$. △

Recall that $C = (v_0, v_1, v_2, v_3, v_0)$ is an arbitrary facial 4-cycle of G, and both $P = (v_0, u, w, v_2)$ and $P' = (v_1, u, v, v_3)$ are paths of G with $u, v, w \notin V(C)$. By Claim 9.19.4, we have $v \neq w$. By Claim 9.19.1, $f_u = v_0v_1u$ is a 3-face of G. If $vv_0 \in E(G)$, then $f_1 = v_0vv_3$ and $f_2 = v_0vu$ are 3-faces of G, too. Then, $G - v_0$ is a plane triangle-free graph (by (1)) and $d_G(v_0) = 4$ implying that $\chi(G) \leq 3$ (by Theorem 9.17). This contradiction to (2) shows that $vv_0 \notin E(G)$. By symmetry, we also have $v_1w \notin E(G)$. Furthermore, $C_1 = (u, v_1, v_2, w, u)$ and $C_2 = (u, v_0, v_3, v, u)$ are 4-cycles of G (see Figure 9.4).

Claim 9.19.5 *Neither C_1 nor C_2 is a facial cycle of G.*

Proof: Suppose this is false. By symmetry, we may assume that C_2 is a facial 4-cycle of G. Then the graph G' obtained from $G - v_0$ by adding the edge vv_1 is a planar graph and $s(G') < s(G)$. If G' has at most three triangles, then G' has a 3-coloring φ' (by (3)). Then φ' is a 3-coloring of $G - v_0$ with $\varphi(v) \neq \varphi(v_1)$. By assigning color $\varphi(v)$ to v_0, we then obtain a 3-coloring of G, a contradiction to (2). Hence, the planar graph G' has at least four triangles. By (1), this implies that there is a vertex v' in G such that (v, v_1, v', v) is a triangle of G', and so $v'v_1, v'v \in E(G)$. Since $v_1w \notin E(G)$, we obtain that $v' = v_2$. Consequently, $C_2 = (u, v_0, v_3, v, u)$ is a facial 4-cycle of G, and G has two paths $Q = (u, v_1, v_2, v)$ and $Q' = (v_0, v_1, v_2, v_3)$ with $v_1, v_2 \notin V(C_2)$. This contradiction to Claim 9.19.4 proves the claim. △

For $i \in \{1,2\}$, let G_i be the plane subgraph of G consisting of the cycle C_i and the part of G inside C_i. Clearly, G_i is a plane graph and C_i is the outer cycle of G_i. By Claim 9.19.5 neither $C_1 = (u, v_1, v_2, w, u)$ nor $C_2 = (u, v_0, v_3, v, u)$ is a facial cycle of G. Hence Claim 9.19.1 implies that G_i has a triangle,. Since G has at most three triangles and (u, v_0, v_1, u) is a triangle of G, we obtain that G_i has exactly one triangle. Furthermore, Claim 9.19.1 implies $C = (u, w, v_2, v_3, v, u)$ is a facial 5-cycle (we may assume that C is the outer cycle of G), and $N_G(u) \cap N_G(v_2) = \{v_1, w\}$ and $N_G(u) \cap N_G(v_3) = \{v_0, v\}$ (by Claim 9.19.4). Then $G_i' = G_i + uv_{i+1}$ is a planar graph with three triangles. Note that $T_1 = (u, v_1, v_2, u)$ is a triangle of G_1' and $T_2 = (u, v_0, v_3, u)$ is a triangle of G_2'. From (3) it follows that G_i' has a 3-coloring φ_i. Then φ_i is a 3-coloring of G_i and $|\varphi_i(V(T_i))| = 3$. By permuting colors if necessary, we may assume that $\varphi_1(u) = \varphi_2(u)$, $\varphi_1(v_1) = \varphi_2(v_3)$ and $\varphi_1(v_2) = \varphi_2(v_0)$. Then $\varphi_1 \cup \varphi_2$ is a 3-coloring of G. This contradiction to (2) completes the proof. □

The following result belongs to a long list of similar results that are motivated by Grötzsch's theorem and by a conjecture due to Steinberg [961] from 1975, see the discussion in Section 9.5. The proof of this result again demonstrates the usefulness of the Kostochka–Yancey density result for 4-critical graphs.

Theorem 9.20 *Every planar graph without cycles of length* 4 *through* 8 *is* 3-*colorable.*

Proof Suppose this is false. Then there exists a plane graph G such that $G \in \mathrm{Crit}(4)$ and G has no cycles whose length belongs to the interval $[4,8]$. Let n, m and ℓ denote the number of vertices, the number of edges and the number of faces of G, respectively. Furthermore, let n_3, m_3 and ℓ_3 denote the number of vertices having degree 3 in G, the number of edges contained in some 3-face of G, and the number of 3-faces of G, respectively. By Euler's formula, we have

$$(1)\ n - m + \ell = 2.$$

Since $G \in \mathrm{Crit}(4)$, we obtain that $\delta(G) \geq 3$ and each face is bounded by a cycle. By summing up the vertex degrees, we obtain $2m \geq 3n_3 + 4(n - n_3) = 4n - n_3$ which leads to

$$(2)\ n_3 \geq 4n - 2m.$$

Since G has no 4-cycle, no edge of G is contained in two triangles. Consequently, we obtain that $3\ell_3 = m_3$. Furthermore, if v is a vertex of degree 3 in G, then at least one edge in $E_G(v)$ is not contained in a triangle. This implies that $m - m_3 \geq n_3/2$. Combining this with (1) and (2), we obtain that

$$(3)\ \ell_3 = \frac{m_3}{3} \leq \frac{m - n_3/2}{3} \leq \frac{2m - 2n}{3} = \frac{2\ell - 4}{3} < \frac{2\ell}{3}.$$

By summing up the face degrees, we obtain that $2m \geq 3\ell_3 + 9(\ell - \ell_3) = 9\ell - 6\ell_3$, since every face is bounded by a 3-cycle or a p-cycle with $p \geq 9$. By using (3), this leads to $9\ell \leq 2m + 6\ell_3 < 2m + 4\ell$, and hence to $\ell < \frac{2}{5}m$. Using (1), we then obtain that

$$m = n + \ell - 2 < n + \tfrac{2}{5}m - 2,$$

and hence $m < \frac{1}{3}(5n - 10)$. On the other hand, since $G \in \text{Crit}(4)$, the Kostochka–Yancey bound (see Theorem 5.10) leads to $m \geq \frac{1}{3}(5n - 2)$; which is impossible. □

In view of the four color theorem, the three color problem has naturally gained in importance. However, we know from complexity theory (see Theorem 7.38) that a good characterization of the class of 3-colorable planar graphs is unlikely unless P=NP. That (vertex) 3-colorability of planar graphs is NP-complete was proved by Stockmeyer [981] in 1973. Surprisingly, the 3-coloring problem for the class of maximal planar graphs can be solved efficiently, see Theorem 9.21. This result was already formulated by Heawood [487], but without proof; the name three color theorem was first used by Franklin [389]. As Heawood pointed out in his follow-up paper of 1898 [488], the three color theorem leads to a necessary and sufficient 3-color criterion for planar graphs (see Corollary 9.24); however, this criterion does not lead to a polynomial algorithm for deciding whether a planar graph is 3-colorable.

Theorem 9.21 (THREE COLOR THEOREM) *A maximal planar graph with at least three vertices is 3-colorable if and only if all its vertices have even degree.*

The proof of Theorem 9.21 is based on the fact that if a graph $G \in \text{Crit}(4)$ contains a wheel W as a subgraph, then $G = W$ and W is an odd wheel (see Lemma 4.23). Recall that a **wheel** is a graph of the form $W_n = K_1 \boxplus C_n$ with $n \geq 3$; the wheel W_n is said to be **odd** or **even** depending on whether n is odd or even.

Let G be a maximal planar graph with $|G| \geq 3$. We may fix an embedding of G; any such embedding is a plane triangulation (by Corollary C.28). If G has a vertex of odd degree, say d, then $d \geq 3$ and G contains an odd wheel W as a subgraph, and so $\chi(G) \geq \chi(W) = 4$. This proves the "only if" direction of Theorem 9.21. For the proof of the "if" direction we consider plane near-triangulations. Let G be a plane near-triangulation whose outer cycle is C; we say that G is **internally even** if every vertex inside C has even degree in G. The "if" direction of Theorem 9.21 is an immediate consequence of the following result.

Theorem 9.22 *Every internally even plane near-triangulation is 3-colorable.*

Proof Let G be plane near-triangulation that is internally even, and let C be the outer cycle of G. We prove by induction on $e(G)$ that G has a coloring with color set $S = \{1,2,3\}$. If $e(G) = 3$, then $G = C = K_3$ and the statement is evident. So assume that $e(G) \geq 4$. First suppose that there is a chord of C in G, that is, an edge $uv \in E(G) \setminus E(C)$ such that $u, v \in V(C)$. Then G is the union of two internally even plane near-triangulations, say G_1 and G_2, which have only the edge uv in common. By the induction hypothesis, for $i \in \{1,2\}$, there is a coloring φ_i of G_i with color set S and, by permuting colors if necessary, we may assume that $\varphi_i(u) = 1$ and $\varphi_i(v) = 2$. Then $\varphi_1 \cup \varphi_2$ is a coloring of G with color set S, and we are done. Now assume that C has no chord in G. Let uv be an edge of C. Then uv belongs to a facial triangle, say $C' = (u, v, w, u)$, where $w \in V(G) \setminus V(C)$. Since G is an internally even near-triangulation, $d_G(w) \geq 4$ is even and $G[N_G(w)]$ contains an even cycle

C_1 with $uv \in E(C_1)$. Then $G' = G - uv$ is a plane near triangulation whose outer cycle is $C - uv + uw + wv$, and G' is internally even. By the induction hypothesis, G' has a coloring φ' with color set S. In G' the vertex w is joined to all vertices of the u-v path $C_1 - uv$ whose length is odd, implying that $\varphi'(u) \neq \varphi'(v)$. Hence φ' is a coloring of G with color set S. This completes the proof. □

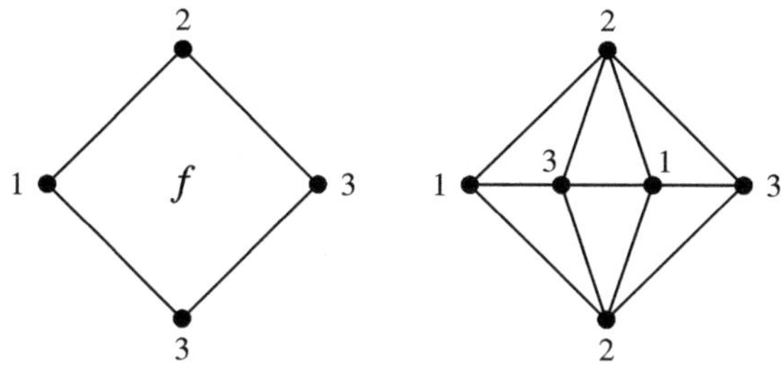

Fig. 9.5 Illustration to Lemma 9.23.

Lemma 9.23 *Let H be a planar graph with $\chi(H) \leq 3$. Then there is a maximal planar graph G such that $|G| \geq 3$, H is an induced subgraph of G and $\chi(G) \leq 3$.*

Proof We may add isolated vertices to H so that $|H| \geq 3$ and $\chi(H) \leq 3$. Let $\mathcal{X}$ be the set of all graphs G such that G is a 2-connected planar graph, H is an induced subgraph of G and $\chi(G) \leq 3$. First we show that $\mathcal{X} \neq \varnothing$. There is a maximal planar graph G' such that $H \subseteq G'$ and $|H| = |G'|$. Let $E' = E(G') \setminus E(H)$. For every edge $e = uv \in E'$ add a new vertex $v(e)$ and replace the edge e by the two edges $uv(e)$ and $v(e)v$. Denote the resulting graph by G. Clearly, $G \in \mathcal{X}$.

To complete the proof, it suffices to show that $\mathcal{X}$ contains a maximal planar graph. For a graph $G \in \mathcal{X}$, define $p(G) = 3|G| - 6 - e(G)$. Since every graph in $\mathcal{X}$ is planar and $|G| \geq 3$, we have $p(G) \geq 0$ for all $G \in \mathcal{X}$ (by Corollary C.27). Hence, there is a graph G in $\mathcal{X}$ for which $p(G)$ is minimum. If $p(G) = 0$, then $e(G) = 3|G| - 6$ and G is a maximal planar graph (by Corollary C.27). It remains to consider the case that $p(G) > 0$. To arrive at a contradiction, we fix a coloring φ of G with color set $S = \{1,2,3\}$, and we fix an embedding G' of G in the plane. Since G is 2-connected, every face is bounded by a cycle. Furthermore, G' is not a triangulation (by Corollary C.28). Hence, there is a facial cycle, say C, whose length is at least 4. If $V(C)$ contains a set U such that $|U| \geq 4$ and $|\varphi(U)| = 2$, then let G' be the graph obtained from G by adding a new vertex v and joining v to all vertices of U. Clearly, $G' \in \mathcal{X}$ and $p(G') = p(G) + 3 - |U| \leq p(G) - 1$, contradicting the choice of G. Otherwise, $|C| = 4$ and $|\varphi(V(C))| = 3$. By permuting colors if necessary, we may assume that $C = (u_1, u_2, u_3, u_4, u_1)$, $\varphi(u_i) = i$ for $i \in \{1,2,3\}$ and $\varphi(u_4) = 2$. Let G' be the graph obtained from G by adding a two new vertices u and v and the new edges $uu_4, uu_1, uu_2, vu_2, vu_3, vu_4$ and uv (see Figure 9.5). Clearly, $G' \in \mathcal{X}$ and $p(G') = p(G) + 6 - 7 < p(G)$, contradiction the choice of G. □

Corollary 9.24 *A planar graph is 3-colorable if and only if it is a subgraph of a maximal planar graph in which all vertices have even degree.*

Corollary 9.25 (Fisk) *Let G be a maximal planar graph with $|G| \geq 3$, and let u, v be the only two vertices having odd degree in G. Then $uv \notin E(G)$.*

Proof Suppose this is false, i.e. $uv \in E(G)$. Then there is an embedding of G in the plane such that G is a triangulation (by Corollary C.28) and uv belongs to the outer face of this embedding (by Proposition C.25). Then $\chi(G) \leq 3$ (by Theorem 9.22) and $\chi(G) \geq 4$ (by Theorem 9.21), which is impossible. □

Fisk did not formulate the statement in Corollary 9.25 in explicit form, but he proved in [376] that if a plane triangulation G has exactly two vertices of odd degree, and they are adjacent, then G has no 4-coloring (see Exercise 9.8)). By the four color theorem, this result immediately implies Corollary 9.25. However, this corollary can be also deduced from the three-color theorem. As pointed out by Král' et al. [644, Proposition 1], Corollary 9.25 can be also proved by using the dual version of the two-color theorem. The two-color theorem is a simple consequence of König's characterization of bipartite graphs, which says that a graph is nonbipartite if and only if it contains an odd cycle; and of (C.5) which implies that a plane graph has an even number of faces of odd degree.

Theorem 9.26 (Two Color Theorem) *A plane graph is bipartite if and only if each of its faces has even degree.*

Proof Let G be a plane graph. First assume that G has a face f such that $d_G(f)$ is odd. Then the boundary subgraph $\mathrm{Bd}(f)$ of G has a closed walk of odd length ≥ 3. Then G contains an odd cycle (see Proposition C.4), and so G is not bipartite. If G is not bipartite, then G contains an odd cycle, say C. Let G' denote the subgraph of G consisting of C and the part of G inside C. Clearly, G' is a plane graph and C is the boundary cycle of the outer face f' of G'. Since $d_{G'}(f') = |C|$ is odd, it follows from (C.5) that there is a face $f \in F(G') \setminus \{f'\}$ such that $d_{G'}(f)$ is odd, too. But then $f \in F(G)$ and $d_G(f) = d_{G'}(f)$ is odd. This completes the proof. □

9.3 Map Color Theorems for the Projective Plane

In this section we study coloring problems of graphs embedded in the projective plane $\mathbf{N}_1$. A **projective planar graph** is a graph that belongs to the class $\mathcal{G}(\hookrightarrow \mathbf{N}_1)$; a **projective plane graph** is a graph that is embedded in $\mathbf{N}_1$. Our aim is to prove a map color theorem for projective planar graphs with respect to the DP-chromatic number. The corresponding result with respect to the list-chromatic number was proved by Böhme, Mohar, and Stiebitz [125].

Theorem 9.27 (Böhme, Mohar, and Stiebitz) *Every graph $G \in \mathcal{G}(\hookrightarrow \mathbf{N}_1)$ satisfies $\chi_{\mathrm{DP}}(G) < H(\mathbf{N}_1) = 6$, unless G contains K_6 as a subgraph.*

For a surface $\mathbf{S}$ and an integer k, let $\text{Crit}_{\text{DP}}(\mathbf{S},k) = \text{Crit}_{\text{DP}}(k) \cap \mathcal{G}(\hookrightarrow \mathbf{S})$. As an immediate corollary of the above theorem we obtain that $\text{Crit}_{\text{DP}}(\mathbf{N}_1,6) = \langle K_6 \rangle$. That K_6 is the only 6-critical graph in $\mathbf{N}_1$ was proved by Albertson and Hutchinson [34]; that this also holds for χ_ℓ-critical graphs was proved by Böhme, Mohar, and Stiebitz [125].

For the proof of Theorem 9.27 we need an auxiliary result about the extension of a given DP-coloring of the outer cycle of a plane graph G to the entire graph G; see Theorem 9.30. Let G be a graph, let (X,H) be a cover of G, let G' be a subgraph of G, and let φ' be an (X,H)-coloring of G'. We say that φ' can be **extended** to an (X,H)-coloring of G if there is an (X,H)-coloring φ of G such that $\varphi(v) = \varphi'(v)$ for all $v \in V(G')$; in this case we call φ an **extension** of φ'. A graph G is **nontrivial** if $|G| \geq 2$.

Theorem 9.28 *Let G be a plane graph, let G' be the boundary subgraph of the outer face of G, and let $uu' \in E(G')$. Furthermore, let (X,H) be a cover of G such that $|X(v)| \geq 3$ for all $v \in V(G')$ else $|X(v)| \geq 5$, and let $x,x' \in V(H)$ be two colors such that $x \in X(u)$, $x' \in X(u')$ and $xx' \notin E(H)$. Then there exists an (X,H)-coloring φ of G such that $\varphi(u) = x$ and $\varphi(u') = x'$.*

Proof The proof is by induction on the order of G. If $|G| = 2$, the statement is obvious; so assume that $|G| \geq 3$. First assume that G is the union of two nontrivial induced subgraphs, say G_1 and G_2, which have at most one vertex in common. By symmetry, we may assume that uu' is an edge of G_1. Then, by the induction hypothesis, there is an (X,H)-coloring φ_1 of G_1 such that $\varphi_1(u) = x$ and $\varphi_1(u') = x'$. Again, by the induction hypothesis, there is an (X,H)-coloring φ_2 of G_2; and if v is the only common vertex of G_1 and G_2, then v belongs to the outer face of G_2 and we may choose φ_2 such that $\varphi_2(v) = \varphi_1(v)$. Then $\varphi = \varphi_1 \cup \varphi_2$ is an (X,H)-coloring of G such that $\varphi(u) = x$ and $\varphi(u') = x'$. It remains to consider the case that G is 2-connected. Then G is a plane graph whose outer face is bounded by a cycle, say C. Then there is a near-triangulation G^* with outer cycle C such that G is a spanning subgraph of G^*. Then there is an (X,H)-coloring φ of G^* such that $\varphi(u) = x$ and $\varphi(u') = x'$ (by Theorem 9.12). Clearly, φ is an (X,H)-coloring of G and we are done. □

Theorem 9.29 *Let G be a plane graph, let G' be the boundary subgraph of the outer face of G, and let $P = (u_1,u_2,\ldots,u_p)$ be a path of G'. Furthermore, let (X,H) be a cover of G such that $|X(v)| \geq 5$ for all $v \in V(G) \setminus V(G')$, $|X(v)| \geq 4$ for all $v \in V(P) \setminus \{u_1,u_p\}$, $|X(v)| \geq 2$ for all $v \in \{u_1,u_p\}$, and $|X(v)| \geq 3$ for all $v \in V(G') \setminus V(P)$. Then G has an (X,H)-coloring.*

Proof For $p \leq 2$, the result follows from Theorem 9.28 and the fact that $|X(v)| \geq 2$ for $v \in \{u_1,u_p\}$. Now assume that $p \geq 3$. To show that G has an (X,H)-coloring, we use induction on the order of G.

First consider the case that G is the union of two nontrivial induced subgraphs, say G_1 and G_2, which have at most one vertex in common and with $P \subseteq G_1$. Then, by the induction hypothesis, there is an (X,H)-coloring φ_1 of G_1. By Theorem 9.28, there is an (X,H)-coloring φ_2 of G_2; and if v is the only common vertex of G_1 and G_2, then v belongs to the outer face of G_2 and we may choose φ_2 such that $\varphi_2(v) = \varphi_1(v)$. Then $\varphi = \varphi_1 \cup \varphi_2$ is an (X,H)-coloring φ of G, and we are done.

It remains to consider the case that G is connected and every block of G is either an edge of P, or a 2-connected plane graph with an outer cycle C' such that $C' \cap P$ a subpath of P with at least two vertices, where for distinct 2-connected blocks of G, these subpaths are edge-disjoint. Then there is a near-triangulation G^* with outer cycle C such that G is a spanning subgraph of G^*, $V(C) = V(G')$, and P is a subpath of C. Then $C = (u_1, u_2, \ldots, u_q, u_1)$ with $q \geq p \geq 3$. If $q = p$, then it follows from Theorem 9.28 with the precolored edge $u_1 u_p$, that there is an (X,H)-coloring of G^* and hence also of G. It remains to consider that case when $q \geq p+1$. To see that there is an (X,H)-coloring of G^* and hence of G, we distinguish two cases.

Case 1: *C has an chord in G^* incident with u_p, say $u_p u_i$.* If $i \in [1, p-2]$, then we apply the induction hypothesis to the cycle $C_1 = (u_1, u_2, \ldots, u_i, u_p, \ldots u_q, u_1)$ and the part of G^* inside C_1 with respect to the path $P' = (u_1, u_2, \ldots, u_i, u_p)$, and then we apply Theorem 9.12 to the cycle $C_2 = (u_i, u_{i+1}, \ldots, u_p, u_i)$ and the part of G^* inside C_2 where $u_p u_i$ is the precolored edge. This results in an (X,H)-coloring of G^*. Otherwise $i \in [p+2, q]$. Then we apply the induction hypothesis to the cycle $C_1 = (u_1, u_2, \ldots u_p, u_i, u_{i+1}, \ldots, u_q)$ and the part of G^* inside C_1 with respect to the path P, and then we apply Theorem 9.12 to the cycle $C_2 = (u_p, u_{p+1}, \ldots, u_i, u_p)$ and the part of G^* inside C_2 where $u_p u_i$ is the precolored edge. This results in an (X,H)-coloring of G^*, too.

Case 2: *C has an no chord in G^* incident with u_p.* Since G^* is a near-triangulation whose outer cycle is C, the neighbors of u_p in G belong to a path P' whose ends are u_{p-1} and u_{p+1}, and whose inner vertices lie inside C. Then $C' = (C - u_p) \cup P$ is a cycle in G, and $G^- = G^* - u_p$ is a plane near-triangulation whose outer cycle is C'. Let $H' = H - X(u_p)$. Since $|X(u_p)| \geq 2$, there are two (distinct) colors $y, y' \in X(u_p)$. Let $X' : V(G') \to 2^{V(H')}$ be the map with $X'(v) = X(v) \setminus (N_H(y) \cup N_H(y'))$ if v is a vertex of $P' - u_{p+1}$ else $X'(v) = X(v)$. By (C2), this implies that $|X'(v)| \geq |X(v)| - 2$ for every vertex v of $P' - u_{p+1}$; in particular, we have $|X'(u_{p-1})| \geq 2$. Then we apply the induction hypothesis to the plane graph G^-, his outer cycle C', the path $P - u_p$, and the cover (X', H'). This results in an (X', H')-coloring φ' of $G^- = G - u_p$. If $v \in N_{G^*}(u_p) \setminus \{u_{p+1}\}$, then $\varphi'(v) \in X'(v) = X(v) \setminus (N_H(y) \cup N_H(y'))$. Furthermore, by (C2), one of the colors y and y' is not adjacent to the color $\varphi'(u_{p+1})$ in H. By assigning that color to u_p, we get an (X,H)-coloring φ of G. This completes the proof. □

Theorem 9.30 *Let G be a plane graph with outer cycle C, where $|C| = p$ and $3 \leq p \leq 6$, let (X,H) be a cover of G such that $|X(v)| \geq 5$ for all $v \in V(G)$, and let φ' be an (X,H)-coloring of $G[V(C)]$. Then φ' can be extended to an (X,H)-coloring of the entire graph G, unless $p \geq 5$ and the notation may be chosen so that $C = (v_1, v_2, \ldots, v_p, v_1)$, $\varphi'(v_i) = y_i \in X(v_i)$ for $i \in [1,p]$, and one of the following four conditions hold, where all indices are computed modulo p:*

(a) *$p \leq 6$ and there is a vertex u inside C such that $u \in N_G(v_i)$ for $i \in [1,5]$, $X(u) = \{x_i \mid i \in [1,5]\}$, and $x_i y_i \in E(H)$ for $i \in [1,5]$.*

(b) *$p = 6$ and there is a vertex u inside C such that $u \in N_G(v_i)$ for $i \in [1,6]$, $X(u) = \{x_i \mid i \in [1,6]\}$, and $x_i y_i \in E(H)$ for $i \in [1,6]$.*

(c) *$p = 6$ and there is an edge u_0u_1 inside C such that, for $i \in \{0,1\}$, we have $v_{3i+1}, v_{3i+2}, v_{3i+3}, v_{3i+4} \in N_G(u_i)$, $X(u_i) = \{x_{i,3i+j}, z_i \mid j \in [1,4]\}$, $z_0z_1 \in E(H)$ and $x_{i,3i+j} y_{3i+j} \in E(H)$ for $i \in \{0,1\}$ and $j \in [1,4]$.*

(d) *$p = 6$ and there is a triangle (u_0,u_1,u_2,u_0) inside C such that, for $i \in [0,2]$, we have $v_{2i+1}, v_{2i+2}, v_{2i+3} \in N_G(u_i)$, $X(u_i) = \{x_{i,2i+j}, z_i, z_i' \mid j \in [1,3]\}$, and $z_0z_1, z_1z_2, z_2z_0, z_0'z_1', z_1'z_2', z_2'z_0' \in E(H)$ and $x_{i,3i+j} y_{3i+j} \in E(H)$ for $i \in [0,2]$ and $j \in [1,3]$.*

Proof The proof is by induction on the order of G. If (G,X,H,φ') satisfies one of the four conditions (a), (b), (c), or (d), we say that (G,X,H,φ') is a *bad configuration.* For $G' \subseteq G$ and $u \in V(G)$, let $d(u:G') = |N_G(u) \cap V(G')|$. For the proof we consider two cases.

Case 1: *There is an edge $vw \in E(C)$ such that $d(u : C-v-w) \le 2$ for every vertex u inside C.* Let $W = V(C-v-w)$ and let $G^- = G-W$. Then G^- is a plane graph. Let G' be the boundary subgraph of the outer face of G^-. Clearly, vw belongs to $E(G')$. For $u \in V(G^-) \setminus \{v,w\}$, let

$$X'(u) = X(u) \setminus \bigcup_{v' \in W \cap N_G(u)} N_H(\varphi'(v')),$$

and, for $u \in \{v,w\}$, let $X'(u) = X(u)$. Let U be the union of the sets $X'(u)$ with $u \in V(G^-)$, and let H'=H[U]. Then (X',H') is a cover of G^- such that $|X'(u)| \ge |X(u)| - 2 \ge 3$ if $u \in V(G')$ else $|X'(u)| \ge 5$. From Theorem 9.28 it then follows that there is an (X',H')-coloring φ^- of G^- such that $\varphi^-(u) = \varphi'(u)$ if $u \in \{v,w\}$. Then $\varphi = \varphi' \cup \varphi^-$ is an (X,H)-coloring of G, that is, φ' can be extended to an (X,H)-coloring of G.

Case 2: *For every edge $vw \in E(C)$ there exists a vertex u inside C such that $d(u : C-v-w) \ge 3$.* Clearly, this implies that $|C| = p \ge 5$. We say that G is *reducible* if G has a separating cycle (i.e., there are vertices inside and outside of this cycle) of length at most four.

First assume that G is reducible, and let C' be a separating cycle of G with $|C'| \le 4$. Clearly, $C' \ne C$. Let G' be the graph obtained from G by deleting all vertices inside C'. Then G' is a plane graph and C is the outer cycle of G'. If φ' can be extended to an (X,H)-coloring of G', then, by using the arguments from Case 1, we can extend this coloring to the vertices inside C', and, therefore, φ' can be extended to an (X,H)-coloring of G. Otherwise, by the induction hypothesis, we obtain that (G',X,H,φ') is a bad configuration, implying that (G,X,H,φ') is a bad configuration, too.

Now assume that G is not reducible. Let u be a vertex inside C such that $d = d(u:C)$ is maximum. By the assumption of Case 2, we have $3 \le d \le |C| = p \le 6$. If $d \ge 5$, the u is the only vertex inside C, since otherwise G would be reducible. Then it is easy to see that φ' can be extended to an (X,H)-coloring of G, unless (a) or (b) holds.

Next, consider the case when $d = 4$. Since G is not reducible, it follows from the assumption of Case 2 that $p = 6$ and the four neighbors of u in C are consecutive on C. So we may assume that u is in G adjacent to v_1, v_2, v_3 and v_4, but not to v_5 and v_6. Furthermore, $C' = (v_1, u, v_4, v_5, v_6, v_1)$ is a 5-cycle of G and all vertices of $G - (V(C) \cup \{u\})$ are inside C'. Let $X'(u)$ be the set of colors $z \in X(u)$ such that $N_H(z) \cap \{y_1, y_2, y_3, y_4\} = \emptyset$. Since $|X(u)| \geq 5$, it follows from (C2) that $|X'(u)| \geq 1$. Let $z \in X'(u)$ be an arbitrary color. Then we can extend φ' to an (X,H)-coloring φ^+ of $G[V(C) \cup \{u\}]$ with $\varphi^+(u) = z$. If φ^+ can be extended to an (X,H)-coloring of G, we are done. Otherwise, it follows from the the induction hypothesis (applied to the planar graph consisting of the cycle C' and the part of G inside C' with the precoloring φ^+) that there is a vertex u' inside C' such that u' is in G adjacent to all vertices of C', $X(u') = \{x_1, x_4, x_5, x_6, z'\}$ and $x_1y_1, x_4y_4, x_5y_5, x_6y_6, zz' \in E(H)$. Then u and u' are the only vertices inside C, and it follows from (C1) that $|X'(u)| = 1$ and (c) holds with $u = u_0$ and $u' = u_1$.

Finally, consider the case when $d = 3$. Since G is not reducible, it follows from the assumption of Case 2 that every vertex u inside C with $d(u : C) = 3$ has three consecutive neighbors on C. Furthermore, we obtain that $p = 6$ and there are three vertices inside C, say u_0, u_1 and u_2, such that $N_G(u_i) \cap V(C) = \{v_{2i+1}, v_{2i+2}, v_{2i+3}\}$ for $i \in [0,2]$. Again, since G is not reducible, this implies that all vertices of $G - (V(C) \cup \{u_0, u_1, u_2\})$ are inside the 6-cycle $(v_1, u_0, v_3, u_1, v_5, u_2, v_1)$.

For $i \in [0,2]$, let $X'(u_i)$ be the set of color $z \in X(u_i)$ such that $N_H(z) \cap \{y_{2i+1}, y_{2i+2}, y_{2i+3}\} = \emptyset$. Since $|X(u)| \geq 5$, it follows from (C2) that $|X'(u_i)| \geq 2$ for $i \in [0,2]$. First, assume that $T = (u_0, u_1, u_2, u_0)$ is a triangle in G. Since G is not reducible, this implies that $V(G) = V(C) \cup V(T)$ and G is a plane near-triangulation with outer cycle C. Let $H^T = H[X'(u_0) \cup X'(u_1) \cup X'(u_2)]$ and let X^T be the map with $X^T(u_i) = X'(u_i)$ for $i \in [0,2]$. Then (X^T, H^T) is a cover of T. If there exists an (X^T, H^T)-coloring φ^T of T, then $\varphi = \varphi' \cup \varphi^T$ is an (X,H)-coloring of G, and we are done. Otherwise, we easily conclude that, for $i \in [0,2]$, the set $X'(u_i)$ consists of precisely two colors, say z_i and z_i', such that $z_0z_1, z_1z_2, z_2z_0, z_0'z_1', z_1'z_2', z_2'z_0' \in E(H^T)$. Then it follows that (d) holds, and we are done, too.

So it remains to consider the case that T is no triangle of G. By symmetry, we may assume that $u_0u_2 \notin E(G)$. Then $G' = G - v_2$ is a plane graph with outer cycle $C' = (v_1, u_0, v_3, v_4, v_5, v_6, v_1)$. There is a color $z \in X'(u_0)$ and φ' can be extended to a (X,H)-coloring φ^+ of $G[V(C) \cup \{u_0\}]$ with $\varphi^+(u_0) = z$. If this coloring can be extended to an (X,H)-coloring of G, we are done. Otherwise, it follows from the induction hypothesis that (G', X, H, φ^+) is a bad configuration. Note that the vertices u_1 and u_2 lie inside C', $N_G(u_2) \cap V(C') = \{v_1, v_5, v_6\}$ and $\{v_3, v_4, v_5\} \subseteq N_G(u_1) \cap V(C') \subseteq \{v_3, v_4, v_5, u_0\}$. This implies that there is a vertex inside C' distinct from u_1 and u_2 which has two neighbors in the set $\{v_1, v_3, v_5\}$. But then G is reducible, a contradiction. This completes the proof. □

Let G be a complete graph K_n with $n \geq 2$, and let (X,H) be a cover of G. Recall from Section 1.4, that (G,X,H) is a K-configuration if there is a partition $(X_1(v), X_2(v), \ldots, X_{n-1}(v))$ of $X(v)$ for every $v \in V(G)$ such that

$$H_i = H[\bigcup_{v \in V(G)} X_i(v)] \text{ is a } K_n \text{ for } i \in [1, n-1] \text{ and } H = \bigcup_{i=1}^{n-1} H_i.$$

Then it is easy to prove by induction on n, that if (X,H) is a cover of G such that $|X(v)| \geq n-1$ for all $v \in V(G)$, then G has no (X,H)-coloring if and only if (G,X,H) is a K-configuration. This is a very special case of a more general result (Theorem 1.17) proved in Section 1.4. Theorem 9.27 is an immediate consequence of the following result.

Theorem 9.31 *Let G be a projective plane graph, and let (X,H) be a cover of G. Suppose that $|X(v)| = 5$ for every vertex $v \in V(G)$ and, for every complete subgraph G' on six vertices of G, $(G',(X,H)/G')$ is not a K-configuration. Then G has an (X,H)-coloring.*

Proof The proof is by contradiction. So let (G,X,H) be a smallest counterexample, that is, G is a projective plane graph and (X,H) is a cover of G such that the following conditions hold:

(1) $|X(v)| = 5$ for all $v \in V(G)$,
(2) for every complete subgraph G' on six vertices of G, $(G',(X,H)/G')$ is not a K-configuration,
(3) G has no (X,H)-coloring, and
(4) $|G|$ is minimum subject to (1), (2) and (3).

Claim 9.31.1 *The following statements hold:*

(a) $\delta(G) \geq 5$,
(b) *each contractible triangle of G is a facial cycle of G, and*
(c) *G is not a K_6.*

Proof: For the proof of (a), suppose there is a vertex $v \in V(G)$ such that $d_G(v) \leq 4$. Since $G' = G - v$ is a projective plane graph, G' has an (X,H)-coloring φ' (by (4)). Since $|X(v)| = 5$ (by (1)), there is a color $x \in X(v)$ such that $N_H(x) \cap \varphi'(N_G(v)) = \varnothing$ (by (C2)). Hence, by coloring v with x, we can extend φ' to an (X,H)-coloring of G, a contradiction to (3). This proves (a). For the proof of (b), let C be a contractible triangle of G. Let G_1 denote the subgraph of G that consists of the triangle C and the part of G inside C; and let $G_2 = G - (V(G_1) \setminus V(C))$. If C is a nonfacial cycle of G, then G_2 is a projective plane graph with $|G_2| < |G|$, implying that there is an (X,H)-coloring φ_2 of G_2 (by (4)). Furthermore, G_1 is a plane graph whose outer cycle is the triangle C. Then it follows from Theorem 9.30 that there is an (X,H)-coloring φ_1 of G_1 such that $\varphi_1(v) = \varphi_2(v)$ for all $v \in V(C)$. Then $\varphi = \varphi_1 \cup \varphi_2$ is an (X,H)-coloring of G. This contradiction to (3) proves (b). Statement (c) follows immediately from (2) (see Theorem 1.17). △

Recall from Appendix C.13 that a projective plane graph has two types of cycles: contractible cycles and noncontractible cycles. If all cycles of G are contractible, then G is a plane graph, and there is an (X,H)-coloring of G (by Theorem 9.11), a contradiction to (3). So we may assume that G contains a noncontractible cycle. Let

p be the minimum length of a noncontractible cycle of G, and let $\mathcal{NC}$ denote the set of all noncontractible cycles of G of length p. Note that $p \ge 3$.

Let $C = (v_1, v_2, \ldots, v_p, v_1) \in \mathcal{NC}$ be an arbitrary noncontractible cycle of length p. Then C has no chords in G and, by cutting the projective plane $\mathbf{N}_1$ along C, we obtain a plane graph G_C with outer cycle $O_C = (v_1^1, v_2^1, \ldots, v_p^1, v_1^2, v_2^2, \ldots, v_p^2, v_1^1)$. The plane graph G_C can be thought of as a representation of G on a closed disc, where antipodal points are identified on the boundary, see Figure 9.6. The plane graphs $G - V(C)$ and $G_C - V(O_C)$ are identical, and, for $u \in V(G) \setminus V(C)$ and $i \in [1, p]$, $uv_i \in E(G)$ if and only if uv_i^1 or uv_i^2 belongs to $E(G_C)$. Furthermore, a path $P = (v_i^1, u_1, u_2, \ldots, u_q, v_i^2)$ of G_C with $u_1, u_2, \ldots u_q \in V(G) \setminus V(C)$ corresponds to a noncontractible cycle $(v_i, u_1, u_2, \ldots, u_q, v_i)$ of G implying that $q + 1 \ge p$. In particular, for every $u \in V(G) \setminus V(C)$, the edges uv_i^1 and uv_i^2 are not both in $E(G_C)$. We denote by A_C the set of vertices of $G - V(C)$ that are in G adjacent to some vertex of C. For a vertex $u \in A_C$, let $N_C(u) = N_G(u) \cap V(C)$.

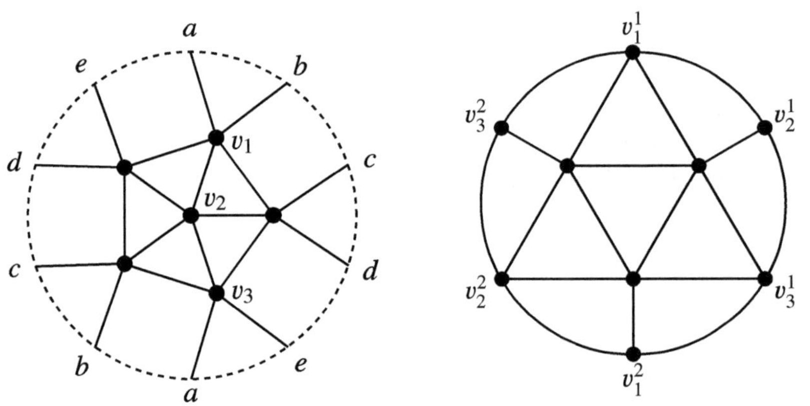

Fig. 9.6 A projective plane graph $G = K_6$ with a noncontractible triangle $C = (v_1, v_2, v_3, v_1)$ on the left side and its plane representation (G_C, O_C) on the right side.

Claim 9.31.2 $p \ge 4$.

Proof: Suppose this is false. Then $p = 3$ and there is a noncontractible cycle $C = (v_1, v_2, v_3, v_1) \in \mathcal{NC}$. Let (G_C, O_C) be the corresponding plane representation of (G, C), with $O_C = (v_1^1, v_2^1, v_3^1, v_1^2, v_2^2, v_3^2, v_1^1)$. Let φ' be an arbitrary (X, H)-coloring of C. To arrive at a contradiction, we show that φ' can be extended to an (X, H)-coloring of G.

First, we construct a cover (X', H') for the plane graph G_C with outer cycle O_C as follows. The cover (X', H') is obtained from (X, H) by splitting each set $X(v_i)$ into two copies $X'(v_i^1)$ and $X'(v_i^2)$. In particular, for the plane graph $G^- = G - V(C) = G_C - V(O_C)$ we have $(X', H')/G^- = (X, H)/G^-$. Furthermore, for $u \in A_C$, we replace an edge yx of H, where $y \in X(u)$ and $x \in X(v_i)$, by the edge yx^1 or by the edge yx^2 in H' depending on whether uv_i^1 or uv_i^2 belongs to G_C, where $x^j \in X(v_i^j)$ is the copy of $x \in X(v_i)$.

Now let φ_C be the (X',H')-coloring of O_C such that the color $\varphi_C(v_i^j) \in X'(v_i^j)$ is the copy of the color $\varphi'(v_i) \in X(v_i)$, where $i \in [1,3]$ and $j \in \{1,2\}$. If φ_C can be extended to an (X',H')-coloring of G_C, then φ' can be extended to an (X,H)-coloring of G, a contradiction to (3). If φ_C cannot be extended to an (X',H')-coloring of G_C, then we conclude from Theorem 9.30, that in the plane graph G_C there is a triangle C' inside O_C, such that each vertex of C' is in G_C adjacent to three vertices that are consecutive in O_C. This implies that in G each vertex of C' is adjacent to all vertices of C implying that $G[V(C) \cup V(C')]$ is a K_6. Since every contractible triangle is a facial triangle (by Claim 9.31.1(b)), we obtain that $V(G) = V(C) \cup V(C')$, and so G is a K_6, a contradiction to Claim 9.31.1(c). This proves the claim. △

In the following we have $p \geq 4$ (by Claim 9.31.2) and all indices are computed modulo p.

Claim 9.31.3 *If $C \in \mathcal{NC}$, then $|N_C(u)| \leq 3$ for all $u \in A_C$.*

Proof: Let $C = (v_1, v_2, \ldots, v_p, v_2) \in \mathcal{NC}$ be an arbitrary noncontractible cycle of G. If some vertex $u \in A_C$ is adjacent in G to v_i and v_j with $i < j$, then exactly one of the two cycles $(v_i, v_{i+1}, \ldots, v_j, u, v_i)$ and $(v_j, v_{j+1}, \ldots, v_i, u, v_j)$ is a noncontractible cycle of G. Therefore, $|N_C(u)| \leq 3$ if $p \geq 5$, for otherwise there would exists a noncontractible cycle in G whose length is at most $p-1$, a contradiction. If $p = 4$ and $N_C(u) = V(C)$ for some vertex $u \in A_C$, then all four triangles (u, v_i, v_{i+1}, u) with $i \in [1,4]$ are contractible and hence facial triangles in G (by Claim 9.31.1(b)). This, however, implies that $d_G(u) = 4$, a contradiction to Claim 9.31.1(b). This completes the proof. △

For a cycle $C \in \mathcal{NC}$, let T_C denote the set of vertices $u \in A_C$ such that $|N_C(u)| = 3$. Furthermore, for $v \in V(C)$, let $T_C^v = \{u \in T_C \mid uv \in E(G)\}$.

Claim 9.31.4 *Let $C = (v_1, v_2, \ldots, v_p, v_2) \in \mathcal{NC}$, let $u \in T_C$, and let $v \in V(C)$. Then the following statements hold:*

(a) *$N_C(u) = \{v_i, v_{i+1}, v_{i+2}\}$ for some $i \in [1,p]$, and both triangles (u, v_i, v_{i+1}, u) and (u, v_{i+1}, v_{i+2}, u) are contractible and hence facial triangles of G.*
(b) *The three neighbors of u in G_C that belong to O_C are consecutive on O_C.*
(c) *$N_C(w) \neq N_C(u)$ for all $w \in T_C \setminus \{u\}$*
(d) *$|T_C^v| \leq 3$ for all $v \in V(C)$.*

Proof: Statement (a) is an immediate consequence of Claim 9.31.1(b) and the fact that C is a shortest noncontractible cycle of G. Statement (b) follows immediately from (a). For the proof of (c), suppose this is false. Then, by (a), $N_C(w) = N_C(u) = \{v_i, v_{i+1}, v_{i+2}\}$ for some $i \in [1,p]$ and the four triangles (v, v_i, v_{i+1}, v) and (v, v_{i+1}, v_{i+2}, v) with $v \in \{u, w\}$ are facial triangles of G. This implies that $d_G(v_{i+1}) = 4$, a contradiction to Claim 9.31.1(a). Statement (d) follows from (a), (b) and (c). △

Let $C \in \mathcal{NC}$ be an arbitrary cycle. Two vertices $u, w \in T_C$ are *C-conform* if there is a vertex in O_C that is adjacent in G_C to both vertices u and w. If $v \in V(C)$ and $|T_C^v| = 3$, then it follows from Claim 9.31.4 that there are exactly two vertices in T_C^v that are C-conform.

Let φ be an arbitrary (X,H)-coloring of C. Then define the reduced cover (X',H') of the plane graph $G' = G - V(C)$ as follows. For a vertex $u \in V(G')$, let $X'(u) = X(u) \setminus N_H(\varphi(N_C(u)))$. So we delete from $X(u)$ all colors that are in H adjacent to a color $\varphi(v)$ with $v \in N_C(u)$. Furthermore, let U be the union of the sets $X'(u)$ with $u \in V(G')$ and let $H' = H[U]$. Clearly, then (X',H') is a cover of G' and we write $(X',H') = (X,H)/(\varphi,C)$. Note that if $u \in V(G') \setminus A_C$, then $X'(u) = X(u)$. Furthermore, every vertex $u \in A_C$ belongs to the boundary subgraph of the outer face of the plane graph G', and $|X'(u)| \geq |X(u)| - |\varphi(N_C(u))| \geq 2$ (by (C2) and Claim 9.31.3). If $A \subseteq A_C$, we say that the (X,H)-coloring φ of C is *A-good* if $|X'(u)| \geq 3$ for all $u \in A_C \setminus A$.

Claim 9.31.5 *Let $C \in \mathcal{NC}$, let φ be an (X,H)-coloring of C, and let $(X',H') = (X,H)/(\varphi,C)$ be the reduced cover of $G' = G - V(C)$. Then G' has no (X',H')-coloring.*

Proof: Suppose this is false, that is, there is an (X',H')-coloring φ' of G'. Since C is an induced cycle of G, it then follows that $\varphi_1 = \varphi \cup \varphi'$ is an (X,H)-coloring of G, a contradiction to (3). △

Claim 9.31.6 *Let $C \in \mathcal{NC}$, let $v \in V(C)$, and let $A = T_C^v$. Then there exists an (X,H)-coloring of C that is A-good.*

Proof: We fix an ordering of C such that $C = (v_1, v_2, \ldots, v_p, v_1)$ and $v_p = v$. Now we construct step by step an (X,H)-coloring φ of C as follows. We choose an arbitrary color $x \in X(v_1)$ and set $\varphi(v_1) = x$. Now assume that $\varphi(v_1), \varphi(v_2), \ldots, \varphi(v_{i-1})$ are already defined where $2 \leq i \leq p$. If $i \leq p-1$, then it follows from Claim 9.31.4 that there is at most one vertex $u \in T_C^{v_i}$ such that $N = N_C(u) \setminus \{v_i\}$ is a subset of $\{v_1, v_2, \ldots, v_{i-1}\}$. Then $N = \{v_{i-2}, v_{i-1}\}$, $\varphi(v_{i-2}) \in X(v_{i-2})$ and $\varphi(v_{i-1}) \in X(v_{i-1})$. Hence, $S = X(u) \setminus (N_H(\varphi(v_{i-1})) \cup N_H(\varphi(v_{i-2})))$ satisfies $|S| \geq 3$ (by (1) and (C2)). Consequently, there is a color $y \in X(v_i) \setminus N_H(\varphi(v_{i-1}))$ such that $|S \setminus N_H(y))| \geq 3$ (by (1) and (C2)). Then we define $\varphi(v_i) = y$. If $i = p$, then there is a color $y \in X(v_p) \setminus (N_H(\varphi(v_1)) \cup N_H(\varphi(v_{p-1})))$ and we define $\varphi(v_p) = y$. Since C is a induced cycle of G, this leads to an (X,H)-coloring φ of C that is A-good. △

Claim 9.31.7 *Let $C \in \mathcal{NC}$ and $v \in V(C)$. Then the following statements hold:*

(a) $|T_C^v| \geq 2$.
(b) *If $T_C^v = \{u,u'\}$ is a 2-set and $p \geq 5$, then u,u' are not C-conform.*

Proof: Let $C \in \mathcal{NC}$, let $v \in V(C)$, and let $A = T_C^v$. Then there is an (X,H)-coloring φ of C such that φ is A-good (by Claim 9.31.6). Let $G' = G - V(C)$ and let $(X'.H') = (X,H)/(\varphi,C)$. Note that G' is a plane graph and (X',H') is a cover of G' such that A_C belongs to the boundary subgraph of the outer face of G', $A \subseteq A_C$, $|X'(u)| \geq 5$ for all $u \in V(G') \setminus A_C$, $|X'(u)| \geq 3$ for all $u \in A_C \setminus A$, and $|X'(u)| \geq 2$ for $u \in A$.

First, assume that $A = T_C^v$ consists of one vertex. Then it follows from Theorem 9.28 that G' has an (X',H')-coloring, a contradiction to Claim 9.31.5.

Now, assume that $A = T_C^v$ consists of two vertices, say u and u', such that $p \geq 5$ and u, u' are C-conform. By Claim 9.31.4, the notation may be chosen so that $C = (v_1, v_2, \dots, v_p, v_1)$, $N_C(u) = \{v_1, v_2, v_3\}$, and $N_C(u') = \{v_3, v_4, v_5\}$, and in G_C the neighbors of u and u' in $O_C = (v_1^1, v_2^1, \dots, v_p^1, v_1^2, \dots, v_p^2, v_1^1)$ are v_1^1, v_2^1, v_3^1 and v_3^1, v_4^1, v_5^1, respectively. Note that $v = v_3$. Let $u, u_1, u_2, \dots, u_q, u'$ be the neighbors of v_3^1 in the plane graph G_C in that clockwise order around v_3^1. By Claim 9.31.1(b), each noncontractible triangle of G is a facial triangle of G. Hence, by adding certain edges, we may assume that $P = (u, u_1, u_2, \dots, u_q, u')$ is a path on the boundary of the outer face of the plane graph $G' = G - V(C) = G_C - V(O_C)$, where each vertex of A_C still belongs to the outer face of G'. Since φ is A-good, we have $|X'(u)| \geq 5$ for all $u \in V(G') \setminus A_C$, $|X'(u)| \geq 3$ for all $u \in A_C \setminus A$ and $|X'(u)| \geq 2$ for $u \in A$. Furthermore, since $p \geq 5$ and $C \in \mathcal{NC}$ is a shortest noncontractible cycle of G, we obtain that $N_C(u_i) = \{v_3\}$ for all $i \in [1,q]$ and, therefore, $|X'(u_i)| \geq 4$. Then Theorem 9.29 implies that G' has a (X', H')-coloring, a contradiction to Claim 9.31.5. This completes the proof of the claim. △

Claim 9.31.8 $p = 4$.

Proof: Suppose this is false. Then $p \geq 5$ (by Claim 9.31.2). Let $C \in \mathcal{NC}$ be an arbitrary cycle. Then $|T_C^v| \leq 3$ for all $v \in V(C)$ (by Claim 9.31.4), and we denote by $t(C)$ the number of vertices $v \in V(C)$ such that $|T_C^v| = 3$. Now chose a cycle $C \in \mathcal{NC}$ for which $t(C)$ is minimum.

First, assume that $t(C) \geq 1$. Then there is a vertex $v \in V(C)$ such that T_C^v is a set of three vertices, say u_1, u_2 and u_3. By Claim 9.31.4, the notation may be chosen so that $C = (v_1, v_2, \dots, v_p, v_1)$ with $v = v_1$, where $N_C(u_1) = \{v_{p-1}, v_p, v_1\}, N_C(u_2) = \{v_1, v_2, v_3\}$ and $N_C(u_3) = \{v_p, v_1, v_2\}$. Since $p \geq 5$, we obtain that $I = \{u_1, u_2, u_3\}$ is an independent set in G. Then the cycle $C' = (v_1, u_2, v_3, \dots, v_p, v_1)$ belongs to $\mathcal{NC}$ such that $N_{C'}(v_2) = \{v_1, u_2, v_3\}$, $N_{C'}(u_1) = \{v_{p-1}, v_p, v_1\}$ and $N_{C'}(u_3) = \{v_p, v_1\}$. From Claim 9.31.4 it then follows that there is a vertex w_1 such that $N_{C'}(w_1) = N_C(w_1) = \{v_{p-2}, v_{p-1}, v_p\}$, since otherwise $T_{C'}(v_p) = \{u_1\}$, a contradiction to Claim 9.31.7(a). By symmetry (and considering the cycle $C' = (v_1, v_2, \dots, v_{p-1}, u_1, v_1)$) we obtain that there is a vertex w_2 such that $N_C(w_2) = \{v_2, v_3, v_4\}$. Now $C' = (u_3, v_2, v_3, \dots, v_p, u_3)$ is a noncontractible cycle in $\mathcal{NC}$. Since I is an independent set of G, it then follows from Claim 9.31.4 that $N_{C'}(u_1) = \{v_{p-2}, v_{p-1}\}$, $N_{C'}(u_2) = \{v_2, v_3\}$ and $N_{C'}(v_1) = \{v_p, u_3, v_2\}$. Hence, $T_{C'}(u_3) = \{v_1\}$, a contradiction to Claim 9.31.7(a).

Now, assume that $t(C) = 0$. Then $|T_C^v| \leq 2$ for all $v \in V(C)$. From Claim 9.31.7 it then follows that, for every vertex $v \in V(C)$, the set T_C^v consists of two vertices and these two vertices are not C-conform. This implies that in G_C each vertex of the cycle O_C has exactly one neighbor in T_C. Since each vertex in T_C has three neighbors in O_C and $|O_C| = 2p$, we obtain that $2p \equiv 0 \pmod 3$ and hence $p \equiv 0 \pmod 3$. Furthermore, for every vertex $u \in T_C$, the three neighbors of u in G_C that belong to O_C are consecutive on O_C (by Claim 9.31.4). Then we obtain that there are two distinct vertices $u, w \in T_C$ such that $N_C(u) = N_C(w)$, a contradiction to Claim 9.31.4(c). Thus the claim is proved. △

Let $C \in \mathcal{NC}$ be an arbitrary cycle. For a vertex $u \in A_C$, we denote by $N_{O_C}(u)$ the set of neighbors of u in G_C that belongs to O_C. Note that if $u \in T_C$, then the three vertices in $N_{O_C}(u)$ are consecutive in O_C (by Claim 9.31.4). We have $|C| = p = 4$ (by Claim 9.31.8) and $|T_C^v| \geq 2$ for all $v \in V(C)$ (by Claim 9.31.7(a)). By Claim 9.31.4, this implies that T_C consists of three vertices, say $T_C = \{u_2, u_3, u_4\}$, two of which, say u_2 and u_4, are C-conform. Hence, up to rotational symmetry, we have $C = (v_1, v_2, v_3, v_4, v_1)$, $N_C(u_2) = \{v_1, v_2, v_3\}$, $N_C(u_4) = \{v_3, v_4, v_1\}$ and $N_C(u_3) = \{v_2, v_3, v_4\}$. For the cycle $O_C = (v_1^1, v_2^1, v_3^1, v_4^1, v_1^2, v_2^2, v_3^2, v_4^2, v_1^1)$, we then have

(A) $N_{O_C}(u_2) = \{v_1^1, v_2^1, v_3^1\}$, $N_{O_C}(u_4) = \{v_3^1, v_4^1, v_1^2\}$, $N_{O_C}(u_3) = \{v_2^2, v_3^2, v_4^2\}$.

Clearly, $C' = (v_1, u_2, v_3, v_4, v_1)$ is a cycle belonging to $\mathcal{NC}$, too, and

$$O_{C'} = (v_1^1, u_2^1, v_3^1, v_4^1, v_1^2, u_2^2, v_3^2, v_4^2, v_1^1).$$

Furthermore, we have $N_{O_{C'}}(u_4) = \{v_3^1, v_4^1, v_1^2\}$, $N_{O_{C'}}(v_2) = \{v_1^2, u_2^2, v_3^2\}$ and $N_{O_{C'}}(u_3) = \{v_3^2, v_4^2\}$. Then we obtain that there is a vertex u_1 in $V(G) \setminus V(C')$ such that $N_{O_{C'}}(u_1) = \{v_4^2, v_1^1, u_2^1\}$. Then $u_1 \notin T_C = \{u_2, u_3, u_4\}$, and we have

(B) $N_{O_C}(u_1) = \{v_4^2, v_1^1\}$ and $u_1u_2 \in E(G)$.

By symmetry (and considering the cycle $C'' = (v_1, v_2, v_3, u_4, v_1)$) we obtain that there is a vertex u_1' such that $u_1' \notin \{u_1, u_2, u_3, u_4\}$ and

(C) $N_{O_C}(u_1') = \{v_1^2, v_2^2\}$ and $u_1'u_4 \in E(G)$.

Finally, we consider the cycle $C^* = (v_1, v_2, u_3, v_4, v_1)$. Clearly, $C^* \in \mathcal{NC}$ and for

$$O_{C^*} = (v_1^1, v_2^1, u_3^1, v_4^1, v_1^2, v_2^2, u_3^2, v_4^2, v_1^1)$$

we obtain that $N_{O_{C^*}}(v_3) = \{v_2^1, u_3^1, v_4^1\}$, $N_{O_{C^*}}(u_2) = \{v_1^1, v_2^1\}$, $N_{O_{C^*}}(u_4) = \{v_4^1, v_1^2\}$, $v_4^2, v_1^1 \in N_{O_{C^*}}(u_1)$ and $v_1^2, v_2^2 \in N_{O_{C^*}}(u_1')$. Since $|T_{C^*}| = 3$, this implies that we have

(D) $u_3u_1, u_3u_1' \in E(G)$.

Now consider the plane graph G_C and his outer cycle O_C. Let $w_1, w_2, \ldots, w_q$ be the neighbors of v_3^1 in the plane graph G_C in that clockwise order around v_3^1. Then $w_1 = u_2$ and $w_q = u_4$. Furthermore, $\{w_2, w_3, \ldots, w_{q-1}\} \cap \{u_1, u_1', u_3\} = \varnothing$, since otherwise there would exists a noncontractible triangle, contradicting the fact that $p = 4$. From (A), (B), (C), and (D) it then follows that $A_C = \{u_1, u_1', u_2, u_3, u_4\} \cup \{w_2, w_3, \ldots, w_{q-1}\}$. Furthermore, we have $N_C(w_i) = \{v_3\}$ for $i \in [2, q-1]$. By Claim 9.31.1(b), each contractible triangle of G is a facial triangle of G. Hence, by adding certain edges, we may assume that $P = (w_1, w_2, \ldots, w_q)$ is a path on the boundary of the outer face of the plane graph $G' = G - V(C) = G_C - V(O_C)$ with $w_1 = u_2$ and $w_q = u_4$, where each vertex of A_C still belongs to the outer face of G'. The vertex set $A = T_C^{v_1}$ satisfies $A = \{u_2, u_4\}$. By Claim 9.31.6, there exists an (X, H)-coloring φ of C that is A-good. Let $(X', H') = (X, H)/(\varphi, C)$ be the reduced cover of the plane graph G'. Since φ is A-good, we have $|X'(u)| \geq 5$ for all $u \in V(G') \setminus A_C$, $|X'(u)| \geq 3$ for all $u \in A_C \setminus A$ and $|X'(u)| \geq 2$ for $u \in A$. Since $N_C(w_i) = \{v_3\}$ for $i \in [2, q-1]$, we have $|X'(w_i)| \geq 4$ for $i \in [2, q-1]$. Then Theorem 9.29 implies that G' has an (X', H')-coloring, a contradiction to Claim 9.31.5. Hence the proof of Theorem 9.31 is complete. □

9.4 Exercises

9.1 Show that if $\mathbf{S}$ is a surface with $\varepsilon(\mathbf{S}) \leq -8$, then $N(\mathbf{S}, H(\mathbf{S}) - 1) \leq H(\mathbf{S}) + p$ for an absolute constant $p \in \mathbb{N}$. (Škrekovski [946])

9.2 Show that the class $\mathrm{Crit}_{\mathrm{DP}}(\mathbf{S}, k) = \mathrm{Crit}_{\mathrm{DP}}(k) \cap \mathcal{G}(\hookrightarrow \mathbf{S})$ is finite for all $k \geq 7$ and all surfaces $\mathbf{S}$. (Hint: Use (9.4) and Corollary C.19.)

9.3 Let $\mathrm{Crit}_{\mathrm{DP}}(\mathbf{S}, k, g)$ the class of graphs $G \in \mathrm{Crit}_{\mathrm{DP}}(\mathbf{S}, k))$ whose girth is at least g. Show that for every surface $\mathbf{S}$, the class $\mathrm{Crit}_{\mathrm{DP}}(\mathbf{S}, k, 4)$ is finite for all $k \geq 5$, and the class $\mathrm{Crit}_{\mathrm{DP}}(\mathbf{S}, k, 6)$ is finite for all $k \geq 4$.

9.4 Show that if G is a triangle-free graph embedded in the torus or Klein bottle, then $\chi_{\mathrm{DP}}(G) \leq 4$. (Kronk and White [657])

9.5 Show that if a graph G has girth at least six and can be embedded in the torus or Klein bottle, then $\chi_{\mathrm{DP}}(G) \leq 3$ (Kronk and White [657])

9.6 Let $\mathrm{Crit}(\mathbf{S}, k, g)$ the class of graphs $G \in \mathrm{Crit}(\mathbf{S}, k)$ whose girth is at least g. Use the Kostochka–Yancey bound for $\mathrm{ext}(k, n)$ to show the existence of an integer $N(\varepsilon, k, g)$ such that every graph $G \in \mathrm{Crit}(\mathbf{S}, k, g)$ with $\varepsilon(\mathbf{S}) = \varepsilon$ satisfies $|G| \leq N(\varepsilon, k, g)$, provided that $g = 4$ and $k \geq 5$, or $g = 5$ and $k \geq 5$, or $g = 6$ and $k \geq 4$. (Gimbel and Thomassen [415])

9.7 Let G be a graph which has a set $\mathcal{T}$ of triangles such that each edge of G is contained in precisely two triangles of $\mathcal{T}$, and let φ be a coloring of G with a set S of four colors. Show that, for any two colors $c, c' \in S$, the number of vertices $v \in \varphi^{-1}(c)$ having odd degree in G has the same parity as the number of vertices $w \in \varphi^{-1}(c')$ having odd degree in G. In particular, if G has precisely two vertices of odd degree, then they have the same color. (Fisk [375, p. 327], Thomassen [1012, Proposition 3.1])

9.8 Let G be a graph which has a 2-cell embedding into some surface such that every face is bounded by a triangle. Show that if G has exactly two vertices of odd degree, and these two vertices are adjacent, then G is not 4-colorable. (Fisk [376]).

9.9 Let G be a graph embedded in the projective plane. Suppose that G has at most two contractible cycles of length four, or one contractible cycle of length three such that all other 4-cycles are noncontractible. Show that G is 3-colorable. (Hint: Use Euler's formula and the Kostochka–Yancey bound for 4-critical graphs.) (Thomassen [1009] and Borodin, Kostochka, Lidický, and Yancey [158])

9.10 Show that if G is a projective planar graph such that every vertex is in at most one triangle and G contains no k-cycle for $k \in \{4, 5, 6\}$, then G is 3-colorable. (Hint: Use Euler's formula and the Kostochka–Yancey bound for 4-critical graphs.) (Borodin, Kostochka, Lidický, and Yancey [158])

9.11 Show that every planar graph without cycles of length 4 through 11 has coloring number at most 3. (Hint: Use similar arguments as in the proof of Theorem 9.20.) (Abbott and Zhou [10])

9.12 A graph is is called **outerplanar** if it has an embedding in the plane such that each vertex lies on the boundary of the outer face. Show that every outerplanar graph G satisfies $\operatorname{col}(G) \leq 3$. (König [608])

9.13 Show that if G is a 2-connected plane graph such that $d_G(f) \equiv 0 \pmod 3$ for all $f \in F(G)$ and $d_G(v) \equiv 0 \pmod 2$ for all $v \in V(G)$, then $\chi(G) \leq 3$. (Hint: Use Corollary 9.24, induction on the number of faces f with $d_G(f) \geq 9$, and the graph $K_3 \boxplus \overline{K}_3$.) (Ore [794, Remark 3.1])

9.14 Let G be a plane triangulation all of whose vertices have even degree. Show that we can color the faces of G such that every face receives a color and faces sharing a common boundary edge get different colors. (Hint: Define an appropriate graph whose vertices are the faces of G and apply the two color theorem to that graph.)

9.15 Show that the following two statements are equivalent:

1. Every plane triangulation all of whose vertices have even degree is 3-colorable
2. Every internally even plane near-triangulation is 3-colorable.

9.16 Let G be a plane graph such that $\delta(G) \geq 3$, G is connected and no two triangles of G have an edge in common. Define an initial charge distribution ch on $W = V(G) \cup F(G)$ by $ch(x) = d_G(x) - 4$ for all $x \in W$. Then, by Euler's formula (see (C.10)), we have

$$\sum_{x \in W} ch(x) = -8.$$

Starting with the initial charge distribution ch, we redistribute the charge according to the following five rules:

1. Transfer a charge of $-\frac{1}{3}$ from every 3-face to each of the three vertices on its boundary.
2. Transfer a charge of $\frac{2}{3}$ from every p-face with $p > 3$ to each boundary vertex of degree three contained in a 3-face.
3. Transfer a charge of $\frac{1}{3}$ from every p-face with $p > 3$ to each boundary vertex of degree 3 not contained in a 3-face.
4. Transfer a charge of $\frac{1}{3}$ from every p-face with $p > 3$ to each boundary vertex of degree 4 contained in two 3-faces.
5. Transfer a charge of $\frac{1}{3}$ from every p-face with $p > 3$ to each boundary vertex of degree 4 contained in a nonadjacent 3-faces. Here a p-face with $p > 3$ and a 3-face are adjacent if their boundaries have an edge in common.

Let ch' denote the resulting new charge distribution. Show that there is a $x \in W$ such that $ch'(x) < 0$ and x is a p-face with $4 \leq p \leq 9$, or x is a 10-face such that all of its boundary vertices have degree three in G and are contained in an adjacent 3-faces. (Borodin [150])

9.17 Show that every planar graph without cycles of length 4 through 9 has list chromatic number at most three. (Hint: Use the previous exercise and either Proposition 5.5 or a direct list coloring argument for even cycles.) (Borodin [150], Sanders and Zhao [889]).

9.18 Show that every plane triangulation G with $\delta(G) \geq 5$ contains an edge uv with $d_G(u) + d_G(v) \leq 11$. (Hint: Use the discharging method as in Exercise 9.16); the initial charge is $ch(v) = 6 - d_G(v)$ for $v \in V(G)$ and $ch(f) = 6 - 2d_G(f)$ for $f \in F(G)$, transfer a charge of $\frac{1}{5}$ from every vertex of degree 5 to each of its neighbors.) (Wernicke [1064])

9.19 Show that, for every surface $\mathbf{S}$ and $4 \leq k \leq H(\mathbf{S})$, the following two statements are equivalent:

1. Every k-chromatic graph that can be embedded on $\mathbf{S}$ has K_k as a minor.
2. No k-contraction-critical graph $G \neq K_k$ can be embedded on $\mathbf{S}$.

9.20 Show that every k-contraction-critical graph $G \neq K_k$ satisfies $\delta(G) \geq k$ and $\omega(G) \leq k - 2$, where $k \geq 5$. (Hint: Use contradiction and contract an appropriate induced path of length two.) (Dirac [299] and [302])

9.21 Show that, if $\mathbf{S}$ with $\varepsilon(\mathbf{S}) \leq -3$ and $k = H(\mathbf{S}) - 1$, then every k-chromatic graph that can be embedded on $\mathbf{S}$ has K_k as a minor. (Dirac [295] and [296])

9.22 Let G be a planar graph, and let $P \subseteq V(G)$ be a vertex set such that $\text{dist}_G(u, v) \geq 3$ whenever u and v are distinct vertices of P. Show that any coloring of $G[P]$ with a set S of six colors can be extended to a coloring of G with color set S. (Hint: Use the Five Color Theorem of Thomassen.) (Albertson [31])

9.5 Notes

The 2001 book *Graphs on Surfaces* by Mohar and Thomassen [752] provides a rigorous and concise introduction to graphs on surfaces. The book concludes with an overview of recent developments on coloring graphs on surfaces achieved at the end of the millennium, later updated by Mohar [751]. There are several articles and books about the history and the proof of the four color theorem. The article by Thomas [995] provides a very readable survey on the four color theorem, discussing in particular different equivalent formulations as well as progress on some generalizations of the four color theorem; the book by Rudolf Fritsch and Gerda Fritsch [393] contains interesting biographical information about the lead players in the four color story. Recommended is also the nontechnical lucid history by Wilson [1070].

Results about (vertex) coloring of graphs embedded in a given surface are often called **map color theorems**. However, when we think of maps, we first think of political maps, such as those found in an atlas; in such maps, neighboring regions are often colored with different colors. Maps may be depicted by graphs embedded in the plane or sphere. An embedded graph (or multigraph) has three basic types of objects that we may color, either separately for each type or simultaneously for two or three types, namely vertices, edges and faces. The 2013 survey by Borodin [152] discusses results about simultaneous colorings of vertices, edges, and faces (in all possible combinations) of plane graphs. The **faces** of an embedded multigraph are also called **regions** (see e.g. Diestel [286]) and **countries** (see e.g. Ringel [865]).

A natural requirement for a **face coloring** of an embedded multigraph G is that each face receives exactly one color, so that the two colors on either side of each edge are distinct. However, this is only possible if G is bridgeless. So a bridgeless graph (multigraph) embedded on a surface **S** is also called a **map** on **S**; often it is required that a map is connected or even 2-connected, and that the embedding in **S** is a 2-cell embedding. Let G be a bridgeless multigraph embedded on a surface **S**. Then the **face graph** (respectively **region graph**) of G, denoted by $\mathrm{R}(G)$, is the graph whose vertices are the faces of G and in which two faces are joined by an edge if their boundaries have a common edge in G. The face graph was introduced by Ringel [865] under the name **country graph**. Note that the face graph of G is closely related to the **dual map** of G, a concept from topological graph theory (for a formal definition see White [1067] or Mohar and Thomassen [752]); since G is embedded on **S**, $\mathrm{R}(G)$ can be embedded on **S**, too. A coloring of $\mathrm{R}(G)$ is called a **face coloring** of G; and the chromatic number of $\mathrm{R}(G)$ is called the **face chromatic number** of G. So face coloring problems for bridgeless multigraphs embedded on **S** can be easily transformed into (vertex) coloring problems for graphs embedded on **S**, and vice versa. For the sphere, this was already observed by Kempe [568]. The following statements are equivalent formulations of the four color theorem:

(F1) Every planar graph has chromatic number at most four.
(F2) Every bridgeless plane multigraph graph has face chromatic number at most four.
(F3) Every bridgeless cubic planar graph has chromatic index three.

That (F1) implies (F2) follows from the fact that if G is a bridgeless plane multigraph, then $\mathrm{R}(G)$ is a planar graph. To see that (F2) implies (F1), we first observe that it suffices to prove (F2) for 2-connected plane multigraphs. Furthermore, it suffices to prove (F2) for 2-connected cubic plane graphs. Given a 2-connected plane multigraph G, replace every vertex of degree 2 by a K_4^- and every vertex of degree $k \geq 3$ by a C_k (see Figure 9.7), and let G' denote the resulting graph. Then G' is a 2-connected cubic plane graph, and a face coloring of G' with four colors induces a face coloring of G with four colors. Hence (F2) is equivalent to the following statement.

(F2') Every 2-connected cubic plane graph has face chromatic number at most four.

To see that (F2') implies (F1), first observe that it suffices to prove (F1) for maximal planar graphs whose order is at least five. So let G be such a graph. We may assume that G is embedded in the plane. Then G is a plane triangulation (see Corollary C.28). Then $G' = \mathrm{R}(G)$ is a 2-connected cubic graph which has an embedding in the plane such that $\mathrm{R}(G') = G$. Then a face coloring of G' with four colors leads to a (vertex) coloring of G with four colors; hence (F2') implies (F1). That the statements (F2') and (F3) are equivalent was proved by Tait [986] in 1878 (see e.g. White [1067, Theorem. 8-6.]).

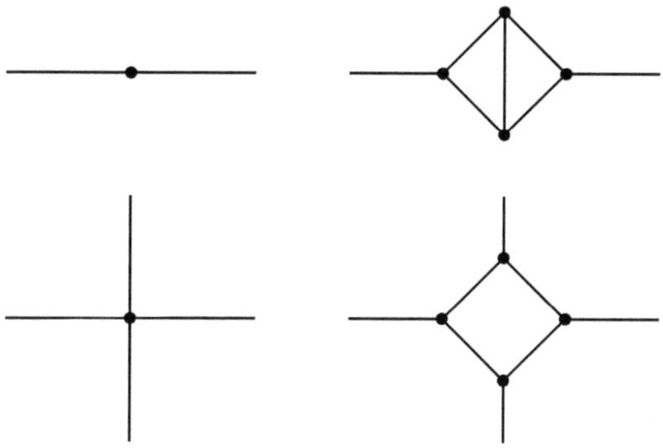

Fig. 9.7 Reduction to cubic graphs.

A standard approach to proving the four color theorem is a proof by induction or a proof by contradiction. To this end, we analyze a **minimal counterexample**, hereafter referred to as **mce-graph**, that is, a graph G such that (1) G is planar, (2) $\chi(G) \geq 5$, and (3) $|V(G)| + |E(G)|$ is minimum subject to (1) and (2). Since $\mathcal{G}(\hookrightarrow \mathbf{S})$ is a monotone graph property for every surface $\mathbf{S}$, (2) and (3) imply that every mce-graph G belongs to Crit(5) and, hence, $\delta(G) \geq 4$ and G is 2-connected. If G is an mce-graph and $uv \in E(\overline{G})$, then the graph $G' = G/\{u,v\}$ obtained from G by identifying u and v satisfies $\chi(G') \geq \chi(G) \geq 5$ (by (2)), implying that G' is not planar (by (3)). From this it follows that every embedding of G in the plane is a triangulation, and so G is a maximal planar graph. Let $\mathcal{T}$ denote the family of plane triangulations G with $\delta(G) \geq 4$. Then every mce-graph belongs to $\mathcal{T}$. To reach a contradiction it would be sufficient to find a family $\mathcal{F}$ of configurations (plane near-triangulation) satisfying the following two conditions:

- Every configuration of $\mathcal{F}$ is **reducible**, that is, a triangulation containing this configuration cannot be an mce-graph (any 4-coloring of the vertices outside the configuration may be extended to a 4-coloring of the whole graph, perhaps after suitable changes of the given 4-coloring).
- $\mathcal{F}$ is **unavoidable** with respect to $\mathcal{T}$, that is, every graph in $\mathcal{T}$ contains an element from $\mathcal{F}$.

To prove the four color theorem, it suffices to establish an unavoidable family of reducible configurations. The first who made a proposal for such a family was Kempe [568]. Kempe claimed that the family $\mathcal{F} = \{W_4 = K_1 \boxplus C_4, W_5 = K_1 \boxplus C_5\}$ is unavoidable with respect to $\mathcal{T}$ and every configuration in $\mathcal{F}$ is reducible. That the family is unavoidable is a trivial consequence of Euler's formula, which implies that every graph $G \in \mathcal{T}$ satisfies $4 \leq \delta(G) \leq \mathrm{col}(G) - 1 \leq 5$. To show that both configurations are reducible, Kempe used the recoloring technique now called Kempe

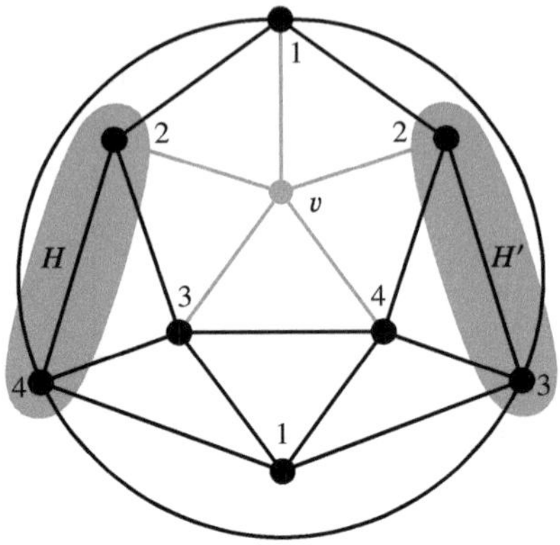

Fig. 9.8 Recoloring two chains and Kempe's argument.

changes (see Section 1.2). First assume that an mce-graph G contains a vertex v having degree 4 in G. We may fix an embedding of G in the plane. Since G is a plane triangulation, the four neighbors of v in G belong in the plane graph $G' = G - v$ to a 4-face, say $f = v_1v_2v_3v_4$. Since G is an mce-graph, there is a coloring φ of $G - v$ with color set $S = [1,4]$. Clearly, $S = \varphi(N_G(v))$ since otherwise we may color v with a color $c \in S \setminus \varphi(N_G(v))$ to get a coloring of G with color set S, which is impossible. So $S = \varphi(N_G(v))$ and, by symmetry, we may assume that $\varphi(v_i) = i$ for $i \in S$. Now let H be the $(1,3)$-chain with respect to φ containing v_1, and let H' be the $(2,4)$-chain with respect to φ containing v_2. If $v_3 \notin V(H)$, then the coloring $\varphi' = \varphi/H$ obtained from H by recoloring H satisfies $\varphi'(N_G(v)) \neq S$ and so we get a coloring of G with color set S; a contradiction. Otherwise, there is a v_1-v_3 path in G whose vertices are colored by 1 and 3. Since f is a face of the plane graph $G - v$, this implies that v_4 does not belong to the $(2,4)$-chain H' containing v_2. But then recoloring H' and coloring v with color 2 results in a coloring of G with color set S, a contradiction. This completes the proof of the claim that W_4 is reducible. To show that W_5 is reducible, Kempe argued in a similar way. So if v has degree 5 in G, then the five neighbors of v belong to a 5-face $f = v_1v_2v_3v_4v_5$ of $G' = G - v$, and, for the coloring $\varphi \in CO(G - v, 4)$, we may assume that $\varphi(v_i) = i$ for $i \in [1,4]$ and $\varphi(v_5) = 2$. Then there is a $(1,3)$-chain with respect to φ containing v_1 and v_3 as well as a $(1,4)$-chain with respect to φ containing v_1 and v_4, since otherwise we can use a Kempe change to get a coloring of G with color set S, a contradiction. Since f is a face of the plane graph $G - v$, this implies that the $(2,4)$-chain H with respect to φ containing v_2 does not contain v_4, and the $(2,3)$-chain H' with respect to φ containing v_5 does not contain v_3. Now recoloring H and H' and coloring v with 2, results in a coloring of G with color set $S = [1,4]$, a contradiction.

Heawood [487] observed that the final conclusion in Kempe's above argument is incorrect. If we recolor two chains, we need to do this one after another, but after recoloring the first chain the second chain may change and not have the right property, see Figure 9.8.

That every mce-graph is a triangulation was first mentioned by Birkhoff in his 1913 paper [118], he also proved that every mce-graph G is **internally 6-connected**, that is, G is 5-connected and for every set $U \subseteq V(G)$ of cardinality 5, $G - U$ is either connected or consists of two components, one of which is just a vertex. A proof of Birkhoff's result may be found in the book by Mohar and Thomassen [752, Lemma 8.2.4]. So we may replace $\mathcal{T}$ by the family $\mathcal{T}^*$ of internally 6-connected plane triangulations. Then $\mathcal{F} = \{W_5\}$ is unavoidable with respect to $\mathcal{T}^*$, but no recoloring argument is known showing that a degree five vertex is reducible. However, Birkhoff began a renewed search for reducible configurations, and he also discovered several such configurations; the most famous among them is the so-called **Birkhoff diamond**: no mce-graph G contains a set X of four vertices all of degree 5 and such that $G[X] = K_4^-$ (a K_4 minus an edge).

It was not until the early 1960's that Heinrich Heesch systematized the study of reducibility. He was also the first who developed an algorithm for the proof of so-called D-reducibility, a concept introduced by himself and which lent itself to investigation with the aid of computers. Heesch also developed a clever method for proving unavoidability results; in German he used the name "Entladung" for his method, which translates to "discharging" in English. In implicit form, the discharging method was already used in a 1904 paper by Wernicke [1064] (see Exercise 9.19), in which the problematic configuration W_5 was replaced by two configurations, one consisting of two adjacent vertices of degree 5, the other of two adjacent vertices of degrees 5 and 6, and their surrounding triangular faces. This result motivated Wernicke's supervisor Hermann Minkowski to belief that a proof of the four color theorem was within reach.

Over the years, the discharging method has become a powerful tool in graph theory. Heesch published his results only in the late 1960s, see [491], but in German, which limited the influence of his research. In the late 1960s he began his collaboration with Karl P. Dürre, a German mathematician with an interest in computer science. Both propagated the idea of proving the existence of an unavoidable set of reducible configurations by using a computer. However, they lacked both financial support and the necessary computing power. In 1976, Appel, Haken, and Koch [68], [69] presented an unavoidable set of 1834 reducible configurations, using extensive computer time; no doubt their research was inspired by the pioneering work of Heesch. Hans-Günther Bigalke published a book [116] on Heesch and his achievements. More than 20 years later, Robertson, Sanders, Seymour, and Thomas [868] provided a much simplified proof of the four color theorem, by determining (also with the help of a computer) an unavoidable set (with respect to $\mathcal{T}^*$) of 633 reducible configurations (see [752, Appendix B]), avoiding some of the more problematic details in the proof by Appel, Haken, and Koch.

The class of planar graphs is minor closed (see Section 4.7) implying that every mce-graph G is 5-contraction critical (that is, $\chi(H) < \chi(G) = 5$ for every proper minor H of G), and $G \neq K_5$ (by (1)). Hadwiger's conjecture for $k = 5$ says that such a graph does not exist. As explained in Section 4.7 Hadwiger's conjecture for $k = 5$ is indeed equivalent to the four color theorem. Hadwiger's conjecture, however, is wide open for all $k \geq 7$. Very little is known about the structure of possible counterexamples.

Theorem 9.32 *Let $G \neq K_k$ be a k-contraction-critical graph, where $k \geq 7$. Then the following statements hold:*

(a) $\delta(G) \geq k$, $\omega(G) \leq k-2$, *and* $\alpha(G[N_G(v)]) = 2$ *for every vertex* v *satisfying* $d_G(v) = k$.
(b) *G is 7-connected.*
(c) *G is k-edge-connected.*
(d) *No four vertices all of degree k in G induces a K_4^- in G.*
(e) $2|E(G)| \geq k|G| + k - 4$ *and* $|G| \geq k + 4$.

Statement (a) was proved by Dirac [299] [302], statement (b) is due to Mader [710], statement (c) was obtained by Toft [1022], statement (d) and (e) are due to Stiebitz and Toft [975]; statement (d) is an abstract generalization of Birkhoff's result that the Birkhoff diamond is reducible. Statement (e) generalizes the theorem of Mayer [726] that there is always a vertex of degree $> k$, the first Brooks type result obtained for contraction critical graphs. For further results see the surveys by Toft [1029], Seymour [931], Kawarabayashi [559], and the paper by Rolek, Song, and Thomas [876]. Combining Corollary C.19 and Theorem 9.32(e), we obtain the following result from [975].

Corollary 9.33 *Let* **S** *be a surface satisfying* $\varepsilon(S) \leq -2$ *and let* $7 \leq k \leq H(S)$. *If* $G \neq K_k$ *is a k-contraction-critical graph and* $G \in \mathcal{G}(\hookrightarrow \mathbf{S})$, *then*

$$|G| \leq \frac{4 - k - 6\varepsilon(\mathbf{S})}{k - 6}.$$

The bound in Corollary 9.33 is a slight improvement of a result by Dirac [295] who instead of Theorem 9.32(e) used only the trivial lower bound $\delta(G) \geq k$ (see Theorem 9.32(a)). Corollary 9.33 combined with the lower bound $|G| \geq k + 4$ (see Theorem 9.32(e)) can be used to exclude the existence of a k-contraction-critical graph $G \neq K_k$ on a surface **S** of Euler characteristic ε for certain pairs (k, ε), and thus proving Hadwiger's conjecture for k-chromatic graphs on **S** (see Exercise 9.19). For instance, a graph G embedded on the torus $\mathbf{S}_1$ satisfies $2|E(G)| \leq 6|G|$, and a 6-contraction-critical graph $G \neq K_6$ satisfies $2|E(G)| \geq 6|G| + 2$. Hence every 6-chromatic graph on $\mathbf{S}_1$ has K_6 as a minor. This result was obtained by Albertson and Hutchinson [35], see also [36]; however, they used the four color theorem. As proved by Robertson, Seymour, and Thomas in the deep paper [871], the four color theorem implies that every 6-chromatic graph has K_6 as a minor, and vice versa.

In his 1890 paper, pointing out the flaw in Kempe's purported proof of the four color theorem, Percy Heawood began the study of maps on orientable surfaces other than the plane. In particular, he showed that some maps on the torus require seven colors but that no map needs more. Heawood actually did more, although less than he claimed. He found a remarkable bound for the maximum number of colors that are needed to color any map on a particular orientable surface so that neighboring regions are colored differently. By considering the minimum degree of a graph embeddable on $\mathbf{S}_h$, for $h \geq 1$, he showed that every graph embeddable on $\mathbf{S}_h$ can be (vertex) colored with at most

$$H(\mathbf{S}_h) = \left\lfloor \frac{7+\sqrt{48h+1}}{2} \right\rfloor$$

colors. An easy calculation shows the important fact, used below, that

$$7 \leq k \leq \frac{7+\sqrt{48h+1}}{2} \quad \text{is equivalent to} \quad h \geq \frac{(k-3)(k-4)}{12} \geq 1.$$

But although the number of colors given by Heawood's formula is indeed sufficient for each surface $\mathbf{S}_h$ with $h \geq 1$, it is not obvious in general that there must be maps that require this number of colors, as happens for the torus. Heawood believed this to be the case, but his proof was deficient, and his belief became known as the Heawood conjecture; it remained unproved until the 1960s. Heawood admitted the deficiency in his arguments and went on to write further papers on the problem, the last being in his 90th year (a complete list of his papers may be found in Dirac's biographical paper [301]).

Heawood Conjecture: *For $h \geq 1$, $\chi(\mathbf{S}_h) = H(\mathbf{S}_h)$, that is, every map that is embedded on the surface $\mathbf{S}_h$ can be colored with $\left\lfloor \frac{1}{2}(7+\sqrt{48h+1}) \right\rfloor$ colors, and there are maps on $\mathbf{S}_h$ that require this number of colors.*

In 1891, the German mathematician Lothar Heffter pointed out the gap in Heawood's argument for the second assertion of this conjecture, see his paper [493]. He then asked for the smallest possible genus of an orientable surface on which we can draw n mutually neighboring regions; for example, on the sphere we can draw four neighboring regions, but not five, and on the torus we can draw seven neighboring regions, but not eight. Heffter next dualized the problem into that of finding the smallest $h \geq 0$ for which the complete graph K_n can be embedded on $\mathbf{S}_h$. For this (orientable) **genus** of K_n, denoted by $\mathrm{og}(K_n)$ (see Section C.13), Heffter noticed that

$$\mathrm{og}(K_n) \geq \left\lceil \frac{(n-3)(n-4)}{12} \right\rceil,$$

which follows directly from the Heawood formula, as we noted above. He then proved that this lower bound for $\mathrm{og}(K_n)$ gives the correct value for $n \leq 12$; for example, when $n = 7$ the formula yields $h = 1$, the torus.

To prove that Heawood's formula gives the exact value for the number of colors, it is sufficient to prove equality in Heffter's inequality; thus, for the torus, it is sufficient to note that K_7 can be embedded there. But this equality in Heffter's formula is also necessary; any graph on the torus needing seven colors must contain K_7, as first realized by Peter Ungar in the 1940s. Subsequently, Gabriel Dirac proved that this is so in general (except for the sphere): Heffter's lower bound for the genus of a complete graph is achieved if and only if Heawood's upper bound for the number of colors needed for embedded graphs is achieved. It is notable that Dirac [290] proved his map color result before the Heawood conjecture was confirmed. To determine the genus of complete graphs is thus precisely what is needed to prove the Heawood conjecture that $\chi(\mathbf{S}_h) = H(\mathbf{S}_h)$ for all $h \geq 1$. The solution was finally completed by the German mathematician Gerhard Ringel and the American J. W. T. Youngs in 1968, while Ringel was on sabbatical leave in California with Youngs, see the paper [866] by Ringel and Youngs, or Ringel's book [865].

Ringel–Youngs Theorem *For $n \geq 3$, the genus of the complete graph K_n is*

$$\mathrm{og}(K_n) = \left\lceil \frac{(n-3)(n-4)}{12} \right\rceil.$$

Because of the form of the answer, it transpired that the problem split naturally into smaller pieces corresponding to the various values of n (modulo 12), and in particular, those complete graphs for which the inside expression is an integer – that is, when $n \equiv 0, 3, 4,$ or $7 \pmod{12}$. In order to facilitate the construction of the embeddings, Ringel found it helpful to create **current graphs**: these are directed graphs with labels on the arcs (directed edges) that satisfy Kirchhoff's current law from the theory of electrical networks. These ingenious devices provided what was needed. For a further discussion of Ringel and Youngs' proof of the Heawood conjecture, see the survey by White [1067].

Subsequently, substantial simplifications of the solution arose when Jonathan Gross and Thomas Tucker developed dual mechanisms called **voltage graphs**. To achieve complicated graph embeddings one can use simpler ones with voltages assigned to the edges and satisfying Kirchhoff's voltage law, see [1067].

Embedding a graph on an orientable surface may be described purely combinatorially by giving the clockwise ordering of the edges around each vertex. There is a one-to-one correspondence between cellular embeddings and such **rotation systems**, consisting for each vertex of a circular permutation of the edges that are incident to the vertex. This had been implicit in Heffter's work, but was treated explicitly in 1960 by Jack Edmonds, and described in his master's thesis at the University of Maryland, see also [320]. It was also used by Ringel and later by Ringel and Youngs. Edmonds provided a method to find the genus of the surface of the embedding from the rotation system. Finding the minimum h for which the graph can be embedded on $\mathbf{S}_h$ is thus equivalent to finding certain such rotation systems, and current and voltage graphs are useful devices for describing these systems. There is however no efficient algorithm known for determining the genus of a given graph, not even for cubic graphs.

Until 1910, all attempts to extend map coloring to surfaces had concentrated on orientable surfaces. But in that year the Austrian mathematician Heinrich Tietze wrote a note [1018] on coloring maps on the Möbius band and the projective plane, showing that every map on either of them can be colored with six colors. He also obtained analogs for nonorientable surfaces of the formulas of Heawood and Heffter. In particular, Euler's formula leads to

$$\chi(\mathbf{N}_h) \leq H(\mathbf{N}_h) = \left\lfloor \frac{7+\sqrt{24h+1}}{2} \right\rfloor.$$

Note that every surface has its own version of Euler's formula (see Theorem C.16); the version for orientable surfaces was established by L'Huilier [672] in 1812–13. Recall from Appendix C.13, that the **non orientable genus** of a graph G, denoted by $\mathrm{ng}(G)$, is the least integer $h \geq 1$ such that G can be embedded in the nonorientable surface $\mathbf{N}_h$. Then the bound for the chromatic number of $\mathbf{N}_h$ yields

$$\mathrm{ng}(K_n) \geq \left\lceil \frac{(n-3)(n-4)}{6} \right\rceil$$

for all $n \geq 3$. It follows from the embeddability of K_6 on the projective plane that $\mathrm{ng}(K_6) = 1$. However, although the Klein bottle has the same Euler characteristic as the torus and K_7 is embeddable on the torus, it cannot be embedded on the Klein bottle: this was first proved by Philip Franklin in 1934. Sixteen years before the orientable genus of the complete graphs was found, Ringel [863] [864] had already established the nonorientable genus of all complete graphs. For a simpler new proof of Ringel's theorem, see Korzhik [612]. It turns out that K_7 is exceptional, being the only complete graph whose nonorientable genus does not equal the bound arising from Euler's formula.

Ringel's Theorem *For $n \geq 3$, the nonorientable genus of the complete graph K_n is*

$$\mathrm{ng}(K_n) = \left\lceil \frac{(n-3)(n-4)}{6} \right\rceil$$

except that $\mathrm{ng}(K_7) = 3$.

In addition to showing that K_7 cannot be drawn on the Klein bottle, Franklin [388] showed that every graph that can be embedded there is 6-colorable. Further, in the same way that the Ringel–Youngs theorem provided what was needed to determine the chromatic number of all orientable surfaces other than the sphere, this earlier theorem of Ringel yields the chromatic number of all nonorientable surfaces, again with the notable exceptional case of the Klein bottle, that is, $\chi(\mathbf{N}_h) = H(\mathbf{S}_h)$ for all $h \geq 3$ and $h = 1$, and $\chi(\mathbf{N}_2) = 6$. The above description of the development from Heawood and Hefter to Ringel and Youngs follows the one presented by Beineke, Wilson, and Toft [92].

For any fixed surface $\mathbf{S}$, the graph property $\mathcal{G}(\hookrightarrow \mathbf{S})$ is hereditary with respect to the subgraph relation, the induced subgraph relation and the minor relation. This has several interesting implications. Since $\mathcal{G}(\hookrightarrow \mathbf{S})$ is minor closed, the Graph Minor

Theorem of Robertson and Seymour (see Theorem 4.58) implies that the graph property $\mathcal{G}(\hookrightarrow \mathbf{S})$ can be characterized by a finite set of forbidden minors. From this it follows that the recognition problem for $\mathcal{G}(\hookrightarrow \mathbf{S})$ is in P (see also Section 2.8). Mohar [747] proved that even an embedding of a given graph G in $\mathbf{S}$ can be found in linear time (linear in $|G|$). This is in sharp contrast to the fact that determining the genus, or nonorientable genus, of graphs is NP-hard, as proved by Thomassen, see [1004] and [1006].

Since $\mathcal{G}(\hookrightarrow \mathbf{S})$ is monotone, we obtain that, whenever $k \geq 1$ is fixed, every graph $G \in \mathcal{G}(\hookrightarrow \mathbf{S})$ satisfies $\chi(G) \leq k-1$ if and only if no subgraph of G belongs to $\mathrm{Crit}(\mathbf{S},k)$, that is,

$$\mathcal{G}(\hookrightarrow \mathbf{S}) \cap \mathcal{G}(\chi \leq k-1) = \mathrm{Forb}_{\subseteq}(\mathrm{Crit}(\mathbf{S},k)).$$

If $\mathrm{Crit}(\mathbf{S},k)$ is finite, then we can decide in polynomial time whether a graph embedded on $\mathbf{S}$ has a coloring with $k-1$ colors, and even the recognition problem for the graph property $\mathcal{G}(\hookrightarrow \mathbf{S}) \cap (\chi \leq k-1)$ belongs to P. That $\mathrm{Crit}(\mathbf{S},k)$ is finite, for every fixed $k \geq 7$, can be proved easily by combining Eulers's formula with the known bounds for $\mathrm{ext}(k,n)$. Proposition 9.7 implies that if $\varepsilon(\mathbf{S}) \leq 0$, $\mathbf{S} \neq \mathbf{N}_3$ and $k = H(\mathbf{S})$, then $N(\mathbf{S},k) \leq k$ and hence $\mathrm{Crit}(\mathbf{S},k) = \langle K_k \rangle$. This result became known as Dirac's map color theorem. For $\mathbf{N}_3$, we have $\varepsilon(\mathbf{N}_3) = -1$, $k = H(\mathbf{N}_3) = 7$ and $N(\mathbf{N}_3,7) \leq 16$. That K_7 is the only 7-critical graph on $\mathbf{N}_3$ was proved by Albertson and Hutchinson [34] using coloring arguments; our proof of this result (see Theorem 9.4) is based on an improved density bound (see Theorem 9.3). However, for $k = 6$ the density argument fails. Thomassen's proof in [1012] that $\mathrm{Crit}(\mathbf{S},6)$ is always finite has not only solved a long-standing open problem (see Jensen and Toft [530, Problem 3.5]), but also led to deeper insights into the theory of map colorings. So for every fixed surface $\mathbf{S}$, there is a polynomial time algorithm deciding whether a given graph on $\mathbf{S}$ has a 5-coloring. Such algorithm can be constructed if an explicit full list of 6-critical graphs on $\mathbf{S}$ is provided. The list of 6-critical graphs are explicitly known for the projective plane, the torus and the Klein bottle. Albertson and Hutchinson [34] proved that $\mathrm{Crit}(\mathbf{N}_1,6) = \langle K_6 \rangle$, thereby solving one of the three missing cases in Dirac's map color theorem. The next two surfaces are the torus and the Klein bottle, both have Euler characteristic $\varepsilon = 0$ and hence Heawood number $H(\mathbf{S}_1) = H(\mathbf{N}_2) = 7$.

Let $H_7 = K_4 \nabla K_4$, and let T_{11} be the graph obtained from an 11-cycle by adding edges joining all pairs of vertices of distance two or three in the cycle. Albertson and Hutchinson [35] proved (using the four color theorem) that T_{11} is the only 6-chromatic 6-regular graph on the torus; note that every embedding of T_{11} on the torus is a triangulation of the torus. In 1994, Thomassen [1007] proved (without using the four color theorem) that

$$\mathrm{Crit}(\mathbf{S}_1,6) = \langle K_6, C_3 \boxplus C_5, K_2 \boxplus H_7, T_{11} \rangle \tag{9.7}$$

and he also discussed some interesting implications of this characterization. The first three graphs in Thomassen's list can be embedded on the Klein bottle, but the last one T_{11} cannot. For the Klein bottle we need six further graphs. The graphs $L_1 = K_6 \nabla K_6$ and $L_2 = K_1 \boxplus (K_5 \nabla K_5)$ are both 6-critical graphs on the Klein bottle.

Let $L_3' = K_5 \blacktriangledown K_5 = ((K_5, v_1, v_2) \blacktriangledown (K_5, w, A_1, A_2))$ be the Gallai join (see Section 4.2), where $|A_1| = |A_2| = 2$, so we split a vertex w of one K_5 in a uniform way into an edge v_1v_2 of the other K_5. Then L_3' is 5-critical and hence $L_3 = K_1 \boxplus L_1'$ is 6-critical. Furthermore $L_4 = K_6 \blacktriangledown K_6 = ((K_6, v_1, v_2) \blacktriangledown (K_6, w, A_1, A_2))$, where $|A_1| = 2$ and $|A_2| = 3$, is 6-critical, too. The last two graphs have the form $L_5 = K_1 \boxplus L_5'$ and $L_6 = K_1 \boxplus L_6'$, where L_5' and L_6' are two specific graphs belonging to $\text{Crit}(5, 9)$. Then we have

$$\text{Crit}(\mathbf{N}_2, 6) = \langle K_6, C_3 \boxplus C_5, K_2 \boxplus H_7, L_1, L_2, L_3, L_4, L_5, L_6 \rangle.$$

The list of 6-critical graphs on the Klein bottle was established, independently, by two teams: Kawarabayashi et al. [560] and Chenette et al. [214].

Since the problem of testing 2-colorability is polynomial time solvable for graphs in general, and the problem of testing 3-colorability is NP-complete even for planar graphs [407], the only interesting remaining case is 4-colorability. On the one hand, it is an open problem whether there exists a polynomial time algorithm for testing 4-colorability of graphs embedded in a fixed surface $\mathbf{S}$ other than the sphere. On the other hand, it is known, by a clever construction due to Fisk [376], that $\text{Crit}(\mathbf{S}, 5)$ is not finite. However, the situation changes when we restrict the girth of the graphs under consideration. Recall from Exercise 9.6 that $\text{Crit}(\mathbf{S}, k, g)$ denotes the class of graphs in $\text{Crit}(\mathbf{S}, k)$ whose girth is at least g, in particular $\text{Crit}(\mathbf{S}, k, 3) = \text{Crit}(\mathbf{S}, k)$. As already noted by Gimbel and Thomassen [415], the Gallai bound for $\text{ext}(k, n)$ is sufficient to show that $\text{Crit}(\mathbf{S}, k, g)$ is finite provided that $g \geq 6$ and $k \geq 4$, or $4 \leq g \leq 5$ and $k \geq 5$. Density arguments are of little use to decide whether $\text{Crit}(\mathbf{S}, 4, 5)$ is finite. That this is indeed the case for all surfaces $\mathbf{S}$ was first proved in 2003 by Thomassen [1014]. In 2020, Dvořák, Kráľ, and Thomas [312] proved that every graph $G \in \text{Crit}(\mathbf{S}, 4, 5)$ satisfies $|G| \leq c \cdot \text{eg}(\mathbf{S})$ where c is an absolute constant. As a consequence of Thomassen's result, the chromatic number of graphs of girth at least five on $\mathbf{S}$ can be found in polynomial time.

As discussed in Section 5.8 the Kostochka–Yancey bound for $\text{ext}(4, n)$ implies that $\text{Crit}(\mathbf{S}_0, 4, 5) = \varnothing$, that is, every planar graph of girth at least five is 3-colorable. Suppose that $G \in \text{Crit}(\mathbf{S}, 4, 5)$ has n vertices and m edges. Then $m \leq 5(n - \varepsilon(\mathbf{S}))/3$ (by Theorem C.18) and $m \geq (5n+2)/3$ (by a result of Liu and Postle [681]), which leads to $\varepsilon(\mathbf{S}) \leq -1$. Consequently, for $\mathbf{S} \in \{\mathbf{S}_0, \mathbf{S}_1, \mathbf{N}_1, \mathbf{N}_2\}$ we obtain that $\text{Crit}(\mathbf{S}, 4, 5) = \varnothing$, that is, every graph on $\mathbf{S}$ with girth at least five is 3-colorable (see also Corollary 5.36).

The Kostochka–Yancey bound implies not only that $\text{Crit}(\mathbf{S}_0, 4, 5) = \varnothing$, but the stronger result that no plane graph $G \in \text{Crit}(4)$ satisfies $d_G(f) \geq 5$ for all $f \in F(G)$. By Lemma 9.15, this easily implies that $\text{Crit}(\mathbf{S}_0, 4, 4) = \varnothing$, which is equivalent to Grötzsch's famous theorem that every triangle-free planar graph is 3-colorable. Even if we consider the proof of the Kostochka–Yancey bound for $k = 4$, this leads to a short and elegant proof of Grötzsch's theorem. This proof was obtained in 2014 by Kostochka and Yancey [639]. Other short proofs of Grötzsch's theorem were given earlier by Thomassen, see [1009], [1010] and [1015]. All three proofs by Thomassen are based on precoloring extension arguments; this kind of argument very often involves list coloring results. In the proofs given in [1010] and [1015] it is shown that every planar graph of girth at least 5 satisfies $\chi_\ell(G) \le 3$, that is, $\mathrm{Crit}_\ell(\mathbf{S}_0,4,5) = \varnothing$, see Theorem 9.14. Here $\mathrm{Crit}_\ell(\mathbf{S},k) = \mathrm{Crit}_\ell(k) \cap \mathcal{G}(\hookrightarrow \mathbf{S})$ is the family of graphs on $\mathbf{S}$ that are χ_ℓ-critical with $\chi_\ell(G) = k$; and $\mathrm{Crit}_\ell(\mathbf{S},k,g)$ is the family of graphs in $\mathrm{Crit}_\ell(\mathbf{S},k)$ whose girth is at least g. Similar we define $\mathrm{Crit}_{\mathrm{DP}}(\mathbf{S},k)$ and $\mathrm{Crit}_{\mathrm{DP}}(\mathbf{S},k,g)$ with respect to the DP-chromatic number χ_{DP}, see Exercise 9.3. As proved by Voigt [1053], Grötzsch's theorem has no counterpart for the list-chromatic number, that is, $\mathrm{Crit}_\ell(\mathbf{S}_0,4,4) \neq \varnothing$.

Lemma 9.15 is a very special case of a stronger result obtained by Klostermeyer and Zhang [603]. This so-called **folding lemma** provides a reduction method which maintains the odd girth of planar graphs.

Lemma 9.34 (Folding Lemma) *Let G be a planar graph with odd girth $p \ge 3$, and let $C = (v_0, v_1, \ldots, v_r, v_0)$ be a facial cycle of G with $r \neq p$. Then there is an index $i \in [0, r-1]$ such that the graph G' obtained from G by identifying v_{i-1} and v_{i+1} (where indices are modulo r) is still of odd girth p.*

Klostermeyer and Zhang used this lemma to show that every planar graph with odd girth at least $10k-3$ has an homomorphism into the cycle C_{2k+1} and hence $\chi^\circ(G) \le 2 + \frac{1}{k}$ (see also Chapter 8). Dvořák, Sereni, and Volec [319] used the folding lemma to show that every triangle-free graph G of order n has fractional chromatic number $\chi^*(G) \le 3 - \frac{3}{3n+1}$.

Grötzsch's result has been the subject of extensive research. Over the years, various proofs of Grötzsch's theorem and its extensions have been presented, including some erroneous proofs. The proof of many of these results can be simplified by using the Kostochka–Yancey bound for $\mathrm{ext}(4,n)$. The proofs of Theorems 9.16, 9.17, 9.18 and 9.19 essentially follows those of Borodin, Kostochka, Lidický, and Yancey [158]. Note that Theorem 9.17 was first obtained by Jensen and Thomassen [529] and Theorem 9.18 is due to Grötzsch [434]. Further applications of the Kostochka-Yancey bound to prove extensions of Grötzsch's theorem are discussed in the paper by La, Lužar, and Štorgel [662]. The most popular extension of Grötzsch's theorem is the Grünbaum–Aksionov theorem (Theorem 9.19). Grünbaum [435] noticed that a planar graph with at most three triangles remains 3-colorable; however the original proof was incorrect, as first noticed by Gallai, who presented a counterexample to the extension Grünbaum has suggested to make induction work. A first correct version was presented by Aksionov [27]. On the one hand, K_4 shows that the result is best possible. There are infinitely many 4-critical planar graphs with four triangles; a constructive characterization of this family was obtained by Borodin et al. [153]; thereby answering a question of Erdős from 1990. On the other hand, Havel [479] conjectured that there is an absolute constant d such that if G is a planar graph and every two distinct triangles in G have distance at least d, then G is 3-colorable. This conjecture was recently confirmed by Dvořák, Král', and Thomas [313] (with d near 10^{100}).

Recall from Section 2.4 that for a graph G we denote by $CL(G)$ the set of cycle lengths of G, i.e. $CL(G) = \{|C| \mid C \text{ is a cycle of } G\}$. Given a set $S \subseteq \mathbb{N} \setminus \{1,2\}$ of possible cycle length, we say that S is **χ-unavoidable** if every graph $G \in \mathrm{Crit}(\mathbf{S}_0, 4)$

satisfies $CL(G) \cap S \neq \emptyset$, otherwise S is **χ-avoidable**. Accordingly, we define that a set S is **χ_ℓ-unavoidable**, respectively **χ_{DP}-unavoidable**. Note that a set S is χ-unavoidable if and only if every planar graph G with $CL(G) \cap S = \emptyset$ is 3-colorable. Since $K_4 \in \mathrm{Crit}(\mathbf{S}_0, 4)$, every χ-unavoidable set S satisfies $3 \in S$ or $4 \in S$. We are interested in χ-unavoidable sets that are minimal (with respect to the subset relation). Obviously, Grötzsch's theorem is equivalent to the statement that $S = \{3\}$ is χ-unavoidable; this set is even minimally χ-unavoidable. However, both $S = \{3\}$ and $S = \{4,5\}$ are χ_ℓ-avoidable (by constructions of Voigt [1053] and [1054]), and $S = \{3,4\}$ is χ_ℓ-unavoidable (by Thomassen's result Theorem 9.14).

Steinberg [961] asked in a 1975 letter to Aksionov and Melnikov whether the set $S = \{4,5\}$ is χ-unavoidable, that is, whether every graph $G \in \mathrm{Crit}(\mathbf{S}_0, 4)$ contains a 4-cycle or a 5-cycle. Steinberg's problem was published by Aksionov and Melnikov in the 1978 proceedings of the graph theory conference, held in Keszthely (Hungary) in 1976, and again in a 1980 journal paper [29]. In the early 1990s Erdős proposed to simplify Steinberg's question and to search for a k for which $S = [4,k]$ is χ-unavoidable. This simplification has greatly stimulated research, and results on unavoidable sets of cycle lengths (with respect to χ, χ_ℓ and χ_{DP}) are now multiple, especially when related results are taken into account. It is therefore difficult to keep up to date.

To recognize avoidable sets we may try to use the construction methods for critical graphs known from Sections 4.1 and 4.2. The Gallai join $G \blacktriangledown G'$ (see Theorem 4.7) seems to be particularly suitable. Let $G \in \mathrm{Crit}(\mathbf{S}_0, 4)$. Note that both K_4 and the Hajós join $H_7 = K_4 \nabla K_4$ belong to $\mathrm{Crit}(\mathbf{S}_0, 4)$. Splitting a vertex of K_4 into an edge of G obviously results in a graph $G' = G \blacktriangledown K_4$ that belongs to $\mathrm{Crit}(\mathbf{S}_0, 4)$. Furthermore, splitting the only high vertex of H_7 in a uniform way into an edge of G results in a graph $G' = G \blacktriangledown H_7$ that also belongs to $\mathrm{Crit}(\mathbf{S}_0, 4)$. By repeated application of the first splitting operation, we can destroy all 5-cycles in G; and by repeated application of the second splitting operation, we can destroy all 4-cycles in G. This observation is due to Aksionov and Melnikov [29] (see also Steinberg's paper [961]). Surprisingly, almost forty years passed before Cohen-Addad, Hebdige, Král', Li, and Salgado [252] found a plane non-3-colorable graph that has neither a 4-cycle nor a 5-cycle, thus showing that $S = \{4,5\}$ is χ-avoidable.

While Steinberg's original question thus has no positive answer, Erdős' relaxation yields many positive results. The first positive result was obtained in 1991 by Abott and Zhou [10]. They proved that if a planar graph G satisfies $CL(G) \cap [4,11] = \emptyset$, then $\mathrm{col}(G) \leq 3$ (see Exercise 9.11). This implies that $S = [4,11]$ is unavoidable with respect to χ, χ_ℓ, and also χ_{DP}. If a planar graph G satisfies $CL(G) \cap [4,8] = \emptyset$, then the argument of Abott and Zhou yields $|E(G)| \leq (5|G| - 10)/3$, see the proof of Theorem 9.20. Then the Kostochka–Yancey bound for $\mathrm{ext}(4,n)$ implies that G cannot be 4-critical. This shows that $S = [4,8]$ is χ-unavoidable; this was observed by Hamaker and Vatter [465]. Unfortunately, this density argument cannot be used for either χ_ℓ or χ_{DP}, since the bounds are too weak. The main proof methods in this area are discharging and precoloring extension; some easy examples of the discharging method are discussed in Exercises 9.16 and 9.18. In 1996 Borodin [150] proved that $S = [4,9]$ is χ_ℓ-anavoidable (see Exercise 9.16). Borodin proved his result by a simple discharging argument (see Exercise 9.16). The result was originally formulated for the chromatic number, but it also applies to the list-chromatic number; this is due to the fact that even cycles cannot occur as blocks of the low vertex subgraph for graphs in $\text{Crit}(k) \cup \text{Crit}_\ell(k)$. However, the low vertex subgraphs of χ_{DP}-critical graphs may have even cycles as blocks; so Borodin's argument is not useable for χ_{DP}. Using advanced discharging arguments Borodin, Glebov, Raspaud, and Salavatipour [155] proved that $S = [4,7]$ is χ-unavoidable. Yet, the question whether $S = [4,7]$ is minimally χ-unavoidable remains open; in particular, it is not known if $S' = [4,6]$ is χ-unavoidable. On the other hand, Borodin, Glebov, Montassier, and Raspaud [154] proved that every planar graph G with $CL(G) \cap \{5,7\} = \varnothing$ and without adjacent triangles is 3-colorable; two subgraphs of a given graph are adjacent if they have an edge in common. The proof of this result in [154] uses precoloring extension arguments and the results from [155]; as pointed out in [154] an earlier similar proof of this result by Boagang Xu uses a wrong precoloring extension lemma. Rao and Wang [851] found a simple discharging argument to show that every planar graph G with $CL(G) \cap \{4,5,7\} = \varnothing$ and without triangles at distance less than two satisfies $\text{col}(G) \leq 3$ and hence $\chi_{\text{DP}}(G) \leq 3$. The paper by Rao and Wang surveys results about unavoidable sets with respect to χ_ℓ and χ_{DP} including related results. The best known result about list-coloring was obtained by Dvořák and Postle [318]; they proved that $S = [4,8]$ is χ_ℓ-unavoidable. It is notable that the proof uses the DP-coloring concept introduced in [318]; yet, the question whether $S = [4,8]$ is χ_{DP}-unavoidable remains unanswered. For a review of further results related to the Steinberg problem we refer the reader to the papers [219], [555] and [683].

The 1993 paper by Steinberg [961] discusses the state of the three-color problem for planar graphs and contains most of the relevant references up to that time. However, even Steinberg was not sure who gave the first complete proof of the three-color theorem (Theorem 9.21). A connected graph whose vertices all have an even degree is often referred to as an **Eulerian graph**. Our proof that each Eulerian triangulation of the plane is 3-colorable might belong to the folklore; however, we have not found any relevant reference. The most famous proof uses the fact that a plane Eulerian triangulation has a face coloring with two colors (see Exercise 9.14) from which a vertex coloring with three colors can then be constructed; the details can be found in Lovász's book [698, Problems 5.26 and 9.56]. The three-color theorem has been rediscovered many times, and thus there are a variety of different proofs. Steinberg [961] provides an easy proof of this result using flow theory; two more recent proofs can be found in the paper by Tsai and West [1030] and in the paper by Diks, Kowalik, and Kurowski [287]. The 2012 paper by Král' et al. [644] contains an overview of coloring results about Eulerian triangulations of surfaces other than the sphere; the main result in this paper says that an Eulerian triangulation G of the Klein bottle satisfies $\chi(G) \leq 6$, where equality holds if and only if G contains a K_6 as a subgraph.

The study of the class $\text{Crit}(\mathbf{S},4,4)$ for surfaces $\mathbf{S}$ different from the sphere has recently attracted increased interest when Dvořák, Král', and Thomas published a series of seven papers on this subject, see [312], [313] and [314]. Further relevant references may be found in the paper by Dvořák and Pekárek [316]. By a result of Youngs [1085] (see Theorem 3.48) it follows that $\mathrm{Crit}(\mathbf{N}_1, 4, 4)$ is infinite. The generalized Mycielski graphs $\mathrm{M}_r(C_{2s+1})$ with $r, s \geq 2$ form an infinite family of projective planar graphs that are 4-critical and triangle-free, see also Section 8.9. On the other hand, Dvořák, Král', and Thomas [314] proved that 3-colorability of triangle-free graphs embedded on a fixed surface can be decided in linear time. Youngs [1085] proved that every nonbipartite quadrangulation of the projective plane has chromatic number 4 (see Theorem 3.48 and the final remark in Section C.13). Gimbel and Thomassen [415] proved the following extension of this result, thereby solving a problem raised by Youngs.

Theorem 9.35 (Gimbel and Thomassen) *Let G be a graph that is embedded in the projective plane $\mathbf{N}_1$ such that all contractible cycle have length at least* 4. *Then $\chi(G) \leq 3$ if and only if G does not contain a nonbipartite quadrangulation of $\mathbf{N}_1$ as a subgraph.*

In 1976, Hutchinson [510] found an appropriate extension of the two-color theorem for planar graphs to graphs on arbitrary surfaces; an improvement was obtained by Liu et al. [685] in 2019. For a surface $\mathbf{S} \neq \mathbf{S}_0$, define

$$H_e(\mathbf{S}) = \left\lfloor \frac{5 + \sqrt{25 - 16\,\varepsilon(\mathbf{S})}}{2} \right\rfloor,$$

and for an arbitrary surface $\mathbf{S}$, define

$$H_{\text{even}}(\mathbf{S}) = \begin{cases} 2 & \text{if } \mathbf{S} = \mathbf{S}_0, \\ H_e(\mathbf{S}) - 1 & \text{if } \mathbf{S} = \mathbf{N}_2 \text{ or } \mathbf{S} = \mathbf{S}_2, \\ H_e(\mathbf{S}) & \text{otherwise.} \end{cases}$$

Theorem 9.36 (Even Map Color Theorem) *Let $\mathbf{S}$ be an arbitrary surface. If a graph G is embedded on $\mathbf{S}$ such that $d_G(f) \equiv 0 \pmod 2$ for all $f \in F(G)$, then $\chi(G) \leq H_{\text{even}}(\mathbf{S})$. Furthermore, there exists a quadrangulation G of $\mathbf{S}$ such that $\chi(G) = H_{\text{even}}(\mathbf{S})$.*

Hutchinson's proof in [510] that $H_e(\mathbf{S})$ is an upper bound for the chromatic number of a graph embedded on $\mathbf{S}$ with all faces of even degree requires that the embedding is cellular. However, working with general embeddings, rather than cellular embeddings, simplifies the proof as shown in [685]. As pointed out by Liu et al. [685], every graph can be embedded in an orientable surface such that all faces have even degree.

It is tempting to suggest that locally plane graphs are 5-colorable, and locally bipartite plane graphs are 3-colorable. Albertson and Stromquist [40] made a first proposal of what locally plane means. Thomassen's result that $\mathrm{Crit}(\mathbf{S}, 6)$ is finite for every surface $\mathbf{S}$ (Theorem 9.8) implies that there is a smallest integer $f(\mathbf{S})$ such that every graph G embedded on $\mathbf{S}$ with all noncontractible cycle having length greater than $f(\mathbf{S})$ satisfies $\chi(G) \leq 5$. Clearly, we have $f(\mathbf{S}) \leq N(\mathbf{S}, 6) = \Theta(\mathrm{eg}(\mathbf{S}))$, where the equality was proved by Postle and Thomas [824], see also [819]. Using different arguments Postle [819, Thoerem 5.8.9] proved that $f(\mathbf{S}) = \Theta(\log \mathrm{eg}(\mathbf{S}))$ for cellular embedded graphs on $\mathbf{S}$. Thomassen's lists (9.7) of 6-critical toroidal graphs implies that $f(\mathbf{S}_1) = 3$, thus improving an earlier bound established by Albertson and Stromquist [40]. Hutchinson [511] proved that locally plane bipartite graphs on orientable surfaces are 3-colorable.

Theorem 9.37 (Hutchinson) *If a graph G is cellular embedded on a surface $\mathbf{S}_g$, $g \geq 1$, with all faces of even degree and with all noncontractible cycle having length at least 2^{3g+5}, then $\chi(G) \leq 3$.*

In 2012, Postle submitted a thesis about 5-list-coloring graphs on surfaces; in this thesis Postle developed a theory of linear isoperimetric inequalities for graphs on surfaces and applied it successfully to map coloring problems. The results were published in a sequence of papers by Postle and Thomas, see the references in [824]. Three of the main results say that, for every surface $\mathbf{S} \neq \mathbf{S}_0$, the graph families $\mathrm{Crit}_\ell(\mathbf{S}, 6)$, $\mathrm{Crit}_\ell(\mathbf{S}, 5, 4)$, and $\mathrm{Crit}_\ell(\mathbf{S}, 4, 5)$ are finite. Recall from Section 5.2 that every graph $G \in \mathrm{Crit}_\ell(k)$ is k-list-critical, that is, there is a $(k-1)$-list-assignment L for which G is L-critical (i.e., every proper subgraph of G has an L-coloring, but G itself does not). In [824] the following stronger result is proved.

Theorem 9.38 (Postle and Thomas) *There is an absolute constant c such that, for every surface $\mathbf{S} \neq \mathbf{S}_0$, if a graph $G \in \mathcal{G}(\hookrightarrow \mathbf{S})$ is either 6-list-critical, or 5-list-critical and triangle-free, or 4-list-critical with girth at least 5, then $|G| \leq c \cdot \mathrm{eg}(\mathbf{S})$.*

Note that $\mathrm{Crit}_\ell(\mathbf{S}_0, 6) = \varnothing$ and $\mathrm{Crit}_\ell(\mathbf{S}_0, 4, 5) = \varnothing$; the first equation follows from Thomassen's 5-list-color theorem, and the second equation follows from Thomassen's Theorem 9.14. As emphasized by Dvořák and Postle [318] both results can be extended to the DP-chromatic number. A plane triangle-free graph G satisfies $\mathrm{col}(G) \leq 4$ by Euler's formula, and so $\mathrm{Crit}_\ell(\mathbf{S}, 5, 4) = \mathrm{Crit}_{\mathrm{DP}}(\mathbf{S}_0, 5, 4) = \varnothing$. Let $G \in \mathrm{Crit}_{\mathrm{DP}}(\mathbf{S}, 5, 4)$ be a graph with n vertices and m edges. Then Euler's formula yields $2m \leq 4(n - \varepsilon(\mathbf{S})) = 4n + 4\mathrm{eg}(\mathbf{S}) - 8$, and the Gallai-bound for χ_{DP}-critical graphs yields $2m \geq 2\mathrm{ext}_{\mathrm{DP}}(5, n) \geq (4 + \frac{1}{11})n$ (see (9.4)). Consequently, we obtain that $n \leq 44 \cdot \mathrm{eg}(\mathbf{S}) - 88$, that is, $\mathrm{Crit}_{\mathrm{DP}}(\mathbf{S}, 5, 4)$ is finite for all surfaces $\mathbf{S}$. To decide whether $\mathrm{Crit}_{\mathrm{DP}}(\mathbf{S}, 6)$ or $\mathrm{Crit}_{\mathrm{DP}}(\mathbf{S}, 4, 5)$ is finite, even the best possible bounds for $\mathrm{ext}_{\mathrm{DP}}(n, k)$ are not sufficient.

An important proof technique is to extend a coloring of a subgraph to the entire graph. The following result by Postle [819, Theorem 1.8.8] (see also [824]) is an extension of earlier results, see, in particular, the papers by Albertson [31], by Albertson and Hutchinson [37], and by Dvořák, Lidický, Mohar, and Postle [315].

Theorem 9.39 (Postle) *Let G be a graph that is cellular embedded in a surface $\mathbf{S}$ such that every noncontractible cycle has length at least $\Omega(\log\ \mathrm{eg}(\mathbf{S}))$. Let L be a 5-list-assignment for G, and let $X \subseteq V(G)$ such that $\mathrm{dist}_G(u, v) \geq \Omega(\log\ \mathrm{eg}(\mathbf{S}))$ whenever $u, v \in X$ and $u \neq v$. Then every L-coloring of $G[X]$ can be extended to an L-coloring of G.*

Because 3-coloring of planar graphs is an NP-complete decision problem (see Theorem 7.38), a good characterization of the class $\mathrm{Crit}(\mathbf{S}_0, 4)$ is unlikely, unless

$\mathsf{P} = \mathsf{NP}$. On the other hand, the structure of graphs in $\text{Crit}(\mathbf{S}_0, 4)$ is more restrictive than that of graphs in $\text{Crit}(4)$. The Toft graphs show that, for every $t \geq 3$, there exists a graph $G \in \text{Crit}(4)$ containing $K_{t,t}$ as a subgraph implying that K_t is a minor of G. Another parameter is the average degree. Let $S = \{\text{ad}(G) \mid G \in \text{Crit}(\mathbf{S}_0, 4)\}$ and $S' = \{\text{ad}(G) \mid G \in \text{Crit}(4)\}$. By the result of Kostochka and Yancey, we have $\text{ext}(4,n) = \lceil \frac{1}{3}(5n-2) \rceil$ and this bound is attained by infinitely many graphs of $\text{Crit}(\mathbf{S}_0, 4)$. Thus, we obtain that $\liminf S = \liminf S' = \frac{10}{3}$. However, for the supremum we have that $\sup\ S \leq 6$ and $\sup\ S' = \infty$. Every graph $G \in \text{Crit}(\mathbf{S}_0, 4)$ satisfies $3 \leq \delta(G) \leq 5$, which is in sharp contrast to Theorem 4.40. Dirac (see [399]) made the conjecture that every graph $G \in \text{Crit}(\mathbf{S}_0, 4)$ has $\delta(G) = 3$, and Gallai [399] made the stronger conjecture that such a graph satisfies $e(G) \leq 2|G| - 2$. In 1985, Köster [613] used an elegant construction to obtain a 4-regular graph $G \in \text{Crit}(\mathbf{S}_0, 4)$ having order $n = 40$, thus disproving both conjectures. Köster's graph shows that $\sup\ S \geq 4$. The first improvement of this lower bound was obtained by Grünbaum [436]; he proved that $\sup\ S = \lim\sup\ S \geq 158/39$. The next improvement, namely $\sup\ S \geq 78/19$, was obtained independently by Abbott and Zhou [9] and by Köster [615]. In 2017, Yao and Zhou [1079] obtained the lower bound $\sup\ S \geq 14/3$. Combining the fact that every planar 4-critical graph of order $n \geq 5$ has at most $n-1$ triangles with Euler's formula, we obtain that every such graph G satisfies $2e(G) \leq 5n - 9$ which yields $\delta(G) \leq 4$ and $\sup\ S \leq 5$; this was proved by Köster [615]. Dvořák and Feghali [310] proved that Köster's bound is best possible, i.e. $\lim\sup\ S = 5$. Using Köster's graph of order $n = 40$, one can obtain an infinite family of 4-regular graphs belonging to $\text{Crit}(\mathbf{S}_0, 4)$, for the details see Abbott and Zhou [9], Köster [614] and [615], and Chiu, Felsner, Scheucher, Schröder, Steiner, and Vogtenhuber [218]. Abbott and Zhou observed that if G is a 4-critical plane graph and f is a 3-face of G, then we obtain a new 4-critical plane graph by subdividing the three edges of f and joining the three new vertices by three edges forming a 3-face.

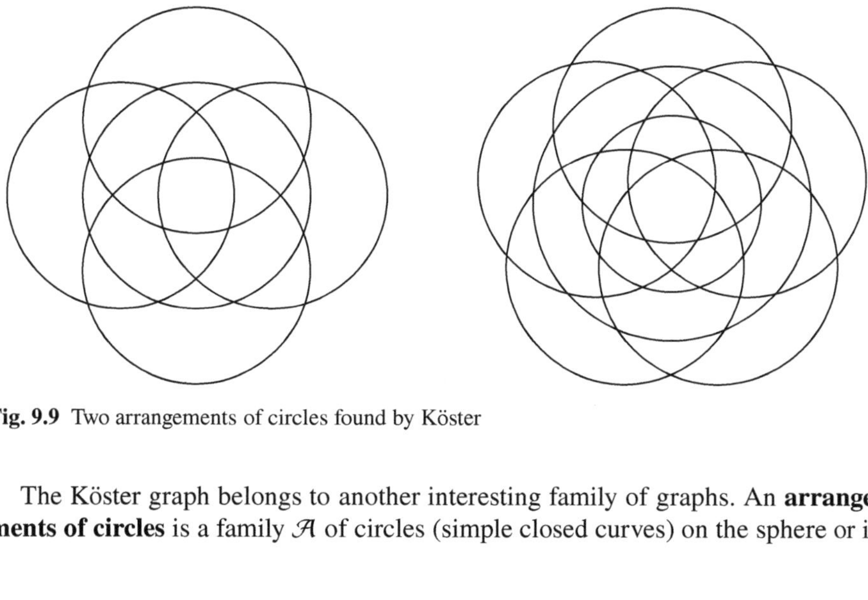

Fig. 9.9 Two arrangements of circles found by Köster

The Köster graph belongs to another interesting family of graphs. An **arrangements of circles** is a family $\mathcal{A}$ of circles (simple closed curves) on the sphere or in the plane no two of which are tangent, no three of which meet at a point, and every pair of which has zero or two crossing points. What we call a circle is often called a pseudocircle to distinguish it from a geometric circle. With an arrangements $\mathcal{A}$ of circles, we can associate the following two graphs. The intersection graph $\mathrm{IG}(\mathcal{A})$ has vertex set $\mathcal{A}$ and two distinct members of $\mathcal{A}$ are adjacent in $\mathrm{IG}(\mathcal{A})$ if and only if they have a nonempty intersection. The **arrangement graph** $\mathrm{AG}(\mathcal{A})$ is the plane graph obtained by placing vertices at the crossing points of $\mathcal{A}$ and thereby subdividing the circles into edges; we call such graphs **circle graphs**. Note that a circle graph may have multiple edges; take for instance two circles having two crossing points resulting in $4K_2$. Clearly, any circle graph G is empty or planar and 4-regular, and thus $\chi(G) \leq 4$ (by Brooks' theorem). Circle graphs were first considered by Grötzsch and Sachs; they conjectured that every circle graph is 3-colorable, see Sachs [883]. Further information can be found in the paper by Dobrynin and Mel'nikov [305] and in the paper by Steinberg [961]. Köster's graph, however, disproves this conjecture; the graph is the arrangement graph $\mathrm{AG}(\mathcal{A})$ belonging to a family $\mathcal{A}$ of seven (geometric) circles shown in Figure 9.9 on the right hand side, note that $\mathrm{IG}(\mathcal{A}) = K_7^-$. The left hand side of Figure 9.9 shows an arrangement $\mathcal{A}'$ of five (geometric) circles, found by Köster, such that $\mathrm{IG}(\mathcal{A}') = K_5$ and $\mathrm{AG}(\mathcal{A}')$ is a 4-chromatic (but not 4-critical) planar graph. Jaeger [513] proved that an arrangement $\mathcal{A}$ of circles satisfies $\chi(\mathrm{AG}(\mathcal{A})) \leq 3$, provided that $\chi(\mathrm{IG}(\mathcal{A})) \leq 3$; this condition cannot be replaced by $\chi(\mathrm{IG}(\mathcal{A})) \leq 4$ as proved in [305]. Figure 9.10 shows two 4-chromatic circle graphs G found by Dobrynin and Mel'nikov; these graphs have the form $G = \mathrm{AG}(\mathcal{A})$ where $\mathcal{A}$ is an arrangement of four circles such that $\mathrm{IG}(\mathcal{A}) = K_4$. Jaeger [513] proposed the conjecture that every circle graph has chromatic index 4; this conjecture seems to be open. Chiu et al. proved in [218] that a circle graph is 3-colorable, provided that every edge belongs to exactly one 3-face; they also constructed an infinite family of 4-regular graphs $G \in \mathrm{Crit}(\mathbf{S}_0, 4)$ having fractional chromatic number $\chi^*(G) = 3$. Felsner, Hurtado, Noy, and Streinu [370] proposed the conjecture that if $\mathcal{A}$ is an arrangement of (geometric) big-circles on the (2-dimensional) sphere, then $\mathrm{AG}(\mathcal{A})$ is 3-colorable.

Fig. 9.10 Two arrangement of four circles found by Dobrynin and Mel'nikov

Extending map color results to coloring parameters other than the chromatic number has attracted a lot of attention over the years. Here we focus on the list-chromatic number and the DP-chromatic number. In what follows let ξ denote an arbitrary monotone graph parameter such that $\omega(G) \leq \xi(G) \leq \mathrm{col}(G)$ for every graph G. These inequalities immediately imply that $\xi(\mathbf{S}) = H(\mathbf{S})$ for every surface $\mathbf{S}$, except possibly for the sphere and the Klein bottle. For the sphere we obtain that $4 \leq \xi(\mathbf{S}_0) \leq 6$. The four color theorem shows that it might be very difficult to determine $\xi(\mathbf{S}_0)$. Thomassen's outstanding short proof in [1008] that $\chi_\ell(\mathbf{S}_0) \leq 5$ can be applied directly to the DP-chromatic number as noted by Dvořák and Postle [318]. Together with the 1993 construction of a non-4-choosable planar graph by Voigt [1052] this shows that $\chi_\ell(\mathbf{S}_0) = \chi_{\mathrm{DP}}(\mathbf{S}_0) = 5$. Mirzakhani's construction from 1996 in [743] and her proof of its correctness are of exquisite simplicity, everything is to the point. In 2014, Maryam Mirzakhani (1977–2017) was honored with the Fields Medal for her outstanding contribution to hyperbolic geometry; becoming the first Iranian to be honored with the award and the first woman.

Since we assume that ξ is monotone, we have $\xi(G) \leq k-1$ if and only if no subgraph G' of G is ξ-critical and has $\xi(G') = k$. So let $\mathrm{Crit}_\xi(\mathbf{S},k)$ denote the family of ξ-critical graphs $G \in \mathcal{G}(\hookrightarrow \mathbf{S})$ with $\xi(G) = k$; and let $\mathrm{Crit}_\xi(\mathbf{S},k,g)$ be the family of graphs $G \in \mathrm{Crit}_\xi(\mathbf{S},k)$ with girth at least g. To decide whether $\mathrm{Crit}_\xi(\mathbf{S},k,g)$ is finite, we can use information about the density of ξ-critical graphs as far as available. So let $\mathrm{ext}_\xi(k,n)$ denote the minimum number of edges possible in a ξ-critical graph G satisfying $\xi(G) = k$ and $|G| = n$ (usually, we have $n \geq k$). As pointed out by Gimbel and Thomassen, if we have a Gallai bound for ξ-critical graphs, i.e.

$$2\mathrm{ext}_\xi(k,n) \geq \big(k-1+\frac{k-3}{(k^2-3)}\big)n \text{ for } n > k \geq 4,$$

then $\mathrm{Crit}_\xi(\mathbf{S},k,g)$ is finite provided that $g \geq 6$ and $k \geq 4$, or $4 \leq g \leq 5$ and $k \geq 5$, or $g = 3$ and $k \geq 7$. In particular, for every surface $\mathbf{S}$, the family $\mathrm{Crit}_\xi(\mathbf{S},k)$ is finite for all $k \geq 7$ (see also Exercise 9.2); this particularly holds for $\xi = \chi_\ell$ and $\xi = \chi_{\mathrm{DP}}$, so both families $\mathrm{Crit}_\ell(\mathbf{S},k)$ and $\mathrm{Crit}_{\mathrm{DP}}(\mathbf{S},k)$ are finite for all $k \geq 7$. The Gallai bound for ξ-critical graphs leads to a bound for the maximum order $N_\xi(\mathbf{S},k)$ for graphs in $\mathrm{Crit}_\xi(\mathbf{S},k)$ for $k \geq 7$. If $\mathbf{S}$ is neither $\mathbf{S}_0$, $\mathbf{N}_1$ nor $\mathbf{N}_3$ and $k = H(\mathbf{S})$, then the Gallai bound for ξ-critical graphs leads to $N_\xi(\mathbf{S},k) \leq k+3$ (see [125, Section 2]). If a Dirac bound holds for ξ-critical graphs, that is, $2\mathrm{ext}_\xi(k,n) \geq (k-1)n + (k-3)$ for $n > k \geq 4$, then we obtain $N_\xi(\mathbf{S},k) \leq k+1$ (see the proof of Theorem 9.2). For $\xi = \chi, \chi_\ell, \chi_{\mathrm{DP}}$, this suffices to show that

$$\mathrm{Crit}(\mathbf{S},k) = \mathrm{Crit}_\ell(\mathbf{S},k) = \mathrm{Crit}_{\mathrm{DP}}(\mathbf{S},k) = \langle K_k \rangle,$$

provided that $\mathbf{S} \neq \mathbf{N}_2$. For the Klein bottle we have $H(\mathbf{N}_2) = 7$, but K_7 cannot be embedded on $\mathbf{N}_2$. From this it follows that $\chi(\mathbf{N}_2) = \chi_\ell(\mathbf{N}_2) = \chi_{\mathrm{DP}}(\mathbf{N}_2) = 6$. So Dirac's map color theorem can be extended to χ_ℓ and χ_{DP} and to all surfaces except possibly $\mathbf{S}_0$, $\mathbf{N}_1$ or $\mathbf{N}_3$. For the projective plane we have $H(\mathbf{N}_1) = 6$. That K_6 is the only ξ-critical graph $G \in \mathcal{G}(\hookrightarrow \mathbf{N}_1)$ with $\xi(G) = 6$ was proved by Albertson and Hutchinson [34] for $\xi = \chi$ and by Böhme et al. [125] for $\xi = \chi_\ell$. While the proof of Albertson and Hutchinson is one page long, the proof of Böhme et al. requires several precoloring extension results for plane graphs. These theorems and their proofs can be extended to DP-colorings without difficulty, see Theorems 9.28, 9.29 and 9.30. The choosability version of Theorem 9.28 was already used by Thomassen [1012]. The condition (b) in Theorem 9.30 is missing in the choosability version in [125], but it is needed as we only require that the lists have size at least 5 (and not equal to 5). For the application of this result in the proof of the main result, this case is not relevant (here we assume that all list have size equal to 5). That this case was missing in [125] was pointed out by Joan Hutchinson in an email to the authors. However, the proof of the main result (Theorem 9.31) differs in the case $p = 4$ (that is, the case when the shortest noncontractible cycle has length 4) from the original proof in [125]. This is due to the fact that $\chi_\ell(C_{2n}) = 2$ but $\chi_{\mathrm{DP}}(C_{2n}) = 3$. For $\mathbf{N}_3$, we have $H(\mathbf{N}_3) = 7$. That K_7 is the only ξ-critical graph $G \in \mathcal{G}(\hookrightarrow \mathbf{N}_3)$ with $\xi(G) = 7$ was proved for $\xi = \chi$ by Albertson and Hutchinson [34], and for $\xi = \chi_\ell$ by Král' and Škrekovski [646]. We were able to simplify the argument for the chromatic number by using a strengthening of the Dirac bound for χ-critical graphs, see Theorem 9.3. Such a result, however, is not known for either χ_ℓ or χ_{DP}. The known proof of Theorem 9.3 uses the simple fact that a graph G' obtained from a graph G by contracting two nonadjacent vertices satisfies $\chi(G) \leq \chi(G')$. Whether the long and sophisticated proof of Král' and Škrekovski can be extended to the DP-chromatic number is unclear, but not impossible.

Let us briefly discuss four other very popular coloring parameters: the **group chromatic number** χ_{gr}, the **online list chromatic number** χ_{ol}, the **Alon-Tarsi number** $\chi_{\mathrm{at}} = \Delta^+_{\mathrm{at}} + 1$, and the **game chromatic number** χ_{ga}. All four coloring parameters are monotone and bounded from below by the clique number. While the first three coloring parameters are bounded from above by the coloring number, this is not the case for the game chromatic number. Hence Heawood's map color theorem only holds for the first three parameters, i.e. for $\xi \in \{\chi_{\mathrm{gr}}, \chi_{\mathrm{ol}}, \chi_{\mathrm{at}}\}$; we have $\xi(\mathbf{S}) = H(\mathbf{S})$ for every surface $\mathbf{S}$, except possibly for the sphere $\mathbf{S}_0$ and and the Klein bottle $\mathbf{N}_2$.

The **graph coloring game** is a two-person game in which, given a graph G and a color set C, the two players, Alice and Bob, take turns coloring an uncolored vertex of G properly using colors from C, Alice having the first move. Alice wins if all vertices of G are colored; Bob wins if a player has no legal move before G is completely colored. The **game chromatic number** $\chi_{\mathrm{ga}}(G)$ is the minimum cardinality of a color set C for which Alice has a winning strategy. This graph coloring game and the corresponding game chromatic number was introduced, independently, by Brams (see Gardner [406]) and Bodlaender [123]); there are other types of graph coloring games (see the paper by Jakovac and Štesl [521]). Brams and Gardner, in their correspondence around 1980, contemplated that maybe there would be no upper limit to the game chromatic number for planar graphs, and the existence of such an upper bound seems first proved by Kierstead and Trotter [932]. In a letter to Gardner from Lloyd Shapley in October 1980 it was pointed out that Bob can win on the dodecahedron with five colors available. Thus 6 as a lower bound for planar graphs was established. In his column in Scientific American in April

1981 Gardner mentioned the following intriguing problem: If Alice has a winning strategy for G using k colors, does she also have a winning strategy with k+1 colors? This problem is still completely open, even if it at a first glance seems trivial. Bartnicki, Grytczuk, Kierstead, and Zu [83] published a delightful paper about the game chromatic number. For the sphere it is known that $8 \leq \chi_{\mathrm{ga}}(\mathbf{S}_0) \leq 17$; the lower bound was obtained by Kierstead and Trotter [586], and the upper bound was obtained by Zhu [1098]. As pointed out by Bodlaender, for the class of trees the maximum game chromatic number is 4.

Let A be an abelian group. We say that a graph G is **A-colorable**, if for every orientation D of G and every map $f : E(D) \to A$, there is a map $\varphi : V(D) \to A$, such that for every edge $e \in E(D)$ we have $\varphi(v_D^-(e)) - \varphi(v_D^+(e)) \neq f(e)$; we then say that φ is an **A-coloring** of (D, f). Note that the choice of the orientation D is unimportant. The **group chromatic number** $\chi_{\mathrm{gr}}(G)$ of G is the least integer k such that G has an A-coloring for every abelian group A of order at least k. Group colorings of graphs were first introduced in 1992 by Jaeger, Linial, Payan, and Tarsi [517]. As an immediate consequence of the definition, we obtain that every graph G satisfies

$$\omega(G) \leq \chi(G) \leq \chi_{\mathrm{gr}}(G) \leq \chi_{\mathrm{DP}}(G) \leq \mathrm{col}(G). \tag{9.8}$$

To see that $\chi_{\mathrm{gr}}(G) \leq \chi_{\mathrm{DP}}(G)$, let A be an abelian group, let D be an orientation of G, and let $f : E(D) \to A$ be a map. Let (X, H) be a cover of G such that $X_v = \{v\} \times A$ for all $v \in V(G)$, and two distinct vertices (v, c) and (v', c') are adjacent in H if and only if vv' is an edge of D directed from v to v' and $c' = c + f(vv')$ (in A). Then (D, f) has an A-coloring if and only if G has an (X, H)-coloring. This immediately implies that $\chi_{\mathrm{gr}}(G) \leq \chi_{\mathrm{DP}}(G)$. That this inequality holds was observed by Lai and Mazza [664] in 2021; this clearly has several interesting implications. First, by Theorem 9.11, we obtain that $\chi_{\mathrm{gr}}(\mathbf{S}_0) \leq 5$; this result was proved by Lai and Zhang [665] in 2002. That $\chi_{\mathrm{gr}}(\mathbf{S}_0) \geq 5$ was proved by Král', Pangrác, and Voss [645] in 2004. Consequently, we have $\chi_{\mathrm{gr}}(\mathbf{S}_0) = \chi_{\mathrm{DP}}(\mathbf{S}_0) = 5$. Another remarkable result proved in [645] says that every graph G with minimum degree $d \geq 2$ satisfies $\chi_{\mathrm{gr}}(G) \geq d/(2 \ln d)$; see also Theorem 5.23. Second, Dirac's map color theorem can be extended to χ_{gr} and to all surfaces except possibly $\mathbf{S}_0$, $\mathbf{N}_1$ and $\mathbf{N}_3$. As noted in [664], the group coloring concept can easily be extended to multigraphs, and the inequality $\chi_{\mathrm{gr}} \leq \chi_{\mathrm{DP}}$ also holds for multigraphs. Because of Theorem 1.20, this implies the following version of Brooks' theorem; for simple graph this result was obtained by Richter, Thomassen, and Younger [862] in 2016.

Theorem 9.40 *Let G be a connected multigraph. Then* $\chi_{\mathrm{gr}}(G) \leq \Delta(G) + 1$, *where equality holds only if G is a DP-brick.*

If G is a complete graph or a cycle, then it is easy to see that $\chi_{\mathrm{gr}}(G) = \Delta(G) + 1$. Clearly, there is also a choosability version of the group coloring concept (see [664] for the details), and the above theorem can be extended to the group choice number. As a consequence of Corollary 1.18, we obtain the following result.

Theorem 9.41 *Let G be a connected multigraph, let D be an an orientation of G, let A be an abelian group, and let* $f : E(D) \to A$ *be a map. For each vertex* $v \in V(G)$, *let $L(v) \subseteq A$ such that $|L(v)| \geq d_G(v)$. Then there is an A-coloring φ of (D, f) such that $\varphi(v) \in L(v)$ for all $v \in V(G)$, unless each block of G is a DB-brick.*

It is very likely that one can establish both a Gallai bound and a Dirac bound for the extremal function $\text{ext}_{\chi_{\text{gr}}}(k,n)$, but we have not checked the details.

The online list chromatic number was introduced independently in 2009 by Schauz [894] and Zhu [1099]. Let G be a graph, and let $f : V(G) \to \mathbb{N}_0$ be a vertex function for G. The graph G is said to be **f-paintable** if either $G = \varnothing$ or $f(v) \geq 1$ for all $v \in V(G)$ and for each nonempty subset $S \subseteq V(G)$ there exists an independent set I of G such that $I \subseteq S$ and $G' = G - I$ is f'-paintable for the restricted vertex function $f' = f(S, I)$ of G' satisfying $f'(v) = f(v)$ for all $v \in V(G') \setminus S$ and $f'(v) = f(v) - 1$ for all $v \in V(G') \cap S$. When $f(v) = k$ for all $V \in V(G)$, the corresponding term becomes **k-paintable**. The **online list chromatic number** or **paintability number** of G, denoted by $\chi_{\text{ol}}(G)$, is the least number $k \geq 0$ such that G is k-paintable. The term paintable was used by Schauz in [894]; he gave an equivalent definition based on a two-person game between Mr. Paint and Mrs. Correct. From the definition it follows immediately that if G is f-paintable, then G is f-choosable. To see this, let L be an arbitrary f-assignment for G. To construct an L-coloring of G, we can proceed as follows. Either $G = \varnothing$ and we are done, or we consider a nonempty set S consisting of all vertices having a certain color c in their list. Since G is f-paintable, there is an independent set $I \subseteq S$ such that $G' = G - I$ is f'-paintable for $f' = f(S, I)$. Now, we color all vertices of S with color c and let L' be the f'-assignment of G' obtained from L by deleting color c from all the lists. Continuing this procedure with G', f' and L' results in an L-coloring of G. This implies, in particular, that the list chromatic number is an lower bound of the online list chromatic number.

The **Alon-Tarsi number** $\chi_{\text{at}}(G) = \Delta^+_{\text{at}}(G) + 1$ was introduced by Jensen and Toft [530], see also Section 3.8. Schauz [895] extended the Alon-Tarsi result in Theorem 3.24 and proved that the Alon-Tarsi number is an upper bound for the online list chromatic number. Hence we obtain the following sequence of inequalities for an arbitrary graph G:

$$\omega(G) \leq \chi(G) \leq \chi_\ell(G) \leq \chi_{\text{ol}}(G) \leq \chi_{\text{at}}(G) \leq \Delta^+_{\text{ke}}(G) + 1 \leq \text{col}(G). \tag{9.9}$$

Zhu [1102] proved that every planar graph has Alon-Tarsi number at most 5; together with the construction of Voigt [1052] this implies that $\chi_\ell(\mathbf{S}_0) = \chi_{\text{ol}}(\mathbf{S}_0) = \chi_{\text{at}}(\mathbf{S}_0) = 5$. A Gallai bound for the function $\text{ext}_{\chi_{\text{ol}}}(k,n)$ was established by Riasat and Schauz [860]; they also proved that $\chi_{\text{ol}}(\mathbf{N}_2) = 6$. Whether Dirac's map color theorem can be extended to the online list chromatic remains open, but seems to be very likely. Improved Gallai-type bounds for both functions $\text{ext}_{\chi_{\text{ol}}}(k,n)$ and $\text{ext}_{\chi_{\text{at}}}(k,n)$ were obtained by Cranston and Rabern [262] as well as by Kierstead and Rabern [581]. These bounds may help to extend Dirac's map color theorem to the coloring parameters $\chi_{\text{ol}}(G)$ and $\chi_{\text{at}}(G)$ and to all surfaces, except $\mathbf{S}_0$, $\mathbf{N}_1$ and $\mathbf{N}_3$; however we did not check the details. As proved by Culver and Hartke [269], the difference $\chi_{\text{DP}} - \chi_{\text{at}}$ can be both positive and negative.

Appendix A
Brooks' Fundamental Paper

A.1 On Colouring the Nodes of a Network

R. L. Brooks

Communicated by W. T. Tutte

Received 15 November 1940

Proceedings of the Cambridge Philosophical Society 37 (1941), 194–197.

The purpose of this note is to prove the following theorem.

Let N be a network (or linear graph) such that at each node not more than n lines meet (where $n > 2$), and no line has both ends at the same node. Suppose also that no connected component of N is an n-simplex. Then it is possible to colour the nodes of N with n colours so that no two nodes of the same colour are joined.

An n-simplex is a network with $n+1$ nodes, every pair of which are joined by one line.

N may be infinite, and need not lie in a plane.

A network in which not more than n lines meet at any node is said to be *of degree not greater than n*. The colouring of its nodes with n colours so that no two nodes of the same colour are joined is called an "n-coloring".

Without loss of generality we may suppose that N is connected, for otherwise the theorem can be proved for each connected component; and that it is not a simplex. With these suppositions, N is finite or enumerable, both as regards nodes and lines.

Now we can $(n+1)$-colour N with colours $c_0, c_1, \ldots, c_n$, by giving to each node in turn a colour different from all those already assigned to nodes to which it is

M. Stiebitz et al., *Brooks' Theorem*, Springer Monographs in Mathematics,
https://doi.org/10.1007/978-3-031-50065-7

directly joined. We can then apply the following operations, in which the colours of directly joined nodes remain distinct.

(1) A node directly joined to not more than $n-1$ colours can be recoloured not-c_0. (In the term "recolouring" we include for convenience the case in which no colour is altered.) In particular, a node directly joined to two nodes of the same colour may be recoloured not-c_0.

(2) If P and Q are directly joined they can be recoloured without altering any other nodes, so that P is not-c_0. For neglecting the join PQ, we may recolour P not-c_0, by (1); and Q can then be recoloured (possibly c_0).

(3) Let $P, P', P'', \ldots, Q$ be a path, i.e. suppose every consecutive pair of nodes directly joined. Then we can recolour $P, P', \ldots, Q$ successively, without altering any other nodes, so that at most Q has finally the colour c_0.

Corollary 1. If N is finite, choose Q arbitrarily in N. Since there is a path joining Q to every node P in N, we can recolour N with at most the node Q coloured c_0.

Corollary 2. If N is infinite, let F be any connected finite part, and Q a node directly joined to, but not in, F. Then we can recolour F and Q so that no node of F is coloured c_0.

PROOF OF THE MAIN THEOREM FOR A FINITE NETWORK

Case 1. If any node X meets fewer than n lines, we can n-colour N. For let N be $(n+1)$-coloured, with at most the node X coloured c_0. Then by (1), X also can be recoloured not-c_0.

Case 2. Suppose that if P, Q, A, B are any four distinct nodes, there is a path from P to Q not including A or B

Since N is not a simplex, we can find nodes P, Q not directly joined. Let N be $(n+1)$-coloured so that only Q is coloured c_0. Then P and all nodes directly joined to it are not-c_0. Either P meets fewer than n lines, when N may be n-coloured by case 1, or there are two nodes A, B directly joined to P, which have the same colour. But there is a path joining P to Q, not including A or B. Hence, by (3), N can be recoloured, without altering A or B, so that at most P is coloured c_0. Since A and B have the same colour, P can be recoloured not-c_0, by (1). Thus N is n-coloured.

Case 3. Suppose there exist distinct nodes P, Q, A, B, such that every path from P to Q passes through A or B.

Then consider the networks, contained in N, with the following specifications:

N_1. *Nodes:* P, and all nodes joined to P by some path not passing through A or B as an intermediate point.
Lines: all lines connecting the above nodes in N.

N_2. *Nodes:* A, B, and all nodes not in N_1.
Lines: all lines connecting the above nodes in N.

Thus N_1 and N_2 are connected non-null networks, together making up the whole network N, and having in common at least one of the nodes A, B, and at most A, B, and any lines AB. Therefore, if m_i is the number of lines in N_i, and m_0 is the number of lines in AB,

$$m_1 + m_2 \leq m_0 + n. \tag{A.1}$$

Clearly, N_1 and N_2 have degree not exceeding n.

There are three subcases, 3.1, 3.2 and 3.3.

Case 3.1. Suppose N_1 and N_2 have only one node, say A, in common. Then in each of them, the node A meets fewer than n lines. Thus by case 1, N_1 and N_2 may be n-coloured; and if we permit the colours of N_2 so that the colours of A in N_1 and N_2 become the same, the whole network N is n-coloured.

Case 3.2. One of N_1 and N_2 (say N_1), is such that when the line AB is added it becomes an n-simplex.

N_1 can be n-coloured by assigning arbitrary colours to the $n-1$ nodes other than A and B, and the remaining colour to A and B. By (A.1) there is just one line in N_2 meeting A and just just one meeting B or case 3.1 holds. Thus if A and B were identified in N_2, there would still (since $n > 2$) be fewer than n lines meeting A $(= B)$. Hence by case 1 the resulting network can be n-coloured.

Case 3.3. Neither N_1 nor N_2 becomes an n-simplex on adding a join AB.

Suppose they become M_1 and M_2 respectively by this addition. Then M_1 and M_2 each contain fewer nodes than N, and by (A.1) they are of degree not greater than n. (If not we should have case 3.1.)

If M_1 and M_2 are n colourable so is N. For since both contain a line AB, in any n-colouring A must have a different colour from B in each network. We can permute the colours of M_1 so that the colours of A and B are the same as in M_2, and then by combination obtain an n-colouring of N.

Thus if N is a finite network of degree not exceeding n, and is not an n-simplex, either it is n-colourable, or it is n-colourable if two networks which satisfy the same conditions, but have fewer nodes, are n-colourable. Now it is obvious that the theorem is true for a network with less than four nodes. Therefore, by induction over the number of nodes, N is always n-colourable.

INFINITE NETWORKS

If F is a network or a set of nodes in N, we denote by $N - F$ the network composed of the nodes of N not in F, and the lines of N neither end of which is in F.

LEMMA. For each positive r we can find a connected finite network F_r such that

(i) the connected component of $N - F_1 - F_2 - \cdots - F_r$ are infinite for all r.
(ii) every node of N lies in one and only one F_r.

Let the nodes of N be enumerated as $P_1, P_2, P_3, \ldots$, and let F_r be defined inductively as follows.

When F_s has been chosen, for $s < r$ (or without any choice when $r = 1$), let R_r be the first P_m not in any F_s. Suppose further that the connected components of $N - F_1 - F_2 - \cdots - F_{r-1}$ are infinite. Then all the finite connected components of $N - F_1 - F_2 - \cdots - F_{r-1} - R_r$ must contain a node directly joined to R_r in N. Therefore the number of these components cannot exceed n, and at least one infinite component of $N - F_1 - F_2 - \cdots - F_{r-1} - R_r$ contains a node joined by a line to R_r.

Take F_r to be the "logical sum" of R_r, all finite connected components of $N - F_1 - F_2 - \cdots - F_{r-1} - R_r$, and all lines joining them in N. Thus F_r is a finite connected network, and has no node in common with F_s, for $s < r$. Further, the connected components of $N - F_1 - F_2 - \cdots - F_r$ are simply the infinite components of $N - F_1 - F_2 - \cdots - F_{r-1} - R_r$. The inductive construction is therefore complete.

By the method of choosing R_r, P_m must lie in some F_s ($s \leq m$).

A node Q_r can be chosen in an infinite connected component of

$$N - F_1 - F_2 - \cdots - F_{r-1} - R_r,$$

so that Q_r is directly joined in N to R_r, which lies in F_r. Thus Q_r does not lie in F_s, if $s \leq r$.

Now N can be $(n+1)$-coloured; and by (3), corollary 2, we can recolour F_r in n colours, altering only F_r and Q_r, i.e. not altering F_s for $s < r$. Thus we can recolour $F_1, F_2, \ldots,$ in turn in n colours, each recolouring not affecting the nodes already recoloured: that is, we can n-colour N.

TRINITY COLLEGE
CAMBRIDGE

A.2 Rowland Leonard Brooks 1916–1993

R. L. Brooks was born on 6 February 1916 in Lincolnshire, England, and he died 18 June 1993 in London. He studied mathematics at the University of Cambridge from 1935. In Cambridge he met Cedric A. B. Smith, Arthur H. Stone and William T. Tutte, who all became life-long friends. Also, the four were close friends of Blanche Descartes. In November 1940 Tutte communicated Brooks' paper [173], containing what later became famous as Brooks' Theorem, to the Cambridge Philosophical Society. After finishing his studies at Cambridge, Brooks worked as an income-tax inspector in London. He kept a strong interest in mathematics throughout his life. A joint paper [177] by the Trinity 4, Brooks, Smith, Stone and Tutte, on determinants and current flows in electrical networks was published in the Julius Petersen volume 100 of Discrete Mathematics in 1992. In 1940 the four had published their paper [175] on the dissection of rectangles into squares in volume 7 of Duke Mathematical Journal, see also [176]. The 1992 paper [177] may be seen as a sequel.

Cedric A. B. Smith sent each year a Christmas letter to his friends, containing thoughts, history, happenings and interesting quotations from the year that had passed. We shall take the liberty to quote from these letters.

From Cedric A. B. Smith's Christmas letter 1992:

Some 56 years ago there were three mathematics students, Arthur, Leonard and Cedric, who were friends. They were joined by a chemist, Bill, when he was not doing experiments. Arthur said, "I've heard about a problem. Show that you can not divide a (geometric) square into squares of all different sizes." Sounds easy. Doesn't it? They quickly discovered that if you take 9 squares, with sides 1, 4, 7, 8, 9, 10, 14, 15, 18, you can fit them together, like a jigsaw, to make what looks exactly like a square. Try it for yourself. But if you add up the sides, alas, they are just slightly different – it is a rectangle, pretending to be a square. They did not know it, but it had been found many years before by a Pole, named Moron. Just as many of the most famous musicians are German or Italian, so, for no obvious reason, many distinguished mathematicians are Polish or Hungarian. But to return to the point. For 3 years Arthur, Leonard, Cedric and Bill worked hard, producing many rectangles filled with unequal squares, but never one which looked remotely like a square, and never any hint as to why a square was impossible. Leonard tried a jigsaw on his mother, who put it together unexpectedly differently. Surprising, but not solving our problem. But one day Arthur suggested to Cedric, "if only we make it sufficiently complicated (following a certain plan) we might succeed." There were no nice pocket calculaters then, but after 3 days hard calculations on paper, there really and truly was before them a square divided into unequal squares. Arthur and Cedric invited Leonard and Bill to coffee, saying "we have something to show you". But Leonard insisted we should have coffee with him. So Arthur and Cedric carefully drew the divided square on paper, went over to Leonard's rooms, walked in and said "look at this". But Leonard came towards them, holding a piece of paper, and said "look at this". And on it was another square divided into unequal squares in a different way. Cheers!

From Cedric A. B. Smith's Christmas letter 1993:

Cedric's great ambition was to go to Cambridge to study mathematics. He was thrilled to go to his very first lecture, on geometry. Perhaps you think that geometry is about squares and triangles, circles and spheres, cones, and all that sort of thing. Cedric thought so too. But the lecturer talked about adding points together, and multiplying points by numbers. At the end of the lecture, Cedric turned to the young man next to him, and said "That was very confusing". "No", replied his companion, "I thought it was a very good lecture. When is the next lecture?" – "at 11" – "No, it's at 10". – They went together into the lecture room. But by a misprint in the timetable, they had walked into an advanced lecture. But, happily, they were blissfully unaware of that, because the lecturer had so thick a Russian accent that for half an hour they literally did not understand one word he said. Later on in the term, when the class roared with laughter at some awful distortion of some word, the lecturer sternly explained, "Yew may laff. Baat feefty meelion people speak yore kind of Eengleesh. Fife haandred meelion people speak my kind of Eengleesh". Anyway, Cedric found that his new friend was called Leonard Brooks. And that was the start of a friendship to last over 57 years. Recently Leonard had been developing on his computer some ideas about the representation of numbers which could be made to produce some dazzling colored patterns. On June 15, this summer, he and Cedric were to have lunched together. But at the last minute he found that it was not possible, and it was put off a week, to June 22. Then it suddenly occurred to Cedric how one could produce a lot of unexpected Brooks patterns, and he worked hard to get them ready to show Leonard on the 22nd. But at the Saturday before that, Leonard's wife Gladys rang up to say that he had quite unexpectedly had a bad heart attack. And he had not survived to see the new pattern.

In a private letter Cedric A. B. Smith, July 2001, said:

Leonard was rather a shy person, and he did not want his photograph published, which is why it is difficult to find a photo of him.

A.3 Brooks' Theorem and Beyond

It took several years for Brooks' theorem to become widely known. In the late 1940s Gabriel Andrew Dirac, influenced by Peter Ungar and supervised by Richard Rado, took up graph coloring theory in his doctoral studies at King's College, University of London. The title of his thesis, submitted in 1951, is *On the Colouring of Graphs. Combinatorial Topology of Linear Complexes*. In the thesis Dirac defined critical graphs and rediscovered Brooks' theorem. The external examiner Cedric A. B. Smith made Dirac aware of Brooks' 1941 paper, which led to some small changes in the final version (Senate House Library, University of London). In particular, Dirac's proof of Brooks' theorem was deleted from the thesis. Thereafter, through Dirac's papers in the 1950s, Brooks' theorem became well known and the cornerstone of abstract graph coloring. To this day, Brooks' 1941 paper [173] is stimulating for many mathematicians; see the two surveys of Brooks' theorem, one by Stiebitz and Toft [976] and another by Cranston and Rabern [260].

Over time, several authors have published alternative proofs of Brooks' theorem that have some advantages compared to the original proof and to those published by other authors. The first new proof of Brooks' theorem was published by Gerencsér [413] in 1965, but only in Hungarian. Gerencsér's proof is by contradiction; so let G be a counterexample to Brooks' theorem whose order is minimum. Then G is Δ-regular, $K_{\Delta+1} \nsubseteq G$, and every induced subgraph of G except G itself has a Δ-coloring, where $\Delta \geq 3$. First, note that G does not contain a $K^-_{\Delta+1}$ (a $K_{\Delta+1}$ minus an edge) as a subgraph. For otherwise, if $D = K_{\Delta+1} - vw \subseteq G$, then $G' = G - V(D)$ has a Δ-coloring. Since both v and w have only one neighbor in G', we can extend this Δ-coloring by first coloring v and w the same color and then coloring the remaining $\Delta - 1$ vertices of D. This results in a Δ-coloring of G, giving a contradiction. Let I be a maximum independent set of G. Then $G - I$ has maximum degree at most $\Delta - 1$. If no subgraph of $G - I$ is a K_Δ if $\Delta \geq 4$ and an odd cycle if $\Delta = 3$, then the minimality of G implies that $G-I$ has a $(\Delta-1)$-coloring and so G has a Δ-coloring, a contradiction. It remains to consider the case that G contains an induced subgraph K, where $K=K_\Delta$ if $\Delta \geq 4$ and K is an odd cycle if $\Delta = 3$. Since $K^-_{\Delta+1} \nsubseteq G$ and G is Δ-regular, we obtain that each vertex of K has exactly one neighbor in $G' = G - V(K)$ and there are at least two different such neighbors, say u and w. Then $G'+uw$ has a Δ-coloring φ (since G is a minimum counterexample). By using a simple list coloring argument, it is easy to show that φ can be extended to a Δ-coloring of G, a contradiction. Gerencsér's proof was rediscovered by Rabern [842]; the case $\Delta = 3$ of Gerencér's proof was also obtained by Dirac [298, Theorem 2]. Rabern [839] gave a refinement of Gerencér's proof presented in Section 1.2.

The list of papers giving alternative proofs of Brooks' theorem keeps growing, see the journal papers [75], [141], [183], [195], [367], [377], [413], [500], [694], [731], [746], [817], [839], [842], [888], [945], [979], [1043], and [1088]. A second group of papers provide proofs of results that directly imply Brooks' theorem, see e.g. [148], [355], [397], [636], [653], [948], [1012], and [1051]. For instance, Gallai's result in [397], saying that the low vertex subgraph of a critical graph is a Gallai forest, immediately implies Brooks' theorem. Borodin [148, 149] as well as Erdős, Rubin, and Taylor [355] proved a choosability version of Brooks' theorem.

The algorithmic aspects of Brooks' theorem have also been studied extensively, for classical algorithms, but also for PRAM algorithms and LOCAL algorithms; see the following papers for corresponding results and further references: Panconesi and Srinivasan [796], Assadi, Kumar, and Mittal [72], and Fischer, Halldórsson, and Maus [373].

Brooks also proved his theorem for locally finite graphs. Schmerl [900] gave a different proof, by constructing a recursive set hitting all maximum cliques; a shorter proof of Schmerl's result was given by Tverberg [1044]. Conley, Marks, and Tucker-Drop [254] examines measurable generalizations of Brooks' theoem.

To date, there are probably several hundred papers related to Brooks' theorem; most of these papers deal with generalizations of Brooks' theorem to other coloring parameters or with strengthening of Brooks' theorem. Already in the 1950s, Dirac noticed that the inequality $2\text{ext}(k,n) \geq (k-1)n+1$ for $n > k \geq 4$ is equivalent to Brooks' theorem, but far from optimal. However, more than half a century passed before Kostochka and Yancey asymptotically determined the function $\text{ext}(k,n)$ for any fixed $k \geq 4$, see Chapter 5. What remains is a complete proof of Ore's conjecture that $\text{ext}(k,n) = \text{ore}(k,n)$.

Let us briefly discuss a precoloring extension version of Brooks' theorem. For a graph G and a vertex set $P \subseteq V(G)$, let

$$\text{mpd}(P:G) = \min\{\text{dist}_G(u,v) \mid u,v \in P,\ u \neq v\}.$$

Note that $\text{mpd}(P:G) \geq 2$ if and only if P is an independent vertex set in G. Let S be a set of colors. Following Brooks, a coloring $\varphi : V(G) \to S$ of the vertices of G with colors from S so that no two vertices of the same color are joined by an edge is called an **S-coloring** of G. A mapping $L: V(G) \to 2^S$ is called a k-assignment for

G (with color set S) if $|L(v)| \geq k$ for every vertex v of G. An L**-coloring** of G is an S-coloring of G such that $\varphi(v) \in L(v)$ for every vertex $v \in V(G)$.

The following result was obtained by Albertson, Kostochka, and West [39] and, independently, by Axenovich [73], answering a question proposed by Albertson at the Southeastern Conference on Graph Theory, Combinatorics and Computing, Boca Raton, March 2002 (see [73, Reference 1]).

Theorem A.1 (Axenovich / Albertson, Kostochka, West) *Let G be a graph with maximum degree $\Delta \geq 3$ not containing $K_{\Delta+1}$ as a subgraph, let P be a vertex set in G such that* $\text{mpd}(P:G) \geq 8$, *and let L be a Δ-assignment for G. Then every L-coloring of $G[P]$ can be extended to an L-coloring of G.*

As pointed out in [39] the distance threshold 8 in Theorem A.1 is sharp even for the constant list assignments. Let $k \geq 3$, and let H be a graph obtained from $K_{k+1} - uv$ by adding a vertex w and joining w to v; note that w is the only vertex having degree 1 in H and u is the only vertex having degree $k-1$ in H. Let G be the graph obtained from k (disjoint) copies of H by adding all possible edges between the vertices of degree 1, let P be the set of of vertices having degree $k-1$, and let $S = \{1, 2, \ldots, k\}$ be a set of k colors. Then it is easy to check that G has maximum degree $\Delta = k$, $\text{mpd}(P:G) = 7$, and there is no coloring φ of G with color set S such that $\varphi(u) = 1$ for all vertices $u \in P$.

The following strengthening of Theorem A.1 was obtained by Rackham [843]; again the new distance threshold is sharp.

Theorem A.2 (Rackham) *Let G be a connected graph with maximum degree $\Delta \geq 3$ not containing $K_{\Delta+1}^-$ as a subgraph, let P be a vertex set in G such that* $\text{mpd}(P:G) \geq 4$, *and let L be a Δ-assignment for G. Then every L-coloring of $G[P]$ can be extended to an L-coloring of G.*

Appendix B
Tutte's Lecture from 1992

B.1 Fifty Years of Graph Colouring

W. T. Tutte

Friday, June 26, 1992

I learned the Four Colour Problem from school, when I borrowed Rouse Ball's "Mathematical Recreations and Essays" from the school library. At the University, in the late Thirties I sought more information. I learned that workers on the problem were studying irreducible configurations. Every so often a paper would appear giving new reducible configurations and raising the minimum number of faces for a 5-chromatic planar cubic map.

I learned also that Hassler Whitney had posed colouring problems for graphs in general. He sought colourings of the vertices of a graph in n colours so that no two of the same colour were adjacent [here Tutte refers to the paper [1068]].

I played with colouring problems myself, with no significant result. But my fellow-student, Leonard Brooks, asked me to check the proof of his memorable theorem, and I had the honour of transmitting his paper to the Cambridge Philosophical Society. It was published in their mathematical Proceedings in 1941. Let me remind you of what this theorem says. If a [connected] graph G, with no loops or multiple joins, cannot be coloured in n colours, where $n > 2$, then either G is a complete graph of $n+1$ vertices or it has a vertex whose valency exceeds n. All graphs of this lecture are to be without loops or multiple joins.

Some time before I began my Ph. D. work in 1945 I learned of two conjectures that would, if valid, generalize Brooks' Theorem. There was the conjecture of Hajós, that every graph not colourable in n colours must contain a subdivision of a $K(n+1)$, that is of a complete graph on $n+1$ vertices. But as we know this conjecture was disproved by P. A. Catlin in a paper of 1979 (J. Combinatorial Theory). Then there was a conjecture of H. Hadwiger, that every graph not colourable in n colours could

M. Stiebitz et al., *Brooks' Theorem*, Springer Monographs in Mathematics,
https://doi.org/10.1007/978-3-031-50065-7

be changed, by deleting some edges and contracting others, into a $K(n+1)$. Nowadays we would express this condition by saying that the graph must have $K(n+1)$ as a "minor". Another way of putting it is to say that the graph must have $n+1$ disjoint connected subgraphs, each being joined to each of the others to an edge. As far as I know Hadwiger's Conjecture still stands.

The Conjectures of Hajós and Hadwiger were of interest because each implied the Four Colour Conjecture.

Some time in the Fifties I learned of Hajós' Theorem. Then I had it explained to me by Hajós himself. That would have been at a Hungarian Conference in 1959. Hajós Theorem was a disappointment to its discover, it seemed to promise so much and yet achieved so little. It had little publicity, and yet I want to expand upon it in this lecture. I will give you a version of the Theorem that I believe to be a little stronger than the one first presented. Let n be a fixed integer greater than 2. Let us define an "achrome" as a graph that cannot be coloured in n colours., and a "minimal achrome" as an achrome that does not contain another as a proper subgraph. The complete graph $K(n+1)$ is the simplest achrome, and it is clearly minimal. It is easy to show that a minimal achrome must be connected and non-separable. Hajós was interested in operations that changed an achrome, or a pair of achromes , into a new achrome.

For example let G be an achrome in which there are two non-adjacent vertices p and q. Let H be the graph got from G by identifying p and q to make a new vertex r. Any digons introduced by this operation are to be replaced by single edges. Then H is another achrome, for any n-colouring of H would determine one of G in which p and q had the same colour. We say that H is formed from G by the Hajós operation of "vertex–identification".

Now consider two disjoint achromes J and L. Let E be an edge of J with ends p and q. Deleting E from J we obtain a graph $J(1)$. Of this we can assert that in any n-colouring p and q must have the same colour. Now let us construct a graph $L(1)$ by splitting a vertex v of L into two new vertices x and y. Each edge originally incident with v is to be incident with one of the new vertices x and y, but not with both. Incidences of edges with other vertices are remain unchanged. We can assert that in any n-colouring of $L(1)$ the vertices x and y must have different colours. We now unite $J(1)$ and $L(1)$ to form a single graph M by identifying p with x and q with y, no other vertex–identifications being made. Evidently M is an achrome. We refer to the construction of M from J and L as the Hajós operation of "achrome–junction". Hajós was interested in the achromes that could be constructed from copies of $K(n+1)$ by repeated application of his two operations. In his Theorem he asserts that every minimal achroma could be constructed in this way.

Actually that is a weakened form of the Theorem. Hajós required, for the operation of achrome–junction, that neither x nor y should be isolated in L. But now I will state a version that I believe to be stronger than Hajós' original.

Let us say that an achrome G is "H–decomposable" if it is the union of two subgraphs $J(1)$ and $L(1)$ having two common vertices x and y such that the following conditions are fulfilled.

(i) J(1) becomes an achrome J when its vertices x and y are joined by a new edge E.
(ii) L(1) becomes a minimal achrome L when its vertices x and y are identified.

We then say that $\{x,y\}$ is the "junctive pair" of the H–decomposition, that $J(1)$ is the first "part" and $L(1)$ its second, and that J is its first "constituent" and L its second.

If G satisfies these conditions it is obtainable by Hajós' operations from J and L, first an achrome–junction with joining vertices x and y and then appropriate vertex–identifications, each of a vertex of $J(1)$ with a vertex of $L(1)$.

Next we define recursively a class of "H–constructible" graphs. The rules are as follows.

(iii) The achrome $K(n+1)$ is H–constructible.
(iv) An achrome M is H-constructible if it is non-separable and has a H–decomposition whose constituent achromes are H–constructible.

This recursively definition makes it clear that every H–constructible achrome is non-separable, but not that every such achrome is minimal.

So a graph is H–constructible if and only if it can be reduced to a set of copies of $K(n+1)$ by a number of H–decompositions into constituent achromes. Let us now attempt a proof of the following version of Hajós' Theorem.

Theorem I. *Every achrome G has as a subgraph an H-constructible achrome Q with the following property. Either Q is a $K(n+1)$ or it has an H–decomposition whose second part does not include all the edges of G.*

Proof. Assume the Theorem is false. Then there is an achrome G for which the Theorem fails, which has the least number of vertices consistent with this and, subject to these two conditions has the greatest possible number of edges.

We note first that any minimal achrome M with fewer vertices than G must be H–constructible. For it contains an H–constructible achrome R by the choice of G, and it is identical with R by minimality. We note also that G must be non-separable since otherwise it would contain such an M.

Suppose that G is a complete graph $K(m)$. Then $m > n$, for otherwise G would be n-colourable. So G has an H–constructible achroma $K(n+1)$ as a subgraph. From this contradiction we deduce that G is not a $K(m)$, that is it has two non-adjacent vertices p and q. Since G is connected we can choose these so that they are joined by an arc of length 2, that is so that each is joined to some third vertex r.

Let us join p and q by a new edge E to get a new achrome U. By the choice of G it contains an H–constructible achrome J, and J canot be a subgraph of G. Hence J contains the edge E joining p and q. Form $J(1)$ by deleting E from J.

In another construction we identify p and q to get a new achrome V. This contains a minimal achrome L. But L has fewer vertices than G and is therefore H–constructible. By an appropriate splitting of the new vertex in L, into the vertices p and q, we get a subgraph $L(1)$ of G. Note that in the construction of V one of the edges pr and qr is erased, by the rule for eliminating digons. Hence there is an edge of G that does not belong to $L(1)$.

Let M be the union of $J(1)$ and $L(1)$, a subgraph of G. Then Conditions (i) and (ii) are satisfied. Evidently J and L are the constituent achromes of an H–decomposition of M. So if M is non-separable it must be H–constructible, by the H–constructibility of J and L and the minimality of L. But this contradicts the choice of G. On the other hand if M is separable it contains a minimal achrome R with fewer vertices than G. By the choice of G the theorem is true for R and therefore it is true for G. This contradiction establishes the Theorem. □

Corollary. *Let G be a minimal achrome. Then G is H–constructible. Moreover if G is not a $K(n+1)$ it has an H–decomposition whose constituent achromes are H–constructible, and whose second part is a proper subgraph of G.*

Theorem II. *Let G be a minimal achrome and let x and y be two non-adjacent vertices of G. Let G be the union of $k > 1$ subgraphs $G(1), G(2), \ldots, G(k)$ such that the intersection of any two of them consists solely of the vertices x and y. Then $k = 2$. Moreover we can adjust the notation so that $\{G(1), G(2)\}$ is an H–decomposition of G whose constituent achromes are both minimal.*

Proof. Form a new achroma P from G by adjoining a new edge E with ends x and y. Let the adjunction of E to $G(i)$ change it into $P(i)$, where i ranges from 1 to k. Then one of the $P(i)$, let us say $P(1)$, is an achrome; otherwise n-colourings of $P(i)$ could be combined to give an n-colouring of P. Let J be a minimal achrome contained in $P(1)$. Since it is not a subgraph of G it includes E, with its ends x and y. Define $J(1)$ as before, the result of deleting E from J.

In an alternative construction we identify x and y in G, as a new vertex z, to get a graph Q. Let this identification change $G(i)$ into $Q(i)$ for each i. Clearly there is a $Q(j)$ that is achrome. It contains a minimal achrome L. Since no isomorph of L can be a subgraph of G the graph L must include z, and give rise under a vertex-splitting to a subgraph $L(1)$ of G that includes x and y.

The union of $J(1)$ and $L(1)$ is an achroma, identical with G since G is minimal. Hence $k = 2$, $J(1) = G(1)$ and $L(1) = G(2)$. We observe that J and L are the constituent achromes of an H–decomposition of G. They are both H–constructible by the Corollary to Theorem I. □

I noticed a few weeks ago that the strengthened version of Hajós' Theorem can actually be used to prove something. We can prove Brooks' Theorem from the above result.

Theorem III. (Brooks' Theorem). *If a minimal achrome G is not a $K(n+1)$ it has a vertex whose valency exceeds n.*

Proof. We note first that the valency of any vertex of a minimal achrome must be at least n. Otherwise we could n-color the rest of the graph, by minimality, and then we could find a color for v.

If G is not a $K(n+1)$ it has a H–decomposition whose first constituent achrome J is H–constructible, whose second constituent achrome L is minimal, and whose secons part $L(1)$ is a proper subgraph of G. We note that $J(1)$ has an edge E not belonging to $L(1)$.

Now E does not join x and y, by the definition of an H–decomposition. So it has an end z that is neither x nor y. If z is a vertex of $L(1)$ its valency in G exceeds its valency in $L(1)$ and is therefore at least $n+1$, by the minimality of L.

If z is not a vertex of $L(1)$ we can represent G as the union of two proper subgraphs P and Q having only vertices of L in common, where P contains L and z is a vertex of Q. The number k of common vertices is at least 2, since G is non-separable. If one of the common vertices is not x or y we can argue as before. So we have only to consider the case in which x and y are the only common vertices of P and Q.

In that case we use Theorem II. By minimality of constituent achromes we infer that in one of P and Q each of x and y has valency at least $n-1$, and in the other their combined valency is at least n. Since n is at least 3 it follows that one of x and y has valency exceeding n in G. □

We still do not see how to use Hajós' Theorem, even in a strengthened version, to prove Hadwiger's Conjecture. But at least we have made it prove one well-known theorem.

Progress on the planar Four Colour Problem in the last fifty years can be summed up quite briefly. People went on collecting reducible configurations until they believed they had enough. But now the wish they did not need quite so many.

There is also a Three Colour Problem for the plane, the problem of finding necessary and sufficient conditions for a planar graph G to be vertex-colourable in 3 colours. Some progress has been made with it; H. Grötzsch proved that G can be 3-coloured if it has no triangle. An extension by B. Grünbaum asserts that G can be 3-coloured if it has fewer than 4 triangles.

I wondered if it would be possible to settle the Three Colour Problem by applying a planar version of Hajós' Theorem. Now Hajós' Theorem for n colours reduces any given achrome, through a sequence of H–decompositions, to a set of copies of $K(n+1)$. But $K(n+1)$ is non-planar when $n > 3$, so then no planar version of the Theorem is to be expected. So let us now discuss the possibility of a planar version for $n = 3$, the coloring problem for $n = 1$ and $n = 2$ being adequately solved already.

I have found no difficulty in putting forward a conjectural planar form of Hajós' Theorem, but I have no proof of it. It is based on the notion that each of the Hajós operations must be carried out in such a way as to preserve planarity, as follows.

The splitting of a vertex v of L into x and y must be carried out planely. That is the edges left incident with x must form a consecutive block in the cyclic sequence of edges round v, and those left incident with y must make up the complementary consecutive block. And in each of these blocks the original cyclic order of the edges must be preserved.

The operation of achrome-junction, in which achromes J and L are combined to form M, does make M planar when J and L are planar. Note that $L(1)$ is confined to the face of $J(1)$ that contained the deleted edge, and $J(1)$ to the face of $L(1)$ that is formed from the two faces of L in the vertex splitting.

After such an achrome-junction some operations of vertex identification are permitted, each is to identify a vertex of $J(1)$ with a vertex of $L(1)$, neither being x or y, and no vertex of one part is to be identified with more than one vertex of the other. Moreover each identification is to be between two vertices with a common incident face, to preserve planarity, and to be made across that face. There are only two possible faces for this operation, each having one bounding arc in $J(1)$ and the complementary bounding arc in $L(1)$.

Of course two vertex-identifications across the same face must not interfere with one another. One pair of vertices must not separate the other in the boundary of the face.

The student is invited to construct some planar achromes for $n = 3$ out of tetrahedra, by Hajós operations restricted as above. We can conjecture that every such achrome that is minimal can be constructed in this way. If true this proposition is a planar version of Hajós' Theorem. But then it can equally well be called a solution of the Three Colour Problem. There is, I think, one encouraging thing about the conjecture. You can prove that every constructible achrome has at least four triangles, a result that Grünbaum proved for all achromes.

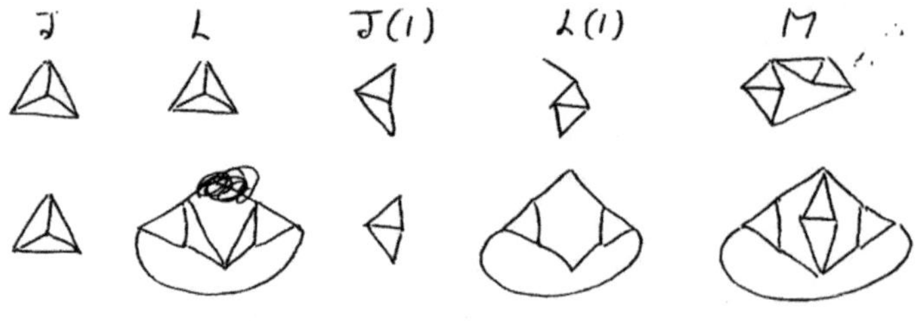

Fig. B.1 Some drawings by Tutte.

B.2 William Thomas Tutte 1917–2002

Bill Tutte was a leading graph and matroid theorist in the second half of the 1900s, but his greatest intellectual feat was probably achieved during World War Two, breaking the German Army High Command's extremely complex Lorenz code (that the British called FISH), uncovering the structure of the machine that generated it, without ever seeing it. In an interview [669] Donald Michie, founder of the Center for Machine Intelligence at the University of Edinburgh, said: *Well, the giant on whose shoulders that everybody was standing, including Allan Turing, was Bill Tutte – he got the original break.*

After studies in chemistry in the second half of the 1930s at Trinity College Cambridge Tutte was called up for national service, and he joined Bletchley Park in 1941 and worked there for the rest of the war. He regarded signing the Official Secrets Act as a lifelong obligation, and only very late in life he felt able to talk about Bletchley Park, when at least some of the secrets were no longer official.

In 1948 Tutte took up a post at the University of Toronto, and later in 1962 he moved to the University of Waterloo. He retired in 1985, but continued to play an important role as Professor Emeritus until his death in 2002. During the many years in Canada he often travelled to England and met there with colleagues and friends, in particular C.A.B. Smith and R.L. Brooks. He was a fellow of the Royal Society of Canada and also of the Royal Society of London. In 2001 he was awarded the Order of Canada.

Tutte's contributions to graph and matroid theory are immense. It is difficult in a short summary to describe all the branches of research pursued by Tutte and the results obtained. He himself gave such summaries in [1036] and [1037]. Also the obituaries [117], [1081] and [1082] give fine descriptions.

B.3 Tutte's Flow Conjectures

One of Tutte's accomplishments was to establish a relationship between colorings and flows of plane graphs. This led him to formulate three flow conjectures for graphs in general; these conjectures are among the most interesting graph problems. Although the conjectures are still unsolved, they have led to a variety of interesting graph-theoretical investigations and results. To formulate the conjectures we need some notation.

Let G be a graph, let Γ be an abelian group, and let $k \in \mathbb{N}$. A **Γ-flow** of G is an assignment to each edge of G of a direction and a value from Γ such that for each vertex v the sum (with respect to the group operation in Γ) of the values of the edges directed out of v is equal to the sum of the values of the edges directed into v. More formally, a Γ-flow is a pair (D, ϕ) such that D is an orientation of G (see Section 3.1) and $\phi : E(D) \to \Gamma$ is a mapping such that for every vertex $v \in V(D)$, we have

$$\sum_{e \in E_D^+(v)} \phi(e) = \sum_{e \in E_D^-(v)} \phi(e),$$

where the sum stands for the addition in the group Γ. A **nowhere zero Γ-flow** of G is a Γ-flow (D, ϕ) of G such that $\phi(e) \neq 0$ for every $e \in E(D)$. A **k-flow** of G is a $\mathbb{Z}$-flow (D, ϕ) of G such that $|\phi(e)| \leq k-1$ for all $e \in E(D)$. A **nowhere zero k-flow** of G is a k-flow (D, ϕ) of G such that $\phi(e) \neq 0$ for all $e \in E(D)$. Note that if D is an orientation of G, then $V(D) = V(G)$ and $E(D) = E(G)$. Suppose that (D, ϕ) is a

Γ-flow of G and D' is an orientation of G. Then there is a mapping $\phi' : E(D') \to \Gamma$ such that (D', ϕ') is a Γ-flow of G (replace $\phi(e)$ by $-\phi(e)$ if the direction of e is changed). Furthermore, (D, ϕ) is nowhere zero if and only if (D', ϕ') is nowhere zero. Hence, the problem whether G has a nowhere zero Γ-flow is a problem about graphs and not digraphs; see e.g. Jaeger [514], [516], Jensen and Toft [530, Chapter 13], and Seymour [930]. A fine general exposition of flows was given by Younger [1080]. Tutte [1032] made two important observations.

Theorem B.1 (TUTTE) *Let G be a graph, let Γ be an abelian group of order k. Then G has a nowhere zero Γ-flow if and only if G has a nowhere zero k-flow.*

Theorem B.2 (TUTTE) *Let G be a 2-edge-connected plane graph, and let $k \in \mathbb{N}$. Then G has a face coloring with a set of k colors if and only if G has a nowhere zero k-flow.*

The proof of Theorem B.1 is by a nonconstructive counting argument using the principle of inclusion and exclusion. Combining Theorem B.2 with the four color theorem, it follows that every 2-edge connected planar graph has a nowhere zero 4-flow. The Petersen graph shows that not every 2-edge-connected graph has a nowhere zero 4-flow, but Tutte proposed the following two conjectures; the first was published in [1032] and the second in [1035].

Conjecture B.3 (TUTTE'S 5-FLOW CONJECTURE) *Every 2-edge-connected graph admits a nowhere zero 5-flow.*

Conjecture B.4 (TUTTE'S 4-FLOW CONJECTURE) *Every 2-edge-connected graph containing no Petersen graph as a minor admits a nowhere zero 4-flow.*

In 1981, Seymour [929] proved that every 2-edge-connected graph admits a nowhere zero 6-flow using Theorem B.1 with the abelian group $\Gamma = \mathbb{Z}_3 \times \mathbb{Z}_2$; an alternative short proof of this result was given by DeVos and Nurse [283]. In 2005, Lai, Li, and Poon [663] proved that every 2-edge-connected graph containing no K_5 as a minor admits a nowhere zero 4-flow, thus solving a conjecture proposed by Jensen and Toft [530, Problem 13.2]. For the history of the next flow conjecture, we refer the reader to the paper by Steinberg [961]. That the conjecture holds for the class of planar graphs follows from Theorem B.2 and the dual version of Grötzsch's Theorem (Theorem 5.13).

Conjecture B.5 (TUTTE'S 3-FLOW CONJECTURE) *Every 4-edge-connected graph admits a nowhere zero 3-flow.*

By Theorem B.1, a graph G has a nowhere zero 3-flow if and only if G has a nowhere zero $\mathbb{Z}_3$-flow. By reversing the direction of the edges of value 2 in a $\mathbb{Z}_3$-flow, it follows that a graph G has a nowhere zero 3-flow if and only if G has an orientation D such that $d_D^+(v) \equiv d_D^-(v) \pmod 3$ for all $v \in V(D)$. For further information about Tutte's 3-flow conjecture, we refer the reader to the paper by Li, Luo, and Wang [673] and the paper by Thomassen, Wu, and Zhang [1017].

Appendix C
Basic Graph Theory Concepts

This appendix presents basic definitions, terminology and notation of graph theory.

C.1 Sets, Mappings, and Permutations

As usual, we denote by $\mathbb{R}$, $\mathbb{Z}$ and $\mathbb{N}$ the set of real numbers, the set of integers and the set of positive integers, respectively. Furthermore, let $\mathbb{N}_0 = \mathbb{N} \cup \{0\}$ and $\mathbb{R}_{\geq 0} = \{x \in \mathbb{R} \mid x \geq 0\}$. For $a, b \in \mathbb{R}$, let $\mathbb{R}[a,b] = \{x \in \mathbb{R} \mid a \leq x \leq b\}$. For $p, q \in \mathbb{Z}$, let $[p,q] = \{h \in \mathbb{Z} \mid p \leq h \leq q\}$. For $x \in \mathbb{R}$, let $\lfloor x \rfloor$ denote the **lower integer part** of x, and $\lceil x \rceil$ the **upper integer part** of x.

If A and B are two sets, we write $A \subseteq B$ if A is a **subset** of B; and we write $A \subset B$ if A is a **proper subset** of B, i.e., $A \subseteq B$ and $A \neq B$. The **set difference** of A and B is denoted by $A \setminus B$, so $A \setminus B = \{x \mid x \in A \text{ and } x \notin B\}$. For a set A and an integer $p \geq 0$, we denote by $[A]^p$ the set of all subsets X of A of **cardinality** (or **size**) $|X| = p$; these sets are called the p**-element subsets** of A. The **power set** of A, written as 2^A, is the set of all subsets of A. A **partition** or k**-partition** of a nonempty set A is a set $\mathcal{A}$ of k pairwise disjoint nonempty subsets of A whose union $\bigcup \mathcal{A}$ is A.

Let A and B be two sets. If f is **function** from A to B, we write $f : A \to B$. A function is also called a **mapping**, or briefly a **map**. Two functions $f, g : A \to B$ are **equal**, written $f = g$, if $f(a) = g(a)$ for all $a \in A$.

Given a function $f : A \to B$, we use the following standard notation. If $A' \subseteq A$, then $f(A') = \{f(a) \mid a \in A'\}$ is the **image** of A' by f; and $\text{im}(f) = f(A)$ is the **image** or **range** of f. The function $f : A \to B$ is **surjective** if $\text{im}(f) = B$. If f is surjective we also say that f is a function from A **onto** B. The **restriction** of f to A' (or f **restricted** to A') is the function $f' : A' \to B$ with $f'(a) = f(a)$ for all $a \in A'$, and we write $f' = f|_{A'}$. If $B' \subseteq B$, then $f^{-1}(B') = \{a \in A \mid f(a) \in B'\}$ is the **preimage** or **inverse image** of B' by f. If $B' = \{b_1, b_2, \ldots, b_n\}$, we also write $f^{-1}(b_1, b_2, \ldots, b_n)$ instead of $f^{-1}(B')$. In particular, $f^{-1}(b) = \{a \in A \mid f(a) = b\}$ for $b \in B$. The function f is **injective** if, for every $a, a' \in A$, $f(a) = f(a')$ implies $a = a'$. So f is injective if and only if $|f^{-1}(b)| \leq 1$ for all $b \in B$. The function f

M. Stiebitz et al., *Brooks' Theorem*, Springer Monographs in Mathematics,
https://doi.org/10.1007/978-3-031-50065-7

is **bijective** or **one-to-one** if f is surjective and injective. A **bijection** from A to B is just a bijective function from A onto B. For a set X, let $\mathrm{id}_X : X \to X$ denote the **identity function** of X, defined by $\mathrm{id}_X(x) = x$ for all $x \in X$; clearly id_X is bijective.

Given two functions $f : A \to B$ and $g : B \to C$, then the function $h : A \to C$ with $h(a) = g(f(a))$ for all $a \in A$ is called the **composition** of g and f, written as $h = g \circ f$. A function $f' : B \to A$ is called the **inverse function** of $f : A \to B$ if $f \circ f' = \mathrm{id}_B$ and $f' \circ f = \mathrm{id}_A$. If f has an inverse function, this function is unique and denoted by f^{-1}. Furthermore, f has an inverse function if and only if f is bijective.

For a set A, a bijection from A to A is called a **permutation** of A. The set of all permutations of A forms a group, under composition of functions, called the **symmetric group** on A, written as $\mathrm{Sym}(A)$. So

$$\mathrm{Sym}(A) = \{\pi \mid \pi : A \to A \text{ is bijective}\}.$$

When $A = [1,n]$ we write S_n instead of $\mathrm{Sym}(A)$. A **permutation group** is just a subgroup of a symmetric group. A permutation group $\Gamma \subseteq \mathrm{Sym}(A)$ is called **transitive** if for every pair $(a,a') \in A \times A$ there is a permutation $\gamma \in \Gamma$ such that $\gamma(a) = a'$; in this case we also say that Γ **acts transitively** on A.

C.2 Graphs and Subgraphs

A **graph** G is a pair of sets, $V(G)$ and $E(G)$, where $V(G)$ is finite and $E(G)$ is a set of 2-element subsets of $V(G)$. The set $V(G)$ is the **vertex set** of G and its elements are the **vertices** of G. The set $E(G)$ is the **edge set** of G and its elements are the **edges** of G. For convenience, we allow a graph G to be **empty**, i.e., $V(G) = E(G) = \varnothing$. In this case we write briefly $G = \varnothing$. The empty graph is sometimes called the **null graph**.

Let G be a graph. The number of vertices of G is its **order**, denoted by $|G|$. A vertex v is **incident** with an edge e if $v \in e$. The two vertices incident with an edge e are the **ends** of e; and e is said to **join** its ends. An edge $\{u,v\}$ is also written as uv or vu. Two distinct vertices u,v of G with $uv \in E(G)$ are called **adjacent vertices** and **neighbors**. The set of all neighbors of v in G is denoted by $N_G(v)$, i.e., $N_G(v) = \{u \in V(G) \mid uv \in E(G)\}$. If $X \subseteq V(G)$, then $N_G(X) = \bigcup_{v \in X} N_G(v)$ is the **neighborhood** of X in G. For a vertex $v \in V(G)$, let $E_G(v) = \{e \in E(G) \mid v \in e\}$. Furthermore, for $X,Y \subseteq V(G)$, let $E_G(X,Y)$ denote the set of all edges of G joining a vertex of X with a vertex of Y. When $Y = V(G) \setminus X$, then $E_G(X,Y)$ is called the **coboundary** of X in G and is denoted by $\partial_G(X)$. The **degree** of a vertex $v \in V(G)$ is $d_G(v) = |E_G(v)|$. If $G \neq \varnothing$, then let

$$\delta(G) = \min_{v \in V(G)} d_G(v) \text{ and } \Delta(G) = \max_{v \in V(G)} d_G(v)$$

denote the **minimum degree** and the **maximum degree** of G, respectively. For $G = \varnothing$, we define $\delta(G) = \Delta(G) = 0$. A vertex of degree 0 is an **isolated vertex**, and a vertex of degree 1 is an **endvertex**. For an integer $p \geq 0$, let $V^{(p)}(G)$ denote the set of all vertices v in G with $d_G(v) = p$. So $V^{(0)}(G)$ is the set of isolated vertices and $V^{(1)}(G)$ is the set of endvertices of G. A nonempty graph G is called **regular** and r**-regular** if all vertices of G have the same degree r. A 3-regular graph is also referred to as a **cubic graph**.

Since the ends of each edge are distinct vertices, it follows by a simple counting argument that every nonempty graph G with n vertices and m edges satisfies

$$\sum_{v \in V(G)} d_G(v) = \sum_{v \in V(G)} |E_G(v)| = 2m;$$

so $\frac{2m}{n}$ is the **average degree** of G and we obtain that

$$\delta(G) \leq \frac{2m}{n} \leq \Delta(G).$$

Let H and G be two graphs. If $V(H) \subseteq V(G)$ and $E(H) \subseteq E(G)$, we write briefly $H \subseteq G$ and call H a **subgraph** of G. Obviously, $H = G$ if and only if $H \subseteq G$ and $G \subseteq H$. A subgraph H of G with $H \neq G$ is said to be a **proper subgraph** of G, written as $H \subset G$. A **spanning subgraph** or **factor** of G is a subgraph of G that contains every vertex of G. An r**-factor** of G is a factor of G that is r-regular.

If there exists a bijection $\phi : V(G) \to V(H)$ such that $vw \in E(G)$ if and only if $\phi(v)\phi(w) \in E(H)$ for all $u, v \in V(G)$, then we write briefly $H \cong G$ and say that H is **isomorphic** to G. A graph isomorphic to a graph G is also said to be a **copy** of G.

Let G be a graph and $X \subseteq V(G)$. Then $G[X]$ is the subgraph of G with $V(G[X]) = X$ and $E(G[X]) = E_G(X,X)$. We call $G[X]$ the subgraph of G **induced** by X; and $H \subseteq G$ is an **induced subgraph** of G if $H = G[V(H)]$. If H is an induced subgraph of G, we write $H \leqslant G$, and $H < G$ if H is a proper induced subgraph of G. So $H < G$ if and only if $H \leqslant G$ and $H \neq G$. Clearly, $H \leqslant G$ implies $H \subseteq G$. Furthermore, let $G - X = G[V(G) \setminus X]$. We write $G - v$ instead of $G - \{v\}$. If $H \subseteq G$, $G - H$ means $G - V(H)$. For an edge set $F \subseteq E(G)$, let $G - F = (V(G), E(G) \setminus F)$. If $F = \{e\}$ is a singleton, we write $G - e$ rather than $G - \{e\}$. We call $G - v$ a **vertex-deleted subgraph** of G, and $G - e$ an **edge-deleted subgraph** of G; the two operations are called **vertex deletion** and **edge deletion**, respectively. If $u, v \in V(G)$ are nonadjacent vertices of G, then we denote by $G + uv$ the graph obtained from G by adding the edge joining u and v.

C.3 Path, Cycles and Complete Graphs

A **path** is a nonempty graph whose vertices can be arranged in a sequence with no repetitions such that two vertices are adjacent if and only if they are consecutive in the sequence. It is common to refer to a path by the sequence of its vertices. So we say that $P = (v_1, v_2, \ldots, v_n)$ is a path if

$$V(P) = \{v_1, v_2, \ldots, v_n\} \text{ and } E(P) = \{v_1v_2, v_2v_3, \ldots, v_{n-1}v_n\},$$

where the vertices $v_1, v_2, \ldots, v_n$ are distinct. In this case we say that v_1 and v_n are **joined** by P, and that P is a path **between** v_1 and v_n (respectively, **from** v_1 **to** v_n), and that P is a v_1-v_n **path**. The vertices v_1 and v_n are the **ends** of P; and the remaining vertices $v_2, \ldots, v_{n-1}$ are the **inner vertices** of P. Note that if $n \geq 2$, the ends of P are the endvertices of P.

If $P = (v_1, v_2, \ldots, v_n)$ is a path and $n \geq 3$, then $C = P + v_nv_1$ is a **cycle**. As for graphs, we refer to a cycle by the cyclic sequence of its vertices, so for C we also write $C = (v_1, v_2, \ldots, v_n, v_1)$.

The number of edges of a path (or cycle) G is its **length**, written $\ell(G)$. So $\ell(P) = |P| - 1$ if P is a path; and $\ell(C) = |C|$ if C is a cycle. A path or cycle is **odd** or **even**, depending on whether its length is odd or even. A cycle of length n is also called an n**-cycle**. If a path (or cycle) H is a subgraph of a graph G, then we also say that H is a path (or cycle) in G (respectively, of G). A **Hamilton path** of a graph G is a path in G that contains every vertex of G; and a **Hamilton cycle** of G is a cycle in G that contains every vertex of G.

A graph in which any two distinct vertices are adjacent is **complete**. If G is a complete graph on n vertices, then it is common to write $G = K_n$ (and neither $G \cong K_n$ nor $G \in K_n$). Note that up to isomorphism, there is a unique complete graph on n vertices; and K_n represents the isomorphism class and not a graph with a specific vertex set. In the same sense, we use P_n for a path on n vertices and C_n for a cycle on n vertices. A K_3 (isomorphic to C_3) is often called a **triangle**.

A **vertex order** or **vertex enumeration** of a graph G of order $n \geq 1$ is a sequence $\ell = (v_1, v_2, \ldots, v_n)$ of all its vertices, that is, $V(G) = \{v_1, v_2, \ldots, v_n\}$ and all the v_i are distinct. Each vertex order ℓ of G induces a **linear order** $\leq$ on its vertex set, where $u \leq v$ if $u = v$ or u is before v in the sequence ℓ.

C.4 Connectivity and Blocks

A nonempty graph G is **connected** if there is a path in G between any two of its vertices, and otherwise G is **disconnected**. A (connected) **component** of a nonempty graph G is a maximal connected subgraph of G. Alternatively, the vertex sets of the connected components of G are the equivalence classes of $V(G)$ under the relation *to be equal or to be joined by a path in G*. Let $\text{com}(G)$ denote the number of components of G. By convention, we put $\text{com}(\varnothing) = 0$. A nonempty graph G is disconnected if and only if there exists a 2-partition $\{X, Y\}$ of $V(G)$ such that $E_G(X, Y) = \varnothing$. A set $S \subseteq V(G) \cup E(G)$ is defined to be a **separating set** of G if $\text{com}(G - S) \geq 2$. If u and w are vertices belonging to different components of $G - S$, then we also say that S **separates** u and w in G. A vertex v of G such that $S = \{v\}$ separates two other vertices of the same component of G is a **separating vertex** of G. So v is a separating vertex of G if and only if $\text{com}(G - v) > \text{com}(G)$. An edge e of G such that $S = \{e\}$ separates the two ends of e in G is a **bridge** of G. So the bridges in a graph G are precisely those edges that do not belong to any cycle of G. A separating set of k vertices of G is called a **separator** or **k-separator** of G. A separator S of G is **minimal** if no proper subset of S is a separator of G. By a **cut** or **edge cut** of a graph G we mean a triple (X, Y, F) such that X is a nonempty proper subset of $V(G)$, $Y = V(G) \setminus X$, and $F = \partial_G(X) = \partial_G(Y)$. Clearly, any edge set of a cut is a separating edge set. The converse is not true. However, if F is a minimal separating edge set of a connected graph G, then $G - F$ has precisely two components with vertex sets X and Y, implying that $F = E_G(X, Y)$, and thus (X, Y, F) is a cut.

A graph G is called **k-connected** ($k \in \mathbb{N}_0$) if $|G| \geq k+1$ and $G - S$ is connected for every set $S \subseteq V(G)$ with $|S| \leq k-1$. So every nonempty graph is 0-connected, and every connected graph is 1-connected except K_1. Note that if G is k-connected, then G is ℓ-connected for $0 \leq \ell \leq k$. For a nonempty graph G, the **connectivity** $\kappa(G)$ is defined to be the largest integer k such that G is k-connected. By convention, $\kappa(\varnothing) = 0$. Note that $\kappa(K_n) = n-1$ for all $n \geq 1$. If G is not complete, then

$$\kappa(G) = \min\{|X| \mid X \subseteq V(G), \mathrm{com}(G - X) \geq 2\}.$$

A graph G is called **k-edge-connected** ($k \in \mathbb{N}_0$) if $|G| \geq 2$ and $G - F$ is connected for every set $F \subseteq E(G)$ with $|F| \leq k-1$. For a graph G with $|G| \geq 2$, the **edge-connectivity** $\kappa'(G)$ is defined to be the largest integer k such that G is k-edge-connected. By convention, $\kappa(\varnothing) = \kappa(K_1) = 0$. If $|G| \geq 2$, then

$$\kappa'(G) = \min\{|\partial_G(X)| \mid \varnothing \neq X \subset V(G)\}.$$

Proposition C.1 *Every graph G satisfies $\kappa(G) \leq \kappa'(G) \leq \delta(G)$.*

Proof If G is a complete graph K_n with $n \geq 0$, all three numbers are equal. So assume that G is not complete. Then $\kappa'(G) \leq \min\{|\partial_G(v)| \mid v \in V(G)\} = \delta(G)$. To prove the first inequality, choose a cut $F = \partial_G(X)$ with $\varnothing \neq X \subset V(G)$ such that $|F| = \kappa'(G)$. Let X_F denote the set of vertices in X incident with some edge in F; and let Y_F denote the set of vertices in $Y = V(G) \setminus X$ incident with some edge in F. If $X_F \neq X$, then X_F is a separating vertex set in G and hence $\kappa(G) \leq |X_F| \leq |F| = \kappa'(G)$, so we are done. By symmetry, we are also done if $Y_F \neq Y$. Hence it remains to consider the case that $X_F = X$ and $Y_F = Y$. Since G is not complete, there is a vertex v of G such that $N_G(v)$ is a separating vertex set of G. By symmetry, we may assume that $v \in X_F$. If $w \in N_G(v)$, then either $vw \in F$ or $w \in X_F$ and hence incident with an edge of F. This implies that $\kappa(G) \leq |U| = d_G(v) \leq |F| = \kappa'(G)$. Thus the proof is complete. □

The following result, known as Menger's theorem [732], is a basic tool in graph theory. We state the global version of Menger's theorem. A proof may be found, for example, in the book by Diestel [286, Theorem 3.3.6]. A set $\mathcal{X}$ of paths in a graph is said to be **internally disjoint** if no inner vertex of a path of $\mathcal{X}$ lies on another path of $\mathcal{X}$.

Theorem C.2 (Menger's Theorem) *For any graph G with $|G| \geq 2$ and any integer $k \geq 0$ the following statements hold:*

(a) *G is k-connected if and only if any two distinct vertices of G are joined by k internally disjoint paths.*
(b) *G is k-edge-connected if and only if any two distinct vertices of G are joined by k edge-disjoint paths.*

Let G be an arbitrary graph. A **block** of G is a maximal connected subgraph of G that has no separating vertex. Alternatively, the edges of the blocks are the equivalence classes under the relation *to be equal or to lie in a common cycle*. The set of blocks of G is denoted by $\mathfrak{B}(G)$, and for $v \in V(G)$, let $\mathfrak{B}_v(G) = \{B \in \mathfrak{B}(G) \mid v \in V(B)\}$. Note that each block of G is a connected induced subgraph of G, and so $\mathfrak{B}(\varnothing) = \varnothing$. If G is disconnected, then $\mathfrak{B}(G)$ is the union of $\mathfrak{B}(H)$ taken over all components H of G. If G is connected and has no separating vertex, then $\mathfrak{B}(G) = \{G\}$ and we say that G is a **block**; in particular, we have $\mathfrak{B}(K_1) = \{K_1\}$. If G is connected and has a separating vertex v, then G is the union of two proper induced subgraphs, say G_1 and G_2, having only v in common. Clearly, both subgraphs G_1 and G_2 are connected, and we have $\mathfrak{B}(G) = \mathfrak{B}(G_1) \cup \mathfrak{B}(G_2)$ and $\mathfrak{B}(G_1) \cap \mathfrak{B}(G_2) = \varnothing$. Any two distinct blocks of G have at most one vertex in common; and a vertex v is a separating vertex of G if and only if it belongs to more than one block, i.e. $|\mathfrak{B}_v(G)| \geq 2$. A block of G which contains at most one separating vertex of G is called an **end-block** of G. If G has a separating vertex, then G has at least two end-blocks. So either G itself is a block or G has at least two end-blocks.

The following proposition characterizes 2-connected graphs; the proof is left to the reader as an exercise.

Proposition C.3 *Let G be a connected graph with $|G| \geq 3$. Then the following statements are equivalent:*

(a) *G is 2-connected.*
(b) *Any two vertices are on a common cycle.*
(c) *Any two edges are on a common cycle.*
(d) *G has no separating vertex.*
(e) *For every vertex $v \in V(G)$, $G - v$ is connected.*
(f) *$\mathfrak{B}(G) = \{G\}$.*

C.5 Distance, Girth and Odd Girth

Let G be a graph and let $v, w \in V(G)$. If there exist a path in G between v and w, define the **distance** $\mathrm{dist}_G(v, w)$ of v nd w in G to be the length of a shortest v-w path contained in G, that is,

$$\mathrm{dist}_G(v, w) = \min\{\ell(P) \mid P \subseteq G \text{ is an } v\text{-}w \text{ path}\}.$$

If no such path exists, define $\mathrm{dist}_G(v, w) = \infty$. If G is a connected graph, then the distance is a metric, and we can use the classical concepts for metric spaces, as ball, sphere, diameter, and radius. For instance, the **radius** of a connected graph G is defined by

$$\mathrm{rad}(G) = \min_{v \in V(G)} \max_{w \in V(G)} \mathrm{dist}_G(v, w).$$

If a graph G contains a cycle, define the **girth** $g(G)$ of G to be the length of a shortest cycle contained in G, that is,

$$g(G) = \min\{\ell(C) \mid C \subseteq G \text{ is a cycle}\}.$$

If G contains no cycle, then define $g(G) = \infty$. If G contains an odd cycle, define the **odd girth** $g_o(G)$ of G to be the length of a shortest odd cycle in G, that is,

$$g_o(G) = \min\{\ell(C) \mid C \subseteq G \text{ is an odd cycle}\}.$$

If G contains no odd cycle, then define $g_o(G) = \infty$.

Let G be a graph and let $W = (v_0, v_1, \ldots, v_n)$ be a nonempty sequence of vertices of G. We call W a **walk** of G if $v_i v_{i+1} \in E(G)$ for $0 \le i \le n-1$. We say that W **connects** v_0 and v_n, and that W **starts** in v_0, and **terminates** in v_n, and that W is a v_0-v_n walk. The vertices v_0 and v_n are called the **ends** of the walk W, v_0 being its **initial vertex** and v_n its **terminal vertex**; the remaining vertices $v_1, v_2, \ldots, v_{n-1}$ are its **internal vertices**. The integer n is called the **length** of W, written as $\ell(W) = n$. The reverse sequence $W^{-1} = (v_n, v_{n-1}, \ldots, v_0)$ is also a walk in G. A walk is **odd** or **even**, depending on whether its length is odd or even. A walk is **closed** if its initial and terminal vertices are identical; otherwise it is **open**. If the edges $v_i v_{i+1}$ of a walk are all different it is called a **tour**. If the vertices of a walk are distinct, it defines a path in G; two vertices of G are connected by a walk if and only if they are joined by a path in G. A closed walk $W = (v_0, v_1, \ldots, v_n)$ with $n \ge 3$ whose vertices $v_0, v_1, \ldots, v_{n-1}$ are distinct defines a cycle; then the closed walk $W' = (v_i, v_{i+1}, \ldots, v_n, v_1, v_2, \ldots, v_{i-1}, v_i)$ (with $1 \le i \le n-1$) is different from W, but it defines the same cycle as W. So a path of length $n \ge 1$ gives rise to two different open walks; and a cycle of length $n \ge 3$ gives rise to $2n$ different closed walks.

Proposition C.4 *Every closed odd walk W in a graph G satisfies $g_o(G) \le \ell(W)$.*

Proof The proof is by induction on the length of W. Since W is a closed walk whose length is odd, $\ell(W) \ge 3$. If $\ell(W) = 3$, then W defines a cycle and $g_o(G) = 3$. So assume that $\ell(W) = n > 3$, say $W = (v_0, v_1, \ldots, v_n)$. If W defines a cycle, we have $g_o(G) = n = \ell(W)$. Otherwise, $v_i = v_j$ for $0 \le i < j \le n-1$. Then we can split W into two closed walks, namely $W_1 = (v_0, v_1, \ldots, v_i, v_{j+1}, \ldots, v_n)$ and $W_2 = (v_i, v_{i+1}, \ldots, v_j)$. Since $\ell(W) = \ell(W_1) + \ell(W_2)$ and $\ell(W) = n$ is odd, either W_1 or W_2 is odd, say W_1. Then $\ell(W_1) < \ell(W)$ and the induction hypothesis implies that $g_o(G) \le \ell(W_1)$, which gives $g_o(G) \le \ell(W)$. □

Proposition C.5 *Let G be a graph and suppose that $g_o(G) \ge 2p+1$ with $p \ge 1$. Then for two vertices $v \ne w$ of G the following conditions are equivalent:*

(a) *There exists an odd v-w walk W in G with $\ell(W) \le 2p+1$.*

(b) *There exists an odd v-w path P in G with $\ell(P) \leq 2p+1$.*

Proof Since each path defines an obvious walk of the same length, (b) implies (a). To show the converse implication, let $W = (v_0, v_1, \ldots, v_{2p}, v_{2q+1})$ be an odd v-w walk, whose length is minimum. By (a), such a walk exists and $q \leq p$. We claim that the vertices of W are distinct and W therefore defines a path, which proves (b). For otherwise, $v_i = v_j$ for $0 \leq i < j \leq 2q$, since $v_0 = v \neq w = v_{2q+1}$. Then $W_1 = (v_i, v_{i+1}, \ldots, v_j)$ is a closed walk with $2 \leq \ell(W_1) < 2q+1 \leq 2p+1$. Since $g_o(G) \geq 2p+1$, it then follows from Proposition C.4 that W_1 is even. But then $W_2 = (v_0, v_1, \ldots, v_i, v_{j+1}, \ldots, v_{2q+1})$ is an odd v-w walk with $\ell(W_2) \leq \ell(W) - 2$, contradicting the choice of W. □

C.6 Trees and Bipartite Graphs

An **acyclic graph**, that is, a graph not containing any cycle, is a **forest**, and it is a **tree** when it is connected. So any component of a forest is a tree. A graph is a forrest if and only if each of its edges is a bridge. An endvertex of a tree is also called a **leaf**. If T is a tree of order $n \geq 2$, then T has at least two leaves. This follows from the simple fact that the two ends of a longest path in T are leaves of T. If T is a tree and v is an leaf, then $T - v$ is a tree. It follows, by induction on its order, that every tree of order n has $n-1$ edges. If T is a tree and v, w are two vertices of T, then there is a unique v-w path in T; this path is denoted by vTw or wTv. Note that $\text{dist}_T(v, w) = \ell(vTw)$. Clearly, every connected graph G contains a **spanning tree**, that is, a tree T such that $V(T) = V(G)$ (take a minimal connected subgraph of G with vertex set $V(G)$, or take a maximal acyclic subgraph of G with vertex set $V(G)$).

Depth-first search (**DFS**) is a fundamental graph searching technique developed by Hopcroft and Tarjan [505] and by Tarjan [992]. The structure of DFS enables efficient graph algorithm for many graph problems. A **DFS tree** T of a given graph G of order $n \geq 1$ can be constructed as follows. First we choose a vertex r of G and a vertex order $\ell = (v_1, v_2, \ldots, v_n)$ of G with $v_1 = r$. Next we construct recursively vertices u_i and trees T_i, where we start with the vertex $u_1 = r$ and the tree T_1 consisting of the single vertex r. If in step $i \geq 1$ the vertex u_i and the tree T_i has been obtained, we proceed as follows. If no vertex in $U = \{u_1, u_2, \ldots, u_i\}$ has a neighbor in G belonging to $U' = V(G) \setminus U$, then we stop and put $T = T_i$. Otherwise, we choose the largest integer $j \in [1, i]$ such that u_j has a neighbor in U', and then we choose the neighbor $u \in U'$ of u_j which comes first in the vertex order ℓ. Then $u_{i+1} = u$ and T_{i+1} is the tree obtained from T_i by adding the vertex u and the edge $u_j u$. Since $V(G)$ is a finite set, the procedure terminates after a finite number of steps with a tree T. The vertex $u_1 = r$ is called the **root** of the DFS tree T. If G is disconnected, then the procedure stops with $T = T_i$ for some $i < n$ and $V(T) = \{u_1, u_2, \ldots, u_i\}$ is the vertex set of a component of G. If G is connected, the procedure stops with $T = T_n$ and T is a spanning tree of G with $V(G) = V(T) = \{u_1, u_2, \ldots, u_n\}$.

Proposition C.6 *Let G be a connected graph with order n. Then the following statements hold:*

(a) *Let T be a DFS tree of G with root r, and let vw be an edge of G. Then either v belongs to the path rTw, or w belongs to the path rTv.*
(b) *Let T be a DFS tree of G with root r, and let $V^{(1)}(T)$ be the set of leaves of T. Then $V^{(1)}(T) \setminus \{r\}$ is an independent set of G.*
(c) *Let $P \subseteq G$ be a path between r and v. Then there exists a DFS tree T of G with root r such that $P = rTv$.*

Proof To prove (a), let u_i be the vertex of G added in step i to T, so $V(G) = V(T) = \{u_1, u_2, \ldots, u_n\}$ and $u_1 = r$. If $vw \in E(G)$, then $v = u_i$ and $u = u_j$, where we may assume that $i < j$. Let U be the set of vertices $u \in V(G)$ for which v belongs to the path rTu. An inductive argument shows that $u_i, u_{i+1}, \ldots, u_j \in U$. Hence $w \in U$, which proves (a). Statement (b) is an immediate consequence of (a). To prove (c), suppose that $P = (v_1, v_2, \ldots, v_p)$ is a path with $v_1 = r$ and $v_p = v$. Then $(v_1, v_2, \ldots, v_p)$ can be extended to a vertex order $\ell = (v_1, v_2, \ldots, v_n)$ of G. If T is the DFS tree corresponding to the vertex order ℓ, then $u_i = v_i$ for $i \in [1, p]$. Hence T has root r and $rTv = v_1 T v_p = P$. □

A **rooted tree** (T, v_T) is a tree T with a designated vertex $v_T \in V(T)$ called its **root**; we then also say that T is a tree **rooted** at v_T. Designating a root imposes a hierarchy on the vertices of a rooted tree, according to their distance from that root. For a rooted tree (T, v_T) we use the following terminology. The **depth** or **level** of a vertex v of T is its distance $d_T(v_T, v)$ from the root in T; thus the root has depth 0. The **height** of (T, v_T) is the length of a longest path in T from the root, or equivalently the greatest depth in (T, v_T). If w is a vertex of T and v is the neighbor of w on the path $v_T T w$, then v is a **parent** of w and w is a **child** of v. If $e = vw$ is an edge of T such that v is a parent of w, then e is called an **outgoing edge** of v (or a **descending edge** of v) and an **incoming edge** of w. A vertex v is called an **ancestor** of w if v belongs to the path $v_T T w$; if, in addition, $v \neq w$, then v is a **proper ancestor** of w. A **leaf** of (T, v_T) is any vertex having no children. An **internal vertex** of (T, v_T) is any vertex that has at least one children. Note that the root v_T is an internal vertex of (T, v_T), unless it is the only vertex of T. If the root v_T has exactly one child, then it is a leaf of T, but not of (T, v_T); otherwise, the leaves of (T, v_T) are the same as the leaves of T.

Breadth-first search (**BFS**) is another natural way of searching a graph; this leads to a **BFS tree** of a given graph G. It seems that BFS and its application in finding components of a graph G were invented in 1945 by Konrad Zuse and Michael Burke. The method and its application in finding shortest path of a graph or digraph were reinvented in the 1950s by various researchers, including Bellman, Dijkstra, Dantzig, Ford, Moore, and possibly others. Those who are interested in the history of the shortest path problem, should read the article by Schrijver [904]. A BFS tree of a connected graph G is a rooted tree (T, v_T) such that T is a spanning subgraph of G and every vertex v of G satisfies $\mathrm{dist}_G(v_T, v) = \mathrm{dist}_T(v_T, v) = \ell(v_T T v)$.

A graph G is called **bipartite** if there are two disjoint sets X, Y such that $V(G) = X \cup Y$ and $E(G) = E_G(X, Y)$; the sets X and Y are called **partite sets** or **parts** of G, and (X, Y) is called a **bipartition** of G. A **complete bipartite** graph is a bipartite graph G with parts X and Y such that $E(G) = \{xy \mid x \in X, y \in Y\}$; if $|X| = p$ and $|Y| = q$ we briefly write $G = K_{p,q}$ $(p, q \geq 0)$. A graph isomorphic to $K_{1,n}$ is called a **star**; if $n = 3$, such a star is also called a **claw**. Clearly, any subgraph of a bipartite graph is bipartite, too. Furthermore, it is easy to see that every tree T is bipartite and has a bipartition (X, Y) such that the path vTw has even length whenever v and w belong to the same part X or Y (up to permuting the parts X and Y, this bipartition of T is unique). The sets X and Y may be described as follows. Fix a vertex r of T as a root of T and put $X = \{v \in V(T) \mid \text{dist}_T(r, v) \text{ is even}\}$ and $Y = \{v \in V(T) \mid \text{dist}_T(r, v) \text{ is odd}\}$. In 1916 Dénes König [609] provided a characterization of bipartite graphs by forbidden subgraphs.

Theorem C.7 (KÖNIG) *A graph is bipartite if and only if it contains no odd cycle.*

Proof Since an odd cycle has no bipartition, no bipartite graph contains an odd cycle. Conversely, we shall show that if a graph G contains no odd cycle, then G has a bipartition. To do this, we may assume that G is connected and $|G| \geq 2$. Then G has a spanning tree T, and T has a bipartition $\{X, Y\}$. If this is a bipartition of G, we are done. Otherwise, there is an edge vw with both ends in one of the two partition classes, say in X. Then the path vTw has even length and hence $vTw + vw$ is an odd cycle, giving a contradiction. □

C.7 Graph Operations and Graph Modifications

Let H and G be two graphs. The **union** $H \cup G$ is the graph with $V(H \cup G) = V(H) \cup V(G)$ and $E(H \cup G) = E(H) \cup E(G)$; and the **intersection** $H \cap G$ is the graph with $V(H \cap G) = V(H) \cap V(G)$ and $E(H \cap G) = E(H) \cap E(G)$. We say that H and G are **disjoint** if $H \cap G = \varnothing$. The **complement** $\overline{G}$ of G is the graph with $V(\overline{G}) = V(G)$ and $E(\overline{G}) = [V(G)]^2 \setminus E(G)$. So if G has order $n \geq 0$, then $G \cup \overline{G} = K_n$ and $G \cap \overline{G} = \overline{K}_n$.

Let G be a graph and let X be a set of vertices of G. We call X an **independent set** or p**-independent set** of G if $G[X] = \overline{K}_p$, that is, $|X| = p$ and no two vertices of X are adjacent in G. The **independence number** of G, written $\alpha(G)$, is the largest integer p for which G contains a p-independent set. We call X a **clique** or p**-clique** of G if $G[X] = K_p$, that is, $|X| = p$ and any two vertices of X are adjacent in G. The **clique number** of G, written $\omega(G)$, is the largest integer p for which G contains a p-clique. Clearly, $\alpha(\overline{G}) = \omega(G)$ and $\omega(\overline{G}) = \alpha(G)$

Join Operation. If H and G are disjoint graphs, we define the **join** $H \boxplus G$ of these two graphs to be the graph obtained from $H \cup G$ by adding all edges uv with $u \in V(H)$ and $w \in V(G)$. Assuming that all graphs P_n, C_n and K_n are disjoint, we have $K_{p+q} = K_p \boxplus K_q$ and $K_{p,q} = \overline{K}_p \boxplus \overline{K}_q$. A **wheel** is a graph isomorphic to

$W_n = K_1 \boxplus C_n$. The wheel W_n is said to be **odd** or **even** depending on whether n is odd or even.

Vertex Splitting. Let H and G be disjoint graphs, let $X \subseteq V(H)$ and let $v \in V(G)$. Let G' denote a graph obtained from $H \cup (G - v)$ by joining each vertex of $N_G(v)$ to exactly one vertex of X such that each vertex in X has a at least one neighbor in $N_G(v)$. This is possible if $d_G(v) \geq |X|$. We say that G' is obtained from G and H by (proper) **splitting** v into X.

Vertex Identification. Let G be a graph and $X \subseteq V(G)$. We say that graph H is obtained from G by **identifying** X to a new vertex v, if $V(H) = V(G - X) \cup \{v\}$ and $E(H) = E(G - X) \cup \{vu \mid u \in V(G) \setminus X,\ N_G(u) \cap X \neq \emptyset\}$.

Vertex Duplication. Duplicating a vertex v of a graph G produces a new graph G' by adding a new vertex v' and joining v' to all neighbors of v in G (but not to v).

Edge Contraction. Let G be a graph and let e be an edge of G. By G/e we denote the graph obtained from G by identifying the ends of e; we then say that G/e is obtained from G by **contracting** the edge e.

Edge Subdivision. Let G be a graph and let e be an edge of G. We say that H is obtained from G by **subdividing** the edge e, if H is obtained from $G - e$ by adding a new vertex v_e and by joining v_e to the ends of e.

Hajós Construction. Suppose H and G are disjoint graphs, $x_1y_1 \in E(H)$ and $x_2y_2 \in E(G)$. The **Hajós join** of G and H with respect to (x_1,y_1) and (x_2,y_2) is the graph G' obtained from $H \cup G$ by deleting the edges x_1y_2 and x_2y_2, identifying x_1,x_2 to a new vertex x and adding the edge y_1y_2.

Graph Products. Let G and H be graphs. The **Cartesian product** $G \square H$ has vertex set $V(G) \times V(H)$, and (v,w) is adjacent to (v',w') if and only if either (1) v is adjacent to v' in G and $w = w'$, or (2) $v = v'$ and w is adjacent to w' in H. The **lexicographic product** (or **composition**) $G[H]$ is the graph with vertex set $V(G) \times V(H)$, in which (v,w) is adjacent to (v',w') if and only if either (1) $vv' \in E(G)$, or (2) $v = v'$ and $ww' \in E(H)$. The **direct product** (or **categorial product**) $G \times H$ has vertex set $V(G) \times V(H)$, and (v,w) is adjacent to (v',w') if and only if v is adjacent to v' in G and w is adjacent to w' in H.

C.8 Minors and Subdivisions

We have already discussed some simple ways of modifying a graph by means of local graph operations, such as vertex deletion, edge deletion, edge contraction and edge subdivision.

Let G be a graph. Any subgraph of G can be obtained from G by repeated application of the operations vertex deletion and edge deletion; any induced subgraph of G can be obtained from G by repeated application of the operation vertex deletion; and any spanning subgraph of G can be obtained from G by repeated application of the operation edge deletion.

A graph H is a **minor** of G if an isomorphic copy of H can be obtained from G by repeated application of the operations of vertex deletion, edge deletion and edge contraction, where these operation can be performed in any order. If H is a minor of G, we write $H \preceq G$. It follows from the definition that if H and H' are isomorphic graphs, then $H \preceq G$ if and only if $H' \preceq G$. Furthermore, every subgraph of G is a minor of G; and every minor of G can be obtained from a subgraph of G by a sequence of edge contractions. It is easy to prove that $H \preceq G$ if and only if there is a set $\{X_v \mid v \in V(H)\}$ of pairwise disjoint subsets of $V(G)$ such that $G[X_v]$ is connected for all $v \in V(H)$ and $E_G(X_v, X_w) \neq \varnothing$ whenever $vw \in E(H)$. The minor relation $\preceq$ is an partial order on the class of graphs, that is, the relation is reflexive, transitive, and antisymmetric in the sense that $H \preceq G$ and $G \preceq H$ implies $H \cong G$.

A graph H is a **subdivision** of G if an isomorphic copy of H can be obtained from G by successive edge subdivision; or equivalently, by replacing edges with pairwise internally disjoint paths (so an edge vw is replaced by an v-w path). A graph H is a **topological minor** of G if a subgraph of G is (isomorphic to) a subdivision of H; in this case we write $H \preceq_t G$. So $H \preceq_t G$ if and only if there is an injective mapping $\varphi : V(H) \to V(G)$ such that for every edge vw of H there is a $\varphi(v)$-$\varphi(w)$ path in G and all these paths are internally vertex disjoint. Clearly, $H \preceq_t G$ implies $H \preceq G$, but not conversely. However, if $\Delta(H) \leq 3$, then $H \preceq G$ and $H \preceq_t G$ are equivalent (see [286, Proposition 1.7.2])

C.9 Multipartite Graphs and Turán Graphs

A p**-partite graph** is a graph G whose vertex set is the disjoint union of p sets (called **partite sets** or **parts**), in such a way that no edge of G has both ends in the same part. If any two vertices in different partite sets are adjacent, then G is called a **complete multipartite graph**, respectively a **complete** p**-partite graph**completeppartitegraph. Thus a graph G is complete p-partite if and only if $\overline{G}$ is the disjoint union of p complete graphs (where one or more of the complete graphs may be empty).

Let n and p be positive integers. The **Turán graph** $TG(n, p)$ is the unique complete p-partite graph with n vertices whose partite sets differ in size by at most 1. Note that if n has the form $n = hp + r$ with $0 \leq r < p$, then

$$TG(n, p) \cong \overline{K}_{n_1} \boxplus \overline{K}_{n_2} \boxplus \cdots \boxplus \overline{K}_{n_p},$$

where $n_i = h+1$ for $i \in [1, r]$ and $n_i = h$ for $i \in [r+1, p]$ (see Figure C.1). In particular, $TG(n, p) = K_n$ if $n \leq p$. The Turán graph $G = TG(n, p)$ is an edge-maximal K_{p+1}-

free graph, that is, $\omega(G) \leq p$ but $\omega(G+vw) = p+1$ for every edge $vw \in E(\overline{G})$; where $G+vw$ is the graph obtained from G by adding the edge vw. The number of edges of $TG(n,p)$, denoted by $t(n,p)$, is called the **Turán number**. The next result, which was proved in 1941 by Turán [1031], opened a new branch in discrete mathematics, known as extremal graph theory.

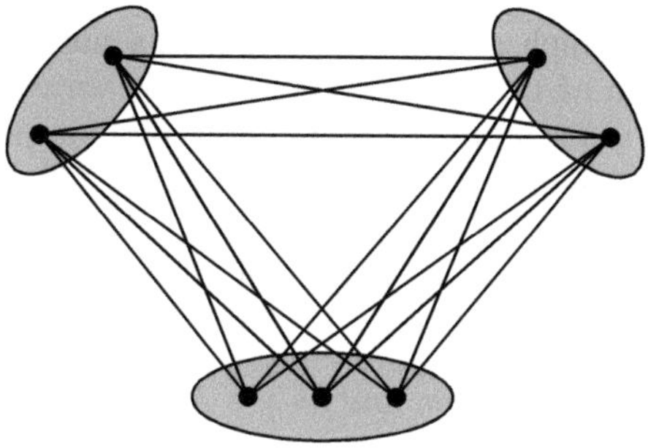

Fig. C.1 The Turán graph $TG(7,3)$.

Theorem C.8 (TURÁN'S THEOREM) *Let n and p be positive integers. If G is an edge-maximal K_{p+1}-free graph of order n, then $G \cong TG(n,p)$.*

Proof First we claim that the degrees of any two nonadjacent vertices of G are equal, and the degrees of any two adjacent vertices of G differ by at most one. For otherwise, deleting the vertex whose degree is smaller and duplicating the other vertex results in K_{p+1}-free graph of order n having more edges than G, giving a contradiction. Next we claim that G is a complete multipartite graph. For otherwise, G contains three vertices v, v_1, v_2 such that $vv_1, vv_2 \notin E(G)$ but $v_1v_2 \in E(G)$. By the first claim, all three vertices have the same degree. But then deleting both v_1 and v_2 and duplicating v twice results in a K_{p+1}-free graph of order n having more edges than G, giving a contradiction. Thus G is a complete p-partite graph and the sizes of any two nonempty parts differ by at most one. Consequently, $G \cong TG(n,p)$ as required. □

The above proof of Turán's theorem is due to Zykov [1105]. Turán's theorem implies, in particular, that

$$t(n,p) = \max\{|E(G)| \mid \omega(G) \leq p \wedge |G| = n\}. \tag{C.1}$$

for all positive integers n and p.

C.10 Automorphism Group of Graphs

Given two graphs G and H, an **isomorphism** from G to H is a bijective mapping $\phi : V(G) \to V(H)$ such that, for all vertices $v, w \in V(G)$,

$$vw \in E(G) \Longleftrightarrow \phi(v)\phi(w) \in E(H).$$

An isomorphism of a graph G to itself is called an **automorphism** of G. Note that any automorphism of G is a permutation on its vertex set V. The set of all automorphisms of G forms a group under composition of mappings; this group is called the **automorphism group** of G written as $\mathrm{Aut}(G)$. So $\mathrm{Aut}(G)$ is a subgroup of the symmetric group $\mathrm{Sym}(V)$. The graph G is said to be **transitive** if $\mathrm{Aut}(G)$ acts transitively on its vertex set V, that is, for all vertices $v, w \in V$ there exists an automorphism $\gamma \in \mathrm{Aut}(G)$ such that $\gamma(v) = w$.

Proposition C.9 *Let G be a graph, and let $\gamma \in \mathrm{Aut}(G)$ be an automorphism of G Then the following statements hold:*

(a) $d_G(\gamma(v)) = d_G(v)$ *for all* $v \in V(G)$.
(b) $\mathrm{dist}_G(\gamma(v), \gamma(w)) = \mathrm{dist}_G(v, w)$ *for all* $v, w \in V(G)$.

Proof Note that γ is bijective and, for $v, w \in V(G)$, we have $vw \in E(G)$ if and only if $\gamma(v)\gamma(w) \in E(G)$. Consequently, $N_G(\gamma(v)) = \gamma(N_G(v))$ and so $d_G(v) = |N_G(v)| = |N_G(\gamma(v))| = d_G(\gamma(v))$, which proves (a). If $P = (v_1, v_2, \ldots, v_n)$ is a v-w path in G, then $P' = (\gamma(v_1), \gamma(v_2), \ldots, \gamma(v_n))$ is a $\gamma(v)$-$\gamma(w)$ path in G (and vice versa) with $\ell(P') = \ell(P)$. Since $\gamma^{-1} \in \mathrm{Aut}(G)$, this implies (b). □

Let Γ be a permutation group acting on a finite nonempty set V. Then Γ is a finite set and $|\Gamma|$ is its **order**. Given a subgroup Γ' of Γ and a permutation $\gamma \in \Gamma$, the set

$$\gamma \circ \Gamma' = \{\gamma \circ \gamma' \mid \gamma' \in \Gamma'\}$$

is a **left coset** of Γ' in Γ, and Γ/Γ' denotes the set of all left cosets. It is well-known, that Γ/Γ' forms a partition of Γ and Lagrange's theorem says that

$$|\Gamma| = |\Gamma/\Gamma'||\Gamma'|. \tag{C.2}$$

For elements $v, w \in V$, define

$$v \sim_\Gamma w \Longleftrightarrow \exists \gamma \in \Gamma : w = \gamma(v).$$

Then $\sim_\Gamma$ is an equivalence relation on V, and an equivalence class is called an **orbit**. So $O \subseteq V$ is an orbit if and only if there is an $v \in V$ and $O = \{\gamma(v) \mid \gamma \in \Gamma\}$. Given an element $v \in V$, the set

$$\Gamma_v = \{\gamma \in \Gamma \mid \gamma(v) = v\}$$

is called the **stabilizer** of v in Γ; the stabilizer is an subgroup of Γ. There is a simple relation between the number m of orbits of Γ and the order of its stabilizers, namely

$$m|\Gamma| = \sum_{v \in V} |\Gamma_v|.$$

This result is due to Cauchy and Frobenius, but it is often referred to as the Burnside lemma. If $\emptyset \neq U \subseteq V$, then define

$$\Gamma_U = \bigcap_{v \in U} \Gamma_v \text{ and } \Gamma|_U = \{\gamma|_U \mid \gamma \in \Gamma\}.$$

Proposition C.10 *Let Γ be a permutation group on a finite nonempty set V, and let $\emptyset \neq U \subseteq V$. Then $|\Gamma| = |\Gamma|_U| \cdot |\Gamma_U|$.*

Proof Clearly, Γ_U is a subgroup of Γ. For $\gamma, \gamma' \in \Gamma$ we then deduce that

$$\begin{aligned} \gamma \circ \Gamma_U = \gamma' \circ \Gamma_U &\iff \gamma^{-1} \circ \gamma' \in \Gamma_U \\ &\iff (\gamma^{-1} \circ \gamma')(v) = v \ \forall v \in U \\ &\iff \gamma(v) = \gamma'(v) \ \forall v \in U \\ &\iff \gamma|_U = \gamma'|_U. \end{aligned}$$

Thus $|\Gamma|_U| = |\Gamma/\Gamma_U|$ and the result follows from equation (C.2). □

C.11 Graph Properties and Graph Invariants

Let $\mathcal{G}$ denote the class of all graphs. A **graph property** $\mathcal{P}$ is a subclass of $\mathcal{G}$ that is closed under isomorphism, that is, if a graph G belongs to $\mathcal{P}$, then so does every graph isomorphic to G. A graph property $\mathcal{P}$ is **trivial** if $\mathcal{P} = \mathcal{G}$, or $\mathcal{P} = \emptyset$, or $\mathcal{P}$ consists only of the empty graph. Thus a graph property is nontrivial if and only if it contains a nonempty graph, but not all graphs. A graph property is **monotone** if it is closed under taking subgraphs, and it is **hereditary** if it is closed under taking induced subgraphs. Obviously, every monotone graph property is a hereditary graph property, but not vice versa.

Every hereditary graph property can be defined in terms of forbidden induced subgraphs. Given a class or set $\mathcal{X}$ of graphs, let $\text{Forb}(\mathcal{X})$ be the class of graphs $G \in \mathcal{G}$ such that no induced subgraph of G is isomorphic to a graph of $\mathcal{X}$. Clearly, $\text{Forb}(\mathcal{X})$ is a hereditary graph property, and $\text{Forb}(\mathcal{X}) = \mathcal{G}$ if and only if $\mathcal{X} = \emptyset$. If $\mathcal{X} = \{H_1, H_2, \ldots, H_n\}$, then we abbreviate $\text{Forb}(\mathcal{X})$ by $\text{Forb}(H_1, H_2, \ldots, H_n)$. A graph G is said to be **$\mathcal{X}$-free** if $G \in \text{Forb}(\mathcal{X})$. If $\mathcal{X}$ consists of a single graph H, then the term becomes **H-free**; a K_3-free graph is also called **triangle-free**. The class $\text{Forb}(H)$ is empty if and only if $H = \emptyset$; and the class $\text{Forb}(H)$ consists only of the empty graph if and only if $H = K_1$. The class $\text{Forb}(K_2)$ of K_2-free graphs consists of all edgeless graphs. Figure C.2 below shows a list of popular forbidden subgraphs; these small graphs have special names. For example, the class $\text{Forb}(K_{1,3})$ of claw-free graphs has attracted particular attention, see the 2005 survey paper by Chudnowsky and Seymour [240].

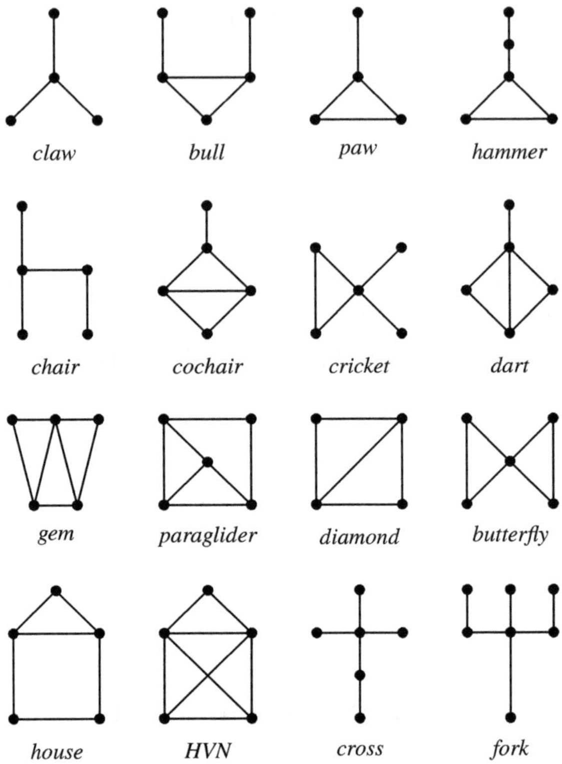

Fig. C.2 Small graphs with names.

In the class of graphs $\mathcal{G}$ we have the subgraph relation $\subseteq$, the induced subgraph relation $\leqslant$, the minor relation $\preceq$, and the topological minor relation $\preceq_t$. In what follows let $\trianglelefteq$ be one of these relations. If $H \trianglelefteq G$ and $H \ncong G$, then we write $H \triangleleft G$. The relation $\trianglelefteq$ is reflexive, transitive, and antisymmetric in the sense that $H \trianglelefteq G$ and $G \trianglelefteq H$ implies $H \cong G$. Furthermore, every strictly descending chain in $(\mathcal{G}, \trianglelefteq)$ is finite. A graph property $\mathcal{P}$ is **$\trianglelefteq$-hereditary** if it is closed with respect to the relation $\trianglelefteq$, that is, if G belongs to $\mathcal{P}$, then so does every graph H with $H \trianglelefteq G$. Thus $\mathcal{P}$ is monotone if and only if it is $\subseteq$-hereditary; and $\mathcal{P}$ is hereditary if and only if it is $\leqslant$-hereditary. For a class $\mathcal{X}$ of graphs, let

$$\mathrm{Forb}_{\trianglelefteq}(\mathcal{X}) = \{G \in \mathcal{G} \mid \forall H \in \mathcal{G}\, \forall H' \in \mathcal{X} : H \trianglelefteq G \Rightarrow H \ncong H'\}$$

Clearly, $\mathrm{Forb}_{\trianglelefteq}(\mathcal{X})$ is a $\trianglelefteq$-hereditary graph property and $\mathrm{Forb}(\mathcal{X}) = \mathrm{Forb}_{\leqslant}(\mathcal{X})$. For a graph property $\mathcal{P}$, let

$$\mathrm{Crit}_{\trianglelefteq}(\mathcal{P}) = \{G \in \mathcal{G} \mid G \notin \mathcal{P} \wedge \forall H (H \triangleleft G \Rightarrow H \in \mathcal{P})\}.$$

Note that $\mathrm{Crit}_{\trianglelefteq}(\mathcal{P})$ is a graph property, which is empty if and only if $\mathcal{P} = \mathcal{G}$. Since every strictly descending chain in $(\mathcal{G}, \trianglelefteq)$ is finite, it follows that for every graph

$G \notin \mathcal{P}$ there exists a graph $G' \in \mathrm{Crit}_{\trianglelefteq}(\mathcal{P})$ such that $G' \trianglelefteq G$. If $\mathcal{P}$ is a $\trianglelefteq$-hereditary graph property, then

$$\mathcal{P} = \mathrm{Forb}_{\trianglelefteq}(\mathrm{Crit}_{\trianglelefteq}(\mathcal{P})). \tag{C.3}$$

A **graph parameter** (or **graph invariant**) is a map ρ that assigns to each graph $G \in \mathcal{G}$ a real number $\rho(G)$ such that $\rho(H) = \rho(G)$ for any two isomorphic graphs $H, G \in \mathcal{G}$. A graph parameter ρ is called **$\trianglelefteq$-monotone** if $H \trianglelefteq G$ implies $\rho(H) \leq \rho(G)$, for any graphs $G, H \in \mathcal{G}$. A graph parameter is **monotone** if it is $\subseteq$-monotone. For a graph parameter ρ and a real number x, let

$$\mathcal{G}(\rho \leq x) = \{G \in \mathcal{G} \mid \rho(G) \leq x\}.$$

Clearly, $\mathcal{G}(\rho \leq x)$ is a graph property. If ρ is a $\trianglelefteq$-monotone graph parameter, then the graph property $\mathcal{G}(\rho \leq x)$ is $\trianglelefteq$-hereditary for all $x \in \mathbb{R}$, and vice versa. For $\mathrm{Forb}_{\trianglelefteq}(\mathcal{G}(\rho \leq x))$ we also write $\mathrm{Forb}_{\trianglelefteq}(\rho \leq x)$, and for $\mathrm{Crit}_{\trianglelefteq}(\mathcal{G}(\rho \leq x))$ we also write $\mathrm{Crit}_{\trianglelefteq}(\rho \leq x)$. Note that if ρ is a graph parameter, then

$$\mathrm{Crit}_{\trianglelefteq}(\rho \leq x) = \{G \in \mathcal{G} \mid \rho(G) > x \wedge \forall H (H \triangleleft G \Rightarrow \rho(H) \leq x)\}.$$

A graph property $\mathcal{P}$ is **additive** if $\mathcal{P}$ is closed under taking vertex disjoint unions. So if $\mathcal{P}$ is an additive graph property, then a nonempty graph belongs to $\mathcal{P}$ if and only if each of its components belong to $\mathcal{P}$. A graph parameter ρ is **additive** if $\rho(H \cup G) = \max\{\rho(G), \rho(H)\}$ whenever G and H are disjoint graphs. Clearly, if ρ is an additive graph parameter, then $\mathcal{G}(\rho \leq x)$ is an additive graph property for every $x \in \mathbb{R}$. Furthermore, if ρ is a monotone graph parameter, then ρ is additive if and only if $\rho(H \cup G) \leq \max\{\rho(G), \rho(H)\}$ whenever G and H are disjoint nonempty graphs. The maximum degree Δ, the clique number ω and the independence number α are examples of monotone graph parameters; while the minimum degree δ and the connectivity κ are nonmonotone graph parameters. For a nonempty graph G, let

$$\mathrm{ad}(G) = \frac{2|E(G)|}{|G|}$$

be the **average degree** of G, and let $\mathrm{mad}(G) = \max\{\mathrm{ad}(H) \mid \varnothing \neq H \subseteq G\}$ be the **maximum average degree**. Clearly, both the average degree and the maximum average degree are monotone graph parameters. To see that the average degree is an additive graph parameter, let G_1 and G_1 be disjoint nonempty graphs, and let $a_i = 2|E(G_i)|$ and $b_i = |G_i|$ $(i = 1,2)$. Then

$$\mathrm{ad}(G_1 \cup G_2) = \frac{a_1 + a_2}{b_1 + b_2} \leq \max\{\frac{a_1}{b_1}, \frac{a_2}{b_2}\} = \max\{\mathrm{ad}(G_1), \mathrm{ad}(G_2)\},$$

where the inequality holds, since if $B = b_1 + b_2$ and $M = \max\{\frac{a_1}{b_1}, \frac{a_2}{b_2}\}$, then we have

$$\frac{a_1 + a_2}{b_1 + b_2} = \frac{a_1}{b_1}\frac{b_1}{B} + \frac{a_2}{b_2}\frac{b_2}{B} \leq M\frac{b_1}{B} + M\frac{b_2}{B} = M.$$

As a consequence, we obtain that the maximum average degree is additive, too. Thus for every nonempty graph G, we have

$$\text{mad}(G) = \max\{\text{ad}(H) \mid H \subseteq G, H \text{ is connected}\}. \tag{C.4}$$

For the empty graph, we define $\text{ad}(\varnothing) = 0$ and $\text{mad}(\varnothing) = 0$.

C.12 Multigraphs and Directed Multigraphs

A **multigraph** is a triple $G = (V, E, i)$, where V and E are two finite disjoint sets, and $i : E \to [V]^2$. As for graphs, $V = V(G)$ is the **vertex set** of G and $E = E(G)$ is the **edge set** of G; moreover, $i = i_G$ is called the **incidence function** of G. A vertex v is **incident** with an edge e if $v \in i_G(e)$. The two vertices incident with an edge e are the **ends** of e; and e is said to **join** its ends. Terminology introduced for graphs will be used correspondingly. For instance, for $v \in V(G)$, we put $E_G(v) = \{e \in E(G) \mid v \in i_G(e)\}$, and for $X, Y \subseteq V(G)$, we denote by $E_G(X, Y)$ the set of edges of G joining a vertex of X with an vertex of Y. The **degree** of a vertex v is $d_G(v) = |E_G(v)|$, and the **multiplicity** of two distinct vertices v and w of G is $\mu_G(v, w) = |E_G(v, w)|$; distinct edges in $E_G(v, w)$ are called **multiple edges** or **parallel edges**. The **maximum multiplicity** of G is denoted by $\mu(G)$. Thus, a graph is essentially the same as a multigraph without multiple edges, that is, with $\mu(G) \leq 1$. Note that our multigraphs have no loops. If G is a multigraph without parallel edges, then for an edge e with $i_G(e) = \{v, w\}$ we also write $e = vw$ or $e = wv$, so we can identify e with the 2-subset $\{u, v\}$. Conversely, every graph G may be viewed as a multigraph whose incidence function is the identity function of its edge set, that is, $i_G(e) = e$ for every edge $e \in E(G)$.

Let G and H be two multigraphs. Then H is a **submultigraph** of G, written $H \subseteq G$, if $V(H) \subseteq V(G)$, $E(H) \subseteq E(G)$, and $i_G(e) = i_H(e)$ for all edges $e \in E(H)$. We say that G and H are **isomorphic** if there is a bijective map $\phi : V(G) \to V(H)$ such $\mu_G(v, w) = \mu_H(\phi(v), \phi(w))$ whenever v and w are distinct vertices of G. If H_1 and H_2 are two submultigraphs of G, then the **union** $H = H_1 \cup H_2$ is the submultigraph of G with $V(H) = V(H_1) \cup V(H_2)$ and $E(H) = E(H_1) \cup E(H_2)$, and the **intersection** $H' = H_1 \cap H_2$ is the submultigraph of G with $V(H) = V(H_1) \cap V(H_2)$ and $E(H) = E(H_1) \cap E(H_2)$. A **decomposition** of a multigraph G is a list of submultigraphs of G such that each edge of G appears in exactly one subgraph in the list.

The **underlying graph** of a multigraph G is the graph H with $V(H) = V(G)$ and $E(H) = \{i_G(e) \mid e \in E(G)\}$; then for $X \subseteq V(G)$, we have $N_G(X) = N_H(X)$. Clearly, such G and H are isomorphic if and only if G has no parallel edges.

Let G be a multigraph. Two distinct edges of G which are incident with a common vertex are called **adjacent edges** of G. A **matching** of G is a set of pairwise nonadjacent edges, so no two edges of a matching have a common end. For a set $M \subseteq E(G)$, we denote by $V(M)$ the set of vertices which are incident to some edge of M; a vertex in $V(M)$ is said to be **covered** by M. In 1935 Hall [463] established a necessary and sufficient condition for the existence of a matching in a bipartite multigraph that covers every vertex in one of its part.

Theorem C.11 (Hall's Theorem) *Let G be a bipartite multigraph with parts X and Y. Then G has a matching M with $V(M) \supseteq X$ if and only if $|N_G(S)| \geq |S|$ for all $S \subseteq X$.*

Let G be a graph. The maximum number of edges in a matching of G is called the **matching number** of G, written $\alpha'(G)$. Furthermore, we denote by $\mathrm{odd}(G)$ the number of components of G having odd order. The following result is due to Tutte and Berge (see [144, Corollary16.12]).

Theorem C.12 (The Berge-Tutte Formula) *Every graph G satisfies $\alpha'(G) = \max\{\frac{1}{2}(|G - \mathrm{odd}(G-P) + |P||) \mid P \subset V(G)\}$.*

Note that our multigraphs may have parallel edges but no loops. A **pseudomultigraph** is a triple $G = (V, E, i_G)$, where V and E are two finite disjoint sets, and $i_G : E \longrightarrow [V]^1 \cup [V]^2$. As for graphs, the elements of V are the vertices and the elements of E are the edges. An edge e of a pseudomultigraph with $i_G(e) = \{v\}$ for a vertex $v \in V$ is called a **loop** at v. Similarly, a **pseudograph** is a pair $G = (V, E)$ with $E \subseteq [V]^1 \cup [V]^2$. Note that a vertex v of a pseudograph is **isolated** if there is no edge $\{v, w\}$ of the pseudograph with $v \neq w$.

A **multidigraph** (or **directed multigraph**) D is a pair of two finite disjoint sets, the **vertex set** $V(D)$ and the **edge set** $E(D)$, together with two maps

$$v_D^+ : E(D) \longrightarrow V(D) \text{ and } v_D^- : E(D) \longrightarrow V(D)$$

assigning to every edge e an **initial vertex** $v_D^+(v)$ and a **terminal vertex** $v_D^-(v)$ such that $v_D^+(e) \neq v_D^-(e)$. If it is clear that we refer to the digraph D we write $v^+(e)$ and $v^-(e)$ rather than $v_D^+(e)$ and $v_D^-(e)$, respectively. The edge e is said to be **directed from** $v^+(e)$ **to** $v^-(e)$. If there exists an edge in D directed from v to w, then w is called a **successor** or **out-neighbor** of v, and v is called a **predecessor** or **in-neighbor** of w. For a vertex v of a multidigraph D, let $N_D^+(v)$ be the set of all out-neighbors of v in D, and let $N_D^-(v)$ be the set of all in-neighbors of v in D. The number of vertices of a multidigraph D is its **order**, denoted by $|D|$.

A multidigraph may have several edges between the same two vertices. Such edges are called **multiple edges**; if they have the same direction, say from u to v, they are **parallel**. Note that our digraphs have no loops. A multidigraph without multiple edges is called a **simple digraph**. A **strict digraph** is a multidigraph without parallel edges. So if D is a strict digraph, then between any two distinct vertices v and w there is at most one edge directed from v to w and at most one edge directed from w to v. If e is an edge of a strict digraph D directed from u to v, we also write $e = (u, v)$, so the edge set of D may be regarded as a subset of $V(D)^2 = V(D) \times V(D)$.

Let D be a multidigraph. For a vertex $v \in V(D)$, define

$$E_D^+(v) = \{e \in E(D) \mid v_D^+(e) = v\} \text{ and } E_D^-(v) = \{e \in E(D) \mid v_D^-(e) = v\}.$$

Then $d_D^+(v) = |E_D^+(v)|$ is the **out-degree** of v in D, and $d_D^-(v) = |E_D^-(v)|$ is the **in-degree** of v in D. Furthermore, $\Delta^+(D) = \max_{v \in V(D)} d_D^+(v)$ is the **maximum out-degree** of D, and $\delta^+(D) = \min_{v \in V(D)} d_D^+(v)$ is the **minimum out-degree** of D. Similarly, we define the **maximum in-degree** $\Delta^-(D)$ and the **minimum in-degree** $\delta^-(D)$ of D. Note that if D is a strict digraph, then $d_D^+(v) = |N_D^+(v)|$ and $d_D^-(v) = |N_D^-(v)|$ for every vertex v of D.

Let D, D' be two multidigraphs. We call D' a **subdigraph** of D if $V(D') \subseteq V(D)$, $E(D') \subseteq E(D)$, and for all $e \in E(D')$ we have $v_{D'}^+(e) = v_D^+(e)$ as well as $v_{D'}^-(e) = v_D^-(e)$. If $X \subseteq V(G)$, then $D[X]$ is the **subdigraph induced by** X, that is, the subdigraph D' of D satisfying

$$V(D') = X \text{ and } E(D') = \{e \in E(D) \mid \{v^+(e), v^-(e)\} \subseteq X\}.$$

So D' is an **induced subdigraph** of G if $D' = D[V(D')]$. A **decomposition** of D is a list of subdigraphs of D such that each edge of D belongs to exactly one subdigraph in the list.

Let D be a multidigraph. The **underlying graph** of D, denoted by G_D, is the unique graph with vertex set $V(G_D) = V(D)$ and $E(G_D) = \{\{v_D^+(e), v_D^-(e)\} \mid e \in E(D)\}$. The multidigraph D is said to be **connected** if G_D is connected. A set $X \subseteq V(D)$ is called an **independent set** of D if X is an independent set of G_D, that is, $D[X]$ is a multidigraph without edges. A set $X \subseteq V(D)$ is called an **clique** of D if X is a clique of G_D. A **kernel** of D is an independent set S of D such that for every vertex $u \in V(D) \setminus S$ there is an edge in D directed from u to some vertex $v \in S$. We say that D is **kernel-perfect** if every induced subdigraph of D has a kernel.

A **directed path** P is a simple directed graph whose underlying graph is a path $(v_1, v_2, \dots, v_n)$ such that $v_D^+(v_i v_{i+1}) = v_i$ for $i \in [1, n-1]$. In this case we also say that P is a **directed path from** v_1 **to** v_n. A **directed cycle** is a simple directed graph whose underlying graph is a cycle $(v_1, v_2, \dots, v_n, v_1)$ such that $v_D^+(v_i v_{i+1}) = v_i$ for $i \in [1, n-1]$ and $v_D^+(v_n v_1) = v_n$. As for graphs, a directed cycle is called **odd** or **even** depending on whether its order is odd or even. A digraph is **acyclic** if it has no directed cycle. Note that if D is an acyclic digraph, then its underlying graph may have cycles. It is easy to show that every nonempty acyclic digraph D contains a vertex v with $d_D^+(v) = 0$ as well as a vertex w with $d_D^-(w) = 0$ (take a directed path in D whose order is maximum). As for graphs, a directed path or directed cycle of a digraph D is called a **directed Hamilton path** respectively a **directed Hamilton cycle** of D if it contains all vertices of D.

Let G be a graph or multigraph. An **orientation** of G is a multidigraph D with vertex set $V(D) = V(G)$ and edge set $E(D) = E(G)$ such that every edge $e \in E(D)$ satisfies $i_G(e) = \{v_D^+(e), v_D^-(e)\}$. So an orientation of G is a multidigraph obtained from G by assigning to each edge of G a direction from one of its ends to the other.

C.13 Graphs on Surfaces

Figure C.3 shows three different drawings of the Petersen graph in the plane, but none of the three drawings is crossing-free. The Petersen graph can be embedded on the torus, but not in the plane. In this section we review some basic facts for graphs embedded in the plane or in other (closed) surfaces. Here a **surface** is a connected compact 2-dimensional manifold without boundary. Such surfaces can be classified according to their Euler characteristic and orientability. The **orientable surfaces** are the surfaces $\mathbf{S}_g$, for $g \geq 0$, obtained from the (2-dimensional) sphere by attaching g handles. So $\mathbf{S}_0$ is the sphere and $\mathbf{S}_1$ is the torus. The **nonorientable surfaces** are the surfaces $\mathbf{N}_h$, for $h \geq 1$, obtained by taking the sphere with h holes and attaching h Möbius bands along their boundary to the boundary of their holes. So $\mathbf{N}_1$ is the projective plane and $\mathbf{N}_2$ is the Klein bottle. The **Euler characteristic** $\varepsilon(\mathbf{S})$ of the surface $\mathbf{S}$ is $2-2g$ if $\mathbf{S} = \mathbf{S}_g$ and $2-h$ if $\mathbf{S} = \mathbf{N}_h$. The derived invariant $\mathrm{eg}(\mathbf{S}) = 2 - \varepsilon(\mathbf{S})$ is the **Euler genus** of $\mathbf{S}$. The famous classification theorem due to Brahana [166] states that every (closed) surface is homeomorphic either to $\mathbf{S}_g$ for some $g \geq 0$, or to $\mathbf{N}_h$ for some $h \geq 1$. Each surface may be viewed as a topological subspace of either $\mathbb{R}^3$ or $\mathbb{R}^4$. By Brahana's theorem, two surfaces are topologically the same if and only if both are either orientable or nonorientable and both have the same Euler characteristic. The Euler characteristic of the sphere with g handles and h Möbius bands is $2-2g-h$, and it is nonorientable when $h \geq 1$.

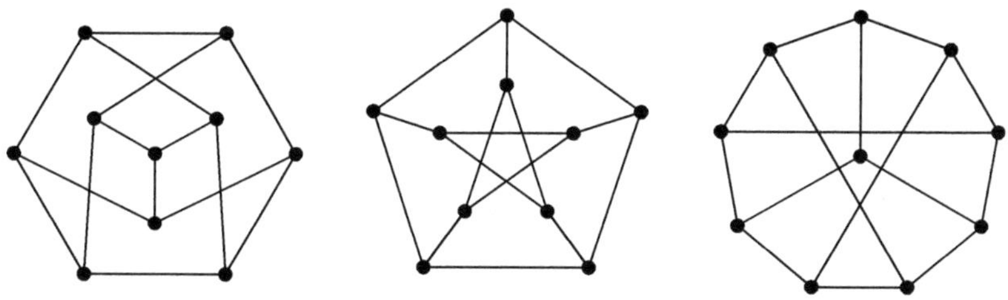

Fig. C.3 Three drawings of the Petersen graph.

Let $\mathbf{T} \subseteq \mathbb{R}^n$ be a topological subspace with the usual induced topology. If $\mathbf{X}$ is a subset of $\mathbf{T}$, then we denote its **boundary** by $\mathrm{bd}(\mathbf{X})$, its **interior** by $\mathrm{int}(\mathbf{X})$, and its **closure** by $\mathrm{cl}(\mathbf{X})$. An **arc**, a **circle**, a **closed disc**, and an **open disc** in $\mathbf{T}$ are subsets that are homeomorphic (in the subspace topology) to the real interval $\mathbb{R}[0,1]$, to the 1-sphere $\mathbf{S}^1 = \{\mathbf{x} \in \mathbb{R}^2 \mid \|\mathbf{x}\| = 1\}$, to the 2-ball $\mathbf{B}^2 = \{\mathbf{x} \in \mathbb{R}^2 \mid \|\mathbf{x}\| \leq 1\}$, and to the disc $\{\mathbf{x} \in \mathbb{R}^2 \mid \|x\| < 1\}$, respectively. If $\mathbf{a}$ is an arc in $\mathbf{T}$ and $g : \mathbb{R}[0,1] \to \mathbf{a}$ is the corresponding homomorphism, then the **endpoints** of $\mathbf{a}$ are $g(0)$ and $g(1)$; we also say that $\mathbf{a}$ is an **arc between** $g(0)$ and $g(1)$ or an $g(0)$-$g(1)$ **arc**; further we call the elements of $\mathbf{a} \setminus \{g(0), g(1)\}$ the **internal points** of $\mathbf{a}$. The **components** of a nonempty subset $\mathbf{X}$ of $\mathbf{T}$ are the equivalence classes of $\mathbf{X}$, where two points of $\mathbf{X}$ are **equivalent** if they are the same or if there exists an arc $\mathbf{a} \subseteq \mathbf{X}$ between these two points. A nonempty subset $\mathbf{X}$ of $\mathbf{T}$ is **connected** if it is its only component. Let

$\mathbf{C}$ be a circle in $\mathbf{T}$. Then $\mathbf{C}$ is a **nonseparating circle** if $\mathbf{T} \setminus \mathbf{C}$ is connected, and a **separating circle** otherwise. The circle $\mathbf{C}$ is **contractible** if it is the boundary of a closed disk in $\mathbf{T}$, and **noncontractible** otherwise. If $\mathbf{S}$ is a surface and $\mathbf{C}$ is a separating circle in $\mathbf{S}$, then $\mathbf{S} \setminus \mathbf{C}$ has exactly two components, both with boundary $\mathbf{C}$. If $\mathbf{S}$ is a surface different from the sphere and $\mathbf{C}$ is a contractible circle in $\mathbf{S}$, then a disk $\mathbf{D}$ such that $\mathbf{C} = \mathrm{bd}(\mathbf{D})$ is uniquely determined; in this case we call the open disk $\mathrm{int}(\mathbf{D})$ the **inside** of $\mathbf{C}$, denoted by $\mathrm{ins}(\mathbf{C})$, and we call $\mathbf{S} \setminus \mathbf{D}$ the **outside** of D. The Jordan curve theorem says that if $\mathbf{C}$ is a circle in the plane $\mathbb{R}^2$, then $\mathbb{R}^2 \setminus \mathbf{C}$ has exactly two components $\mathbf{X}_1$ and $\mathbf{X}_2$, and $\mathrm{bd}(\mathbf{X}_1) = \mathrm{bd}(\mathbf{X}_2) = \mathbf{C}$. Exactly one of the two components, say $\mathbf{X}_1$ is bounded and the other one is unbounded. Then the **inside** of $\mathbf{C}$ is $\mathrm{int}(\mathbf{X}_1)$ and the **outside** of $\mathbf{C}$ is $\mathrm{int}(\mathbf{X}_2)$. Furthermore, every surface different from the sphere contains a nonseparating circle.

Let $\mathbf{T} \subseteq \mathbb{R}^n$ be a topological space. An **embedded multigraph** in $\mathbf{T}$ (or a **T-multigraph**) is a multigraph G such that every member of $V(G)$ is a point of $\mathbf{T}$, every member e of $E(G)$ is an arc of $\mathbf{T}$, such that both endpoints of an arc $e \in E(G)$, but no interior point of e, belong to $V(G)$, and such that any two different arcs of $E(G)$ have at most two endpoints and no interior point in common. Furthermore, for every arc $e \in E(G)$, $i_G(e)$ is the set consisting of the two endpoints of e. Then we denote by $\mathbf{G}$ the subspace of $\mathbf{T}$ induced by G, that is,

$$\mathbf{G} = V(G) \cup \bigcup_{e \in E(G)} e.$$

If $H \subseteq G$ is a subgraph of G, then $\mathbf{H}$ is a subspace of $\mathbf{G}$ and hence of $\mathbf{T}$. The **faces** of G are the components of $\mathbf{T} \setminus \mathbf{G}$; we denote by $F(G)$ the set of all faces of G. A vertex $v \in V(G)$ is **incident** to a face $f \in F(G)$ if $v \in \mathrm{bd}(f)$; an edge $e \in E(G)$ is **incident** to f if $\mathrm{bd}(f)$ contains an interior point of e. We say that G is **2-cell embedded** in $\mathbf{T}$ if all faces of G are open disks. Let G be an (abstract) graph or multigraph. We say that G is **embeddable** in $\mathbf{T}$ if G is isomorphic to a $\mathbf{T}$-multigraph G'; in this case we call G' an **embedding** of G in $\mathbf{T}$. The class of all graphs embeddable in $\mathbf{T}$ is denoted by $\mathcal{G}(\hookrightarrow \mathbf{T})$; clearly, this class is a graph property. A **2-cell embedding** of G is an embedding of G which is 2-cell embedded; a 2-cell embedding is also referred to as a **cellular embedding**. If $\mathbf{T}$ is the plane $\mathbb{R}^2$ or a surface, then a multigraph is embeddable in $\mathbf{T}$ if and only if its underlying graph is embeddable in $\mathbf{T}$. The same statement holds for pseudographs, that is, loops do not affect the embeddability.

Let $\mathbf{T}$ be a surface or the plane, and let G be a graph embedded in $\mathbf{T}$. Then every edge is incident either to exactly one face or to two different faces of G. Furthermore, an edge $e \in E(G)$ is incident to a face $f \in F(G)$ if and only if $e \subseteq \mathrm{bd}(f)$. Thus for every face $f \in F(G)$, there is a unique graph $H \subseteq G$ such that $\mathbf{H} = \mathrm{bd}(f)$. This graph H is denoted by $\mathrm{Bd}(f)$; it is called the **boundary subgraph** of f. If $f \in F(G)$ and H is a subgraph of G, we say that f is **bounded** by H if $H = \mathrm{Bd}(f)$. A subgraph of G is called a **facial subgraph** if it is the boundary subgraph of some face of G. A **facial cycle** is a facial subgraph that is a cycle. Every vertex of $\mathbf{T}$ has a (topological) neighborhood which is homeomorphic to an open disk. If $v \in V(G)$ is a vertex, then the edges incident with v are arranged in a cyclic order, which can be described by a cyclic permutation $\pi_v \in \mathrm{Sym}(E_G(v))$ (respectively by the inverse permutation π_v^{-1}). Then for all edges $e \in E(G)$ there is a face $f \in F(G)$ such that both e and $\pi_v(e)$ are incident with f. This face is not unique if $d_G(v) = 2$. Furthermore, if a vertex v is incident with an edge e and this edge is incident with a face f, then either $\pi_v(e)$ is incident with f or $\pi_v^{-1}(e)$ is incident with f. These observations can be used to define **facial walks** (see Thomassen [1011] or Mohar and Thomassen [752]). Let $f \in F(G)$ be a face, let $H = \mathrm{Bd}(f)$ the subgraph bounded by f, and let H' be a component of H. Then there is a closed walk $W = (v_1, v_2, \dots, v_m, v_{m+1})$ with $v_{m+1} = v_1$ such that $V(H') = \{v_1, \dots, v_m\}$, $E(H') = \{v_i v_{i+1} \mid i \in [1,m]\}$, every edge appears at most twice in this walk (if an edge appears twice in the walk, then it appears in both directions), and an edge appears twice if and only if f is the only face incident with that edge. If the boundary subgraph of f is connected, then modulo orientation and starting vertex, there is exactly one facial walk associated with f. The **degree** of a face $f \in F(G)$, denoted by $d_G(f)$, is the number of edges in the boundary subgraph of f plus the number of bridges of G belonging to the boundary subgraph of f. Note that if e is a bridge of G belonging to the boundary subgraph of f, then e does not belong to the boundary subgraph of any other face. Hence, each edge of G appears twice as an edge of the boundary subgraph of a face, implying that

$$\sum_{f \in F(G)} d_G(f) = 2|E(G)|. \tag{C.5}$$

If G is a connected graph and $|G| \geq 3$, then, for every face $f \in F(G)$, we have $d_G(f) \geq 3$ and equality hods if and only if f is bounded by a triangle K_3. If $C \subseteq G$ is a cycle, then $\mathbf{C}$ is a circle in $\mathbf{T}$. We say that C is a **contractible cycle**, if the circle $\mathbf{T}$ is contractible in $\mathbf{S}$; and a **noncontractible cycle** otherwise.

Proposition C.13 *Let* $\mathbf{T}$ *be a surface or the plane, let* G *be a graph embedded in* $\mathbf{T}$, *and let* $C \subseteq G$ *a contractible cycle, such that* C *is an induced subgraph of* G *and* $G - V(C)$ *is connected. Then* C *is a facial cycle of* G.

Proof As C is contractible, $\mathbf{T} \setminus \mathbf{C}$ has exactly two components $\mathbf{X}_1$ and $\mathbf{X}_2$ such that $\mathrm{bd}(\mathbf{X}_1) = \mathrm{bd}(\mathbf{X}_1) = \mathbf{C}$. Since $G - V(C)$ is a connected graph embedded in $\mathbf{S} \setminus \mathbf{C}$, we deduce that that either $V(G) \cap \mathbf{X}_1 = \varnothing$ or $V(G) \cap \mathbf{X}_2 = \varnothing$; say $V(G) \cap \mathbf{X}_2 = \varnothing$. This implies that $V(G) \subseteq \mathrm{cl}(\mathbf{X}_1)$. Let $e \in E(G)$. If both ends of e are in C, then $e \in E(C)$ (since C is an induced cycle of G) and so $e \subseteq \mathbf{C} \subseteq \mathrm{cl}(\mathbf{X}_1)$. If one endpoint of e lies in $\mathbf{X}_1$, then $e \subseteq \mathrm{cl}(\mathbf{X}_1)$. It follows that $\mathbf{G} \subseteq \mathrm{cl}(\mathbf{X}_1)$, which implies that $\mathbf{X}_2$ is a face of G whose boundary subgraph is C. □

A proof of the following well known results can be found in Diestel's book [286, Proposition 4.2.6 and Proposition 4.2.7]. Let G be an arbitrary graph, and let $C \subseteq G$ be a cycle. We call C a **separating cycle** of G if $G - V(C)$ is disconnected, and a **nonseparating cycle** otherwise.

Proposition C.14 *Let* G *be a 2-connected graph embedded in the plane or the sphere. Then for all faces* $f \in F(G)$, *the closure* $\mathrm{cl}(f)$ *is a closed disk and so* $\mathrm{Bd}(f)$ *is a cycle.*

Proposition C.15 *The facial subgraphs in a 3-connected graph embedded in the plane or the sphere are precisely its nonseparating induced cycles.*

In order to view an embedding of a graph or multigraph on a surface, we may use a polygonal representation of that surface. A torus can be obtained from a rectangle $ABCD$ by identifying the side AB with the side DC and the side AD with the side BC. A projective plane may be represented by the disk $\mathbf{B}^2$ in which every point on the boundary is identified with its antipodal point. Figure C.4(a) shows an 2-cell embedding of K_4 in the torus, it has two faces bounded by the cycle $(1,2,3,4,1)$ and the closed walk $(1,2,4,1,3,4,2,3,1)$, respectively. Figure C.4(b) shows a non-2-cell embedding of K_4 in the torus, it has three faces and one of its faces is homeomorphic to a cylinder and is bounded by the facial walk $(1,2,3,1,4,3,1)$. Figure C.5 shows a 2-cell embedding of the Petersen graph in the projective plane. Note that 2-cell embedded graphs are connected.

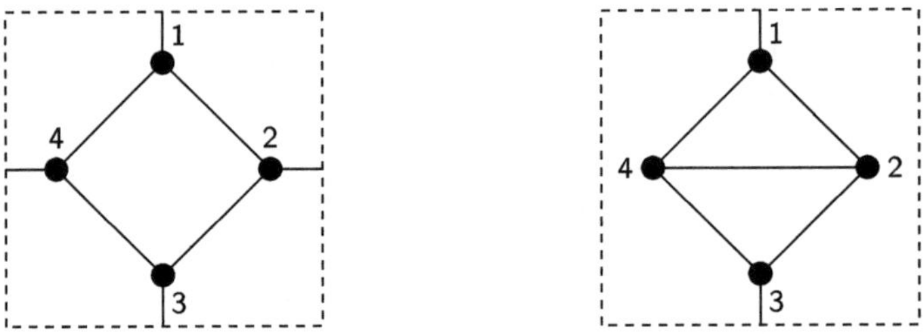

(a) (b)

Fig. C.4 Two embeddings of K_4 in the torus.

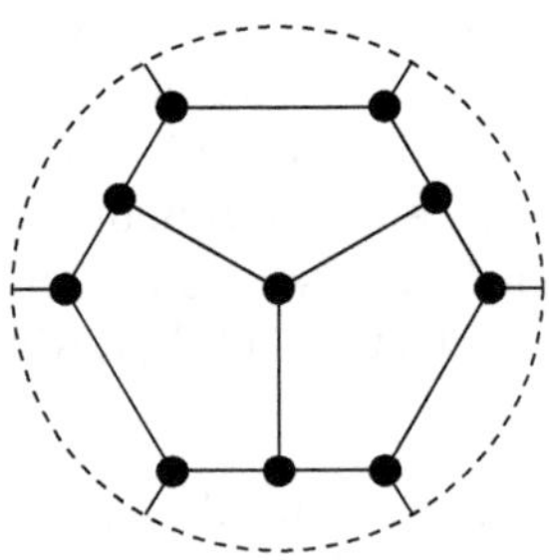

Fig. C.5 An embedding of the Petersen graph in the projective plane.

Many results about graphs embedded on surfaces are just consequences of the following result, which is known as Euler's Formula. A proof of this result may be found in [865] or in [752].

Theorem C.16 (Euler's Formula) *If a connected graph G is 2-cell embedded in a surface* $\mathbf{S}$*, then*

$$|V(G)| - |E(G)| + |F(G)| = \varepsilon(\mathbf{S}).$$

If a connected graph G has an embedding in a surface $\mathbf{S}$, but not a 2-cell embedding in $\mathbf{S}$, then we cannot apply Euler's formula directly. However in this case we can apply the following result obtained by Kagno [544] (see also [752] or [1086]).

Theorem C.17 *Let* $\mathbf{S}$ *be a surface, and let $G \in \mathcal{G}(\hookrightarrow \mathbf{S})$ be a connected graph embeddable in* $\mathbf{S}$*. Then there is a surface* $\mathbf{S}'$ *with $\varepsilon(\mathbf{S}') \geq \varepsilon(\mathbf{S})$ such that G has a 2-cell embedding in* $\mathbf{S}'$*.*

Theorem C.18 *Let* $\mathbf{S}$ *be a surface. If $G \in \mathcal{G}(\hookrightarrow \mathbf{S})$ is a connected graph with n vertices, m edges and girth at least g, where $n \geq g \geq 3$, then*

$$m \leq \frac{g}{g-2}(n - \varepsilon(S)).$$

Proof By Theorem C.17 there is a surface $\mathbf{S}'$ such that $\varepsilon(\mathbf{S}') \geq \varepsilon(\mathbf{S})$ and G has a 2-cell embedding G' in $\mathbf{S}$. Clearly, G' has n vertices and m edges. Let $\ell = |F(G')|$. Since G' is 2-cell embedded in $\mathbf{S}'$, for every face $f \in F(G')$ the facial subgraph $\mathrm{Bd}(f)$ is connected and gives rise to exactly one facial walk, which is a closed walk in G'. Since G' has girth at least g and $n \geq g$, we deduce that every facial walk in G' has at least g edges. Since each edge occurs exactly twice in facial walks, we conclude that $g\ell \leq 2m$. Then, according to Theorem C.16, we obtain that

$$\varepsilon(\mathbf{S}) \leq \varepsilon(\mathbf{S}') = n - m + \ell \leq n - m + \frac{2m}{g},$$

from which the result follows. □

Note that the condition $n \geq g$ in the above theorem is necessary, since we consider graphs with girth at least g and not graphs with girth exactly g, and any tree has girth at least g. Since any graph has girth at least three, Theorem C.18 immediately implies the following result.

Corollary C.19 *Let* $\mathbf{S}$ *be a surface. If $G \in \mathcal{G}(\hookrightarrow \mathbf{S})$ is a connected graph with n vertices and m edges, where $n \geq 3$, then $m \leq 3(n - \varepsilon(S))$.*

Corollary C.20 *Let* $\mathbf{S}$ *be a surface. Suppose that G is connected graph that is 2-cell embedded in* $\mathbf{S}$*, and G has n vertices and m edges, where $n \geq 3$. Then $m \leq 3(n - \varepsilon(\mathbf{S}))$ and equality holds if and only if $d_G(f) = 3$ for all faces $f \in F(G)$.*

Proof Let $\ell = |F(G)|$. Since G is connected and 2-cell embedded in $\mathbf{S}$, we have $m - n + \ell = \varepsilon(\mathbf{S})$ (by Theorem C.16). Since G is connected and $n \geq 3$, we have $d_G(f) \geq 3$ for all $f \in F(G)$. By (C.5), this leads to

$$2m = \sum_{f \in F(G)} d_G(f) \geq 3\ell = 3(m - n + \varepsilon(\mathbf{S})), \tag{C.6}$$

which is equivalent to

$$m \leq 3(n - \varepsilon(\mathbf{S})). \tag{C.7}$$

Equality holds in (C.7) if and only if it holds in (C.6), that is, if and only if $d_G(f) = 3$ for all $f \in F(G)$. □

Since the graph property $\mathcal{G}(\hookrightarrow \mathbf{S})$ is monotone, Corollary C.19 leads to an upper bound for the maximum average degree of graphs embeddable in $\mathbf{S}$ in terms of its Euler characteristic. Such a bound was established by Heawood [487]. For a surface $\mathbf{S}$, we call

$$H(\mathbf{S}) = \left\lfloor \frac{7 + \sqrt{49 - 24\,\varepsilon(\mathbf{S})}}{2} \right\rfloor \tag{C.8}$$

the **Heawood number** of $\mathbf{S}$. We may also express the Heawood number of $\mathbf{S}$ in terms of its Euler genus $\operatorname{eg}(\mathbf{S}) = 2 - \varepsilon(\mathbf{S})$, so

$$H(\mathbf{S}) = \left\lfloor \frac{7 + \sqrt{24\,\operatorname{eg}(\mathbf{S}) + 1}}{2} \right\rfloor.$$

For instance, the Heawood number of the sphere is $H(\mathbf{S}_0) = 4$. In his fundamental paper [487] published in 1890 Heawood proved the following result.

Theorem C.21 *Let* $\mathbf{S}$ *be a surface, and let* G *be a graph embeddable in* $\mathbf{S}$*. If* $\varepsilon(\mathbf{S}) \geq 1$*, then* $\operatorname{mad}(G) < 6$ *and so* $\delta(G) \leq 5$*; and if* $\varepsilon(\mathbf{S}) \leq 0$*, then*

$$\operatorname{mad}(G) \leq \frac{5 + \sqrt{49 - 24\varepsilon(\mathbf{S})}}{2}$$

and so $\delta(G) \leq H(\mathbf{S}) - 1$.

Proof Note that $\delta(G) \leq \lfloor \operatorname{mad}(G) \rfloor$. Thus, because of (C.4) and since $\mathcal{G}(\hookrightarrow \mathbf{S})$ is monotone, it suffices to show that every connected graph $G \in \mathcal{G}(\hookrightarrow \mathbf{S})$ satisfies $\operatorname{ad}(G) < 6$ if $\varepsilon(\mathbf{S}) \geq 1$, and

$$\operatorname{ad}(G) \leq h(\mathbf{S}) = \frac{5 + \sqrt{49 - 24\varepsilon(\mathbf{S})}}{2}$$

otherwise. So let $G \in \mathcal{G}(\hookrightarrow \mathbf{S})$ be a connected graph. Suppose that G has n vertices, m edges and average degree D. If $n \leq 2$, then $D \leq 1$ and there is nothing to prove. So assume that $n \geq 3$. Then it follows from Corollary C.19 that $m \leq 3(n - \varepsilon(\mathbf{S}))$, and since $D = \frac{2m}{n}$ this gives $nD \leq 6(n - \varepsilon(\mathbf{S}))$. If $\varepsilon(\mathbf{S}) \geq 1$, this implies $D < 6$ and we are done. So assume that $\varepsilon(\mathbf{S}) \leq 0$. Then $h(\mathbf{S}) \geq 6$ and there is nothing to prove if $D \leq 6$. So assume that $D > 6$. Since $m \leq 3(n - \varepsilon(\mathbf{S}))$ and $D = 2m/n$, we get again $Dn \leq 6(n - \varepsilon(\mathbf{S}))$, which is equivalent to $n(D-6) + 6\varepsilon(\mathbf{S}) \leq 0$. Since $D > 6$ and $n \geq \Delta(G) + 1 \geq D + 1$, we get $(D+1)(D-6) + 6\varepsilon(\mathbf{S}) \leq 0$, from which we obtain that

$$D \leq \frac{5+\sqrt{49-24\varepsilon(\mathbf{S})}}{2}$$

and so $\delta(G) \leq \lfloor D \rfloor \leq H(\mathbf{S})-1$. Thus the proof is complete. □

For a surface $\mathbf{S}$ and a graph parameter ρ, let

$$\rho(\mathbf{S}) = \max\{\rho(G) \mid G \in \mathcal{G}(\hookrightarrow \mathbf{S})\}.$$

Theorem C.21 then says that every surface $\mathbf{S}$ different from the sphere satisfies $\delta(\mathbf{S}) \leq H(\mathbf{S})-1$ and so $\omega(\mathbf{S}) \leq H(\mathbf{S})$. For the sphere we have $\delta(\mathbf{S}_0) \leq 5$ and $\omega(\mathbf{S}_0) = H(\mathbf{S}_0) = 4$. For every surface $\mathbf{S}$ distinct from the Klein bottle, the Heawood number $H(\mathbf{S})$ is, in fact, the maximum order of a complete graph enbeddable in $\mathbf{S}$, that is, $\omega(\mathbf{S}) = H(\mathbf{S})$. This landmark result, that was conjectured by Heawood [487], is due to Ringel [863] and Ringel and Youngs [866]. For the Klein bottle, we have $\omega(\mathbf{N}_2) = 6$, but $H(\mathbf{N}_2) = 7$.

The **(orientable) genus** $\mathrm{og}(G)$ and the **nonorientable genus** $\mathrm{ng}(G)$ of a graph G is the minimum g and the minimum h, respectively, for which G is embeddable in the orientable surface $\mathbf{S}_g$, respectively, in the nonorientable surface $\mathbf{N}_h$. So if $\mathrm{og}(G) = g$, then G is embeddable in an orientable surface $\mathbf{S}$ if and only if $\mathrm{eg}(G) \geq 2g$. It is known, see [752], that $\mathrm{ng}(G) \leq 2\mathrm{og}(G)+1$ for every graph G. If $\mathrm{ng}(G) = 2\mathrm{og}(G)+1$, then G is said to be **orientably simple**. For instance, every planar graph G is orientably simple, since $\mathrm{og}(G) = 0$ and $\mathrm{ng}(G) = 1$. Note that we consider trees as orientably simple graphs, although some authors exclude them. An embedding of G in the surface $\mathbf{S}_g$ with $g = \mathrm{og}(G)$ is called a **minimum genus embedding** of G. Similarly, a **nonorientable minimum genus embedding** of G is an embedding of G in the surface $\mathbf{N}_h$ with $h = \mathrm{ng}(G)$. For the next theorem the reader is referred to Mohar and Thomassen [752] and to Parson et al. [798].

Theorem C.22 *Let G be a connected graph. Then the following statements hold:*

(a) *Every minimum genus embedding of G is a 2-cell embedding.*
(b) *If $\mathrm{ng}(G) \leq 2\mathrm{og}(G)$, then every nonorientable minimum genus embedding of G is a 2-cell embedding.*

Remark C.23 Let G be a connected graph with n vertices and m edges, where $m \geq 3n+2$ and $n \geq 4$. If $G \in \mathcal{G}(\hookrightarrow \mathbf{S})$, then $3n+2 \leq m \leq 3(n-\varepsilon(\mathbf{S}))$ (by Corollary C.19), which leads to $\varepsilon(\mathbf{S}) \leq -1$. Hence $\mathrm{ng}(G) \geq 3$ and $\mathrm{og}(G) \geq 2$. Suppose that G can be embedded in $\mathbf{N}_3$. Then $\mathrm{ng}(G) = 3 \leq 2\mathrm{og}(G)$, and Theorem C.22(b) implies that every embedding of G in $\mathbf{N}_3$ is a 2-cell embedding. Now we fix such an embedding G in $\mathbf{N}_3$, which we may also denote by G. Then $m \leq 3(n+1)$ (by Corollary C.19), and so $3n+2 \leq m \leq 3n+3$. If $m = 3n+3$, then $d_G(f) = 3$ for all $f \in F(G)$ (by Corollary C.20). If $m = 3n+2$, then it follows from the proof of Corollary C.20 that there is a face $f' \in F(G)$ such that $d_G(f) = 3$ for all $f \in F(G) \setminus \{f'\}$ and $d_G(f') = 4$. Since $n \geq 4$ and G is connected, this implies that each face of the embedding G is bounded by a 3-cycle except for at most one face which is bounded by a 4-cycle.

The genus of the complete graphs was established by Ringel [863] and by Ringel and Youngs [866]; they used this result to give a complete answer to the Heawood problem concerning the clique number of surfaces.

Theorem C.24 *The following statements hold:*

(a) $\mathrm{og}(K_n) = \left\lceil \frac{(n-3)(n-4)}{12} \right\rceil$ *for* $n \geq 3$.

(b) $\mathrm{ng}(K_n) = \left\lceil \frac{(n-3)(n-4)}{6} \right\rceil$ *for* $n \geq 3$ *and* $n \neq 7$.

(c) *If* $\mathbf{S}$ *is a surface distinct from the Klein bottle, then* $\omega(\mathbf{S}) = H(\mathbf{S})$*; moreover,* $\omega(\mathbf{N}_2) = 6$ *and* $\mathrm{ng}(K_7) = 3$.

A **planar graph** is a graph embeddable in the plane; and a **plane graph** is a graph embedded in the plane. So each planar graph is isomorphic to a plane graph, and vice versa. If G is a plane graph, then exactly one of its faces is unbounded. This face is the **outer face** of G, the other faces are its **inner faces**.

A graph is embeddable in the plane $\mathbb{R}^2$ if and only if it is embeddable in the sphere $\mathbf{S}_0$, that is, the two classes $\mathcal{G}(\hookrightarrow \mathbb{R}^2)$ and $\mathcal{G}(\hookrightarrow \mathbf{S}_0)$ are the same. This follows from the fact that if x is any point of the sphere, then $\mathbf{S}_0 \setminus \{x\}$ is homeomorphic to $\mathbb{R}^2$; and the stereographic projection $\pi_x : \mathbf{S}_0 \setminus \{x\} \to \mathbb{R}^2$ is a homeomorphism between these two spaces. Let G be a planar graph, let G' be an embedding of G in the sphere, let $f \in F(G')$ be an arbitrary face, and let $x \in f$ be an arbitrary point. Then the image G^* of G' under the stereographic projection π_x is an embedding of G in the plane, and there is a one-to-one correspondence between the faces of G' and G^*, where the image of the face $f \setminus \{x\}$ with respect to π_x is the outer face of G^*. Conversely, if G^* is an embedding of G in the plane, then its preimage with respect to π_x is an embedding of G on the sphere. This can be used to prove the following result (see also Bondy and Murty [144, Proposition 10.5]).

Proposition C.25 *Let G be a planar graph, and let f be a face in some embedding of G in the plane. Then there is an embedding of G in the plane whose outer face has the same boundary subgraph as f.*

By Theorem C.22(a) it follows that every embedding of a planar connected graph in the sphere is a 2-cell embedding; and the same holds for every embedding of a planar connected graph in the plane. Furthermore, if G is a plane connected graph, then

$$|V(G)| - |E(G)| + |F(G)| = 2, \tag{C.9}$$

implying, by (C.5), that

$$\sum_{x \in F(G) \cup V(G)} (d_G(x) - 4) = -8. \tag{C.10}$$

Moreover, it follows from the proof of Theorem C.18 that every connected plane graph G with $n \geq 3$ vertices has at most $3n - 6$ edges, where equality hods if and only if $d_G(f) = 3$ for all $f \in F(G)$, that is, each face is bounded by a triangle. A plane graph in which each face is bounded by a triangle is called a **plane triangulation**. A **maximal planar graph** is a planar graph G such that adding any new edge (from its complement $\overline{G}$) to G results in a nonplanar graph. Hence, each planar graph is a spanning subgraph of a maximal planar graph. If G is a planar graph of order $n \leq 4$, then G is maximal planar if and only if G is a K_n. A proof of the following proposition can be found in [286, Proposition 4.2.8].

Proposition C.26 *If G is a maximal planar graph with $|G| \geq 3$, and G' is an embedding of G in the plane, then G' is a triangulation.*

Summarizing all results about planar and plane graphs mentioned in this section, we obtain the following result.

Corollary C.27 *Let G be a planar graph with n vertices and m edges, where $n \geq 3$. Then $m \leq 3n - 6$ and equality holds if and only if G is maximal planar.*

Proof There is a maximal planar graph $\tilde{G}$ containing G as a subgraph and with $|\tilde{G}| = |G|$. Let G' be an embedding of $\tilde{G}$ in the plane and let $\tilde{m}$ be the number of edges of $\tilde{G}$. Then G' is a triangulation (by Proposition C.26), and so $\tilde{G}$ is connected. Hence, $m \leq \tilde{m} = 3n - 6$ (by Corollary C.20). Furthermore, $m = 3n - 6$ if and only if $G = \tilde{G}$, that is, if and only if G is maximal planar. □

Corollary C.28 *Let G be a planar graph with $|G| \geq 3$, and let G' be an embedding of G in the plane. Then G is maximal planar if and only if G' is a plane triangulation.*

A **triangulation** (respectively, **quadrangulation**) of a surface $\mathbf{S}$ is a graph G that is cellular embedded in $\mathbf{S}$ such that each facial subgraph of G is a C_4 (respectively, a C_3); the corresponding embedding of G in $\mathbf{S}$ is said to be **triangular** (respectively **quadrangular**). A graph G embedded in the sphere is a triangulation of the sphere if and only if G is a maximal planar graph. Let G be an embedding of K_3 in the torus, such that the triangle is a contractible cycle of the torus. Then G has two faces both of degree 3, but G is not 2-cell embedded in the torus and hence G is not a triangulation of the torus. On the other hand, every embedding of K_7 in the torus is a triangular embedding. There is an quadrangular embedding of K_4 in the projective plane $\mathbf{N}_1$ with three facial 4-cycles, namely (u,v,w,x,u), (u,v,x,w,u) and (u,x,v,w,u).

Suppose that G is a graph embedded in a surface or the plane and $\delta(G) \geq 2$. Then every face of G of degree at most five is bounded by a cycle.

Remark C.29 Let G be a nonbipartite graph that is embedded in the projective plane such that each facial subgraph of G is a C_4. Then G is a quadrangulation of $\mathbf{N}_1$. To prove this it suffices to show that the embedding of G in $\mathbf{N}_1$ is a cellular embedding. Let $n = |V(G)|$, $m = |E(G)|$ and $\ell = |F(G)|$. Let C be a shortest odd cycle of G, and let $p = |C|$. Since G is nonbipartite such a cycle exists. Since every facial subgraph is a C_4, we obtain that G is connected and C is a noncontractible cycle of $\mathbf{N}_1$. Clearly, C has no chords in G. Then, by cutting the projective plane $\mathbf{N}_1$ along C, we obtain a plane graph G_C with outer cycle O_C, where $|O_C| = 2p$ and G can be obtained from

G_C by identifying opposite vertices of O_C. In particular, G_C has order $n_C = n+p$ and $m_C = |E(G_C)| = m+p$. Since every facial subgraph of G is a C_4, we obtain that G_C is connected and $\ell_C = |F(G_C)| = \ell+1$. Clearly, G_C is 2-cell embedded in the plane and so $n_C - m_C + \ell_C = 2$ which implies that $n - m + \ell = 1 = \varepsilon(\mathbf{N}_1)$. From this it follows that G is 2-cell embedded in $\mathbf{N}_1$.

C.14 Hypergraphs

A **hypergraph** H is a pair of sets, $V(H)$ and $E(H)$, where $V(H)$ is finite and $E(H)$ is a subset of $2^{V(H)}$ such that $|e| \geq 2$ for all $e \in E(H)$. The set $V(H)$ is the **vertex set** of H and its elements are the **vertices** of H. The set $E(H)$ is the **edge set** of H and its elements are the **edges** of H. Thus graphs are special hypergraphs and we adopt the notation for graphs. A hypergraph H is **empty** if $V(H) = E(H) = \varnothing$; in this case we write $H = \varnothing$.

Let H be a hypergraph. The number of vertices of H is its **order**, denoted by $|H|$. An edge e with $|e| \geq 3$ is called a **hyperedge**, and an edge e with $|e| = 2$ is an **ordinary edge**. If all edges of H have the same size p, then H is said to be p-**uniform**. So a graph is a 2-uniform hypergraph, that is, a hypergraph in which each edge is ordinary. A vertex v is **incident** with an edge e if $v \in e$. For a vertex v of H, let $E_H(v) = \{e \in E(H) \mid v \in e\}$. The **degree** of v in H is $d_H(v) = |E_H(v)|$. Let $\delta(H)$ and $\Delta(H)$ denote the **minimum degree** and the **maximum degree** of H, respectively. A nonempty hypergraph H is said to be **regular** and r-**regular** if all vertices of H have the same degree r, that is, if $\delta(G) = \Delta(G) = r$. By double counting, we deduce that

$$\sum_{v \in V(H)} d_H(v) = \sum_{e \in E(H)} |e|, \tag{C.11}$$

which yields $r|H| = p|E(H)|$ if the hypergraph H is p-uniform and r-regular.

Let H and H' be hypergraphs. If $V(H') \subseteq V(H)$ and $E(H') \subseteq E(H)$, then H' is called a **subhypergraph** of H and we write $H' \subseteq H$. Clearly, $H = H'$ if and only if $H' \subseteq H$ and $H \subseteq H'$. A subhypergraph H' of H with $H' \neq H$ is said to be a **proper subhypergraph** of H, written as $H' \subset H$. As for graphs, $H' \subseteq H$ is a **spanning subhypergraph** if $V(H') = V(H)$. Let $X \subseteq V(H)$. We define two hypergraphs $H[X]$ and $H\langle X\rangle$ by

$$V(H[X]) = X \text{ and } E(H[X]) = E(H) \cap 2^X,$$

and

$$V(H\langle X\rangle) = X \text{ and } E(H\langle X\rangle) = \{e \cap X \mid e \in E(H) \text{ and } |e \cap X| \geq 2\}.$$

While $H[X]$ is a subhypergraph of H (the subhypergraph of H **induced** by X), $H\langle X\rangle$ need not be a subhypergraph of H. Clearly, if H is a graph, then $H[X] = H\langle X\rangle$. Furthermore, let $H - X = H[V(H) \setminus X]$ and $H \div X = H\langle V(H) \setminus X\rangle$. As usual, we abbreviate $H - \{v\}$ by $H - v$ and $H \div \{v\}$ by $H \div v$. A subhypergraph H' of H is called an **induced subhypergraph** of H if $H' = H[V(H')]$. For a set $F \subseteq 2^{V(H)}$, we define $H - F = (V(H), E(H) \setminus F)$ and $H + F = (V(H), E(H) \cup F)$. When $F = \{e\}$, we abbreviate $H - F$ by $H - e$ and $H + F$ by $H + e$.

The hypergraphs H and H' are **isomorphic**, written $H \cong H'$, if there is a bijection $\phi : V(H) \to V(H')$ such that, for all sets $e \subseteq V(H)$, we have $e \in E(H)$ if and only if $\phi(e) \in E(H')$.

For hypergraphs H and H' let $H \cup H' = (V(H) \cup V(H'), E(H) \cup E(H'))$ be the **union** of H and H', and let $H \cap H' = (V(H) \cap V(H'), E(H) \cap E(H'))$ be the **intersection** of H and H'. We say that H and H' are **disjoint** if $H \cap H' = \varnothing$.

Let H be a hypergraph and let $X \subseteq V(H)$ be a vertex set. We call X an **independent set** or **p-independent set** of H if $|X| = p$ and $H[X]$ has no edges. The **independence number** of H, written $\alpha(H)$, is the largest integer p for which H contains a p-independent set. We call X a **clique** or **p-clique** of H if $|X| = p$ and $[X]^2 \subseteq E(H)$. The **clique number** of H, written $\omega(G)$, is the largest integer p for which H contains a p-clique. If X is a p-clique of H, then $H[X]$ contains K_p as a spanning subhypergraph, but not necessarily as an induced subhypergraph.

Let $q \geq 2$ be an integer. For a set X, let $K^q(X)$ denote the hypergraph with $V(K^q(X)) = X$ and $E(K^q(X)) = [X]^q$. We call $K^q(X)$ an **complete q-uniform hypergraph**. So $K^2(X)$ is a complete graph with vertex set X, and $K^q(X)$ is edgeless if $|X| < r$. If $|X| = q$, then we write $K(X)$ instead of $K^q(X)$. So $K(X)$ is the hypergraph with $V(K(X)) = X$ and $E(K(X)) = \{X\}$, where $|X| \geq 2$. If H is a complete q-uniform hypergraph of order n, then we also write $H = K_n^q$ and say that H is a K_n^q.

A **bipartite hypergraph** is a hypergraph H whose vertex set is the disjoint union of two sets (called **partite sets** or **parts**), each of which is an independent set of H. So if H is a bipartite hypergraph with parts X and Y, then both hypergraphs $H[X]$ and $H[Y]$ are edgeless, $V(H) = X \cup Y$ and $X \cap Y = \varnothing$.

Let H be a hypergraph and let $W = (v_0, e_1, v_1, \ldots, v_{n-1}, e_n, v_n)$ be a sequence with $n \geq 0$ such that $v_0, v_1, \ldots, v_n$ are vertices of H and $e_1, e_2, \ldots, e_n$ are edges of H. We call W a **hyperwalk** of H if $\{v_{i-1}, v_i\} \subseteq e_i$ for $1 \leq i \leq n$. We say that W **connects** or **joins** v_0 and v_n, and that W **starts** in v_0, and **terminates** in v_n, and that W is a v_0-v_n hyperwalk. The vertices v_0 and v_n are called the **ends** of the hyperwalk W, v_0 being its **initial vertex** and v_n its **terminal vertex**; the remaining vertices $v_1, v_2, \ldots, v_{n-1}$ are its **internal vertices**. The integer n is called the **length** of W, written as $\ell(W) = n$. The reverse sequence $W^{-1} = (v_n, e_n, v_{n-1}, \ldots, v_1, e_1, v_0)$ is also a hyperwalk in G. A walk is **odd** or **even**, depending on whether its length is odd or even. The hyperwalk W is called a **hyperpath** if all the vertices v_i and all the edges e_j are distinct. The hyperwalk W with $n \geq 2$ is called a **hypercycle** if the the vertices $v_0, v_1, \ldots, v_{n-1}$ are distinct, $v_0 = v_n$ and all the edges e_j are distinct. As for graphs, the **girth** $g(G)$ of a hypergraph is the length of a shortest hypercycle of G; if G has no hypercycle we define $g(G) = \infty$.

Proposition C.30 *Let H be a hypergraph and let v, w be vertices of H. Then the following statements are equivalent:*

(a) *There exists a v-w hyperwalk in H.*

(b) *There exists a v-w hyperpath in H.*

Proof Since every hyperpath is a hyperwalk, (b) implies (a). To see that the reverse implication is true let $W = (v_0, e_1, v_1, \dots, v_{n-1}, e_n, v_n)$ be a v-w hyperwalk, whose length is minimum. By (a) such a hyperwalk exists. If W is no hyperpath, then $n \geq 1$ and either two vertices of W or two edges of W are equal. If $v_i = v_j$ with $0 \leq i < j \leq n$, then $W' = (v_0, e_1, v_1, \dots, e_{i-1}, v_j, e_j, \dots, v_{n-1}, e_n, v_n)$ is a v-w hyperwalk of H with $\ell(W') < \ell(W)$. If $e_i = e_j$ with $0 \leq i < j \leq n-1$, then $W' = (v_0, e_1, v_1, \dots, e_{i-1}, v_i, e_j, v_{j+1} \dots, v_{n-1}, e_n, v_n)$ is a v-w hyperwalk of H with $\ell(W') < \ell(W)$. So in booth cases we obtain a contradiction. □

A nonempty hypergraph H is **connected** if there exists a hyperpath between any two of its vertices, and otherwise it is **disconnected**. A (connected) **component** of a nonempty hypergraph H is a maximal connected subhypergraph of H. As for graphs, let $\text{com}(H)$ denote the number of components of H. By convention, we put $\text{com}(\varnothing) = 0$.

Let H be a nonempty hypergraph. For vertices v and w of H, let $v \sim w$ if there exists a v-w hyperwalk in H. Then it is easy to check that $\sim$ is an equivalence relation on $V(H)$. By Proposition C.30 it follows that H' is a component of H if and only if H' is an induced subhypergraph of H and $V(H')$ is an equivalence class of $V(H)$ with respect to the relation $\sim$.

Proposition C.31 *Let H be a nonempty hypergraph, let U be the vertex set of a component of H, and let e be an edge of H. Then either $e \subseteq U$ or $e \subseteq V(G) \setminus U$.*

Proof Suppose, to the contrary, that there is an edge $e \in E(H)$ such that e contains a vertex u of U as well as a vertex w of $V(G) \setminus U$. The set U is an equivalence class with respect to the above defined relation $\sim$. Since $F = (u, e, w)$ is a hyperwalk, we obtain $u \sim w$ implying that $w \in U$, a contradiction. □

Let H be a hypergraph, and let $e \in E(H)$. Then it is easy to see that $\text{com}(H) \leq \text{com}(H-e) \leq \text{com}(H) + |e| - 1$. We call e a **bridge** of H if $\text{com}(H-e) > \text{com}(H)$; and we call e a **strong bridge** of H if $\text{com}(H-e) = \text{com}(H) + |e| - 1$. Obviously, e is a bridge of H if and only if two vertices of e belong to different components of $H-e$; and e is a strong bridge of H if and only if no two vertices of e belong to the same component of $H-e$. The proof of the following proposition is straight forward.

Proposition C.32 *Let H be a connected hypergraph and let e be an edge of H. Then the following statements are equivalent:*

(a) *e is a strong bridge of H.*
(b) *e contains exactly one vertex from each component of $H-e$.*
(c) *e lies in no hypercycle of H.*

Let H be a hypergraph. We call $F \subseteq E(H)$ a **separating edge set** of H if we have $\mathrm{com}(H-F) \geq 2$. If $X \subseteq V(H)$, then we denote by $\partial_H(X)$ the set of edges of H containing a vertex of X as well as a vertex of $V(H) \setminus X$. By a **cut** or **edge cut** of H we mean a triple (X,Y,F) such that X is a nonempty proper subset of $V(H)$, $Y = V(H) \setminus X$, and $F = \partial_H(X) = \partial_H(Y)$. Clearly, if H is connected and $\varnothing \neq X \subset V(H)$, then $F = \partial_H(X)$ is a separating edge set of H, but not necessarily a minimal one.

Proposition C.33 *Let H be a connected hypergraph and let $F \subseteq E(H)$ be a minimal separating edge set. Then $F = \partial_H(X)$ for a set X with $\varnothing \neq X \subset V(H)$, and so $(X, V(H) \setminus X, F)$ is a cut.*

Proof Since H is connected and F is a separating edge set of H, $H - F$ is the union of two nonempty disjoint hypergraphs, say H_1 and H_2. Then $F_1 = \partial_H(V(H_1))$ is a separating edge set of H and $F_1 \subseteq F$. Since F is a minimal separating edge set, $F = F_1$ and we are done. □

A hypergraph H is called k**-edge-connected** ($k \in \mathbb{N}_0$) if $|H| \geq 2$ and $H - F$ is connected for every set $F \subseteq E(H)$ with $|F| \leq k-1$. For a hypergraph H with $|H| \geq 2$, the **edge-connectivity** $\kappa'(H)$ is defined to be the largest integer k such that H is k-edge-connected. By Proposition C.33 it follows that if $|H| \geq 2$, then

$$\kappa'(H) = \min\{|\partial_H(X)| \mid \varnothing \neq X \subset V(H)\}. \tag{C.12}$$

Let H be a hypergraph. A **separation** of H is a triple (S, H_1, H_2) consisting of a set $S \subseteq V(H)$ and two induced subhypergraphs H_1 and H_2 of H such that $H = H_1 \cup H_2$, $|H_i| > |S|$ $(i = 1,2)$ and $V(H_1) \cap V(H_2) = S$. If (S, H_1, H_2) is a separation of H, then S is called a **separation set** of H. A separation set S of H is **minimal** if no proper subset of S is a separation set of H. Clearly, the empty hypergraph and the hypergraph consisting of one vertex have no separation set and the only minimal separation set of a disconnected hypergraph is the empty set.

Proposition C.34 *Let H be a connected hypergraph and let $S \subseteq V(H)$. Then S is a separation set of H if and only if $H \div S$ is disconnected.*

Proof If S is a separation set of H, then there is a separation (S, H_1, H_2) of H. Then there is a vertex $v \in V(H_1) \setminus S$ as well as a vertex $w \in V(H_2) \setminus S$. We claim that v and w belong to different components of $H' = H \div S$. If not, then there is a v-w hyperwalk W of H'. Then W contains a subhyperwalk $W' = (v', e', w')$ with $v' \in V(H_1) \setminus S$, $w' \in V(H_2) \setminus S$ and $e' \in E(H')$. But then there is an edge $e \in E(H)$ such that $\{v', w'\} \subseteq e \subseteq S \cup \{v', w'\}$ and so $e \notin E(H_1) \cup E(H_2)$, a contradiction. Hence v and w belong to different components of H' as claimed and so H' is disconnected.

If $H' = H \div S$ is disconnected, then $\mathrm{com}(H') = p \geq 2$. Let $H'_1, H'_2, \ldots, H'_p$ be the components of H'. Let $H_1 = H[V(H'_1) \cup V(H'_2) \cup \cdots \cup V(H'_{p-1}) \cup S]$ and let $H_2 = H[V(H'_p) \cup S]$. We claim that (S, H_1, H_2) is a separation of H and so S is a separation set of H. If not then one edge e of H does not belong to $H_1 \cup H_2$. So e contains a vertex of $V(H'_p)$ as well as a vertex of $V(H'_1) \cup V(H'_2) \cup \cdots \cup V(H'_{p-1})$. But then H' contains also such an edge, which is impossible. □

A hypergraph H is k-**connected** ($k \in \mathbb{N}_0$) if $|H| \geq k+1$ and every separation set S of H satisfies $|S| \geq k$. For a nonempty hypergraph H the **connectivity** $\kappa(H)$ is the largest integer k such that H is k-connected. Note that Proposition C.1 is not true for hypergraphs. It is not difficult to show that a hypergraph H is k-connected if $|H| \geq k+1$ and $H \div S$ is connected for every set $S \subseteq V(H)$ with $|S| \leq k-1$. Note that if H is a graph, then $H \div S = H - S$.

Let H be a hypergraph. A set $S \subseteq V(H)$ is called a **separating vertex set** of H if $\text{com}(H \div S) \geq 2$. Then Proposition C.34 says that if H is a connected hypergraph and $S \subseteq V(H)$, then S is a separating vertex set of H if and only if S is a separation set of H. A **separating vertex** of H is a vertex v of H such that $\text{com}(H \div v) > \text{com}(H)$. This obviously implies that the separating vertices of a disconnected hypergraph are those of its components. If H is a connected hypergraph, then v is a separating vertex of H if and only if $S = \{v\}$ is a separation set of H (by Proposition C.34), that is, H is the union of two (induced) subhypergraphs having only vertex v in common and having both at least two vertices.

Let H be a hypergraph. A **block** of H is a maximal connected subhypergraph of H that has no separating vertex. Let $\mathfrak{B}(H)$ denote the set of all blocks of H. For a vertex $v \in V(H)$, let $\mathfrak{B}_v(H) = \{B \in \mathfrak{B}(H) \mid v \in V(B)\}$. Note that $\mathfrak{B}(\varnothing) = \varnothing$ and that every block of H is a connected induced subhypergraph of H. If H is disconnected, then $\mathfrak{B}(H)$ is the union of $\mathfrak{B}(H')$ taken over all components H' of H. If H is connected and has no separating vertex, then $\mathfrak{B}(H) = \{H\}$ and we say that H is a **block**. If H is connected and $(\{v\}, H_1, H_2)$ is a separation of H, $\mathfrak{B}(H) = \mathfrak{B}(H_1) \cup \mathfrak{B}(H_2)$ and $\mathfrak{B}(H_1) \cap \mathfrak{B}(H_2) = \varnothing$. As for graphs it is not difficult to show that any two distinct blocks of H have at most one vertex in common, and a vertex of H is a separating vertex of H if and only if it belongs to more that one block. A block of H, which contains at most one separating vertex of H is called an **end-block** of H. If H contains a separating vertex then H has at least two end-blocks. Note that for a hypergraph H the relation on $E(H)$ *to be equal or to lie on a common cycle* is not necessarily an equivalence relation.

C.15 Lovász Local Lemma

A (finite) **probability space** is a pair (Ω, Pr), where Ω is a finite set, called the **sample space**, and $\text{Pr} : \Omega \to \mathbb{R}$ is a **probability function**, that is, $0 \leq \text{Pr}(\omega) \leq 1$ for all $\omega \in \Omega$ and $\sum_{\omega \in \Omega} \text{Pr}(\omega) = 1$.

Let (Ω, Pr) be a probability space. Any subset A of Ω is called an **event** and the **probability** of A is defined by:

$$\text{Pr}(A) = \sum_{\omega \in A} \text{Pr}(\omega).$$

Two events A and B are **independent** if $\text{Pr}(A \cap B) = \text{Pr}(A)\text{Pr}(B)$; otherwise, they are **dependent**. Clearly, the independence relation is symmetric and if $\text{Pr}(B) > 0$, then A and B are independent if and only if $\Pr(A|B) = \Pr(A)$, where $\Pr(A|B)$ is the **conditional probability** $\Pr(A \cap B)/\Pr(B)$ that the event A occurs given that event B occurs. If A and B are independent events, then so are A and the negation $\overline{B}$ of B. If $A_1, A_2, \ldots, A_n$ are events and $S \subseteq [1,n]$, we put $A_S = \bigcup_{i \in S} A_i$. Events $A_1, A_2, \ldots, A_n$ are **mutually independent** if

$$\Pr(A_S) = \prod_{i \in S} \Pr(A_i)$$

for all $S \subseteq [1,n]$. If $A_1, A_2, \ldots, A_n$ are mutually independent events, then so are $\overline{A}_1, \overline{A}_2, \ldots, \overline{A}_n$. An event A is mutually independent of the events $A_1, A_2, \ldots, A_n$ if

$$\Pr(A \cap A_S) = \Pr(A)\Pr(A_S)$$

for all $S \subseteq [1,n]$. A **dependency digraph** for events $A_1, A_2, \ldots, A_n$ is a strict digraph D with vertex set $V(D) = [1,n]$ such that for each $i \in I$, the event A_i is mutually independent of the events A_j with $(i,j) \notin E(D)$. Note that the dependency digraph is not unique. The following result, first stated, proved and used in Erdős and Lovász [354], is known as the Local Lemma (see the book of Alon and Spencer [57]); it is a powerful tool in combinatorics and graph theory.

Lemma C.35 (The Local Lemma; General Case) *Let $A_1, A_2, \ldots, A_n$ be events in an arbitrary probability space, and let D be a dependency digraph for these events. Suppose that there exist real numbers $x_1, x_2, \ldots, x_n$ such that $0 \le x_i < 1$ and*

$$\Pr(A_i) \le x_i \cdot \prod_{(i,j) \in E(D)} (1 - x_j) \tag{C.13}$$

for all $i \in [1,n]$. Then

$$\Pr\left(\bigcap_{i=1}^{n} \overline{A}_i\right) \ge \prod_{(i,j) \in E(D)} (1 - x_j).$$

In particular, with positive probability no event A_i ($i \in [1,n]$) holds.

Corollary C.36 (The Local Lemma; Symmetric Case) *Let $A_1, A_2, \ldots, A_n$ be events in an arbitrary probability space. Suppose that each event A_i is mutually independent of all the other events A_j except at most d, and that $\Pr(A_i) \le p$ for all $i \in [1,n]$. If*

$$ep(d+1) \le 1,$$

then $\Pr\left(\bigcap_{i=1}^{n} \overline{A}_i\right) > 0$ (the constant e is Euler's number 2.718..).

Proof If $d = 0$, then the events $A_1, A_2, \ldots, A_n$ are mutually independent and the result is trivial. So assume that $d \ge 1$. By assumption, there is a dependency digraph D for the events $A_1, A_2, \ldots, A_n$ with $\Delta^+(D) \le d$. We now apply Lemma C.35 with

$x_i = \frac{1}{d+1}$ ($i \in [1,n]$). Clearly, $0 \leq x_i < 1$ and the proof is completed by showing that (C.13) holds. Since $ep(d+1) \leq 1$ and $\Delta^+(D) \leq d$ we deduce that

$$\Pr(A_i) \leq p \leq \frac{1}{d+1} \cdot \frac{1}{e} \leq \frac{1}{d+1} \cdot \left(1 - \frac{1}{1+d}\right)^d \leq x_i \prod_{(i,j) \in E(D)} (1 - x_j).$$

Thus Corollary C.36 is proved. □

Meanwhile, an incredible number of applications of the Lovás Local Lemma (LLL) are known, some interesting applications were discovered in 1994 by Alon [46]. For later use we want to discuss one of his results here in detail, see [46, Proposition 5.3].

Lemma C.37 (Alon) *Let H be a graph, and let $\mathcal{X}$ be a partition of the vertex set of H. Suppose there is a positive integer k such that $|X| \geq 2ek$ for all $X \in \mathcal{X}$ and $\Delta(H) \leq k$. Then H has an independent set I such that $|I \cap X| = 1$ for all $X \in \mathcal{X}$.*

Proof For the proof it suffices to consider the case that each set of $\mathcal{X}$ has cardinality $\ell = \lceil 2ek \rceil$ and is an independent set of H. For each set $X \in \mathcal{X}$, choose independently and uniformly a single vertex of X. Let I denote the resulting random subset of $V(H)$. Clearly, $|I \cap X| = 1$ for all $X \in \mathcal{X}$. To complete the proof it suffices to show that with positive probability I is an independent set. To do this, we shall use Corollary C.36. For an edge uv of H, let A_{uv} denote the event that $\{u, v\} \subseteq I$. Since each set in $\mathcal{X}$ has size ℓ, we have $\Pr(A_{uv}) = \ell^{-2}$. Now let uv be an edge of H. Then there are two sets in $\mathcal{X}$, say U and V, such that $u \in U$ and $v \in V$. Let $F(uv)$ be the set of edges of H such that one of its end belongs to $U \cup V$. Note that $uv \in F(uv)$ and

$$|F(uv)| \leq \sum_{u' \in U} d_H(u') + \sum_{v' \in V} d_H(v') \leq 2\ell k.$$

An event A_{uv} is mutually independent of all the events $A_{u'v'}$, provided that $u'v' \notin F(uv)$. So an event A_{uv} is mutually independent of all other events $A_{u'v'}$ except at most $d = 2\ell k - 1$. Since

$$e\,\ell^{-2}(d+1) = \frac{2ek}{\ell} = \frac{2ek}{\lceil 2ek \rceil} \leq 1,$$

Corollary C.36 implies that I is an independent set of H with positive probability, as desired. □

In 1975, Bollobás, Erdős, and Spencer [130] conjectured that the quantity $2ek$ in the above lemma can be replaced by $2k$. This is best possible in the sense that $2k$ could not be replaced by ck for any constant $c < 2$. This was first proved in [130]; further constructions were later given by Yuster [1087]. Alon's proof of his lemma indicates that $2ek$ might be best value that can be obtained by applying the LLL. A proof that the conjectured value $2k$ is the right one was obtained in 2001 by Aharoni and Haxell as a special case of a more general result (unpublished); the theorem was first stated in a short paper by Haxell [482], but without a proof. Meanwhile several extensions of the Aharoni–Haxell result have been obtained by applying methods from topology (see [482]), but a direct proof would still be of interest.

Theorem C.38 (Aharoni and Haxell) *Let H be a graph, and let $\mathcal{X}$ be a partition of the vertex set of H. Suppose there is a positive integer k such that $|X| \geq 2k$ for all $X \in \mathcal{X}$ and $\Delta(H) \leq k$. Then H has an independent set I such that $|I \cap X| = 1$ for all $X \in \mathcal{X}$.*

Alon's decomposition lemma and its proof is a typical example how the LLL is used in order to prove the existence of combinatorial objects satisfying certain constraints. In 2010 Moser and Tardos [761] obtained a fully constructive proof of the LLL, thereby extending an earlier approach of Beck [88]. They also developed a general framework, the so-called *entropy compression method*, that can be used to provide an efficient algorithms for finding the combinatorial objects whose existence are guaranteed by the LLL. It tur ned out that one could obtain better combinatorial results by a direct application of the entropy compression method instead of applying the LLL in its original form. The LLL requires most events to be independent of each other. Erdős and Spencer [357] presented another version of the LLL, the lopsided LLL, which relaxes the requirement of independence to negative dependence, which is more general but also more difficult to identify. We state here the symmetric version, see, for example, [57, Theorem x.x].

Lemma C.39 (The Lopsided Local Lemma) *Let $A_1, A_2, \ldots, A_n$ be events in an arbitrary probability space. Suppose that for every $i \in [1,n]$ there is a set $N_i \subseteq [1,n]$ such that $|N_i| \leq d$ and for all $M \subseteq [1,n] \setminus N_i$,*

$$\Pr\left(A_i \mid \bigcap_{j \in M} \overline{A}_j\right) \leq p$$

If

$$4pd \leq 1,$$

then $\Pr\left(\bigcap_{i=1}^{n} \overline{A}_i\right) > 0$

A collection $X_1, X_2, \ldots, X_n$ of $\{0,1\}$-valued random variables is called **negatively correlated** if for all $I \subseteq [1,n]$,

$$\Pr(X_i = 1 \text{ for all } i \in I) \leq \prod_{i \in I} \Pr(X_i = 1). \tag{C.14}$$

Panconesi and Srinivasan [797] noted that several Chernoff-type bounds for mutually independent variables also hold on negatively correlated variables. For the following result see also [754, Lemma 3]. Here $\mathbb{E}(X)$ denotes the **expectation** of X.

Lemma C.40 (Chernoff bounds) *Let $X_1, X_2, \ldots, X_n$ be $\{0,1\}$-valued random variables, let $Y_i = 1 - X_i$ for $i \in [1,n]$, let $X = \sum_{i=1}^{n} X_i$, and let $0 < t < \mathbb{E}(X)$. Then the following statements hold:*

(a) *If the variables $X_1, X_2, \ldots, X_n$ are negatively correlated, then*

$$\Pr(X > \mathbb{E}(X) + t) < \exp(-t^2/3\mathbb{E}(X))$$

(b) *If the variables $Y_1, Y_2, \ldots, Y_n$ are negatively correlated, then*

$$\Pr(X < \mathbb{E}(X) - t) < \exp(-t^2/2\mathbb{E}(X)).$$

If a random variable is the sum of mutually independent $\{0,1\}$-valued random variables, the following standard concentration bound can be used, see for instance [729, Theorem 2.3] or [306, Theorem 1.10.1].

Lemma C.41 *Let X be the sum of n mutually independent $\{0,1\}$-valued random variables, and let $\vartheta > 0$. Then*

$$\Pr(X \geq (1+\vartheta)\mathbb{E}(X)) \leq \exp\left(-\frac{\vartheta^2\mathbb{E}(X)}{2+\frac{2}{3}\vartheta}\right) \leq \exp\left(-\frac{\min\{\vartheta^2,\vartheta\}\mathbb{E}(X)}{3}\right)$$

Let $((\Omega_i, \Pr_i))_{i=1}^n$ be a family of finite probability spaces, and let $(\Omega, \Pr)$ be their product space. Let $X : \Omega \to \mathbb{R}$ be a nonnegative random variable, and let $\ell, r > 0$. We say that X is (r,ℓ)**-certifiable** if for every $s > 0$ and every $\omega = (\omega_1, \omega_2, \ldots, \omega_n)$ such that $\omega \in \Omega$ and $X(\omega) \geq s$, there exists an index set $I \subseteq [1,n]$ with $|I| \leq rs$ such that, for all $k \geq 0$, we have $X(\omega') \geq s - k\ell$, for all $\omega' = (\omega'_1, \omega'_2, \ldots, \omega'_n) \in \Omega$ such that $\omega_i \neq \omega'_i$ for at most k indices $i \in I$. For the following combinatorial version of Talagrand's inequality see Kelly and Postle [566, Theorem 6.3 with $\Omega^* = \varnothing$].

Theorem C.42 (Talagrand's Inequality) *Let $((\Omega_i, \Pr_i))_{i=1}^n$ be a family of finite probability spaces, and let $(\Omega, \Pr)$ be their product space. Let $X : \Omega \to \mathbb{R}$ be a nonnegative random variable, not identical 0, which is (r,ℓ)-certifiable, where $\ell, r >$ 0. Then*

$$\Pr\left(|X - \mathbb{E}(X)| > t)\right) \leq 4\exp\left(-\tfrac{t^2}{8r\ell^2(4\mathbb{E}(X)+t)}\right)$$

for any real number t with $t > 96\ell\sqrt{r\mathbb{E}(X)} + 128r\ell^2$.

C.16 Fekete's Lemma

The following result is due to Pólya and Szegő [816]; in the literature it is also known as Fekete's lemma [369].

Lemma C.43 (Fekete's lemma) *Let $f : \mathbb{N} \to \mathbb{R}$ be a function such that*

$$f(m+n) \leq f(m) + f(n)$$

for all integers $m, n \geq 1$. Then the limit $\lim_{n\to\infty} \frac{f(n)}{n}$ exists and

$$\lim_{n\to\infty}\frac{f(n)}{n}=\inf_{n\in\mathbb{N}}\frac{f(n)}{n}.$$

Proof Let c be an arbitrary real number satisfying $c>\inf_n \frac{f(n)}{n}$. Then there exists an integer $p\in\mathbb{N}$ with $\frac{f(p)}{p}<c$. Now let $n\in\mathbb{N}$. Then $n=gp+r$ with $0\le r<p$. From the hypothesis of the lemma we deduce that

$$f(n)=f(gp+r)\le f(gp)+f(r)\le gf(p)+rf(1),$$

which gives

$$\frac{f(n)}{n}\le\frac{gf(p)+rf(1)}{gp+r}=\frac{f(p)+\frac{r}{g}f(1)}{p+\frac{r}{g}}.$$

If n tends to infinity, then $\frac{r}{g}$ tends to zero. Hence the above inequality implies that

$$\limsup_{n\to\infty}\frac{f(n)}{n}\le\frac{f(p)}{p}<c.$$

By the choice of c, we then we deduce that

$$\limsup_{n\to\infty}\frac{f(n)}{n}\le\inf_n\frac{f(n)}{n},$$

which yields $\lim_{n\to\infty}\frac{f(n)}{n}=\inf_n\frac{f(n)}{n}$. □

A function $f:\mathbb{N}\to\mathbb{R}$ satisfying the hypothesis of Fekete's lemma is called **subadditive**. We need the following strengthening of Fekete's lemma.

Lemma C.44 *Let $f:\mathbb{N}\to\mathbb{R}_{\ge 0}$ be a function and let $k,h\in\mathbb{N}$. Suppose that*

$$f(m+n)\le f(m)+f(n+h) \tag{C.15}$$

for all integers $m,n\ge k$. Then the limit $\lim_{n\to\infty}\frac{f(n)}{n}$ exists and is finite.

Proof First we claim that $f(n)\le cn$ for all $n\in\mathbb{N}$, where

$$c=\max\{\frac{f(k)}{k},\frac{f(k+h)}{k},\frac{f(k+1+h)}{k+1},\ldots,\frac{f(2k-1+h)}{2k-1}\}.$$

To see this, let $n\in\mathbb{N}$. Then $n=gk+r$ with $g\ge 0$ and $0\le r\le k-1$. By repeated application of (C.15), we deduce that

$$\begin{aligned}f(n)&=f(gk+r)\\&\le f(k)+(g-2)f(k+h)+f(k+r+h)\\&\le ck+(g-2)ck+c(k+r)=c(gk+r)=cn.\end{aligned}$$

Since $0\le f(n)\le cn$ for all $n\in\mathbb{N}$, there are real numbers a and b such that

$$a = \liminf_{n\to\infty} \frac{f(n)}{n} \le \limsup_{n\to\infty} \frac{f(n)}{n} = b$$

To show that $a = b$, let $\varepsilon > 0$ be given. Then there exists an integer $p \ge \max\{\frac{c}{\varepsilon}, k\}$ for which

$$\frac{f(p)}{p} \le a + \varepsilon.$$

Now let n be some suitable large integer. Then $n = gp + r$ for some integers g, r satisfying

$$\max\{k, \frac{c(h+k+p)}{\varepsilon p}\} \le g \text{ and } k \le r < k + p \tag{C.16}$$

Then by repeated application of (C.15) we obtain

$$f(n) = f(gp+r) \le gf(p) + f(gh+r) \le gf(p) + f(gh) + f(h+r).$$

Using (C.16) and the fact that $f(n) \le cn$ for all $n \in \mathbb{N}$, we deduce from the above inequality that

$$\begin{aligned}
\frac{f(n)}{n} &\le \frac{gf(p) + f(gh) + f(h+r)}{gp+r} \\
&\le \frac{gf(p) + cgh + c(h+r)}{gp} \\
&\le \frac{f(p)}{p} + \frac{c}{p} + \frac{c(h+k+p)}{gp} \\
&\le (a+\varepsilon) + \varepsilon + \varepsilon = a + 3\varepsilon.
\end{aligned}$$

Since this holds for all $\varepsilon > 0$ and all sufficiently large n, we deduce that $b \le a$, so $a = b$, and the desired limit exists and is finite. □

References

1. Abbott, H.L. and Hanson, D. (1969). On a combinatorial problem of Erdős. *Can. Math. Bull.,* 12:823–829.
2. Abbott, H.L. and Hare, D.R. (1989). Sparse color-critical hypergraphs. *Combinatorica,* 9:233–243.
3. Abbott, H.L., Hare, D.R. and Zhou, B. (1994). Sparse color-critical graphs and hypergraphs with no short cycles. *J. Graph Theory,* 18:373–388.
4. Abbott, H.L., Hare, D.R. and Zhou, B. (1998). Color-critical graphs and hypergraphs with few edges and no short cycles. *Discrete Math.,* 182:3–11.
5. Abbott, H.L. and Liu, A. (1978). The existence problem for color critical linear hypergraphs. *Acta Math. Hungar.,* 32:273–282.
6. Abbott, H.L. and Liu, A. (1980). On property B of families of sets. *Can. Math. Bull.,* 23:429–435.
7. Abbott, H.L., Liu, A. and Toft, B. (1980). The enumeration problem for color critical linear hypergraphs. *J. Combin. Theory Ser. B,* 29:106–115.
8. Abbott, H.L. and Moser, L. (1964). On a combinatorial Problem of Erdős and Hajnal. *Can. Math. Bull.,* 7:177–181.
9. Abbott, H.L. and Zhou, B. (1991). The edge density of 4-critical planar graphs. *Combinatorica,* 11:185–189.
10. Abbott, H.L. and Zhou, B. (1991). On small faces in 4-critical planar graphs. *Ars Combin.,* 32:203–207.
11. Abbott, H. L. and Zhou, B. (1992). On a conjecture of Gallai concerning complete subgraphs of k-critical graphs. *Discrete Math.,* 100:223–228.
12. Abbott, H.L. and Zhou, B. (1995). Some remarks on $(k-1)$-critical subgraphs of k-critical graphs. *Combinatorica,* 15:469–474.
13. Aboulker, P., Aubian, G. and Charbit, P. (2023). Digraph colouring and arc-connectivity. Manuscript arXive:2304.04690
14. Aboulker, P., Bellitto, T., Havet, F. and Rambaud, C. (2022). On the minimum number of arcs in k-critical oriented graphs. Manuscript arXive:2207.01051
15. Aboulker, P., Brettell, N., Havet, F., Marx, D. and Trotignon, N. (2017). Colouring graphs with constraints on connectivity. *J. Graph Theory,* 85:814–838.
16. Aboulker, P., Havet, F., Knauer, K. and Rambaud, C. (2022). On the dichromatic number of surfaces. *Electron. J. Combin.,* 29:#P1.30.
17. Aboulker, P. and Vermande, Q. (2022). Various bounds on the minimum number of arcs in a k-dicritical digraph. Manuscript arXive:2208.02112
18. Abu-Khzam, F.N., Feghali, C. and Heggernes, P. (2020). Partitioning a graph into degenerate subgraphs. *European J. Combin.,* 83:103015.
19. Addario-Berry, L., Havet, F. and Thomassé, S. (2007). Paths with two blocks in n-chromatic digraphs. *J. Combin. Theory Ser. B,* 97:620–626.

M. Stiebitz et al., *Brooks' Theorem*, Springer Monographs in Mathematics,
https://doi.org/10.1007/978-3-031-50065-7

20. Aglave, S., Amarnath, V.A., Shannigrahi, S. and Singh, S. (2020). Improved bounds for uniform hypergraphs without property B. *Australas. J. Combin.,* 76:73–86.
21. Aharoni, R., Ben-Arroyo Hartman, I., and Hofmann, A.J. (1985). Path partitions and packs of acyclic digraphs. *Pacific J. Math.,* 118:249–259.
22. Aharoni, R., Berger, E. and Kfir, O. (2008). Acyclic systems of representatives and acyclic colorings of digraphs. *J. Graph Theory,* 59:177–189.
23. Aharoni, R. and Holzman, R. (1998). Fractional kernels in digraphs. *J. Combin. Theory Ser. B,* 73:1–6.
24. Ajtai, M., Komlós, J. and Szemerédi, E. (1980). A note on Ramsey numbers. *J. Combin. Theory Ser. A,* 3:354–360.
25. Akbari, S., Mirrokni, V. S. and Sadjad, B.S. (2006). A relation between choosability and uniquely list colorability. *J. Combin. Theory Ser. B,* 96:577–583.
26. Akolzin, I. and Shabanov, D. (2016). Colorings of hypergraphs with large number of colors. *Discrete Math.,* 339:3020–3031.
27. Aksionov, V.A. (1974). The extension of a 3-colouring on planar graphs (in Russian). *Diskret. Analiz,* 26:3–19.
28. Aksionov, V.A. (1977). Chromatic connected vertices in planar graphs (in Russian). *Diskret. Analiz,* 31:5–16.
29. Aksionov, V.A. and Mel'nikov, L.S. (1980). Some counterexamples associated with the three-color problem. *J. Combin. Theory Ser. B,* 28:1–9.
30. Albar, B. and Gonçalves, D. (2018). On triangles in K_r-minor free graphs. *J. Graph Theory,* 88:154–173.
31. Albertson, M.O. (1998). You can't paint yourself into a corner. *J. Combin. Theory Ser. B,* 73:189–194.
32. Albertson, M.O., Bollobás, B. and Tucker, S. (1976). The independence ratio and maximum degree of a graph. *Congr. Numer.,* 17:43–50.
33. Albertson, M.O. and Collins, K.L. (1985). Homomorphisms of 3-chromatic graphs. *Discrete Math.,* 54:127–132.
34. Albertson, M.O. and Hutchinson, J.P. (1979). The three excluded cases of Dirac's map-color theorem. (1979). *Ann. New York Acad. Sci.,* 319:7–17.
35. Albertson, M.O. and Hutchinson, J.P. (1980). On six-chromatic toroidal graphs. *Proc. London Math. Soc.,* 3:533–556.
36. Albertson, M.O. and Hutchinson, J.P. (1980). Hadwiger's conjecture for graphs on the Klein bottle. *Discrete Math.,* 29:1–11.
37. Albertson, M.O. and Hutchinson, J.P. (2004). Extending precoloring of subgraphs of locally planar graphs. *European J. Combin.,* 25:863–871.
38. Albertson, M.O., Jamison, R.E., Hedetniemi, S.T. and Locke, S.C. (1989). The subchromatic number of a graph. *Discrete Math.,* 74:33–49.
39. Albertson, M.O., Kostochka, A.V. and West, D. (2005). Precoloring extensions of Brooks' theorem. *SIAM J. Discrete Math.,* 18:542–553.
40. Albertson, M.O. and Stromquist, W.R. (1982). Locally planar toroidal graphs are 5-colorable. *Proc. Amer. Math. Soc.,* 84:449–457.
41. Alishahi, M., Hajiabolhassan, H. and Meunier, F. (2017). Strengthening topological colorful results for graphs. *European J. Combin.,* 64:27–44.
42. Alon, N. (1985). Hypergraphs with high chromatic number. *Graphs Combin.,* 1:387–389.
43. Alon, N. (1992). The strong chromatic number of a graph. *Random Struct. Alg.,* 3:1–7.
44. Alon, N. (1992). Choice numbers of graphs: a probabilistic approach. *Combin. Probab. Comput.,* 1:107–114.
45. Alon, N. (1993). Restricted colorings of graphs. In: *Surveys in Combinatorics, 1993,* pages 1–33. London Math. Soc. Lecture Notes Series, Vol. 187, Cambridge University Press.
46. Alon, N. (1994). Probabilistic methods in coloring and decomposition problems. *Discrete Math.,* 127:31–46.
47. Alon, N. (1996). Independence numbers of locally sparse graphs and a Ramsey type problem. *Random Struct. Alg.,* 9:271–278.

48. Alon, N. (1999). Combinatiorial Nullstellensatz. *Combin. Probab. Comput.,* 8:7–29.
49. Alon, N. (2000). Degrees and choice numbers. *Random Struct. Alg.,* 16:364–368.
50. Alon, N. and Bregman, Z. (1988). Every 8-uniform 8-regular hypergraph is 2-colorable. *Graphs Combin.,* 4:303–306.
51. Alon, N., Frankl, P. and Lovász, L. (1986). The chromatic number of Kneser hypergraphs. *Trans. Amer. Math. Soc.,* 298:359–370.
52. Alon, N., Kostochka, A.V., Reiniger, B., West, D.B. and Zhu, X. (2016). Coloring, sparseness and girth. *Israel J. Math.,* 214:315–331.
53. Alon, N., Krivelevich, M. and Seymour, P. (2000). Long cycles in critical graphs. *J. Graph Theory,* 35:193–196.
54. Alon, N., Krivelevich, M. and Sudakov, B. (1999). Coloring graphs with sparse neighborhoods. *J. Combin. Theory Ser. B,* 77:73–82.
55. Alon, N., Krivelevich, M. and Sudakov, B. (2023). Complete minors and average degree: A short proof. *J. Graph Theory,* 103:599–602.
56. Alon, N., Pach, J. and Solymosi, J. (2001). Ramsey-type theorems with forbidden subgraphs. *Combinatorica,* 21:155–170.
57. Alon, N. and Spencer, J. (1992). *The Probabilistic Method.* Wiley-Interscience Series in Discrete Mathematics and Optimization, John Wiley & Sons.
58. Alon, N. and Tarsi, M. (1992). Colorings and orientations of graphs. *Combinatorica,* 12:125–134.
59. Alon, N. and Tarsi, M. (1997). A note on graph colorings and graph polynomials. *J. Combin. Theory Ser. B,* 70:197–201.
60. Alon, N., Voigt, M. and Tuza, Zs. (1997). Choosability and fractional chromatic number. *Discrete Math.,* 165:31–38.
61. Andersen, L.D. (1977). On edge-colourings of graphs. *Math. Scand.,* 40:161–175.
62. Anderson, J., Bernshteyn, A., and Dhawan, A. (2023). Coloring graphs with forbidden almost bipartite subgraphs. *Combin. Probab. Comput.,* 32:45–67.
63. Andrásfai, B. (1964). Graphentheoretische Extremalprobleme. *Acta Math. Hungar.,* 15:413–438.
64. Andrásfai, B., Erdős, P. and Sós, V. (1974). On the connection between chromatic number, maximal clique and minimal degree of a graph. *Discrete Math.,* 8:205–218.
65. Andres, S.D. and Hochstättler, W. (2015). Perfect digraphs. *J. Graph Theory,* 79:21–29.
66. Andrews, J.A. and Jacobsen, M.S. (1985). On a generalization of chromatic number. *Congr. Numer.,* 47:33-48.
67. Aouchiche, M. and Hansen, P. (2013). A survey of Nordhaus-Gaddum type relations. *Discrete Math.,* 161:466-546.
68. Appel, K. and Haken, W. (1977). Every planar map is 4-colorable. Part I: Discharging. *Illinois J. Math,* 21:429–490.
69. Appel, K., Haken, W. and Koch, J. (1977). Every planar map is 4-colorable. Part II: Reducibility. *Illinois J. Math,* 21:491–567.
70. Aravind, N.R., Karthick, T., and Subramanian, C.R. (2011). Bounding χ in terms of ω and Δ for some classes of graphs. *Discrete Math.,* 311:911–920.
71. Asplund, E. and Grünbaum, B. (1960). On a coloring problem. *Math. Scand.,* 8:181–188.
72. Assadi, S., Kumar, P. and Mittal, P. (2022). Brooks' theorem in graph streams: a single-pass semi-streaming algorithm for Δ-coloring. In: *Prooceedings of the 54th Annual ACM SIGACT Symposium on Theory of Computing, STOC,* pages 234–247.
73. Axenovich, M. (2003). A note on graph coloring extensions and list-colorings. *Electron. J. Combin.,* 10:#N1.
74. Babson, E. and Kozlov, D.N. (2006). Complexes of graph homomorphisms. *Israel J. Math.,* 152:285–312.
75. Baetz, B. and Wood, D.R. (2014). Brooks' vertex-colouring theorem in linear time. Manuscript arXive:1401.8023
76. Balogh, J., Kostochka, A., Prince, N. and Stiebitz, M. (2009). The Erdős–Lovász Tihany conjecture for quasi-line graphs. *Discrete Math.,* 309:3985–3991.

77. Bang-Jensen, J., Bellitto, T., Schweser, T. and Stiebitz, M. (2020). On DP-coloring of digraphs. *J. Graph Theory,* 95:76–98.
78. Bang-Jensen, J., Bellitto, T., Schweser, T. and Stiebitz, M. (2020). Hajós and Ore construction for digraphs. *Electron. J. Combin.,* 27:#P1.63.
79. Bang-Jensen, J., Reed, B., Schacht, M., Šámal, R., Toft, B. and Wagner, U. (2006). On six problems posed by Jarik Nešetřil. In: *Topics in Discrete Mathematics,* pages 613–627. Algorithms and Combinatorics, Vol. 26, Springer.
80. Bang-Jensen, J., Schweser, T. and Stiebitz, M. (2022). Digraphs and variable degeneracy. *SIAM J. Discrete Math.,* 36:578–595.
81. Bárány, I. (1978). A short proof of Kneser's conjecture. *J. Combin. Theory Ser. A,* 25:325–326.
82. Barát, J., Joret, G. and Wood, D.R. (2011). Disproof of the list Hadwiger conjecture. *Electron. J. Combin.,* 18:#P232.
83. Bartnicki, T., Grytczuk, J., Kierstead, H.A. and Zhu, X. (2007). The map-coloring game. *Amer. Math. Monthly,* 114:793–803.
84. Baum, S. (2005). *Färbungen und Homomorphismen von Graphen.* Diploma thesis, Technical University Ilmenau.
85. Baum, S. and Stiebitz, M. (2005). Coloring of graphs without short odd paths between vertices of the same color. Preprint University of Southern Denmark, Department of Mathematics and Computer Science.
86. Beaudou, L., Naserasr, R. and Tardif, C. (2015). Homomorphisms of binary Cayley graphs. *Discrete Math.,* 338:2539–2544.
87. Beck, J. (1978). On 3-chromatic hypergraphs. *Discrete Math.,* 24:127–137.
88. Beck, J. (1991). An algorithmic approach to the Lovász local lemma. *Random Struct. Alg.,* 2:343–365.
89. Behzad, M. (1965). *Graphs and Their Chromatic Numbers.* Ph.D. Thesis, Michigan State University.
90. Beineke, L.W. (1968). On derived graphs and digraphs. In: *Beiträge zur Graphentheorie* (Manebach, Germany, 1967), pages 17–23. B. G. Teubner-Verlagsgesellschaft.
91. Beineke, L.W. (1970). Characterizations of derived graphs. *J. Combin. Theory,* 9:129–135.
92. Beineke, L.W., Wilson, R. and Toft, B. (2023). *Milestones in Graph Theory.* Manuscript submitted for publication
93. Bellitto, T., Bousquet, N., Kabela, A, and Pierron, T. (2023). The smallest 5-chromatic tournament. Manuscript arXive:2210.09936
94. Benedict, J.M. and Chinn, P.Z. (1978). On graphs having prescribed clique number, chromatic number, and maximum degree. In: *Theory and Applications of Graphs* (Michigan, 1976), pages 132–140. Lecture Notes in Math., Vol. 642, Springer.
95. Bensmail, J., Harutyunyan, A. and Le, N. (2018). List coloring digraphs. *J. Graph Theory,* 87:492–508.
96. Benzer, S. (1959). On the topology of the genetic fine structure. *Proc. Nat. Acad. Sci. U.S.A.,* 45:1607–1620.
97. Berge, C. (1961). Färbung von Graphen, deren sämtliche bzw. deren ungerade Kreise starr sind. *Wiss. Z. Martin-Luther-Univ. Halle-Wittenberg Math.-Natur. Reihe,* 10:114.
98. Berge, C. (1963). Some classes of perfect graphs. In: *Six Papers on Graph Theory,* pages 1–21. Indian Statistical Institute, Calcutta.
99. Berge, C. (1978). *Fractional Graph Theory.* Lecture Notes, Vol. 1, Indian Statistical Institute, Mc Millan, Calcutta.
100. Berge, C. (1982). k-optimal partitions of a directed graph. *European J. Combin.,* 3:97–101.
101. Berge, C. (1994). About vertex-critical non-bicolorable hypergraphs. *Australas. J. Combin.,* 9:211–220.
102. Berge, C. (1997). Motivations and history of some of my conjectures. *Discrete Math.,* 165/166:61–70.
103. Berge, C. and Duchet, P. (1983). *Problème,* Séminaire MSH, Paris.
104. Bernshteyn, A. (2016). The asymptotic behavior of the correspondence chromatic number. *Discrete Math.,* 339:2680–2692.

105. Bernshteyn, A. (2017). The local cut lemma. *European J. Combin.,* 63:95–114.
106. Bernshteyn, A. (2019). The Johannson-Molloy theorem for DP-coloring. *Random Struct. Alg.,* 54:653–664.
107. Bernshteyn, A., Brazelton, T., Cao, R. and Kang, A. (2023). Counting colorings of triangle-free graphs. *J. Combin. Theory Ser. B,* 161:86–108.
108. Bernshteyn, A. and Kostochka, A.V. (2018). Sharp Dirac's theorem for DP-critical graphs. *J. Graph Theory,* 88:521–546.
109. Bernshteyn, A. and Kostochka, A.V. (2019). DP-colorings of hypergraphs. *European J. Combin.,* 78:134–146.
110. Bernshteyn, A., Kostochka, A.V., and Pron, S. (2017). On DP-coloring of graphs and multigraphs. *Sib. Math. J.,* 58:28–36.
111. Bernshteyn, A., Kostochka, A.V., and Zhu, X. (2020). Fractional DP-colorings of sparse graphs. *J. Graph Theory,* 93:203–221.
112. Bernshteyn, A. and Lee, E. (2023). Weak degeneracy of graphs. *J. Graph Theory,* 103:607–634.
113. Bernstein, F. (1908). Zur Theorie der trigonometrischen Reihen. *Leipz. Ber.,* 60:325–338.
114. Beutelspacher, A. and Hering, P.-R. (1984). Minimal graphs for which the chromatic number equals the maximal degree. *Ars Combin.,* 18:201–216.
115. Bielecki, A. (1948). Problem 56. *Colloq. Math.,* 1:333.
116. Bigalke, H.-G. (1988). *Heinrich Heesch: Kristallgeometrie, Parkettierungen, Vierfarbenforschung*. Vita Mathematica Band 3, Birkhäuser Verlag.
117. Biggs, N. (2003). W.T. Tutte 1917-2002. In: *Surveys in Combinatorics, 2003*, pages 1–6. London Math. Soc. Lecture Notes Series, Vol. 307, Cambridge University Press.
118. Birkhoff, G.D. (1913). The reducibility of maps. *Amer. J. Math.,* 35:114–128.
119. Biró, M., Hujter, M. and Tuza, Zs. (1992). Precoloring extension. I. Interval graphs. *Discrete Math.,* 100:267–279.
120. Björner, A. (1995). Topological metheods. In: *Handbook of Combinatorics,* Vol. 2, pages 1819–1872. Elsevier.
121. Björner, A. and de Longueville, M. (2003). Neighborhood complexes of stable Kneser graphs. *Combinatorica,* 23:23–34.
122. Blasiak, J. (2007). A special case of Hadwiger's conjecture. *J. Combin. Theory Ser. B,* 97:1056–1073.
123. Bodlaender, H.L. (1991). On the complexity of some coloring games. *Intern. J. Found. Comput. Sci.,* 2:133–147.
124. Böhme, T., Broersma, H.J., Göbel, F., Kostochka, A.V. and Stiebitz, M. (1997). Spanning trees with pairwise nonadjacent endvertices. *Discrete Math.,* 170:219–222.
125. Böhme, T., Mohar, B. and Stiebitz, M. (1999). Dirac's map-color theorem for choosability. *J. Graph Theory,* 32:327–339.
126. Bokal, D., Fijavž, G., Juvan, M., Kayll, P.M. and Mohar, B. (2004). The circular chromatic number of a digraph. *J. Graph Theory,* 46:227–240.
127. Bollobás, B. (1978). Chromatic number, girth and maximal degree. *Discrete Math.,* 24:311-314.
128. Bollobás, B. (1984). The evolution of sparse graphs. In: *Graph Theory and Combinatorics* (Conference in honor of Paul Erdős, Cambridge, 1983), pages 36–58. Academic Press.
129. Bollobás, B., Catlin, P.A. and Erdős, P. (1980). Hadwiger's conjecture is true for almost every graph. *European J. Combin.,* 1:195–199.
130. Bollobás, B., Erdős, P. and Szemerédi, E. (1975). On complete subgraphs of r-chromatic graphs. *Discrete Math.,* 13:97-107.
131. Bollobás, B. and Harary, F. (1975). Point arboricity critical graphs exist. *J. London Math. Soc.,* 2:97–102.
132. Bollobás, B. and Harris, A.J. (1985). List-colourings of graphs. *Graphs Combin.,* 1:115–127.
133. Bollobás, B. and Hind, H.R. (1989). A new upper bound for the list chromatic number. *Discrete Math.,* 74:65–75.

134. Bollobás, B. and Manvel, B. (1979). Optimal vertex partitions. *Bull. London Math. Soc.,* 11:113–116.
135. Bollobás, B. and Thomason, A. (1979/2006). Set colourings of graphs. *Discrete Math.,* 25:21–26 and 306:948-952.
136. Bonamy, M., Charbit, P., Defrain, O., Joret, G., Lagoutte, V., Limouzy, V., Pastor, L. and Sereni, J.S. (2020). Revisiting a theorem by Folkman on graph colouring. *Electron. J. Combin.,* 27:#P1.56.
137. Bonamy, M., Dabrowski, K.K., Feghali, C., Johnson, M. and Paulusma, D. (2021). Recognizing graphs close to bipartite graphs with an application to colouring reconfiguration. *J. Graph Theory,* 98:81–109.
138. Bonamy, M., Kelly, T., Nelson, P. and Postle, L. (2022). Bounding χ by a fraction of Δ for graphs without large cliques. *J. Combin. Theory Ser. B,* 157:263–282.
139. Bonamy, M., Perrett, T. and Postle, L. (2022). Coloring graphs with sparse neighborhoods: bounds and applications. *J. Combin. Theory Ser. B,* 155:278–317.
140. Bondy, J.A. (1976). Disconnected orientations and a conjecture of Las Vergnas. *J. London Math. Soc.,* 2:277–282.
141. Bondy, J.A. (2003). Short proofs of classical theorems. *J. Graph Theory,* 44:159–165.
142. Bondy, J.A. and Hell, P. (1990). A note on the star chromatic number. *J. Graph Theory,* 14:479–482.
143. Bondy, J.A. and Locke, S.C. (1981). Relative lengths of paths and cycles in 3-connected graphs. *Discrete Math.,* 33:111–122.
144. Bondy, J.A. and Murty, U.S.R. (2008). *Graph Theory*. Graduate Texts in Mathematics, Vol. 244, Springer.
145. Bonnet, É., Bourneuf, R., Duron, J., Geniet, C., Thomassé, S. and Trotignon, N. (2023). A tamed family of triangle-free graphs with unbounded chromatic number. Manuscript arXive:2304.04296
146. Bonomo, F., Chudnovsky, M., Maceli, P., Schaudt, O., Stein, M. and Zhong, M. (2018). Three-coloring and list three-coloring of graphs without induced paths on seven vertices. *Combinatorica,* 38:779–801.
147. Borodin, O.V. (1976). On decomposition of graphs into degenerate subgraphs (in Russian). *Diskret. Analiz,* 28:3–12.
148. Borodin, O.V. (1977). A criterion of choosability of graphs with degree size list assignments (in Russian). In: Abstracts of IV. All-Union Conference on Theoretical Cybernetics, Novosibirsk, 1977, pages 127–128.
149. Borodin, O.V. (1979). *Problems of Colouring and of Covering the Vertex Set of a Graph by Induced Subgraphs (in Russian).* Ph.D. Thesis, Novosibirsk State University, Novosibirsk.
150. Borodin, O.V. (1996). Structural properties of plane graphs without adjacent triangles and an application to 3-colorings. *J. Graph Theory,* 21:183–186.
151. Borodin, O.V. (1997). A new proof of Grünbaum's 3-color theorem. *Discrete Math.,* 169:177–183.
152. Borodin, O.V. (2013). Colorings of plane graphs: A survey. *Discrete Math.,* 313:517–539.
153. Borodin, O.V., Dvořák, Z., Kostochka, A. V., Lidický, B. and Yancey, M. (2014). Planar 4-critical graphs with four triangles. *European J. Combin.,* 41:138–151.
154. Borodin, O.V., Glebov, A.N., Montassier, M. and Raspaud, A. (2009). Planar graphs without 5- and 7-cycles and without adjacent triangle are 3-colorable. *J. Combin. Theory Ser. B,* 99:668–673.
155. Borodin, O.V., Glebov, A.N., Raspaud, A. and Salavatipour, M.R. (2005). Planar graphs without cycles of length from 4 to 7 are 3-colorable. *J. Combin. Theory Ser. B,* 93:303–311.
156. Borodin, O., Kim, S-J., Kostochka, A.V. and West, D.B. (2004). Homomorphisms from sparse graphs with large girth. *J. Combin. Theory Ser. B,* 90:147–159.
157. Borodin, O.V. and Kostochka, A.V. (1977). On an upper bound of a graph's chromatic number, depending on the graph's degree and density. *J. Combin. Theory Ser. B,* 23:247–250.
158. Borodin, O.V., Kostochka, A.V., Lidický, B. and Yancey, M. (2014). Short proofs of coloring theorems on planar graphs. *European J. Combin.,* 36:314–321.

159. Borodin, O.V., Kostochka, A.V. and Toft, B. (2000). Variable degeneracy: extensions of Brooks' and Gallai's theorems. *Discrete Math.,* 214:101–112.
160. Borodin, O.V., Kostochka, A.V. and Woodall, D.R. (1997). List edge and list total colourings of multigraphs. *J. Combin. Theory Ser. B,* 71:184–204.
161. Boros, E. and Gurvich, V. (1996). Perfect graphs are kernel-solvable. *Discrete Math.,* 159:35–55.
162. Boros, E. and Gurvich, V. (2006). Perfect graphs, kernels, and cores of cooperative games. *Discrete Math.,* 306:2336–2354.
163. Borowiecki, M., Drgas-Burchardt, E. and Mihók, P. (1995). Generalized list colourings of graphs. *Discuss. Math., Graph Theory,* 15:185–193.
164. Borowiecki, M. and Mihók, P. (1991). Hereditary properties of graphs. In: *Advances in Graph Theory*, pages 41–68. Vishwa International Publisher.
165. Bosse, C. (2019). A note on Hadwiger's conjecture for W_5-free graphs with independence number two. *Discrete Math.,* 342:111595.
166. Brahana, H.R. (1921). Systems of circuits of two-dimensional manifolds. *Ann. of Math.,* 23:144-168.
167. Brandt, S. and Thomassé, S. (2005). Dense triangle-free graphs are four-colorable: A solution to the Erdős–Simonovits problem. Manuscript unpublished
168. Briański, M., Davies, J. and Walczak, B. (2023). Separating polynomial χ-boundedness from χ-boundedness. *Combinatorica,* in press.
169. Broere, I. and Mynhardt, C.M. (1985). Generalized colorings of outerplanar graphs and planar graphs. In: *Graph Theory with Applications to Algorithms and Computer Science* (Kalamazoo. Michigan, 1984), pages 151–161. John Wiley & Sons.
170. Broersma, H., Fomin, F.V., Golovach, P.A. and Paulusma, D. (2013). Three complexity results on coloring P_k-free graphs. *European J. Combin.,* 34:609–619.
171. Broersma, H., Golovach, P.A., Paulusma, D. and Song, J. (2012). Updating the complexity status of coloring graphs without a fixed induced linear forest. *Theoret. Comput. Sci.,* 414:9–19.
172. Broersma, H., Golovach, P.A., Paulusma, D. and Song, J. (2012). Determining the chromatic number of triangle-free $2P_3$-free graphs in polynomial time. *Theoret. Comput. Sci.,* 423:1–10.
173. Brooks, R.L. (1941). On colouring the nodes of a network. *Proc. Cambridge Philos. Soc.,* 37:194–197.
174. Brooks, R.L. (1970). A procedure for dissecting a rectangle into squares, and an example for the rectangle whose sides are in the ratio 2:1. *J. Combin. Theory,* 8:232–243.
175. Brooks, R.L., Smith, C.A.B., Stone, A.H. and Tutte, W.T. (1940). The dissection of rectangles into squares. *Duke Math. J.,* 7:312–340.
176. Brooks, R.L., Smith, C.A.B., Stone, A.H. and Tutte, W.T. (1947). A simple perfect square. *Nederl. Akad. Wetensch. Proc.,* 50:1300–1301.
177. Brooks, R.L., Smith, C.A.B., Stone, A.H. and Tutte, W.T. (1992). Determinants and current flows in electric networks. *Discrete Math.,* 100:291–301.
178. Brown, J.I. (1992). A vertex critical graph without critical edges. *Discrete Math.,* 102:99–101.
179. Brown, J.I. and Corneil, D.G. (1987). On generalized graph colorings. *J. Graph Theory,* 11:87–99.
180. Brown, J.I. and Corneil, D.G. (1991). Graph properties and hypergraph colourings. *Discrete Math.,* 98:81–93.
181. Brown, W.G. and Jung, H.A. (1969). On odd circuits in chromatic graphs. *Acta Math. Hungar.,* 20:129–134.
182. Brown, W.G. and Moon, J.W. (1969). Sur les ensembles de sommets indépendants dans les graphes chromatiques minimaux. *Canad. J. Math.,* 21:274–278.
183. Bryant, V. (1996). A characterisation of some 2-connected graphs and a comment on an algorithmic proof of Brooks' theorem. *Discrete Math.,* 158:279–281.
184. Burling, J.P. (1965). *On coloring problems of families of prototypes.* Ph.D. Thesis, University of Colorado.

185. Burr, S.A., Erdős, P. and Lovász, L. (1976). On graphs of Ramsey type. *Ars Combin.,* 1:167–190.
186. Burstein, M.I. (1976). Critical hypergraphs with minimal number of edges (in Russian). *Bull. Acad. Sci. Georgian SSR,* 83:285–288.
187. Cambie, S. (2021). Hadwiger's conjecture implies a conjecture of Füredi-Gyárfás-Simonyi. Manuscript arXive:2108.10303
188. Cameron, B., Hoàng, C.T. and Sawada, J. (2022). Dichotomizing k-vertex-critical H-free graphs for H of order four. *Discrete Appl. Math.,* 312:106–115.
189. Cameron, K., Goedgebeur, J., Huang, S. and Shi, Y. (2021). k-critical graphs in P_5-free graphs. *Theoret. Comput. Sci.,* 864:80–91.
190. Cameron, K. and Vušković, K. (2020). Hadwiger's conjecture for some hereditary classes of graphs: A survey. *Bulletin EATCS,* 131:pp. 19.
191. Carter, D. (2022). Hadwiger's conjecture with certain forbidden induced subgraphs. (2022). Manuscript arXive:2211.00259
192. Catlin, P.A. (1976). *Embedding Subgraphs and Coloring Graphs Under Extremal Degree Conditions.* Ph.D. Thesis, The Ohio State University.
193. Catlin, P.A. (1978). A bound on the chromatic number of a graph. *Discrete Math.,* 22:81–83.
194. Catlin, P.A. (1979). Hajós' graph-coloring conjecture: variations and counterexamples. *J. Combin. Theory Ser. B,* 26:268–274.
195. Catlin, P.A. (1979). Brooks' graph-coloring theorem and the independence number. *J. Combin. Theory Ser. B,* 27:42–48.
196. Catlin, P.A. and Lai, H.-J. (1995). Vertex arboricity and maximum degree. *Discrete Math.,* 141:37–46.
197. Chandran, L.S., Issac, D., and Zhou, S. (2019). Hadwiger's conjecture is true for squares of 2-trees. *European J. Combin.,* 76:159–174.
198. Chang, G.J., Liu, D.D.-F., Zhu, X. (2013). A short proof for Chen's alternative Kneser coloring lemma. *J. Combin. Theory Ser. A,* 120:159–163.
199. Chang, G.J., Tong, L.-D., Yan, J.-H. and Yeh, H.-G. (2002). A note on the Gallai-Roy-Vitaver theorem. *Discrete Math.,* 256:441–444.
200. Chang, M.S., Kloks, T. and Lee, C.M. (2001). Maximum clique transversal. In: *Graph-Theoretic Concepts in Computer Science* (Boltenhagen, 2001), pages 32–43. Lecture Notes in Comput. Sci., Vol. 2204, Springer.
201. Char, A. and Karthick, T. (2022). Coloring (P_5,4-wheel)-free graphs. *Discrete Math.,* 345:112795.
202. Char, A. and Karthick, T. (2023). On graphs with no induced P_5 and $K_5 - e$. Manuscript arXive:2308.08166
203. Chartrand, G., Geller, D. and Hedetniemi, S. (1968). A generalization of the chromatic number. *Proc. Cambridge Philos. Soc.,* 64:265–271.
204. Chartrand, G., Geller, D. and Hedetniemi, S. (1971). Graphs with forbidden subgraphs. *J. Combin. Theory,* 10:12–41.
205. Chartrand, G. and Kronk, H.V. (1968). Randomly traceable graphs. *SIAM J. Appl. Math.,* 16:696–700.
206. Chartrand, G., Kronk, H.V. and Wall, C.E. (1968). The point-arboricity of a graph. *Israel J. Math.,* 6:169–175.
207. Chen, B.-L., Lih, K.-W. and Wu, P.-L. (1994). Equitable coloring and the maximum degree. *European J. Combin.,* 15:443–447.
208. Chen, C.C., Jin, G. and Koh, K.M. (1997). Triangle-free graphs with large degree. *Combin. Probab. Comput.,* 6:381–396.
209. Chen, G., Jing, G. and Zang, W. (2019). Proof of the Goldberg-Seymour conjecture on edge-colorings of multigraphs. Manuscript arXive:1901.10316.
210. Chen, M. and Raspaud, A. (2010). Homomorphisms from sparse graphs to the Petersen graph. *Discrete Math.,* 310:2705–2713.
211. Chen, P.-A. (2011). A new coloring theorem of Kneser graphs. *J. Combin. Theory Ser. A,* 118:1062–1071.

212. Chen, R. (2022). Graphs with girth at least $2\ell + 1$ and without longer odd holes are 3-colorable. Manuscript arXive:2301.00112
213. Chen, Z., Ma, J. and Zang, W. (2015). Coloring digraphs with forbidden cycles. *J. Combin. Theory Ser. B,* 115:210–223.
214. Chenette, N., Postle, L., Streib, N., Thomas, R. and Yerger, C. (2012). Five-coloring graphs on the Klein bottle. *J. Combin. Theory Ser. B,* 102:1067–1098.
215. Cherkashin, D. and Kozik, J. (2015). A note on random greedy coloring of uniform hypergraphs. *Random Struct. Alg.,* 47:407–413.
216. Cherkashin, D. and Petrov, F. (2019). Regular behaviour of the maximal hypergraph chromatic number. *SIAM J. Discrete Math.,* 34:1326–1333.
217. Chetwynd, A.G. and Häggkvist, R. (1989). A note on list-colorings. *J. Graph Theory,* 13:87–95.
218. Chiu, M.-K., Felsner, S., Scheucher, M., Schröder, F., Steiner, R. and Vogtenhuber, B. (2023). Coloring circle arrangements: New 4-chromatic planar graphs. *European J. Combin.,* 103839.
219. Choi, I., Liu, C.-H. and Oum, S.-I. (2018). Characterization of cycle obstruction sets for improper coloring planar graphs. *SIAM J. Discrete Math.,* 32:1209–1228.
220. Choudum, S.A. (1977). Chromatic bounds for a class of graphs. *Quart. J. Math. Oxford,* 28:257–270.
221. Chudnovsky, M. (2014). The Erdős–Hajnal conjecture– a survey. *J. Graph Theory,* 75:178–190.
222. Chudnovsky, M., Cook, L., Davies, J. and Oum, S. (2023). Reuniting χ-boundedness with polynomial χ-boundedness. Manuscript arXive:2310.11167
223. Chudnovsky, M., Cornuéjols, G., Liu, X., Seymour, P. and Vušković, K. (2005). Recognizing Berge graphs. *Combinatorica,* 25:143–186.
224. Chudnovsky, M. and Fradkin A.O. (2008). Hadwiger's conjecture for quasi-line graphs. *J. Graph Theory,* 59:17–33.
225. Chudnovsky, M. and Fradkin, A.O. (2010). An approximate version of Hadwiger's conjecture for claw-free graphs. *J. Graph Theory,* 63:259–278.
226. Chudnovsky, M., Fradkin, A. and Plumettaz, M. (2013). On the Erdős–Lovász Tihany conjecture for claw-free graphs. Manuscript arXive:1309.1020.
227. Chudnovsky, M., Goedgebeur, J., Schaudt, O. and Zhong, M. (2020). Obstructions for three-coloring and list three-coloring H-free graphs. *SIAM J. Discrete Math.,* 34:431–469.
228. Chudnovsky, M., Karthick, T., Maceli, P. and Maffray, F. (2020). Coloring graphs with no induced five-vertex path or gem. *J. Graph Theory,* 95:527–542.
229. Chudnovsky, M., King, A.D., Plumettaz, M. and Seymour, P. (2013). A local strengthening of Reed's ω, Δ, χ conjecture for quasi-line graphs. *SIAM J. Discrete Math.,* 27:95–108.
230. Chudnovsky, M., Robertson, N., Seymour, P. and Thomas, R. (2006). The strong perfect graph theorem. *Ann. of Math.,* 164:51–229.
231. Chudnovsky, M. and Safra, S. (2008). The Erdős–Hajnal conjecture for bull-free graphs. *J. Combin. Theory Ser. B,* 98:1301–1310.
232. Chudnovsky, M., Schaudt, O., Spirkl, S., Stein, M. and Zhong, M. (2019). Approximately coloring graphs without long induced paths. *Algorithmica,* 81:3186–3199.
233. Chudnovsky, M., Scott, A. and Seymour, P. (2016). Induced subgraphs of graphs with large chromatic number. II. Three steps towards Gyárfás' conjectures. *J. Combin. Theory Ser. B,* 118:109–128.
234. Chudnovsky, M., Scott, A. and Seymour, P. (2017). Induced subgraphs of graphs with large chromatic number. III. Long holes. *Combinatorica,* 37:1057–1072.
235. Chudnovsky, M., Scott, A. and Seymour, P. (2019). Induced subgraphs of graphs with large chromatic number. XI. Orientations. *European J. Combin.,* 76:53–61.
236. Chudnovsky, M., Scott, A. and Seymour, P. (2019). Induced subgraphs of graphs with large chromatic number. XII. Distant stars. *J. Graph Theory,* 92:237–254.
237. Chudnovsky, M., Scott, A. and Seymour, P. (2021). Induced subgraphs of graphs with large chromatic number. V. Chandeliers and strings. *J. Combin. Theory Ser. B,* 150:195–243.

238. Chudnovsky, M., Scott, A., Seymour, P. and Spirkl, S. (2020). Induced subgraphs of graphs with large chromatic number. VIII. Long odd holes. *J. Combin. Theory Ser. B,* 140:84–97.
239. Chudnovsky, M., Scott, A., Seymour, P. and Spirkl, S. (2023). Erdős–Hajnal for graphs with no 5-hole. *Proc. London Math. Soc.,* 126:997–1014.
240. Chudnovsky, M. and Seymour (2005). The structure of claw-free graphs. In: *Surveys in Combinatorics, 2005*, pages 153–171. London Math. Soc. Lecture Notes Series, Vol. 327, Cambridge University Press.
241. Chudnovsky, M. and Seymour (2012). Packing seagulls. *Combinatorica,* 32:251–282.
242. Chudnovsky, M. and Seymour (2023). Proof of a conjecture of Plummer and Zha. *J. Graph Theory,* 103:437–450.
243. Chudnovsky, M., Spirkl, S. and Zhong, M. (2018). Four-coloring P_6-free graphs. I. Extending an excellent precoloring. Manuscript arXive:1802.02282.
244. Chudnovsky, M., Spirkl, S. and Zhong, M (2018). Four-coloring P_6-free graphs. II. Finding an excellent precoloring. Manuscript arXive:1802.02283.
245. Chvátal, V. (1970). The smallest triangle-free 4-chromatic 4-regular graph. *J. Combin. Theory,* 9:93–94.
246. Chvátal, V. (1973). On the computational complexity of finding a kernel. Report No. CRM-300, Centre de Recherches Mathématiques, Université de Montréal.
247. Chvátal, V. (1974). The minimality of the Mycielski graph. In: *Graphs and Combinatorics*, pages 243–247. Lecture Notes in Math., Vol. 406, Springer.
248. Chvátal, V. (1975). On certain polytopes associated with graphs. *J. Combin. Theory Ser. B,* 18:138–154.
249. Chvátal, V., Garey, M.R. and Johnson, D.S. (1978). Two results concerning multicolorings. *Ann. Discrete Math.,* 2:151–154.
250. Clarke, F.H. and Jamison, R.E. (1976). Multicolorings, measures and games on graphs. *Discrete Math.,* 14:241–245.
251. Cockayne, E.J. (1972). Colour classes for r-graphs. *Canad. Math. Bull.,* 15:349–354.
252. Cohen-Addad, V., Hebdige, M., Kráľ, D., Li, Z. and Salgado, E. (2017). Steinberg's conjecture is false. *J. Combin. Theory Ser. B,* 122:452–456.
253. Collins, K.L., Heenehan, M.E. and McDonald, J. (2021). A note on the immersion number of generalized Mycielski graphs. Manuscript: arXive:2105.05724
254. Conley, C., Marks, A. and Tucker-Drop, R. (2016). Brooks' theorem for measurable colorings. *Forum Math. Sigma,* 4: E16.
255. Corsini, T., Deschamps, Q., Feghali, C., Goncalves, D. Langlois, H. and Talon, A. (2023). Partitioning into degenerate graphs in linear time. *European J. Combin.,* 114:103771.
256. Cranston, D.W., Lafayette, H. and Rabern, L. (2020). Coloring (P_5, gem)-free graphs with $\Delta - 1$ colors. *J. Graph Theory,* 101:633–642.
257. Cranston, D.W. and Rabern, L. (2012). Conjectures equivalent to the Borodin-Kostochka conjecture that are a priori weaker. *European J. Combin.,* 44:23–42.
258. Cranston, D.W. and Rabern, L. (2013). Coloring claw-free graphs with $\Delta - 1$ colors. *SIAM J. Discrete Math.,* 27:534–549.
259. Cranston, D.W. and Rabern, L. (2015). Graphs with $\chi = \Delta$ have big cliques. *SIAM J. Discrete Math.,* 29:1792–1814.
260. Cranston, D.W. and Rabern, L. (2015). Brooks' theorem and beyond. *J. Graph Theory,* 80:199–225.
261. Cranston, D.W. and Rabern, L. (2017). The fractional chromatic number of the plane. *Combinatorica,* 37:837–861.
262. Cranston, D.W. and Rabern, L. (2018). Edge lower bounds for list critical graphs, via discharging. *Combinatorica,* 38:1045–1065.
263. Cranston, D.W. and Rabern, L. (2018). Planar graphs are 9/2-colorable. *J. Combin. Theory Ser. B,* 133:32–45.
264. Cristofides, D., Edwards; K. and King, A.D. (2013). A note on hitting maximum and maximal cliques with a stable set. *J. Graph Theory,* 73:354–360.

265. Csiszár, I. and Körner, J. (1981). Graph decomposition: a new key for coding theorems. *IEEE Trans. Inform. Theory,* 27:5–12.
266. Csorba, P. (2005). *Non-tidy Spaces and Graph Colorings.* Doctoral Thesis, ETH Zürich.
267. Csorba, P. (2008). Fold and Mycielskian on homomorphism complexes. *Conributions Discrete Math.,* 3:1–8.
268. Csorba, P. and Osztényi, J. (2010). On the topological lower bound for the multichromatic number. *Discrete Math.,* 310:1334–1339.
269. Culver, E. and Hartke, S.G. (2023). Relation between the correspondence chromatic number and the Alon–Tarsi number. *Discrete Math.,* 346:113347.
270. Davies, E., de Joannis de Verclos, R., Kang, R.J. and Pirot, F. (2020). Coloring triangle-free graphs with local list sizes. *Random Struct. Alg.,* 57:730–744.
271. Davies, E., Kang, R.J. Pirot, F. and Sereni, J.-S. (2020). Graph structure via local occupancy. Manuscript arXive:2003.14361
272. De Bruijn, N.G. and Erdős, P. (1951). A colour problem for infinite graphs and a problem in the theory of infinite relations. *Indag. Math.,* 13:371–373.
273. De Grey, A.D.N.J. (2018). The chromatic number of the plane is at least 5. *Geombinatorics,* 28:18–31.
274. Delcourt, M. and Postle, L. (2022). Reducing linear Hadwiger's conjecture to coloring small graphs. Manuscript arXiv: 2108.01633
275. Deligkas, A., Eiben, E., Gutin, G. Neary, P.R., and Yeo, A. (2023). Complexity Dichotomies for the maximum weighted digraph partition problem. Manuscript arXive: 2307.01109
276. De Loera, J.A., Goaoc, X., Meunier, F., and Mustafa, N.H. (2019). The discrete yet ubiquitous theorems of Carathéodory, Helly, Sperner, Tucker, and Tverberg. *Bull. Am. Math. Soc.,* 56:415–511.
277. De Longueville, M. (2004). 25 years proof of the Kneser conjecture. The advent of topological combinatorics. *EMS Newsletter,* 53:16–19.
278. De Longueville, M. (2012). *A Course in Topological Combinatorics.* Springer Science & Business Media.
279. Demidovich, Y.A. (2022). On some generalizations of the property B-problem of an n-uniform hypergraph. *J. Math. Sci.,* 262:457–475.
280. Descartes, B. (1947). A three colour problem. *Eureka,* 9:24–25.
281. Deuber, W.A., Kostochka, A.V. and Sachs, H. (1996). A shorter proof of Dirac's theorem on the number of edges in chromatically critical graphs (in Russian). *Diskret. Anal. Issled. Oper.,* 3:28–34.
282. DeVos, M. (2007). Mapping planar graphs to odd cycles. Open Problem Garden, Graph Theory, Coloring, Homomorphisms.
283. DeVos, M. and Nurse, K. (2023). A short proof of Seymour's 6-flow theorem. Manuscript arXive:2302.08625
284. De Vries, H.L. (1977). On property B and on Steiner systems. *Math. Z.,* 153:155–159.
285. Dhurandhar, M (1982). Improvement on Brooks' chromatic bound for a class of graphs. *Discrete Math.,* 42:51–56.
286. Diestel, R. (2017). *Graph Theory.* Fifth Edition. Graduate Texts in Mathematics, Vol. 173, Springer.
287. Diks, K., Kowalik, L. and Kurowski, M. (2002). A new 3-color criterion for planar graphs. In: *Graph-Theoretic Concepts in Computer Science, WG 2002,* pages 138–149. Lecture Notes in Comput. Sci., Vol. 2573, Springer.
288. Dilworth, R.P. (1950). A decomposition theorem for partially ordered sets. *Ann. of Math.,* 51:161–166.
289. Dirac, G.A. (1951). Note on the colouring of graphs. *Math. Z.,* 54:347–353.
290. Dirac, G.A. (1952). Map-colour theorems. *Canad. J. Math.,* 4:480–490.
291. Dirac, G.A. (1952). A property of 4-chromatic graphs and some remarks on critical graphs. *J. London Math. Soc.,* 27:85–92.
292. Dirac, G.A. (1952). Some theorems on abstract graphs. *Proc. London Math. Soc.,* 3:69–81.

293. Dirac, G.A. (1953). The structure of k-chromatic graphs. *Fund. Math.,* 40:42–55.
294. Dirac, G.A. (1955). Circuits in critical graphs. *Monatsh. Math.,* 59:178–187.
295. Dirac, G.A. (1956). Map colour theorems related to the Heawood colour formula. *J. London Math. Soc.,* 31:460–471.
296. Dirac, G.A. (1957). Map colour theorems related to the Heawood colour formula (II). *J. London Math. Soc.,* 32:436–455.
297. Dirac, G.A. (1957). Short proof of a map-colour theorem. *Canad. J. Math.,* 9:225–226.
298. Dirac, G.A. (1957). A theorem of R. L. Brooks and a conjecture of H. Hadwiger. *Proc. London Math. Soc.,* 3:161–195.
299. Dirac, G.A. (1960). Trennende Knotenpunktmengen und Reduzibilität abstrakter Graphen mit Anwendung auf das Vierfarbenproblem. *J. Reine Angew. Math.,* 204:116–131.
300. Dirac, G.A. (1961). On rigid circuit graphs. *Abh. Math. Sem. Univ. Hamburg,* 25:71–76.
301. Dirac, G.A. (1963). Percy John Heawood. *J. London Math. Soc.,* 38:263–277.
302. Dirac, G.A. (1964). On the structure of 5- and 6-chromatic abstract graphs. *J. Reine Angew. Math.,* 214/215:43–52.
303. Dirac, G.A. (1974). The number of edges in critical graphs. *J. Reine Angew. Math.,* 268:150–164.
304. Diwan, A.A., Kenkre, S., and Vishwanathan, S. (2013). Circumference, chromatic number and online coloring. *Combinatorica,* 33:319–334.
305. Dobrynin, A.A. and Mel'nikov, L.S. (2006). Counterexamples to Grötzsch–Sachs–Köster's conjecture. *Discrete Math.,* 306:591–594.
306. Doerr, B. (2020). Probabilistic tools for the analysis of randomized optimization heuristics. In: *Theory of Evolutionary Computation: Recent Developments in Discrete Optimization*, pages 1–87. Springer. Also available arXive:1801.06733
307. Dol'nikov, V.L. (1988). A certain combinatorial inequality. *Sib. Math. J.,* 29:375–397.
308. Du, D.Z., Hsu, D.F. and Hwang, F.K. (1993). The Hamiltonian property of consecutive-d digraphs. *Math. Comput. Modelling,* 17:61–63.
309. Dvořák, Z. (2022). Coloring of triangle-free graphs and the Rosenfeld counting method. Lecture Notes, Computer Science Institute of Charles University, Prague.
310. Dvořák, Z. and Feghali, C. (2023). Solution to a problem of Grünbaum on the edge density of 4-critical planar graphs. Manuscript arXive:2311.03132
311. Dvořák, Z., Hu, X. and Sereni, J.-S. (2019). A 4-choosable graph that is not $(8:2)$-choosable. *Advances Combin.,* 5:9 pp..
312. Dvořák, Z., Král', D. and Thomas, R. (2020). Three-coloring triangle-free graphs on surfaces III. Graphs of girth five. *J. Combin. Theory Ser. B,* 145:376–432.
313. Dvořák, Z., Král', D. and Thomas, R. (2021). Three-coloring triangle-free graphs on surfaces V. Coloring planar graphs with distant anomalies. *J. Combin. Theory Ser. B,* 150:244–269.
314. Dvořák, Z., Král', D. and Thomas, R. (2022). Three-coloring triangle-free graphs on surfaces VII. A linear-time algorithm. *J. Combin. Theory Ser. B,* 152:483–504.
315. Dvořák, Z., Lidický, B., Mohar, B. and Postle, L (2017). 5-list-coloring planar graphs with distant precolored vertices. *J. Combin. Theory Ser. B,* 122:311–552.
316. Dvořák, Z. and Pekárek, J. (2022). Characterization of 4-critical triangle-free toroidal graphs. *J. Combin. Theory Ser. B,* 154:336–369.
317. Dvořák, Z. and Postle, L. (2017). Density of 5/2-critical graphs. *Combinatorica,* 37:863–886.
318. Dvořák, Z. and Postle, L. (2018). Correspondence coloring and its application to list-coloring planar graphs without cycles of lengths 4 to 8. *J. Combin. Theory Ser. B,* 129:38–54.
319. Dvořák, Z., Sereni, J.-S. and Volec, J. (2015). Fractional coloring of triangle-free planar graphs. *Electron. J. Combin.,* 22:#P4.11.
320. Edmonds, J. (1960). A combinatorial representation for polyhedral surfaces. *Notices Amer. Math. Soc.,* 7:646.
321. Edmonds, J. (1965). Minimum partition of a matroid into independent subsets. *J. Res. Nat. Bur. Standards Sect. B,* 69:67–72.

322. Edwards, K. and King, A.D. (2014). A superlocal version of Reed's conjecture. *Electron. J. Combin.,* 21:#P4.48.
323. Ehrlich, G., Even, S. and Tarjan, R.E. (1976). Intersection graphs of curves in the plane. *J. Combin. Theory Ser. B,* 21:8–20.
324. Ellingham, M.N. and Goddyn, L. (1996). List edge colourings of some 1-factorable multigraphs. *Combinatorica,* 16:343–352.
325. El-Zahar, M. and Sauer, N.W. (1985). The chromatic number of the product of two 4-chromatic graphs is 4. *Combinatorica,* 5:121–126.
326. Emden-Weinert, T., Hougardy, S. and Kreuter, B. (1998). Uniquely colourable graphs and the hardness of colouring graphs of large girth. *Combin. Probab. Comput.,* 7:375–386.
327. Erdős, P. (1947). Some remarks on the theory of graphs. *Bull. Am. Math. Soc.,* 53:292–294.
328. Erdős, P. (1959). Graph theory and probability. *Canad. J. Math.,* 11:34–38.
329. Erdős, P. (1961). Graph theory and probability II. *Canad. J. Math.,* 13:346–352.
330. Erdős, P. (1962). On circuits and subgraphs of chromatic graphs. *Mathematika,* 9:170–175.
331. Erdős, P. (1963). On a combinatorial problem. *Nordisk Mat. Tidskr.,* 11:5–10.
332. Erdős, P. (1964). On a combinatorial problem II. *Acta Math. Hungar.,* 15:445–447.
333. Erdős, P. (1964). Problem 9. In: *Theory of Graphs and Its Application* (Smolenice, 1963), page 159. Academia Praha and Academic Press.
334. Erdős, P. (1968). Problem 2. In: *Theory of Graphs* (Tihany, Hungary, 1966), page 361. Academic Press.
335. Erdős, P. (1971). Some unsolved problems in graph theory and combinatorial analysis. In: *Combinatorial Mathematics and Its Applications* (Oxford, 1969), pages 97–109. Academic Press.
336. Erdős, P. (1979). Problems and result in graph theory and combinatorial analysis. In: *Graph Theory and Related Topics* (Waterloo, 1977), pages 153–163. Academic Press.
337. Erdős, P. (1980). Problems and results in number theory and graphs theory. *Congr. Numer.,* 27:3–21.
338. Erdős, P. (1985). Problems and results on chromatic numbers in finite and infinite graphs. In: *Graph Theory with Applications to Algorithms and Computer Science*, pages 201–213. John Wiley & Sons.
339. Erdős, P. (1989). On some aspects of my work with Gabriel Dirac. In: *Graph Theory in Memory of G. A. Dirac*, pages 111–116. Ann. Discrete Math., Vol. 41, North-Holland.
340. Erdős, P. (1990). Some of my favourite unsolved problems. In: *A Tribute to Paul Erdős*, pages 467–468. Cambridge University Press.
341. Erdős, P. (1992). Some of my favourite problems in various branches of combinatorics. *Le Matematiche,* 47:231–240.
342. Erdős, P. and Fajtlowicz, S. (1981). On the conjecture of Hajós. *Combinatorica,* 1:141–143.
343. Erdős, P., Faudree, R.J., Rousseau, C.C. and Schelp, R.H. (1999). The number of cycle lengths in graphs of given minimum degree and girth. *Discrete Math.,* 200:55–60.
344. Erdős, P., Füredi, Z., Hajnal. A., Komjáth, P., Rödl, V. and Seress, Á. (1986). Coloring graphs with locally few colors. *Discrete Math.,* 59:21–34.
345. Erdős, P. and Hajnal, A. (1961). On a property of families of sets. *Acta Math. Hungar.,* 12:87–123.
346. Erdős, P. and Hajnal, A. (1964). Some remarks on set theory. IX. Combinatorial problems in measure theory and set theory. *Michigan Math. J.,* 11:107–127.
347. Erdős, P. and Hajnal, A. (1966). On chromatic number of graphs and set-systems. *Acta Math. Hungar.,* 17:61–99.
348. Erdős, P. and Hajnal, A. (1967). On chromatic graphs (Hungarian). *Mat. Lapok.,* 18:1–4.
349. Erdős, P. and Hajnal, A. (1977). On spanned subgraphs of graphs. In: Beiträge zur Graphentheorie und deren Anwendungen (Int. Kolloquium in Oberhof, DDR, 1977), pages 80–96. Ilmenau: Techn. Hochschule.
350. Erdős, P. and Hajnal, A. (1985). On the chromatic number of finite and infinite graphs and hypergraphs. *Discrete Math.,* 53:281–285.

351. Erdős, P. and Hajnal, A. (1989). Ramsey-types theorems. *Discrete Appl. Math.,* 25:37–52.
352. Erdős, P. and Hanani, H. (1963). On a limit theorem in combinatorial analysis. *Publ. Math. Debrecen,* 10:10–13.
353. Erdős, P., Ko, C. and Rado, R. (1961). Intersection theorems for systems of finite sets. *Quart. J. Math. Oxford,* 12:313–320.
354. Erdős, P. and Lovász, L. (1973). Problems and results on 3-chromatic hypergraphs and some related questions. In: *Infinite and Finite Sets* (Keszthely, Hungary, 1973), pages 609–627. Coll. Math. Soc. János Bolyai , Vol. 10, North-Holland.
355. Erdős, P., Rubin, A.L. and Taylor, H. (1979). Choosability in graphs. *Congr. Numer.,* 26:125–157.
356. Erdős, P. and Simonovits, M. (1973). On a valence problem in extremal graph theory. *Discrete Math.,* 5:323–334.
357. Erdős, P. and Spencer, J. (1991). Lopsided Lovász Local Lemma and Latin transversals. *Discrete Appl. Math.,* 30:151–154.
358. Erdős, P. and Stone, A.H. (1946). On the structure of linear graphs. *Bull. Am. Math. Soc.,* 52:1087–1091.
359. Erdős, P. and Szekeres, G. (1935). A combinatorial problem in geometry. *Compos. Math.,* 2:463–470.
360. Esperet, L. (2017). *Graph Colorings, Flows and Perfect Matchings.* Habilitation thesis, Université Greneoble Alpes.
361. Exoo, G. and Goedgebeur, J. (2019). Bounds for the smallest k-chromatic graphs of given girth. *Discrete Math. Theor. Comput. Sci.,* 21:#9.
362. Fajtlowicz, S. (1978). On the size of independent sets in graphs. *Congr. Numer.,* 21:269–274.
363. Fan, K. (1952). A generalization of Tucker's combinatorial lemma with topological applications. *Ann. of Math.,* 56:431–437.
364. Farrugia, A. (2004). Vertex-partitioning into fixed additive induced-hereditary properties is NP-hard. *Electron. J. Combin.,* 11:#R46.
365. Farzad, B. and Molloy, M. (2009). On the edge-density of 4-critical graphs. *Combinatorica,* 29:665–689.
366. Farzad, B., Molloy, M. and Reed, B. (2005). $(\Delta - k)$-critical graphs. *J. Combin. Theory Ser. B,* 93:173–185.
367. Feghali, C. (2024). Dirac's theorem on chordal graphs implies Brooks' theorem. *Discrete Math.,* 347:113791.
368. Feige, U. and Kilian, J. (1998). Zero knowledge and the chromatic number. *J. Comput. System Sci.,* 57:187–199.
369. Fekete, M. (1923). Über die Verteilung der Wurzeln bei gewissen algebraischen Gleichungen mit ganzzahligen Koeffizienten. *Math. Z.,* 17:228–249.
370. Felsner, S., Hurtado, F., Noy, M. and Streinu, I. (2006). Hamiltonicity and colorings of arrangement graphs. *Discrete Appl. Math.,* 154:2470–2483.
371. Fernandez de la Vega, W. (1983). On the maximum density of graphs which have no subcontraction to K^s. *Discrete Math.,* 46:109–110.
372. Finck, H.J. and Sachs, H. (1969). Über eine von H. S. Wilf angegebene Schranke für die chromatische Zahl endlicher Graphen. *Math. Nachr.,* 39:373–386.
373. Fischer, M., Halldórsson, M.M. and Maus, Y. (2023). Fast distributed Brooks' theorem. In: Proceedings of the 2023 Annual ACM-SIAM Symposium on Discrete Algorithms (SODA), pages 2567–2588.
374. Fisher, D. and Ullman, D. (1992). The fractional chromatic number of the plane. *Geombinatorics,* 2:8–12.
375. Fisk, S. (1973). Combinatorial structure on triangulations. I. The structure of four colorings. *Advances Math.,* 11:326–338.
376. Fisk, S. (1978). The nonexistence of colorings. *J. Combin. Theory Ser. B,* 24:247–248.
377. Fleiner, T. (2014). Yet another proof of Brooks' theorem. EGRES, Quick Proof No. 2014-01
378. Fleiner, T. (2018). A note on restricted list edge-colourings. *Combinatorica,* 38:1265–1267.

379. Fleischner, H. and Stiebitz, M. (1992). A solution to a colouring problem of P. Erdős. *Discrete Math.,* 101:39–48.
380. Fleischner, H. and Stiebitz, M. (1997). Some remarks on the cycle plus triangles problem. In: *The Mathematics of Paul Erdős II,* pages 136–142. Algorithms and Combinatorics, Vol. 14, Springer.
381. Folkman, J.H. (1970). An upper bound on the chromatic number of a graph. In: *Combinatorial Theory and its Application* (Balatonfüred, Hungary, 1969), Vol II, pages 437–457. Coll. Math. Soc. János Bolyai, Vol. 4, North-Holland.
382. Foniok, J. and Tardif, C. (2017). Hedetniemi's conjecture and adjoint functors in thin categories. *Appl. Categor. Struct.,* 26:113–128.
383. Fouquet, J.L. and Vanherpe, J.M. (2012). Reed's conjecture on some special classes of graphs. Manuscript arXive:1205.0730
384. Fox, J. and Pach, J. (2012). Coloring K_k-free intersection graphs of geometric objects in the plane. *European J. Combin.,* 33:853–866.
385. Frank, A. (1995). Connectivity and network flows. In: *Handbook of Combinatorics*, Vol. I, pages 111–178. North-Holland.
386. Frank, A. (2011). *Connections in Combinatorial Optimization.* Oxford Lecture Series in Mathematics and its Application 38, Oxford University Press.
387. Frank, A. and Gyárfás, A. (1978). How to orient the edges of a graph? In: *Combinatorics* (Keszthely, Hungary, 1976), Vol 1, pages 353–364. Coll. Math. Soc. János Bolyai, Vol. 18, North-Holland.
388. Franklin, P. (1934). A six color problem. *J. Math. Phys.,* 13:363–369.
389. Franklin, P. (1939). The four color problem. *Scripta Mathematica,* 6:149–156, 197–210.
390. Freund, R.M. and Todd, M.J. (1981). A constructive proof of Tucker's combinatorial lemma. *J. Combin. Theory Ser. A,* 30:321–325.
391. Frieze, A. and Luczak, T. (1992). On the independence and chromatic numbers of random regular graphs. *J. Combin. Theory Ser. B,* 54:123-132.
392. Frieze, A. and Mubayi, D. (2013). Coloring simple hypergraphs. *J. Combin. Theory Ser. B,* 103:767-794.
393. Fritsch, R. and Fritsch, G. (1998). *The Four-Color Theorem: History, Topological Foundations, and Idea of Proof.* Springer, Translated from the 1994 German original by Julie Peschke.
394. Fulkerson, D.R. (1971). Blocking and anti-blocking pairs of polyhedra. *Math Program.,* 1:168-194.
395. Fulkerson, D.R. (1973). On the perfect graph theorem. In: *Mathematical Programming*, pages 69–76. Academic Press.
396. Gale, D. and Shapley, L.S. (1962). College admissions and the stability of marriage. *Amer. Math. Monthly,* 69:9–15.
397. Gallai, T. (1963). Kritische Graphen I. *Magyar Tud. Akad. Mat. Kutató Int. Közl.,* 8:165–192.
398. Gallai, T. (1963). Kritische Graphen II. *Magyar Tud. Akad. Mat. Kutató Int. Közl.,* 8:373–395.
399. Gallai, T. (1964). Critical graphs. In: *Theory of Graphs and its Applications* (Smolenice, 1963), pages 43–45. Academia Praha and Academic Press.
400. Gallai, T. (1968). On directed paths and circuits. In: *Theory of Graphs* (Tihany, Hungary, 1966), pages 115–118. Academic Press.
401. Gallai, T. and Milgram, A.N. (1960). Verallgemeinerung eines graphentheoretischen Satzes von Rédei. *Acta Sci. Math.,* 21:181–186.
402. Galvin, F. (1995). The list chromatic index of a bipartite multigraph. *J. Combin. Theory Ser. B,* 63:153–158.
403. Gao, J. and Ma, J. (2023). Tight bounds towards a conjecture of Gallai. *Combinatorica,* 43:447–453.
404. Gao, W. and Postle, L. (2018). On the minimal edge density of K_4-free 6-critical graphs. Manuscript arXiv:1811.02940

405. Gardner, M. (1960). Mathematical games. *Sci. Amer.,* 202:174–188.
406. Gardner, M. (1981). Mathematical games. *Sci. Amer.,* 244:18.
407. Garey, M.R. and Johnson, D.S. (1979). *Computers and Intractability: A Guide to the Theory of* NP*-completeness.* Freeman.
408. Garey, M.R., Johnson, D.S. and Stockmeyer, L. (1976). Some simplified NP-complete graph problems. *Theoret. Comput. Sci.,* 1:237–267.
409. Gasparian, G.S. (1996). Minimal imperfect graphs: a simple approach. *Combinatorica,* 16:209–212.
410. Gebauer, H. (2013). On the construction of 3-chromatic hypergraphs with few edges. *J. Combin. Theory Ser. A,* 120:1483–1490.
411. Geetha, J., Narayanan, N. and Somasundaram, K. (2023). Total colourings–a survey. *AKCE Int. J. Graphs Combin.,* in press.
412. Geller, D. (1976). r-tuple colorings of uniquely colorable graphs. *Discrete Math.,* 16:9–12.
413. Gerencsér, L. (1965). Szinezési problémákrol (On coloring problems). *Mat. Lapok.,* 16:274–277.
414. Gimbel, J., Kündgen, A., Li, B. and Thomassen, C. (2019). Fractional coloring methods with applications to degenerate graphs and graphs on surfaces. *SIAM J. Discrete Math.,* 33:1415-1430.
415. Gimbel, J. and Thomassen, C. (1997). Coloring graphs with fixed genus and girth. *Trans. Amer. Math. Soc.,* 349:4555–4564.
416. Gimbel, J. and Thomassen, C. (2000). Coloring triangle-free graphs with fixed size. *Discrete Math.,* 219:275–277.
417. Goddard, W. and Lyle, J. (2011). Dense graphs with small clique number. *J. Graph Theory,* 66:319–331.
418. Godsil, C. and Royle, G. (2001). *Algebraic Graph Theory.* Graduate Texts in Mathematics, Vol. 207, Springer.
419. Goedgebeur, J., Huang, S., Ju, Y. and Merkel, O. (2023). Colouring graphs with no induced six-vertex path or diamond. *Theoret. Comput. Sci.,* 941:278–299.
420. Goldberg, M.K. (1973). On multigraphs of almost maximal chromatic class (in Russian). *Diskret. Analiz,* 23:3–7.
421. Goldberg, M.K. (1984). Edge-coloring of multigraphs: Recoloring technique. *J. Graph Theory,* 8:123–137.
422. Golovach, P., Johnson, M., Paulusma, D. and Song, J. (2017). A survey on the computational complexity of coloring graphs with forbidden subgraphs. *J. Graph Theory,* 84:331–363.
423. Golowich, N. (2016). The m-degenerate chromatic number of a digraph. *Discrete Math.,* 339:1734–1743.
424. Golumbic, M.C. (2004). *Algorithmic Graph Theory and Perfect Graphs.* Elsevier.
425. Gould, R., Larsen, V. and Postle, L. (2022). Structure in sparse k-critical graphs. *J. Combin. Theory Ser. B,* 156:194–222.
426. Graham, R.L. and Nešetřil, J. (1997). Ramsey theory in the work of Paul Erdős. In: *The Mathematics of Paul Erdős II,* pages 193–209. Algorithms and Combinatorics, Vol. 14, Springer.
427. Gravier, S. (1996). A Hajós-like theorem for list coloring. *Discrete Math.,* 152:299–302.
428. Gravier, S., Hoàng, C.T. and Maffray, F. (2003). Coloring the hypergraph of maximal cliques of a graph with no long path. *Discrete Math.,* 272:285–290.
429. Gravin, N.V. and Karpov, D.V. (2012). On proper colorings of hypergraphs. *J. Math. Sci.,* 184:595–600.
430. Greene, J.E. (2002). A new short proof of Kneser's conjecture. *Amer. Math. Monthly,* 109:918–920.
431. Greenwell, D. and Lovász, L. (1974). Applications of product colouring. *Acta Math. Hungar.,* 25:335–340.
432. Grötschel, M., Lovász, L. and Schrijver, A. (1981). The ellipsoid method and its consequences in combinatorial optimization. *Combinatorica,* 1:169–197.
433. Grötschel, M., Lovász, L. and Schrijver, A. (1984). Polynomial algorithms for perfect graphs. *Ann. Discrete Math.,* 21:325–356.

434. Grötzsch, H. (1958/59). Ein Dreifarbensatz für dreikreisfreie Netze auf der Kugel. *Wiss. Z. Martin-Luther Univ. Halle-Wittenberg, Math.-Nat. Reihe,* 8:109–120.
435. Grünbaum, B. (1963). Grötzsch's theorem on 3-colorings. *Michigan Math. J.,* 10:303–310.
436. Grünbaum, B. (1988). The edge-density of 4-critical planar graphs. *Combinatorica,* 8:137–139.
437. Gupta, R.P. (1963). A note on a theorem on k-coloring the vertices of a graph. In: *Six Papers on Graph Theory,* pages 30–32. Indian Statistical Institute, Calcutta.
438. Gupta, R.P. (1967). *Studies in the Theory of Graphs.* Ph.D. Thesis, Tata Institute of Fundamental Research, Bombay.
439. Gupta, R.P. (1978). On the chromatic index and the cover index of a multigraph. In: *Theory and Applications of Graphs* (Michigan, 1976), pages 204–215. Lecture Notes in Math., Vol. 642, Springer.
440. Gupta, U.K. and Pradhan, D. (2021). Borodin–Kostochka's conjecture for (P_5, C_4)-free graphs. *J. Appl. Math. Comp.,* 65:877–884.
441. Guthrie, F. (1880). Note on the colouring of maps. *Proc. Roy. Soc. Edinburgh,* 10:727–728.
442. Gyárfás, A. (1975). On Ramsey covering-numbers. In: *Infinite and Finite Sets* (Keszthely, Hungary, 1973), pages 801–816. Coll. Math. Soc. János Bolyai Vol. 10, North-Holland.
443. Gyárfás, A. (1987). Problems from the world surrounding perfect graphs. *Zastos. Mat.,* 19:413–441.
444. Gyárfás, A. (1992). Graphs with k odd cycle lengths. *Discrete Math.,* 103:41–48.
445. Gyárfás, A., Jensen, T. and Stiebitz, M. (2004). On graphs with strongly independent color-classes. *J. Graph Theory,* 46:1–14.
446. Gyárfás, A. and Lehel, J. (1988). On-line and first fit colorings of graphs. *J. Graph Theory,* 12:217–227.
447. Gyárfás, A., Szemerédi, E. and Tuza, Zs. (1980). Induced subtrees in graphs of large chromatic number. *Discrete Math.,* 30:235–244.
448. Hadwiger, H. (1943). Über eine Klassifikation der Streckenkomplexe. *Vierteljschr. Naturforsch. Ges. Zürich,* 88:133–142.
449. Häggkvist, R. (1989). Towards a solution of the Dinitz problem? *Discrete Math.,* 75:247–251.
450. Häggkvist, R., Hell, P., Miller, D.J. and Neumann Lara, V. (1988). On multiplicative graphs and the product conjecture. *Combinatorica,* 8:63–74.
451. Häggkvist, R. and Janssen, J. (1997). New bounds on the list-chromatic index of the complete graph and other simple graphs. *Combin. Probab. Comput.,* 6:295–313.
452. Hahn, G. and Tardif, C. (1997). Graph homomorphisms: structure and symmetry. In: *Graph Symmetry,* pages 107–166. NATO ASI Series (Series C: Mathematical and Physical Sciences), Vol. 497, Springer.
453. Hajebi, S., Li, Y. and Spirkl, S. (2021). List-5-coloring graphs with forbidden induced subgraphs. Manuscript arXive:2105.01787
454. Hajiabolhassan, H. (2009). On colorings of graph powers. *Discrete Math.,* 309:4299–4305.
455. Hajiabolhassan, H. and Taherkhani, A. (2010). Graph powers and graph homomorphisms. *Electron. J. Combin.,* 17:#R17.
456. Hajnal, A. A theorem on k-saturated graphs. (1965). *Can J. Math.,* 10:720–724.
457. Hajnal, A. and Surányi, J. (1958). Über die Auflösung von Graphen in vollständige Teilgraphen. *Ann. Univ. Sci. Budapest Eötvös Sect. Math.,* 1:113–121.
458. Hajnal, A. and Szemerédi, E. (1970). Proof of a conjecture of P. Erdős. In: *Combinatorial Theory and its Application* (Balatonfüred, Hungary, 1969), Vol. II, pages 601–623. Coll. Math. Soc. János Bolyai, Vol. 4, North-Holland.
459. Hajnal, P. (1983). Partition of graphs with condition on the connectivity and minimum degree. *Combinatorica,* 3:95–99.
460. Hajós, G. (1956). Über eine Art von Graphen. *Intern. Math. Nachr.,* 47/48:65.
461. Hajós, G. (1961). Über eine Konstruktion nicht n-färbbarer Graphen. *Wiss. Z. Martin Luther Univ. Halle-Wittenberg, Math.-Natur. Reihe,* 10:116–117.
462. Hakimi, S.L. (1965). On the degrees of the vertices of a directed graph. *J. Franklin Inst.,* 279:290–308.

463. Hall, P. (1935). On representatives of subsets. *J. London Math. Soc.,* 10:26–30.
464. Halldórsson, M.M. and Lau, H.C. (1997). Low-degree graph partitioning via local search with application to constraint satisfaction, max cut, and coloring. *J. Graph Algorithms and Applications,* 1:1–13.
465. Hamaker, Z. and Vatter, V. (2022). Three coloring via triangle counting. Manuscript arXiv:2203.08136
466. Han, M., Li, J., Wu, Y. and Zhang, C. (2018). Counterexamples to Jaeger's circular flow conjecture. *J. Combin. Theory Ser. B,* 131:1–11.
467. Hanson, D., Robinson, G.C. and Toft, B. (1986). Remarks on the graph colour theorem of Hajós. *Congr. Numer.,* 55:69–76.
468. Harary, F. (1953–54). On the notion of balance of a signed graph. *Michigan Math. J,* 2:143–146. Addendum ibid., preceding p. 1.
469. Harary, F. (1985). Conditional colorability in graphs. In: *Graphs and Applications*, pages 127–136. John Wiley & Sons.
470. Harary, F. and Tutte, W.T. (1965). A dual form of Kuratowski's theorem. *Canad. Math. Bull.,* 8:17–20.
471. Hare, D.R. (2019). Tools for counting odd cycles in graphs. *J. Combin. Theory Ser. B,* 139:163–192.
472. Harris, D.G. (2019). Some results on chromatic number as a function of triangle count. *SIAM J. Discrete Math.,* 33:546–563.
473. Harutyunyan, A., Le, T.-N., Newmann, A. and Thomassé, S. (2019). Coloring dense digraphs. *Combinatorica,* 39:1021–1053.
474. Harutyunyan, A. and Mohar, B. (2011). Gallai's theorem for list coloring of digraphs. *SIAM J. Discrete Math.,* 25:170–180.
475. Harutyunyan, A. and Mohar, B. (2011). Strengthened Brooks' theorem for digraphs of girth at least three. *Electron. J. Combin.,* 18:#P195.
476. Harutyunyan, A. and Mohar, B. (2012). Two results on the digraph chromatic number. *Discrete Math.,* 312:1823–1826.
477. Harutyunyan, A. and Mohar, B. (2017). Planar digraphs of digirth five are 2-colorable. *J. Graph Theory,* 84:408–427.
478. Hasse, M. (1965). Zur algebraischen Begründung der Graphentheorie I. *Math. Nachr.,* 28:275–290.
479. Havel, I. (1969). On a conjecture of B. Grünbaum. *J. Combin. Theory Ser. B,* 7:184–186.
480. Havet, F., Picasarri-Arrieta, L. and Rambaud, C. (2023). On the minimum number of arcs in 4-dicritical oriented graphs. In: *Graph-Theoretic Concepts in Computer Science, WG 2023*, pages 376–387. Lecture Notes in Comput. Sci., Vol. 14093, Springer.
481. Havet, F. and Thomassé, S. (2000). Oriented Hamiltonian paths in tournaments: A proof of Rosenfeld's conjecture. *J. Combin. Theory Ser. B,* 78:243–273.
482. Haxell, P. (2001). A note on vertex list colouring. *Combin. Probab. Comput.,* 10:345–347.
483. Haxell, P. and MacDonald, C. (2023). Large cliques in graphs with high chromatic number. *Discrete Math.,* 346:113181.
484. Haxell, P. and McDonald, J. (2012). On characterizing Vizing's edge colouring bound. *J. Graph Theory,* 69:160–168.
485. Haxell, P. and Naserasr, R. (2023). A note on Δ-critical graphs. *Graphs Combin.,* 39:#101.
486. He, X. and Wigderson, Y. (2021). Hedetniemi's conjecture is asymptotically false. *J. Combin. Theory Ser. B,* 146:485–494.
487. Heawood, P.J. (1890). Map colour theorem. *Quart. J. Pure Appl. Math.,* 24:332–338.
488. Heawood, P.J. (1898). On the four-color map theorem. *Quart. J. Pure Appl. Math.,* 29:270–285.
489. Hedetniemi, S.T. (1966). Homomorphisms of graphs and automata., University of Michigan, Technical Report 03105-44-T.
490. Hedetniemi, S.T. (1968). On partitioning planar graphs. *Canad. Math. Bull.,* 11:203–211.
491. Heesch, H. (1969). *Untersuchungen zum Vierfarbenproblem,* BI Hochschulskripten Band 810.

492. Hefetz, D. (2011). On two generalizations of the Alon-Tarsi polynomial method. *J. Combin. Theory Ser. B,* 101:403–414.
493. Heffter, L. (1891). Über das Probem der Nachbargebiete. *Math. Ann.,* 38:477–508.
494. Hell, P. and Nešetřil, J. (1990). On the complexity of H-coloring. *J. Combin. Theory Ser. B,* 48:92–110.
495. Hell, P. and Nešetřil, J. (2004). *Graphs and Homomorphisms*. Oxford Lecture Series in Mathematics and its Application, Vol. 28, Oxford University Press.
496. Henning, M.A. and Yeo, A. (2015). On 2-colorings of hypergraphs. *J. Graph Theory,* 80:112–135.
497. Henning, M.A. and Yeo, A. (2018). Not-all-equal 3-SAT and 2-colorings of 4-regular 4-uniform hypergraphs. *Discrete Math.,* 341:2285–2292.
498. Hilton, A.J.W. and Johnson, P D. (1999). The Hall number, the Hall index, and the total Hall number of a graph. *Discrete Appl. Math.,* 94:227–245.
499. Hilton, A.J.W., Rado, R. and Scott, S.H. (1973). A (¡5)-colour theorem for planar graphs. *Bull. London Math. Soc.,* 5:302–306.
500. Hladký, J., Král', D. and Schauz, U. (2010). Brooks' theorem via the Alon-Tarsi theorem. *Discrete Math.,* 310:3426–3428.
501. Hoàng, C.T., Kamiński, M., Sawada, J. and Shu, X. (2010). Deciding k-colorability of P_5-free graphs in polynomial time. *Algorithmica,* 57:74–81.
502. Hochberg, R. and O'Donnell, P. (1993). A large independent set in the unit distance graph. *Geombinatorics,* 3:83–84.
503. Hochstättler, W., Schröder, F. and Steiner, R. (2019). On the complexity of digraph colourings and vertex arboricity. *Discrete Math. Theor. Comput. Sci.,* 22:#4.
504. Holyer, I. (1981). The NP-completeness of edge-coloring. *SIAM J. Comput.,* 10:718–720.
505. Hopcroft, J.E. and Tarjan, R.E. (1973). Efficient algorithms for graph manipulation. *Commun. ACM,* 16:372–378.
506. Hoshino, R. and Kawarabayashi, K. (2015). The edge density of critical digraphs. *Combinatorica,* 35:619–631.
507. Huang, S. (2016). Improved complexity results on k-coloring P_t-free graphs. *European J. Combin.,* 51:336–346.
508. Hungerford, T.W. (1974). *Algebra*. Graduate Texts in Mathematics, Vol. 73, Springer.
509. Hurley, E. and Pirot, F. (2020). Coloring locally sparse graphs with the first moment method. Manuscript arXive:2109.15215
510. Hutchinson, J.P. (1976). On coloring maps made from Eulerian graphs. *Congr. Numer.,* 15:343–354.
511. Hutchinson, J.P. (1995). Three-coloring graphs embedded on surfaces with all faces even-sided. *J. Combin. Theory Ser. B,* 65:139–155.
512. Imrich, W. and Klavžar, S. (2000). *Product Graphs: Structure and Recognition*. Wiley-Interscience Series in Discrete Mathematics and Optimization, John Wiley & Sons.
513. Jaeger, F. (1978). Sur les graphes couverts par leurs bicycles et la conjecture des quatre couleurs. In: *Problémes Combinatoires et Theorie des Graphes,* pages 243–247. Editions du Centre National de la Recherche Scientifique.
514. Jaeger, F. (1979). Flows and generalized coloring theorems in graphs. *J. Combin. Theory Ser. B,* 26:205–216.
515. Jaeger, F. (1981). A constructive approach to the critical problem for matroids. *European J. Combin.,* 2:137–144.
516. Jaeger, F. (1984). On circular flows in graphs. In: *Finite and Infinite Sets* (Eger, Hungary, 1981), pages 391–402. Coll. Math. Soc. János Bolyai, Vol. 37, North-Holland.
517. Jaeger, F., Linial, N., Payan, C. and Tarsi, M. (1992). Group connectivity of graphs – A non-homogenous analogue of nowhere-zero flow properties. *J. Combin. Theory Ser. B,* 56:165–182.
518. Jakobsen, I.T. (1971). *On Certain Homomorphism-Properties of Graphs With Applications to the Conjecture of Hadwiger.* Ph.D. Thesis, University of London. Various Publ. Ser. 15 (Aarhus University).

519. Jakobsen, I.T. (1971). A homomorphism theorem with an application to the conjecture of Hadwiger. *Studia Sci. Math. Hungar.,* 6:151–160.
520. Jakobsen, I.T. (1973). Some remarks on the chromatic index of a graph. *Arch. Math. (Basel),* 24:440–448.
521. Jakovac, M. and Štesl, D. (2023). On game chromatic vertex-critical graphs. *Bull. Malays. Math. Sci. Soc,* 46:#27.
522. Jamall, M.S. (2011). A Brooks' theorem for triangle-free graphs. Manuscript arXive:1106.1958
523. Jamall, M.S. (2011). A coloring algorithm for triangle-free graphs. Manuscript arXive:1101.5721
524. Javdekar, M. (1980). Note on Choudum's "chromatic bounds for a class of graphs". *J. Graph Theory,* 4:265–267.
525. Jensen, T.R. (2002). Dense critical and vertex-critical graphs. *Discrete Math.,* 258:63–84.
526. Jensen, T.R. (2017). Grassmann homomorphism and Hajós-type theorems. *Linear Algebra Appl.,* 522:140–152.
527. Jensen, T.R. and Royle, G.F. (1995). Small graphs with chromatic number 5: a computer search. *J. Graph Theory,* 19:107–116.
528. Jensen, T.R. and Royle, G.F. (1999). Hajós constructions of critical graphs. *J. Graph Theory,* 30:37–50.
529. Jensen, T.R. and Thomassen, C. (2000). The color space of a graph. *J. Graph Theory,* 34:234–245.
530. Jensen, T.R. and Toft, B. (1995). *Graph Coloring Problems.* Wiley-Interscience Series in Discrete Mathematics and Optimization, John Wiley & Sons.
531. Jensen, T.R, and Toft, B. (1995). Choosability versus chromaticity – the plane unit distance graph has a 2-chromatic subgraph of infinite list-chromatic number. *Geombinatorics,* 5:45–64.
532. Jiang, T. (2001). Small odd cycles in 4-chromatic graphs. *J. Graph Theory,* 37:115–117.
533. Jin, G. (1995). Triangle-free four-chromatic graphs. *Discrete Math.,* 145:151–170.
534. Johansson, A. (1996). Asymptotic choice number for triangle free graphs. Unpublished Manuscript.
535. Johansson, A. (1996). The choice number of sparse graphs. Unpublished Manuscript.
536. Johnson, A., Holroyd, F.C. and Stahl, S. (1997). Multichromatic numbers, star chromatic numbers and Kneser graphs. *J. Graph Theory,* 26:137–145.
537. Johnson, D.S. (1974). Worst case behavior of graph colouring algorithms. *Congr. Numer.,* 10:513–527.
538. Johnson, D.S. (1974). Approximation algorithms for combinatorial problems. *J. Comp. System Sci.,* 9:256–278.
539. Johnson, D.S. (1976). On property B_r. *J. Combin. Theory Ser. B,* 20:64–66.
540. Johnson, P.D. (1994). The choice number of the plane. *Geombinatorics,* 3:122–128.
541. Jones, R.P. (1975). Brooks' theorem for hypergraphs. *Congr. Numer.,* 15:379–384.
542. Juvan, M., Mohar, B. and Škrekovski, R. (1998). List-total colourings of graphs. *Combin. Probab. Comput.,* 7:181–188.
543. Juvan, M., Mohar, B. and Thomas, R. (1999). List edge-colorings of series-parallel graphs. *Electron. J. Combin.,* 6:#R42.
544. Kagno, I.N.. (1936). The triangulation of surfaces and the Heawood color formula. *J. Math. Phys.,* 15:179-186.
545. Kahle, M. and Martinez-Figueroa, F. (2020). The chromatic number of random Borsuk graphs. *Random Struct. Alg.,* 56:838–850.
546. Kahn, J. and Seymour, P.D. (1992). A fractional version of the Erdős-Faber-Lovász conjecture. *Combinatorica,* 12:155–160.
547. Kaiser, T. and Stehlík, M. (2015). Colouring quadrangulations of projective spaces. *J. Combin. Theory Ser. B,* 113:1–17.
548. Kaiser, T. and Stehlík, M. (2019). Generalised Mycielski graphs and the Borsuk–Ulam theorem. *Electron. J. Combin.,* 26:#P4.1.
549. Kaiser, T. and Stehlík, M. (2020). Edge-critical subgraphs of Schrijver graphs. *J. Combin. Theory Ser. B,* 144:191–196.

550. Kaiser, T. and Stehlík, M. (2020). Edge-critical subgraphs of Schrijver graphs II: The general case. *J. Combin. Theory Ser. B,* 152:453–482.
551. Kamiński, M. and Lozin, V.V. (2007). Coloring edges and vertices of graphs without short or long cycles. *Contrib. Discrete Math.,* 2:61–66.
552. Kang, D., Kelly, T., Kühn, D., Methuku, A. and Osthus, D. (2023). A proof of the Erdős–Faber–Lovász conjecture. *Ann. of Math.,* 189:537–618.
553. Kang, Y. (2018). *Coloring of signed graphs.* Dissertation, Universität Paderborn.
554. Kang, Y. (2018). Hajós-like theorem for signed graphs. *European J. Combin.,* 67:199–207.
555. Kang, Y., Jin, L. and Wang, Y. (2016). The 3-colorability of planar graphs without cycles of length 4, 6 and 9. *Discrete Math.,* 339:299–307.
556. Karp, R. M. (1972). Reducibility among combinatorial problems. In: *Complexity of Computer Computation*, pages 85–103. Plenum Press.
557. Karthick, T. and Maffray, F. (2018). Coloring (gem,co-gem)-free graphs. *J. Graph Theory,* 89:283–303.
558. Kaul, H. and Mudrock, J.A. (2019). A note on fractional DP-coloring of graphs. Manuscript arXive:1910:03416
559. Kawarabayashi, K. (2015). Hadwiger's conjecture. In: *Topics in Chromatic Graph Theory*, pages 73–93. Cambridge University Press.
560. Kawarabayashi, K., Kráľ, D., Kynčl, J. and Lidický, B (2009). 6-critical graphs on the Klein bottle. *SIAM J. Discrete Math.,* 23:372–383.
561. Kawarabayashi, K, Pedersen, A.S., and Toft, B. (2010). Double-critical graphs and complete minors. *Electron. J. Combin.,* 17:#R87.
562. Kawarabayashi, K. and Toft, B. (2005). Any 7-chromatic graph has K_7 or $K_{4,4}$ as a minor. *Combinatorica,* 25:327–353.
563. Keevash, P., Li, Z., Mohar, B. and Reed, B. (2013). Digraph girth via chromatic number. *SIAM J. Discrete Math.,* 27:693–696.
564. Kelly, J.B. and Kelly, L.M. (1954). Path and circuits in critical graphs. *Amer. J. Math.,* 76:786–792.
565. Kelly, T. (2019). *Cliques, Degrees, and Coloring: Expanding the ω, Δ, χ Paradigm.* Ph.D. Thesis, University of Waterloo (Canada).
566. Kelly, T. and Postle, L. (2020). A local epsilon version of Reed's Conjecture. *J. Combin. Theory Ser. B,* 141:181–222.
567. Kelly, T. and Postle, L. (2020). Corrigendum to "A local epsilon version of Reed's Conjecture". *J. Combin. Theory Ser. B,* 151:509–512.
568. Kempe, A.B. (1879). On the geographical problem of four colours. *Amer. J. Math.,* 2:193–200.
569. Kenkre, S. and Vishwanathan, S. (2017). A bound on the chromatic number using the longest odd cycle length. *J. Graph Theory,* 54:267–276.
570. Kézdy, A.E. and Snevily, H.S. (1997). On extensions of a conjecture of Gallai. *J. Combin. Theory Ser. B,* 70:317–324.
571. Kierstead, H.A. (1984). On the chromatic index of multigraphs without large triangles. *J. Combin. Theory Ser. B,* 36:156-160.
572. Kierstead, H.A. (1989). Applications of edge coloring of multigraphs to vertex coloring of graphs. *Discrete Math.,* 74:117–124.
573. Kierstead, H.A. (2000). On the choosability of complete multipartite graphs with part size three. *Discrete Math.,* 211:255–259.
574. Kierstead, H.A. and Kostochka, A.V. (2008). A short proof of the Hajnal-Szemerédi theorem on equitable coloring. *Combin. Probab. Comput.,* 17:265–270.
575. Kierstead, H.A. and Kostochka, A.V. (2009). Ore-type version of Brooks' theorem. *J. Combin. Theory Ser. B,* 99:298–305.
576. Kierstead, H.A. and Kostochka, A.V. (2010). Equitable versus nearly equitable coloring and the Chen-Lih-Wu conjecture. *Combinatorica,* 30:201–216.
577. Kierstead, H.A., Kostochka, A.V., Mydlarz, M. and Szemerédi, E. (2010). A fast algorithm for equitable coloring. *Combinatorica,* 30:217–224.

578. Kierstead, H.A., Kostochka, A.V. and Yu, G. (2009). Extremal graph packing problems: Ore-type versus Dirac-type. In: *Surveys in Combinatorics, 2009,* pages 113–136. London Math. Soc. Lecture Notes Series, Vol. 365, Cambridge University Press.
579. Kierstead, H.A. and Penrice, S. (1994). Radius two trees specify χ-bounded classes. *J. Graph Theory,* 18:119–129.
580. Kierstead, H.A. and Rabern, L. (2017). Extracting list colorings from large independent sets. *J. Graph Theory,* 86:315–328.
581. Kierstead, H.A. and Rabern, L. (2020). Improved lower bounds on the number of edges in list critical and online list critical graphs. *J. Combin. Theory Ser. B,* 140:147–170.
582. Kierstead, H.A. and Rödl, V. (1996). Applications of hypergraph coloring to coloring graphs not inducing certain trees. *Discrete Math.,* 150:187–193.
583. Kierstead, H.A. and Schmerl, J.H. (1983). Some applications of Vizing's theorem to vertex colorings of graphs. *Discrete Math.,* 45:277–285.
584. Kierstead, H.A. and Schmerl, J.H. (1986). The chromatic number of graphs which induce neither $K_{1,3}$ nor $K_5 - e$. *Discrete Math.,* 58:253–262.
585. Kierstead, H.A., Szemerédi, E. and Trotter, W.T. (1984). On coloring graphs with locally small chromatic number. *Combinatorica,* 4:183–185.
586. Kierstead, H.A. and Trotter, W.T. (1994). Planar graph coloring with an uncooperative partner. *J. Graph Theory,* 18:569–584.
587. Kierstead, H.A. and Zhu, Y. (2004). Radius three trees in graphs with large chromatic number. *SIAM J. Discrete Math.,* 17:571–581.
588. Kilakos, K. and Reed, B. (1993). Fractionally colouring total graphs. *Combinatorica,* 13:435–440.
589. Kim, J.H. (1995). On Brooks' Theorem for sparse graphs. *Combin. Probab. Comput.,* 4:97–13.
590. Kim, J.H. (1995). The Ramsey number $R(3,t)$ has order of magnitude $t^2/\log t$. *Random Struct. Alg.,* 7:173–207.
591. Kim, S.-J. and Ozeki, K. (2019). A note on a Brooks' type theorem for DP-coloring. *J. Graph Theory,* 91:148–161.
592. Kim, S.-J. and Park, B. (2015). Counterexamples to the list square coloring conjecture. *J. Graph Theory,* 78:239–247.
593. King, A.D. (2009). *Claw-free Graphs and two Conjectures on omega, Delta and chi.* Ph.D. Thesis, McGill University (Canada).
594. King, A.D. (2011). Hitting all maximum cliques with a stable set using lopsided independent transversals. *J. Graph Theory,* 67:300–305.
595. King, A.D. and Reed, B. (2008). Bounding χ in terms of ω and Δ for quasi-line graphs. *J. Graph Theory,* 59:215–228.
596. King, A.D. and Reed, B. (2016). A short proof that χ can be bounded ε away from $\Delta+1$ towards ω. *J. Graph Theory,* 81:30–34.
597. King, A.D., Reed, B. and Vetta, A. (2007). An upper bound for the chromatic number of line graphs. *European J. Combin.,* 28:2182–2187.
598. Király, T. (2003). *Edge-connectivity of undirected and directed hypergraphs.* Dissertation, Eötvös Loránd University, Budapest.
599. Király, T. and Pap, J. (2009). A note on kernels and Sperner's lemma. *Discrete Appl. Math.,* 157:3327–3331.
600. Klein, R. (1992). *On $(m_1, m_2, \ldots, m_s)$-composed graphs.* Doctoral Thesis, Tel Aviv University.
601. Klein, R. and Schönheim, J. (1995). On the colorability of graphs decomposable into degenerate graphs with specified degeneracy. *Australas. J. Combin.,* 12:201–208.
602. Klimošová, T., Malík, J., Masařík, T., Novotná, J., Paulusma, D. and Slívová, V. (2020). Colouring $(P_r + P_s)$-free graphs. *Algorithmica,* 82:1833–1858.
603. Klostermeyer, W. and Zhang, C.Q. (2000). $(2+\epsilon)$-Coloring of planar graphs with large odd-girth. *J. Graph Theory,* 33:109–119.
604. Klostermeyer, W. and Zhang, C.Q. (2002). n-tuple coloring of planar graphs with large odd girth. *Graphs Combin.,* 18:119–131.

605. Kneser, M. (1955/56). Aufgabe 360. *Jahresber. Dtsch. Math.-Ver.,* 58 (2): 27.
606. Kohl, A. and Schiermeyer, I. (2010). Some results on Reed's conjecture about ω, Δ, and χ with respect to α. *Discrete Math.,* 310:1429–1438.
607. Komjáth, P. (2011). The list-chromatic number of infinite graphs defined on Euclidean spaces. *Discrete Comput. Geom.,* 45:497–502.
608. König, D. (1905). On map colouring (in Hungarian). *Mat. Fiz. Lapok,* 14:193–200.
609. König, D. (1916). Über Graphen und ihre Anwendung auf Determinantentheorie und Mengenlehre. *Math. Ann.,* 77:453–465.
610. König, D. (1936). *Theorie der Endlichen und Unendlichen Graphen.* Akademische Verlagsgesellschaft M.B.H., Leipzig. Reprinted by Chealsea 1950 and by B. G. Teubner Verlagsgesellschaft 1986. English translation published by Birkhäuser 1990.
611. Körner, J., Pilotto, C. and Simonyi, G. (2005). Local chromatic number and Sperner capacity. *J. Combin. Theory Ser. B,* 95:101–117.
612. Korzhik, V.P. (2022). A simpler proof of the Map Color Theorem for nonorientable surfaces. *J. Combin. Theory Ser. B,* 156:1–17.
613. Köster, G. (1985). Note to a problem of T. Gallai and G. A. Dirac. *Combinatorica,* 5:227–228.
614. Köster, G. (1990). 4-critical 4-valent planar graphs constructed with crowns. *Math. Scand.,* 67:15–22.
615. Köster, G. (1991). On 4-critical planar graphs with high edge density. *Discrete Math.,* 98:147–151.
616. Kostochka, A.V. (1978). Degree, girth and chromatic number. In: *Combinatorics* (Keszthely, Hungary, 1976), Vol II, pages 679–696., Coll. Math. Soc. János Bolyai, Vol. 18, North-Holland.
617. Kostochka, A.V. (1980). Degree, density, and chromatic number of graphs (in Russian). *Metody Diskret. Analiz.,* 35:45–70.
618. Kostochka, A.V. (1982). A modification of Catlin's algorithm (in Russian). *Methods and Programs of Solutions Optimization Problems on Graphs and Networks,* 2:75–79.
619. Kostochka, A.V. (1984). Lower bound of the Hadwiger number of graphs by their average degree. *Combinatorica,* 4:307–316.
620. Kostochka, A.V. (2004). Coloring uniform hypergraphs with few colors. *Random Struct. Alg.,* 24:1–10.
621. Kostochka, A.V. (2004). Coloring intersection graphs of geometric figures with a given clique number. In: *Towards a Theory of Geometric Graphs,* pages 127–138. Contemporary Mathematics, Vol. 342, American Mathematical Society.
622. Kostochka, A.V. (2006). Color-critical graphs and hypergraphs with few edges: A survey. In: *More Sets, Graphs and Numbers* (A Salute to Vera Sós and András Hajnal), pages 175–197. Bolyai Soc. Math. Stud., Vol. 15, Springer.
623. Kostochka, A.V. and Mazurova, N.P. (1977). An inequality in the theory of graph coloring (in Russian). *Metody Diskret. Analiz.,* 30:23–29.
624. Kostochka, A.V., Mubayi, D., Rödl, V. and Tetali, P. (2001). On the chromatic number of set systems. *Random Struct. Alg.,* 19:87–98.
625. Kostochka, A.V. and Nakprasit, K. (2005). On equitable Δ-coloring of graphs with low average degree. *Theoret. Comput. Sci.,* 349:82–91.
626. Kostochka, A.V. and Nešetřil, J (1999). Properties of Descartes' contruction of triangle-free graphs with high chromatic number. *Combin. Probab. Comput.,* 8:467–472.
627. Kostochka, A.V., Rabern, L. and Stiebitz, M. (2012). Graphs with chromatic number close to maximum degree. *Discrete Math.,* 312:1273–1281.
628. Kostochka, A.V., Schweser, T. and Stiebitz, M. (2023). Generalized DP-colorings of graphs. *Discrete Math.,* 346:113186.
629. Kostochka, A.V. and Stiebitz, M. (1998). Colour-critical graphs with few edges. *Discrete Math.,* 191:125–137.
630. Kostochka, A. . and Stiebitz, M. (1999). Excess in colour-critical graphs. In: *Graph Theory and Combinatorial Biology* (Balatonlelle, Hungary, 1996), pages 87–99. Bolyai Soc. Math. Stud., Vol. 7, Springer.

631. Kostochka, A.V. and Stiebitz, M. (2000). On the number of edges in colour-critical graphs and hypergraphs. *Combinatorica,* 20:521–530.
632. Kostochka, A.V. and Stiebitz, M. (2002). A list version of Dirac's theorem on the number of edges in colour-critical graphs. *J. Graph Theory,* 39:165–177.
633. Kostochka, A.V. and Stiebitz, M. (2003). A new lower bound on the number of edges in colour-critical graphs and hypergraphs. *J. Combin. Theory Ser. B,* 87:374–402.
634. Kostochka, A.V. and Stiebitz, M. (2008). Partitions and edge colourings of multigraphs. *Electron. J. Combin.,* 15:#N25.
635. Kostochka, A.V. and Stiebitz, M. (2020). The minimum number of edges in 4-critical digraphs of given order. *Graphs Combin.,* 36:703–718.
636. Kostochka, A.V., Stiebitz, M. and Wirth, B. (1996). The colour theorems of Brooks and Gallai extended. *Discrete Math.,* 162:299–303.
637. Kostochka, A.V., Sudakov, B. and Verstraëte, J. (2017). Cycles in triangle-free graphs of large chromatic number. *Combinatorica,* 37:481–494.
638. Kostochka, A.V. and Woodall, D.R. (2001). Choosability conjectures and multicircuits. *Discrete Math.,* 240:123–143.
639. Kostochka, A.V. and Yancey, M. (2014). Ore's conjecture for $k = 4$ and Grötzsch's theorem. *Combinatorica,* 34:323–329.
640. Kostochka, A.V. and Yancey, M. (2014). Ore's conjecture on color-critical graphs is almost true. *J. Combin. Theory Ser. B,* 109:73–101.
641. Kostochka, A.V. and Yancey, M. (2018). A Brooks-type result for sparse critical graphs. *Combinatorica,* 38:887–934.
642. Kráľ, D. (2004). Hajós' theorem for list coloring. *Discrete Math.,* 287:161–163.
643. Kráľ, D., Kratochvíl, J., Tuza, Zs. and Woeginger, G.J. (2001). Complexity of coloring graphs without forbidden induced subgraphs. In: *Graph-Theoretic Concepts in Computer Science, WG 2001,* pages 254–262. Lecture Notes in Comput. Sci., Vol. 2204, Springer.
644. Kráľ, D., Mohar, B., Nakamoto, A., Pangrác, O. and Suzuki, Y. (2012). Coloring Eulerian triangulations of the Klein bottle. *Graphs Combin.,* 28:499–530.
645. Kráľ, D., Pangrác, O. and Voss, H.-J. (2005). A note on group colorings. *J. Graph Theory,* 50:123–129.
646. Kráľ, D. and Škrekovski, R. (2006). The last excluded case of Dirac's map-color theorem for choosability. *J. Graph Theory,* 51:319–354.
647. Kratochvíl, J. and Tuza, Zs. (1994). Algorithmic complexity of list colorings. *Discrete Appl. Math.,* 50:297–302.
648. Kratochvíl, J., Tuza, Zs. and Voigt, V. (1999). New trends in the theory of graph colorings: choosability and list colorings. In: *Contemporary Trends in Mathematics,* pages 183–198. DIMACS, Series in Discrete Mathematics and Theoretical Computer Science, Vol. 49.
649. Kriesell, M. (2010). On Seymour's strengthening of Hadwiger's conjecture for graphs with certain forbidden subgraphs. *Discrete Math.,* 310:2714–2724.
650. Kriesell, M. (2017). Unique colorability and clique minors. *J. Graph Theory,* 85:207–216.
651. Krivelevich, M. (1997). On the minimal number of edges in color-critical graphs. *Combinatorica,* 17:401–426.
652. Krivelevich, M. (1998). An improved bound on the minimal number of edges in color-critical graphs. *Electron. J. Combin.,* 5:#P4.
653. Krivelevich, M. (2022). The choosability version of Brooks' theorem - a short proof. Manuscript arXive:2205.08326
654. Kříž (1989). A hypergraph-free construction of highly chromatic graphs without short cycles. *Combinatorica,* 9:227–229.
655. Kronk, H.V. and Mitchem, J. (1972). On Dirac's generalization of Brooks' theorem. *Canad. J. Math.,* 24:805–807.
656. Kronk, H.V. and Mitchem, J. (1975). Critical point-arboritic graphs. *J. London Math. Soc.,* 2:459–466.
657. Kronk, H.V. and White, A.T. (1972). A 4-color theorem for toroidal graphs. *Proc. Amer. Math. Soc.,* 34:83–86.

658. Krul, M. and Thoma, L. (2019). 2-colorability of r-uniform hypergraphs. *Electron. J. Combin.,* 26:#P3.30.
659. Krusenstjerna-Hafstrøm, U. and Toft, B. (1980). Special subdivisions of K_4 and 4-chromatic graphs. *Monatsh. Math.,* 89:101–110.
660. Kühn, D. and Osthus, D. (2002). Topological minors in graphs of larger girth. *J. Combin. Theory Ser. B,* 86:364–380.
661. Kuratowski, C. (1930). Sur le problème des courbes gauches en Topologie. *Fund. Math.,* 15:271–283.
662. La, H., Lužar, B. and Štorgel, K. (2022). Further extensions of the Grötzsch theorem. *Discrete Math.,* 345:112849.
663. Lai, H.-J., Li, X. and Poon, H. (2005). Nowhere zero 4-flows in regular matroids. *J. Graph Theory,* 49:196–204.
664. Lai, H.-J. and Mazza, L. (2021). Group colorings and DP-colorings of multigraphs using edge-disjoint decompositions. *Graphs Combin.,* 37:2227–2243.
665. Lai, H.-J. and Zhang, X. (2002). Group chromatic number of graphs without K_5-minors. *Graphs Combin.,* 18:147–154.
666. Larsen, M., Propp, J. and Ullmann, D.H. (1995). The fractional chromatic number of Mycielski's graphs. *J. Graph Theory,* 19:411–416.
667. Larsen, V. (2015). *An Epsilon Improvement to the Asymptotic Density of k-Critical Graphs.* Ph.D. Thesis, Emory University.
668. Lawrence, J. (1978). Covering the vertex set of a graph with subgraphs of smaller degree. *Discrete Math.,* 21:61–68.
669. Lee, J.A.N. and Holtzman, G. (1995). 50 Years after breaking the codes: interviews with two of the Bletchley Park scientists. *IEEE Annals History Comp.,* 17:32–43.
670. Lekkerkerker, C. and Boland, J.Ch. (1962). Representation of a finite graph by a set of intervals on the real line. *Fund. Math.,* 51:45–64.
671. Leven, D. and Galil, Z. (1983). NP-completeness of finding the chromatic index of regular graphs. *J. Algorithms,* 4:35–44.
672. L'Huilier, S. (1812/1813). Géométrie. *Annales de Mathématiques, pures et appliquées,* 3:169–189.
673. Li, J., Luo, R. and Wang, Y. (2018). Nowhere-zero 3-flow of graphs with small independence number. *Discrete Math.,* 341:42–50.
674. Li, Z. and Mohar, B. (2017). Planar digraphs of digirth four are 2-colorable. *SIAM J. Discrete Math.,* 31:2201–2205.
675. Lick, D.R. and White, A. (1970). k-degenerate graphs. *Can. J. Math,* 22:1082–1096.
676. Lih, K.-W. (1998). The equitable coloring of graphs. In: *Handbook of Combinatorial Optimization*, Vol. 3, pages 543–566. Kluwer Academic Publisher.
677. Lih, K.-W. and Wu, P.-L. (1996). On equitable coloring of bipartite graphs. *Discrete Math.,* 151:155–160.
678. Lin, W., Wu, J., Lam, P.C.B. and Gu, G. (2006). Several parameters of generalized Mycielskians. *Discrete Appl. Math.,* 154:1173–1182.
679. Liu, A. (1976). *Some results on hypergraphs.* Ph.D. Thesis, University of Alberta (Canada).
680. Liu, C.-H. and Ma, J. (2018). Cycle lengths and minimum degree of graphs. *J. Combin. Theory Ser. B,* 128:66–95.
681. Liu, C.-H. and Postle, L. (2017). On the minimum edge-density of 4-critical graphs of girth five. *J. Graph Theory,* 86:387–405.
682. Liu, D.D.-F. and Zhu, X. (2016). A combinatorial proof for the circular chromatic number of Kneser graphs. *J. Comb. Optim.,* 32:765–774.
683. Liu, R., Loeb, S., Rolek, M., Yin, Y. and Yu, G. (2019). DP-3-coloring of planar graphs without 4,9-cycles and cycles of two length from $\{6,7,8\}$. *Graphs Combin.,* 35:695–705.
684. Liu, S. and Zhang, J. (2006). Using Hajós' construction to generate hard graph 3-colorability instances. In: *Artificial Intelligence and Symbolic Computation, AISC 2006,* pages 211–225. Lecture Notes in Comput. Sci., Vol. 4120, Springer.

685. Liu, W., Lawrencenko, S., Chen, B., Ellingham, M.N., Hartsfield, N. Yang, H., Ye, D. and Zha, X. (2019). Quadrangular embeddings of complete graphs and the even map color theorem. *J. Combin. Theory Ser. B,* 139:1–26.
686. Lovász, L. (1966). On decomposition of graphs. *Studia Sci. Math. Hungar.,* 1:237–238.
687. Lovász, L. (1967). Operations with structures. *Acta Math. Hungar.,* 18:321–328.
688. Lovász, L. (1968). On chromatic number of finite set-systems. *Acta Math. Hungar.,* 19:59–67.
689. Lovász, L. (1970). A generalization of König's theorem. *Acta Math. Hungar.,* 21:443–446.
690. Lovász, L. (1972). A characterization of perfect graphs. *J. Combin. Theory Ser. B,* 13:95–98.
691. Lovász, L. (1972). Normal hypergraphs and the perfect graph conjecture. *Discrete Math.,* 2:253–267.
692. Lovász, L. (1973). Coverings and colorings of hypergraphs. *Congr. Numer.,* 8:3–12.
693. Lovász, L. (1973). Independent sets in critical chromatic graphs. *Studia Sci. Math. Hungar.,* 8:165–168.
694. Lovász, L. (1975). Three short proofs in graph theory. *J. Combin. Theory Ser. B,* 19:269–271.
695. Lovász, L. (1975). On combinatorial min-max theorems (Hungarian). *Mat. Lapok,* 26:209–264.
696. Lovász, L. (1976). Chromatic number of hypergraphs and linear algebra. *Studia Sci. Math. Hungar.,* 11:113–114.
697. Lovász, L. (1978). Kneser's conjecture, chromatic number, and homotopy. *J. Combin. Theory Ser. A,* 25:319–324.
698. Lovász, L. (1979). *Combinatorial Problems and Exercises*. Second Edition 1993, North-Holland.
699. Lovász, L (1982). Bounding the independence number of a graph. In: *Bonn Workshop on Combinatorial Optimization,* pages 213–223. Ann. Discrete Math., Vol. 16, North-Holland.
700. Lovász, L (1983). Self-dual polytopes and the chromatic number of distance graphs on the sphere. *Acta Sci. Math.,* 45:317–323.
701. Lovász, L. (1994). Stable sets and polynomials. *Discrete Math.,* 124:137–153.
702. Lovász, L. (2012). *Large Networks and Graph Limits.* American Mathematical Society.
703. Lovász, L.M., Thomassen, C., Wu, Y. and Zhang, C.-Q. (2013). Nowhere-zero 3-flows and modulo k-orientations. *J. Combin. Theory Ser. B,* 103:587–598.
704. Lubotzky, A., Phillips, R. and Sarnak, P. (1988). Ramanujan graphs. *Combinatorica,* , 8:261–277.
705. Luczak, T. (2006). On the structure of triangle-free graphs of large minimum degree. *Combinatorica,* 26:489–493.
706. Lund, C. and Yannakakis, M. (1994). On the hardness of approximating minimization problems. *J. ACM,* 41:960–981.
707. Luo, C., Ma, J. and Yang, T. (2023). On the maximum number of edges in k-critical graphs. *Combin. Probab. Comput.,* 1–12.
708. Lyle, J. (2014). A not on independent sets in graphs with large minimum degree and small cliques. *Electron. J. Combin.,* 21:#P2.38.
709. Ma, J. and Yang, T. (2021). Counting critical subgraphs in k-critical graphs. *Combinatorica,* 41:669–694.
710. Mader, W. (1967). Über trennende Eckenmengen in homomorphiekritischen Graphen. *Math. Ann.,* 175:243–252.
711. Mader, W. (1968). Homomorphiesätze für Graphen. *Math. Ann.,* 178:154–168.
712. Mader, W. (1973). Grad und lokaler Zusammenhang in endlichen Graphen. *Math. Ann.,* 205:9–11.
713. Maffray, F. (1992). Kernels in perfect line-graphs. *J. Combin. Theory Ser. B,* 55:1–8.
714. Maffray, F. and Preissmann, M. (1996). On NP-completeness of k-colorability problem for triangle-free graphs. *Discrete Math.,* 162:313–317.
715. Mansfield, A.J. and Welsh, D.J.A. (1982). Some coloring problems and their complexity. *Ann. Discrete Math.,* 13:159–170.
716. Marshall, T., Naserasr, R. and Nešetřil, J. (2006). Homomorphism bounded classes of graphs. *European J. Combin.,* 27:592–600.

717. Martinsson, A. (2021). A simplified proof of the Johansson–Molly theorem using Rosenfeld counting method. Manuscript arXive:2111.06214
718. Martinsson, A. and Steiner, R. (2023). Vertex critical graphs far from edge-criticality. Manuscript arXive:2310.12891
719. Martinsson, A. and Steiner, R. (2024). Strengthening Hadwiger's conjecture for 4- and 5-chromatic graphs. *J. Combin. Theory Ser. B,* 164:1–16.
720. Matamala, M. (2007). Vertex partition and maximum degenerate subgraphs. *J. Graph Theory,* 55:227–232.
721. Mathews, J., Panda, M.K. and Shannigrahi, S. (2015). On the construction of non-2-colorable uniform hypergraphs. *Discrete Appl. Math.,* 180:181–187.
722. Matoušek, J. (2003). *Using the Borsuk–Ulam Theorem: Lectures on Topological Methods in Combinatorics and Geometry*, Springer.
723. Matoušek, J. (2004). A combinatorial proof of Kneser's conjecture. *Combinatorica,* 24:163–170.
724. Matoušek, J. and Ziegler, G. (2004). Topological lower bounds for the chromatic number: A hierarchy. *Jahresber. Dtsch. Math.-Ver.,* 106:71–90.
725. Matula, D.W. (1968). A min-max theorem for graphs with application to graph coloring. *SIAM Rev.,* 10:481–482.
726. Mayer, J. (1989). Conjecture de Hadwiger: Un graphe k-chromatique contraction-critique n'est pas k-régulier. *Ann. Discrete Math.,* 41:341–346.
727. McConnell, R.M., Mehlhorn, K., Näher, S. and Schweitzer, P. (2011). Certifying algorithms. *Comput. Sci. Rev.,* 5:119–161.
728. McDiarmid, C. (1997). Hypergraph colouring and the Lovász Local Lemma. *Discrete Math.,* 167/168:481–486.
729. McDiarmid, C. (1998). Concentration. In: *Probabilistic Methods for Algorithmic Discrete Mathematics*, pages 195–248. Springer.
730. McDonald, J. (2015). On Galvin orientations of line graphs and list-edge-colouring. Manuscript arXive:1508.01820.
731. Melnikov, L.S. and Vizing, V.G. (1969). New proof of Brooks' theorem. *J. Combin. Theory,* 7:289–290.
732. Menger, K. (1927). Zur allgemeinen Kurventheorie. *Fund. Math.,* 10:96–115.
733. Meyer, W. (1972). Five-coloring planar maps. *J. Combin. Theory Ser. B,* 13:72–82.
734. Meyer, W. (1973). Equitable coloring. *Amer. Math. Monthly,* 80:920–922.
735. Mihók, P. (1981). On the structure of the point arboricity critical graphs. *Math. Slovaca,* 31:101–106.
736. Mihók, P. (1992). An extension of Brooks' theorem. *Ann. Discrete Math.,* 51:235–236.
737. Mihók, P. and Schiermeyer, I. (2004). Cycle lengths and chromatic number of graphs. *Discrete Math.,* 286:147–149.
738. Mihók, P. and Škrekovsky, R. (2001). Gallai's inequality for critical graphs of reducible hereditary properties. *Discuss. Math., Graph Theory,* 21:167–177.
739. Miller, E.W. (1937). On a property of families of sets. *Comptes Rendus Varsovie,* 30:31–38.
740. Miller, Z. and Müller, H. (1981). Chromatic numbers of hypergraphs and coverings of graphs. *J. Graph Theory,* 5:299–305.
741. Minty, G.J. (1962). A theorem on n-colouring the points of a linear graph. *Amer. Math. Monthly,* 69:623–624.
742. Mirsky, L. (1971). A dual of Dilworth's decomposition theorem. *Amer. Math. Monthly,* 78:876–877.
743. Mirzakhani, M. (1996). A small non-4-choosable planar graph. *Bull. Inst. Combin. Appl.,* 17:15–18.
744. Mitchem, J. (1977). An extension of Brooks' theorem to n-degenerate graphs. *Discrete Math.,* 17:291–298.
745. Mitchem, J. (1978). A new proof of a theorem of Dirac on the number of edges in critical graphs. *J. Reine Angew. Math.,* 299/300:84–91.
746. Mitchem, J. (1978). A short proof of Catlin's extension of Brooks' theorem. *Discrete Math.,* 21:213–214.

747. Mohar, B. (1999). A linear time algorithm for embedding graphs in an arbitrary surface. *SIAM J. Discrete Math.,* 12:6–26.
748. Mohar, B. (2003). Circular colorings of edge-weighted graphs. *J. Graph Theory,* 43:107–116.
749. Mohar, B. (2004). Hajós theorem for colorings of edge-weighted graphs. *Combinatorica,* 25:65–76.
750. Mohar, B. (2010). Eigenvalues and colorings of digraphs. *Linear Algebra Appl.,* 432:2273–2277.
751. Mohar, B. (2015). Colouring graphs on surfaces. In: *Topics in Chromatic Graph Theory,* pages 13–35. Cambridge University Press.
752. Mohar, B. and Thomassen, C. (2001). *Graphs on Surfaces.* The John Hopkins University Press.
753. Molloy, M. (1999). Chromatic neighborhood sets. *J. Graph Theory,* 31:303–311.
754. Molloy, M. (2019). The list chromatic number of graphs with small clique number. *J. Combin. Theory Ser. B,* 134:264–284.
755. Molloy, M. and Reed, B. (1998). Coloring graphs whose chromatic number is almost their maximum degree. In: *LATIN'98: Theoretical Informatics* (Campinas, 1998), pages 216–225. Lecture Notes in Comput. Sci., Vol. 1380, Springer.
756. Molloy, M. and Reed, B. (2001). *Graph Colouring and the Probabilistic Method.* Algorithms and Combinatorics, Vol. 23, Springer.
757. Molloy, M. and Reed, B. (2014). Coloring graphs when the number of colours is almost the maximum degree. *J. Combin. Theory Ser. B,* 109:134–195.
758. Montassier, M., Ossona de Mendez, P., Raspaud, A. and Zhu, X. (2012). Decomposing a graph into forests. *J. Combin. Theory Ser. B,* 102:38–52.
759. Moore, B. and Smith-Roberge, E. (2023). A density bound for triangle-free 4-critical graphs. *J. Graph Theory,* 103:66–111.
760. Moser, L. and Moser, W. (1961). Solution to problem 10. *Can. Math. Bull.,* 4:187–189.
761. Moser, R.A. and Tardos, G. (2010). A constructive proof of the general Lovász local lemma. *J. ACM,* 57:1–15.
762. Mozhan, N.N. (1983). Chromatic number of graphs with a density that does not exceed two-thirds of the maximal degree (in Russian). *Metody Diskret. Analiz.,* 39:52-65.
763. Mozhan, N.N. (1987). On doubly critical graphs with chromatic number five. *Metody Diskret. Analiz.,* 46:50–59.
764. Müller, V. (1975). On colorable critical and uniquely colorable critical graphs. In: *Recent Advances in Graph Theory* (Prague, 1974), pages 385–386. Academia Praha.
765. Müller, V. (1979). On colorings of graphs without short cycles. *Discrete Math.,* 26:165–176.
766. Mycielski, J. (1955). Sur le coloriage des graphes. *Colloq. Math.,* 3:161–162.
767. Mynhardt, C.M. and Broere, I. (1985). Generalized colorings of graphs. In: *Graph Theory with Applications to Algorithms and Computer Science* (Kalamazoo, Michigan, 1984), pages 583–594. John Wiley & Sons.
768. Nash-Williams, C.St.J.A. (1961). Edge-disjoint spanning trees of finite graphs. *J. London Math. Soc.,* 36:445–450.
769. Nash-Williams, C.St.J.A. (1964). Decomposition of finite graphs into forests. *J. Combin. Theory Ser. B,* 39:12.
770. Năstase, E., Rödl, V. and Siggers, M. (2010). Note on robust critical graphs with large odd girth. *Discrete Math.,* 310:499–504.
771. Nešetřil, J. (1966). K-chromatic graphs without cycles of length ≤ 7 (in Russian). *Comment. Math. Univ. Carol.,* 7:373–376.
772. Nešetřil, J. (1999). The homomorphism structure of classes of graphs. *Combin. Probab. Comput.,* 8:177–184.
773. Nešetřil, J. (2013). A combinatorial classic - sparse graphs with high chromatic number. In: *Erdős Centennial,* pages 383–407. Bolyai Soc. Math. Stud., Vol 25, Springer.
774. Nešetřil, J. and Nigussie, Y. (2012). Finite dualities and map-critical graphs on a fixed surface. *J. Combin. Theory Ser. B,* 102:131–152.

775. Nešetřil, J. and Rödl, V. (1979). A short proof of the existence of highly chromatic hypergraphs without short cycles. *J. Combin. Theory Ser. B,* 27:225–227.
776. Nešetřil, J. and Rödl, V. (1985). Three remarks on dimensions of graphs. *Ann. Discrete Math.,* 28:199–207.
777. Nešetřil, J. and Rödl, V. (1989). Chromatically optimal rigid graphs. *J. Combin. Theory Ser. B,* 46:133–141.
778. Nešetřil, J. and Zhu, X. (1996). On bounded treewidth duality of graphs. *J. Graph Theory,* 23:151–162.
779. Neumann, J. von and Morgenstern, O. (1944). *Theory of Games and Economic Behavior.* Princeton University Press.
780. Neumann-Lara, V. (1982). The dichromatic number of a digraph. *J. Combin. Theory Ser. B,* 33:265–270.
781. Neumann-Lara, V. (1985). Vertex colourings in digraphs - some problems. Technical Report, University of Waterloo.
782. Neumann-Lara, V. (1994). The 3- and 4-chromatic tournaments of minimum order. *Discrete Math.,* 135:233–243.
783. Ngoc, N. Van and Tuza, Zs. (1995). 4-chromatic graphs with large odd girth. *Discrete Math.,* 138:387–392.
784. Nielsen, F. and Toft, B. (1975). On a class of planar 4-chromatic graphs due to T. Gallai. In: *Recent Advances in Graph Theory* (Prague, June 1974), pages 425–430. Academia Praha
785. Nigussie, Y. (2009). Extended Gallai's theorem. *Electron. Notes Discrete Math.,* 34:399–403.
786. Nikoforov, V. (2010). Chromatic number and minimum degree of K_r-free graphs. Manuscript arXive:1001.2070
787. Nilli, A. (1999). Short odd cycles in 4-chromatic graphs. *J. Graph Theory,* 31:145–147.
788. Nilli, A. (2000). Triangle-free graphs with large chromatic numbers. *Discrete Math.,* 211:261–262.
789. Noel, J.A., Reed, B.A. and Wu, H. (2014). A proof of a conjecture of Ohba. *J. Graph Theory,* 79:86–102.
790. Nordhaus, E.A. and Gaddum, J.W. (1956). On complementary graphs. *Amer. Math. Monthly,* 63:175–177.
791. Oberkampf, H. and Schacht, M. (2020). On the structure of dense graphs with bounded clique number. *Combin. Probab. Comput.,* 29:641–649.
792. Ohba, K. (2002). On chromatic-choosable graphs. *J. Graph Theory,* 40:130–135.
793. Ore, O. (1960). A note on hamiltonian circuits. *Amer. Math. Monthly,* 67:55.
794. Ore, O. (1967). *The Four-Color Problem.* Academic Press.
795. Östergård, P.R.J. (2014). On the minimum size of 4-uniform hypergraphs without property B. *Discrete Appl. Math.,* 163:199–204.
796. Panconesi, A. and Srinivasan, A. (1995). The local nature of Δ-coloring and its algorithmic applications. *Combinatorica,* 15:255–280.
797. Panconesi, A. and Srinivasan, A. (1997). Randomized distributed edge coloring via an extension of the Chernoff–Hoeffding bounds. *SIAM J. Comput.,* 26:350–368.
798. Parsons, T.D., Pica, G., Pisanski, T. and Ventre, A.G.S. (1987). Orientably simple graphs. *Math. Slovaca,* 37:391–394.
799. Pawlik, A., Kozik, J., Krawczyk, T., Lasoń, M., Micek, P., Trotter, W.T. and Walczak, B. (2014). Triangle-free intersection graphs of line segments with large chromatic number. *J. Combin. Theory Ser. B,* 105:6–10.
800. Payan, C. (1992). On the chromatic number of cube-like graphs. *Discrete Math.,* 103:271–277.
801. Pedersen, A.S., Plummer, M. and Toft, B. (2016). Inflations of anti-cycles and Hadwiger's conjecture. J. Combin.. 7:413–421.
802. Pedgen, W. (2017). Critical graphs without triangles: An optimum density construction. *Combinatorica,* 33:495–512.
803. Petersen, J. (1891). Die Theorie der regulären graphs. *Acta Math.,* 15:193–220.

804. Peterson, D. and Woodall, D.R. (1999). Edge-choosability in line perfect multigraphs. *Discrete Math.,* 202:191–199.
805. Pettie, S. and Su, H.-H. (2015). Distributed coloring algorithms for triangle-free graphs. *Inform. Comput.,* 243:263–280.
806. Picassari-Arrieta, L. and Stiebitz, M. (2023). Minimum number of arcs in k-critical digraphs with order at most $2k - 1$. Manuscript arXive:2310.03584
807. Pirnazar, A. and Ullman, D.H. (2002). Girth and fractional chromatic number of planar graphs. *J. Graph Theory,* 39:201–217.
808. Pitassi, T. and Urquhart, A. (1995). The complexity of the Hajós calculus. *SIAM J. Discrete Math.,* 8:464–483.
809. Plantholt, M.J. and Tipnis, S.K. (1999). On the list chromatic index of nearly bipartite multigraphs. *Australas. J. Combin.,* 19:157–170.
810. Plesnevič, G.S. and Vizing, V.G. (1965). On the problem of the minimal coloring of the vertices of a graph (in Russian). *Sibirsk. Mat. Zh.,* 6:234–236.
811. Pluhár, A. (2009). Greedy colorings of uniform hypergraphs. *Random Struct. Alg.,* 35:216–221.
812. Plummer, M. and Zha, X. (2014). On a conjecture concerning the Petersen graph: part II. *Electron. J. Combin.,* 21:#P1.34.
813. Plummer, M., Stiebitz, M. and Toft, B. (2003). On a special case of Hadwiger's conjecture. *Discuss. Math., Graph Theory,* 23:333-363.
814. Poljak, S. and Roberts, F.S. (2009). An application of Stahl's conjecture about the k-tuple chromatic numbers of Kneser graphs. In: *The Mathematics of Preference, Choice and Order,* pages 345–352. Studies in Choice and Welfare, Springer.
815. Poljak, S. and Rödl, V. (1981). On the arc-chromatic number of a digraph. *J. Combin. Theory Ser. B,* 31:190–198.
816. Pólya, G. and Szegő, G. (1925). *Aufgaben und Lehrsätze aus der Analysis*, Springer.
817. Ponstein, J. (1969). A new proof of Brooks's chromatic number theorem for graphs. *J. Combin. Theory,* 7:255–257.
818. Postel, J. von, Schweser, T. and Stiebitz, M. (2022). Point partition numbers: decomposable and indecomposable critical graphs. *Discrete Math.,* 345:112903.
819. Postle, L. (2012). *5-List-Coloring Graphs on Surfaces.* Ph.D. Thesis, Georgia Institute of Technology.
820. Postle, L. (2017). On the minimum number of edges in triangle-free 5-critical graphs. *European J. Combin.,* 66:264–280.
821. Postle, L. (2020). Further progress towards Hadwiger's conjecture. Manuscript arXive:2006.11798
822. Postle, L. (2021). 3-List-coloring graphs of girth at least five on surfaces. *J. Combin. Theory Ser. B,* 147:1–36.
823. Postle, L. and Smith-Roberge, E. (2022). On the density of C_7-critical graphs. *Combinatorica,* 42:253–300.
824. Postle, L. and Thomas, R. (2018). Hyperbolic families and coloring graphs on surfaces. *Trans. Amer. Math. Soc., Ser. B,* 5:167–221.
825. Potapov, V.N. (2022). DP-colorings of uniform hypergraphs and splitting boolean hypercubes into faces. *Electron. J. Combin.,* 29:#P3.37.
826. Pretzel, O.R.L. and Youngs, D. (1990). Cycle lengths and graph orientations. *SIAM J. Discrete Math.,* 3:544–553.
827. Pretzel, O.R.L. and Youngs, D. (1991). Balanced graphs and noncovering graphs. *Discrete Math.,* 88:279–287.
828. Prowse, A. and Woodall, D.R. (2003). Choosability of powers of circuits. *Graphs Combin.,* 19:137–144.
829. Przybylo, J. (2024). On triangle-free list assignments. *Discrete Math.,* 347:113779.
830. Puš, V. (1988). Chromatic number of products of graphs. *Comment. Math. Univ. Carolin.,* 29:457–463.
831. Rabern, L. (2006). On graph associations. *SIAM J. Discrete Math.,* 20:529–535.

832. Rabern, L. (2008). A note on Reed's conjecture. *SIAM J. Discrete Math.,* 22:820–827.
833. Rabern, L. (2011). The Borodin–Kostchka conjecture for graphs with a doubly critical edge. *Electron. J. Combin.,* 14:#N22.
834. Rabern, L. (2011). On hitting all maximum cliques with an independent set. *J. Graph Theory,* 66:32–37.
835. Rabern, L. (2011). A strengthening of Brooks' theorem for line graphs. *Electron. J. Combin.,* 18:#P145.
836. Rabern, L. (2012). Δ-Critical graphs with small high vertex cliques. *J. Combin. Theory Ser. B,* 102:126–130.
837. Rabern, L. (2013). Destroying noncomplete regular components in graph partitions. *J. Graph Theory,* 72:123–127.
838. Rabern, L. (2013). Partitioning and coloring graphs with degree constraints. *Discrete Math.,* 313:1028–1034.
839. Rabern, L. (2014). A different short proof of Brooks' theorem. *Discuss. Math., Graph Theory,* 34:633–634.
840. Rabern, L. (2016). A better lower bound on average degree of 4-list-critical graphs. *Electron. J. Combin.,* 23:#P3.37.
841. Rabern, L. (2016). A better lower bound on average degree of online k-list-critical graphs. *Electron. J. Combin.,* 25:#P1.51.
842. Rabern, L. (2023). Yet another proof of Brooks' theorem. *Discrete Math.,* 346:113261.
843. Rackham, T. (2009). A note on $K^-_{\Delta+1}$-free precolouring with Δ colours. *Electron. J. Combin.,* 16:#N28.
844. Radhakrishnan, J. and Srinivasan, A. (2000). Improved bounds and algorithms for hypergraph 2-coloring. *Random Struct. Alg.,* 16:4–32.
845. Raigorodskii, A.M. and Shabanov, D. (2011). The Erdős–Hajnal problem of hypergraph colouring, its generalization, and related problems (in Russian). *Russ. Math. Surv.,* 66:933–1002.
846. Ramirez-Alfonsin, J. and Reed, B. (eds.) (2001). *Perfect Graphs.* John Wiley & Sons.
847. Ramsey, F.P. (1930). On a problem of formal logic. *Proc. London Math. Soc.,* 30:264–286.
848. Randerath, B. (1998). *The Vizing Bound for the Chromatic Number Based on Forbidden Pairs.* Dissertation, RWTH Aachen, Shaker Verlag.
849. Randerath, B. and Schiermeyer, I. (2004). 3-Colorability ∈ P for P_6-free graphs. *Discrete Appl. Math.,* 136:299–313.
850. Randerath, B. and Schiermeyer, I. (2004). Vertex colouring and forbidden subgraphs – a survey. *Graphs Combin.,* 20:1–40.
851. Rao, M. and Wang, T. (2021). DP-3-coloring of planar graphs without certain cycles. *Discrete Appl. Math.,* 297:35–45.
852. Read, R.C. (1957). Maximal circuits in critical graphs. *J. London Math. Soc.,* 32:456–466.
853. Rédei, L. (1934). Ein kombinatorischer Satz. *Acta. Litt. Sci. Szeged,* 7:39–43.
854. Reed, B. (1998). ω, Δ, and χ. *J. Graph Theory,* 27:177–212.
855. Reed, B. (1999). A strengthening of Brooks' theorem. *J. Combin. Theory Ser. B,* 76:136–149.
856. Reed, B. (2001). A gentle introduction to semi-definite programming. In *Perfect graphs,* Chapter 11, John Wiley & Sons
857. Reed, B. and Seymour, P. (1998). Fractional colouring and Hadwiger's conjecture. *J. Combin. Theory Ser. B,* 74:147–152.
858. Reed, B. and Seymour, P. (2004). Hadwiger's conjecture for line graphs. *European J. Combin.,* 25:873–876.
859. Reed, B. and Sudakov, B. (2002). Asymptotically the list coloring constants are 1. *J. Combin. Theory Ser. B,* 86:27–37.
860. Riasat, A. and Schauz, U. (2012). Critically paintable, choosable or colorable graphs. *Discrete Math.,* 312:3373-3383.
861. Richardson, M. (1953). Solutions of irreflexive relations. *Ann. of Math.,* 58:573–590.
862. Richter, R.B., Thomassen, C. and Younger, D.H. (2016). Group-colouring, group-connectivity, claw-decompositions, and orientations in 5-edge-connected planar graphs. *J. Combin,* 7:219–232.

863. Ringel, G. (1954). Bestimmung der Maximalzahl der Nachbargebiete auf nichtorientierbaren Flächen. *Math. Ann.,* 127:181–214.
864. Ringel, G. (1959). *Färbungsprobleme auf Flächen und Graphen.* VEB Deutscher Verlag der Wissenschaften.
865. Ringel, G. (1974). *Map Color Theorem.* Springer.
866. Ringel, G. and Youngs, J.W.T. (1968). Solution of the Heawood map-coloring problem. *Proc. Nat. Acad. Sci. U.S.A.,* 60:438–445.
867. Ritter, E. (1892). Die eindeutigen automorphen Formen vom Geschlecht Null, eine Revision und Erweiterung der Poincaréschen Sätze. *Math. Ann.,* 41:1-82.
868. Robertson, N., Sanders, D.P., Seymour, P. and Thomas, R. (1997). The four-colour theorem. *J. Combin. Theory Ser. B,* 70:2–44.
869. Robertson, N. and Seymour, P. (1983). Graph minors I. Excluding a forest. *J. Combin. Theory Ser. B,* 35:39–61.
870. Robertson, N. and Seymour, P. (2004). Graph minors XX. Wagner's conjecture. *J. Combin. Theory Ser. B,* 92:325–357.
871. Robertson, N., Seymour, P. and Thomas, R. (1993). Hadwiger's conjecture for K_6-free graphs. *Combinatorica,* 13:279–361.
872. Rödl, V. (1977). On the chromatic number of subgraphs of a given graph. *Proc. Amer. Math. Soc.,* 64:370–371.
873. Rödl, V. (1985). On a packing and covering problem. *European J. Combin.,* 6:69–78.
874. Rödl, V. and Siggers, M. (2006). Color critical hypergraphs with many edges. *J. Graph Theory,* 53:56–74.
875. Rödl, V. and Tuza, Zs. (1985). On colour critical graphs. *J. Combin. Theory Ser. B,* 38:204–213.
876. Rolek, M., Song, Z.-X. and Thomas, R. (2023). Properties of 8-contraction-critical graphs with no K_7 minor. *European J. Combin.,* 110:103711.
877. Rosenfeld, M. (2020). Another approach to non-repetitive colorings of graphs of bounded degree. *Electron. J. Combin.,* 27:#P3.43.
878. Roussel, F., Rusu, I. and Thuillier, H. (2009). The strong perfect graph conjecture: 40 years of attempts, and its resolution. *Discrete Math.,* 309:6092–6113.
879. Roy, B. (1967). Nombre chromatique et plus longs chemins d'un graphe. *ESAIM Math. Model. Numer. Anal.,* 1:129–132.
880. Saaty, T.L. and Kainen, P.C. (1977). *The Four-Color Problem.* McGraw-Hill.
881. Sabidussi, G. (1961). Graph derivatives. *Math. Z.,* 76:385–401.
882. Sachs, H. (1969). Finite graphs (investigations and generalizations concerning the construction of finite graphs having given chromatic number and no triangles). In: *Recent Progress in Combinatorics* (Proc. Third Waterloo Conf. on Combinatorics, 1968), pages 175–184. Academic Press.
883. Sachs, H. (1978). A three-colour conjecture of Grötzsch. In: *Problémes Combinatoires et Theorie des Graphes,* page 471. Editions du Centre National de la Recherche Scientifique.
884. Sachs, H. (1993). Elementary proof of the cycle-plus-triangles theorem. In: *Combinatorics, Paul Erdős is Eighty,* (Kesthely (Hungary), 1993), Vol. 1, pages 347–359. Bolyai Soc. Math. Stud., Vol. 2, János Bolyai Math. Soc.
885. Sachs, H. and Stiebitz, M. (1983). Construction of colour-critical graphs with given major-vertex subgraph. In: *Combinatorial Mathematics* (Marseille-Luminy, 1981), pages 581–598. Ann. Discrete Math., Vol. 17, North-Holland.
886. Sachs, H. and Stiebitz, M. (1988). Colour-critical graphs with vertices of low valency. In: *Graph Theory in Memory of G. A. Dirac*, pages 371–396. Ann. Discrete Math., Vol. 41, North-Holland.
887. Sachs, H. and Stiebitz, M. (1989). On constructive methods in the theory of colour-critical graphs. *Discrete Math.,* 74:201–226.
888. Sajith, G. and Saxena, S. (2022). On Brooks' theorem. Manuscript: arXive:2208.02186
889. Sanders, D.P. and Zhao, Y. (1995). A note on the three color problem. *Graphs Combin.,* 11:91–94.

890. Sauer, N. (1993). Problem 18 in "Open Problems"; International Colloquium on Combinatorics (Paul Erdős is Eighty), Keszthely (Hungary), July, 19–24, 1993. Unpublished.
891. Sauer, N. (2001). Hedetniemi's conjecture – a survey. *Discrete Math.,* 229:261–292.
892. Scarf, H.E. (1967). The core of an n person game. *Econometrica,* 35:50–69.
893. Schäuble, M. (1969). Bemerkungen zur Konstruktion dreikreisfreier k-chromatischer Graphen. *Wiss. Z. TH Ilmenau,* 15 (Heft 2) :59–63.
894. Schauz, U. (2009). Mr. Paint and Mrs. Correct. *Electron. J. Combin.,* 16:#R77.
895. Schauz, U. (2010). Flexible color lists in Alon and Tarsi's theorem, and time scheduling with unreliable participants. *Electron. J. Combin.,* 17:#R13.
896. Schauz, U. (2014). Proof of the list edge coloring conjecture for complete graphs of prime degree. *Electron. J. Combin.,* 21:#P3.43.
897. Scheide, D. (2010). Graph edge colouring: Tashkinov trees and Goldberg's conjecture. *J. Combin. Theory Ser. B,* 100:68–96.
898. Scheinerman, E.R. and Ullman, D.H. (1997). *Fractional Graph Theory, A Rational Approach to the Theory of Graphs.* Wiley-Interscience Series in Discrete Mathematics and Optimization, John Wiley & Sons.
899. Schiermeyer, I. (2016). Chromatic number of P_5-free graphs: Reed's conjecture. *Discrete Math.,* 339:1940–1943.
900. Schmerl, J.H. (1982). The efective version of Brooks' theorem. *Canad. J. Math.,* 34:1036–1046.
901. Schmerl, J.H. (1995). The list-chromatic number of Euclidean space. *Geombinatorics,* 5:65–68.
902. Schrijver, A. (1978). Vertex-critical subgraphs of Kneser-graphs. *Nieuw Arch. Wiskd.,* 26:454–461.
903. Schrijver, A. (2003). *Combinatorial Optimization: Polyhedra and Efficiency.* Algorithms and Combinatorics, Vol. 24, Springer.
904. Schrijver, A. (2012). On the history of the shortest path problem. *Documenta Math.,* 17:155–167.
905. Schweser, T. (2019). DP-degree colorable hypergraphs. *Theoret. Comput. Sci.,* 796:196–206.
906. Schweser, T. (2020). *Colorings of Graphs, Digraphs, and Hypergraphs.* Dissertation, TU Ilmenau (Germany).
907. Schweser, T. (2021). Generalized hypergraph coloring. *Discuss. Math., Graph Theory,* 41:103–121.
908. Schweser, T. and Stiebitz, M. (2017). Degree choosable signed graphs. *Discrete Math.,* 340:882–891.
909. Schweser, T. and Stiebitz, M. (2021). Partition of hypergraphs under variable degeneracy constraints. *J. Graph Theory,* 96:7–33.
910. Schweser, T. and Stiebitz, M. (2021). Vertex partition of hypergraphs and maximum degenerate subhypergraphs. *Electr. J. Graph Theory Appl.,* 9:1–9.
911. Schweser, T., Stiebitz, M., and Toft, B. (2022). Coloring hypergraphs of low connectivity. *J. Combin.,* 13:1–21.
912. Scott, A. (1997). Induced trees in graphs of large chromatic number. *J. Graph Theory,* 24:297–311.
913. Scott, A. and Seymour, P. (2016). Induced subgraphs of graphs with large chromatic number. I. Odd holes. *J. Combin. Theory Ser. B,* 121:68–84.
914. Scott, A. and Seymour, P. (2017). Induced subgraphs of graphs with large chromatic number. IX. Rainbow path. *Electron. J. Combin.,* 24:#P2.53.
915. Scott, A. and Seymour, P. (2018). Induced subgraphs of graphs with large chromatic number. IV. Consecutive holes. *J. Combin. Theory Ser. B,* 132:180–235.
916. Scott, A. and Seymour, P. (2019). Induced subgraphs of graphs with large chromatic number. X. Holes of specific residue. *Combinatorica,* 39:1105–1132.
917. Scott, A. and Seymour, P. (2020). Induced subgraphs of graphs with large chromatic number. VI. Banana trees. *J. Combin. Theory Ser. B,* 145:487–510.

918. Scott, A. and Seymour, P. (2020). Induced subgraphs of graphs with large chromatic number. VII. Gyárfás' complementation conjecture. *J. Combin. Theory Ser. B,* 142:43–55.
919. Scott, A. and Seymour, P. (2020). A survey of χ-boundedness. *J. Graph Theory,* 95:473–504.
920. Scott, A., Seymour, P. and Spirkl, S. (2022). Polynomial bounds for chromatic number. III. Excluding a double star. *J. Graph Theory,* 101:323–340.
921. Sebő, A. (2009). Path partitions, cycle covers and integer decomposition. In: *Graph Theory, Computational Intelligence and Thought,* pages 183–199. Lecture Notes in Comput. Sci., Vol. 5420, Springer.
922. Sebő, A. (2019). Color-critical graphs and hereditary hypergraphs. Manuscript arXive: 1910.11302
923. Seidman, S.B. (1983). Network structure and minimum degree. *Social Networks,* 5:269–287.
924. Seinsche, D. (1974). On a property of the class of n-colorable graphs. *J. Combin. Theory Ser. B,* 16:191–193.
925. Seymour, P. (1974). On the two-colouring of hypergraphs. *Quart. J. Math. Oxford,* 25:303–312.
926. Seymour, P. (1974). A note on a combinatorial problem of Erdős and Hajnal. *J. London Math. Soc.,* 8:681–682.
927. Seymour, P. (1979). Some unsolved problems on one-factorizations of graphs. In: *Graph Theory and Related Topics* (Waterloo, 1977), pages 367–368. Academic Press.
928. Seymour, P. (1979). On multi-colourings of cubic graphs, and conjectures of Fulkerson and Tutte. *Proc. London Math. Soc.,* 38:423–460.
929. Seymour, P. (1981). Nowhere zero 6-flows. *J. Combin. Theory Ser. B,* 30:130–135.
930. Seymour, P. (1995). Nowhere zero flows. In: *Handbook of Combinatorics,* Vol. 1, pages 290–299. Elsevier.
931. Seymour, P. (2016). Hadwiger's conjecture. In: *Open Problems in Mathematics*, pages 417–437. Springer.
932. Shabanov, D.A. (2007). On some extremal properties of hypergraph colorings. *Electron. Notes Discrete Math.,* 29:97–100.
933. Shabanov, D.A. (2012). On r-chromatic hypergraphs. *Discrete Math.,* 312:441–458.
934. Shannon, C.E. (1949). A theorem on coloring the lines of a network. *J. Math. Phys.,* 28:148–152.
935. Shapira, A. and Thomas, R. (2011). Color-critical graphs have logarithmic circumference. *Adv. Math.,* 227:2309–2326.
936. Shearer, J.B. (1983). A note on the independence number of triangle-free graphs. *Discrete Math.,* 46:83–87.
937. Shearer, J.B. (1995). A note on the independence number of sparse graphs. *Random Struct. Alg.,* 7:269–271.
938. Shitov, Y. (2019). Counterexamples to Hedetniemi's conjecture. *Ann. of Math.,* 190:663–667.
939. Simonovits, M. (1972). On colour-critical graphs. *Studia Sci. Math. Hungar.,* 7:67–81.
940. Simons, G., Tardif, C. and Wehlau, D. (2017). Generalised Mycielski graphs, signature systems, and bounds on chromatic numbers. *J. Combin. Theory Ser. B,* 122:776–793.
941. Simonyi, G. and Tardos, G. (2006). Local chromatic number, Ky Fan's theorem, and circular colorings. *Combinatorica,* 26:587–626.
942. Simonyi, G. and Tardos, G. (2007). Colorful subgraphs in Kneser-like graphs. *European J. Combin.,* 28:2188–2200.
943. Simonyi, G. and Tardos, G. (2011). On directed local chromatic number, shift graphs, and Borsuk-like graphs. *J. Graph Theory,* 66:65–82.
944. Simonyi, G., Tardos, G. and Vrećica, S. (2009). Local chromatic number and distinguishing the strength of topological obstruction. *Trans. Amer. Math. Soc.,* 361:889–908.
945. Sivaraman, V. (2015). A unified proof of Brooks' theorem and Catlin's theorem. *Discrete Math.,* 338:272–273.
946. Škrekovsky, R. (2002). A theorem on map colorings. *Bull. Inst. Combin. Appl.,* 35:53–60.

947. Škrekovsky, R.(2002). On the critical point-arboricity graphs. *J. Graph Theory,* 39:50–61.
948. Skulrattanakulchai, S. (2006). Δ-List vertex coloring in linear time. *Inform. Process. Lett.,* 98:101–106.
949. Slivnik, T. (1996). Short proof of Galvin's theorem on the list-chromatic index of a bipartite multigraph. *Combin. Probab. Comput.,* 5:91–94.
950. Soifer, A. (2009). *The Mathematical Coloring Book.* Springer
951. Song, Z.-X. (2019). Erdős-Lovász Tihany conjecture for graphs with forbidden holes. *Discrete Math.,* 342:2632–2635.
952. Song, Z.-X. and Thomas, B. (2017). Hadwiger's conjecture for graphs with forbidden holes. *SIAM J. Discrete Math.,* 31:1572–1580.
953. Spanier, E. H. (1966). *Algebraic Topology.* Springer.
954. Spencer, J. (1981). Coloring n-sets red and blue. *J. Combin. Theory Ser. A,* 30:112–113.
955. Stacho, L. (2001). New upper bounds for the chromatic number of a graph. *J. Graph Theory,* 36:117–120.
956. Stahl, S. (1976). n-tuple colorings and associated graphs. *J. Combin. Theory Ser. B,* 20:185–203.
957. Stahl, S. (1998). The multichromatic numbers of some Kneser graphs. *Discrete Math.,* 185:287–291.
958. Stehlík, M. (2003). Critical graphs with connected complements. *J. Combin. Theory Ser. B,* 89:189–194.
959. Stehlík, M. (2006). Minimal connected τ-critical hypergraphs. *Graphs Combin.,* 22:421–426.
960. Stehlík, M. (2019). Critical digraphs with few vertices. Manuscr. arXive:1910.02454
961. Steinberg, R. (1993). The state of the three color problem. In: *Quo Vadis Graph Theory,* pages 211–248. Ann. Discrete Math., Vol. 55, North-Holland.
962. Steiner, R. (2020). A note on coloring digraphs of large girth. *Discrete Appl. Math.,* 287:62–64.
963. Steiner, R. (2022). Asymptotic equivalence of Hadwiger's conjecture and its odd minor-variant. *J. Combin. Theory Ser. B,* 155:45–51.
964. Steiner, R. (2022). Improved lower bound for the list chromatic number of graphs with no K_t minor. *Combin. Probab. Comput.,* 31:1070–1075.
965. Stiebitz, M. (1982). Proof of a conjecture of T. Gallai concerning connectivity properties of colour-critical graphs. *Combinatorica,* 2:315–323.
966. Stiebitz, M. (1985). *Beiträge zur Theorie der Färbungskritischen Graphen.* Habilitation, TU Ilmenau (Germany).
967. Stiebitz, M. (1987). Subgraphs of colour-critical graphs. *Combinatorica,* 7:303–312.
968. Stiebitz, M. (1987). K_5 is the only double-critical 5-chromatic graph. *Discrete Math.,* 64:91–93.
969. Stiebitz, M. (1988). On k-critical n-chromatic graphs. In: *Combinatorics* (Eger, Hungary, 1987), pages 509–514. Coll. Math. Soc. János Bolyai, Vol. 52, North-Holland.
970. Stiebitz, M. (1994). The forest plus stars colouring problem. *Discrete Math.,* 126:385–389.
971. Stiebitz, M. (1996). Decomposing graphs under degree constraints. *J. Graph Theory,* 23:321–324.
972. Stiebitz, M. (2017). A relaxed version of the Erdős-Lovász Tihany conjecture. *J. Graph Theory,* 85:278–287.
973. Stiebitz, M., Scheide, D., Toft, B. and Favrholdt, L. (2012). *Graph Edge Coloring: Vizing's Theorem and Goldberg's Conjecture.* Wiley-Interscience Series in Discrete Mathematics and Optimization, John Wiley & Sons.
974. Stiebitz, M., Storch, P. and Toft, B. (2016). Decomposable and indecomposable critical hypergraphs. *J. Combin.,* 7:423–451.
975. Stiebitz, M. and Toft, B. (1995). An abstract generalization of a map reduction theorem of Birkhoff. *J. Combin. Theory Ser. B,* 65:165–185.
976. Stiebitz, M. and Toft, B. (2015). Brooks's theorem. In: *Topics in Chromatic Graph Theory,* pages 36–55. Cambridge University Press.

977. Stiebitz, M. and Toft, B. (2018). A Brooks type theorem for the maximum local edge connectivity. *Electron. J. Combin.,* 25:#P1.50.
978. Stiebitz, M., Tuza, Zs. and Voigt, M. (2009). On list critical graphs. *Discrete Math.,* 309:4931–4941.
979. Stiebitz, M., Tuza, Zs. and Voigt, M. (2015). Orientations of graphs with prescribed weighted out-degrees. *Graphs Combin.,* 31:265–280.
980. Stiebitz, M. and Voigt, M. (2015). List-colourings. In: *Topics in Chromatic Graph Theory*, pages 114–136. Cambridge University Press.
981. Stockmeyer, L. (1973). Planar 3-colorability is polynomial complete. *ACM SIGACT News,* 5(3):19–25.
982. Su, X.Y. (1995). On complete subgraphs of color-critical graphs. *Discrete Math.,* 143:243–249.
983. Sumner, D.P. (1981). Subtrees of a graph and chromatic number. In: *The Theory and Application of Graphs*, pages 557–576. John Wiley & Sons.
984. Sylvester, J.J. (1878). On an application of the new atomic theory to the graphical representation of the invariants and covariants of binary quantics, with three appendices. *Amer. J. Math.,* 1:64–104.
985. Szekeres, G. and Wilf, H.S. (1968). An inequality for the chromatic number of a graph. *J. Combin. Theory,* 4:1–3.
986. Tait, P.G. (1878-1880). On the colouring of maps. *Proc. Roy. Soc. Edinburgh,* 10:501–503,729.
987. Tardif, C. (2001). Fractional chromatic numbers of cones over graphs. *J. Graph Theory,* 38:87–94.
988. Tardif, C. (2021). The chromatic number of the product of 14-chromatic graphs can be 13. *Combinatorica,* 1–8.
989. Tardif, C. (2023). The chromatic number of the product of 5-chromatic graphs can be 4. *Combinatorica,*
990. Tardif, C. and Zhu, X. (2002). The level of nonmultiplicativity of graphs. *Discrete Math.,* 244:461-471.
991. Tardif, C. and Zhu, X. (2019). A note on Hedetniemi's conjecture, Stahl's conjecture and the Poljak-Rödl function. *Electron. J. Combin.,* 26:#P4.32.
992. Tarjan, R. (1972). Depth-first search and linear graph algorithms. *SIAM J. Comput.,* 1:146–160.
993. Tarsi, M. (1999). The graph polynomial and the number of proper vertex coloring. *Annales de l'institut Fourier,* 49:1089–1093.
994. Tashkinov, V.A. (2000). On an algorithm for the edge coloring of multigraphs (in Russian). *Diskretn. Anal. Issled. Oper. Ser. 1,* 7:72–85.
995. Thomas, R. (1998). An update on the four-color theorem. *Notices Amer. Math. Soc.,* 45:848–859.
996. Thomas, R. and Walls, B. (2004). Three-coloring Klein bottle graphs of girth five. *J. Combin. Theory Ser. B,* 92:115–135.
997. Thomason, A. (1979). Critically partitionable graphs, I.. *J. Combin. Theory Ser. B,* 27:254–259.
998. Thomason, A. (1982). Critically partitionable graphs II.. *Discrete Math.,* 41:67–77.
999. Thomason, A. (1984). An extremal function for contractions of graphs. *Math. Proc. Cambridge Philos. Soc.,* 95:261–265.
1000. Thomassen, C. (1974). Graphs in which every path is contained in a Hamilton path. *J. Reine Angew. Math.,* 268:271–281.
1001. Thomassen, C. (1983). Graph decomposition with constraints on the connectivity and minimum degree. *J. Graph Theory,* 7:165–167.
1002. Thomassen, C. (1983). Disjoint cycles in digraphs. *Combinatorica,* 3:393–396.
1003. Thomassen, C. (1988). Paths, circuits and subdivisions. In: *Selected Topics in Graph Theory*, Vol. III, pages 97–133. Academic Press.
1004. Thomassen, C. (1989). The graph genus problem is NP-complete. *J. Algorithms,* 10:568–576.

1005. Thomassen, C. (1992). The even cycle problem for directed graphs. *J. Amer. Math. Soc.,* 5:217–229.
1006. Thomassen, C. (1993). Triangulating a surface with a prescribed graph. *J. Combin. Theory Ser. B,* 57:196–206.
1007. Thomassen, C. (1994). Five-coloring graphs on the torus. *J. Combin. Theory Ser. B,* 62:11–33.
1008. Thomassen, C. (1994). Every planar graph is 5-choosable. *J. Combin. Theory Ser. B,* 62:180–181.
1009. Thomassen, C. (1994). Grötzsch's 3-color theorem and its counterparts for the torus and the projective plane. *J. Combin. Theory Ser. B,* 62:268–279.
1010. Thomassen, C. (1995). 3-list-coloring planar graphs of girth 5. *J. Combin. Theory Ser. B,* 64:101–107.
1011. Thomassen, C. (1995). Embeddings and minors. In: *Handbook of Combinatorics*, Vol. I, pages 301–349. North-Holland.
1012. Thomassen, C. (1997). Color-critical graphs on a fixed surface. *J. Combin. Theory Ser. B,* 70:67–100.
1013. Thomassen, C. (2002). On the chromatic number of triangle-free graphs of large minimum degree. *Combinatorica,* 22:591–596.
1014. Thomassen, C. (2003). The chromatic number of a graph of girth 5 on a fixed surface. *J. Combin. Theory Ser. B,* 87:38–71.
1015. Thomassen, C. (2003). A short list color proof of Grötzsch's theorem. *J. Combin. Theory Ser. B,* 88:189–192.
1016. Thomassen, C. (2005). Some remarks on Hajós' conjecture. *J. Combin. Theory Ser. B,* 93:95–105.
1017. Thomassen, C., Wu, Y. and Zhang, C.-Q. (2016). The 3-flow conjecture, factors modulo k, and the 1-2-3-conjecture. *J. Combin. Theory Ser. B,* 121:308–325.
1018. Tietze, H. (1910). Einige Bemerkungen über das Problem des Kartenfärbens auf einseitigen Flächen. *Jahresber. Dtsch. Math.-Ver.,* 19:155–159.
1019. Toft, B. (1970). *Some Contributions to the Theory of Colour-critical Graphs*. Ph.D. Thesis, University of London. Published as No. 14 in Various Publication Series, Mathematisk Institut, Aarhus Universitet.
1020. Toft, B. (1970). On the maximal number of edges of critical k-chromatic graphs. *Studia Sci. Math. Hungar.,* 5:461–470.
1021. Toft, B. (1972). Two theorems on critical 4-chromatic graphs. *Studia Sci. Math. Hungar.,* 7:83–89.
1022. Toft, B. (1972). On separating set of edges in contraction-critical graphs. *Math. Ann.,* 196:129–147.
1023. Toft, B. (1974). Color-critical graphs and hypergraphs. *J. Combin. Theory Ser. B,* 16:145–161.
1024. Toft, B. (1974). On critical subgraphs of colour-critical graphs. *Discrete Math.,* 7:377–392.
1025. Toft, B. (1975). On colour-critical hypergraphs. In: *Infinite and Finite Sets* (Keszthely, Hungary, 1973), pages 1445–1457. Coll. Math. Soc. János Bolyai, Vol. 10, North-Holland.
1026. Toft, B. (1978). An investigation of colour-critical graphs with complements of low connectivity. In: *Advances in Graph Theory,* pages 279–287. Ann. Discrete Math., Vol. 3, North-Holland.
1027. Toft, B. (1985). Some problems and results related to subgraphs of colour critical graphs. In: *Graphen in Forschung und Unterricht: Festschrift K. Wagner*, pages 178–186., Barbara Franzbecker Verlag.
1028. Toft, B. (1990). 75 graph coloring problems. In: *Graph Colorings*, pages 9–35. Pitman Res. Notes, Vol. 218, Longman.
1029. Toft, B. (1996). A survey of Hadwiger's conjecture. *Congr. Numer.,* 115:249–283.
1030. Tsai, M.-T. and West, D.B. (2011). A new proof of 3-colorability of Eulerian triangulations. *Ars Math. Contemp.,* 4:73–77.
1031. Turan, P. (1941). Eine Extremalaufgabe aus der Graphentheorie. *Mat. Fiz. Lapok,* 48:435–452.

1032. Tutte, W.T. (1956). A contribution to the theory of chromatic polynomials. *Canad. J. Math.,* 6:80–91.
1033. Tutte, W.T. (1960). A colloquiun on graph theory. *Canad. Math. Bull.,* 3:102–103.
1034. Tutte, W.T. (1961). On the problem of decomposing a graph into n connected factors. *J. London Math. Soc.,* 36:221–230.
1035. Tutte, W.T. (1966). On the algebraic theory of graph colorings. *J. Combin. Theory,* 1:15–50.
1036. Tutte, W.T. (1979). *Selected Papers by W. T. Tutte, Volumes I and II.* (edited by McCarthy, D. and Stanton, R. G.), Charles Babbage Research Center.
1037. Tutte, W.T. (1998). *Graph Theory As I Have Known It.* Oxford University Press.
1038. Tuza, Zs. (1992). Theorem proving through depth-first test. *Acta Univ. Carol., Math. Phys.,* 33:135–141.
1039. Tuza, Zs. (1992). Graph coloring in linear time. *J. Combin. Theory Ser. B,* 55:236–243.
1040. Tuza, Zs. (1997). Graph coloring with local constraints – a survey. *Discuss. Math., Graph Theory,* 17:161–228.
1041. Tuza, Zs. (2013). Problems on cycle and colorings. *Discrete Math.,* 313:2007–2013.
1042. Tuza, Zs. and Voigt, M. (1996). On a conjecture of Erdős, Rubin and Taylor. *Tatra Mt. Math. Publ.,* 9:69–82.
1043. Tverberg, H. (1983). On Brooks' theorem and some related results. *Math. Scand.,* 52:37–40.
1044. Tverberg, H. (1984). On Schmerl's efecctive version of Brooks' theorem. *J. Combin. Theory Ser. B,* 37:27–30.
1045. Urquhart, A. (1997). The graph constructions of Hajós and Ore. *J. Graph Theory,* 26:211–215.
1046. Vince, A. (1988). Star chromatic number. *J. Graph Theory,* 12:551–559.
1047. Vishwanathan, S (2003). On 2-coloring certain k-uniform hypergraphs. *J. Combin. Theory Ser. A,* 101:168–172.
1048. Vitaver, L.M. (1962). Determination of minimal coloring of vertices of a graph by means of Boolean powers of the incidence matrix (in Russian). *Dokl. Akad. Nauk. SSSR,* 147:758–759.
1049. Vizing, V.G. (1964). On an estimate of the chromatic class of a p-graph (in Russian). *Diskret. Analiz,* 3:25–30.
1050. Vizing, V. G. (1968). Some unsolved problems in graph theory (in Russian). *Uspekhi Mat. Nauk,* 23:117–134. English translation in: *Russian Mathematical Surveys,* 23:125–141.
1051. Vizing, V.G. (1976). Coloring the vertices of a graph in prescribed colors (in Russian). *Diskret. Analiz,* 29:3–10.
1052. Voigt, M. (1993). List colourings of planar graphs. *Discrete Math.,* 120:215–219.
1053. Voigt, M. (1995). A not 3-choosable planar graph without 3-cycles. *Discrete Math.,* 146:325–328.
1054. Voigt, M. (2007). A non-3-choosable planar graph without cycles of length 4 and 5. *Discrete Math.,* 307:1013–1015.
1055. Voss, H.-J. (1977). Graphs with prescribed maximal subgraphs and critical chromatic graphs. *Comment. Math. Univ. Carolinae,* 18:129–142.
1056. Vu, V. (2002). A general upper bound on the list chromatic number of locally sparse graphs. *Combin. Probab. Comput.,* 11:103–111.
1057. Wagner, K. (1937). Über eine Eigenschaft der ebenen Komplexe. *Math. Ann.,* 114:570–590.
1058. Wagner, K. (1964). Beweis einer Abschwächung der Hadwiger-Vermutung. *Math. Ann.,* 153:139–141.
1059. Wagner, K. (1970). *Graphentheorie.* BI Hochschultaschenbücher, Band 248.
1060. Wang, Y. and Wu, R. (2023). Graphs with girth 9 and without larger odd holes are 3-colorable. Manuscritpt arXive:2307.01460
1061. Wanless, I.M. and Wood, D.R. (2022). A general framework for hypergraph coloring. *SIAM J. Discrete Math.,* 36:1663–1677.
1062. Weichsel, P. (1962). The Kronecker product of graphs. *Proc. Amer. Math. Soc.,* 13:47–52.

1063. Weinstein, J (1975). Excess in critical graphs. *J. Combin. Theory Ser. B,* 18:24–31.
1064. Wernicke, P. (1904). Über den katographischen Vierfarbensatz. *Math. Ann.,* 58:413–426.
1065. Wessel, W. (1981). *Critical Lines, Critical Graphs, and Odd Cycles.* Report, Akademie der Wissenschaften der DDR, Institut für Mathematik.
1066. West, D.B. (2001). *Introduction to Graph Theory.* Second Edition. Prentice-Hall.
1067. White, A.T. (1973/1984). *Graphs, Groups, and Surfaces.* North-Holland 1973; Mathematics Studies 8, Revised Edition 1984.
1068. Whitney, H. (1932). The coloring of graphs. *Ann. of Math.,* 33:688–718.
1069. Wilf, H.S. (1967). The eigenvalues of a graph and its chromatic number. *J. London Math. Soc.,* 42:330–332.
1070. Wilson, R. (2002). *Four Color Suffice: How the Map Problem Was Solved.* Revised Color Edition, 2014, Princeton Science Library 128.
1071. Woeginger, G.J. and Sgall, J. (2001). The complexity of coloring graphs without long induced paths. *Acta Cybernet.,* 15:107–117.
1072. Woodall, D.R. (1972). Property B and the four-color problem. In: *Combinatorics* (Proc. Oxford Combinatorial Conference, Institute of Mathematics and its Applications, Southend-on-Sea, 1972), pages 322–340.
1073. Woodall, D.R. (1999). Edge-choosability of multicircuits. *Discrete Math.,* 202:271–277.
1074. Woodall, D.R. (2001). List colourings of graphs. In: *Surveys in Combinatorics, 2001,* pages 269–301. London Math. Soc. Lecture Notes Series, Vol. 288, Cambridge University Press.
1075. Wrochna, M. (2019). On inverse powers of graphs and topological implications of Hedetniemi's conjecture. *J. Combin. Theory Ser. B,* 139:267–295.
1076. Wrochna, M. (2020). Smaller counterexamples to Hedetniemi's conjecture. Manuscript arXiv:2012.13558
1077. Wu, Y.Q., Yuan, J.J. and Zhao, Y.C. (1996). Partition a graph into two induced forests. *J. Math. Study,* 29:1–6.
1078. Yang, A. and Yuan, J. (2006). Partition the vertices of a graph into one independent set and one acyclic set. *Discrete Math.,* 306:1207–1216.
1079. Yao, T. and Zhou, G. (2017). Constructing a family of 4-critical planar graphs with high edge density. *J. Graph Theory,* 86:244–249.
1080. Younger, D.H. (1983). Integer flows. *J. Graph Theory,* 7:349–357.
1081. Younger, D.H. (2004). Dedication W. T. Tutte. *J. Combin. Theory Ser. B,* 92:193–198.
1082. Younger, D.H. (2012). Williams Thomas Tutte. *Biogr. Mems Fell. R. Soc.,* 58:283–297.
1083. Youngs, D.A. (1992). Gallai's problem on Dirac's construction. *Discrete Math.,* 101:343–350.
1084. Youngs, D. A. (1995). Minimal orientations of colour critical graphs. *Combinatorica,* 15:289–295.
1085. Youngs, D.A. (1996). 4-chromatic projective graphs. *J. Graph Theory,* 21:219–227.
1086. Youngs, J.W.T. (1963). Minimal imbeddings and the genus of a graph. *J. Math. Mech.,* 12:303–315.
1087. Yuster, R. (1997). Independent transversals in r-partite graphs. *Discrete Math.,* 176:255–261.
1088. Zajac, M. (2018). A short proof of Brooks' theorem. Manuscript arXiv:1805.11176
1089. Zaker, M. (2008). New bounds for the chromatic number of graphs. *J. Graph Theory,* 58:110–122.
1090. Zaslavsky, T. (1982). Signed graphs. *Discrete Applied Math.,* 4:47–74.
1091. Zhou, Q. (2023). Hadwiger's conjecture for some graphs with independence number two. Manuscript arXive:2305.05868
1092. Zhu, X. (1992). Star chromatic numbers and products of graphs. *J. Graph Theory,* 16:557–569.
1093. Zhu, X. (1998). A survey on Hedetniemi's conjecture. *Taiwanese J. Math.,* 2:1–24.
1094. Zhu, X. (2001). Circular chromatic number: a survey. *Discrete Math.,* 229:371–410.
1095. Zhu, X. (2001). An analogue of Hajós' theorem for the circular chromatic number. *Proc. Amer. Math. Soc.,* 129:2845–2852.

1096. Zhu, X. (2003). An analogue of Hajós' theorem for the circular chromatic number. II.. *Graphs Combin.,* 19:419–432.
1097. Zhu, X. (2006). Recent developments in circular colouring of graphs. In: *Topics in Discrete Mathematics*. Algorithms and Combinatorics, Vol 26, pages 497–550. Springer.
1098. Zhu, X. (2008). Refined activation strategy for the marking game. *J. Combin. Theory Ser. B,* 98:1–18.
1099. Zhu, X. (2009). On-line list colouring of graphs. *Electron. J. Combin.,* 16:#R127.
1100. Zhu, X.. (2011). The fractional version of Hedetniemi's conjecture is true. *European J. Combin.,* 32:1168–1175.
1101. Zhu, X. (2020). A note on the Poljak-Rödl function. *Electron. J. Combin.,* 27:#P3.2.
1102. Zhu, X. (2019). The Alon-Tarsi number of planar graphs. *J. Combin. Theory Ser. B,* 134:354–358.
1103. Zhu, X. (2021). Relatively small counterexamples to Hedetniemi's conjecture. *J. Combin. Theory Ser. B,* 146:141–150.
1104. Ziegler, G.M. (2002). Generalized Kneser coloring theorems with combinatorial proofs. *Invent. Math.,* 147:671–691.
1105. Zykov, A.A. (1949). On some properties of linear complexes (in Russian). *Mat. Sbornik N. S.,* 24:163–188. English translation in: *Amer. Math. Soc. Transl.,* (1952), 79:418–449.

Graph and Hypergraph Parameters

$\alpha(G)$	independence number of G	2, 562
$\alpha'(G)$	matching number of G	571
$\mathrm{ad}(G)$	average degree of G	569
$\chi(G)$	chromatic number of G	1, 356
$\chi(G,g)$	chromatic number of (G,g)	40
$\chi(G:\mathcal{P})$	$\mathcal{P}$-chromatic number of G	90
$\chi'(G)$	$=\chi(\mathrm{L}(G))$, chromatic index of G	37
$\chi''(G)$	$=\chi(\mathrm{T}(G))$, total chromatic number of G	39
$\chi^{=}(G)$	equitable chromatic number of G	31
$\chi_{\mathrm{DP}}(G)$	DP-chromatic number of G	16
$\chi_\ell(G)$	list chromatic number of G	10, 357
$\chi_\ell(G,g)$	list chromatic number of (G,g)	41
$\chi_\ell(G:\mathcal{P})$	$\mathcal{P}$-list chromatic number of G	90
$\chi'_\ell(G)$	$=\chi_\ell(\mathrm{L}(G))$, list chromatic index of G	38
$\chi''_\ell(G)$	$=\chi_\ell(\mathrm{T}(G))$, total list chromatic number of G	39
$\chi^*(G)$	fractional chromatic number of G	404
$\chi^\circ(G)$	circular chromatic number of G	413
$\vec{\chi}(D)$	dichromatic number of D	161
$\vec{\chi}_\ell(D)$	list dichromatic number of D	161
$\mathrm{c}(G)$	$=\max CL(G)$, circumference of G	78
$\mathrm{c_o}(G)$	$=\max CL_o(G)$, odd circumference of G	78
$\mathrm{c_e}(G)$	$=\max CL_e(G)$, even circumference of G	78
$\mathrm{col}(G)$	coloring number of G	11
$\delta(G)$	minimum degree of G	554
$\Delta(G)$	maximum degree of G	554
$\Delta^+_{\mathrm{fr}}(G)$	fractional maximum out-degree of G	123
$\delta^-(D)$	minimum in-degree of D	572

M. Stiebitz et al., *Brooks' Theorem*, Springer Monographs in Mathematics,
https://doi.org/10.1007/978-3-031-50065-7

$\delta^+(D)$	minimum out-degree of D	572
$\Delta^-(D)$	maximum in-degree of D	572
$\Delta^+(D)$	maximum out-degree of D	572
$\Delta^+_{\mathcal{D}}(G)$	$= \min\{\Delta^+(D) \mid D \in \mathcal{D}, G = G_D\}$	121
$\Delta^+_{\text{all}}(G)$	$= \Delta^+_{\mathcal{D}}(G)$ where $\mathcal{D}$ are all multidigraphs	121
$\Delta^+_{\text{ac}}(G)$	$= \Delta^+_{\mathcal{D}}(G)$ where $\mathcal{D}$ are all acyclic digraphs	121
$\Delta^+_{\text{ke}}(G)$	$= \Delta^+_{\mathcal{D}}(G)$ where $\mathcal{D} = \{D \mid D$ kernel perfect$\}$	125
$\Delta^+_{\text{at}}(G)$	$= \Delta^+_{\mathcal{D}}(G)$ where $\mathcal{D} = \{D \mid \varepsilon_e(D) \neq \varepsilon_o(D)\}$	147
$\varepsilon(D)$	number of spanning Eulerian subdigraphs of D	134
$\varepsilon_o(D)$	number of odd spanning Eulerian subdigraphs of D	134
$\varepsilon_e(D)$	number of even spanning Eulerian subdigraphs of D	134
$\text{exc}_k(G)$	$= 2\lvert E(G)\rvert - (k-1)\lvert G\rvert$, k-excess of G	232
$g(G)$	$= \min CL(G)$, girth of G	559
$g_o(G)$	$= \min CL_o(G)$, odd girth of G	559
$\kappa(G)$	connectivity of G	557
$\kappa_G(v,w)$	local connectivity of vertex pair (v,w) in G	178
$\bar{\kappa}(G)$	maximum local connectivity of G	178
$\kappa'(G)$	edge-connectivity of G	557
$\kappa'_G(v,w)$	local edge connectivity of vertex pair (v,w) in G	178
$\bar{\kappa}'(G)$	maximum local edge-connectivity of G	178
$\Bbbk_p(G)$	$= \lvert\mathcal{K}_p(G)\rvert$	188
$\ell(P), \ell(C)$	length of a path P and a cycle C	556
$\text{mad}(G)$	maximum average degree of G	12
$\mu(G)$	maximum multiplicity of G	570
$\text{odd}(G)$	number of components G having odd order	571
$\omega(G)$	clique number of G	2, 562
$\omega_{\preceq}(G)$	Hadwiger number of G	229
$\omega^*(G)$	fractional clique number of G	405
$\text{rad}(G)$	radius of G	559

Graph and Hypergraph Families

$BG(n,\alpha)$	Borsuk graph, also $BG'(n,\alpha)$	454
$\mathcal{BF}(n)$	$= \{BG(n,\alpha) \mid 0 < \alpha < 2\}$	457
$\mathcal{BF}'(n)$	$= \{BG'(n,\alpha) \mid 0 < \alpha < 2\}$	457
$\text{Crit}(k)$	k-critical graphs	33
$\text{Crit}(k,n)$	$= \{G \in \text{Crit}(k) \mid \lvert G\rvert = n\}$	167
$\text{Crit}_{\trianglelefteq}(\mathcal{P})$	$\trianglelefteq$-minimal graphs not in $\mathcal{P}$	568
$\text{Crit}^*(k,n)$	graphs in $\text{Crit}(k,n)$ with no dominating vertex	186
$\text{Crit}_\ell(k)$	χ_ℓ-critical graphs with $\chi_\ell = k$	281
$\text{Crit}_\ell(k,n)$	$= \{G \in \text{Crit}_\ell(k) \mid \lvert G\rvert = n\}$	281
$\text{Crit}_{\text{DP}}(k)$	χ_{DP}-critical graphs with $\chi_{\text{DP}} = k$	489
$\text{Crit}_{\text{DP}}(k,n)$	$= \{G \in \text{Crit}_{\text{DP}}(k) \mid \lvert G\rvert = n\}$	490
$\text{Crit}(\mathbf{S},k)$	$= \text{Crit}(k) \cap \mathcal{G}(\hookrightarrow \mathbf{S})$	487
$\text{Crit}(\mathbf{S},k,g)$	$= \{G \in \text{Crit}(\mathbf{S},k) \mid g(G) \geq g\}$	513
$\text{Crit}_\ell(\mathbf{S},k)$	$= \text{Crit}_\ell(k) \cap \mathcal{G}(\hookrightarrow \mathbf{S})$	526
$\text{Crit}_\ell(\mathbf{S},k,g)$	$= \{G \in \text{Crit}_\ell(\mathbf{S},k) \mid g(G) \geq g\}$	526
$\text{Crit}_{\text{DP}}(\mathbf{S},k)$	$= \text{Crit}_{\text{DP}}(k) \cap \mathcal{G}(\hookrightarrow \mathbf{S})$	503
$\text{Crit}_{\text{DP}}(\mathbf{S},k,g)$	$= \{G \in \text{Crit}_{\text{DP}}(\mathbf{S},k) \mid g(G) \geq g\}$	513
$\text{Crih}(k)$	k-critical hypergraphs	362
$\text{Crih}(k,n)$	k-critical hypergraphs of order n	362
$CG(n,d)$	circular graph	414
$C\mathcal{T}(n)$	$= \{G \in \mathcal{G} \mid \text{coind}(\mathbf{T}_H(G)) \geq n\}$	461
$\mathcal{D}$	digraph property	121
$\mathcal{DG}(k)$	subfamily of $\text{Crit}(k,2k-1)$	51
$\mathcal{DG}'(k)$	subfamily of $\text{Crit}(k,2k-1)$	232
$\mathcal{DG}_k$	k-degenerate graphs	99
$\mathcal{EG}(k)$	subfamily of $\text{Crit}(k,2k)$	51
$\text{Ext}(k,n)$	$= \{G \in \text{Crit}(k,n) \mid \lvert E(G)\rvert = \text{ext}(k,n)\}$	231

M. Stiebitz et al., *Brooks' Theorem*, Springer Monographs in Mathematics,
https://doi.org/10.1007/978-3-031-50065-7

$\mathrm{Exth}(k,n)$	$= \{H \in \mathrm{Crih}(k,n) \mid \lvert E(H)\rvert = \mathrm{exth}(k,n)\}$	380
$\mathrm{Forb}_{\trianglelefteq}(\mathcal{X})$	$\mathcal{X}$-free graphs w.r.t. the graph relation $\trianglelefteq$	568
$\mathrm{Forb}(\mathcal{X})$	$\mathcal{X}$-free graphs w.r.t. the induced subgraph relation	567
$\mathcal{G}$	class of all graphs	567
$\mathcal{G}(\rho \le x)$	$= \{G \in \mathcal{G} \mid \rho(G) \le x\}$	569
$\mathcal{G}(\hookrightarrow \mathbf{S})$	$= \{G \in \mathcal{G} \mid G$ is embeddable on $\mathbf{S}\}$	574
$GY(k)$	Gyárfás graph	428
$GY(k,\ell)$	Gyárfás–Simonyi–Tardos graph	429
$\mathcal{GF}(n)$	$= \{GY(n+2,\ell) \mid \ell \in \mathbb{N}_0\}$	453
$\mathcal{H}$	hypergraph property	360
K_n, C_n, P_n	complete graph, cycle and path of order n	556
$K_{p,q}$	complete bipartite graph with parts of size p and q	562
$K_{p,q,r}$	complete tripartite graph with parts of size p,q,r	564
K_n^q	complete q-uniform hypergraph of order n	583
$KG(n,p)$	Kneser graph	407
$\mathcal{KF}(n)$	$= \{KG(n+2p,p) \mid p \in \mathbb{N}\}$	453
$\mathcal{LG}$	line graphs	102
$LB(n,\alpha)$	Lovász–Borsuk graph	455
$\mathcal{LF}(n)$	$= \{LB(n,\alpha) \mid \alpha_n' < \alpha < 2\}$	457
$\mathcal{MG}(2)$	$= \{G \in \mathcal{G} \mid G = K_2\}$	422
$\mathcal{MG}(k+2)$	$= \{\mathrm{M}_{r_1,r_2,\ldots,r_k}(K_2) \mid r_1,r_2,\ldots,r_k \in \mathbb{N}\}$ with $k \ge 1$	421
$\mathcal{MF}(n)$	$= \mathcal{MG}(n+2)$	453
$\mathcal{P}$	graph property	567
$\mathcal{PG}$	perfect graphs	101
P_n^o	path P_n with a loop at one end	441
$SG(n,p)$	Schrijver graph	412
$\mathcal{SF}(n)$	$= \{SG(n+2p,p) \mid p \in \mathbb{N}\}$	453
TG_p	Toft graph	200
W_n	$= K_1 \boxplus C_n$, a wheel	563
$W(\ell,d)$	$= C_\ell \boxplus K_d$, a d-wheel	188

Operations and Relations

$G-F$	$=(V(G),E(G)\setminus F)$ with $F\subseteq E(G)$	555
$G-e$	$G-\{e\}$ with $e\in E(G)$, edge-deleted subgraph	555
$G+uv$	$=(V(G),E(G)\cup\{uv\})$ with $u,v\in V(G)$, $u\neq v$	555
$G-X$	$G[V(G)\setminus X]$ with $X\subseteq V(G)$	555
$G-v$	$=G-\{v\}$ with $v\in V(G)$, vertex-deleted subgraph	555
G/I	identifying the independent set I of G	170
G/e	contracting the edge e of G	187
$\overline{G}$	complement of G	562
$G\Box H$	cartesian product of G and H	563
$G[H]$	composition of G and H	563
$G\times H$	direct product of G and H	563
$G[g]$	inflation of G with $g: V(G)\to\mathbb{N}_0$	41
$G_1\boxplus G_2$	Dirac join of G_1 and G_2	168
$G_1\nabla G_2$	Hajós join of G_1 and G_2	168
$G_1\nabla^{o}G_2$	Ore join of G_1 and G_2	171
$G_1\blacktriangledown G_2$	Gallai join of G_1 and G_2	176
$G_1\boxminus G_2$	Toft join of G_1 and G_2	210
$G\to H$	G is homomorphic to H	401
$G\nrightarrow H$	G is not homomorphic to H	401
$H\cong G$	H is isomorph to G	555
$H\subseteq G$	H is subgraph of G	555
$H\subset G$	H is proper subgraph of G	555
$H\leqslant G$	H is induced subgraph of G	555
$H<G$	H is proper induced subgraph of G	555
$H\preceq G$	H is minor of G	564
$H\preceq_t G$	H is topological minor of G	564
$H\cap G$	intersection of H and G	562
$H\cup G$	union of H and G	562
$H-X$	$=H[V(H)\setminus X]$ with $X\subseteq V(H)$	582

M. Stiebitz et al., *Brooks' Theorem*, Springer Monographs in Mathematics,
https://doi.org/10.1007/978-3-031-50065-7

$H \div X$	$= H\langle V(H) \setminus X\rangle$ with $X \subseteq V(H)$	582
$H - F$	$= (V(H), E(G) \setminus F)$ with $F \subseteq 2^{V(H)}$	583
$H + F$	$= (V(H), E(G) \cup F)$ with $F \subseteq 2^{V(H)}$	583
$\mathrm{L}(G)$	line graph of G	37
$\mathrm{M}_r(G)$	generalized Mycielski graph of G	418
$\mathrm{M}_{r_1,r_2,\ldots,r_k}(G)$	iterated Mycielski graph of G	418
$\mathcal{P} \to \mathcal{P}'$	$\mathcal{P}'$ hom-bounded from below by $\mathcal{P}$	453
$\mathcal{P} \leftrightarrow \mathcal{P}'$	$\mathcal{P}$ hom-equivalent to $\mathcal{P}'$	453
$\mathrm{T}(G)$	total graph of G	39
tH	multigraph G with $\mu_G(v,w) = t\mu_H(v,w)$	17

Other Notation

$\mathrm{Aut}(G)$	automorphism group of G	566
$\mathfrak{B}(G)$	set of blocks of G	558
$\mathfrak{B}_v(G)$	set of blocks B of G with $v \in V(B)$	558
$\mathrm{bro}(\Delta,\omega)$	$= \max\{\chi(G) \mid \Delta(G) = \Delta$ and $\omega(G) = \omega\}$	293
$C(G)$	maximum cliques of G	295
$\mathrm{CL}(G)$	cycle lengths in G	76
$\mathrm{CL}_e(G)$	even cycle lengths in G	76
$\mathrm{CL}_o(G)$	odd cycle lengths in G	76
$\mathcal{CO}(G,C)$	colorings of G with color set C	61
$\mathcal{CO}(G,k)$	colorings of G with color set $C = [1,k]$	61
$d_G(v)$	degree of v in G	554
$d_G(f)$	degree of f in G	575
$\mathrm{dist}_G(v,w)$	distance of v and w in G	558
$d_D^-(v)$	in-degree of v in D	572
$d_D^+(v)$	out-degree of v in D	572
$D[X]$	subdigraph of D induced by $X \subseteq V(D)$	572
D/F	reverse the orientation of $e \in F \subseteq E(D)$	132
$D(F)$	$= (V(D),F)$, subdigraph of D with $F \subseteq E(D)$	132
$E(G)$	edge set of G	554
$E_G(v)$	set of edges incident with vertex v in G	554
$E_G(X,Y)$	set of edges incident joining X and Y in G	554
$E_D^-(v)$	edges e of D with $v_D^-(e) = v$	571
$E_D^+(v)$	edges e of D with $v_D^+(e) = v$	571
$\varepsilon(\mathbf{S})$	Euler characteristic of $\mathbf{S}$	573
$\mathrm{eg}(\mathbf{S})$	$= 2 - \varepsilon(\mathbf{S})$, Euler genus of $\mathbf{S}$	573
$\mathrm{ext}(k,n)$	$= \min\{\lvert E(G)\rvert \mid G \in \mathrm{Crit}(k,n)\}$	231

M. Stiebitz et al., *Brooks' Theorem*, Springer Monographs in Mathematics,
https://doi.org/10.1007/978-3-031-50065-7

$\text{ext}_\ell(k,n)$	$= \min\{\lvert E(G)\rvert \mid G \in \text{Crit}_\ell(k,n)\}$	281
$\text{ext}_{\text{DP}}(k,n)$	$= \min\{\lvert E(G)\rvert \mid G \in \text{Crit}_{\text{DP}}(k,n)\}$	490
$\text{exth}(k,n)$	$= \min\{\lvert E(H)\rvert \mid H \in \text{Crih}(k,n)\}$	380
$\lvert G\rvert$	$= \lvert V(G)\rvert$, order of G	554
$G[X]$	subgraph of G induced by $X \subseteq V(G)$	555
G_D	underlying graph of a multidigraph D	121
G_H	high vertex subgraph of a critical graph G	35
G_L	low vertex subgraph of a critical graph G	35
(G,g)	weighted graph with $g: V(G) \to \mathbb{N}_0$	40
$g_k(n,c)$	Gallai-function	237
$\text{Hom}(G,H)$	homomorphisms from G to H	439
$H[X]$	subhypergraph of H induced by $X \subseteq V(H)$	582
$H\langle X\rangle$	hypergraph H shrinked to $X \subseteq V(H)$	582
$H(\mathbf{S})$	Heawood number of $\mathbf{S}$	578
h_Δ	$= \max\{h \in \mathbb{N}_0 \mid (h+2)(h+1) \leq \Delta\}$	328
$\mathcal{I}(G)$	independent sets of G	335
$\mathcal{I}(G,v)$	independent sets of G containing v	404
$\mathcal{K}_p(G)$	subgraphs K of G isomorphic to K_p	188
K^G	the map graph of G w.r.t. K	441
$\text{koy}(k,n)$	Kostochka–Yancey function	247
$L_k(n)$	$= \min\{\text{c}(G) \mid G \in \text{Crit}(k,n)\}$	206
$\mu_G(v,w)$	multiplicity of two distinct vertex v and w in G	570
$N_G(v)$	set of neighbors of v in G	554
$N_G(X)$	set of neighbors of X in G	554
$N_G^p(X)$	pth distance class of X in G	426
$N_D^-(v)$	set of in-neighbors of v in D	571
$N_D^+(v)$	set of out-neighbors of v in D	571
$N(\mathbf{S},k)$	$= \max\{\lvert G\rvert \mid G \in \text{Crit}(\mathbf{S},k)\}$	487
$\mathbf{N}_h$	nonorientable surface with $h \geq 1$	573
$\text{ore}(k,n)$	Ore-function	233
$OC(k,\ell)$	odd (k,ℓ)-colorings of G	426
(Ω,Pr)	probability space	586
P_G, P_D	polynomials associated with G and D, respectively	128
$\partial_G(X)$	$= E_G(X, V(G)\setminus X)$ coboundary of X in G	554
$\rho(\mathbf{S})$	$= \max\{\rho(G) \mid G \in \mathcal{G}(\hookrightarrow \mathbf{S})\}$	579

$R(k,\ell)$	Ramsey number	96
$\mathrm{Sym}(A)$	symmetric group of a set A	554
$\mathbf{S}_g$	orientable surface with $g \geq 0$	573
vTw	the unique v-w path in a tree T	560
$t(n,p)$	Turan number	565
$\mathbf{T}$	topological space	574
$v_D^-(v)$	terminal vertex of e in D	571
$v_D^+(v)$	initial vertex of e in D	571
$V(G)$	vertex set of G	554
$\mathcal{V}_p(G)$	functions from $V(G)$ to $\mathbb{N}_0^p$	83

Index

active component, 319
acyclic
  acyclic digraph, 572
  acyclic graph, 560
acyclic coloring, 161
acyclic L-coloring, 161
adjacency polynomial, 159
adjacent edges, 37
adjacent vertices, 554
algorithm
  certifying algorithm, 106
  graph coloring algorithm, 53
  optimal, 53
  performance guarantee, 54
  polynomial, 53
  suboptimal, 53
  t-absolute approximation algorithm, 54
  t-relative approximation algorithm, 54
  worst-case running time, 54
Alon-Tarsi number, 160, 536
assignment
  f-assignment, 10
  k-assignment, 10, 357
  list-assignment, 10
augmented tree
  r-augmented tree, 71
augmenting edges, 71
automorphism, 566
automorphism group, 566
average degree, 555, 569
avoidable, 28

bad list-assignment, 56
ball
  d-ball, 410
barycentric subdivision, 458
BFS, *see* breadth-first search
bijection, 554
bipartite graph, 562
  bipartition, 562
  partite sets, 562
  parts, 562
bipartite hypergraph, 583
  partite sets, 583
  parts, 583
bipartition, 562
block, 357, 558, 586
body, *see* simplicial complex
Borsuk graph, 454
Borsuk–Ulam theorem, 453
boundary subgraph, 574
bounded graph property
  χ-bounded, 69
  polynomially χ-bounded, 104
bounding function for a graph property
  χ-bounding function, 69
box
  d-dimensional box, 105
box-complexes, 460
breadth-first search, 561
brick, 7
  DP-brick, 17
  k-brick, 7
bridge, 556, 584
broom, 101

cardinality, 553
Cartesian product, 563
categorial product, *see* direct product, 563
Catlin graph, 52
Cayley graph, 473
cellular embedding, 574
certifying algorithm, 106
chain: (c, c')-chain, 5

M. Stiebitz et al., *Brooks' Theorem*, Springer Monographs in Mathematics,
https://doi.org/10.1007/978-3-031-50065-7

characteristic function, 126, 189
Chernoff bounds, 589
choice number, 10
choice ratio, 471
choosability, 55
$(a : b)$-choosability, 60
choosable
degree-choosable, 15
f-choosable, 10
k-choosable, 10
$(n : p)$-choosable, 60
chord, 36
chordal graph, 47
chordless, 36
chromatic
k-chromatic, 1
chromatic index, 37
chromatic number, 1, 40, 356
circular chromatic number, 413
$\mathcal{D}$-chromatic number, 161
dichromatic number, 161
$\mathcal{D}$-list-chromatic number, 161
DP-chromatic number, 16, 60
equitable chromatic number, 31
face chromatic number, 516
fractional chromatic number, 59, 404
game chromatic number, 534
group chromatic number, 535
k-chromatic number, 111
list chromatic number, 10, 41, 357
local chromatic number, 477
odd chromatic number, 438
online list chromatic number, 536
$\mathcal{P}$-chromatic number, 90
$\mathcal{P}$-list-chromatic number, 90
set chromatic number, 59
signed chromatic number, 51
total chromatic number, 39
total list-chromatic number, 39
chromatic-choosable graph, 160
circular chromatic number, 413
circular graph, 414
circular (n, d)-coloring, 413
circumference, 79
claw, 562
clique, 357, 562, 572, 583
p-clique, 562
v-clique, 41
fractional clique, 405
maximum clique, 295
clique number, 2, 357, 562, 583
fractional clique number, 406
clique sum, 47
clique-acyclic, *see* normal digraph
closed out-neighborhood, 126
club, 322
cluster
k-cluster, 190
coboundary, 554
color, 1, 16
color class, 2, 356
color set, 1, 40, 356
colorable
DP-degree-colorable, 17
DP-f-colorable, 16
DP-k-colorable, 16
f-list-colorable, 10, 41
H-colorable, 466
k-colorable, 1, 40, 356
k-list-colorable, 10, 41, 357
L-colorable, 10, 41
list-colorable, 10
$(\mathcal{P}, L)$-colorable, 90
uniquely odd (k, ℓ)-colorable, 436
(X, H)-colorable, 16, 60
coloring, 1, 40, 161, 356
acyclic coloring, 161
acyclic L-coloring, 161
circular (n, d)-coloring, 413
correspondence coloring, 57
$\mathcal{D}$-coloring, 161
$(\mathcal{D}, L)$-coloring, 161
edge coloring, 37
equitable k-coloring, 27
face coloring, 516
fractional coloring, 404
H-coloring, 465
improper coloring, 90
k-coloring, 1, 40, 356
L-coloring, 10, 40
local (m, r)-coloring, 477
multi-coloring, 59
nearly p-uniform C-coloring, 28
odd (k, ℓ)-coloring, 426
optimal coloring, 53
partial coloring, 306
$\mathcal{P}$-coloring, 90
$(\mathcal{P}, L)$-coloring, 90
precoloring, 55
proper coloring, 90
p-uniform C-coloring, 28
set-coloring, 59
(s, t)-wide coloring, 477
total coloring, 39
tuple-coloring, 59
(X, H)-coloring, 16, 60
coloring graph
n-coloring graph, 442

coloring number, 11, 16
coloring of digraphs, 161
coloring of graphs, 1
coloring of hypergraphs, 356
coloring of weighted graphs, 40
coloring problem, 106
  k-coloring problem, 106
common neighborhood, 458
comparability graph, 149
complement, 562
complete
  H is complete to H' in G, 319
  v is complete to H' in G, 319
  two subsets are complete in G, 319
complete d-ary tree of heigth m, 71
complete q-uniform hypergraph, 357, 583
complete bipartite graph, 562
complete club, 322
complete graph, 556
complete multipartite graph, 564
component, 556, 584
composition, 554, 563
conditional colorings, 112
conditional probability, 587
cone, *see* simplicial complex
configuration
  C-configuration, 19
  colorable, 17
  constructible configurations, 19
  degree-feasible, 17
  disjoint, 19
  even C-configuration, 18
  feasible configuration, 17
  K-configuration, 18
  merging, 19
  minimal uncolorable, 17
  odd C-configuration, 18
  uncolorable, 17
Conjecture
  $(a{:}b)$-choosability, 44
  Alon–Krivelevich–Sudakov, 334
  Borodin–Kostochka, 293, 345
  Chen–Raspaud, 470
  Cranston–Rabern, 303
  Digraph, 166
  Double-Critical Graph, 212
  Equitable Δ-Coloring, 32
  Erdős–Hajnal, 104
  Erdős–Lovász Tihany, 228
  Five Flow, 552
  Forest-Plus-Starforest, 67
  Four Flow, 552
  Gyárfás / Sumner, 69
  Gyárfás / Sumner (polynomial version), 104
  Hadwiger, 215
  Heawood, 521
  Jaeger, 467
  Kierstead–Kostochka, 32
  King, 345
  Klein–Schönheim, 68
  Kostochka–Stiebitz, 284
  List-Edge-Coloring, 39
  List-Total-Coloring, 40
  Local DP, 345
  Molloy–Reed, 351
  Ngoc–Tuza, 473
  Ore (hypergraph version), 382
  Planar 2-Coloring, 164
  Postle, 280
  Product, 440
  Reed, 290, 294
  Saaty–Kainen, 229
  Stahl, 469
  Su, 225
  Three Flow, 552
  Total Graph, 40
connected, 44, 572
  k-connected, 557
connected graph, 556
connected hypergraph, 584
connectivity, 557
constructible, 173
constructible configurations, 19
Construction
  Descartes' construction, 108
  Dirac's construction, 168, 364
  Dirac–Gallai construction, 176
  Enlarging operation, 366
  Hajós' construction, 169, 365
  Liu's construction, 395
  Mycielski's construction, 418
  Ore's construction, 171
  Toft's construction, 210
  Toft's splitting operation, 366
  Urquhart's construction, 171
  Zykov's construction, 97
  Zykov-Schäuble-Lovász construction, 385
construction sequence, 173
construction tree, 173
contractible cycle, 575
contracting an edge, 563
contraction-critical, 214
  k-contraction-critical, 214
copy of a graph, 555
core, 56
  k-core, 100
  L-core, 234
core number, 100

core of graph, 463
Corollary
  Ajtai–Komlós–Szemerédi, 270
  Ajtai–Komlós–Szemerédi / Shearer, 269, 270
  Bernshteyn, 267, 340
  Bernshteyn–Kostochka–Pron, 27
  Bonamy–Kelly–Nelson–Postle, 352
  Bondy–Boppana–Siegel, 118
  Borodin / Erdős–Rubin–Taylor, 15
  Cranston–Rabern, 302, 327
  Farzad–Malloy–Reed, 334
  Fisk, 502
  Gallai, 181, 207
  Galvin, 120
  Gimbel–Thomassen / Nilli, 271
  Jensen–Toft, 271
  Kostochka–Stiebitz, 269
  Kostochka–Yancey, 261
  Local (symmetric case), 587
  Molloy–Reed, 332
  Rabern, 313, 314
  Reed, 345, 346
  Simonyi–Tardos, 479
correspondence coloring, 57
cover, 15
  k-cover, 262
cover-critical
  k-cover-critical, 274
  (X, H)-critical, 273
covered, 570
critical
  contraction-critical, 214
  critical graph, 32
  critical hypergraph, 360
  critical vertex/edge, 32
  double-critical, 212
  H-critical, 466
  k-contraction-critical, 214
  k-cover-critical, 274
  k-critical, 32, 360
  k-list-critical, 235, 362
  L-critical, 235, 362
  list-critical, 235
  strong k-critical, 273
  vertex-critical, 107, 362
  (X, H)-critical, 273
  χ_{DP}-critical, 489
  $(\chi_\ell, \mathcal{P})$-vertex-critical, 93
  $(\mathcal{P}, L)$-vertex-critical, 92
  ξ-critical, 280
critical with respect to, 280
cubic graph, 555
cut, *see* edge cut
cycle, 556
cycle-plus-triangles graph, 136

decomposable graph, 181
decomposable hypergraph, 375
decomposition, 570, 572
degeneracy, 100
degenerate
  f-degenerate, 152
  k-degenerate, 61
  strictly h-degenerate, 82
degree
  degree of a face, 575
  degree of a vertex, 355, 554, 570, 582
degree-choosable, 15, 357
degree-feasible, 17
dependency digraph, 587
depth-first search, 560
Descartes' construction, 108
DFS, *see* depth-first search
diamond, 9
dichromatic number, 161
digon, 165
digraph, *see* multidigraph
digraph property, 121
Dirac join, 168, 364
Dirac's bound, 273
Dirac's construction, 168, 364
Dirac's map color theorem, 274
Dirac–Gallai construction, 176
direct product, 439, 563
directed cycle, 572
directed from, 571
directed Hamilton path, 572
directed multigraph, *see* multidigraph
directed path, 572
disconnected graph, 556
disjoint, 562
distance, 426, 558
  pth distance class, 426
distance graph, 455
domain, 262
dominating edge, 378
dominating vertex, 185, 378
double-critical, 212
DP-coloring problem, 15
DP-forest, 273
DP-tree, 273
DP-brick, 17
DP-chromatic number, 16, 60
DP-degree-colorable, 17
DP-k-colorable, 16
duplicating a vertex, 52, 563

edge
edge of a graph, 554
edge of a hypergraph, 582
hyperedge, 355
ordinary edge, 355
edge of a multidigraph, 571
edge coloring, 37
edge coloring problem, 58
edge cut, 373, 557, 585
edge deletion, 555
edge density, 38
edge set
edge set of a graph, 554
edge set of a hypergraph, 355, 582
edge set of a multidigraph, 571
edge set of a multigraph, 570
edge-connected
k-edge-connected, 557, 585
edge-connectivity, 557, 585
edge-deleted subgraph, 555
efficiently realizable upper bound, 53
embeddable, 574
embedded multigraph, 574
T-multigraph, 574
embedding, 574
2-cell embedding, 574
empty, 355, 582
empty graph, 554
end-block, 558, 586
endomorphism, 463
ends
ends of a path, 556
ends of an edge, 554
endvertex, 555
enlarging, 366
Enlarging operation, 366
equitable chromatic number, 31
equitable k-coloring, 27
equivalent colorings, 61, 436
Erdős–Hajnal property, 104
Euler characteristic, 483, 573
Euler genus, 483, 573
Euler's Formula, 577
Eulerian multidigraph, 133
Eulerian orientation, 136
even, 556, 559, 563, 572
even circumference, 79
events, 586
dependent, 586
independent, 586
exact p-partition, 315
excess
k-excess, 232
expectation, 589
extended, 503
extension of a coloring, 62, 503

face chromatic number, 516
face coloring, 516
face graph, 516
facial cycle, 574
facial subgraph, 574
facial walk, 575
factor, 555
feasible configuration, 17
feasible move sequence, 324
FF, *see* first-fit
finite, *see* simplicial complex
first-fit, 79
flow
Γ-flow, 551
k-flow, 551
forest, 560
fractional chromatic number, 59, 404
fractional clique, 405
fractional clique number, 406
fractional coloring, 404
fractional DP-chromatic number, 471
fractional list chromatic number, 470
fractional orientation, 123
fractional out-degree, 123
fractionally absorbing, 126
fractionally independent, 126
free
H-free, 567
triangle-free, 567
$\mathcal{X}$-free, 69, 567
full component, 318
function
bijective, 554
image, 553
injective, 553
inverse image, 553
one-to-one, 554
onto, 553
preimage, 553
range, 553
restricted, 553
restriction, 553
surjective, 553

Gallai forest, 7
Gallai hyperforest, 358
Gallai hypertree, 358
Gallai join, 176
Gallai tree, 7
Gallai–Roy theorem, 115
game chromatic number, 534

generalized graph coloring, 112
generalized Kneser graph, 464
genus
  (orientable) genus, 579
  nonorientable genus, 579
girth, 559, 583
good move, 321
good move sequence, 322
  club, 322
  complete club, 322
  nonrepetitive, 322
  nontraveler, 322
  traveler, 322
graph, 554
  acyclic graph, 560
  bipartite graph, 562
  Borsuk graph, 454
  Catlin graph, 52
  chordal graph, 47
  circular graph, 414
  complete bipartite graph, 562
  complete graph, 556
  connected graph, 556
  cubic graph, 555
  disconnected graph, 556
  distance graph, 455
  (d,r,ℓ)-graph, 71
  empty graph, 554
  ε_k-graph, 7
  generalized Kneser graph, 464
  Gyárfás graph, 428
  Gyárfás–Simonyi–Tardos graph, 430
  imperfect graph, 44
  intersection graph, 105
  interval graph, 52, 105
  Kneser graph, 407
  line graph, 37
  Lovász–Borsuk graph, 455
  minimally imperfect graph, 45
  Mirzakhani graph, 493
  multiplicative graph, 440
  Mycielski graph, 418
  nontrivial graph, 503
  perfect graph, 44
  Petersen graph, 52, 408
  planar graph, 484
  plane graph, 485
  regular graph, 555
  Schrijver graph, 412
  shift graph, 98
  split graph, 221
  Toft graph, 200
  total graph, 39
  Turán graph, 564
  weighted graph, 40
graph calculus, 173
graph coloring algorithm, 53
graph coloring game, 534
graph coloring problem, 53
graph construction, 471
graph invariant, *see* graph parameter
graph parameter, 569
  additive, 569
  monotone, 569
  $\trianglelefteq$-monotone, 569
graph product, 479
graph property, 567
  additive, 569
  χ-bounded, 69
  hereditary, 567
  irreducible, 112
  minor-closed, 214
  monotone, 567
  polynomially χ-bounded, 104
  smooth, 91
  trivial, 567
  $\trianglelefteq$-hereditary, 568
greedy coloring, 53
group chromatic number, 535
Gyárfás graph, 428
Gyárfás–Simonyi–Tardos graph, 430

Hadwiger number, 229
Hadwiger–Nelson problem, 455
Hajós calculus, 174
Hajós join, 168, 364, 563
Hajós number
  k-Hajós number, 175
Hajós' construction, 169, 365
Hajós-k-constructible graphs, 170
Hamilton cycle, 556
Hamilton path, 556
hard pairs, 83
  f-hard, 83
  merging, 84
Heawood number, 484, 578
hereditary, *see* graph property
high vertex, 34, 368
high vertex subgraph, 35
hitting every maximum clique, 295
hitting set for a set system, 349
hole, 101
hom-bounded from above, 453
hom-bounded from below, 453
hom-equivalent, 453, 463
homomorphic, 401
homomorphism, 401
hyperbrick, 357

hypercycle, 583
hyperedge, 582
hypergraph, 355, 582
  empty, 355
  linear, 361
  nontrivial, 355
  q-uniform, 355
  simple, 361
hypergraph parameter, 360
  monotone, 360
hypergraph property, 360
  monotone, 360
hyperpath, 583
hyperwalk, 583

identifying independent vertices, 170
identifying vertices, 563
identity function, 554
imperfect graph, 44
improper coloring, 90
in-degree, 572
in-neighbor, 571
incidence function, 570
incident, 554, 570, 582
indecomposable graph, 181
indecomposable hypergraph, 375
independence number, 2, 562, 583
independence tree, 36
independent partial transversal, 262, 296
independent set, 2, 356, 562, 572, 583
  p-independent set, 562
independent transversal, 15, 262, 296
index of a club, 322
induced subdigraph, 572
induced subgraph, 555
induced subhypergraph, 583
inflation of a graph, 41
initial vertex, 571
inner faces, 580
inner vertices of a path, 556
inside, 490
interior, 490
internally disjoint, 557
intersection, 562, 570
intersection graph, 105
interval graph, 52, 105
inverse function, 554
irreversible, 125
isolated vertex, 554
isomorphic graphs, 555
isomorphism, 566

join
  Dirac join, 168, 364
  Gallai join, 176
  Hajós join, 168, 364
  join of two graphs, 562
  join of two vertices, 554, 570
  Ore join, 171
  Toft join, 210
joined by a path, 556

Kempe-change, 5
kernel, 118, 572
kernel-perfect, 118, 572
kernel-solvable, 125
Kneser graph, 407

leaf, 560
left coset, 566
Lemma
  Alon, 588
  Chernoff, 589
  Cranston–Rabern, 325
  El-Zahar–Sauer, 444, 445
  Fekete, 590
  Folding, 526
  Hajnal, 298
  Kernel Method, 118
  Local (general case), 587
  Local (lopsided), 589
  Mihók–Škrekovsky, 280
  Mozhan, 317
  Rabern, 290, 313
  Simonovits, 197
  Simonyi–Tardos, 462
  Small Pot, 342
  Stiebitz, 188
  Toft, 195
  Tucker (Octahedral), 411
length, 173, 556, 559
lexicographic product, 563
line graph, 37
line perfect multigraph, 159
linear degree, *see* star degree
linear forest, 107
linear hypergraph, 361
linear order, 556
list chromatic number, 10, 357
list coloring constant, 341
list-assignment, 10
  bad list-assignment, 56
  uncolorable list-assignment, 56
list-chromatic index, 38
list-colorable, 10
  f-list-colorable, 10, 41
  k-list-colorable, 10, 41, 357
list-critical, 235

k-list-critical, 235, 362
minimal k-list-critical, 235
Liu's construction, 395
local chromatic number, 477
local connectivity, 178
local edge connectivity, 178
loop, 571
looped path, 441
Lovász–Borsuk graph, 455
low vertex, 34, 92, 368
low vertex subgraph, 35, 92

map, 516, *see* function
map color theorems, 515
map graph, 441
mapping, *see* function
matching, 38, 570
matching number, 38, 571
maximal independent set coloring, 53
maximal planar graph, 581
maximum average degree, 12, 569
maximum clique, 295
maximum degree, 554, 582
maximum fractional out-degree, 123
maximum in-degree, 572
maximum local connectivity, 178
maximum local edge connectivity, 178
maximum multiplicity, 570
maximum out-degree, 572
maximum star degree, 361
maximum triangle degree, 292
minimal separation set, 585
minimal separator, 557
minimally imperfect graph, 45
minimum degree, 554, 582
minimum genus embedding, 579
minimum in-degree, 572
minimum out-degree, 572
minimum star degree, 361
minor, 564
minor-closed, 214
Mirzakhani graph, 493
modulo $(2p+1)$-orientation, 467
mono-block, 83
monotone graph parameter, 33
monotone graph property, 33
monotone orientation, 128
Moser spindle, 455
movable
(c, c')-movable, 28
move, 315
moving a vertex, 315
Mozhan partition, 315
multi-coloring, 59
multicycle, 23
(s,t)-multicycle, 23
multidigraph, 571
multigraph, 15, 570
multicycle, 23
multipath, 23
signed multigraph, 50
multipartite graph
complete p-partite graph, 564
complete multipartite graph, 564
partite sets, 564
parts, 564
p-partite graph, 564
multipath, 23
multiple edges, 570, 571
multiplicative graph, 440
multiplicity, 15, 38, 570
mutually independent events, 587
Mycielski graph, 418
cone, 472
ith level, 418
r-Mycielskian, 472
top vertex, 418
Mycielski's construction, 418
Mycielski–Grötzsch graph, *see* Mycielski graph

negative color, 28
negative pair, 448
negatively correlated, 589
neighborhood, 554
neighborhood complex, 410
neighbors, 554
noncontractible cycle, 575
nonorientable minimum genus embedding, 579
nonorientable surfaces, 573
nonpartitionable pair of dimension p, 85
nonrepetitive, 322
nonseparating cycle, 575
nontraveler, 322
nontrivial, 355, *see* graph and hypergraph
Nordhaus–Gaddum inequalities, 96
normal digraph, 124
nowhere zero Γ-flow, 551
nowhere zero k-flow, 551
null graph, *see* empty graph

odd, 101, 556, 559, 563, 572
odd antihole, 101
odd chromatic number, 438
odd circumference, 79
odd girth, 559
online coloring algorithm, 79
online graph, 79

online list chromatic number, 160, 536
optimal coloring, 53
optimal partition, 317
orbit, 566
order, 554, 566, 571, 582
ordinary edge, 582
Ore join, 171
Ore's construction, 171
Ore-k-constructible graphs, 171
Ore-degree of a graph, 276
Ore-degree of an edge, 276
Ore-function, 234
orientable surfaces, 573
orientably simple, 579
orientation, 115, 572
oriented graph, 165
out-degree, 572
out-going edge, 74
out-neighbor, 571
outer cycle, 490
outer face, 580
outerplanar graph, 514

paintability number, 160, 536
paintable
  f-paintable, 536
parallel, 571
parallel edges, 570
partial (X, H)-coloring, 262
partial coloring, 306
partial transversal, 262, 296
partially ordered set, 149
partionable multigraph
  f-partitionable, 83
  (s, t)-partitionable, 111
partite sets, 562, 564
partition of a multigraph, 81
  active component, 319
  component, 315
  critical vertex, 315
  exact p-partition, 315
  feasible move sequence, 324
  f-partition, 83
  full component, 318
  good move, 321
  good move sequence, 322
  $(k_1, k_2, \ldots, k_p)$-M-partition, 315
  move, 315
  moving a vertex, 315
  optimal, 317
  ith part, 315
  partition, 315
  p-partition, 81, 315
  reverse move, 315
  weight, 317
partition of a set, 553
path, 555
perfect graph, 44
performance guarantee, 54
permutation, 554
permutation group
  acts transitively, 554
  transitive, 554
Petersen graph, 52, 408
planar graph, 484, 580
plane graph, 485, 580
plane near-triangulation, 490
plane triangulation, 581
point aboricity, 112
point partition number, 112
Poljak–Rödl function, 481
positive color, 28
power
  pth partial power, 427
  pth power, 427
power set, 553
precoloring, 55
predecessor, 571
probability, 586
probability function, 586
probability space, 586
Problem
  Erdős–Hajnal, 197
  Gallai, 195
  Gao–Ma, 225
  Haxell–Naserasr, 344
  Nešetřil–Rödl, 196
  Stiebitz, 224
  Toft, 196, 210
projections, 439
projective planar graph, 502
projective plane graph, 502
proof sequence, 174
proper coloring, 90
proper cycle, 125
proper splitting, 204
proper subgraph, 555
proper subhypergraph, 582
proper subset, 553
property B, 396
pseudograph, 571
pseudomultigraph, 571

quadrangular embedding, 581
quadrangulation, 581
quasi-order, 465

radius, 424, 559

Ramsey number, 96
rank number, 96
recognition problem, 106
recoloring, 5, 28
reducer
k-reducer, 329
reducible
f-reducible, 153
weakly f-reducible, 153
reduction
k-reduction, 329
regular, 582
regular coloring, *see* Borsuk graph
regular graph, 555
relative complement of a hypergraph, 375
remainder, 132
restriction of a coloring, 62
retract of a graph, 463
reverse move, 315
root, 560, 561
rooted tree, 561
ancestor, 561
child, 561
depth, 561
descending edge, 561
height, 561
incoming edge, 561
internal vertex, 561
leaf, 561
level, 561
outgoing edge, 561
parent, 561
proper ancestor, 561
Rubin's Block Lemma, 58

sample space, 586
Schrijver graph, 412
send, 324
separating cycle, 575
separating edge set, 585
separating set, 556
separating vertex, 357, 556
separation, 585
separation set, 585
separator, 557
sequential coloring, 53
sequential coloring algorithm, 4
set chromatic number, 59
set difference, 553
set-coloring, 59
shift graph, 98
shrinking, 363
shrinking operation, 356
signed chromatic number, 51
signed multigraph, 50
simple digraph, 571
simple hypergraph, 361
simple splitting, 366
simplices, 410
simplicial complex, 410
body, 410
cone, 422
finite, 410
suspension, 422
simplicial map, 458
simplicial vertex, 47
size, 553
smallest last order, 11
spanning subgraph, 555
spanning subhypergraph, 582
spanning tree, 36, 560
SPGT, *see* Strong Perfect Graph Theorem
sphere
$(d-1)$-sphere, 410
split graph, 221
split-up, 176
splitting
proper splitting, 204
simple splitting, 366
splitting a vertex, 204, 366
splitting a vertex into an independent set, 367
splitting operation for graphs, 204
splitting operation for hypergraphs, 366
spread of a complete club, 323
spread of a move sequence, 323
square of a graph, 49
stabilizer, 566
stable Kneser graph, *see* Schrijver graph
stable matching, 152
star, 562
star degree, 361
star forest, 64
strict digraph, 125, 571
strictly h-degenerate, 82
strong bridge, 358, 584
strong fractional kernel, 126
strong product, 480
strong k-critical, 273
subadditive function, 591
subdigraph, 572
subdividing an edge, 563
subdivision, 564
subgraph, 555
subhypergraph, 582
submultigraph, 570
subset, 553
p-element subsets, 553
successor, 571

super-orientation, 125
surface, 573
suspension, *see* simplicial complex
symmetric difference, 136
symmetric group, 554

Talagrand's inequality, 590
terminal vertex, 571
Theorem
  Abbott–Zhou, 189
  Aharoni / Haxell, 298, 589
  Akbari–Mirrokni–Sadjad, 134
  Albertson–Hutchinson, 486
  Alon–Tarsi, 133, 134
  Axenovich / Albertson–Kostochka–West, 544
  Böhme–Mohar–Stiebitz, 502
  Berge–Tutte, 571
  Bernshteyn, 263, 267, 335
  Bernshteyn / Kostochka–Stiebitz, 396
  Bernshteyn–Kostochka, 282
  Bernshteyn–Kostochka–Pron, 27
  Bernshteyn–Lee, 160
  Beutelspacher–Hering, 344
  Bonamy–Kelly–Nelson–Postle, 352
  Bondy–Hell, 416
  Borodin–Kostochka–Toft, 84
  Borodin–Kostochka–Woodall, 119, 120
  Brooks, 4
  Chen–Jing–Zang, 39
  Chen–Ma–Zang, 165
  Chudnovsky–Goedgebeur–Schaudt–Zhong, 108
  Chudnovsky–King–Plumettaz–Seymour, 347
  Chudnovsky–Robertson–Seymour–Thomas, 210
  Combinatorial Nullstellensatz, 130
  Cranston–Rabern, 326
  Cycle-Plus-Triangles, 136
  Dilworth, 150
  Dirac, 34, 485
  Dirac / Gallai, 175, 176
  Dirac Construction, 168, 364
  Dol'nikov, 464
  Dvořák–Postle, 57
  El-Zahar–Sauer, 443, 446
  Emden-Weinert–Hougardy–Kreuter, 107, 328
  Enlarging Operation, 366
  Erdős, 69
  Erdős–Rubin–Taylor, 56
  Euler, 577
  Even Map Color, 529
  Fajtlowicz, 310
  Farzad–Malloy–Reed, 332
  Fischer–Hörsch, 349
  Five Color, 489
  Four Color, 485
  Frieze–Mubayi, 394
  Gallai, 35, 36, 155, 181, 207, 223, 239
  Gao–Ma, 189
  Gimbel–Thomassen, 529
  Grötzsch, 246
  Grünbaum / Aksionov, 497
  Graph Minor, 214
  Gravin–Karpov, 394
  Greenwell–Lovász, 188
  Gyárfás, 70, 76
  Gyárfás–Jensen–Stiebitz, 427
  Hadwiger, 215
  Hadwiger / Dirac, 216
  Hajós, 170
  Hajós Construction, 169, 365
  Hajnal–Szemerédi, 27
  Hall, 571
  Harutyunyan–Mohar, 161
  Haxell–McDonald, 350
  Heawood, 484
  Hedetniemi, 444
  Hutchinson, 530
  Jensen–Toft, 123
  Jiang, 425
  König, 38, 562
  Kaiser–Stehlík, 473
  Kelly–Postle, 346
  Kernel Solvable Graph, 126
  Kierstead–Szemerédi–Trotter, 424
  King, 296
  King–Reed, 303
  King–Reed–Vetta, 348
  Kleitman–Lovász, 129
  Kostochka, 58
  Kostochka–Nakprasit, 54
  Kostochka–Nešetřil, 396
  Kostochka–Stiebitz, 486
  Kostochka–Stiebitz–Wirth, 358, 359
  Kostochka–Yancey, 243, 244, 251
  Král'–Kratochvíl–Tuza–Woeginger, 106
  Lovász, 46, 82, 198, 399
  Lovász–Kneser, 409
  Lund–Yanakakis, 55
  Müller, 224
  Map Color, 487
  Martinsson / Dvorák, 287
  McDiarmid, 347
  Menger, 557
  Mihók, 68

Mihók–Schiermeyer, 78
Minty, 116, 155
Mohar, 284
Molloy–Reed, 303, 306, 328, 331, 351
Mozhan / Stiebitz, 213
Ore, 171
Postel–Schweser–Stiebitz, 283
Postle, 530
Postle–Thomas, 530
Pretzel–Youngs, 157
Przybylo, 340, 341
Rödl, 196, 205
Rödl–Siggers, 400
Rödl–Tuza, 421
Rabern, 311, 313
Rackham, 544
Reducibility, 388
Reed, 295, 345
Richardson, 122
Ringel, 523
Ringel–Youngs, 522
Robertson–Seymour–Thomas, 216
Schweser, 162
Scott–Seymour, 102
Shannon, 38
Shitov / Tardiv / Wrochna, 449
Simonyi–Tardos, 478
Splitting Operation, 366
Stable Marriage, 151
Stehlík, 283, 377
Stiebitz, 224, 422
Stiebitz–Storch–Toft, 375, 378, 382
Stiebitz–Tuza–Voigt, 44
Strong Perfect Graph, 48
Tallagrand, 590
Thomassen, 281, 488–490, 494
Three Color, 500
Toft, 177, 200, 201, 211, 371–374, 392
Turán, 565
Tutte, 552
Tuza, 116
Two Color, 502
Urquhart, 171
Vizing / Erdős–Rubin–Taylor, 15
Vizing / Gupta, 38
Wanless–Wood, 393
Weak Perfect Graph, 46
Wrochna, 449
Youngs, 157
Zig–Zag, 478
Toft graph, 200
Toft join, 210
Toft's construction, 210
Toft's splitting operation, 366
topological minor, 564
topological space
antipodal, 454
arc, 573
circle, 573
closed disc, 573
d-ball, 410
diameter, 454
$(d-1)$-sphere, 410
k-connected, 410
open disc, 573
total chromatic number, 39
total coloring, 39
total coloring problem, 59
total graph, 39
total list-chromatic number, 39
totally dominating set, 296
tour, 559
tournament, 149
transitive, 566
transitive digraph, 150
transversal
transversal of a cover, 15, 262
transversal of a set system, 296
travel, 324
traveler, 322
tree, 560
BFS tree, 561
broom, 101
DFS tree, 560
independence tree, 36
rooted tree, 561
spanning tree, 36, 560
tree decomposition, 226
triangle, 556
triangle-free, 567
triangular embedding, 581
triangulation, 581
tuple-coloring, 59
Turán graph, 564
Turán number, 565

uncolorable hyperpair, 358
uncolorable list-assignment, 56
uncolorable pair, 13
underlying graph, 17, 118, 570, 572
underlying rooted tree, 71
uniform coloring
nearly p-uniform C-coloring, 28
negative color, 28
positive color, 28
p-uniform C-coloring, 28
uniform hypergraph
q-uniform, 355

union, 562, 570
uniquely C-colorable, 61
uniquely k-colorable, 61
universal vertex, 367
Urquhart's construction, 171

value of a fractional coloring, 404
vector function, 82
vertex deletion, 555
vertex enumeration, 556
vertex order, 556
vertex set
  vertex set of a graph, 554
  vertex set of a hypergraph, 355, 582
  vertex set of a multidigraph, 571
  vertex set of a multigraph, 570
  vertex set of a simplicial complex, 410
vertex-critical, 107, 362
  k-vertex-critical, 107, 362
vertex-deleted subgraph, 555
Vizing property, 103

walk, 559
  closed, 559
  open, 559
weak degeneracy, 113, 154, 160
weighted graph, 40
  connected, 44
wheel, 562
  d-wheel, 188
worst-case running time, 54

Zykov's construction, 97
Zykov-Schäuble-Lovász construction, 385

MIX
Papier aus verantwortungsvollen Quellen
Paper from responsible sources
FSC® C105338

If you have any concerns about our products,
you can contact us on
ProductSafety@springernature.com

In case Publisher is established outside the EU,
the EU authorized representative is:
Springer Nature Customer Service Center GmbH
Europaplatz 3, 69115 Heidelberg, Germany

Printed by Libri Plureos GmbH
in Hamburg, Germany